CONTROL DATA AND MONUMENTS

Aerial photograph roll and frame number*	3 - 20

Horizontal control

Third order or better, permanent mark	Neace △ Neace ⊕
With third order or better elevation	BM △ 45.1 Pike BM ⊕ 45.1
Checked spot elevation	△ 19.5
Coincident with section corner	Cactus △ Cactus ⊕
Unmonumented*	+

Vertical control

Third order or better, with tablet	BM × 16.3
Third order or better, recoverable mark	× 120.0
Bench mark at found section corner	BM + 18.6
Spot elevation	× 5.3

Boundary monument

With tablet	BM □ 21.6 BM ⊕ 71
Without tablet	□ 171.3
With number and elevation	67 □ 301.1
U.S. mineral or location monument	▲

CONTOURS

Topographic

Intermediate	
Index	
Supplementary	
Depression	
Cut; fill	

Bathymetric

Intermediate	
Index	
Primary	
Index Primary	
Supplementary	

BOUNDARIES

National	
State or territorial	
County or equivalent	
Civil township or equivalent	
Incorporated city or equivalent	
Park, reservation, or monument	
Small park	

*Provisional Edition maps only
Provisional Edition maps were established to expedite completion of the remaining large scale topographic quadrangles of the conterminous United States. They contain essentially the same level of information as the standard series maps. This series can be easily recognized by the title "Provisional Edition" in the lower right hand corner.

LAND SURVEY

U.S. Public Land

Township or range	
Location doubtful	- - - -
Section line	
Location doubtful	- - - -
Found section corner; found closing corner	+ · +
Witness corner; meander corner	WC ⊕ MC ⊕

Other land surveys

Township or range line	··········
Section line	··········
Land grant or mining claim; monument	- · — □
Fence line	- - - - -

SURFACE FEATURES

Levee	—	Levee
Sand or mud area, dunes, or shifting sand		Sand
Intricate surface area		Strip Mine
Gravel beach or glacial moraine		Gravel
Tailings pond		Tailings Pond

MINES AND CAVES

Quarry or open pit mine	✕
Gravel, sand, clay, or borrow pit	⋏
Mine tunnel or cave entrance	⊣
Prospect; mine shaft	✕ ▪
Mine dump	/ Mine Dump
Tailings	Tailings

VEGETATION

Woods	
Scrub	
Orchard	
Vineyard	
Mangrove	Mangrove

GLACIERS AND PERMANENT SNOWFIELDS

Contours and limits	
Form lines	

MARINE SHORELINE

Topographic maps

Approximate mean high water	
Indefinite or unsurveyed	

Topographic-bathymetric maps

Mean high water	
Apparent (edge of vegetation)	

Topographic map symbols (Courtesy U.S. Geological Survey).

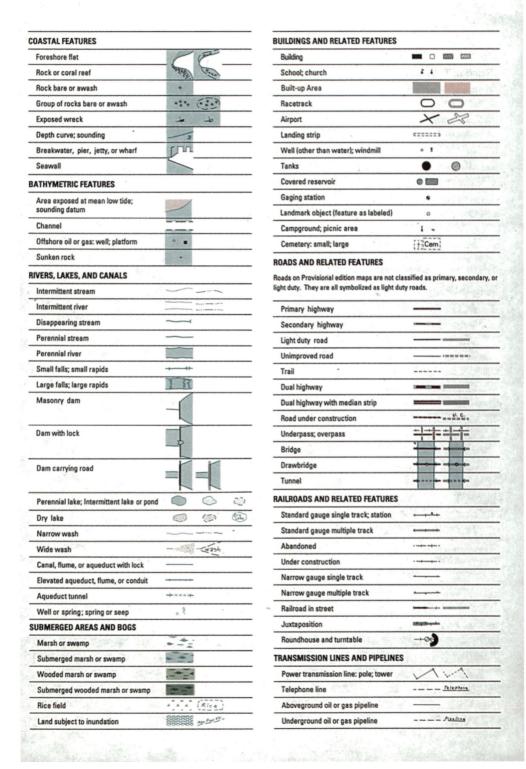

COASTAL FEATURES

Foreshore flat

Rock or coral reef

Rock bare or awash

Group of rocks bare or awash

Exposed wreck

Depth curve; sounding

Breakwater, pier, jetty, or wharf

Seawall

BATHYMETRIC FEATURES

Area exposed at mean low tide; sounding datum

Channel

Offshore oil or gas: well; platform

Sunken rock

RIVERS, LAKES, AND CANALS

Intermittent stream

Intermittent river

Disappearing stream

Perennial stream

Perennial river

Small falls; small rapids

Large falls; large rapids

Masonry dam

Dam with lock

Dam carrying road

Perennial lake; Intermittent lake or pond

Dry lake

Narrow wash

Wide wash

Canal, flume, or aqueduct with lock

Elevated aqueduct, flume, or conduit

Aqueduct tunnel

Well or spring; spring or seep

SUBMERGED AREAS AND BOGS

Marsh or swamp

Submerged marsh or swamp

Wooded marsh or swamp

Submerged wooded marsh or swamp

Rice field

Land subject to inundation

BUILDINGS AND RELATED FEATURES

Building

School; church

Built-up Area

Racetrack

Airport

Landing strip

Well (other than water); windmill

Tanks

Covered reservoir

Gaging station

Landmark object (feature as labeled)

Campground; picnic area

Cemetery: small; large

ROADS AND RELATED FEATURES

Roads on Provisional edition maps are not classified as primary, secondary, or light duty. They are all symbolized as light duty roads.

Primary highway

Secondary highway

Light duty road

Unimproved road

Trail

Dual highway

Dual highway with median strip

Road under construction

Underpass; overpass

Bridge

Drawbridge

Tunnel

RAILROADS AND RELATED FEATURES

Standard gauge single track; station

Standard gauge multiple track

Abandoned

Under construction

Narrow gauge single track

Narrow gauge multiple track

Railroad in street

Juxtaposition

Roundhouse and turntable

TRANSMISSION LINES AND PIPELINES

Power transmission line: pole; tower

Telephone line

Aboveground oil or gas pipeline

Underground oil or gas pipeline

Topographic map symbols.

Surveying
Theory and Practice

Surveying

Theory and Practice

SEVENTH EDITION

James M. Anderson
University of California, Berkeley

Edward M. Mikhail
Purdue University

Boston   Burr Ridge, IL   Dubuque, IA   Madison, WI   New York   San Francisco   St. Louis
Bangkok   Bogotá   Caracas   Lisbon   London   Madrid
Mexico City   Milan   New Delhi   Seoul   Singapore   Sydney   Taipei   Toronto

WCB/McGraw-Hill

*A Division of The **McGraw·Hill** Companies*

Previously published as Davis, F. E., Foote, F. S., Anderson, J. M., and Mikhail, E. M., *Surveying: Theory and Practice*.

SURVEYING: THEORY AND PRACTICE

This book is printed on acid-free paper.

5 6 7 8 9 0 DOC/DOC 0 9 8 7 6 5 4 3 2 1

ISBN 0-07-015914-9

Vice president and editorial director: *Kevin T. Kane*
Publisher: *Tom Casson*
Executive editor: *Eric Munson*
Marketing manager: *John T. Wannemacher*
Project manager: *Paula M. Buschman*
Production supervisor: *Michael McCormick*
Designer: *Kiera Cunningham*
Cover design: *Good Studio*
Senior photo research coordinator: *Keri Johnson*
Compositor: *Interactive Composition Corporation*
Typeface: *10/12 Times Roman*
Printer: *R. R. Donnelley & Sons Company*

Library of Congress Cataloging-in-Publication Data

Anderson, J. M. (James McMurry) (date)
   Surveying, theory and practice / James M. Anderson. Edward M.
Mikhail.—7th ed.
     p. cm.
   Rev. ed. of: Surveying, theory and practice / Raymond E. Davis . . .
[et al.]. 1981.
   Includes bibliographical references and index.
   ISBN 0-07-015914-9
   1. Surveying. I. Mikhail, Edward M. II. Title.
TA545.D45   1998
526.9—dc21
                                                             97-28584

http://www.mhhe.com

PREFACE

This new edition continues the complete and extensive coverage found in the sixth edition, which will be suitable as a text for elementary and advanced college courses without losing its value as a reference for the practicing professional. The philosophy behind this revision is based on a desire to bring all materials up to date and introduce new methods and procedures that have been developed in the past 10 years as a result of extremely rapid technological advances in positioning equipment and data acquisition and processing procedures. In a major organizational change, appropriate chapters are separated into two sections: Part A on elementary topics; and Part B, advanced topics. This innovation facilitates use of the textbook for level one and level two courses.

Technological advances that have had substantial impact on surveying include: the development and refinement of the Global Positioning System (GPS) as the pre-eminent method for determining basic control point positions; potential applications of GPS and almost complete use of total station systems and data collectors in other types of surveys; and the natural relationship which exists among the data and products of surveying and mapping and Geographical Information Systems (GIS) and Land Information Systems (LIS). In addition to these factors, the North American Datum of 1983 (NAD 83) is now in place and the North American Vertical Datum of 1988 (NAVD 88) is in the process of being adopted.

As in the previous edition, this book consists of five distinct parts. Part I consists of three chapters. Chapter 1 is introductory and contains basic definitions and differences between surveying and mapping. Information about the NAD 1983 with more emphasis on the geoid and ellipsoid are now included in this chapter. Chapter 2 describes the basic concepts of the theory of errors as related to survey measurements and is divided into Parts A and B. A major change in this chapter is the inclusion of a consolidated development of the least squares adjustment, moved from an Appendix to Part B of this chapter. Chapter 3, which covers field and office work, has been altered to include introductions to GPS, GIS, LIS, new computational aids, and use of data collectors as electronic field books.

The basic surveying measurements of distance, vertical distance or leveling, directions and angles, and position in three-dimensional space by total station systems and stadia and tacheometry are covered in Part II, Chapters 4 through 7, respectively. Note that these topics are covered in a very logical order for presentation in an elementary course. Although uses of traditional types of equipment (tape, level, rod, etc.) are still covered as basic methods for obtaining distances and elevations (Chapters 4 and 5), applications of the most modern Electronic Distance Measurement (EDM) instruments, automated self-reading levels, and GPS systems are included in each case. In Chapter 6, directions and angles, the emphasis is now shifted to the use of optical reading and/or digitized, three-screw repeating and single-center theodolites. A major change has been made in Chapter 7, where total station systems are covered in detail and the material on stadia and tacheometry, although still included, has been shortened.

Part III, Survey Operations, follows details of basic measurements quite naturally and is composed of Chapters 8 through 10 on traverse, standard methods for horizontal positioning (intersection, resection, triangulation, trilateration, and associated error propagation and adjustment), and astronomy. Chapter 8 on traverse now includes a section on two-dimensional coordinate transformations and a completely revised section on three-dimensional traverse. The treatment of triangulation and trilateration, in Chapter 9, has been altered to put more emphasis on trilateration and combined triangulation/ trilateration in three-dimensional control networks. Astronomy in Chapter 10 now includes a section on azimuth by the hour angle method and all examples are modified to accommodate currently available ephemerides.

Part IV, now titled *Modern Surveying and Mapping,* has been substantially rearranged and revised and represents another major change in the textbook. Chapter 11 (old Chapter 13), on map projections, has been modified to include coordinate reference frameworks and to accommodate NAD 1983 parameters in the procedures and equations for computing state plane coordinates. Chapter 12 contains a complete treatment of the Global Positioning System (GPS), and has been shifted from Part III into Part IV. Chapter 13 (old Chapter 14) covers photogrammetric surveying and mapping and includes sections on the analytical plotter, softcopy photogrammetry, and digital data and image processing. Finally, in Chapter 14, *Mapping, Digital Mapping, and Spatial Information Systems* are covered. Sufficient material has been retained from old Chapter 13 on mapping so as to develop the concept of mapping and what elements are found in a topographic map. This basic information is supported by articles on digital mapping, digital terrain models, computer aided drafting and design (CADD), geographic and land information systems (GIS) and (LIS), and automated mapping systems.

In the last section, Part V, *Types of Surveys*, topographic, route, construction, and land surveys are covered, each in its own chapter. The common elements involved in revisions for these chapters include use of total station systems and GPS for data acquisition in topographic, route, construction and land surveys, and applications of GIS and LIS in all types of surveys.

The Appendices are also reorganized and supplemented. A new Appendix A, *Elementary Mathematical Concepts* is added, followed by Appendix B, *Introduction to Matrix and Vector Algebra.* Another new section on *Coordinate Transformations* is found in Appendix C. This is followed by a third new section, Appendix D, *Introduction to Probability and Statistics* that provides the theoretical background for all of the error propagation, adjustments, and analysis of results in the main part of the textbook. The last appendix, E, contains the necessary trigonometric formulas and statistical tables.

Thus, this edition preserves the depth and breadth of the previous edition and is organized according to the same format. However, all coverage is brought up to date and includes the latest technological developments in equipment and procedures. As in the past, each chapter is followed by a list of problems and references for additional reading. Throughout, error and measurement analyses based on modern statistical techniques are included in each chapter. This very important aspect of surveying and mapping gains increasing importance with the rapidly developing technology that involves complicated, sophisticated systems, which produce large amounts of data that must then be analyzed. It is absolutely essential to prepare the future surveyor to handle the analysis of the system and resulting data in a statistically valid manner. After all, the reader should recognize that surveying is essentially the science of metrology or measurement.

ACKNOWLEDGMENTS

The authors would like to acknowledge the assistance offered by their colleagues and users of the text and authors of contributed chapters. Dr. Clyde C. Goad, Topcon-GeoComp, Ltd., Professors Sayed R. Hashimi, Ferris State University, Steven D. Johnson, Purdue University, James P. Reilly, New Mexico State University, Carl Shangraw, Ferris State University, and Howard Turner, California Polytechnic, Pomona, CA, Messrs James Appleton, Richard Burns, Richard Davis, Lawrence Fenske, Michael S. Honnold, Leland D. Ho, Mark O'Dowd, John Rogers, and Hugh Schultz, all of the California Department of Transportation, Mr. Hans Haselbach Jr., Haselbach Surveying Instruments, and Dr. George Y. G. Lee, U.S. Geological Survey, all provided useful information and rendered valuable suggestions and advice. Mr. Bryn Fosburgh, Trimble Navigation Limited, compiled the material for Chapter 12, Global Positioning System, and Professors James Bethel and Steven Lambert, Purdue University and Dr. Jolyon D. Thurgood, Space Imaging, cooperated in the writing of Chapter 14, Mapping, Digital Mapping, and Spatial Information Systems. Their efforts are very much appreciated.

Many of the illustrations and tables have been from or adapted from publications or furnished directly by federal or state agencies, specifically the U.S. Bureau of Land Management, U.S. Geological Survey, U.S. National Ocean Survey, National Geodetic Survey, and the California Department of Transportation. Use has also been made of photographs and tabular information from the American Railway Engineering Association (AREA), American Society of Civil Engineers (ASCE), Institute for Photogrammetry at the University of Stuttgart (Germany), U.S. Air Force, U.S. Army Corps of Engineers, and the U.S. Naval Observatory.

Illustrative material was also furnished by the following manufacturers and distributors of surveying, photogrammetric, and electronic equipment and consulting services: Chicago Steel Tape, The Cooper Group (Lufkin), Geotronics of North America, Intergraph Corporation, Laserline Manufacturing, Inc., Leica, Inc., Nikon, Inc., Pentax Corporation, Sokkia Corporation, Spectra-Physics Laserplane, Inc., Topcon America, Inc., Topcon Laser Systems, Inc., Trimble Navigation Limited, Raymond Vail & Associates, Warren-Knight Instrument Co., and Carl Zeiss, Inc.

During the checking of page proofs, substantial editing assistance was provided by: Hazem Barakat, Chris Dietsch, Jim Elithorpe, Andrea Johnson, John Marshall, Kent Ross, Sarabjit Singh, and Hank Theiss, all of Purdue University. The authors are grateful to these individuals for their contributions of time and expertise. Professor Mikhail's secretary, Cheryl Kerker, provided valuable typing support in the early stages; her effort is acknowledged and appreciated.

Finally, a very special note of gratitude is expressed to our families, Ruth and Connie Anderson and LaVerne Mikhail and the Mikhail children, who persevered with us through the preparation of the manuscript and editing of the manuscript and page proofs. Their support during this time is deeply appreciated.

James M. Anderson
Edward M. Mikhail

CONTENTS

Appendixes

PART I

Concepts

CHAPTER 1

Surveying and Mapping

1.1
SURVEYING

Surveying has to do with the determination of the relative spatial location of points on or near the surface of the earth. It is the art of measuring slope and horizontal and vertical distances between objects, of measuring angles between lines, of determining the directions of lines, and of establishing point locations by predetermined angular and linear measurements.

Accompanying the actual measurements of surveying are the mathematical calculations. Distances, angles, directions, locations, elevations, areas, and volumes are thus determined from the data of the survey. Also much of the information of the survey is portrayed graphically, either by the construction of maps, profiles, cross sections, and diagrams on map sheets or by viewing a video screen using digitized data stored in an electronic computer.

The equipment available and the methods applicable for measurement, storage, calculation, and compilation of the data have changed tremendously in the past decade, mainly due to the growth of the electronics industry and development of the microprocessor. The Global Positioning System (GPS, a positioning method based on measurements to orbiting satellites), total station systems in which the distance and direction can be observed with one compact instrument, Geographic and Land Information Systems (GIS and LIS), digital photogrammetry, and inertial surveying are examples of current systems to measure, collect, and display information usable in the surveying procedure. The relatively easy access to electronic computers of all sizes and capabilities facilitates the rigorous processing and storage of large volumes of data in the field and in the office.

With the development of these sophisticated data acquisition and processing systems, the duties of the surveyor have expanded beyond the traditional tasks of the fieldwork of taking the measurements and the office work of computing and drawing. Surveying not only is required for conventional engineering construction projects but also increasingly is used in manufacturing for product quality control as well as in the other physical sciences, such as geology and geophysics; biology, including agriculture, forestry, grasslands, wetlands, and wildlife; hydrology and oceanography; and geography, including human and cultural resources. The tasks in these operations need to be redefined to include design of the surveying procedure and selection of the equipment and system(s) appropriate to the project, acquisition of data in the field by a combination of electronic storage devices and manual recording, reduction or analysis of data in the field or office, storage of data in a form compatible with future retrieval and use in GIS or LIS, preparation of the maps in numerical form (for display on a video screen) needed for the survey, and setting monuments and boundaries in the field as well as control for construction layout. Even with the wide array of sophisticated hardware and software now available, performance of these tasks still requires familiarity with the uses of surveying, knowledge of the fundamentals of the surveying process, and a knowledge of the various means by which data can be prepared for presentation.

1.2
USES OF SURVEYS

The earliest surveys known were to establish the boundaries of land. Such surveys are still the important work of many surveyors.

Every construction project of any magnitude is based on measurements taken during the progress of a survey and constructed about lines and points established by the surveyor. Aside from land surveys, practically all surveys of a private nature and most of those conducted by public agencies are of assistance in the conception, design, and execution of engineering works.

For many years the federal government, and in some instances individual states, have conducted surveys over large areas for a variety of purposes. The principal work so far accomplished consists of fixing the national and state boundaries, charting coastlines and navigable streams and lakes, precisely locating reference points throughout the country, collecting valuable facts concerning the earth's magnetism at widely scattered stations, establishing and observing a greater network of gravity stations throughout the world, establishing and operating tidal and water level stations, extending the area covered by hydrographic and oceanographic charts and maps, and extending the area covered by topographic maps of the earth's land surfaces. The United States has achieved almost complete coverage of its surface area to map scales as large as 1:24,000 or 1 inch to 2000 feet.

Observations of a worldwide net of satellite triangulation stations were made during the decade 1964–1974. Reduction of these measurements led to an improved determination of the shape of the earth, one to two magnitudes better than previously known. Consequently, surveys of a global extent have been performed, and with the advent of GPS, surveys of a global nature will become more common in the future.

Surveys are divided into three classes: (1) those for the primary purpose of establishing the boundaries of land, (2) those providing information necessary for the planning and construction of public and private works, and (3) those large, high-precision surveys conducted by the government and, to some extent, by the states. No hard and fast line of demarcation separates one class of survey from another in the methods employed, the results obtained, or the use of the data.

1.3
THE EARTH AS A SPHEROID

The earth has the approximate shape of an *oblate spheroid* (Figure 1.1(a)) or ellipsoid of revolution that, on its outer crust, is composed of irregular land masses and uniform surfaces of oceans and seas. This irregular solid has a polar axis somewhat less than its

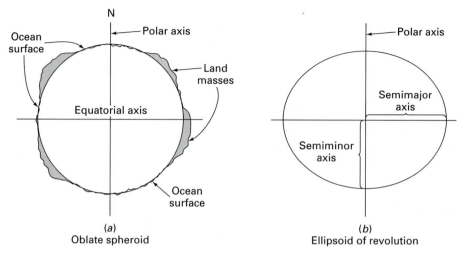

(a)
Oblate spheroid

(b)
Ellipsoid of revolution

FIGURE 1.1
Shape of the earth.

equatorial axis and, for mathematical purposes, is approximated by an *ellipsoid of revolution,* generated by rotating an ellipse about its shorter axis (Figure 1.1(b)). The lengths of the major and minor semiaxes of this ellipsoid are variously computed as follows:

Reference	Equatorial	Polar
	$a =$	$b =$
	(semimajor axis), m	(semiminor axis), m
Clarke (1866)	6,378,206	6,356,584
GRS 80	6,378,137	6,356,752.314

The lengths computed by Clarke were generally accepted in the United States and used in government surveys until 1986, when the Geodetic Reference System of 1980 (GRS 80) was adopted.

Note that the polar semiminor axis for GRS 80 is shorter than the equatorial semimajor axis by about 13 mi, or about 21 km. Relative to the radius of the earth this is a very small quantity, less than 0.34 percent. Imagine the earth shrunk to the size of a billiard ball, still retaining the same shape. In this condition, it would appear to the eye as a smooth sphere, and only by precise measurement could its lack of true sphericity be detected.

Now suppose the irregular land masses of the earth were removed. The surface of this imaginary spheroid is a curved surface, every element of which is normal to the force of gravity (the plumb line or vertical line, a very important reference line in surveying). Such an *equipotential* surface is termed a *level surface.* The particular surface at the average sea level is called the *mean sea level.* When the mean sea-level surface encompasses the true figure of the earth, the resulting surface is referred to as the *geoid,* an equipotential surface whose potential is the same as mean sea level. Figure 1.2 shows a cross section of an ellipsoid of revolution superimposed on a cross section of the geoid. Note that the geoid is a smoothly curving, undulating line that either crosses or is definitely separated from the ellipsoid of revolution. These undulations, which are much exaggerated in the sketch, are due to the uneven distribution of the earth's mass.

The reader's understanding of the various concepts regarding the earth is greatly facilitated by assuming it to be a sphere. Imagine a plane passing through the center of the

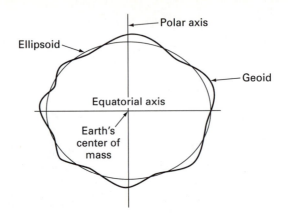

FIGURE 1.2
The geoid.

earth, which is assumed to be spherical, as in Figure 1.3. Its intersection with the level surface forms a continuous line around the earth. Any portion of such a line is termed a *level line,* and the circle defined by the intersection of such a plane with the mean level of the earth is termed a *great circle* of the earth. The distance between two points on the earth, as *A* and *B* (Figure 1.3), is the length of the arc of the great circle passing through the points and is always more than the chord intercepted by this arc. The arc is a level line; the chord is a mathematically straight line.

If a plane is passed through the poles of the earth and any other point on the earth's surface (as *A* in Figure 1.4), the line defined by the intersection of the level surface and plane is called a *meridian.* Imagine two such planes as passing through two points as *A* and *B* (Figure 1.4) on the earth, and the section between the two planes removed like the slice of an orange, as in Figure 1.5. At the equator the two meridians are parallel; above and below the equator they converge, and the angle of convergency increase as the poles are approached. No two meridians are parallel except at the equator.

Imagine lines, normal to the meridians, drawn on the two cut surfaces of the slice. If the earth is a perfect sphere, these lines would converge at a point at the center of the earth. Considering the lines on either or both of the cut surfaces, no two are parallel. The radial lines may be considered vertical or plumb lines; hence, we arrive at the deductions that all plumb lines converge at the earth's center and that no two are parallel. (Strictly speaking, this is not quite true, owing to the unequal distribution of the earth mass and to the fact that normals to an oblate spheroid do not all meet at a common point.)

Consider three points on the mean surface of the earth. Let us make these three points the vertices of a triangle, as in Figure 1.6. The surface within the triangle *ABC* is a curved surface, and the lines forming its sides are arcs of great circles. The figure is a spherical triangle. In the figure, the dashed lines represent the plane triangle whose vertices are points *A*, *B*, and *C*. (Actually, the "auxiliary plane triangle" has sides equal in length to the *arcs* of the corresponding spherical triangle.) Lines drawn tangent to the sides of the spherical triangle at its vertices are shown. The angles α, β, and γ of the spherical triangle are seen to be greater than the corresponding angles α', β', and γ' of the plane triangle. The amount of this excess would be small if the points were close together, and the surface forming the triangle would not depart far from a plane passing through the three points. If the points were far apart, the difference would be considerable. Evidently, the same conditions would exist for a figure of any number of sides. Hence, we see that angles on the surface of the earth are spherical angles.

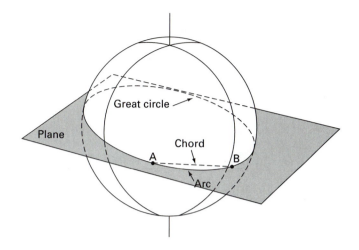

FIGURE 1.3

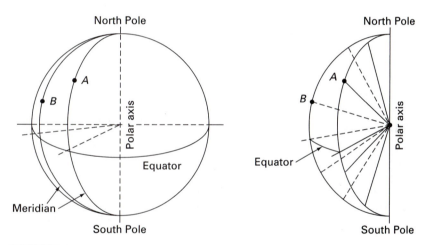

FIGURE 1.4   **FIGURE 1.5**

In everyday life these facts are of no concern, primarily because only a small portion of the earth's surface is involved. A line passing along the surface of the earth directly between two points is considered a straight line, plumb lines are parallel, a level surface is a flat surface, and angles between lines in such a surface are plane angles.

As to whether the surveyor must regard the earth's surface as curved or a plane (a much simpler premise) depends on the character and magnitude of the survey and the precision required.

In either case, to provide a suitable framework to which all surveys are referenced, it is necessary to establish a horizontal datum and a vertical datum. A *horizontal datum* is the surface to which horizontal distances are referred and consists of an ellipsoid of revolution approximating the figure of the earth and a set of constants or constraints that specify the size, position, and orientation of the ellipsoid. The horizontal datum now being used in the United States is the North American datum of 1983 (NAD 83), which is referenced to GRS 80. A *vertical datum* is a surface to which all elevations and depths are referred. The vertical datum currently being used in the United States is the North American Vertical

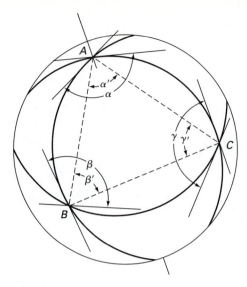

FIGURE 1.6

Datum (NAVD) of 1988. Additional details concerning the datum can be found in Section 14.2.

1.4
GEODETIC SURVEYING

The type of surveying that takes into account the true shape of the earth is called *geodetic surveying.* Surveys employing the principles of geodesy are of high precision and generally extend over large areas. Surveys of this character have been conducted primarily through government agencies. In the United States, such surveys have been performed principally by the U.S. National Geodetic Survey and the U.S. Geological Survey. Geodetic surveys also have been conducted by the Great Lakes Survey, National Imagery and Mapping Agency (NIMA), Department of Defense (DOD), U.S. Corps of Engineers, Mississippi River Commission, several boundary commissions, and others.

Although relatively few engineers and surveyors are employed full-time in geodetic work, the data of the various geodetic surveys are of great importance, because they furnish precise positions for points of reference to which a multitude of surveys of lower precision may be tied. For each state, a system of plane coordinates has been devised to which all points in the state can be referred without an error of more than 1 part in 10,000 in distance or direction arising from the difference between the reference surface and the actual mean surface of the earth.

The Global Positioning System, by virtue of the system configuration and data reduction procedure (Chapter 12), automatically provides geodetic positions referred to in the GRS 80 when proper network adjustment methods are performed. Since GPS equipment and total station systems are now in common use by most practitioners, the majority of the surveyors in private practice, as well as those working for government agencies, have the technical capability to perform geodetic surveys. Even so, it cannot be sufficiently overemphasized that, in spite of the availability of this very sophisticated equipment, a thorough knowledge of the principles of geodesy is an absolute prerequisite for the proper planning and execution of geodetic surveys.

1.5
PLANE SURVEYING

That type of surveying in which the mean surface of the earth is considered a plane, or in which its ellipsoidal shape is neglected, generally is called *plane surveying*. With regard to horizontal distances and directions, a level line is considered mathematically straight, the direction of the plumb line is considered to be the same at all points within the limits of the survey, and all angles are considered to be plane angles.

By far most surveys are of this type. When it is considered that the length of an arc 11.5 mi or 18.5 km long lying in the earth's surface is only 0.02 ft or 0.007 m greater than the subtended chord and, further, that the difference between the sum of the angles in a plane triangle and the sum of those in a spherical triangle is only 1 second for a triangle at the earth's surface having an area of 75.5 mi² or 196 km², it will be appreciated that the shape of the earth must be taken into consideration only in surveys of precision covering large areas. However, with the increasing size and sophistication of engineering and other scientific projects, surveyors who restrict their practice to plane surveying are severely limited in the types of surveys in which they can be engaged.

Surveys for the location and construction of highways, railroads, canals, and in general, the surveys necessary for the works of human beings are plane surveys, as are the surveys made to establish boundaries, except state and national. The United States system of subdividing the public lands employs the methods of plane surveying but takes into account the shape of the earth in the location of certain of the primary lines of division.

The operation of determining *elevation* usually is considered a division of plane surveying. Elevations are referred to an ellipsoidal surface, a tangent at any point in the surface being normal to the plumb line at that point. The curved surface of reference, usually mean sea level, is called a *datum* or, curiously and incorrectly, a *datum plane*. The procedure ordinarily used in determining elevations automatically takes into account the curvature of the earth, and elevations referred to the curved surface of reference are secured without extra effort by the surveyor. In fact, it would be more difficult for the surveyor to refer elevations to a true plane than to the imaginary ellipsoidal surface that was selected. Imagine a true plane tangent to the surface of mean sea level at a given point. At horizontal distances of 10 mi and 10 km from the point of tangency, the vertical distances (or elevations) of the plane above the surface represented by mean sea level are 67 ft and 7.9 m and at horizontal distances of 100 mi and 100 km from the point of tangency, the elevations of the plane are 6670 ft and 785 m, respectively. Obviously, curvature of the earth's surface is a factor that cannot be neglected in obtaining even rough values of elevations.

Heights referred to the ellipsoid of revolution also can be determined directly, using GPS. As noted previously (Figure 1.2) there are separations between the geoid and the ellipsoid. Therefore, corrections must be applied to GPS-derived heights before they can be used as elevations or heights referred to the geoid, frequently called *orthometric elevations*. This is an extremely important aspect of surveying with GPS that cannot be neglected. See Chapter 12 (Section 12.13, subsection 5) for details.

1.6
OPERATIONS IN SURVEYING

A *control survey* consists of establishing the horizontal and vertical positions of arbitrary points. A *land, boundary,* or *property survey,* is performed to determine the length and direction of land lines and to establish the position of these lines on the ground. A *topographic survey* is made to secure data from which a *topographic map* indicating the configuration of the terrain and the location of natural and human-made objects may be made. *Hydrographic surveying* refers to surveying bodies of water for the purposes of navigation, water supply, or subaqueous construction. *Mine surveying* utilizes the principles for control, land, geologic, and topographic surveying to control, locate, and map underground and surface works related to mining operations. *Construction surveys* are performed to lay out, locate, and monitor public and private engineering works. *Route surveying* refers to those control, topographic, and construction surveys necessary for the location and construction of lines of transportation or communication, such as highways, railroads, canals, transmission lines, and pipelines. *Photogrammetric surveys* utilize the principles of aerial and terrestrial photogrammetry, in which measurements made on photographs are used to determine the positions of photographed objects. Photogrammetric surveys are applicable in practically all the operations of surveying and in a great number of other sciences.

1.7
SUMMARY OF DEFINITIONS

A *level surface* is a curved continuous surface every element of which is normal to a plumb line. Disregarding local deviations of the plumb line from vertical, a level surface is parallel with the surface of the geoid. A body of still water provides the best example of a level surface.

An *ellipsoid of revolution* is a figure generated when an ellipse is rotated about its shorter axis and used to approximate the surface of the earth for geodetic calculations.

The *geoid* is an equipotential surface (a surface everywhere perpendicular to gravity) with a potential equal to the mean sea level surface that encompasses the earth.

A *horizontal plane* is a plane tangent to a level surface at a particular point.

A *horizontal line* is a line tangent to a level surface. In surveying, it is commonly understood that a horizontal line is straight.

A *horizontal angle* is an angle formed by the intersection of two lines in a horizontal plane.

A *vertical line* is a line perpendicular to the horizontal plane. A plumb line is an example. A vertical line in the direction toward the center of the earth is said to be in the direction of the nadir. A vertical line in the direction away from the center of the earth and above the observer's head is said to be directed toward the zenith.

A *vertical plane* is a plane in which a vertical line is an element.

A *vertical angle* is an angle between two intersecting lines in a vertical plane. In surveying, it is commonly understood that one of these lines will be horizontal, and a vertical angle to a point is understood to be the angle in a vertical plane between a line to that point and the horizontal plane.

A *zenith angle* is an angle between two lines in a vertical plane, where it is understood that one of the lines is directed toward the zenith. A *nadir angle* is an angle between two lines in a vertical plane, where it is understood that one of the lines is directed toward the nadir.

In plane surveying, distances measured along a level line are termed *horizontal distances*. The distance between two points commonly is understood to be the horizontal distance from the plumb line through one point to the plumb line through the other. Measured distances may be either horizontal or inclined, but in practically all cases the inclined distances are reduced to equivalent horizontal lengths.

The *elevation* of a point is its vertical distance above (or below) some arbitrarily assumed level surface or datum.

An elevation *contour* is an imaginary line of constant elevation on the ground surface. The corresponding line on the map is called an elevation *contour line.*

The vertical distance between two points is termed the *difference in elevation.* It is the distance between an imaginary level surface containing the high point and a similar surface containing the low point. The operation of measuring difference in elevation is called *leveling.*

The *grade,* or *gradient,* of a line is its slope or rate of ascent or descent.

1.8
UNITS OF MEASUREMENT

The operations of surveying entail both angular and linear measurements.

The sexagesimal units of angular measurement are the *degree, minute,* and *second.* A plane angle extending completely around a point equals 360 degrees; 1 degree = 60 minutes; 1 minute = 60 seconds. In Europe, the centesimal unit, the gon, is the angular unit; 400 gons[g] equal 360°; 1 gon[g] = 100 centesimal minutes[c] = 0.9°; 1 centesimal minute[c] = 100 centesimal seconds[cc] = 0°00'32.4" (see Section A.1 in Appendix A). Gons usually are expressed in decimals. For example, 100[g]42[c]88[cc] is expressed as 100.4288 gons.

The international unit of linear measure is the meter. Originally, the meter was defined as 1/10,000,00 of the earth's meridional quadrant. The meter, as used in the United States, where it has always been employed for geodetic work, initially was based on an iron meter bar standardized at Paris in 1799. In 1866, by act of Congress, the use of metric weights and measures was legalized in the United States and English equivalents were specified. Among these equivalents, the meter was given as 39.37 in, or the ratio of the foot (12 in) to the meter was specified as 1200/3937. This ratio yields 1 in = 0.025400051 m, 1 ft = 0.304800610 m, and 1 yd (3 ft) = 0.914401829 m. In the United States, the foot as defined by this ratio is called the *U.S. survey foot.*

In 1875, a treaty was signed in Paris by representatives of 17 nations (the United States included) to establish a permanent International Bureau of Weights and Measures. As a direct result of this treaty, the standard for linear measure was established as the International Meter, defined by two marks on a bar composed of 90 percent platinum and 10 percent iridium. The original International Meter Bar was deposited at Sèvres, near Paris. Two copies of this meter bar (called *prototype meters*) were sent to the United States in 1889. These bars were first acquired by the United States Coast and Geodetic Survey (now called the National Ocean Survey, or NOS) and later were transferred to the National Bureau of Standards in Washington, DC. By executive order, in 1893, these new standards were established as *fundamental standards* for the nation. In this executive order, the foot-to-meter ratio was defined to be 1200/3937, the same as specified by congressional action in 1866.

Also established by the Treaty of 1875 was a General Conference on Weights and Measures (Conference General de Poids et Mesures, abbreviated to CGPM) that was to meet every six years. In 1960, the CGPM modernized the metric system and established the *Système International d'Unités* (SI). The standard unit for linear measure, as fixed by this conference, was the meter, defined as a length equal to 1,650,763.73 wavelengths of the orange-red light produced by burning the element krypton at a specified energy level. Such a standard has the advantages of being reproducible and immune to inadvertent or purposeful damage.

The meter is subdivided into the following units:

$$1 \text{ decimeter (dm)} = 0.1 \text{ meter (m)}$$

$$1 \text{ centimeter (cm)} = 0.01 \text{ meter}$$

$$1 \text{ millimeter (mm)} = 0.001 \text{ meter}$$

$$1 \text{ micrometer } (\mu m) = 0.001 \text{ mm} = 10^{-6} \text{ m}$$

$$1 \text{ nanometer (nm)} = 0.001 \ \mu m = 10^{-9} \text{ m}$$

In the United States, by executive order on July 1, 1959, the yard (3 ft) was redefined to be 0.9144 m. This order changed the foot-to-meter ratio as previously established to shorten the foot by 1 part per 500,000. The relationships between the English system of linear units of measure (the foot and the inch) as agreed upon in 1959 are

$$1 \text{ International in} = 1 \text{ British in} = 2.54 \text{ cm}$$

$$1 \text{ International ft} = 30.48 \text{ cm}$$

The English system was used by the United States and the United Kingdom and is now employed officially only in the United States, where conversion to the international system is slowly but surely occurring.

The difference between the U.S. *survey foot* (1200/3937) and the International foot is very small and amounts to 0.2 m/100,000 m or 0.66 ft/60 mi. When six digits or fewer are used in surveying computations, no significant difference will result. However, note that many handheld calculators contain automatic conversion factors based on the International System. The surveyor must be aware of this possibility and recognize that when 6–10 digits are used in computations, the proper conversion factor of 1200/3937 must be calculated or significant differences will occur.

In the United States, the *rod* and *Gunter's chain* were units used in land surveying. Gunter's chain is still employed in the subdivision of U.S. public lands. The Gunter's chain is 66 ft long and divided into 100 *links,* each 7.92 in in length. One mile = 80 chains = 320 *rods, perches,* or *poles* = 5280 ft.

The *vara* is a Spanish unit of measure used in Mexico and several other countries that were under early Spanish influence. In portions of the United States formerly belonging to Spain or Mexico, the surveyor frequently will have occasion to rerun property lines in which lengths are given in terms of the vara. Commonly, 1 vara equals 32.993 in (Mexico), 33 in (California), or $33\frac{1}{3}$ in (Texas); but other somewhat different values have been used for many surveys.

The units of area, as used in the United States, are the *square foot, acre,* and *square mile.* Formerly, the *square rod* and *square Gunter's chain* also were used. One acre = 10 square Gunter's chains = 160 square rods = 43,560 square feet. One square mile equals 640 acres.

In the metric system the unit of area is the square meter. A European metric unit of area is 1 hectare = 10,000 m^2 = 107,639.10 ft^2 = 2.471 acres. One acre equals 0.4047 hectares. The unit of volume is the *cubic meter,* where 1 cubic meter = 35.315 ft^3 = 1.308 yd^3.

The units of volume measurement in the United States have been the *cubic foot* and the *cubic yard.* One cubic foot = 0.0283 m^3 and 1 cubic yard = 0.7646 m^3.

1.9
THE DRAWINGS OF SURVEYING

The drawings of surveying consist of maps, profiles, cross sections, and to a certain extent graphical calculations. In many of these products, few dimensions are shown, and the person who uses the drawings must rely on distances measured with a scale and angles measured with a protractor. Therefore, the usefulness of these drawings largely is dependent on the accuracy with which the points and lines are projected on the drawing surface, which may be paper, plastic film, or a high-resolution graphics monitor. Quite frequently, the drawings of surveying are so irregular and the data on which the drawings are based are of such a nature that use of conventional drafting tools to construct these drawings is the exception rather than the rule. Although the drawings may be constructed manually by traditional drafting methods, most of the drawings of surveying now are compiled from numerical data stored in the central memory of an electronic computer, on diskette, on magnetic tape, or in some other peripheral storage unit and then automatically plotted by a computer-controlled plotter. In this latter process, the computer, peripheral units, and plotter are called *hardware;* the collection of programs that run the computer and allow operator interface with the system is referred to as *software.* This overall system is called Computer-Aided Design and Drafting (CADD). Details concerning the hardware and software for CADD can be found in Chapter 14, Section 14.14.

An important concept involved in the construction of these drawings is scale. *Scale* is defined as the ratio of a distance on the drawing to the corresponding distance on the object. Scale can be expressed as a ratio; for example, 1:1000, where the units of measure are the same or by stating that 1 cm on the drawing represents 1000 cm on the object. Another way is to use different units; for example, 1 in on the drawing representing 100 ft on the object, which is equivalent to the scale ratio of 1:1200.

1.10
MAP PROJECTIONS

A map shows graphically the location of certain features on or near the surface of the earth. The surface of the earth is curved and the surface of the map is flat, so no map can be made to represent a given territory without some distortion. If the area considered is small, the earth's surface may be regarded as a plane, and a map constructed by orthographic projection, as in mechanical drawing, will represent the relative locations of points without measurable distortion. The maps of plane surveying are constructed in this manner, points being plotted by rectangular coordinates or by horizontal angles and distances.

As the size of the territory increases, this method becomes inadequate, and various forms of projection are employed to minimize the effect of map distortion. Control points are plotted by spherical coordinates or the latitude and longitude of the points. *Latitude* is the angular distance above or below the plane of the equator, and *longitude* is the angular distance in the plane of the equator east or west of the Greenwich meridian (see Figure 6.3, Section 6.3). It also is customary to show meridians and parallels of latitude on the finished map. The maps of states and countries, as well as those of some smaller areas such as the U.S. Geological Survey topographic quadrangle maps, are constructed in this manner. Map projection, or the projection of the curved surface of the earth onto the plane of the map sheet, usually is *conformal,* which means that angles measured on the earth's surface are preserved on the map. There are several different conformal map projection methods, each suited for a particular purpose or area of specific configuration. Each method is based on minimizing one or more of the distortions inherent in map production, which include distortions in length, distortions in azimuth and angle, and distortions in area. The various methods of map projection are discussed in Chapter 11.

State plane coordinate systems have been designed for each state in the United States, whereby, even over large areas, points can be mapped accurately using a simple orthogonal coordinate system without the direct use of spherical coordinates.

On a global basis, the Universal Transverse Mercator projection system (Chapter 11) is defined according to parameters of an international ellipsoid. This system can be used throughout the world for plane coordinate computations.

1.11
MAPS

Maps may be divided into two classes: those that become a part of public records of land division and those that form the basis of studies for private and public works. The best examples of the former are the plans filed as parts of deeds in the county registry of deeds in most states of the United States. Good examples of the latter are the preliminary maps along the proposed route of a highway. It is evident that any dividing line between these two classes is indistinct, because many maps might serve both purposes.

Maps that form the basis for studies may be divided into two types: *line-drawn maps,* drawn automatically by CADD systems or by a draftsperson using data from field surveys and those compiled photogrammetrically using stereo pairs of vertical aerial photographs (Chapter 13), and *orthophoto maps,* also produced by photogrammetric methods from stereo-aerial coverage (Chapter 13).

Maps are further divided into the following two general categories: *planimetric maps,* which graphically represent in plan such natural and artificial features as streams, lakes, boundaries, condition and culture of the land, and public and private works; and *topographic maps,* which include not only some or all of the preceding features but also represent the relief or contour of the ground.

Maps of large areas, such as a state or country, which show the locations of cities, towns, streams, lakes, and the boundary lines of principal civil divisions, are called *geographic maps.* Maps of this character, which show the general location of some kind of the works of human beings, are designated by the name of the works represented. Thus we have a *railroad map of the United States* or an *irrigation map of California.* Maps of this type which emphasize a single topic and where the entire map is devoted to showing the geographic distribution or concentration of a specific subject, are called *thematic maps.*

Topographic maps indicate the relief of the ground in such a way that elevations may be determined by inspection. Relief is usually shown by irregular lines, called contour lines, drawn through points of equal elevation (Chapter 14). General topographic maps represent topographic and geographic features, public and private works, and usually are drawn to a small scale. The quadrangle maps of the U.S. Geological Survey are good examples.

Hydrographic maps show shorelines, locations and depths of soundings or lines of equal depth, bottom conditions, and sufficient planimetric or topographic features of lands adjacent to the shores to interrelate the positions of the surface with the underwater features. Charts of the U.S. National Ocean Survey are good examples of hydrographic maps. Such maps form an important part of the basic information required for the production of environmental impact statements.

Maps are constructed using data acquired in the field by field survey methods, such as radial surveys with a total station system or kinematic GPS surveys (Chapter 12). Information from these surveys, stored in electronic field data collectors supplemented by notes in conventional field notebooks, is electronically transferred or keyed into an electronic computer in the office for processing and plotting using CADD. Maps more frequently are compiled photogrammetrically using data from aerial photography or other remote sensors carried by either aircraft or satellite. Note that, for a map compiled by photogrammetric methods, a sufficient number of ground control points of known positions must be identifiable in the photographic or sensed record to allow scaling the map and selecting the proper datum.

For small jobs (e.g., lot surveys, land developments, and construction jobs), traditional methods and manual drawing still may be employed. From an educational standpoint, one cannot underestimate the value of the experience of manually plotting a map as it leads to a much-enhanced ability to use and evaluate properly the many features found in maps.

An increasing number of maps also is produced from map data obtained from a wide variety of sources. Data from old maps, from aerial photography, and data acquired by field methods can be digitized and stored in the memory of an electronic computer, on diskette, or in some other storage device. Information stored electronically in this way frequently is referred to as a *database*. The database so formed is called a *digital map*. When desired, these data can be retrieved and plotted automatically using a CADD system (Chapter 14). This database, with its digital maps and CADD plotting capability, then becomes an integral part of a larger electronic controlling system called a *Geographic or Land Information System*. Use of a GIS or LIS permits the operator to retrieve and overlay maps with an extensive variety of other types of data such as utility maps, land use portrayals, census information, or vehicle traffic distributions and provides an extremely powerful tool for the planning and design involved in a wide range of projects.

1.12
PRECISION OF MEASUREMENTS

In dealing with abstract quantities, we have become accustomed to thinking in terms of exact values. The student of surveying should appreciate that the physical measurements acquired in the process of surveying are correct only within certain limits because of errors that cannot be eliminated. The degree of precision of a given measurement depends on the methods and instruments employed and on other conditions surrounding the survey. It is desirable that all measurements be made with high precision, but unfortunately a given increase in precision usually is accompanied by more than a directly proportionate increase in the time and effort of the surveyor. It therefore becomes the duty of the surveyor to maintain as high a degree of precision as can be justified by the purpose of the survey but no higher. It is important, then, that the surveyor have a thorough knowledge of (1) the sources and types of errors, (2) the effect of errors on field measurements, (3) the instruments and methods to be employed to keep the magnitude of the errors within allowable limits, and (4) the intended use of the survey data.

A complete discussion of error analysis as related to surveying procedure is presented in Chapter 2.

1.13
PRINCIPLES INVOLVED IN SURVEYING

Plane surveying requires a knowledge of geometry and trigonometry, to a lesser degree physics and astronomy, and the behavior of random variables and the adjustment of survey data. Some knowledge of mathematical statistics is invaluable for understanding error propagation, which, in turn, is needed for the design of survey procedures and selection of equipment. In this respect, a knowledge of differential calculus also is necessary. Applicable portions of physics, astronomy, and statistics are developed in succeeding chapters when the need arises.

Data processing with high-speed electronic computers, minicomputers, desktop electronic calculators, and laptop computers is so much a part of the recording, reduction, storage, and retrieval of survey data that familiarity with these systems is absolutely essential to the surveyor. No attempt is made to cover these subjects in this book because such tools are being changed continuously and newer models introduced. It is assumed that the students will have acquired these skills in other courses and from other textbooks.

1.14
THE PRACTICE OF SURVEYING

The practice of surveying is complex. No amount of theory will make a good surveyor unless the requisite skill in the art and science of measuring and in field and office procedures is obtained. The importance of the practical phases of the subject cannot be overemphasized.

Surveying frequently is one of the first professional subjects studied by civil engineering students. Even though these students may not expect to major in surveying, they should understand that the education received in the art and science of measuring and computing and in the practice of mapping will contribute directly to success in other subjects, regardless of the branch of engineering ultimately chosen. The same principles are equally important in the geosciences.

REFERENCES

American Congress on Surveying and Mapping (ACSM). *Metric Practice Guide for Surveying and Mapping.* Bethesda, MD: Metric System Committee, ACSM, 1990.

Bomford, G. *Geodesy.* 4th ed. London: Oxford Clarendon Press, 1980.

Bossler, J. D. "New Adjustment of the North American Datum." *ASCE Journal of Surveying and Mapping* 108, SU2 (1982), p. 47.

Gilson, H. L. "Cadastral Survey 2000." *Surveying and Land Information Systems* 51, no. 4 (December 1991), pp. 269–71.

Hamilton, A. C. "Surveying: 1976–2176" *Surveying and Mapping* 32, no. 2 (June 1977), pp. 119–26.

Korte, George B. "How a GIS Relates to CADD, CAM, and AM/FM." *POB, Point of Beginning* 16, no. 5 (June–July 1991), p. 56.

Lane, Anna. "Metric Mandate." *Professional Surveyor* 12, no. 2 (March–April 1992), pp. 13–14.

Moffitt, F. H., and H. B. Bouchard. *Surveying.* 9th ed. New York: HarperCollins Publishers, 1992.

National Bureau of Standards. *The International System of Units (SI).* Washington DC: U.S. Government Printing Office, 1974.

North American Vertical Datum (NAVD 88). "Final Report of the Ad Hoc Committee on the Proposed NAVD 88." *Surveying and Land Information Systems* 50, no. 3 (September 1990) pp. 225–28.

Petrie, G., and T. J. M. Kennie. *Engineering Surveying Technology.* New York: Halsted Press, a division of John Wiley & Sons, 1990.

Strasser, Georg. "The Toise, the Yard and the Metre—The Struggle for a Universal Unit of Length." *Surveying and Mapping* 35, no. 1 (March 1975), pp. 25–46.

Toscano, P. "The Gunter's Chain" *Surveying and Land Information Systems* 51, no. 3 (September 1991), pp. 155–61.

Vaníček, P. "Vertical Datum and NAVD 88." *Surveying and Land Information Systems* 51, no. 2 (1991), pp. 83–86.

Wade, Elizabeth B. "Impact of the North American Datum of 1983." *ASCE Journal of Surveying Engineering* 112, no. 1 (June 1986), p. 49.

Survey Measurements and Adjustments

PART A
ELEMENTARY TOPICS

The sections in this part of the chapter are intended for the students in a first course in surveying. They cover the general concepts of measurements and their errors and the need for adjustment due to taking more measurements than the minimum necessary. Simple methods of adjustment are also introduced in this part. In Part B of the chapter, we will introduce more advanced topics, which are suitable for second and third courses in modern surveying. It is important to note that the amount of material selected for a first course depends entirely on the instructors, who may certainly choose somewhat more or less than indicated here based on their preference. This is only the authors' recommendation. Of course, both Parts A and B combined provide sufficient breadth and depth to retain the reference value of this work.

2.1
INTRODUCTION

Measurements are essential to the functions of the surveyor. The surveyor's task is to design the survey; plan its field operations; designate the amount, type, and acquisition techniques of the measurements; and then adjust and analyze these measurements to arrive at the required survey results. It is important, then, that the individual studying surveying under-stand the basic idea of a measurement or an observation (here, the two terms are used interchangeably).

Except for counting, measuring entails a physical operation that usually consists of several more elementary ones, such as preparation (instrument setup, calibration, or both), pointing, matching, and comparing. Yet, the result of these physical operations is assigned a numerical value and called the *measurement*. Therefore, it is important to note that a measurement really is an indirect thing, even though in some simple instances it may appear

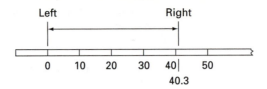

FIGURE 2.1
Length of line using a tape.

to be direct. Consider, for example, the simple task of determining the length of a line using a measuring tape. This operation involves several steps: setting up the tape and stretching it, aligning the zero mark to the left end of the line, and observing the reading on the tape at the right end. The value of the distance as obtained by subtracting 0 from the second reading (40.3 in Figure 2.1) is what we call the *measurement,* although actually two alignments have been made. To make this point clear, visualize the tape simply aligned next to the line and the tape read opposite to its end. In this case, the reading on the left end may be 11.9 and the one on the right end 52.2, with the net length measurement of 40.3.

A more common situation regarding observations involves the determination of an angle. This raises the question of distinguishing between directions and angles. Directions are more fundamental than angles, since an angle can be derived from two directions. Where both directions and angles may be used, care must be exercised in treating each group, because an angle is the difference between two directions. As in measuring a line, an angle is obtained directly if the first pointing (or direction) is taken specifically at 0; then the reading at the second direction will be the value of the angle and may be considered the observation. However, if the surveying process is formulated to be fundamentally in terms of directions, the values read for these directions then should be the observations. In this case, any angle determination will be a simple linear function of the observed directions, and its properties can be evaluated from the properties of the directions and the known function, as will be shown later.

2.2
OBSERVATIONS AND ERRORS

Let us consider the measurement of the length of a tract of land. Begin by assuming that a standardized tape (Section 4.15) is available (which may not be true!). To make a length measurement, we have to align the zero mark and note that reading at the other end of the line. Being human beings working with a manufactured tool (the tape), we have no assurance that our determination of length is the best value we can obtain. We at least are *uncertain* about whether this *one measurement* in fact will be the best we can do. Perhaps, if we repeat the measurement twice, we would feel more confident. And if we do, we probably will end up with two different values for the length. We have no reason to accept one measurement over the other; both appear equally reasonable.

Return to the assumption that the tape is standardized. What if the tape is too short by a certain amount and we do not know it? Would not all our measurements with that tape be too large by factors of the amount by which the tape is short? When the measured length is used directly in computing an area, it obviously will be incorrect.

It is clear, then, that variability in repeated measurements (under similar conditions) is an inherent quality of physical processes and must be accepted as a basic property of observations. Therefore, observations or measurements are numerical values for *random variables,* which are subject to random fluctuations (see Appendix D). Once this is recognized, we may proceed to treat the observations employing established statistical techniques to derive estimates or make appropriate inferences. The term *error* in general, can be considered to refer to the difference between a given measurement and the "true" or "exact" value of the measured quantity. It will always exist, because the repeated measurements will vary naturally, whereas the "true" value remains a constant.

Classically, errors are considered to be of three types: blunders or mistakes, systematic errors, and random errors. Each type will be discussed in turn.

2.3
BLUNDERS OR MISTAKES

Blunders or mistakes actually are not errors because they usually are so gross in magnitude compared to the other two types of errors. One of the most common reasons for mistakes is simple carelessness on the part of the observer, who may take the wrong reading of a scale or a dial or, if the proper value was read, may record the wrong one by transposing numbers, for example. If the operation of collecting the observations is performed through an automatic recording technique, mistakes still may occur through failure of the equipment, although they may be less frequent in this case. Another cause of blunders is failure in the technique, such as reading the fraction on a tape on the wrong side of the zero mark or selecting the wrong whole degree in the measurement of angles that are very close to an integer of degrees (e.g., $71°59'58''$ instead of $72°00'02''$). Finally, a mistake may occur due to misinterpretation, such as sighting to the wrong target.

Once the possibility is recognized that mistakes and blunders can occur, observational procedures and methodology must be designed to allow for their detection and elimination. Here, a variety of ways can be employed, such as taking multiple independent readings (not mere replications) and checking for reasonable consistency; careful checking of both sighting on targets and recording; using simple and quick techniques for verification, applying logic and common sense; checking and verifying the performance of equipment, particularly that with an automatic readout; repeating the experiment with perhaps slightly different techniques or adopting a different datum or indexes; increasing the number of observations; in relatively complex models, applying simplified geometric or algebraic checks to detect mistakes; and simply noting that most mistakes have a large magnitude, which may lead directly to their detection. For example,

1. Measuring an angle or a distance several times and computing the average. Any single measurement deviating from that average by an amount that is larger than a present value can be assumed to contain a blunder.
2. Taking two readings on the horizontal circle of a transit differing by 180°.
3. Using two types of units, such as meters and feet, on a level rod and converting one into the other for comparison.
4. Realizing simple facts, such as that a tape is usually 30 meters long or vertical angles must lie between ±90°.
5. Making a quick check on a spherical triangle using plane trigonometry.
6. Using small portable calculators to check surveying computations to a low number of significant digits.

Modern statistical concepts designate observations as samples from probability distributions, and their variability therefore is governed by the rules of probability (see Appendix D). Consequently, observations containing mistakes or blunders are considered not to belong to the same distributions from which the observational samples are drawn. In other words, an observation with a mistake is not useful unless the mistake is removed; otherwise, that observation must be discarded.

2.4
SYSTEMATIC ERRORS

Systematic errors or effects occur according to a *system* that, if known, always can be expressed by mathematical formulation. They follow a defined pattern, and if the experiment is repeated while maintaining the same conditions, the same pattern will be duplicated and the systematic errors will reoccur. The system causing the pattern can be due to the observer, the instrument used, or the physical environmental conditions of the observational experiment. Any change in one or more of the elements of the system will cause a change in the character of the systematic effect if the observational process is repeated. It must be emphasized, however, that repetition of the measurement under the same conditions will not result in the elimination of the systematic error. One must find the system causing these errors before they can be eliminated from the observations.

Systematic effects take on different forms depending on the value and sign of each of the effects. If the value and sign remain the same throughout the measuring process, so-called *constant* error is present. Making distance measurements with a tape that is either too short or too long by a constant value is an example of a constant error. All lengths measured by that tape will undergo the same systematic effect due to the tape alone. If the sign of the systematic effect changes, perhaps due to personal bias of an observer, the resulting systematic errors are often called *counteracting*.

In surveying, systematic errors occur due to natural causes, instrumental factors, and the observer's human limitations. Temperature, humidity, and barometric pressure are examples of natural sources that will affect angle and distance measurements either by tapes or electronic distance measuring equipment. Instrumental factors are caused by either imperfections in construction or lack of adequate adjustment of equipment before their use in data acquisition. Examples include unequal graduations on linear and circular scales; lack of centering of different components of the instrument; compromise in optical design, thus leaving certain amounts of distortions and aberrations; and physical limitations in machining parts, such as straight ways and pitch of screws.

Although automation has been considered and in some cases introduced (with its own sources of systematic effects) to several tasks, the human observer remains an important element in the activities of surveying. The observer relies mostly on the natural senses of vision and hearing, both of which have limitations and vary due to circumstances and from one individual to another. Although some personal systematic errors are constant and some are counteracting, many others may be erratic. In a way, the set of errors committed by an observer will depend on the precise physical, psychological, and environmental conditions during the particular observation experiment.

Although the sources of systematic effects just discussed pertain to the experiment as such, other systematic errors may be a result of the choice of geometric or mathematical model used to treat the measurements. For example, three options exist in treating a triangle on the earth's surface—plane, spherical, or ellipsoidal—and the choice of one rather than another may result in systematic error.

2.5
SYSTEMATIC ERROR CORRECTIONS AND RESIDUALS

Classically, if an "error" is removed from a measurement, the value of that measurement should improve. This idea is applicable only to *systematic errors*. Thus, if x is an observation and e_s is a systematic error in that observation

$$x_c = x - e_s \qquad (2.1)$$

is the value of the observation "corrected" for the systematic error. Again, in the classical error theory, a *correction* c_s for systematic error was taken equal in magnitude but opposite in sign to the systematic error; therefore,

$$x_c = x + c_s \qquad (2.2)$$

In fact, the concept of error and correction is not limited to the systematic type but is regarded as general in nature, with x_c termed the "true value," which is never known. More realistically, the observations are sample values from a random probability distribution with mean μ and some specified variation (this concept is explained in Section 2.7). Like the true value, the distribution mean μ usually is unknown, and we can make only an improved estimate for the observation. The improved estimate for an observation x is designated $\hat{x}$, and instead of error or correction, we use the term *residual* v, such as

$$\hat{x} = x + v \qquad (2.3)$$

The residual, which will be used throughout this development, although having the same sign sense as the "correction," is not a hypothetical concept, because it has value in the reduction of redundant observational data.

2.6
PREPROCESSING (EXAMPLES OF COMPENSATION FOR SYSTEMATIC ERRORS)

Preprocessing of survey measurements involves both the elimination of blunders (see Section 2.3) and correction for all known systematic errors. Examples of a variety of systematic errors abound in geodetic and surveying operations. Consider the operation of measuring by tape to determine distances between points on the earth's surface. The length of a given tape may be physically different from the values indicated by the numbers written on its graduations, owing to some or all of the following factors:

1. The temperature has changed between that used for tape standardization (calibration) and the temperature actually recorded in the field during the observation (Equation (4.6), Section 4.17).
2. The tension or pull applied to the tape during measurement is different from that used during calibration (Equation (4.7), Section 4.18).
3. The method of tape support is different during measurement from that used during calibration (Equation (4.8), Section 4.19).
4. The endpoints of the distance to be measured are at different elevations and the horizontal distance is desired. In this case a correction is needed because one would be measuring the slope distance instead of the horizontal distance (Equations (4.1) to (4.4), Section 4.13).

Electronic (and electro-optical) distance measuring techniques also are subject to a number of systematic effects whose sources and characteristics must be determined and alleviated. Of these sources the following are mentioned here: the density of air through which the signal travels may change and cause a change in the signal frequency (due to variations in wave propagation velocity); the instrument (and sometimes the reflector or remote unit) may not be properly centered on the ends of the line to be measured; and the path of propagation of the signal may not conform to the straight-line assumption and may be deviated due to environmental and other factors.

Another surveying operation for which systematic effects must be determined is leveling, or the observation of differences in elevation between points on the earth's surface. A level is the instrument used to acquire data by placing it at one point and sighting on a graduated rod on another point. In the instrument is a spirit bubble that must be centered to ensure that the line of sight is horizontal. But if the axis of the telescope is not parallel to the axis of the bubble, the line of sight will be inclined to the horizontal even when the bubble is centered. This error, however, may be compensated for by a relatively simple observational procedure of selecting equal backsight and foresight distances. In this fashion, the systematic effects will be counteracting and cancel each other.

In addition to the level itself, the rod also may cause systematic errors. For example, the rod may not be held vertical or may change in dimensions due to thermal expansion (being in the sun as opposed to being in the shade). The environment also contributes its share of systematic effects. The geodetic concept of the *deflection of the vertical* due to local gravity anomalies is one source. In addition, the earth is not a plane and the line of sight is not a straight line, because of atmospheric refraction.

Another instrument used extensively in surveying and geodesy to measure horizontal and vertical angles is the theodolite or transit. The following are some of the sources of systematic effects that may be associated with such instruments:

1. The horizontal circle may be off center.
2. Graduation on either or both horizontal and vertical circles may not be uniform.
3. The horizontal axis of the telescope (about which it rotates) may not be perpendicular to the vertical axis of the instrument.
4. The longitudinal (or optical) axis of the telescope may not be normal to the horizontal axis of rotation. (In this case, the axis of the telescope would describe a cone instead of a plane as it rotates through a complete revolution.)
5. When the optical axis of the telescope is horizontal, the reading on the vertical circle is different from 0.
6. The optical axis of the telescope and the axis of the leveling bubble may not be parallel.
7. This source does not pertain to the transit but to the target on which sightings are taken. If the natural illuminating conditions are such that part of the target is in shadow, the observer will tend to center the transit's cross hair so as to bisect the illuminated portion of the target. This type of error is often referred to as *phase error*.

These are only some examples of systematic sources in surveying and geodesy. There are numerous others, such as spherical and spheroidal excess, gravimeter and other instrument errors, and timing and other errors in astrogeodetic work. Once the systematic errors are recognized, they often can be compensated for by

1. Actual formulation and computation of corrections, which are then applied to the raw observations.
2. Careful calibration and adjustment of equipment and measuring under the same conditions specified by calibration results.
3. Devising observational procedures that will eliminate the systematic errors that otherwise would occur.
4. Extending the functional model to include the effect of the systematic errors in the adjustment of data.

2.7
RANDOM ERRORS

After mistakes are detected and eliminated and all sources of systematic errors are identified and corrected, the values of the observations will be free of any bias and regarded as sample values for random variables. A *random variable* (see also Section D.1 in Appendix D) may be defined as a variable that takes on *several possible values* and a probability is associated with each value. *Probability* may be defined as the number of chances for success divided by the total number of chances or as the limit value to which the relative frequency of occurrence tends as the number of repetitions is increased indefinitely. As a sample illustration, consider the experiment of throwing a die and noting *the number of dots on the top face.* That number is a random variable, because there are six possible values, from 1 to 6. For example, the probability that three dots will occur tends to be $\frac{1}{6}$ as the number of throws gets to be very large. This probability value of $\frac{1}{6}$ or 0.166 is the limit of the relative frequency, or the ratio between the number of times three dots show up and the total number of throws.

A survey measurement, such as a distance or angle, after mistakes are eliminated and systematic errors corrected, is a random variable such as the number of dots in the die example. If the nominal value of an angle is $41°13'36''$ and the angle is measured 20 times, it is not unusual to get values for each of the measurements that differ slightly from the nominal angle. Each of these values has a probability that it will occur. The closer the value approaches $41°13'36''$, the higher is the probability; and the farther away it is, the lower is the probability.

In the past, the value $41°13'36''$ was designated the "true value," which was never known. Then, when an observation was given that, owing to random variability, was different from the true value, an "error" was defined as

$$\text{error} = \text{measured value} - \text{true value} \qquad (2.4)$$

and a correction, which is the negative of the error, was defined as

$$\text{correction} = \text{true value} - \text{measured value} \qquad (2.5)$$

Whereas for systematic effects, the concept of error and correction is reasonable, in the case of random variation it is not, since there is no reason to say that "errors" have been committed. The so-called random error actually is nothing but a random variable, because it represents the difference between a random variable, the measurement, and a constant, the "true value." The ideal value of the error is 0 (which in statistics is called the *expected value*); that for the observation is the true value (in statistics called the *expectation* or *distribution mean*, μ, as will be explained later). The variation of the random errors around 0 is identical to the variation of the observations around the expectation μ—or the true value. Thus, it is better to talk about the observations themselves and seek better estimates for these observations than discuss errors, because, strictly speaking, the values being analyzed are not errors.

Another classically used term is *discrepancy,* which is the difference between two measurements of the same variable. *Best value, most probable value,* and *corrected value* are all terms that refer to a new estimate of a random variable in the presence of *redundancy,* or having more measurements than the minimum necessary. Such an estimate usually is obtained by some adjustment technique, such as least squares. If the random variable is x, the estimate from the adjustment is called the *least-squares estimate* or sometimes the *adjusted value,* denoted by $\hat{x}$. The *residual* has been defined by Equation (2.3) and will be used in the same sense as it has been in the past. The *deviation* is simply the negative of the residual and may be used on occasion in the course of statistical computation.

2.8
SAMPLING

The statistical term *population* refers to the totality of all possible values of a random variable (see also Section D.3, in Appendix D). Because of the large size of the population, it is either impossible or totally impractical to seek all of its elements to evaluate its characteristics. Instead, only a certain number of observations, called a *sample,* is selected

TABLE 2.1

All 127 micrometer readings taken with the Wild T-2 theodolite

n	Value, of arc	n	Value, of arc	n	Value, of arc
1	27.2	44	30.5	87	36.2
2	29.0	45	35.7	88	33.9
3	37.4	46	30.0	89	28.2
4	37.0	47	36.7	90	33.0
5	32.0	48	35.6	91	31.2
6	32.5	49	33.8	92	32.1
7	36.0	50	30.8	93	34.0
8	35.0	51	37.1	94	33.0
9	33.8	52	33.9	95	33.6
10	34.0	53	32.4	96	33.5
11	35.0	54	33.9	97	33.8
12	30.0	55	33.9	98	36.2
13	34.7	56	34.5	99	37.3
14	35.5	57	36.7	100	33.9
15	36.0	58	32.9	101	35.8
16	37.0	59	32.1	102	36.2
17	35.0	60	36.3	103	38.3
18	36.0	61	30.1	104	39.9
19	34.8	62	32.8	105	37.4
20	33.5	63	31.8	106	37.1
21	35.0	64	33.1	107	37.9
22	35.7	65	30.8	108	33.2
23	38.0	66	35.7	109	35.8
24	31.3	67	34.5	110	39.1
25	33.0	68	30.6	111	31.0
26	31.0	69	35.6	112	34.0
27	31.2	70	37.2	113	34.0
28	31.5	71	31.7	114	31.0
29	32.0	72	32.0	115	33.1
30	36.1	73	35.7	116	32.0
31	37.5	74	35.5	117	34.1
32	36.1	75	32.1	118	36.5
33	33.0	76	35.2	119	39.1
34	36.0	77	34.5	120	36.0
35	28.8	78	34.9	121	41.1
36	28.8	79	33.9	122	35.6
37	37.4	80	32.7	123	38.7
38	27.0	81	35.6	124	40.7
39	31.2	82	34.3	125	35.2
40	31.1	83	34.2	126	40.6
41	31.2	84	35.0	127	33.1
42	34.0	85	34.1		
43	31.9	86	32.9		

First interval, 26.7–27.7″.

Last interval, 40.8–41.7″.

$n = 127$.

Mean Value = 34.2″.

and studied. From the results of the sample study, inferences and statistical statements regarding the population can be made. As an example, consider a horizontal angle. The population in this case may be composed of essentially an infinite number of measurements. In practice, a few measurements of the angle usually are considered sufficient. Specifically, assume a sample of size 10, and use the 10 measurements to estimate parameters that belong to the population. One such parameter is the population mean, μ (the classical "true value"), for which the sample mean, or the mean of the 10 observations, is an estimate. These estimates are usually called *sample statistics*.

From the sample data, some or all of the following can usually be computed:

1. Frequency diagrams (histograms and stereograms).
2. Sample statistics for location (mean, median, mode, midrange).
3. Sample statistics for dispersion (variances and covariances).

Histograms and Stereograms

When a sample of repeated measurements for the same random variable is given, a *histogram* may be constructed to represent the probability density function. The sample values are divided into equal intervals of appropriate size, and the numbers of observations in each interval are determined. For example, Table 2.1 contains 127 micrometer readings observed using a Wild T-2 theodolite. If an interval range of 1" is selected, the number of micrometer observations falling into each interval can be determined. Table 2.2 shows the interval ranges, number of observations in each interval, and relative frequencies for each interval. The first interval is from 26.7 to 27.7" inclusive; the second, from 27.8 to 28.7" inclusive; and so on. The number of measurements falling in each interval is determined and divided by 127 (the total number of observations) to give the corresponding relative frequencies that represent probabilities. Rectangles are constructed over the intervals, each having an area equal to the relative frequency, resulting in the histogram depicted in Figure 2.2. For this example, the number of observations in the interval 33.8 to 34.7" inclusive is 22, yielding a relative frequency of 22/127 = 0.17. Because the interval is 1", the heights of rectangles in the histogram for respective intervals equal the relative frequencies. Note that the sum of the relative frequencies equal 1.

A histogram is constructed for a single random variable, but for two random variables, the frequency diagram is called a *stereogram*. The base would be a plane with square units when equal intervals are used for both variables, and square columns erected on them whose volumes correspond to the relative frequencies. Beyond two random variables, frequency diagrams become impractical.

In practice, histograms frequently are not used because their construction requires large sample sizes, which often are expensive due to the large amount of data necessary. Furthermore, the type of distribution (such as the normal distribution, see Section D.2 in Appendix D) often is known or assumed and estimates for its parameters (for example, the mean and variance) are sought using the sample data. In the following two sections the most commonly used sample statistics are discussed.

TABLE 2.2

Interval range, " of arc	26.7– 27.7	27.8– 28.7	28.8– 29.7	29.8– 30.7	30.8– 31.7	31.8– 32.7	32.8– 33.7	33.8– 34.7	34.8– 35.7	35.8– 36.7	36.8– 37.7	37.8– 38.7	38.8– 39.7	39.8– 40.7	40.8– 41.7	Σ
Number of observations in each interval	2	1	3	5	13	12	14	22	19	16	10	4	2	3	1	127
Relative frequency	0.02	0.01	0.02	0.04	0.10	0.09	0.11	0.17	0.15	0.13	0.08	0.03	0.02	0.02	0.01	1.0

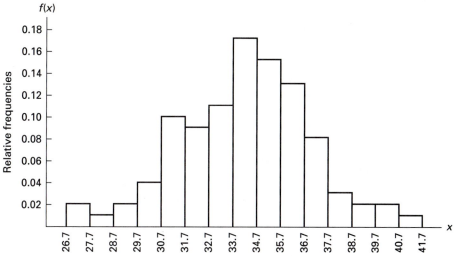

(a) Histogram of micrometer observations.

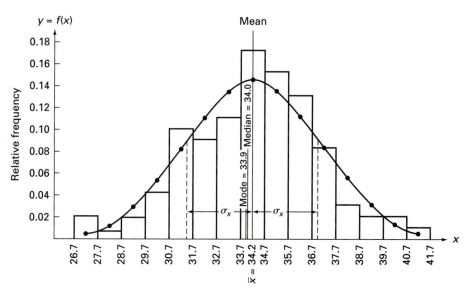

(b) Histogram and density distribution curve for micrometer readings.

FIGURE 2.2
Histograms and density distribution curve for micrometer observations.

2.9
SAMPLE STATISTICS FOR LOCATION

The Sample Mean

The first and most commonly used measure of location is the *mean* of a sample, which is defined as

$$\bar{x} = \frac{1}{n} \sum_{i=1}^{n} x_i \qquad (2.6)$$

where $\bar{x}$ is the mean, x_i are the observations, and n is the sample size, or total number of observations in the sample. It can be shown that the arithmetic mean of a set of independent observations is an unbiased estimate of the mean of the population, describing the random variable for which observations are made. The sample mean of the 127 micrometer readings in Table 2.1 is $\bar{x} = 4339.9/127 = 34.2''$.

The Sample Median

Another measure of location is the median, x_m, which is obtained by arranging the values in the sample in their order of magnitude. The median is the value in the middle if the number of observations n is odd and the mean of the middle two values if n is even. Thus, the number of observations larger than the median equals the number smaller than the median.

The sample median from Table 2.1 is 34.0''.

The Sample Mode

The sample mode is the value that occurs most often in the sample. Therefore, if a histogram is developed from the sample data, it is the value at which the highest rectangle is constructed.

The sample mode from Table 2.1 is 33.9''.

Midrange

If the observation of smallest magnitude is subtracted from that of the largest magnitude, a value called the *range* is obtained. The value of the observation that is midway along the range, called the *midrange,* may be used, although infrequently, as a measure of position for a given sample. It is simply the arithmetic mean of the largest and smallest observations.

The midrange of the 127 micrometer readings is $27.0 + (41.1 - 27.0)/2 = 27.0 + 14.1/2 = 34.05''$.

2.10
SAMPLE STATISTICS FOR DISPERSION

The Range

The simplest of dispersion (scatter or spread) measures of the given observations is the range, as defined in the preceding section; however, it is not as useful a measure as others.

The Mean (or Average) Deviation

The mean deviation is a more useful measure of dispersion and has been conventionally called the *average error*. It is the arithmetic mean of the absolute values of the deviations from any measure of position (usually the mean). Thus, the mean deviation from the mean for a sample of n observations would be given by

$$\text{mean deviation} = \frac{1}{n} \sum_{i=1}^{n} |(x_1 - \bar{x})| \tag{2.7}$$

Sample Variance and Standard Deviation

The mean deviation, although useful in certain cases, does not reflect the dispersion or scatter of the measured values as effectively as the *standard deviation,* which is defined as the positive square root of the *variance* (see also Section D.2, in Appendix D). The variance of a sample is

$$\hat{\sigma}_x^2 = \frac{1}{n - 1} \sum_{i=1}^{n} (x_i - \bar{x})^2 \tag{2.8}$$

where $\bar{x}$ is the sample mean and n the sample size. The sample variance $\hat{\sigma}_x^2$, is an unbiased estimate of the population variance σ^2. It is often called the *estimate of the variance* of any one observation. On the other hand, the estimate of the variance of the mean $\bar{x}$ is given by

$$\hat{\sigma}_{\bar{x}}^2 = \frac{\hat{\sigma}_x^2}{n} \tag{2.9}$$

for n independent observations. Equation (2.9) can be derived by error propagation, as developed in Section 2.15.

EXAMPLE 2.1. Twenty measurements of an angle are given in Table 2.3. Compute the mean of the angle $\bar{x}$, the standard deviation σ_x, and the standard deviation of the mean $\sigma_{\bar{x}}$.

Solution. Applying Equation (2.6), the sample mean is

$$\bar{x} = 31°02' \frac{532.4''}{20} = 31°02'26.6''$$

TABLE 2.3

Observation number	Value of angle	Observation number	Value of angle
1	31°02′29.3″	11	31°02′24.1″
2	31°02′24.0″	12	31°02′26.2″
3	31°02′27.9″	13	31°02′30.1″
4	31°02′26.8″	14	31°02′29.7″
5	31°02′26.1″	15	31°02′24.1″
6	31°02′25.9″	16	31°02′26.2″
7	31°02′26.1″	17	31°02′27.1″
8	31°02′27.8″	18	31°02′24.9″
9	31°02′27.2″	19	31°02′25.7″
10	31°02′28.0″	20	31°02′25.2″

The squared deviations, v_i^2, from $\bar{x}$ are calculated and the variance $\hat{\sigma}_x^2$ is equal to the sum of v_i^2 divided by $(n - 1)$ (Equation (2.8)). Thus,

$$\hat{\sigma}_x^2 = \frac{61.12}{19} = 3.22 \qquad \text{or} \qquad \hat{\sigma}_x = 1.8''$$

Also, the standard deviation of the mean may be obtained from Equation (2.9) as

$$\hat{\sigma}_{\bar{x}} = \frac{\hat{\sigma}_x}{\sqrt{n}} = \frac{1.8}{\sqrt{20}} = 0.4''$$

EXAMPLE 2.2. Refer to the sample of 127 micrometer observations illustrated in Table 2.1. Compute the mean of observations, $\bar{x}$; the estimated standard deviation, $\hat{\sigma}_x$; the estimated standard deviation of the mean, $\hat{\sigma}_{\bar{x}}$; and the ordinates for the normal density distribution. Then, plot the curve superimposed on the histogram.

Solution. By Equation (2.6), the sample mean is

$$\bar{x} = \frac{4339.9}{127} = 34.2''$$

The variance can be obtained using Equation (2.8):

$$\hat{\sigma}_x^2 = \frac{1}{(127 - 1)} \sum_1^{127} (x_i - 34.2)^2 = 7.552$$

$$\hat{\sigma}_x = 2.75''$$

The standard deviation of the mean is calculated using Equation (2.9):

$$\hat{\sigma}_x = \frac{\hat{\sigma}_x}{\sqrt{n}} = \frac{2.75}{\sqrt{127}} = 0.24''$$

Because the mean $\bar{x}$ and standard deviation $\hat{\sigma}_x$ are estimated from the sample, if one assumes a normal distribution, then by Equation (D.8) in Appendix D

$$y = f(x) = \frac{1}{\hat{\sigma}_x\sqrt{2\pi}} \exp\left[-\frac{(x - \bar{x})^2}{2\hat{\sigma}_x^2}\right]$$

or

$$y = A_1 e^{-A_2(x-\bar{x})^2} = \frac{A_1}{e^{A_2(x-\bar{x})^2}}$$

in which

$$A_1 = \frac{1}{\hat{\sigma}_x\sqrt{2\pi}} \qquad \text{and} \qquad A_2 = \frac{1}{2\hat{\sigma}_x^2}$$

In this way the density distribution curve can be plotted to the scale of the histogram of the relative frequencies illustrated in Figure 2.2(a). For this example,

$$A_1 = \frac{1}{(2.75)\sqrt{2\pi}} = 0.145 \qquad \text{and} \qquad A_2 = \frac{1}{(2)(7.552)} = 0.066$$

The remainder of the calculations are best performed in tabular form as follows:

$x,$ $"$ of arc	$(x - \bar{x})^2,$ $("$ of arc$)^2$	$A_2(x - \bar{x})^2$	$e^{A_2(x-\bar{x})^2}$	$y = \dfrac{A_1}{e^{A_2(x-\bar{x})^2}}$
34.2	0	0	1.000	0.145
35.2	1	0.0662	1.068	0.136
36.2	4	0.2648	1.303	0.111
37.2	9	0.5959	1.815	0.080
38.2	16	1.059	2.88	0.050
39.2	25	1.655	5.23	0.028
40.2	36	2.384	10.84	0.013
41.2	49	3.244	25.63	0.006

Figure 2.2(b) shows the ordinates y just listed plotted over the histogram. The density distribution curve is the continuous smooth curve formed by connecting these plotted points. Note that this curve has the characteristic bell shape of the normal density function previously discussed and that the area under the curve is the same as that bounded by the histogram and is equal to 1.

The values of standard deviation obtained in Examples 2.1 and 2.2 actually are a measure of the "quality" or precision of the estimate to which they refer. This concept is explained further in the next section.

2.11
MEASURES OF QUALITY

Several terms are used to describe the quality of measurements and the quantities derived from them. The following are those most commonly used in surveying:

Accuracy. The term *accuracy* refers to the closeness between measurements and their true values. The further a measurement is from its true value, the less accurate it is. (The mathematical terms *expectation* or *expected value* are used more often at present than *true value;* these will be introduced in Part B of this chapter, in Section 2.15.)

Precision. As opposed to accuracy, the term *precision* pertains to the closeness to one another of a set of repeated observations of a random variable. If such observations are closely clustered together, then they are said to have been obtained with high precision. It should be apparent, then, that observations may be precise but not accurate, if they are closely grouped together but about a value that is different from the true value by a significant amount. Also, observations may be accurate but not precise if they are well distributed about the true value but dispersed significantly from one another. Finally, observations will be both precise and accurate if they are closely grouped around the expected value (or the distribution mean).

An example often used to demonstrate the difference between the two concepts of accuracy and precision is the grouping of rifle shots on a target. Figure 2.3 shows three

(a) (b) (c)

FIGURE 2.3
Rifle shot groupings.

different groupings that are possible to obtain. From the preceding discussion, group (a) is both accurate and precise, group (b) is precise but not accurate, and group (c) is accurate but not precise. A hard notion to accept is that case (c) in fact *is accurate,* even though the scatter between the different shots is rather large. A justification that may help is that we can visualize that the center of mass (which is equivalent to the expected or true value of the different shots) turns out to be very close to the target center (which is the true value).

Precision is measured by any of the statistics for dispersion given in Section 2.10. The most commonly used measure is the variance or its positive square root, the standard deviation.

Weight. It is easy to see that the *higher* is the precision, the *smaller* is the variance. To avoid the apparent reversal in meaning, the term *weight* is used to express a quantity that is proportional to the reciprocal of the variance (for uncorrelated or independent random variables). It is denoted by w.

Relative precision. The term *relative precision* refers to the ratio of the measure of precision, usually the standard deviation, to the quantity measured or estimated. For example, if a distance s is measured with a standard deviation σ_s, the relative precision is σ_s/s.

Ratio of misclosure. In traverse computations (Chapter 8, Section 8.14), *the ratio of misclosure* (RoM) is given by

$$\text{RoM} = d_c/S \qquad (2.10)$$

where d_c is the misclosure in m or ft and S is the length of traverse in m or ft. Although the ratio of misclosure has been used by some practicing surveyors to classify the quality of traverses performed by traditional methods (Chapter 8, Section 8.21), it has no statistical basis and its use therefore is not recommended.

Relative line accuracy. When the propagated standard deviation, σ_{ij}, is given for a line of length s_{ij}, between points i and j, then the *relative line accuracy* (RLA) is given by

$$\text{RLA} = (\sigma_{ij})/(s_{ij}) \qquad (2.11)$$

Note that the RLA is just a special example of the general concept of relative precision.

Mean square error. When the so-called true values are available to compare with calculated values, the *mean square error* (MSE) is given by

$$\text{MSE} = [\Sigma \, (x - \tau)^2]/n \qquad (2.12)$$

in which x is the measured value, τ is the true value, and n is the number of measurements.

Root mean square positional error. Horizontal positions are frequently specified by X and Y rectangular coordinates. The *root mean square positional error* (RMSPE) is

$$\text{RMSPE} = (\text{MSE}_X + \text{MSE}_Y)^{1/2} \qquad (2.13)$$

in which $\text{MSE}_X = 1/n \, \Sigma \, (X_{\text{meas}} - X_{\text{true}})^2$ and $\text{MSE}_Y = 1/n \, \Sigma \, (Y_{\text{meas}} - Y_{\text{true}})^2$ each is computed with Equation (2.12). The RMSPE is often used in determining map accuracy (Chapter 14, Section 14.26).

2.12
SIGNIFICANT FIGURES

The expression *significant figures* is used to designate those digits in a number that have meaning. A significant figure can be any one of the digits 1, 2, 3, . . . , 9; and 0 is a significant figure except when used to fix a decimal point. Thus, the number 0.00456 has three significant figures and the number 45.601 has five significant figures.

In making computations, you will have occasion to use numbers that are mathematically exact and numbers determined from observations (directly or indirectly), which inevitably contain observational errors. Mathematically exact numbers are absolute and have as many significant figures as are required. For example, if you multiply a given quantity by 2, the number 2 is absolute and exact.

An observed value is *never* exact. A distance measured with a steel tape is an example of a directly observed quantity. If the distance is measured roughly, it may be recorded as 52 ft or as 16 m. If greater precision is required, the distance is observed as 52.3 ft (15.94 m), or if still more refinement is desired, 52.33 ft (15.950 m) may be the value recorded. Note that none of these three numbers expresses the correct distance exactly but they contain two, three, four, and five significant figures, respectively. Thus, the number of significant figures in a directly observed quantity is related to the precision of refinement employed in the observation.

An indirectly measured quantity is obtained when the observed value is determined from several related, dependent observations. For example, a total distance determined by a summation of a number of directly measured quantities is an indirectly observed distance. The number of significant figures in a directly observed quantity is evident. If you measure the width of a room using a 50-ft steel tape (graduated to hundredths of a foot), your answer can contain up to four significant figures. Assume that the width is recorded as 32.46 ft. This is an example of a directly observed measurement in which the number of significant figures is related to the precision of the equipment and the method of observation. Note that the *precision* of the method of observation is a function of the *skill* and *experience* of the observer as well as the *least count* of the instrument being used.

When numbers are determined indirectly from observed quantities, the number of significant figures is less easily determined. Suppose that the length of the same room is slightly over 50 ft and that two directly observed quantities 25.12 and 27.56 ft, yield a sum of 52.68 ft. This distance also could have been found by using the full 50-ft length of tape and a short increment, resulting in measurements of 50.00 ft and 2.68 ft, to yield a total of 52.68 ft. The total distance is expressed correctly to four significant figures, even though one distance contained only three significant digits. This example illustrates that the number of significant figures in individual directly observed quantities does not control the precision of the quantity calculated by the addition of these quantities.

Consistency in computations requires that one distinguish between exact and observed quantities and that the rules of significant figures be followed. First consider the act of rounding numbers.

Numerical Rounding Off

Assume that 27 and 13.1 are exact numbers. The quotient of $27/13.1 = 2.061068702 \ldots$ never terminates. To use this quotient, superfluous digits are removed from the right and the quotient is retained as 2.06 or 2.0611. This process is called *rounding off*.

To round a number, retain a certain number of digits counted from the left and drop the others. If it is desired to round π to three, four, and five significant figures, the results are 3.14, 3.142, and 3.1416, respectively. Rounding is performed to cause the least possible error and should be done according to the following rule:

To round a number to n significant figures, discard all digits to the right of the nth place. If the discarded digit in the $(n + 1)$th place is less than one-half a unit in the nth place, leave the nth digit unchanged; if the discarded digit is greater than one-half a unit in the nth place, add 1 to the nth digit. If the discarded digit is exactly one-half a unit in the nth place, leave the nth digit unaltered if it is even and increase the nth digit by 1 if it is odd.

A number rounded according to this rule is said to be correct to n significant figures. Examples of numbers rounded to four significant figures are

$$31.68234 \text{ becomes } 31.68$$

$$45.6874 \text{ becomes } 45.69$$

$$2.3453 \text{ becomes } 2.345$$

$$10.475 \text{ becomes } 10.48$$

Rules of Significant Numbers Applied to Arithmetic Operations

Addition

Find the sum of approximate numbers 561.32, 491.6, 86.954, and 3.9462, where each number is correct to its last digit. Round all the numbers to one more decimal than the least significant number and add.

$$
\begin{array}{r}
561.32 \\
491.6 \\
86.95 \\
\underline{3.95} \\
1143.82 \quad \text{or } 1143.8
\end{array}
$$

Retention of the extra significance in the more accurate numbers eliminates the errors inherent in these numbers and reduces the total errors in the sum. Round the final answer to tenths. The average of a set of measurements provides a more reliable result than any of the individual measurements from which the average is computed. Retention of one more figure in the average than in the numbers themselves is justified.

Subtraction

To subtract one approximate number from another, first round each number to the same decimal place before subtracting. Consider the difference between 821.8 and 10.464:

$$821.8 - 10.5 = 811.3$$

Errors resulting from subtraction of approximate numbers are most significant and serious when the numbers are very nearly equal and the leading significant figures are lost. Such errors in subtraction of nearly equal figures can cause computations to be worthless.

Multiplication

Round the more accurate numbers to one more significant figure than the least accurate number. The answer should be given to the same number of significant figures as found in the least accurate factor. For example,

$$(349.1)(863.4) = 301,412.94$$

which should be rounded to

$$301,400 = 3.014 \times 10^5$$

which has four significant figures.

Division

The same rules apply as for multiplication. For example, assume that 5 is an exact number. Then,

$$\frac{56.3}{\sqrt{5}} = \frac{56.3}{2.236}$$

where $\sqrt{5} = 2.236$ is rounded to one more significant figure than the numerator and the quotient, $56.3/2.236 = 25.2$, is rounded to the same number of significant figures as found in the least significant number in the calculation.

Powers (see Scarborough 1966)

If k is the value of the first significant figure of a number having n significant digits, its pth power, is correct to

$$n - 1 \text{ significant figures if } p \leq k$$

$$n - 2 \text{ significant figures if } p \leq 10k$$

For example, raise 0.3862 to the fourth power. Thus, $k = 3$, $n = 4$, and $p = 4$, so that $p < 10k$ or $p < 30$ and $(0.3862)^4 = 0.02225$ should be rounded to $n - 2$ significant figures, or to 0.022.

Roots (see Scarborough 1966)

The rth root of a number having n significant digits and in which k is the first significant digit is correct to

$$n \text{ significant figures if } rk \geq 10$$

$$n - 1 \text{ significant figures if } rk < 10$$

For example, consider the roots of the following numbers, where the given numbers are correct to the last digit:

$(25)^{1/2} = 5$ where $r = 2$, $k = 2$, $n = 2$, and $rk = 4 < 10$

$(615)^{1/4} = 5.00$ where $r = 4$, $k = 6$, $n = 3$, and $rk = 24 > 10$

$(32,768)^{1/5} = 8.0000$ where $r = 5$, $k = 3$, $n = 5$, and $rk = 15 > 10$

2.13
THE CONCEPT OF ADJUSTMENT

In surveying, the measurements or observations rarely are used directly as the required information. In general, they are used in subsequent operations to derive, often computationally, other quantities, such as directions, lengths, relative positions, areas, shapes, and volumes. The relationships applied in the computational effort are the mathematical representations of the geometric and/or physical conditions of the problem. Such a representation, in the context of the present activities, is referred to as the *mathematical model*. The moment you have observations that must be reduced in some fashion to yield the required useful information, you use a model in such a reduction process. Suppose we are interested in the elevation of two points, B and C, given an elevation E of a third point A. We usually do not measure the elevations directly. Instead, we measure differences in elevation such as

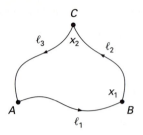

FIGURE 2.4
Single-loop level network.

l_1 from A to B, l_2 from B to C, and l_3 from C to A, as shown in Figure 2.4. The model in this case is relatively simple; it consists of the determination of the elevations X_1, X_2 of points B, C relative to a reference level surface (the vertical datum). It is clear that, to determine the model uniquely, we need to measure *only* two differences in elevation. However, because we expect the measurements to contain random errors, we measure more than the necessary minimum, in this case three, l_1, l_2, l_3. The given elevation E, the three measurements (and their quality as may be expressed by the standard deviations), and the two unknown elevations X_1, X_2, constitute the mathematical model of this problem. To compute X_1, X_2, we need to describe the relationships between the elements of the mathematical model in the form of equations:

$$E + l_1 = X_1 \tag{2.14a}$$

$$X_1 + l_2 = X_2 \tag{2.14b}$$

$$X_2 + l_3 = E \tag{2.14c}$$

We need to calculate only two quantities, X_1, X_2, so not all three equations are required to describe the *functional model* of the problem; two will suffice. In this situation, we have three choices: (2.14a) and (2.14b), or (2.14a) and (2.14c), or (2.14b) and (2.14c). Because of random errors, each of these choices is likely to yield different values for X_1, X_2. The selection of an appropriate choice cannot be done arbitrarily, because the surveyor needs to be confident of the results. Therefore, when *redundant* measurements exist, a means must be found to arrive at a set of values for the *unknown parameters* that are unique and consistent with the mathematical model. The procedure used to obtain unique values of the unknown quantities no matter which sufficient subset of the observations is used to compute them is called *data adjustment*. The term *adjustment* is used to imply that the given values of the observations must be altered, or adjusted, to make them consistent with the model, leading to the uniqueness of the estimated unknowns.

Several techniques can be used to adjust redundant measurements. The most rigorous and commonly used is the method of *least squares,* which will be discussed in detail in Part B of this chapter. Other approximate adjustment techniques have been used by the surveyor. As an example, consider the closure of the level loop in Figure 2.4. If all three measurements are consistent, their algebraic sum should be 0. Usually, this would not be the case and a closure error would result, or

$$l_1 + l_2 + l_3 = c$$

where c is the closure error. If all three observed differences in elevation are of equal quality, or weight, then each observation would have a residual equal to $(-c/3)$. Thus,

$$(l_1 + v_1) + (l_2 + v_2) + (l_3 + v_3) = \left(l_1 - \frac{c}{3}\right) + \left(l_2 - \frac{c}{3}\right) + \left(l_3 - \frac{c}{3}\right)$$

$$= \hat{l}_1 + \hat{l}_2 + \hat{l}_3 = 0$$

in which v_1, v_2, v_3 are the three residuals and $\hat{l}_1$, $\hat{l}_2$, $\hat{l}_3$ are the three *adjusted* observations.

Now, because the $\hat{l}_i$ are consistent with the model, any two would lead to the same values for X_1 and X_2.

2.14
SIMPLE ADJUSTMENT METHODS

In the preceding section we gave a relatively simple example of a single leveling loop. The error of closure was divided equally among the three measured differences in elevation so that the sum of the adjusted (or corrected) observations became 0, thus eliminating the closure error. The assumption was made in that example that all three differences in elevation were measured along courses of approximately the same length, using the same instruments and observers, and so forth. In short, it was assumed that the three measurements are of equal quality, or *weight*. If this were not true and the different weights of the observations could be estimated, then each observation would be assigned a residual that is somewhat different from the others.

To illustrate the use of weights, assume that the three differences in elevations, l_1, l_2, l_3, have the weights $w_1 = 1$, $w_2 = 2$, $w_3 = 2$, respectively. These can be loosely interpreted as that we are twice as confident in the quality of l_2 and l_3 than we are in l_1. It follows that l_2 and l_3 would receive corrections each of which is half that for l_1. Thus, the proportion of the closure error c assigned to each residual is in the ratio $v_1 : v_2 : v_3 = 1/w_1 : 1/w_2 : 1/w_3 = 1 : 0.5 : 05$. If the closure error is, say, 12 cm, then

$$v_1 = \frac{1}{1 + 0.5 + 0.5} = 6 \text{ cm}$$

$$v_2 = \frac{0.5}{1 + 0.5 + 0.5} = 3 \text{ cm}$$

$$v_3 = \frac{0.5}{1 + 0.5 + 0.5} = 3 \text{ cm}$$

The adjusted observation therefore would be $\hat{l}_1 = l_1 - 6$ cm, $\hat{l}_2 = l_2 - 3$ cm, and $\hat{l}_3 = l_3 - 3$ cm.

The procedure of distributing closure error, either by simple proportion or by weighted proportion, is used, at least as an approximate adjustment, in different surveying situations. For example, if the three interior angles of a plane triangle are measured and do not add to 180°, the difference is apportioned to the three angles either equally or inversely proportional to their weights, if such weights are available. Traverse closure errors also are distributed according to simple rules, as will be described in Chapter 8. A single line or loop of levels is adjusted in proportion to the number of instrument setups required or the distance leveled.

PART B
ADVANCED TOPICS

In Part A we introduced the elementary topics to be used in a first course in surveying. In this part, we continue this important subject of survey measurements and their adjustments, by discussing topics on special and general propagation, various least squares adjustment techniques, and pre- and postadjustment analysis.

2.15
PROPAGATION OF RANDOM ERRORS

In surveying, as in other science and engineering disciplines, the quantities we measure frequently are used to compute other quantities that are of interest. The new quantities are expressed as functions of the survey measurements. Because, as we have shown, the measurements contain random errors, the computed quantities will also. Evaluation of the errors in those computed quantities from the known functions and the errors in the measurements is called *propagation of random errors* or simply *error propagation*. The errors in the measurements usually are represented by the standard deviations, which, when propagated through the functions, yield standard deviations of the computed quantities. As an example, in trigonometric leveling, one measures the slope distance s and the inclination angle β and computes the height h from $h = s \sin \beta$ and its standard deviation σ_h given σ_s and σ_β.

The relation used to evaluate σ of the computed quantity from the values of σ of the measurements is formulated as follows. Let y represent a quantity computed from several measurements (random variables) represented by $x_1, x_2, \ldots, x_n$, in general using a nonlinear function, or

$$y = y(x_1, x_2, \ldots, x_n) \tag{2.15}$$

Further, let $\sigma_1, \sigma_2, \ldots, \sigma_n$ represent the standard deviations of the measurements x_1, $x_2, \ldots, x_n$, *which are assumed to be independent*. Then, σ_y is computed by first computing the variance

$$\sigma_y^2 = \left(\frac{\partial y}{\partial x_1}\right)^2 \sigma_1^2 + \left(\frac{\partial y}{\partial x_2}\right)^2 \sigma_2^2 + \cdots + \left(\frac{\partial y}{\partial x_n}\right)^2 \sigma_n^2 \tag{2.16}$$

and then taking its positive square root.

EXAMPLE 2.3. In trigonometric leveling, the slope distance is $s = 50.00$ m with $\sigma_s = 0.05$ m and $\beta = 30°00'$ with $\sigma_\beta = 00°30'$. Compute h and σ_h. (Assume s and β to be uncorrelated.)

Solution

$$h = s \sin \beta = (50.00)(0.5) = 25.00 \text{ m}$$

$$\sigma_h^2 = \left(\frac{\partial h}{\partial s}\right)^2 \sigma_s^2 + \left(\frac{\partial h}{\partial \beta}\right)^2 \sigma_\beta^2$$

$$= (\sin \beta)^2 (0.05)^2 + (s \cos \beta)^2 (0.0087)^2 = 0.1425 \text{ m}^2$$

$$\sigma_h = 0.38 \text{ m}$$

In this example, note that σ_β was converted to radians to balance the dimensions in the relationship.

EXAMPLE 2.4. The area of a rectangular parcel of land is required together with its standard deviation. The length is $a = 100$ m with $\sigma_a = 0.10$ m, and the width is $b = 40$ m with $\sigma_b = 0.08$ m. (Assume the mutual variation between a and b is 0 or they are uncorrelated; see also Section 2.18.)

Solution

$$A = ab = 4000 \text{ m}^2$$

$$\sigma_A^2 = b^2 \sigma_a^2 + a^2 \sigma_b^2$$

$$= (40)^2 (0.1)^2 + (100)^2 (0.08)^2 = 80 \text{ m}^4$$

$$\sigma_A = 8.94 \text{ m}^2$$

A simplification for the nonlinear case, given by Equation (2.15), is to consider y a *linear* function of x_i, or

$$y = a_1 x_1 + a_2 x_2 + \cdots + a_n x_n \tag{2.17}$$

in which case Equation (2.16) becomes

$$\sigma_y^2 = a_1^2 \sigma_1^2 + a_2^2 \sigma_2^2 + \cdots + a_n^2 \sigma_n^2 \tag{2.18}$$

If all a_i in Equation (2.17) are equal to either $+1$ or -1 (addition or subtraction of the random variables), that is,

$$y = x_1 \pm x_2 \pm \cdots \pm x_n \tag{2.19}$$

then the variance of y is simply the *sum* of all the variances of x_i, or

$$\sigma_y^2 = \sigma_1^2 + \sigma_2^2 + \cdots + \sigma_n^2 \tag{2.20}$$

Note that the variances are *added* regardless whether the values of x_i are added *or* subtracted. Finally, if, in addition to being uncorrelated, all values of x_i have the same precision with a variance σ_x^2 for each, then σ_y^2 for the function in Equation (2.19) will be

$$\sigma_y^2 = n \sigma_x^2 \tag{2.21}$$

EXAMPLE 2.5. Three adjacent distances along the same line were measured independently with the following results: $x_1 = 51.00$ m with $\sigma_1 = 0.05$ m; $x_2 = 36.50$ m with $\sigma_2 = 0.04$ m; and $x_3 = 26.75$ m with $\sigma_3 = 0.03$ m. Compute the total distance and its standard deviation.

Solution

$$y = x_1 + x_2 + x_3 = 114.25 \text{ m}$$

and, from Equation (2.20),

$$\sigma_y^2 = \sigma_1^2 + \sigma_2^2 + \sigma_3^2 = (0.05)^2 + (0.04)^2 + (0.03)^2 = 0.005 \text{ m}^2$$

$$\sigma_y = 0.07 \text{ m}$$

EXAMPLE 2.6. If, in Example 2.5, all three distances were measured with a standard deviation $\sigma_x = 0.05$ m, what would σ_y be?

Solution. From Equation (2.21),

$$\sigma_y^2 = 3(0.05)^2 = 0.0075 \text{ m}^2$$

$$\sigma_y = 0.09 \text{ m}$$

EXAMPLE 2.7. A distance is measured independently by two observers to be $x_1 = 110.00$ m and $x_2 = 110.80$ m. If the weights of these two measurements are $w_1 = 2$ and $w_2 = 3$, respectively, and its best estimate is the weighted mean given by $\hat{x} = (w_1 x_1 + w_2 x_2)/(w_1 + w_2)$, compute the standard deviation of $\hat{x}$.

Solution. Taking the weights as the reciprocals of the variances, $w_1 = 1/\sigma_1^2$ and $w_2 = 1/\sigma_2^2$. Next,

$$\sigma_{\hat{x}}^2 = \left(\frac{\partial \hat{x}}{\partial x_1}\right)^2 \sigma_1^2 + \left(\frac{\partial \hat{x}}{\partial x_2}\right)^2 \sigma_2^2$$

$$= \left(\frac{w_1}{w_1 + w_2}\right)^2 \left(\frac{1}{w_1}\right) + \left(\frac{w_2}{w_1 + w_2}\right)^2 \left(\frac{1}{w_2}\right)$$

$$= \frac{w_1 + w_2}{(w_1 + w_2)^2} = \frac{1}{w_1 + w_2}$$

which means that the weight $w_{\hat{x}}$ of the weighted mean of two observations is equal to the sum of their weights, or $w_{\hat{x}} = w_1 + w_2$. Finally,

$$\sigma_{\hat{x}} = \left(\frac{1}{w_1 + w_2}\right)^{1/2} = \left(\frac{1}{2 + 3}\right)^{1/2} = \left(\frac{1}{5}\right)^{1/2} = 0.45$$

2.16
EXPECTATION

A number of properties that relate a random variable and its probability density function are useful in our understanding of its behavior. The first of these expresses our intuitive concept of the mean as being the most probable value of the random variable. It is referred to by any of several terms—*expectation, expected value, mean,* or *average*—and will be denoted $E(x)$.

The *expected value*, $E(x)$, of a random variable x is defined as the weighted sum μ_x of all possible values, where the weights are the corresponding probabilities. In mathematical terms

$$E(x) = \mu_x = \sum_{i=1}^{n} x_i P(x_i) \text{ for a discrete distribution} \qquad (2.22)$$

and

$$E(x) = \mu_x = \int_{-\infty}^{\infty} xf(x)\, dx \text{ for a continuous distribution} \qquad (2.23)$$

where $f(x)$ is the density function. Expectation may also be defined for a function $g(x)$ of a random variable x having a density function $f(x)$ as (assuming continuous function)

$$E[g(x)] = \int_{-\infty}^{\infty} g(x)f(x)\, dx \qquad (2.24)$$

In the discrete case (2.22), if there are n values of equal probability, then $P(x_i) = 1/n$ and $E(x) = \sum_{i=1}^{n} x_i/n = \mu_x$ or the mean, as previously stated.

If the probabilities are not equal, the weighted mean would result. As an example, consider a random variable y equal to the sum of two fair dice yielding the following data:

y_i	$P(y_i)$	$y_i P(y_i)$
2	1/36	2/36
3	2/36	6/36
4	3/36	12/36
5	4/36	20/36
6	5/36	30/36
7	6/36	42/36
8	5/36	40/36
9	4/36	36/36
10	3/36	30/36
11	2/36	22/36
12	1/36	12/36
		$\Sigma = 252/36$

so that

$$E(y) = \sum_{i=2}^{12} [y_i P(y_i)] = 7 = \mu_y$$

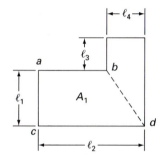

FIGURE 2.5
Expected value of an area.

41

CHAPTER 2:
Survey
Measurements
and Adjustments

Several useful properties of expectation include

$$E(c) = c \qquad (c = \text{a constant}) \qquad (2.25a)$$

$$E(cx) = cE(x) \qquad (2.25b)$$

$$E[E(x)] = E(\mu_x) = \mu_x \qquad (2.25c)$$

$$E(x + y) = E(x) + E(y) \qquad (2.25d)$$

$$E(xy) \neq E(x)\,E(y) \qquad (2.25e)$$

The relation in (2.25e) in general is true unless x and y are independent random variables, in which case $E(xy) = E(x)\,E(y)$.

As an example of the application of the properties of expectation, consider the expected value of the area A_1 of the trapezoid $abcd$ in Figure 2.5 assuming that l_1, l_2, l_3, and l_4 are not correlated.

$$E(A_1) = E\left[l_1\left(\frac{2l_2 - l_4}{2}\right)\right]$$

$$E(A_1) = \frac{1}{2}\{E(l_1)[2E(l_2) - E(l_4)]\}$$

and assuming that each of the values l_1, l_2, l_4 has respective means of μ_1, μ_2, μ_4, then

$$E(A) = \frac{1}{2}(2\mu_1\mu_2 - \mu_1\mu_4)$$

2.17
VARIANCE, COVARIANCE, AND CORRELATION

If the function $g(x)$ in Equation (2.24) is specifically $(x - \mu_x)^2$, then the expectation is called the *variance*, designated by either var(x) or σ_x^2; therefore,

$$\text{var}(x) = \sigma_x^2 = E[(x - \mu_x)^2] = \int_{-\infty}^{\infty} (x - \mu_x)^2 f(x)\,dx \qquad (2.26)$$

The *positive square root* of the variance is the standard deviation, σ_x, which is a measure of the dispersion or spread of the random variable.

Similarly, if in Equation (2.24) the specific function $(x - \mu_x)(y - \mu_y)$ relating to two random variables x and y is used, the expectation is called the *covariance*, referred to by cov (x, y) or σ_{xy}:

$$\text{cov}(x, y) = \sigma_{xy} = E[(x - \mu_x)(y - \mu_y)] \qquad (2.27)$$

The covariance expresses the mutual interrelation between the two random variables. (Unlike the standard deviation, the square root of the covariance, if σ_{xy} is positive, has no meaning and therefore is not used.) Another term, the *correlation coefficient,* ρ_{xy}, is defined as

$$\rho_{xy} = \frac{\sigma_{xy}}{\sigma_x \sigma_y} = E\left[\left(\frac{x - \mu_x}{\sigma_x}\right)\left(\frac{y - \mu_y}{\sigma_y}\right)\right] \tag{2.28}$$

where σ_x and σ_y are the standard deviations of x and y, respectively.

2.18
COVARIANCE, COFACTOR, AND WEIGHT MATRICES

The one-dimensional case contains one random variable, x, with mean or expectation μ_x and a variance σ_x^2. The two-dimensional case has two random variables, x and y, with means μ_x and μ_y and variances σ_x^2 and σ_y^2, respectively, and covariance σ_{xy}. These three parameters can be collected in a *square symmetric* matrix, Σ, of order 2, and called the *variance-covariance matrix* or simply the *covariance matrix.* It is constructed as

$$\Sigma = \begin{bmatrix} \sigma_x^2 & \sigma_{xy} \\ \sigma_{xy} & \sigma_y^2 \end{bmatrix} \tag{2.29}$$

where the variances are along the main diagonal and the covariance is off the diagonal. The concept of the covariance matrix can be extended to the multidimensional case by considering n random variables $x_1, x_2, \ldots, x_n$ and writing

$$\Sigma_{xx} = \begin{bmatrix} \sigma_1^2 & \sigma_{12} & \cdots & \sigma_{1n} \\ \sigma_{12} & \sigma_2^2 & \cdots & \sigma_{2n} \\ \cdots & & & \\ \sigma_{1n} & \sigma_{2n} & \cdots & \sigma_n^2 \end{bmatrix} \tag{2.30}$$

which is an $\mathbf{n} \times \mathbf{n}$ square symmetric matrix.

Often in practice, the variances and covariances are not known in absolute terms but only to a scale factor. The scale factor, given the symbol σ_0^2, is termed the *reference variance,* although other names, such as *variance factor* and *variance associated with weight unity,* also have been used. The square root σ_0, of σ_0^2, is called the *reference standard deviation,* classically known as the *standard error of unit weight.* The relative variances and covariances, called *cofactors,* are given by

$$q_{ii} = \frac{\sigma_i^2}{\sigma_0^2} \qquad \text{and} \qquad q_{ij} = \frac{\sigma_{ij}}{\sigma_0^2} \tag{2.31}$$

Collecting the cofactors in a square symmetric matrix produces the *cofactor matrix,* $\mathbf{Q}$, with the obvious relationship with covariance matrix

$$\mathbf{Q} = \frac{1}{\sigma_0^2}\Sigma \tag{2.32}$$

When $\mathbf{Q}$ is nonsingular, its inverse is called the *weight matrix* and designated $\mathbf{W}$; thus,

$$\mathbf{W} = \mathbf{Q}^{-1} = \sigma_0^2 \Sigma^{-1} \tag{2.33}$$

If σ_0^2 is equal to 1 or, in other words, if the covariance matrix is known, the weight matrix becomes its inverse.

Equation (2.33) should be carefully understood, particularly in view of the *classical* definition of weights as being inversely proportional to variances. This clearly is not true *unless all covariances are equal to 0,* which means that all the random variables are mutually uncorrelated. Only then would Σ (and $\mathbf{Q}$) become a diagonal matrix and the weight of the random variable become equal to σ_0^2 divided by its variance.

2.19
VARIANCE-COVARIANCE PROPAGATION

In Section 2.15, we introduced the concept of propagation of errors, where consideration was given to the case of uncorrelated random errors. In this section, we generalize it to both correlated quantities and multiple functions of such quantities. This general case of propagation can be expressed as follows.

Let y be a set (vector) of m quantities, each of which is a function of another set (vector) x of n random variables. Given the covariance matrix Σ_{xx} (or the cofactor matrix $\mathbf{Q}_{xx}$) for the variables x, the covariance matrix Σ_{yy} (or cofactor matrix $\mathbf{Q}_{yy}$) for the new quantities y may be evaluated from

$$\Sigma_{yy} = \mathbf{J}_{yx} \Sigma_{xx} \mathbf{J}_{yx}^T \tag{2.34}$$

or

$$\mathbf{Q}_{yy} = \mathbf{J}_{yx} \mathbf{Q}_{xx} \mathbf{J}_{yx}^T \tag{2.35}$$

where $\mathbf{J}_{yx}$ is $m \times n$ and called the Jacobian matrix, or the partial derivative of y with respect to x, with the following elements (see Section B.10, Appendix B):

$$\mathbf{J}_{yx} = \begin{bmatrix} \dfrac{\partial y_1}{\partial x_1} & \dfrac{\partial y_1}{\partial x_2} & \cdots & \dfrac{\partial y_1}{\partial x_n} \\[2mm] \dfrac{\partial y_2}{\partial x_1} & \dfrac{\partial y_2}{\partial x_2} & \cdots & \dfrac{\partial y_2}{\partial x_n} \\[2mm] \cdots & & & \\[2mm] \dfrac{\partial y_m}{\partial x_1} & \dfrac{\partial y_m}{\partial x_2} & \cdots & \dfrac{\partial y_m}{\partial x_n} \end{bmatrix} \tag{2.36}$$

Equation (2.34) or (2.35) is quite general inasmuch as multiple functions in terms of several variables are considered and, more important, no restrictions are imposed on the structure of the given covariance matrix Σ_{xx}. Therefore, the given random variables in general could be of unequal precision and correlated, so that Σ_{xx} would no longer be a diagonal matrix. From the general propagation relationships of Equation (2.34), several relationships could be obtained. First, consider the case of a single function y of several (n) variables x_1, $x_2, \ldots, x_n$, that are *uncorrelated* and with variances $\sigma_1^2, \sigma_2^2, \ldots, \sigma_n^2$, respectively. Equation (2.34) would become

$$\sigma_y^2 = \begin{bmatrix} \dfrac{\partial y}{\partial x_1} & \dfrac{\partial y}{\partial x_2} & \cdots & \dfrac{\partial y}{\partial x_n} \end{bmatrix} \begin{bmatrix} \sigma_1^2 & & & 0 \\ & \sigma_2^2 & & \\ & & \ddots & \\ 0 & & & \sigma_n^2 \end{bmatrix} \begin{bmatrix} \partial y/\partial x_1 \\ \partial y/\partial x_2 \\ \cdots \\ \partial y/\partial x_n \end{bmatrix} \tag{2.37}$$

or

$$\sigma_y^2 = \left(\dfrac{\partial y}{\partial x_1}\right)^2 \sigma_1^2 + \left(\dfrac{\partial y}{\partial x_2}\right)^2 \sigma_2^2 + \cdots + \left(\dfrac{\partial y}{\partial x_n}\right)^2 \sigma_n^2 \tag{2.38}$$

Of course, if the variables x_i were correlated, $\mathbf{\Sigma}_{xx}$ in Equation (2.37) would not be a diagonal matrix and Equation (2.38) would include cross-product terms in all combinations. In such a case, it would be unwise to write the expanded form in Equation (2.38) but instead work directly with the matrix form.

Note that Equations (2.34) through (2.38) are given in terms of variances and covariances of the distributions. However, because such parameters are rarely known in practice, the equations apply equally using sample variances and covariances.

2.20
MATHEMATICAL MODEL FOR ADJUSTMENT

In Section 2.13 we introduced the notion that measurements are obtained for some, but not all, of the elements of a *mathematical model* describing the geometric or physical situation at hand. It is important that the reader appreciate the importance of the concept of the mathematical model as well as its adequacy for the surveying problem. As an example, suppose we are interested in the area of a rectangular tract of land (assumed reasonably small to use computations in a plane). To compute that area, we need the length a and the width b. The area will naturally be computed from $A = ab$, which is the "functional" model. Now, it appears that all we have to do is go out with a tape, measure the length and width, each once, and apply the two values into the formula to compute the area. This sounds simple enough until we begin taking a closer look at the problem and analyzing the factors involved in its solution.

In addition to the random and systematic errors in tape measurements as discussed in Section 2.2 (see also Chapter 4), let us also examine the adequacy of the model $A = ab$. This model is formulated on the basic assumption that the parcel of land is exactly rectangular. What if one, or more, of the four angles is checked and differences from the right-angle assumption are discovered? The simple model given no longer could be used; instead, another that more properly reflects the *geometric* conditions should be utilized.

As another example, suppose that we are interested in the *shape* of a plane triangle. All that is required for this operation is to measure two of its angles, and the shape of the triangle will be uniquely determined. However, if we were to decide, for safety's sake, to measure all three angles, any attempt to construct such a triangle will immediately show *inconsistencies* among the three observed angles. In this case the model simply is that the sum of the three angles must equal 180°. If three observations are used in this model, it is highly unlikely that the sum will equal exactly 180°. Therefore, when *redundant* observations, or more observations than are absolutely necessary, are acquired, these observations will rarely fit the *model* exactly. Intuitively, and relying on our previous discussion, this results from something characteristic to the observations and makes them inconsistent in the case of redundancy. Of course, we first need to be sure of the adequacy of the model (it is a plane triangle and not spherical or spheroidal, for example). Then, we need to express the quality of the measurements before we seek to *adjust* the observations to fit the model. These concepts are elaborated on in the following sections.

2.21
FUNCTIONAL AND STOCHASTIC MODELS

We indicated previously that survey measurements are planned with a mathematical model in mind that describes the physical situation or set of events for which the survey is designed. This mathematical model is composed of two parts: a functional model and a stochastic model. The *functional model* is the more obvious part, because it usually describes the geometric or physical characteristics of the survey problem. Thus, the functional model for the example concerning the triangle discussed at the end of the preceding section *involves the determination of the shape* of a plane triangle through the measurement of interior angles. If three angles in a triangle are available, redundant measurements are present with respect to the functional model. This does not say anything about the properties of the measured angles. For example, in one case, each angle may be measured by the same observer, using the same instrument, applying the same measuring technique, and performing the measurements under very similar environmental conditions. In such a case, the three measured angles are said to be equally "reliable." But certainly in other cases the resulting measurements will not be of equal quality. In fact, as most practicing surveyors know from experience, measurements always are subject to unaccountable influences that result in variability when observations are repeated. Such statistical variations in the observations are important and must be taken into consideration when using the survey measurements to derive the required information. The *stochastic model* is the part of the mathematical model that describes the statistical properties of all the elements involved in the functional model. For example, in the case of the plane triangle, to say that each interior angle was measured and to give its value is insufficient. Additional information should also be included as to how well each angle was measured and if there is reason to believe in a statistical "correlation" (or interaction) among the angles, and if so, how much. The outcome of having a unique shape for the triangle from the redundant measurements depends both on knowing that the sum of its internal angles is 180° (the functional model) and knowing the statistical properties of the three observed angles (the stochastic model).

2.22
LEAST-SQUARES ADJUSTMENT

The concept of adjustment was introduced in Section 2.13 and some simple adjustment methods in Section 2.14. A more systematic procedure of adjustment is least squares, which is most commonly used in surveying and geodesy. Most people refer to least squares as an adjustment technique equivalent to *estimation* in statistics. Although *adjustment* is not the most precise term, it is appropriate because adjustment is needed when there are redundant observations (i.e., more observations than are necessary to specify the model). In this case the observations given are not consistent with the model and are replaced by another set of estimates, classically called *adjusted observations* (which also is not a precise term), that satisfy the model. As an example, consider the determination of a distance between two points. The distance may be considered a random variable and, if measured once, would have one estimate, so that no adjustment is needed. On the other hand, if the distance were measured three times, there would likely be three slightly different values, x_1, x_2, and x_3. Since the model concerns a single distance that would be *uniquely* specified by one measurement, it is obvious that there are two *redundant* measurements. It also is clear that the situation requires adjustment to have a unique solution. Otherwise, there are several different possibilities for the required distance. We can take any one of x_1, x_2, x_3 or a combination of x_1 and x_2, or x_2 and x_3, or x_1 and x_3, or x_1, x_2, and x_3. Therefore, having redundant observations makes it possible to have numerous ways of computing the desired values. The multiplicity of possibilities and arbitrariness of choice in obtaining the required information obviously is undesirable. Instead, a process or technique must be found so that one always would get one unique answer, which is derived from the data and is the "best" that can be obtained. This is why the relative confidence in, or merit of, the different observations should be taken into account when computing the best estimate, which is defined as the estimate that deviates least from all the observations while considering their relative reliability. This is basically the role of least-squares adjustment.

As another example, consider the case of a plane triangle in which the three angles must add up to 180°. That the three angles must add to 180° represents a functional relationship that reflects the geometrical system involved in the problem. If the shape and not the size of the triangle is of interest, it is unnecessary to observe the magnitudes of three angles, because two angles will be sufficient to determine the third from the functional relationship just mentioned. However, in practice, the three angles α, β, and γ are measured whenever possible, and their sum likely will be different from 180°. Suppose that the sum of the angles exceeds 180° by 3″ of arc. Any two of the three measured angles would give the shape of the triangle, but all three possibilities, in general, will be different. Therefore, to satisfy the condition that the sum of the angles must be 180°, the values of the observed angles must be altered. Here, there are numerous possibilities: 3″ may be subtracted from any one of the three angles, perhaps 2″ could be subtracted from the largest angle, 1″ from the second largest, and nothing from the third angle; or it may appear to be more satisfactory if 1″ were subtracted from each angle, assuming that they are equally reliable, taken by the same instrument and observer under quite similar conditions. An alternative criterion to those just given may be to apply alterations that are proportionate to the relative magnitudes of the angles, or the magnitudes of their complements or supplements, or even inversely proportionate to such magnitudes. It is clear, then, that although adjustment is necessary, the large number of possibilities given are quite arbitrary and a *criterion* is required in addition to the satisfaction of the functional model of summing the angle to 180°. Such is the *least-squares* criterion, which is introduced in the next section.

2.23
THE LEAST-SQUARES CRITERION

Let l designate the vector of given observations and v the vector of *residuals* (or alterations), which when added to l yields a set of new estimates, $\hat{l}$, that is consistent with the model:

$$\hat{l} = l + v \tag{2.39}$$

The statistical or stochastic properties of the observations are expressed by either the covariance or cofactor matrix Σ or Q, respectively, or by the weight matrix W. (Note that $W = Q^{-1}$ and $W = \Sigma^{-1}$ if the reference variance σ_0^2 is equal to unity.) With these variables, the general form of the least-squares criterion is given by

$$\phi = v^T W v \rightarrow \text{minimum} \tag{2.40}$$

Note that ϕ is a scalar, for which a minimum is obtained by equating to 0 its partial derivative with respect to v. In Equation (2.40), the weight matrix of the observations W is not necessarily a diagonal matrix, implying that the observations may be correlated. If the observations are uncorrelated, W will be a diagonal matrix and the criterion will simplify to

$$\phi = \sum_{i=1}^{n} w_i v_i^2 = w_1 v_1^2 + w_2 v_2^2 + \cdots + w_n v_n^2 \rightarrow \text{minimum} \tag{2.41}$$

which says that the sum of the weighted squares of the residuals is a minimum. Another, simpler case involves observations that are uncorrelated and of equal weight (precision), for which $W = I$ and ϕ becomes

$$\phi = \sum_{i=1}^{n} v_i^2 = v_1^2 + v_2^2 + \cdots + v_n^2 \rightarrow \text{minimum} \tag{2.42}$$

The case covered by Equation (2.42) is the oldest and may have accounted for the name *least squares,* because it seeks the "least" sum of the squares of the residuals.

If we refer back to the example of measuring a distance three times and assume that x_1, x_2, and x_3 are of equal precision (weight) and uncorrelated, it can be shown that the Σv_i^2 is a minimum if the best estimate $\hat{x}$ is taken as the arithmetic mean of the three observations. Similarly, if the three interior angles α, β, and γ in the plane triangle example have a unit weight matrix, the method of least squares will yield all three residuals equal to $-1''$. Therefore, when each angle is reduced by $1''$, their sum will be $180°$ and the functional model will be satisfied. These two examples, as well as several others, are worked out in detail in the following section.

2.24
REDUNDANCY AND THE MODEL

Before planning the acquisition phase of surveying data, a general model usually is specified either explicitly or implicitly. Such a model is determined by a certain number of variables and a possible set of relationships among them. Whether or not an adjustment of the survey data is necessary depends on the amount of observational data acquired. A *minimum number* of independent variables always is needed to determine the selected model uniquely. Such a minimum number is designated n_0. If n measurements are acquired ($n > n_0$) with respect to the specified model, then the *redundancy,* or (statistical) degrees of freedom, is specified as the amount by which n exceeds n_0. Denoting the redundancy by r,

$$r = n - n_0 \qquad (2.43)$$

As illustrations, consider the following examples:

1. The shape of a plane triangle is uniquely determined by a minimum of two interior angles, or $n_0 = 2$. If three interior angles are measured, then with $n = 3$, the redundancy is $r = 1$.
2. The size and shape of a plane triangle require a minimum of three observations, at least one of which is the length of one side; or $n_0 = 3$. If three interior angles and all three lengths are available, then with $n = 6$, the redundancy is $r = 3$.
3. In addition to the size and shape of the plane triangle, its location and orientation with respect to a specified Cartesian coordinate system xy also are of interest (Figure 2.6). In this case, the minimum number of variables necessary to determine the model is $n_0 = 6$, which can be explained in one of two ways. From example 2, the size and shape requires that $n_0 = 3$, then the location of one point (e.g., x_1 and y_1 in Figure 2.6) and the orientation of one side (e.g., α in the figure) add three more to make a minimum

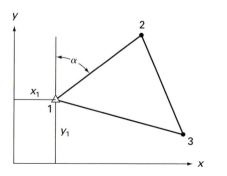

FIGURE 2.6

total of six. Another way to determine n_0 is to express the model as simply locating three points, (1, 2, 3 in the figure) in the two-dimensional coordinate system xy, which obviously requires *six* coordinates. If observations x_1, y_1, α are known in addition to the three interior angles and three sides, then with $n = 9$, the redundancy is $r = 3$.

The success of a survey adjustment depends to a large measure on the proper definition of the model and the correct determination of n_0. Next, the acquired measurements must relate to the specified model and have a set that is sufficient to determine the model. If this is not the case, the adjustment would not be meaningful. This can be illustrated by having three different measurements of *one* interior angle in a plane triangle. In this case, even though $n = 3$ and $n_0 = 2$, it is clear that the shape of the triangle cannot be determined from these data.

2.25
CONDITION EQUATIONS—PARAMETERS

After the redundancy r is determined, the adjustment proceeds by writing equations that relate the model variables to reflect the existing redundancy. Such equations will be referred to either as *condition equations* or simply as *conditions*. The number of conditions to be formulated for a given problem will depend on whether only observational variables are involved or other unknown variables as well. To illustrate this point, consider having two measurements α_1 and α_2 for the angle α. If no additional unknown variables are introduced, there will be only *one condition equation* corresponding to the one redundancy, or $\hat{\alpha}_1 - \hat{\alpha}_2 = 0$. Once the adjustment is performed, the least-squares estimate of the angle $\hat{\alpha}$ is obtained from another relationship; namely, $\hat{\alpha} = \hat{\alpha}_1 = \alpha_1 + v_1$ (or $\hat{\alpha} = \hat{\alpha}_2$). Note that this relationship is almost self-evident. Nevertheless, such additional relations are required to evaluate other variables, as will be shown in the following example. As an alternative, $\hat{\alpha}$ could be carried in the adjustment as an additional unknown variable. In such a case, one more condition must be written in addition to the one corresponding to $r = 1$ (i.e., there must be two conditions). These may be written as

$$\hat{\alpha} - \hat{\alpha}_1 = 0$$
$$\hat{\alpha} - \hat{\alpha}_2 = 0$$

The additional unknown variable, which is a random variable like the observations, will be called a *parameter*. The one thing that distinguishes a parameter from an observation is that the parameter has no a priori sample value but the observation does. After the adjustment, both the observations and the parameters will have new least-squares estimates, as well as estimates for their cofactor or covariance matrices, as will be explained in later sections of this chapter.

To summarize, if the redundancy is r, there exist r independent condition equations, which can be written in terms of the given n observations. If u additional unknown parameters are included in the adjustment, a total of

$$c = r + u \qquad (2.44)$$

independent condition equations in terms of both the n observations and u parameters must be written. For the parameters to be functionally independent, their number, u, should not exceed the minimum number of variables, n_0, necessary to specify the model. Hence, the following relation must be satisfied:

$$0 \le u \le n_0 \qquad (2.45)$$

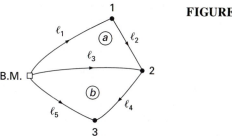

FIGURE 2.7

Similarly, for the formulated condition equations to be independent, their number c should be no larger than the total number of observations, n. Hence,

$$r \le c \le n \tag{2.46}$$

To demonstrate these relations as well as elaborate further on the concept of a parameter, consider another example.

> **EXAMPLE 2.8.** Figure 2.7 is a sketch of a small level network that contains a bench mark (B.M.) and three points, the elevations of which are needed.

Solution. To determine these three elevations, five differences in elevation, $l_1, l_2, \ldots, l_5$, are measured. The arrow along a line in Figure 2.7 (by convention) leads from a low point to a higher point.

The model involves the elevations of four points, of which one is a bench mark having a known elevation. Therefore, the minimum number of variables needed to fully specify the model is $n_0 = 3$. Given $n = 5$ observations, it follows that the redundancy is $r = 2$. Consequently, for the least-squares adjustment, two independent condition equations are written in terms of the five observations. One possibility is to write one condition for each of the two loops, a and b, shown in Figure 2.7 as follows:

$$\hat{l}_1 + \hat{l}_2 - \hat{l}_3 = 0$$
$$\hat{l}_3 + \hat{l}_4 - \hat{l}_5 = 0 \tag{2.47}$$

After the adjustment, new estimates, $\hat{l}_1, \hat{l}_2, \ldots, \hat{l}_5$, for the five observations are obtained. From these new values the elevations of points, 1, 2, and 3 can be computed uniquely no matter which combination of estimated observations is used. For instance, the elevation of point 2 may be computed in any one of the following ways and its value would be identical:

$$\text{elevation of point 2} = \text{B.M.} + \hat{l}_1 + \hat{l}_2$$
$$= \text{B.M.} + \hat{l}_3$$
$$= \text{B.M.} + \hat{l}_5 - \hat{l}_4$$

An alternative least-squares procedure is possible if the elevations of points 1, 2, 3 are carried as parameters in the adjustment. In such a case, $u = 3$ and, according to Equation (2.44), the number of conditions would be $c = 2 + 3 = 5$. Hence, u takes on its upper limit of $n_0 = 3$, and likewise c is equal to n (see Equations (2.45) and (2.46)). Denoting the parameters by x_1, x_2, x_3, the five condition equations may be written as

$$\text{B.M.} + \hat{l}_1 - x_1 = 0$$
$$x_1 + \hat{l}_2 - x_2 = 0$$

$$\text{B.M.} + \hat{l}_3 - x_2 = 0 \qquad (2.48)$$

$$x_2 + \hat{l}_4 - x_3 = 0$$

$$\text{B.M.} + \hat{l}_5 - x_3 = 0$$

The estimates $\hat{x}_1$, $\hat{x}_2$, $\hat{x}_3$ computed from the direct least-squares adjustment would be identical to those computed from $\hat{l}_1$, $\hat{l}_2$, ..., $\hat{l}_5$ in the previous technique.

This example also demonstrates that least-squares adjustments can be performed by at least one of two techniques: involving both observations and parameters, or involving only observations and no parameters in the condition equations. The two techniques are discussed in greater detail in the following section.

2.26
TECHNIQUES OF LEAST SQUARES

Although there are many techniques for least-squares adjustment, we will consider the three most commonly used. It is important to point out, however, that for any given survey adjustment problem, *all* least-squares techniques yield *identical* results. Therefore, the choice of technique depends mainly on the type of problem, the required information, and to some extent the computing equipment available. Further discussion of this point will be given after the actual techniques are presented.

The first technique, called *adjustment of indirect observations,* is characterized by the following properties:

1. The condition equations include both observations and parameters.
2. There are as many conditions as the number of observations, or $c = n$.
3. Each condition equation contains *only one observation,* with the specific stipulation that its coefficient is unity.

With these properties the condition equations take the functional form

$$v_1 + b_{11}\delta_1 + b_{12}\delta_2 + \cdots + b_{1u}\delta_u = f_1$$

$$v_2 + b_{21}\delta_1 + b_{22}\delta_2 + \cdots + b_{2u}\delta_u = f_2$$

$$\cdots \qquad (2.49)$$

$$v_n + b_{n1}\delta_1 + b_{n2}\delta_2 + \cdots + b_{nu}\delta_u = f_n$$

where
$v_1, v_2, \ldots, v_n$ = residuals for the n observations
$\delta_1, \delta_2, \ldots, \delta_u$ = u unknown parameters
$b_{11}, b_{12}, \ldots, b_{nu}$ = numerical coefficients of the parameters
$f_1, f_2, \ldots, f_n$ = constant terms for the n conditions, which usually will contain the a priori numerical values of the observations

The set of equations in (2.49) can be collected into matrix form as

$$\begin{bmatrix} v_1 \\ v_2 \\ \vdots \\ v_n \end{bmatrix} + \begin{bmatrix} b_{11} & b_{12} & \cdots & b_{1u} \\ b_{21} & b_{22} & \cdots & b_{2u} \\ \cdots & & & \\ b_{n1} & b_{n2} & \cdots & n_{nu} \end{bmatrix} \begin{bmatrix} \delta_1 \\ \delta_2 \\ \vdots \\ \delta_u \end{bmatrix} = \begin{bmatrix} f_1 \\ f_2 \\ \vdots \\ f_n \end{bmatrix} \qquad (2.50)$$

or, more concisely,

$$\underset{n,1}{v} + \underset{n,u}{B} \underset{u,1}{\Delta} = \underset{n,1}{f} \qquad (2.51)$$

In linear problems, the constant-term vector f in Equation (2.51) usually is the vector of given observations l subtracted from a vector of numerical constants d,

$$f = d - l \tag{2.52}$$

and Equation (2.51) takes the form

$$v + B\Delta = d - l \tag{2.53}$$

EXAMPLE 2.9. To demonstrate Equations (2.49), write the conditions in Equation (2.48) for the level network of Example 2.8 as follows:

$$
\begin{aligned}
v_1 - x_1 \qquad\qquad &= -l_1 - \text{B.M.} = f_1 \\
v_2 + x_1 - x_2 \qquad &= -l_2 \qquad\qquad = f_2 \\
v_3 \qquad - x_2 \qquad &= -l_3 - \text{B.M.} = f_3 \\
v_4 \qquad + x_2 - x_3 &= -l_4 \qquad\qquad = f_4 \\
v_5 \qquad\qquad - x_3 &= -l_5 - \text{B.M.} = f_5
\end{aligned}
\tag{2.54}
$$

or, in matrix form,

$$
\underset{5,1}{v} + \underset{5,3}{B} \, \underset{3,1}{\Delta} = \underset{5,1}{f} \tag{2.55}
$$

with

$$
v = \begin{bmatrix} v_1 \\ v_2 \\ \vdots \\ \vdots \\ v_5 \end{bmatrix}; \quad
B = \begin{bmatrix} -1 & 0 & 0 \\ 1 & -1 & 0 \\ 0 & -1 & 0 \\ 0 & 1 & -1 \\ 0 & 0 & -1 \end{bmatrix}; \quad
\Delta = \begin{bmatrix} x_1 \\ x_2 \\ x_3 \end{bmatrix}; \quad
f = \begin{bmatrix} f_1 \\ f_2 \\ \vdots \\ f_5 \end{bmatrix}
\tag{2.56}
$$

It is clear from Equation (2.54) that the vector f is

$$
f = \begin{bmatrix} -\text{B.M.} \\ 0 \\ -\text{B.M.} \\ 0 \\ -\text{B.M.} \end{bmatrix} - \begin{bmatrix} l_1 \\ l_2 \\ l_3 \\ l_4 \\ l_5 \end{bmatrix} = d - l
$$

which demonstrates Equation (2.52).

The second least-squares technique to be considered is called *adjustment of observations only*. As its name implies, no parameters are included in the condition equations, which therefore must be equal in number to the redundancy r. They take the general functional form

$$
\begin{aligned}
a_{11}v_1 + a_{12}v_2 + \cdots + a_{1n}v_n &= f_1 \\
a_{21}v_1 + a_{22}v_2 + \cdots + a_{2n}v_n &= f_2 \\
&\cdots \\
a_{r1}v_1 + a_{r2}v_2 + \cdots + a_{rn}v_n &= f_r
\end{aligned}
\tag{2.57}
$$

which in matrix notation becomes

$$
\begin{bmatrix}
a_{11} & a_{12} & \cdots & a_{1n} \\
a_{21} & a_{22} & \cdots & a_{2n} \\
\vdots & & & \\
a_{r1} & a_{r2} & \cdots & a_{rn}
\end{bmatrix}
\begin{bmatrix}
v_1 \\ v_2 \\ \vdots \\ v_n
\end{bmatrix}
=
\begin{bmatrix}
f_1 \\ f_2 \\ \vdots \\ f_r
\end{bmatrix}
\tag{2.58}
$$

or, more concisely,

$$
\underset{r,n}{\mathbf{A}} \; \underset{n,1}{v} = \underset{r,1}{f}
\tag{2.59}
$$

For linear adjustment problems, the constant-term vector f in Equation (2.59) is given by

$$
f = d - \mathbf{A}l
\tag{2.60}
$$

in which d is a vector of numerical constants. Then, Equation (2.59) becomes

$$
\mathbf{A}v = d - \mathbf{A}l
\tag{2.61}
$$

EXAMPLE 2.10. As an illustration, rewrite the two conditions in Equation (2.47) for the level network in Example 2.8.

Solution

$$
\begin{aligned}
v_1 + v_2 - v_3 &= -l_1 - l_2 + l_3 = f_1 \\
v_3 + v_4 - v_5 &= -l_3 - l_4 + l_5 = f_2
\end{aligned}
\tag{2.62}
$$

or

$$
\underset{2,5}{\mathbf{A}} \; \underset{5,1}{v} = \underset{2,1}{f}
\tag{2.63}
$$

in which

$$
\mathbf{A} =
\begin{bmatrix}
1 & 1 & -1 & 0 & 0 \\
0 & 0 & 1 & 1 & -1
\end{bmatrix};
\qquad
v = [v_1 \quad v_2 \quad \cdots \quad v_5]^t;
\qquad
f = \begin{bmatrix} f_1 \\ f_2 \end{bmatrix}
\tag{2.64}
$$

Each of these two techniques of least-squares adjustment is discussed separately in the following two sections.

2.27
ADJUSTMENT OF INDIRECT OBSERVATIONS (PARAMETRIC ADJUSTMENT)

To appreciate the matrix derivation to follow, consider the level network in Figure 2.7. Also, to keep the introductory treatment uncomplicated, assume that the five observed differences in elevation are uncorrelated and have the weights $w_1, w_2, \ldots, w_5$, respectively. Under this assumption, the minimum criterion from Equation (2.41) applies:

$$
\phi = w_1 v_1^2 + w_2 v_2^2 + w_3 v_3^2 + w_4 v_4^2 + w_5 v_5^5
$$

which, in view of the equations in (2.54), becomes

$$
\phi = w_1(f_1 + x_1)^2 + w_2(f_2 - x_1 + x_2)^2 + w_3(f_3 + x_2)^2 + w_4(f_4 - x_2 + x_3)^2
$$
$$
+ w_5(f_5 + x_3)^2
\tag{2.65}
$$

For ϕ in Equation (2.65) to be a minimum, its partial derivative with respect to each unknown variable (i.e., with respect to x_1, x_2, x_3) must equal 0. Hence,

$$\frac{\partial \phi}{\partial x_1} = 2w_1(f_1 + x_1) - 2w_2(f_2 - x_1 + x_2) \qquad\qquad = 0$$

$$\frac{\partial \phi}{\partial x_2} = 2w_2(f_2 - x_1 + x_2) + 2w_3(f_3 + x_2) - 2w_4(f_4 - x_2 + x_3) = 0 \qquad (2.66)$$

$$\frac{\partial \phi}{\partial x_3} = 2w_4(f_4 - x_2 + x_3) + 2w_5(f_5 + x_3) \qquad\qquad = 0$$

Clearing and rearranging, the set of equations in (2.66) becomes

$$(w_1 + w_2)x_1 - w_2 x_2 = -w_1 f_1 + w_2 f_2$$

$$-w_2 x_1 + (w_2 + w_3 + w_4)x_2 - w_4 x_3 = -w_2 f_2 - w_3 f_3 + w_4 f_4 \qquad (2.67)$$

$$-w_4 x_2 + (w_4 + w_5)x_3 = -w_4 f_4 - w_5 f_5$$

To express Equation (2.67) in matrix form, recall that $\mathbf{\Delta}$ and f are vectors given in Equation (2.56) and that the weight matrix of the observations is a diagonal matrix.

$$\mathbf{W} = \begin{bmatrix} mw_1 & & & \mathbf{0} \\ & w_2 & & \\ & & \ddots & \\ \mathbf{0} & & & w_5 \end{bmatrix} \qquad (2.68)$$

Then, the equations in (2.67) can be written in matrix form as

$$\begin{bmatrix} -1 & 1 & 0 & 0 & 0 \\ 0 & -1 & -1 & 1 & 0 \\ 0 & 0 & 0 & -1 & -1 \end{bmatrix} \begin{bmatrix} w_1 & & & \mathbf{0} \\ & w_2 & & \\ & & \ddots & \\ \mathbf{0} & & & w_5 \end{bmatrix} \begin{bmatrix} -1 & 0 & 0 \\ 1 & -1 & 0 \\ 0 & -1 & 0 \\ 0 & 1 & -1 \\ 0 & 0 & -1 \end{bmatrix} \begin{bmatrix} x_1 \\ x_2 \\ x_3 \end{bmatrix}$$

$$= \begin{bmatrix} -1 & 1 & 0 & 0 & 0 \\ 0 & -1 & -1 & 1 & 0 \\ 0 & 0 & 0 & -1 & -1 \end{bmatrix} \begin{bmatrix} w_1 & & & \mathbf{0} \\ & w_2 & & \\ & & \ddots & \\ \mathbf{0} & & & w_5 \end{bmatrix} \begin{bmatrix} f_1 \\ f_2 \\ \vdots \\ f_5 \end{bmatrix} \qquad (2.69)$$

Multiply the matrices in Equation (2.69) to ascertain that the equations in (2.67) will result. Recalling matrix $\mathbf{B}$ from Equation (2.56), Equation (2.69) can be written more concisely as

$$\underset{3,5}{(\mathbf{B}^T} \underset{5,5}{\mathbf{W}} \underset{5,3}{\mathbf{B})} \underset{3,1}{\mathbf{\Delta}} = \underset{3,5}{\mathbf{B}^T} \underset{5,5}{\mathbf{W}} \underset{5,1}{\mathbf{f}} \qquad (2.70)$$

The relation in Equation (2.70), although derived for an example, is in fact general and applies to any problem for which the condition equations are of the general form given in Equation (2.51). Furthermore, there is no restriction on the structure of the weight matrix $\mathbf{W}$. The following derivation shows that this is true.

The least-squares criterion is

$$\phi = \mathbf{v}^T \mathbf{W} \mathbf{v}$$

which, upon substituting for v from Equation (2.51) becomes

$$\phi = (\mathbf{f} - \mathbf{B}\Delta)^T \mathbf{W}(\mathbf{f} - \mathbf{B}\Delta)$$
$$\phi = \mathbf{f}^T\mathbf{W}\mathbf{f} - \Delta^T\mathbf{B}^T\mathbf{W}\mathbf{f} - \mathbf{f}^T\mathbf{W}\mathbf{B}\Delta + \Delta^T\mathbf{B}^T\mathbf{W}\mathbf{B}\Delta \tag{2.71}$$

Because all terms on the right side of Equation (2.71) are scalars, the second and third terms are equal and thus

$$\phi = \mathbf{f}^T\mathbf{W}\mathbf{f} - 2\mathbf{f}^T\mathbf{W}\mathbf{B}\Delta + \Delta^T\mathbf{B}^T\mathbf{W}\mathbf{B}\Delta \tag{2.72}$$

For ϕ to be a minimum, $\partial\phi/\partial\Delta$ must be zero, or

$$-2\mathbf{f}^T\mathbf{W}\mathbf{B} + 2\Delta^T\mathbf{B}^T\mathbf{W}\mathbf{B} = 0$$

or

$$\underset{u,n\ n,n\ n,u\ \ u,1}{(\mathbf{B}^T\ \mathbf{W}\ \mathbf{B})\ \Delta} = \underset{u,n\ n,n\ n,1}{\mathbf{B}^T\ \mathbf{W}\ \mathbf{f}} \tag{2.73}$$

which is identical to Equation (2.70). Note that $\mathbf{W}$ could be a full matrix and therefore no restriction is placed on the statistical properties of the observations. Of course, certain assumptions about the structure of $\mathbf{W}$ may be made in practice. If the auxiliaries

$$\underset{u,u}{\mathbf{N}} = \mathbf{B}^T\mathbf{W}\mathbf{B}$$
$$\underset{u,1}{\mathbf{t}} = \mathbf{B}^T\mathbf{W}\mathbf{f} \tag{2.74}$$

are used, Equation (2.73) becomes

$$\mathbf{N}\Delta = t \tag{2.75}$$

the solution of which (assuming a nonsingular $\mathbf{N}$) is

$$\Delta = \mathbf{N}^{-1}t \tag{2.76}$$

The set of equations in (2.73) (or (2.75)) usually is called the *reduced normal equations,* or simply the *normal equations* in the parameters. $\mathbf{N}$ is a nonsingular matrix when $\mathbf{B}$ is of full rank u, which is usually the case. The matrices $\mathbf{N}$ and $\mathbf{t}$ (in (2.74)) are called the *normal equations coefficient matrix* and the *normal equations constant term vector,* respectively.

The precision of the estimated parameters is the cofactor matrix $\mathbf{Q}_{\Delta\Delta}$. This matrix is obtained by applying the relationships of error propagation developed in Section 2.19 to Equation (2.76). Using Equation (2.74) for t and Equation (2.52) for f, Equation (2.76) becomes

$$\Delta = \mathbf{N}^{-1}\mathbf{B}^T\mathbf{W}(d - l) \tag{2.77}$$

The only vector of random variables on the right side of Equation (2.27) is l, since d is a vector of numerical constants. The matrix $\mathbf{Q}$ is used to designate the cofactor of the observations (i.e., in place of $\mathbf{Q}_{ll}$), and therefore when Equation (2.34) is applied to Equation (2.77), the following results:

$$\mathbf{Q}_{\Delta\Delta} = \mathbf{J}_{\Delta l}\mathbf{Q}_{ll}\mathbf{J}_{\Delta l}^T = \mathbf{J}_{\Delta l}\mathbf{Q}\mathbf{J}_{\Delta l}^T$$
$$= (-\mathbf{N}^{-1}\mathbf{B}^T\mathbf{W})\mathbf{Q}(-\mathbf{N}^{-1}\mathbf{B}^T\mathbf{W})^T$$

or

$$\mathbf{Q}_{\Delta\Delta} = \mathbf{N}^{-1}\mathbf{B}^T\mathbf{W}\mathbf{Q}\mathbf{W}\mathbf{B}\mathbf{N}^{-1}$$

noting that $\mathbf{N}^{-1}$ and $\mathbf{W}$ are symmetric matrices. Because $\mathbf{W} = \mathbf{Q}^{-1}$, using Equation (2.74) for $\mathbf{N}$,

$$\mathbf{Q}_{\Delta\Delta} = \mathbf{N}^{-1}(\mathbf{B}^T\mathbf{W}\mathbf{B})\mathbf{N}^{-1} = \mathbf{N}^{-1}\mathbf{N}\mathbf{N}^{-1}$$

or

$$\mathbf{Q}_{\Delta\Delta} = \mathbf{N}^{-1} \tag{2.78}$$

Once $\mathbf{\Delta}$ is computed (from Equation (2.75)), the observational residuals may be computed using Equation (2.51) or

$$v = f - \mathbf{B}\mathbf{\Delta} \tag{2.79}$$

and the least-squares estimate of the observations, $\hat{l}$, is evaluated from Equation (2.72):

$$\hat{l} = l + v \tag{2.80}$$

In a manner similar to that used to obtain $\mathbf{Q}_{\Delta\Delta}$, a relationship for the cofactor matrix $\mathbf{Q}_{\hat{l}\hat{l}}$ may be derived as

$$\mathbf{Q}_{\hat{l}\hat{l}} = \mathbf{B}\mathbf{N}^{-1}\mathbf{B}^T \tag{2.81}$$

If, originally, the covariance matrix of the observations $\mathbf{\Sigma}$ was given and used in the least-squares solution instead of the cofactor matrix $\mathbf{Q}$, then, $\mathbf{\Sigma}_{\Delta\Delta} = \mathbf{N}^{-1}$ and $\mathbf{\Sigma}_{\hat{l}\hat{l}} = \mathbf{B}\mathbf{N}^{-1}\mathbf{B}^T$, instead of Equations (2.78) and (2.81). On the other hand, if only relative variances and covariances were given a priori, Equations (2.78) and (2.81) may be used to compute $\mathbf{Q}_{\Delta\Delta}$ and $\mathbf{Q}_{\hat{l}\hat{l}}$. Then, to get $\mathbf{\Sigma}_{\Delta\Delta}$ and $\mathbf{\Sigma}_{\hat{l}\hat{l}}$, an estimate $\hat{\sigma}_0^2$ of the reference variance may be computed from the adjustment using the relationship

$$\hat{\sigma}_0^2 = \frac{v'\mathbf{W}v}{r} \tag{2.82}$$

in which r is the redundancy (see Equation (2.43)), v is the vector of residuals computed from Equation (2.79), and $\mathbf{W}$ is the a priori weight matrix of the observations. Then, according to Equation (2.32),

$$\mathbf{\Sigma}_{\Delta\Delta} = \hat{\sigma}_0^2 \mathbf{Q}_{\Delta\Delta} \tag{2.83}$$

$$\mathbf{\Sigma}_{\hat{l}\hat{l}} = \hat{\sigma}_0^2 \mathbf{Q}_{\hat{l}\hat{l}} \tag{2.84}$$

EXAMPLE 2.11. Suppose that l_1, l_2, l_3 are three different measurements of a distance (which is a random variable) and that these measurements are uncorrelated and of equal precision; find the least squares estimate of the distance $\bar{l}$.

Solution. If $\bar{l}$ is the final estimate of the distance, it will be equal to the sum of each distance and its corresponding residual, or

$$l_1 + v_1 = \bar{l}$$
$$l_2 + v_2 = \bar{l}$$
$$l_3 + v_3 = \bar{l}$$

These three condition equations may be rearranged and put in matrix form:

$$\begin{bmatrix} v_1 \\ v_2 \\ v_3 \end{bmatrix} + \begin{bmatrix} -1 \\ -1 \\ -1 \end{bmatrix} \bar{l} = \begin{bmatrix} -l_1 \\ -l_2 \\ -l_3 \end{bmatrix}$$

which is of the general form $v + \mathbf{B}\mathbf{\Delta} = \mathbf{f}$, with

$$\mathbf{B} = \begin{bmatrix} -1 \\ -1 \\ -1 \end{bmatrix}; \quad \mathbf{\Delta} = \bar{l}; \quad \text{and} \quad \mathbf{f} = \begin{bmatrix} -l_1 \\ -l_2 \\ -l_3 \end{bmatrix}$$

Because the observations are uncorrelated and of equal precision, $\mathbf{W} = \mathbf{I}$ and the normal equations coefficient matrix $\mathbf{N}$ and constant-term vector t are computed from Equation (2.74) as

$$\mathbf{N} = \mathbf{B}^T\mathbf{B} = [-1 \quad -1 \quad -1]\begin{bmatrix} -1 \\ -1 \\ -1 \end{bmatrix} = 3$$

$$t = \mathbf{B}^T f = [-1 \quad -1 \quad -1]\begin{bmatrix} -l_1 \\ -l_2 \\ -l_3 \end{bmatrix} = l_1 + l_2 + l_3$$

According to Equation (2.76), the least-squares estimate of the distance is given by $\bar{l} = \Delta = \mathbf{N}^{-1}t = (3)^{-1}(l_1 + l_2 + l_3) = (l_1 + l_2 + l_3)/3$.

This result is rather interesting, because it shows that the least-squares estimate of the distance using three direct and uncorrelated observations of equal weights is their arithmetic mean, which is what one would intuitively have taken as the "best estimate" of the distance. Although only three observations are considered, the result is general and may be applied to any number of observations. Thus, for n observations that are uncorrelated and of equal precision (weight),

$$\mathbf{N} = n \tag{2.85}$$

$$t = l_1 + l_2 + \cdots + l_n = \sum_{i=1}^{n} l_i \tag{2.86}$$

$$\bar{l} = \frac{1}{n}\sum_{i=1}^{n} l_i \tag{2.87}$$

and from Equation (2.78),

$$q_{\bar{l}\bar{l}} = \mathbf{Q}_{\Delta\Delta} = \mathbf{N}^{-1} = \frac{1}{n} \tag{2.88}$$

Therefore, the weight of the arithmetic mean of n uncorrelated observations, each of unit weight, is equal to n, because

$$w_{\bar{l}\bar{l}} = q_{\bar{l}\bar{l}}^{-1} = n \tag{2.89}$$

EXAMPLE 2.12. Compute the least-squares estimate, $\bar{l}$, for the distance measured three times (l_1, l_2, l_3) of Example 2.11, if the observations are uncorrelated but have different weights, w_1, w_2, w_3, respectively.

Solution. The condition equations are the same as those given in Example 2.11. However, the normal equations coefficient matrix in this case is

$$\mathbf{N} = \mathbf{B}^T\mathbf{W}\mathbf{B} = [-1 \quad -1 \quad -1]\begin{bmatrix} w_1 & 0 & 0 \\ 0 & w_2 & 0 \\ 0 & 0 & w_3 \end{bmatrix}\begin{bmatrix} -1 \\ -1 \\ -1 \end{bmatrix} = w_1 + w_2 + w_3$$

recognizing of course that the weight matrix $\mathbf{W}$ is a diagonal matrix. The constant-term vector t is computed as

$$t = \mathbf{B}^T\mathbf{W}f = [-1 \quad -1 \quad -1]\begin{bmatrix} w_1 & 0 & 0 \\ 0 & w_2 & 0 \\ 0 & 0 & w_3 \end{bmatrix}\begin{bmatrix} -l_1 \\ -l_2 \\ -l_3 \end{bmatrix} = w_1 l_1 + w_2 l_2 + w_3 l_3$$

and thus the least-squares estimate is

$$\bar{l} = \mathbf{N}^{-1}t = \frac{w_1 l_1 + w_2 l_2 + w_3 l_3}{w_1 + w_2 + w_3}$$

which is the weighted mean of the given observations.

As before, the result from Example 2.12 is general and may be extended to any number of observations, n:

$$\mathbf{N} = w_1 + w_2 + \cdots + w_n = \sum_{i=1}^{n} w_i \tag{2.90}$$

$$t = w_1 l_1 + w_2 l_2 \cdots + w_n l_n = \sum_{i=1}^{n} w_i l_i \tag{2.91}$$

$$\bar{l} = \frac{\sum\limits_{i=1}^{n} w_i l_1}{\sum\limits_{i=1}^{n} w_i} \tag{2.92}$$

$$q_{\bar{l}\bar{l}} = \mathbf{Q}_{\Delta\Delta} = \mathbf{N}^{-1} = \left(\sum_{i=1}^{n} w_i \right)^{-1} \tag{2.93}$$

$$w_{\bar{l}\bar{l}} = q_{\bar{l}\bar{l}}^{-1} = \sum_{i=1}^{n} w_i \tag{2.94}$$

Equation (2.94) implies that the weight of the weighted mean of a set of uncorrelated observations (with different weights) is equal to the sum of their weights. It can be seen then that Equation (2.89) is the special case of Equation (2.94), in which the weights are all equal to unity.

EXAMPLE 2.13. The three measured interior angles of a plane triangle are $l_1 = 45°25'01''$, $l_2 = 65°20'00''$, and $l_3 = 69°15'02''$. Compute the least-squares estimates for the three angles, assuming that the measurements are uncorrelated and of equal precision.

Solution. Let x_1 and x_2 be the adjusted measurements for angles l_1 and l_2. Three condition equations can be formed as follows:

$$v_1 + l_1 = x_1$$
$$v_2 + l_2 = x_2$$
$$v_3 + l_3 = 180° - x_1 - x_2$$

which become

$$v_1 - x_1 = -l_1$$
$$v_2 - x_2 = -l_2$$
$$v_3 + x_1 + x_2 = 180° - l_3$$

which may be written in matrix form as

$$v + \mathbf{B}\Delta = f$$

where

$$v = \begin{bmatrix} v_1 \\ v_2 \\ v_3 \end{bmatrix}; \quad \mathbf{B} = \begin{bmatrix} -1 & 0 \\ 0 & -1 \\ 1 & 1 \end{bmatrix}; \quad \Delta = \begin{bmatrix} x_1 \\ x_2 \end{bmatrix}; \quad f = \begin{bmatrix} -45°25'01'' \\ -65°20'00'' \\ 110°44'58'' \end{bmatrix}$$

Because uncorrelated observations of equal precision have been assumed, $\mathbf{W} = \mathbf{I}$, and the normal equations (2.73) become $(\mathbf{B}^T\mathbf{B})\mathbf{\Delta} = \mathbf{B}^T f$ or

$$\begin{bmatrix} -1 & 0 & 1 \\ 0 & -1 & 1 \end{bmatrix} \begin{bmatrix} -1 & 0 \\ 0 & -1 \\ 1 & 1 \end{bmatrix} \begin{bmatrix} x_1 \\ x_2 \end{bmatrix} = \begin{bmatrix} -1 & 0 & 1 \\ 0 & -1 & 1 \end{bmatrix} \begin{bmatrix} -45°25'01'' \\ -65°20'00'' \\ 110°44'58'' \end{bmatrix}$$

$$\begin{bmatrix} 2 & 1 \\ 1 & 2 \end{bmatrix} \begin{bmatrix} x_1 \\ x_2 \end{bmatrix} = \begin{bmatrix} 156.166389° \\ 176.082778° \end{bmatrix}$$

The solution of these equations yields

$$x_1 = 45°25'00''$$

$$x_2 = 65°19'59''$$

Substitution of these values into the original condition equations gives

$$v_1 = 45°25'00'' - 45°25'01'' = -01''$$

$$v_2 = 65°19'59'' - 65°20'00'' = -01''$$

$$v_3 = 180° - (45°25'00'' + 65°19'59'' + 69°15'02'') = -01''$$

The propagated cofactor matrix for the adjusted observations by Equation (2.81) is

$$\mathbf{Q}_{\hat{\ell}\hat{\ell}} = \mathbf{B}(\mathbf{B}^T\mathbf{B})^{-1}\mathbf{B}^T = \mathbf{B}\mathbf{N}^{-1}\mathbf{B}^T$$

Thus

$$\mathbf{Q}_{\hat{\ell}\hat{\ell}} = \begin{bmatrix} -1 & 0 \\ 0 & -1 \\ 1 & 1 \end{bmatrix} \begin{bmatrix} \frac{2}{3} & -\frac{1}{3} \\ -\frac{1}{3} & \frac{2}{3} \end{bmatrix} \begin{bmatrix} -1 & 0 & 1 \\ 0 & -1 & 1 \end{bmatrix}$$

or

$$\mathbf{Q}_{\hat{\ell}\hat{\ell}} = \begin{bmatrix} \frac{2}{3} & -\frac{1}{3} & -\frac{1}{3} \\ -\frac{1}{3} & \frac{2}{3} & -\frac{1}{3} \\ -\frac{1}{3} & -\frac{1}{3} & \frac{2}{3} \end{bmatrix}$$

Note that the precision of each adjusted angle is higher than that for the unadjusted, original observation, because $\hat{q} = \frac{2}{3}$ as compared to $q = 1.0$. In addition, although the original observations were uncorrelated, the adjusted observations are correlated as indicated by the off-diagonal elements in $\mathbf{Q}_{\hat{\ell}\hat{\ell}}$.

2.28
ADJUSTMENT OF OBSERVATIONS ONLY (CONDITIONAL ADJUSTMENT)

Similar to the case of indirect observations, this technique is introduced by working the level-network problem discussed in Example 2.8 and assuming that the five observations are uncorrelated and have the weights w_1, w_2, w_3, w_4, and w_5. With this information the minimum criterion from Equation (2.41) is

$$\phi = w_1v_1^2 + w_2v_2^2 + w_3v_3^2 + w_4v_4^2 + w_5v_5^2 \rightarrow \text{minimum}$$

Unlike the adjustment of indirect observations, it is not possible here to substitute for the residuals from the condition equations because there are five residuals and only two condition equations (in (2.62)). Therefore, in this case, a minimum for the function ϕ is sought under the constraint imposed by the condition Equations (2.62). This makes the problem that of seeking a *constrained minimum* instead of a free minimum, as it is termed in mathematics. Such a constrained minimum is obtained most conveniently by adding (algebraically) to ϕ each of the condition equations multiplied by a factor λ. These factors are called *Lagrange multipliers* after the great French analyst Lagrange (see Leick, 1982). It is numerically more convenient for our later development to use $2k$ instead of λ; therefore, the function to be minimized becomes

$$\phi' = w_1 v_1^2 + w_2 v_2^2 + w_3 v_3^2 + w_4 v_4^2 + w_5 v_5^2 - 2k_1(v_1 + v_2 - v_3 - f_1)$$
$$- 2k_2(v_3 + v_4 - v_5 - f_2) \qquad (2.95)$$

Note that, after the adjustment, the quantities within parentheses in (2.95) vanish because the two condition Equations (2.62) are fully satisfied after the adjustment. Consequently, the minimum of ϕ' corresponds to the minimum of the original function ϕ. Taking the partial derivatives of ϕ' with respect to each of the five residuals and equating to 0 leads to

$$\frac{\partial \phi'}{\partial v_1} = 2w_1 v_1 - 2k_1 = 0$$

$$\frac{\partial \phi'}{\partial v_2} = 2w_2 v_2 - 2k_1 = 0$$

$$\frac{\partial \phi'}{\partial v_3} = 2w_3 v_3 + 2k_1 - 2k_2 = 0 \qquad (2.96)$$

$$\frac{\partial \phi'}{\partial v_4} = 2w_4 v_4 - 2k_2 = 0$$

$$\frac{\partial \phi'}{\partial v_5} = 2w_5 v_5 + 2k_2 = 0$$

Partial differentiation of ϕ' with respect to k_1 and k_2 and equating the result to 0 yields the two condition equations in (2.62). Therefore, combining Equations (2.96) and (2.62) results in seven linear equations with seven unknowns: $v_1, v_2, \cdots, v_5, k_1, k_2$. Solving Equation (2.96) for the five residuals yields

$$v_1 = \frac{1}{w_1} k_1$$

$$v_2 = \frac{1}{w_2} k_1$$

$$v_3 = \frac{1}{w_3}(-k_1 + k_2)$$

$$v_4 = \frac{1}{w_4} k_2$$

$$v_5 = \frac{1}{w_5}(-k_2)$$

which in matrix form becomes

$$\begin{bmatrix} v_1 \\ v_2 \\ v_3 \\ v_4 \\ v_5 \end{bmatrix} = \begin{bmatrix} 1/w_1 & & & & 0 \\ & 1/w_2 & & & \\ & & 1/w_3 & & \\ & & & 1/w_4 & \\ 0 & & & & 1/w_5 \end{bmatrix} \begin{bmatrix} 1 & 0 \\ 1 & 0 \\ -1 & 1 \\ 0 & 1 \\ 0 & -1 \end{bmatrix} \begin{bmatrix} k_1 \\ k_2 \end{bmatrix} \qquad (2.97)$$

The first (diagonal) matrix on the right side of Equation (2.97) is the inverse of the weight matrix, or $\mathbf{W}^{-1}$, which is equal to the cofactor matrix of the observations, $\mathbf{Q}$. The second matrix is $\mathbf{A}^T$, as can be seen by reference to Equation (2.64). Therefore, Equation (2.97) may be written more concisely as

$$v = \mathbf{QA}^T k \qquad (2.98)$$

in which all the terms are as defined and k is the vector of Lagrange multipliers, or $k = [k_1 \quad k_2]^T$. Substituting for v from Equation (2.98) into the condition equations of (2.63) gives

$$\mathbf{AQA}^T k = f \qquad (2.99)$$

which may by solved for k as

$$k = (\mathbf{AQA}^T)^{-1} f \qquad (2.100)$$

Finally, substitute the value of k computed from Equation (2.100) into Equation (2.98) to get values for the residuals.

The relations (2.99) and (2.100) are not specific for this particular example but are general for this technique of least-squares *adjustment of observations only* as shown in the following derivation. Let k be the vector of r Lagrange multipliers, one for each of the r condition equations in (2.59). Then the function to be minimized is

$$\phi' = v^T \mathbf{W} v - 2k^T(\mathbf{A}v - f)$$

For ϕ' to be a minimum, $\partial\phi'/\partial v$ must be 0, or

$$\frac{\partial\phi'}{\partial v} = 2v^T \mathbf{W} - 2k^T A = 0$$

which, after transposing and rearranging, becomes

$$\mathbf{W}v = \mathbf{A}^T k$$

and solving for v yields

$$v = \mathbf{W}^{-1}\mathbf{A}^T k = \mathbf{Q}\,\mathbf{A}^T\,k \qquad (2.101)$$

Substituting Equation (2.101) into Equation (2.59) yields

$$(\mathbf{AQA}^T)k = f \qquad (2.101a)$$

which when using the auxiliary

$$\mathbf{Q}_e = \mathbf{A}\,\mathbf{Q}\,\mathbf{A}^T \qquad (2.102)$$

leads to

$$\mathbf{Q}_e k = f$$

or

$$k = \mathbf{Q}_e^{-1} f = \mathbf{W}_e f \qquad (2.103)$$

The matrix $\mathbf{Q}_e$ can be considered the cofactor matrix for an equivalent set of observations, l_e, containing as many observations as there are condition equations. Because $r < n$, the number of equivalent observations always is less than the number of original observations. Each equivalent observation is a linear combination of the original observations, as can be seen from Equation (2.61) or (2.62). The linear relations are expressed by the matrix $\mathbf{A}$; therefore

$$l_e = \mathbf{A}l \qquad (2.104)$$

By error propagation, then, the cofactor matrix $\mathbf{Q}_e$ for l_e may be evaluated as

$$\mathbf{Q}_e = \mathbf{J}_{l_e l}\mathbf{Q}\mathbf{J}_{l_e l}^T = \mathbf{A}\mathbf{Q}\mathbf{A}^T$$

which is identical to Equation (2.102). The inverse of $\mathbf{Q}_e$ is designated $\mathbf{W}_e$, as shown in Equation (2.103).

The final relation for v is obtained by substituting for k from Equation (2.103) into Equation (2.101):

$$v = \mathbf{Q}\mathbf{A}^T\mathbf{W}_e f \qquad (2.105)$$

Precision estimation after the adjustment may be performed using the rules of error propagation developed in Section 2.19. The estimated observations, $\hat{l}$, are given by Equation (2.80):

$$\hat{l} = l + v = l + \mathbf{Q}\mathbf{A}'\mathbf{W}_e f$$

which, from Equation (2.52), becomes

$$\hat{l} = l + \mathbf{Q}\mathbf{A}'\mathbf{W}_e(d - \mathbf{A}l) \qquad (2.106)$$

Applying Equation (2.34) to Equation (2.106) results in

$$\mathbf{Q}_{\hat{l}\hat{l}} = \mathbf{J}_{\hat{l}l}\mathbf{Q}\mathbf{J}_{\hat{l}l}^T$$
$$= (\mathbf{I} - \mathbf{Q}\mathbf{A}^T\mathbf{W}_e\mathbf{A})\,\mathbf{Q}(\mathbf{I} - \mathbf{Q}\mathbf{A}^T\mathbf{W}_e\mathbf{A})^T$$
$$= (\mathbf{Q} - \mathbf{Q}\mathbf{A}^T\mathbf{W}_e\mathbf{A}\mathbf{Q})(\mathbf{I} - \mathbf{A}^T\mathbf{W}_e\mathbf{A}\mathbf{Q})$$

or

$$\mathbf{Q}_{\hat{l}\hat{l}} = \mathbf{Q} - \mathbf{Q}\mathbf{A}^T\mathbf{W}_e\mathbf{A}\mathbf{Q} - \mathbf{Q}\mathbf{A}^T\mathbf{W}_e\mathbf{A}\mathbf{Q} + \mathbf{Q}\mathbf{A}^T\mathbf{W}_e\mathbf{A}\mathbf{Q}\mathbf{A}^T\mathbf{W}_e\mathbf{A}\mathbf{Q}$$

From the definition of $\mathbf{Q}_e$ in Equation (2.102) and the fact that $\mathbf{W}_e = \mathbf{Q}_e^{-1}$, the last term reduces to the negative of the third term, and therefore the two cancel out and the final expression for the cofactor matrix of the estimated observations becomes

$$\mathbf{Q}_{\hat{l}\hat{l}} = \mathbf{Q} - \mathbf{Q}\mathbf{A}^T\mathbf{W}_e\mathbf{A}\mathbf{Q} \qquad (2.107)$$

As an exercise, the reader should evaluate the cofactor matrix of the residuals $\mathbf{Q}_{vv}$ from Equation (2.105) and verify that it is the negative of the last term in Equation (2.107). So, an alternative to Equation (2.107) is

$$\mathbf{Q}_{\hat{l}\hat{l}} = \mathbf{Q} - \mathbf{Q}_{vv} \qquad (2.108)$$

EXAMPLE 2.14. The interior angles of a plane triangle are measured to be $l_1 = 45°25'01''$, $l_2 = 65°20'00''$, and $l_3 = 69°15'02''$ (the same data as used for Example 2.13). Compute the least-squares estimates of the three angles if the measurements are uncorrelated and of equal precision.

Solution. Because it takes a minimum of two angles to fix the shape of a plane triangle, given three measured angles, the redundancy according to Equation (2.43) is

$$r = n - n_0 = 3 - 2 = 1$$

Therefore, one condition equation relates the adjusted observations:

$$\hat{l}_1 + \hat{l}_2 + \hat{l}_3 = 180°$$

meaning that the sum of the interior angles in a plane triangle must equal $180°$. This equation may be rewritten in the form of $\mathbf{A}v = f$:

$$v_1 + v_2 + v_3 = 180° - l_1 - l_2 - l_3 = -3''$$

or

$$[1 \quad 1 \quad 1] \begin{bmatrix} v_1 \\ v_2 \\ v_3 \end{bmatrix} = -3''$$

Because the observations are uncorrelated and of equal weight, $\mathbf{Q} = \mathbf{W} = \mathbf{I}$, and from Equations (2.102), (2.103), and (2.105), we obtain

$$\mathbf{W}_e = (\mathbf{A}\mathbf{A}^T)^{-1} = \frac{1}{3}$$

$$v = \mathbf{A}^T \mathbf{W}_e f = \begin{bmatrix} 1 \\ 1 \\ 1 \end{bmatrix} \frac{1}{3}(-3'') = \begin{bmatrix} -1'' \\ -1'' \\ -1'' \end{bmatrix}$$

The least-squares estimates of the observations are

$$\hat{l} = l + v = \begin{bmatrix} 45°25'01'' \\ 65°20'00'' \\ 69°15'02'' \end{bmatrix} + \begin{bmatrix} -1'' \\ -1'' \\ -1'' \end{bmatrix} = \begin{bmatrix} 45°25'00'' \\ 65°19'59'' \\ 69°15'01'' \end{bmatrix}$$

which when added together yield $180°$, thus satisfying the required condition.

As expected, these adjusted angles correspond to the angles obtained by the least-squares adjustment by indirect observations in Example 2.13. The propagated cofactor matrix for adjusted observations is obtained as follows by Equation (2.108):

$$\mathbf{Q}_{\hat{l}\hat{l}} = \mathbf{Q} - \mathbf{Q}\mathbf{A}^T\mathbf{W}_e\mathbf{A}\mathbf{Q} =$$

$$\begin{bmatrix} 1 & 0 & 0 \\ 0 & 1 & 0 \\ 0 & 0 & 1 \end{bmatrix} - \begin{bmatrix} 1 & 0 & 0 \\ 0 & 1 & 0 \\ 0 & 0 & 1 \end{bmatrix} \begin{bmatrix} 1 \\ 1 \\ 1 \end{bmatrix} [\tfrac{1}{3}][1 \quad 1 \quad 1] \begin{bmatrix} 1 & 0 & 0 \\ 0 & 1 & 0 \\ 0 & 0 & 1 \end{bmatrix} = \begin{bmatrix} \frac{2}{3} & -\frac{1}{3} & -\frac{1}{3} \\ -\frac{1}{3} & \frac{2}{3} & -\frac{1}{3} \\ -\frac{1}{3} & -\frac{1}{3} & \frac{2}{3} \end{bmatrix}$$

which is identical to the propagated cofactor matrix for adjusted angles calculated in Example 2.13.

2.29
COMBINED ADJUSTMENT OF OBSERVATIONS AND PARAMETERS

A more general least-squares technique than those in the preceding two sections arises when the condition equations contain parameters but the number of conditions is less than the number of observations n. In the adjustment of indirect observations, the number of parameters u must be equal to n_0 so that the number of conditions becomes n. In the present case, u usually is less than n_0 and hence c is less than n. Some or all of the conditions then will contain more than one measurement. In this case, the general form of condition equations becomes

$$\underset{c,n\ \ n,1}{\mathbf{A}\ \boldsymbol{v}}\ +\ \underset{c,u\ \ u,1}{\mathbf{B}\ \boldsymbol{\Delta}}\ =\ \underset{c,1}{\boldsymbol{f}} \qquad (2.109)$$

For linear problems, the vector f would be given by Equation (2.60).

Equation (2.109) is a representation of the form of the conditions when observations and parameters are combined. Under these conditions, the normal equations are given by (their derivation is beyond the scope of this book)

$$[\mathbf{B}^T(\mathbf{AQA}^T)^{-1}\mathbf{B}]\ \boldsymbol{\Delta}\ =\ \mathbf{B}^T(\mathbf{AQA}^T)^{-1}\boldsymbol{f} \qquad (2.110)$$

The auxiliaries of Equation (2.110) are

$$\mathbf{N}\ =\ \mathbf{B}^T(\mathbf{AQA}^T)^{-1}\mathbf{B}$$

$$\boldsymbol{t}\ =\ \mathbf{B}^T(\mathbf{AQA}^T)^{-1}\boldsymbol{f} \qquad (2.111)$$

in which $\mathbf{Q}$ is the cofactor matrix of the observations.

2.30
POST-ADJUSTMENT ANALYSIS

After the least-squares adjustment, it is quite important in surveying to analyze the results and provide a statement regarding the quality of the estimates. This operation, referred to as *post-adjustment analysis,* applies various statistical techniques that are given in Appendix D.

Test on the Reference Variance

The first test is on the estimated reference variance, $\hat{\sigma}_0^2$. Let the a priori reference variance be σ_0^2, r is the degrees of freedom (redundancy) in the adjustment, and assume that the residuals v_i are normally distributed. The statistic $r\,\hat{\sigma}_0^2/\sigma_0$ has a χ^2 probability distribution with r degrees of freedom (see Section D.5, Appendix D). The two-tailed $100(1 - \alpha)$ confidence region for σ_0^2 is given by

$$(r\hat{\sigma}_0^2/\chi_{r,\alpha/2}^2) < \sigma_0^2 < (r\hat{\sigma}_0^2/\chi_{r,1-\alpha/2}^2) \qquad (2.112)$$

If σ_0^2 is incorrect or the mathematical model used is improper or incomplete (i.e., does not adequately account for systematic errors), then σ_0^2 will fall outside this interval.

Basically, two broad categories are to be investigated when the test on $\hat{\sigma}_0^2$ fails: one corresponding to the functional model and the other to the stochastic model. In the first category one must ascertain that (1) the computations are correctly performed; (2) the mathematical model is properly formulated, the correct equations are written and linearized (if nonlinear) correctly, and so on; (3) all possible systematic errors (model deficiencies) have been adequately corrected for; and (4) any blunders or outliers have been identified, located, and eliminated. Once these matters have been done, consideration is given next to the a priori value σ_0^2. Note that this essentially is a scale factor that indicates how realistic were the chosen values of variances and covariances of the observations used in the adjustment. If we are unable to ascertain the adequacy of σ_0^2 or when no a priori value for σ_0^2 is available, such as for level networks, the rest of the postadjustment statistical evaluations will be performed using the a posteriori reference variance $\hat{\sigma}_0^2$ because it becomes the only available information to use. This value of $\hat{\sigma}_0^2$, is computed from the calculated observational residuals, v, which are estimates of the true errors, ϵ, in the observations and show only part of the true errors (see Mikhail 1979). Furthermore, $\hat{\sigma}_0^2$ combines all error sources that cannot be separated, and therefore, it is a very limited statistic, so its corresponding global statistical test is not very effective.

Test for Blunders or Outliers

If v_i is the ith residual and σ_{v_i} is its standard deviation, then

$$\bar{v}_i = v_i/\sigma_{v_i} \tag{2.113}$$

is called the *standardized residual*. Frequently, the effort involved in computing Σ_{vv} is quite extensive and therefore an approximate estimate of σ_{v_i} may be obtained from

$$\hat{\sigma}_{v_i} = [(n-u)/n]^{1/2}\hat{\sigma}_0\sigma_{l_i}/\sigma_0 = \left[\frac{n-u}{n}\right]^{1/2}\hat{\sigma}_0\,q_{l_i} \tag{2.114}$$

in which n is the number of observations, u is the number of parameters (thus, $n - u = r$, the redundancy), σ_{l_i} is the a priori standard deviation of observation l_i, and q_{l_i}, is the a priori cofactor or relative standard deviation of l_i. When σ_0^2 is known, $\bar{v}_i$ has a probability density function (pdf) $N(0,1)$ and

$$\bar{v}_i = |v_i/\sigma_{v_i}| < N_{1-\alpha/2} \tag{2.115}$$

If σ_0^2 is not known,

$$\bar{v}_i = |v_i/\hat{\sigma}_{v_i}| < \tau_{r,1-\alpha/2} \tag{2.116}$$

in which $\hat{\sigma}_{v_i}$ is computed from (2.114), and τ_r has a τ pdf with r degrees of freedom. If r is large, as in surveying, photogrammetric, or geodetic networks with extensive observations, τ_r may be replaced by the student t_r pdf or even the normal pdf.

Confidence Region for Estimated Parameters

The covariance matrix for the parameters as evaluated from the least squares is given by

$$\Sigma_{\hat{x}\hat{x}} = \sigma_0^2 N^{-1} \tag{2.117}$$

It can be shown that a region of constant probability is bounded by a u-dimensional hyperellipsoid centered at $\hat{x}$, if the parameters are assumed to have a multivariate normal probability density function. The function

$$k^2 = (x - \hat{x})^T \Sigma_{\hat{x}\hat{x}} (x - \hat{x}) \tag{2.118}$$

describes the hyperellipsoid. The quadratic k^2 has a χ_u^2 distribution (see Section D.5), with the probability for a point estimate being

$$P(\chi_u^2 < k^2) = 1 - \alpha$$

For the two-dimensional case (error ellipses, see the next section), the typical values are given on the left side of Table 2.4 and for the three-dimensional case, they are given on the right side. When $k = 1$, we usually call it the *standard region* or *standard error ellipse* (for two dimensions) and *standard error ellipsoid* (for three dimensions).

TABLE 2.4
**Point estimate probability
values**

2 dimensions		3 dimensions	
P	**k**	**P**	**k**
0.394	1.000	0.199	1.000
0.500	1.177	0.500	1.538
0.632	1.414	0.900	2.500
0.900	2.146	0.950	2.700
0.950	2.447	0.990	3.368
0.990	3.035		

Given $\boldsymbol{\Sigma}$, for example, for a point in a plane, the semimajor axis, a, and semiminor axis, b, of the standard error ellipse are computed from the eigenvalues and eigenvectors (see Section B.8, in Appendix B). If one is interested in the 90 percent confidence region (i.e., a significance level of $\alpha = 0.10$), and a, b are multiplied by 2.146. For the standard regions, there is a 0.683 probability that an adjusted point falls in a one-dimensional interval, a 0.394 probability that it falls inside the standard error ellipse, and only a 0.199 probability that it falls within the standard error ellipsoid.

If the a posteriori reference variance, $\hat{\sigma}_0^2$, is used,

$$\hat{\boldsymbol{\Sigma}}_{xx} = \hat{\sigma}_0^2 \mathbf{Q}_{xx} = \hat{\sigma}_0^2 \mathbf{N}_{xx}^{-1} \tag{2.119}$$

Whereas $\boldsymbol{\Sigma}_{xx}$ or $\hat{\boldsymbol{\Sigma}}_{xx}$ provide the most complete description of the quality of the estimated values, such as survey point coordinates, quite frequently there is a need for one representative assessment quantity. Such a quantity is often taken as the mean of the variances:

$$\bar{\sigma}_x^2 = \frac{1}{u} \operatorname{tr}(\boldsymbol{\Sigma}_{xx})$$

$$\hat{\bar{\sigma}}_x^2 = \frac{1}{u} \operatorname{tr}(\hat{\boldsymbol{\Sigma}}_{xx}) \tag{2.120}$$

in which $\bar{\sigma}_x^2$ and $\hat{\bar{\sigma}}_x^2$ are estimates of the "mean" accuracy of the coordinates, and u is the dimension of the covariance matrix or the number of the coordinates with which it is associated. Note that the trace of the covariance matrix also is equal to the sum of its eigenvalues or the sum of the variances of the transformed matrix such that all covariances are 0 (i.e., all correlations have been eliminated).

2.31
ERROR ELLIPSES

The variance and standard deviation are measures of precision of the one-dimensional case of an angle or a distance. In two-dimensional problems, such as the horizontal position of a point, error ellipses may be established around the point to designate precision regions of different probabilities. The orientation of the ellipse relative to the x, y axes system (Figure 2.8) depends on the correlation between x and y. If they are uncorrelated, the ellipse

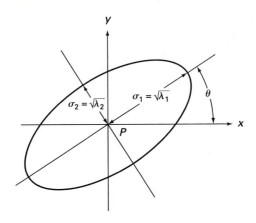

FIGURE 2.8
Error ellipse.

axes will be parallel to x and y. If the two coordinates are of equal precision, or $\sigma_x = \sigma_y$, the ellipse becomes a circle.

Consider the general case where the covariance matrix for the position of point P is given as

$$\Sigma = \begin{bmatrix} \sigma_x^2 & \sigma_{xy} \\ \sigma_{xy} & \sigma_y^2 \end{bmatrix} \tag{2.121}$$

The semimajor and semiminor axes of the corresponding ellipse are computed in the following manner. First, a second-degree polynomial (called the *characteristic polynomial*) is set up using the elements of Σ as

$$\lambda^2 - (\sigma_x^2 + \sigma_y^2)\lambda + (\sigma_x^2\sigma_y^2 - \sigma_{xy}^2) = 0 \tag{2.122}$$

The two roots, λ_1 and λ_2, of Equation (2.122) (which are called the *eigenvalues of* Σ) are computed, and their square roots are the semimajor and semiminor axes of the *standard error ellipse,* as shown in Figure 2.8. The orientation of the ellipse is determined by computing θ between the x axis and the semimajor axis from

$$\tan 2\theta = \frac{2\sigma_{xy}}{\sigma_x^2 - \sigma_y^2} \tag{2.123}$$

The quadrant of 2θ is determined from the fact that the sign of $\sin 2\theta$ is the same as the sign of σ_{xy} and $\cos 2\theta$ has the same sign as $(\sigma_x^2 - \sigma_y^2)$. Whereas, in the one-dimensional case, the probability of falling within $+\sigma$ and $-\sigma$ is 0.683, the probability of falling on or inside the standard error ellipse is 0.394. In a manner similar to constructing intervals with given probabilities for the one-dimensional case (Section 2.30), different size ellipses may be established, each with a given probability. It should be obvious that the larger the size of the error ellipse, the larger is the probability. Using the standard ellipse as a base, Table 2.4 (in the previous section) gives the scale multiplier k to enlarge the ellipse and the corresponding probability (see Mikhail 1979).

As an example, for an ellipse with axes $a = 2.447a_s$ and $b = 2.447b_s$ where a_s and b_s are the semimajor and semiminor axes, respectively, of the standard ellipse, the probability that the point falls inside the ellipse is 0.95.

In surveying, one frequently is interested in the relative accuracy between two points, 1 and 2, in a horizontal network. Then, the coordinate differences are

$$d_x = x_2 - x_1$$

$$d_y = y_2 - y_1$$

and the total covariance matrix for the coordinates is

$$\Sigma_{1,2} = \begin{bmatrix} \sigma_{x_1}^2 & \sigma_{x_1 y_1} & \sigma_{x_1 x_2} & \sigma_{x_1 y_2} \\ & \sigma_{y_1}^2 & \sigma_{y_1 x_2} & \sigma_{y_1 y_2} \\ & & \sigma_{x_2}^2 & \sigma_{x_2 y_2} \\ \text{symm.} & & & \sigma_{y_2}^2 \end{bmatrix}$$

Then,

$$d = \begin{bmatrix} d_x \\ d_y \end{bmatrix} = \begin{bmatrix} -1 & 0 & 1 & 0 \\ 0 & -1 & 0 & 1 \end{bmatrix} \begin{bmatrix} x_1 \\ y_1 \\ x_2 \\ y_2 \end{bmatrix} = \mathbf{JP}_{1,2}$$

and

$$\Sigma_{dd} = \mathbf{J}\Sigma_{1,2}\mathbf{J}^T$$

from which

$$\sigma_{dx}^2 = \sigma_{x_1}^2 - 2\sigma_{x_1 x_2} + \sigma_{x_2}^2$$
$$\sigma_{dy}^2 = \sigma_{y_1}^2 - 2\sigma_{y_1 y_2} + \sigma_{y_2}^2 \tag{2.124}$$
$$\sigma_{dxdy} = \sigma_{x_1 y_1} - \sigma_{x_1 y_2} - \sigma_{x_2 y_1} + \sigma_{x_2 y_2}$$

Introducing the variances, σ_{dx}^2 and σ_{dy}^2, and covariance, σ_{dxdy}, in Equations (2.122) and (2.123), the elements of a relative error ellipse can be computed.

In the three-dimensional case, where the horizontal position as well as the elevation of the point is involved, the precision region becomes an ellipsoid. Table 2.4 (Section 2.30) gives the corresponding multipliers. For more details on error ellipsoids, the reader may consult Mikhail (1979).

The concepts of error ellipse and error ellipsoid are quite useful in establishing confidence regions about points determined by surveying techniques. These regions are measures of the reliability of the positional determination of such points. They could also be specified in advance as a means of establishing specifications.

Although both absolute error ellipses (for points) and relative error ellipses (for lines) are used to evaluate adjustment quality, it is frequently more convenient to replace the two-dimensional representation by a one-dimensional single quantity (similar to σ in Section 2.30). In this case, a circular probability distribution is substituted for the elliptical probability distribution. Consequently, a single circular standard deviation, σ_c, is calculated from the two semiaxes of the error ellipse. The value of σ_c depends on the relative magnitudes of these axes.

Let

$$\Sigma_{xx} = \begin{bmatrix} \sigma_x^2 & \sigma_{xy} \\ \sigma_{xy} & \sigma_y^2 \end{bmatrix}$$

represent the covariance matrix for the x, y coordinates of a point. Then, $\lambda_1 = a^2$ and $\lambda_2 = b^2$ are the eigenvalues ($\lambda_1 > \lambda_2$) of Σ_{xx}, and a and b are the semimajor and semiminor axes of the error ellipse, respectively. Note that

$$\text{tr}(\Sigma_{xx}) = \sigma_x^2 + \sigma_y^2 = \lambda_1 + \lambda_2 = a^2 + b^2$$

If $\sigma_{min} = b = \sqrt{\lambda_2}$ and $\sigma_{max} = a = \sqrt{\lambda_1}$, then the value of the ratio $\sigma_{min}/\sigma_{max}$ determines the relationship used to calculate σ_c.

When $\sigma_{min}/\sigma_{max}$ is between 1.0 and 0.6,

$$\sigma_c \simeq (0.5222\sigma_{min} + 0.4778\sigma_{max}) \tag{2.125}$$

A good approximation that yields a slightly larger σ_c (i.e., on the safe side) is given by

$$\sigma_c \simeq 0.5(a + b) \tag{2.126}$$

which may be extended to the limit of $\sigma_{min}/\sigma_{max} \leq 0.2$.

The "mean" accuracy measure, $\overline{\sigma}$ (see Equation 2.120), is

$$\overline{\sigma}^2 = \frac{1}{2}(\sigma_x^2 + \sigma_y^2) = \frac{1}{2}(a^2 + b^2) = \frac{1}{2}(\lambda_1 + \lambda_2)$$

and

$$\overline{\sigma}_c = \left[\frac{1}{2}(\sigma_x^2 + \sigma_y^2)\right]^{1/2} \tag{2.127}$$

actually is applicable only when $\sigma_{min}/\sigma_{max}$ is between 1.0 and 0.8, in which case it yields essentially the same value of σ_c as in Equation (2.126). As the ratio of $\sigma_{min}/\sigma_{max}$ decreases, σ_c from Equation (2.127) gets progressively larger than that from Equation (2.126) with a maximum increase of about 20 percent at $\sigma_{min}/\sigma_{max} = 0.2$.

Of course, the probability associated with the standard error circle is the same as for the standard error ellipse, 0.394. The multipliers given in Table 2.4 also still apply for circular errors of different probabilities. Figure 2.9 shows several standard error ellipses and their corresponding circles for several $\sigma_{min}/\sigma_{max}$ ratios.

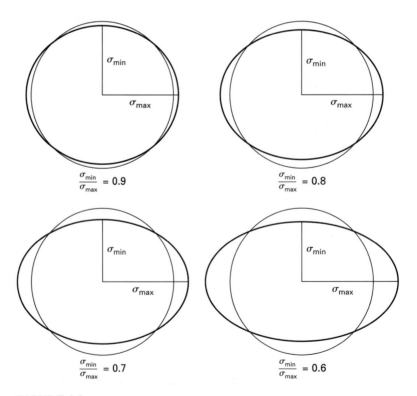

FIGURE 2.9
Error ellipses and their corresponding error circles (from "*ACIC Technical Report No. 96*," United States Air Force, February 1962).

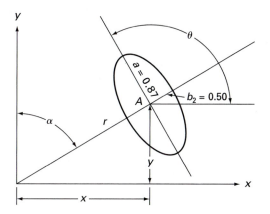

FIGURE 2.10
Error ellipse for rectangular coordinates.

EXAMPLE 2.15. The position of point A in Figure 2.10 is determined by the radial distance $r = 100$ m, with $\sigma_r = 0.5$ m, and the azimuth angle $\alpha = 60°$, with $\sigma_\alpha = 0°30'$. Compute the rectangular coordinates x and y and the associated covariance matrix for point A. (Assume r and α to be uncorrelated.) Then calculate the semimajor and semiminor axes of the standard error ellipse and its orientation.

Solution

$$x = r \sin \alpha = 86.60$$

$$y = r \cos \alpha = 50.00$$

$$\mathbf{J} = \begin{bmatrix} \dfrac{\partial x}{\partial \alpha} & \dfrac{\partial x}{\partial r} \\ \dfrac{\partial y}{\partial \alpha} & \dfrac{\partial y}{\partial r} \end{bmatrix} = \begin{bmatrix} r \cos \alpha & \sin \alpha \\ -r \sin \alpha & \cos \alpha \end{bmatrix} = \begin{bmatrix} 50 & 0.866 \\ -86.6 & 0.5 \end{bmatrix}$$

The covariance matrix of the known variables, r and α, is a diagonal matrix because they are uncorrelated. It is equal to

$$\mathbf{\Sigma} = \begin{bmatrix} \sigma_\alpha^2 & 0 \\ 0 & \sigma_r^2 \end{bmatrix} = \begin{bmatrix} (0.0087)^2 & 0 \\ 0 & (0.5)^2 \end{bmatrix}$$

Applying Equation (2.34), the covariance matrix of the Cartesian coordinates is

$$\mathbf{\Sigma}_{coord} = \begin{bmatrix} 50 & 0.866 \\ -86.6 & 0.5 \end{bmatrix} \begin{bmatrix} 0.7569 \times 10^{-4} & 0 \\ 0 & 0.25 \end{bmatrix} \begin{bmatrix} 50 & -86.6 \\ 0.866 & 0.5 \end{bmatrix}$$

$$= \begin{bmatrix} 0.3767 & -0.2195 \\ -0.2195 & 0.6300 \end{bmatrix}$$

This matrix expresses the reliability of the Cartesian coordinates of point A. According to Equation (2.122), the characteristic polynomial is

$$\lambda^2 - (0.3767 + 0.6300)\lambda + (0.3767)(0.6300) - (-0.2195)^2 = 0$$

$$\lambda^2 - 1.0067\lambda + 0.189141 = 0$$

from which the two roots (eigenvalues) are

$$\lambda_1 = 0.7567; \qquad \lambda_2 = 0.25$$

The semimajor axis is $a = \sqrt{\lambda_1} = 0.87$ m and the semiminor axis is $b = \sqrt{\lambda_2} = 0.50$ m. To get the orientation of the semimajor axis, use Equation (2.123):

$$\tan 2\theta = \frac{2\sigma_{xy}}{\sigma_x^2 - \sigma_y^2} = \frac{-2(0.2195)}{0.3767 - 0.6300} = \frac{-0.4390}{-0.2533} = 1.732$$

Because both $\sin 2\theta$ and $\cos 2\theta$ are negative, 2θ is in the third quadrant, or $2\theta = 240°$, and $\theta = 120°$.

The results obtained here are rather important. They show, as demonstrated in Figure 2.10, that the semiminor axis is along the extension of r and equal to $\sigma_r = 0.5$ m. Similarly, the semimajor axis is oriented normal to r and is equal to $r\sigma_\theta = (100)(0.0087) = 0.87$ m. Therefore, the error ellipse axes always are oriented in the directions where the variables are uncorrelated, as in this case r and α. Along any other pair of directions, the variables are correlated, as in the case of x and y, for which the correlation coefficient ρ_{xy} may be computed from Equation (2.28) as

$$\rho_{xy} = \frac{-0.2195}{(0.6138)(0.7937)} = 0.45$$

For this example $\sigma_{min} = b = 0.5$ m and $\sigma_{max} = a = 0.87$ m, and the ratio is $\sigma_{min}/\sigma_{max} = 0.57$. The radius of the standard error circle, according to Equation (2.126), is

$$\sigma_c \cong 0.5(a + b) \simeq 0.5(0.5 + 0.87) = 0.685 \text{ m}$$

If Equation (2.127) is used,

$$\sigma_c = [0.5(0.3767 + 0.6300)]^{1/2} = 0.709 \text{ m}$$

which is about 9 percent larger than σ_c. If one can ascertain that the ratio of $\sigma_{min}/\sigma_{max}$ is not below 0.2, then the advantages of using Equation (2.127) are that there is no need to compute the eigenvalues, and the equation produces a slightly larger error circle.

As with circular errors for two-dimensional locations, a standard error sphere with a radius σ_s may replace the standard error ellipsoid around a survey point in three-dimensional space. If $a = \sqrt{\lambda_1}, b = \sqrt{\lambda_2}, c = \sqrt{\lambda_3}$, where $\lambda_1 \geq \lambda_2 \geq \lambda_3$ are the eigenvalues of the covariance matrix Σ_{xx} of the coordinates X, Y, Z, and $a \geq b \geq c$ are the semiaxes of the error ellipsoid, then

$$\sigma_s = \frac{1}{3}(a + b + c) \qquad (2.128)$$

is used when $\sigma_{min}/\sigma_{max}$ is between 1.0 and 0.35, where σ_{min} and σ_{max} are the smallest and largest semiaxes of the ellipsoid, respectively. When the ratio of $\sigma_{min}/\sigma_{max}$ is between 1.0 and 0.9, the mean of the trace, or

$$\bar{\sigma}_s = \left[\frac{1}{3}(\sigma_x^2 + \sigma_y^2 + \sigma_x^2) \right]^{1/2} \qquad (2.129)$$

essentially will give the same value as in Equation (2.128). As this ratio gets smaller, $\bar{\sigma}_s$ will get progressively larger than σ_s, with an increase of about 16 percent as this ratio approaches 0.35 (depending, of course, on the value of the middle semiaxis, b). The spherical concept is not recommended when $\sigma_{min}/\sigma_{max}$ is less than 0.35.

2.32
DESIGN OR PREANALYSIS

Engineering *design* more frequently is known as *preanalysis* in surveying engineering. It refers to the task of determining the observations to be made and their required accuracy so that the accuracy of the final product, as given in the project specifications, is met. This usually is done by an iterative procedure, often using interactive graphics and applying the established mathematical model of the problem. The physical limitations of the project, such as visibility and accessibility problems, first are imposed on the design. Next, what is considered to be an adequate set of measurements with suitable accuracy estimates is entered into a design program to estimate the required unknown parameters and their expected accuracy. The overall accuracy of the estimated parameters is reflected by the posterior covariance matrix Σ. (This is why preanalysis is equivalent to *covariance analysis* used in the mathematical literature.) From Σ, individual confidence measures, such as error ellipses and ellipsoids, are computed and compared to the design requirements. If the confidence measures are too large, additional observations or better-quality measurements are attempted and the process is repeated. On the other hand, if they are too small (i.e., too good), reduced observations or lower-quality observations (using less expensive equipment or techniques) are attempted instead. The procedure is iterated until an optimum design results.

The covariance matrix, $\Sigma_{\Delta\Delta}$, of the estimated parameters, such as survey point coordinates, is directly related to $\mathbf{N}^{-1}$ (see Equations (2.78) and (2.83)). The value of $\mathbf{N}$ is given by either $\mathbf{N} = \mathbf{B}^T\mathbf{W}\mathbf{B}$ (Equation (2.74)) or $\mathbf{N} = \mathbf{B}^T(\mathbf{A}\mathbf{Q}\mathbf{A}^T)^{-1}\mathbf{B}$ (Equation (2.111)), depending on the technique of least squares applied. It is clear that $\mathbf{N}$, and consequently $\Sigma_{\Delta\Delta}$, do not depend on the values of the observations, l. They are functions of only the coefficient matrices, $\mathbf{A}$ and $\mathbf{B}$, and the a priori quality of the observations specified by either $\mathbf{Q}$ or its inverse $\mathbf{W}$. The coefficient matrices are Jacobians of the condition equations evaluated at approximate values for the parameters x° and observations l°. Variations of these approximations will have some effect on $\Sigma_{\Delta\Delta}$, but the effect usually is minor in survey networks.

From $\Sigma_{\Delta\Delta}$, the quality of each survey station position estimate can be evaluated by determining its error ellipse as described in the preceding section. Similarly, relative error ellipses between various pairs of survey stations also can be evaluated.

Either standard error ellipses or, more practically, those corresponding to 90 percent or 95 percent confidence, are used in the evaluation of the design. The evaluation could be for each point and each line in the network or for an overall accuracy measure. Both the number of observations and their quality (in Σ_{ll} or $\mathbf{W}$) may be changed, the simulation program rerun, and new quality measures evaluated. The procedure is iterated until one reaches the combination required to meet the stated specifications. This could all be done on-line, taking advantage of computer graphics for examining not only the overall network but each segment to avoid any geometric weaknesses in the design.

PROBLEMS

2.1. Three angles α, δ, and τ, are measured in a plane triangle. The respective standard deviations estimated for the angles are σ_α, σ_δ, and σ_τ. Form the mathematical model and the stochastic model for this problem.

2.2. Describe and give examples of three types of errors and their influence on survey measurements.

2.3. Explain and give examples of the differences between systematic and random errors in surveying measurements.

2.4. What are some common ways of detecting blunders in measurement?

2.5. Describe how one can compensate for systematic errors in the measurement of (a) distance and (b) angles.

2.6. An angle is assumed to be normally distributed with a mean $\mu = 40°30'32.5''$ and variance of $\sigma^2 = 4.0 \, sec^2$. Sketch the distribution showing the 0.6827 probability region. Give the range in the angle corresponding to this probability. Provide numerical values with a graphical interpretation.

2.7. Using the concept of expectation, define the mean, standard deviation, covariance, and correlation coefficient.

2.8. Explain what is meant by reference variance and describe its significance with respect to surveying measurements.

2.9. Three independent angle measurements have standard deviations of $02''$, $06''$ and $12''$. Form the corresponding covariance matrix. Determine the correlation coefficient between the second and third angles.

2.10. If the reference variance in Problem 2.9 is $\sigma_{01}^2 = 24''^2$, calculate the weight matrix $\mathbf{W}$ for the angles. Using another value, $\sigma_{02}^2 = 48''^2$, what is the new weight matrix?

2.11. Estimates of the X and Y coordinates of a survey point have standard deviations of $\sigma_X = 5$ mm and $\sigma_Y = 10$ mm. If the covariance term, $\sigma_{XY} = 0.600$, form the covariance matrix and calculate the correlation coefficient between X and Y.

2.12. The random errors in the planimetric position of a survey point have standard deviations of $\sigma_X = \sigma_Y = 0.1$ m in the X and Y directions and a correlation coefficient of $\rho_1 = 0.6$. Evaluate the standard error ellipse and the 95 percent error ellipse. Sketch both ellipses.

2.13. Given the same standard deviations as in Problem 2.12, evaluate and sketch error ellipses for $\rho_2 = -0.6$, $\rho_3 = 0$, and $\rho_4 = 1$.

2.14. Assume that the following 25 independent measurements of a distance have been corrected for systematic errors:

206.159	206.139	206.143	206.131	206.129
206.161	206.137	206.137	206.133	206.150
206.145	206.135	206.136	206.140	206.149
206.140	206.144	206.147	206.140	206.152
206.143	206.148	206.142	206.132	206.120

(a) Construct a histogram from these data, using a class interval of 2 mm, with the lowest and highest intervals being $206.120 - 206.121$ and $206.160 - 206.161$ m, respectively.
(b) Calculate the sample mean, sample median, sample mode, and sample midrange.
(c) Calculate the sample variance and standard deviation and the standard deviation of the sample mean.

2.15. Using the sample data in Problem 2.14, calculate the parameters of a normal density curve and plot this curve superimposed on the histogram.

2.16. Assuming that the calculated standard deviation for the curve in Problem 2.15 in fact is the population parameter, evaluate three ranges within which the population mean μ falls, corresponding to the three probabilities 0.6827, 0.95, and 0.99.

2.17. Angles measured by repetition by five different groups are $33°42'10''$, $33°42'00''$, $33°42'15''$, $33°42'20''$, and $33°42'30''$. Compute the standard deviation of (a) a single value and (b) the mean angle. Determine the confidence interval for $33°42'10''$, turned in by the first group, with a probability of 0.95.

2.18. Uncorrelated angles a and b in triangle ABC are measured with estimated values of $\sigma_a = 15''$ and $\sigma_b = 30''$. Compute the estimated value of σ_c. Form the covariance matrix for angles a, b, and c.

2.19. The estimated standard deviation for measurement of a distance, determined by Electronic Distance Measurement (EDM) is 0.005 m + 10 ppm (parts per million) of the distance measured. Calculate estimated standard deviations for distances 300.05, 30,100.60, and 2500.45 m measured with this EDM instrument. Assuming a reference variance of 1.00, form the covariance and weight matrices for the distances, which are assumed to be uncorrelated.

2.20. Plane rectangular coordinates are calculated using these equations:

$$X = X_0 + d \sin (A) \qquad \text{and} \qquad Y = Y_0 + d \cos (A)$$

If X_0 and Y_0 are assumed errorless and the respective standard deviations for d, the measured distance, and A, the measured azimuth, are 0.005 m and $05''$, then form the covariance matrix for calculated coordinates X and Y for $d = 500.000$ m and $A = 42°30'30''$ and calculate the correlation coefficient between the X and Y coordinates.

2.21 Three adjacent angles have the following mean values, standard deviations, and correlation coefficients.

Angle	Mean value	σ	ρ
1	30°10'30"	30"	
			0.6
2	10°20'10"	10"	
			0.6
3	25°35'15"	20"	
			0.6
1			

Calculate the mean value and standard deviation of the total angle (sum of the three angles).

2.22. Two sides and the included angle of a plane triangle are measured to give the following values:

$$a = 100.000 \text{ m}, \qquad \sigma_a = 0.030 \text{ m}; \qquad b = 200.000 \text{ m}, \qquad \sigma_b = 0.040 \text{ m};$$

$$C = 45°00'00'', \qquad \sigma_C = 30''$$

The measurements are uncorrelated. Calculate the area of the triangle and standard deviation of the area.

2.23. A line was measured by three parties in three sections that have respective lengths and standard deviations as follows:

Party	Distance, m	Estimated σ, m
151	220.451	0.030
152	150.550	0.050
153	300.250	0.090

Calculate the estimated standard deviation of the total length of line.

2.24. A tract of land has a trapezoidal shape with two parallel sides a and b and the distance between them (i.e., height of trapezoid) is h. You are given the following data:

$$a = 150.12 \text{ m} \qquad \sigma_a = 0.20 \text{ m}$$
$$b = 225.64 \text{ m} \qquad \sigma_b = 0.10 \text{ m}$$
$$h = 30.00 \text{ m} \qquad \sigma_h = 0.05 \text{ m}$$

The three measurements are uncorrelated. Calculate the area of the tract and its standard deviation.

2.25 A sample is given of $x_1, x_2, \ldots, x_n$ measurements, with $\sigma_{x1} = \sigma_{x2} = \cdots = \sigma_{xn} = \sigma$ also known. Assuming uncorrelated measurements and using the special law of error propagation, show the derivation of an expression for the standard deviation of the mean value, $\bar{x}$.

2.26. A side and two internal angles of a plane triangle are measured as follows:

$$\text{side } a = 124.25 \text{ m} \qquad \text{with } \sigma_a = 2 \text{ cm}$$

$$\text{angle } B = 90°00' \qquad \text{with } \sigma_B = 2'$$

$$\text{angle } C = 35°00' \qquad \text{with } \sigma_C = 2'$$

The three measurements are uncorrelated. In keeping with the general notation of geometry, the two angles B and C are at the extremities of side a.

(a) Calculate the two sides b and c and their covariance matrix.
(b) Calculate the third angle A and its standard deviation.
(c) Calculate the area of the triangle and its standard deviation.

2.27. An angle was measured 16 times by repetition and the standard deviation of a *single* angle was found to be 04". If the reference variance $\sigma_0^2 = 16''^2$, calculate the weight for the *average* angle.

2.28. Explain the difference between accuracy and precision.

2.29. When does adjustment become necessary in a survey network?

2.30. What is the basic objective of a survey adjustment?

2.31. What is the least squares criterion and why is it needed?

2.32. Why do we need to add residuals to the observations?

2.33. Is the covariance matrix of the survey measurements used in least squares? Show where it is used and why.

REFERENCES

Aguilar, A. M. "Principles of Survey Error Analysis and Adjustment." *Surveying and Mapping,* September 1973.

Allman, J. S. "Angular Measurement in Traverses." *Australian Surveyor,* December 1973.

Ananga, N. "Least Squares Adjustment of Seasonal Leveling." *Journal of Surveying Engineering, ASCE* 117, no. 2 (May 1991), pp. 66–76.

Barry, A. *Errors in Practical Measurement in Engineering, Science, and Technology.* New York: John Wiley & Sons, 1978.

Cooper, M. A. R. *Fundamentals of Survey Measurements and Analysis.* London: Crosby Lookwood, 1974.

Cross, P. A. "The Effects of Errors in Weights." *Survey Review,* July 1972.

Cross, P. A. "Ellipse Problem—More Solutions." *Survey Review,* January 1972.

Cross, P. A. "Error Ellipse: A Simple Example." *Survey Review,* October 1971.

Gao, Y.; E. J. Krakiwsky; and J. Czompo. "Robust Testing Procedure for Detection of Multiple Blunders." *Journal of Surveying Engineering, ASCE* 118, no. 1 (February 1992), pp. 11–23.

Haug, M. D. "Application of Mhor's Circle Technique in Error Analysis." *Surveying and Mapping* 44, no. 4 (December 1984), pp. 309–21.

Janes, H. W. "An Error Budget for GPS Relative Positioning." *Surveying and Land Information Systems* 51, no. 3 (September 1991), pp. 133–37.

Leick, A. "Minimal constraints in Two-Dimensional Networks." *Journal of the Surveying and Mapping Division, ASCE* 108, no. SU2 (August 1982), pp. 53–68.

Mikhail, E. M. *Observations and Least Squares.* New York: Harper and Row, Publishers, 1976.

Mikhail, E. M. "Review and Some Thoughts on the Assessment of Aerial Triangulation." Aerial Triangulation Symposium, University of Queensland, Australia, October 1979.

Mikhail, E. M., and G. Gracie. *Analysis and Adjustment of Survey Measurements.* New York: Van Nostrand Reinhold Company, 1980.

Richardus, P. *Project Surveying.* Amsterdam: North-Holland Publishing Company, 1966.

Scarbourough, J. B. *Numerical Mathematical Analysis.* 6th ed. Baltimore: Johns Hopkins Press, 1966.

Tan, W. "Free Net Analysis Under Squared Error Loss." *Journal of Surveying Engineering, ASCE* 155, no. 4 (November 1989), pp. 373–79.

Thompson, E. H. "A Note on Systematic Errors." *Photogrammetria* 10 (1953–1954).

Uotila, U. A. "Useful Statistics for Land Surveyors." *Surveying and Mapping,* March 1973.

CHAPTER 3

Field and Office Work

3.1
GENERAL

The nature of surveying measurements already has been indicated. Much of the field and office work involved in the acquisition and processing of measurement is performed concurrently. Field and office work for a complete survey consists of

1. Planning and design of the survey; adoption of specifications; adoption of a map projection and coordinate system and of a proper datum; selection of equipment and procedures.
2. Care, handling, and adjustment of the instruments.
3. Fixing the horizontal location of objects or points by horizontal angles and distances by one of the surveying procedures to be developed.
4. Determining the elevations of objects or points by one of the methods of leveling.
5. Recording field measurements.
6. Field computations to verify the data.
7. Office computations in which data are reduced, adjusted, and filed or stored for current or future use.
8. Setting points in the field to display land property location and to control construction layout (as may be necessary).
9. Performing the final *as-built survey,* in which all structures built as a part of the project are located with respect to the basic control network or established property lines.

Discussions in this chapter are restricted to general comments related to planning and designing the survey, care and adjustment of instruments, methods for recording data, computational methods, and computational aids.

3.2
PLANNING AND DESIGN OF THE SURVEY

There are many types of surveys related to an almost infinite variety of projects. Therefore, the planning of the survey can be discussed only in very general terms at this point. Assuming that the nature of the project is established and the results desired from the survey are known, the steps involved in planning the survey are these:

1. Establish specifications for horizontal and vertical control accuracies.
2. Locate and analyze all existing control, maps, photographs, and other survey data.
3. Do a preliminary examination of the site in the office, using existing maps and photographs, and in the field to locate existing and set new control points.
4. Select the equipment and surveying procedures appropriate for the task.
5. Select the computational procedures and the method to present the data in a final form.

To understand the procedure, consider an example. Assume that a private surveying firm has been engaged to prepare a topographic map of a 2000 acre (809.4 ha) tract at a scale of 1:1500 (see Section 1.9 for a definition of scale) and with a contour interval of 2 m.* Control surveys are to be performed using a combination of theodolite and EDM traverse (see Section 8.38, Chapter 8) and the Global Positioning System (GPS; see Chapter 12); the major portion of the topographic mapping is to be compiled photogrammetrically. The tasks to be performed by the surveying firm are to select a proper coordinate grid and datum plane and to establish a horizontal and vertical control network throughout the area including sufficient points to control aerial photographs for the photogrammetric mapping, and to map heavily wooded valleys not accessible to the photogrammetric methods. Another firm has been engaged to do the photogrammetric work.

The map is to be used by the client for presenting plans of a proposed industrial park to the county planning commission. If plans are approved, earthwork quantities will be calculated using measurements from the map, and construction surveys will be made from control points set in the survey. Consider some of the problems involved in planning for this type of a survey.

3.3
SPECIFICATIONS

The accuracy of the measurements should be consistent with the purpose of the survey. Each survey is a problem in itself, for which the surveyor must establish the limits of error using a knowledge of the equipment involved, the procedure to be employed, error propagation, and judgment based on practical experience. The best survey is one that provides data of adequate accuracy without wasting time or money.

The Federal Geodetic Control Subcommittee (FGCS) publishes specifications for first-, second-, and third-order horizontal and vertical control surveys and for the GPS relative positioning techniques (see Chapters 5 and 9). These specifications provide a starting point for establishing standards on most jobs that require basic control surveys.

For the example given, the work can be divided into three categories: establishing basic control, establishing supplementary control, and performing topographic surveys.

---

*A contour is the line of intersection of a level surface with the terrain surface; a contour interval is the vertical distance between two successive such level surfaces (see Section 14.6 for more details).

Basic control consists of a primary network of rather widely spaced horizontal and vertical control points that will coordinate with local, state, and national controls and be used to control all other surveys throughout the entire area to be studied. Second-order specifications generally are recommended for basic control networks. For this phase, GPS methods most often are used. Because photogrammetric surveys will follow, particular attention should be paid to establishing basic horizontal control points around the perimeter of the area to be mapped, plus several vertical control points near the center of the project. Construction surveys are definitely a possibility, so that a sufficient number of basic control points should be established in the interior of the area to allow tying all subsequent construction surveys to second-order control points.

Supplementary control is set to provide additional control points for photogrammetric mapping and for topographic mapping by field methods. Third-order specifications or lower (down to a ratio of misclosure (RoM; see Section 2.11, Chapter 2) of 1 part in 3000) usually are considered adequate for supplementary horizontal control. Total station system traverses generally are used for this category of survey. On the example project, most of the supplementary control will be used for controlling the photogrammetric mapping, and a specification for error in position is appropriate. Therefore, it could be specified that supplementary control points should be to within $\pm$ a tolerance based on the requirements of the photogrammetric mapping system. Vertical control usually is established to satisfy third-order specifications. Close cooperation is required between the surveyor and photogrammetrist in this phase of the work. Under certain circumstances, supplementary control can be established by photogrammetric methods, which could be a topic for discussion between the surveyor and the photogrammetrist.

Topographic measurements in the field are made from supplementary control points. Accuracies in these measurements usually are based on map accuracies. For the example, if the map scale is 1:1500, then 0.5 mm on the map is equivalent to 0.75 m on the ground or $\frac{1}{50}$ in represents 2.5 ft. Thus, the accuracy in position for mapping is ± 0.8 m or ± 2.5 ft. Elevations are governed by the contour interval. National map accuracy standards (Section 14.26) require that 90 percent of all spot elevations determined from the map be within $\pm$ one-half contour interval of the correct elevation. Experience has shown that this accuracy of contour interpolation normally can be achieved for maps if spot elevations in the field are determined to within one-tenth of the contour interval. Heights of topographic points should be located to within ± 0.2 m for the example project. Because elevations within this interval can be well-determined by both field and photogrammetric methods, this specification presents no problem for either the field or the photogrammetric survey method.

The specifications for photogrammetric mapping are largely a function of the map scale and contour interval. As the contour interval decreases, the cost of the photogrammetric map increases at an exponential rate. Therefore, it is worth studying the desired contour interval to be sure that this interval is *really* necessary. For this example, earthwork will be calculated from measurements made on the map so that both contour interval and map scale represent optimum values that must be maintained. This being the case, aerial photography must be flown at an altitude and to specifications compatible with the stated scale and contour interval.

Unique situations on the project may require special treatment. If construction is to be concentrated in a particular area of the site, second-order control can be clustered in that region and mapping of that portion can be at a larger scale and with a smaller contour interval. For a relatively small area, such as the example problem of 2000 acres, mapping at two scales probably would not be practical. Such an approach could be feasible under special circumstances in larger areas. The requirements for construction surveys may be the limiting factor. When structures for the transport of water or other liquids by gravity flow are involved, first-order levels are required. Bridges, complicated steel structures, tunnels, and city surveys are a few examples where first-order control surveys may be necessary.

3.4
EXISTING MAPS AND CONTROL POINTS

A thorough search should be performed to locate all survey data in all existing Geographic or Land Information Systems or other data bank sources, existing maps, and aerial photographs of the area to be surveyed.

The National Geodetic Survey Division (NGSD) of the National Ocean and Atmospheric Administration (NOAA) in Silver Springs, Maryland 20910, maintains records of all first-, second-, third-, and selected fourth-order and photogrammetrically derived control point locations in the horizontal and vertical control networks of the United States. Control point data plus descriptions will be provided for a fee when requested. The Bureau of Land Management keeps a file on property survey data related to public lands survey. State, county, city, and town engineering and surveyor's offices also should be consulted for useful survey data.

The U.S. Geological Survey (USGS) is responsible for compiling maps, at various scales, of the United States. The USGS *quadrangle* topographic maps at a scale of 1 : 24,000 or 1 : 25,000 in the metric series are the maps most commonly used for engineering studies. Information can be obtained from the National Cartographic Information Centre (NCIC) in Reston, Virginia 22092. The Nautical Charting Division, also of the NOAA, compiles and sells nautical and aeronautical charts. State, county, city, and town engineering and survey offices should also be checked for maps at scales larger than the 1:24,000.

Aerial photography is very useful for preliminary planning purposes and such photographs can be acquired from the following government agencies: U.S. Geological Survey, U.S. Department of Agriculture, Stabilization and Conservation Service, and Soil Conservation Service; U.S. Forest Service; Bureau of Land Management; National Ocean Survey; and the Tennessee Valley Authority (TVA). Many of these agencies have local or regional offices that maintain files of photographs of that particular area. State highway departments generally have photographs taken along the right of way for highways. Private photogrammetric firms usually keep files of photographs for the areas that they have mapped.

Images from satellite-based multispectral scanners (MSS) provide broad, synoptic coverage that is very useful for planning, especially on large projects. The USGS Landsat System collects data of this type, which can be obtained from the USGS EROS Data Center, Sioux Falls, South Dakota 57198. Similar coverage, with higher resolution and provided also in stereo, is furnished by the French satellite SPOT (Le Système Probatoire d'Observation de la Terre) and available in the United States through the SPOT Image Corporation, Reston, Virginia 22091.

Data from all sources should be gathered and studied to exploit all previous survey work in the area. Preliminary studies can be made in the office using existing maps or aerial photographs to choose possible locations for control points. These studies are followed by field reconnaissance to locate all existing control points and set new horizontal and vertical control points in the basic and supplementary networks to be established. At this point, close cooperation between the surveyor and photogrammetrist is absolutely essential when photogrammetric mapping is involved.

Whenever possible, the basic control network should be tied to at least two existing government monuments (one for a GPS network) to provide a basis for performing the entire survey in a proper state or national coordinate system.

3.5
SELECTION OF EQUIPMENT AND PROCEDURES

The specifications for horizontal and vertical control provide an allowable tolerance for horizontal and vertical positions. Using estimates for the capabilities inherent in various types of instruments, errors in distance and direction can be propagated for a given procedure. These propagated errors then are compared with the allowable tolerance and the instrument or procedure can be modified if necessary. Various instruments, procedures for use of these instruments, and error propagation methods are described in subsequent chapters. Considerable judgment and practical experience in surveying are required in this phase of the survey design.

3.6
SELECTION OF COMPUTATIONAL PROCEDURES AND THE METHOD FOR DATA PRESENTATION

Systematic procedures for gathering, filing, processing, and disseminating of the survey data should be developed. Field notes and methods for recording data are described in Section 3.12.

The prevalence of modern, efficient electronic computing equipment virtually dictates that rigorous methods and adjustment techniques be utilized in all but the smallest of surveys.

Methods for filing or storing survey information should be standardized for all jobs and ought to allow current or future retrieval with no complications. The use of a Geographic or Land Information System (GIS or LIS) that has an interface with a Computer-Aided Drafting and Design (CADD) system (see Chapter 14) is highly recommended as an efficient way to exploit current electronic technology for storing and filing data.

The method for presenting the data must be carefully considered. For the example case, a conventional line map or an orthophoto map (see Chapter 14) with an overlay of contours would be appropriate for presentation to a planning commission. Selection of the method for presenting the data underscores the value of using GIS or LIS and CADD. For example, if a line map with an overlay of contours were chosen, then most surveying firms with GIS/CADD capability would be able to provide this product quickly at a minimum cost.

3.7
RELATION BETWEEN ANGLES AND DISTANCES

In Section 3.5 reference was made to the propagated errors in distance and in direction. These errors are used to determine the uncertainty in position for a point. Assume that estimated standard deviations in a distance r and direction α are σ_r and σ_α, respectively. As developed in Section 2.31 and Example 2.15, the position of a point determined using r and α has an uncertainty region defined by an ellipse centered about a point located by r and α as shown in Figure 3.1. In Example 2.15, $\sigma_r = 0.5$ m, $r = 100$ m, $\alpha = 60°$, and $\sigma_\alpha = 30'$, which produce an ellipse having a semiminor axis of $\sigma_r = 0.5$ m that is parallel to r. The semimajor axis of the ellipse $= r\sigma_\alpha = 0.87$ m and is normal to line r. Note that $\sigma_x = 0.61$ m and $\sigma_y = 0.79$ m, both of which are less than the maximum uncertainty in the point $= r\sigma_\alpha = 0.87$ m.

Suppose that σ_α is chosen so that $r\sigma_\alpha = \sigma_r$ or $\sigma_\alpha = 0.5/100 = 0.00500$ rad. In this case, the two axes of the region of uncertainty are equal and the ellipse becomes a circle, as illustrated in Figure 3.2. Therefore, to have the same contribution from distance and angle errors, σ_α should be about $0°17'$.

The preceding analysis illustrates the relationship between uncertainties in direction and distance and emphasizes the desirability of maintaining consistent accuracy in the two measurements. The error in distance is normally expressed as a relative precision of ratio

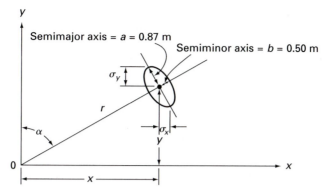

FIGURE 3.1
Error ellipse for $r = 100$ m, $\alpha = 60°$, $\sigma_r = 0.50$ m, and $\sigma_\alpha = 30'$.

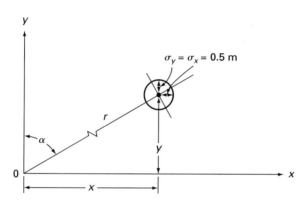

FIGURE 3.2
Error ellipse for $r = 100$ m, $\alpha = 30°$, and $\sigma_r = r\sigma_\alpha$.

of the error to the distance (see Section 2.11). In the example, the relative precision is 0.5/100, or 1 part in 200. Similarly, the linear distance subtended by σ_α in a distance r equals 0.5 and the tangent or sine of the error or its value in radians is 1 part in 200. Accordingly, a consistent relation between accuracies in angles and distances will be maintained if the estimated standard deviation in direction equals σ_r/r radians or the relative precision in the distance.

It is impossible to maintain an exact equality between these two relative accuracies; but with some exceptions, to be considered presently, surveys should be conducted so that the difference between angular and distance accuracies is not great. Table 3.1 shows, for various angular standard deviations, the corresponding relative precision and the linear errors for lengths of 1000 ft and 300 m. For a length other than 1000 ft or 300 m, the linear error is in direct proportion. A convenient relation to remember is that an angular error of $01'$ corresponds to a linear error of about 0.3 ft in 1000 ft or 3 cm in 100 m.

To illustrate the use of the table, suppose that distances are to be measured with a precision of 1/10,000. From the table the corresponding permissible angular error is $20''$. As another example, suppose that the distance from the instrument to a desired point is determined as 250 m with a standard deviation of 0.8 m. For an angular error of $10'$ the

TABLE 3.1
Corresponding angular and linear errors

Standard deviation in angular measurement	Linear error in		Relative precision
	1000 ft	300 m	
10′	2.9089	0.87267	1/344
5′	1.4544	0.43633	1/688
1′	0.2909	0.08727	1/3440
30″	0.1454	0.04363	1/6880
20″	0.0970	0.02909	1/10,300
10″	0.0485	0.01454	1/20,600
5″	0.0242	0.00727	1/41,200
1″	0.004848	0.00145	1/206,000

corresponding linear error is $(250/300)(0.87) = 0.73$ m. Therefore, the angle needs to be determined only to the nearest 10′.

The prevalence of Electronic Distance Measurement (EDM) equipment for measuring distances creates a situation where an exception occurs. Distances can be measured using EDM with a very good relative precision without additional effort. For example, suppose that a distance of 3000 m is observed with an estimated standard deviation of 0.015 m, producing a relative precision of 1/200,000. The corresponding angular standard deviation is 01″. For an ordinary survey this degree of angular accuracy would be entirely unnecessary. At the other end of the spectrum, distances may be determined roughly by taping or pacing and angles measured with more than the required accuracy. For example, in rough taping, the relative precision in distance might be 1/1000, corresponding to a standard deviation in angles of 03′. Even using an ordinary 01′ transit, angles could be observed as easily to the nearest 01′ as to the nearest 03′.

Often, field measurements are made on the basis of computations involving the trigonometric functions, and it is necessary that the computed results be of a required precision. If the values of these functions were exactly proportional to the size of the angles—in other words, if any increase in the size of an angle were accompanied by a proportional increase or decrease in the value of a function—the problem of determining the precision of angular measurements would resolve itself into that explained in the preceding section. However, because the rates of change of the sines of small angles, of the cosines of angles near 90°, and the tangents and cotangents of small and large angles are relatively large, it is evident that the degree of precision with which an angle is determined should be made to depend on the size of the angle and the function to be used in the computations. It is not practical to measure each angle with exactly the precision necessary to ensure sufficiently accurate computed values, but at least the surveyor should have a sufficiently comprehensive knowledge of the purpose of the survey and the properties of the trigonometric functions to keep the angles within the required precision.

The curves of Figures 3.3 and 3.4 show the relative precision corresponding to various standard deviations in angles from 05″ to 01′ for sines, cosines, tangents, and cotangents. For the function under consideration these curves may be used as follows:

1. To determine the relative precision corresponding to a given angle and error.
2. To determine the maximum or minimum angle that for a given angular error will furnish the required relative precision.
3. To determine the precision with which angles of a given size must be measured to maintain a required relative precision in computations.

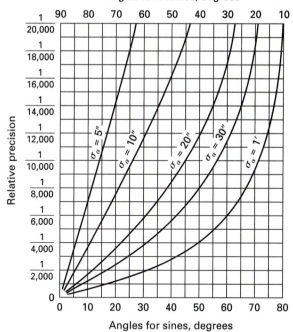

FIGURE 3.3
Relative precision for sines and cosines.

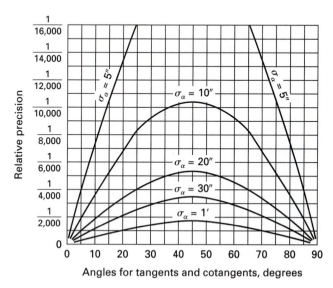

FIGURE 3.4
Relative precision for tangents and cotangents.

The following examples illustrate the use of the curves:

1. An angle measured with a 1' transit is recorded as 32°00'. It is desired to know the relative precision of a computation involving the tangent of the angle if the error of the angle is 30". In Figure 3.4, it will be seen that the relative precision opposite the intersection of the curve $\sigma_\alpha = 30"$ and a line corresponding to 32° is 1/3000.

2. In a triangulation system, the angles can be measured with σ_α not exceeding 05". Computations involving the use of sines must maintain a precision no lower than 1/20,000. It is desired to determine the minimum allowable angle. In Figure 3.3 (sines) the angle corresponding to a relative precision of 1/20,000 and σ_α of 05" is about 26°.

3. In computations involving the use of cosines a relative precision of 1/10,000 is to be maintained. It is desired to know with what precision angles must be measured. In Figure 3.3 (cosines) opposite 1/10,000, it will be seen that, for angles of about 76°, the σ_α cannot exceed 05"; for angles of about 64°, the σ_α cannot exceed 10"; and so on.

3.8
DEFINITIONS

For a better understanding of the following sections, brief definitions of a few of the terms of surveying are appropriate.

Taping. The operation of measuring horizontal or inclined distances with a tape. The persons who make such measurements are called *tape persons* or *tapers* as used hereafter in this book. Historically, this procedure has been called *chaining.*

Flagger. A person whose duty it is to hold the flagpole, or range pole, at selected points, as directed by the transit operator or other person in charge.

Rodperson. A person whose duty it is to hold the rod and to assist level operator or topographer.

Backsight. (1) A sight taken with the level to a point of known elevation. (2) A sight or observation taken with the transit along a previous line of known direction to a reference point.

Foresight. (1) A sight taken with the level to a point the elevation of which is to be determined. (2) A sight taken with the transit to a point (usually in advance), along a line whose direction is to be determined.

Grade or gradient. The slope, or rate of regular ascent or descent, of a line. It usually is expressed in percent; for example, a 4 percent grade is one that rises or falls 4 ft in a horizontal distance of 100 ft or 4 m in 100 m. The term *grade* is also used to denote an established line on the profile of an existing or a proposed roadway. In such expressions as *at grade* or *to grade,* it denotes the elevation of a point either on a grade line or at some established elevation as in construction work.

Hub. A transit station, or point over which the transit is set, in the form of a heavy stake set nearly flush with the ground, with a tack in the top marking the point.

Line. The path or route between points of control along which measurements are taken to determine distance or angle. To *give line* is to direct the placing of a flagpole, pin, or other object on line.

Turning point. A fixed point or object, often temporary in character, used in leveling where the rod is held first for a foresight, then for a backsight.

Bench mark. A fixed reference point or object, more or less permanent in character, the elevation of which is known. A bench mark also may be used as a turning point.

3.9
SIGNALS

Although field surveys can be performed by one person using the latest in robotic surveying systems (see Chapter 7), many surveys are still done by *field parties* composed of two or three people. Except for short distances, a good system of hand signals among different members of the party makes a more efficient means of communication than is possible by word of mouth. A few of the more common hand signals follow:

Right or left. The corresponding arm is extended in the direction of the desired movement. A long, slow, sweeping motion of the hand indicates a long movement; a short, quick motion indicates a short movement. This signal may be given by the transit operator in directing the tapeperson on line, by the level operator in directing the rodperson for a turning point, by the chief of the party to any member, or by one tapeperson to another.

Up or down. The arm is extended upward or downward, with wrist straight. When the desired movement is nearly completed, the arm is moved toward the horizontal. The signal is given by the level operator.

All right. Both arms are extended horizontally and the forearms waved vertically. The signal may be given by any member of any party.

Plumb the flagpole or plumb the rod. The arm is held vertically and moved in the direction in which the flagpole or rod is to be plumbed. The signal is given by the transit operator or level operator.

Give a foresight. The instrument operator holds one arm vertically above the head.

Establish a turning point or set a hub. The instrument operator holds one arm above the head and waves it in a circle.

Turning point or bench mark. In profile leveling, the rodperson holds the rod horizontally above the head and then brings it down on the point.

Give line. The flagger holds the flagpole horizontally in both hands above the head and then brings it down and turns it to a vertical position. If a hub is to be set, the flagpole is waved (with one end of it on the ground) from side to side.

Wave the rod. The level operator holds one arm vertically and moves it from side to side.

Pick up the instrument. Both arms are extended outward and downward, then inward and upward, as they would be in grasping the legs of the tripod and shouldering the instrument. The signal is given by the chief of the party or by the head tapeperson when the transit is to be moved to another point.

When distances are sufficiently long to render hand signals inadequate, the use of a citizen's band radio provides an excellent method for communication among members of a field party.

3.10
THE CARE AND HANDLING OF INSTRUMENTS

As the use of the various surveying instruments is discussed in the following chapters, suggestions for the care and manipulation of these instruments are given.

Surveying Instruments

The following suggestions apply to instruments such as the optical reading theodolite, total station systems, American standard style transit, self-leveling level, and Dumpy level:

Care of surveying instruments

1. Handle the instrument with care, especially when removing it from or replacing it in its case.
2. See that the instrument is securely fastened to the tripod head.
3. Whenever the instrument is being carried or handled, the clamp screws should be clamped very lightly to allow the parts to move if the instrument is struck.
4. Protect the instrument from impact and vibration.
5. If the instrument is to be shipped, pack paper, cloth, or styrofoam padding around it in the case; pack the case, well padded, in a larger box.
6. Never leave the instrument while it is set up in the street, on the sidewalk, near construction work, in the fields where there are livestock, or in any other place where there is possibility of an accident or theft.
7. Just before setting up the instrument, adjust the mechanism controlling the friction between tripod legs and head so that each leg, when placed horizontally, will barely fall under its own weight.
8. Do not set the tripod legs too close together, and see that they are firmly planted. Push *along* the leg, not vertically downward. As far as possible, select solid ground for instrument stations. On soft or yielding ground, do not step near the feet of the tripod.
9. While an observation is being made, do not touch the instrument except as necessary to make a setting; and do not move about.
10. In tightening the various clamp screws, adjusting screws, and leveling screws, bring them only to a firm bearing. *The general tendency is to tighten these screws far more than necessary.* Such a practice may strip the threads, twist off the screw, bend the connecting parts, or place undue stress on the instrument, so that the setting may not be stable. Special care should be taken not to strain the small screws that hold the cross-hair ring.
11. For the plumb-bob string, learn to make a sliding bowknot that can be undone easily. Hard knots in the string indicate an inexperienced or careless instrument operator.
12. Before observations are begun, focus the eyepiece on the cross hairs and (by moving the eye slightly from side to side) see that no parallax is present.
13. *When the magnetic needle is not in use, see that it is raised off the pivot.* While the needle is resting on the pivot, impact is apt to blunt the point of the pivot or to chip the jewel, thus causing the needle to be sluggish.
14. Always use the sunshade. Attach or remove it by a clockwise motion, in order not to unscrew the objective.
15. If the instrument is to be returned to its box, put on the dust cap if one was supplied by the manufacturer and wipe the instrument clean and dry.

Transport and Mounting of Surveying Instruments

Optical reading theodolites and total station systems

1. Always transport the theodolite or total station instrument to the observation site in its carrying case. Specially constructed backpacks are available for this purpose.
2. Do not carry the theodolite or total station instrument mounted on the tripod between stations. They should be moved in the carrying case.
3. When mounting the theodolite on the tripod always grasp the standard, which does not contain the vertical circle and compensator, with one hand and the bottom plate with the other hand. Never hold the instrument by telescope, plate level bubble tube, micrometer knob, tangent screws, clamps, or optical plummet.
4. When mounting a total station instrument on the tripod grasp it by the handle or yoke at the top of the standards and use both hands. Never hold it by the lens barrel.

1. Normally, carry the instrument mounted on the tripod and over one shoulder with the tripod legs forward and held together by that hand and forearm.
2. Avoid carrying the instrument on the shoulder while passing through doorways or beneath low-hanging branches; carry it under the arm, with the head of the instrument in front.
3. When walking along a sidehill, always carry the instrument on the downhill shoulder to leave the uphill arm free to catch the body in case of tripping or stumbling and to prevent the instrument from slipping off the shoulder.
4. Before climbing over a fence or similar obstacle, place the instrument on the other side, with the tripod legs well spread.
5. When carrying a self-leveling or automatic level that is mounted on a tripod, grasp the tripod with both hands and carry it in front of you with the level in a near-vertical position to avoid damage to the compensator.

Taping equipment

Keep the tape straight when in use; any tape will break when kinked and subjected to a strong pull. Steel tapes rust readily and for this reason should be wiped dry after use.

Use special care when working near electric power lines. Fatal accidents have resulted from throwing a metallic tape over a power line.

Do not use the flagpole as a bar to loosen stakes or stones, such use bends the steel point and soon renders the point unfit for lining purposes.

To avoid losing pins, tie a piece of colored cloth (preferably bright red) through the ring of each.

Leveling rod

Do not allow the metal shoe on the foot of the rod to strike against hard objects, as this, if continued, will round off the foot of the rod and thus introduce a possible error in leveling. Keep the foot of the rod free of dirt. When not in use, long rods should be either placed upright or supported for their entire length; otherwise, they are likely to warp. When not in use, jointed rods should have all clamps loosened to allow for possible expansion of the wood.

EDM equipment

EDM instruments are precision surveying devices that require the same care and handling as a theodolite or level. These instruments always should be transported in the carrying case especially designed for that purpose. All EDM devices have a power unit. Excessive shock can damage this unit. Care must be exercised when transporting the power unit. Most EDM instruments require little service other than the normal cleaning of transmitter lens and other parts. The power unit does require periodic recharging and checking to ensure that power is not lost at a critical moment. EDM instruments should be calibrated and tested periodically to verify instrument constants and check for frequency drift (Section 4.38).

Total Station Systems

These systems are battery operated and consist of a theodolite (either optical reading or digital) and an EDM unit; hence, they require the same care and handling as a theodolite and EDM. The following additional precautions are suggested.

1. Never leave the instrument unprotected in a high temperature, as this may increase the interior temperature excessively, reducing its service life.
2. Avoid sudden change in temperature (e.g., taking from a heated vehicle), as this may reduce the range in measuring distances.
3. When changing setups, remove the instrument from the tripod and place it in the carrying case for transport. Under no circumstances should the tripod be carried over the shoulder or under the arm with the instrument attached, as damage to the instrument centers may result.

GPS Equipment

Current models of GPS receivers and antennas are battery operated, compact, rugged units built to withstand extremes in weather, temperature, and humidity. These units respond well with reasonable care and handling. However, the antenna usually is attached to a tripod that is centered over the survey point. Consequently, this tripod and the related centering device and adapter must be handled and maintained like any other piece of precision surveying equipment.

3.11
ADJUSTMENT OF INSTRUMENTS

Surveying instrument *adjustment* means bringing the various fixed parts into proper relation with one another, as distinguished from the ordinary operations of leveling the instrument, aligning the telescope, and so on.

The ability to adjust ordinary surveying instruments is an important qualification of the surveyor. Although the effect of instrumental errors largely may be eliminated by proper field methods, instruments in good adjustment greatly expedite the fieldwork. It is important that the surveyor

1. Understand the principles on which the adjustments are based.
2. Learn the method by which nonadjustment is discovered.
3. Know how to make adjustments.
4. Appreciate the effect of one adjustment on another.
5. Know the effect of each adjustment on the use of the instrument.
6. Learn the order in which adjustments may be performed most expeditiously.

The frequency with which adjustments are required depends on the particular adjustment, the instrument and its care, and the precision with which measurements are to be taken. Often, in a good instrument, well cared for, the adjustments will be maintained with sufficient precision for ordinary surveys over a period of months or even years. On the other hand, blows that may pass unnoticed are likely to disarrange the adjustments at any time. On ordinary surveys, it is good practice to test the critical adjustments once each day, especially on long surveys, where frequent checks on the accuracy of the field data are impossible. Failure to observe this simple practice sometimes results in the necessity of retracing lines, which may represent the work of several days. Testing the adjustments with reasonable frequency lends confidence to the work and is a practice to be strongly commended. The instrument operator should make the necessary tests, if possible, at a time that will not interfere with the general progress of the survey party. Some adjustments may be made with little or no loss of time during the regular progress of the work.

The adjustments are made by tightening or loosening certain screws. Usually, these screws have capstan heads that may be turned by a pin called an *adjusting pin*. The following are some general suggestions:

1. The adjustng pin should be carried in the pocket and not left in the instrument box. Disregard of this rule frequently leads to loss of valuable time.
2. The adjusting pin should fit the hole in the capstan head. If the pin is too small, the head of the screw soon will be ruined.
3. Preferably make the adjustments with the instrument in the shade.
4. Before adjusting the instrument, see that no parts (including the objective) are loose. When an adjustment is completed, always check it before using the instrument.
5. When several interrelated adjustments are necessary, time will be saved by first making an approximate or rough series of adjustments and then by repeating the series to make finer adjustments. In this way, the several disarranged parts are gradually brought to their correct positions. This practice does not refer to those adjustments which are in no way influenced by others.
6. Most of the more precise instruments, such as theodolites, precise levels, and EDM equipment, require more complicated tools, more stable or constant environmental conditions for the test and adjustment phase, and more specialized personnel to achieve the more refined adjustments. Consequently, such instruments rarely are adjusted in the field, but the operator does need to test the instrument periodically to determine if adjustment is necessary.

3.12
DATA RECORDING

Information acquired in field surveys is recorded in the standard field notebook in the form of hand lettered notes and sketches or by using an automatic data collector, sometimes referred to as an *electronic field book*. First, consider notes taken in a standard field notebook.

No part of the operations of surveying is of greater importance than the field notes. The competency of the surveyor is reflected with great fidelity by the character of the notes recorded in the field. These notes should constitute a permanent record of the data in such a form as to be interpreted with ease by anyone having a knowledge of surveying. Unfortunately, this often is not the case. Many surveyors seem to think that their work is well done if the field record, reinforced by their own memory, is sufficiently comprehensive to make the field data of immediate use for whatever purpose the survey may have. On most surveys, however, it is impossible to predict to what extent the information gathered may become of value in the remote future. Often court proceedings involve surveys made long before. Often, it is desirable to return, extend, or otherwise use surveys made years previously. In such cases it is quite likely that the old field notes will be the only visible evidence, and their value will depend largely on the clarity and completeness with which they are recorded.

The notes consist of numerical data, explanatory notes, and sketches. Also, the record of every survey or student problem should include a title indicating the location of the survey and its nature or purpose; the equipment used, including manufacturer and number for major instruments; the data; the weather conditions; and the names and duties of the members of the party.

All field notes should be recorded *in the field book* at the time the work is being done. Notes made later, from memory or copied from temporary notes, may be useful, but they are not field notes. Notes should be neat. They generally are recorded in pencil, but they should be regarded as a permanent record and not as memoranda to be used only in the immediate future.

It is not easy to take good notes. The recorder should realize that the notes may be used by persons not familiar with the locality, who must rely entirely on what has been recorded. Not only should the notebook contain all necessary information, but the data should be recorded in a form that will allow only the correct interpretation. A good sketch will help to convey a correct impression, and sketches should be used freely. The use to be made of the notes will guide the recorder in deciding which data are necessary and which are not. To make the notes clear, recorders should put themselves in the place of one who is not in the field at the time the survey is made. Before any survey is made, the necessary data to be collected should be considered carefully, and in the field all such data should be obtained, but no more.

Although convenient forms of notes such as those shown in this book are in common use, it generally will be necessary to supplement these, and in many cases it will be necessary for surveyors to devise their own form of record. A code of symbols is desirable.

In some cases, as in locating details for mapping, the field notes may be supplemented by photographs taken with an ordinary camera. Frequently, in field completion surveys for photogrammetric mapping, an enlargement of the aerial photograph that covers the area to be mapped is annotated in the field, thus providing a record of additional information.

Other methods for recording field data are described in the subsection titled "Data Collectors."

Notebook

In practice, the field notebook should be of good-quality rag paper, with a stiff board or leather cover, made to withstand hard use, and of pocket size. Treated papers are available that will shed rain; some of these can be written on when wet.

Special field notebooks are sold by engineering supply companies for particular kinds of notes, such as cross sections or earthwork. For general surveying or for students in fieldwork, where the problems to be done are general in character, an excellent form of notebook has the right-hand page divided into small rectangles with a red line running up the middle, and has the left-hand page divided into six columns; both pages have the same horizontal ruling. In general, tabulated numerical values are written on the left-hand page, sketches and explanatory notes on the right. This type is called a *field book* (see, e.g., Figure 8.12). Another common form, used in leveling, has both pages ruled in columns and has wider horizontal spacing than the field book; this is called a *level book*.

The field notebook may be bound in any of three ways: conventional, spiral ring, or loose leaf. The ring type, which consists of many metal rings passing through perforations in the pages, is not loose-leaf; it has the advantage over the conventional binding that the book opens quite flat and that the covers can be folded back against each other.

Loose-leaf notebooks have the following advantages:

1. Only one book need be carried, as in it may be inserted blank pages of various rulings, together with notes and data relating to the current fieldwork.
2. Sheets can be withdrawn for use in the field office while the survey is being continued.
3. Carbon copies can be made in the field, for use in the field or headquarters office. Carbon copies are also a protection against loss of data. Duplicating books are available.
4. Notes of a particular survey can be filed together. Files can be made consecutive and are less bulky than for bound books.
5. The cost of binders is less than that for bound books.

Its disadvantages are

1. Sheets may be lost or misplaced.
2. Sheets may be substituted for other sheets—an undesirable practice.
3. There may be difficulty in establishing the identity of the data in court, as compared with a bound book. (This feature is important in land surveying.) When loose-leaf books are used, *each sheet* should be fully identified by date, serial number, and location.

Loose leaves are furnished in either single or double sheets. Single sheets are ruled on both sides and are used consecutively. Double sheets comprising a left-hand and a right-hand page joined together are ruled on one side only. Sheets for carbon copies need not be ruled.

Manually Recording Data

A 4H pencil, well pointed, should be used. Lines made with a harder pencil are not as distinct; lines made with a softer pencil may become smeared. Reinhardt slope lettering (Figure 14.15) is commonly considered to be the best form of lettering for taking notes rapidly and neatly. Office entries of reduced or corrected values should be made in red ink, to avoid confusion with the original data.

The figures used should be plain; one figure should never be written over another. In general, *numerical data should not be erased*; if a number is in error, a line should be drawn through it, and the corrected value written above.

In tabulating numbers, the recorder should place all figures of the tens column, and so forth, in the same vertical line. Where decimals are used, the decimal point should never be omitted. The number should always show with what degree of precision the measurement was taken; thus, a rod reading taken to the nearest 0.01 ft should be recorded not as 7.4 ft but as 7.40 ft. Notes should not be made to appear either more precise or less precise than they really are.

Sketches are rarely made to exact scale, but in most cases they are made approximately to scale. They are made freehand and of liberal size. The recorder should decide in advance just what the sketch is to show. A sketch crowded with unnecessary data often is confusing, even though all necessary features are included. Large detailed sketches may be made of portions having much detail. Many features may be shown most readily by conventional symbols (Section 14.24); special symbols may be adopted for the particular organization or job.

Explanatory notes are employed to make clear what the numerical data and sketches fail to do. Usually, they are placed on the right-hand page in the same line with the numerical data that they explain. If sketches are used, the explanatory notes are placed where they will not interfere with other data and as close as possible to that which they explain.

If a page of notes is abandoned, either because it is illegible or because it contains erroneous or useless data, it should be retained and the word "void" written in large letters diagonally across the page. The page number of the continuation of the notes should be indicated.

Data Collectors

Data collectors essentially are handheld calculators that have an interface with electronic surveying equipment (e.g., digital theodolites, total station systems, and GPS receivers) and allow automatic real-time storage or keyed-in recording of observed data, point identification, and other attributes associated with the survey. When the data collector is combined with the necessary software modules that permit processing and adjusting raw survey data and the download of these data to office computers for transferral to CADD and GIS for plotting and design, the system, composed of data collector and software, is called an *electronic field book*.

Numerous manufacturers produce data collectors, all of which are of about the size of a pocket calculator and come in a wide range of prices ($150 to $6,000), storage capacities (64 kB to 16 MB),* and other capabilities. In addition to data storage, some collectors also provide for computation and adjustment of the data in the field and display of the results.

Some survey systems (e.g., total station systems) are equipped with on-board data storage units that can be used to download data to off-line storage devices for subsequent calculation and processing. This eliminates the cables that tend to interfere with rotation of the instrument. However, current practice seems to be oriented toward use of a separate handheld unit, attached to the instrument by a cable. This unit is compatible with several, and in some cases all, surveying systems, so as to provide additional flexibility in operations with a wide variety of field equipment.

Data collectors usually are programmed to provide a main menu with options for selecting different screen displays designed to allow recording data using the various surveying procedures (e.g., distance, horizontal angle, or vertical circle measurement). When one of these screens is selected, prompts are displayed on the screen to guide the operator through the sequence of operations required for a given process. Observed data such as angles, distances, rod readings, and the like are automatically recorded in the collector's memory and the surveyor can enter comments and attributes related to the objects located in the survey being performed.

The procedure for recording data follows the rules of note keeping as previously discussed. Thus, the operator, using a menu displaying various options, can enter the date, time, weather, location, instrument number, and calibration constants. Then menu entry is chosen and, following prompts displayed on the screen, the procedure is completed and data are stored along with necessary comments, and so on. When the task is finished, data can be evaluated and checked. Any values not satisfying the preprogrammed criteria can be rejected and repeated immediately.

When the entire job is completed, preliminary calculations may be performed in the field to test the validity of the survey. This option permits detecting mistakes and correcting them before leaving the site. At this stage, the data collector contains a complete record of all data measured, descriptive comments and attributes related to these measurements, and the results of the preliminary computations performed.

In the office, the records can be downloaded to a PC, mini, or mainframe computer for subsequent computations, adjustment, analysis, plotting, and graphical display. Also, the files can be printed to furnish a hard-copy record of the survey.

Data collectors and associated software provide an extremely powerful tool for the surveyor, which opens up the possibility of automated surveying from the acquisition of the data to the plotted map or plan. The electronic field book approach has many advantages but some disadvantages are present as well. Data stored in memory can be accidentally lost due to operator mistakes or software malfunction. Strict attention to careful procedures and care in the software design can minimize the chance of this type of blunder but the possibility does exist. One very practical procedure to protect against this type of disaster would be to keep minimal notes and sketches in a standard field notebook as a backup record.

Other aspects requiring precautions are the liability and legality aspects of digital data stored in memory or on diskettes. Liability can occur when digital data obtained by the surveyor is entered into a GIS that is accessible to parties other than the original client and then is used for purposes other than originally intended.

The rules with respect to liability and legality of field notes in the standard field notebook are well established in the courts. Such is not the case concerning liability and legality of digital records which are in a state of evolution. To be on the safe side, the

---

*A kB = 1000 bytes, MB = 1×10^6 bytes, byte = 8 bits (8 b).

surveyor who uses a data collector or an electronic field book should exercise the following recommendations to provide authentication of the data and prevent potential tampering with the digital record:

1. Always archive the actual observations in condensed binary form.
2. This binary file also should contain all measurement and keyboard blunders.
3. Any mistake in the binary data should be "crossed out," which means the software must be designed to permit flagging bad measurements.
4. The binary file, when converted to ASCII, should have time tags to the nearest second for each data field entry and it should be indicated whether the entry was automatic or via the keyboard.

The references at the end of this chapter contain more detail on this subject. Readers interested in this continuing evolution of liability and legal aspects of digital records also must participate in the activities of their professional societies and read the latest society proceedings and periodicals.

3.13
COMPUTATIONS AND CHECKING

Calculations of one kind or another form a large part of the work of surveying, so that familiarity with computational processes is essential for the surveyor. To achieve this familiarity, one must (1) possess a knowledge of the precision of measurements and the effect of errors in the given data on the precision of values calculated from these data and (2) know the algebraic and graphical processes and the computational equipment to be used in the computation.

Computations are made *algebraically* by the use of arithmetical procedures and trigonometery, *graphically* by accurately scaled drawings, or *electronically* using data collectors, pocket calculators, personal computers (PC), and mini and mainframe computers.

Before making calculations of importance, the surveyor should plan a clear and orderly arrangement, using tabular forms wherever possible. Such steps save time, prevent mistakes, and facilitate the work of the checker.

Because the bulk of the computations in surveying now are done using software programmed for an electronic computer, particular care should be exercised in preparing data for computation by a program on one of the various types of electronic computers. The presence of undetected blunders or data out of order that then are processed by an unsuspecting terminal operator can lead to incorrect results and the consequent waste of inestimable amounts of time.

All computations should be preserved in a loose-leaf notebook maintained for that purpose. A record of electronic notebook data, calculations, and adjusted results should be kept on a backup diskette and a hardcopy listing of these data, *signed and dated,* ought to be kept in a the project notebook.

Office computations are a continuation of some fieldwork. They should be easily accessible for future reference; for this reason, pages should be numbered and a table of contents included. Parts of problems separated by other calculations should be cross-referenced. Each problem must have a clear heading that should include the name of the survey, the kind of computations, field book number or file name and page of original notes, the name of the computer and name of checker, and the dates of computing and checking. Usually enough of the original notes should be transcribed to make computations possible without reference to the field notebook or original file of data.

With respect to checking when using traditional field notes and manual computations, *no confidence is placed in results that have not been checked,* and important results preferably are checked by more than one method. A good habit to develop is that of checking one's work until certain that the results are correct. These checks should be performed independently.

Many problems can be solved by more than one method. Because the use of the same method in checking may cause the same error to occur, results should be checked by a different method when this is feasible. Approximate checks to discover large arithmetical mistakes may be obtained by applying approximate solutions to the problem with a pocket calculator or checking the answer by scaling from a map or drawing with protractor and scale. Graphical methods may be used as an approximate check; they take less time than arithmetical solutions and possible incorrect assumptions in the precise solution may be detected.

Each step in a long computation that cannot be verified otherwise should be checked by repeating the computation.

When work is being checked and a difference is found, the computation should be repeated before a correction is made, because the check itself may be incorrect.

In many cases, large mistakes, such as faulty placement of a decimal point, can be located by inspection of the value to see if it looks reasonable.

Electronic computer programs, either purchased from a vendor or developed in-house, always should be tested before used in production. Such tests are best performed by comparing results from the same problem calculated independently with another program or manually with a pocket calculator.

When software packages, known to be operating properly, are used for the solution, checking should concentrate on the input data and output results to and from the computer. The input data to any computer program should be checked exhaustively to eliminate tabulation and keyboard mistakes.

If data collectors or the electronic notebook are used, the checking routines need to be built into the software at each stage of the automatic data recording and computational procedure. In this case, field and office personnel must be thoroughly familiar with the entire computational process to be able to evaluate and analyze the quality of results at each step of the procedure.

PROBLEMS

3.1. The planning of a survey involves what major steps?

3.2. What factors should be considered by the surveyor when setting the specifications for accuracy on a given project?

3.3. On a typical mapping project what types of surveys must be considered?

3.4. You are responsible for obtaining existing maps of a region proposed as right of way for a highway. What government agencies would you contact?

3.5. Small-scale planimetric maps or photographs covering a very large area are needed for planning on a dam project. What are the most likely sources of this type of information?

3.6. Filing and presentation of the processed surveying data are important parts of the overall survey. Discuss four systems that can be used to perform these functions.

3.7. A point is to be established on the ground at a distance of 300.00 m from a given point by means of one linear and one angular measurement. This point is to be established to within 5 cm of its true location. What angular error could be allowed to achieve this accuracy in position? Express the answer in seconds of arc.

3.8. A horizontal distance has been measured with a relative precision of 1 part in 25,000. This specification for accuracy in distance of 500.000 m corresponds to what tolerance in horizontal angular measurement?

3.9. An angle can be measured with a standard deviation of 30″ on a line 1000.00 m long. Determine the corresponding accuracy of this distance, expressed as a relative precision.

3.10. An angle of 50°00′00″ is measured with a theodolite having a least count of 20″. Assume the maximum error is one-half the least count of the theodolite. What is the relative precision if the sine of the angle is to be used in computations? The cosine? The tangent?

3.11. A horizontal angle of 25°00′ is measured with a transit having a least count of 01′. If the maximum error to be expected is one-half the least count, what will be the relative precision if the sine of the angle is used in the computations? The cosine? The tangent?

3.12. If tangents or cotangents are involved, what is the highest precision corresponding to single measurements (of angles of any size) with the (a) theodolite in Problem 3.10, and (b) transit in Problem 3.11.

3.13. If the angular error is 20″, what is the minimum allowable angle that, for computations involving sines, will permit a relative precision of 1/20,000 to be maintained?

3.14. What is the range of allowable angles that, for computations involving tangents and cotangents, will permit a relative precision of 1/5000 to be maintained if the angular error is 20″?

3.15. Field notes, in either the standard field notebooks or using the electronic notebook, consist of what primary components?

3.16. Describe the elements that make up the electronic notebook.

3.17. Discuss the advantages and disadvantages of data collectors and the electronic notebook concept for collecting and processing survey data.

3.18. Discuss the legal and liability aspects of the electronic notebook and suggest precautions that can be taken to minimize potential problems.

REFERENCES

Colvocoresses, A. P. "The Case for the High Altitude Version of Landsat-7." *Photogrammetric Engineering and Remote Sensing* 58, no. 4 (April 1992), p. 454.

Crawford, W. G., "Shopping for a Total Station." *P.O.B., Point of Beginning* 22, no. 7 (April 1997), pp. 22–47.

Hermansen, K. E., and G. Wigal. "Who Owns the Records?" *ACSM Bulletin*, no. 143 (May–June 1993), pp. 52, 53.

Onsrud, H. J., and J. H. Hintz. "Evidentiary Admissibility and Reliability of Automated Field Recorder Data." *Surveying and Land Information Systems* 51, no. 1 (1991), pp. 23–28.

Roth, A. W. "P.O.B. 1992 Data Collector Survey." *P.O.B., Point of Beginning* 17, no. 6 (August– September 1992), pp. 30–44.

Shrestha, R. L. "An Overview of the Development of an Electronic Field Book and Data Processing System." *Surveying and Land Information Systems* 50, no. 3 (1990), pp. 209–14.

Shrestha, R. L. "Formats and Specifications for an Electronic Field Book." *Surveying and Land Information Systems* 51, no. 3 (1990), pp. 215–24.

Sollers, G. C. *Geodetic and Charting Publications.* Silver Spring, MD: Coast and Geodetic Survey, April 1992.

Walkowski, W. "Data Collector Guide." *P.O.B., Point of Beginning* 22, no. 3 (December 1996), pp. 33–46.

Welch, R. "Cartographic Potential of SPOT Image Data." *Photogrammetric Engineering and Remote Sensing* 51, no. 8 (August 1985), pp. 1085–91.

PART II

Basic Survey Measurements

CHAPTER 4

Distance Measurement

PART A
ELEMENTARY TOPICS

4.1
DISTANCE

The distance between any two random points in three-dimensional space is a spatial distance. This concept is particularly pertinent considering the advanced stage of electronic distance measurement (EDM) techniques, which automatically provide spatial or slope distances in the range of from 15 m or 50 ft to 60 km or 36 mi. Although slope distances frequently are observed in the surveying operation, these distances then are reduced to a horizontal projection for more convenient use in subsequent calculations and field layout. Horizontal distance is further reduced to its equivalent distance on the ellipsoid for geodetic surveys (Section 9.11). In plane surveying, horizontal distances are reduced to the ellipsoid only when it is desired to convert them into equivalent distances at another elevation, such as that of the state plane coordinate system or the average elevation of a survey for which the variation in elevation over the area is large.

There are several methods of determining distance, the choice of which depends on the accuracy required, the cost, and other conditions. The methods, in ascending order of accuracy, are estimation, scaling from a map, pacing, odometer, tacheometry, taping, photogrammetry, inertial systems, and electronic distance measurement. For example, on rough reconnaissance, a relative precision of 1 part in 100 (i.e., a standard deviation of 1/100 of the distance) or less may be adequate for the survey; on the other hand, specifications for first-order surveys (refer to Section 9.23) call for base-line measurement to have a standard deviation of 1 part in 1 million of the distance measured. On certain surveys, a combination of methods may be warranted.

Table 4.1 classifies the principal methods of measuring distance according to the usual relative precisions obtained. The various methods are discussed further in the following sections.

TABLE 4.1
Methods of measuring distance

Method	Relative precision*	Use	Instruments required	Instruments for measuring angles with corresponding accuracy
Pacing odometer or mileage recorder	1/100	Reconnaissance, small-scale mapping, checking tape measurement, quantity surveys	Pedometer, odometer	Hand compass, peepsight alidade
Tacheometry Stadia	1/300– 1/1000	Location of details for topographic mapping, rough traverse, checking more accurate measurements	Level rod or stadia board, calibrated optical line of sight with stand	Transit, telescopic alidade and plane table, surveyor's compass
Distance wedge	1/5000– 1/10,000	Traverse for land surveys, control of route and topographic surveys and construction work	Horizontal graduated rod and support; calibrated optical line of sight equipped with a distance wedge	Transit or theodolite
Subtense bar	1/1000– 1/9000	Hydrographic surveys, traverse	Calibrated subtense bar and tripod; 1″ theodolite	Transit or theodolite
Ordinary taping	1/3000– 1/5000	Traverse for land surveys and for control of route and topographic surveys and construction	Steel tape, chaining pins, plumb bobs	Transit
Precise taping	1/10,000– 1/30,000	Traverses for city surveys, base lines for triangulation of low accuracy, and construction surveys requiring high accuracy	Calibrated steel tape, thermometer, tension handle, hand level, plumb bobs	Transit or theodolite
Photogrammetry	up to 1/50,000	Location of detail for topographic mapping, second- and third-order ground control surveys	Stereoplotters, mono and stereo comparators, electronic computer	
Inertial systems	up to 1/50,000	Rapid, reconnaissance surveying: large area surveys, second- and third-order ground control surveys	Inertial positioning system	
Base-line taping	1/100,000 1/1,000,000	First-, second-, and third-order triangulation for large areas, city surveys, long bridges, and tunnels	Calibrated steel tape, thermometers, tension handle, taping supports, level, level rod	Theodolite (1″)
EDM	1 mm + 2 ppm to 5 mm + 5 ppm	Traverse, triangulation, and trileration for control surveys of all types and for construction surveys	EDM equipment	Theodolite (1″) which can be built into the EDM

*Relative precision may be defined as the ratio of the allowed standard deviation to the distance measured.

100

4.2
PACING

Pacing furnishes a rapid means of approximately checking more precise measurements of distance. It is used on reconnaissance surveys and, in small-scale mapping, for locating details and traversing with the plane table. Pacing over rough country has furnished a relative precision of 1/100; under average conditions, a person of experience will have little difficulty in pacing with a relative precision of 1/200.

Each two paces or double step is called a *stride*. Thus, for a pace the stride would be 5 ft, or there would be roughly 1000 strides per mile.

Paces or strides are usually counted by means of a tally register operated by hand or by means of a pedometer attached to the leg. Digital pedometers, which permit entering a person's average pace, are available. In hilly country, rough corrections for slope can be applied.

Beginners should standardize their pace by walking over known distances on level, sloping, and uneven ground.

4.3
MILEAGE RECORDER, ODOMETER, AND OTHER MEASUREMENT METHODS

Distance may be measured by observing the number of revolutions of the wheel of a vehicle. The *mileage recorder* attached to the ordinary automobile speedometer registers distance to 0.1 mi and may be read by estimation to 0.01 mi. Special speedometers can read to 0.01 or 0.002 mi. The *odometer* is a simple device that can be attached to any vehicle and directly registers the number of revolutions of a wheel. With the circumference of the wheel known, the relation between revolutions and distance is fixed. The distance indicated by either the mileage recorder or the odometer is somewhat greater than the true horizontal distance, but in hilly country a rough correction based on the estimated average slope may be applied.

Distances sometimes are estimated roughly by the *time interval of travel* for a person at walk, a saddle animal at walk, or a saddle animal at gallop.

By *mathematical* or *graphical* methods, unknown distances may be determined through their relation to one or more known distances. These methods are used in triangulation and plane-table work.

4.4
TACHEOMETRY

Tacheometry includes stadia with transit or theodolite and stadia rod, distance wedge and horizontal rod, and subtense bar and theodolite. The stadia method, described in detail in Chapter 7, offers a rapid indirect means of determining distances. The telescope of the theodolite transit, level, or plane-table alidade is equipped with two horizontal hairs, one above and the other an equal distance below the horizontal cross hair. The distance from the instrument to a given point is indicated by the interval between these stadia hairs, as shown on a graduated rod held vertically at the point. The precision of the stadia method depends on the instrument, the observer, the atmospheric conditions, and the length of sights. Under average conditions, the stadia method will yield a relative precision between 1/300 and 1/1000. The use of a self-reducing tacheometer or self-reducing alidade for the stadia method simplifies the operation substantially.

The distance wedge is used in conjunction with a horizontally supported rod placed perpendicular to the optical line of sight.

Another method for determining distance by tacheometry uses a subtense bar and precise transit or theodolite. The subtense bar is a 2-m calibrated bar mounted horizontally on a tripod and placed perpendicular to the line of sight. The subtended angle is then measured with a 1″ theodolite placed at the other end of the line to be measured so that the horizontal distance between the points can be calculated.

4.5
TAPING

Taping involves direct measurement of the distance with steel tapes varying in length from 3 ft to 300 ft. Metric tapes vary in length from 1 m to 60 m. Graduations are in feet, tenths, and hundredths, or meters, decimeters, centimeters, and millimeters. Formerly, on surveys of ordinary precision, it was the practice to measure the length of lines with the *engineer's chain* or *Gunter's chain;* for measurements of the highest precision, special bars were used.

The *engineer's chain* is 100 ft long and composed of 100 links each 1 ft long. At every 10 links, brass tags are fastened, notches on the tags indicating the number of 10-link segments between the tag and the end of the tape. Distances measured with the engineer's chain are recorded in feet and decimals.

The *surveyor's* or *Gunter's chain* is 66 ft long and divided into 100 links each 0.66 ft = 7.92 in long; distances are recorded in chains and links. Thus,

$$1 \text{ (Gunter's) chain} = 66 \text{ ft} = 100 \text{ links} = 4 \text{ rods}$$

$$80 \text{ (Gunter's) chains} = 1 \text{ mi}$$

$$10 \text{ square (Gunter's) chains} = 1 \text{ acre} = 43{,}560 \text{ ft}^2$$

Measuring with chains was originally called *chaining.* The term has survived and often is applied to measuring lines with tapes.

The precision of distance measured with tapes depends on the degree of refinement with which measurements are taken. On the one hand, rough taping through broken country may be less accurate than the stadia. On the other hand, when extreme care is taken to eliminate all possible errors, measurements have been taken with a relative precision of less than 1/1 million. In ordinary taping over flat, smooth ground, the relative precision is about 1/3000 to 1/5000.

4.6
ELECTRONIC DISTANCE MEASUREMENT

Electro-optical and electromagnetic instruments are of great value to the surveyor for accurate measurements of distances. Measurement of distance with electronic distance measuring (EDM) equipment is based on the invariant speed of light or electromagnetic waves in a vacuum. EDM equipment that can be used for traverse, triangulation, and trilateration as well as for construction layout has supplanted taping for modern surveying operations except for short distances and certain types of construction layout. A detailed explanation of EDM systems and procedures is given in Sections 4.27 to 4.32.

4.7
CHOICE OF METHODS

In general, boundary, control, and construction surveys involving long lines and large areas can be performed most accurately and economically using modern EDM equipment. Where the distances involved are relatively short or specific construction layout requirements are present, taping the distances can be more practical.

Each of these methods mentioned has a field of usefulness. On the survey for a single enterprise, the surveyor may find occasion to employ a combination of methods to advantage.

4.8
TAPES

Tapes are made in a variety of materials, lengths, and weights. Those more commonly used by the surveyor and for engineering measurements are the steel tapes, sometimes called the *engineer's* or *surveyor's tape,* and fiberglass tapes.

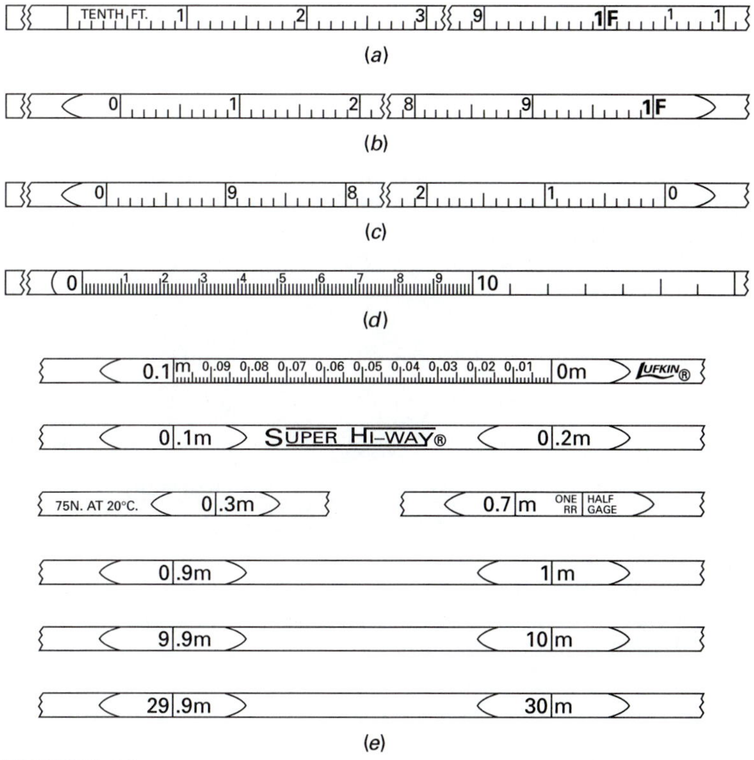

FIGURE 4.1
Various styles of steel tapes. (a) Tape graduated in feet, tenths, and hundredths throughout. (b) First foot graduated to hundredths, tape graduated throughout to feet. A "cut" tape. (c) Added foot graduated to hundredths of a foot, tape graduated to feet throughout. An "add" tape. (d) Metric tape. First decimeter graduated to millimeters, first meter graduated to centimeters, graduated throughout to decimeters. A "cut" tape. (e) Metric tape. Added decimeter graduated to millimeters. Tape graduated throughout to decimeters. An "add" tape. (*Courtesy the Cooper Group (Lufkin).*)

Metric steel tapes are 15, 30, or 50 m long. The light box tapes and some medium weight tapes are graduated throughout in meters, decimeters, and centimeters. Heavier tapes are graduated as follows: throughout to meters, and decimeters; first and last meter in centimeters; and first and last decimeters in millimeters; or throughout to meters and half-meters with end meters to decimeters.

Steel tapes for which the feet is the unit of length are graduated as follows: lightweight tapes and some engineer's tapes are graduated to hundredths of feet throughout the length; normally, the heavier tapes have graduations with numbers every foot, with only the end feet graduated in tenths or hundredths of feet. Various styles of steel tapes graduated in feet and meters are shown in Figure 4.1 on page 103.

Fiberglass tapes also are available in units of meters and feet and may be obtained in lengths of 15, 20, 30, and 50 m and 50, 100, 200, and 300 ft.

Tapes for which the Gunter's chain is the unit of length are 1 and 2 chains in length and graduated in links throughout, with the end links in tenths.

Ordinarily, rawhide thongs serving as handles are fastened to the rings at each end of the chain tape. Wire handles sometimes are used, but they are objectionable when the tape must be dragged through grass or brush. Detachable clamp handles are available for grasping the tape at any point. Figure 4.2 illustrates the use of a tape clamp.

The tape should be kept straight when in use; any tape will break when kinked and subjected to a straight pull. Steel tapes rust readily and should be wiped dry after use. Special care is required when working near power lines. Fatal accidents have occurred from throwing a steel or metallic tape over a power line.

FIGURE 4.2
Use of a tape clamp.

4.9
EQUIPMENT FOR TAPING

Additional equipment employed for determining the length of lines by direct measurement with a tape consists of plumb bobs, the hand level, the tension handle, chaining pins, and range poles.

The plumb bob is a pointed metal weight used to project the horizontal location of a point from one elevation to another.

The hand level, described in Section 5.21, can be used to keep the two ends of the tape at the same elevation when measuring over irregular terrain. A tension handle is a spring scale that can be attached to the end of the tape and allows applying the proper tension.

Steel taping pins, also called *chaining pins, taping arrows,* or *surveyor's arrows,* commonly are employed to mark the ends of the tape during taping between two points more than a tape length apart. They are usually 10 to 14 in (25 to 35 cm) long. A set consists of 11 pins (see Figure 4.3a).

For more precise taping or for future reference, nails may be driven into the earth. Marking tags, which can be stamped with identification as desired, may be threaded over the nails before they are driven. On paved surfaces, tape lengths and other points may be marked with keel (carpenter's chalk), pencil, or spray paint; for precise work a short piece of opaque adhesive tape may be stuck on the pavement and the point marked with a pencil or ballpoint pen.

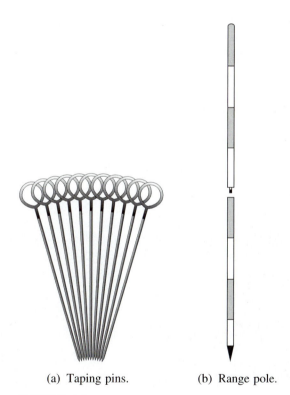

(a) Taping pins.   (b) Range pole.

FIGURE 4.3
Taping equipment.

Wooden stakes or hubs are driven into the ground to mark the significant points of profile-level and transit-traverse lines. Taping tripods and wooden or concrete posts are used to mark tape ends in base-line taping.

Metal, wooden, or fiberglass range or prism poles, also called *flags, flagpoles,* or *lining rods,* are used as temporary signals to indicate the location of points or the direction of lines. They are of octagonal or circular cross section, pointed at the lower end. Wooden and fiberglass range poles are shod with a steel point. The common length is 2 or 3 m or 6 or 8 ft. Usually, the pole is painted with alternate bands of red and white 1 ft or $\frac{1}{2}$ m long (see Figure 4.3b).

4.10
TAPING ON SMOOTH, LEVEL GROUND

The procedure followed in measuring distances with the tape depends to some extent upon the required accuracy and the purpose of the survey. The following description represents the usual practice when the measurements are of ordinary relative precision of, say, 1/5000. Errors in taping are discussed in Sections 4.14 to 4.25 and mistakes in Section 4.26.

The tape is supported throughout its length. If only the distance between two fixed points (as the corners of a parcel of land) is to be determined, the equipment will consist of one or more range poles, 11 chaining pins, and a 30-m or 100-ft heavy steel tape graduated as described in Section 4.8. One range pole is placed behind the distant point to indicate its location.

The rear taper *with one pin* is stationed at the point of beginning. The head taper, with the zero (graduated) end of the tape and 10 pins, advances toward the distant point. When the head taper has gone nearly 100 ft, or 30 m, the rear taper calls "chain" or "tape," a signal for the head taper to halt. The rear taper holds the 100-ft or 30-m mark at the point of beginning and, by hand signals or voice, lines in a taping pin (held by the head taper) with the range pole (or other signal) marking the distant point. During the lining-in process, the rear taper kneels on the line and faces the distant point; the head taper kneels to one side of and faces the line so that the tape can be held steady and the rear taper will have a clear view of the signal marking the distant point. The head taper with one hand, sets the pin vertically on line and a short distance to the rear of the zero mark. With the other hand, the taper then pulls the tape taut and, making sure that it is straight, brings it in contact with the pin. The rear taper, on observing that the 30-m or 100-ft mark is at the point of

FIGURE 4.4
Taping on smooth, level ground.

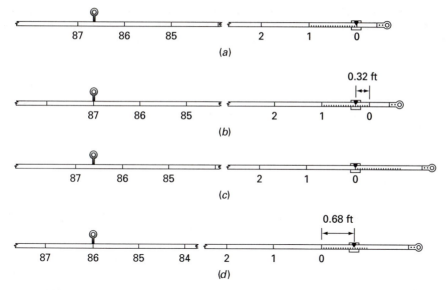

FIGURE 4.5
Plan view of a tape graduated in feet. (a) and (b) Measuring with a "cut" tape. (c) and (d) Measuring with an "add" tape.

beginning, calls "good" or "mark." The head taper pulls the pin and sticks it at the zero mark of the tape, with the pin sloping away from the line (Figure 4.4). As a check, the head taper again pulls the tape taut and notes that the zero point coincides with the pin at its intersection with the ground. The head taper then calls "good," the rear taper releases the tape, the head taper moves forward as before, and so the process is repeated.

As the rear taper leaves each intermediate point, the pin is pulled. Thus, one pin always is in the ground, and the number of pins held by the rear taper at any time indicates the number of multiples of 30 m or hundreds of feet, or *stations,* or from the point of beginning to the pin in the ground.

At the end of each (10 full stations), 300 m, or 1000 ft the head taper places the last pin in the ground. Then, the head taper signals for pins, the rear taper comes forward and gives 10 pins to the head taper, and both check the total, which is recorded. The procedure then is repeated. The count of pins is important, as the number of tape lengths is easily forgotten, owing to distractions.

When the end of the course is reached, the head taper halts and the rear taper comes forward to the last pin set. The head taper holds the zero mark at the terminal point. The rear taper pulls the tape taut and observes the number of meters, or decimeters or *whole feet* between the last pin and the end of the line (Figure 4.5a). The rear taper then holds the next *larger* meter, or decimeter foot, mark at the pin, and the head taper pulls the tape taut and reads the decimal by means of the finer graduations at the end of the tape (Figure 4.5b). For tapes graduated in feet and having fine graduations between the 0 and 1-ft points, (a "cut" tape), the distance is the whole number of feet held by the rear taper minus the fractional increment observed by the head taper. For example, if the rear taper holds 87 ft and the head taper observes 0.32 ft directly, the distance is $87 - 0.32 = 86.68$ ft as illustrated in Figure 4.5b.

For tapes having an extra graduated foot beyond the zero point (an "add" tape), the rear taper holds the next *smaller* foot mark at the pin, and the head taper reads the decimal. Figure 4.5c and d illustrates this procedure for the same distance of 86.68 ft.

Metric tapes have the first decimeter graduated to millimeters and the first meter to centimeters (a "cut" tape; see Figure 4.1d) or a decimeter graduated to millimeters is added to the zero end of the tape (an "add" tape; see Figure 4.1e). Both styles usually are graduated throughout to decimeters.

Cut and add metric tapes are used in the same way as tapes graduated in feet. Suppose that, when the zero end of a metric cut tape is at the terminal point, the rear taper observes that the rear pin falls just past 25.3 m. The rear taper "gives" a decimeter and holds 25.400 m at the rear taping pin. If the head taper then reads 0.075 m on the first graduated decimeter at the terminal point, the distance is $25.400 - 0.075 = 25.325$ m.

In the same situation, using a metric add tape, the rear taper would hold 25.300 m at the rear taping pin. The head taper then would read 0.025 m on the added decimeter at the terminal point, yielding a distance of $25.300 + 0.025 = 25.325$ m.

When the theodolite is set up on the line to be measured, the theodolite operator usually directs the head taper in placing the pins on line.

On some surveys, stakes must be set on line at short intervals, usually 30 m or 100 ft. Sometimes stakes are driven by the rear taper after the pin is pulled. On surveys of low precision, no pins are used and the head taper sets the stakes in the manner previously described for pins, measuring the distance between the centers of the stakes at their junction with the ground. On more precise surveys, measurements are carried forward by setting a tack in the head of each stake. When the tack has been driven, it is tested for line and distance.

4.11
HORIZONTAL TAPING OVER SLOPING GROUND

The process of taping over uneven or sloping ground or over grass and brush is much the same as that just described for smooth, level ground, except that a plumb bob is used. The tape is held horizontal, and a plumb bob is used by either or at times by both tapers for projecting from tape to pin, or vice versa. For rough work, plumbing can be accomplished with the range pole.

To secure accuracy comparable to taping over level ground, considerable skill is required. Some experience is necessary to determine when the tape is nearly horizontal; the tendency is to hold the downhill end of the tape too low. A hand level is useful to estimate the proper height to hold the tape ends so as to have a horizontal tape. The tape is unsupported between its two ends, and either the pull must be increased to eliminate the sag or a correction for the sag must be applied. A firm stance is important; the tapers should place the planes of their bodies parallel to the tape, with legs well apart. The forearms should be in line with the tape and steadied or snubbed against the respective legs or bodies, as required by the height at which the tape must be held.

Where the slope is less than 1.5 m in 30 m or 5 ft in 100 ft, the head taper advances a full tape length at a time. If the course is downhill, the head taper holds the plumb line at the zero point of the tape, with the tape horizontal and the plumb bob a few inches from the ground, then pulls the tape taut and is directed to line by the rear taper. When the plumb bob comes to rest, it is lowered carefully to the ground (see Figure 4.6) and a pin is set in its place. As a check, the measurement is repeated. If the course is uphill, the rear taper holds a plumb line suspended from the 30-m or 100-ft mark and signals the head taper to give or take until the rear taper's plumb bob comes to rest over the pin. The head taper sets a pin, and the measurement is repeated.

Where the course is steeper and 30-m or 100-ft stations still are desired, as for a route survey, the following procedure is recommended. Assuming that the slope is downhill, the head taper advances a full tape length and then returns to an intermediate point from which

FIGURE 4.6
Plumbing at the downhill end of a horizontal tape.

the tape can be held horizontal. The head taper suspends the plumb line at a foot mark, is lined in by the rear taper, and sets a pin at the indicated point. The rear taper comes forward, *gives the head taper a pin,* and at the pin in the ground holds the tape at the foot mark, from which the plumb line was previously suspended. The head taper proceeds to another point from which the tape can be held horizontal, and so the process is repeated until the head taper reaches the zero mark on the tape. At each *intermediate* point of a tape length the rear taper gives the head taper a pin, but not at the point marking the full tape length. In this manner the tape is always advanced a full length at a time; the number of pins held by the rear taper at each 30-m or 100-ft point indicates the number of multiples of 30 m hundreds of feet or from the last tally, and the rear taper's count of pins is not confused. The process is called *breaking tape.*

To illustrate, Figure 4.7 represents the profile of a line to be measured in the direction of *A* to *D*, and *A* represents a pin marking the end of a 30-m interval from the point of beginning. The head taper goes forward until the 30-m mark is at *A*, where the rear taper is stationed. The head taper then returns to *B*, where the tape is held horizontal and plumbed from the 25-m mark to set a pin at *B*. The rear taper gives the head taper a pin and holds the 25-m mark at *B*. The head taper plumbs from the 15-m mark and sets a pin at *C*. The rear taper gives the head taper a pin and holds the 15-m mark at *C*. The head taper plumbs from the zero mark to set a pin at *D* at the end of the full tape length. The rear taper goes forward but keeps the pin that was pulled at *C*.

Some surveyors prefer to measure distances less than the tape length individually and add these measurements. However, this practice requires recording and may lead to mistakes in addition.

Usually, the tape is estimated to be horizontal by eye. This practice commonly results in the downhill end being too low, sometimes causing a significant error in horizontal measurement. The safe procedure in rough country is to use a hand level.

In horizontal measurements over uneven or sloping ground, the tape sags between supports and becomes effectively shorter. The effect of the sag can be eliminated by standardizing tape, by applying a computed correction, or by using the normal tension (Section 4.20). In breaking the tape or when the tape is supported for part of its length, the difference in effect of sag as between a full tape length and the unsupported length can be taken into account roughly by varying the pull on the tape.

In measurements over low obstructions, if the tape is partially supported between the tapers, account should be taken of the effect this support makes on sag corrections.

4.12
SLOPE TAPING

Where the ground is fairly smooth, slope measurements generally are preferred because they usually are made more accurately and quickly than horizontal measurements. Some means of determining either the slope or the difference in elevation between successive tape ends and breaks in slope is required. For surveys of ordinary accuracy, either the clinometer or transit (for measuring vertical angles) or the hand level (for measuring difference in elevation) may be used to advantage. If only the distance between the ends of the line is required, the procedure of taping is the same as on level ground, but a record is kept of either the slope or the difference in elevation of each tape length (or less at breaks in slope). The horizontal distances then are computed from the distances measured on the slope.

Where stakes are to be placed at regular intervals (e.g., 100 ft, 20 m, 30 m), corrections to the slope distances may be applied as the taping progresses.

4.13
CORRECTION FOR SLOPE

Where the slopes are measured with sufficient accuracy to warrant, the corresponding horizontal distances may be computed by exact trigonometric relations. In Figure 4.8, let s represent the slope distance between two points A and B, h is the difference in elevation, and H is the horizontal distance, all in the same units. The angles θ and z are the vertical and zenith angles, respectively, observed with a clinometer, theodolite, or total station system. If θ has been observed, then

$$H = s \cos \theta \qquad (4.1)$$

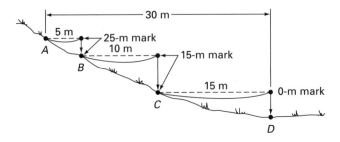

FIGURE 4.7
Horizontal measurements on a steep slope.

When z has been measured,

$$H = s \sin z \qquad (4.2)$$

EXAMPLE 4.1. A slope measurement of 29.954 m was made between two points where θ is 4°30′. Determine the horizontal distance.

Solution. Using Equation (4.1) yields

$$H = (29.954)(\cos 4°30′) = (29.954)(0.996917) = 29.862 \text{ m}$$

EXAMPLE 4.2. It is desired to set point D a horizontal distance of 195.00 ft from point E along a line that has a slope angle of 5°30′. What slope distance should be measured in the field?

Solution. Using Equation (4.1), we obtain

$$S = \frac{H}{\cos \theta} = \frac{195.00}{0.995396} = 195.90 \text{ ft}$$

EXAMPLE 4.3. A slope measurement of 250.542 m was made between two points for which the measured zenith angle is $z = 85°45′30″$. Calculate the horizontal distance.

Solution. Using Equation (4.2) yields

$$H = S \sin z = (250.542)(0.9972610) = 249.856 \text{ m}$$

EXAMPLE 4.4. A horizontal distance of 150.00 ft is to be set from point E to set point F along a line for which the zenith angle is $z = 94°25′00″$. Determine the slope distance that should be measured in the field.

Solution. Using Equation (4.2) we have

$$S = H/(\sin z) = 150.00/(0.997030) = 150.45 \text{ ft}$$

Reduction of the slope to horizontal distances also can be determined by using the difference in elevation between the two ends of the line. Referring to Figure 4.8 yields

$$H^2 = s^2 - h^2$$

from which
$$s = (H^2 + h^2)^{1/2} \qquad (4.3)$$

and
$$H = (s^2 - h^2)^{1/2} \qquad (4.4)$$

which are the direct relations for the slope and horizontal distances, respectively.

EXAMPLE 4.5. A distance of 130.508 m was measured over terrain with a constant slope along a sloping line that has a difference in elevation between the two ends of 5.56 m. Calculate the horizontal distance between the two points.

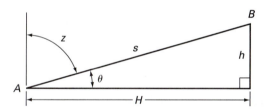

FIGURE 4.8
Slope correction.

Solution. By Equation (4.4),

$$H = [(130.508)^2 - (5.56)^2]^{1/2} = 130.390 \text{ m}$$

EXAMPLE 4.6. Point R is to be set at a horizontal distance of 98.25 ft from point Q along a sloping line where the difference in elevation between R and Q is 4.35 ft. Calculate the slope distance to be measured in the field.

Solution. Using Equation (4.3) yields

$$s = [(98.25)^2 + (4.35)^2]^{1/2} = 98.35 \text{ ft}$$

EXAMPLE 4.7. A distance was measured over irregularly sloping terrain. Slope distances and differences in elevation are tabulated in the two columns on the left of the following table. Calculate the horizontal distance.

Slope distance, ft	Difference in elevation, ft	Horizontal distance, ft
100.00	3.50	99.94
100.00	5.30	99.86
80.50	4.20	80.39
100.00	8.05	99.68
62.35	5.25	62.13
442.85		442.00

Solution. H is calculated using Equation (4.4), individually for each length and tabulated in the third column. The total horizontal distance is the sum of the calculated values of H.

4.14
SYSTEMATIC ERRORS IN TAPING

Systematic errors in taping linear distances are those attributable to the following causes: (1) The tape is not of standard length, (2) the tape is not horizontal, (3) variations in temperature, (4) variations in tension, (5) sag, (6) incorrect alignment of tape, and (7) the tape is not straight.

4.15
TAPE IS NOT OF STANDARD LENGTH

The nominal length of a tape, as stated by the manufacturer, rarely corresponds exactly with the true length. The true length is taken as the length determined by comparison with a known standard length under given conditions of temperature, tension, and support. The National Institute of Standards and Technology (NIST), Gaithersburg, Maryland, will standardize tapes and provide a certificate of standardization upon payment of a fee. The absolute value for the tape correction C_d is

$$C_d = \text{true length} - \text{nominal length} \tag{4.5}$$

This discrepancy normally is assumed to be distributed uniformly throughout the tape and directly proportional to the fractional portion of the tape used for a measurement.

Tapes are usually standardized at 20°C (68°F) and with tensions varying from 4 to 14 kg or 10 to 30 lb. Tapes may be supported throughout, at specified points, or at the two ends only. The desired conditions of standardization are specified when requesting the standardization.

The normal use and abuse of a tape in the field over a period of time can invalidate the standard length. Consequently, the standard tape should be kept in the office and used only for comparison. At regular intervals, field tapes should be compared with the standard tape or to a standard length measured with the standardized tape. Such comparisons are extremely important if specified accuracies are to be maintained in the field.

4.16
TAPE IS NOT HORIZONTAL

When horizontal taping is being performed over sloping ground, discrepancies frequently occur because the tape is not truly horizontal. Slopes often are deceptive, even to experienced surveyors; the tendency is to hold the downhill end of the tape too low. Inexperienced tapers have been observed keeping very careful alignment, yet taping what they thought were horizontal distances on a slope of perhaps 10 percent. The resulting discrepancy is 0.5 ft/100 ft or 26 ft/mi (0.15 m/30 m or ≈ 5 m/km). In ordinary taping, this is one of the largest of contributing errors. It will not be eliminated by repeated measurements, but it can be reduced to a negligible amount by leveling the tape with either a hand level or a clinometer.

4.17
VARIATIONS IN TEMPERATURE

The tape expands as the temperature rises and contracts as the temperature falls. Therefore, if the tape is standardized at a given temperature and measurements are taken at a higher temperature, the tape will be too long. For a change in temperature of 15°F, a 100-ft steel tape will undergo a change in length of about 0.01 ft, introducing an error of about 0.5 ft/mi. Similarly, a change of 15°C produces a change in length of 5 mm/30 m or 0.17 m/km. Under a change of 50°F the error would be 1.5 ft/mi or 0.58 m/km. It is seen that, even for measurements of ordinary precision, the error due to thermal expansion becomes of consequence when the measurements are taken during cold or hot weather.

The coefficient of thermal expansion of steel is approximately 0.00000645/1°F or 0.0000116/1°C. If the tape is standard at a temperature of T_0 degrees and measurements are taken at a temperature of T degrees, the correction C_t for change in length is given by the formula

$$C_t = \alpha L(T - T_0) \tag{4.6}$$

in which L is the measured length and α the coefficient of thermal expansion.

Errors due to variations in temperature are greatly reduced by attaching one or more tape thermometers to it and observing the temperature for each tape length.

4.18
VARIATIONS IN TENSION

If the tension or pull is greater or less than that for which the tape is verified, the tape will be elongated or shortened accordingly. The correction for variation in tension in a steel tape is given by the formula

$$C_p = \frac{(P - P_0)L}{aE} \tag{4.7}$$

where C_p = correction per distance L, ft or m
P = applied tension, lb or kg
P_0 = tension for which the tape is standardized, lb or kg
L = length, ft or m
a = cross-sectional area, in^2 or cm^2
E = elastic modulus of the steel, lb/in^2 or kg/cm^2

The modulus of elasticity is taken as 28 to 30 million lb/in^2 or 2.1×10^6 kg/cm^2. The cross-sectional area of the tape can be computed from the weight and dimensions, because steel weighs approximately 490 lb/ft^3 or 7.85×10^{-3} kg/cm^3. Light (1-lb) and heavy (3-lb) 100-ft tapes have cross-sectional areas of approximately 0.003 and 0.009 in^2, respectively. Light and heavy 30-m tapes have respective cross-sectional areas of about 0.019 and 0.058 cm^2.

Some idea of the effect of variation in tension can be obtained from the following examples.

EXAMPLE 4.8. Assume that a light 100-ft tape is standard with full support under a tension of 10 lb, $E = 30,000,000$ lb/in^2, and the cross-sectional area of the tape is 0.003 in^2. Determine the elongation for an increase in tension from 10 to 30 lb. Assume the tape is used with full support.

Solution. Using Equation (4.7), we obtain

$$C_p = \frac{(30 - 10)(100)}{(30,000,000)(0.003)} = 0.0222 \text{ ft}$$

EXAMPLE 4.9. A heavy 30-m tape having a cross-sectional area of 0.06 cm^2 has been standardized with full support at a tension of 5 kg. If $E = 2.1 \times 10^6$ kg/cm^2, calculate the elongation of the tape for an increase in tension from 5 to 15 kg. Assume the tape used is fully supported.

Solution. Using Equation (4.7) yields

$$C_p = \frac{(15 - 5)(30)}{(2,100,000)(0.06)} = 0.0024 \text{ m}$$

Corrections for variation in tension are seen to be significant for ordinary taping when light tapes are used and the differences in tension are substantial. For measurements of high accuracy such as base lines, tension variations are significant regardless of the weight of tape.

4.19
CORRECTION FOR SAG

When the tape sags between the points of support, it takes the form of a catenary. The correction to be applied is the difference in length between the arc and the subtending chord. To determine the correction, the arc may be assumed to be a parabola. The correction then is given with sufficient precision for most purposes by the formula

$$C_s = \frac{w^2 L^3}{24 P^2} = \frac{W^2 L}{24 P^2} \tag{4.8}$$

where C_s = correction between points of support, ft or m
w = weight of tape, lb/ft or kg/m
W = total weight of tape between supports, lb or kg
L = distance *between supports,* ft or m
P = applied tension, lb or kg

The correction is seen to vary directly as the cube of the unsupported length and inversely as the square of the pull. Although the equation is intended for use with a *level* tape, it may be applied without error of consequence to a tape held on a slope up to approximately 10°.

The changes in correction for the sag due to a variation in tension, the weight of the tape, and the distance between supports are illustrated in Figure 4.9. for 100-ft and 25-ft spans and in Figure 4.10. for 10-m and 30-m spans.

Analysis of Figures 4.9 and 4.10 reveals the hazard of measuring with a heavy tape without using a tension handle. A variation of −0.5 kg, from a standard tension of 8 kg, with a very heavy tape over a 30-m span results in a discrepancy of 8 mm in the sag correction or a relative precision of 1 part in 3750. A similar variation in tension using a medium-weight 30-m tape causes discrepancies of 1.5 mm for a relative precision of 1 in 20,000.

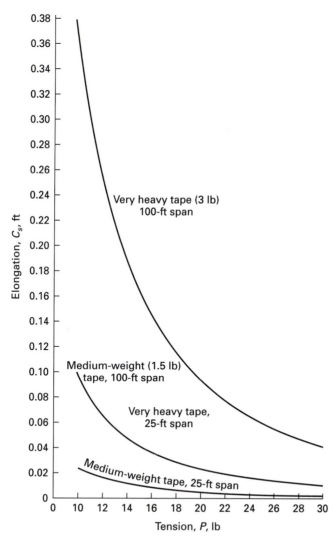

FIGURE 4.9
Sag correction versus tension, 100- and 25-ft spans.

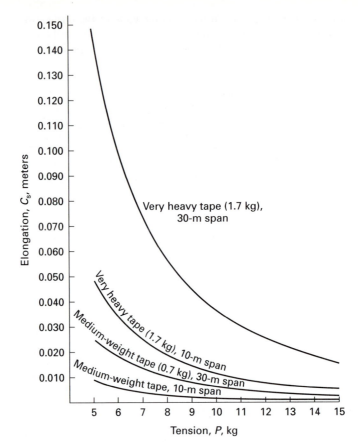

FIGURE 4.10
Sag correction versus tension, 30- and 10-m spans.

Estimation of tension to within ± 0.5 kg or about ± 1 lb is very difficult, especially with a heavy tape. Consequently, even for ordinary taping where a relative precision of 1 part in 5000 is sought, a tension handle should be used for measurements with heavy tapes.

The effect of reducing span length also is illustrated in Figures 4.9 and 4.10 where discrepancies in C_s are reduced substantially for the shorter 10-m and 25-ft spans. This factor should not be interpreted to indicate that a shorter tape length should be used, as the error propagation due to setting and measuring additional points probably would offset any reduction in sag correction errors. Reduction of span length is most useful when using long, heavy tapes for precise measurement. Note that a 25-ft unsupported span of a very heavy 100-ft tape has the same sag corrections as a medium-weight 100-ft unsupported span.

4.20
NORMAL TENSION

By equating the right-hand members of Equations (4.7) and (4.8), the elongation due to increase in tension is made equal to the shortening due to sag; thus, the effect of the sag can be eliminated. The pull that will produce this condition, called *normal tension, P_n*, is given by the formula

$$P_n = \frac{0.204 \, W\sqrt{aE}}{\sqrt{P_n - P_0}} \tag{4.9}$$

This equation can be solved using successive approximations for P_n. Normal tensions for very heavy ($W = 3$ lb or 1.7 kg; $a = 0.01$ in^2 or 0.06 cm^2) and medium-weight ($W = 1.5$ lb or 0.7 kg; $a = 0.005$ in.^{2} or 0.03 cm^2) 100-ft and 30-m tapes, assuming that $P_0 = (10$ lb, 5 kg) and $E = (30,000,000$ lb/in^2, 2,100,000 kg/cm^2), respectively, are as follows:

	Very heavy tape		Medium-weight tape	
	100 ft	**30 m**	**100 ft**	**30 m**
Normal tension	51.8 lb	26.5 kg	28.0 lb	12.8 kg

It can be seen that the normal tension for the very heavy tape over the long span is not practical.

4.21
IMPERFECT TAPE ALIGNMENT

The head taper is likely to set the pin sometimes on one side and sometimes on the other side of the correct line. This produces a variable systematic error, because the horizontal angle that the tape makes with the line is not the same for one tape length as for the next. The error cannot be eliminated, but it can be reduced to a negligible quantity by care in lining. Generally, it is the least important of the errors in taping, and extreme care in lining is not justified. The linear error when one end of the tape is off-line a given amount can be computed by Equation (4.4). For a 100-ft tape, the error amounts to 0.005 ft when one end is off-line 1 ft with respect to the other and to only 0.001 ft when the error in alignment is 0.5 ft. Displacements at the ends of a 30-m tape of 3 dm and 1.5 dm produce discrepancies of 1.5 mm and 0.4 mm, respectively. Many surveyors use unnecessary care in securing good alignment without paying much attention to other, more important sources of error. Errors in alignment tend to make the measured length between fixed points greater than the true length and hence are positive.

4.22
TAPE IS NOT STRAIGHT

In taping through grass and brush or when the wind is blowing, it is impossible to have all parts of the tape in perfect alignment with its ends. The error arising from this cause is systematic and variable and of the same sign (positive) as that from measuring with a tape that is too short. Care must be exercised to stretch the tape taut and to observe that it is straight by sighting over it, so that the error is not of consequence.

4.23
SUMMARY OF SYSTEMATIC ERRORS IN TAPING

The various systematic errors discussed in Sections 4.12 through 4.22 are summarized in Table 4.2. The apparently simple task of linear measurement is affected by a remarkable number of factors.

TABLE 4.2
Systematic errors in taping

Source	Amount	Error of 0.01 ft/ 100 ft or 3 mm/ 30 m tape length caused by	Makes tape too	Importance in 1:5000 taping	Procedure to eliminate or reduce
Tape not of standard length	—	—	Long or short	Usually small but must be checked	Standardize tape and apply computed correction
Temperature	$C_t = \alpha L(T - T_0)$	15°F or 9°C	Long or short	Of consequence only in hot or cold weather	Measure temperature and apply computed correction; for precise work, tape at favorable times
Change in pull or tension	$C_p = \dfrac{(P - P_0)L}{aE}$	15 lb or 4.2 kg	Long or short	Negligible	Apply computed correction; in precise work use a spring balance
Sag	$C_s = \dfrac{w^2 L^3}{24P^2} = \dfrac{W^2 L}{24P^2}$	$\Delta P = 0.6$ lb or* 1 kg too small	Short	Large, especially with heavy tape	Apply computed correction; use tape fully supported
Slope	$H = (s^2 - h^2)^{1/2}$ $H = s\cos\theta$	1.4 ft or† 0.42 m in h, 0°48′ in θ (both from horizontal)	Short	—	At breaks in slope determine differences in elevation or slope angle; apply computed correction
Imperfect horizontal alignment	Same as slope	1.4 ft or 0.42 m	Short	Not serious	Use reasonable care in aligning tape; keep tape taut and reasonably straight

For the pull, sag, and change in tension rows:

Use normal tension

$$P_n = \frac{0.204\, W\sqrt{aE}}{\sqrt{P_n - P_0}}$$

or use standard tension P_0

Notes: In measuring a distance between fixed points with a tape that is too long, add the correction (assume a tape of medium weight).
*Error affected nonlinearly by the tension P.
† Nonlinear relationship, error increases nonlinearly as the slope increases.

118

TABLE 4.3

119

CHAPTER 4:
Distance
Measurement

Taping operation	Sign of correction to be applied	
	Tape too long	Tape too short
Measurement	+	−
Layout	−	+

When corrections are applied to the observed length of a line measured between fixed points (designated *measurement*) with a tape that is too long, the correction is *added*. When the *layout* of a distance is performed (i.e., a second point is established at a specified distance from a starting point) with a tape that is too long, the correction is *subtracted* from the specified distance to determine the distance to be read on the tape. For a tape that is too short, the corrections are opposite in direction to those just stated. Table 4.3 provides a convenient summary of the signs for these corrections.

Refer to the sample problems in Section 4.24 for examples of how corrections are applied.

4.24
COMBINED CORRECTIONS

Whenever corrections for several effects, such as slope, tension, temperature, and sag, are to be applied, for convenience they may be combined as a single net correction per tape length. Since the corrections are relatively small, the value of each is not affected appreciably by the others and each may be computed on the basis of the nominal tape length. For example, even if the verified length of a tape were 100.21 ft, the correction for temperature (within the required precision) would be found to be the same whether computed for the exact length or for a nominal length of 100 ft. The combined correction per tape length can be used as long as conditions remain constant; however, the temperature is likely to vary considerably during the day, and the ground may be such that it may be necessary to support the tape at various intervals of length and with different slopes so that corrections must be calculated for individual tape lengths (see Example 4.7). Several examples to illustrate applying corrections follow.

EXAMPLE 4.10. A 30-m tape weighing 0.55 kg and with a cross-sectional area of 0.02 cm^2 was standardized and found to be 30.005 m at 20°C, with 5-kg tension and supported at the 0- and 30-m points. This tape was used to measure a distance of about 89 m over terrain of a uniform 5 percent slope. The temperature was constant at 30°C, the tape was fully supported throughout, and a tension of 5 kg was applied to each tape length. The observed distances were 30.000 m, 30.000 m, and 29.500 m. Calculate the horizontal distance between the points.

Solution. Sag, temperature, and standardization are in direct proportion to the length and can be calculated for the total distance.

The sag correction (Equation (4.8)) for two 30.000-m and one 29.500-m tape lengths is

$$C_s = (2)\frac{(0.55)^2(30.005)}{(24)(5)^2} + \frac{[0.55(29.5/30)]^2(29.500)(30.005/30.000)}{(24)(5)^5} = 0.045 \text{ m}$$

The tape is long.

The temperature correction (Equation (4.6)) is

$$C_t = (0.0000116)(89.50)(30 - 20) = 0.0104 \text{ m} \qquad \text{The tape is long.}$$

The standardization correction (Equation (4.5)) is

$$C_d = (30.005 - 30.000)\left(\frac{89.50}{30}\right) = 0.0149 \text{ m} \qquad \text{The tape is long.}$$

Corrected slope distance = $89.500 + 0.045 + 0.010 + 0.015 = 89.570$ m

The slope is uniform, so the entire distance can be reduced to the horizontal distance using Equation (4.4):

$$H = (s^2 - h^2)^{1/2} = \{(89.570)^2 - [(0.05)(89.570)]^2\}^{1/2} = 89.458 \text{ m}$$

EXAMPLE 4.11. A 100-ft tape has a standardized length of 100.000 ft at 68°F, with 20-lb tension and with the tape fully supported. The observed distance between two points, with the tape held horizontal, supported at the 0- and 100-ft points only, and 30-lb tension is 100.452 ft. The tape weighs 3.0 lb and has a cross-sectional area of 0.01 in². Calculate the true distance between the two points. Assume that $E = 3 \times 10^7$ lb in².

Solution. First calculate the sag correction at the standard tension with Equation (4.8):

$$C_s = \frac{(3)^2(100.452)}{(24)(30)^2} = -0.042 \text{ ft}$$

The tape is shorter, so the correction is negative when measuring between two fixed points. Next, compute the correction due to the increase in tension using Equation (4.7):

$$C_p = \frac{(30 - 20)(100.452)}{(0.01)(3)(10^7)} = 0.003 \text{ ft}$$

The increased tension makes the tape long, so that the correction is positive. The corrected distance is

$$100.452 - 0.042 + 0.003 = 100.413 \text{ ft}$$

Note that the foregoing corrections can be calculated with adequate accuracy with a calculator and using $L = 100$ ft without introducing significant errors:

$$C_s = \frac{(3)^2(100)}{(24)(30)^2} = -0.042 \text{ ft}$$

$$C_p = \frac{(30 - 20)(100)}{(0.01)(3)(10^7)} = 0.003 \text{ ft}$$

EXAMPLE 4.12. A slope distance is measured from A on the west side of a railroad underpass to B on the east side of the tracks, using a 200-ft steel tape having a standardized length of 200.005 ft, with the tape fully supported, a tension of 18 lb, and at a temperature of 68°F. The recorded distance of about 251 ft was measured in two segments: (1) an unsupported slope distance of 199.96 ft, with 18 lb tension, at a temperature of 40°F, and a zenith angle of 94°30′30″; and (2) an unsupported, horizontal length of 55.45 ft at 40°F and 5 lb tension, which is the correct tension for that length with tape fully supported. Calculate the corrected horizontal distance A–B. The tape weighs 3 lb and has a temperature coefficient of 0.00000645/°F.

Solution. Calculate the standardization, temperature, and sag corrections separately for the sloping (199.96 ft) and horizontal (55.45 ft) segments of the total distance. Use Equations (4.5), (4.6), and (4.8), respectively.

$$C_{d1} = (199.96/200)(200.005 - 200.000) = 0.005 \text{ ft}$$
$$C_{d2} = (55.45/200)(200.005 - 200.000) = 0.001 \text{ ft}$$
$$C_{t1} = \alpha L(T - T_0) = 0.00000645(199.96)(40 - 68) = -0.036 \text{ ft}$$
$$C_{t2} = (0.00000645)(55.45)(40 - 68) = -0.010 \text{ ft}$$

$$C_{s1} = \frac{W^2 L}{24P^2} = \frac{(3)^2(199.96)}{(24)(18)^2} = 0.231 \text{ ft}$$

$$C_{s2} = \frac{\left[\left(\dfrac{3}{200}\right)(55.45)\right]^2 (55.45)}{(24)(5)^2} = 0.064 \text{ ft}$$

For C_{s1} and C_{s2}, the tape is short so the respective signs will be negative for a measurement (Table 4.3) when corrections are applied.

Corrected sloping segment, $s = 199.96 + 0.005 - 0.036 - 0.231 = 199.698$ ft
Sloping segment reduced to horizontal distance using Equation (4.2):
$H = s \sin z = (199.698)(0.9969059)$ = 199.080 ft
Corrected horizontal segment = $55.45 + 0.001 - 0.010 - 0.064$ = 55.377 ft
Total corrected distance between A and B = 254.46 ft

4.25
RANDOM ERRORS IN TAPING

Random errors in taping occur as a result of human limitations in observing measurements and manipulating equipment. Examples of specific causes include marking tape ends with the tape fully supported, plumbing to mark tape ends, small variations in applying tension, determining slope angle or differences in elevation between tape ends, and temperature measurements.

Note that random errors occur as a result of (1) direct observations, such as marking tape ends or reading graduations, or (2) indirect observation of slope angles, applying tension, and the like. These random errors are much smaller than the systematic errors and tend to be of opposite signs so that they often cancel each other in the long run.

When every device is employed to eliminate systematic errors using the methods described in Sections 4.14 through 4.24, random errors of observation become more significant; for this reason, long tapes (200 ft, 50 m) frequently are used where very precise measurement is required.

A more detailed analysis of the sources and effects of random errors in taping and error propagation in taping can be found in Section 4.34 in Part B of this chapter.

4.26
MISTAKES IN TAPING

Some of the mistakes commonly made by individuals inexperienced in taping are

1. *Adding or dropping a full tape length.* This is not likely to occur if both tapers count the pins or when numbered stakes are used, if the rear taper calls out the station number of

the rear stake in response to which the head taper calls out the number of the forward stake while marking it. A tape length may be added through failure of the rear taper to give the head taper a pin at breaks marking fractional tape lengths. A tape length may be dropped through failure of the rear taper to take a pin at the point of beginning.

2. *Adding a foot or a decimeter.* This usually happens in measuring the fractional part of a tape length at the end of the line. This distance should be checked by the head taper holding the zero mark on the tape at the terminal point and the rear taper noting the number of feet or decimeters and approximate fraction at the last pin set.

3. *Other points incorrectly taken as 0- or 100-ft (or 0- or 30-m) marks on tape.* The taper should note whether these marks are at the end of rings or on the tape itself, also whether there is an extra graduated unit at one end of the tape.

4. *Reading numbers incorrectly.* Frequently "68" is read "89" or "6" is read as "9." As a check, a good practice is to observe the number of the foot or decimeter marks on each side of the one indicating the measurement, especially if the numbers are dirty or worn. Also, the tape should be read with the numbers right side up.

5. *Calling numbers incorrectly or not clearly.* For example, 50.3 might be called "fifty, three" and recorded as 53.0. If called as "fifty, point, three" or preferably "five, zero, point, three," the mistake would not be likely to occur. When a decimal point or a zero occurs in a number, it should be called. When numbers are called to a recorder, the notekeeper should repeat them as they are recorded.

Often, large mistakes will be prevented or discovered if those involved in the work form the habit of pacing distances or of estimating them by eye. If a transit is being used to give line, distances can be checked by reading the approximate stadia interval on a range pole.

4.27
ELECTRONIC DISTANCE MEASUREMENT

Electronic distance measuring (EDM) equipment includes electro-optical (lightwaves) and electromagnetic (microwaves) instruments. The first generation of electro-optical instruments, typified by the Geodimeter, originally was developed in Sweden in the early 1950s. The basic principle of electro-optical devices is the indirect determination of the time required for a light beam to travel between two stations. The instrument, set up on one station, emits a modulated light beam to a passive reflector set up on the other end of the line to be measured. The reflector, acting as a mirror, returns the light pulse to the instrument, where a phase comparison is made between the projected and reflected pulses. The velocity of light is the basis for computation of the distance. A clear line of sight is required, and observations cannot be made if conditions do not permit intervisibility between the two stations.

Electromagnetic EDM devices, first developed in South Africa during the 1950s, utilize high-frequency microwave transmission. A typical electromagnetic device, such as the Tellurometer, consists of two interchangeable instruments, one being set up on each end of the line to be measured. The sending instrument transmits a series of microwaves that are run through the circuitry of the receiving unit and retransmitted to the original sending unit, which measures the time required. Distances are computed on the basis of the velocity of the radio waves. An unobstructed measuring path between the two instruments is necessary. However, intervisibility is not required, and therefore observations can be made in fog and other unfavorable weather conditions. Distances up to 66 to 80 km (40 to 50 mi) can be measured under favorable conditions.

Development of electro-optical EDM equipment has gone through several stages since it was first introduced in the 1950s. Miniaturization of electronics and development of small light-emitting diodes resulted in the design and production of very small, light, compact EDM instruments that can operate on low power. These devices, which have short to moderate to long ranges in distance, can be operated in a stand-alone mode when mounted on a tribrach. More important, they also can be attached to the standards of a theodolite or integrated into the framework of a theodolite to form a *total station system,* which has the advantage of permitting measurement of distance and direction with a single instrument (see Chapter 7). Included in this group are instruments that have ranges of from 0.2 m to 14 km (1 ft to 8+ mi) with inherent mean square errors $\pm(0.2$ mm + 0.2 ppm) to $\pm(5$ mm + 5 ppm) or $\pm(0.001$ ft + 0.2 ppm) to (0.02 ft + 5 ppm), where ppm is parts per million.* These instruments are well adapted to the needs of the practicing surveyor,

TABLE 4.4
Selected electronic distance measuring instruments (EDM)

Instrument short range	Manufacturer	Emission source	Range, m (single prism)	Accuracy (mean square error)
DI1001	Leica	Infrared	1–800	$\pm(5$ mm + 5 ppm)
RED Mini 2	Sokkia	Infrared	800	$\pm(5$ mm + 3 ppm)
DM-H1	Topcon	Infrared	0.15–800	$\pm(1$ mm + 2 ppm)
DM-A5	Topcon	Infrared	0.15–800	$\pm(5$ mm + 3 ppm)
ND20/21	Nikon	Infrared	N/A–700/1000	$\pm(5$ mm + 5 ppm)
MD-14/MD-20	Pentax	Infrared	1–1,000/1,600	$\pm(5$ mm + 5 ppm)
MA200	Navigation Electronics	Infrared	–1,600	$\pm(0.25$ mm + 0.5 ppm)
ND-26	Nikon	Infrared	N/A–2,000	$\pm(5$ mm + 5 ppm)
DI1600	Leica	Infrared	1–3,000	$\pm(3$ mm + 2 ppm)
Intermediate				
Geodimeter 220	Geotronics	Infrared	0.2–2,300	$\pm(5$ mm + 3 ppm)
DM-S2/DM-S3L	Topcon	Infrared	0.15–2,400	$\pm(5$ mm + 3 ppm)
DI2002	Leica	Infrared	1–2,500	$\pm(1$ mm + 1 ppm)
RED 2A/RED 2L	Sokkia	Infrared	2,000/3,800	$\pm(5$ mm + 5 ppm)
Leica/Kern ME5000	Leica	Laser	20–5,000	$\pm(0.2$ mm + 0.2 ppm)
DIOR 3002S	Leica	Infrared	0–6,000 No prism, 300	$\pm(3.5$ mm + 0.2 ppm)
RED 2LV	Sokkia	Infrared	6,000	$\pm(5$ mm + 5 ppm)
Eldi 10	Zeiss	Infrared	0.2–7,000	$\pm(5$ mm + 3 ppm)
Pulsar 50	Geo-Fennel	Infrared	2–8,000	$\pm(5$ mm + 5 ppm)
DI 3000S	Leica	Infrared	1–9,000	$\pm(3$ mm + 1 ppm)
Criterion 100	Laser Technology	Laser	1.5–8,000 No prism, 457	$\pm(90$ mm + 50 ppm)
Long Range				
Pro Survey 1000	Laser Atlanta	Laser	1–10,000 No prism, 850	±100 mm ±100 mm
Atlas 2000	Laser Atlanta	Laser	1–10,000 No prism, 1,500	±100 mm ±100 mm
Geodimeter	Geotronics	Infrared	0.2–14,000	$\pm(5$ mm + 1 ppm)
MRA7	Navigation	Microwave	10–50,000	$\pm(15$ mm + 3 ppm)

* Thus, if a distance of 1000 m is measured, the standard deviation would be ±0.2 to ±5 mm + 0.2 to 5 mm.

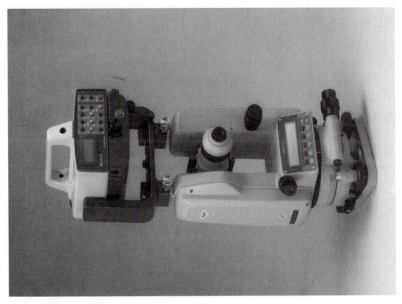

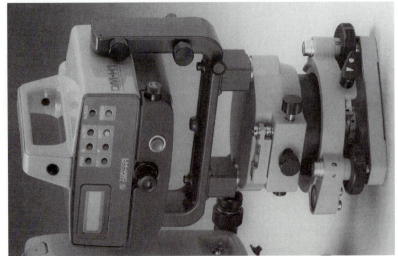

(a)

(b)

FIGURE 4.11

(a) EDM in a stand-alone mode, mounted on the tribrach; (b) EDM mounted on the standards of an electronic theodolite. (*Courtesy of TOPCON America Corp.*)

and those with ranges of up to 14 km (8+ mi) also are applicable to higher-order control surveys of fairly large extent, such as city or county surveys. An intermediate-range EDM is illustrated in Figure 4.11 in (a) the stand-alone mode fastened to a tribrach and (b) mounted on the standards of a theodolite.

Incorporation of visible laser light into light-wave EDM devices provides a group of short- to long-range electro-optical instruments that can be used with or without a reflector (see Section 4.31 for details). Microwave equipment also is available with a range of from 10 m to 50 km and having accuracies of ±(15 mm + 3 ppm). A listing of selected short-, intermediate-, and long-range EDM instruments is given in Table 4.4 on page 123.

Much of the EDM equipment now being used in surveying is part of a total station system. All of the short-range and most of the intermediate-range EDM units are adaptable as add-on devices or occur as an integral part of such a system. Some stand-alone EDM units still are used for special purpose surveys, where very long lines are measured for high-accuracy surveys, but it is reasonable to assume that in the future practically all of the EDM equipment will be contained in the more comprehensive total station surveying systems.

Topics concerning EDM distance measurement in this chapter necessarily are related to operations with stand-alone instruments. However, the basic principles are the same no matter what framework supports the EDM. Therefore, the concepts developed in the sections that follow are equally applicable to the EDM components for total station systems, which are discussed in detail in Chapter 7.

4.28
BASIC PRINCIPLES OF ELECTRO-OPTICAL EDM INSTRUMENTS (EDM)

A continuous-wave carrier beam of light, generated in the transmitter, is modulated by an electronic shutter before entering the aiming optics and being transmitted to the other end of the line, where a reflector is placed. The original light-wave instruments, such as the Geodimeter, utilized incandescent or mercury light and a Kerr cell modulator to achieve the high-frequency modulation of the beam. More recent instruments use infrared or laser light as carrier beams and ultrosonic modulators to impose the intensity modulation on the continuous-wave beam. This intensity modulation is analogous to turning a light off and on using a switch. In the EDM device, the modulator chops the beam into wavelenths that are proportional to the modulating frequency, where the wavelength is given by

$$\lambda = V/f \tag{4.10}$$

where λ = wave length,

V = velocity of light through the atmosphere, m/sec

f = modulating frequency, Hz (hertz = cycles per second)

The velocity of light in the atmosphere varies with temperature, humidity, and the partial pressure of water vapor. Corrections for these factors are described in Section 4.36. The intensity of the modulated light varies from zero or no light at 0° to maximum light at 90° to a second zero at 180°, to a second maximum at 270° and back to zero at 360°, as illustrated graphically in Figure 4.12. Measurement of a distance is accomplished by placing the electro-optical transmitter at one end and a reflector at the other end of the line to be measured. The reflector is a corner cube of glass in which the sides are perpendicular to within a very close tolerance. It has the characteristic that incident light is reflected parallel to itself, and it is called a *retrodirective prism* or *retro-reflector*. The EDM instrument transmits an intensity-modulated light beam to the reflector, which reflects the light beam back to the transmitter, where the incoming light is converted to an electrical signal, allowing a phase comparison between transmitted and received signals. The amount by

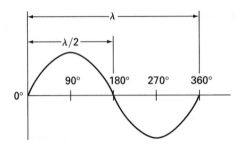

FIGURE 4.12
Wavelength of modulated light.

which transmitted and received wavelengths are out of phase can be measured electonically to within a few millimeters or hundredths of feet.

To visualize the operation, consider an electro-optical EDM instrument, operating on one modulating frequency and with a null meter to record positive or negative maximum or zero intensity of modulation. This relationship is illustrated schematically in Figure 4.13. Assume that the instrument centered over A in Figure 4.13 is an even number of wavelengths, λ, from the reflector at C and the meter registers 0. Now, move the instrument slowly toward C so that the meter will reach a maximum on the right at 90°, return to zero at 180°, a second maximum on the left at 270°, and a second zero at 360°. This procedure is repeated as the instrument is moved toward C, with the needle on the meter moving from 0 to 1 maximum to 0 to another maximum until the instrument is at a distance of $\lambda/2$ from the reflector. Note that the constant or uniform "tape length" between successive zero points is $\lambda/2$ for the double distance.

In practice, the instrument rarely is an even number of wavelengths from the reflector and not moved along the line of sight toward the reflector. Assume that distance AC is not an integral number of wavelengths. Figure 4.14 shows a section through AC with

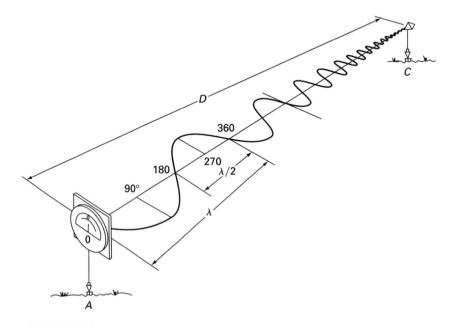

FIGURE 4.13
Modulation frequency of EDM. (*Courtesy of Geotronics from the AGA Operating Manual,* 1965.)

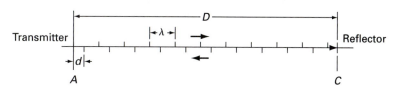

FIGURE 4.14
Measurement principle of EDM.

the transmitter at A and the reflector at C. The modulated beam transmitted at A travels the double distance from A to C and back, where the received beam is out of phase with the transmitted signal by an amount d. The increment d is measured electornically by adjusting some type of phase meter in the instrument, and the distance from A to C is

$$D = \frac{1}{2}(m\lambda + d) \qquad (4.11)$$

in which λ is the wavelength of the modulated beam and m the integer number of wavelengths in the double path of the light. Note that if the reflector were moved $\lambda/2$ or any full number of $\lambda/2$s, the increment d would not change. Therefore, if one knows the double distance to within $\lambda/2$ or the distance to within $\lambda/4$, the total distance can be resolved.

Knowledge of the distance to within $\pm\lambda/4$ is an impractical constraint on an EDM device. The ambiguity in m is resolved by using some type of multiple-frequency technique. Light is transmitted on one frequency and d_1 is observed. Light then is transmitted on a second but slightly different frequency and d_2 is observed. Two equations (4.11) are formed and solved for m. The multiple-frequency technique is built into the EDM device, permitting electronic determination of m and a direct readout of the distance. Decade modulation is one of several techniques applicable and is used for the following example.

EXAMPLE 4.13. Assume that a modulation frequency of 15 MHz (megahertz = 10^6 cps), giving $\lambda/2$ of 10 m, is used.* A full sweep of the phase meter represents this 10-m distance and allows reading the unit and decimal parts of a meter from 0 to 9.999 m. If a given distance is 5495.800 m, the phase meter records 5.800 m on this frequency. This operation is analogous to measuring the distance with a steel tape graduated at every 10-m point (but not labeled), and with one added 10-m section graduated to meters and millimeters, where the increment 5.800 is observed. Now switch to a second frequency with $f = 1.5$ MHz, yielding $\lambda/2 = 100$ m. A full sweep of the phase meter allows reading 10 to 90 m and permits resolving the number of tens of meters, or 90 m for this example. Next, switch to 0.15 MHz, yielding $\lambda/2 = 1000$ m, permitting resolution of the number of hundreds of meters, in this case 400 m. Finally, turn to a frequency of 15 kHz (kilohertz = 10^3 cps) with $\lambda/2 = 10,000$ m, which will allow resolving the number of 1000 m in the distance, 5000 m for the example. The total distance is $5.800 + 90 + 400 + 5000 = 5495.800$ m.

4.29
ELECTROMAGNETIC OR MICROWAVE EDM PRINCIPLES

Microwave EDM transmitters generate electromagnetic carrier beams having frequencies from 3 to 35 GHz (gigahertz = 10^9 cps) and modulated with frequencies varying from

---

*Assume that $V \simeq 300,000$ km/sec.

10 to 75 MHz. The wavelength is given by Equation (4.10), in which λ is the wavelength of modulation in meters or feet; V the velocity of microwaves through the atmosphere in meters, m/sec or ft/sec; and f the modulating frequency in Hz. The velocity V is a function of temperature, atmospheric pressure, and the partial pressure of water vapor. Details of corrections for these factors are given in Section 4.36.

The fundamental concepts of microwave instrument operations are similar to those involved in electro-optical EDM devices. The major difference is that two similar instruments and two operators, one for each end of the line, are required to measure a line using microwave equipments. The master unit transmits to the remote unit, which acts as a reflector for the transmitted microwaves. Voice communication is available for the operators at each unit. The operator selects a frequency at the master unit and by voice signal transmits this information to the operator at the other station, who sets the remote unit to correspond to the transmitted frequency. The signal is received at the remote unit and retransmitted to the master station, where the odd increment is measured with a phase meter.

The number of integer measuring units m between the two ends of the line is determined by the method of multiple frequencies similar to that employed in the electro-optical EDM. The multiple frequencies employed in a typical microwave transmitter are

A pattern:	7.492 MHz
B pattern:	7.490 MHz
C pattern:	7.477 MHz
D pattern:	7.432 MHz
E pattern:	5.993 MHz

The A pattern resolves to the nearest centimeter and combination of the B, C, D and E patterns enable resolving to the nearst 1, 10, 100 and 1000 m, respectively. A detailed description of microwave systems can be found in Burnside (1991).

4.30
SYSTEMATIC ERRORS IN EDM AND REDUCTION OF SLOPE DISTANCES

The effects of atmospheric conditions on wave velocity, uncertainties in the position of the electrical center of the transmitter, uncertainties in the effective center(s) of reflector(s), and transmitter nonlinearity all contribute to the systematic errors found in light-wave and microwave systems. The magnitudes and signs of each of these errors must be evaluated and applied to measured distances to achieve optimum results. Details concerning these systematic errors and their correction can be found in Sections 4.35 through 4.40.

All distances measured by EDM systems represent *slope distances*. Reduction of these measurements to horizontal distance means additional data, such as the slope angle of the line or difference in elevation between the electrical center of the transmitter and center of the reflector or remote unit, also must be obtained. Methods for reducing short and long lines measured by EDM are given in Sections 4.41 and 4.43 through 4.47.

4.31
ELECTRO-OPTICAL EDM WITHOUT REFLECTORS

Several manufacturers market EDM systems that transmit an infrared-timed pulse or a visible laser beam that reflects from a variety of solid surfaces such as wood, brick, stone, or metal. These EDM instruments usually operate with or without a reflector. Without the reflector, ranges vary from 450 to 1500 m with accuracies of from $\pm(90 \text{ mm} + 50 \text{ ppm})$

to ±100 mm. With a single reflector, the range increases to 8,000–10,000 m and accuracies remain the same. The projected red laser beam or, in the case of those systems using an infrared-timed pulse, an attached laser beam projects a red spot on the object sighted to avoid misidentification of the point to which the measurement is made.

These systems usually are modular, that is, they are top mounted on a conventional theodolite, but some can be used in a stand-alone mode on a tripod, on a handheld staff, or simply be held by hand. Such instruments are especially useful where quick acquisition of lower accuracy distances is adequate (e.g., GPS controlled surveys, see Foster, 1997). Applications include hydrographic, topographic, and route surveys as well as in the timber industry and forestry for timber cruising and determination of tree heights and diameters.

4.32
ACCURACY POSSIBLE WITH EDM SYSTEMS

Short-range EDM systems (maximum range of 600–7000 m) have constant uncertainties that vary from 1 to 5 mm +1 to 5 ppm (Crawford, 1997). Long-range systems (maximum range of 15–50 km) have constant uncertainties of from 0.5 to 1.5 cm + 1 to 3 ppm. It is reasonable to assume that these figures represent legitimate estimates of the standard deviation of a measurement when corrections for meteorological conditions and instrumental systematic errors have been corrected as completely as possible. Then, the worst of the short-range instruments would provide a relative precision of 1 part in 40,000 or better for lines equal to or in excess of 250 m. The worst of the long-range systems would provide similar relative precisions for lines equal to or in excess of 680 m. Achievement of relative precisions of 1 part in 40,000 over comparable distances by taping would require the use of taping tripods in precise taping procedures. Hence, EDM equipment has a clear superiority over traditional taping for surveys with distances of these magnitudes. For longer distances, relative precisions of 1 part in 100,000 to 1 part in 1 million are attainable. It is important to recognize that relative precisions of these magnitudes are possible only when corrections for meteorological conditions are evaluated very carefully and the EDM system is in good adjustment and calibrated against a legal standard.

Another point to consider is the use of EDM for measuring distances of very short lines, equal to or less than 50 m. Again, using the worst case of 5 mm + 5 ppm as the uncertainty for most short-range instruments leads to relative precisions for distances of 50, 35, and 15 m (about 160, 115, and 50 ft) of approximately 1 part in 9500, 1 part in 6800, and 1 part in 3000. Clearly, if one desires high accuracy over very short distances, precise taping using a steel tape is advisable.

PART B
ADVANCED TOPICS

4.33
SLOPE CORRECTION, PRECISION REQUIRED
FOR θ OR z AND h

The relative precision of a measured line usually is expressed as the ratio of the allowable discrepancy to the distance measured. Therefore, a relative precision of 1 in 10,000 implies a discrepancy of 1 unit in 10,000 or 0.01 unit in 100 units. Required relative precisions in measured distances for various classes or work are given in Chapter 8, Section 8.21 and Chapter 9, Section 9.23. Ordinary taping generally is said to have a relative precision of 1 part in 3000 to 1 part in 5000.

The final precision in H will depend on the precision with which s and z, θ, or h are measured. In other words, the standard deviation σ_H may be evaluated in terms of σ_s and σ_z, or σ_θ, or σ_h, applying the principle of error propagation as developed in Chapter 2. The contribution of σ_θ alone is (see Equation (2.16))

$$\sigma_H^2 = \left(\frac{dH}{d\theta}\right)^2 \sigma_\theta^2 \tag{4.12}$$

in which σ_H and σ_θ are estimated standard deviations in H and θ, respectively, and $dH/d\theta$ is found by differentiating Eq. (4.1) with respect to θ to yield

$$\frac{dH}{d\theta} = -s \sin \theta \tag{4.13}$$

so that

$$\sigma_H = s \sin \theta \sigma_\theta \tag{4.14}$$

The relative precision in H due to σ_θ thus is

$$\frac{\sigma H}{H} = \frac{s \sin \theta}{s \cos \theta} \sigma_\theta = \sigma_\theta \tan \theta \tag{4.15}$$

Similarly, differentiating Equation (4.2) with respect to H, and using error propagation from Equation (2.16) gives the standard deviation in H due to z, which is

$$\sigma_H = s \cos z \, \sigma_z \tag{4.16}$$

and the relative precision of H due to z is

$$\frac{\sigma_H}{H} = \left(\frac{s \cos z}{s \sin z}\right) \sigma_z = \sigma_z \cot z \tag{4.17}$$

Therefore, if the standard deviation or relative precision in the distance measurment is given, the precision necessary in obtaining θ or z can be calculated.

EXAMPLE 4.14. In Example 4.1, assume that the contribution to σ_H due to θ is to be within ± 0.005 m. Determine the allowable tolerance in θ to meet this specification.

Solution. Using Equation (4.14), we obtain

$$\sigma_\theta = \frac{\sigma_H}{s \sin \theta} = \frac{(0.005)(206,265)}{(29.954)(0.07846)} = 439'' = 7'19''$$

Note that σ_θ in radians was converted to seconds of arc through multiplication by 206,265. Consequently, the vertical angle needs to be read to the nearest $\pm 5'$ to satisfy the stated specification.

EXAMPLE 4.15. Assume that a relative precision of 1 part in 25,000 is due to the measurement of θ in Example 4.2. Compute the tolerance required in the measurement of θ.

Solution. Using Equation (4.15), we obtain

$$\sigma_\theta \tan 5°30' = \frac{1}{25,000}$$

$$\sigma_\theta = \frac{206,265}{(25,000)(0.09629)} = 85.7'' = 1'26''$$

and θ should be read to the nearest minute to maintain the stated relative precision.

EXAMPLE 4.16. In Example 4.3, the distance is to be determined with a standard deviation of 0.005 m. Compute the tolerance in the measured zenith angle to satisfy this specification.

Solution. Using Equation (4.16), we obtain

$$\sigma_z = \frac{\sigma_H}{s \cos z} = \frac{(0.005)(206,265)}{(250.542)(0.0739634)} = 55.6''$$

and z should be measured to the nearest $30''$ to satisfy the given specification for H.

A similar analysis, utilizing Equation (4.3), yields the estimated precision required in h to satisfy its contribution to horizontal distance measurement specifications. In this case,

$$\sigma_H^2 = \left(\frac{dH}{dh}\right)^2 \sigma_h^2 \tag{4.18}$$

in which σ_h is the estimated standard deviation in h, the difference in elevation, and dH/dh is found by differentiating Equation (4.4) with respect to h. Thus,

$$\frac{dH}{dh} = (1/2)(s^2 - h^2)^{-1/2}(-2h) = -\frac{h}{(s^2 - h^2)^{1/2}}$$

or

$$\frac{dH}{dh} = -\frac{h}{H} \approx -\frac{h}{s}$$

Thus, from Equation (4.17),

$$\sigma_H = \frac{h}{s}\sigma_h \tag{4.19}$$

and the relative precision relationship will be

$$\frac{\sigma_H}{s} = \frac{h}{s^2}\sigma_h \tag{4.20}$$

EXAMPLE 4.17. The tolerance in the distance in Example 4.5 due to error in h is ±0.005 m. Compute the allowable tolerance in the determination of the elevation difference.

Solution. Using Equation (4.19) yields

$$\sigma_h = \frac{(130.508)(0.005)}{5.56} = 0.12 \text{ m}$$

so that in this case elevations to the nearest 10 cm would be adequate.

Where the total measurement consists of a series of full or partial tape lengths having different elevation differences, σ_h may be evaluated with either Equation (4.19) or (4.20), using the line having the maximum change in elevation.

The final precision σ_H in the horizontal distance H must take into account the precision σ_s of the actual field taping to be theoretically correct. This point will be discussed in detail and illustrated with examples in Section 4.34.

4.34
ERROR PROPAGATION IN TAPED DISTANCES

Even though all systematic errors are eliminated from the distance measurements, random errors due to various causes remain. General comments about the causes of random errors in taping can be found in Section 4.25. The effects of these random errors on the final reduced measurements can be evaluated by *error propagation*. In this section, we examine error propagation in distances determined by taping.

Random errors in taping occur due to human limitations in observing measurements and manipulating the equipment. Specific causes of random errors in taping are listed in Table 4.5, together with corresponding estimates of the errors per tape length, assuming experienced personnel, conditions as given in the table, and medium-weight (1.5 lb, 0.7 kg) 100-ft and 30-m tapes having cross-sectional areas of 0.006 in.^{2} and 0.040 cm^2, respectively.

Note that random errors can occur as a result of (1) direct observations, such as marking tape ends, reading graduations, and so on; or (2) indirect observation of the difference in elevation or slope between the tape ends, applying tension, and so on. In these latter cases, the observational error must be propagated through the appropriate equations before arriving at an estimate for the error in taping. For example, the error when observing elevation differences is propagated through Equation (4.3) (Section 4.13) by using Equation (4.19) in Section 4.33.

A comparison of Tables 4.2 and 4.5 reveals that systematic errors are of much greater magnitude than random errors. When every device is employed to eliminate systematic errors using the methods described in Sections 4.13 through 4.24, random errors of observation become more significant; for this reason long tapes (200 ft, 50 m) frequently are employed if long lines are to be measured.

It would be valuable if a definite outline of procedure could be established to produce any desired degree of precision in taping. Unfortunately, the conditions are so varied and so much depends on the skill of the individual that the surveyor must be guided largely by experience and knowledge of the errors involved. However, by making certain assumptions concerning the errors involved and procedures employed, a reasonable methodical approach can be developed that is very useful in establishing procedures related to measurements required for specific tasks.

The usual practice in rough taping through broken country (maximum slopes 6 percent) is to take measurements with the tape horizontal, plumbing at the downhill end, breaking tape where necessary, applying tension by estimation, and making no corrections for sag, temperature, or tension. The tape usually is 100 ft or 30 m long and weighs about 2 lb or 1 kg. For this the most significant errors and assumed magnitudes are

| | Assigned error (σ) | | Effect on |
Cause of error	ft	m	measured line
σ_h, tape not level	0.05	0.015	Systematic
σ_s, sag in tape	0.03	0.009	Random
σ_v, plumbing	0.05	0.015	Random

TABLE 4.5
Random errors in taping

Designation	Source	Governing conditions and causes	Estimated value per tape length
σ_v	Plumbing to mark tape ends	Rugged terrain, breaking tape frequently	0.05–0.10 ft (15–30 mm)
σ_m	Marking tape ends with tape fully supported	Tape graduated to hundredths of ft or mm	0.01 ft(3 mm)
σ_p	Applying tension	Change in sag correction due to variations in tension of ±2 lb or 0.9 kg from standard tension	0.01 ft(3 mm)
σ_h	Determining elevation difference or slope angle (assume a maximum 6 percent slope)	$\ln h = \dfrac{\pm 0.8 \text{ ft}}{\pm 0.25 \text{ m}}$ $\ln \theta = \pm 0°28'$	0.50 ft(15 mm)
σ_d	Standardization	Field tapes compared to standardized tape kept in office	0.005 ft(1.5 mm)

For a distance of 1000 ft (10 tape lengths) the total error would be $10 \times 0.05 = 0.50$ ft due to the tape not being level plus $(0.03 + 0.05)\sqrt{10} = 0.25$ ft due to sag and plumbing, for a total of 0.75 ft. Therefore, under these assumptions the relative accuracy is only about 1 part in 1300.

To achieve a relative accuracy of 1 part in 5000, more control must be exercised over the taping procedure. The following magnitudes for the most significant errors are

Cause of error	Assigned errors		Effect on measured line
	ft	m	
Plumbing, σ_v	0.02	0.006	Random
Sag, σ_s	0.01	0.003	Random
Temperature, σ_t	0.004	0.001	Systematic
Tape not level, σ_h	0.003	0.001	Systematic
Standardization, σ_d	0.004	0.001	Systematic

Using the error propagation rules, the total error is 0.205 ft for a 1000-ft distance and 0.058 m for a 300-m distance, yielding the desired relative accuracy of 1 part in 5000.

To meet the stated conditions, the following specifications should govern the taping operation (assume a 100-ft medium-weight tape having cross-sectional area of 0.006 in^2 and maximum slopes of 6 percent):

1. The tape shall be standardized to within ±0.004 ft ($\sigma_d = 0.004$ ft) with the tape supported at both ends at 68°F and with 12 lb of tension.
2. A tension handle shall be used with tension applied to within ±2 lb of the standardized tension.

3. Endpoints of each tape length shall be marked to within ± 0.02 ft ($\sigma_v + \sigma_m = 0.02$ ft).
4. A hand level shall be used to ensure that both tape ends of a full tape length are within ± 0.5 ft of horizontal, assuming the maximum 6 percent slope.
5. Temperatures at the site of the measurement shall be observed to within $\pm 7°F$.

Note that some rather stringent specifications are necessary to ensure a relative accuracy of 1 part in 5000. Note also that more than half the total errors occur in plumbing or marking the endpoints of each tape length. Results can be improved by using a longer tape with fewer tape lengths to be marked.

4.35
SYSTEMATIC ERRORS IN EDM EQUIPMENT

General comments on the effects of atmospheric conditions on wave velocity, uncertainties in the position of the electrical center of the transmitter, uncertainties in the effective center(s) of the reflectors(s), and transmitter nonlinearity were made in Section 4.30. Details of these systematic errors and their compensation are given in the sections that follow.

4.36
EFFECTS OF ATMOSPHERIC CONDITIONS ON WAVE VELOCITY

The velocity V of electromagnetic waves in air is a function of the speed of light in a vacuum V_0 and the refractive index of air n, yielding

$$V = \frac{V_0}{n} \tag{4.21}$$

The constant V_0 is 299,792.5 km sec. The refractive indexes of light waves and microwaves in air are functions of air temperature, atmospheric pressure, and the partial pressure of water vapor, which in turn depend on temperature and relative humidity. A knowledge of these atmospheric conditions is required to determine the refractive index and the consequent effects on the velocity of the propagated waves in air. Because light waves and microwaves react somewhat differently to varying atmospheric conditions, each is treated separately.

For light waves, it is first necessary to calculate the index of refraction n_g of standard air given by the Barrell and Sears equation for an atmosphere at 0°C, 760 mm Hg pressure, and 0.03 percent carbon dioxide.

$$n_g = 1 + \left(287.604 + \frac{4.8864}{\lambda_1^2} + \frac{0.068}{\lambda_1^4} \right) 10^{-6} \tag{4.22}$$

in which λ_1 is the wavelength of the carrier beam of light in micrometers. Values of λ_1 for various sources of light are

Carrier	$\lambda_1, \mu m$
Mercury vapor	0.5500
Standard lamp	0.5650
Red laser	0.6328
Infrared	0.900–0.9300

Owing to changes in temperature, pressure, and humidity, the refractive index of air becomes n_a, given by

$$n_a = 1 + \frac{0.359474(n_g - 1)p}{273.2 + t} - \frac{1.5026e(10^{-5})}{273.2 + t} \quad (4.23)$$

where p = atmospheric pressure, mmHg
     t = air temperature, °C
     e = vapor pressure, mmHg

The humidity represented by water vapor pressure in the second term of Equation (4.23), has little effect on light waves and usually can be ignored. For measurements of high precision an average value of 0.5 ppm can be assumed as a humidity correction and substituted for the second term in Equation (4.23).

EXAMPLE 4.18. An electro-optical instrument utilizing infrared light with a wavelength of 0.9100 μm has a modulation frequency at 24.5 MHz. At the time of measurement, the temperature was 27.0°C and atmospheric pressure was 755.1 mmHg. Calculate the modulated wavelength of the light under the given atmospheric conditions.

Solution. Using Equation (4.22), the refractive index of air under standard conditions is

$$n_g = 1 + \left[287.604 + \frac{4.8864}{(0.91)^2} + \frac{0.068}{(0.91)^4} \right] 10^{-6} = 1.0002936$$

The refractive index of air under the given conditions, by Equation (4.23), neglecting the second term, is

$$n_a = 1 + \frac{(0.359474)(1.0002936 - 1)(755.1)}{273.2 + 27.0} = 1.00026547$$

The velocity of the infrared light through the atmosphere by Equation (4.21) where $n = n_a$, is

$$V_a = \frac{299,792.5}{1.00026547} = 299,712.9 \text{ km/sec}$$

and by Equation (4.10), the modulated wavelength in the given atmosphere is

$$\lambda = \frac{299,712.9}{(24.5)(10^6)} = 0.01223318 \text{ km} = 12.23318 \text{ m}$$

Microwaves are more sensitive than light waves of humidity. The refractive index of the atmosphere for microwaves is

$$(n_r - 1)10^6 = \frac{103.49}{273.2 + t}(p - e) + \frac{86.26}{273.2 + t}\left(1 + \frac{5748}{273.2 + t}\right)e \quad (4.24)$$

where n_r = refractive index
     p = atmospheric pressure, mmHg
     e = vapor pressure, mmHg
     t = temperature, °C

The velocity of microwaves in air is given by Equation (4.21) where $n = n_r$.

EXAMPLE 4.19. The modulation frequency for microwave transmission is exactly 10 MHz at a temperature of 15.4°C, atmospheric pressure of 645 mmHg, and vapor pressure of 3.8 mmHg. Calculate the modulated wavelength under these conditions.

Solution. Using Equation (4.24), we obtain

$$(n_r - 1)(10^6) = \frac{103.49}{288.6}(645 - 3.8) + \frac{86.26}{288.6}\left(1 + \frac{5748}{288.6}\right)(3.8)$$

$$(n_r - 1)(10^6) = 253.69$$

$$n_r = 1.00025369$$

By Equation (4.21),

$$V = \frac{299,792.5}{1.00025369} = 299,716.5 \text{ km/sec}$$

and by Equation (4.10), the modulated wavelength in the atmosphere is

$$\lambda = \frac{299,716.5}{(10)(10^6)} = 0.02997165 \text{ km} = 29.97165 \text{ m}$$

The examples indicate that, as atmospheric conditions vary, the velocity of the modulated wave is altered, resulting in a corresponding change in the modulated wavelength and hence in the basic measuring unit of the EDM instrument.

4.37
CORRECTION OF DISTANCES FOR ATMOSPHERIC CONDITIONS

To correct the observed slope distances for varying atmospheric conditions for light-wave instruments, the temperature, t, and atmospheric pressure, p, are recorded at each end of the line. Some models of EDM systems allow entry of the mean values of t and p using the keyboard. The microcomputer in the EDM then automatically computes and applies a correction factor to the measured slope distances. In other EDM systems, the correction term, in parts per million, is determined using the mean observed values of t and p from a chart or nomogram furnished by the manufacturer of the EDM. This correction term then is entered into the EDM via the control panel or a special dial. Some practitioners using these latter models, prefer to calculate a correction factor (see the next paragraph), which then is applied to the measured slope distance. In this case, 0 ppm would be dialed into the EDM prior to measuring the slope distance.

To determine these correction factors, a reference value for the index of refraction for the instrument, n_0, using a chosen standard atmosphere must be given or calculated for each EDM device. From Equations (4.10) and (4.21), we can form the following equation:

$$n_0 = \frac{V_0}{\lambda_0 f_0} \tag{4.25}$$

in which $V_0 = 299,792.458$ km/sec, the velocity of light in a vacuum
λ_0 = the modulation wave length of the EDM in m
f_0 = the modulation frequency for the EDM in kHz

The manufacturers of EDM instruments use a *reference index of refraction, N_0,* expressed in ppm where

$$N_0 = (n_0 - 1)10^6 \tag{4.26}$$

Also, a refractive index number can be computed for conditions of observations, using

$$N_a = (n_a - 1)10^6 \tag{4.27}$$

where n_a is calculated by Equation (4.23). Thus, a reference index of refraction and a refractive index number, based on the values for λ_0 and f_0 given in the manufacturer's specifications for a specified EDM and conditions of observation, can be calculated with Equations (4.25) and (4.26) and (4.23) and (4.27), respectively. The correction to the slope distance in ppm is given by

$$\text{Correction (ppm)} = (N_0 - N_a) \qquad (4.28)$$

This correction term, in ppm, corresponds to the value obtained using the nomogram or chart provided with the EDM and can be dialed into the EDM to provide for automatic compensation for atmospheric conditions.

Obviously, the simplest approach is to observe temperature and atmospheric pressure and either enter these values into the EDM by the keyboard or control panel, if possible, or use the nomogram or chart furnished by the manufacturer for the EDM and dial in the correction.

For some high-precision surveys, the ppm correction dial is set to 0 and the slope distance is corrected using the correction (ppm) found in Equation (4.28). An alternate approach is to use the ratio of n_0 to n_a to correct the slope distance directly, using

$$s_{\text{corr}} = \frac{n_0}{n_a} s \qquad (4.29)$$

in which s_{corr} = slope distance corrected for atmospheric conditions
$\qquad n_0$ and n_a are calculated using Equations (4.25) and (4.23), respectively
$\qquad s$ = the observed slope distance

Some examples illustrate the application of these equations.

EXAMPLE 4.20. An EDM has the following specifications: $f_0 = 14.9854$ MHz, $\lambda_1 = 0.820$ μm, and $\lambda_0 = 20$ m. This EDM was used to measure a slope distance at 25°C and atmospheric pressure of 750 mm Hg. Determine the ppm correction to be used either to enter into the EDM or for correcting the slope distance.

Solution. Using Equation (4.22), the refractive index of air under standard conditions is

$$n_g = 1 + \left[287.604 + \frac{4.8864}{(0.820)^2} + \frac{0.068}{(0.820)^4} \right] 10^{-6} = 1.000295022$$

The refractive index of air under the given conditions by Equation (4.23), neglecting the second term, is

$$n_a = 1 + \frac{(0.359474)(1.000295022 - 1)(750)}{(273.2 + 25)} = 1.000266732$$

The reference index of refraction for the instrument, by Equation (4.25), is

$$n_0 = \frac{299{,}792.5}{(20)(14{,}985.4)} = 1.000281941$$

Using these values, the refractive index numbers, N_0, for the EDM and N_a for operating conditions, are calculated by Equations (4.26) and (4.27), respectively, to give

$$N_0 = (1.000281941 - 1)(10^6) = 281.9 \text{ ppm}$$

$$N_a = (1.000266732 - 1)(10^6) = 266.7 \text{ ppm}$$

which allow calculation of the correction by Equation (4.28) to yield

$$\text{Correction (ppm)} = (281.9 - 266.7) = 15.2 \text{ ppm or 15 ppm}$$

This correction is the value automatically calculated and applied to the measured slope distance by the microcomputer of the EDM system using temperature and atmospheric pressure input via the keyboard or control panel. For EDM systems not possessing this feature, the correction represents the value obtained using temperature and atmospheric pressure as arguments in a nomogram or chart furnished with the EDM. The example serves to show the relationship between the equations developed and the number obtained either by the microcomputer or by the analog method.

Next consider an example to illustrate numerical correction of an uncorrected slope measurement.

EXAMPLE 4.21. The EDM in Example 4.20 was used under the given atmospheric conditions to measure slope distances recorded as $s_1 = 2341.454$ m and $s_2 = 345.123$ m. Each distance was measured with the ppm correction dial set to 0.

Solution. From Example 4.21, the correction (ppm) = 15.2 ppm, so the respective corrections are

$$\text{Correction } S_1 = (2{,}341.454)(10^{-6})(15.2) = 0.036 \text{ m}$$

$$S_{1,\text{corr}} = 2{,}341.490 \text{ m}$$

$$\text{Correction } s_2 = (345.123)(10^{-6})(15.2) = 0.005 \text{ m}$$

$$s_{2,\text{corr}} = 345.128 \text{ m}$$

In an alternate solution, n_0 and n_a are used in Equation (4.29) to compute the corrected slope distance

$$s_{1,\text{corr}} = \left(\frac{1.0002819}{1.0002667}\right)(2{,}341.454) = 2{,}341.490$$

$$s_{2,\text{corr}} = \left(\frac{1.0002819}{1.0002667}\right)(345.123) = 345.128 \text{ m}$$

In recording meteorological data, care should be exercised in obtaining air temperatures. Estimating temperature to $\pm 10°F(\approx \pm 6°C)$ and atmospheric pressure to $\approx \pm 1$ in or 25 mmHg will introduce a relative error of 10 ppm in light-wave equipment. Thermometers should be placed above the ground and in the shade. Atmospheric pressure can be determined from readings on an aneroid barometer (Section 5.6) converted to the appropriate pressure reading in mmHg or inHg using a chart provided by the instrument manufacturer.

The partial pressure of water vapor as determined by wet- and dry-bulb thermometer readings must be determined for microwave instruments. An error of 1°C between wet- and dry-bulb temperatures at normal conditions (temperature of 20°C, atmospheric pressure of 760 mmHg) produces a relative error of 7 ppm in the distance. This relative error at 45°C is 17 ppm. Consequently, wet- and dry-bulb temperatures must be carefully observed using calibrated equipment. Once the meteorological data are recorded, corrections to observed slope distances are made using charts and nomographs provided with the instrument or may be calculated using methods similar to those outlined for electro-optical measuring systems.

4.38
SYSTEMATIC INSTRUMENTAL ERRORS IN EDM SYSTEMS

Systematic instrumental errors occurring in electro-optical systems include uncertainties in the position of the electrical center of the transmitter, uncertainties in the effective center

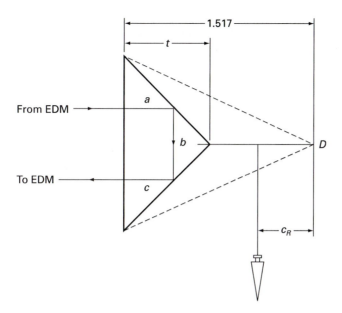

FIGURE 4.15
Reflector offset.

of the reflectors, frequency drift, and instrument nonlinearity. The first two sources of error must be taken into account in all survey measurements, the third requires constant monitoring, and the fourth is critical only for measurements of high precision.

Microwave systems are affected by uncertainties in the electrical centers of the master and remote units and by a phenomenon called *ground swing* or *reflection.*

In EDM systems properly adjusted at the factory, these errors will be very small and, in a practical sense, may be insignificant. However, it is important that users of these systems realize that periodic calibration of the instruments against a known distance is absolutely necessary to assure consistent results. First, consider calibration of an electro-optical system.

Uncertainties in the effective center of the reflector are illustrated by reference to Figure 4.15, which shows a cross section through a corner-cube retroreflector. The distance from the face of the cube to the back corner is t. The ray path within the cube is $a + b + c = 2t$. Owing to the refractive properties of the glass in the cube, the equivalent travel in air of the ray path within the cube is $1.517 \times 2t$. Point D represents the effective center of the corner cube and actually is the end of the line being measured. If it were possible to mount the reflector so that D coincided with the plumb line, the reflector offset would be 0. However, D is so far behind the face of the prism that such an arrangement would be unbalanced and difficult to manage in the field. Prisms usually are mounted so that c_R is from 28 to 40 mm. Some manufacturers eliminate C_R by making an adjustment in the transmitter to absorb the offset.

When slope distances are measured, light rays striking the reflector are not perpendicular to the front face of the reflector, as indicated in Figure 4.15, altering the path of the rays within the reflector and changing the position of the effective center (Kivioja, 1978). The amount of this change is a function of the degree of slope and varies from a few tenths of a millimeter at slopes of 4° to 6° to 7 to 14 mm at slopes from 30° to 40°. Given the angle of slope, corrections can be calculated to compensate for this error. Another option is to design the reflector with an adjustment allowing the front fact of the reflector to be placed perpendicular to the incoming rays of light, compensating for the error instrumentally.

Details concerning the formulas for making corrections can be found in Kivioja (1978). Corrections of this type would be necessary on surveys of high precision where slope angles are consistently large.

At the other end of the line, if the plumb line does not coincide with the exact electrical center of the electro-optical transmitter, an instrument offset exists.

Similar errors exist in microwave systems, where the instrument offsets in the master and remote transmitters are analogous to the instrument and reflector offsets of the light-wave system.

A base line measured with very high precision is required to calibrate an EDM system. To establish such a base line, the procedures and specifications as set forth by the National Geodetic Survey Division (NGSD) of the National Oceanic and Atmospheric Administration (NOAA) *must be followed* (see Fronzcek, 1980). Following these procedures, the NGSD has established calibration base lines in each state in the United States. These base lines contain from 4 to 10 or more monumented stations and vary in length from 1000 to 2000 m with distances being given to all stations set on the line. Unfortunately, these base lines have not always been adequately maintained. Another option is for local surveyors' professional organizations and societies to establish their own base lines for a given region. Since special procedures and equipment are required, such efforts are usually undertaken in cooperation with the NGSD (see Prescott, 1996).

The calibration procedure is the same for both the electro-optical and microwave systems. A series of EDM distances should be recorded (10, for example). Meteorological data should be gathered with extreme care. The average of the EDM distances corrected for meteorological conditions and for the slope of the line should agree with the taped distance for the base line. The difference between the two measurements represents the system constant, Δ_s. In electro-optical systems, a system constant should be determined for each reflector or group of reflectors to be used. Reflectors should be numbered and constants recorded for use in subsequent surveys.

Once a base line has been established for calibration, the instrument should be recalibrated by remeasuring the base line at periodic intervals to guard against frequency drift. The time between check calibrations is a function of the age of the equipment and the user's confidence in it. With new equipment, a check every week would not be unreasonable until confidence in the system stability is established. Thereafter, one per month would be sufficient. The same equipment should be used for each check calibration. The mean of the observations should be plotted versus the date of measurement and a record of meteorological conditions should be maintained. Ideally, the graph should be a horizontal straight line within the specification range for the system.

If a known base line is not available, calibration on a line of unknown length is possible. This type of calibration is accomplished by measuring a line a of unknown length in several sections. Let the distance D in Figure 4.16 be divided into arbitrary increments d_1, d_2, ..., d_n. Measure the total length of line using the EDM system and then measure the n sections separately, where $n = 2$ represents the minimum number of sections. For electro-optical systems, the same reflector should be used for the entire set of measurements. All distances should be corrected for meteorological conditions and slope. Let the system

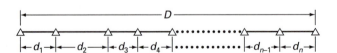

FIGURE 4.16
Base-line calibration using a line of unknown length.

constant be Δ_s so that the total distance is $D + \Delta_s$, each of the increments is $d_1 + \Delta_s$, $d_2 + \Delta_s, \ldots, d_n + \Delta_s$ and the following equation can be formed:

$$D + \Delta_s = (d_1 + \Delta_s) + (d_2 + \Delta_s) + \cdots + (d_n + \Delta_s)$$

from which

$$\Delta_s = \frac{-(d_1 + d_2 + \cdots + d_n) + D}{n - 1} \tag{4.30}$$

If the distance is divided into sections such that each section contains a distance with a different unit digit (e.g., for five sections $d_1 = 152$ m, $d_2 = 153$ m, $d_3 = 241$ m, $d_4 = 305$ m, $d_5 = 206$ m, and $D = 1057$ m), then errors due to nonlinearity (discussed in the next section) will be averaged and the net effect of this error on Δ_s will be small.

EXAMPLE 4.22. EDM No. T-61034 and corner-cube reflector No. 2 were calibrated on a base line of unknown length composed of three collinear points, A, B, and C. Ten slope distances and zenith angles were measured and temperatures and atmospheric pressures observed at both ends of each line for three line segments, AC, AB, and BC. Reduced horizontal distances for these segments, corrected for atmospheric conditions by dialing the ppm correction into the EDM for each measurement are

$$AC = 148.987 \text{ m} = D$$

$$AB = 52.149 \text{ m} = d_1$$

$$BC = 96.149 \text{ m} = d_2$$

Solution. This situation represents the simplest case for which Equation (4.30) is applicable, as follows:

$$\Delta_s = [148.987 - (52.149 + 96.832)]/(1) = +0.006 \text{ m}$$

A slope distance of 200.546 m was measured with EDM No. T-61034 and reflector No. 2. The corrected slope distance is

$$s_{\text{corr}} = 200.546 + 0.006 = 200.552 \text{ m}$$

EXAMPLE 4.23. A base line was established over uniformly, moderately sloping terrain with a total distance of about 1400 m between the two endpoints, A and G, with distances also being measured between four intermediate stations. Each distance was measured 10 times and corrected for atmospheric conditions and slope. The mean corrected distances of EDM No. 06M8208, reflector No. 5, are

From	To	Distance, m	From	To	Distance, m
A	G	1,406.311	D	E	184.558
A	B	156.064	E	F	503.079
B	C	229.999	F	G	97.369
C	D	185.562			

Solution. The system correction, Δ_s, using Equation (4.30), is

$$\Delta_s = \frac{1403.311 - (156.064 + 279.999 + 185.562 + 184.558 + 503.079 + 97.36)}{(6 - 1)}$$

$$\Delta_s = -0.064 \text{ m}$$

A slope distance measured with EDM No. 06M8208 and reflector No. 5 was recorded as 2,003.216 m. The corrected slope distance is

$$s_{corr} = 2,003.216 - 0.064 = 2,003.152 \text{ m}$$

4.39
NONLINEARITY IN ELECTRO-OPTICAL INSTRUMENTS

A nonlinear periodic bounded error occurs in all electro-optical instruments. As the distance between the transmitter and reflector is changed, the error will rise to a maximum, fall to a minimum, and so forth, repeating over and over again within the same boundary. To achieve maximum accuracy from a given EDM system, this cyclic error must be evaluated and the error corrected.

The nonlinearity of the instrument can be detected by taking a series of readings from a single instrument set up to the same reflector set, successively, on a series of points located at intervals, the sum of which should equal one full period of the nonlinearity. It is not necessary to know the total distance from the transmitter to this span of intervals with high precision, but the distances between the points must be measured with a standardized steel tape under standard conditions of support and tension and making corrections for temperature. Preferably, the calibration range should be inside and on a firm, level, uncluttered surface. At least 10 measurements should be made to each of 10 points with an allowable standard deviation of 0.0003 to 0.004 m.

For example, assume that the instrument to be calibrated has a basic measuring unit ($\lambda/2$) of 10 m. Figure 4.17 illustrates the configuration for a calibration base line, where the points are set at 1-m intervals over a span of 10 m, resulting in 11 stations. Note that these intervals may be somewhat more or less than 1.000 m (e.g., 0.999, 0.998, 1.001) but they do need to be known, precisely, as noted previously. Let $d_i(i = 1, 2, \ldots, 10)$ be these intervals, D_i be the distances measured by the EDM to points 0, 1, 2, ..., 10. Then, e_i, the errors in each reading due to nonlinearity, will be given by

$$e_i = D_i - (D_0 + \Sigma d_i) \tag{4.31}$$

Corrections for nonlinear errors will be

$$c_i = -e_i + (\Sigma e_i)/10 \tag{4.32}$$

in which $i = 1, 2, \ldots, 10$, the number of calibration test intervals.

EXAMPLE 4.24. A test range was established for calibrating the short-period nonlinear errors in an EDM with $\lambda/2 = 10$ m. To do this, points 0, 1, ..., 10 (Figure 4.17)

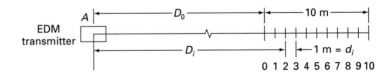

FIGURE 4.17
Setup for calculating nonlinear errors in EDM.

were set on a line at 1 m intervals and then measured very carefully with a standardized steel tape held horizontally, using standard support and tension and making corrections for temperature. The mean of 10 measurements to these stations had a standard deviation of 0.003 m and yielded the intervals listed in the second column of Table 4.6.

To complete the calibration, the EDM was set on a point on the line with the calibration test line and about 100 m from station 0. Distances were then measured by EDM (corrections for atmospheric conditions were dialed into the instrument) to the same reflector set successively at stations 0, 1, . . . , 10. These distances (reduced to a horizontal), which represent the first term of Equation (4.31), are listed in the third column of the table. The first distance, $D_0 = 100.912$ m, successively added to the cumulated, taped intervals for each station, is given in the fourth column and represents the second term in Equation (4.31). The errors calculated by Equation (4.31) and the average error are shown in the fifth column. Finally, the negative values of these errors or the *corrections,* translated to an origin at the average of the errors and calculated using Equation (4.32), are given in the last column. As an example, consider computations for station 4: $D_i = D_4 = 104.907$ m, $D_0 = 100.912$ m, $\Sigma d_i = (1.000 + 1.000 + 1.000 + 0.999) = 3.999$ m. Using Equation (4.31), $e_4 = 104.907 - (100.912 + 3.999) = -0.004$ m. Because the average error was -0.004 m (column 5), by Equation (4.32), the correction, $c_4 = -(-0.004) + (-0.004) = 0.000$ m. Similarly, $c_1, c_2, c_3, c_5, . . . , c_{10}$ can be calculated and are listed in the sixth column of the table. These corrections, plotted versus distance on a graph, produce a calibration curve for nonlinearity, which is illustrated in Figure 4.18.

TABLE 4.6

Station	Taped interval, d_i, m	EDM measured, D_i, m	$D_0 + \Sigma d_i$, m	$e_i = D_i - (D_0 + \Sigma d_i)$, m	$c_i = -e_i + (\Sigma e_i)/10$, m
0		100.912			
	1.000				
1		101.913	101.912	0.001	−0.005
	1.000				
2		102.913	102.912	0.001	−0.005
	1.000				
3		103.910	103.912	−0.002	−0.002
	0.999				
4		104.907	104.911	−0.004	0.000
	1.002				
5		105.906	105.913	−0.007	0.003
	1.000				
6		106.903	106.913	−0.010	0.006
	0.999				
7		107.901	106.912	−0.011	0.007
	1.002				
8		108.909	108.914	−0.005	0.001
	1.002				
9		109.914	109.916	−0.002	−0.002
	0.997				
10		110.911	110.913	−0.002	−0.002
			Average error =	−0.004	

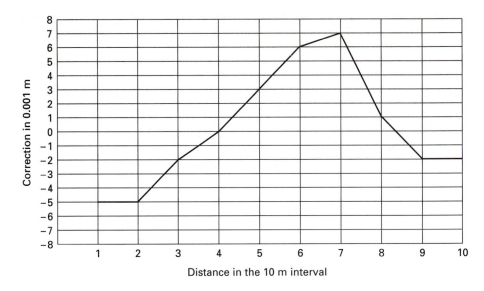

FIGURE 4.18
Typical calibration curve for nonlinearity. (*After Moffitt, 1975.*)

If the system constant, as determined by one of the methods outlined in Section 4.38 for the EDM used in Example 4.24, is −0.005 m and the observed slope distance is 342.541 m, the corrected slope distance = 342.541 − 0.005 − 0.003 = 342.533 m. The correction, −0.003, is taken from the intersection of the ordinate for a distance of 2.5 m with the calibration curve from the graph in Figure 4.18. This value is determined for 2.5 m because it represents the units and decimal parts of a meter that were measured.

The nonlinear errors that have been discussed up to now are referred to as *short-period errors. Long-period* nonlinear *errors* also may exist in EDM measurements. Long-period errors have been shown to exist in certain short-range EDM with periods of 40 and 50 m (see Santaala, 1983, 1984). Calibration of this type of error would require a base line several hundred meters in length with stations at 40- or 50-m intervals, where the distances are measured with extremely high precision.

Calibration for most electro-optical instruments is performed in the factory, so that the average nonlinearity and estimated true distance coincide. In other words, errors due to nonlinearity are uniformly distributed about the calibration point so the total peak-to-peak error is within specified limits. This instrumental adjustment is analogous to the distribution of corrections in the calibration chart in Figure 4.18. From a practical standpoint, calibration for nonlinearity rarely is needed unless the instrument is to be used for projects in which very high accuracy is required, such as earth crustal movement and dam or structural deformation studies.

4.40
GROUND REFLECTIONS OF MICROWAVE EDM INSTRUMENTS

The beam transmitted by microwave instruments is relatively wide. Therefore, when transmitting over smooth, level terrain or water, reflections can occur that yield erroneous distances. Occurrence of errors due to reflections, or ground swing, as it is called, can be reduced by elevating the master and remote units above the surface over which the line is to be measured. Microwave instruments of recent design operate on higher frequencies with a narrower beamwidth ($1\frac{1}{2}°$), so that errors due to ground swing have been reduced. It is recommended that a series of fine readings (the A pattern for the Tellurometer) be taken, each at a different frequency, from both ends of the line at the beginning and end of each measurement. Strong reflections will cause these readings to vary in a cyclic manner, and if the readings are plotted versus frequency, the resulting curve ideally will take the shape of a sine wave. When swing is present, considerable experience and judgment are required to interpret the results properly. Usually a straight-line average of the fine readings will be adequate. The National Ocean Survey recommends that the two measurements from the two ends of the line agree to within 1 part in 100,000 and that the spread of the fine readings not exceed 4 cm.

4.41
REDUCTION OF SLOPE DISTANCES OBTAINED BY EDM

For short lines, less than 2 mi or 3.3 km in length or with zenith angles between 85° and 90° or vertical angles of less than 5°, EDM slope distances corrected for meteorological conditions and system constants can be reduced to horizontal with the usual slope correction equations as outlined in Section 4.13, and still maintain a relative accuracy of 1 part in 10,000. Thus, the zenith or vertical angle, or the difference in elevation between the two ends of every measured line must be obtained. When the zenith or vertical angle is observed, using a stand-alone EDM separate from the theodolite or top-mounted (on a theodolite) EDM, additional data required for reduction include the heights of the EDM transmitter and the reflector or remote unit above the ground at respective stations and the heights of the theodolite used to measure the angle and the target sighted in this measurement, above the ground at their respective stations.

Figure 4.19 illustrates a typical situation for distance measurement by separate EDM equipment, accompanied by observation of the zenith or vertical angle. The EDM transmitter is at E while the reflector or remote unit is at E', with the respective heights above the ground of EA and $E'B$. The theodolite and target, located at T and T', have heights above the ground (h.i.'s) of AT and BT', respectively. Angle z_T or α_T is the observed zenith or vertical angle, respectively. The correction Δz or $\Delta \alpha$ must be calculated to determine the

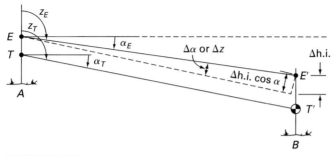

FIGURE 4.19
Difference between vertical angle and line along which EDM slope distance is measured.

corrected zenith angle, z_E, or vertical angle, α_E, of the measured line EE'. The difference in heights of instruments required for this calculation is

$$\Delta \text{h.i.} = (\text{h.i. reflector} - \text{h.i. target}) - (\text{h.i. EDM} - \text{h.i. theodolite}) \quad (4.33)$$

and

$$\Delta z(\text{sec. of arc}) = -\left[\frac{(\Delta \text{h.i. sin } z)}{EE'} \right] \rho \quad (4.34a)$$

$$\Delta \alpha(\text{sec. of arc}) = \left[\frac{(\Delta \text{h.i. cos } \alpha)}{EE'} \right] \rho \quad (4.34b)$$

in which $\rho = 1/(\text{arc } 01'') = 206{,}265$ (sec of arc)/rad. Equations (4.34) lead to

$$z_E = z_T + \Delta z \quad (4.35a)$$

$$\alpha_E = \alpha_T + \Delta \alpha \quad (4.35b)$$

The sign of Δz is a function of $\Delta \text{h.i.}$ Of course, the zenith angle is unambiguous with respect to sign. The sign of $\Delta \alpha$ depends on the signs of α and $\Delta \text{h.i.}$ Care should be exercised in calculating z_E and α_E to reflect the proper signs of $\Delta \text{h.i.}$, α, and Δz or $\Delta \alpha$.

EXAMPLE 4.25. The slope distance from A to B corrected for meteorological conditions and EDM system constants is 920.850 m. The EDM transmitter is 1.840 m above the ground and the reflector is 2.000 m above the ground. The observed zenith angle is $94°30'00''$ with theodolite and target 1.740 m and 1.800 m above the ground, respectively. Calculate the horizontal distance.

Solution. From Equation (4.33), calculate $\Delta \text{h.i.}$:

$$\Delta \text{h.i.} = (2.000 - 1.800) - (1.840 - 1.740) = 0.100$$

Using Equation (4.34a), determine Δz.

$$\Delta z = -\frac{(0.100)(\sin 94°30'00'')}{920.850} 206{,}265 = -22''$$

and the corrected vertical angle by Equation (4.35a) is

$$z_E = z_T + \Delta z = 94°30'00'' - 0°00'22'' = 94°29'38''$$

so that the horizontal distance by Equation (4.2) is

$$H = s \sin z_E = (920.850)(0.996926) = 918.019 \text{ m}$$

EXAMPLE 4.26. A slope distance from C to D corrected for EDM constants and meteorological conditions is 5005.45 ft with a vertical angle of $+3°30'00''$ observed from a theodolite having an h.i. of 4.40 ft to a target with h.i. of 4.60 ft. The h.i.'s of the EDM and reflector are 5.00 and 4.80 ft, respectively. Compute the horizontal distance.

Solution. By Equation (4.33),

$$\Delta \text{h.i.} = (4.80 - 4.60) - (5.00 - 4.40) = -0.40 \text{ ft}$$

Using Equation (4.34b) we obtain

$$\Delta \alpha = \frac{(-0.40)(\cos 3°30'00'')}{5005.45} 206{,}265 = -16''$$

so the corrected angle using Equation (4.35b) is

$$\alpha_E = \alpha_T + \Delta \alpha = 3°29'44''$$

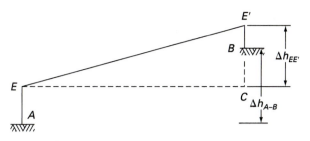

FIGURE 4.20
Reduction of short slope distances by EDM using
differences in elevation.

and the horizontal distance by Equation (4.1) is

$$H = s \cos \alpha_E = 4996.14 \text{ ft}$$

Vertical circle readings should be observed from both ends of the line using an instrument with a least count in the vertical circle of from 1 to 6″ (see Section 6.26). Note that, if first-order horizontal distances are desired or long lines are measured, then the method of reduction from slope to horizontal should follow methods detailed in Sections 4.43 to 4.47.

Most total station systems now are designed with *collinear optics* to determine distance and direction. Therefore, the second term in Equation (4.33) becomes 0. In addition, sighting targets that can be fitted symmetrically about the reflector mount are available. In this case, the h.i. of the reflector equals the h.i. of the target, eliminating the first term of Equation (4.33) and the need to compute a correction to the zenith or vertical angle.

When slope distances are to be reduced using differences in elevation between the two ends of the line, the heights of the EDM instrument and the reflector (or remote unit) above the ground at their respective stations must be observed and recorded. In Figure 4.20, the EDM instrument is at E with a height above the ground of AE and the reflector or remote unit is at E' at a height of BE' above the ground. Then the elevation difference between E and E' is

$$\Delta h_{EE'} = \Delta h_{A-B} - AE + BE'$$

or

$$\Delta h_{EE'} = \Delta h_{A-B} - \text{h.i. EDM} + \text{h.i. reflector or remote unit} \qquad (4.36)$$

4.42
ERROR PROPAGATION IN DISTANCES BY EDM

Systematic errors in EDM systems were covered in Sections 4.37 and 4.38. To recapitulate, systematic errors in distances measured by EDM are caused by (a) the effects of atmospheric conditions on propagated wave velocities of light waves and electromagnetic waves in air; (b) the differences between the effective centers of the transmitter and reflector and their respective plumb lines; and (c) transmitter nonlinearity. Atmospheric conditions are observed allowing corrections to be made for (a) and the system is calibrated to permit determination of constants to correct for (b) and (c).

In this section, comments will focus on short-range, light-wave instruments, where the errors that remain in distances measured by EDM are the small uncertainties in the (1) modulation frequency, f, that differs from its theoretical value; (2) index of refraction caused by variations in atmospheric conditions along the line measured; (3) determination of partial wavelengths by phase comparison; (4) calibration of the system constant to account for systematic errors from (b) and (c) above; and (5) centering of the EDM transmitter and reflector over their respective stations. Please note that the developments that follow are based primarily on work by Laurilla (1983).

Effects of the errors due to uncertainties (1) and (2) are obtained by propagating them through an equation for distance derived from Equations (4.10) and (4.11). First, modify Equation (4.11) as follows:

$$D = \frac{1}{2}\left(\frac{mV_0}{nf} + d\right) \tag{4.37}$$

Applying the law of error propagation to Equation (4.37), neglecting the second term, and taking $D \approx (mV_0)/(2nf)$ yields

$$\sigma_{D_1} = \sqrt{\left(\frac{(\sigma_r)^2}{(n)^2} + \frac{(\sigma_f)^2}{(f)^2}\right)D^2} \tag{4.38}$$

in which σ_{D_1} equals the standard deviation in the measured distance due to uncertainties in the refractive index, σ_r, and the modulation frequency, σ_f. Note that these two sources of error are proportional to the distance measured.

Independent errors in phase comparison, σ_ϕ, determination of the system constant, σ_{sys}, and centering of the EDM and reflector, σ_{cent} propagate to form

$$\sigma_{D_2} = \sqrt{\sigma_{C sys}^2 + \sigma_\phi^2 + \sigma_{cent}^2} \tag{4.39}$$

which, when combined with D_1, yields the total standard deviation for the distance measurement of

$$\sigma_{D, total} = \sqrt{\sigma_{D_1}^2 + \sigma_{D_2}^2} \tag{4.40}$$

Equation (4.40) is a rigorous expression for propagating the error in a distance measured by EDM. The manufacturers of EDM equipment use a similar form to describe the accuracy of their systems where their expression is 3–5 mm + 1–5 ppm (see Table 4.4, Section 4.27) that results from a simplification of Equation (4.40). The 3–5 mm is independent of distance and corresponds to D_2, whereas the 1–5 ppm depends on distance and is analogous to D_1, both as in Equation (4.40).

EXAMPLE 4.27. The standard deviation for distances measured by a short-range EDM is to be evaluated for the instrument and, specifically, the standard deviation for a slope distance of 5000.550 m is to be determined. The estimated standard deviations in modulation frequency, index of refraction, phase comparison, system constant, and centering of EDM instrument and reflector are

$$\sigma_f = 4 \times 10^{-6}$$

$$\sigma_r = 3 \times 10^{-6}$$

$$\sigma_\phi = 0.003 \text{ m}$$

$$\sigma_{c sys} = 0.004 \text{ m}$$

$$\sigma_{cent} = 0.003 \text{ m}$$

Assume $n = 1$ and σ_f determined using a frequency demodulator and frequency counter, is in units of $\Delta f/f$ (Santaala, 1984).

Solution. Using Equation (4.38), calculate the effects of uncertainties in refraction and modulation frequency that depend on distance

$$\sigma_{D_1} = \sqrt{[(3 \times 10^{-6})^2 + (4 \times 10^{-6})^2]D^2} = 5 \times 10^{-6}D = 5 \text{ ppm of } D$$

Next, compute the effects of errors in phase comparison that are independent of distance, using Equation (4.39):

$$\sigma_{D_2} = \sqrt{(0.003 \text{ m})^2 + (0.004 \text{ m})^2 + (0.003 \text{ m})^2} = 0.006 \text{ m}$$

Then, using Equation (4.40), the total standard deviation in a distance measurement is

$$\sigma_{D, \text{total}} = \sqrt{(0.006)^2 + (5 \times 10^{-6} \, D)^2}$$

Thus, the slope distance of 5000.550 m would have a propagated standard deviation of

$$\sigma_{D, \text{total}} = \sqrt{(0.006)^2 + (5 \times 10^{-6} \times 5000.550)^2} = 0.026 \text{ m}$$

Now, the manufacturer of the EDM would use the information obtained with Equation (4.40), in the preceding step, as the basis for describing the instrument as having an accuracy of 5 mm + 5 ppm. Applying this rule to the given distance yields a standard deviation of 0.030 m, so the simplification provides an estimate for the error that is somewhat on the safe side.

Propagation of errors in reduction of short slope distances (s < 2 mi or 3 km and/or $85° < z < 95°$ or $\theta < 5°$) by EDM to horizontal distances can be accomplished using the same procedures outlined in Section 4.33. For long slope distances, propagation through equations for reducing long lines in Sections 4.43 to 4.47 would be necessary.

4.43
SLOPE REDUCTIONS FOR LONG LINES

In the previous section, a short line was defined as less than 2 mi or 3 km in length or with a zenith angle between 85° and 95° or a vertical angle less than 5°. If a long line is composed of short sloping sections and a ratio of precision of 1 part in 10,000 is adequate, one of Equations (4.1), (4.2), or (4.4) can be used to reduce the sections individually to a horizontal so that the sum of the these distances follows the curvature, of the earth, as illustrated in Figure 4.21. When such a distance is reduced to the ellipsoid,* as described in Chapter 9,

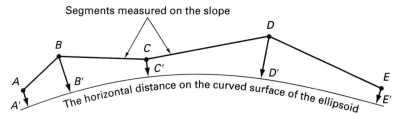

FIGURE 4.21
Sections (*ABCDE*) reduced to horizontal distance (*A'B'C'D'E'*) on the ellipsoid.

---

* The ellipsoid is the mathematical figure used to approximate the figure of the earth for geodetic computations (Section 1.3, Figure 1.1b) and is the reference surface for Global Positioning System (GPS) positions (Chapter 12, Section 12.13). The ellipsoid height, $h = H + N$, where H = the elevation above the vertical datum and N is the separation between the ellipsoid and the vertical datum. Values for N can be determined from an appropriate mathematical model; for example, Geoid 96, available on diskette from the National Geodetic Information Center, Silver Spring, Maryland.

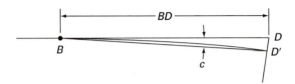

FIGURE 4.22
Curvature expressed as an angle.

Section 9.11, it is called a *geodetic distance*. This logic also applies to taped distances, where the individual increments are tape lengths.

If this same distance is measured in a single observation, the assumption that the slope distance is the hypotenuse of a right triangle no longer is satisfied. In this case, Equations (4.1), (4.2), and (4.4) are not adequate for slope correction, and a different procedure must be developed for slope reduction of lines longer than 2 mi or 3.3 km, or that are 2 mi long and have zenith angles less than 85° or greater than 95°, or vertical angles greater than 5°.

Reduction of long slope lines involves use of zenith and vertical angles affected by curvature and refraction. The formulation can be simplified by determining the effect of earth curvature as an angular value per unit of distance and using this value in the slope reduction. The fundamental equations for earth curvature and refraction are derived in Chapter 5. Assuming an average radius for the earth of 6371 km or 3959 mi, curvature is 0.667 ft/mi or 0.0785 m/km. So, in Figure 4.22, the curvature DD' is 0.667 ft when BD is 1 mi and 0.0785 m when $BD = 1$ km. The curvature c, expressed as an angle in seconds, is

$$c = \frac{0.0785 \text{ m/km}}{1000 \text{ m arc } 1''} = 16.192''/1 \text{ km} \tag{4.41}$$

$$c = \frac{0.667 \text{ ft/mi}}{5280 \text{ ft arc } 1''} = 26.06''/\text{mi}$$

or

$$c = \frac{26.06''/\text{mi}}{5.28} = 4.935''/1000 \text{ ft} \tag{4.42}$$

4.44
SLOPE REDUCTION WITH RECIPROCAL ZENITH AND VERTICAL ANGLES

Figure 4.23 illustrates the geometry involved when a long slope distance s is measured from O', at elevation $O'L$ above the ellipsoid to P at elevation QP above the ellipsoid. The curved path of the light or electromagnetic waves from O' to P is approximated by the straight-line distance $O'P$. As illustrated in Section 4.47, the difference is insignificant; for all practical purposes (in 90 mi or 150 km) the differences are less than 0.7 ft or 0.2 m for EDM instruments. The zenith and elevation vertical angles (Section 6.37) at O' to P are angles $z_{O'}$ and α. From P at the other end of the observed line PO', a horizontal line will intersect the plumb line through O' at V, and point O' itself will be observed with zenith angle z_P or depression angle β(Section 6.37). When both z_P and $z_{O'}$ or α and β are measured, they are called *reciprocal angles*.

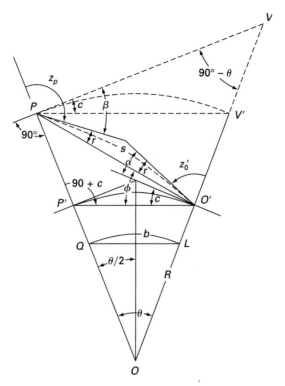

FIGURE 4.23
Reduction of long slope distances to horizontal
distance ellipsoid.

In triangle OPV, θ is the angle subtended by O' and P' at the center of the ellipsoid and angle PVO equals $90 - \theta$.

In the figure, $O'P'$ is the chord distance at elevation $Q'L$ above the ellipsoid, QL is the ellipsoid chord distance, b is the ellipsoid arc distance, r is the angle of refraction, c is the angle of curvature, and by the geometry of the figure $c = \theta/2$.

Assume that reciprocal angles $z_{O'}$ and z_P or α and β have been observed at each end of the measured line $O'P$. In the triangle $O'PV$, using zenith angles $z_{O'}$ and z_P,

$$90° - \theta + z_P - 90° + r + z_{O'} + r = 180°$$

so that

$$r = 90° - \frac{z_P}{2} - \frac{z_{O'}}{2} + \frac{\theta}{2}$$

In the triangle $PO'P'$,

$$\phi = 90° - z_{O'} - r + \frac{\theta}{2} = 90° - z_{O'} - \left(90° - \frac{z_P}{2} - \frac{z_{O'}}{2} + \frac{\theta}{2}\right) + \frac{\theta}{2}$$

or

$$\phi = \frac{z_P - z_{O'}}{2} \tag{4.43}$$

Using the vertical angles in triangle $O'PV$

$$90 - \theta + \beta + r + 90 - \alpha + r = 180°$$

so that

$$r = \frac{\theta + \alpha - \beta}{2}$$

In the triangle $PO'P'$,

$$\text{angle } P'O'P = \phi = \alpha - r + c = \alpha - \left(\frac{\theta + \alpha - \beta}{2}\right) + c$$

or

$$\theta = \frac{\alpha + \beta}{2} \tag{4.44}$$

Therefore, when reciprocal vertical angles are observed, the curvature and refraction cancel.

Also in the triangle $PO'P'$,

$$\gamma = 180° - \left(\phi + 90 + \frac{\theta}{2}\right) = 180° - (\phi + 90 + c) \tag{4.45}$$

Because

$$O'P' \simeq O'P \cos \phi \tag{4.46}$$

then

$$\sin \frac{\theta}{2} = \frac{O'P'}{2R} \tag{4.47}$$

is a sufficiently close approximation for θ. Here, R is the radius of the ellipsoid for the average latitude of the observation in the direction of the line. An alternative approach is to calculate $c = \theta/2$ by either Equation (4.41) or (4.42).

With the angles ϕ, γ, and c evaluated, solve the triangle $PO'P'$ by the law of sines for $O'P$:

$$O'P' = \frac{O'P \sin \gamma}{\sin(90° + c)} \tag{4.48}$$

to give the horizontal distance at elevation $O'L'$ above the ellipsoid.

Reduce $O'P'$ to an ellipsoid chord QL using Equation (9.11) (Section 9.11):

$$QL = \frac{(R)(O'P')}{R + h_{o'}} \tag{4.49}$$

in which R is the radius of the earth ellipsoid for the average latitude of the area in the direction of the line measured and $h_{o'} = O'L$ is the elevation of O' above the ellipsoid.

For the case where a single vertical angle is measured, an angle of refraction r must be estimated. Assuming that refraction reduces the curvature correction by 14 percent, Equations (4.41) and (4.42) become

$$(c \ \& \ r) = (4.935''/1000 \text{ ft})(0.86) = 4.244''/1000 \text{ ft} \tag{4.50}$$

and

$$(c \ \& \ r) = (16.192''/\text{km})(0.86) = 13.925''/\text{km} \tag{4.51}$$

If $z_{O'}$ or the elevation angle α is observed in triangle $PO'P'$,

$$\phi = 90° - z_{O'} + (c \ \& \ r) = \alpha + (c \ \& \ r) \tag{4.52}$$

When z_P or β, the depression angle, is observed:

$$\phi = 90° - z_P + (c \ \& \ r) = \beta + (c \ \& \ r) \tag{4.53}$$

in which β always has a negative sign. Thus, ϕ is calculated by Equation (4.52) or (4.53); γ, with Equation (4.45), and $O'P'$ is computed using Equation (4.48).

EXAMPLE 4.28. A slope distance of 4994.481 m, corrected for system constants and atmospheric conditions, is measured at station O' that has an elevation of 146.852 m above the ellipsoid to station P. Reciprocal zenith angles, $z_{O'}$ and z_p are 79°00'55" and 100°59'55", respectively. Calculate the ellipsoid chord distance between these two stations.

Solution. Calculate ϕ with Equation (4.43):

$$\phi = \frac{z_P - z_{O'}}{2} = \frac{(100°59'55") - (79°00'55")}{2} = 10°59'30"$$

Compute curvature, c, using Equation (4.42):

$$c" = (16.192"/\text{km})(4.9945 \text{ km}) = 80.9" = 0°01'20.9"$$

Determine γ using Equation (4.45):

$$\gamma = 180° - (\phi + 90° + c) = 180° - [(10°59'30") + 90° + 0°01'20.9")]$$
$$= 78°59'09.1"$$

Solve the oblique triangle for the horizontal distance $P'O'$ Using Equation (4.48):

$$P'O' = \frac{O'P \sin \gamma}{\sin(90° + C)} = \frac{(4,994.481) \sin(78°59'09.1")}{\sin(90°01'20.9")}$$
$$= 4,902.483 \text{ m at } 146.852 \text{ m above the ellipsoid}$$

Reduce to the ellipsoid with Equation (4.49) using $R = 6,371,000$ m:

$$QL = \frac{(R)(O"P")}{(R + h_o')} = \frac{(6,371,000)(4,902.483)}{(6,371,000 + 146.852)}$$
$$= \text{ellipsoid chord distance} = 4,902.370 \text{ m}$$

EXAMPLE 4.29. A slope distance of 23,457.500 m is measured with an electro-optical EDM system at station O', 323.00 m above the ellipsoid, to station P. Reciprocal vertical angles $\alpha_{O'}$ and β_P are 3°03'00" and −3°13'10", respectively. Calculate the ellipsoid distance for $O'P$. Refer to Figure 4.23.

Solution. Calculate ϕ by Equation (4.44):

$$\phi = \frac{\alpha_{O'} + \beta_P}{2} = \frac{(3°03'00") + (3°13'10")}{2} = 3°08'05"$$

Compute c by Equation (4.41):

$$c" = 16.192"/\text{km})(23.457 \text{ km}) = 379.8" = 6'19.8"$$
$$\phi + c = 3°14'24.8"$$

Determine γ by Equation (4.45):

$$\gamma = 180° - (\phi + 90 + c) = 180° - (93°14'24.8'') = 86°45'35.2''$$

Solve the oblique triangle for the horizontal distance $P'O'$ with Equation (4.48):

$$P'O' = \frac{O'P \sin \gamma}{\sin (90° + c)} = \frac{(23,457.500)(\sin 86°45'35.2'')}{\sin 90°06'19.8''}$$

$$= 23,420.039 \text{ m at elevation } 323.00 \text{ m above the ellipsoid}$$

Reduce to the ellipsoid by Equation (4.49), using an average radius for the earth of 6,371,000 m:

$$QL = \frac{(R)(O'P')}{R + h_{o'}} = \frac{(6,371,000)(23,420.039)}{6,371,000 + 323.00}$$

$$= \text{ellipsoid chord distance} = 23,418.852 \text{ m}$$

EXAMPLE 4.30. A slope distance of 58,050.30 ft is measured from station O' in Figure 4.23, which has an elevation of 2422.85 ft above the ellipsoid. A single vertical angle $\alpha = 2°30'00''$ is observed from O' to P. Calculate the ellipsoid chord distance. Assume $R = 20,902,190$ ft.

Solution. Use Equation (4.50) to evaluate (c & r).

$$(c \text{ \& } r) = (4.244'')(58.050) = 246.4'' = 0°04'06.4''$$

Evaluate c by Equation (4.42):

$$c = 4.935''/1000 \text{ ft}$$

$$c = (4.935)(58.050) = 286.5'' = 0°04'46.5''$$

By Equation (4.52)

$$\phi = \alpha + (c \text{ \& } r) = (2°30'00'') + (0°04'06.4'') = 2°34'06.4''$$

$$\phi + c = 2°38'52.9''$$

and by Equation (4.45),

$$\gamma = 180° - (90 + \phi + c) = 87°21'07.1''$$

Solve for $O'P'$ by Equation (4.48):

$$O'P' = \frac{O'P \sin \gamma}{\sin (90 + c)} = \frac{(58,050.30)(\sin 87°21'07.1'')}{\sin 90°04'46.5''}$$

$$= 57,988.37 \text{ ft}$$

and reduce to the chord distance with Equation (4.49):

$$QL = \frac{(20,902,190)(57,988.37)}{20,902,190 + 2423} = 57,981.65 \text{ ft}$$

4.45
REDUCTION OF LONG LINES USING ELEVATION DIFFERENCES

In Figure 4.24 the slope distance s is measured between points O' and P for which respective elevations equal to LO' and QP above the ellipsoid are known. The difference in elevation between O' and $P = PP' = h_P - h_{O'} = \Delta h$. As in the previous example, s is approximated by the straight-line distance $O'P'$; θ is the angle subtended by O' and P; and the angle of curvature $c = \theta/2$.

In the triangle PFP', $\theta/2 = c$ can be calculated using Equation (4.41) or (4.42), assuming that $O'P \simeq O'P'$. Then the horizontal straight-line distance $O'P'$ is given by

$$O'P' = [(O'P)^2 - (PF)^2]^{1/2} - PP' \sin c \qquad (4.54)$$

where $PF = P'P \cos c = \Delta h \cos c$.

Because c is calculated assuming that $O'P = O'P'$, the solution may need to be iterated once more using an improved value of c to achieve a sufficiently accurate value of $P'O'$. The horizontal distance $P'O'$ should then be reduced to the ellipsoid by Equation (4.49).

An exact solution for a direct reduction to the ellipsoid also is possible. In the triangle OQL, by the law of cosines,

$$(QL)^2 = 2R^2 - 2R^2 \cos \theta$$

from which

$$\cos \theta = 1 - \frac{(QL)^2}{2R^2}$$

In the triangle $OO'P$, by the law of cosines,

$$(O'P)^2 = (R + h_P)^2 + (R + h_{O'})^2 - 2(R + h_P)(R + h_{O'}) \cos \theta$$

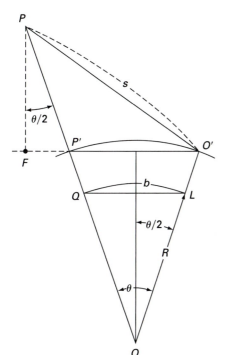

FIGURE 4.24
Reduction of long lines using elevations.

where $h_P = QP$, and $h_{O'} = LO'$. Substitution of the value found for $\cos \theta$ yields

$$(O'P)^2 = (h_P - h_{O'})^2 + \frac{(QL)^2(R + h_P)(R + h_{O'})}{R^2}$$

which can be further simplified by setting $\Delta h = h_P - h_{O'}$ and solving for $(QL)^2$.

$$(QL)^2 = \frac{R^2[(O'P)^2 - \Delta h^2]}{(R + h_P)(R + h_{O'})}$$

or

$$QL = \left\{ \frac{R^2[(O'P)^2 - \Delta h^2]}{(R + h_P)(R + h_{O'})} \right\}^{1/2} \tag{4.55}$$

QL is the ellipsoid chord distance and needs no further correction unless a chord to arc correction is warranted. Extreme care should be excercised when using Equation (4.55) with pocket calculators, as significant figures may be lost.

EXAMPLE 4.31. Stations O' and P (Figure 4.24) have elevations above the ellipsoid of 414.52 m and 1517.91 m. The slope distance corrected for meteorological conditions and instrumental constants from O' to P is 28,545.600 m. Calculate the ellipsoid chord distance. Assume $R = 6,371,000$ m.

Solution. Calculate c by Equation (4.41) and the horizontal distance with Equation (4.54).

$$c = (16.192)(28.546) = 462.2'' = 0°07'42.2''$$

The difference in elevation between O' and P is 1103.39 m:

$$PF = P'P \cos c = \Delta h \cos c = (1103.39) \cos 0°07'42.2'' = 1103.387 \text{ m}$$

so that, by Equation (4.54),

$$O'P' = [(28,545.600)^2 - (1103.387)^2]^{1/2} - (1103.39)(\sin 0°07'42.2'')$$

$$= (28,524.267) - 2.472 = 28,521.795 \text{ m}$$

Now recalculate c with the improved estimate for the distance:

$$c = (16.192)(28.522) = 461.8'' = 0°07'41.8''$$

The second iteration involves evaluation of only the second term of Equation (4.54), so that

$$O'P' = 28,524.267 - (1103.39)(\sin 0°07'41.8'')$$

$$= 28,524.267 - 2.470 = 28,521.797 \text{ m}$$

Reduce to the ellipsoid chord distance by Equation (4.49):

$$QL = \frac{(28,521.797)(6,371,000)}{(6,371,000) + (414.52)} = 28,519.941 \text{ m}$$

As an alternative solution, the direct reduction to the ellipsoid is by Equation (4.55), where

$$QL = \left\{ \frac{(6,371,000)^2[(28,545.600)^2 - (1103.39)^2]}{(6,371,000 + 1517.91)(6,371,000 + 414.52)} \right\}^{1/2}$$

$$= 28,519.942 \text{ m}$$

The difference of 1 mm can be attributed to a round-off error in the second solution.

4.46
CHORD-TO-ARC CORRECTION

In Figure 4.24 the ellipsoid distance b exceeds the chord QL by a very small amount. In 100-mi and 200-km distances the respective differences are 11 ft and 3.5 m. The chord-to-arc correction is calculated as follows:

$$QL = 2R \sin\left(\frac{\theta}{2}\right)$$

Expanding $\sin \theta/2$ in series expansion yields

$$QL = 2R\left(\frac{\theta}{2} - \frac{\theta^3}{48} + \frac{\theta^5}{3840} - \cdots + \cdots\right)$$

Also

$$b = R\theta = \frac{2R\theta}{2}$$

$$\theta = \frac{b}{R} \approx \frac{QL}{R}$$

chord-to-arc correction $= b - QL = \dfrac{2R\theta}{2} - 2R\left(\dfrac{\theta}{2} - \dfrac{\theta^3}{48} + \dfrac{\theta^5}{3840}\right)$, which reduces to $b - QL = R\theta^3/24$, and because $\theta \approx QL/R$,

$$b - QL = \frac{(QL)^3}{24R^2} = \text{arc} - \text{chord} \qquad (4.56)$$

The chord-to-arc correction always is positive.

In 20 mi or 33 km, this correction is 1 ppm and, for lines this length or less, is insignificant.

EXAMPLE 4.32. In Example 4.31, the ellipsoid chord distance is 28,519.941 m. What is the chord-to-arc correction? As before, $R = 6,371,000$ m and $QL = 28,519.941$ m.

Solution. By Equation (4.56),

$$b - QL = \text{arc to chord} = \frac{(28,519.941)^3}{24(6,371,000)^2} = 0.024 \text{ m}$$

So, the geodetic distance from S to L is

$$b = 28,519.941 + 0.024 \text{ m} = 28,519.965$$

Note that the correction for this length of line is 1 ppm and is insignificant except for very special cases.

4.47
CORRECTION FOR RAY PATH

Determination of the correction due to curvature of the ray path for propagated electronic waves is insignificant except for surveys with very long lines, where extraordinary precision is required. The radius of the ray path is a function of the coefficient of refraction in air, k, which has values of approximately 0.13 and 0.20, respectively, for daytime and night

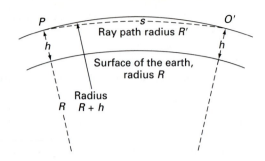

FIGURE 4.25
Curvature of ray path.

observations with electro-optical systems. For microwave instruments, k is approximately 0.25 (Mead, 1972).

The equation for the correction is the same as for the chord-to-arc correction (Equation 4.56) but with the opposite sign. Referring to Figure 4.25 and using $R' = R/k$,

$$PO' - s = -\frac{s^3}{24(R')^2} = -\frac{k^2 s^3}{24R^2} \tag{4.57}$$

Assuming a radius of 6,371,000 m and the previously cited values for k, $k^2/24R^2$ takes the following values:

| | k | $k^2/24R^2$ | Correction in meters for a line of | | |
			100 km	150 km	200 km
Electro-optical, day	0.13	1.74×10^{-17}	0.017	0.059	0.139
Electro-optical, night	0.20	4.11×10^{-17}	0.041	0.139	0.329
Microwave	0.25	6.42×10^{-17}	0.064	0.217	0.513

Therefore, for lines less than 150 km, the correction is insignificant, so that the correction for ray path curvature primarily is of academic interest.

4.48
ACCURACY OF VERY LONG LINES WITH EDM

When very long lines and high precision are involved, acquisition and correction for meteorological conditions represent the limiting factors in relative precisions possible with EDM systems. For maximum accuracy, meteorological conditions always should be recorded at both ends of the line. For best results, temperatures should be recorded 30 ft or about 10 m above terminal stations. The next step would be to acquire meteorological data at specified intervals along the line, to obtain a more realistic estimate for the index of refraction. An approach in this direction has been taken on long lines measured in conjunction with fault-line movements in California. In this case, an aircraft was flown between the two terminal points of the line to acquire additional meteorological data along the line while observations were being performed.

The ultimate solution probably lies in the instrumental determination of the index of refraction simultaneously with the observations for measurement by utilizing a combination of light and microwaves. Such a system would use modulated red and blue laser light in addition to simultaneous modulated microwave transmission. The difference in velocities of the two light-wave signals permits a real-time determination of the dry air density, leading to an average air density over the path. From the average air density, the average group index of refraction can be computed. The microwave index of refraction is much more sensitive to water vapor pressure than light waves, so that simultaneous transmission of microwaves permits determination at the average water vapor pressure from the optical-microwave dispersion. A microcomputer built into the instrument would allow real-time processing of the data and output consisting of slope distances corrected for meteorological conditions.

A system utilizing these ideas was developed and a prototype built for distance determinations in studies of earth crustal movement (see Huggett and Slater, 1974). To date, no such instrument is available in the commercial market. Development of the GPS (Chapter 12), which permits measuring very long lines with extremely high accuracy, has reduced the need for continued improvement of accuracies in EDM measurements of very long lines except in special cases where sub-mm accuracies are required for long lines.

PROBLEMS

4.1. The length of a line measured with a 30-m tape is 310.550 m. When the tape is compared to the standard it is found to be 0.010 m too short under the same conditions of support, tension, and temperature as existed during measurement of the line. Compute the length of the line.

4.2. A building 80.00 ft by 160.00 ft is to be laid out with a 50-ft tape that is 0.050 ft too short. What ground measurements should be made?

4.3. The slope measurement of a line is 245.840 m. The differences in elevation between successive 30-m points are 0.50, 0.46, 0.75, 1.20, 1.40, 1.50, 2.00, and 0.20 m. Calculate the horizontal distance along the line.

4.4. The slope measurement of a line is 1240.4 ft. Slope angles measured with a clinometer are as follows. Determine the horizontal distance.

Cumulated distance, ft	0		300.0		800.0		1000.0		1240.4
Slope angle, deg		$\frac{1}{2}$		$1\frac{1}{4}$		$2\frac{1}{2}$		4	

4.5. A 30-m tape having a standard length of 30.005 m is to be used to lay out a building that has plan dimensions of 150.000 by 270.000 m. What horizontal measurements must be made on the ground in the field to perform this layout?

4.6. The following slope distances and differences in elevations between tape ends were recorded for a measurement:

Slope distance	Difference in elevation, m
30.005	1.452
29.950	3.500
30.000	0.505
25.989	2.445
10.345	1.595

Calculate the horizontal distance between the two endpoints of this line.

4.7. Two points at a slope distance of 100.00 m apart have a difference in elevation of 8.00 m. Calculate the slope distance to be set off to establish a horizontal distance of 100.00 m.

4.8. Two points at a slope distance of 25.000 m apart have a difference in elevation of 5.000 m. Determine the slope distance to be laid off to set the second point a horizontal distance of 25.000 m from the first.

4.9. The distance measured over smooth level ground between two monuments was recorded as 105.00 m. Measurements consisted of three full tape lengths and one partial tape length. If the first taping point was misaligned 0.4 dm left, the second 9.2 dm right, and the third 5.0 dm left, compute the correct distance between the points.

4.10. A tape that has a standardized length of 99.900 at 68°F with the tape fully supported is to be used to set the corner points of a foundation for a building that is 100.00 m by 200.00 m. What distances should be measured between the building corners with the tape fully supported over level terrain when the temperature is 40°F?

4.11. The tape in Problem 4.10 was used to measure the distance between two monuments set at the corners of a city block. The tape was fully supported over level terrain and the temperature was 88°F. The distance recorded was 610.86 ft. What is the correct horizontal distance between the two monuments?

4.12. A horizontal distance of 200.000 m is to be set out over uniformly sloping terrain that has a slope of 6.000 percent. Calculate the slope distance that must be measured to lay out this distance.

4.13. Vertical angles, measured with an engineer's transit, and slope distances for a line measured over irregular terrain are as follows:

Slope distance, m	Vertical angle
30.000	3°25′
20.268	2°32′
29.567	1°55′
5.230	0°00′

Calculate the horizontal distance.

4.14. The organizing committee for a "fun run" wants a 10-km course laid out over a clear smooth path passing over undulating terrain. The distance is to have an accuracy of ±5 m (1 part in 2000). Describe the most efficient way to measure this distance so as to set start and finish marks. Assume that neither EDM or GPS equipment is available.

4.15. A distance measured over smooth, level terrain was recorded as 99.5 m. Measurements consisted of three full and one partial tape lengths using a 30.000 tape. If the first tape length was misaligned 0.50 m to the right, the second 0.60 m right, and the third 0.10 m to the left, all offsets referred to the true alignment, compute the correct distance.

4.16. A steel tape having a standardized length of 29.990 m, with 8 kg tension, at 20°C, and supported at the two ends, was used to measure a distance over smooth, level terrain using *standard support and tension* and at an average recorded temperature of 30°C. The recorded distance was 915.258 m. Determine the correct horizontal distance. The weight of the tape is 0.60 kg.

4.17. The tape described in Problem 4.16 was used to measure a single tape length with the tape *fully supported and with the temperature at 20°C*. Compute the distance that was laid out. $P = 8$ kg.

4.18. The tape described in Problem 4.16 was used to measure a horizontal distance of 310.455 m between two monuments with the tape fully supported, with 8 kg tension on all full tape lengths, and at an average temperature of 5°C. A tension of 3.5 kg (taken from the chart for this tape and is the correct tension) was applied to the last partial increment taped. Determine the correct horizontal distance.

4.19. A horizontal distance of 165.299 m is to be measured from a monument to set a point using the tape in Problem 4.16. The tape was *fully supported,* with 8 kg tension on full tape lengths (5 kg, taken from the table and correct for the last partial length), and at a temperature of 45°C. Determine the distance that must be measured in the field to set this point.

4.20. A standardized 30-m steel tape weighing 0.7 kg measures 29.9935 m supported at the two ends only, with 8.0 kg of tension and at 20°C.

(a) A horizontal distance of 3248.835 m is measured with the tape fully supported, with 8.0 kg of tension on full tape lengths (3.0 kg on the partial length) and at a temperature recorded at 8°C. Determine the correct horizontal distance.

(b) A horizontal distance of 200.843 m is to be measured from a monument to set an adjacent property corner. What horizontal distance must be measured under standardization conditions of support and tension to set this property corner where the temperature is 21°C?

4.21. A 100-ft tape weighing 2 lb is of standard length under a tension of 15 lb, supported for the full length of the tape. A line on smooth level ground is measured with the tape fully supported under a tension of 30 lb and found to be 4863.5 ft long. $E = 30,000,000$ lb/in.^{2} and 3.53 in.^{3} of steel weighs 1 lb. Make the correction for the increase in tension.

4.22. A second line is measured with the tape of Problem 4.21, the tape being supported at intervals of 50 ft and the pull being 25 lb on full tape lengths. The measured length is 1823.00 ft. Compute the corrections for sag and variation in tension and determine the corrected length of the line. Assume 6-lb tension on the 23-ft increment (the correct amount).

4.23. A 30-m steel tape weighing 0.7 kg has a cross-sectional area of 0.03 cm^2 and a standard length of 29.995 m when fully supported, at 20°C, and with 6 kg of tension.

(a) A line on a smooth level ground is measured with the tape under a tension of 12 kg at a temperature of 20°C and is recorded as 1242.823 m long. Compute the correction for the increase in tension. $E = 2.100,000$ kg/cm^2. Tension on the 12.823 increment was 4 kg, the correct amount for that span.

(b) This same tape is used to measure a distance recorded as 985.423 m over level terrain with the tape supported at the two ends and 6 kg of tension (temperature $= 20$°C). Calculate the correction due to sag for this distance.

4.24. Compute the normal tension for the tape in (a) Problem 4.21; (b) Problem 4.23.

4.25. For the purpose of establishing monuments in a city, a line along a paved street having a grade of 2.5 percent is measured on the slope. The applied tension is 7.5 kg, and observations of temperature are made at each application of the tape. The measured length on the slope is 390.351 m, and the mean of the observed temperatures is 29.3°C. The 30-m steel tape used for the measurements is standardized at 20°C, supported for its full length, and is found to be 0.001 m too short under a tension of 7.5 kg. Determine the horizontal length of the line.

4.26. A line roughly 2 mi long along a railroad track is measured with a steel tape, and corrections are made for observed temperatures. What error will be introduced if the actual temperature of the tape is 4°F higher than the observed temperature? State the error in fractional form with 1 as the numerator.

4.27. Assume that an invar tape (a tape made of a special alloy) having a coefficient of thermal expansion of 0.00000083/1°F is used under the conditions of Problem 4.26. Compute the error introduced.

4.28. A hedge along the line AB makes direct measurement impossible. A point C is established at an offset distance of 3.95 m from the line AB and roughly equidistant from A and B. The distances AC and CB then are taped; $AC = 392.34$ m and $CB = 412.39$ m. Compute the length of the line AB.

4.29. Describe and explain the fundamental differences between electro-optical and electromagnetic distance measuring instruments.

4.30. Describe and explain the major systematic errors that affect EDM systems.

4.31. An electro-optical EDM using infrared light with a wavelength of 0.9300 μm has a modulation frequency of 30 MHz. Determine the modulated wavelength of light at 30°C and an atmospheric pressure of 755.9 mmHg. Compare this wavelength with the wavelength of the modulation frequency in a standard atmosphere at 0°C and 760 mmHg.

4.32. An electro-optical EDM transmits a modulated beam with carrier wavelength of 0.910 μm and modulation frequency of 24.582 MHz. Calculate the wavelength at 24°C and 752.5 mmHg.

4.33. Compute the velocity of red laser light at a temperature of 25°C and atmospheric pressure of 752.8 mmHg.

4.34. Compute the modulation wavelength of microwaves having a modulation frequency of 25 MHz at a temperature of 25°C, an atmospheric pressure of 755.5 mmHg, and a vapor pressure of 6.5 mmHg.

4.35. A microwave EDM has a modulation frequency of 7.500 MHz. Determine the modulation wavelength for this frequency at 15°C, atmospheric pressure of 755.6 mmHg, and vapor pressure of 5.0 mmHg.

4.36. An electro-optical EDM has f_0 = 14.9850 MHz, carrier wave frequency, λ_1 = 0.900 μm, and λ_0 = 20.00 m. A slope distance was measured using this instrument at a temperature of 18°C and an atmospheric pressure of 755.0 mmHg. Calculate the ppm correction that should be entered into the instrument prior to making this measurement.

4.37. An electro-optical EDM has a modulation frequency of 7.4927 MHz with a modulation wavelength of 40.0000 m and carrier wavelength of 0.850 μm. On the site, the temperature is 24°C and atmospheric pressure registers 752.6 mmHg. Determine the ppm correction to be entered into the EDM before making measurements.

4.38. The EDM described in Problem 4.36 was used under the given atmospheric conditions and with the ppm correction dial set to 0 to measure a slope distance of 2,356.781 m. Calculate the slope distance corrected for environmental conditions.

4.39. The EDM described in Problem 4.37 was used to measure a slope distance of 1555.467 m. If the atmospheric conditions are as given in the problem and the environmental correction dial in ppm was set to 0 before measurement, calculate the slope distance corrected for environmental conditions.

4.40. EDM No. 8208 and reflector No. 3 were calibrated by measuring base line 1-2-3-4 in several segments, where averaged distances for each segment, corrected for slope and atmospheric conditions, are

From–To	Distance, m
1–2	150.009
2–3	280.073
3–4	770.014
1–4	1200.032

Calculate the system constant for EDM No. 8208 and reflector No. 3.

4.41. Describe nonlinearity in electro-optical EDM and outline the conditions under which errors due to nonlinearity should be corrected.

4.42. A slope distance of 327.500 m is measured using an EDM for which the system constant is −0.002 m and the calibration chart for nonlinearity is given in Figure 4.18. If the two stations between which the distance is measured have a difference in elevation of 9.055 m, compute the horizontal distance.

4.43. The slope distance from E to G, measured by EDM and corrected for atmospheric conditions and system constants, is 520.625 m with $z = 87°00'00''$ measured with a theodolite at E, where the h.i. of the instrument $= 1.350$ m above E, to a target at G, where the h.i. $= 1.425$ m above station G. The h.i.'s of the EDM above E and the reflector above G are 1.607 m and 1.839 m, respectively. Compute the horizontal distance between points E and G.

4.44. The vertical angle from station 1 to station 2 is $+2°27'10''$ and the slope distance, corrected for atmospheric conditions and systematic errors, is 710.099 m measured with an EDM. The h.i.'s of EDM and reflector above stations 1 and 2 were 1.623 m and 1.850 m, respectively. The h.i.'s of theodolite and target above 1 and 2 were 1.350 m and 1.245 m, respectively. Determine the horizontal distance from 1 to 2.

4.45. Specifications state that the horizontal distance calculated in Problem 4.42 be determined with a standard deviation of 0.001 m. Determine the standard deviation of the difference in elevation between the two ends of the line required to satisfy this specification.

4.46. In Problem 4.43, the horizontal distance is to be found with a standard deviation of 0.005 m. Calculate the standard deviation in the zenith angle needed to satisfy this requirement.

4.47. For rough taping with a 30-m steel tape over irregular terrain, the following standard deviations in taping operations are assumed: (1) tape is not level, $\sigma_h = 0.03$ m; (2) sag in the tape is caused by estimating tension, $\sigma_s = 0.02$ m; (3) plumbing the points, $\sigma_v = 0.03$ m.
(a) Calculate the standard deviation in measuring a tape length due to these estimated errors.
(b) A line 1200 m long is to be measured with this tape using procedures that satisfy the estimates just given. Compute the standard deviation for the length of the measured line.

4.48. The vertical angle from point 1 to 2 is $+3°27'01''$ and the slope distance (corrected for atmospheric conditions and system constants) is 17,728.974 m, as measured by an EDM. If the elevation of point 1 is 1000.55 m above the ellipsoid, compute the horizontal ellipsoid distance from 1 to 2.

4.49. The vertical angles between stations A and B are $4°02'05''$ from A to B and $-4°12'55''$ from B to A. The slope distance AB corrected for atmospheric conditions and system constants is 25,500.123 m and A is 522.073 m above the ellipsoid. Compute the horizontal distance AB reduced to the ellipsoid.

4.50. The zenith angle from C to D is $86°30'50''$ and the slope distance corrected for atmospheric conditions and EDM system constants is 60,200.458 ft. The elevation of C is 2500.44 ft above the ellipsoid. Calculate the horizontal distance from C to D and reduce to the ellipsoid.

4.51. The difference in elevation between points A and B is 928.540 m and the slope distance (corrected for atmospheric conditions and EDM system constants) is 20,400.851 m. Determine the horizontal distance reduced to the ellipsoid. The elevation of A is 200.82 m above the ellipsoid.

4.52. The standard deviations for measuring the zenith angle and corrected slope distance in Problem 4.50 are $\sigma_z = 10''$ and $\sigma_{\text{slope distance}} = 0.040$ m. Calculate the standard deviation in the horizontal distance from C to D.

REFERENCES

AGA Operating Manual. AGA Corporation Geodimeter Division. Sweden, 1965.

Burnside, C. D. *Electromagnetic Distance Measurement,* 3rd ed. Cambridge, MA: Blackwell Scientific Publications, 1991.

Crawford, W. G. "Shopping for a Total Station?" *P.O.B., Point of Beginning* 22, no. 7 (April 1997), pp. 22–47.

Dracup, J. F.; C. F. Kelley; G. B. Lesley; and R. W. Tomlinson. *National Geodetic Survey Lecture Notes for Use at Workshops on Surveying Instrumentation and Coordinate Computation.* Silver Springs, MD: Coast and Geodetic Survey, 1972.

Foster, N. "In the Heart of the Big Thicket." *Professional Surveyor* 17, no. 3 (April 1997), pp. 8–12.

Fronczek, C. J. NOAA Technical Memorandum NOS NGS-10. *Use of Calibration Base Lines.* National Geodetic Information Center, Rockville, MD, December 1977, revised 1980.

Gossett, F. R. *Manual of Geodetic Triangulation.* C&GS, NGSD, Special Publication 247, U.S. Government Printing Office, Washington DC. January 1950.

Greene, J. "Accuracy Evaluation of Electro-Optic Distance Measuring Instruments." *Surveying and Mapping* 37, no. 3 (September 1977), pp. 247–56.

Huggett, G. R. "High Accuracy Techniques Using the HP 3800 Distance Meter." *Insight,* Hewlett-Packard Engineering Products 2, no. 1 (November 1973).

Huggett, G. R. "Instrument Non-linearity." *Insight,* Hewlett-Packard Engineering Products 2, no. 4 (August 1974).

Huggett, G. R., and S. L. Slater. "Precision Electromagnetic Distance Measuring Instrument for Determining Secular Strain and Fault Movement." Paper presented at the International Symposium on Recent Crustal Movements, Eidgonössische Technische Hochschule, Zurich, Switzerland, August 26–31, 1974.

Kester, J. M. "EDM Slope Reduction and Trigonometric Leveling." *Surveying and Mapping* 33, no. 1 (March 1973), p. 61.

Kivioja L. A. "The EDM Corner Reflector Is Not Constant." *Surveying and Mapping* 38, no. 2 (June 1978), pp. 142–55.

Laurilla, S. H. *Electronic Surveying.* New York: John Wiley & Sons, 1983.

Mead, B. D. "Precision in Electronic Distance Measuring." *Surveying and Mapping* 32, no. 1 (March 1972), pp. 69–78.

Moffitt, F. H. "Field Evaluation of the Hewlett-Packard Model 3800 Distance Meter." *Surveying and Mapping* 31, no. 1 (March 1971).

Mofitt, F. H. "Calibration of EDM's for Precision Measurement." *Surveying and Mapping* 35, no. 2 (June 1975), pp. 147–154.

Moffitt, F. H., and H. B. Bouchard. *Surveying,* 9th ed. New York: Harper & Row, 1992.

Prescott, R. R. "EDM Calibration Base Lines: Do Them Yourself." *ACSM Bulletin,* no. 164 (November/December 1996), pp. 23–28.

Reilly, J. P., "EDM Survey, 1993." *P.O.B., Point of Beginning* 19, no. 2 (December–January 1993), pp. 78–83.

Santaala, J. "New Tools for Determination of Instrumental Errors in EDM." *Surveying Science in Finland,* no. 1 (1983), pp. 9–21.

Santaala, J. "On Determination of Long-Period Instrumental Errors in EDM Using Field Observations." *Surveying Science in Finland,* no. 1 (1984), pp. 11–16.

Stoughton, H. W. "Development and Application of Refractivity Correction: Formula for Optical and Infrared Observations." *Surveying and Land Information Systems* 53, no. 52 (June 1993), pp. 79–84.

Zilkosky, D. B.; J. H. Richards; and G. M. Young. "Results of the General Adjustment of the North American Vertical Datum of 1988." *Surveying and Land Information Systems* 52, no. 3 (September 1992).

Vertical Distance Measurement: Leveling

PART A
ELEMENTARY TOPICS

5.1
DEFINITIONS

The *elevation* or *height* of a point near the surface of the earth is its vertical distance above or below an arbitrarily assumed *level surface* or curved surface every element of which is normal to the plumb line. The level surface (real or imaginary) used for reference is called the *datum*. A *level line* is a line in a level surface.

The *difference in elevation* or height between two points is the vertical distance between the two level surfaces in which the points lie. *Leveling* is the operation of measuring vertical distances, either directly or indirectly, to determine differences in elevation.

A *horizontal line* is a line, in surveying taken as straight, tangent to a level surface.

A *vertical angle* is an angle between two intersecting lines in a vertical plane. In surveying, it is commonly understood that one of these lines is horizontal.

Of special interest is the vertical datum (see also Section 14.2) used as a reference surface. Two different kinds of vertical datums are considered in this chapter: the *geoid* (Section 1.3), and the *reference ellipsoid* (Section 1.3).

The geoid is a horizontal surface shaped by the gravity field of the earth, a concept ideally suited for a vertical datum. However, because an infinite number of these surfaces exists, specification of the desired surface as a datum is necessary. In North America, the vertical datum most commonly used has been mean sea level, particularly the mean sea level datum established by the United States government. Until 1991, the official datum in the United States was the *National Geodetic Vertical Datum of 1929* (NGVD 29). This datum was the result of a general adjustment, in 1929, of about 107,000 km of leveling in the United States and Canada to fit 26 mean sea level stations (21 in the United States and 5 in Canada).

Between 1977 and 1991, a new adjustment was undertaken, in which the approximately 600,000 km of leveling added since 1929 and, in addition, another 81,500 km of releveled lines were adjusted to create the *North American Vertical Datum 1988* (NAVD 88). This adjustment contains leveling from the United States, Canada, and Mexico and includes the *International Great Lakes Datum of 1985* (IGLD 85). NAVD 88 became effective in September 1991, when 80 percent of the total number of bench marks were made available to the public. The remaining bench marks are being added to the network in a long-term project, scheduled to be completed in 1999. During this period of transition to the new datum (until approximately 1999), the National Geodetic Survey is committed to maintaining two datums. Differences between the two datums within the continental United States range from −0.040 to 1.50 m.

Strictly speaking, NAVD 88 is not a mean sea level surface since the 26 mean sea level stations were not used as constraints in this latest adjustment and the height of the primary tidal bench mark at Father Point/Rimouski, Quebee, Canada, was held fixed as the constraint (see Zilkoski, Richards, and Young, 1992). Elevations referred to this datum are called *orthometric heights* (elevations), where the orthometric height of a point is the geometrical distance between the geoid and the point, measured along the plumb line passing through the point (Vaníček and Krakiwsky, 1986).

When elevations or heights are discussed in this text, it will be assumed that they are referred to NAVD 88 or simply to the datum. Additional details on NAVD 88 can be found in Zilkoski et al. (1992).

Ellipsoidal elevations (an indirect by-product of GPS surveys) are referred to the reference ellipsoid (GRS 80, see Section 1.3 and Figure 1.1b), a mathematical surface of reference* for geodetic surveys that is used in reduction and adjustment of GPS data (see Chapter 12). A separation exists between this ellipsoid and the geoid as defined by NAVD 88. Because elevations used for mapping, engineering planning and design, and scientific applications invariably are referred to the datum (i.e., NAVD 88), this separation between the geoid and the ellipsoid must be taken into account before ellipsoidal elevations are useful to the engineering and scientific communities.

5.2
CURVATURE AND REFRACTION

In leveling, it is necessary to consider the affects of (1) the curvature of the earth and (2) atmospheric refraction, which affects the line of sight. Usually, these two effects are considered together.

Figure 5.1 shows a horizontal line tangent at A to a level line near the surface of the earth. The vertical distance between the horizontal line and the level line is a measure of the earth's curvature. It varies approximately as the square of the distance from the point of tangency. In Figure 5.1, let $OA = r$, the average radius of the earth. Also let $c = ED$, the correction for each curvature. Then,

$$r^2 + \overline{AE}^2 = (r + c)^2 = r^2 + 2rc + c^2$$

$$\overline{AE}^2 = c(2r + c)$$

$$c = \frac{\overline{AE}^2}{2r + c} \tag{5.1}$$

---

* A PC program on a diskette can be obtained from NOAA, NGS, N/C 6174, 1315 East-West Highway, Station 9202, Silver Spring, MD 20910-3282.

167

CHAPTER 5:
Vertical
Distance
Measurement:
Leveling

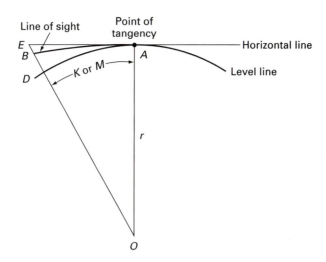

FIGURE 5.1
Earth curvature and refraction.

Because c is very small compared to r, a reasonable approximation for earth curvature is

$$c = \frac{\overline{AE}^2}{2r} \tag{5.2}$$

Assuming a mean radius of the earth of 3959 mi or 6371 km, the curvature correction is

$$c = 0.667M^2 \text{ feet} \tag{5.3}$$

in which M is the distance from the point of tangency (station of the observer) in miles. In the metric system, the earth curvature correction c_m, in meters, is

$$c_m = 0.0785K^2 \tag{5.4}$$

in which K is the distance from the point of tangency in kilometers. Thus, the curvature correction is 0.67 ft/mi and 7.9 cm/km. For distances of 100 ft or 30 m the respective corrections would be 0.00024 ft and 0.07 mm.

Owing to the phenomenon of atmospheric refraction, rays of light are refracted, or bent, downward slightly. This bending of the rays of light toward the center of the earth tends to diminish the effect of earth curvature by approximately 14 percent. In Figure 5.1, AB is the refracted line of sight and the distance BD represents the combined effect of curvature and refraction. Let $(c\&r) = BD$ be computed by the following equations:

$$(c\&r) = 0.574M^2 \text{ feet} \tag{5.5}$$

$$(c\&r) = (2.06) \, 10^{-2}D^2 \text{ feet} \tag{5.6}$$

$$(c\&r) = 0.0675K^2 \text{ meters} \tag{5.7}$$

in which M is in miles, K is in kilometers as previously stated, and D is in thousands of feet.

In most ordinary spirit leveling operations, the line of sight is rarely more than 6 ft or approximately 2 m above the ground, where variations in temperature cause substantial uncertainties in the refractive index of air. Fortunately, most lines of sight in leveling are relatively short (about 100 ft or 30 m) and backsight and foresight distances are balanced (see Section 5.4). Consequently, curvature and refraction corrections are rarely significant except for precise leveling (see Section 5.50).

5.3
METHODS

Difference in elevation may be measured by the following methods:

1. *Direct* or *spirit leveling,* by measuring vertical distances directly. Direct leveling is the most precise method of determining elevations and the one commonly used (Section 5.4).
2. *Indirect* or *trigonometric leveling,* by measuring vertical angles and horizontal or slope distances (Section 5.5).
3. *Stadia leveling,* in which vertical distances are determined by tacheometry using the engineer's transit and level rod; plane table and alidade and level rod; or self-reducing tacheometer and level rod. Details for these procedures are given in Chapter 7.
4. *Barometric leveling,* by measuring the differences in atmospheric pressure at various stations by means of a barometer (Section 5.6).
5. *Gravimetric leveling,* by measuring the differences in gravity at various stations by means of a gravimeter for geodetic purposes.
6. *Inertial positioning system,* in which an inertial platform has three mutually perpendicular axes, one of which is "up," so that the system yields elevation as one of the outputs. Vertical accuracies of from 15 to 50 cm in distances of 60 and 100 km, respectively, have been reported. The equipment cost is extremely high and applications are restricted to very large projects where terrain, weather, time, and access impose special constraints on traditional methods.
7. *GPS survey* elevations are referenced to the ellipsoid but can be corrected to the datum if a sufficient number of points with datum elevations are located in the region surveyed. Standard deviations in elevation differences of 0.053 to 0.094 m are possible under these conditions (Fiedler, 1992; also see Chapter 12, Section 12.13).

Differential leveling is the operation of determining differences in elevation of points some distance apart or of establishing bench marks. Usually, differential leveling is accomplished by direct leveling. *Precise leveling* is a precise form of differential leveling.

Profile leveling is the operation—usually by direct leveling—of determining elevations of points at short measured intervals along a definitely located line, such as the center line for a highway or a sewer.

Direct leveling also is employed for determining elevations for cross sections, grades, and contours.

5.4
DIRECT LEVELING

In Figure 5.2, A represents a point of known elevation and B represents a point the elevation of which is desired. In the method of direct or spirit leveling, the level is set up at some intermediate point as L, and the vertical distances AC and BD are observed by holding a leveling rod first at A and then at B, the line of sight of the instrument being horizontal. (Owing to refraction, the line of sight is slightly curved, as explained previously.)

If the difference in elevation between the points A and E is designated as H_a and between E and B as H_b,

$$H_a = h_a - h'_a \quad \text{and} \quad H_b = h_b - h'_b$$

in which h_a and h_b are the vertical distances read at A and B, respectively, and $h'_a = (c\&r)_a$ and $h'_b = (c\&r)_b$ are the effects of curvature and refraction for the horizontal distances LA and LB, respectively, calculated using one of Equations (5.5), (5.6), or (5.7).

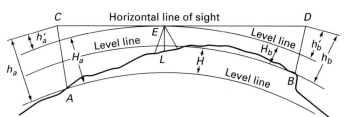

169

CHAPTER 5:
Vertical
Distance
Measurement:
Leveling

FIGURE 5.2
Direct leveling. (Owing to refraction, the line of sight is slightly
curved.)

The difference in elevation H between A and B then is

$$H = H_a - H_b = (h_a - h'_a) - (h_b - h'_b)$$

$$= h_a - h_b - h'_a + h'_b \tag{5.8}$$

If the backsight distance LA is equal to the foresight distance LB, then $h'_a = h'_b$ and

$$H = h_a - h_b \tag{5.9}$$

Therefore, if backsight and foresight distances are balanced, the difference in elevation
between two points is equal to the difference between the rod readings taken to the two
points, and no correction for curvature and refraction is necessary. In direct leveling, the
work usually is conducted so that the effect of curvature and refraction is reduced to a
negligible amount (Section 5.37).

On lines of direct levels the usual procedure is as follows. Let A and M be two
established points some distance apart whose difference in elevation is desired. With the
level in some convenient location, not necessarily on a line joining A and M, a backsight is
taken to point A and a foresight is taken to some convenient point B. The level then is moved
ahead, a backsight is taken to B, and a foresight is taken to some accessible point as C. And
so the process is repeated until the terminal point M is reached.

5.5
TRIGONOMETRIC LEVELING FOR SHORT LINES

Trigonometric leveling involves observing the zenith or vertical angle and either the horizontal or slope distance between two points. The difference in elevation then can be
calculated. In Figure 5.3, A represents a point of known elevation and B a point the elevation
of which is desired. For short lines (lines ≤ 3 km or 2 mi, $85° \leq z \leq 95°$ or $\alpha \leq 5°$),
triangle ECD can be assumed to be a right triangle and the distance $ED = EF$. If the
horizontal distance is known, then

$$DC = ED \cot z = ED \tan \alpha$$

$$DF = (c\&r)_{ED}$$

in which the correction for the earth's curvature and refraction $(c\&r)$ can be calculated
using one of Equations (5.5), (5.6), or (5.7). The difference in elevation, ΔH_{AB},

$$\Delta H_{AB} = AE + DF + DE \cot z - BC \tag{5.10a}$$

$$= AE + DF + DE \tan \alpha - BC \tag{5.10b}$$

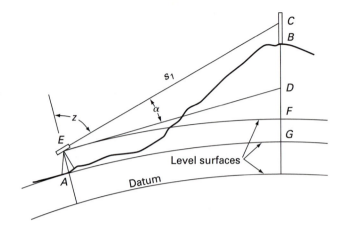

FIGURE 5.3
Trigonometric leveling.

where AE and BC are the instrument height above A and the target height above B, respectively. These heights are measured using a level rod (Section 5.23), tape, or by means of a telescopic plumbing rod if the tripod is so equipped.

A more common situation occurs when the slope distance is measured very accurately using EDM equipment (Sections 4.28–4.32 and 4.35–4.40). In this case $DC = EC \cos z = EC \sin \alpha$ and ΔH_{AB} becomes

$$\Delta H_{AB} = AE + DF + EC \cos z - BC \tag{5.11a}$$

$$\Delta H_{AB} = AE + DF + EC \sin \alpha - BC \tag{5.11b}$$

A major source of error in determining the difference in elevation by this method is the uncertainty in the curvature and refraction correction caused by variations in atmospheric conditions. To reduce the effects of this uncertainty, zenith or vertical angles are observed from both ends of the line, at the same time if possible. In Figure 5.4, z_1 and α are the zenith angle and *elevation* vertical angle, respectively, from the instrument line of sight E over point A to target C at B and z_2 and β are the zenith angle and *depression* vertical angle, respectively, from instrument line of sight C' over B to target E' at A. The difference in elevation, ΔH_{AB} is calculated using the appropriate Equation (5.11). The difference in elevation from A to B using z_2 or the absolute value of β from the other end of the line at B is

$$\Delta H_{ABz_2} = -HK - C'E' \cos z_2 - C'B + E'A \tag{5.12a}$$

$$\Delta H_{AB\beta} = C'E' \sin \beta - HK - C'B + E'A \tag{5.12b}$$

If zenith or vertical angles, z_1, z_2 or α and β are measured simultaneously, they are said to be *reciprocal* zenith or vertical angles (see also Section 4.44). Because the atmospheric conditions will be the same, $HK = DF$ and the addition of Equations (5.11a) and (5.12a) or (5.11b) and (5.12b) yields

$$\Delta H_{ABz_1} + \Delta H_{ABz_2} = EC \cos z_1 - C'E' \cos z_2 + AE + E'A - BC - C'B$$

or

$$\Delta H_{\alpha} + \Delta H_{\beta} = EC \sin \alpha + C'E' \sin \beta + AE + AE' - BC - C'B$$

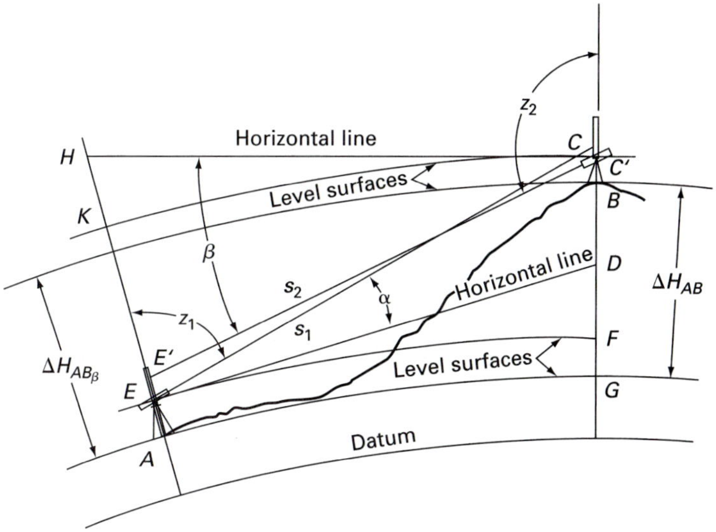

171

CHAPTER 5:
Vertical
Distance
Measurement:
Leveling

FIGURE 5.4
Trigonometric leveling with reciprocal zenith vertical angles.

so that

$$\Delta H_z = \frac{\Delta H_{ABz_1} + \Delta H_{ABz_2}}{2}$$

$$\Delta H_z = \frac{EC \cos z_1 - C'E' \cos z_2}{2} + \frac{(AE - BC) + (E'A - C'B)}{2} \qquad (5.13a)$$

and

$$\Delta H_{\alpha\beta} = \frac{\Delta H_\alpha + \Delta H_\beta}{2}$$

$$\Delta H_{\alpha\beta} = \frac{EC \sin \alpha + C'E' \sin \beta}{2} + \frac{(AE - BC) + (E'A - C'B)}{2} \qquad (5.13b)$$

Therefore, when reciprocal zenith or vertical angles are observed, the difference in elevation is the average of the elevations obtained from the two ends of the line and the correction for earth curvature and refraction cancels.

When reciprocal zenith or vertical angles are employed with horizontal distances (as, for example, when using a total station system where the reduced horizontal distance is available or when employing stadia), ΔH_z and ΔH_α are

$$\Delta H_z = \frac{DE \cot z_1 + C'E' \cot z_2}{2} + \frac{(AE - BC) + (E'A - C'B)}{2} \qquad (5.14a)$$

$$\Delta H_\alpha = \frac{DE \tan \alpha + C'H \tan \beta}{2} + \frac{(AE - BC) + (E'A - C'B)}{2} \qquad (5.14b)$$

Normally, when using total station systems and in stadia operations (see Chapter 7) $AE = BC$ and $E'A = C'B$, so that Equations (5.14) become

$$\Delta H_z = \frac{DE \cot z_1 + C'H \cot z_2}{2} \tag{5.15a}$$

$$\Delta H_\alpha = \frac{DE \tan \alpha + C'H \tan \beta}{2} \tag{5.15b}$$

Uses

Indirect trigonometric leveling furnishes a rapid means of determining elevations of points in rolling or rough terrain. Using a total station system with vertical circle accuracy of from 1 to 3 seconds of arc and slope distances measured to mm by EDM, accuracies in elevations nearly comparable to those obtained by differential leveling are possible. This can be used to establish vertical control for photogrammetric, topographic, and construction surveys and also in the determination of structural deformations at points in a building, bridge, dam, and so forth, virtually inaccessible to differential leveling. In this last case, extreme care must be exercised in the observational and reduction procedures (see Teskey, 1992). On less precise surveys, angles are measured by theodolite and distances by stadia. For reconnaissance surveys, angles may be observed with a clinometer and distances determined by pacing.

Procedure

On lines of indirect levels where zenith angles and slope distances are measured with a total station system, the usual procedure is as follows. Let A and D be two points whose difference in elevation is desired. With the instrument at some intermediate location, T_1, not necessarily on line between A and D, the slope distance and vertical circle reading to A are determined by a *backsight* and similar quantities are measured by taking a *foresight* to another intermediate point B at a convenient location not necessarily on the line between A and D. The system then is moved ahead to T_2 and similar observations are taken to B and C. And so the process is repeated until the end of the line is reached at D. All heights of instrument and target must be observed, curvature and refraction calculated, and differences in elevation computed using Equation (5.11a or 5.12a). To increase the precision of the results, occupy each station (e.g., $A, T_1, B, T_2, C, \ldots, D$), measuring slope distances and reciprocal zenith angles to the nearest second with the telescope direct and reversed (see Section 5.55). Average angles and slope distances then are used in Equation (5.13) to compute differences in elevation between each pair of occupied points. Ideally, two total station systems should be used for this last method, allowing simultaneous observation of reciprocal zenith angles, thus more nearly ensuring that curvature and refraction indeed will cancel.

When a theodolite and level rod are used in the stadia method (Chapter 7), the procedure as initially outlined is followed and either the zenith or vertical angles are measured. If the instrument is equidistant from the points on either side of it to which sights are taken, the effect of curvature and refraction will be eliminated and the appropriate Equation (5.10) is used to reduce the data.

In practice when using stadia, little attention is paid to equalizing backsight and foresight distances, because the effect of curvature and refraction is negligible (6 mm in 300 m and 0.02 ft in 100 ft) compared to the precision with which elevation can be determined by this method (generally, 5 cm to 1 dm or several tenths of a foot).

173

CHAPTER 5:
Vertical
Distance
Measurement:
Leveling

EXAMPLE 5.1. Referring to Figure 5.3, the slope distance s_1 from E to C is 332.791 m with an average zenith angle of $80°13'31''$ from a height of instrument at E of 1.558 m above A to a target height C of 1.372 m above B. If the elevation at A is 21.935 m above the datum, calculate the difference in elevation from A to B and the elevation of B.

Solution. Use Equation (5.11a), where the correction for curvature and refraction is calculated with Equation (5.7):

$$\Delta H_{AB} = 1.558 + (0.067)(0.332791)^2 + (332.791)(\cos 80°13'31'') - 1.372$$

$$= 56.693 \text{ m}$$

elevation of $B = 21.935 + 56.693 = 78.628$ m

EXAMPLE 5.2. Referring to Figure 5.4, the slope distance s_1 from E to C is 404.163 m with a vertical angle of $1°48'26''$ from an instrument height at E of 1.558 m above A to a target height at C of 1.521 m above B. A reciprocal vertical angle of $-1°48'38''$ is observed from an instrument height of 1.560 m at C' above B to a target height of 1.587 m at E' above A. If the observed slope distance s_2 from C' to E' is 404.161 m and the elevation of A is 29.935 m above the datum, calculate the difference in elevation from A to B and the elevation of B.

Solution. Use Equation (5.13b) to compute ΔH_{AB}:

$$\Delta H_{AB} = \tfrac{1}{2}[(404.163)(\sin 1°48'26'') + (404.161)(\sin 1°48'38'')]$$
$$+ \tfrac{1}{2}[(1.558 - 1.521) + (1.587 - 1.560)]$$

$$= 12.758 + 0.032$$

$$= 12.790 \text{ m}$$

elevation of $B = 29.935 + 12.790 = 42.725$ m above the datum.

EXAMPLE 5.3. The slope distance s_1 from station 49 to station 24 is 122.265 m with a zenith angle of $92°21'40''$ from an instrument height at 49 of 1.642 m above point 49 to a combined reflector and target height of 1.573 m above the station at 24. From station 24, a reciprocal zenith angle of $87°46'00''$ and slope distance of 122.254 m were measured from respective h.i.'s of 1.712 m at 24 to a reflector and target 1.503 above 24. If the elevation at 49 is 89.733 m above the datum, calculate the difference in elevation from station 49 to 24 and the elevation of station 24.

Solution. Use Equation (5.13a) to compute ΔH_{49-24}:

$$\Delta H_{49-24} = \tfrac{1}{2}[(122.265)(\cos 92°21'40'') - (122.254)(\cos 87°46'00'')]$$
$$+ \tfrac{1}{2}[(1.642 - 1.573) + (1.712 - 1.503)]$$

$$= -4.9006 + 0.1390$$

$$= -4.762 \text{ m}$$

elevation of $24 = 89.733 - 4.762 = 84.971$ m above the datum.

5.6
BAROMETRIC LEVELING

Because the pressure of the earth's atmosphere varies inversely with the elevation, the barometer may be employed for observations of difference in elevation. Barometric leveling is used principally on exploratory or reconnaissance surveys where differences in elevation are large, as in hilly or mountainous country. Atmospheric pressure may vary over a considerable range in the course of a day or even an hour, so elevations determined by one ordinary barometer carried from one elevation to another may be several feet in error. However, by means of sensitive barometers and special techniques, elevations can be determined within a foot or about 3 dm.

Usually, barometric observations are taken at a fixed station during the same period that observations are made on a second barometer, which is carried from point to point in the field. The procedure makes it possible to correct the readings of the portable barometer for atmospheric disturbances.

Instruments and Methods

The mercurial barometer is accurate, but it is cumbersome and suitable only for observations at a fixed station. For field use, an aneroid barometer commonly is used because it is light and easily transported. The usual type has a dial about 3 in. in diameter, graduated both in inches of mercury and in feet of altitude (elevation); it is compensated for temperature. At a point of known altitude, the pointer can be set at the corresponding reading on the scale to index readings. The aneroid barometer can be calibrated against a mercurial barometer by comparing values at a given station over a range of temperatures.

In use, the barometer should be given time to reach the temperature of the air before an observation is made. In most instruments today, this time is much less than 1 min.

A single aneroid barometer sometimes is used by topographers on small-scale surveys where the contour interval is large. Stops are made at frequent intervals during the day at a single point of assumed elevation or at various points of known elevation, instrument readings are made, and the rate of change in atmospheric conditions is calculated; suitable corrections thus are determined and applied to the observed values. Where distances permit, it is preferable to return to the starting point and correct the intermediate readings in proportion to the change in atmospheric pressure during the interval between observations.

Sensitive Barometers

Extremely sensitive barometers known as *altimeters,* have been developed, with which elevations can be determined within a foot or so. In one procedure used in topographic surveying, two of the instruments are employed at fixed bases and one or more instruments are carried from point to point over the area being surveyed. One fixed instrument is located at a point of known elevation near the highest elevation of the area, and one near the lowest elevation; these instrument stations are called the *upper base* and *lower base,* respectively. Other altimeters are carried to points whose elevations are desired, and readings are taken. Readings on the fixed instruments are taken either simultaneously (as determined by signaling) or at fixed intervals of time; in the latter case the readings at the desired instant are determined by proportion. The elevation of a portable instrument is then determined by interpolation. The horizontal location of each point at which a reading is taken is determined by conventional methods.

> **EXAMPLE 5.4.** Given the elevation of an upper base of 275 ft, of a lower base of 56 ft; the difference in elevation between the bases, therefore, is $275 - 56 = 219$ ft. At a given instant, the three altimeter readings indicate that the difference in elevation of an intermediate point from the upper base is 209 ft and from the lower base is 25 ft; therefore, the indicated total difference in elevation between bases is 234 ft. The corrected differences in elevation are proportionately $(219/234)(209) = 196$ ft (from the upper base) and $(219/234)(25) = 23$ ft (from the lower base); as a check, the total computed difference in elevation between bases is now $196 + 23 = 219$ ft. The

elevation of the point is 79 ft, computed by difference from either base (275 − 196 = 79; or 56 + 23 = 79).

175

CHAPTER 5:
Vertical
Distance
Measurement:
Leveling

Another procedure is to employ one barometer at a fixed base and one barometer that is carried to points whose elevations are desired, simultaneous readings being taken. The carried barometer finally is brought either back to the starting point or to another point of known elevation; the computation of the elevation of each point then takes account of the corresponding change in atmospheric pressure at the fixed base.

Next, we discuss instruments for direct leveling.

5.7
KINDS OF LEVELS

The types of levels used for direct leveling are (1) the traditional *dumpy level,* (2) the *self-leveling* or *automatic level,* (3) the *tilting level,* and (4) the *electronic laser* or *infrared level.* Dumpy and automatic levels are general purpose instruments, commonly referred to as *engineer's levels;* tilting levels are used for precise leveling (see Section 5.50); and electronic laser levels are employed primarily for construction surveys (Chapter 17).

The fundamental characteristics of levels are best described by referring to the engineer's level, which has as its essential features a *line of sight* and *spirit level tube* (as on the dumpy level) or some other means of making the line of sight horizontal such as an *optical compensator* (as on the self-leveling level). Figure 5.5a is a diagram of the principal parts of an engineer's level equipped with a spirit level. The level consists of telescope *A* mounted on the level bar *B* that is rigidly fastened to the spindle *C*. Attached to the telescope or the level bar and parallel to the telescope is the level tube *D*. The spindle fits into the cone-shaped bearing of the leveling head *E*, so that the level is free to revolve about the spindle *C* as an axis. The leveling head is attached to a tripod *F*. In the tube of the telescope are cross hairs at *G*, which appear on the image viewed through the telescope, as illustrated in Figure 5.5b. The bubble of the level is centered by means of leveling screws *H*. The *architect's* or *construction level,* a modified form of the dumpy level but with a telescope of lower magnifying power and a less sensitive spirit level, sometimes is used in establishing grades for buildings. The *hand level* (Sections 5.21 and 5.22) is a simple handheld device for roughly determining differences in elevation. An instrument frequently used for direct leveling but not primarily designed for this purpose is the *engineer's theodolite* (Chapter 6).

The measurements of difference in elevation are made by sighting on a graduated rod, called a *leveling rod* (Sections 5.23–5.25). Other accessories sometimes used are a *rod level* (Section 5.29) and *turning points* (Section 5.30).

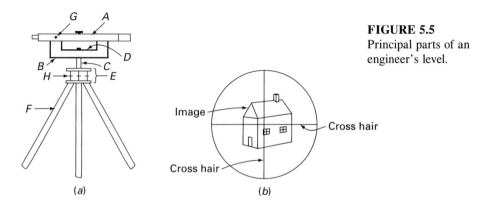

FIGURE 5.5
Principal parts of an engineer's level.

The leveling head of the engineer's level usually is equipped with three leveling screws, but the four-screw type sometimes is found on the older, traditional-style dumpy levels. The three-screw type can be leveled rapidly, requires the use of only one hand, and is relatively stable compared with the four-screw type when the latter type is not perfectly set. A disadvantage of a three-screw leveling head is that, when the level screws are turned, the elevation of the telescope may change slightly because the telescope is not supported by a center bearing as on a four-screw leveling head. Therefore, one should avoid adjusting the level screws in the midst of a setup when precise leveling is being performed with a three-screw instrument.

5.8
TELESCOPE

Modern levels are equipped with internally focusing telescopes that have features as illustrated in Figure 5.6a. Rays of light emanating from an object point are caught by the objective lens A brought to a focus, and form an image in the plane of cross hairs B. The lenses of the eyepiece C form a microscope that is focused on the image at the cross hairs. The objective lens is fixed in the end of the telescope tube and a negative lens L is attached to a slide D within the telescope tube. This negative lens can be moved parallel to the line of sight by rotating the focusing knob F and allows focusing the telescope on objects at different distances from the instrument. The eyepiece C is held in position by threaded sleeve G, which may be moved in a longitudinal direction for focusing. By means of adjusting screws, the cross-hair reticule may be moved transversely so that the intersection of the cross hairs will appear in the center of the field of view.

The line of sight is defined by the intersection of the cross hairs and the optical center of the objective lens. The instrument is so constructed that the optical axis of the objective lens coincides (or practically coincides) with the axis of the negative or objective slide (D, Figure 5.6a); in other words, a given ray of light passing through the optical center of

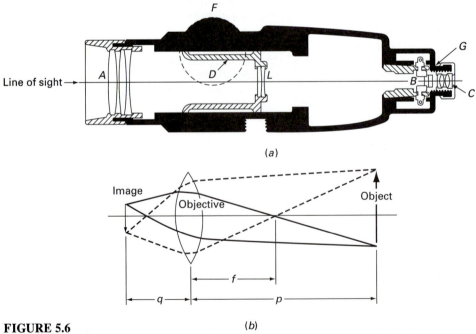

FIGURE 5.6
(a) Longitudinal section of Internal-focusing telescope. (*Courtesy the Sokkia Corporation*).
(b) How an image is focused.

the objective always occupies the same position in the telescope tube regardless of the longitudinal position of the negative lens. The cross hairs can be adjusted so that the line of sight and the optical axis coincide.

177

CHAPTER 5:
Vertical
Distance
Measurement:
Leveling

All modern surveying instruments are manufactured with internally focusing telescopes equipped with coated lenses to reduce light reflections from each glass surface and thus provide better illumination for the images.

5.9
OBJECTIVE LENS

The principal function of the telescope objective lens is to form an image for sighting purposes. For the accuracy of measurements, the objective should produce an image that is well lighted, accurate in form, distinct in outline, and free of discoloration. A single biconvex lens meets the first two of these requirements but is faulty in regard to the other two, for the following reasons:

1. Rays entering the lens near its edge come to a focus nearer the objective than do those entering near its center. The image does not lie in a plane, but on the surface of a sphere. Hence, as viewed through the telescope, portions of the object are blurred. This defect is called *spherical aberration.*
2. Rays of the various colors of the spectrum are deviated by different amounts as they pass through the lens; hence, the field of view appears discolored by lights of various hues. This is called *chromatic aberration.*

These two objectionable features of the single lens are nearly eliminated in most surveying instruments by providing an outer double-convex lens of crown glass and an inner concavo-convex lens of flint glass. The two lenses usually are cemented together with balsam but are sometimes separated by a thin spacer ring.

The *optical center* of the objective is that point in the lens through which a ray of light will pass without permanent deviation, regardless of the direction of the object from which the light emanates. In other words, the direction of the ray is the same after leaving the lens as before entering it. In a biconvex lens with faces of equal curvature, the optical center and geometrical center coincide.

The *optical axis* is the line taken by a light ray that experiences no deviation on either entering or leaving the objective. It passes through the optical center and the centers of curvature of the lens.

The *principal focus* is a point on the optical axis back of the objective where rays entering the telescope parallel with the optical axis are brought to a focus; or it is a point in front of the objective from which diverging light rays entering the lens emerge from it parallel with the optical axis. Stated in another form, the image of a point on the optical axis and an infinite distance away is at the principal focus back of the objective. If a point is at the principal focus in front of the objective, however, it will have no image.

The *focal length* of the objective is the distance from its optical center to the principal focus. When the telescope is focused on a distant point, the focal length is very nearly the distance from the optical center of the objective to the plane of the cross hairs, for reasons that the preceding paragraph makes clear.

The basic principle (lens equation), which relates object distance p, image distance q, and focal length f (for externally focusing telescopes), is

$$\frac{1}{p} + \frac{1}{q} = \frac{1}{f} \tag{5.16}$$

in which f is a constant for the telescope. Figure 5.6b illustrates the manner in which rays from an object are deviated by the objective and brought to a focus to form the image. Note that the image is inverted.

5.10
FOCUSING

When the telescope is to be used, the eyepiece is first moved in or out until the cross hairs appear sharp and distinct. This adjustment of the eyepiece should be tested frequently, as the observer's eye becomes tired.

When an object is sighted, the negative lens or objective slide is moved in or out until the image appears clear. At this point, the image should be in the plane of the cross hairs. If a slight movement of the eye from side to side or up or down produces an apparent movement of the cross hairs over the image, the plane of the image and the plane of the cross hairs do not coincide, and *parallax* is said to exist. Parallax is a source of error in observations and should be eliminated by refocusing the objective, the eyepiece, or both until further trial shows no apparent movement. Figure 5.7 provides a graphic illustration of the concept of parallax. The negative or objective lens must be focused for each distance sighted.

The telescope cannot be focused on objects closer than about 2 m or 6 ft from the center of the instrument, unless special short-focus lenses are employed.

Most telescopes are equipped with peep sights or simple sighting collimators on top or alongside the telescope. The observer should point the telescope approximately at the object by first sighting along the peep sights at the target. Then, looking through the telescope, the observer should focus the objective lens so that there is a clear, distinct image of the target at which the telescope is pointed. A subsequent movement of the pointing of the telescope by a degree or two one way or another should point the line of collimation (cross hairs) precisely on the target.

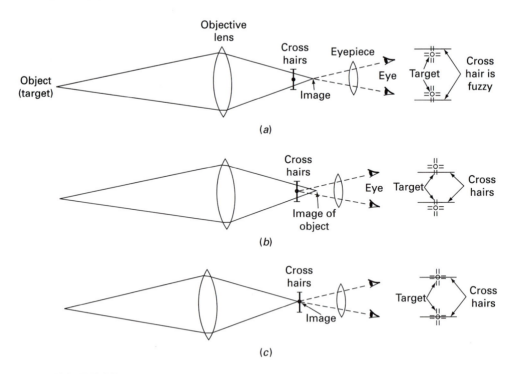

FIGURE 5.7
Presence of parallax. (a) Objective not focused, eyepiece not focused, parallax is present. (b) Objective not focused, eyepiece focused on cross hairs, parallax present. (c) Objective focused, eyepiece focused, no parallax.

179

CHAPTER 5:
Vertical
Distance
Measurement:
Leveling

The strain on the eyes will be reduced if the observer tries keeping both eyes open while sighting.

Any lateral movement of the negative lens in an internally focusing telescope or the objective at the externally focusing telescope will cause a deviation in the line of sight, introducing errors in measurements. The quality of work on any good instrument is sufficiently precise to ensure practical elimination of errors of this sort when the telescope is new, but in the course of long use, wear develops between the sliding parts and the mechanism becomes loose. This condition produces uncertainties in observation that no amount of adjustment can overcome. In this respect, the internally focusing telescope, in which the focusing lens and mechanism are sealed within the telescope tube protected from dust, water, and other foreign matter, is definitely superior. Errors caused by changing focus rarely are significant for ordinary leveling but can be of consequence in precise leveling (Section 5.50).

5.11
CROSS HAIRS

The cross hairs, which define the line of sight, are supported in the *cross-hair ring* or *reticule,* which consists of a glass plate on which are etched fine vertical and horizontal lines that serve as cross hairs. Special patterns of additional lines are used on some instruments; for example, stadia hairs, or double horizontal and vertical lines closely spaced for precise sighting between them. The spacing between a double line can be pointed more precisely than a single line.

As shown in Figure 5.8, the cross-hair ring is held in position by four capstan-headed screws that pass through the telescope tube and tap into the ring. The holes in the telescope

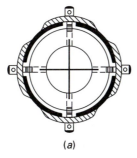

(a)

FIGURE 5.8
Cross-hair reticule.

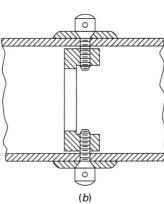

(b)

tube are slotted so that, when the screws are loosened, the ring may be rotated through a small angle about its own axis. To rotate the ring without disturbing its centering, two adjacent screws are loosened; and the same two screws are tightened after the ring has been rotated. The ring is smaller than the inside of the tube, and it may be moved either horizontally or vertically by means of the screws. Thus, to move it to the left, the right-hand screw is loosened and the left-hand screw is tightened. If the movement is to be large, first the top or bottom screw is loosened slightly; and after the movement to the left has been completed, the same (top or bottom) screw is tightened again.

5.12
EYEPIECE

Attention has previously been drawn to the fact that the image formed by the objective is inverted. Eyepieces are of two general types.

The *erecting* or *terrestrial* eyepiece, the more common of the two types, reinverts the image so that the object appears to the eye in its normal position. Usually, it consists of four plano-convex lenses placed in a metal tube called the *eyepiece slide* (Figure 5.9a). In the figure, A represents the object, B is the inverted image in the plane of the cross hairs, C is the image magnified by the lens nearest the eye, and D is the magnified image as it appears to the eye.

The *inverting* or *astronomical* eyepiece simply magnifies the image without reinverting it. It is composed of two plano-convex lenses generally arranged as shown in Figure 5.9b. The arrangement is seen to be identical with that of the two lenses farthest apart in the erecting eyepiece. The magnified image D is inverted, and the object, as viewed through the telescope, is upside down.

For either eyepiece, the ratio of the angle at the eye subtended by the magnified image to that subtended by the object itself is the magnifying power of the telescope. If, in either Figure 5.9a or 5.9b, D is the apparent length of the magnified image and E is the apparent length of the object as seen by the naked eye, the ratio of D to E is the magnifying power.

Each lens interposed between the object and the eye absorbs some of the light that strikes it and each lens surface reflects some of the entering and departing light. Hence, other things being equal, the object is more brilliantly illuminated when viewed through the inverting eyepiece; this is a great advantage, particularly when observations are made during cloudy days or near nightfall. Another important advantage of the inverting eyepiece is that the telescope is shorter and the instrument lighter in weight. The beginner experiences some inconvenience on viewing things apparently upside down, but this difficulty is overcome with a little practice. The single advantage of the erecting eyepiece is that objects appear in their natural position, and this is why its use is so common.

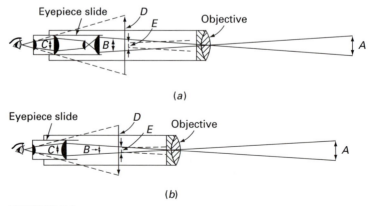

FIGURE 5.9
Eyepieces: (a) erecting, (b) inverting.

5.13
PROPERTIES OF THE TELESCOPE

181

CHAPTER 5:
Vertical
Distance
Measurement:
Leveling

The *illumination* of the image depends on the effective size of the objective, the quality and number of lenses, and the magnifying power. Other conditions being the same, either the larger the objective or the smaller the magnifying power, the better is the illumination; that is, the better lighted appears the object.

Distortion of the field of view so that it does not appear flat is caused mainly by what is termed the *spherical aberration* of the eyepiece. Although this introduces no appreciable error in ordinary measurements, it is not desirable when two points in the field are to be observed at the same time, as in stadia measurements.

The *definition* of a telescope is its power to produce a sharp image, which depends on the quality of the glass, the accuracy with which the lenses are ground and polished, and the precision with which they are spaced and centered. Light rays passing through the lenses near their edges are particularly troublesome. To improve the definition, these rays are intercepted by diaphragms or screens placed between the lenses of the eyepiece and in the rear of the objective. These screens decrease the illumination somewhat.

The angular width of the field of view is the angle subtended by the arc, whose center is nearly at the eye and whose length is the distance between opposite points of the field viewed through the telescope. For a particular instrument, this angle may be readily determined by observation. It is independent of the size of the objective. In general, the larger the telescope and the greater the magnifying power, the less is the angle of the field of view. For most surveying of moderate precision, the work is greatly retarded if the instrument lacks a fairly large field of view, and this is one reason why the telescopes are not usually made of high magnifying power. Generally, the angular width of the field ranges from about 1°30′ for a magnifying power of 20 to 45′ for a magnifying power of 40.

The magnifying power of the telescope for the better grade of engineer's level is about 30 diameters. A precise level has a magnification of about 40 diameters.

5.14
SPIRIT LEVEL TUBE

The level tube used in surveying instruments (Figure 5.10) is a glass vial with the inside ground barrel-shaped, so that a longitudinal line on its inner surface is the arc of a circle (Figure 5.11). The tube is nearly filled with sulfuric ether or alcohol. The remaining space is occupied by a bubble of air that takes up a location at the high point in the tube. The tube usually is graduated in both directions from the middle; thus, by observing the ends of the bubble it may be "centered," or its center brought to the midpoint of the tube.

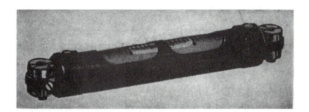

FIGURE 5.10
Level tube.

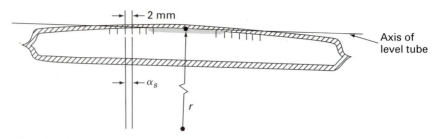

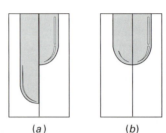

FIGURE 5.11
Cross section of a spirit level tube

FIGURE 5.12
Coincidence bubble: (a) bubble not centered, (b) bubble centered. (*Courtesy of Leica, Inc.*)

(*a*)   (*b*)

A longitudinal line tangent to the curved inside surface at its upper midpoint is called the *axis of the level tube* or *axis of the level*. When the bubble is centered, the axis of the level tube is horizontal.

The tube is set in a protective metal housing, usually with plaster of paris. The housing is attached to the instrument by means of screws that permit vertical adjustment at one end and lateral movement at the other end of the tube, as shown in Figure 5.10.

Some leveling instruments are equipped with a prismatic viewing device by means of which one end of the bubble appears reversed in direction and alongside the other end. Figure 5.12a illustrates such a *coincidence* bubble when it is not centered. The bubble is centered by matching its ends as shown in Figure 5.12b. Settings can be made more accurately by such coincidence than by matching the ends of the bubble with respect to the graduations on the level tube. Tilting levels (Section 5.18) usually are equipped with coincidence viewing of the spirit level tube.

5.15
SENSITIVITY OF THE LEVEL TUBE

If the radius of the circle to which the inner chamber level tube is ground is large, a small vertical movement of one end of the tube will cause a large displacement of the bubble; if the radius is small, the displacement will be small. Thus, the radius of the tube is a measure of its sensitivity. The sensitivity generally is expressed in seconds of the central angle, whose arc is one division of the tube. For most instruments, the length of a division is 2 mm. The sensitivity expressed in seconds of arc is not a definite measure unless the spacing of graduations is known.

183

CHAPTER 5:
Vertical
Distance
Measurement:
Leveling

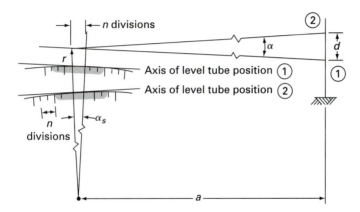

FIGURE 5.13
Determination of level tube sensitivity.

Sensitivities of bubble tubes, expressed in seconds per 2-mm division, vary from 1 to 2" for precise levels up to 10 to 30" for engineer's levels.

Should determination of bubble tube sensitivity be necessary, proceed as follows: (1) align the bubble tube axis with a pair of diagonally opposite level screws; (2) hold a rod in a vertical position at a measured distance from the level; (3) observe the rod reading; (4) tilt the telescope by manipulating the level screws, moving the bubble tube through n divisions; and (5) observe the rod reading. Repeat steps (3) and (4) at least five times and calculate the average increment on the rod subtended by the angle caused by movement of the bubble through n divisions. From Figure 5.13, the sensitivity α_s in seconds per division is

$$\alpha_s = \frac{d}{(a)(n)} \rho \tag{5.17}$$

in which d is the average increment on the rod, a is the distance from level to rod in compatible units, n is the number of divisions the bubble tube moved, and $\rho = 206{,}264.8''/\text{rad}$.

5.16
RELATION BETWEEN MAGNIFYING POWER AND SENSITIVITY

It is desirable that the sensitivity of the level tube be such that, for the smallest noticeable movement of the bubble, there is an apparent movement of the cross hairs on a level rod held at an average sighting distance from the instrument; and likewise, for the smallest notice-able movement of the cross hairs, there should be an observable movement of the bubble. The least noticeable movement of the cross hairs depends to some extent on the definition and illumination of the image but principally on the magnification of the telescope.

If the level tube is more sensitive than is necessary, time is wasted in centering the bubble. If the magnifying power is higher than it need be, unnecessary labor is expended due to the more limited field of view and the increased difficulties of focusing the objective properly. A satisfactory test may be conducted by one person sighting at a rod while a second person bears down slightly on one end of the telescope and at the same time observes the level tube. If the first noticeable movement of the bubble is accompanied by an apparent movement of the cross hairs, there is a satisfactory balance between sensitivity and magnification. If the cross hairs move first, a level tube of greater radius might be properly employed.

5.17
DUMPY LEVELS

The conventional dumpy level provides an excellent introduction to levels in general, so although this type level largely has been superseded by self-leveling instruments (Section 5.19), the dumpy level shown in Figure 5.14 is used here to illustrate the basic construction and properties of engineer's levels. The telescope A is rigidly attached to the level bar B, and the instrument is constructed so that the optical axis of the telescope is perpendicular to the axis of the center spindle. The level tube, C, is permanently placed so that its axis lies in the same vertical plane as the optical axis, but it is adjustable in altitude by means of a capstan-headed screw at one end. The spindle revolves in the socket of the leveling head, D, which is controlled in position by the four leveling screws, E. At the lower end of the spindle is a ball-and-socket joint that makes a flexible connection between the instrument proper and the foot plate, F. When the leveling screws are turned, the level is moved about this joint as a center. The sunshade, G, protects the objective from the direct rays of the sun. The adjusting screws, H, for the cross-hair ring are near the eyepiece end of the telescope.

The telescope of the dumpy level shown has a magnifying power of 36 diameters, and the level tube has a sensitivity of $20''$ of arc per graduation (2 mm).

The current primary advantage of the dumpy-style level is that its weight lends stability under conditions where vibration or wind render lighter automatic levels less stable.

5.18
TILTING LEVELS

The distinctive feature of a tilting level (Figure 5.15) is that the telescope is mounted on a transverse fulcrum, A, near the vertical axis and toward the objective end of the telescope, and on micrometer screw, B (not visible), at the eyepiece end of the telescope. After the instrument has been approximately leveled by use of the level screws and centering the circular spirit level bubble, the telescope is pointed in the desired direction and then "tilted," or rotated slightly in the vertical plane of its axis by turning the micrometer screw knob C, until the sensitive telescope level bubble, enclosed under housing D, is centered. The line of sight then is horizontal, although the instrument as a whole is not exactly level.

The telescope bubble is centered by bringing the images of the ends of the bubble into coincidence as shown in Figure 5.12. These coincidence bubbles are viewed through a system of prisms from the eyepiece end of the telescope E, either through a separate parallel eyepiece (F in Figure 5.15) or off to one side of the field of view of the main telescope. Figure 5.16 illustrates an optical train for viewing the telescope bubble where the images of the ends of the bubble are projected into the field of view of the main telescope. The instrument shown in Figure 5.15 has a level bubble with a sensitivity of $10''$ per 2-mm division and the images of the bubble ends are viewed through the eyepiece, E.

The coincidence bubble permits more accurate centering of the spirit level bubble than is possible by observing the bubble directly. Consequently, tilting levels, of geodetic quality, equipped with coincidence bubbles are used for precise leveling (see Sections 5.47 and 5.50). On older models of engineer's and geodetic tilting levels, the tilting knob is

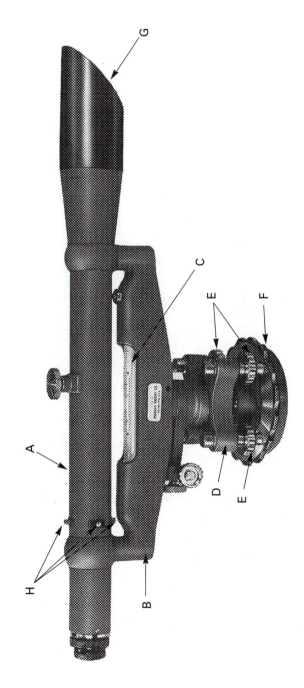

FIGURE 5.14
Engineer's dumpy level. (*Courtesy of Warren Knight Instrument Company.*)

185

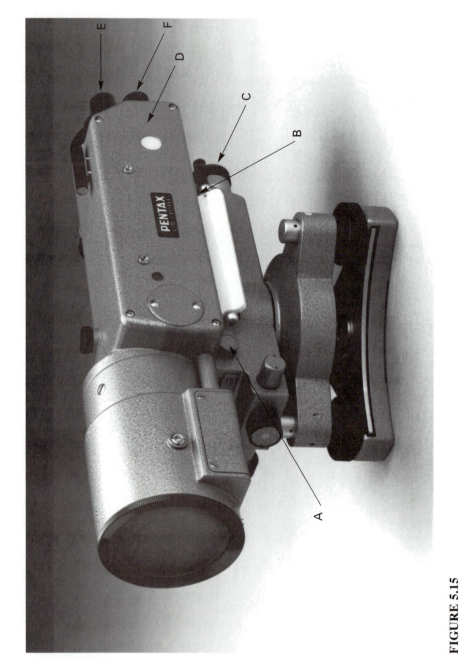

FIGURE 5.15
Tilting level. (*Courtesy of Pentax, ASAHI Precision Co., Ltd.*)

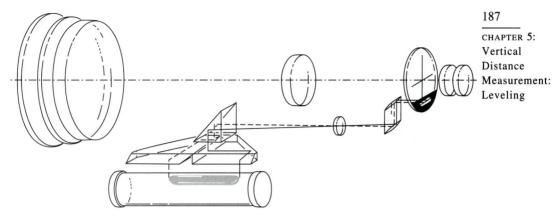

187

CHAPTER 5:
Vertical
Distance
Measurement:
Leveling

FIGURE 5.16
Optical train for viewing coincidence bubble. (*Courtesy of Sokkia Corporation.*)

fitted with a graduated drum so that the slope angle of the line of sight can be measured. This device is called a *gradienter* and is especially useful in reciprocal leveling (see Section 5.54).

5.19
SELF-LEVELING LEVELS

A self-leveling level allows establishing a horizontal line of sight by means of a compensator, consisting of a system of prisms and mirrors supported by wires as in a pendulum. Annotated Figure 5.17 shows one form of a self-leveling level and Figure 5.18 illustrates a typical cross-sectional view of the optics involved. Light enters at the objective lens at A, passes through focusing lens, B, is reflected by the optical compensator, C, which is suspended by wires at D and clamped magnetically at E. When the compensator C is freely supported, the line of sight defined by the objective lens, compensator, and eyepiece is automatically horizontal. Most automatic levels are designed so the line of sight is horizontal when the telescope barrel is within $\pm 10'$ of being horizontal. At each setup, the instrument is leveled by use of a circular spirit or bull's-eye level (Figure 5.17) and the compensator maintains a horizontal line of sight. The instrument is light, easy to handle, and its operation is quick and accurate. Note that random error in centering the bubble is absent in this instrument.

5.20
DIGITAL SELF-LEVELING LEVELS

The attributes of a self-leveling level combined with a solid-state camera and electronic image processing have resulted in a *digital, self-leveling level* that comes very close to being a truly automatic level.

Figure 5.19 shows the Leica NA2000 and its associated level rod with bar-code graduations. This compensator-equipped, self-leveling instrument can be used in the conventional leveling mode if desired and is furnished with the usual eyepiece, A; focusing knob, B; circular level bubble, C; tangent motion, D; three level screws, E; and objective lens, F. In addition, this level has a built-in solid-state camera, a microcomputer, a storage module, G, a control panel, H, and display register, K.

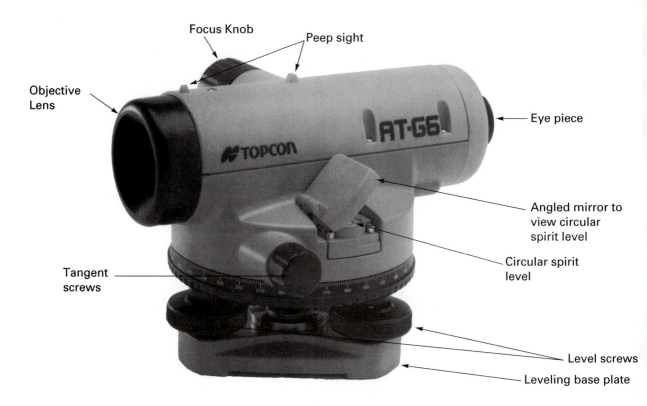

FIGURE 5.17
Topcon AT-G6 self-leveling level. (*Courtesy of Topcon America Corporation.*)

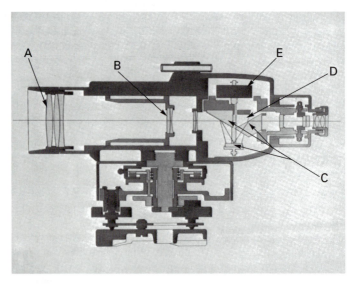

FIGURE 5.18
Sectional view of a Lietz/Sokkisha B-1 self-leveling level.
(*Courtesy of Sokkia Corporation.*)

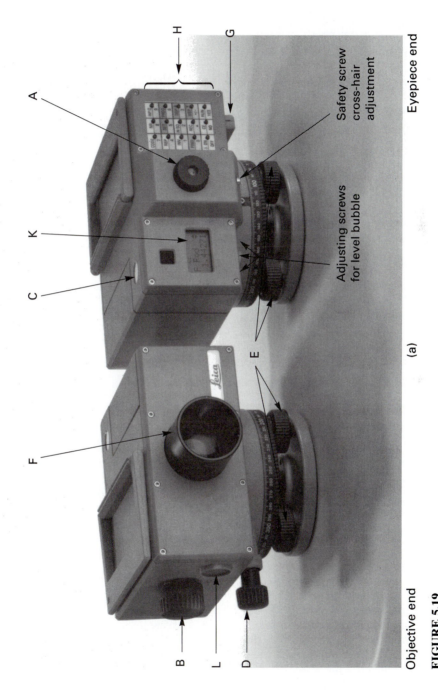

Objective end

(a)

Eyepiece end

FIGURE 5.19

(a) Electronic digital level NA2000. (*Courtesy of Leica, Inc.*)

189

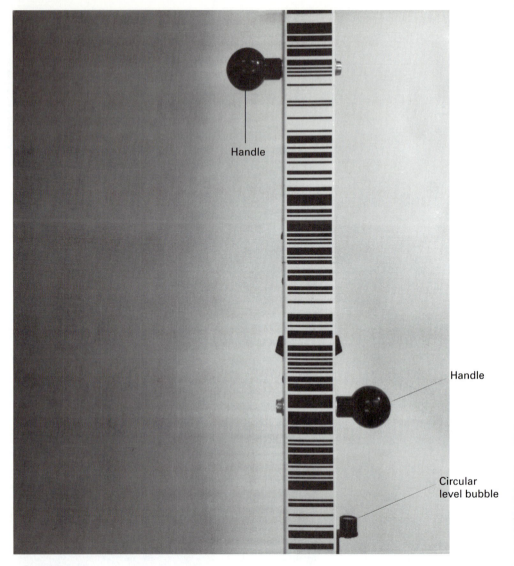

FIGURE 5.19
(b) Level rod with bar-coded graduations.

To operate the NA2000, the level is set up, leveled, pointed toward, and focused on the rod with bar-code graduations. When the on-off key on the control panel is pressed, instructions directing the next operation or asking for information appear on the display register. For example, when the level is in the "measuring without recording" mode, the display in the register will be "meas. only" followed by "Rod." At this time, the operator would press the measure button (K in Figure 5.19a) and in a few seconds the rod reading would be shown in the display register. In the interim, the solid-state camera will have recorded a digital image of the rod with the middle and two stadia cross hairs superimposed over the bar-coded graduations. This image then was compared to a master image of the entire rod, stored in memory of the on-board microcomputer, and the rod reading was determined to five decimal places by image correlation and averaging the readings of the

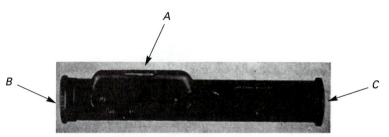

FIGURE 5.20
Hand level.

191

CHAPTER 5:
Vertical
Distance
Measurement:
Leveling

three cross hairs. This rod reading was displayed on register K (e.g., 1.4271; Figure 5.19a) and could be stored in the memory of the storage module, if desired. In addition, by pressing the proper key on the control panel, the distance from level to rod determined by stadia (see Chapter 7) also would be displayed. When in the "measuring without recording" mode, these readings may be recorded manually in a field notebook as directed in Section 5.38 for differential leveling with a dumpy or a self-leveling instrument.

If desired, the level can be put into the "measuring and recording" mode. Then, by proper manipulation of keys on the control panel, the elevation of the initial bench mark can be entered and the instrument's software will display, calculate, and store rod readings, heights of instrument, heights of turning points, and distances to turning points. At the end of a job, the removable memory module (G in Figure 5.19a) can be interfaced to a PC and the data downloaded to provide a hard-copy listing of the data and adjusted elevations.

5.21
HAND LEVELS

The hand level (Figure 5.20) is widely used for rough leveling. It consists of a metal sighting tube about 6 in. long on which is mounted a level vial, A. On the bottom of the level vial is etched a black graduation placed at the midpoint of the vial, perpendicular to the axis of the tube. In the tube beneath the vial is a prism that reflects the image of the bubble and the black graduation to the eyepiece end, C, of the level. The reflected image of the black graduation serves as the middle cross hair of the hand level. The eyepiece, C, consists of a peephole held in place by a threaded sleeve. The objective, B, consists of a plain piece of glass also held in place by a threaded sleeve. Mounted on the left inner side of the tube is a semicircular convex lens that magnifies the image of the bubble and cross hair as reflected by the prism. The hand level shown in Figure 5.20 has a fixed focus. Magnifying hand levels also are available.

In using the level, the object is viewed directly through one half of the sighting tube and brought into focus if a magnifying level is being used. At the same time and with the same eye, the position of the bubble with respect to the cross hair is observed in the other half of the field of view. The level is held with the bubble vial uppermost and tipped up or down until the cross hair bisects the bubble, when the line of sight is horizontal. After a little practice, one can make observations with more facility by keeping both eyes open. Some observers steady the hand level by holding it against, or fastening it to, a staff.

5.22
ABNEY HAND LEVEL AND CLINOMETER

As its name indicates, this level is suitable both for direct leveling and for measuring the angles of slopes. The instrument shown in Figure 5.21 is graduated in both degrees and

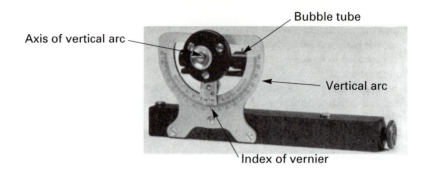

FIGURE 5.21
Abney hand level and clinometer. (*Courtesy of Sokkia Corporation.*)

percentage of slope, or grade. When it is used as a level, the index of the vernier is set at 0, and it is then used in the same way as the hand level in Figure 5.20. When it is used as a clinometer, the object is sighted and the level tube is caused to rotate about the axis of the vertical arc until the cross wire bisects the bubble as viewed through the eyepiece. Either the slope angle or the slope percentage then is read on the vertical arc.

5.23
LEVELING RODS

These are graduated wooden rods of rectangular cross section by means of which difference in elevation is measured. The lower or ground end of the rod is shod with metal to protect it from wear and usually is the point of zero measurement from which the graduations are numbered. Aluminum alloy and fiberglass rods also are available in certain specific styles.

The rod is held vertically; hence, the reading of the rod as indicated by the horizontal cross hair of the level is a measure of the vertical distance between the point on which the rod is held and the line of sight.

Rods are obtainable in a variety of types, patterns, and graduations, and either in single pieces or in sections that are jointed together or slide past each other and are clamped. Common lengths are 12 and 13 ft or 3.65 and 3.7m. In the United States, rods are graduated in hundredths of a foot and in metric units.

The two general classes of leveling rods are (1) *self-reading* rods, which may be read directly by the leveler while looking through the telescope of the level, and (2) *target* rods, for which a target sliding on the rod is set by the rod person as directed by the leveler. Under ordinary conditions, observations with the self-reading rod can be made with nearly the same precision and much more rapidly. The self-reading rod is the one commonly employed, even for precise leveling.

5.24
SELF-READING RODS

The self-reading rod is held vertically; the leveler observes the graduation at which the line of sight intersects the rod and records the reading. Observations closer than the smallest division on the rod are made by estimation.

193

CHAPTER 5:
Vertical
Distance
Measurement:
Leveling

The self-reading rod should be marked so that the graduations appear sharp and distinct for any normal distance between level and rod. Normally, the background is white or yellow with graduations 0.01 ft or 0.010 m wide painted in black as shown in Figure 5.22. The readings to 0.01 ft and to 0.010 m are made on the edges of the graduations. Usually, the rod is read to half-hundredths of a foot or to millimeters, by estimation. The numbers indicating feet or meters are in red, and those indicating tenths of feet or decimeters are in black. This style of graduation is satisfactory for self-reading when the length of sight is less than 400 to 500 ft or 120 to 150 m.

The *Philadelphia* rod is the most widely used rod. It is made in two sliding sections, held in contact by two brass sleeves. A screw attached to the upper sleeve permits clamping the two sections together in any desired relative position. For readings of 7 ft (2 m on the metric rod) or less, the back section is clamped in its normal collapsed position. For greater readings, the rod is extended to its full length so that graduations on the front face of the

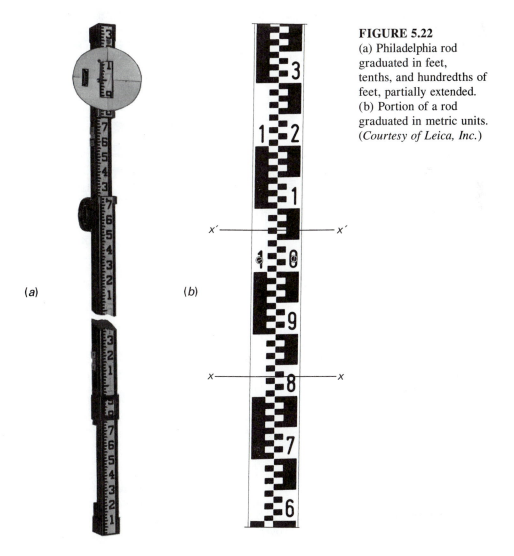

(a)

(b)

FIGURE 5.22
(a) Philadelphia rod graduated in feet, tenths, and hundredths of feet, partially extended.
(b) Portion of a rod graduated in metric units.
(*Courtesy of Leica, Inc.*)

back section are a continuation of those on the lower front strip. When thus extended, the rod is called a *high* rod. Figure 5.22a shows a partially extended Philadelphia rod graduated in feet, tenths, and hundredths of feet. A portion of the face of a European-style rod graduated in metric units is shown in Figure 5.22b. This rod comes in 3- and 4-m lengths, is hinged for folding to simplify carrying, and contains a built-in circular level bubble used to keep the rod vertical.

Direct observation of a self-reading rod involves estimation to obtain the third decimal place in meters and feet. In Figure 5.22b, graduations are in meters numbered on the left, decimeters numbered on the right, and centimeters designated by alternating black and white squares. These squares are arranged in a checkerboard pattern so that the cross hair always will appear on a white square. Again referring to Figure 5.22b, readings at lines *xx*

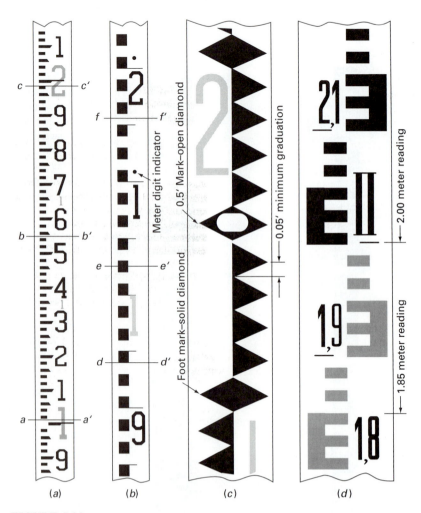

(a)   (b)   (c)   (d)

FIGURE 5.23
Direct reading of (a) rod graduated in feet, tenths, and hundredths of feet; (b) rod graduated in meters, decimeters, and centimeters; (c) stadia rod graduated in feet, tenths, and 0.05 ft; (d) stadia rod graduated in meters, decimeters, and centimeters. (*Courtesy of Sokkia Corporation.*)

and $x'x'$ to the nearest centimeter yields 0.83 m and 1.07 m, respectively. Estimation to the nearest millimeter at these same lines gives readings of 0.831 m and 1.066 m, respectively.

195

CHAPTER 5:
Vertical
Distance
Measurement:
Leveling

Figure 5.23a and b shows the faces of Philadelphia-style rods graduated in feet and meters, respectively. In Figure 5.23a the largest crosshatched (red) numerals indicate the whole numbers of feet, the black numerals are tenths of feet, and the alternating black and white graduations represent hundredths of feet. Thus, the readings at lines aa', bb', and cc' to the nearest hundredth are 1.01, 1.54, and 1.98 ft, respectively. Estimation to the nearest thousandth of a foot yields respective readings of 1.009, 1.545, and 1.980. In Figure 5.23b the light gray numbers are meters, black numerals are decimeters, and alternating black and white graduations are centimeters. Readings to the nearest centimeter at lines dd', ee', and ff' are 0.94, 1.02, and 1.15 m, respectively. If millimeters are desired, they must be estimated yielding respective readings of 0.940, 1.024, and 1.155 m. Note that, in Figure 5.23a, the whole number of feet is indicated by the small light gray (actually red on the rod) number between tenth foot numerals and the whole number of meters in Figure 5.23b is given by the number of dots above each decimeter mark on the metric rod. Consequently, even though the field of view of the telescope may fall within full unit graduations, determination of the full unit is possible.

When using a Philadelphia rod, high rod readings are made with the rod fully extended. These rods have a metal stop against which the lower brass sleeve makes contact when the rod is fully extended. Occasionally, this metal stop becomes loose and out of adjustment, so that the rod may not be extended to the proper length, introducing errors in high rod readings. To guard against the occurrence of this type of error, examine the vernier on the back of the rod (Section 5.26).

5.25
STADIA RODS

This type of rod is used for stadia surveying as described in Chapter 7. Any type of self-reading rod may be used as a stadia rod, but the leveling rod graduated in hundredths of feet or in centimeters as in Figure 5.23a and b is suitable only for sights of less than 400 ft or 120 m. For longer sights, a rod with larger, heavier graduations arranged in a different pattern is necessary. Figure 5.23c and d illustrates two such patterns, graduated in feet, tenths, and 0.05 ft and meters, decimeters, and centimeters, respectively. These graduations are suitable for distances of up to about 700 ft or 210 m. Readings can be estimated to the nearest 0.01 ft and nearest millimeter on these two rods. These rods are equally suitable for use in stadia surveying and leveling.

5.26
VERNIERS

A vernier, or vernier scale, is a short auxiliary scale placed alongside the graduated scale of an instrument, by means of which fractional parts of the least division of the main scale can be measured precisely; the length of one space on the vernier scale differs from that on the main scale by the amount of one fractional part. The precision of the vernier depends on the fact that the eye can determine more closely when two lines coincide than it can estimate the distance between two parallel lines. The scale may be either straight (as on a leveling rod) or curved (as on the horizontal and vertical circles of a transit). The zero of the vernier scale is the index for the main scale.

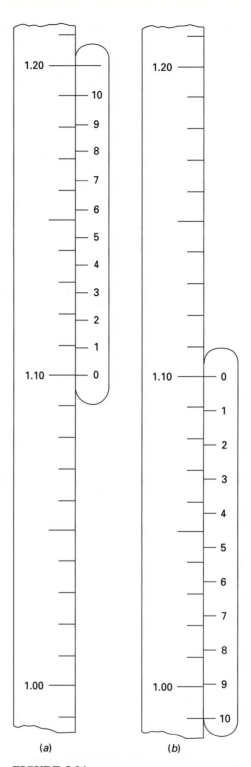

FIGURE 5.24
(a) Direct vernier. (b) Retrograde vernier.

197

CHAPTER 5:
Vertical
Distance
Measurement:
Leveling

Verniers are of two types: (1) the *direct* vernier, which has spaces slightly shorter than those of the main scale; and (2) the *retrograde* vernier, which has spaces slightly longer than those of the main scale. The use of the two types is identical, and they are equally sensitive and equally easy to read. Since they extend in opposite directions, however, one or the other may be preferred because it permits a more advantageous location of the vernier on the instrument. Both types are in common use.

Direct Vernier

Figure 5.24a represents a scale graduated in hundredths of a meter (cm) and a direct vernier having each space 0.001 m shorter than a 0.01-m space on the main scale; thus, each vernier space is equal to 0.009 m, and 10 spaces on the vernier are equal to 9 spaces on the scale. The index, or 0, of the vernier is set at 1.100 m on the scale. If the vernier were moved upward 0.001 m, its graduation numbered 1 would coincide with a graduation (1.11 m) on the scale, and the index would be at 1.111 m, and so on. Thus, the position of the index is determined to thousandths of a meter without estimation, simply by noting which graduation on the vernier coincides with one on the scale. Note that the coinciding graduation on the main scale does *not* indicate the main-scale reading.

The *least count,* or fineness of reading, of the vernier is equal to the difference between a scale space and a vernier space. For a direct vernier, if s is the length of a space on the scale and if n is the number of vernier spaces of total length equal to that of $(n - 1)$ spaces on the scale, then the least count is s/n.

Retrograde Vernier

On the retrograde vernier shown in Figure 5.24b, each space on the vernier is 0.001 m *longer* than a 0.01-m space on the main scale, and 10 spaces on the vernier are equal to 11 spaces on the scale. As before, the index is set at 1.10 m on the scale. If the vernier were moved upward 0.001 m, its graduation numbered 1 would coincide with a graduation (1.09 m) on the scale, the index would be at 1.101, and so on. It is seen that, from the index, the retrograde vernier extends backward along the main scale and that the vernier graduations also are numbered in reverse order; however, the retrograde vernier is read in the same manner as the direct vernier.

The least count of the retrograde vernier is equal to the difference between a scale space and a vernier space, as in the case of the direct vernier. For the retrograde vernier, if s is the length of a space on the scale and if n is the number of vernier spaces of total length equal to that of $(n + 1)$ spaces on the scale, then the least count is s/n.

Reading the Vernier

Figure 5.25 illustrates settings of direct verniers on the target (at the left) and on the back (at right) of a Philadelphia leveling rod (Section 5.24). The rod reading indicated by the target (4.347 ft; Figure 5.25a) is determined by first observing the position of the vernier index on the scale to hundredths of feet (4.34 in the figure), next by observing the number of spaces *on the vernier* from the index to the coinciding graduations (seven spaces in the figure), and finally by adding the vernier reading (0.007 ft in the figure) to the scale reading (4.34 ft). On the back of the rod (Figure 5.25b) both the main scale and the direct

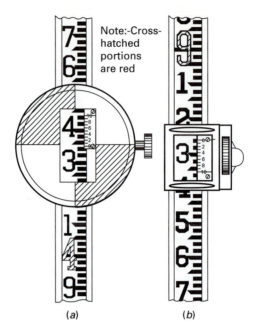

Note:-Cross-
hatched
portions
are red

(a)   (b)

FIGURE 5.25
Direct vernier settings.

vernier read *down* the rod. The scale reading is 9.26 ft and the vernier reading is 0.004 ft; hence, the rod reading is 9.264 ft.

A helpful check in reading the vernier is to note that the lines on either side of the coinciding line should depart from coincidence by the same amount, in opposite directions. As a check against possible mistakes, it is advisable to estimate the fractional part of the main-scale division by reading the index directly.

Figure 5.26 illustrates the setting and reading of an auxiliary scale on the back of a metric rod. (Figure 5.23b) where one reads from top to bottom of the rod (Section 5.38). For this example, the rod reading is 2.15 m, read from the scale on the rod at the 0 index on the auxiliary scale, plus 0.007 m observed on the auxiliary scale, to give a reading of 2.157 m. Note that this is not a vernier as just defined but simply is a scale graduated to millimeters and set so that 0 on the scale is referred to the bottom of the rod. The size of the least graduation (cm) on the metric rod makes this arrangement the most practical for this case.

5.27
TARGETS

The usual target (Figures 5.22a and 5.25a) is a circular or elliptical disk about 5 in. in diameter, with horizontal and verticals lines formed by the junction of alternate quadrants of white and red. A rectangular opening in the front of the target exposes a portion of the rod to view so that readings can be taken. The attached vernier (Section 5.26) fits closely

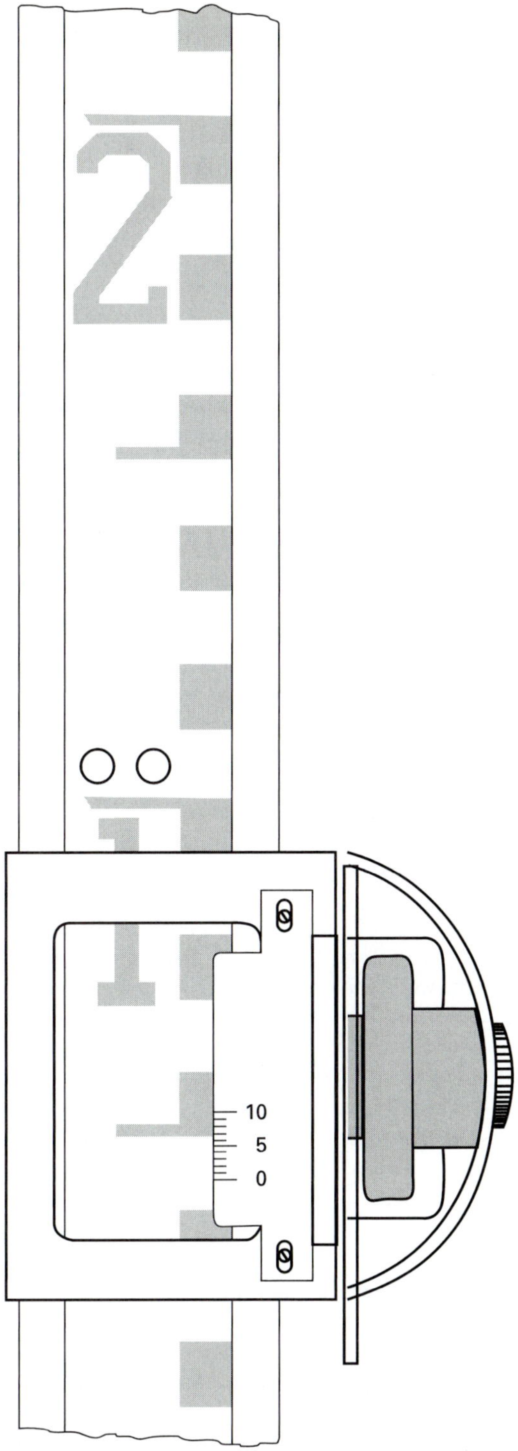

199

CHAPTER 5:
Vertical
Distance
Measurement:
Leveling

FIGURE 5.26
Auxiliary scale on the back of metric rod.

to the rod, its 0 point or index being at the horizontal line of the target. In Figure 5.25a a direct vernier is shown, but both retrograde and direct verniers are in common use.

5.28
TARGET RODS

With the target rod, the leveler signals the rodperson to slide the target up or down until it is bisected by the line of sight. The target then is clamped, and the rodperson, leveler, or both observe the indicated reading. Usually, the target is equipped with a vernier or other device by which fractional measurements of the rod graduations can be read without estimation.

The Philadelphia rod (Figure 5.22a), previously described, is designed as a self-reading rod but also may be used as a target rod. Lugs on the target engage in a groove on either side of the front strip. For readings on the lower half of the rod, the reading is made to thousandths of feet by means of a vernier attached to the target. For example, the reading in Figure 5.25a is 4.347 ft using the target.

Graduations on the back of the rear strip are a continuation of those on the front strip and read downward. On the back of the top sleeve is a vernier employed for observations with the rod extended. For readings greater than can be taken with a "short" rod, the target in Figure 5.22a is clamped on the 7.000-ft graduation (the 2.000-m graduation on the metric rod shown in Figure 5.23b) on the front face of the upper section. The vertically held rod then is extended until the target is bisected by the line of sight. The vertical distance from the foot of the rod is indicated by the reading of the vernier on the back of the rod (Figure 5.25b). Note that the back of the rod is a continuation of graduations on the front, with numbers increasing from top to bottom. For the example illustrated in Figure 5.25b with the target set at 7.00 ft on the front of the rod, the vertical distance from the bottom of the rod to the target is 9.264 ft. For the metric rod shown in Figure 5.26 with the target set at 2.000 m on the front of the rod, the vertical distance from the bottom of the rod to the target is 2.157 m.

The principal advantage of the target rod is that mistakes are less likely to occur, particularly if both rodperson and leveler read the rod. Under certain conditions its use materially facilitates the work; for example, where very long sights are taken, where the rod is partly obscured from view, or where it is necessary to establish a number of points all at the same elevation. However, under ordinary conditions its use retards progress without adding much, if anything, to the precision.

5.29
ROD LEVELS

The rod level is an attachment for indicating the verticality of the leveling rod. One type consists of a circular or "bull's-eye" level vial mounted on a metal angle or bracket, which is either attached by screws to the side of the rod or held against the rod, as desired. As noted for the level rod shown in Figure 5.22b, the rod level may be built into the rod.

Another type consists of a hinged casting on each wing of which is mounted a level tube that is held parallel to a face of the rod. When both the bubbles are centered, the rod is plumb. The hinge makes it possible to fold the level compactly when it is not in use.

201

CHAPTER 5:
Vertical
Distance
Measurement:
Leveling

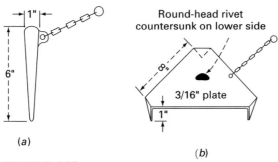

FIGURE 5.27
(a) Turning point. (b) Turning plate.

5.30
TURNING POINTS

A metal plate or pin that will serve temporarily as a stable object on which the leveling rod may be held at turning points is a useful part of the leveling equipment for lines of differential levels. The iron pin shown in Figure 5.27a is adapted for use in firm ground. Often a railroad spike or rivet pin is used.

In soft ground, the steel plate of Figure 5.27b makes a satisfactory turning point for ordinary leveling. The plate also is adapted for use where the ground is so solid as to make driving the pin impossible or at least impractical, as along highways. Under these conditions, the plate with the dogs at its corners acts as a tripod, no special attempt being made to secure bearing between the lower surface of the plate and the ground.

5.31
SETTING UP ENGINEER'S LEVELS

The engineer's level is placed in a desired location, with the tripod legs well spread and firmly pressed into the ground, with the tripod head nearly level, and with the telescope at a convenient height for sighting. If the setup is on a slope, it is preferable to orient the tripod so that one of its legs extends up the slope.

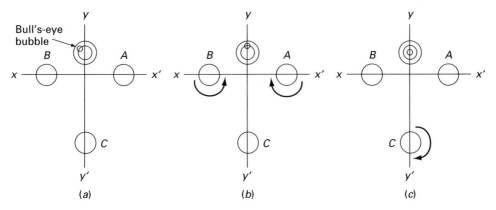

FIGURE 5.28
Leveling a three-screw instrument.

When operating three-screw tilting or automatic levels that have bull's-eye bubbles, as illustrated in Figure 5.28a, opposite rotation of level screws A and B moves the bubble along axis xx', as shown in Figure 5.28b, whereas rotation of screw C moves the bubble in the direction of axis yy' (Figure 5.28c). When the bubble is centered within the circle, the instrument is sufficiently leveled to be within the range of the spirit level in a tilting instrument or the pendulum device in the automatic level.

If the leveling head has four screws, the telescope is brought over one pair of opposite leveling screws and the bubble is centered approximately by turning the level screws in *opposite directions,* then the process is repeated with the telescope over the other pair. By repetition of this procedure the leveling screws are manipulated until the bubble remains centered, or nearly so, for any direction in which the telescope is pointed. If the instrument is in adjustment, the line of sight will be horizontal.

5.32
READING THE ROD

For observations to millimeters and to hundredths or thousandths of feet, the rod is held on some well-defined point of a stable object. The rodperson holds the rod vertical, either by observing the rod level or by estimation. First, the cross hairs are carefully focused to suit the eye of the observer. This can be done by holding an open field book in front of the objective lens of the telescope and rotating the eyepiece adjustment knob until the cross hairs are clear, sharp, and black. Next, the level operator revolves the telescope about the vertical axis until the rod is about in the middle of the field of view, focuses the objective for distinct vision, checks for presence of parallax (Sections 5.10 and 5.40), and carefully centers the bubble. If a self-reading rod is used, the leveler observes and records the reading indicated by the line of sight, that is, the apparent position of the horizontal cross hair on the rod. Checking, the leveler again observes the bubble and the rod. If the target rod is used, the procedure is identical except that the target is set by the rodperson as directed by the leveler.

For leveling of lower precision, as when rod readings for points on the ground are determined to the nearest 0.1 ft or nearest cm, the observations usually are not checked and proportionally less care is exercised in keeping the rod vertical and the bubble centered, always bearing in mind the errors involved and the precision with which measurements are desired.

If no rod level is used, in calm air the rodperson can plumb the rod accurately by balancing it on the point on which it is held. By means of the vertical cross hair the leveler can determine when the rod is held in a vertical plane passing through the instrument but

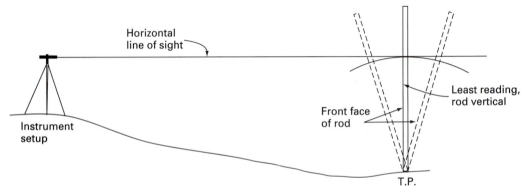

FIGURE 5.29
Swinging the rod.

203

CHAPTER 5:
Vertical
Distance
Measurement:
Leveling

cannot tell whether it is tipped forward or backward in this plane. If it is in either of these positions, the rod reading will be greater than the true vertical distance, as illustrated by Figure 5.29. To eliminate this error, the rodperson *swings the rod,* or tilts it forward and backward as indicated by the figure, and the leveler takes the least reading, which occurs when the rod is vertical. The larger the rod reading, the larger is the error due to the rod being held at a given inclination; hence, it is more important to swing the rod for large readings than for small readings. Further, whenever the rod is tipped backward about any support other than the front edge of its base, the graduated face rises and an error is introduced; for small readings this error is likely to be greater than that caused by not swinging the rod.

5.33
DIFFERENTIAL LEVELING

Differential leveling is the operation of determining the elevations of points some distance apart. Usually, this is accomplished by direct leveling. Differential leveling requires a series of setups of the instrument along the general route and, for each setup, a rod reading back to a point of known elevation and forward to a point of unknown elevation.

5.34
BENCH MARKS

A *bench mark* (B.M.) is a definite point on an object, the elevation and location of which are known. It may be permanent (P.B.M.) or temporary (T.B.M.). Bench marks serve as points of reference for levels in a given locality. Their elevations are established by differential leveling, except that the elevation of the initial bench mark of a local project may be assumed.

   Throughout the United States permanent bench marks are established by the U.S. Geological Survey and the U.S. National Geodetic Survey. Similarly, bench marks have been established by various other federal, state, and municipal agencies and by such private interests as railroads and water companies, so that the surveyor has not far to go before finding some point of known elevation.

   The National Geodetic Survey bench marks consist of metal alloy plates having a semispherical upper surface, set in stone or concrete and marked with the elevation above mean sea level. Those of the other agencies are similar. Other objects frequently used as bench marks are stones, pegs or pipes driven in the ground, nails or spikes driven horizontally in trees or vertically in pavements, and marks painted or chiseled on street curbs.

   For any survey or construction enterprise, levels are run from some initial bench mark of known or assumed elevation to scattered points in desirable locations for future reference as bench marks.

   In some areas, the elevation of bench marks may be altered by earth movements such as those caused by earthquakes, slides, lowering of water tables, pumping from oil fields, mining, or construction.

5.35
DIFFERENTIAL LEVELING DEFINITIONS

A *turning point* (T.P.) is an intervening point between two bench marks on which point foresight and backsight rod readings are taken. It may be a pin or plate (see Section 5.30) that is carried forward by the rodperson after observations have been made, or it may be any stable object such as a street curb, railroad rail, or stone. The nature of the turning point usually is indicated in the notes, but no record is made of its location unless it is to be reused. A bench mark may be used as a turning point.

A *backsight* (B.S.) is a rod reading taken on a point of known elevation such as a bench mark or a turning point. Usually, it will be taken with the level sighting back along the line, hence, the name. A backsight is sometimes called a *plus sight*. The horizontal distance from level to rod on a B.S. is called the *backsight distance*.

A *foresight* (F.S.) is a rod reading taken on a point the elevation of which is to be determined, such as on a turning point or on a bench mark that is to be established. A foresight is sometimes called a *minus sight*. The horizontal distance from the level to the rod on a F.S. is called the *foresight distance*.

The *height of instrument* (H.I.) is the elevation of the line of sight of the telescope above the datum when the instrument is leveled.

In surveying with the a theodolite or a total station system, the terms *backsight*, *foresight*, and *height of instrument* have meanings different from those defined here.

5.36
PROCEDURE

In Figure 5.30, B.M.$_1$ represents a point of known elevation (bench mark), and B.M.$_2$ represents a bench mark to be established some distance away. It is desired to determine the elevation of B.M.$_2$. The rod is held at B.M.$_1$, and the level is set up in some convenient

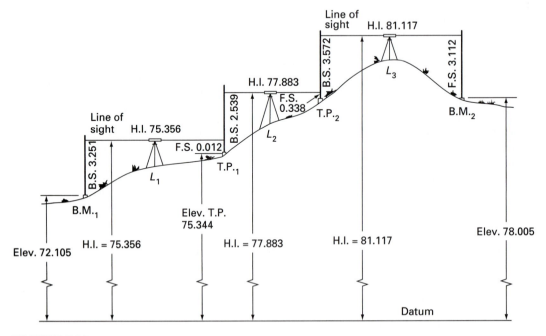

FIGURE 5.30
Differential leveling.

location, such as L_1, along the general route B.M.$_1$ to B.M.$_2$. The level is placed in such a location that a clear rod reading is obtainable, but no attempt is made to keep on the direct line joining B.M.$_1$ and B.M.$_2$. A backsight is taken on B.M.$_1$. The rodperson then goes forward and, as directed by the leveler, chooses a turning point, T.P.$_1$, at some convenient spot within the range of the telescope along the general route B.M.$_1$ to B.M.$_2$. It is desirable, but not necessary, that each foresight distance, such as L_1–T.P.$_1$, be approximately equal to its corresponding backsight distance, such as B.M.$_1$–L_1. The chief requirement is that the turning point shall be a stable object at an elevation and in a location favorable to a rod reading of the required precision. The rod is held on the turning point, and a foresight is taken. The leveler than sets up the instrument at some favorable point, as L_2, and takes a backsight to the rod held on the turning point; the rodperson goes forward to establish a second turning point T.P.$_2$, and so the process is repeated until finally a foresight is taken on the terminal point B.M.$_2$.

Figure 5.30 shows that a *backsight added to the elevation* of a point on which the backsight is taken gives the height of instrument and that a *foresight subtracted from the height of instrument* determines the elevation of the point on which the foresight is taken. Thus, if the elevation of B.M.$_1$ is 72.105 m and the B.S. is 3.251 m, the H.I. with the instrument set at L_1 is $72.105 + 3.251 = 75.356$ m. If the following F.S. is 0.012 m, the elevation of T.P.$_1$, is $75.356 - 0.012 = 75.344$ m. Also, the difference between the backsight taken on a given point and the foresight taken on the following point is equal to the difference in elevation between the two points. It follows that the difference between the sum of all backsight and the sum of all foresights gives the difference in elevation between the bench marks.

Because the level normally is at a higher elevation than that of the points on which rod readings are taken, the backsights are often called *plus* sights and the foresights *minus* sights and are so recorded in the field notes. Sometimes, however, in leveling for a tunnel or a building, it is necessary to take rod readings on points that are at a higher elevation than that of the H.I. In such cases the rod is held inverted, and in the field notes each such backsight is indicated with a minus sign and each foresight with a plus sign.

When several bench marks are to be established along a given route, each intermediate bench mark is made a turning point in the line of levels. Elevations of bench marks sometimes are checked by rerunning levels over the same route but more often by "tying on" to a previously established bench mark near the end of the line or by returning to the initial bench mark. A line of levels that ends at the point of beginning is called a *level circuit*. The final observation in a level circuit therefore is a foresight on the initial bench mark. If each bench mark in a level circuit also is a turning point and the circuit checks within the prescribed limits of error, it is regarded as conclusive evidence that the elevations of all turning points in the circuit, including all bench marks used as turning points, are correct within prescribed limits.

5.37
BALANCING BACKSIGHT AND FORESIGHT DISTANCES

In Section 5.4 it was shown that, if a foresight distance were equal to the corresponding backsight distance, any error in readings due to earth's curvature and to atmospheric refraction (under uniform conditions) would be eliminated. In ordinary leveling, no special attempt is made to balance *each* foresight distance against the preceding backsight distance. Whether or not such distances are approximately balanced between bench marks will depend on the desired precision. The effect of the earth's curvature and atmospheric

205

CHAPTER 5:
Vertical
Distance
Measurement:
Leveling

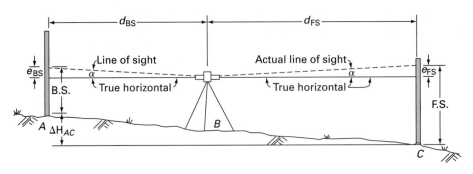

FIGURE 5.31
Instrumental error in line of sight.

refraction is slight unless there is an abnormal difference between backsight and foresight distances. The effect of *instrumental errors* is likely to be of considerably greater consequence with regard to the balancing of these distances. The chances are that the line of sight will not be exactly horizontal when (1) the circular bubble is centered and the compensator is free in a self-leveling instrument or (2) the level bubble on a dumpy level is centered. As a result, even if the instrument were perfectly leveled, the line of sight always would be inclined slightly upward or slightly downward. Figure 5.31 illustrates the geometry involved where a level is set at B, a B.S. rod reading is taken on a rod held at A at a distance d_{BS}, an F.S. rod reading is observed on a rod at C at a distance d_{FS}, and the actual line of sight is inclined upward by an angle α in each direction. For this example, the respective errors in B.S. and F.S. are $e_{BS} = d_{BS}\alpha$ and $e_{FS} = d_{FS}\alpha$, where α is in radians. Corrected B.S. and F.S. rod readings are B.S.$_{\text{corr.}}$ = B.S $- d_{bs}\alpha$, and F.S.$_{\text{corr.}}$ = F.S. $- d_{FS}\alpha$. Therefore, the true difference in elevation between A and C, $\Delta H_{AC} =$ (B.S. $-$ F.S.) $- (d_{BS}\alpha - d_{FS}\alpha)$. Consequently, if $d_{BS} = d_{FS}$, the errors would cancel; but if $d_{BS} \neq d_{FS}$, an error would be introduced into each setup that is directly proportional to the difference between the foresight distance and the backsight distance. In a line or circuit of levels, consisting of several to many instrument setups, the *sum* of the backsight distances must be made equal to the *sum* of the foresight distances to eliminate this instrumental error.

In ordinary leveling no special attempt is made to equalize these distances *if there is assurance that the instrument is in good adjustment.* Normally, for levels run over flat or gently rolling ground, the line of sight will fall within the length of the rod regardless of the position of the instrument, and the distance between instrument and rod is governed by the optical qualities of the telescope. While moving forward, the leveler generally paces or estimates by eye a distance about the proper maximum length of sight; the rodperson similarly estimates the proper distance from the instrument.

For leveling of moderately high precision, it is necessary to equalize backsight and foresight distances between bench marks; hence, these distances are recorded. In less-refined leveling, distances usually are determined by pacing; in precise leveling, they usually are measured by stadia or the gradienter.

In leveling uphill or downhill, the length of sight usually is governed by the slope of the ground. So that maximum distances between turning points may be obtained and progress be most rapid, the leveler sets up the instrument in a position such that the line of sight will intersect the rod near its top if the route is uphill or near its bottom if the route is downhill; the leveler directs the rodperson to a similarly favorable location for the turning point. In leveling uphill or downhill, a balance between foresight and backsight distances can be obtained with a minimum number of setups by following a zigzag course.

When bench marks are at roughly the same elevation, the backsight and foresight distances will tend to balance in the long run, regardless of the character of the terrain. However, for levels between two points having a large difference in elevation, a very small inclination of the line of sight will produce a marked error unless some attempt is made to equalize backsight and foresight distances.

207

CHAPTER 5:
Vertical
Distance
Measurement:
Leveling

5.38
DIFFERENTIAL-LEVEL NOTES

For ordinary differential leveling, when no special effort is made to equalize backsight and foresight distances between bench marks, usually the record of fieldwork is kept in the form indicated by Figure 5.32, in which the levels from B.M.$_1$ to B.M.$_2$ are the same as shown by Figure 5.30 and the line of levels closes on known B.M.$_3$. The left page is divided into columns for numerical data, and the right page is reserved for descriptive notes concerning bench marks and turning points. In the same horizontal line with each turning point or bench mark shown in the first column are all data concerning that point. The heights of instrument and the elevations are computed as the work progresses. Thus, when the backsight (3.251) has been taken on B.M.$_1$ it is added to the elevation (72.105) to determine the

Sta.	B.S.	H.I.	F.S.	Elev.		
	LEVELS FOR BENCH MARKS ALONG RIDGE ROAD					June 20, 1997 / Fair 70° F — J.G. Sutter / W.R. Knowles Rod — Level #42, Rod #12
B.M.$_1$	3.251	75.356		72.105		Top of Hydrant Corner of Oak & Ridge Elev. in meters above sea level.
T.P.$_1$	2.539	77.883	0.012	75.344		
T.P.$_2$	3.572	81.117	0.338	77.545		
B.M.$_2$	0.933	78.938	3.112	78.005		Spike in Pole North of Williams House No. 260 Oak St. Marked B.M. 75.902 m
T.P.$_3$	0.317	75.949	3.306	75.632		
T.P.$_4$	0.835	74.068	2.716	73.233		Stone
T.P.$_5$	0.247	70.773	3.542	70.526		
B.M.$_3$			3.786	66.987 [66.980]		Top of Brass Plate in Concrete Monument 4 m North of N. Edge Pvt. Oak St. at County Line
ΣB.S. =	11.694	ΣF.S. =−	16,812.		} Check	
			11.694			
	B.M.$_1$	72.105−	5.118 =	66.987		
Error of closure = 66.987−66.980 = + 0.007m						

FIGURE 5.32
Differential-level notes.

H.I. (75.356). The height of instrument is recorded on the same line with the backsight by means of which it is determined. When the first foresight (0.012) is observed, it is recorded on the line below and subtracted from the preceding H.I. (75.356) to determine the elevation of T.P.$_1$ (75.344). And so the notes are continued. Note that B.M.$_2$ is included as a T.P. in the line of levels. Usually, at the foot of each page of level notes, the *computations* are checked by comparing the difference between the sum of the backsights and the sum of the foresights with the difference between the initial and the final elevation, as illustrated at the bottom of Figure 5.32. Agreement between these two differences signifies that the additions and subtractions are correct but does not check against mistakes in observing or recording. For this example, the line of levels closes on known B.M.$_3$, where an F.S. of 3.786 subtracted from the H.I. of 70.773 yields a calculated elevation of 66.987 m. The difference between this calculated elevation and the known elevation of B.M.$_3$ (66.980 m, column five) is called the *error of closure* for the level net and listed at the bottom of the level notes (Figure 5.32). The error of closure for this case is 0.007 m, not particularly good for a line of ordinary levels containing only seven instrument setups. This magnitude of error of closure indicates that probably no mistakes were made and the error resulted from accumulation of systematic instrumental and random errors (Sections 5.37 and 5.40). Error of closure will be discussed further in Section 5.42.

Bench marks should be briefly but definitely described and should be so marked in the field that they readily can be identified. They usually are marked with paint or with crayon that will withstand the effects of the weather. When the bench mark is on stone or concrete, the position often is indicated by a cross cut with a chisel. A bench mark may or may not be marked with its elevation. Whenever any question might arise as to the exact location of the point on which the rod was held, its nature should be clearly indicated in the notes. A description of turning points is of no particular importance unless the points are on objects that can be identified and therefore might become of some value in future leveling operations. Such points usually are marked with crayon and briefly described in the notes.

5.39
MISTAKES IN LEVELING

Some of the mistakes commonly made in leveling are

1. Confusion of numbers in reading the rod, as, for example, reading and recording 2.345 when it should be 3.345. The mistake is not likely to occur if the observer notes the numbers on both sides of the observed reading.
2. Recording backsights in foresight column and vice versa.
3. Faulty addition and subtraction; adding foresights and subtracting backsights. As a check, the difference between the sum of the backsights and the sum of the foresights should be computed for each page or between bench marks.
4. Rod not held on same point for both the foresight and backsight. This is not likely to occur if the turning points are marked or otherwise clearly defined.
5. Not having the Philadelphia rod fully extended when reading the long rod. Before a reading on a turning point is taken, the clamp should be inspected to see that it has not slipped.
6. Wrong reading of vernier when the target rod is used.
7. When the long target rod is used, not having the vernier on the target set to read exactly the same as the vernier on the back of the rod when the rod is short.

5.40
ERRORS IN LEVELING

209

CHAPTER 5:
Vertical
Distance
Measurement:
Leveling

In leveling, errors are due to some or all of the following causes (see also Table 5.1):

1. *Imperfect adjustment of the instrument.* The essential relations are that the optical line of sight should be level when the circular bubble is centered on self-leveling levels and the compensator is free or the attached spirit level bubble is centered on dumpy-style levels. In self-leveling instruments, this error is due to imperfect adjustment of the circular level bubble; in dumpy levels, it occurs because the optical line of sight is not parallel to the axis of the attached spirit level. Any inclination of the actual line of sight from true horizontal leads to an error proportional to the distance from the level to the rod and, for a given distance, would be the same magnitude and sign for a backsight as for a foresight (see Section 5.37 and Figure 5.31). Because backsights are added and foresights are subtracted, it is clear that the error in elevations will be eliminated to the extent that, between bench marks, the sum of the backsight distances is made equal to the sum of the foresight distances. Conversely, a systematic error will result to the extent that these distances are not equalized between any two bench marks. Often, these distances will be sufficiently balanced in the long run, regardless of the terrain, to yield a satisfactory final result; but that does not ensure a corresponding accuracy for the bench marks established along the line. The effect of imperfect adjustment of the instrument is minimized by adjusting the instrument and balancing backsight and foresight distances. In precise leveling, this error also is reduced further by computing a collimation correction, as shown in Section 5.51.

2. *Parallax.* This condition is present when either or both of the following occur: (1) the objective lens is not focused on the object; (2) the observer's eye is not focused on the plane of the cross hairs (see Figure 5.7). The effect of parallax is to cause relative movement between the image of the cross hairs and image of the object when the eye is moved up and down. Parallax causes a random error and in practice can be eliminated by careful focusing as described in Section 5.10.

3. *Earth's curvature.* This produces an error only when backsight and foresight distances are not balanced. Under ordinary conditions these distances do not tend to vary greatly, and whatever resultant error arises from this source in ordinary leveling is so small as to be of no consequence. When backsight distances consistently are made greater than foresight distances or vice versa, a systematic error of considerable magnitude is produced, particularly when the sights are long. The effect is the same as that due to the line of sight being inclined. The error varies as the square of the distance from instrument to rod and hence will be eliminated not merely by equalizing the sum total of backsight and foresight distances between bench marks but rather by balancing *each* length of backsight by a corresponding length of foresight.

4. *Atmospheric refraction.* This varies as the square of the distance but, under normal conditions, is only about one-seventh of that due to the earth's curvature and its effect is opposite in sign. It usually is considered together with the earth's curvature, but although the effect of the latter will be entirely eliminated if each backsight distance is made equal to the following foresight distance, the atmospheric refraction often changes rapidly and greatly in a short distance. It is particularly uncertain when the line of sight passes close to the ground. Hence, it is impossible to eliminate entirely the effect of refraction even though the backsight and foresight distances are balanced. In ordinary leveling, its effect is negligible. In leveling of greater precision, the change in refraction can be minimized by keeping the line of sight well above the ground (at least 2 ft or ~ 0.7 m) and by taking the backsight and foresight readings in quick succession. In

the long run, the error is random, but over a short period, as a day, it may be systematic. So-called heat waves are evidence of rapidly fluctuating refraction. Errors from this source can be reduced by shortening the length of sight until the rod appears steady. In precise geodetic leveling, corrections for earth curvature and refraction are calculated using the equations given in Section 5.2.

5. *Variations in temperature.* The sun's rays falling on top of the telescope, or on one end and not on the other, will produce a warping or twisting of its parts and hence may influence rod readings through temporarily disturbing the adjustments. Although this effect is not of much consequence in leveling or ordinary precision, it may produce an appreciable error in more refined work. The error usually is random, but under certain conditions it may become systematic. In practice, it is eliminated by shielding the instrument from the rays of the sun.

6. *Rod not of standard length.* If the error is distributed over the length of the rod, a systematic error is produced that varies directly as the difference in elevation and bears no relation to the length of the line over which levels are run. The error can be eliminated by comparing the rod with one of standard length and applying the necessary corrections. The case is analogous to measurement of distance with a tape that is too long or too short. If the rod is too long, the correction is added to a measured difference in elevation; if the rod is too short, the correction is subtracted. Most manufactured rods are nearly of standard length, but where large differences in elevation are to be determined, few rods are near enough to the standard that corrections can be ignored in precise work. If the rod is worn uniformly at the bottom, an erroneous height of instrument is shown at each setup, but the error in backsight is balanced by that in the following foresight, and no error results in the elevation of the foresight point.

7. *Expansion or contraction of the rod.* Owing to a change in moisture content or a change in temperature, the leveling rod may expand or contract. The resultant error is systematic. Wood when well-seasoned and painted will shrink or swell but little in the direction of the grain. Also, its coefficient of thermal expansion is small. The error is of no particular consequence in ordinary leveling. For precise leveling, gauge points may be established by inserting metal plugs in the rod, and corrections for shrinkage may be determined by observing any change in distance between the gauge points. Corrections for thermal expansion may be based on observed temperatures of the rod, as indicated by an attached thermometer, the temperature being recorded in the notes. An invar tape may be used on the rod.

8. *Rod is not held plumb.* This condition produces rod readings that are too large. In running a line of levels uphill or downhill, it becomes a systematic error, inasmuch as the backsights are larger than the foresights or *vice versa.* Over rolling or level ground, the resultant error is random because the backsights, on the average, are about equal to the foresights. The error varies directly with the first power of the rod reading and directly as the square of the inclination. Thus, if a 3 m rod is 0.06 m out of plumb, the error amounts to 0.0006 m for a 3 m reading and 0.0002 m for a 1 m reading; but if the rod is 0.12 m out of plumb, the corresponding errors are 0.002 and 0.0008 m, respectively. It therefore is evident that appreciable inclinations of the rod must be avoided, particularly for high rod readings. The error can be eliminated by swinging the rod or by using a rod level.

9. *Faulty turning points.* A random error results when turning points are not well defined. Even a flat, rough stone, for example, does not make a good turning point for precise leveling because no definite point exists on which to hold the rod, which is not likely to be held in the same position for both backsight and foresight.

10. *Settling of tripod or turning points.* If the tripod settles in the interval that elapses between taking a backsight and the following foresight, the foresight will be too small and the observed elevation of the forward turning point will be too large. Similarly, if a turning point settles in the interval between foresight and backsight readings, the

height of instrument as computed from the backsight reading will be too great. It thus is seen that, by the normal leveling procedure, if either the level tripod or the turning point settles, as may occur to some extent when leveling over soft ground, the error will be systematic and the resulting elevations always will be too high.

Few occasions arise when turning points cannot be so selected or established as to eliminate the possibility of settling, but care should be taken not to strike the bottom of the rod against the turning point between sights or exert any pressure whatsoever on the turning point.

On the other hand, some settling of the instrument is nearly certain to occur when leveling over muddy, swampy, or thawing ground or over melting snow. The errors due to such settling can be greatly reduced by employing two rods and two persons, one setting the turning point ahead while the other remains at the turning point in the rear. Backsight and foresight readings then can be made in quick succession. Small errors remaining from this source can be minimized by reversing the order of sights at alternate setups, as described in Section 5.50.

11. *Bubble is not exactly centered at instant of sighting.* This condition produces a random error that tends to vary as the distance from instrument to rod. Hence, the longer the sight, the greater is the care that should be observed in leveling the instrument.

12. *Inability of observer to read the rod exactly or to set the target exactly on the line of sight.* This inability causes a random error of a magnitude depending on the instrument, weather conditions, length of sight, and observer. It can be confined within reasonable limits through proper choice of length of sight.

5.41
ERROR PROPAGATION IN LEVELING

The sources of errors discussed in Section 5.40 are summarized in Table 5.1. Obviously, a great many factors affect the differential leveling operation. Although the final accuracy achieved is influenced by the instrument used, a great deal depends on the skill of the leveler and degree of refinement with which the work is executed. Assuming the proper leveling procedures are employed and care is given to detail, systematic errors can be nearly eliminated. Thus, the remaining errors are random and can be attributed to centering the spirit level bubble, reading the rod, and variations in refractions in air. The first two errors are proportional to the length of sight. Although refraction varies as the square of the distance, the exact relationship cannot be determined; these variations may occur anywhere between the instrument and rod. Consequently, refraction also is assumed proportional to the length of sight, and these three sources of random error act together. One can say that, in ordinary differential leveling, the error propagates approximately in proportion to the distance leveled or, lacking the distance, in proportion to the number of instrument setups in the level circuit or line of levels. Although this concept is not rigorous, a statistical basis underlies the idea and it is adequate to use in simple adjustment of level circuits or lines, as discussed in Section 5.42. For rigorous examination of error propagation in differential leveling, refer to Part B, Section 5.46.

5.42
SIMPLE ADJUSTMENT OF LEVEL CIRCUITS OR LINES

When a line of levels makes a complete circuit or closes on another bench mark of known elevation, determined with a higher order of accuracy, almost invariably the final elevation of the initial or closing bench mark will not agree with the initial or closing elevation of this

211

CHAPTER 5:
Vertical
Distance
Measurement:
Leveling

TABLE 5.1
Errors in leveling

Source	Type	Cause	Remarks	Procedure to eliminate or reduce
Instrumental	Systematic	Line of sight not parallel to axis of level tube	Error of each sight proportional to distance*	Adjust instrument; balance sum of backsight and foresight distances
	Random	Rod not of standard length (throughout length)†	May be due to manufacture, moisture, or temperature; error usually small	Standardize rod and apply corrections, same as for tape
Personal	Random	Parallax Bubble is not centered at instant of sighting		Focus carefully Check bubble before making each sight
		Rod is not held plumb	Error varies as length of sight	Wave the rod, or use rod level
			Readings are too large; error of each sight proportional to square of inclination*	
		Faulty reading of rod or setting of target		Check each reading before recording; for self-reading rod, use fairly short sights
		Faulty turning points		Choose definite and stable points
Natural	Random	Temperature	May disturb adjustment of level	Shield level from sun
	Systematic	Earth's curvature	Error of each sight proportional to square of distance*	Balance *each* backsight and foresight distance or apply computed correction
	Random	Variations in atmospheric refraction	Error of each sight proportional to square of distance*	Same as for earth's curvature; also take short sights, well above ground, and take backsight and foresight readings in quick succession
	Systematic	Settling of tripod or turning points	Observed elevations are too high	Choose stable locations; take backsight and foresight readings in quick succession, preferably alternating order of sights

*The error of *each sight* is systematic, but the resultant error is the difference between the systematic error for foresights and that for backsights; hence, the resultant error tends to be random.
†Uniform wear of the bottom of the rod causes no error.

212

point. The difference is the error of running the circuit or line of levels, called the *error of closure* (see also Section 5.38 and Figure 5.32). It is evident that elevations of intermediate bench marks established while running the levels also will be in error, and the problem arises of determining errors for these intermediate points and adjusting their elevations accordingly.

213

CHAPTER 5:
Vertical
Distance
Measurement:
Leveling

In Section 5.41 the concept was stated that errors in leveling accumulate approximately in proportion to the distance leveled or the number of instrument setups. It follows that the appropriate correction to the observed bench mark is directly proportional to the distance of the bench mark from the point of beginning or to the number of instrument setups since the point of beginning. Therefore, if E_c is the error of closure of a level circuit of length L, and if $C_a, C_b, \ldots, C_n$ are the respective corrections to the applied to observed elevations of bench marks $A, B, \ldots, N$, whose respective distances from the point of beginning are $a, b, \ldots, n$, then

$$C_a = -\frac{a}{L} E_c; \qquad C_b = -\frac{b}{L} E_c; \qquad \ldots; \qquad C_n = -\frac{n}{L} E_c \qquad (5.18)$$

EXAMPLE 5.5 The accepted elevation of the initial bench mark, B.M.$_i$ of a level circuit is 470.46 ft. The length of the circuit is 10 mi. The final elevation of the initial bench mark as calculated from the level notes is 470.76. The observed elevations of bench marks established along the route and the distances to the bench marks from B.M.$_i$ are as shown in the third and second columns of the following table. The elevations of these intermediate points are required:

$$E_c = 470.76 - 470.46 = +0.30 \text{ ft}$$

Point	Distance from B.M.$_i$, mi	Observed elevation, ft	Correction, ft	Adjusted elevation, ft
B.M.$_i$	0	470.46	0.0	
B.M.$_a$	2	780.09	−0.06	780.03
B.M.$_b$	5	667.41	−0.15	667.46
B.M.$_c$	7	544.32	−0.21	544.11
B.M.$_i$	10	470.76	−0.30	470.46

Solution. By Equation (5.18),

$$C_a = -\tfrac{2}{10} \times 0.30 = -0.06 \text{ ft}$$

$$C_b = -\tfrac{5}{10} \times 0.30 = -0.15 \text{ ft}$$

$$C_c = -\tfrac{7}{10} \times 0.30 = -0.21 \text{ ft}$$

These corrections subtracted from the corresponding observed elevations give the adjusted elevations as tabulated. Note that, if the error of closure is positive, all corrections are to be subtracted and *vice versa*.

If desired, corrections may be made directly proportional to the number of setups instead of the distances.

The principles described in this section apply also to the adjustment of elevations of bench marks on a line of levels run between two points whose difference in elevation has previously been determined by more accurate methods and is assumed to be correct.

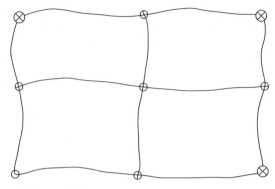

⊗ Known B.M.

○ B.M. to be established

FIGURE 5.33
Network of differential levels.

EXAMPLE 5.6. In Figure 5.30, levels were run from $B.M._1$ to a known $B.M._3$, establishing $B.M._2$ on the way. Compute the adjusted elevation of $B.M._2$. For this example, distances are not known, so the correction to $B.M._2$ is made in proportion to the number of instrument setups. From Figure 5.30, the total number of setups equals the number of h.i.'s $= 7$, the number of setups from $B.M._1$ to $B.M._2 = 3$, and the error of closure $= E_c = +0.007$ m.

Solution. Substituting number of setups for distances n and L in Equation (5.18), the correction to $B.M._2$ is

$$C_{B.M._1} = -(3/7)(0.007) = -0.003 \text{ m}$$

and the adjusted elevation for $B.M._2 = 78.005 - 0.003 = 78.002$ m.

Note that simple adjustment of levels is appropriate for single circuits or lines of levels. When several circuits and lines intersect, forming a network (Figure 5.33) of differential levels, a rigorous *least-squares adjustment* is required. Details of this procedure and its application can be found in Chapter 2, Sections 2.22 to 2.29, and in Part B of this chapter, Sections 5.57 and 5.58.

5.43
PROFILE LEVELING

The process of determining the elevations of points at short measured intervals along a fixed line is called *profile leveling*. During the location and construction of highways, railroads, canals, and sewers, stakes or other marks are placed at regular intervals along an established line, usually the center line. Ordinarily, the interval between stakes is 100 ft, 50 ft, or 25 ft, with intervals of 100 m, 50 m, 20 m, and 10 m being utilized in the metric system. The 100-ft or 100-m points, reckoned from the beginning of the line, are *full stations,* and all other points are called *plus stations.* Each stake is marked with its station and plus. Therefore, a stake set at 1600 ft from the point of beginning is numbered 16 or 16+00, and one set at 1625 ft from the point of beginning is numbered 16+25. Similarly, a point 450 m

215

CHAPTER 5:
Vertical
Distance
Measurement:
Leveling

from the origin is station 4+50 and a point set 1410 m from the origin is number 14+10 (a more detailed explanation of stationing can be found in Section 16.4). Elevations by means of which the profile may be constructed are obtained by taking level-rod readings on the ground at each stake and at intermediate points where marked changes in slope occur.

Figure 5.34 illustrates in plan (a) and elevation (b) the steps in leveling for profile. In this case stakes are set every 20 m or $\frac{1}{5}$ th station. The instrument is set at some convenient location, not necessarily on line (as at L_1), the rod is held on a bench mark (B.M. 10, elevation 55.805 m above the datum), a backsight (1.254 m) is taken, and the height of instrument (57.059) is obtained as in differential leveling. Readings in meters then are taken with the rod held on the ground, hubs, or other features at successive stations along the line. These rod readings, being for points of unknown elevation, are foresights regardless of whether they are back or ahead of the level. They frequently are designated *intermediate foresights* (IFS) to distinguish them from foresights taken on turning points or bench marks. The intermediate foresights (1.050, 1.50, 2.65, 3.050, . . . , 0.80), subtracted from the H.I. (57.059), give elevations of the stations. The readings on 0+00, 0+45 and 0+47 are

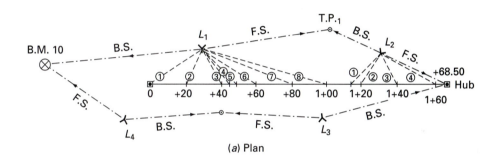

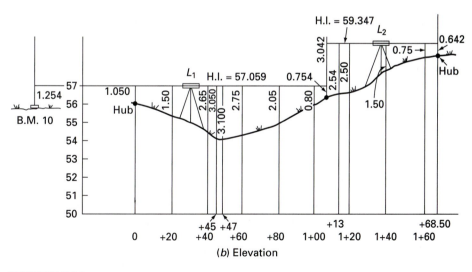

FIGURE 5.34
Profile leveling.

to the nearest millimeter, because they are taken on a hub and two edges of a sidewalk, respectively, and warrant this precision. Readings on the ground are to the nearest centimeter. When the rod has been advanced to a point beyond which further readings to stations cannot be observed, a turning point (T.P.$_1$) is selected and a foresight (0.754) is taken to establish its elevation. The level is set up in an advanced position (L_2), and a backsight (3.042) is taken on T.P.$_1$, just established. Rod readings at subsequent stations then are continued as before. Note that, throughout, the rodperson observes where changes in slope occur (as at 0+45, 0+47, 1+13), and readings are taken at these intermediate stations. The "plus," or distance from the preceding station, is measured by pacing or with a tape or rod according to the precision required. When the end of the line is reached at station 0+68.50, a reading of 0.642 is taken on the hub set at that point. The reading on this station is treated as a foresight and included as a turning point in the circuit, which is closed back on B.M. 10 by making two additional setups at L_3 and L_4, thus, providing a check on the leveling.

The care exercised in taking observations on turning points depends on the distance between bench marks, the elevations of which have been determined previously, and on the required precision of the profile. For a ground profile, the backsights and foresights usually are read to mm (hundredths of feet) and no particular attention is paid to balancing backsight and foresight distances; the intermediate foresights to ground points are read to cm or tenths of feet only. Occasions arise when it is desirable or necessary to determine intermediate foresights to within 3 mm or hundredths of feet, for example, in securing the profile of a railroad track or of the water grade in a canal; rod readings on turning points then, generally are taken to mm or thousandths of feet, and backsight and foresight distances often are balanced.

As the work of leveling for profile progresses, bench marks generally are established to facilitate later work; these are made turning points wherever possible. To check the elevation of turning points, it is necessary either to run levels back to the point of beginning or to run short lines of differential levels connecting with bench marks previously established by some other survey. The effect of an occasional error in the elevations of intermediate ground points on the profile usually is not of sufficient importance to justify the additional work that checking of rod readings on intermediate foresights would make necessary, and if turning points are checked, it is regarded as a sufficient.

5.44
PROFILE-LEVEL NOTES

The notes for profile leveling may be recorded as shown in Figure 5.35, in which foresights to turning points and bench marks are in a separate column from intermediate foresights to ground points. The notes for turning points are kept in the same manner as for differential leveling.

The computations shown at the foot of the notes of Figure 5.35 check all computations for H.I.'s and elevations of T.P.'s on the page; and thus for the notes shown the difference between the sum of all backsights and the sum of all foresights is equal to the error of closure on B.M. 30. Elevations of ground points are recorded only to the *number of decimal places contained in the intermediate foresights,* regardless of the number of places in the H.I.

The right-hand page is reserved for concise descriptions of bench marks and other pertinent items. Occasionally, as shown in Figure 5.35, simple sketches are employed in conjunction with the explanatory notes.

217

CHAPTER 5:
Vertical
Distance
Measurement:
Leveling

	PROFILE LEVELS FOR STORM					SEWER LOCATION STATE UNIV.	
Sta.	B.S.	H.I.	F.S.	I.F.S.	Elev.,m	Lietz Level No. 21, Rod No. 12	April 27, 1995
							Clear, windy 70°F
B.M.30	3.478	33.478			30.000	Intersection West Rd., Gayley Ave. Top of fireplug N.W. Corner	Jensen
0+00				3.617	29.861	Top manhole	Laird, Notes
0+00				5.141	28.337	Flow line manhole	Lucas, Rod
+10				1.72	31.76		
T.P.1	3.134	36.419	0.193		33.285		0+53.3 ⊙ Exist. M.H.
+20				2.86	33.56		
+29.5				1.852	34.567	Edge of asphalt sidewalk	+46.5 Sidewalk
+30				1.805	34.614	Asphalt walk	
+31.9				1.738	34.681	Edge of asphalt sidewalk	+43.8
+34.7				1.250	35.169	Edge of asphalt sidewalk	+34.7 +38
+38.0				0.951	35.468	Edge of asphalt sidewalk	+31.9 Sidewalk
+40				0.65	35.77		Sidewalk
+43.8				0.054	36.365	Edge of asphalt sidewalk	Stakes
T.P.2	3.551	39.844	0.126		36.293		set at +29.5
+46.5				3.178	36.666	Edge of asphalt sidewalk	10-m
+48.1				2.87	36.97		intervals N
+50				2.74	37.10		on slope
+53.3				2.289	37.555	Top of manhole	South of
+53.3				3.508	36.336	Flow line manhole	Allison Hall
T.P.3	0.081	37.636	2.289		37.555		
T.P.4	0.333	34.308	3.661		33.975		
T.P.5	0.515	31.972	2.851		31.457		Exist. M.H. ⊙
B.M.30			1.974		29.998		
	11.092		−11.094 11.092				
		30.000	−0.002 = 29.998 Check				

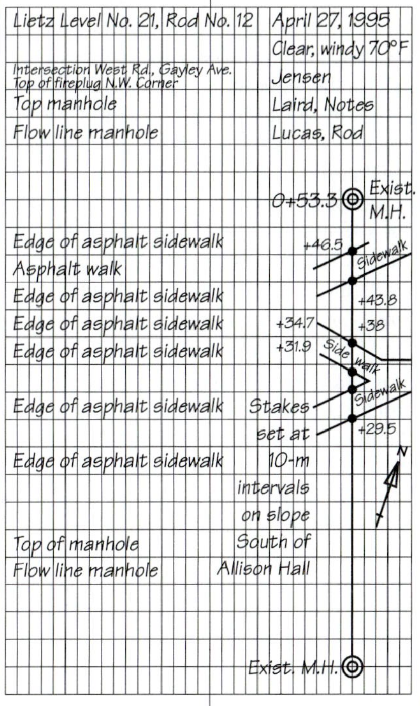

FIGURE 5.35
Profile notes in metric units.

5.45
PLOTTING THE PROFILE

To provide a useful product for the designer, the profile is plotted on profile paper for specified horizontal and vertical scales. Profile paper is a cross-sectioned paper of good quality with a specified number of lines per horizontal and vertical unit. Common scales are 1:600 (50 ft/in.) horizontally and 1:120 (10 ft/in.) vertically. Profile paper has 20 lines per inch vertically and 10 lines per inch horizontally. In the metric system comparable horizontal and vertical scales are 1:500 and 1:100 in the horizontal and vertical scales, respectively. Figure 5.36 illustrates a profile plotted using the elevations in Figure 5.35. The horizontal and vertical scales in Figure 5.36 are 1:500 and 1:100, respectively. Note that a certain balance between horizontal and vertical scale is necessary to portray the ground profile in a realistic manner. The plotted profile is utilized by the designer to establish grade lines for highways, roads, sewer lines, canals, and so on. Frequently, a plan view of the area through which the profile runs is placed above or below the profile on the same sheet. Further details concerning profiles and establishing grade lines can be found in Section 16.28. A plan and profile sheet for a route alignment is shown in Figure 16.35.

Use of an electronic digital level, similar to the one described in Section 5.20, would expedite the entire profile leveling process tremendously. A function of the onboard software of a digital level is a routine that includes taking intermediate foresights on center line stations along with backsights on turning points and bench marks. Therefore, when operating in the "measure and record" mode, intermediate foresights can be stored in the memory module along with the elevations of these points and all other data pertinent to the line of profile levels such as B.M.s and T.P.s. When the profile is completed, the memory module will contain all of the information related to the profile, as illustrated in Figure 5.36. The module can be removed and the data downloaded into a personal or mainframe computer in the office for computation and adjustment. These data then can be printed out as a hard-copy listing of the profile levels. Another option is to store the results in the computer memory or on diskette for automatic plotting of the profile using a computer-aided design program (Chapter 14, Section 14.14) or for use in a geographic information system (Chapter 14, Section 14.18). In the latter case, the profile could be displayed on a video screen for interactive design purposes. For example, on a route location job, this design would include establishing grade lines and designing vertical curve alignments on the project related to the profile levels.

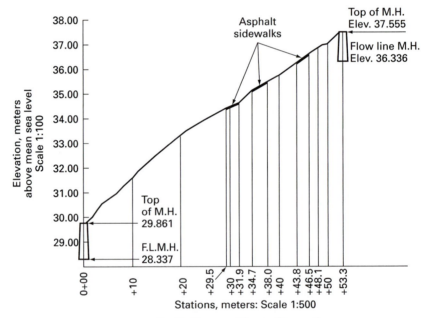

Profile Proposed Sewer Connection
South of Allison Hall
State University Campus Treadway, Oregon
Drawn by: R.Combes, September 15, 1995

FIGURE 5.36
Plotted profile.

PART B
ADVANCED TOPICS

219

CHAPTER 5:
Vertical
Distance
Measurement:
Leveling

5.46
ERROR PROPAGATION IN DIFFERENTIAL LEVELING

Errors in leveling were discussed in Section 5.40 and summarized in Table 5.1. Nonrigorous error propagation was developed in Section 5.41, where it was shown that the primary errors which propagate consist of random errors due to centering the spirit level bubble, reading the rod, and variations in refraction in air. In this section, rigorous error propagation is developed.

   To evaluate the propagated error in a line of levels, the random errors in a sight from an instrument setup are combined into a single value and the backsights are assumed equal to the foresights.

Let

$\hat{\sigma}_s$ = estimated standard deviation of the combined random errors in a single sight from
   one instrument setup expressed in units/unit of length of sight
l = length of a single sight

Because Δ elevation = B.S. $-$ F.S. (by Equation (2.16), Section. 2.15),

$$\hat{\sigma}_\Delta = l\hat{\sigma}_s\sqrt{2} \tag{5.19}$$

in which $\hat{\sigma}_\Delta$ is the estimated standard deviation for a complete instrument setup.

Next let

$\hat{\sigma}_d$ = estimated standard deviation in a line of levels
d = length of the line of levels
n = number of instrument setups required for closure

Then, also by Equation (2.16),

$$\hat{\sigma}_d = \hat{\sigma}_\Delta\sqrt{n} \tag{5.20}$$

Because $n = d/2l$,

$$\hat{\sigma}_d = \hat{\sigma}_\Delta\sqrt{\frac{d}{2l}} = \hat{\sigma}_s\sqrt{ld} \tag{5.21}$$

   Equations (5.20) and (5.21) show that the propagated error in a line of levels varies as the square root of the number of instrument setups or as the square root of the distance leveled and the square root of the lengths of sights. Therefore, most specifications governing closures in lines or circuits of levels are expressed as a function of the square of the distance leveled and the maximum length of sights is specified.

   Levels to establish vertical control are classified as first, second, or third order, depending on the equipment and procedures employed. Table 5.2 shows the classification and general specifications for first-, second-, and third-order vertical control. The specifications given in Table 5.2 were prepared by the Federal Geodetic Control Committee (FGCC), now known as the Federal Geodetic Control Subcommittee (FGCS). Standards of accuracy are discussed in Section 5.59.

5.47
GEODETIC LEVELS

The high order of accuracy needed in geodetic leveling requires an extremely sensitive level of special design. Automatic compensator-equipped levels, designed according to the principles described in Section 5.19, now are used for most geodetic leveling. They are heavier,

TABLE 5.2

Classification and general specifications for vertical control (abstracted from FGCC, *Standards and Specifications for Geodetic Control Networks*, September 1984)

Order	First		Second		Third
Class	I	II	I	II	—
Network geometry					
Bench mark spacing not to exceed (km)	3	3	3	3	3
Average B.M. spacing not to exceed (km)	1.6	1.6	1.6	3.0	3.0
Distance between network control points not to exceed (km)	300	100	50	50	25
Instrumentation					
Leveling instrument					
Minimum repeatability of line of sight	0.25"	0.25"	0.50"	0.50"	1.00"
Leveling rod construction	IDS	IDS	IDS[a] or ISS	ISS	Wood, metal
Instrument and rod resolution (combined)					
Least count (mm)	0.1	0.1	0.5–1.0[b]	1.0	1.0
Calibration procedures					
Leveling instrument					
Maximum collimation error, single line of sight, mm/m	0.05	0.05	0.05	0.05	0.10
Reversible compensator, mean of two values, mm/m	0.02	0.02	0.02	0.02	0.04
Time interval between compensator checks not to exceed (days), reversible compensator	7	7	7	7	7
Other types (days)	1	1	1	1	1
Maximum angular difference between two lines of sight, compensator	40"	40"	40"	40"	60"
Leveling rod					
Minimum scale calibration	N	N	N	M	M
Time interval between scale calibrations, years	1	1	—	—	—
Leveling rod verticality to within	10'	10'	10'	10'	10'
Field procedures					
Minimal observation method	Micrometer	Micrometer	Micrometer or 3-wire	3-wire	Center wire
Section running	SRDS, DR, or SP	SRDS, DR, or SP	SRDS, DR, or SP	SRDS or DR[d]	SRDS or DR[e]
Minimal observation method	Micrometer	Micrometer	Micrometer or 3-wire	3-wire	Center wire

Difference between forward and backward sights not to exceed

Per setup (m)	2	5	5	10	10
Per section (m)	4	10	10	10	10
Maximum sight length (m)	50	60	60	70	90
Minimum ground clearance, line of sight (m)	0.5	0.5	0.5	0.5	0.5
Even number setups when using rod without detailed calibration	Yes	Yes	Yes	Yes	—
Determine temperature gradient for vertical range of line of sight at each setup	Yes	Yes	Yes	—	—
Maximum section misclosure (mm)	$3\sqrt{D}$	$4\sqrt{D}$	$6\sqrt{D}$	$8\sqrt{D}$	$12\sqrt{D}$
Maximum loop misclosures (mm)	$4\sqrt{D}$	$5\sqrt{D}$	$6\sqrt{D}$	$8\sqrt{D}$	$12\sqrt{D}$
Single-run methods					
Reverse directions of single run every half	Yes	Yes	Yes	—	—
Nonreversible compensator levels					
Off-level/relevel between observing high and low rod scales	Yes	Yes	Yes	—	—
3-wire method					
Reading check (difference between top and bottom intervals) for one setup not to exceed (tenths of rod units)	—	—	2	2	3
Read rod 1 first in alternate setup method	—	—	Yes	Yes	Yes
Double-scale rods					
Low-high scale elevation difference for one setup not to exceed (mm)					
With reversible compensator	0.4	1.00	1.00	2.00	2.00
Other types of levels					
Half-centimeter rods	0.25	0.30	0.60	0.70	1.30
Full-centimeter rods	0.30	0.30	0.60	0.70	1.30

Notes: Code: IDS-Invar, double scale; ISS-Invar, single scale; N-national standard; M-manufacturer's standard; SRDS-single-run double simultaneous procedure; DR-double run; SP-spur, less than 25 km, double run; D-shortest length of section, one way, km; E-perimeter of loop, km.

[a] If optional micrometer is used.

[b] 1.0 mm if 3-wire method, 0.5 mm if optical micrometer.

[c] Must be double run when using 3-wire method.

[d] May be single run if length between network control points is less than 25 km.

[e] May be single run if length between network control points is less than 10 km.

more stable, have more powerful optics (magnification of about 40 diameters), and are equipped with optical micrometers. Figure 5.37 shows a self-leveling instrument suitable for geodetic leveling. This instrument has the added feature of a reversible compensator to help reduce effects of collimation errors.

A digital level similar to the one illustrated in Figure 5.19 has been approved for certain classes of geodetic leveling. This geodetic digital level is heavier and has a more precise compensator than the one shown in Figure in 5.19 but is the same in other respects and is used with a rod having bar-coded graduations.

Tilting levels also are appropriate for geodetic leveling. The geodetic tilting level is similar to the engineer's tilting level illustrated in Figure 5.15 but is heavier, more stable, and has higher-quality optics. The geodetic model also usually is equipped with an optical micrometer and has coincidence viewing (Section 5.14, Figure 5.12) of the bubble of a spirit level, which has a sensitivity of one second per 2 mm division (Section 5.15). Note that use of this style of instrument for geodetic leveling has decreased substantially in recent

FIGURE 5.37
Automatic level suitable for geodetic leveling.
(*Courtesy of Carl Zeiss, Inc.*)

years due to the speed with which comparable accuracies can be obtained with self-leveling and digital levels.

223

CHAPTER 5:
Vertical
Distance
Measurement:
Leveling

5.48
PRECISE LEVEL RODS

Precise or geodetic levels (Sections 5.47–5.52) require a specially constructed self-reading rod, which is a little over 3 m long and has a hardened steel foot 25 mm in diameter with a slightly convex bottom face. The rod is made of a single piece of extruded aluminum alloy or well-seasoned wood. The front of the rod is graduated in meters. Along the face of the rod runs a strip of invar metal graduated with a least division of centimeters. The invar strip is fastened rigidly only at the bottom of the rod and kept taut by a spring at the top; thus; it is free to expand or contract independent of the main rod. Precise level rods usually are manufactured in pairs, because the procedure for precise levels requires the use of two rods (Section 5.50).

Precise level rods have a double set of graduations with one scale offset from the other by known amount. This arrangement permits two independent readings with a single observation using a single cross hair.

One style of precise level rod, constructed of a light metal extruded T-section and equipped with a graduated invar strip and built-in circular level bubble, is illustrated in Figure 5.38a. This rod has a double scale of the line graduation type. The least division is 0.005 m, with every tenth graduation numbered. There are two vertical rows of graduations. Note that numerals and graduations on this rod are upside down and reverse, to provide an upright, right-reading image with an inverting telescope. The main scale on the right (on the observer's left using an inverting telescope) has 0 at the foot of the rod whereas the check scale on the left (the observer's right) is offset exactly 0.025 m below the main scale and numbered with a higher series of numbers. The first number on the check scale

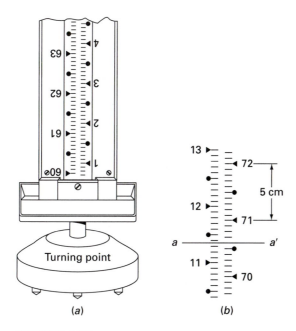

FIGURE 5.38
Double-scale precise level rod.

is 60 units, placed so that the corresponding reading on the main scale is 59.250 units less. For example, in Figure 5.38b, the scale is shown as visible to an observer through an inverting telescope. The reading on the main scale at cross hair aa' is 11.350 and on the check scale is 70.600, so that the difference is $70.600 - 11.350 = 59.250$. Thus, different readings are made on two different scales, providing two independent readings and a positive numerical check against blunders. This rod can be used with the special turning point shown in Figure 5.38a or with any other turning point suitable for precise leveling (Section 5.50). Rods of this style having erect numerals for use with levels equipped with erecting telescopes also are available.

Some older styles of precise level rods are graduated in meters on the front side of the rod and have graduations in feet and tenths of feet on the back of the rod to provide an independent check reading.

5.49
OPTICAL MICROMETER

Estimation to the nearest millimeter or to thousandths of a foot is inadequate in leveling operations where extremely high accuracy is required. In such cases, the engineer's precise level, tilting level, or self-leveling level is equipped with an optical micrometer mounted in front of the objective lens of the telescope. Figure 5.39a shows a cross-sectional view of a telescope equipped with an optical micrometer.

The basic element in an optical micrometer is a plane-parallel glass plate or optical flat that can be rotated about an axis perpendicular to the optical axis and in a horizontal plane. The principle of the optical flat is illustrated by Figure 5.40. The light ray, obliquely incident to side AB at an angle α with the normal, is refracted toward the normal (of the more dense

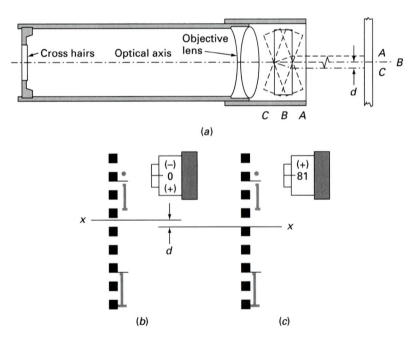

FIGURE 5.39
Optical micrometer and micrometer settings.

225

CHAPTER 5:
Vertical
Distance
Measurement:
Leveling

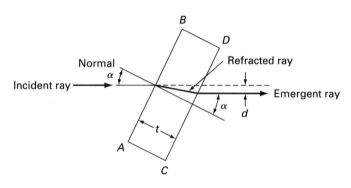

FIGURE 5.40
Principle of the optical micrometer.

medium) as it passes through the glass. This refracted ray is bent in the opposite direction by the same angle α on emerging in air (the less dense medium) at side CD, which is parallel to AB. Thus, the emergent ray is parallel to the incident ray and it can be shown that the displacement d is

$$d = t \tan \alpha \left(\frac{n - 1}{n} \right) \quad \text{(approximately)} \quad (5.22)$$

where d = displacement between incident and emergent rays
$\quad t$ = thickness of the optical flat
$\quad \alpha$ = angle of incidence with the normal to the optical flat
$\quad n$ = refractive index of the glass in the optical flat

Rotation of the plane parallel plate in the optical micrometer produces an apparent vertical shift d of the field of view (Figure 5.39a), so that the image of the cross hair x in Figure 5.39b can be brought into coincidence with a graduation on the scale being observed (Figure 5.39c).

From Equation (5.22), the displacement d varies as the tangent of the rotation angle α. The knob that rotates the optical flat is linked to the plate, so that uniform rotation of the knob generates a tangent function of the tilting plate. A drum scale attached to the knob permits observing the amount of rotation in units of the vertical scale being observed.

The range on an optical micrometer must be sufficient to include one full least division plus one-tenth of this division on the scale to be observed. Figure 5.39 illustrates a cross section of the micrometer attachment and example readings on a metric scale with a least division of 10 mm. In this case the range of the micrometer should be 11 mm with a 0 reading on the drum occurring with the optical flat is position B or normal to the line of sight (Figure 5.39a). Maximum upward (negative) displacement is produced by a counterclockwise rotation of the knob and occurs when the flat is in position A. Maximum downward (positive) displacement results from a clockwise rotation and occurs with the flat in position C. The estimated reading at cross hair x in Figure 5.39b is 1.058 m. A clockwise rotation of the knob displaces the image of the graduations to yield coincidence between the cross hair x and 1.050, where the increment is $+81 = +0.0081$, giving a reading of 1.0581 m. As a check, a counterclockwise rotation of the knob permits displacement of the image and coincidence of cross hair x with 1.060 (not illustrated), yielding a micrometer reading of $-15 = -0.0015$. This gives a reading of $1.060 - 0.0015 = 1.0585$, so the final observation is the average of the two readings, or 1.0583 m. Extreme care must be exercised prior to each observation to ensure proper focusing and coincidence of the level bubble.

The preceding example is for a hypothetical instrument, to illustrate the principle of an optical micrometer applied to a rod reading. For geodetic levels, the range of the micrometer is exactly one rod unit and micrometer units usually are numbered with 0 at A, the position of maximum upward displacement (Figure 5.39a), to 10 at C the position of maximum downward displacement. Thus, when the line of sight is horizontal (B in Figure 5.39a), the micrometer reading is 5. Consequently, all micrometer readings are positive, only one graduation can be intercepted, and all readings are 5 micrometer units too large. The first two factors eliminate two sources of blunder, and the last factor is of no consequence because the 5 micrometer units is a constant for all readings that cancels, since the difference in elevation equals the difference between backsight and foresight rod readings

5.50
PRECISE, THREE-WIRE LEVELING

The subject of precise leveling as practiced on government surveys is not to be considered in detail, as it is thoroughly covered in the U.S. National Geodetic Survey Manual (see Geodetic Leveling, 1984). However, it is appropriate to call attention to certain refinements by means of which a relatively high degree of accuracy may be obtained with the ordinary dumpy, self-leveling, or tilting levels and the self-reading rod.

The rod should be calibrated at frequent intervals by comparison with a standard length. Rods constructed with the graduations on a strip of invar metal are preferable. The rod should have an attached rod level for plumbing. Turning points *must* be on solid objects with rounded tops so that the base of the rod can be held in the same position for backsights and foresights.

The level must be equipped with stadia hairs in addition to the regular cross hairs (Section 5.11). Engineers' levels in the United States have stadia cross hairs spaced so that a reading interval between the upper and lower cross hairs of 1 ft or 1 m is equivalent to 100 ft or 100 m, respectively, of horizontal distance. A tilting level with a coincidence bubble or a self-leveling instrument may be used. To prevent unequal thermal expansion, the level should be protected from the sun by an umbrella. The level tripod should be set very firmly to prevent settlement. To eliminate, as nearly as possible, the effects of variation in atmospheric refraction, settlement of the tripod, or warping of the level, the shortest possible time elapse between backsight and succeeding foresight is desirable. For each sight, the three cross hairs are read by estimation to thousandths of feet or millimeters and recorded. The mean of the readings is taken as the correct rod reading for each sight. The interval between the reading on the upper cross hair and the reading on the lower cross hair is a measure of the distance from level to rod.

A suitable form for notes is shown in Figure 5.41. Note that the entire two-page spread is used for the data, with backsights on the left and foresights on the right page. The terms *thread* and *wire* are used interchangeably to designate cross hair. The station refers to the instrument setup and not to the turning point number. In the example of Figure 5.41, levels are carried in the forward direction form B.M.$_{25}$ to B.M.$_{26}$.

At the beginning of each day the collimation correction or C factor for the level should be determined, as outlined in Section 5.51. When this has been done, rod A is held on B.M.$_{25}$ and a turning point is set in a forward position for rod B. The level is set up to balance the backsight and foresight distances. A backsight is taken on rod A held at B.M.$_{25}$, and the three thread readings 2037, 1843, and 1648 are recorded in the second column. On tilting levels, the level operator must exercise care to be sure the bubble is centered at the instant of reading the rod. When a self-leveling instrument is used, the observer should tap the

227

CHAPTER 5:
Vertical
Distance
Measurement:
Leveling

THREE WIRE LEVELING Date: Oct. 10, 1997 Sun: 8			Forward			U.S. 24, LAKE CO., COLO. (NORTH) From: BM 25 Level No. 1623 Wind: 1				To: BM 26 Time: 0900MST	
No. of Station	Thread Reading, mm	Mean, mm	Middle Thread, ft	Thread Interval	Σ Intervals	Rod No. & Temp.	Thread Reading, mm	Mean, mm	Middle Thread, ft	Thread Interval, mm	Σ Intervals, mm
(1)	(2)	(3)	(4)	(5)	(6)	(7)	(8)	(9)	(10)	(11)	(12)
	2037			194		A	0850			0209	
1	1843	18427	6.04	195		22°C	0641	06403	209	0211	
	1648			389	389		0430			0420	0420
	2446			222		B	0712			0245	
2	2224	22240	7.30	222		21°C	0467	04677	1.53	0243	
	2002			444	833		0224			0488	0908
	1725			114		A	1324			055	
3	1611	16110	5.29	114		22°C	1269	12687	4.16	056	
	1497			228	1061		1213			111	1019
	2089			198		B	0658			167	
4	1891	18913	6.21	197		22°C	0491	04917	1.61	165	
	1694			395	1456		0326			332	1351
	2240			089		A	1116			171	
5	2151	21527	7.07	084		20°C	0945	09453	3.10	170	
	2067			173	1629		0775			341	1692
	29165	97217	31.91ft				11441	38137	12.49ft		1629
	−11441	−38137	(9.73m)						3.81m		63
	3)17724	59080									
	5908										

FIGURE 5.41
Three-wire level notes.

housing or leg lightly during each pointing. The image should oscillate, indicating the compensator is free. If precise level rods graduated on the back side in feet (Section 5.48) are being used, rod *A* is turned and a reading is taken in feet and recorded in the fourth column. The notekeeper immediately calculates and records the half-intervals 194, 195 mm in column 5. If these intervals do not agree to within ± 3 mm (third order, see Table 5.2), the observations should be repeated. The mean thread reading, 1842.7 mm, is calculated and recorded in column 3. This mean should compare closely with the middle thread reading. The sum of half-intervals, 389, is recorded in column 6. The temperature at rod *A* should be recorded.

The telescope now is sighted on rod *B* at the forward turning point and the foresight thread readings 850, 641, and 430 mm are observed and recorded in column 8 of the notes. the middle thread is read on the back side of rod *B* as 2.09 ft and recorded in column 10. The half-intervals 209, 211 mm agree satisfactorily and are recorded in column 11. The mean thread reading 640.3 mm is calculated and recorded in column 9. At this point, instrument setup at station 1 is complete.

Rod *A* is moved ahead and a turning point is established in the direction of B.M.$_{26}$. The level is set at station 2 and the first sight is a foresight on rod *A*, with the data being recorded on the right page as before. This step is followed by a backsight on rod *B* on the previous turning point, with data being recorded on the left page. This completes instrument setup 2. Rod *B* is moved ahead to a forward turning point and the level is moved to station 3. At station 3, the backsight is taken first and the foresight second. This routine of observing backsight-foresight, foresight-backsight, and so on, is alternated at each successive setup to minimize the effects of instrumental settlement (Section 5.40). The backsight and foresight distances should balance to within the specifications given in Table 5.2. For third-order levels, the difference between backsight and foresight thread intervals should not exceed 100 mm or 10 m horizontal distance. Similarly, the difference between cumulated backsight and foresight thread intervals should not exceed 100 mm or 10 m on the ground. For example, the first specification is exceeded at station 3, where the difference is $228 - 111 = 117$ mm, which is equal to a horizontal distance of 11.7 m. However, the difference between the cumulated B.S. and F.S. distances between B.M.$_{25}$ and B.M.$_{26}$ is 63 mm (or 6.3 m), well within the latter specifications. Levels are run forward and backward in sections, the lengths of which are not to exceed the specification given in Table 5.2. Thus, even on long lines, each day's work provides a closed loop.

When the leveling is completed, thread readings, mean thread readings, and middle thread readings (on back of rod) are cumulated and listed in their respective columns at the bottom of the note sheet. The sum of the thread readings divided by 3 should equal the sum of the mean readings, and the sum of the middle thread readings converted to meters should check closely with the sum of the mean values, as shown in Figure 5.41. The difference in elevation between B.M.$_{25}$ and B.M.$_{26}$, uncorrected for collimation error or curvature and refraction, equals the difference between the sum of the mean backsights and the sum of the mean foresights.

5.51
COLLIMATION CORRECTION

In precise three-wire and geodetic leveling, when the sum of backsights and sum of foresights are unbalanced, a correction can be applied if the slope of the line of sight is known. This slope of the optical line of sight occurs in a tilting level when the axis of the spirit level tube and optical axis are not parallel or in a self-leveling instrument if the line of sight is not truly horizontal when the compensator is free. The amount of slope in units per unit of length of sight is the collimation correction, referred to as the C factor.

The effects of an inclined line of sight were discussed in Sections 5.37 and 5.40 and are further illustrated in Figure 5.42, where the dashed line of sight is inclined by angle α with the true horizontal line. In Figure 5.42a, the line of sight deviates downward from true horizontal, causing the backsight and foresight rod readings N_{s1} and N_{l1} to be too small. Due to this condition, the sign of α (and C) is taken as positive. Conversely, when the line of sight deviates upward from true horizontal, the sign of C is negative.

The determination of C involves a procedure similar to the *peg test* described in Section 5.62 for adjustment of the level. Set two points, A and B, about 200 ft or 60 m apart (Figure 5.42a). Set up the level to be checked at station 1 about 20 ft or 6 m from A. Read and record the backsight thread readings, half-intervals, sum of intervals, and backsight mean in columns 2, 4, 5, and 3, as indicated in Figure 5.43. The backsight mean 1356.0 mm $= N_{s1}$ and the sum of intervals 92 mm $= d_{s1}$ in Figure 5.42a. Next, read and record foresight thread readings, half-intervals, sum of intervals, and foresight mean in columns 6, 8, 9, and 7 of Figure 5.43. The foresight mean 1070.7 mm $= N_{l1}$ and the sum of intervals 441 mm $= d_{l1}$ in Figure 5.42a. The level is then moved to station 2 (Figure 5.42b) and the entire procedure as outlined for station 1 is repeated. From station 2, $N_{s2} = 1281.7$ mm, $d_{s2} = 119$ mm, $N_{l2} = 1574.0$ mm, and $d_{l2} = 426$ mm.

The true differences in elevation between A and B from stations 1 and 2 are

$$\Delta H_{AB} = (N_{s1} + Cd_{s1}) - (N_{l1} + Cd_{l1}) \qquad (5.23a)$$

$$\Delta H_{AB} = (N_{l2} + Cd_{l2}) - (N_{s2} + Cd_{s2}) \qquad (5.23b)$$

229

CHAPTER 5:
Vertical
Distance
Measurement:
Leveling

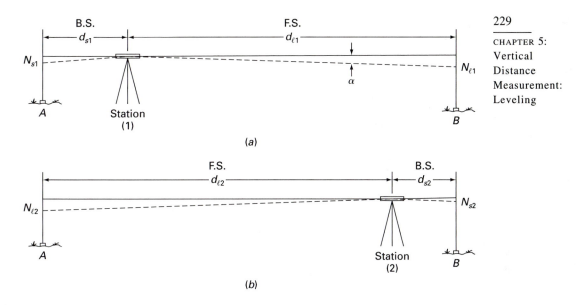

FIGURE 5.42
Determination of the C factor.

Sta. (1)	Thread Reading (2)	B.S. Mean (3)	Thread Interval (4)	Sum Intervals (5)		Thread Reading (6)	F.S. Mean (7)	Thread Interval (8)	Sum Intervals (9)
	1402		46			1291		220	
1	1356	1356.0	46			1071	1070.7	221	
	1310	$= N_{s1}$	92	$92 = d_{s1}$		850	$= N_{\ell1}$	441	$441 = d_{\ell1}$
	1341		59			1787		213	
2	1282	1281.7	60			1574	1574.0	213	
	1222	$= N_{s2}$	119	$211 = d_{s1} + d_{s2}$		1361	$= N_{\ell2}$	426 $= d_{\ell2}$	$867 = d_{\ell1} + d_{\ell2}$
			$= d_{s2}$						
	7913	2637.7	$= N_{s1} + N_{s2}$			7934	2644.7	$= N_{\ell1} + N_{\ell2}$	$-211 = d_{s1} + d_{s2}$
	$-$	2644.7	$= N_{\ell1} + N_{\ell2}$						
							$(d_{\ell1} + d_{\ell2}) - (d_{s1} + d_{s2}) = 656$		
	$-$	7.0	$= (N_{s1} + N_{s2}) - (N_{\ell1} + N_{\ell2})$						
	$C =$	$\dfrac{-7.0}{656}$	$= -0.011$ mm /unit of stadia			interval			

C FACTOR DETERMINATION

Level No. 438

Oct 10, 1006 Temp. 80°
Clear

FIGURE 5.43
Notes of C factor determination.

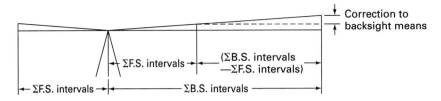

FIGURE 5.44
Application of C factor to imbalance in Σ F.S. and Σ B.S. intervals.

where C = correction per unit of stadia interval and is positive when the line of sight is
below horizontal
N_{s1}, N_{s2}, d_{s1}, and d_{s2} are defined in Figure 5.42.

Equating Equation (5.23a) with (5.23b) yields

$$(N_{s1} + Cd_{s1}) - (N_{l1} + Cd_{l1}) = (N_{l2} + Cd_{l2}) - (N_{s2} + Cd_{s2}) \qquad (5.24)$$

Solving Equation (5.24) for C gives

$$C = \frac{(N_{s1} + N_{s2}) - (N_{l1} + N_{l2})}{(d_{l1} + d_{l2}) - (d_{s1} + d_{s2})} \qquad (5.25)$$

Therefore, the C factor equals the difference between the sum of the short backsights minus
the sum of the long foresights, the quantity divided by the difference between the sum of
the long foresight distances and short backsight distances. For the example given in
Figure 5.43, C is -0.011 mm/mm of stadia interval. If rods graduated in feet were used,
the units would be feet per foot of stadia interval. The minus sign indicates that the line of
sight is above the horizontal.

The value for C is applied to the lack of balance between cumulated foresight and
backsight distances. The difference in elevation corrected for inclination of the line of
sight is

$$\Delta H \text{ corrected} = \Delta H \text{ observed} + C(\Sigma \text{ B.S. intervals} - \Sigma \text{ F.S. intervals}) \qquad (5.26)$$

This correction is shown graphically in Figure 5.44. If the C factor calculated in Figure 5.43 were applied to the change in elevation between bench marks 25 and 26 in
Figure 5.41, the elevation corrected for inclination of the line of sight would be

$$\Delta H_{25-26} = 5.9080 \text{ m} + (-0.011 \text{ mm/mm})(1629 \text{ mm} - 1692 \text{ mm})$$

$$= 5.9080 \text{ m} + 0.0007 \text{ m} = 5.9087 \text{ m}$$

When the collimation test is performed for a tilting level or automatic level equipped
with an optical micrometer, the general field procedure is similar but the sequence of
observations and the note form are different. The sequence of readings follows the instructions given in Section 5.52 under "Geodetic Leveling Procedure." Figure 5.45 will show
the note form used. A detailed example of this type of C-factor determination can be found
in Schomaker and Berry (1981).

5.52
GEODETIC LEVELING

Geodetic leveling is direct leveling of a high order of accuracy and precision. It is based on
the irregular surfaces associated with the geoid rather than the regular surface of a spheroid
or an ellipsoid representing the surface of the earth. The geoid is a surface shaped by gravity

that closely conforms to a sea level surface extended continuously through the continents (Section 1.3, Figure 1.2). At every point, the geoid's surface is perpendicular to the plumb line; it is determined by measuring the acceleration of gravity.

231

CHAPTER 5:
Vertical
Distance
Measurement:
Leveling

Geodetic leveling must comply with the general specifications as set forth in Table 5.2 for the various orders and classifications for leveling (e.g., first order, Classes I and II). It normally is conducted in connection with control surveys over large areas to furnish vertical control for the national network, for extensive engineering surveys, and crustal movement studies, to mention a few application areas.

The methods employed generally are the same as described in Section 5.50 for precise three-wire levels. However, the procedure and equipment are more refined and more attention is paid to details. For example, an automatic self-leveling level or a tilting level, each equipped with an optical micrometer, is specified. In addition, digital bar-code systems (Section 5.20, Figure 5.19) currently are being evaluated and, in all probability, will be found adequate for certain classes of vertical control surveys (see Table 5.2). Calibrated matched pairs of precise level rods as described in Section 5.50 must be used and a double-scale rod similar to the one illustrated in Figure 5.38 is preferred. The level must be protected from the sun's rays by an umbrella at all times.

For first-order, Class I leveling, the maximum length of sight is 50 m, the maximum unbalance between cumulated backsight and foresight distances is 4 m, and the allowable closure error forward and backward per section is 3 mm $\sqrt{D}$, where D is the one-way length of the section in km. These and other specifications are listed in Table 5.2. Details concerning instruments, procedures, and note forms can be found in the National Geodetic Survey publication (Schomaker and Berry 1981).

Bench marks are spaced at intervals not to exceed 3 km. The level routes almost invariably are along railroads or highways; therefore, the bench marks of the system of vertical control usually are near no horizontal control points, many of which are elevated to provide visibility for the GPS satellite network (Chapter 12) or were in positions suitable for older triangulation-trilateration networks. The usual form of a bench mark is an inscribed metal disk set solidly into a concrete post, masonry structure, or rock.

The procedure currently preferred by the National Geodetic Survey is to use an automatic level equipped with a reversible compensator and optical micrometer (Figure 5.37) with a pair of double-scale rods (Figure 5.38), each rod having a different constant offset. Use of the reversible compensator and double-scale rods with different constants permits minimization of systematic instrumental errors and detection of blunders such as transposition of backsight and foresight rod readings. Therefore, when checking against existing elevations is possible, single-run, double-simultaneous (SRDS) leveling can be performed.

A definite sequence for leveling the instrument and observing the low and high scales is specified to reduce the accumulation of small systematic errors that occur in the instrument and rods. Details of these systematic errors and recommended compensatory steps can be found in Schomaker and Berry (1981).

Observations may be (1) keyed into a portable, battery-powered, programmable computer that serves as an electronic notebook and field calculator or (2) recorded on a standard form. The field computer is programmed to provide a warning signal when discrepancies occur in the observations. For example, if backsight and foresight distances do not balance or elevations determined by the left and right scale readings do not agree, a warning signal is displayed. This feature permits an immediate check on the quality of the measurements and correction of the irregularities before the level is moved to the next setup. This aspect of the electronic notebook is a definite advantage. Note that the procedure and form for the written record is designed to provide the same checks but at the expense of considerable time and accompanied by the inevitable human mistakes in recording and calculation.

At the end of a day's work, data stored in the field computer or recorded on the standard forms are either downloaded or keyed into a personal computer in the field office. This computer record (a file in computer memory or on diskette) then is transmitted by telephone line to the central office for further analysis of the data, correction for all systematic errors, and eventual adjustment of the level net when the field operations are completed. In addition to the computer record, the results of the day's work are added to the abstract of elevations for the entire project, a running record, maintained in the field office, of all the work done between control points.

NOAA FORM 76—191 (8—77)

U.S. DEPARTMENT OF COMMERCE — NATIONAL OCEANIC AND ATMOSPHERIC ADMINISTRATION

GEODETIC LEVELING
MICROMETER OBSERVATIONS (Δh)

Example, Second Order, Class I

PAGE 1 OF 1

· 4 0 ·	YR. / MO. / DAY	CODE	INSTRUMENT SERIAL NO.	M	CODE	ROD SERIAL NO.	CODE	ROD SERIAL NO.		Z	TIME
· 4 1 ·	940216		101288		1	1045324	2	1034525	$C = +0.020\ ^{mm}/_{m}$		

FROM	BM DESIGNATION.	TO	BM DESIGNATION.	Z	TIME BEGIN	TIME END	F/C	TEMPERATURE BEGIN	TEMPERATURE END	W	S	OBSERVER
	BM 25		JP 1		1400	1715	F	55	50			JMA

① SET UP#	② STADIA BACK	③ SB Σ	④ STADIA FORE	⑤ SF Σ		⑥ LOW SCALE BACK/FORE	⑦ ΔhL Σ		⑧ HIGH SCALE BACK/FORE	⑨ ΔhH±d Σ	⑩ ΔhL−ΔhH	REMARKS
1	184	210	424	212	B	18427	−24048	B	77677	−25045	−03	*Constant difference between high and low scales:*
	1630		4028		F	42475		F	102722			
		$\Delta s = S_B - S_F = -0.2$								+10		*scales:*
2	222	222	573	224	B	22240	−35126	B	82490	−34131	+05	*Rod 1- 592.5*
	1998	432	5506	434	F	57366	−59174	F	116621	−59176	+02	*Rod2- 602.5*
				+0.2						−10		
3	161	114	229	118	B	16110	− 6846	B	75360	− 7845	−01	
	1496	546	2172	552	F	22956	−66020	F	83205	−67021	+01	
				−0.4						+10		
4	189	198	469	201	B	18913	−27992	B	79163	−26988	−04	
	1692	744	4759	753	F	46905	−94012	F	106151	−94009	−03	
				−0.3						−10		
5	215	89	249	86	B	21527	−24148	B	80777	−25154	+06	
	2061	833	2404	839	F	45675	118160	F	105931	−119163	+03	
				+0.3						+.10		
6	210	104	220	99	B	21055	− 1028	B	81305	− 0033	+05	
	1996	937	2101	938	F	22083	−119188	F	81338	−119196	+08	
		+93.8		+0.5						−10	*Mean ΔH =*	*1191.920 hcm*
		1875			B			B				*÷ 200 hcm/m*
		×.001			F			F				*−5.9596 m*
		S = 0.188km										
					B			B				
					F			F				
					B			B				*Rod 1: J. Campbell*
					F			F				*Rod 2: R. Kovaks*
					B			B				*Recorder: K. McIntyre*
					F			F				

FIGURE 5.45
Notes for geodetic leveling, micrometer observations.

In the main office, field data and computations are checked and recomputed. Corrections for collimation error, rod calibration and temperature, earth curvature and refraction, and tidal acceleration are computed and applied. An *orthometric* correction (Section 5.53), which is a function of latitude and elevation, is made to account for the irregular shape of the geoid. Circuit closures are made to detect and rectify serious misclosures. When all mistakes and systematic errors have been detected and removed, a least-squares adjustment of the entire network (Sections 2.22–2.28 and 5.58) is performed to provide adjusted elevations of all benchmarks in the network. Elevations are provided in meters and feet and descriptions of bench marks are prepared for publication.

233

CHAPTER 5:
Vertical
Distance
Measurement:
Leveling

Geodetic Leveling Procedure

Figure 5.45 shows the notes, on a NOAA standard form, for a line of geodetic levels run using double-scale, half-centimeter rods and an automatic level equipped with a micrometer and with a nonreversible compensator. With this equipment, two independent elevation differences are taken from each setup. This procedure is referred to as *single run, double simultaneous* (SRDS) in Table 5.2.

First, one records equipment numbers, date and time zone, temperature units, collimation factor, and names or initials of the observer, recorder, and rodpersons. With the level set on the line between the initial control point and first foresight (equalize foresight and backsight distances), check to make sure the rods are plumb and then perform the following sequence of operations for instrument setup number 1.

1. *Backsight, low scale.* Sight telescope on the backsight rod and read the center cross hair on the low scale of 184.27 (three- or four-digit rod reading plus a two-digit micrometer reading) and record this in column 6, line 1. Record the rod reading only of 184 in column 2, line 1. Read the lower stadia hair of 163.0 to tenths of rod units and record in column 2, line 2. Compute the half-interval of 21.0 and record in column 3, line 1. The rod is graduated in half-centimeters, so the backsight distance is 21.0 m.
2. *Foresight, low scale.* Point toward the foresight rod and read the center cross hair of 424.75, which is recorded in column 6, line 2, with 424 also being recorded in column 4, line 1. Read and record the lower stadia hair observation of 402.8 in column 4, line 2. Compute the foresight distance $= 424 - 402.8 = 21.2$ and record this in column 5, line 1. Check the backsight distance $-$ foresight distance $= 21.0 - 21.2 = -0.2 <$ 5 m, the specification given for this value in Table 5.2. Calculate the low-scale difference in elevation, $\Delta h_L = 184.27 - 424.75 = -240.48$ and record this in column 7, line 1. Because the level has a nonreversible compensator, delevel the instrument by rotating the level screw toward the foresight and relevel it using the other two screws. When the instrument has been leveled, high-scale observations are made as follows.
3. *Foresight, high scale.* With the telescope still focused on the foresight, read the center cross hair intercept on the high scale of 1027.22 and record this in column 8, line 2.
4. *Backsight, high scale.* Point toward the backsight rod, focus, and read the center cross hair on the high scale of 776.77, which is recorded in column 8, line 1. Compute the high-scale difference in elevation, $\Delta h_H \pm d = 776.77 - (1027.22 \pm d) = -250.45 \pm d$. Rod 1 is the backsight, so $d = +10$ (when rod 2 is the backsight, $d = -10$), $\Delta h_H = -250.45 + 10 = 240.45$, and $\Delta h_L - \Delta h_H = -.03$ rod units (half-centimeters), recorded in column 10. This value in millimeters equals $0.03 \times 5 = 0.15$ mm, less than the specified amount of 0.60 mm given for this parameter in Table 5.2 for second-order, Class-I levels.

Next, the level is moved to instrument setup 2, rod 2 remains at the old foresight position and becomes the backsight, rod 1 is moved ahead to equalize backsight and foresight distances, and the procedure from steps 1 through 4 is repeated for the next setup and then for all subsequent setups in the line of levels.

Referring to Figure 5.45, note that backsight and foresight distances, low-scale and high-scale elevation differences (Δh_L and Δh_H), and the differences between low-scale and high-scale elevations ($\Delta h_L - \Delta h_H$) are cumulated for each setup in columns 3, 5, 7, 9, and 10, respectively. For this example, the sum of cumulated backsight and foresight distances = 93.7 + 93.8 = 187.5 m or 0.188 km, the distance leveled. The mean of the cumulated sum of low-scale and high-scale elevation differences is −1191.92 half-cm = (−1191.92) half-cm ÷ 200 (half-cm)/m = − 5.9596 m. The cumulated differences between low- and high-scale readings is 0.08 rod units or 0.40 mm. This value has no meaning in this context because the example line of levels is only a part of several lines that make up a section, and even though it has been run SRDS, it must close on a known control point to be used as part of a section of second-order, Class-I levels. Otherwise, the line must be run in the opposite direction to be included as part of a section that provides a closure which can be checked against the specified loop closure for this order and class of leveling.

When the observations are keyed into a field data collector, the procedure is the same except that all entries are stored in computer memory. All specified checks are shown on the data collector display screen with warnings included when a check is not satisfied. The entire process is speeded up substantially and human errors in calculation are eliminated. Extreme care must be exercised in keying in data to avoid mistakes in data recording and the recommended backup forms (see Schomaker and Berry 1981) also must be carefully maintained.

The example cited is only a bare bones description of the overall procedure. Those who are interested in geodetic leveling must obtain Manual NOS NGS 13, *Geodetic Leveling* (1981) and follow the detailed instructions meticulously to achieve results that satisfy the specifications and standards.

Procedure with Digital Levels

As noted in Section 5.47, a heavier, more precise version of the digital level shown in Figure 5.19 has been approved for certain types of geodetic leveling. When a digital level and the associated bar-code level rod are used for this purpose, it is referred to as an *electronic digital/bar-code leveling system*. The use of such a system for geodetic leveling is governed by special conditions and specifications (in addition to those in Table 5.2), which are currently evolving and being given interim approval, contingent on evaluation of results obtained with these systems. Therefore, organizations and individuals desiring to use digital systems for geodetic leveling must contact the Vertical Control Branch of the National Geodetic Survey* to inquire about the latest interim specifications and procedures set forth by the FGCS for the use of electronic digital/bar-code leveling systems for geodetic leveling.

5.53
ORTHOMETRIC CORRECTION

As noted in Section 5.1, the datum for leveling is *level surface* everywhere perpendicular to gravity. The most commonly used datum is the mean sea level surface. The force of gravity is less at the equator than at the poles, owing primarily to the action of centrifugal force. Because the level surface is a function of gravity, a given level surface is farther from

---

* NOAA, National Geodetic Survey, 1315 East-West Highway, Station 9202, Silver Spring, MD 20910–3282.

N

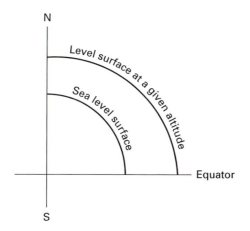

Level surface at a given altitude

Sea level surface

Equator

S

235

CHAPTER 5:
Vertical
Distance
Measurement:
Leveling

FIGURE 5.46
Divergence of a level surface from sea level.

sea level at the equator than at a point nearer the pole. What this means in terms of leveling is that level lines at different elevations that progress along northerly or southerly directions are referenced to different level surfaces. Figure 5.46 illustrates the divergence of a level surface from the mean sea level surface.

For example, assume that levels referred to the datum are run along an inland route from San Diego, California, to Seattle, Washington, where closure in both cases is on a bench mark at or near the datum. The level surface at San Diego is farther from the datum than the level surface at Seattle, so the error due to this convergence of level surfaces is about 1.3 m. This amount in that distance is significant for first-order levels.

The orthometric correction is applied to account for this convergence of level surfaces at different elevations and can be calculated using the equation

$$\text{Correction} = -0.005288 \sin 2\phi h \Delta\phi \text{ arc } 1' \tag{5.27}$$

where ϕ = latitude at the starting point
h = datum elevation, m or ft, at the starting point
$\Delta\phi$ = change in latitude in minutes between the two points (+ in the direction of increasing latitude or toward the pole)

EXAMPLE 5.7. The elevation of a lake running in a north-south direction is 200 m above the datum at the northern end, where $\phi = 42°00'$N. A line of levels is run along the lake to the south end, where $\phi = 40°40'$N. Calculate the orthometric correction.

Solution. By Equation (5.27),

$$\text{Orthometric correction} = (-0.005288)(\sin 84°)(200 \text{ m})(-80')(0.000290)$$

$$= 0.024 \text{ m}$$

Therefore, the elevation at the south end of the lake = 200.000 m + 0.024 = 200.0244 m.

EXAMPLE 5.8. A level line is run from latitude 38°00'N along a south to north line to latitude 38°05'N at an average elevation of 7000 ft above the datum. Compute the orthometric correction.

Solution. By Equation (5.27),

$$\text{Orthometric correction} = (-0.005288)(\sin 76°)(7000)(05')(0.000290)$$

$$= -0.052 \text{ ft}$$

The elevation at 38°05'N latitude = 7000.00 − 0.052 = 6999.948 ft above the datum.

5.54
RECIPROCAL LEVELING

Occasionally, it becomes necessary to determine the relative elevations of two widely separated intervisible points between which levels cannot be run in the ordinary manner. For example, it may be desired to transfer levels from one side to the other of a deep canyon or from bank to bank of a wide stream.

If *A* and *B* are two such points, then the level is set up near *A* and one or more rod readings are taken on both *A* and *B* (Figure 5.47a). Then the level is set up in a similar location near *B*, and rod readings to near and distant points are taken as before. The mean of the two differences in elevation thus determined is taken to be the difference between the two points. Usually, the distance between points is large (often a half mile or a kilometer), so that it is necessary to use a target on the distant rod. If more accurate results are desired, a series of foresights is taken on the distant rod and sometimes also a series of backsights on the near rod, the bubble being recentered and the target reset after each observation. The difference in elevation then is computed by using the mean of the backsights and the mean of the foresights. The three-wire method described in Section 5.50 can be used here.

More refined results can be obtained if a tilting level equipped with a gradienter is available (Section 5.18). In this case, two targets are placed on the distant rod at *B*, one above and one below the approximate point where the horizontal line of sight strikes the rod. The optical line of sight is tilted using the micrometer screw and a series of gradienter readings is observed and recorded to each of the targets on the distant rod and also when the spirit level bubble is centered. Readings for each target position on the rod are recorded and the position of the line of sight is calculated by proportion using the gradienter readings. The entire operation is repeated with a setup near *B*.

Factors that may appreciably affect the results are variations in temperature, causing unequal expansion of parts of the instrument, variations in atmospheric refraction, and the collimation error of the level. To minimize these effects, the instrument should be shaded and corrections for curvature and refraction applied (Section 5.2). A more satisfactory approach is to use two levels, each having about the same magnifying power and sensitiveness of level bubble. With one level set up near *A* and the other near *B*, sets of observations

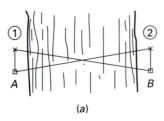

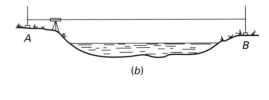

(b)

(a)

FIGURE 5.47
Reciprocal leveling: (a) plan view, (b) section.

are obtained simultaneously from both locations. Thus, the effects of curvature and refraction cancel. Then, to limit the effects of collimation errors, the two instruments are interchanged and the operation is repeated. The mean of the differences in elevations is taken as the best estimate of actual difference in elevation.

237

CHAPTER 5:
Vertical
Distance
Measurement:
Leveling

When self-leveling instruments are used for reciprocal leveling, they must be equipped with rotary wedge attachments. A *rotary wedge* is an optical device, attached to the objective end of the telescope, that permits determining the slope of the line of sight. Thus, the self-leveling instrument can be used in the same way as the tilting level.

When the highest precision is required, special river-crossing equipment is available. This system consists of two self-leveling instruments attached to an elongated, rigid base plate that is then securely screwed to the top of a tripod. Each level is equipped with a rotary wedge attachment. Two of these systems are employed, one on each side of the river or canyon, so there is no need to transport instruments from one side to the other. Reciprocal collimation is performed between each level in each system to minimize effects of collimation errors. Repeated measurements then are made, simultaneously on both sides, from the nearby point to the opposite side. The average of the changes in elevations, thus obtained, is used as the best estimate of the difference in elevation. A detailed description of the procedure for using this special equipment may be found in Schomaker and Berry (1981).

5.55
TRIGONOMETRIC LEVELING FOR LONG LINES

Trigonometric leveling using short lines was described in Section 5.5 and illustrated in Figures 5.3 and 5.4. For these cases, the triangle used in calculating the height difference was assumed to be a right triangle. When long slope lines (lines ≥ 3 km or 2 mi, $85° \geq z \geq 95°$ or $\alpha \geq 5°$) are measured, the assumption of a right triangle no longer is adequate. In trigonometric leveling with long lines (Figure 5.48), the slope distance $s = O'P$ and zenith angles $z_{O'}$ or z_P or vertical angles α and β are measured using total station systems (Chapter 7) or an EDM and theodolites at stations O' and P. In Figure 5.48, point O' is the center of an EDM transmitter and P represents the effective center of a reflector at heights of h.i.$_{O'}$ and h.i.$_P$ above their respective stations. Note, that if the h.i.'s of theodolites and targets employed to measure $z_{O'}$ and z_P or α and β are different, corrections to these angles must be computed and applied (see Section 4.41). Then obtuse triangle $O'PP'$ is solved for vertical side $P'P$ that is used to calculate the difference in elevation, ΔH_{AB}.

Solution of triangle $O'PP'$ is similar to the procedure developed in Section 4.43 for reduction of long slope lines to horizontal distances. As in that reduction, simultaneous measurement of reciprocal zenith or vertical angles causes the effects of curvature and refraction (*c&r*; Section 5.2) to cancel. Because variations in refraction constitute a significant source of uncertainty in trigonometric leveling, observation of reciprocal angles is almost mandatory if high precision is sought. Section 4.43 also contains derivations for the basic equations required to solve obtuse triangle $O'PP'$.

Consider two cases: Case I where reciprocal angles are measured, and Case II in which angular measurements are made from only one end of the line. In Case I (Figure 5.48), look at the general situation where a stand-alone or top-mounted EDM occupies station A, a reflector is set up at station B, at respective heights of instrument of h.i.$_A$ and h.i.$_B$, and the slope distance $O'P$ is measured. The reflector at B then is replaced by a target allowing measurement of $z_{O'}$ or α, and a theodolite is set on the tripod at B and a target set at A, permitting measurement of z_P or β in as nearly a simultaneous mode as possible. Near simultaneous observation of angular values is necessary to achieve similar atmospheric conditions for both sets of measurements and avoid the uncertainties caused by variations in refraction. Note that, for the described situation, $z_{O'}$ and z_P or α and β must be corrected for any differences in heights of instrument between theodolites and EDM and reflectors and targets (Section 4.41). Use of total station systems with coaxial optics and a combined target-reflector, with the target mounted symmetrically about the reflector, eliminates the need for certain corrections. In addition, if a target can be mounted on the standards of total station instruments or theodolites so the center of the target is aligned with the horizontal axes of the instruments at either end of the line, true simultaneous angular measurements are possible.

When the data are acquired, the slope distance, s, is corrected for atmospheric conditions (Sections 4.36 and 4.37) and all system constants (Sections 4.38 and 4.39), and angles $z_{O'}$ and z_P or α and β are corrected for differences in h.i.'s. Angle ϕ, in triangle $O'PP'$ (Figure 5.48), can be computed by Equation (4.43), and then, using this value of ϕ, a reasonable approximation is calculated for the horizontal distance PP' by Equation (4.46), from which the angle of curvature, c, is computed by Equation (4.41) or (4.42). Triangle $O'PP'$ then can be solved by the law of sines for the vertical distance PP', using

$$PP' = \frac{O'P \sin \phi}{\sin(90 + c)} \tag{5.28}$$

and the difference in elevation from A to B is

$$\Delta H_{AB} = \text{h.i.}_A + PP' - \text{h.i.}_B \tag{5.29}$$

If two total station systems occupy stations A and B and distances are measured in each direction in addition to reciprocal zenith or vertical angles, then two separate obtuse

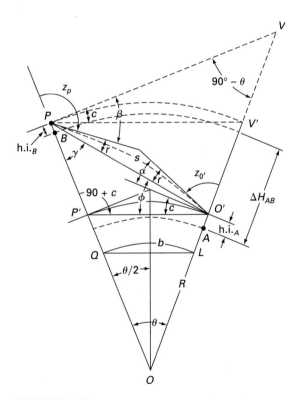

FIGURE 5.48
Trigonometric leveling.

triangles are involved. The computational procedure just outlined is applied to each triangle to give two values for the difference in elevation, which are averaged.

239

CHAPTER 5:
Vertical
Distance
Measurement:
Leveling

For Case II, when observations are made from only one end of the line, the measured distance and angles are corrected as in Case I. Next, an estimate for $(c\&r)$ is calculated using the slope distance in Equation (4.50) or (4.51). With this value for $(c\&r)$, angle ϕ is found by Equation (4.52) or (4.53) and used in Equation (4.46) to get a reasonable estimate for the horizontal distance $O'P'$, which can be employed to recompute improved estimates for $(c\&r)$ and ϕ and a value for c using Equation (4.41) or (4.42). Two iterations usually are sufficient for convergence to stable value of $(c\&r)$ and ϕ. Then, these improved estimates are used in Equations (5.28) and (5.29) to yield the vertical difference PP' and the difference in elevation.

EXAMPLE 5.9. A slope distance of 5,457.500 was measured from an add-on EDM at station G, 323.000 m above the datum, to a reflector at station H. Reciprocal zenith angles were then measured at G and H. Additional data are as follows:

		h.i.'s at G, m	h.i.'s at H, m
$z_G = 86°57'00''$	EDM	1.615	
$z_H = 93°13'10''$	Theodolite	1.431	1.440
	Reflector		1.275
	Target	1.300	1.465

Calculate the elevation of H.

Solution. Let the centers of the EDM at G and the reflector at H be labeled O' and P, respectively, to correspond to Figure 5.48. Compute Δz_G and Δz_H by Equations (4.33) and (4.34a):

$$\Delta\text{h.i.}_G = (1.275 - 1.465) - (1.615 - 1.431) = -0.374 \text{ m}$$

$$\Delta z_G = -\left(\frac{\Delta\text{h.i. } \sin z_G}{S}\right)\rho = -\left(\frac{-0.374 \text{ m} \sin 86°57'00''}{5,457.500 \text{ m}}\right) 206,265''/\text{rad}$$

$$\Delta z_G = +14.1'' \quad \text{so that } z'_G = z_G + \Delta z_G = 86°57'14.1''$$

$$\Delta\text{h.i.}_H = (1.615 - 1.300) - (1.275 - 1.440) = 0.480 \text{ m}$$

$$\Delta z_H = -\left(\frac{0.480 \text{ m} \sin 93°13'10''}{5,457.500 \text{ m}}\right)\rho = -18.1''$$

$$z'_H = z_H + \Delta z_H = 93°13'10'' - 18.1'' = 93°12'51.9''$$

Compute angle ϕ by Equation (4.43):

$$\phi = \frac{z'_H - z'_G}{2} = \frac{93°12'51.9'' - 86°57'14.1''}{2} = 3.130250° = 3°07'48.9''$$

Calculate a reasonable estimate for the horizontal distance between G and H.

$$S_G \text{ (horizontal)} \approx S_{\text{slope}} \cos \phi \approx 5,457.500 \cos 3°07'48.9'' \approx 5,449.357 \text{ m}$$

and then c may be obtained with sufficient precision from Equation (4.42).

$$c = (16.192''/\text{km})(5.449 \text{ km}) = 88.23'' = 0°01'28.2''$$

Next, compute the vertical difference between the horizontal axis of the EDM at G to the center of the reflector at H (PP', Figure 5.48). Use Equation (5.28):

$$PP' = \frac{s \sin \phi}{\sin (90° + c)} = \frac{5,457.500 \sin 3°07'48.9''}{\sin 90°01'28.2''} = 298.012 \text{ m}$$

Finally, determine ΔH_{GH} by Equation (5.29):

$$\Delta H_{GH} = \text{h.i.}_{\text{EDM}@G} + PP' - \text{h.i.}_{\text{reflector}@H} = 1.615 + 298.012 - 1.275$$

$$\Delta_{GH} = 298.352 \text{ m above the datum}$$

EXAMPLE 5.10. A slope distance of 6,489.254 m, corrected for atmospheric conditions and all system constants, was measured from a total station system set over station A, which has an elevation of 100.525 m above the datum, to a combined target-reflector at station B. The zenith angle at A is 82°20′30″, measured from an $\text{h.i.}_A = 1.243$ m, to the center of the combined target-reflector where the $\text{h.i.}_B = 1.428$ m. Compute the elevation of B. Let the horizontal axis of the total station system at A and the center of the reflector at B be labeled O' and P, respectively, to correspond to Figure 5.48.

Solution. Calculate $(c\&r)$ with Equation (4.51), approximating the horizontal distance by the slope distance, $O'P$:

$$(c\&r) = \frac{(13.925''/\text{km}) (6489.254 \text{ m})}{(1000 \text{ m/km})} = 90.4'' = 0°01'30.4''$$

Compute a first approximation for ϕ using Equation (4.52):

$$\phi = 90° - z_A + (c\&r) = 90° - (82°20'30'') + 0°01'30.4'' = 7°41'00.4''$$

Compute an estimate for the horizontal distance from A to B or $O'P'$ in Figure 5.48. Use Equation (4.46):

$$O'P' \approx O'P \cos \phi = (6489.254) \cos 7°41'00.4'' = 6430.993 \text{ m}$$

Now, recompute $(c\&r)$ and ϕ and calculate c using this estimate for $O'P'$ in Equations (4.51), (4.52), and (4.42).

$$(c\&r) = (13.925''/\text{km})(6431 \text{ km})/(1000 \text{ m/km}) = 89.6'' = 00°01'29.6''$$

Another iteration would reveal no significant change in $(c\&r)$, so now recalculate ϕ,

$$\phi = 90° - 82°20'30'' + 0°01'29.6'' = 7.68322° = 7°40'59.6''$$

and determine c with Equation (4.42),

$$c = (16.192''/\text{km}) (6431 \text{ m})/(1000 \text{ m/km}) = 104.1'' = 00°01'04.1''$$

The vertical distance PP' is given by Equation (5.28):

$$P'P = \frac{O'P \sin \phi}{\sin (90° + c)} = \frac{6489.254 \sin (7°40'59.6'')}{\sin (90°01'44.1'')} = 867.587 \text{ m}$$

The difference in elevation between A and B is given by Equation (5.29):

$$\Delta H_{AB} = \text{h.i.}_A + PP' - \text{h.i.}_B = 1.243 + 867.587 - 1.428 = 867.402 \text{ m above the datum.}$$

5.56
ERROR PROPAGATION IN TRIGONOMETRIC LEVELING

Error propagation in trigonometric leveling is an application of the general case and special law of error propagation (Section 2.19, Equations 2.34 through 2.38) to the equations developed in Sections 5.5 and 5.55 for trigonometric leveling with short and long lines, respectively. In this procedure, standard deviations are estimated for the measured values and the error is propagated through the equations used to calculate the difference in elevation. Consider the error propagation involved for trigonometric leveling using short and long lines as given in Examples 5.1 and 5.2 in Section 5.5 and also for Example 5.9 in Section 5.55.

241

CHAPTER 5:
Vertical
Distance
Measurement:
Leveling

Error Propagation for Examples 5.1 and 5.2

Assume the following estimated standard deviations for Example 5.1: $\hat{\sigma}_{s_1} = 0.020$ m, $\hat{\sigma}_\alpha = 04''$, $\hat{\sigma}_{\text{h.i.}} = \hat{\sigma}_{\text{target height}} = 0.005$ m, and $\hat{\sigma}_{(c\&r)} = 0.002$ m. Referring to Equation (5.11) and Figure 5.3 and applying Equation (2.38), the estimated standard deviation of ΔH from A to B is (assuming no correlation between the different variables in Equation (5.11) and letting $s = EC$, $c \& r = DF$, h.i. $= AE$, h.i. $= BC = $ target height)

$$\hat{\sigma}^2_{\Delta H_{(A-B)}} = \left(\frac{\partial \Delta H}{\partial AE}\right)^2 \hat{\sigma}^2_{\text{h.i.}} + \left(\frac{\partial \Delta H}{\partial BC}\right)^2 \hat{\sigma}^2_{\text{h.i.}} + \left(\frac{\partial \Delta H}{\partial s}\right)^2 \hat{\sigma}^2_s + \left(\frac{\partial \Delta H}{\partial \alpha}\right)^2 \hat{\sigma}^2_\alpha$$
$$+ \left(\frac{\partial \Delta H}{\partial (c\&r)}\right)^2 \hat{\sigma}^2_{(c\&r)}$$

$$= 2(0.005)^2 + (\sin \alpha)^2 (0.02)^2 + [(332.791)\cos \alpha]^2 [(4)(4.848)(10^{-6})]^2$$
$$+ (0.002)^2$$

$$= 0.000050 + 0.000012 + 0.000043 + 0.000004 = 0.000109$$

$$\hat{\sigma}_{\Delta H_{(A-B)}} = 0.010 \text{ m} = 10 \text{ mm}$$

The uncertainty in $(c\&r)$ was estimated by neglecting refraction completely. Note that, for short lines, the effect is not significant. For long lines, the uncertainty in $(c\&r)$ is substantially larger and has a pronounced effect on the error propagation.

For Example 5.2, the error must be propagated using Equation (5.13). Assume that $\hat{\sigma}_\beta = \hat{\sigma}_\alpha = 04''$, $\hat{\sigma}_{s_1} = \hat{\sigma}_{s_2} = 0.020$ m, $\sigma_{AE} = \sigma_{E'A} = \sigma_{BC} = \sigma_{C'B} = 0.005$ m, and let $s_1 = EC$, $s_2 = E'C'$. Applying Equation (2.38) to Equation (5.13), the estimated standard deviation for ΔH_{AB} is

$$\hat{\sigma}^2_{\Delta H_{(A-B)}} = \frac{1}{2}[4\hat{\sigma}^2_{\text{h.i.}} + (\sin^2\alpha + \sin^2\beta)\hat{\sigma}^2_s + (s_1^2 \cos^2\alpha + s_2^2 \cos^2\beta)\sigma^2_\alpha]$$
$$= \frac{1}{2}[(4)(0.005)^2 + (\sin^2 1°48'26'' + \sin^2 1°48'38'')(0.02)^2 + \{(404.163)^2$$
$$(\cos 1°48'26'')^2 + (404.161)^2(\cos 1°48'38'')^2\}[(4)(4.848)(10^{-6})]^2]$$

$$= \frac{1}{2}(0.0001 + 0.000000797 + 0.000122731)$$

$$= 0.0001118$$

$$\hat{\sigma}_{\Delta H_{(A-B)}} = 0.011 \text{ m} = 11 \text{ mm}$$

Error Propagation in Example 5.9

Assume estimated standard deviations for the measured values as follows: $\sigma_s = 5$ mm + 5 ppm $= 0.005$ m + 0.027 m $= 0.032$ m; $\sigma_z = 05''$; and $\sigma_{\text{h.i.}} = 0.005$ m for the EDM, reflector, target and theodolite, labeled $\sigma_{\text{h.i.}_E}$, $\sigma_{\text{h.i.}_R}$, $\sigma_{\text{h.i.}_t}$, and $\sigma_{\text{h.i.}_T}$, respectively. First propagate errors in the respective h.i.'s through Equations (4.33), (4.34a), and (4.35a) to evaluate $\sigma_{z'}$. Starting with Equation (4.33), we have

$$\Delta h.i._G = (h.i._R - h.i._t)_H - (h.i._E - h.i._T)_G$$

$$\Delta h.i._H = (h.i._E - h.i._t)_G - (h.i._R - h.i._T)_H$$

Note that $\Delta h.i._G$ and $\Delta h.i._H$ are functions of some of the same values, $h.i._R$ at H and $h.i._E$ at G. This being the case, correlation will occur and Equation. (2.37), for the general case, is used. Rearranging the preceding equations, we have

$$\Delta h.i._G = h.i._{RH} - h.i._{EG} - h.i._{tH} + h.i._{TG}$$

$$\Delta h.i._H = -h.i._{RH} + h.i._{EG} - h.i._{tG} + h.i._{TH}$$

that can be expressed in matrix form as

$$\begin{bmatrix} \Delta h.i._G \\ \Delta h.i._H \end{bmatrix} = \begin{bmatrix} +1 & -1 & -1 & +1 & 0 & 0 \\ -1 & +1 & 0 & 0 & -1 & +1 \end{bmatrix} \begin{bmatrix} h.i._{R_H} \\ h.i._{E_G} \\ h.i._{t_H} \\ h.i._{T_G} \\ h.i._{t_G} \\ h.i._{T_H} \end{bmatrix}$$

or, in more compact form, as

$$\Delta_H = \mathbf{H}\,\mathbf{X}_{h.i.}$$

The application of Equation (2.37) to this relationship gives

$$\Sigma_{\Delta_{h.i.}\Delta_{h.i.}} = \mathbf{J}_H \Sigma_{h.i.} \mathbf{J}_H^T$$

in which

$$\mathbf{J}_H = \mathbf{H}, \qquad \Sigma_H = \begin{bmatrix} \sigma_{RH}^2 & 0 & 0 & 0 & 0 & 0 \\ 0 & \sigma_{EG}^2 & 0 & 0 & 0 & 0 \\ 0 & 0 & \sigma_{tH}^2 & 0 & 0 & 0 \\ 0 & 0 & 0 & \sigma_{TG}^2 & 0 & 0 \\ 0 & 0 & 0 & 0 & \sigma_{tG}^2 & 0 \\ 0 & 0 & 0 & 0 & 0 & \sigma_{TH}^2 \end{bmatrix}$$

Where $\mathbf{J}$ is the matrix $\mathbf{H}$ and substitution of given variances into Σ_H results in

$$\Sigma_H = \begin{bmatrix} (0.005)^2 & 0 & 0 & 0 & 0 & 0 \\ 0 & (0.005)^2 & 0 & 0 & 0 & 0 \\ 0 & 0 & (0.005)^2 & 0 & 0 & 0 \\ 0 & 0 & 0 & (0.005)^2 & 0 & 0 \\ 0 & 0 & 0 & 0 & (0.005)^2 & 0 \\ 0 & 0 & 0 & 0 & 0 & (0.005)^2 \end{bmatrix}$$

$$= (0.005)^2 \mathbf{I}_6$$

where $\mathbf{I}_6$ is the identity matrix or unit matrix of order 6; that is the covariance matrix for the measured h.i.'s.

Substitution of the matrices $\mathbf{J}_H$ and Σ_H into the general Equation (2.37) for error propagation gives

243

CHAPTER 5:
Vertical
Distance
Measurement:
Leveling

$$\Sigma_{\Delta_{h.i.}} = \begin{bmatrix} 1.0 \times 10^{-4} & -1.0 \times 10^{-5} \\ -1.0 \times 10^{-5} & 1.0 \times 10^{-4} \end{bmatrix} = \begin{bmatrix} \sigma^2_{\Delta_{h.i.G}} & \sigma_{\Delta_{h.i.GH}} \\ \sigma_{\Delta_{h.i.HG}} & \sigma^2_{\Delta_{h.i.H}} \end{bmatrix}$$

which is the propagated covariance matrix for $\Delta_{h.i.H}$ and $\Delta_{h.i.G}$. Examination of this covariance matrix shows that the covariance terms are very small and the correlation coefficient, $\rho = \sigma_{\Delta_{h.i.GH}}/(\sigma_{\Delta_{h.i.G}})(\sigma_{\Delta_{h.i.H}}) = 0.10$ (Section 2.17, Equation (2.28)), indicating that, although correlation exists, it is so small as to be insignificant.* Consequently, error propagation will proceed from this point, assuming uncorrelated values of $\Delta_{h.i.}$ and using Equations (2.37) and (2.38), the special law of error propagation for uncorrelated measurements.

The next step is to use these propagated variances for $\Delta_{h.i.G}$ and $\Delta_{h.i.H}$, along with variances in zenith angles and the distance, to propagate the errors in Δz_G and Δz_H, which are calculated using Equation (4.34a):

$$\Delta z_G = -\left(\frac{\Delta_{h.i.G} \sin z_G}{s}\right)\rho, \qquad \Delta z_H = -\left(\frac{\Delta_{h.i.H} \sin z_H}{s}\right)\rho$$

Applying error propagation Equation (2.38) to the first of these equations leads to

$$\sigma^2_{\Delta z_G} = \left(\frac{\partial \Delta z_G}{\partial \Delta_{h.i.G}}\right)^2 \sigma^2_{\Delta_{h.i.G}} + \left(\frac{\partial \Delta z_G}{\partial z_G}\right)^2 \sigma^2_{\Delta z_G} + \left(\frac{\partial \Delta z_G}{\partial s}\right)^2 \sigma^2_s$$

$$\sigma^2_{\Delta z_G} = \left(\frac{-\sin z_G \rho}{s}\right)^2 \sigma^2_{\Delta_{h.i.G}} + \left(\frac{-\Delta_{h.i.G} \cos z_G \rho}{s}\right)^2 \sigma^2_z + \left(\frac{\Delta_{h.i.G} \sin z_G \rho}{s^2}\right)^2 \sigma^2_s$$

$$\sigma^2_{\Delta z_G} = \left(\frac{-\sin 86°57'00''\rho}{5457.500}\right)^2 0.0001 + \left(\frac{-0.374 \cos 86°57'00''\rho}{5457.500}\right)^2 (5.88 \times 10^{-10})$$
$$+ \left(\frac{0.374 \sin 86°57'00''\rho}{s^2}\right)^2 (1.02 \times 10^{-4})$$

in which $\rho = 206,264.8''/\text{rad}$ so that

$$\sigma^2_{\Delta z_G} = 0.142 + (3.33 \times 10^{-10}) + (6.82 \times 10^{-10}) = 0.142^{\text{sec}^2}$$

Similarly, $\sigma^2_{\Delta z_H}$ can be evaluated by substituting Z_H for Z_G and $\Delta_{h.i.H}$ for $\Delta_{h.i.G}$ in the preceding equation to give

$$\sigma^2_{\Delta z_H} = \left(\frac{-\sin 93°13'10''\rho}{5457.5}\right)^2 0.0001 + \left(\frac{-0.480 \cos 93°13'10''\rho}{5457.500}\right)^2 (5.88 \times 10^{-10})$$
$$+ \left(\frac{0.480 \sin 93°13'10''\rho}{(5457.500)^2}\right)^2 (1.02 \times 10^{-4})$$

$$\sigma^2_{\Delta z_H} = 0.142 + (6.104 \times 10^{-10}) + (1.124 \times 10^{-9}) = 0.142 \text{ sec}^2$$

Note that variances in corrections to zenith angles are extremely small and due entirely to the variances in the h.i.

Next propagate variances in the corrected zenith angles through Equation (4.35a), where

$$z'_G = z_G + \Delta z_G, \qquad z'_H = z_H + \Delta z_H$$

---

*A complete, rigorous error propagation applied to this problem revealed that the propagated standard deviation changed only a few tenths of a millemeter.

Propagated variances are

$$\sigma^2_{\Delta z'_G} = \left(\frac{\partial z'_G}{\partial z_G}\right)^2 \sigma^2_{z_G} + \left(\frac{\partial z'_G}{\partial \Delta z_G}\right)^2 \sigma^2_{\Delta z_G} = \sigma^2_{z_G} + \sigma^2_{\Delta z_G} = (05)^2 + (0.142)$$

$$\sigma^2_{z'_G} = \sigma^2_{z'_H} = 25.142 \text{ sec}^2$$

With $\sigma^2_{z'_G}$ and $\sigma^2_{z'_H}$ evaluated, propagate the variance in angle ϕ using Equation (4.43), where

$$\phi = \frac{z'_G - z'_H}{2}$$

so that

$$\sigma^2_\phi = \left(\frac{\partial\phi}{\partial z'_G}\right)^2 \sigma^2_{z'_G} + \left(\frac{\partial\phi}{\partial z'_H}\right)^2 \sigma^2_{z'_H} = \left(\frac{1}{2}\right)^2 25.142 + \left(\frac{1}{2}\right)^2 25.142 = \left(\frac{1}{2}\right) 25.142$$

$$\sigma^2_\phi = 12.6 \text{ sec}^2$$

One more value, earth curvature, c, needs to be considered. Propagation of the variance in c through Equation (5.2), where $c = (AE)^2/2R$ and $AE = s \cos \phi$ is an acceptable approximation for calculating c, reveals that the variance in c is 8.24×10^{-10} m^2 and can be ignored.

Next, propagate the error through Equation (5.28) used to calculate the vertical difference in elevation PP', where

$$PP' = \frac{s \sin \phi}{\sin (90° + c)}$$

$$\sigma^2_{PP'} = \left(\frac{\partial PP'}{\partial s}\right)^2 \sigma^2_s + \left(\frac{\partial PP'}{\partial \phi}\right)^2 \sigma^2_\phi + \left(\frac{\partial PP'}{\partial c}\right)^2 \sigma^2_c$$

Because σ^2_c is being neglected, only the first two terms are evaluated, to yield

$$\sigma^2_{PP'} = \left(\frac{\sin \phi}{\sin (90° + c)}\right)^2 \sigma^2_s + \left(\frac{s \cos \phi}{\sin (90° + \phi)}\right)^2 \sigma^2_\phi$$

in which $\sigma^2_\phi = 12.6''^2/(206,265''/\text{rad})^2 = 2.96 \times 10^{-10}$ rad^2 to give

$$\sigma^2_{PP'} = \left(\frac{\sin 3°07'48.9''}{\sin (90° + c)}\right)^2 (0.032)^2 + \left(\frac{5457.500 \cos 3°07'48.9''}{\sin (90°01'28.2'')}\right)^2 (2.96 \times 10^{-10})$$

$$\sigma^2_{PP'} = 3.05 \times 10^{-6} + 0.00878 = 0.0088 \text{ m}^2$$

Finally, the difference in elevation between G and H, by Equation (5.29), is

$$\Delta_{GH} = \text{h.i.}_{EG} + PP' - \text{h.i.}_{RH}$$

which has respective variance and standard deviation of

$$\sigma^2_{\Delta H} = \left(\frac{\partial \Delta H}{\partial \text{h.i.}_{EG}}\right)^2 \sigma^2_{\text{h.i.}_{EG}} + \left(\frac{\partial \Delta H}{\partial PP'}\right)^2 \sigma^2_{PP'} + \left(\frac{\partial \Delta H}{\partial \text{h.i.}_{RH}}\right)^2 \sigma^2_{\text{h.i.}_{RH}}$$

$$\sigma^2_{\Delta H} = (0.005)^2 + (0.0088) + (0.005)^2 = 0.08850 \text{ m}^2$$

and

$$\sigma_{\Delta H} = 0.094 \text{ m}$$

Some comments about this example are useful. The assumption of uncorrelated values of Δh.i. was justified and, in fact, error propagation through Equations (4.43), (5.28), and (5.29), the last three steps of the example, would yield an acceptable standard deviation of 0.093 m. Note that this conclusion was clear only after rigorous error propagation verified the simplification. Also note that the uncertainty in vertical circle readings contributes the largest portion of the propagated error (the second term in propagation through Equation (5.28)). Therefore, if optimum accuracy is to be achieved with trigonometric leveling, simultaneous, reciprocal zenith or vertical angles must be observed with the telescope direct and reversed (Section 6.37 and 6.38) using an instrument having a least count (Section 6.26) of 5 sec on the vertical circle.

245

CHAPTER 5:
Vertical
Distance
Measurement:
Leveling

5.57
ADJUSTMENT OF LEVELS OVER DIFFERENT ROUTES

When differential levels are run over several different routes from a fixed bench mark to establish a bench mark, there will be as many observed elevations for the bench mark as there are lines terminating at that point. In other words, redundant data are available that must be resolved. As indicated in Chapter 2, the best answer in this case is the *least-squares estimate*. To make such an adjustment, weights must be established for each line.

Assume that the standard deviation for each individual observed difference in elevation, σ_l varies as the square root of the distance leveled, d_i, or the number of instrument setups, n (Section 5.46). Consequently, $\sigma_{li} = \sqrt{d_i}\, \sigma_s$, where σ_s is the estimated standard deviation in units per unit of distance leveled. If $i = 1, 2, \ldots, n$ lines are involved and noncorrelation is assumed between observed differences in elevation, then according to Equation (2.30), Section 2.18, the covariance matrix for observations is

$$\Sigma_{li} = \begin{bmatrix} \sigma_s^2\, d_1 & & & \mathbf{0} \\ & \sigma_s^2\, d_2 & & \\ & & \ddots & \\ \mathbf{0} & & & \sigma_s^2\, d_n \end{bmatrix} \tag{5.30}$$

and, according to Equations (2.32) and (2.33), the cofactor matrix for the observations is

$$Q_{li} = \frac{1}{\sigma_0^2}\, \Sigma_{li} \tag{5.31}$$

By definition (Eq. (2.33)), the matrix of weights is

$$\mathbf{W}_{li} = \mathbf{Q}_{li}^{-1} = \sigma_0^2 \Sigma_{li}^{-1} = \sigma_0^2 \begin{bmatrix} 1/\sigma_s^2\, d_1 & & & \mathbf{0} \\ & 1/\sigma_s^2\, d_2 & & \\ & & \ddots & \\ \mathbf{0} & & & 1/\sigma_s^2\, d_n \end{bmatrix} \tag{5.32}$$

Now, σ_0 is the standard deviation of unit weight, an arbitrary constant that can be taken equal to σ_s so that the weight matrix becomes

$$\mathbf{W}_{li} = \begin{bmatrix} 1/d_1 & & & \mathbf{0} \\ & 1/d_2 & & \\ & & \ddots & \\ \mathbf{0} & & & 1/d_n \end{bmatrix} \tag{5.33}$$

Therefore, the weights of individual lines are inversely proportional to the distance leveled or the number of instrument setups. Consider an example to illustrate the use of such a weight matrix in the least-squares adjustment of a leveling problem with redundant data.

EXAMPLE 5.11. Lines of levels between $B.M._A$ and $B.M._B$ are run over four different routes. The fixed elevation of $B.M._A$ is 640.000 m. The lengths of lines, corresponding weights (according to Equation (5.33)), and observed differences in elevation are shown in the following table. Determine the least-squares estimate for $B.M._B$.

Route	Length d_i, km	Weight, w_i	Observed D.E., m	Observed elevation, m
1	2	$\frac{1}{2}$	0.720	640.720
2	4	$\frac{1}{4}$	0.560	640.560
3	10	$\frac{1}{10}$	1.080	641.080
4	20	$\frac{1}{20}$	0.260	640.260

Solution. To perform the adjustment, a mathematical model must be formed. In this case the model consists of condition equations of the following general form for each line:

(Observed difference in elevation + residual)$_i$

$$= \text{(adjusted elevation of } B.M._B - \text{elevation of } B.M._A) \quad (5.34)$$

For the example problem, these equations can be written for the four lines:

$$0.720 + v_1 = \Delta - 640.000$$

$$0.560 + v_2 = \Delta - 640.000 \quad (5.35)$$

$$1.080 + v_3 = \Delta - 640.000$$

$$0.260 + v_4 = \Delta - 640.000$$

in which Δ is the adjusted elevation of $B.M._B$. Equations (5.35) become

$$v_1 - \Delta = -640.720$$

$$v_2 - \Delta = -640.560$$

$$v_3 - \Delta = -641.080 \quad (5.36)$$

$$v_4 - \Delta = -640.260$$

which can be stated in matrix form as (see Equation (2.55))

$$\underset{4,1}{\mathbf{v}} + \underset{(4,1)}{\mathbf{B}} \underset{(1,1)}{\Delta} = \underset{4,1}{\mathbf{f}} \quad (5.37)$$

The criterion for the least-squares adjustment is (Section 2.27, Equation (2.65))

$$w_1 v_1^2 + w_2 v_2^2 + w_3 v_3^2 + w_4 v_4^2 = \text{a minimum} \quad (5.38)$$

from which the normal equations are (see Equation (2.70))

$$\underset{1,4 \ 4,4 \ 4,1}{(\mathbf{B}^T \mathbf{W} \mathbf{B})} \underset{1,1}{\Delta} = \underset{1,4 \ 4,4 \ 4,1}{\mathbf{B}^T \mathbf{W} \mathbf{f}} \quad (5.39)$$

so that

$$\Delta = (\mathbf{B}^T \mathbf{W} \mathbf{B})^{-1} \mathbf{B}^T \mathbf{W} \mathbf{f} \quad (5.40)$$

247

CHAPTER 5:
Vertical
Distance
Measurement:
Leveling

in which

$$
\mathbf{B} = \begin{bmatrix} -1 \\ -1 \\ -1 \\ -1 \end{bmatrix}
$$

and Δ is the adjusted elevation of point B, $\mathbf{W}$ is defined by Equation (5.33), and the values d_i in the table, and $\mathbf{f}$ is defined in Equations (5.36) and (5.37).

Substitution of these values into (5.40) yields

$$
\Delta = \left\{ \begin{bmatrix} -1 & -1 & -1 & -1 \end{bmatrix} \begin{bmatrix} \tfrac{1}{2} & & & 0 \\ & \tfrac{1}{4} & & \\ & & \tfrac{1}{10} & \\ 0 & & & \tfrac{1}{20} \end{bmatrix} \begin{bmatrix} -1 \\ -1 \\ -1 \\ -1 \end{bmatrix} \right\}^{-1}
$$

$$
\begin{bmatrix} -1 & -1 & -1 & -1 \end{bmatrix} \begin{bmatrix} \tfrac{1}{2} & & & 0 \\ & \tfrac{1}{4} & & \\ & & \tfrac{1}{10} & \\ 0 & & & \tfrac{1}{20} \end{bmatrix} \begin{bmatrix} -640.720 \\ -640.560 \\ -641.080 \\ -640.260 \end{bmatrix}
$$

which gives the solution for Δ. Therefore,

$$
\Delta = \frac{(640.720)(\tfrac{1}{2}) + (640.560)(\tfrac{1}{4}) + (641.080)(\tfrac{1}{10}) + (640.260)(\tfrac{1}{20})}{\tfrac{1}{2} + \tfrac{1}{4} + \tfrac{1}{10} + \tfrac{1}{20}}
$$

$$
= 640.690 \text{ m} \tag{5.41}
$$

is the least-squares estimate for the elevation of B.M.$_B$.

The astute reader already would have noted that the solution for Δ is actually nothing more than the weighted mean of the given four observed values, which in this case provides the least-squares estimate. In fact, for the example provided, the least-squares solution is trivial. The logical approach would be to simply form the weights and calculate the *weighted mean*, of course, with the knowledge that such a procedure provides a least-squares estimate for this type of problem, because *no correlation* exists between the observed differences in elevation.

EXAMPLE 5.12. Lines of levels are run from bench marks A, B, and C to establish elevations at junction point E. The number of instrument setups per line, weights, and observed elevations are listed in the following table:

Route	Number of setups, n	Weight w_i	Observed elevation, ft
1	9	1/9	320.48
2	12	1/12	320.32
3	4	1/4	320.89

Calculate the least-squares estimate for the elevation of junction point E.

Solution. Weights are taken as inversely proportional to the number of setups, as shown in the third column of the table. As in the previous example, the least-squares adjustment is given by the weighted mean as follows:

$$\text{Adjusted elevation of } E = \frac{(320.48)(\frac{1}{9}) + (320.32)(\frac{1}{12}) + (320.89)(\frac{1}{4})}{\frac{1}{9} + \frac{1}{12} + \frac{1}{4}}$$

$$= 320.68 \text{ ft}$$

5.58
ADJUSTMENT OF LEVEL NETWORKS

A level network consists of a number of intersecting lines of levels tied into known bench marks. Figure 5.49 shows a simple level network that contains known bench marks A and D. The adjusted elevations of bench marks B, C, and E are to be found. Lengths and differences in elevations for each line are indicated, with the arrow showing the direction in which levels were run. The errors of closure and total lengths of loops $ABEA$, $AEDA$, and $DEBCD$ also are given (Figure 5.49). Least-squares adjustments by the techniques of indirect observations and observations only provide the most rigorous and expedient solutions to this problem (Sections 2.27 and 2.28). Both techniques as used in level-network adjustments are summarized.

Let l_{ij} represent the measured difference in elevation between two points i and j and in the particular direction i to j (see Figure 5.50). If x_i and x_j represent the elevations of points i and j, respectively (either or both known or unknown), *the condition equation for the adjustment of indirect observations* is

$$x_i + l_{ij} + v_{ij} = x_j$$

or

$$v_{ij} + x_i - x_j = -l_{ij} \tag{5.42}$$

in which x_i and x_j are the parameters.

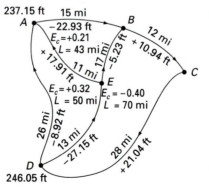

237.15 ft   15 mi
A
−22.93 ft
E_c=+0.21
L = 43 mi

B

12 mi
+10.94 ft

E
E_c=+0.32
L = 50 mi
E_c = −0.40
L = 70 mi

C

26 mi
−8.92 ft
13 mi
−27.15 ft

28 mi
+21.04 ft

D
246.05 ft

FIGURE 5.49
Simple level network.

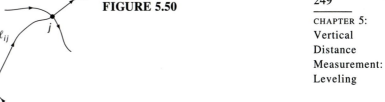

FIGURE 5.50

249

CHAPTER 5:
Vertical
Distance
Measurement:
Leveling

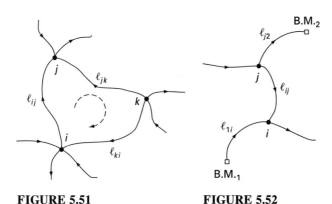

FIGURE 5.51   **FIGURE 5.52**

In the *adjustment of observations only,* the elevations of the different points do not appear in the condition equations. For every closed *loop* composed of points for which elevations are to be determined (see Figure 5.50), one condition equation is written. For example, going from point i to j to k and back to i (Figure 5.51), the sum of the differences in elevation must add up to 0. Noting that the arrow implies the direction in which the elevation increases from a lower point to a higher point (by convention) the condition is

$$l_{ij} + v_{ij} - l_{jk} - v_{jk} + l_{ki} + v_{ki} = 0$$

or

$$v_{ij} - v_{jk} + v_{ki} = -l_{ij} + l_{jk} - l_{ki} \tag{5.43}$$

In some cases, the loop is not closed on the original bench mark but rather begins and ends at bench marks for which the elevations are known. Let x_1 and x_2 represent the elevations of the two bench marks in Figure 5.52. The condition equation for the adjustment of observations only is

$$x_1 + l_{1i} + v_{1i} - l_{ij} - v_{ij} + l_{j2} + v_{j2} = x_2$$

or

$$v_{1i} - v_{ij} + v_{j2} = x_2 - l_{1i} + l_{ij} - l_{j2} - x_1 \tag{5.44}$$

EXAMPLE 5.13. ADJUSTMENT OF A LEVEL NETWORK. Figure 5.49 shows a simple level network that contains two known bench marks A and D, whose elevations are 237.15 ft and 246.05 ft, respectively. The method of least squares is to be used to determine the adjusted elevations of points B, C, and E. Lengths and differences in elevations of each line are indicated on the figure (with the arrows showing the direction in which levels are run) and summarized in the following table:

From	To	Length, mi	Observation	Difference in elevation, ft	Weight
A	B	15	l_1	−22.93	$w_1 = \frac{1}{15}$
B	C	12	l_2	+10.94	$w_2 = \frac{1}{12}$
C	D	28	l_3	+21.04	$w_3 = \frac{1}{28}$
D	A	26	l_4	−8.92	$w_4 = \frac{1}{26}$
E	B	17	l_5	−5.23	$w_5 = \frac{1}{17}$
E	A	11	l_6	+17.91	$w_6 = \frac{1}{11}$
D	E	13	l_7	−27.15	$w_7 = \frac{1}{13}$

The seven observations are assumed to be *uncorrelated*. Because it was not possible to determine variances, each observation is assumed to have an empirical weight that is inversely proportional to the length of line in miles. Under the assumption of lack of correlation, the weights are computed directly as given in the last column in the preceding table of data. The weight matrix then is the following diagonal matrix, containing the elements in that last column:

$$\mathbf{W} = \begin{bmatrix} \frac{1}{15} & & & & & & \mathbf{0} \\ & \frac{1}{12} & & & & & \\ & & \frac{1}{28} & & & & \\ & & & \frac{1}{26} & & & \\ & & & & \frac{1}{17} & & \\ & & & & & \frac{1}{11} & \\ \mathbf{0} & & & & & & \frac{1}{13} \end{bmatrix}$$

Solution. *Adjustment by the method of indirect observations*
Denoting the elevations of points, B, C, and E by x_1, x_2 and x_3, respectively, the seven condition equations, according to Equation (5.42), are

$$v_1 + 237.15 - x_1 \quad = -l_1 \quad \text{or} \quad v_1 - x_1 \quad\quad = -214.22$$
$$v_2 + x_1 - x_2 \quad = -l_2 \quad \text{or} \quad v_2 + x_1 - x_2 \quad = -10.94$$
$$v_3 + x_2 - 246.05 \quad = -l_3 \quad \text{or} \quad v_3 + \quad x_2 \quad = 225.01$$
$$v_4 + 246.05 - 237.15 = -l_4 \quad \text{or} \quad v_4 \quad\quad = 0.02 \quad \text{(ft)}$$
$$v_5 + x_3 - x_1 \quad = -l_5 \quad \text{or} \quad v_5 - x_1 + x_3 = 5.23$$
$$v_6 + x_3 - 237.15 \quad = -l_6 \quad \text{or} \quad v_6 + \quad x_3 = 219.24$$
$$v_7 + 246.05 - x_3 \quad = -l_7 \quad \text{or} \quad v_7 - \quad x_3 = -218.90$$

that in matrix form become

$$\begin{bmatrix} v_1 \\ v_2 \\ v_3 \\ v_4 \\ v_5 \\ v_6 \\ v_7 \end{bmatrix} + \begin{bmatrix} -1 & 0 & 0 \\ 1 & -1 & 0 \\ 0 & 1 & 0 \\ 0 & 0 & 0 \\ -1 & 0 & 1 \\ 0 & 0 & 1 \\ 0 & 0 & -1 \end{bmatrix} \begin{bmatrix} x_1 \\ x_2 \\ x_3 \end{bmatrix} = \begin{bmatrix} -214.22 \\ -10.94 \\ 225.01 \\ 0.02 \\ 5.23 \\ 219.24 \\ -218.90 \end{bmatrix} \quad \text{(ft)}$$

that is of the general form $\mathbf{v} + \mathbf{B\Delta} = \mathbf{f}$.

With the given diagonal weight matrix, the normal equations are $(\mathbf{B'WB})\ \Delta = (\mathbf{B'Wf})$

251

CHAPTER 5:
Vertical
Distance
Measurement:
Leveling

$$\begin{bmatrix} 0.209 & -0.083 & -0.059 \\ -0.083 & 0.119 & 0 \\ -0.059 & 0 & 0.227 \end{bmatrix} \begin{bmatrix} x_1 \\ x_2 \\ x_3 \end{bmatrix} = \begin{bmatrix} 13.062 \\ 8.948 \\ 37.077 \end{bmatrix}$$

The solution of these equations yields the three adjusted elevations:

$$\hat{x}_1 = \text{elevation of point } B = 214.07 \text{ ft}$$

$$\hat{x}_2 = \text{elevation of point } C = 225.01 \text{ ft}$$

$$\hat{x}_3 = \text{elevation of point } E = 219.14 \text{ ft}$$

For purposes of comparison with results from adjustment by observations only, the vector of residuals is computed from $\mathbf{v} = \mathbf{f} - \mathbf{B\Delta}$ and the adjusted observations from $\hat{l} = l + \mathbf{v}$ (see Equations (2.55) and (2.80)):

$$\mathbf{v} = \begin{bmatrix} -0.146 \\ -0.001 \\ -0.003 \\ 0.020 \\ 0.163 \\ 0.099 \\ 0.241 \end{bmatrix} \quad \text{and} \quad \hat{l} = \begin{bmatrix} -23.076 \\ 10.939 \\ 21.037 \\ -8.900 \\ -5.067 \\ 18.009 \\ -26.909 \end{bmatrix} \quad \text{(ft)}$$

According to Equation (2.78), the cofactor matrix for the adjusted elevations of the three points is

$$\mathbf{Q}_{\Delta\Delta} = \mathbf{Q}_{xx} = (\mathbf{B'WB})^{-1} = \mathbf{N}^{-1} = \begin{bmatrix} 7.395 & 5.176 & 1.919 \\ & 12.024 & 1.343 \\ \text{symmetric} & & 4.910 \end{bmatrix}$$

Because only relative precisions for the observations were given in the form of weights, evaluation of the a posteriori reference variance, $\hat{\sigma}_0^2$, is necessary to calculate the corresponding covariance matrix, $\mathbf{\Sigma}_{xx}$. Therefore, from Equation (2.82),

$$\hat{\sigma}_0^2 = \frac{\mathbf{v'Wv}}{r} = \frac{0.00836}{7 - 3} = 0.0021 \text{ ft}^2$$

in which $r = n - n_0$ by Equation (2.43). Finally, according to Equation (2.83), the covariance matrix for adjusted observations is

$$\mathbf{\Sigma}_{xx} = \hat{\sigma}_0^2 \mathbf{Q}_{xx} = \begin{bmatrix} 0.0155 & 0.0108 & 0.0040 \\ & 0.0251 & 0.0028 \\ \text{symmetric} & & 0.0103 \end{bmatrix} \quad \text{(ft}^2\text{)}$$

Solution. Adjustment of observations only

Because there are seven observations and a minimum of only three are required for a unique solution, the redundancy (see Equation (2.43)) is

$$r = n - n_0 = 7 - 3 = 4$$

Therefore, four condition equations relate the observations together. In a manner similar to Equation (5.44), the following four conditions are written

Loop $A{\to}B{\to}E{\to}A$: $\quad l_1 + v_1 - l_5 - v_5 + l_6 + v_6 = 0$

Loop $B{\to}C{\to}D{\to}E{\to}B$: $\quad l_2 + v_2 + l_3 + v_3 + l_7 + v_7 + l_5 + v_5 = 0$

Loop $A{\to}E{\to}D{\to}A$: $\quad -l_6 - v_6 - l_7 - v_7 + l_4 + v_4 = 0$

Loop $A{\to}B{\to}C{\to}D$: $\quad 237.15 + l_1 + v_1 + l_2 + v_2 + l_3 + v_3 - 246.05 = 0$

So,

$$v_1 - v_5 + v_6 = -l_1 + l_5 - l_6 = -(-22.93) + (-5.23) - 17.91 = -0.21$$

$$v_2 + v_3 + v_5 + v_7 = -l_2 - l_3 - l_5 - l_7 = -10.94 - 21.04 + 5.23 + 27.15 = 0.40$$

$$v_4 - v_6 - v_7 = -l_4 + l_6 + l_7 = 8.92 + 17.91 - 27.15 = -0.32$$

$$v_1 + v_2 + v_3 = 8.9 - l_1 - l_2 - l_3 = 8.9 + 22.93 - 10.94 - 21.04 = -0.15$$

or

$$\begin{bmatrix} 1 & 0 & 0 & 0 & -1 & 1 & 0 \\ 0 & 1 & 1 & 0 & 1 & 0 & 1 \\ 0 & 0 & 0 & 1 & 0 & -1 & -1 \\ 1 & 1 & 1 & 0 & 0 & 0 & 0 \end{bmatrix} \begin{bmatrix} v_1 \\ v_2 \\ \vdots \\ \vdots \\ v_7 \end{bmatrix} = \begin{bmatrix} -0.21 \\ 0.40 \\ -0.32 \\ -0.15 \end{bmatrix} \text{(ft)}$$

which is of the form $\mathbf{Av} = \mathbf{f}$ (Equation (2.59)) Applying Equation (2.101a), we get $(\mathbf{AQA'})\mathbf{k} = \mathbf{f}$, in which $\mathbf{Q} = \mathbf{W}^{-1}$, or

$$\mathbf{Q} = \begin{bmatrix} 15 & & & & & & \mathbf{0} \\ & 12 & & & & & \\ & & 28 & & & & \\ & & & 26 & & & \\ & & & & 17 & & \\ & & & & & 11 & \\ \mathbf{0} & & & & & & 13 \end{bmatrix}$$

Then,

$$\begin{bmatrix} 43 & -17 & -11 & 15 \\ & 70 & -13 & 40 \\ & & 50 & 0 \\ \text{symmetric} & & & 55 \end{bmatrix} \begin{bmatrix} k_1 \\ k_2 \\ k_3 \\ k_4 \end{bmatrix} = \begin{bmatrix} -0.21 \\ 0.40 \\ -0.32 \\ -0.15 \end{bmatrix} \text{(ft)}$$

and, from Equations (2.101) and (2.103), the residuals are

$$\mathbf{v} = \mathbf{QA'}(\mathbf{AQA'})^{-1}\mathbf{f} = \begin{bmatrix} -0.146 \\ -0.001 \\ -0.003 \\ 0.020 \\ 0.163 \\ 0.099 \\ 0.241 \end{bmatrix}$$

and the adjusted measurements are

253

CHAPTER 5:
Vertical
Distance
Measurement:
Leveling

$$\hat{l} = l + v = \begin{bmatrix} -23.076 \\ 10.939 \\ 21.037 \\ -8.900 \\ -5.067 \\ 18.009 \\ -26.909 \end{bmatrix} \quad \text{(ft)}$$

Both v and $\hat{l}$ are identical to those computed by the preceding technique.

Let the adjusted elevations of points B, C, and E be designated by $\hat{X}^t = [x_B x_C x_E]$ and known elevations of points A and D be indicated by $X_0^t = [x_A x_A x_D]$. Therefore, the adjusted elevations (referring to Figure 5.49) are

$$\begin{bmatrix} x_B \\ x_C \\ x_E \end{bmatrix} = \begin{bmatrix} x_A \\ x_A \\ x_D \end{bmatrix} + \begin{bmatrix} 1 & 0 & 0 & 0 & 0 & 0 & 0 \\ 1 & 1 & 0 & 0 & 0 & 0 & 0 \\ 0 & 0 & 0 & 0 & 0 & 0 & 1 \end{bmatrix} \begin{bmatrix} \hat{l}_1 \\ \hat{l}_2 \\ \hat{l}_3 \\ \hat{l}_4 \\ \hat{l}_5 \\ \hat{l}_6 \\ \hat{l}_7 \end{bmatrix}$$

which can be written more compactly as $\hat{X} = X_0 + J\hat{l}$.

Substitution of known elevations and adjusted differences in elevations into the preceding equations yields the following adjusted elevations:

$$x_A = x_A + \hat{l}_1 = 237.15 - 23.076 \qquad\qquad = 214.07 \text{ ft}$$

$$x_C = x_A + \hat{l}_1 + \hat{l}_2 = 237.15 - 23.076 + 10.939 = 225.01 \text{ ft}$$

$$x_E = x_D + \hat{l}_7 = 246.05 - 26.909 \qquad\qquad = 219.14 \text{ ft}$$

Obviously, results from both techniques of least-squares adjustment are identical.

As indicated, adjusted elevations are functions of the adjusted measurements. Consequently, to determine the cofactor matrix for adjusted elevations, it is first necessary to evaluate the cofactor matrix for the adjusted measurements, $Q_{\hat{l}\hat{l}}$. According to Equations (2.107) and (2.108),

$$Q_{\hat{l}\hat{l}} = Q - QA^tW_eAQ = Q - Q_{vv}$$

in which $W_e = (AQA^t)^{-1}$. For this problem,

$$Q_{vv} = \begin{bmatrix} 7.6049 & 2.2185 & 5.1766 & 0.0 & -5.4759 & 1.9193 & -1.9193 \\ & 2.9344 & 6.8470 & 0.0 & 1.6428 & -0.5758 & 0.5758 \\ & & 15.9764 & 0.0 & 3.8331 & -1.3435 & 1.3435 \\ & & & 26.0000 & 0.0 & 0.0 & 0.0 \\ & & & & 8.5333 & -2.9908 & 2.9908 \\ & & & & & 6.0899 & 4.9101 \\ \text{symmetric} & & & & & & 8.0899 \end{bmatrix} \quad \text{(ft}^2\text{)}$$

and

$$\mathbf{Q}_{\hat{l}\hat{l}} = \begin{bmatrix} 7.3951 & -2.2185 & -5.1766 & 0.0 & 5.4759 & -1.9193 & 1.9193 \\ & 9.0656 & -6.8470 & 0.0 & -1.6428 & 0.5758 & -0.5758 \\ & & 12.0236 & 0.0 & -3.8331 & 1.3435 & -1.3435 \\ & & & (7)10^{-15} & 0.0 & 0.0 & 0.0 \\ & & & & 8.4667 & 2.9908 & -2.9908 \\ & & & & & 4.9101 & -4.9101 \\ \text{symmetric} & & & & & & 4.9101 \end{bmatrix} \quad (\text{ft}^2)$$

Note that the cofactor for the fourth observation is essentially zero (7×10^{-15}). This is because the fourth observation occurs between known bench marks A and D. Therefore, $\hat{l}_4$ will have zero variance. Application of error propagation Equation (2.34), Section 2.18, to the equation used to evaluate adjusted elevations yields

$$\mathbf{Q}_{xx} = \mathbf{JQJ}^t$$

and the cofactor matrix for adjusted elevations is

$$\mathbf{Q}_{xx} = \begin{bmatrix} 7.3952 & 5.1766 & 1.9193 \\ & 12.0236 & 1.3435 \\ \text{symmetric} & & 4.9101 \end{bmatrix}$$

Using Equation (2.82), the a posteriori reference variance is

$$\hat{\sigma}_0^2 = 0.002090 \text{ ft}^2$$

so that the propagated covariance matrix for adjusted elevations (Equation (2.32)) is

$$\mathbf{\Sigma}_{xx} = \sigma_0^2 \mathbf{Q}_{xx} = \begin{bmatrix} 0.0155 & 0.0108 & 0.0040 \\ & 0.0251 & 0.0028 \\ \text{symmetric} & & 0.0103 \end{bmatrix} \quad (\text{ft}^2)$$

Note that the elements in this matrix are identical to the corresponding elements in the covariance matrix propagated in the adjustment of this same level network by indirect observations.

Similarly, the propagated covariance matrix for the adjusted differences in elevation (Equation (2.32)) is

$$\mathbf{\Sigma}_{ll} = \sigma_0^2 \mathbf{Q}_{ll}$$

$$\mathbf{\Sigma}_{ll} = \begin{bmatrix} 0.0154 & -0.0046 & -0.0108 & 0.0000 & 0.0114 & -0.0040 & 0.0040 \\ & 0.0189 & -0.0143 & 0.0000 & -0.0034 & 0.0012 & -0.0012 \\ & & 0.0251 & 0.0000 & -0.0080 & 0.0028 & -0.0028 \\ & & & 0.1 \times 10^{-16} & 0.0000 & 0.0000 & 0.0000 \\ & & & & 0.0177 & 0.0062 & -0.0062 \\ & & & & & 0.0103 & -0.0103 \\ \text{symmetric} & & & & & & 0.0103 \end{bmatrix} \quad (\text{ft}^2)$$

The diagonal elements of this covariance matrix define the propagated variances of the respective lines of the level network and will be used to evaluate the accuracy of the adjustment (Section 5.59). Therefore, this last propagation represents an important and necessary part of the adjustment.

255

CHAPTER 5:
Vertical
Distance
Measurement:
Leveling

5.59
STANDARDS OF ACCURACY FOR LEVELING

In Section 5.46, classification and general specifications for leveling, as outlined by the FGCS, were discussed and presented in tabular form (Table 5.2) for first-, second-, and third-order levels. The FGCS (Bossler, 1984) also sets forth *vertical control network standards* based on an *elevation difference accuracy,* designated b, where

$$b = s/\sqrt{d} \qquad (5.45)$$

in which d = approximate horizontal distance between control point positions, in km
s = estimated standard deviation of elevation difference, in mm, between survey control points propagated from a least-squares adjustment

The units of b are in $(mm)/(\sqrt{(km)})$. Elevation difference accuracy values for b are given for the various orders of leveling in Table 5.3.

Theoretically, the elevation difference accuracy criterion should be applied to all pairs of points in a network. In practice, it is computed for a sample and the largest value for b is taken as the governing amount.

EXAMPLE 5.14. Evaluate the accuracy of the level network adjusted in Example 5.13.

Solution. Propagated standard deviations are calculated for each line by taking the square root of the diagonal elements from the propagated covariance matrix for the measured differences in elevation determined in the last step of the adjustment. Using these values and the lengths of the lines given with the original data from the problem, elevation difference accuracy values are as in the following table:

Line	Standard deviation, s		Length of line, d_i		$b = s/\sqrt{d}$
	ft	mm	mi	km	
1	0.1241	37.9	15	24.1	7.72
2	0.1375	41.9	12	19.3	9.53
3	0.1584	48.2	28	45.1	7.18
4	0.3×10^{-8}	9×10^{-7}	26	41.8	1.4×10^{-7}
5	0.1330	40.5	17	27.4	7.74
6	0.1015	30.9	11	17.7	7.34
7	0.1015	30.9	13	20.9	6.76

Comparison of the maximum value of $b = 9.53$ for line 2 with the specified allowable amounts in Table 5.3 shows that these levels do not qualify even for third-order accuracy, a conclusion one could have reached intuitively by examining the large loop closures in the network (see Figure 5.49).

When a new survey is being integrated into an existing level network, a test for systematic error can be performed. In this test, the new data are combined with existing data (assumed to be correctly weighted and free of systematic errors) and a *variance factor ratio* is computed (see Bossler, 1984). If this factor is substantially greater than unity, the presence of systematic error is possible and both new and existing networks must be examined for the cause.

TABLE 5.3
Elevation accuracy standards

Classification	Maximum elevation difference accuracy, b
First-order, Class I	0.5
First-order, Class II	0.7
Second-order Class II	1.0
Second-order, Class II	1.3
Third-order	2.0

5.60
ADJUSTMENT OF LEVELING INSTRUMENTS

Regardless of the precision of manufacture, levels in process of use require certain field adjustments from time to time. It becomes an important duty of the surveyor to test the instrument at short intervals and to make such adjustments as are found necessary.

In some instances one adjustment is likely to be altered by, or depends on, some other adjustment made subsequently. For example, lateral movement of the cross-hair ring also may produce a small rotation, and the lateral adjustment of the level tube depends on the vertical adjustment. Hence, if an instrument is badly out of adjustment, related adjustments must be repeated until gradually they are perfected.

5.61
DESIRED RELATIONS IN DUMPY LEVELS

For a dumpy level in perfect adjustment the following relations should exist (see Figure 5.53)

1. The axis of the level tube should be perpendicular to the vertical axis.
2. The horizontal cross hair should lie in a plane perpendicular to the vertical axis, so that it will lie in a horizontal plane when the instrument is level.
3. The line of sight should be parallel to the axis of the level tube.

Also the optical axis, the axis of the objective slide, and the line of sight should coincide; however, in the dumpy levels commonly manufactured in the United States, the optical axis and the axis of the objective slide are fixed perpendicular to the vertical axis by the manufacturer and no provision for further adjustment is made.

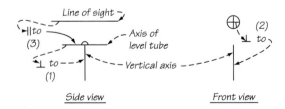

FIGURE 5.53
Desired relations among principal lines of dumpy level.

5.62
ADJUSTMENT OF DUMPY LEVELS

257

CHAPTER 5:
Vertical
Distance
Measurement:
Leveling

The parts capable of and requiring adjustment are the cross hairs and the level tube. The basis for adjustments is the vertical axis. The adjustments for a level with four foot-screws are given in the following:

1. *To make the axis of the level tube perpendicular to the vertical axis.* Approximately center the bubble over each pair of opposite leveling screws; then carefully center the bubble over one pair. Rotate the level end for end about its vertical axis. If the level tube is in adjustment, the bubble will retain its position. If the tube is not in adjustment, the displacement of the bubble will indicate double the actual error, as shown by Figure 5.54. If $(90° - \alpha)$ represents the angle between the vertical axis and axis of level tube, then when the bubble is centered, the vertical axis will make an angle of α with the true vertical. When the level is reversed, the bubble will be displaced through the arc whose angle is 2α. Hence, the correction is the arc whose angle is α. Make the correction by bringing the bubble halfway back to the center by means of the capstan nuts at one end of the tube. Relevel the instrument with the leveling screws, and repeat the process until the adjustment is perfected. Usually, three or four trials are necessary. As a final check, the bubble should remain centered over each pair of opposite leveling screws. This adjustment involves the principle of *reversion*.

2. *To make the horizontal cross hair lie in a plane perpendicular to the vertical axis (and thus horizontal when the instrument is level).* Sight the horizontal cross hair on some clearly defined point (as *A*, Figure 5.55) and rotate the instrument slowly about its vertical axis. If the point appears to travel along the cross hair, no adjustment is needed.

   If the point departs from the cross hair and takes some position as *A'* on the opposite side of the field of view, loosen two adjacent capstan screws and rotate the cross-hair ring until, by further trial, the point appears to travel along the cross hair. Tighten the same two screws. The instrument need not be level when the test is made.

3. *To make the line of sight parallel to the axis of the level tube (two-peg test).* Set two pegs 60 to 90 m (200 to 300 ft) apart on approximately level ground, and designate as *A* the peg near which the second setup will be made (Figure 5.56); call the other peg *B*. Set up and level the instrument at any point *M* equally distant from *A* and *B*, that is, in a vertical plane bisecting the line *AB*. Take rod readings *a* on *A* and *b* on *B*; then $(a - b)$ will be the true difference in elevation, because any error would be the same for the two equal sight distances L_m. Due account must be taken of signs throughout the test.

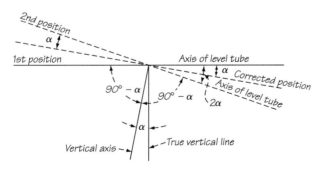

FIGURE 5.54
Adjustment of axis of level tube of dumpy level.

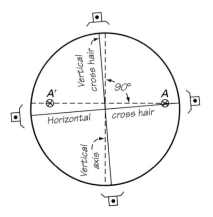

FIGURE 5.55
Adjustment of horizontal cross hair.

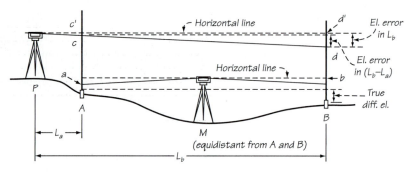

FIGURE 5.56
Two-peg test.

Move the instrument to a point P near A, preferably but not necessarily on line with the pegs; set up as before, and measure the distances L_a to A and L_b to B. Take rod readings c on A and d on B. Then $(c - d)$, taken in the same order as before, will be the indicated difference in elevation; if $(c - d) = (a - b)$, the line of sight is parallel to the axis of the level tube, and the instrument is in adjustment. If not, $(c - d)$ is called the *false* difference in elevation, and the inclination (error) of the line of sight in the net distance $(L_b - L_a)$ is equal to $(a - b) - (c - d)$. By proportion, the error in the reading on the far rod is

$$e_{fr} = \frac{L_b}{L_b - L_a}[(a - b) - (c - d)] \qquad (5.46)$$

Subtract algebraically the amount of this error from the reading d on the far rod to obtain the correct reading d' at B for a horizontal line of sight with the position of the instrument unchanged at P. Set the target at d' and bring the line of sight on the target by moving the cross-hair ring vertically.

EXAMPLE 5.15. With the level at M, the rod reading a is 0.970 and b is 2.986; the true difference in elevation $(a - b)$ is then $0.970 - 2.986 = -2.016$ ft., with B thus

259

CHAPTER 5:
Vertical
Distance
Measurement:
Leveling

indicated as being lower than A. With level at P, the rod reading c is 5.126 and d is 7.018; the false difference in elevation $(c - d)$ is then $5.126 - 7.018 = -1.892$, with B again indicated as being lower than A. The distance L_a is observed to be 30 ft and L_b to be 230 ft. The inclination of the line of sight in $(230 - 30 = 200)$ ft is $(-2.016) - (-1.892) = -0.124$ ft. The error in elevation of the line of of sight at the far rod is $(230/200) \times (-0.124) = -0.143$ ft. The correct rod reading d' for a horizontal line of sight is $7.018 - (-0.143) = 7.161$ ft.

As a partial check on the computations, the correct rod reading c' at A may be computed by proportion; the difference in elevation computed from the two corrected rod readings c' and d' should be equal to the true difference in elevation observed originally at M.

EXAMPLE 5.16. In the preceding example, the error in elevation of the line of sight at the near rod is $(30/200) \times (-0.124) = -0.019$ ft. The correct rod reading c' is $5.126 - (-0.019) = 5.145$ ft. The "false" difference in elevation is $5.145 - 7.161 = -2.016$ ft, which is equal to the true difference in elevation; hence, the computations are checked to this extent.

A sketch always should be drawn. Also, theoretically a correction for the earth's curvature and atmospheric refraction (see Section 5.2) should be added numerically to the final rod reading d', although in practice it is usually considered negligible.

Some surveyors prefer to set up at P within 2 to 3 m or 6 to 10 ft of A and consider $[(a - b) - (c - d)]$ the *total* error in elevation, to be subtracted directly from d. This serves as a first approximation; the procedure for the setup at P is then repeated. The amount of computation thus is reduced, but the amount of fieldwork is increased.

5.63
ADJUSTMENT OF AUTOMATIC LEVELS

Prior to leveling, the circular bubble and activity of the compensator, which automatically makes the line of sight horizontal, should be checked. Center the circular bubble in the engraved circle (Figure 5.57a) and rotate the instrument through 180° or to the direction of maximum bubble displacement (Figure 5.57b). When the displacement d exceeds one-half the bubble diameter, adjustment is necessary. Adjustment is performed in two steps by reducing the two components of the displacement, l from left to right and t from top to bottom. First, remove $\frac{1}{2}l$ using the appropriate level screws (Figure 5.57c). Second, remove the remaining $\frac{1}{2}l$ by manipulating the adjusting screws (Figure 5.57d). One-half the top to bottom displacement t (Figure 5.57e) then is removed by the level screws and the remainder $t/2$ is compensated, using the adjusting screws to bring the bubble to the center of the engraved circle. Turn 180° and check the bubble. Repeat the entire procedure until the bubble remains centered in all positions.

To check the compensator, sight a target about 30 m away. Center the bull's-eye bubble. Next, tap one tripod leg with a hand. The image of the target will appear to swing in the field of view but the target will return to its original position. Next, turn a level screw, causing the line of sight to slope slightly. Once again the target will appear to swing but will return to the original position.

The horizontal cross hair can be tested and adjusted following the procedure given in Section 5.62, Problem 2. The standard peg test described in Section 5.62 also can be employed with automatic, self-leveling levels. If the slope of the optical line of sight is excessive, the cross hair can be adjusted. For these last two adjustments, reference should be made to the instrument manual of instructions for the operation, because most automatic

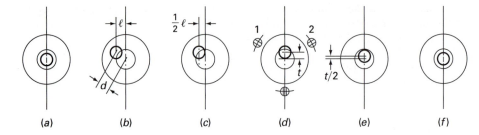

FIGURE 5.57
Adjustment of circular level bubble.

levels do not have exposed adjusting screws for the cross-hair reticule. The cross-hair reticule for the level illustrated in Figure 5.18 is exposed by unscrewing the cover around the eyepiece.

5.64
ADJUSTMENT OF DIGITAL OR BAR-CODED SELF-LEVELING INSTRUMENTS

The circular level bubble can be adjusted as described in Section 5.63 for the automatic level. For the level discussed in Section 5.20, three adjusting screws are located on the underside of the instrument directly beneath the display register (Figure 5.19a).

To test for inclination of the line of sight, the two-peg test given in Section 5.62 is used. Note that the digital level described in Section 5.20 has a program in the onboard software, Check and Adjust, that provides menu instructions displayed on the register to step the operator through this test. At the conclusion of the test, the amount of the deviation of the line of sight is given. Should adjustment be necessary, the safety screw for the cross-hair adjustment is located on the underside of the instrument directly below the eyepiece (Figure 5.19a).

For each of these adjustments, the instrument operator (or owner) should proceed very cautiously. If a person knowledgeable about the instrument and its adjustment is available, proceed according to the methods outlined in the manual of instructions for the system. Otherwise, it would be wise to have the instrument checked and adjusted at the nearest qualified service center or at the manufacturer.

5.65
ADJUSTMENT OF TILTING LEVELS

The circular level bubble is adjusted according to the method outlined for automatic levels in Section 5.63. Also, the horizontal cross hair can be checked by the procedure outlined for dumpy levels in Section 5.62, Problem 2.

Generally, tilting levels are used for precise three-wire or geodetic leveling (Sections 5.50 and 5.52) in which a collimation test (Section 5.51), similar to the two-peg test, is performed and a correction or C factor is calculated. If this C factor indicates excessive inclination of the line of sight, an adjustment is needed to make the line of sight horizontal to the axis of the level bubble. Perform the two-peg test described in Section 5.62, Problem 3, up to the point where the correct reading d' at B is determined. Then sight the middle cross hair on d', at the rod on B, using the tilt knob of the level. At this stage, the line of sight will be horizontal but the level bubble will not be perfectly centered. With the line of sight still horizontal, the spirit level bubble is centered or the two ends of a coincidence bubble can be made to coincide by means of the adjusting screws at one end of the bubble. When the adjustment has been made, again perform the collimation test (Section 5.51) to be certain the inclination of the line of sight is within tolerable limits.

261

CHAPTER 5:
Vertical
Distance
Measurement:
Leveling

PROBLEMS

5.1. What is the combined effect of the earth's curvature and mean atmospheric refraction in a distance of 200.0 ft? In a distance of 4000.0 ft? In a distance of 8.5 mi? In a distance of 30.0 mi?

5.2. Calculate the combined effect of the earth's curvature and refraction for the following distances: (*a*) 150.00 m; (*b*) 400.00 m; (*c*) 1.50 km; (*d*) 15.0 km; (*e*) 150.0 km.

5.3. An observer standing on the shoreline of a lake can just see the top of a tower on an island. If the eye of the observer is 1.35 m above lake level and the top of the tower is 20 m above lake level, how far is the tower from the observer?

5.4. On a clear day, an observer whose eye is 600.0 ft above mean sea level notes that the line of the horizon of the Pacific Ocean and the deck of a suspension bridge are in vertical alignment. The bridge is 13.8 mi from the observer. Determine (*a*) the distance from observer to horizon and (*b*) the height of bridge deck above the water.

5.5. A backsight to a temporary B.M. at a horizontal distance of 1100.5 ft is 1.25 ft and a foresight from the same instrument setup to a turning point is 8.594 ft at a distance of 50.0 ft. Compute the difference in elevation between the temporary B.M. and the T.P.

5.6. The backsight to a bench mark is 2.005 m and the foresight from the same instrument setup is 2.421 m. If the distance from the level to the bench mark is 10 m and from the level to the turning point is 500.0 m, compute the difference in elevation between the bench mark and the turning point.

5.7. Two points, A and B, each are distant 1500 ft from a third point C, from which the measured vertical angle to A is $+4°21'$ and that to B is $+0°32'$. What is the difference in elevation between A and B?

5.8. Let A be a point of elevation 30.000 m, and let B and C be points of unknown elevation. By means of an instrument set 1.300 m above B, zenith angles are observed, that to A being $92°30'00''$ and that to C being $85°30'00''$. If the horizontal distance AB is 500 m and the horizontal distance BC is 1500 m, what are the elevations of B and C, making due allowance for the earth's curvature and atmospheric refraction?

5.9. Two points, A and B, are 1200 ft apart. The elevation of A is 615.03 ft. A level is set up on the line between A and B at a distance of 200 ft from A. The rod reading on A is 10.52 ft and that on B is 1.45 ft. What is the elevation of B?

5.10. The slope distance from E to G by EDM, corrected for atmospheric conditions and system constants, is 625.50 ft with $z = 87°30'00''$ measured with a theodolite at E having an h.i.= 4.40 ft, to a target at G where the h.i. $= 4.75$ ft above G. The respective h.i.'s of the EDM above E and the reflector above G are 5.55 ft and 6.15 ft. Compute the horizontal distance and difference in elevation between points E and G.

5.11. The zenith angles between stations A and B are $86°57'55''$ from A to B and $93°02'55''$ from B to A. The slope distance measured by EDM from A to B is 2458.487 m and the elevation of station *A* above the datum is 150.500 m. The heights of the theodolite and target above stations *A* and *B* are 1.55 m and 1.38 m, respectively. Similarly, the heights of the theodolite and target above *B* and *A* for observing the zenith angle from *B* are 1.43 m and 1.51 m, respectively. The heights of EDM and reflector above stations *A* and *B* are 1.62 m and 1.141 m, respectively. Compute the elevation of station *B*.

5.12. What is the radius of curvature of a level tube graduated to 0.1 in. and having a sensitivity of 20″ per division?

5.13. What is the sensitivity of a level tube graduated to 2 mm having a radius of curvature of 15 m?

5.14. A sight is taken with an engineer's level at a rod held 50.00 m away, and an initial reading of 1.925 m is observed. The bubble then is moved through five spaces on the level tube, when the rod reading is 2.006 m. What is the sensitivity of the level tube in seconds of arc? What is the radius of curvature of the level tube if one space is 2 mm?

5.15. Design (*a*) a direct vernier and (*b*) a retrograde vernier reading to hundredths of feet, each space on the rod being equal to 0.05 ft.

5.16. Design a direct vernier and retrograde vernier, both reading to 1 mm for a rod graduated to centimeters. Draw a neat sketch of each vernier and a portion of the rod for a reading of 2.226 m with graduations shown and labeled.

5.17. On a rod graduated to 1 cm, a direct-reading vernier is to read to millimeters. State the following: (*a*) the length of one vernier space; (*b*) the number of spaces on the vernier; (*c*) the number of spaces on the main scale corresponding to the full length of vernier scale; (*d*) the least count of the vernier.

5.18. Design a direct vernier reading to 1 mm, applied to a scale graduated to 0.5 cm. For the same scale, design a retrograde vernier with least count of 0.2 mm. Sketch the verniers in relation to the scale.

5.19. If the rod were inclined 0.5 ft forward in a length of 12 ft, what error would be introduced in a rod reading of 6.0 ft?

5.20. The rod placed on a pointed turning point is inclined backward 20 cm in a length of 4 m. What error is introduced in a rod reading of (*a*) 0.500 m; (*b*) 2.50 m?

5.21. A rod that is 50 mm square at the base is placed on a flat surface for a turning point. If the top of the 4-m rod is inclined backward 30 cm, what error is introduced in a reading of 0.300 m?

5.22. A line of differential levels was run between two bench marks 20 mi apart, and the measured difference in elevation was found to be 1542.6 ft. Later, the rod whose nominal length was 13 ft was found to be 0.005 ft too short, the error being distributed over its full length. Correct the measured difference in elevation for erroneous length of rod.

5.23. Suppose that the levels of Problem 5.22 had been run using a rod that was 0.005 ft too short, owing to wear on the lower end. What would have been the error?

5.24. Differential levels were run from B.M.$_1$ (elevation 200.125 m) to B.M.$_2$, a distance of 90 km. On the average, the backsight distances were 100 m in length and the foresight distances were 50 m in length. The elevation of B.M.$_2$ as computed from the level notes, was 1113.355 m. Compute the error due to the earth's curvature and atmospheric refraction, and correct the elevation of B.M.$_2$.

5.25. The levels of Problem 5.24 were rerun using an average backsight distance of 60 m and an average foresight distance of 20 m. The elevation of B.M.$_2$, as deduced from the level notes, was

1112.941 m. Compute the error due to curvature and refraction and correct the elevation of B.M.$_2$.

263

CHAPTER 5:
Vertical
Distance
Measurement:
Leveling

5.26. Suppose that the instrument used in running the levels of Problems 5.24 and 5.25 was out of adjustment, so that when the bubble was centered the line of sight was inclined 0.002 m upward in a distance of 100 m. Correct the observed results of Problems 5.24 and 5.25 for inclination of line of sight.

5.27. If in running levels between two points the rod were inclined 15 cm forward in a height of 3.5 m, what error would be introduced per setup when backsight readings averaged 3 m and foresight readings averaged 0.3 m?

5.28. If levels are run from B.M.$_1$ (elevation 600.00 m) to B.M.$_2$ (observed elevation 900.00 m) and the rod is on the average 10 cm out of plumb in a height of 3.5 m, what error is introduced due to the rod's not being plumb? What is the correct elevation of B.M.$_2$?

5.29. What would be the error if, in Problem 5.28, both bench marks were at the same elevation?

5.30. Complete the differential level notes that follow, putting the entire set of notes in proper field note form and making all the customary checks. All units are in meters.

Station	B.S.	h.i.	F.S.	Elevation
B.M.$_A$	0.232		—	55.482
T.P.$_1$	0.503		3.892	
T.P.$_2$	0.212		3.524	
T.P.$_3$	0.246		3.302	
T.P.$_4$	2.169		0.257	
T.P.$_5$	2.895		0.678	
T.P.$_6$	2.912		0.202	
B.M.$_B$	—		1.423	

5.31. Complete the differential-level notes that follow, putting the entire set of notes in proper field note form. Determine the error of closure of the level circuit and adjust the elevations of B.M.$_2$ and B.M.$_3$, assuming that the error is a constant per setup. Units are in meters.

Station	B.S.	h.i.	F.S.	Elevation
B.M.$_1$	3.150		—	225.412
T.P.$_1$	2.831		3.346	
T.P.$_2$	4.104		2.725	
T.P.$_3$	2.654		3.008	
B.M.$_2$	0.368		3.208	
T.P.$_4$	0.089		1.534	
T.P.$_5$	2.863		3.736	
B.M.$_3$	3.356		0.100	
T.P.$_6$	2.781		1.662	
T.P.$_7$	3.365		0.111	
B.M.$_1$	—		6.059	

5.32. The data for differential levels, in the order taken in the field (e.g., B.S., F.S.), for a level circuit originating on B.M. 44 and closing on B.M. 45 are as follows: 0.501, 3.525, 1.279, 3.218, 0.458, 1.789, 3.125, 0.345, 2.368, 0.356, 2.554, 0.125 m. Elevations of B.M. 44 and B.M. 45 are 219.561 and 220.500 m, respectively. Put these data in proper field note form, reduce the data, determine the error of closure, and calculate the adjusted elevation for T.P. 3.

5.33. Differential levels were run from B.M. X to establish B.M. H and then back to B.M. X. Data, in the order taken in the field (e.g., B.S., F.S., and so on) are as follows: 3.735, 0.700, 1.617, 1.899, 1.937, 1.881, 0.186, F.S. on B.M. H = 3.533, 2.302, 3.779, 1.539, 1.857, 1.839, 1.708, 3.558,

1.420 m. The elevation of B.M. *X* is 150.422 m above the datum. Put the data in the proper field note form, calculate the error of closure, and determine the adjusted elevation of B.M. *H*.

5.34. Data for profile notes, in the order taken, are as follows: B.S. on B.M. 13 = 1.931, F.S. on 0+00 (hub) = 0.688, B.S. on 0+00 = 1.967, I.F.S. on 0+10 = 1.98, I.F.S. on 0+20 = 2.30, I.F.S. on 0+30 = 3.12, F.S. on T.P. 1 = 3.245, B.S. on T.P. 1 = 0.010, I.F.S. on 0+40 = 0.45, I.F.S. on 0+50 = 1.62, I.F.S. on 0+60 = 2.78, I.F.S. on 0+70 = 3.00, I.F.S. on 0+80 = 3.173, F.S. on 0+85.2 (hub) = 3.297, B.S. on 0+85.2 = 0.272, F.S. on T.P. 2 = 0.239, B.S. on T.P. 2 = 2.671, F.S. on T.P. 3 = 0.047, B.S. on T.P. 3 = 3.654, F.S. on B.M. 13 = 2.995. B.M. 13 = 306.539 m above the datum. Prepare profile notes for these data, make all the necessary checks, adjust the h.i.'s in proportion to the number of instrument setups, and calculate the elevations for all stations.

5.35. Plot the profile using the data from Problem 5.34 at horizontal and vertical scales of 1 : 500 and 1 : 100, respectively. Plot it on $8\frac{1}{2}$ by 11 in. paper, including all the items necessary on an engineering drawing.

5.36. The data for profile levels, in the order taken, are B.S. on B.M. 30 = 1.980, I.F.S. on 0+00 (top of hub) = 2.754, I.F.S. on 0+20 = 1.50, F.S. on T.P. 1 = 0.175, B.S. on T.P. 1 = 2.875, I.F.S. on 0+40 = 3.29, I.F.S. on 0+60 = 2.10, I.F.S. on 0+80 = 0.98, F.S. on T.P. 2 = 0.188, B.S. on T.P. 2 = 2.098, I.F.S on 1+00 = 2.87, I.F.S. on 1+10 = 1.53, F.S. on T.P. 3 = 0.094, B.S. on T.P. 3 = 2.950, I.F.S. on 1+20 = 1.50, F.S. on 1+24.50 = 1.000 (hub), B.S. on 1+24.50 (top of hub) = 1.500, F.S. on B.M. 40 = 2.365m. The elevations of B.M. 30 and B.M. 40 are 128.301 and 135.892, respectively. Prepare the profile notes, make all the necessary checks, adjust the h.i.'s in proportion to the number of setups, and calculate the elevations for all stations. Units are in meters.

5.37. Plot the profile data from Problem 5.36 at horizontal and vertical scales of 1 : 500 and 1 : 100, respectively. Plot it on $8\frac{1}{2}$ by 11 in. paper, including all the items needed on an engineering drawings.

5.38. If sights average 30 m in length and the standard deviation of a single observation is 0.002 m/ 30 m sight, what is the standard deviation of running a line of levels 15 km long? 60 km long?

5.39. If sights with a level average 100 m in length and the standard deviation per 100 m sight is 0.005 m, what is the standard deviation in running levels 10 km long? 40 km long?

5.40. The error of closure of a level circuit 20 km in length is 0.050 m. The average length of sight is 60 m. Assuming that all systematic errors have been eliminated, what is the estimated error per instrument setup? What is the estimated error for a single observation of the rod?

5.41. Describe the major differences between geodetic leveling and ordinary differential leveling.

5.42. Outline the differences between precise three-wire leveling and precise leveling with an automatic level equipped with an optical micrometer and using double-scale rods.

5.43. The observations for determining the *C* factor for tilting level number 6034 are as follows: (*a*) from instrument position 1, the thread readings on rod *A*, the close backsight, are 1449, 1419, and 1388 mm; thread readings on rod *B*, the distant foresight, are 0764, 0520, and 0277 mm; (*b*) from position 2, thread readings on rod *B*, the close backsight, are 1398, 1370, and 1341 mm; thread readings on distant rod *A* are 2512, 2265, and 2019 mm. Compute the *C* factor for this level.

5.44. Three-wire levels were run from B.M. 17 to B.M. 21. The thread readings in the order taken (e. g., B.S., F.S.) from instrument setups 1, 2, . . . , 5 are (*a*) 2037, 1843, 1648; 0849, 0641, 0430; (*b*) 2446, 2224, 2002; 0712, 0467, 0224; (*c*) 1725, 1611, 1497; 1324, 1269, 1213;

265

CHAPTER 5:
Vertical
Distance
Measurement:
Leveling

(d) 2089, 1891, 1694; 0658, 0491, 0326; (e) 2240, 2153, 2067; 1116, 0945, 0775 mm. Put these observations in the proper three-wire level note form, make all the necessary checks, and determine the difference in elevation. The C factor for the level used was determined to be -0.013 mm/mm of interval.

5.45. A line of levels extends $2°$ northward from a point at latitude $40°$ N having an elevation of 1000 m. Calculate the orthometric correction for this line of levels.

5.46. Reciprocal leveling across a canyon from station 1 to station 2 using one tilting level, moved from one side of the canyon to the other, and two rods on opposite sides of the canyon, yielded the following average readings:

Instrument station	Average near readings, m	Average distant readings, m
1	2.680	3.236
2	2.506	1.966

The distance from station 1 to 2 is 1500 m. Compute the difference in elevation between stations 1 and 2.

5.47. The slope distance from E to G by EDM, corrected for atmospheric conditions and system constants, is 5255.000 m with $z = 87°00'00''$ measured from a theodolite at E having an h.i. $= 1.350$ m to a target at G where the h.i. $= 1.425$ m above G. The respective h.i.'s of the EDM above E and reflector above G are 1.670 and 1.839 m. The elevation of E is 554.250 m above the datum. Compute (a) the horizontal distance from E to G and (b) the elevation of G above the datum.

5.48. Zenith angles, observed simultaneously between stations A and B, are $86°57'55''$ from A to B and $93°12'55''$ from B to A. The slope distance (corrected for all systematic errors) is 4584.871 m and the elevation of A is 322.073 m above the datum. Compute (a) the horizontal distance from A to B and (b) the elevation of station B above the datum. Assume h.i.$_A$ = h.i.$_B$.

5.49. Using the data from Problem 5.10, propagate the error in the difference in elevation between stations A and B. Assume $\sigma_{EG} = 0.02$ ft $+ 5$ ppm, $\sigma_z = 05''$, and $\sigma_{h.i.} = \sigma_{theodolite} = \sigma_{target} = \sigma_{reflector}$ 0.01 ft.

5.50. Using the data from Problem 5.11, propagate the error in the elevation of station B. Assume $\sigma_{AB} = 5$ mm $+ 5$ ppm, $\sigma_z = 02''$, and the estimated standard deviations of all the h.i.'s $= 0.005$ m.

5.51. Using the data in Problems 5.47 and 5.48 and assuming the same estimated standard deviations for distance, angles, and h.i.'s as given in Problem 5.50, propagate the error in the elevations of (a) G in Problem 5.47 and (b) B in Problem 5.48.

5.52. The elevation of junction point X has been determined by differential leveling from bench marks A, B, and C, which have elevations of 91.805, 85.324, and 83.225 m, respectively. The data are as follows:

Line	Number of setups	Difference in elevation, m
AX	4	−0.810
BX	6	5.650
CX	8	−7.745

Compute the best estimate for the elevation of junction point X.

5.53. Lines of differential levels are run from B.M.$_1$ to B.M.$_2$ over three different routes. The following are the lengths of the routes and the observed elevations of B.M.$_2$. Determine the best estimate value of the elevation of B.M.$_2$.

Route	Length km	Elevation of B.M.$_2$ m
a	8	625.91
b	20	625.70
c	60	625.99

5.54. The following data are for a level network whose perimeter (reading clockwise) is *ABCDEFA*. Within the network, a line of levels extends from *B* to *F* and from *C* to *E*. The elevation of *A* is 130.485 m. Adjust the elevations by the method of least squares.

Circuit	From	To	Distance, m	Difference in elevation, km
ABFA	A	B	12	+5.320
	B	F	10	−3.313
	F	A	16	−1.906
BCEFB	B	C	10	+3.620
	C	E	5	−2.587
	E	F	8	−4.272
	F	B	10	+3.310
CDEC	C	D	10	−4.988
	D	E	8	+2.315
	E	C	5	+2.585

5.55. In the two-peg test of a level, the following observations were taken:

	Instrument at *M*, m	Instrument at *P*, m
Rod reading on A	2.101	1.576
Rod reading on B	1.990	1.288

M is equidistant from *A* and *B*; *P* is 12 m from *A* and 75 m from *B*. What is the true difference in elevation between the two points? With the level in the same position at *P*, to what rod reading on *B* should the line of sight be adjusted? What is the corresponding rod reading on *A* for a horizontal line of sight? Check these two rod readings against the true difference in elevation, previously determined.

REFERENCES

Berry, R. M. "History of Geodetic Leveling in the United States." *Surveying and Mapping* 36, no. 2 (June 1976), pp. 174–54.

Berry, R. M. "Observational Techniques for Use with Compensator Leveling Instruments for First-Order Levels." *Surveying and Mapping* 37, no. 1 (March 1977), p. 17.

Bossler, J. D. *Standards and Specifications for Geodetic Control Surveys.* Silver Spring, MD: Federal Geodetic Control Committee, 1984.

Fiedler, J. "Orthometric Heights from Global Positioning System." *ASCE Journal of Surveying Engineering* 118, no. 3 (August 1992), pp. 75, 77.

Holdahl, S. R. *Empirical Calibration of Zeiss NI-1 Level Instruments to Account for Magnetic Errors.* NOAA Technical Memorandum NOS NGS-45. Silver Spring, MD: National Geodetic Information Center, June 1986.

267

CHAPTER 5:
Vertical
Distance
Measurement:
Leveling

Hou, G. Y.; S. A. Veress; and J. E. Colcord. "Refraction in Precise Leveling." *Surveying and Mapping* 32, no. 2 (June 1972), p. 231.

Hradilek, G. "Refraction in Trigonometric and Three-Dimensional Terrestrial Networks." *The Canadian Surveyor* 26, no. 1 (March 1972), pp. 59–70.

Moffitt, F. H., and H. Bouchard. *Surveying,* 9th ed. New York: HarperCollins Publishers, 1992.

Reuger, J. M., and F. K. Brunner. "EDM-Height Traversing versus Geodetic Leveling." *The Canadian Surveyor* 36, no. 1 (March 1982), pp. 69–87.

Schomaker, M. C., and R. M. Berry. *Geodetic Leveling.* NOAA Manual NOS NGS 3. Silver Springs, MD: National Geodetic Information Center, September 1981.

Teskey, W. F. "Trigonometric Leveling in Precise Engineering Surveys." *Surveying and Land Information Systems* 52, no. 1 (March 1992), pp. 46–53.

Vanicek, P. "Vertical Datum and NAVD 88." *Surveying and Land Information Systems* 51, no. 2 (June 1991), pp. 83–86.

Vanicek, P., and E. Krakiwsky. *Geodesy,* 2d ed. New York: North Holland Publishing Company, 1986.

Whalen, C. T., and E. I. Balacz. "Test Results of First-Order Class III Leveling." *Surveying and Mapping* 37, no. 1 (March 1977), pp. 45–58.

Wolf, P. R., and R. C. Brinker. *Elementary Surveying,* 9th ed. New York: HarperCollins Publishers, 1994.

Young, G. M., and E. I. Balacz. *Corrections Applied by the National Geodetic Survey to Precise Leveling Observations.* NOAA Technical Memorandum NOS NGS 34. Silver Spring, MD: National Geodetic Information Center, 1982.

Zilkoski, D. B.; J. H. Richards; and G. M. Young. "Results of the General Adjustment of the North American Adjustment Vertical Datum of 1988." *Surveying and Land Information Systems* 52, no. 3 (September 1992), pp. 133–49.

Angle and Direction Measurement

6.1
LOCATION OF POINTS

As previously stated, the purpose of a survey is to determine the relative locations of points below, on, or above the surface of the earth. Because the earth is three-dimensional, the most logical reference framework with which to locate points is a three-dimensional rectangular coordinate system. Figure 6.1 shows an earth-centered, three-dimensional rectangular coordinate system, $X'Y'Z'$, called the *geocentric coordinate system*. It is a right-handed system, with X' passing through Greenwich in England and Z' passing through the North Pole (Y' is uniquely fixed because $X'Y'Z'$ is specified as right-handed). In addition to the $X'Y'Z'$ coordinates, any point may also be located by a set of spherical coordinates consisting of latitude ϕ, longitude λ, and the distance $\rho + h$ along the normal to the earth ellipsoid. The latter system is used mainly for geodetic surveys and is not discussed in further detail in this text.

A reference more suitable for plane surveying is a local XYZ right-handed rectangular coordinate system similar to the one illustrated in Figure 6.2. The origin of such a system usually is chosen near the center of the area to be surveyed. Its XY plane is tangent to the reference ellipsiod at the point of origin, and the Y axis generally is directed toward the North Pole. Any point P may be located either by XYZ or by the two angles α (in the XY plane) and β, and the distance r from O to P, as shown in Figure 6.2.

Visualization of the local XYZ system is more convenient when the Z axis coincides with the reader's local vertical or the plumb line, as illustrated in Figure 6.3. In this figure, point A, which is assumed to be of a known position, also is introduced. With respect to Figure 6.3, the position of point P in space may be determined in one of the following measurement procedures:

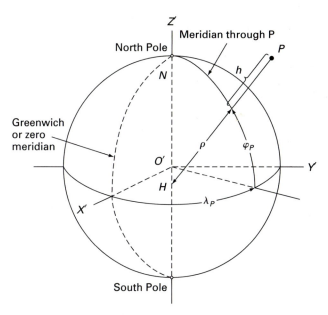

FIGURE 6.1
Position of P in the geocentric system.

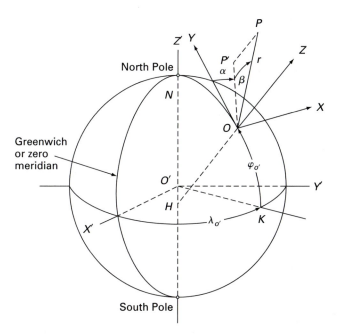

FIGURE 6.2
Local coordinate system.

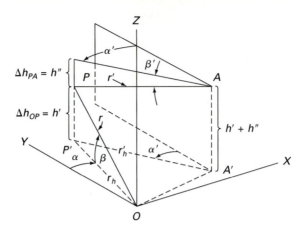

FIGURE 6.3
Position of a point in local coordinate system.

1. The direction of line OP (angles α, β) and the distance $r = OP$.
2. The directions of lines OP (angles α, β) and AP (angles α' in the XY plane and β' in plane $APP'A'$) from two known points O and A.
3. The distances r (from O to P), r' (from A to P), and h' (height of P above the XY plane).

For most surveys of small extent, the positions of P and A and the slope distances r and r' are projected onto the horizontal XY plane. Still referring to Figure 6.3, the position of P' in the horizontal XY plane is defined when measurements in one of the following cases are given:

1. Distance and direction from one known point, such as r_h and α from known point O.
2. Directions from two known points such as α and α' from known points O and A, respectively.
3. Distances from two known points such as r_h and r'_h from points O and A'

Note that the direction of spatial line OP is specified by angles α in the horizontal XY plane and β in the vertical plane OPP'. Similarly, the direction of line AP is determined by α' and β' in the horizontal XY plane and the vertical plane $APP'A'$, respectively. The directions of horizontal lines OP' and $A'P'$ are specified only by horizontal angles α and α', respectively. Therefore, in general, the angular measurements of surveying are in either horizontal or vertical planes.

The angle between two points is understood to mean the horizontal angle or angle between projections in the horizontal plane of two lines passing through the two points and converging at a third point. Thus, at P in Figure 6.3, the angle between A and O is the horizontal angle $OP'A'$.

Angles in the vertical plane are measured (1) from the zenith (the OZ axis), called *zenith angles,* so that in Figure 6.3 the zenith angle from O to P is angle ZOP, or (2) from the horizontal, called *vertical angles.* In Figure 6.3, the elevation vertical angle from O to P is β and the depression vertical angle from A to P is β'.

6.2
MERIDIANS

The relative directions of lines connecting survey points may be obtained in a variety of ways. Figure 6.4 shows lines intersecting at a point. The direction of any line (such as OB)

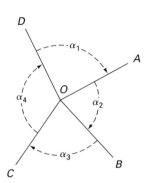

FIGURE 6.4
Directions by angles.

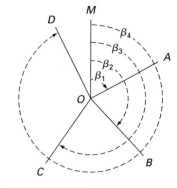

FIGURE 6.5
Directions referred to the meridian.

with respect to an adjacent line (such as *OA*) is given by the horizontal angle between the two lines (such as α_2) and the direction of rotation (such as clockwise). The direction of any line (such as *OC*) with respect to a line not adjacent (such as *OA*) is not given by any of the measured angles but may be computed by adding the intervening angles (such as $\alpha_2 + \alpha_3$).

Figure 6.5 shows the same system of lines but with all angles measured from a line of reference *OM*. The direction of any line (such as *OA*) with respect to the line of reference (such as *OM*) is given by the angle between the lines (such as β_1) and its direction of rotation (such as clockwise). The angle between any two lines (such as *AOC*) is not given directly, but may be computed by taking the difference between the direction angles of the two lines (such as $\angle \beta_3 - \angle \beta_1 = \angle AOC$).

The fixed line of reference may be any line in the survey, or it may be purely imaginary. It is termed a *meridian.* If arbitrarily chosen, it is called an *assumed meridian;* if north-and-south line passing through the geographical poles of the earth, it is called a *true* or *astronomic meridian;* if a line parallel to a central true meridian, it is called a *grid meridian;* or if it lies parallel to the magnetic lines of force of the earth as indicated by the direction of a magnetized needle, it is called a *magnetic meridian.*

6.3
TRUE MERIDIAN

The *true meridian,* or *astronomic meridian,* is determined by astronomical observations as described in Chapter 10. In Figure 6.2, arc *NOK* is a true meridian. For any given point on the earth its direction is always the same, and hence directions referred to the true meridian remain unchanged regardless of time. The lines of most extensive surveys and usually the lines marking the boundaries of landed property are referred to the true meridian.

At many triangulation stations established throughout the United States by the National Ocean Survey, reference lines of known true direction have been established for use by surveyors.

6.4
GRID MERIDIAN

Examination of Figure 6.1 shows that true meridians converge at the poles. In plane surveys of limited extent, it is convenient to perform the work in a rectangular *XY* coordinate system in which one central meridian coincides with true meridian. All remaining meridians are parallel to this central true meridian, eliminating the need to calculate the convergence of meridians when determining positions of points in the system. These parallel meridians are called *grid meridians.*

Each state in the United States has plane coordinate projections established by the National Ocean Survey. These projections are designed so that surveys can be performed as on a horizontal surface within tolerable levels of accuracy. The two projections used for this purpose in the United States are the *Lambert conformal projection* and the *transverse Mercator projection,* which are described in Chapter 11. The oblique Mercator is used for one zone in Alaska.

6.5
MAGNETIC MERIDIAN

The direction of the *magnetic meridian* is that taken by a freely suspended magnetic needle. The magnetic poles are at some distance from the true geographic poles; hence, in general, the magnetic meridian is not parallel to the true meridian. The location of the magnetic poles is constantly changing; therefore, the direction of the magnetic meridian is not constant. However, the magnetic meridian is employed as a line of reference on rough surveys where a magnetic compass is used and often is employed in connection with more precise surveys in which angular measurements are checked approximately by means of the compass. It formerly was used extensively for land surveys.

Details concerning magnetic compasses and meridians can be found in Sections 6.13 to 6.20.

6.6
ANGLES AND DIRECTIONS

Directions and angles may be defined by means of *bearings, azimuths, deflection angles, angles to the right,* or *interior angles,* as described in the following sections. These quantities are said to be *observed* when obtained directly in the field and *calculated* when obtained indirectly by computation. Conversion from one means of expressing angles and directions to another is a simple matter if a sketch is drawn to show the existing relations.

6.7
BEARINGS

The direction of any line with respect to a given meridian may be defined by the *bearing.* Bearings are called *true (astronomic) bearings, magnetic bearings,* or *assumed bearings,* depending on whether the meridian is true, magnetic, or assumed. The bearing of a line is indicated by the quadrant in which the line falls and the acute angle that the line makes with the meridian in that quadrant. Therefore, in Figure 6.6, the bearing of the line *OA* is read north 37° east and written N37°E. The bearings of *OB, OC,* and *OD* are, respectively, S62°E, S50°W, and N20°W. In all cases, values of bearing angles lie between 0° and 90°. If the direction of the line is parallel to the meridian and north, it is written as N0° or *due North;* if perpendicular to the meridian and east, it is written as N90°E or *due East.*

In Figure 6.7, if the observed bearing of *OA* is N37°E and the angle *AOB* = 81°, the calculated bearing of *OB* is S62°E.

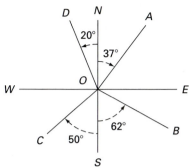

FIGURE 6.6
Bearings.

FIGURE 6.7
Angles and bearings.

6.8
AZIMUTHS

The *azimuth* of a line is its direction as given by the angle between the meridian and the line measured in a clockwise direction usually from the north branch of the meridian. In astronomical observations azimuths generally are reckoned from the true south; in surveying, some surveyors reckon azimuths from the south and some from the north branch of whatever meridian is chosen as a reference, but on any given survey the direction of zero azimuth is either always south or always north. Therefore, it is necessary to designate whether the azimuth is from the north or the south. Azimuths are called *true (astronomic) azimuths, magnetic azimuths,* or *assumed azimuths,* depending on whether the meridian is true, magnetic, or assumed. Azimuths may have values between 0° and 360°.

In Figure 6.8, azimuths measured from the south point are $A_{OA} = 217°$, $A_{OB} = 298°$, $A_{OC} = 50°$, $A_{OD} = 160°$; in Figure 6.9, in which are shown the same lines with azimuths measured from the north point, $A_{OA} = 37°$, $A_{OB} = 118°$, $A_{OC} = 230°$, and $A_{OD} = 340°$. In Figure 6.8 if the observed azimuth of *OA* as reckoned from the south is 217° and the observed angle *AOB* is 81°, the calculated azimuth of *OB* is 298°.

Azimuths may be calculated from bearings or vice versa, preferably with the aid of a sketch. For example, if the bearing of a line is N16°E, its azimuth (from the south) is $180 + 16 = 196°$; and if the azimuth (from the south) of a line is 285°, its bearing is $360 - 285 = S75°E$.

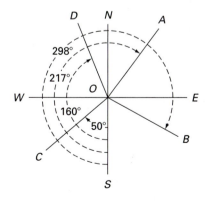

FIGURE 6.8
Azimuths from the south.

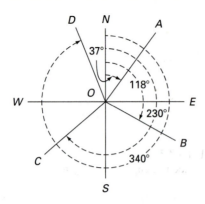

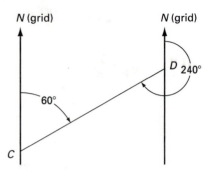

FIGURE 6.9
Azimuths from the north.

FIGURE 6.10
Forward and back azimuths.

In some special cases, the term *azimuth* is used in the sense of a bearing and therefore may be taken either clockwise or counterclockwise, as in "the azimuth of Polaris" (Section 10.38).

It is assumed that, when an azimuth is given as being from O and A, as in Figure 6.9, where the azimuth from north $A_{OA} = 37°$, this specifies the direction of the line from an origin at O to a terminal point A and the azimuth is called a *forward azimuth*. Conversely, the azimuth from A to O is called the *back azimuth* of OA. When the azimuth of a line is less than 180°, the back azimuth of the line is the forward azimuth plus 180°. For example, the back azimuth of OA in Figure 6.9 is $37° + 180° = 217°$. When the forward azimuth of a line is greater than 180°, the back azimuth equals the forward azimuth minus 180°. In Figure 6.9, the back azimuth of line $OC = 230° - 180° = 50°$. The concept of forward and back azimuths is further illustrated in Figure 6.10, where the azimuth from the north of line $CD = 60°$. Now, suppose that the initial point were D. Then the azimuth of DC would be 240° and the back azimuth of DC would be $240° - 180° = 60°$. Note that the idea of forward and back azimuths as developed here is valid only for plane surveys of limited extent where grid north is used as the reference meridian.

6.9
INTERIOR ANGLES

In a closed polygon, angles inside the figure between adjacent lines are called *interior angles*. Figure 6.11 illustrates interior angles $\alpha_1, \alpha_2, \ldots, \alpha_6$ in polygon $ABCDEF$. If n equals the number of sides in a closed polygon, the sum of the interior angles is $(n - 2)(180°)$. In Figure 6.11, $\sum_{i=1}^{6} \alpha_i = (4)(180°) = 720°$.

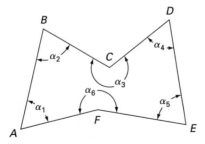

FIGURE 6.11
Interior angles in polygon.

FIGURE 6.12
Deflection angles.

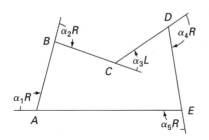

FIGURE 6.13
Deflection angles in polygon.

6.10
DEFLECTION ANGLES

The angle between a line and the prolongation of the preceding line is called *deflection angle*. Deflection angles are recorded as *right* or *left* depending on whether the line to which measurement is taken lies to the right (clockwise) or left (counterclockwise) of the prolongation of the preceding line. In Figure 6.12, the deflection angle at B is 22°R, and at C is 33°L. Deflection angles may have values between 0° and 180°, but usually they are not employed for angles greater than 90°. In any closed polygon, the algebraic sum of the deflection angles (considering right deflections as of a sign opposite to left deflections) is 360°. Figure 6.13 shows the polygon $ABCDE$ with deflection angles $\alpha_1, \alpha_2, \ldots, \alpha_5$. In this example $\alpha_1 + \alpha_2 - \alpha_3 + \alpha_4 + \alpha_5 = 360°$.

6.11
ANGLES TO THE RIGHT

Angles may be determined by clockwise measurements from the preceding to the following line, as illustrated by Figure 6.14. Such angles are called *angles to the right* or *azimuths from the back line*.

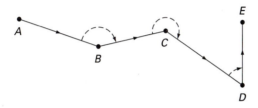

FIGURE 6.14
Angles to the right.

6.12
METHODS OF DETERMINING ANGLES AND DIRECTIONS

Angles normally are measured with a theodolite or total station system but also can be determined by means of a tape or compass. Directions are observed with a direction theodolite or with a magnetic compass.

1. *Theodolite.* The engineer's theodolite is designed to enable observing horizontal and vertical angles. Most engineer's theodolites allow measuring the horizontal angle to the nearest minute of arc, but on some instruments, the horizontal angle can be observed to the nearest 10″ of arc. Zenith and vertical angles can be read to the nearest minute of arc. A detailed description of the engineer's theodolite and its use can be found in Sections 6.21 to 6.38.

2. *Total station systems.* The total station system incorporates angle and distance measurement capability in the framework of a theodolite. Horizontal and zenith or vertical angles are measured digitally and can be observed with least counts from the nearest 20″, in standard models, to 1″ of arc in the most precise systems.

3. *Tape.* If the sides of a triangle are measured, sufficient data are obtained for computing the angles in the triangle. The error in the computed values of the angles depends on the care with which points are established and the accuracy with which measurements are taken. For acute angles on level ground, the error need not exceed 05′ to 10′. For angles greater than 90°, the corresponding acute angle should be observed. The method is slow and generally used only as a check.

4. *Direction theodolite.* In the United States, *direction theodolite* refers to an extremely precise theodolite that has only one horizontal motion. With this type of instrument, directions are observed and the angle is computed as the difference between the two directions. Direction theodolites usually have a nominal accuracy of ± 1″ of arc. The characteristics and use of these instruments are given in Sections 6.21, 6.29, 6.48 and 6.49.

5. *Magnetic compass.* The use of the magnetic compass is described in the following sections. By itself, the compass is useful in making rough surveys and retracing early land surveys. Mounted on a transit, the compass is useful to approximately check horizontal angles measured by more precise methods.

6.13
MAGNETIC COMPASS

Any slender symmetrical bar of magnetized iron when freely suspended at its center of gravity takes up a position parallel to the lines of magnetic force of the earth. In a horizontal projection, these lines define the magnetic meridians. In elevation, the lines are inclined downward toward the north in the northern hemisphere and downward toward the south in the southern hemisphere. Because the bar takes a position parallel to the lines of force, it becomes inclined with the horizontal. This phenomenon is called the *magnetic dip*. The angle of dip varies from 0° at or near the equator to 90° at the magnetic poles. The needle of the magnetic compass rests on a pivot. To counteract the effect of dip, so that the needle will take a horizontal position when directions are observed, a counterweight is attached to one end (the south end in the northern hemisphere). The counterweight usually consists of a short piece of fine brass wire wound around the needle and held in place by spring action. As long as the needle is used in a given locality and loses none of its magnetism, it will remain balanced. When, for any reason, it becomes unbalanced, it is adjusted to the

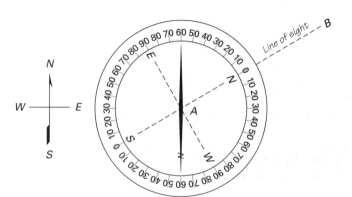

FIGURE 6.15
Features of magnetic compass used in surveying.

horizontal by sliding the counterweight along the needle. At the midpoint of the needle is a jewel that forms a nearly frictionless bearing for the pivot.

The essential features of the magnetic compass used by the surveyor are (1) a compass box with a circle graduated from 0° to 90° in both directions from the N and S points and usually having the E and W points interchanged, as illustrated in Figure 6.15; (2) a line of sight in the direction of the SN points of the compass box; and (3) a magnetic needle. When the line of sight is pointed in a given direction, the compass needle (when pivoted and brought to rest) gives the magnetic bearing. Therefore, in the figure the bearing of *AB* is N60°E. If the N point of the compass box is nearest the object sighted, the bearing is read by observing the north end of the needle.

The varieties of compasses exhibiting the features just mentioned are

1. Various *pocket compasses,* which generally are held in the hand when bearings are observed and used in reconnaisance or other rough surveys. Another style, the Brunton pocket transit, is designed primarily as a hand instrument but may be mounted on a tripod or a Jacob's staff (a pointed stick about 1.5 m (5 ft) long).
2. The *surveyor's compass,* which usually is mounted on a light tripod or sometimes on a Jacob's staff, now is used only for forest surveys or retracing old land surveys.
3. The *transit compass,* a compass box similar to that of the surveyor's compass, mounted on the upper or vernier plate of the engineer's transit (Section 6.21) and often used to check horizontal angles.
4. The *tube* or *trough compass,* in which the magnetized needle is suspended in a tube that can be mounted on the standard of an engineer's theodolite, thus permitting the telescope of the instrument to be oriented toward magnetic north.

6.14
MAGNETIC DECLINATION

The angle between the true meridian and the magnetic meridian is called the *magnetic declination* or *variation.* If the north end of the compass needle points to the east of the true meridian, the declination is said to be *east* (Figure 6.16); if it points to the west of the true meridian, the declination is said to be *west.* Declination may be set off on the compass as shown in Figure 6.17.

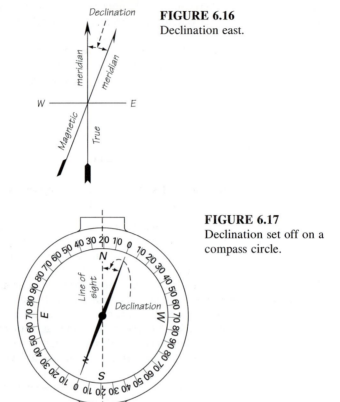

FIGURE 6.16
Declination east.

FIGURE 6.17
Declination set off on a
compass circle.

If a true north-south line is established, the mean declination of the needle for a given locality can be determined by compass observations extending over a period of time. The declination may be estimated from an *isogonic chart,* traditionally published by the U.S. Geologic Survey (U.S.G.C.), Branch of Global Seismology and Geomagnetism.* Specific values for a particular location can be obtained from the Internet via a personal computer. Simply dial 800-358-2663, enter **telnet neis.cr.usgs.gov,** and follow the instructions displayed. To use this program, one needs the latitude, longitude (Section 10.3), and elevation of the location for which the declination is desired. Latitude, longitude, and elevation can be obtained with sufficient accuracy by scaling and interpolating, respectively, from a U.S.G.S. 1 : 24,000 scale topographic map (Section 14.4).

6.15
ISOGONIC CHART

The isogonic chart of the continental United States shown in Figure 6.18 applies to January 1, 1990. It is based on observations made by the U.S. National Ocean Survey at stations widely scattered throughout the country. The solid lines are lines of equal magnetic declination, or *isogonic lines.* East of the heavy solid line of zero declination, or *agonic line,*

* Box 25046, MS968, Denver Federal Center, Denver, CO 80255.

the north end of the compass needle points west of north; west of that line, it points east of north. The north end of the compass needle is moving eastward over the area of eastward annual change and westward elsewhere over the chart at an annual rate indicated by the lines of equal annual change. Note that the U.S.G.S. does not plan to publish additional isogonic charts (Figure 6.18) for dates later than 1990. For declinations and other geomagnetic information related to 1995, 2000, . . . and so forth, the user will have to use the Internet as described already.

6.16
VARIATIONS IN MAGNETIC DECLINATION

The magnetic declination changes more or less systematically in cycles over periods of (1) approximately 300 years, (2) 1 year, and (3) 1 day, as follows:

1. *Secular variation.* Like a pendulum, the magnetic meridian swings in one direction for perhaps a century and a half until it gradually comes to rest and then swings in the other direction, and as with a pendulum, the velocity of movement is greatest at the middle of the swing. The rate of change per year, however, varies irregularly. The causes of this secular variation are not well understood. In the United States, it amounts to several degrees in a half-cycle of approximately 150 years. In Figure 6.18, the annual rates of change in the secular variation for the year 1990 are shown by dashed lines. Because of its magnitude, the secular variation is of considerable importance to the surveyor, particularly in retracing lines the directions of which are referred to the magnetic meridian as it existed years previously. When *variation* is mentioned without further qualification, it is taken to mean the secular variation.
2. *Annual variation.* This is a small annual swing distinct from the secular variation. For most places in the United States, it amounts to less than 01'.
3. *Daily variation.* This variation, also called *solar-diurnal* variation, is a periodic swing of the magnetic needle occurring each day. For points in the United States, the north end of the needle reaches its extreme easterly swing at about 8 or 9 A.M. and its extreme westerly swing at about 1 or 2 P.M. The needle usually reaches its mean position between 10 and 11 A.M. and between 7 and 11 P.M. In general, the higher the latitude, the greater is the range in the daily variation. The average range for points in the United States is less than 08', a quantity so small as to need no consideration for most of the work for which the compass needle is employed. However, in the United States in summer, a line run 1000 ft by compass at 8 A.M. would end as much as 3 ft to the right of the point where it would end if run at 1 P.M.
4. *Irregular variations.* Irregular variations are due to magnetic disturbances usually associated with sunspots. They cannot be predicted but are most likely to occur during magnetic storms, when auroral displays occur and radio transmission is disturbed. They may amount to a degree or more, particularly at high latitudes.

6.17
CORRECTION FOR DECLINATION

When magnetic directions are used to obtain coarse estimates for bearings or when an old survey must be retraced, it is necessary to reduce the magnetic directions to true bearings or azimuths. Conversion from magnetic to true azimuths or vice versa is accomplished most easily by using azimuths. Consider the following examples.

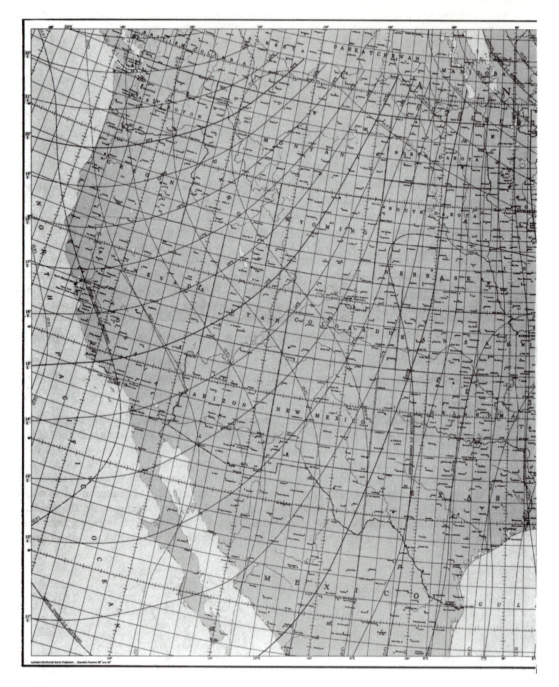

FIGURE 6.18
Isogonic chart for the United States, showing magnetic declination in the United States for 1990.
(*Courtesy of U.S. Geological Survey.*)

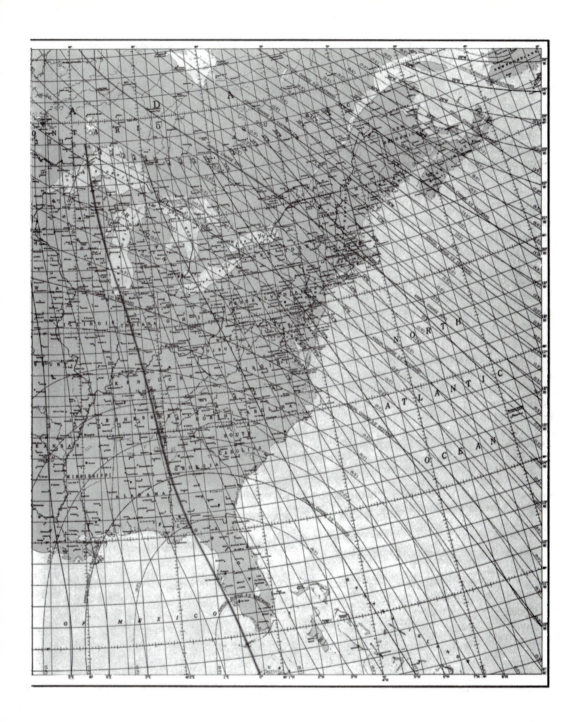

EXAMPLE 6.1. A magnetic azimuth of 54°30′ was observed along line *AD* in June 1994. The declination for the area surveyed is found by interpolation from an isogonic chart dated 1990 to be 17°30′E with an annual change of 3′ westward. Compute the true azimuth of line *AD*.

Solution. First draw a careful sketch of the relationship among true north, magnetic north, and the direction of the line as illustrated by Figure 6.19a.

Magnetic azimuth *AD*	54°30′
Declination 1990	17°30′
Change = (4 yr)(3′/yr)	−0°12′
True azimuth *AD*	71°48′

EXAMPLE 6.2. A magnetic bearing of N34°30′W is recorded on an old survey plan dated August 20, 1910. It is desired to reestablish this direction on the site in 1995. The 1990 isogonic chart shows a declination of 20°W for the area, with an annual change of 2′ eastward. Determine the magnetic bearing that must be used to relocate the direction of the line in the field.

Solution. As before, draw a careful sketch of the lines involved, as shown in Figure 6.19b.

Declination in 1990	20°00′W
Change in 85 yr = 85 × 2′	2°50′
Declination 1910	22°50′W
Magnetic bearing 1910	N34°30′W
True bearing of line	N57°20′W
Declination 1995 = 20°W − (5 yr)(2′E)	19°50′W
Magnetic bearing, 1995	N37°30′W

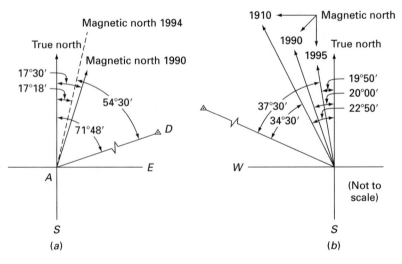

FIGURE 6.19
Corrections for declination.

6.18
LOCAL ATTRACTION

283

CHAPTER 6:
Angle and
Direction
Measurement

Objects of iron or steel, some kinds of iron ore, and currents of direct electricity alter the direction of the lines of magnetic force in their vicinity and hence are likely to cause the compass needle to deviate from the magnetic meridian. The deviation arising from such local sources is called *local attraction* or *local disturbance*. In certain localities, particularly in cities, its effect is so pronounced as to render the magnetic needle of no value for determining directions. It is not likely to be the same at one point as at another, even though the points may be but a short distance apart. It is affected even by such objects as the steel tape, chaining pins, axe, and small objects of iron or steel that are on the person. Usually, its magnitude can be determined, and directions observed with the compass can be corrected accordingly. Local attraction usually can be detected by observing the compass bearing of a line at two or more points on the line.

6.19
USE OF THE COMPASS

In order that *true* bearings may be read directly, some compasses, such as the one shown in Figure 6.17, are designed so that the compass circle may be rotated with respect to the box in which it is mounted. When the circle is in its normal position, the line of sight as defined by the vertical slits in the sight vanes is in line with the N and S points of the compass circle, and the observed bearings are magnetic. If the magnetic declination is set off by means of the circle, the observed bearings will be true, as is evident from Figure 6.17. If the declination is east, as in the figure, the circle is rotated clockwise with respect to the plate; if the declination is west, counterclockwise.

When the direction of a line is to be determined, the compass is set up on a line and leveled. The needle is released, and the compass is rotated about its vertical axis until a range pole or other object on the line is viewed through the slits in the two sight vanes. When the needle comes to rest, the bearing is read. Ordinarily, the sight vane at the end of the compass box marked *S* is held next to the eye; in this case the bearing is given by the north end of the needle.

The following suggestions apply to compass observations. At each observation the compass box should be tapped lightly as the needle comes to rest, so that the needle may swing freely. In order not to confuse the north and south ends of the needle when taking bearings, the observer should always note the position of the counterbalancing wire (which is on the south end in the northern hemisphere). Since the *precision* with which angles may be read depends on the delicacy of the needle, special care should be taken to avoid jarring between the jewel bearing of the needle and the pivot point. *Never move the instrument without making certain that the needle is lifted and clamped.*

Sources of magnetic disturbance such as chaining pins and axes should be kept away from the compass while a reading is being taken. Care should be taken not to produce static charges of electricity by rubbing the glass; a moistened finger pressed against the glass will remove such charges. Ordinarily the amount of metal about the person of the instrument operator is not large enough to deflect the needle appreciably, but a change of position between two readings should be avoided.

Surveying with the compass usually is by traversing. Only alternate stations need be occupied, but a check is secured and local attraction is detected if both a backsight and a foresight are taken from each station. Unlike a transit traverse, in which an error in any angle affects the observed or computed directions of all following lines, an error in the observed bearing of one line in a compass traverse has no effect on the observed *directions* of any of the other lines. This is an important advantage, especially in a traverse having many angles. Another advantage of the compass is that obstacles such as trees can be passed readily by offsetting the instrument a short measured distance from the line. The procedure for performing and adjusting a compass traverse is outlined in Section 8.10.

6.20
SOURCES OF ERROR: ADJUSTMENT OF THE COMPASS

1. *Needle is bent.* If the needle is not perfectly straight, a constant error will be introduced in all observed bearings. As shown by Figure 6.20, one end of the needle will read higher than the correct value and the other end will read lower. For each observation the error can be eliminated by reading both ends of the needle and averaging the two values. The needle can be straightened with pliers.

2. *Pivot is bent.* If the point of the pivot supporting the needle is not at the center of the graduated circle, a variable systematic error will be introduced, the magnitude of which depends on the direction in which the compass is sighted. For one direction, the error is zero; for the normal to this direction, it is a maximum. In this case also, one end of the needle will read higher than the correct value and the other end will read lower (Figure 6.21); for each observation the error can be eliminated by reading both ends of the needle and averaging the two values. The instrument can be corrected by bending the pivot until the end readings of the needle are 180° apart for any direction of pointing.

3. *Plane of sight is not vertical, or graduated circle is not horizontal.* This misalignment introduces a systematic error but it usually is so small as to be of no consequence. However, the sight vanes may become bent so that, even though the instrument is leveled, an appreciable error is introduced, particularly if the line of sight is steeply inclined when taking a bearing. The vanes may be tested by leveling the compass and sighting at a plumb line. The adjustment of the level tubes may be tested by reversal, as described for the theodolite in Section 6.51.

4. *Needle is sluggish.* The needle is not likely to come to rest exactly on the magnetic meridian. This lag produces a random error. As the needle comes nearly to rest, tapping the glass lightly will tend to prevent the needle from sticking to the pivot. If the needle is "weak," it may be remagnetized by drawing its ends over a bar magnet, from the

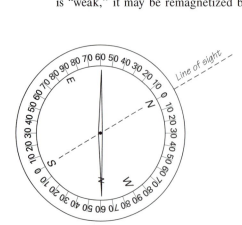

FIGURE 6.20
Bent needle.

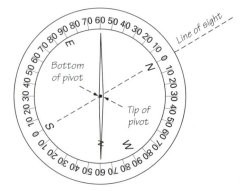

FIGURE 6.21
Bent pivot.

center to the ends of the magnet. The south-seeking end of the compass needle is drawn over the north-seeking half of the bar magnet and vice versa. On each return stroke the needle should be lifted well above the magnet. If the pivot point is blunt, it may be sharpened by rubbing it on a fine-grained oilstone.

5. *Reading the needle.* The inability of the observer to determine exactly the point on the graduated circle at which the needle comes to rest generally is the source of the most important and largest random error in compass work. The needle should be level, and the eye of the observer should be above the coinciding graduation and in line with the needle. If the needle dips perceptibly, its counterweight should be adjusted. Other conditions being equal, the longer the needle, the smaller is the error of observing. With the 6-in. needle used on many surveyor's compasses, the estimated standard deviation need not exceed $\pm 10'$; with the $3\frac{1}{2}$-or 4-in. needle on the engineer's transit, the standard deviation is likely to be as much as $15'$.

6. *Magnetic variations.* Undetected deviations of the magnetic needle from whatever cause are the source of the largest and most important systematic errors in compass work. Largely because of such variations, the compass, no matter how finely constructed, is not a suitable instrument for any except rough surveys. Deviations due to local attraction can be detected and corrections can be applied as described earlier.

6.21
THEODOLITE AND TRANSIT

A theodolite or transit is an instrument designed to measure horizontal and zenith or vertical angles. It consists of a telescope mounted so as to rotate vertically on a horizontal axis supported by a pair of vertical standards attached to a revolvable circular plate containing a graduated circle for reading horizontal angles. Another graduated arc is attached to one standard so that zenith or vertical angles can be observed.

Theodolite was the name given to the earliest version of a device to measure angles in horizontal and vertical planes, designed and built in England in about 1725.* The first *transit* was designed and manufactured by William Young of Philadelphia in 1831.** The transit was distinguished by the ability of its telescope to be "transited" or turned 180° about its horizontal axis, thus providing a means for extending a straight line. The theodolites made at that time were not designed to perform this function, although subsequently the transiting capability was included to yield an instrument called a *transiting theodolite.* In Europe, the name *theodolite* was retained, and in the United States, the term *transit* was kept. Hence, the two names describe essentially similar instruments.

Currently, several styles of theodolites, transits, and theodolite equipped systems (e.g., theodolites with add-on EDM instruments, total station systems) are used to make horizontal and vertical circle measurements. In this chapter, the primary emphasis is on (1) double-center instruments, with three-screw leveling and optical reading or electronic digitized horizontal and vertical scales, called *repeating theodolites,* and (2) single-center instruments with optical reading or digitized horizontal and vertical circles, referred to as *direction theodolites.* These two styles make up the majority of the instruments used by surveyors interested primarily in making only angular measurements. In addition, the *engineer's transits* are designed with the traditional configuration of the instrument, with four-screw leveling and some with three-screw leveling, both equipped for direct reading of horizontal

---

*S. A. Bedini, "Being a History Detective," *Professional Surveyor* 7, no. 5 (1987), p. 6.
**D. Beeks, "Evolution of the Transit," *Professional Surveyor* 7, no. 5 (1987), p. 10.

and vertical angles with a pocket magnifier. Few of these types of instruments are now being manufactured, and they are used mainly in the building construction industry.

The *total station system,* a digital angle and distance measuring device built into the framework of a digitized repeating theodolite, is becoming more and more popular and prevalent in all types of surveying operations. Total station systems are described in detail in Chapter 7.

6.22
REPEATING THEODOLITE

Figure 6.22 shows a repeating, optical reading theodolite with three-screw leveling and Figure 6.23 illustrates a vertical section of a typical optical reading theodolite. It is seen to consist of an *alidade,* which contains two vertical standards that support the horizontal axis to which the telescope is attached; a horizontal circle or *plate,* which is partially enclosed by the lower housing of the alidade; and a *leveling head,* which contains the three leveling screws and is supported on a baseplate attached to the top of the tripod by a hollow-threaded bolt. The leveling head and baseplate together form the *tribrach.*

The alidade, which also contains the microscope for viewing through the optical plummet (see Figure 6.23) and a scale for measuring horizontal angles, is supported on an *inner center,* the axis of which is hollow (for the optical plummet) and concentric to the instrumental vertical axis. The plate that holds the glass horizontal circle is supported by the *outer center,* which is concentric to the inner center and the vertical axis of the instrument. The outer center can be clamped in any position by a *lower clamp screw* that locks the outer center and its horizontal circle to the leveling head. Similarly, the inner center, which carries the alidade with its scale, may be clamped to the outer center by an *upper clamp screw,* which locks the alidade to the outer center. After either clamp has been tightened, small relative movements of the respective centers are made by turning the corresponding *tangent screw.* The axis about which the inner and outer centers revolve is called the *vertical axis* of the theodolite.

A *circular level bubble* is mounted on the leveling head for approximate leveling (Figure 6.22). A *plate level bubble* is fixed to the alidade between the two standards and is used for precise leveling (Figure 6.23). The circular level bubble usually is centered by manipulating the tripod legs. The plate level bubble then is centered precisely by the three leveling screws (see Section 6.32). When the threaded bolt holding the theodolite to the top of the tripod is slightly loosened, the entire theodolite can be shifted laterally 3 to 4 cm in any direction to permit centering the instrumental vertical axis over a point by using the optical plummet (Figure 6.23).

The telescope, fixed to a transverse axis that rests on bearings on the standards, is short and can be rotated completely about the horizontal axis. It can be fixed in any position in a vertical plane by means of the *vertical or telescope clamp screw;* small movements about the horizontal axis then can be performed by turning the *vertical or telescope tangent screw* (see Figure 6.23). The vertical circle, which is made of glass, is supported by a pendulum-type apparatus or compensator (similar to that used in a self-leveling level, Section 5.19) or an optical-mechanical or electro-optical tilt sensor is built into the vertical circle observation system. Thus, when the instrumental vertical axis is made vertical by using the circular and plate level bubbles, readings on the vertical circle will be automatically referred to the true zenith or true horizontal. This process is called *automatic vertical circle indexing.* Most theodolites of advanced design are equipped with this feature.

The glass horizontal and vertical circles are viewed simultaneously by a microscope mounted parallel to the main telescope (Figure 6.23). Images of both horizontal and vertical

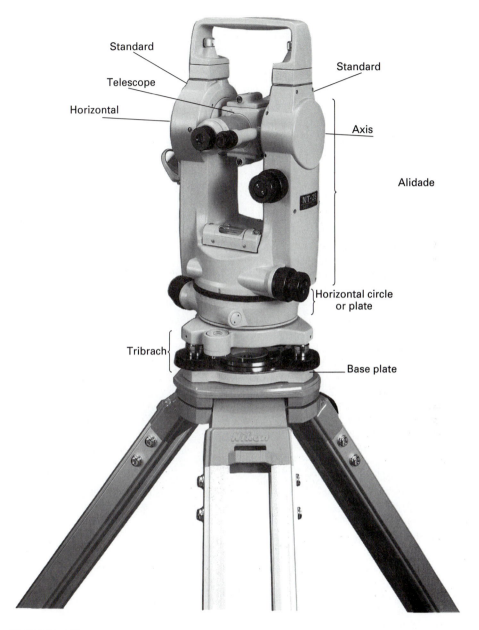

Standard

Telescope

Horizontal

Standard

Axis

Alidade

Horizontal circle
or plate

Tribrach

Base plate

FIGURE 6.22
Optical reading, repeating theodolite. (*Courtesy of the Nikon Corporation.*)

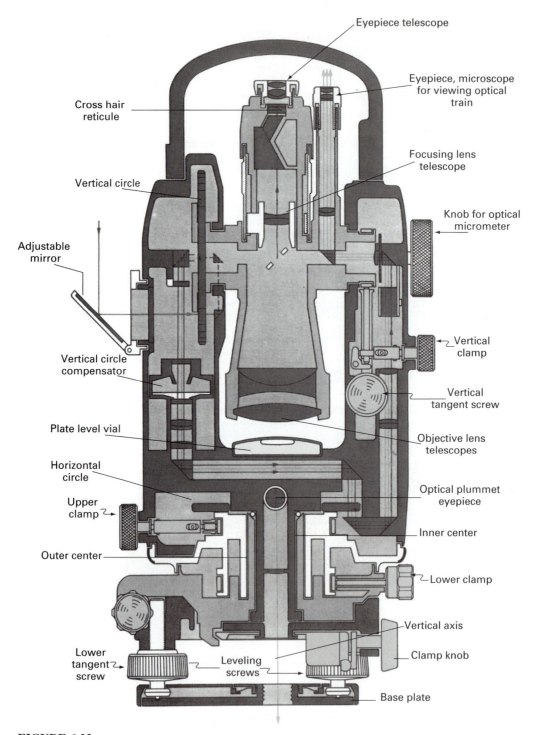

FIGURE 6.23
Vertical cross section of optical reading, repeating theodolite. (*Courtesy of Leica, Inc.*)

circles and their respective indexes are transmitted to the observer's eye by an optical train composed of prisms, mirrors, and lenses. This optical train is illuminated by outside light reflected by a mirror into an aperture on the left standard of the instrument (Figure 6.23). Alternatively, a small light bulb in a battery pack can be mounted over this aperture on the standard to allow night observations.

On some models of repeating theodolites, a *tube or trough compass* can be mounted on a standard of the theodolite to allow orienting the telescope toward the magnetic or true north.

Summarizing the main features,

1. The center of the theodolite can be brought over a given point by viewing through the optical plummet, slightly loosening the clamp that holds the theodolite to the tripod, and then shifting the instrument laterally until the cross or circle in the optical plummet is centered on the point.

2. The instrument can be leveled by adjusting the tripod legs to center the circular level bubble (approximate leveling), followed by centering the plate level bubble using the leveling screws (precise leveling).

3. The telescope can be rotated about either the horizontal or vertical axis.

4. When the upper clamp screw is tightened and the telescope is rotated about the vertical axis, no relative movement will take place between the scale and the horizontal circle.

5. When the lower clamp screw is tightened and the upper one is loose, a rotation of the telescope about the vertical axis will cause the alidade and scale to revolve but will leave the horizontal circle fixed in position; hence, the designation *double-center* instrument.

6. When both the upper and lower clamps are tightened, the telescope cannot be rotated about the vertical axis.

7. The telescope can be rotated about the horizontal axis and can be fixed in a vertical plane by the vertical clamp screw and tangent screw.

8. On most current instruments, the vertical circle is equipped with a compensator that automatically orients 0 toward the zenith or the horizon; thus, the telescope can be leveled by setting the zenith angle on the vertical circle equal to 90° or the vertical angle to 0°, so that the theodolite can be used for direct leveling.

9. By observing the vertical circle, index, and scale, zenith or vertical angles can be determined; hence, the theodolite is suitable for indirect leveling.

10. By observing the horizontal circle, index, and scale, horizontal angles can be measured.

11. When a tube compass is attached, magnetic north can be established and magnetic directions determined.

6.23
LEVEL BUBBLES AND SENSORS

The sensitivity of the spirit levels of the theodolite should be such as to produce a well-balanced instrument and ought to correspond to the fineness of the graduations of the circles and the optical properties of the telescope. If the level bubbles are more sensitive than necessary to maintain this balance, time is wasted in centering the bubbles; if less sensitive than necessary, the precision of measurements is less than it should be for the theodolite as otherwise designed.

The circular level bubble, used to approximately level the instrument, generally has a sensitivity of 8–10'. The plate level bubble for instruments with a least reading on the horizontal and vertical scales of 01' usually has a sensitivity that varies from 30″ to 60″ per

2-mm graduation. Most optical reading and electronic theodolites have plate level bubbles with a sensitivity of 30″ per 2-mm division. When horizontal angles are measured between points nearly in the same horizontal plane, it can be shown that no significant error is introduced even when the bubble is some distance off-center (Section 6.45). On the other hand, when there is a large difference in elevation between the points sighted, a small displacement of the bubble in the plate level tube causes a relatively large error in the horizontal angle. For additional details concerning spirit level tubes, see Sections 5.14 and 5.15.

An electronic sensor, used in place of a plate spirit level bubble, is available on some electronic theodolites. One such device is an electro-optical tilt sensor with a liquid surface that creates an artificial horizon. The angle between the sensor and the artificial horizon is sensed by a position-sensitive detector, and this misalignment is converted to digital or analog signals. For the electronic plate sensor, the digital signals are displayed (on command) on the register of the instrument as horizontal and vertical lines, which can be centered between fixed graduations (also displayed) by manipulating the level screws of the instrument. This type sensor is capable of producing an accuracy of 0.2″ of arc. However, as installed as a plate leveling device, specified accuracies are 5″ to 6″ of arc.

The compensators (similar to the tilt sensor just described), which automatically orient the vertical circle with respect to gravity, have specified accuracies of from 0.3″ to 6″ of arc with ranges of from 2′ to 6′ of arc. This means that, if the vertical axis of the theodolite is within 2′ to 6′ of being vertical, the compensator will be activated to provide vertical circle readings with a precision of from 0.3″ to 6″, depending on the type instrument being used. With a plate spirit level having a sensitivity of 30″ per 2-mm division, and assuming the bubble can be centered to within 0.1 of a division, the vertical axis easily can be vertical with a precision of 3″ of arc. Therefore, even approximate leveling of the instrument will yield the nominal accuracy of the compensator in observed vertical circle readings.

Older models of optical reading theodolites had vertical circles in which the reading scale was equipped with a spirit level bubble of the coincidence type (Section 5.18). These bubbles had sensitivities of from 20″ to 30″ per 2 mm division.

6.24
TELESCOPE

The telescope of the theodolite is similar to that of the engineer's level (Section 5.8). When the theodolite is used as an instrument for establishing lines, measuring angles, or taking azimuths or bearings, any point on the vertical cross hair may be used; when the theodolite is used for either direct or trigonometric leveling, any point on the horizontal cross hair may be used. Most instruments are equipped with stadia hairs (Chapter 7), mounted in the same plane as the cross hairs. The telescope for the theodolite, as for the level, is internally focusing and has a magnifying power of from 23 to 30 diameters. Usually, the erecting eyepiece is employed, but the superior optical properties of the inverting eyepiece make it the favorite of some surveyors and it is the type used in instruments of extremely high precision. The instrument illustrated in Figure 6.22 has an internally focusing telescope with an erecting eyepiece.

6.25
GRADUATED CIRCLES

The graduated horizontal and vertical circles are made of glass and are relatively small (70 mm or 2.8 in. diameter). The graduations are etched on the surface of the glass, a process that produces an extremely fine line. The horizontal circle is graduated in the

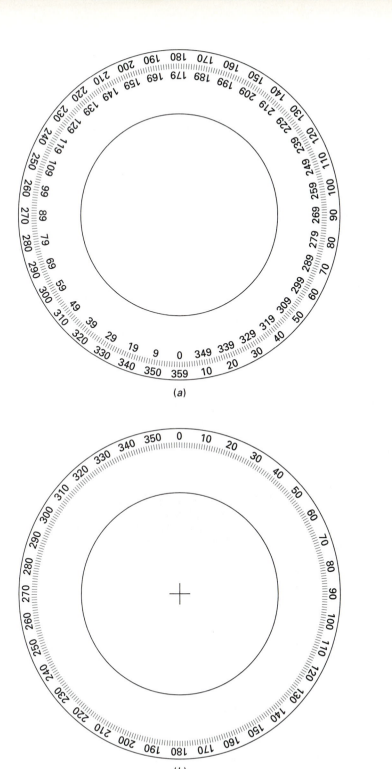

(a)

(b)

FIGURE 6.24
Numbering of circles: (a) horizontal circle, (b) vertical circle.

sexagesimal system (360° per full circle) with a graduation at each full degree or in the centesimal system (400 gons per circle) with a graduation at each full gon. For a repeating theodolite, the horizontal circle is numbered in clockwise and counterclockwise directions, as illustrated in Figure 6.24a, where the circle is graduated in degrees.

The vertical circle also is graduated in full degrees or gons and numbered clockwise from 0° to 360° or from 0^g to 400^g, as shown in Figure 6.24b, in which the circle is subdivided in degrees. Note that Figure 6.24 shows the circle numbering as it would appear when viewed through the optical train.

6.26
SCALES AND MICROMETERS

As noted previously, horizontal and vertical circle readings are observed by viewing through a microscope mounted parallel to the telescope that directs the eye to the scales on the respective circles by means of an optical train composed of a series of mirrors and prisms (Figure 6.23). Two methods are used to read the circles to a higher precision than is possible using the least division on the circles of 1° or 1^g: (1) a direct-reading scale optically superimposed over the horizontal and vertical circles, and (2) an optical micrometer.

Direct-Reading Scale

The direct-reading scale, attached to the alidade so that it rotates with respect to the horizontal circle, is visible through the optical train and covers the least division on the horizontal circle. Figure 6.25a shows a direct-reading scale as viewed through an optical train. Note there are 60 graduations on the scale, which spans the least division of the circle which is 1° or 60'. The *least count* of a theodolite of this type, is the number of units in the least division of the circle divided by the number of divisions on the scale. Therefore, the least count for this instrument equals 60'/60 = 01' and angles can be estimated to the nearest 20". The clockwise and counterclockwise angles for the example in Figure 6.25a are 359°45'30" and 0°14'30", respectively. Note that clockwise angles are observed using the upper numbers and graduations on the circle and scale, both of which are graduated in a clockwise direction. Figure 6.25b illustrates the same scale and circle set to 0 on both the clockwise and counterclockwise circles. Figure 6.25c shows the entire field of view visible when viewing through the optical train. Here, the horizontal angles and vertical circle readings are read in the lower window, labeled *H*, and upper window, labeled *V*. The clockwise and counterclockwise horizontal angles, for this example, are 42°54'00" and 317°06'00", respectively. The vertical circle reading, a zenith angle, is 85°36'00".

Optical Micrometer

An optical micrometer is used in theodolites in which a least count less than 1' is desired. The optical micrometer, as designed for a tilting level, was described in Section 5.49. The same principle is used on a theodolite to measure the increment from the circle index mark to the nearest full graduation of the circle. To accomplish this, the image of the micrometer scale is introduced into the optical train.

The procedure for making horizontal and vertical circle readings with a micrometer varies with the make of the instrument. Figure 6.26 shows the scene visible through the optical train for the instrument shown in Figure 6.23, using the viewing microscope

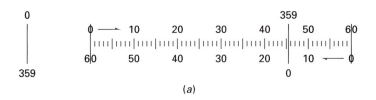

(a)

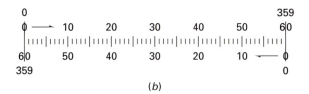

(b)

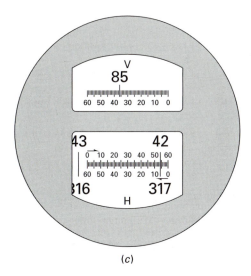

(c)

FIGURE 6.25
Horizontal and vertical circle readings using
a scale in an optical theodolite.

(Figures 6.22 and 6.23). Figure 6.26a illustrates the reading immediately on taking a sight and before measuring the increment for the horizontal circle, labeled *Hz*. Note that the graduation mark for 327° is displaced to the right of the bifilar graduations that constitute a fixed index in the center of the window. By rotating the optical micrometer knob on the right standard (Figure 6.23), the graduation mark is optically moved until it precisely bisects the bifilar graduations, as shown in Figure 6.26b. As this operation occurs, the graduations on the scale, visible in the small window to the right and linked to the micrometer, move and yield the increment measured, which is 59′36″, so the angle reading for this example is 327°59′36″. The procedure for reading the vertical circle is the same, except that the graduation mark for 92° would be centered on the bifilar index in the window labeled *V* and the increment read on the optical micrometer (not illustrated in Figure 6.26). The least counts of optical reading, repeating theodolites vary from 06″ to 20″. The instrument for which readings are illustrated in Figure 6.26 has a least count of 06″.

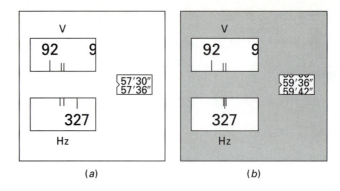

FIGURE 6.26
Horizontal circle reading using optical micrometer, Wild T1.
(*Courtesy of Leica, Inc.*)

The horizontal circle on an instrument equipped with an optical micrometer is numbered only in the clockwise direction. So, when using one of these theodolites, angles to the right are always measured (see Section 6.33).

6.27
ERRORS DUE TO ECCENTRICITY

The mechanical parts of the inner and outer centers may become worn or otherwise damaged, so that the center of rotation of the inner center, supporting the scales and index marks, does not coincide with the geometric center of the graduated circle held by the outer center. As indicated in Figure 6.27, there will be one setting on the graduated circle for

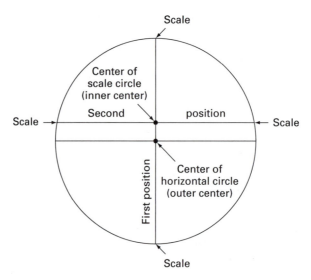

FIGURE 6.27
Ecentricity of centers.

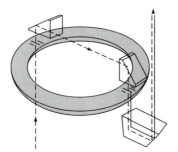

FIGURE 6.28
Compensation for error due to eccentric centers (*Courtesy of Topcon, Inc.*)

which the index marks on the scales are exactly 180° apart (first position) and 90° from there will be another setting for which the scales fail to register 180° apart by a maximum amount (second position).

Correction of this defect mechanically would require the services of an instrument maker, but by observing both scales and taking the mean value, the error can be eliminated. In optical-reading theodolites, this type of error can be automatically compensated for by introducing readings on both sides of the circle into the optical train, as shown in Figure 6.28. In general, this option is not exercised in optical-reading theodolites having least counts of from 06″ to 1′, because errors due to eccentricity are most likely to be less than these amounts. However, some optical repeating instruments have this feature, and it is an integral part of the direction theodolites described in Section 6.29.

6.28
GEOMETRY OF THE THEODOLITE

A knowledge of the geometry of the theodolite is necessary for a thorough understanding of its operation. Figure 6.29 shows this geometry as related to the instrument and in schematic form. The intersecting lines, labeled *A*, define the plane of the plate level tube as determined by two positions of the axis of the level tube, one 90° from the other; the vertical axis is designated *B*, the horizontal trunion axis is *C*; and the optical line of sight is *D*. In a perfectly constructed and adjusted instrument, the following relationships should exist among these axes: *B* is perpendicular to *A*; *C* is perpendicular to *B*; *D* is perpendicular to *C*; *B*, *C*, and *D* pass through a single point; and the zero index of the vertical circle, as corrected by the compensator, falls on line *E* which is parallel to vertical axis *B* and passes through horizontal axis *C*. Achievement of these relationships by adjustment is detailed in Sections 6.51 and 6.52.

6.29
DIRECTION THEODOLITES

Direction theodolites have only a single center and single horizontal motion, in contrast to the double-center repeating instruments. A direction instrument is shown in Figure 6.30 and a vertical cross section of this theodolite is illustrated in Figure 6.31. Note (Figure 6.31), that these instruments have only a single horizontal clamp and tangent screw. There also is a coarse motion for advancing the horizontal circle in large increments (Figures 6.30 and 6.31). *Directions are observed* with these theodolites and the angle is the difference between the two directions (Section 6.48). In general, direction instruments are

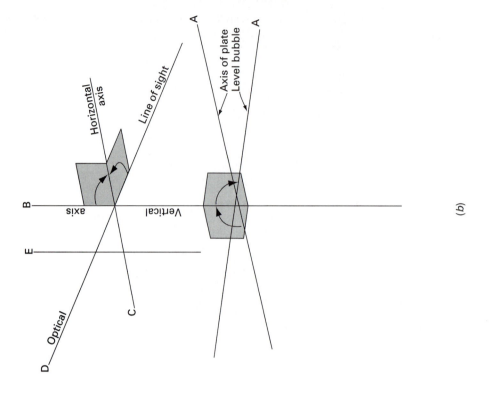

Horizontal axis

Line of sight

Axis of plate

Level bubble

B

E

C

D

Optical

axis

Vertical

(b)

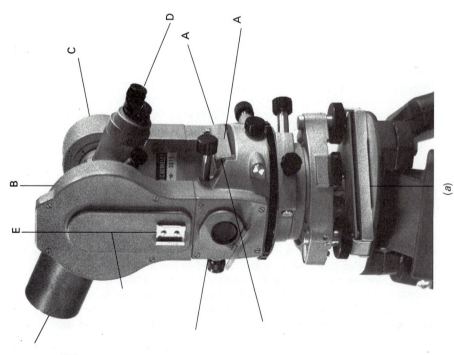

C

D

A

A

B

E

FIGURE 6.29
Geometry of the theodolite.

(a)

Micrometer
knob

Housing coarse
motion horizontal
circle

FIGURE 6.30
Leica theodolite. (*Courtesy of Leica, Inc.*)

more precise than repeating theodolites. The plate level bubble has a sensitivity of 20″ per 2 mm division on the bubble tube, the vertical circle is equipped with an automatic compensator having a setting accuracy of from 0.3″ to 0.5″, and horizontal and vertical circle readings are made using an optical micrometer.

The method of observing horizontal directions and vertical circle readings is the same as for a repeating theodolite and involves viewing through a microscope mounted parallel to the telescope (Figure 6.31). The optical train is illuminated by adjustable mirrors for daylight operation (Figure 6.31). These mirrors can be replaced by attachments with light bulbs for night observation.

As with the repeating theodolite, the procedure for reading angles varies with the make of instrument. The field of view visible, when looking through the reading microscope of the direction theodolite, illustrated in Figure 6.30, is shown in Figure 6.32. In the top diagram, before coincidence, the middle frame contains a scale that shows degrees and minutes to the nearest 10′, and the bottom frame has a scale graduated in minutes and seconds. The horizontal and vertical circles are graduated in 5′ intervals and an optical micrometer is employed to measure directly to seconds. Before coincidence, the horizontal circle reading is estimated as 94°10′. Coincidence is obtained by rotating the micrometer knob (Figure 6.30 and 6.31) until the vertical marks in the topmost scale of Figure 6.32 coincide. This rotation causes the pointer in the middle scale to indicate the nearest 10′ and the bottom scale to provide minutes and seconds. Note that these vertical marks, in the upper frame, are images of graduations on opposite sides of the horizontal circle, so that this

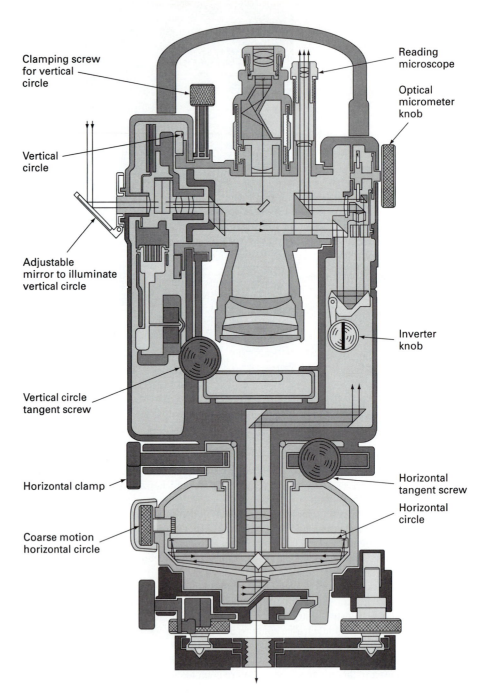

FIGURE 6.31
Optical train of a direction theodolite, Leica T2. (*Courtesy of Leica, Inc.*)

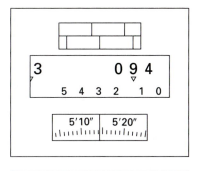

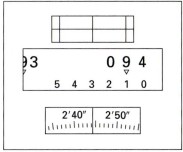

FIGURE 6.32
Horizontal circle reading, Leica
T2. (*Courtesy of Leica, Inc.*)

operation averages two sides of the circle and compensates for eccentricity errors (Section 6.27).

The setting shown on the bottom diagram of Figure 6.32 yields a horizontal circle reading of 94°10′ plus 02′44.4″, or 94°12′44.4″ where the 0.4″ is estimated. Horizontal circle readings are made with the indentation on the inverter knob (Figure 6.31) in a horizontal position. Vertical circle readings are obtained in exactly the same way but with the inverter knob rotated 90° so that the indentation is in a vertical position. Turning this knob rotates a reflective prism so that the optical path is directed toward either the horizontal or vertical circle. In Figure 6.31, illustrating the optical train, this knob is set to permit viewing the vertical circle. Also note that the optical path for viewing the horizontal circle yields readings at two points on the circle, 180° away from each other. This characteristic is found in all precision direction theodolites and provides automatic compensation for eccentricity errors (Section 6.27).

6.30
ELECTRONIC DIGITAL THEODOLITES

Theodolites containing digitized horizontal and vertical scales are readily available and rapidly replacing optical reading instruments. In these devices, the horizontal and vertical circles consist of encoded disks scanned by a sensor containing a source of light and a photo diode. This scan produces electrical pulses that can be converted to angles in digital form to be stored or displayed. In most current models of digital theodolites, decoding is by the *incremental method,* in which no zero mark is physically placed on the encoded circle and horizontal angles are initialized electronically by pressing a zero-set button on the control panel. Vertical circle readings are initialized by moving the telescope through the vertical ($z = 90°$ or $\alpha = 0°$) direction so that the vertical circle sensor detects horizontal or vertical as established automatically by an automatic tilt device (similar to the compensator in an automatic level, Section 5.19) built into the theodolite. Electrical-mechanical and electro-optical liquid tilt devices (Section 6.23) are used for this purpose.

An electronic theodolite is shown in Figure 6.33. This instrument is built into the framework of a single-center direction-type theodolite. Thus, there is only one horizontal clamp and tangent screw. However, it is classified as a repeating instrument because a horizontal angle can be held by pressing the *hold* button (Figure 6.33) so the angle can then be cumulated electronically even though the instrument has only one center and one horizontal motion. Otherwise, this instrument is similar to the repeating optical theodolite with three leveling screws, a circular level bubble, a vertical clamp, and a tangent screw, a plate level bubble, and an optical plummet (Figure 6.33). Vertical and horizontal circle readings are displayed on a register in the sexagesimal or centesimal system or in mils. On the control panel, the R/L toggle switch allows selecting clockwise or counterclockwise horizontal angles, the (RST) or zero-set button is used to initialize a horizontal angle to 0, and the hold button holds a horizontal angle when turning horizontal angles by repetition (Section 6.35).

Most electronic theodolites have automatic vertical circle compensation and several are equipped with *dual-axis compensation,* which includes a device that senses any discrepancy between the horizontal axis of the theodolite (Figure 6.29) and horizontal. This discrepancy then is used to compute a correction that is automatically applied to all measured horizontal angles. The instrument shown in Figure 6.33 has an automatic, electrical-liquid vertical compensator.

6.31
USE OF THE ENGINEER'S THEODOLITE

Succeeding sections describe operations with a theodolite for measuring horizontal and zenith or vertical angles as well as for running lines. Theodolite surveys are described in detail in Chapter 8.

Magnetic directions can be taken with an optical reading theodolite by attaching a tube or trough magnetic compass to the standard. When the tube compass is centered, with zero set on the horizontal circle, the telescope of the theodolite will point toward magnetic north and all directions observed on the horizontal circle will be magnetic directions. If the horizontal circle first is set to the correct declination and the tube compass is centered, then all observed directions will be from true north.

The theodolite also may be employed for running direct differential levels in the same manner as the engineer's level, the telescope level bubble being centered each time a reading is taken when using a theodolite equipped with an attached spirit level bubble. For instruments with automatic vertical circle compensation and no attached level bubble, the telescope is made horizontal by setting the vertical circle to a zenith angle of 90° or a vertical angle of 0°.

The primary function of the theodolite is for observing horizontal and zenith or vertical angles. One of the major assets of the theodolite in this application is that the telescope can be reversed. *Reversing,* or plunging, the telescope consists of rotating it about the horizontal

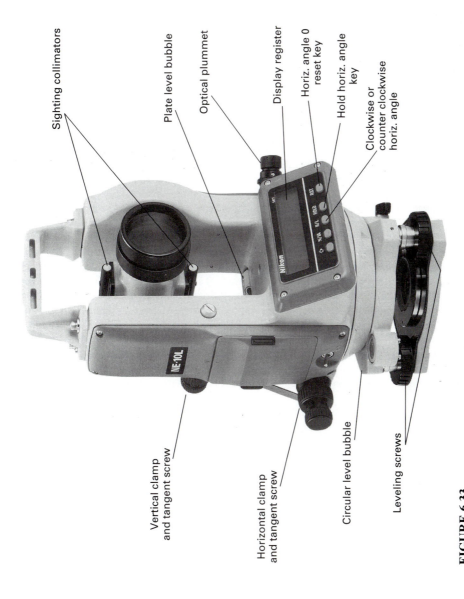

Sighting collimators

Plate level bubble

Optical plummet

Display register

Horiz. angle 0 reset key

Hold horiz. angle key

Clockwise or counter clockwise horiz. angle

Vertical clamp and tangent screw

Horizontal clamp and tangent screw

Circular level bubble

Leveling screws

FIGURE 6.33
Nikon 10 LA electronic digital, electronic theodolite. (*Courtesy of the Nikon Corporation.*)

axis. The telescope is said to be in the *direct* or *normal* position when the vertical clamp and tangent screw and the telescope eyepiece are on the same side of the instrument and in front of the operator (telescope on theodolite shown in Figure 6.22 is in the direct position); the telescope is said to be in the *reversed* or *inverted* position when the vertical clamp and tangent screws and telescope eyepiece are on opposite sides of the theodolite and the eyepiece is in front of the operator. As will be explained subsequently, observations repeated with the telescope direct and reversed permit compensation of most of the systematic instrumental errors.

First, consider the procedure for setting a theodolite over a station point.

6.32
SETTING UP THE THEODOLITE

Ordinarily, the theodolite is set over a definite point, such as a tack in a hub. As noted previously, the theodolite (optical reading, repeating or direction, and electronic digital) is equipped with a circular level bubble, one plate level bubble, an optical plummet, and a tripod with adjustable legs.

First, check to ensure that the theodolite is fastened securely to the top of the tripod; the bottom of the instrument is approximately parallel to and centered on the top of the tripod; there is an equal amount of run, up and down, on the level screws; and the tripod legs are adjusted so the telescope will be near the observer's eye level. Next, the theodolite is placed approximately over the point by estimation and a coin or small stone is dropped from underneath the instrument to observe where it strikes the ground relative to the point. On hillside setups, one leg should be up the hill, with two on the downhill side of the point. Each tripod leg then is moved as required to bring the optical plummet (or the small dropped object) to within 1 to 2 cm (0.03 to .05 ft) of being over the tack, being very careful to ensure that the top of the tripod is nearly level, to simplify succeeding operations. When this has been accomplished, be sure each tripod leg is pressed firmly into the ground, remove the plumb bob (if necessary), and carefully focus the optical plummet on the tack in the hub. At this point, the instrument is not centered by about 1–2 cm and is not leveled.

First, center the cross hairs of the optical plummet exactly on the tack by turning one or two level screws. The theodolite then is approximately leveled by manipulating the tripod legs to center the circular level bubble. To do this, position the circular bubble so that a line between the center of the bubble and the ring on the vial is radially opposite one tripod leg, as illustrated in Figure 6.34a where the line is opposite leg *C*. The bubble will move toward a leg being lengthened and away from a leg being shortened. For the example shown in Figure 6.34a, leg *C* should be lengthened. If the bubble goes to one side or the other during this operation, stop and clamp leg *C* at a point where the bubble can be moved in the direction of one of the other legs, as in Figure 6.34b. Center the bubble by lengthening or shortening the appropriate second leg. For the example in Figure 6.34b, lengthen leg *A*. Repeat this process until the bubble is within the circle on the bubble vial and the vertical axis of the instrument is approximately vertical to within 8–10′.

Checking the optical plummet, you will note that the cross hairs are off-center by only a few millimeters, because centering the circular bubble by using the legs does not seriously decenter the vertical axis. Precise leveling now is performed using the plate level bubble. The instrument is rotated so that the axis of the plate level bubble is aligned with any two level screws, as shown in Figure 6.35a. The bubble is centered by rotating screws 1 and 2 in opposite directions. The rotations indicated in Figure 6.35a will cause the bubble to move from left to right.

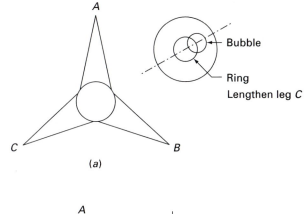

Bubble

Ring

Lengthen leg C

(a)

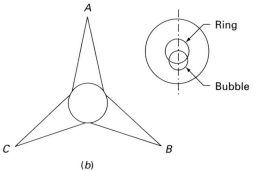

Ring

Stop at this point
so bubble will move
parallel to leg A.
Lengthen leg A.

Bubble

(b)

FIGURE 6.34
Centering the circular level bubble.

Next, rotate the instrument so that one end of the bubble tube is aligned with the remaining level screw (Figure 6.35b). Center the bubble in this position by rotating the remaining screw. The rotation indicated in Figure 6.35b will cause the bubble to move away from level screw 3.

Return to the original position (Figure 6.35a) and check the centering of the bubble.

When the bubble remains centered in both positions and with the tube in the original location (Figure 6.35a), rotate through 180° so bubble end A is opposite level screw 1. If the bubble moves off-center, bring the bubble halfway to center using level screws 1 and 2. Rotate 90° to be in the position of Figure 6.35b, and check the level bubble. Rotate through 180° so end B of the bubble tube coincides with level screw 3. If the bubble moves off-center, bring it halfway to center using level screw 3. If this deviation is large (greater than one 2-mm division), the bubble tube must be adjusted (see Section 6.51).

Now, check the centering of the optical plummet. Centering will probably be off by a few millimeters, so loosen the clamp that secures the instrument to the tripod, very slightly, and gently slide the theodolite on the top of the tripod until the cross hairs in the optical plummet are exactly centered over the tack.

Finally, check the centering of the plate level bubble. If centering by using the optical plummet caused the plate level bubble to decenter, repeat the leveling operation as outlined here until the plate level bubble is precisely centered and the optical plummet is centered on the point.

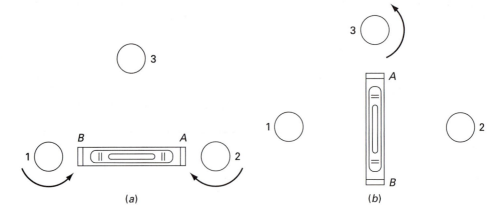

FIGURE 6.35
Leveling the plate level bubble, three-screw instrument.

To summarize the procedure, the theodolite is placed approximately over the station, the optical plummet is centered precisely using the level screws, and the instrument is approximately leveled by manipulating the tripod legs to center the circular level bubble. Precise leveling is accomplished by centering the plate level bubble in two positions and the optical plummet is then recentered by loosening the clamp and shifting the instrument laterally until the cross hairs in the optical plummet are centered over the point. This process of leveling, centering, releveling, and recentering is repeated until the plate level bubble is reasonably centered in all positions of the horizontal plate and the vertical collimation line coincides with the point over which the theodolite is being set. Note that this sequential leveling and centering is made much easier when the top of the tripod is close to being level, initially. Care should be exercised to ensure that this condition is satisfied.

The plate level bubble on most instruments with a least count of 1″ (direction theodolite) is quite sensitive (20″ per 2-mm division) and centering the bubble throughout a 360° revolution is a challenging task unless the theodolite is protected from the rays of the sun.

Just before the theodolite is moved to another setup, the instrument is centered on the top of the tripod, the leveling screws are roughly equalized, the upper motion is clamped, the lower motion is either unclamped or lightly clamped, and the telescope is pointed vertically with the vertical clamp lightly locked.

6.33
HORIZONTAL ANGLES WITH THE THEODOLITE

The basic procedure for turning horizontal angles is the same for optical reading, repeating instruments with direct reading and for repeating theodolites with micrometer scales. The only differences are in the way initial and final readings are observed with the scale or the micrometer. The method for measuring horizontal angles with electronic digital theodolites is somewhat different, due to the characteristics of these instruments. Therefore, they will be discussed separately.

Horizontal Angles with a Repeating Optical Theodolite

If a horizontal angle such as *AOB* in Figure 6.36 is to be measured, the theodolite is set over *O*. The upper motion is clamped with the index mark near 0, and by means of the upper tangent screw, the zero mark on the scale or the index mark of the micrometer is set precisely to 0 on the horizontal scale. The telescope is sighted approximately to *A* (the telescope in the direct position) by first sighting over the top of the telescope to get near the point and then sighting through the telescope.

Most current instruments have a very effective tubular sighting collimator mounted on the top and bottom of the telescope (Figure 6.33). These collimators simplify the approximate sighting immensely and should be used. The lower clamp screw is locked, the telescope is raised or lowered so that the cross-hair intersection is near the point, and the telescope is focused on point *A*. Care should be exercised to detect and remove parallax. By turning the lower tangent screw, the vertical cross hair is set exactly on the plumbed line, range pole, or other target that marks point *A*.

At this stage, the optical line of sight is on the line between *O* and *A* and the index mark on the scale or micrometer is set to 0 on the horizontal circle. This sight is called a *backsight*. Next, the upper clamp is loosened and the telescope is turned until the line of sight is approximately on *B*. The upper clamp is tightened and the line of sight is set exactly on *B* by turning the upper tangent screw. This sight is called a *foresight*.

The reading of the scale on the horizontal circle or the micrometer setting gives the angle. If point *B* falls to the right of point *A*, as shown in Figure 6.36, the objective end of the telescope has moved to the right and the angle is said to be *turned to the right*. Assume a scale-reading instrument is being used and the view of the horizontal circle through the optical train is as shown in Figure 6.25c. The horizontal angle $\alpha = AOB$ turned to the right is observed on the upper or clockwise circle as 42° plus 54′00″ = 42°54′00″, where the 54′00″ is observed on the upper scale. From this same setup, the counterclockwise angle from *A* to *B* or β (Figure 6.36) is observed on the bottom or counterclockwise circle using the bottom scale, which is read from the observer's right to left. This counterclockwise angle $\beta = 317°06′00″$.

If *B* falls to the left of *A*, the objective lens of the telescope is turned to the left and angle $\alpha' = AOB$ is said to be *turned to the left*, as shown in Figure 6.37. Observation of an angle to the left corresponds to the procedure previously described for an angle to the right, noting that the counterclockwise circle is observed to obtain angle α' and the clockwise circle reading yields β'.

When a theodolite equipped with an optical micrometer is used to turn horizontal angles, the procedure is the same with one difference: Because only the clockwise horizontal circle is numbered (Section 6.26), the angle observed always is an angle to the right. For example, suppose a repeating theodolite with an optical micrometer occupies station *O* in

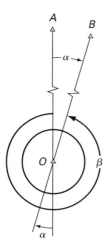

FIGURE 6.36
Horizontal angle to the right.

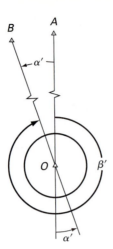

FIGURE 6.37
Horizontal angle to the left.

Figure 6.37 and the angle is turned from A to B. The reading on the horizontal circle when viewed through the optical train, after coincidence (Figure 6.26b), would be 327°59′36″. This horizontal angle is the only one that can be observed with this instrument and is the clockwise angle β' = angle to the right AOB in Figure 6.37. Note that it makes no difference whether the telescope is physically moved to the right or the left from the backsight on A when making this measurement, the answer is the same either way.

Horizontal Angles with an Electronic Digital Theodolite

First assume that a double-center electronic repeating theodolite with lower and upper motions is set over point O in Figure 6.37. A backsight is taken on B using the horizontal clamp and tangent screw following the procedures set forth previously. Next, the zero-set, or equivalent, button on the control panel (the reset button in Figure 6.33) is pressed to automatically set the backsight to 0. The upper clamp is loosened and a foresight taken on B using the upper horizontal clamp and tangent screw. If a clockwise angle is desired, the appropriate button on the control panel is pressed (R/L toggle switch, Figure 6.33) to yield this option and the angle is displayed on the register as HA 358°06′35″, which corresponds to the angle β' in Figure 6.37. For the instrument shown in Figure 6.33, the symbol HA signifies a clockwise angle whereas the symbol HL would indicate a counterclockwise angle. Different makes of instruments have different symbols to indicate the direction of the angle, so it is the responsibility of the operator to study the owner's manual and become informed about the various symbols that identify the output on the register.

If the angle is to be measured from A to B with a single-center electronic theodolite set at station O (Figure 6.36), the procedure would be the same except both backsight and foresight would be taken using the single horizontal clamp and tangent screw. In addition, the *hold* key would be pressed after each foresight to hold the angle for repetition and then again after the next backsight to release the horizontal angle so it could be cumulated. Although there is no chance of using the wrong motion on either backsight or foresight, care must be exercised to engage the hold key in the proper sequence or mistakes will occur.

The following is a list of suggestions for turning horizontal angles with optical reading or electronic repeating theodolites:

1. Be sure the initial backsight is on 0. In optical instruments with direct-scale reading, a mistake of exactly 1° is common. On electronic theodolites, one can easily forget to press the zero-set button, on the initial backsight.

2. When using the optical train, be sure the operator's eye is properly focused on the image of the scales and horizontal circle. This is done by adjusting the eyepiece on the viewing microscope (Figure 6.23).

3. Make reasonably close settings by hand so that the tangent screws need not be turned through more than one or two revolutions.

4. Make the last movement of the tangent screw clockwise, thus compressing the spring.

5. The plate bubble should be centered before measuring an angle, but between the initial and final settings of the line of sight, the leveling screws should not be disturbed. When an angle is being measured by repetition (Section 6.35), the plate may be releveled after each turning of the angle before again sighting on the initial point.

6. The person giving the sight should stand directly behind the range pole, holding it lightly with the fingers of both hands, and balancing it on the tack or other mark indicating the point.

7. In sighting at a range pole the bottom of which is not visible, particular care should be taken to see that it is held vertical. When the view is obstructed for a considerable distance above the point to which the sight is taken, use a plumb line behind which a white card is held. For short sights, a pencil or ruler held on the point makes a satisfactory target. Where the lighting is poor, the sight may be taken on a flashlight.

8. When a number of angles are to be observed from one point without moving the horizontal circle, the instrument operator should sight at some clearly defined object that will serve as a reference mark and should observe the angle. If occasionally the angle to the reference mark is read again, any accidental movement of the horizontal circle will be detected.

9. Whenever an angle is doubled, if the instrument is in adjustment, the two readings should not differ by more than the least count of the scale or vernier. A greater discrepancy, if confirmed by repeating the measurement, will indicate that the instrument is out of adjustment.

6.34
LAYING OFF HORIZONTAL ANGLES

If an angle *AOB* is to be laid off from line *OA*, the theodolite is set up at *O*, the scale or micrometer on optical reading instruments is set to 0, and the line of sight is set on *A* using the lower clamp and tangent screw. The upper clamp is loosened, and the plate is turned until the scale is approximately at the required angle. The upper clamp is tightened, and the scale or micrometer reading is set exactly on the given angle by means of the upper tangent screw. Point *B* then is established on the line of sight.

When an electronic theodolite (double or single center) is set on point *O* to lay off angle *AOB*, the procedure is the same except that 0 on the backsight is set by pressing the zero-set button, and on the foresight, the horizontal tangent screw is used to set the exact required angle, which is displayed on the display register of the theodolite.

6.35
MEASURING HORIZONTAL ANGLES BY REPETITION

By means of the repeating or electronic theodolite, a horizontal angle may be mechanically or electronically cumulated and the sum can be read with the same precision as the single value. When this sum is divided by the number of repetitions, the resulting angle has a precision that exceeds the nominal least count of the instrument. Therefore, with a theodolite having a least count of 1′, an angle for which the true value is between the limits

30°00′30″ and 30°01′30″ will be read as 30°01′ and the limits of possible error will be ±30″. If the angle is cumulated six times on the horizontal circle, the sum, also read to the nearest minute, might be 180°04′, its true value being within the limits 180°03′30″ and 180°04′30″; the limits of possible error, as far as reading the horizontal circle is concerned, also will be ±30″. Dividing the observed sum 180°04′ by 6, the single value becomes 30°00′40″, for which the limits of possible "reading" error are ±30″/6 = ±05″. Similarly, a repeating instrument with a least count of 30″ (estimating to 15″) would yield an angle having a possible reading error of about ±03″ in five repetitions. This method of determining an angle is called *measurement by repetition*. The precision with which an angle can be observed by this method increases directly with the number of times the angle is cumulated or repeated, up to between 6 and 12 repetitions; beyond this number the precision is not appreciably increased by further repetition because of errors in graduations of the horizontal circle, eccentricity of instrument centers, play in the instrument, and random errors such as those due to setting the line of sight, setting the angle on the horizontal circle, and reading angles on the horizontal circle.

When angles are observed by repetition, the principle of *double centering* also is employed. As noted in Section 6.28, on geometry of the theodolites, the lines and planes defined by vertical axis, horizontal axis, optical line of sight, and so on, theoretically have exact relationships with respect to one another. In practice, these relationships are not exact, even in the best-adjusted transit or theodolite, so that systematic instrumental errors are present (see Section 6.45). One of the excellent features of the transit or theodolite is that the telescope can be transited about the horizontal axis so that observations can be made with the telescope in the direct and reversed positions (Section 6.31). When operations on the instrument include a direct and reversed observation or setting, systematic instrumental errors will occur in opposite directions. Thus, if the average of the observations or settings is used, the effect of the systematic errors cancel. Implementation of this process is called *double centering* or *double sighting*.

Another reason for repeating an angle is to decrease the possibility of mistakes in observations. Therefore, the primary reasons for observing an angle by repetition with the telescope direct and reversed are to increase the accuracy of reading the angle, compensate for systematic errors, and eliminate mistakes. Consider some examples.

To repeat an angle once direct and once reversed with an optical-reading theodolite, as for angle *AOB* in Figure 6.36, the theodolite is set up at *O*, where it is very carefully centered and leveled. A sight is taken on point *A* with the telescope direct and a single value of the angle is observed as described in Section 6.33. Next, with this angle still set on the plate, the lower motion is unlocked, the telescope is reversed, the instrument is turned on its lower motion, and a second sight is taken using the lower clamp and lower tangent screw. Note that, on this sighting from *O* to *A*, the value of the angle *AOB* is on the horizontal circle. The upper clamp is loosened and the telescope, now in the reversed position, again is sighted on point *B*. At this stage, approximately twice the original value of the first angle appears at the index of the horizontal circle, and the angle has been repeated with the telescope direct and reversed. The following notes show how data can be recorded for observation of angle *AOB* by repetition once direct and once reversed using a theodolite with a least count of 30″:

Station	From To	Rep.	Tel.	Circle reading	Angle
O	*A*	0	Direct	0°00′00″	
	B	1	Direct	34°20′15″	34°20′15″
	B	2	Reversed	68°41′00″	34°20′45″

The first angle is obtained by subtracting the initial circle reading on *A* (0°00′00″ in this example) from the circle reading on *B*, yielding a first angle of 34°20′15″. The second angle equals the second circle reading minus the first reading on point *B*, or

$(68°41'00'') - (34°20'15'') = 34°20'45''$. This second angle should agree with the first angle to within $\pm$ the least count of the transit, $\pm30''$ in this case. If this difference is greater, the set should be repeated. This set is acceptable. The final angle then is the average of angle 1 and angle 2: $[(34°20'15'') + (34°20'45'')]/2 = 34°20'30''$. The final angle also can be calculated by dividing the cumulated angle minus the initial observation by the number of repetitions; in this case, $[(68°41'00'') - (0°00'00'')]/2$ equals $34°20'30''$.

The preceding example dealt with an interior angle. Next, consider the procedure for turning a deflection angle by repetition (Section 6.10). Assume that the deflection angle at B in Figure 6.12 is to be observed. The theodolite is set at point B and with the horizontal circle set on 0, a backsight is taken on point A with the telescope in the direct position. Next, the telescope is plunged so that the line of sight falls along the extension of line AB and the horizontal circle still is on 0. The upper clamp is loosened and a sight is taken on point C using the upper clamp and tangent screw. The deflection angle is observed on the clockwise circle using the scale or micrometer. Assume that the first deflection angle at B from A to C is $22°35'00''$ to the right. To *double* the angle, the lower clamp is loosened and a second backsight is taken on point A using the lower clamp and lower tangent screw. The telescope is reversed along the backsight and the angle $22°35'00''$ is on the horizontal circle at the zero point on the scale or micrometer. Again the telescope is plunged, the upper clamp loosened, and a sight is taken on point C with the upper clamp and tangent screw. The telescope now is in the direct position and the angle observed on the horizontal circle is the doubled deflection angle. Notes for the example follow:

Station	From To		Rep.	Tel.	Circle reading	Deflection angle
B	A		0	Direct	$0°00'00''$	
		C	1	Reversed	$22°35'00''$	$22°35'00''$ R
		C	2	Direct	$45°10'30''$	$22°35'30''$ R

Note: Theodolite number 37, least count 30".

The procedure for calculating the first and second deflection angles is similar to that described for interior angles in the preceding example. Because the first angle agrees with the second to within the least count of the transit, the set is acceptable. The final average angle is $22°35'15''$, the average of the two angles. The procedure just outlined is called observing angles by *double deflection*. Note that, when deflection angles are observed, the direction of the deflection must be recorded.

Whenever possible, angles should always be turned by repetition the minimum number of times, once direct and once reversed. In this way, systematic instrumental errors are compensated for and blunders are eliminated, even though a high order of accuracy is not achieved.

When an angle is turned more than twice, the number of repetitions always is some multiple of two. Thus, an angle repeated twice direct and twice reversed involves four repetitions. Similarly, angles can be turned three, four, five, and six times direct and reversed involving 6, 8, 10, and 12 repetitions, respectively. There should always be an equal number of direct and reversed observations.

The note form for turning an angle with six repetitions using a 30" repeating theodolite follows:

Station	From To		Rep.	Tel.	Circle reading
O	M		0	Direct	$0°0'00''$
		N	1	Direct	$80°20'00''$
		N	6	Reversed	$122°01'00''$

Note that the circle reading for the first repetition on point N is recorded and then the angles are cumulated without recording until the final repetition is completed. The first angle is the difference between circle observation 1 on point N and circle observation 0 on

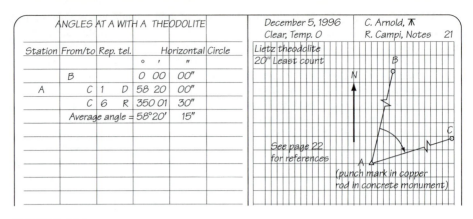

FIGURE 6.38
Field notes for angles by repetition using a repeating instrument.

point M. For this set, the initial angle, used *only* as a check value, is $80°20'00''$. To calculate the average angle, it is necessary to know the number of times the horizontal circle index has passed 0. The total number of degrees passed in six repetitions is the initial approximate angle times the number of repetitions or $(6)(80°20') = 482°$. Because 0 was passed once, $360°$ must be added to the cumulated angle recorded for the sixth repetition before calculating the average angle. Therefore, the average angle for this set is $[(122°01'00'') + 360°] / 6 = 80°20'10''$. This average angle agrees with the initial check angle to within $\pm$ the least count of the transit, so that the set of observations is acceptable. Figure 6.38 shows a field note form for angles by repetition.

The circle reading for the first sight need not be 0, as illustrated by the following notes for six repetitions of angle EFG using a theodolite having a least count of $10''$:

Station	From	To	Rep.	Tel.	Circle reading
F	E		0	Direct	$136°01'50''$
		G	1	Direct	$216°21'00''$
		G	6	Reversed	$257°57'50''$

The approximate angle therefore is $(216°21'00'') - (136°01'50'') = 80°19'10''$. Six repetitions of this angle or $6(80°19'10'') = 481°55''00''$. This value added to $136°01'50''$ equals $617°56'50''$, indicating that 0 was passed once and $360°$ must be added to the final circle reading to calculate the final angle. Thus, $[(257°57'50'' + 360°) - (136°01'50'')]/6 = 80°19'20''$, the final average angle.

In another example, notes for angle SRT observed 12 times with a repeating theodolite having a least count of $10''$ are as follows:

Station	From	To	Rep.	Tel.	Circle reading
R	S		0	Direct	$70°10'30''$
		T	1	Direct	$170°30'30''$
		T	12	Reversed	$194°11'30''$

The approximate angle is $100°20'00''$. Twelve times this approximate angle plus the first circle reading equals $1274°10'30''$. Therefore, the circle index passed 0 three times. The average angle equals $[(194°11'30'') + (1080°) - (70°10'30'')]/12 = 100°20'05''$, which is within the least count or $\pm10''$ of the approximate angle.

The preceding remarks have been directed mainly toward angles by repetition with optical-reading, repeating instruments. When an electronic digital theodolite (Section 6.30) is used for repetition of angles, the procedure as already described is the same with three exceptions. On the initial backsight the horizontal circle is set to 0 by the zero-set key (Figure 6.33) and the desired direction is selected by using the R/L toggle switch. Then, on each foresight, when the sight has been completed, the hold key (a toggle switch that either holds or releases the horizontal angle, Figure 6.33) is pressed to retain the current angle so that it will accumulate in the repetition process. After each successive backsight, following the initial backsight, the hold key is pressed to release the horizontal angle. All other aspects with respect to the operations and notes are the same. As already noted in Section 6.30, the presence of only one horizontal clamp and tangent screw should result in fewer blunders due to use of the wrong motion. However, due care must be exercised in the operation of the hold key or mistakes still will occur.

Some electronic theodolites have a repeat function, implemented by pressing a key on the control panel. This function cumulates the angle electronically and allows displaying the average angle on completion of the repetition.

Angles by repetition should not be attempted unless the instrument operator is experienced both in the instrument setup and observing angles. Speed and accuracy in observation are essential if the repetition procedure is to be successful. Otherwise, considerable time can be wasted without obtaining the desired results.

6.36
LAYING OFF ANGLES BY REPETITION

If it is desired to establish an interior angle with a precision greater than that possible by a single observation, the methods of the preceding section may be employed in the following manner. In Figure 6.39a, OA represents a fixed line and AOB the angle to be laid off to establish the line OB. The repeating theodolite is set up at O, the horizontal circle is set at $0°$, and a sight is taken to A. The given angle is set as closely as possible and a trial point B' is established with the line of sight in its new position. The angle AOB' then is measured by repetition, and the line OB' is measured. The angle AOB' must be corrected by an angular amount $B'OB$ to establish the correct angle AOB. The correction, which is too small to be laid off accurately by angular measurement, is applied by offsetting the distance $B'B = OB'$ tan(or sin) $B'OB$, thus establishing the point B beside B'. It is convenient to remember that the tangent or sine of $1' = 0.0003$ and the tan or sine of $1'' = 4.848 \times 10^{-6}$. As a check, the angle AOB is measured by repetition.

> **EXAMPLE 6.3.** Suppose that an angle of $30°00'$ correct to the nearest $05''$ is to be laid off and that the theodolite to be employed reads to the nearest $01'$. Let the total value of AOB' after six repetitions be $180°02'$, correct to the nearest $30''$. Then, the measured value of AOB' is $180°02'/6 = 30°00'20''$ correct to the nearest $05''$, and the correction to be applied to AOB' is $20''$. Suppose that $OB' = 100$ m. The length of the offset $B'B$ equals tan $20'' \times 100$ m $= (0.0001)(100) = 0.010$ m.

In route surveying, highway center lines frequently are set by deflection angles. Referring to Figure 6.39b, assume that it is desired to establish the direction of the line from $16 + 10.450$ to station $18 + 84.900$ by turning the deflection angle $45°10'00''$ to the right from a backsight on $14 + 55.250$. The instrument is set on station $16 + 10.450$, the horizontal circle is set on 0, and with the telescope in the direct position, a backsight is taken on $14 + 55.250$ using the lower clamp and tangent screw. The telescope is plunged,

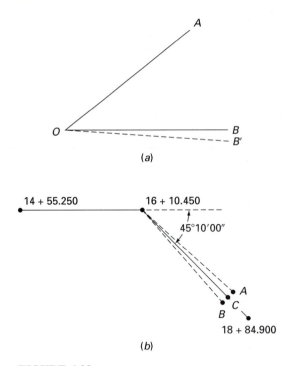

FIGURE 6.39
Laying off horizontal angles: (a) interior angle,
(b) deflection angle.

the upper clamp is loosened, the angle 45°10′00″ is set off on the clockwise circle, and point A is set on the line along the line from 16 + 10.450 to 18 + 84.900. With 45°10′00″ still set on the horizontal circle, release the lower clamp and take a second backsight on 14 + 55.250 using the lower clamp and tangent screw. At this point, the telescope is in the reversed position. Next, the telescope is plunged a second time, the upper clamp is loosened, and twice the deflection angle, or 90°20′00″, is set off on the horizontal circle using the upper clamp and upper tangent screw. This procedure is called setting the line by *double deflection angle.* If the instrument is in excellent adjustment, the last direction will coincide with the first point set at A. However, this rarely occurs, so that a second point is set at B on the top of the same stake set for A. Point C is set such that BC = AC and the line from 16 + 10.450 to C represents the best estimate of the line that deflects 45°10′00″ to the right.

6.37
MEASUREMENT OF ZENITH AND VERTICAL ANGLES

Angles in a vertical plane are designated as (1) the *zenith angle,* an angle that varies from 0° to 360°, is always positive, and is measured from an index with 0 directed toward the zenith; or (2) the *vertical angle,* an angle that varies from 0° to 90°, is measured relative to 0 that coincides with horizontal, and is positive (an elevation vertical angle) above

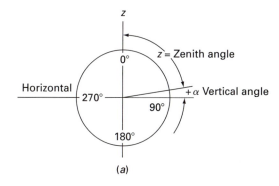

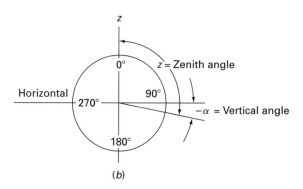

(b)

FIGURE 6.40
Zenith and vertical angles: (a) above horizontal,
(b) below horizontal.

horizontal and negative (a depression vertical angle) below the horizon. Zenith and vertical angles for lines above and below the horizontal plane are illustrated in Figure 6.40. Zenith angles are from 0° to 90° above horizontal (+ vertical angles, Figure 6.40a), and zenith angles below horizontal are from 90° to 180° (− vertical angles, Figure 6.40b). Therefore, the vertical angle $\alpha = 90° - z$.

The method for measuring angles in a vertical plane and determining whether or not zenith or vertical angles are used varies with the style of instrument used. Optical-reading, repeating, and direction theodolites have vertical circles on which 0 corresponds to the zenith. Recent models of these theodolites are equipped with some form of vertical circle index compensator actuated by the force of gravity (Section 6.30), which automatically ensures that 0 on the vertical circle coincides with the zenith. Electronic digital theodolites also have automatic vertical circle compensation. The index of the vertical circle is set to 0 on these instruments by moving the objective end of the telescope in a vertical plane through horizontal, immediately after turning on the power switch. Some electronic theodolites have the option of electronically selecting zenith angles, vertical angles, or a percent slope to fix the direction of an inclined line.

Some older models of optical reading theodolites are equipped with control level bubbles attached to the vertical circle. For this type of instrument, after the theodolite has been set up and leveled using the plate level bubble, the control level bubble must be carefully centered prior to reading the vertical circle. This step ensures that 0 on the vertical circle coincides with the zenith for all subsequent readings of the vertical circle.

Given the zenith angle, observed on a theodolite, the vertical angle can be computed. For example, a zenith angle of $85°10'45''$ yields a vertical angle of $(90°00'00'') - (85°10'45'') = +4°49'15''$. A zenith angle of $94°12'44''$ corresponds to a vertical angle of $(90°00'00'') - (94°12'44'') = -4°12'44''$. Both of these angles are observed with the telescope in the direct position. With the telescope reversed and assuming no vertical collimation error (Section 6.39), corresponding zenith angles are $360° - (85°10'45'') = 274°49'15''$ and $360° - (94°12'44'') = 265°47'16''$. These zenith angles then yield respective vertical angles of $(274°49'15'') - 270° = +4°49'15''$ and $(265°47'16'') - 270° = -4°12'44''$. Note that the zenith angles are unambiguous, because they are positive in all cases.

6.38
ZENITH AND VERTICAL ANGLES BY DOUBLE CENTERING

The method of observing zenith or vertical angles by double centering consists of reading once with the telescope direct and once with it reversed and taking the mean of the two values thus obtained.

Double centering is used in astronomical observations and in the measurement of zenith or vertical angles to distant objects (e.g., in trigonometric leveling, Section 5.5). In traversing, a similar result is obtained by measuring the zenith or vertical angle of each traverse line from each end, with the telescope the same side up for both observations and taking the mean of the two values.

6.39
ERRORS IN VERTICAL CIRCLE MEASUREMENTS

Vertical Collimation Error

A collimation error is present in optical-reading and electronic theodolites when the vertical circle compensator does not make 0 on the vertical circle coincide exactly with the zenith. The same type of error occurs in theodolites equipped with vertical circles having attached control level bubbles when the axis of the level bubble is not exactly parallel to the optical line of sight. In each case, the effect is the same, there is an error, e, between the true vertical and the 0 point on the vertical circle, as illustrated in Figure 6.41. If the instrument were in perfect adjustment and leveled perfectly for each observation, there would be no vertical collimation error; however, in practice these conditions seldom exist.

The effect of collimation errors can be eliminated by either double centering (Section 6.38) for each observation or applying, to each observation, a correction determined (by double centering) for the instrument in its given condition of adjustment. Referring to Figure 6.41, $z_{D\,\text{true}} = z_{D\,\text{meas.}} - e$ and $z_{R\,\text{true}} = z_{R\,\text{meas.}} + e$, so that $z_{\text{aver.}} = [z_{D\,\text{meas.}} + z_{R\,\text{meas.}} - e + e]/2 = (z_D + z_R)/2$ and the error cancels. It also follows from the same formulation that the error can be calculated by observing a single well-defined point with the telescope direct and reversed and then the error, $e = (z_{D\,\text{meas.}} - z_{R\,\text{meas.}})/2$.

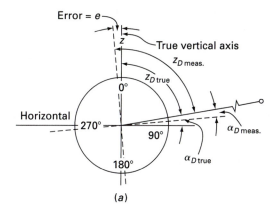

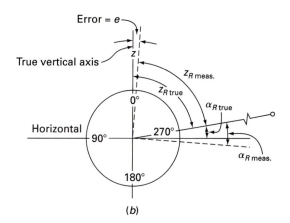

FIGURE 6.41
Vertical collimation error: (a) telescope direct,
(b) telescope reversed.

The vertical collimation correction is equal in amount but opposite in sign to the collimation error (Section 2.5). Therefore, if the observed zenith angle is 87°46′ and the error has been determined to be −02′, the correct value of the angle would be

$$(87°46') + 02 = 87°48'$$

If direct and reversed zenith angles are 94°12′44″ and 265°47′24″, respectively, the average zenith angle and vertical collimation error would be

$$z_{\text{aver.}} = [94°12'44'' + (360° - 265°47'24'')]/2 = 94°12'40''$$

$$e = (94°12'44'' - 360° + 265°47'24'')/2 = 0°00'04''$$

Note that, had the collimation error been known beforehand to be 0°00′04″ and only the direct zenith angle of 94°12′44″ was observed, then the corrected value would be

$$z_{\text{corr.}} = 94°12'44'' - 04'' = 94°12'40''$$

Similarly, if the observed vertical angle to a point is $+2°58'30''$ with the telescope direct and $2°55'30''$ with the telescope reversed, the average vertical angle and collimation error for readings with the telescope in the direct position would be

$$\alpha_{\text{aver.}} = (2°58'30'' + 2°55'30'')/2 = 2°57'00''$$

$$e = (2°58'30'' - 2°55'30'')/2 = +00°01'30''$$

Therefore, if a single vertical angle, with the telescope direct, were $2°41'30''$, then the corrected angle would be

$$\alpha_{\text{corr.}} = 2°41'30'' - 00°01'30'' = 2°40'00''$$

This type of error can be eliminated by double centering or can be made nearly negligible for most ordinary surveys by adjustment of the instrument. For optical-reading, repeating, and direction theodolites with vertical circle tilt compensation, the precision of the vertical circle compensator (0.5 to 06'') is adequate to virtually eliminate the need to apply a correction for ordinary surveys. On some electronic theodolites the adjustment (if needed) can be made electronically by following the directions in the instruction manual for the theodolite.

Inclination of the Vertical Axis

For theodolites equipped with vertical-circle compensators or vertical-circle control spirit levels, moderate inclination of the vertical axis due to erroneous leveling does not introduce appreciable error in zenith or vertical angles. This is true, provided the instrument is in adjustment and assuming that, for instruments so equipped, the control spirit level is in adjustment and centered each time an observation is made. This error is not eliminated by double centering, because the condition causing the error is not changed by reversal of the telescope.

6.40
PRECISE LEVELING OF THE THEODOLITE

When using an optical-reading or electronic theodolite equipped with a vertical circle compensator, for astronomical observations or for measurement of horizontal angles requiring steeply inclined sights, a special, precise leveling procedure is required to ensure the vertical axis is truly vertical. First the theodolite is leveled by the plate level bubble in the usual manner (Section 6.32). With the telescope aligned over a pair of adjacent leveling screws, set a zenith angle of $90°00'00''$ (or a vertical angle of $0°00'00''$) on the vertical circle using the vertical clamp and tangent screw. The telescope is rotated end for end about the vertical axis and the zenith angle is observed again in this position. If there is a discrepancy between this observed value and $90°00'00''$ reduce the discrepancy by one-half, using the two adjacent level screws. Now, turn the telescope so it is at $90°$ to its original position, with the objective of the telescope over the third level screw, and repeat the operation. Repeat this process alternately in the two positions until the zenith angle or vertical angle registers $90°00'00''$ or $0°00'00''$, respectively, for any direction of pointing. Because the vertical circle compensator has a precision of from 0.5 to 6'' of arc compared to the plate level bubble with a sensitivity of 30'' per 2-mm division, leveling the instrument by this method is substantially more precise than simply using the plate level bubble.

If the theodolite (three-screw leveling) is not fitted with a vertical circle compensator but does have a spirit level attached to the telescope, the procedure as just outlined can be employed using the attached spirit level bubble. With the telescope and attached level bubble aligned with any two adjacent leveling screws (Figure 6.35a), the bubble of the telescope level is centered using the vertical tangent screw. The telescope is rotated end for end about the vertical axis; then the bubble is brought back halfway to center by means of the two adjacent leveling screws, the plate level bubble being disregarded. Next turn the telescope so it is 90° to its original position and aligned over the third level screw (Figure 6.35b) and repeat the previous operation. The entire process is repeated alternately, for the two adjacent and third leveling screws, until the bubble of the telescope remains centered for any direction of pointing.

6.41
OPERATIONS WITH THE THEODOLITE: PROLONGING
A STRAIGHT LINE

If a straight line such as *AB* (Figure 6.42) is to be prolonged to *P* (not already defined on the ground), which is beyond the limit of sighting distance or is invisible from *A* and *B*, the line is extended by establishing a succession of stations *C*, *D*, and so forth, each of which is occupied by the theodolite. Two methods may be employed. The second method is the more reliable and is the recommended procedure.

Method 1

The theodolite is set up at *B*, and a backsight is taken to *A*. With both upper and lower motions clamped, the telescope is plunged, and a point *C* is set on the line. If the line of sight is perpendicular to the horizontal axis, as it will be if the instrument is in perfect adjustment, it will generate a vertical plane as the telescope is revolved and point *C* will lie on the prolongation of *AB*. The theodolite is moved to *C*, a backsight is taken to *B*, a point *D* is set in similar manner, and the process is repeated until point *P* is set.

If the line of sight is not perpendicular to the horizontal axis of the theodolite, as the telescope is plunged (say, from the reversed to the direct position), the line of sight will generate a portion of a cone whose vertex is at the center of the instrument and two of whose elements are *AB* and *BC′*, and *C′* will not lie on the true prolongation of *AB*. If the instrument is set up at *C′*, a backsight taken to *B* with the telescope reversed as before, and the telescope plunged to the direct position, a second and similar cone is generated and *D′* will not lie on the prolongation of *BC′*. Thus, if the line is extended by the method outlined and all backsights are taken with the telescope in one position (either direct or reversed), the points established will lie along a curve instead of a straight line, and each segment of the line will be deflected in the same direction (to the right or to the left) by double the error of adjustment of the line of sight. On the other hand, if, say, at the even-numbered stations

FIGURE 6.42
Prolonging a straight line.

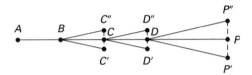

FIGURE 6.43
Double sighting to prolong a line.

B, D, F, and so on, backsights were taken with the telescope reversed and at odd-numbered stations, C, E, and so on, backsights were taken with the telescope direct, a zigzag line would be established with some points on one side of the line joining the terminals and some perhaps on the other. This method should not be used where the adjustment of the instrument is questionable and should never be employed where more than a low order of accuracy is required.

Method 2

This procedure, known as *double centering,* should be employed for all surveys requiring from moderate to high accuracy. If the line AB (Figure 6.43) is to be prolonged to some point P, the theodolite is set up at B and a backsight is taken to A with the telescope in its *direct* position. The telescope is plunged, and a point C' is set on the line. The theodolite then is revolved about its vertical axis, and a second backsight is taken to A with telescope *reversed*. The telescope is plunged, and a point C'' is established on line beside C'. It is evident that C' will be as far on one side of the true prolongation of AB as C'' is on the other. Midway between C' and C'', a point C is set defining a point on the correct prolongation of AB. In a similar manner the next point D is established by setting up at C, double centering to B, and setting points at D', D'', and D. The process is repeated until the desired distance is traversed.

6.42
PROLONGING A LINE PAST AN OBSTACLE

Figure 6.44 illustrates one method of prolonging a line AB past an obstacle where the offset space is limited. The theodolite is set up at A, a right angle is turned, and a point C is established at a convenient distance from A. Similarly, the point D is established, the distance BD being made equal to AC. The line CD, which is parallel to AB, is prolonged; and points E and F are established in convenient locations beyond the obstacle. From E and F right-angle offsets are made, and G and H are set as were C and D; GH then defines the prolongation of AB. The distance AH is determined by measuring the length of the lines AB, DE, and GH. If measurement of distance is to be carried forward with high accuracy, it is necessary to erect the perpendiculars AC, BD, and so forth, with greater than ordinary care; and if the line is to be prolonged with high accuracy, it is essential not only that the offset distances be measured carefully but also that AB and EF, the distances between offsets, be as long as practicable.

Another method of prolonging a line AB past an obstacle is illustrated by Figure 6.45. A small angle α is turned off at B, and the line is prolonged to some convenient point C that

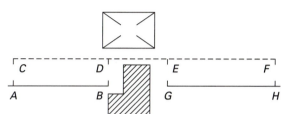

FIGURE 6.44
Prolonging a line past an obstacle by perpendicular
offsets.

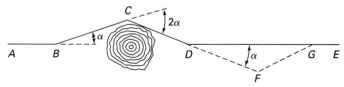

FIGURE 6.45
Prolonging a line past as obstacle by angles.

will enable the obstacle to be cleared. At C, the angle 2α is turned off in the reverse direction, and the line is prolonged to D, with CD made equal to BC. The point D then is on the prolongation of AB; and DE, the further prolongation of AB, is established by turning off the angle α at D. If another obstacle were between D and E, as there often might be in wooded country, the line CD would be prolonged to some point, as F, from which the obstacle could be cleared; and so a zigzag course would be followed until it were possible to resume traversing on the direct prolongation of the main line AB. As compared with the method of perpendicular offsets, this method is more convenient in the field, but it requires computation to determine the length BD. However, if the angle α is small, say, no greater than a degree or so, often it will be sufficiently precise to take the distance along the main line as equal to that along the auxiliary lines.

6.43
RUNNING A STRAIGHT LINE BETWEEN TWO POINTS

If the terminal points A and B of a line are fixed and it is desired to establish intervening points on the straight line joining the terminals, the method to be employed depends on the length of the line and the character of the terrain. Two common cases are considered.

Case 1: Terminals Intervisible

The theodolite is set up at A, a sight is taken to B, and intervening points are established on line (Figure 6.46).

If the intervening points thus established were to lie in the same plane with the center of the instrument and the terminal point B, they would define a truly straight line regardless of whether or not the horizontal axis of the theodolite were truly horizontal. If the horizontal axis is inclined with the horizontal, the line of sight will not generate a vertical plane as the

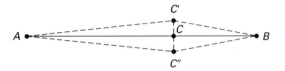

FIGURE 6.46
Balancing in.

telescope is revolved; thus, if it is necessary to rotate the telescope about the horizontal axis to set the intervening points, the points so established will not lie on a truly straight line (as seen in plan view) joining the terminals.

Ordinarily, the vertical angles through which it is necessary to rotate the telescope will be small, and if the horizontal axis is in fair adjustment and the plate bubbles are centered, the error arising from this source is negligible. Occasionally, however, when the intervening points are to be set with high precision or when the adjustment of the instrument is uncertain and the vertical angles are large, the intervening stations are set by double sighting.

Case 2: Terminals Not Intervisible but Visible from an Intervening Point on the Line

The location of the line at the intervening point C is determined by trial, as follows. In Figure 6.46, A and B represent the terminals, both of which can be seen from the vicinity of C. The theodolite is set up on the estimated location of the line near C, a backsight is taken to A, the telescope is plunged, and the location of the line of sight at B is noted. The amount that the theodolite must be shifted laterally is estimated and the process is repeated until, when the telescope is plunged, the line of sight falls on the point at B. This process is known as *balancing in.* The location of the theodolite then should be tested by double centering; the test also will disclose whether or not the line of sight and the horizontal axis are in adjustment.

If the instrument were in perfect adjustment, its center would be on the true line joining AB. On the other hand, if the line of sight were not perpendicular to the horizontal axis, a cone would be generated by the line of sight when the telescope was plunged, as explained in Section 6.41. Also, if the horizontal axis were not truly horizontal, the line of sight in its rotation would not generate a vertical plane, as explained in Case 1. Hence, if the theodolite were not in adjustment, its center might be at C' and still the line of sight would bisect B when the telescope were plunged.

To locate the intermediate point truly on the line, trials are made first with the telescope, say, in its direct position for backsights to A, until an intermediate point, such as C', is determined. Next, a second series of trials is made with the telescope reversed for backsights to A, until the corresponding point C'' is located. Then, for reasons previously explained, the true line would be at C, halfway between C' and C''. Other intermediate points then may be established by setting up the theodolite at C and proceeding as in Case 1.

6.44
GENERAL ERRORS IN DETERMINING ANGLES AND DIRECTIONS

Except in field astronomy, a measured angle always is closely related to a measured distance; and in general, a consistent relation should exist between the precision of measured angles and that of measured distances. From the standpoint of both precision achieved and expediting the work, it is important that the surveyor be able to (1) visualize the effect of errors in terms of both angle and distance, (2) appreciate what degree of care must be exercised to keep certain errors within specified limits, and (3) know under what conditions various instrumental errors can be eliminated.

On surveys of ordinary accuracy, it usually requires much more care to keep *linear* errors within prescribed limits than to maintain a corresponding degree of *angular* precision. Often undue attention is paid to securing precision in angular measurements, while large and important errors in the measurement of distances are overlooked.

Errors in theodolite work may be instrumental (Section 6.45), personal (Section 6.46), or natural (Section 6.47). To enhance visualization of the errors described in the following sections, please refer to Figure 6.29, on the geometry of the theodolite.

6.45
INSTRUMENTAL ERRORS

Errors due to instrumental imperfections or nonadjustment are all systematic, and they can either be eliminated or reduced to a negligible amount by proper methods of procedure.

Errors in Horizontal Angles Caused by Nonadjustment of the Plate Level Bubble

When the plate level bubble in nonadjustment is centered, the vertical axis is inclined, and hence measured angles are not truly horizontal angles. Also the horizontal axis is inclined to a varying degree depending on the direction in which the telescope is sighted. One vertical plane will include the vertical axis in its inclined position; this is illustrated by Figure 6.47a, in which the horizontal axis OH and the vertical axis ON are in the plane of the paper. When the horizontal axis is rotated about the vertical axis until it is normal to the plane of the paper, it becomes truly horizontal and the line of sight will generate a vertical plane when the telescope is reversed.

Figure 6.47b illustrates a three-dimensional portrayal of this condition, in which plane $HNN'OH'$ represents the plane of the paper in Figure 6.47a, OH is the horizontal axis in that plane inclined by angle Ω, and OA is the horizontal axis rotated to be perpendicular to plane $HNN'OH'$ (plane of paper in Figure 6.47a). In this latter position, no error in horizontal direction is introduced regardless of the angle of elevation to the point sighted. Then, as the theodolite is rotated back about its vertical axis (ON') from A to H, the horizontal axis becomes inclined by variable amounts (e.g., by angle Ω' at B) until it reaches H, where the optical line of sight now at OA generates a plane $AN'O$ (Figure 6.47b) that makes an angle with the vertical equal to the error in the position of the vertical axis or angle Ω; and with the line of sight inclined at a given angle, the maximum error in determining the direction is introduced. The larger the vertical angle, the greater is the error in direction. This error cannot be eliminated by double centering.

The diagram in Figure 6.48 shows for various vertical angles (values of α), the errors introduced in horizontal angles due to an inclination of 01′ in the vertical axis or two spaces on the plate level bubble vial of most optical reading and electronic theodolites (Section 6.23). The values of θ are the horizontal angles that the line of sight makes with the vertical plane in which the vertical axis lies in its inclined position (i.e., with the plane of the paper, Figure 6.47a, and with plane $HNN'OH'$, Figure 6.47b). Within reasonable limits, the error in horizontal angle varies directly as the inclination of the vertical axis; hence, a similar diagram for an inclination of 02′ would show ordinates twice as great as those of Figure 6.48.

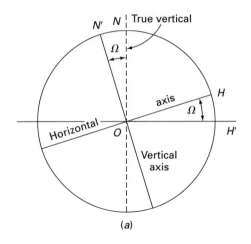

(a)

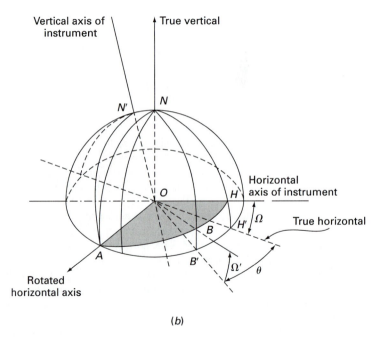

(b)

FIGURE 6.47
Inclination of vertical axis due to maladjustment of plate level
bubbles.

Although the diagram may not be of much practical value, it illustrates some noteworthy facts.

1. For observations of ordinary accuracy taken in flat country, where the vertical angles are rarely greater than 3° and usually much less, the plate bubble may be out several spaces without appreciably affecting the precision of horizontal angular measurements. For example, if an angle were measured between $\theta = 0°$ and $\theta = 90°$, with bubble out two spaces ($\Omega = 01'$) and $\alpha = 3°$, the error would be about 03″; or if in prolonging a straight

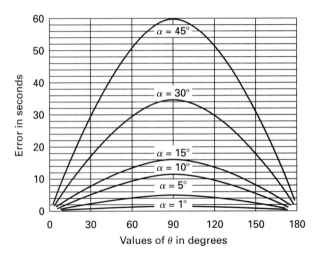

FIGURE 6.48
Errors in horizontal angles for a 1′ inclination of the
vertical axis (α = vertical angle).

line the telescope were reversed from the position $\theta = 90°$ to $\theta = 270°$ with both backsight and foresight taken at a vertical angle of $+3°$, the error would be doubled and for the bubble out two spaces (vertical axis in error 1′), the angular error introduced in the direction of the line would be 06″.

2. For angular measurements of higher precision, such as when measuring an angle by repetition, the plate levels must be in good adjustment and the bubbles must be centered with reasonable care even though the survey is conducted over fairly smooth ground. For example, if a horizontal angle were measured between the positions $\theta = 0°$ and $\theta = 90°$, $\alpha = 5°$, and the vertical axis were inclined 30″, the error in horizontal angle would amount to 02.2″.

3. In rough country where the vertical angles are large, even for surveys of ordinary precision the plate level bubble must be in good adjustment and the bubble must be centered carefully if errors in horizontal angles or in the prolongation of lines are to be kept within negligible limits. For example, if a line were prolonged by reversing the telescope from the position $\theta = 90°$ to $\theta = 270°$, α for both backsight and foresight being $+30°$ and the vertical axis being inclined 30″, the diagram (Figure 6.47a) shows that the error introduced is $2 \times 17.8″ = 34.6″$. In other words, the angle at the station at which the instrument was set instead of being a true 180° would be 180° 00′35″, and beyond the station, the established line would depart from the true prolongation about 17.0 mm in 100 m or 0.05 ft in each 300 ft.

Errors in Vertical Angles Due to Nonadjustment of the Plate Level Bubble

These errors obviously vary with the direction in which the instrument is pointed. With the fixed vertical circle and scale, they are eliminated by observing (for each sighting) the vertical-circle collimation error of the corresponding observed vertical angle (Section 6.39).

Certain styles of repeating and direction theodolites have vertical circle level bubbles that are centered prior to each vertical angle, eliminating the error caused by nonadjustment of plate level bubbles (Section 6.51, test for adjustment 2). Theodolites equipped with automatic vertical circle indexing also are immune to this type of error.

It may be noted further that nonadjustment of the plate level bubble causes an inclination of the plane of the vertical arc. This source of error may be considered negligible.

Line of Sight Is Not Perpendicular to the Horizontal Axis

If the telescope is not reversed between backsight and foresight, if the sights are of the same length so that it is not required to refocus the objective, and if both points sighted are at the same angle of inclination of the line of sight, no error is introduced in the measurement of horizontal angles even though the error due to this lack of adjustment is large. If the instrument is reversed between backsight and foresight, the telescope rotating about the horizontal axis will generate a cone, the apex of which would fall at the intersection of the line of sight and the horizontal axis. The resultant error in the observed angle would be double the error of adjustment.

With the line of sight out of adjustment by a given amount, the effect of the error depends on the vertical angle to the point sighted. In Figure 6.49, OA and DB are horizontal and perpendicular to the horizontal axis OH of the instrument; e is the angle between the nonadjusted line of sight and a vertical plane normal to OH (that is, e is the error in direction for a horizontal sight OB); E is the error in direction for an inclined sight OC; h is the actual vertical angle to C, the point sighted; and OB is made equal to OC. Then,

$$\sin e = \frac{AB}{OB}$$

$$= \frac{AB}{OB}\frac{OD}{OD} = \frac{AB}{OD}\frac{OD}{OB} = \frac{AB}{OD}\frac{OD}{OC}$$

$$= \sin E \cos h$$

or

$$\sin E = \frac{\sin e}{\cos h} \tag{6.1}$$

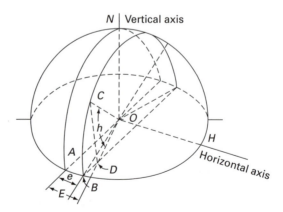

FIGURE 6.49

For all ordinary cases, Equation (6.1) may be taken as

$$E = \frac{e}{\cos h} = \frac{e}{\cos \alpha} \qquad \text{(approximately)} \qquad (6.2)$$

where α is the observed vertical angle.

For example, assume that the optical line of sight makes an angle of 89°58′00″ with the horizontal axis of the instrument. In turning a horizontal angle, if the line of sight on the backsight is horizontal and inclined with a vertical angle of 40°00′ on the foresight, the error in the observed horizontal angle due to the lack of adjustment is

$$E = (02') \cos 40°00' = (02')/(0.76604) = 0°02'37''$$

For two direct pointings (the backsight and the foresight), there will be a value of E for each, and the error in the angle will be the difference between them. In the measurement of a deflection angle by the method in which the telescope is inverted between backsight and foresight, the error in angle is the *sum* of the two values of E.

The error may be eliminated by taking the mean of two angular observations, one with the telescope in the direct position and the other with the telescope reversed. In prolonging a line, errors are avoided by the method of double centering described in Section 6.41.

Horizontal Axis Is Not Perpendicular to the Vertical Axis

No error is introduced in horizontal angles so long as the points sighted are at the same angle of inclination of the line of sight. The angular error in the observed direction of any line depends on both the angle by which the horizontal axis departs from the perpendicular to the vertical axis and the vertical angle to the point sighted. In Figure 6.50, OH is perpendicular to the vertical axis; OH' is the horizontal axis in nonadjustment with the vertical axis by the angular amount e'; OA is horizontal and perpendicular to OH and OH'; α is the observed vertical angle to C, the point sighted; B is directly beneath C and in the same horizontal plane with OA; angle OAB, OAC, and ABC are right angles; and θ is the angular

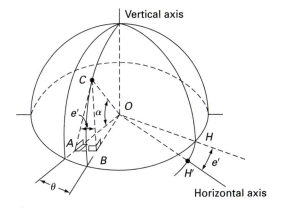

FIGURE 6.50

error in direction. From the figure,

$$\tan \theta = \frac{BA}{OA}$$

$$\tan \theta = \frac{BA}{OA} \frac{CA}{CA} = \frac{BA}{CA} \frac{CA}{OA}$$

$$= \sin e' \tan \alpha \tag{6.3}$$

or, with sufficient precision,

$$\theta = e' \tan \alpha \quad \text{(approximately)} \tag{6.4}$$

For example, if a horizontal angle were measured between D, to which the vertical angle is $-30°$, and E, to which the vertical angle is $+15°$, the horizontal axis being inclined 02', then the error in horizontal angle would be

$$\theta = 02'[\tan 15° - (-\tan 30°)] = 32'' + 01'09'' = 01'41''$$

This example is sufficient to show that the error in an observed horizontal angle may become large. Obviously, the sign of the error in a horizontal angle depends on the direction of displacement of the horizontal axis from its correct position. Hence, if any angle is measured with the telescope first in the direct and then in the reversed position, one value will be too great by the amount of the error and the other will be correspondingly too small; thus, the error is eliminated by taking the mean of the two values.

Effect of the Lack of Coincidence between the Line of Sight and the Optical Axis

Under these conditions, if the line of sight is perpendicular to the horizontal axis for one position of the objective, it will not be perpendicular for other positions but will swing through an angle as the objective is moved in or out. If an angle is measured without disturbing the position of the objective, no error is introduced. For most instruments, the error from this source is not sufficiently large to be of consequence in ordinary transit or theodolite work. It is eliminated by taking the mean of two angles, one observed with the telescope direct and the other with it reversed.

Errors Due to Eccentricity

With a theodolite in good condition, errors due to eccentricity of scales or centers are of no consequence in the ordinary measurement of angles. In any case, when using most optical-reading theodolites and electronic digital instruments, all directions are the mean of readings on opposite sides of the circle, so that errors due to eccentricity are virtually eliminated.

Imperfect Graduations

Errors from this source are of consequence only in work of high precision. They can be reduced to a negligible amount by taking the mean of multiple observations for which the readings are distributed over the circle and direct reading or micrometer scale. Many electronic digital theodolites are designed to use several portions of the circle for each angle measured, thus automatically compensating for this source of error in every measurement.

Lack of Parallelism between the Axis of the Telescope Level and the Line of Sight

For instruments equipped with a level bubble attached to the telescope, this condition introduces an error in leveling that is compensated by equalizing backsight and foresight distances (see Section 5.37). An error also is introduced into the zenith or vertical angles, where compensation is achieved by observing two angles, one with the telescope direct and one with the telescope reversed (Section 6.38). The mean of the two observations is the best estimate of the angle. Most optical-reading and electronic theodolites are not equipped with an attached telescope level bubble but do have automatic vertical circle index compensation. Establishing a horizontal line of sight is achieved by setting the zenith angle to 90° or the vertical angle to 0°. To guard against any potential errors in the compensator, one should apply the same compensatory rules outlined previously for leveling and measure the zenith or vertical angles with the telescope direct and reversed.

Summary

Summing up, it is seen that, with regard to instrumental errors,

1. Errors in horizontal angles due to nonadjustment of the plate level bubble or of the horizontal axis become large as the inclination of sights increases.
2. The maximum error in horizontal errors due to nonadjustment of the line of sight is introduced when the telescope is reversed between backsight and foresight readings. If the telescope is not reversed between backsight and foresight readings when observing horizontal angles, no error is introduced if these distances are equal and the inclination of the line of sight is the same for both the backsight and the foresight.
3. Errors due to instrumental imperfections or nonadjustment all are systematic, and without exception they can be either eliminated or reduced to a negligible amount by proper procedure. In general, this procedure consists of obtaining the mean of two values: one observed before and one after reversal of the horizontal plate by reversing the telescope and rotating it about the vertical axis. One of these values is as much too large as the other is too small. An exception is the error in either horizontal or zenith angle due to inclination of the vertical axis, which cannot be so eliminated. The most effective compensation for this error is to relevel very carefully, using the plate level bubble, following the procedure set forth in Sections 6.32 and 6.40.

Note that some electronic digital theodolites are equipped with a vertical axis or horizontal axis tilt sensor or both. Use of electro-optical liquid or electro-mechanical tilt sensors for vertical circle compensators in theodolites was discussed in Section 6.30. A similar device also can be installed to detect any deviation of the horizontal axis from true horizontal. These detected tilts can be stored for each position of the telescope and then used to compute corrections to horizontal and vertical angles for inclination of the vertical and horizontal axes from true vertical and horizontal, respectively. These corrections are stored, computed, and applied in real time, so that angles displayed on the register or stored in a data collector are the corrected values. Instruments with a tilt sensor on one axis (e.g., the vertical axis) are said to be equipped with *single-axis compensation;* those with tilt sensors on both the horizontal and vertical axes are said to have *dual-axis compensation.*

6.46
PERSONAL ERRORS

Personal errors arise from the limitations of the human eye in setting up and leveling the theodolite and in making observations.

Effect of Not Setting Up Exactly over the Station

This produces an error in all angles measured at a given station, the magnitude of the error varying with the direction of pointing and inversely with the length of sight. It is convenient to remember that an arc of 3 cm with a radius of about 100 m or an arc of 0.1 ft with a radius of 300 ft yields an angle of 01'. Therefore, if a theodolite were offset 1.5 cm or 0.05 ft from the ends of lines 50 m or 150 ft long, respectively, the error in observed directions would be 01'. On the other hand, if the lines were 200 m or 600 ft long, the error would be only 15″. This example emphasizes the need for increased precision in setting up and pointing when the sights are short and the desirability of long sights whenever possible. In general, the error can be kept within negligible limits by reasonable care. Many instrument operators waste time by exercising needless care in setting up when the sights are long.

Centering errors of less than 0.001 m or 0.003 ft are possible with the optical plummet or telescopic centering rod. When using a plumb bob, centering to within 0.006 m or 0.02 ft is possible without difficulty.

Effect of Not Centering the Plate Level Bubble Exactly

This produces an error in horizontal angles after the manner described in Section 6.45 for a plate level bubble out of adjustment. The error from this source is small when the sights are nearly level but may be large for steeply inclined sights (see Figure 6.48). The average theodolite operator does not appreciate the importance of careful leveling for steeply inclined sights; on the other hand, unnecessary care in leveling often is used when sights are nearly horizontal. Because the error in a horizontal angle is caused largely by inclination of the vertical axis, precise leveling (Section 6.40) is a necessity for precise work.

Errors in Setting and Reading Scales and Micrometers

For optical-reading theodolites, these errors are functions of the least count of the scale, the legibility of lines on the horizontal circle and scale, and the magnification of the optical train. Careful adjustment of the focus and illumination in the optical train is essential to obtaining maximum precision.

Horizontal and vertical circle readings with an optical theodolite having a least count of 1″ can be observed with a standard deviation of from 2 to 3″.

Reading and setting angles on electronic theodolites are primarily functions of the resolution of the encoded horizontal and vertical circles, because the angles are either displayed on a register or stored in a data collector and no particular skill is required of the observer.

Effect of Not Sighting Exactly on the Point

This is likely to be a source of rather large error on ordinary surveys, where sights are taken on a range or prism pole of which often only the upper portion is visible from the theodolite.

The effect on direction, of course, is the same as the effect of not setting up exactly over the station. For short sights, greater care should be taken than for long sights and the plumb line should be employed instead of the prism pole. When theodolite and EDM or total station surveys are performed, a target frequently is mounted symmetrically around the prism, which is supported on a prism pole or a tribrach on top of a tripod. This arrangement provides an excellent sight, but one must be sure the target center actually does coincide with the plumb line of the supporting mechanism that is centered over the station.

Imperfect Focusing (Parallax)

The errors due to imperfect focusing always is present to a greater or lesser degree, but with reasonable care it can be reduced to a negligible quantity. The manner of detecting parallax is described in Section 5.10.

Summary

All personal errors are random and hence cannot be eliminated. Of the personal errors, those due to inaccuracies in reading and setting the scale or the optical-reading micrometer (for optical-reading theodolites) and to not sighting exactly on the point are likely to be of greater magnitude. When electronic systems are employed, personal errors in reading and setting directions are somewhat diminished and those personal errors due to the setup and in sighting would predominate.

6.47
NATURAL ERRORS

Sources of natural errors are (1) settlement of the tripod, (2) unequal atmospheric refraction, (3) unequal expansion of parts of the telescope due to temperature changes, and (4) wind, producing vibration of the theodolite or making it difficult to plumb correctly.

In general, the errors resulting from natural causes are not large enough to affect appreciably the measurements of ordinary precision. However, large errors are likely to arise from settlement of the tripod when the theodolite is set up on boggy or thawing ground; in such cases, the instrument may be kept relatively stable through the use of extra-long tripods or of small platforms for the tripod legs. Settlement usually is accompanied by an angular movement about the vertical axis as well as linear movements both vertically and horizontally. When horizontal angles are being measured, usually a larger error is produced by the angular displacement of the circle between backsight and corresponding foresight than by the movement of the transit laterally from the point over which it is set. Errors due to adverse atmospheric conditions usually can be rendered negligible by choosing appropriate times for observing.

For measurements of high precision, the methods of observing are such that instrumental and personal errors are kept within very small limits, and natural errors become of relatively great importance. Natural errors generally are random, but under certain conditions systematic errors may arise from natural causes. On surveys of very high precision, a special attempt is made to establish a procedure that, as nearly as possible, will eliminate natural systematic errors. The instrument may be set up on a masonry pier and protected from sun and wind; also, certain readings may be made at night when temperature and atmospheric conditions are nearly constant.

PART B
ADVANCED TOPICS

6.48
DIRECTIONS AND ANGLES WITH A DIRECTION THEODOLITE

As noted in Section 6.29, the direction theodolite has only a single horizontal motion equipped with one clamp and tangent screw for azimuth. Therefore, it is possible to observe *directions only,* and angles are computed by subtracting one direction from another. For example, assume that the angle *AOB* (Figure 6.36) is to be measured with a direction instrument. The theodolite is set up over point *O*, leveled and centered, and a sight is taken on point *A* using the horizontal clamp and tangent screw for azimuth (Figure 6.31). The horizontal circle then is viewed through the optical-viewing system and the circle reading is observed and recorded (Section 6.29). Assume that this reading is 89°34′22″. The horizontal clamp then is released and the telescope is sighted on point *B*. Once again, the sight is taken with the horizontal clamp and the tangent screw for azimuth. The horizontal circle is observed through the optical-viewing system and this reading is recorded as 102°10′52″. These two observations constitute directions having a common reference direction that is completely arbitrary. The clockwise angle = (102°10′52″) − (89°34′22″) = 12°36′30″. Note that, on a direction theodolite, there is only one horizontal circle graduated in a clockwise direction so only clockwise angles can be derived from observations of directions. In Figure 6.37, point *B* is to the left of point *A*. Suppose that the theodolite is set up on point *O*, the initial sight is on point *A*, and the horizontal circle observation is 10°15′12″. The horizontal clamp is loosened and the telescope is rotated to sight on point *B*, where the circle reading is 351°50′32″. The clockwise angle from *A* to *B* therefore is (351°50′32″) − (10°15′12″) = 341°35′20″. Again, the angle is clockwise from the initial pointing on *A*. Note that in neither case was any attempt made to set the horizontal circle on 0 or on an integer value. A control allows the horizontal circle to be advanced to a desired approximate value (Figure 6.31). This control is a coarse motion only and does not permit fine settings.

6.49
REPEATED DIRECTIONS USING THE DIRECTION THEODOLITE

Measuring angles with the direction theodolite involves observing directions with the telescope direct and reversed, calculating mean directions to the observed points, and determining mean angles from the mean directions. Figure 6.51 shows notes for directions observed from a theodolite set at point *R* to points *A*, *B*, and *C*. Referring to these notes, the procedure for observing the directions is as follows:

1. Make an initial pointing and record the horizontal circle observation. Normally, the telescope is in the direct position and the leftmost point is selected as the initial sighting. The value of the initial circle reading is governed by the number of repetitions to be observed (see Gossett 1950).
2. Release the horizontal clamp and sight each point in succession, recording the circle reading for each sight. These observations constitute the circle readings to *A*, *B*, and *C* with the telescope direct.

Station sighted position	D or R	Circle ° ' "	Mean D & R	Direction "	December 4, 1997 Clear, T = 55°F Kern theodolite 45553 Abstract of directions		L. Patterson, Ⴕ G. Putz, Notes 32 Angles
1, Ⴕ @R					° ' "	° ' "	
A	D	0 00 09					
	R	180 00 15	12	00	0 00 00		
B	D	44 56 36				44 56 28	
	R	224 56 44	40	28	44 56 28		
C	D	50 52 21				5 55 44	
	R	230 52 27	24	12	50 52 12		

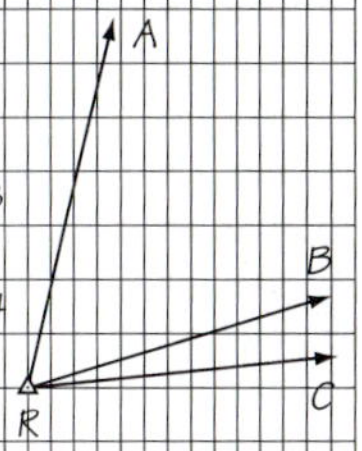

FIGURE 6.51
Field notes for directions with a direction theodolite.

3. Reverse the telescope, sight point *C*, record the circle reading, and then sight on points *B* and *A* in reverse directions, recording the circle observation with the telescope in the reversed position in each case. These reversed readings are different by exactly 180° from the direct observation plus or minus a few seconds that represent the collimation error of the instrument (Section 6.52). The observations taken as described constitute *one position.*

The means of the direct and reversed readings are calculated and recorded in column 4 of the notes (Figure 6.51). Then, an *abstract* of *directions* is calculated and recorded in column 5 of the notes. This abstract is determined by assuming a direction of $0°00'00''$ along the initial sight from *R* to *A*. The direction from *R* to *B* is $(44°56'40'') - (0°00'12'') = 44°56'28''$ and the direction from *R* to $C = (50°52'24'') - (0°00'12'') = 50°52'12''$. Finally, angles *ARB* and *RBC* are computed by subtracting successive directions, as given in the abstract.

Depending on the accuracy desired, anywhere from 1 to 16 positions of directions may be observed. In such cases, the horizontal circle would be advanced for each successive position so as to use the entire circle.* The final abstract of directions and angles would be the mean of all the positions.

The comments in Section 6.48 and the preceding paragraphs are directed toward use of the traditional optical-reading direction theodolite, as illustrated in Figures 6.30 and 6.31, for observing directions. Electronic theodolites (Figure 6.33), many of which are built into the framework of a direction instrument, also can be used to measure directions. This is accomplished very simply, by proper manipulation of the zero-set button (reset (RST) key, Figure 6.33) to set $0°00'00''$ or any other random value on the initial backsight. Once this step is taken, the procedure is exactly the same as outlined for the optical-reading instrument.

---

* The horizontal circle is advanced by an interval *I*, where

$$I = \frac{360°}{mn} + \frac{\text{micrometer range (in minutes)}}{n}$$

in which m = number of positions to be observed and n = number of indexes used to observe the angle.

6.50
ERROR PROPAGATION IN ANGLES

Estimation of attainable precision in observing angular measurements is essential for proper planning and design of a survey. Realistic estimation of the standard deviations achievable in angles observed with the theodolite is possible by analyzing the observational procedure and assigning procedural specifications to eliminate known systematic errors (Section 6.45). Then, estimates can be assigned to major sources of personal and natural random errors (Sections 6.46 and 6.47); the error propagated in the final angle can be evaluated using error propagation procedures detailed in Section 2.15.

As an example, consider the error propagation in a horizontal angle determined by repetition using a repeating theodolite (Section 6.22). As mentioned in Section 6.46, the major sources of random error in angles are those due to reading and setting the index mark, and pointing on signals that mark the backsight and foresight. The final averaged angle for an angle measured by repetition with repeating theodolite is

$$\alpha = \frac{R_2 - R_1}{n} \tag{6.5}$$

where α = average angle
n = number of repetitions
R_1 = initial circle reading on the backsight
R_2 = final circle reading on the foresight after n repetitions of the angle

Assume that the initial circle reading R_1 has an estimated variance

$$\sigma_{R_1}^2 = \sigma_s^2 \tag{6.6}$$

where σ_s is the estimated standard deviation in reading or setting the horizontal circle.

The final circle reading also is affected by $2n$ pointings. Therefore, the estimated variance in R_2 is

$$\sigma_{R_2}^2 = \sigma_s^2 + 2n\sigma_p^2 \tag{6.7}$$

where σ_p is the estimated standard deviation in pointing on a signal.

The final average angle is calculated by Equation (6.5). Propagating the error through Equation (6.5) by Equation (2.16) and using Equations (6.6) and (6.7) yields the estimated variance in the average angle:

$$\sigma_\alpha^2 = \frac{2\sigma_s^2}{n^2} + \frac{2\sigma_p^2}{n} \tag{6.8}$$

Assuming that $\sigma_s = 20''$ and $\sigma_p = 02''$ for a $10''$ repeating theodolite and using Equation (6.8), estimated standard deviations have been calculated according to several different numbers of repetitions. These are listed in Table 6.1. The table offers ample evidence of the value of repeating an angle.

Angles determined from observations using a direction theodolite are calculated by taking the difference between final and initial observed *directions* (Section 6.49). Thus, if D_1 and D_2 are the averaged directions of n backsights and foresights, respectively, the final angle is

$$\alpha = D_2 - D_1 \tag{6.9}$$

in which

TABLE 6.1

333

CHAPTER 6:
Angle and
Direction
Measurement

**Propagated errors in angles by repetition
using a repeating theodolite**

Number of repetitions	Propagated variance σ_α^2 (" of arc)^2	Standard deviation σ_α (" of arc)
2	204.00	14.3
4	52.00	7.2
6	23.56	4.9
8	13.50	3.7
10	8.80	3.0
12	6.22	2.5

Note: 10" repeating theodolite, $\sigma_s = 20"$, $\sigma_p = 02"$.

$$D_1 = \frac{\sum\limits_{i=1}^{n} D_{1i}}{n} \tag{6.10a}$$

$$D_2 = \frac{\sum\limits_{i=1}^{n} D_{2i}}{n} \tag{6.10b}$$

where D_{1i} = ith circle observation on the backsight
D_{2i} = ith circle observation on the foresight
n = number of repetitions

Assume that

σ_m^2 = estimated variance in optical micrometer observations on the horizontal circle
σ_p^2 = estimated variance in pointing on a signal

The error in a single backsight or foresight is a combination of reading and pointing errors. Therefore, the propagated estimated variance for a single backsight or foresight is

$$\sigma_{D_{1i}}^2 = \sigma_{D_{2i}}^2 = \sigma_m^2 + \sigma_p^2 \tag{6.11}$$

The estimated variance in an average of n backsights or foresights can be found by propagating through Equations (6.10a) or (6.10b) by Equation (2.16) and using Equation (6.11) to yield

$$\sigma_{D_1}^2 = \sigma_{D_2}^2 = \frac{\sigma_m^2 + \sigma_p^2}{n} \tag{6.12}$$

Then, the estimated variance in the angle calculated from average directions D_1 and D_2, propagated through Equation (6.9) by Equation (2.16) and using Equation (6.12), is

$$\sigma_\alpha^2 = \frac{2(\sigma_m^2 + \sigma_p^2)}{n} \tag{6.13}$$

Estimated standard deviations for different numbers of repetitions of an angle using a direction theodolite for which $\sigma_m = 02"$ and $\sigma_p = 02"$ are listed in Table 6.2.

Note that an important difference exists between angles by repetition with a repeating theodolite and angles obtained from directions observed with a direction theodolite. When the angle is turned by repetition using a repeating instrument, the angle is the observed quantity. If more than one angle is observed about a point, as shown in Figure 6.52, each

TABLE 6.2

Propagated errors in angles from repeated directions using a direction theodolite

Number of repetitions	σ_α^2 (" of arc)^2	σ_α (" of arc)
2	8.00	2.8
4	4.00	2.0
6	2.67	1.6
8	2.00	1.4

Note: 1" direction, theodolite, $\sigma_m = 02"$, $\sigma_p = 02"$.

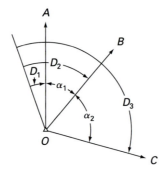

FIGURE 6.52
Angles and directions about a point.

angle can be considered an independent observed value with an estimated standard deviation propagated by Equation (6.8). When this procedure is followed, no correlation exists between α_1 and α_2.

On the other hand, if a direction theodolite is used to observe directions D_1, D_2, and D_3 (Figure 6.52), these directions are the independent, uncorrelated observed values for which respective variances can be calculated by Equation (6.11). The covariance matrix for these directions (Equation (2.29)) is

$$\Sigma_{DD} = \begin{bmatrix} \sigma_{D_1}^2 & 0 & 0 \\ 0 & \sigma_{D_2}^2 & 0 \\ 0 & 0 & \sigma_{D_3}^2 \end{bmatrix} \qquad (6.14)$$

The angles α_1 and α_2 are functions of these observations (Equation (6.9)) as follows:

$$\begin{bmatrix} \alpha_1 \\ \alpha_2 \end{bmatrix} = \begin{bmatrix} D_2 - D_1 \\ D_3 - D_2 \end{bmatrix} = \begin{bmatrix} -1 & 1 & 0 \\ 0 & -1 & 1 \end{bmatrix} \begin{bmatrix} D_1 \\ D_2 \\ D_3 \end{bmatrix} \qquad (6.15)$$

Therefore, the covariance matrix for angles derived from directions must be propagated through Equation (6.15) by Equation (2.34a) to give

$$\Sigma_{\alpha\alpha} = J_{\alpha D} \, \Sigma_{DD} \, J_{\alpha D}^T \qquad (6.16)$$

where Σ_{DD} is defined by Equation (6.14) and

$$J_{\alpha D} = \begin{bmatrix} \partial\alpha_1/\partial D_1 & \partial\alpha_1/\partial D_2 & \partial\alpha_1/\partial D_3 \\ \partial\alpha_2/\partial D_1 & \partial\alpha_2/\partial D_2 & \partial\alpha_2/\partial D_3 \end{bmatrix} = \begin{bmatrix} -1 & 1 & 0 \\ 0 & -1 & 1 \end{bmatrix} \qquad (6.17)$$

is the matrix of partial derivatives of α_1, α_2 with respect to the observed directions. Substitution of Equation (6.14) into Equation (6.16) yields

$$\Sigma_{\alpha\alpha} = \begin{bmatrix} (\sigma_{D_1}^2 + \sigma_{D_2}^2) & -\sigma_{D_2}^2 \\ -\sigma_{D_2}^2 & (\sigma_{D_2}^2 + \sigma_{D_3}^2) \end{bmatrix} \tag{6.18}$$

in which the diagonal terms represent the propagated variances and the off-diagonal elements are covariances of angles α_1, and α_2, respectively. The presence of these covariance elements in $\Sigma_{\alpha\alpha}$ means that the angles α_1 and α_2 definitely are correlated. Consequently, when a direction theodolite is employed to observe directions from which angles are calculated and these angles then are utilized in further adjustment computations, the entire covariance matrix must be used in these computations or erroneous results will be obtained.

EXAMPLE 6.4. Consider the observations of Figure 6.52 taken with a 1″ theodolite using four repetitions of each direction. Assume that estimated standard deviations in reading the horizontal circle and pointing on the signals are $\sigma_m = \sigma_p = 02''$, and signals A, B, and C are of equal quality. Determine covariance and weight matrices for angles α_1 and α_2. Assume that the reference variance $\sigma_0^2 = 1$.

Solution. By Equation (6.12), the variance in the direction D_1 is

$$\hat{\sigma}_{D_1}^2 = \frac{\sigma_m^2 + \sigma_p^2}{n} = \frac{(02)^2 + (02)^2}{4} = 02''^2$$

Because the signals are of equal quality, $\sigma_{D_1}^2 = \sigma_{D_2}^2 = \sigma_{D_3}^2$. Propagation of the covariance matrix by Equation (6.18) yields

$$\Sigma_{\alpha_1\alpha_2} = \begin{bmatrix} 2\sigma_{D_1}^2 & -\sigma_{D_1}^2 \\ -\sigma_{D_1}^2 & 2\sigma_{D_1}^2 \end{bmatrix} = \begin{bmatrix} 4 & -2 \\ -2 & 4 \end{bmatrix}$$

By definition, the weight matrix is (Equation (2.33), Section 2.18)

$$\mathbf{W}_{\alpha_1\alpha_2} = \sigma_0^2 \Sigma_{\alpha_1\alpha_2}^{-1} = \begin{bmatrix} \sigma_{\alpha_1}^2 & \sigma_{\alpha_1\alpha_2}^2 \\ \sigma_{\alpha_2\alpha_1}^2 & \sigma_{\alpha_2}^2 \end{bmatrix}^{-1} = \begin{bmatrix} 4 & -2 \\ -2 & 4 \end{bmatrix}^{-1}$$

which reduces to

$$\mathbf{W}_{\alpha_1\alpha_2} = \begin{bmatrix} \frac{1}{3} & \frac{1}{6} \\ \frac{1}{6} & \frac{1}{3} \end{bmatrix}$$

This is the weight matrix that must be used in all subsequent calculations involving α_1 and α_2.

To ensure the results cited in Tables 6.1 and 6.2, all angles must be observed with an equal number of direct and reversed repetitions to eliminate systematic instrumental errors. The instrument should be in good general adjustment, and in particular, the plate level bubbles must be in adjustment (Section 6.51). Targets to mark backsights and foresights must be selected to ensure the pointing accuracy specification. For example, a range pole is 30 mm or 0.10 ft in diameter. A plumbed range pole can be sighted to within 6 mm or 0.02 ft. Given a length of 600 m or about 2000 ft, a pointing accuracy of ±02″ could be achieved. However, for a sight of 150 m or about 500 ft in length on the same range pole, pointing accuracy becomes ±8 to ±10″. Consequently, for shorter sights a plumbed line or other well-defined target, compatible with the desired pointing accuracy, must be employed. Figure 6.53 shows a target that can be used for daylight operation and also may be

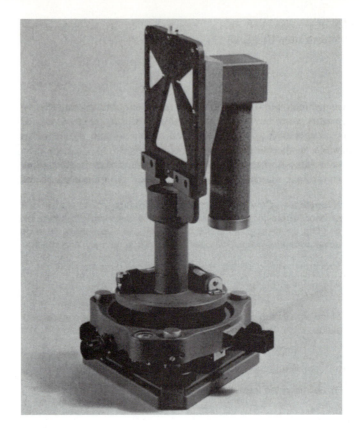

FIGURE 6.53
Target adaptable to tripod. (*Courtesy of Sokkia.*)

illuminated for night operation. This target is mounted on the same tripod as used for the theodolite and is adaptable to forced centering where instrument and target may be interchanged from one tripod to another.

EXAMPLE 6.5. A horizontal control survey, where the angles are to be observed with a standard deviation of 01′, is to be run for a route alignment. Sights are to have a minimum length of 150 m or about 500 ft, and range poles plumbed by eye are to be used to give line. Owing to intervening irregularities in the terrain and brush, many sights will have to be taken well above ground level. A 01′ repeating theodolite is available for observations. An experienced instrument operator is available, but inexperienced personnel must be utilized to give line. The site is wooded and has relatively steep slopes. Evaluate the procedures necessary to provide 01′ accuracy in the angles.

Solution. A range pole held by an inexperienced worker over traverse points on steep terrain in the woods may be out of plumb by $\pm$ the diameter of the pole much of the time. Therefore, a 30-mm or 0.10-ft deviation can be expected in a distance of 150 m or about 500 ft. This deviation yields an angular error of $\sim \pm 40''$, so that σ_p can be estimated as 40″. A 01′ theodolite permits observation of the horizontal circle to within $\pm 30''$ by an experienced operator.

If angles are turned once (Equation (6.8)),

$$\sigma_\alpha^2 = 2\sigma_s^2 + 2\sigma_p^2$$
$$= 2(30)^2 + 2(40)^2 = 5000''^2$$
$$\sigma_\alpha = 71''$$

which exceeds the allowable limit assuming a transit in near perfect adjustment. Good surveying practice would dictate turning the angle at least twice, once direct and once reversed (Section 6.35). Under these conditions the estimated standard deviation in the angles by Equation (6.8) is

$$\sigma_\alpha^2 = \frac{2\sigma_s^2}{n^2} + \frac{2\sigma_p^2}{n}$$
$$= \frac{2(30)^2}{4} + \frac{(2)(40)^2}{2} = 2050''^2$$
$$\sigma_\alpha = 45.3'' \qquad \text{or} \qquad 45''$$

The specifications necessary to guarantee $\sigma_\alpha = 45''$ are as follows:

1. A $01'$ engineer's theodolite in good adjustment should be used. Plate level bubbles must be in good adjustment.
2. Angles must be turned once direct and once reversed. Angles should be observed to the nearest $\pm 30''$.
3. Sights must be no less than 150 m or 500 ft in length.
4. Sights may be taken on range poles plumbed by eye.
5. An experienced theodolite operator must be available.
6. Inexperienced personnel may be utilized to give line.

Consider the effect of random errors in centering the instrument over the point as applied to the preceding example. For a given random error of centering equal to r, the point may occur anywhere on a circle of radius r centered about the point over which the transit is to be set. In Figure 6.54a, the angle $ACB = \theta$ is to be observed from a setup at C. Assume that the random error in centering is $CC' = r$. The angle actually turned is $AC'B = \theta + e_2 - e_1$, where $\tan e_1 = CC'/AC$ and $\tan e_2 \cong r \sin(90° - \theta)/BC$. Assuming that CB and CA are approximately equal, θ is less than $90°$, and the error is approximately normal to line CA, the total angular error introduced is $e_2 - e_1$. For $CC' = 15$ mm or 0.05 ft, $CA = CB = 150$ m or 500 ft, and $\theta = 45°$, $e_1 = 21''$, and $e_2 = 15''$, yielding a total angular error of $06''$. When the centering error falls on the bisector of the angle θ as shown in Figure 6.54b, the observed angle is $AC''B = \theta + e_1' + e_2'$ and the total angular error introduced is $e_1' + e_2'$. In this case, $\tan e_1' \cong (r \sin \theta/2)/AC$ and $\tan e_2' \cong (r \sin \theta/2)/CB$. Under the previous assumptions and given values, $e_1' = e_2' = 08''$, so the total angular error is $16''$. The maximum angular error will occur when the centering error falls on the bisector of the observed angle.

In Example 6.5, the propagated angular error due to reading ($\sigma_s = 30''$) and pointing is $45''$. According to the analysis just completed, the uncertainty in angular observations due to a centering error of 15 mm or 0.05 ft is $\pm 16''$. Therefore, the total estimated standard deviation in the angle due to reading, pointing, and centering errors is

$$\sigma_{\alpha, \text{total}}^2 = (45)^2 + (16)^2 = 2281''^2$$
$$\sigma_{\alpha, \text{total}} = 48''$$

and the effect of centering error is seen to be negligible.

A centering error can be significant if the lengths of sights are short and high accuracy is desired.

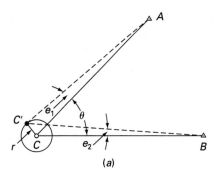

(a)

FIGURE 6.54
Errors due to improper centering.

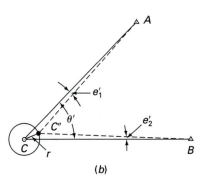

(b)

EXAMPLE 6.6. Relative positions of targets in a camera calibration test array are to be determined by triangulation from a base line 6 m in length using a 01″ theodolite. Lengths of sight, D, are from 6 to 8 m and angles vary from 45 to 90°. Angular measurements are specified to be within ±30″ of arc. Angles are to be observed twice direct and twice reversed, for a total of four repetitions. Determine whether this procedure will produce the desired angular accuracy.

Solution. Assuming that the standard deviation in reading, σ_m = standard deviation in pointing σ_p = 02″, then σ_α = 02″ (Table 6.2). A centering accuracy of 0.001 m or 0.003 ft will lead to

$$e'_1 + e'_2 = \frac{2r \sin(\theta/2)}{D \sin 01''} = \frac{(0.001)(\sin 45°)(2)}{(6)(4.848)(10^{-6})} = 49''$$

so that $\sigma_{\alpha, \text{total}} = [(02)^2 + (49)^2]^{1/2} = 49''$, which exceeds the specified value. Using an optical plummet, centering to within 0.0005 m is possible. This value leads to an angular error of

$$e'_1 + e'_2 = \frac{(0.0005)(\sin 45°)(2)}{(6)(4.848)(10^{-6})} = 24.3''$$

so that

$$\sigma_{\alpha, \text{total}} = [(02)^2 + (24.3)^2]^{1/2} = 24.4''$$

Note that, in this problem, centering is the critical factor and the specifications should be changed to require centering with recently adjusted optical plummet.

EXAMPLE 6.7. Combined triangulation-trilateration (Chapter 9, Section 9.15) is to be performed for a preliminary bridge survey. The shortest length of sight is about

305 m or 1000 ft. The average triangle closure is not to exceed ±05″ with a maximum

339

CHAPTER 6:
Angle and
Direction
Measurement

closure not to exceed ±10″. Observation of each angle with an estimated standard deviation of 02″ will more than satisfy this requirement. Determine equipment and methods required for such accuracy.

Solution. As shown in Table 6.2, Section 6.50, two positions or four repetitions of an angle with a 01″ theodolite will yield an estimated standard deviation of $\sigma_{aD} = 02″$ (based on $\sigma_m = \sigma_p = 02″$). Twelve repetitions using a 30″ repeating theodolite would be required to achieve comparable accuracy (based on $\sigma_s = 20″$, $\sigma_p = 02″$, Table 6.1). From the standpoint of time consumed, use of the 01″ theodolite would be most economical. The following specifications should be written to ensure attaining $\sigma_{aD} = 02″$:

1. A 01″ optical-reading direction theodolite in good adjustment and equipped with an optical plummet or centering tripod shall be used for observing directions. An experienced operator should be available.
2. Two positions of directions shall be observed at each triangulation station. One direct and one reversed direction to each point sighted constitutes one position. The initial circle readings for each position should be distributed about the horizontal circle to eliminate the effects of graduation errors (Gossett, 1950).
3. Targets must be chosen, plumbed, and maintained to provide the 3-mm or 0.01-ft centering accuracy required in a 305-m or 1000-ft sight to ensure a pointing accuracy of ±02″. The use of theodolites and targets compatible (Figure 6.53) with the same tripods to permit interchanging instruments would simplify this task.
4. Observations should be scheduled to take advantage of atmospheric conditions. Observations should not be made looking directly into the sun or when the atmosphere is excessively hazy.

Angles determined from directions observed about a single station according to these specifications will be correlated. This correlation must be taken into account in subsequent calculations with angles in triangulation computations.

6.51
ADJUSTMENT OF THE OPTICAL-READING THEODOLITE: DESIRED RELATIONS

The following relations are desired for a theodolite equipped with a vertical circle compensator and in perfect adjustment. The number of each relation is the same as that of the corresponding adjustment described in the following sections. For adjustments 1 through 4, Figure 6.55 shows the desired relations among the principal lines of the theodolite.

1. The vertical cross hair should lie in a plane perpendicular to the horizontal axis so that any point on the hair may be employed when measuring horizontal lines or when running lines.
2. The axis of the plate level bubble should lie in a plane perpendicular to the vertical axis so that, when the instrument is leveled, the vertical axis will be truly vertical; thus, horizontal angles will be measured in a horizontal plane and vertical angles will be measured without collimation error due to inclination of the vertical axis.
3. The line of sight should be perpendicular to the horizontal axis at its intersection with the vertical axis. Also, the optical axis, the axis of the objective slide, and the line of sight should coincide. If these conditions exist, when the telescope is rotated about the horizontal axis, the line of sight will generate a plane when the object is focused for either a near sight or a far sight, and that plane will pass through the vertical axis.
4. The axis of the circular level bubble should be perpendicular to the vertical axis.

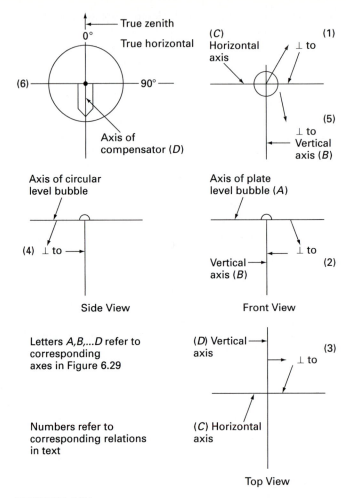

FIGURE 6.55
Desired relations among principal lines of theodolite.

5. The horizontal axis should be perpendicular to the vertical axis so that when the telescope is reversed, the line of sight will generate a vertical plane.
6. When the instrument is approximately leveled, the vertical circle compensator should activate automatically so that 0 on the vertical circle will coincide with the vertical toward the zenith and 90° will coincide with the horizontal.
7. The optical axis and the line of sight should coincide (see relation 3).
8. The line of sight of the telescope of the optical plummet should coincide with the vertical axis of the instrument.

6.52
ADJUSTMENTS

For those adjustments that involve sighting through the telescope, particular attention should be given to proper focusing of both the eyepiece and the objective prior to testing the adjustments.

The adjustments of the theodolite are more or less dependent on one another. For this reason, if the instrument is badly out of adjustment, time will be saved by first making the adjustments roughly for related adjustments and then repeating the tests and corrections in the same order. The plate level bubble will not be disturbed by other adjustments and should be corrected, exactly, before other adjustments are attempted.

Any movement of the screws controlling the cross-hair ring is likely to produce both lateral displacement and rotation of the ring; hence, any considerable adjustment of the line of sight is likely to disturb the vertical cross hair so that it no longer will remain on a point when the telescope is rotated about the horizontal axis.

The theodolite adjustments commonly made are those in relations 1 through 4 and occasionally 7 and 8. Relations 5 and 6 can be checked, but for optical reading instruments the actual adjustment, if needed, should be performed by a qualified instrument repair person.

Relation 1: To Make the Vertical Cross Hair Lie in a Plane Perpendicular to the Horizontal Axis

Test. Sight the vertical cross hair on a well-defined point no less than 60 m or 200 ft away. With both horizontal motions of the instrument clamped, swing the telescope through a small vertical angle so that the point follows the length of the vertical cross hair. If the point appears to move continuously on the cross hair, the cross hair lies in a plane perpendicular to the horizontal axis (see Figure 6.56).

Correction. If the point appears to depart from the cross hair, remove the eyepiece lens holder and reticule adjustment cover to expose the reticule adjustment screws on the eyepiece end of the telescope. Loosen two adjacent capstan screws and rotate the cross-hair ring in the telescope tube until the point traverses the entire length of the hair. Tighten these two screws, being very careful not to overtighten them and warp the reticule. This adjustment is similar to adjustment 2 of the dumpy level (Section 5.62), with the terms *vertical* and *horizontal* interchanged.

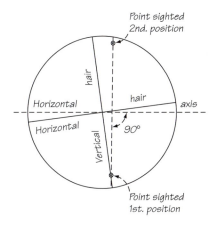

FIGURE 6.56
Adjustment of vertical cross hair of theodolite.

Relation 2: To Make the Axis of the Plate Level Bubble Lie in a Plane Perpendicular to the Vertical Axis

Test. Prelevel the instrument using the circular level bubble. Rotate the instrument about the vertical axis until the axis of the level tube is parallel to a pair of adjacent leveling screws (Figure 6.57a). Center the bubble in the plate level vial using the adjacent level screws 1 and 2 in Figure 6.57a. Rotate the instrument through 180° about the vertical axis. If the bubble remains centered, the axis of the level tube is perpendicular to the vertical axis.

Correction. If the bubble is displaced, bring it halfway to center by means of the leveling screws 1 and 2 (Figure 6.57a). Rotate it back to the original position and repeat the test to verify the results. Now, the vertical axis is vertical but the bubble remains out of adjustment.

The steps involved in adjustment are shown in Figure 6.57b–e. View (b) shows the level tube out of adjustment by the amount of the angle α but with the bubble centered; the support therefore is not level and the vertical axis is not vertical. In view (c) the level tube has been lifted and reversed, end for end, which is effected by rotating the instrument 180° about its vertical axis. The axis of the level tube now departs from the horizontal by 2α, or *double* the error of the setting. In view (d) the bubble has been brought back *halfway* to the middle of the tube by means of the adjusting screw C, without moving the support; the tube now is in adjustment. Finally, in view (e) the bubble again is centered by raising the low end

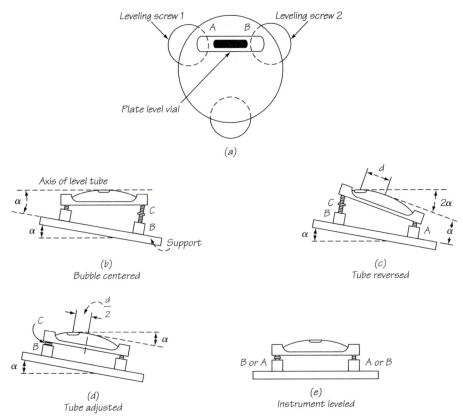

(a)

(b)
Bubble centered

(c)
Tube reversed

(d)
Tube adjusted

(e)
Instrument leveled

FIGURE 6.57
Adjustment of the level tube.

(or lowering the high end) of the support; the support now is level and the adjustment may be checked by reversing the tube again.

In this example, it were desired to level the support in the direction of the tube without taking time to adjust the tube, this could be accomplished by centering the bubble as in view (b), Figure 6.57; reversing the tube as in view (c); and raising the low end (or lowering the high end) of the support until the bubble is brought halfway back to the center of the tube. This position of the bubble corresponds to the error of setting the tube; and whenever the bubble is in this position, the support will be level.

Relation 3: To Make the Line of Sight Perpendicular to the Horizontal Axis

Test. Level the instrument. Sight on a point A (see Figure 6.58) about 150 m or 500 ft away, with telescope in the direct position. With both horizontal motions of the instrument clamped, reverse the telescope and set another point B on the line of sight and about the same distance away on the opposite side of the transit. Unclamp the upper motion, rotate the instrument end for end about the vertical axis, and again sight at A (with the telescope reversed). Clamp the upper motion. Reverse the telescope as before; if B is on the line of sight, the desired relation exists.

Correction. If the line of sight does not fall on B, set a point C on the line of sight beside B. Mark a point D, *one-fourth* the distance from C to B, and adjust the cross-hair ring (by the two opposite horizontal screws) until the line of sight passes through D. The points sighted should be at about the same elevation as the transit.

Relation 4: To Adjust the Circular Level Bubble

Level the theodolite using the plate level bubble (Section 6.32). The circular bubble in the tribrach of the leveling head should be centered. If the circular bubble is not centered, adjust it using the adjusting screws located around the bubble case. The adjustment procedure is the same as described for automatic levels in Section 5.63.

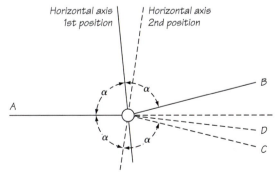

FIGURE 6.58
Adjustment of the line of sight of the theodolite.

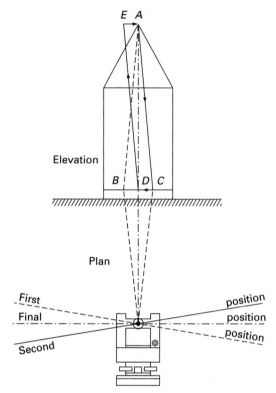

FIGURE 6.59
Adjustment of the horizontal axis of the theodolite.

Relation 5: To Make the Horizontal Axis Perpendicular to the Vertical Axis

Test. Set up the transit near a building or other object on which is some well-defined point A at a considerable vertical angle. Level the instrument very carefully, making the vertical axis truly vertical. Sight at the high point A (see Figure 6.59) and, with the horizontal motions clamped, depress the telescope and set a point B on or near the ground. If the horizontal axis is perpendicular to the vertical axis, A and B will be in the same vertical plane. Reverse the telescope, rotate the instrument end for end about the vertical axis, and again sight on A. Depress the telescope as before; if the line of sight falls on B, the horizontal axis is perpendicular to the vertical axis.

If the line of sight does not fall on B, set point C on the line beside B. A point D, halfway between B and C, will lie in the same vertical plane with A, and the distance BC is an indication of the amount by which the horizontal is out of alignment. Adjustment is performed by raising or lowering the adjustable end of the horizontal axis. On optical-reading and electronic theodolites, consult the owner's manual for instructions on how to proceed. If no instructions are given and adjustment is felt absolutely necessary, it would be wise to take the instrument to a service center where qualified personnel can check the instrument.

Correction (for open-frame instruments). Sight on D, elevate the telescope until the line of sight is beside A, loosen the screws of the bearing cap and raise or lower the adjustable end of the horizontal axis until the line of sight is in the same vertical plane with A.

The high end of the horizontal axis always is on the same side of the vertical plane through the high point as the last point set.

In readjusting the bearing cap, take care not to bind the horizontal axis, but it should not be left so loose as to allow the objective end of the telescope to drop of its own weight when not clamped.

Relation 6: Adjustment of the Vertical Circle Compensator

Test (optical-reading and electronic instruments). Level the instrument and, using the vertical clamp and tangent screw, take a sight on a well-defined point at a distance of about 120 m or 400 ft and observe and record the zenith angle on the vertical circle. Reverse the telescope and sight on the same point, again observing the zenith angle. The sum of the direct and reversed zenith angle should be 360°00′00″. A difference between the sum of these two angles and 360° is an indication of the error in the compensator. Repeat this test several times and calculate an average error. If this error exceeds the least count of the vertical circle, adjustment is necessary.

Mechanical, electrical mechanical, and electro-optical liquid compensators are rugged elements that rarely need adjustment unless the instrument has been severely jarred or dropped. In case adjustment of the compensator of either an optical-reading or digital theodolite is necessary, the instrument should be taken or shipped to the nearest service center or manufacturer, where a qualified person can perform the adjustment.

Some electronic theodolites have a *zero index* setting operation for the vertical circle, which can be done using the zero-set button on the register or some other specified control. Consult the instructional manual for the instrument to see if this is possible for your system.

Relation 7: To Make the Line of Sight, as Defined by the Horizontal Cross Hair, Coincide with the Optical Axis

Test. Set two pegs, one about 8 m or approximately 25 ft and the other 90 to 100 m or 300 to 400 ft from the theodolite. With the vertical motion clamped, take a rod reading on the distant point and, without disturbing the vertical motion, read the rod on the near point. Reverse the telescope, rotate the instrument about the vertical axis, and set the horizontal cross hair at the last rod reading with the rod held at the near point. Now sight on the rod held at the distant point. If the desired relation exists, the first and last readings on the rod will be the same.

Correction. If there is a difference between the rod readings, move the horizontal cross hair by means of the upper and lower adjusting screws until it intercepts the distant rod at the mean of the two readings. Repeat the process until, by successive approximations, the error is reduced to 0. Exercise great care when reading the near rod, because a small difference in the position of the cross hair on the near rod would be enough to cause a considerable error in the reading on the distant rod.

Relation 8: Adjustment of the Optical Plummet

Test. Adjustment of an optical plummet mounted in the alidade of the theodolite (Figures 6.22, 6.33) is performed as follows. Set the instrument over a well-defined point with the plate approximately level and with the optical plumb line within about a centimeter from the point. Center the optical plumb line precisely on the point using the leveling screws (the theodolite need not be precisely leveled). Rotate the instrument 180° about the vertical axis and check the centering of the plumb line over the target. If the cross (or circle) in the optical plummet is properly centered on the point, no adjustment is necessary.

For instruments having an optical plummet mounted in the tribrach, the instrument is leveled and centered over a clearly marked point using a plumb bob suspended from a bracket attached to the clamping screw that secures the theodolite to the tripod. The plumb line is removed and the cross in the optical plummet is very carefully focused on the point. If the cross coincides with the point, no adjustment is needed. For best results, this test should be conducted in an area protected from wind and using a heavy plumb bob (24 to 32 oz).

Adjustment. In either case, any deviations between the cross and the point are due to maladjustment of the optical plummet and can be corrected by the adjusting screws, usually near the eyepiece of the optical plummet. On many instruments, it first is necessary to unscrew a protective cover to expose the four capstan adjusting screws that should be adjusted to remove one-half of the discrepancy. Repeat the test and adjustment until the cross remains centered on the point.

PROBLEMS

6.1. Describe two coordinate systems used to locate points on or near the surface of the earth.

6.2. Define the true, astronomic, and magnetic meridians.

6.3. The following azimuths are from the north: 329°20′, 180°35′, 48°32′, 170°30′, 145°25′, 319°35′, 350°45′, 95°49′, 11°30′, 235°45′. Express these directions as (*a*) azimuths from the south, (*b*) back azimuths, (*c*) bearings.

6.4. The following are bearings: S21°25′W, N88°42′W, N69°52E, N30°50′E, S42°25′E, N15°50′W, N61°32′W, S75°25′W, S42°35′W, N64°29′W. Express these directions as (*a*) azimuths from the north, (*b*) azimuths from the south.

6.5. The following are back azimuths: $A_{BA} = 275°25′$, $A_{BC} = 45°15′$, $A_{CD} = 190°15′$, $A_{CE} = 161°45′$, $A_{CF} = 320°45′$. Convert these directions to forward azimuths.

6.6. In Problem 6.5, determine the counterclockwise horizontal angles between lines *BA* and *BC*, *CB* and *CD*, and *CE* and *CF*.

6.7. The following azimuths are reckoned from the north: $FE = 4°25′$, $ED = 90°15′$, $DC = 271°32′$, $CB = 320°21′$, and $BA = 190°45′$. What are the corresponding bearings? What are the deflection angles between consecutive lines?

6.8. The interior angles of a five-sided closed polygon *ABCDE* are as follows: *A*, 120°24′; *B*, 80°15′; *C*, 132°24′; *D*, 142°20′. The angle at *E* is not measured. Compute the angle at *E*, assuming the given values to be correct.

6.9. The magnetic bearing of a line is S47°30′E and the magnetic declination is 8°20′E. What is the true bearing of the line?

6.10. The true bearing of a line is S25°14′E and the magnetic declination is 5°10′W. What is the magnetic bearing of the line?

6.11. In an old survey made when the declination was 4°15′E, the magnetic bearing of a given line was N35°15′E. The declination in the same locality now is 1°10′W. What are the true bearing and the present magnetic bearing that would be used in retracing the line?

6.12. A theodolite for which the circle is graduated 0° to 360° clockwise and with a least count of 30″ is used to measure an angle by 10 clockwise "repetitions," 5 with the telescope direct and 5 with the telescope reversed. Compute the value of the angle from the following data.

Telescope	Reading	Circle
Direct	Initial backsight	50°45′00″
Direct	After first turning	170°56′
Reversed	After 10th turning	172°38′00″

6.13. An angle is repeated three times direct and three times reversed with a repeating theodolite having a least count of 01′. The circle reading for the initial backsight is 230°35′00″ and after the first repetition is 129°33.0′. The circle reading after the sixth repetition is 344°30′00″. Calculate the average angle. Is this set acceptable?

6.14. The horizontal circle on an optical-reading theodolite is graduated in 1° intervals and the direct-reading scale used to observe angles has 60 equal divisions. What is the least count of this theodolite?

6.15. The optical micrometer in an optical-reading theodolite contains 240 equal divisions. If the horizontal circle is graduated into 1° intervals, what is the least count of this instrument?

6.16. An angle is turned twice direct and twice reversed (2DR) using a theodolite having a least count of 30″. The circle reading for the initial backsight is 260°10′30″ and after the first repetition is 35°51′30″. The circle reading after the fourth repetition is 82°54′00″. Compute the average angle. In this set of angles acceptable? Explain why or why not.

6.17. A theodolite, having a circle graduated 0° to 360° clockwise, is used to measure and angle three times direct and three times reversed (3DR). The data are as follows:

Station	From/To	Rep.	Horizontal circle
	B2	0	00°00′00″
A2	C	1D	145°30′30″
	C	3R	153°04′30″

The theodolite has a least count of 30″. Calculate the average angle. Is the set acceptable? Explain why or why not.

6.18. A direction theodolite with a least count of 1″ is set over station D to measure directions to stations C, B, and A. The observed directions for one position are as follows:

Station	Tel.	Circle
C	Direct	0°44′20″
	Reversed	181°44′24″
B	Direct	46°39′49″
	Reversed	226°39′57″
A	Direct	52°36′11″
	Reversed	232°36′14″

Compile an abstract of average directions and compute the average angles.

6.19. In laying out the lines for a building, a 90° angle was laid off as precisely as possible with a 01′ theodolite. The angle then was measured by repetition and found to be 90°00′35″. What offset should be made at a distance of 400 m from the theodolite to establish the true line?

6.20. The standard procedure for measuring angles in the field is to repeat the angle with the telescope in an equal number of direct and reversed positions. Explain why this procedure is necessary.

6.21. A theodolite is tested by prolonging a line using the method of double centering. In doing so, the theodolite was set at B and two backsights were taken on A, permitting the setting of points C' and C'', 100 m from B on a prolonged line, AB. If the distance between C' and C'' is 4 cm (0.040 m), what is the error between the optical line of sight and the horizontal axis of the theodolite, expressed in seconds of arc? Assume A, B, and C have the same elevations. Note, refer to Section 6.41 in the text.

6.22. The instrument described in Problem 6.21 was used to lay out an angle of $90°00'00''$ by turning the angle once with the telescope in the direct position on the backsight and reversed on the foresight. Thus, if the transit is set over B and a backsight is made on A, then the telescope is reversed and $90°00'00''$ is turned to the left to set C. If the distance BC is 400 m, what error is introduced in point C? If the angle is turned once direct and once reversed, what error occurs in point C? Assume A, B, and C are at the same elevation. Express the magnitude of the error in linear units at right angle to BC (Section 6.36).

6.23. It has been determined that the error between the optical line of sight and the horizontal axis of a transit is $05'$ of arc. What error would be introduced into a horizontal angle (*a*) with sights taken to points at the same elevation as the point occupied? (*b*) with both sights inclined at $+10°$? (*c*) with one sight inclined at $+25°$ and the other at $-35°$?

6.24. Two points, A and B, 1500 m apart, are to be connected by a straight line. A random line run from A in the general direction of B is found by computation to deviate $04'20''$ from the true line. On the random line, at a distance 385.22 m from A, an intermediate point C is established. What must be the offset from C to locate a corresponding point D on the true line?

6.25. What error would be introduced in the measurement of a horizontal angle, with sights taken to points at the same elevation as the theodolite, if, through nonadjustment, the horizontal axis were inclined (*a*) $04'$? (*b*) $3°$? (*c*) If the horizontal axis were inclined $04'$, what error would be introduced if both sights were inclined at angles of $+35°$? (*d*) If one sight was inclined at $+40°$ and the other at $-40°$?

6.26. Vertical angles turned to the same point with the telescope direct and reversed are $+15°25'00''$ and $+15°26'00''$. Calculate the vertical collimation correction. A single direct vertical angle of $+10°20'30''$ then was measured. Determine the corrected vertical angle and the corresponding zenith angle.

6.27. Direct and reversed zenith angles measured to determine the vertical collimation correction are $92°25'30''$ and $266°34'30''$. Compute the vertical collimation correction for instrument. A single zenith angle of $95°24'35''$ with the telescope direct then is measured. Determine (*a*) the corrected zenith angle and (*b*) the corrected vertical angle.

6.28. The following observations were made to determine a vertical collimation correction: Vertical angle to point $A = +9°22'$ with the telescope direct and $+9°24'$ with the telescope inverted. Compute the collimation correction for observations with the telescope direct.

6.29. A vertical angle measured by a single observation is $+15°04'$, and the collimation error is determined to be $-02'$. What is the correct value of the angle?

6.30. A theodolite equipped with a vertical circle level bubble has 0 on the vertical circle oriented toward the zenith. The following direct and reversed zenith angles were observed to determine the vertical collimation correction: $84°21'30''$ and $275°39'00''$. Compute the collimation correction for the instrument. The same instrument is used to measure a zenith angle of $82°14'25''$. Calculate (*a*) the corrected zenith angle and (*b*) the corrected vertical angle.

6.31. The estimated standard deviation in reading and setting the horizontal circle on a 30″ repeating theodolite is $\sigma_s = 20''$. The standard deviation in pointing on a specified target is $\sigma_p = 30''$. Compute the standard deviation for an angle turned (*a*) once, with the telescope in the direct position; (*b*) twice, once with the telescope direct and once with the telescope reversed; (*c*) six times, three direct and three reversed.

6.32. The abstract of directions to triangulation stations *C*, *D*, and *E* observed from station *A* is

Station	Direction	Estimated σ
C	0°00′00.0″	04″
D	44°32′00.8″	04″
E	92°25′23.3″	02″

Calculate angles *CAD* and *DAE* from these directions. Form the covariance matrix for angles *CAD* and *DAE*.

6.33. The angles in a simple triangular figure are to be determined with a standard deviation of 10″. A 06″ theodolite is to be used. The standard deviation in setting and reading the horizontal circle is $\sigma_s = 10''$. The standard deviation in pointing on the targets is estimated to be $\sigma_p = 10''$. Determine the number of repetitions of the angle required to satisfy the specified accuracy.

6.34. Write specifications to ensure that the standard deviation in measured angles as specified in Problem 6.33 is obtained. Assume an average length of sight of 200 m.

6.35. The random error in centering a theodolite over a point is estimated as 0.004 m. What is the maximum angular error introduced by this centering error to an angle of 40°, assuming lengths of sights of (*a*) 60 m? (*b*) 200 m?

6.36. Determine the estimated standard deviation in measuring the angle under the conditions of Problem 6.31 if the random error in centering the instrument over the point is estimated to be 0.002 m, the lengths of sights are 100 m, and the angle is 50°.

REFERENCES

Ali, A. E. "Electronic Theodolites: Comparison Test." *ASCE Journal of Surveying* 117, no. 1 (February 1991), pp. 3–6.

Clark, M. M., and R. B. Buckner. "A Comparison of Precision in Pointing to Various Targets at Different Distances." *Surveying and Land Information Systems* 52, no. 1 (1992), pp. 41–45.

Fialovsky, L. *Surveying Instruments and Their Operational Principles.* Budapest, Hungary: Akadèmia Nyomda, 1991.

Gossett, F. R. *Manual of Geodetic Triangulation.* NOAA C&GS Special Publication 247. Washington, DC: U.S. Government Printing Office, 1950.

Katowski, O., and W. Salzmann. "The Angle Measurement System in the Wild THEOMAT T2000." Heerbrugg, Switzerland: Wild Heerbrugg Ltd., CH-9435, October 1983.

Kavanagh, B. F., and S. J. G. Bird. *Surveying, Principles and Applications,* 3d ed. Englewood Cliffs, NJ: Prentice-Hall, 1992.

Kissam, P. *Surveying for Civil Engineers,* 2d ed. New York: McGraw-Hill Book Co., 1981.

Moffitt, F. H., and H. Bouchard. *Surveying,* 9th ed. New York: HarperCollins Publishers, 1992.

Paiva, J. V. "Evaluating the Accuracy of Electronic Theodolites." *Technical Papers American Congress on Surveying and Mapping* (Fall Technical Meeting, 1985), p. 249.

Pepling, A. "The Wild TC500." *Professional Surveyor* 14, no. 4 (July–August 1994), pp. 38, 39.

Reilly, J. P. "Transit/Theodolite Survey." *P.O.B., Point of Beginning* 18, no. 2 (December 1992–January 1993), pp. 78–83.

Wolf, P. R., and R. C. Brinker, *Elementary Surveying,* 9th ed. New York: HarperCollins Publishers, 1994.

Combined Distance and Angular Measurement Systems

7.1
GENERAL

Rapid determination of distance and direction from a single instrument setup without the need for a tape has always been a top priority in surveying operations. Tacheometry was the traditional procedure by which distances and differences in elevation were determined indirectly using subtended intervals and angles observed with a theodolite on a graduated rod or scale. The distances and elevations thus obtained usually were of a lower order of accuracy than is possible by taping, EDM, or differential leveling. However, the results were very adequate for many purposes. Horizontal distances can be determined with a relative accuracy of 1 part in 300 to 400, and differences in elevation can be determined to within ±0.03 m or ±0.1 ft. The most common tacheometric method practiced in the United States is stadia surveying, in which an engineer's theodolite and graduated level rod are used.

Applications of tacheometry include traverse and leveling for topographic surveys, location of detail for topographic surveys, leveling and field completion surveys for photogrammetric mapping, and hydrographic mapping.

During the past decade, the development of microelectronics and microprocessors and their incorporation into EDM equipment and electronic theodolites, have led to the *total station system,* which yields distance, direction, and elevation, in real time, from a single setup of one compact instrument. Accuracies of 1 to 5 mm + 2 to 5 ppm in distance, 5 arc sec in directions, and a few cm in elevation are possible with such a system. These devices also include onboard computational capacity and are adaptable to data collectors (electronic field book). Therefore, the total station system is an extremely powerful tool for all types of surveys from precise traverse operations (Chapter 8) to data collection for topographic surveys (Chapter 15).

Tacheometry still is a useful procedure for many surveying applications, such as topographic mapping of small areas, keeping track of backsight and foresight distances in precise leveling (Section 5.50), and trigonometric leveling of a low order of accuracy. Although total station systems are replacing tacheometry in many operations of surveying, stadia surveying will continue to be used for some time to come.

In this chapter, stadia surveying using a theodolite and graduated level rod is discussed first. The balance of the chapter is devoted to the characteristics and operation of total station systems.

351

CHAPTER 7:
Combined
Distance and
Angular
Measurement
Systems

7.2
STADIA METHOD

Equipment for stadia measurements consists of a telescope with two horizontal cross hairs called *stadia hairs* and a graduated rod called a *stadia rod*.

The process of taking stadia measurements consists of observing, through the telescope, the apparent locations of the two stadia hairs on the rod, which is held in a vertical position. The interval thus determined, called the *stadia interval* or *stadia reading,* is a direct function of the distance from instrument to rod as developed in Section 7.4. The ratio of distance to stadia interval is 100 for most instruments.

7.3
STADIA HAIRS AND STADIA RODS

The telescopes of transits, theodolites, plane-table alidades, and many levels are furnished with stadia hairs in addition to the regular cross hairs. One stadia hair is above and the other an equal distance below the horizontal cross hair. Stadia hairs usually are mounted on the same reticule and in the same plane as the horizontal and vertical cross hairs. Under these conditions, the stadia hairs are not adjustable and the distance between the hairs remains unchanged. Figure 7.1 shows two styles of stadia cross hairs: the stub cross hair (Figure 7.1a), and the stadia hairs, which go completely across the cross-hair ring (Figure 7.1b).

A conventional rod such as the Philadelphia rod is adequate for stadia work where the sights do not exceed 60 m or 200 ft. For longer sights, stadia rods are advisable and provide more accurate results. Stadia rods are described in Section 5.25 and illustrated in Figure 5.23.

7.4
PRINCIPLE OF STADIA

Figure 7.2a illustrates the principle on which the stadia method is based. In the figure, the line of sight of the telescope is horizontal and the stadia rod is vertical. The stadia hairs are

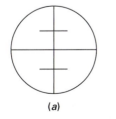
(a)

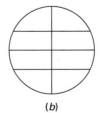
(b)

FIGURE 7.1
Two styles of stadia hairs.

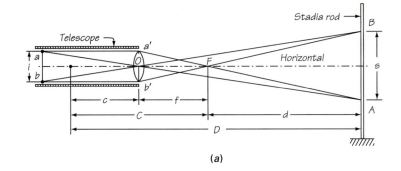

(a)

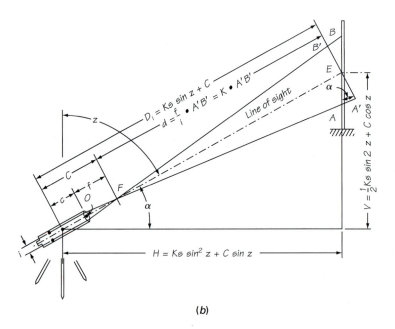

(b)

FIGURE 7.2
(a) Horizontal stadia sight. (b) Inclined stadia sight.

indicated by points a and b. The distance between the stadia hairs is i. The apparent locations of the stadia hairs on the rod are points A and B and the stadia interval is s.

Rays from a passing through the optical center of the lens O and the focal point of the lens F are brought to a focus at A. Similarly, the reverse is true, so that rays from A that pass through F and O are brought to a focus at a.

Because $ab = a'b'$, by similar triangles

$$\frac{f}{i} = \frac{d}{s}$$

Hence, the horizontal distance from the principal focus to the rod is

$$d = (f/i)s = Ks$$

in which $K = f/i$ is a coefficient called the *stadia interval factor,* which for a particular instrument is a constant as long as conditions remain unchanged. So, for a horizontal sight the distance from principal focus to rod is obtained by multiplying the stadia interval factor by the stadia interval. The horizontal distance from center of instrument to rod then is

353

CHAPTER 7:
Combined
Distance and
Angular
Measurement
Systems

$$D = Ks + (f + c) = Ks + C \qquad (7.1)$$

in which C is the distance from center of instrument to principal focus. This formula is employed in computing horizontal distances from stadia intervals when sights are horizontal.

7.5
STADIA CONSTANTS

The focal distance f is a constant for a given instrument. It can be determined with all necessary accuracy by focusing the objective on a distant point and measuring the distance from the cross-hair ring to the objective. The distance c, though a variable depending on the position of the objective, for all practical purposes may be considered a constant. Its mean value can be determined by measuring the distance from the vertical axis to the objective when the objective is focused for an average length of sight.

Usually, the value of $C = f + c$ is determined by the manufacturer and stated on the inside of the instrument box. For external-focusing telescopes, under ordinary conditions, C may be considered as 1 ft without error of consequence. Internal-focusing telescopes (Section 5.8) are so constructed that C is 0 or nearly so; this is an important advantage of internal-focusing telescopes for stadia work. Most instruments now used for stadia are equipped with internal-focusing telescopes.

7.6
STADIA INTERVAL FACTOR

The nominal value of the stadia factor $K = f/i$ usually is 100. The interval factor can be determined by observation. The usual procedure is to set up the instrument in a location where a horizontal sight can be obtained. With a tape lay off, from a point $C = f + c$ in front of the center of the instrument, distances of 30 m, 60 m, and so forth, up to perhaps 300 m, and set stakes at the points established. The stadia rod then is held on each of the stakes and the stadia interval is read. The stadia interval factor is computed for each sight by dividing the distance from the principal focus to the stake by the corresponding stadia interval, and the mean is taken as the best estimate. Owing to errors in observation and perhaps from natural sources, the values of K for the several distances are not likely to agree exactly.

To overcome any prejudical tendencies by the instrument operator, observations may be made on the rod held on stakes at random distances from the instrument; these distances being measured later with the tape.

For use on long sights, where the full stadia interval would exceed the length of the rod, the stadia interval factor may be determined separately for the upper stadia hair and horizontal cross hair and for the lower stadia hair and horizontal cross hair.

7.7
INCLINED SIGHTS

In stadia surveying, most sights are inclined, and usually it is desired to find both the horizontal and the vertical distances from instrument to rod. The problem therefore resolves itself into finding the horizontal and vertical projections of an inclined line of sight. For convenience in field operations, the rod is always held vertical.

Figure 7.2b illustrates an inclined line of sight, AB being the stadia interval on the vertical rod and $A'B'$ being the corresponding projection normal to the line of sight. The length of the inclined line of sight from center of instrument is

$$D_i = \frac{f}{i}(A'B') + C \tag{7.2}$$

For all practical purposes, the angles at A' and B' may be assumed to be 90°. Let $AB = s$; then $A'B' = s \sin z = s \cos \alpha$. Making this substitution in Equation (7.2), and letting $K = f/i$, the inclined distance is

$$D_i = Ks \sin z + C \tag{7.3a}$$

$$D_i = Ks \cos \alpha + C \tag{7.3b}$$

The horizontal component of this inclined distance is

$$H = Ks \sin^2 z + C \sin z \tag{7.4a}$$

$$H = Ks \cos^2 \alpha + C \cos \alpha \tag{7.4b}$$

which is the general equation for determining the horizontal distance from center of instrument to rod, when the line of sight is inclined. In Equations (7.3) and (7.4), z is the zenith angle and α is the vertical angle measured at the instrument.

The vertical component of the inclined distance is

$$V = Ks \sin z \cos z + C \cos z \tag{7.5a}$$

$$V = Ks \cos \alpha \sin \alpha + C \sin \alpha \tag{7.5b}$$

The equivalents of $\sin z \cos z$ and $\cos \alpha \sin \alpha$ are conveniently expressed in terms of double the angles z and α, or

$$V = \tfrac{1}{2} Ks \sin 2z + C \cos z \tag{7.6a}$$

$$V = \tfrac{1}{2} KS \sin 2\alpha + C \sin \alpha \tag{7.6b}$$

which are the general equations for determining the difference in elevation between the center of the instrument and the point where the line of sight cuts the rod. To determine the difference in ground elevations, the height of instrument and the rod reading of the line of sight must be considered.

Equations (7.5) and (7.6) are known as the *stadia formulas for inclined sights*.

7.8
PERMISSIBLE APPROXIMATIONS

More approximate forms of the stadia formulas are sufficiently precise for most stadia work. Usually, distances are computed only to decimeters or feet and elevations to centimeters or tenths of feet. Under these conditions, for side shots where vertical angles are less than 3°, Equation (7.4) for horizontal distances may properly be reduced to the form

$$H = Ks + C \tag{7.7}$$

355

CHAPTER 7:
Combined
Distance and
Angular
Measurement
Systems

which is the same as for horizontal sights (Section 7.4). But for traverses of considerable length, owing to the systematic error introduced, this approximation should not be made for vertical angles greater than perhaps 2°.

Owing to unequal refraction and to accidental inclination of the rod, observed stadia intervals in general are slightly too large. To offest the systematic errors from these sources, frequently on surveys of ordinary precision, the constant C is neglected. Hence, in any ordinary case Equations (7.4) may be expressed, with sufficient precision, in the form

$$H = Ks \sin^2 z = Ks \cos^2 \alpha \qquad \text{(approximately)} \qquad (7.8)$$

Also, Equations (7.6) often may be expressed with sufficient precision for ordinary work in the form

$$V = \tfrac{1}{2} Ks \sin 2z = \tfrac{1}{2} Ks \sin 2\alpha \qquad \text{(approximately)} \qquad (7.9)$$

However, the error in elevation introduced through using Equation (7.9) may not be negligible, as for large vertical angles it amounts to several tenths of a foot or up to a decimeter.

Equations (7.8) and (7.9) are simple in form and most generally are employed.

When K is 100, the common practice is to multiply mentally the stadia interval by 100 at the time of observation and record this value in the field notebook. This distance Ks is often called the *stadia distance*. Therefore, if the stadia interval were 7.375 m, the stadia distance recorded would be 737.5 m.

7.9
STADIA REDUCTIONS

Horizontal distance and difference in elevation can be computed from stadia observations by using the exact Equations (7.4) and (7.5) or (7.6) or approximate Equations (7.8) and (7.9).

EXAMPLE 7.1. The following data were obtained by stadia observation: zenith angle $= 81°50'$, $s = 2.50$ ft. The stadia interval factor is known to be 100 and $C = 0.75$ ft. Calculate H and V.

Solution. By exact Equations (7.4) and (7.6a),

$$H = Ks \sin^2 z + C \sin z$$

$$= (100)(2.50)(\sin^2 81°50') + (0.75)(\sin 81°50')$$

$$= (100)(2.50)(0.9798) + (0.75)(0.9899)$$

$$= 245.0 + 0.7$$

$$= 245.7 = 246 \text{ ft.}$$

$$V = \tfrac{1}{2} Ks \sin 2z + C \cos z$$

$$= \tfrac{1}{2}(100)(2.50)(0.2812) + (0.75)(0.1421)$$

$$= 35.2 + 0.11$$

$$= 35.3 \text{ ft}$$

If approximate Equations (7.8) and (7.9) are used and $C = 1.0$ is assumed, we may add 0.01 ft to the stadia interval. Thus, $s = 2.50 + 0.01 = 2.51$:

$$H = (100)(2.51)(\sin^2 81°50') = (251)(0.9798)$$

$$= 246 \text{ ft}$$

$$V = \tfrac{1}{2}(100)(2.51) \sin(2 \times 81°50') = (125.5)(0.2812)$$

$$= 35.3 \text{ ft}$$

and the approximate equations are seen to yield adequate accuracy for these data.

EXAMPLE 7.2. The stadia interval was 1.372 m at a vertical angle of 20°32'. $C = 0.30$ m and the stadia interval factor is 100. Calculate H and V.

Solution. By Equation (7.4).

$$H = (100)(1.372)(\cos^2 20°32') + (0.30)(\cos 20°32')$$

$$= 120.6 \text{ m}$$

By Equation (7.6b)

$$V = \tfrac{1}{2}(100)(1.372)(\sin 41°04') + (0.30)(\sin 20°32')$$

$$= 45.17 \text{ m}$$

If approximate Equations (7.8) and (7.9) are used for this problem, add 0.003 to the stadia interval so that

$$s = 1.372 + 0.003 = 1.375$$

and

$$H = 100(1.375) \cos^2 \alpha = 120.6 \text{ m}$$

$$V = \tfrac{1}{2}(100)(1.375) \sin 41°04' = 45.16 \text{ m}$$

Use of the approximate equations for a solution with the large vertical angle resulted in an error of 1 cm in the difference in elevation. Had the telescope of the instrument been internally focusing, as in most modern transits and theodolites, C would be equal to 0 and approximate Equations (7.8) and (7.9) could be used with no danger of error.

7.10
OBSERVATION OF STADIA INTERVAL

On transit or plane-table surveys the stadia interval usually is determined by setting the lower stadia hair on a foot or decimeter mark and then reading the location of the upper stadia hair. Thus, the stadia interval is mentally computed more easily and with less chance of mistake than if the lower hair were allowed to take a random position on the rod. When the vertical angle is taken to a given mark on the rod, the corresponding stadia interval is observed with the lower hair on the foot or decimeter mark that renders a minimum displacement of the horizontal cross hair from the mark to which the vertical angle is referred.

357

CHAPTER 7:
Combined
Distance and
Angular
Measurement
Systems

Thus, if a vertical angle were taken with the line of sight cutting the rod at 4.9 ft and the lower stadia hair fell at 2.3 ft, the telescope would be rotated about the horizontal axis until the lower hair was at 2.0 ft. The center horizontal cross hair then would fall at 4.6 ft.

On metric rods it is more convenient to set the bottom cross hair on the nearest even decimeter. For example, if the vertical angle were taken when the middle cross hair is on 1.495 m and the lower stadia hair was at 1.23, rotate the telescope in a vertical plane about the horizontal axis until the lower stadia hair coincides with 1.200 and read the top stadia hair as 1.730 m. The interval then is $1.730 - 1.200 = 0.530$ m.

Whenever the stadia interval is in excess of the length of the rod, the separate half-intervals are observed and their sum is taken.

For precise stadia work, the readings may be made by means of two targets on the rod, one target on, say, the 0.500-m or 2-ft mark and the other set by the rodperson as directed by the instrument operator. To avoid excessive effects of atmospheric refraction, the intercept of the lower cross hair with the rod should fall no nearer the ground than necessary.

Note forms that can be used for recording stadia data in stadia traverses (Section 8.11) and topographic surveys by the stadia method (Section 15.12) are shown in Figures 8.14 and 15.6, respectively.

7.11
DIFFERENCE IN ELEVATION

In Figure 7.3 the instrument is at station A with a height of instrument, or h.i., above A equal to AB and the rod is held in a vertical position at station D. It is desired to determine the difference in elevation between points A and D given stadia observations and the zenith or vertical angles z or α. First, consider the problem when the difference in elevation between points A and D on the ground is desired. This situation occurs in stadia traversing and topographic surveying from a known point. The h.i. $= AB$, the stadia interval s, the zenith angle z or vertical angle α with the middle cross hair set on E, and the rod reading DE are observed and recorded. The difference in elevation between the telescope and E, which is equal to V, can be calculated by using Equation (7.5a), (7.5b), or (7.9). The difference in elevation is Δel. $= AB + V - DE$ or Δ el. $=$ h.i. $+ V -$ rod reading at D. Therefore, the elevation at $D =$ the elevation at $A + (AB + V - DE) =$ h.i. at $B + V - DE$. If the middle cross hair is sighted on the rod such that $DE = AB =$ h.i., then Δ el. $= V$ and the elevation at $D =$ the elevation at $A + V$. Obviously, setting the horizontal cross hair on

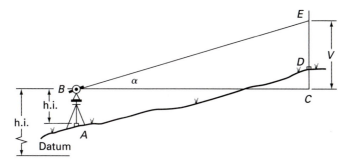

FIGURE 7.3
Difference in elevation between instrument and rod stations.

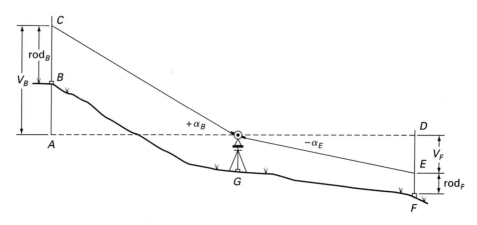

FIGURE 7.4
Difference in elevation between two points.

the rod at a value equal to the h.i. prior to reading the vertical angle simplifies calculating the change in elevation between the transit station and the point on which the rod is held. Also, note that it is important that the rod reading always be recorded.

Figure 7.4 illustrates the situation where the difference in elevation is desired between points B and F utilizing an intermediate instrument set up at G. Assume that a backsight is taken on a rod held at B. The stadia interval, rod reading BC, and α_B are recorded. Using these data, V_B can be calculated. Next, a foresight is taken on the rod held at F. The stadia interval, rod reading FE, and vertical angle $-\alpha_F$ are observed and recorded so that V_F can be computed. A general expression for difference in elevation between two points B (the backsight) and F (the foresight) is

$$\Delta el._{BF} = rod_B \pm V_B \pm V_F - rod_F \qquad (7.10)$$

in which the sign on V_B is opposite the sign of α_B and the sign of V_F corresponds to the sign on α_E. Therefore, in Figure 7.4, $\Delta el._{BF} = BC - V_B - V_F - EF$.

In the reverse direction, if the backsight is on F and the foresight on B, then $\Delta el. = EF + V_F + V_B - BC$.

7.12
INDIRECT LEVELING BY STADIA

Where the required precision is low and the country is rolling or rough, the stadia method of indirect leveling is rapid. The theodolite preferably should be equipped with a vertical circle compensator or a sensitive control level for the vertical circle so that the collimation error can be readily eliminated.

With an optical-reading theodolite having a vertical circle reading to 20″, differences in elevation usually are computed only to the nearest 0.03 m or to the nearest 0.1 ft. The average length of sight in stadia surveying generally is considerably greater than in differential leveling.

In running a line of levels by this method, the transit is set up in a convenient location. A backsight is taken on the rod held at the initial bench mark, first by observing the stadia interval then by measuring the zenith or vertical angle to some arbitrarily chosen mark on

359

CHAPTER 7:
Combined
Distance and
Angular
Measurement
Systems

STADIA LEVELS FOR INDIAN						TRAIL RECONNAISANCE					July 20, 1995 Fair 70°F J.L.Black V.A.Alden, Rod
Station	Backsight					Foresight					Theod. No. 1564
(1)	(2)	(3)	(4)	(5)	(6)	(7)	(8)	(9)	(10)	(11)	
	Inverval, m	Zenith angle	$(rod)_B$, m	V_B	Interval	Zenith angle	$(rod)_F$	V_F	Δ Elev	Elev m	
BM 42	1.286	90°06'00"	0.98	+0.22						1820.39	
TP1	0.963	90°00'00"	2.96	0.00	0.805	85°44'00"	1.34	+5.97	+5.83	1826.22	C = 0
TP2	1.234	92°15'00"	1.83	+4.84	1.344	88°58'00"	2.35	+2.42	+3.03	1829.25	
TP3					1.201	87°40'00"	1.83	+4.88	+9.72		K = 100

FIGURE 7.5
Notes for stadia leveling.

the rod. A turning point then is established in advance of the transit, and similar observations are taken, the vertical angle being measured with the middle cross hair set on the same mark as before. The transit is moved to a new location in advance of the turning point, and the process is repeated. The stadia distances and vertical angles are recorded, as is the rod reading, which is used as an index when vertical angles are measured. If it is impractical to sight at this chosen index reading, the zenith or vertical angle is measured to some other graduation and this rod reading is recorded in the notes.

Figure 7.5 shows a form of notes for stadia leveling. The bench mark sighted for a backsight is recorded in column 1 and its known elevation is shown on the same line in column 11. The stadia interval for the backsight is recorded in column 2, the zenith angle in column 3, and the rod reading, rod_B, where the middle cross hair cuts the rod, is given in column 4. Similar readings for the foresight are recorded one line down in columns 6, 7, and 8. The values V_B and V_F (Figure 7.4) for backsight and foresight are calculated by Equation (7.9) and are recorded in columns 5 and 9, respectively. The sign of V_B is opposite the sign yielded by Equation (7.9), whereas the sign of V_F corresponds to the sign yielded by Equation (7.9). The values of rod_B, V_B, V_F, and rod_F then are used in Equation (7.10) to calculate Δel. between the point on the ground where the rod was held for the backsight and the point on the ground where the rod was held for the foresight. This value is recorded in column 10. Therefore, for the notes in Figure 7.5 for the first instrument setup between B.M.$_{42}$ and T.P.$_{1}$, Δel. $= 0.98 + 0.22 + 5.97 - 1.34 = 5.83$ m. This Δel. (recorded in column 10) is applied to the elevation on the ground at the backsight (B.M.$_{42}$) to yield the elevation on the ground at the foresight (T.P.$_{1}$). This method of recording the data then is continued for subsequent setups, as indicated in Figure 7.5.

7.13
ERRORS IN STADIA

Many of the errors of stadia are those common to all similar operations of measuring horizontal angles (Sections 6.44 to 6.47) and differences in elevation (Section 5.40). Sources of error in horizontal and vertical distances computed from observed stadia intervals are as follows:

1. *Stadia Interval factor is not that assumed.* This condition produces a systematic error in distances, the error being proportional to that in the stadia interval factor. The case is

parallel to that of the tape which is too long or too short. When the value of the interval factor is closely determined by observations as described in Section 7.6 and the stadia measurements are taken under similar conditions, the error from this source may be made negligible.

2. *Rod is not of standard length.* If the spaces on the rod are uniformly too long or too short, a systematic error proportional to the stadia interval is produced in each distance. Errors from this source may be kept within narrow limits if the rod is standardized and corrections for erroneous length are applied to observed stadia intervals. Except for stadia surveys of more than ordinary precision, errors from this source usually are of no consequence.

3. *Incorrect stadia Interval.* The stadia interval varies randomly owing to the inability of the instrument operator to observe the stadia interval exactly. In a series of connected observations (such as a traverse) the error may be expected to vary as the square root of the number of sights. This is the principal error affecting the precision of distances. It can be kept to a minimum by proper focusing to eliminate parallax, by taking observations at favorable times, and by care in observing. Where high precision is required, stadia measurements may be taken by sighting on a rod with two targets, one fixed and the other movable.

4. *Rod is not plumb.* This condition produces a small error in the vertical angle. It also produces an appreciable error in the observed stadia interval and hence in computed distances, the error being greater for large vertical angles than for small angles. It can be eliminated by using a rod level.

5. *Unequal refraction.* Unequal refraction of light rays in layers of air close to the earth's surface affects the sight on the lower stadia hair more than the sight on the upper stadia hair and thus introduces systematic positive errors in stadia measurements. Although errors from this source are of no consequence in ordinary stadia surveying, they may be important on more precise surveys. The periods most favorable for equal refraction are at times when it is cloudy or, if the sun is shining, during the early morning or late afternoon. On precise stadia surveys where it is necessary to work under a variety of atmospheric conditions, it is appropriate to determine the stadia interval factor for each condition and apply the factor to all observations taken under that condition. Whenever atmospheric conditions are unfavorable, the sights should not be taken near the bottom of the rod.

6. *Errors in zenith or vertical angles.* Errors in vertical circle readings are relatively unimportant in their effects on *horizontal distances*. For example, analysis of Equation (7.8) shows that an uncertainty of $01'$ in a zenith or vertical angle of $85°$ or $5°$, respectively, yields discrepancies of 0.005 m in a 100-m sight and 0.02 ft in a 300-ft sight. Similarly, an error of $01'$ in a zenith or vertical angle of $75°$ or $15°$, respectively, produces discrepancies of 0.02 m in a 100-m sight and 0.05 ft in a 300-ft sight. With respect to differences in elevation, analysis of Equation (7.9) reveals that an uncertainty of $01'$ in zenith or vertical angles of $85°$ or $5°$, respectively, results in discrepancies in elevation differences of 0.03 m for a 100-m sight and 0.1 ft for a 300-ft sight. Uncertainty in the stadia interval, especially at higher angles of inclination, will have a more pronounced effect on elevation differences than errors in zenith or vertical angles.

7.14
ERROR PROPAGATION IN STADIA

Estimation of accuracies possible in horizontal distances and differences in elevation calculated from stadia observations are necessary for the planning of stadia surveys. Evaluation of these accuracies is best achieved by error propagation using the basic equations developed in Sections 7.4 and 7.7.

361

CHAPTER 7:
Combined
Distance and
Angular
Measurement
Systems

If the stadia rod is standardized and proper corrections applied for erroneous length, if the stadia interval factor K is carefully determined, if the instrument constant $C = 0.0$ or is carefully determined, and if the rod is plumbed using a rod level, the major source of error affecting both horizontal and vertical distances is that of observing the stadia interval. Errors in observing the vertical angle have an effect but are of secondary importance.

Horizontal Distance Error Propagation

Propagate the error through Equation (7.4a) using Equation (2.16) in Section 2.15. Thus, the estimated variance in the horizontal distance H is

$$\sigma_H^2 = \left(\frac{\partial H}{\partial s}\right)^2 \sigma_s^2 + \left(\frac{\partial H}{\partial z}\right)^2 \sigma_z^2 \tag{7.11}$$

where

$$\frac{\partial H}{\partial s} = K \sin^2 z \quad \text{and} \quad \frac{\partial H}{\partial z} = -(Ks \sin 2z + C \cos z)$$

are obtained by partial differentiation of Equation (7.4a). Assume that $\sigma_s = 0.003$ m or 0.01 ft and $\sigma_z = 20''$ or $\sigma_z = (20'')(\text{arc } 01'') = (20'')(0.000004848) = 0.00009696$ rad. Further, assuming that $s = 1.00$ m, $z = 85°$, $C = 0.0$, and $K = 100$, the estimated variance and standard deviation in H are

$$\sigma_H^2 = (100 \sin^2 85°)^2(0.003)^2 + [(100)(1) \sin 170°]^2(0.00009696)^2$$
$$\sigma_H^2 = 0.0886 + 0.000002 = 0.2977 \text{ m}^2$$
$$\sigma_H = 0.3 \text{ m in a 100-m sight}$$
or
$$\sigma_H = 0.99 \text{ ft} = 1.0 \text{ ft in a 300-ft sight}$$

A similar analysis using vertical angles of $10°$ and $15°$ yields estimated values for σ_H that are virtually the same. The contribution of observational error in the zenith angle (the second term) does become somewhat larger with the higher vertical angles but still is insignificant. Even if $\sigma_z = 05'$, the second term in Equation (7.11) is only 0.00064 m^2 and contributes little to the propagated standard deviation of 0.28 m, assuming a vertical angle of $15°$. Therefore, the error in horizontal distance determined from stadia is primarily dependent on the magnitude of the error in the stadia interval. A value of σ_s of 0.003 m or 0.01 ft is quite reasonable for sights up to 120 m or 400 ft. For this reason, the maximum relative accuracy that can be expected in horizontal stadia distances is 1 part in 300 to 1 part in 400.

Differences in Elevation

The difference in elevation between the center of the telescope and the point on which the rod is held is found by Equation (7.6), Section 7.7. Under similar assumptions as made for horizontal distance, the estimated variance in V by propagating the errors through Equation (7.6) using Equation (2.16) in Section 2.15 is

$$\sigma_V^2 = \left(\frac{\partial V}{\partial s}\right)^2 \sigma_s^2 + \left(\frac{\partial V}{\partial z}\right)^2 \sigma_z^2 \tag{7.12}$$

where

$$\left(\frac{\partial V}{\partial s}\right) = \frac{1}{2}K \sin 2z \quad \text{and} \quad \left(\frac{\partial V}{\partial z}\right) = Ks \cos 2z + C \sin z$$

Assuming that $s = 1.00$ m, $z = 85°$, $K = 100$, $C = 0.0$, $\sigma_s = 0.003$ m or 0.01 ft, and $\sigma_z = 20''$, the estimated variance and standard deviation by Equation (7.12) are

$$\sigma_V^2 = \left[\frac{1}{2}(100) \sin 170°\right]^2 (0.003)^2 + [(100)(1.00)(\cos 170°)]^2(0.00009696)^2$$

$$= 0.000678 + 0.00009 = 0.0007696 \text{ m}^2$$

$$\sigma_V = 0.028 \text{ in a 100-m sight}$$

or

$$\sigma_V = 0.12 \text{ ft in a 300-ft sight}$$

If $z = 75°$ is assumed,

$$\sigma_V^2 = [(50)(\sin 150°)]^2(0.003)^2 + [(100)(\cos 150°)]^2(0.00009696)^2$$

$$= 0.005625 + 0.00007 = 0.005696 \text{ m}^2$$

$$\sigma_V = 0.075 \text{ m in a 100-m sight}$$

or

$$\sigma_V = 0.26 \text{ ft in a 300-ft sight}$$

So, for zenith angles of about 85°, which is normal for most stadia operations, the effects of errors in stadia intervals dominate the total propagated error in V. The effects of errors in zenith or vertical angles are virtually insignificant. For very steep angles, which do occur in rugged terrain, the effects of errors in the stadia interval on V increase and the effects of errors in zenith and vertical angles on V remain about the same.

Consequently, if care is exercised in handling the equipment and observing the stadia interval, differences in elevation between the center of the telescope and the point where the rod is held can be determined by stadia for sights of up to 100 m or 300 ft to within ±0.030 m or 0.1 ft, respectively. However, when the vertical angles are steep, the estimated standard deviation is approximately doubled and, to maintain this accuracy, extreme care must be exercised in observing the stadia interval.

Differences in elevation by stadia leveling are determined with Equation (7.10), using values of V and rod readings from backsights and foresights. The estimated standard deviation in elevation change between two points is found by propagating the errors through Equation (7.10) to give

$$\sigma_{\Delta el.}^2 = \left(\frac{\partial \Delta el.}{\partial V_B}\right)^2 \sigma_{V_B}^2 + \left(\frac{\partial \Delta el.}{\partial V_F}\right)^2 \sigma_{V_F}^2 + \left(\frac{\partial \Delta el.}{\partial rod_B}\right)^2 \sigma_{rod_B}^2 + \left(\frac{\partial \Delta el.}{\partial rod_F}\right)^2 \sigma_{rod_F}^2$$

$$= \sigma_{V_B}^2 = \sigma_{V_F}^2 + \sigma_{rod_B}^2 + \sigma_{rod_F}^2$$

Assuming that $\sigma_{rod_B} = \sigma_{rod_F} = 0.003$ m or 0.01 ft, sights of 100 m or 300 ft, and using values of σ_{V_B} and σ_{V_F} from previous examples, yields

$$\sigma_{\Delta el.}^2 = 2\sigma_V^2 + 2\sigma_{rod}^2$$

$$\sigma_{\Delta el.}^2 = 2(0.0007696) \text{ m}^2 + 2(0.003)^2 = 0.0016 \text{ m}^2$$

$$\sigma_{\Delta el.} = 0.04 \text{ m assuming backsights and foresights of 100 m}$$

or

$$= 0.18 \text{ ft with backsights and foresights of 300 ft}$$

363

CHAPTER 7:
Combined
Distance and
Angular
Measurement
Systems

7.15
TOTAL STATION SYSTEMS

The total station system consists of a short- to medium-range EDM instrument installed in the framework of an electronic theodolite with all components under the control of a built-in microprocessor. The result is a single instrument that permits observing distance and directions from a single setup. Distance measurement by EDM was discussed in Sections 4.29 to 4.32 and theodolites and electronic theodolites were described in Section 6.30. References already have been made to applications of total station systems in Sections 1.1, 3.12, 5.5, 5.55, and 6.21. Obviously, these systems constitute a major component in the surveyor's current arsenal of equipment. In the sections that follow, the features, functions performed, and use and applications of total station systems and their data collectors are discussed.

7.16
FEATURES OF TOTAL STATION SYSTEMS

Total station systems can be classified as modular and self-contained. One form of a modular system is illustrated in Figure 4.11b, where a lightweight EDM unit is attached to the standards of an electronic theodolite. Strictly speaking, current use of the term *modular* means that the EDM optical path is separate from the optical path of the theodolite. In a *self-contained* total station system, a compact EDM device is fitted around the theodolite's telescope so that the EDM and the sighting optics are *coaxial*. Modular systems can be fitted with more powerful EDM devices and have greater ranges in distance. However, they are heavier, bulkier, and less adaptable to automation. Self-contained systems are lighter, more compact, easier to operate, and much more adaptable to automation. In 1996–1997, 99 different models of total station systems were on the market (Crawford, 1997). Of these, four were modular systems. Comments in this chapter are focused primarily on self-contained systems with varying degrees of automation.

All of these systems contain three components: the distance measuring unit or EDM, the angle measuring device or theodolite, and an onboard microprocessor. The EDM units, with infrared carrier beams, have ranges of from 300 to 3300 m (1,000 to 11,000 ft) to a single reflector and 900 to 11,000 m (3,000 to 36,000 ft) to a triple prism. One EDM with a laser carrier beam (a modular system) has a range of 20,000 m (66,000 ft) to one reflector and 40,000 m (131,000 ft) to a triple reflector. Specified accuracies in distance measurement are from 1 mm + 2 ppm (0.003 ft + 2 ppm) to 10 cm + 50 ppm (0.032 ft + 50 ppm). The time required for single measurement of a distance is 0.25 to 5 sec.

The angle measuring unit essentially is an electronic theodolite (Section 6.30). Therefore, the horizontal and vertical circles are encoders scanned by a photoelectric cell to provide an output that can be converted to angular measurements determined by *absolute* or *incremental* decoding. For absolute decoding, a reference point detectable by an electronic sensor is marked on the graduated circle and the angle is determined relative to this *zero mark*. The encoders for incremental decoding are graduated in unnumbered gratings, and the number of gratings involved in an angle is determined from whole circle scanning. The advantages of the incremental method are that its repeatability gives a strong indication of sensor performance and fewer sensors are required. Most total station systems have incremental decoding. In either case, it is possible to assign 0 degrees, or any desired angle, to the pointing after the sight has been taken. The displayed resolutions of the horizontal and vertical circles vary from 0.03 to 20″.

A total station instrument is shown in Figure 7.6. Note that this device is similar to the electronic theodolite (Section 6.30, Figure 6.33) but somewhat heavier (about 5 to 8 kg or

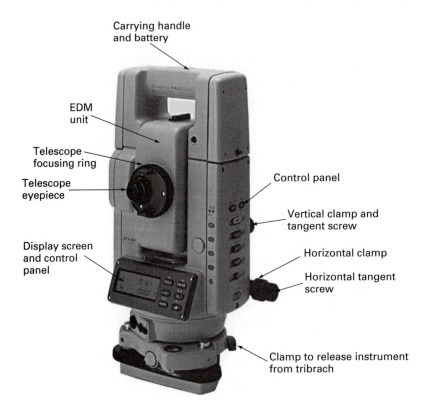

Carrying handle
and battery

EDM
unit

Telescope
focusing ring

Telescope
eyepiece

Control panel

Vertical clamp and
tangent screw

Display screen
and control
panel

Horizontal clamp

Horizontal tangent
screw

Clamp to release instrument
from tribrach

FIGURE 7.6
Total station instrument, Topcon GTS-304. (*Courtesy of Topcon America Corporation.*)

12 to 18 lb) and bulkier, due mainly to the EDM unit that surrounds the telescope and the additional control buttons on the control panel and standard. The instrument illustrated in Figure 7.6 is a direction instrument and has only one horizontal clamp and tangent screw. The basic operations of measurement, zero set, hold, and enter are located on the control panel and keys for additional functions are found on the right-hand standard. Because this is a single-center instrument, angles to be repeated are cumulated electronically by proper manipulation of the hold button and the horizontal clamp and tangent screw. A slightly different style of total station instrument and a data collector are illustrated in Figure 7.7. This instrument is similar to the double-center repeating theodolite (Section 6.22) and has both upper and lower horizontal clamps and tangent screws so that angles to be repeated are cumulated by proper manipulation of these upper and lower motions (Section 6.35).

A sectional view of the instrument pictured in Figure 7.7 is shown in Figure 7.8. Note the vertical and horizontal circles, upper and lower horizontal clamps and tangent screws, vertical clamp and tangent screw, circular and plate-level bubbles, inner and outer centers, optical plummet, and leveling screws, all similar to the repeating optical theodolite (Figure 6.23). Absent are the optical trains for viewing angles on vertical and horizontal circles,

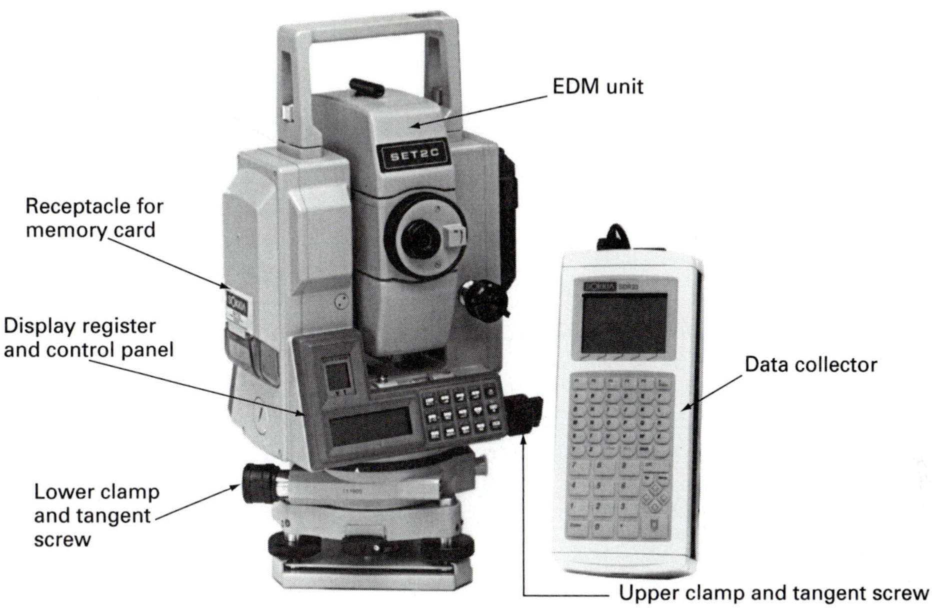

365

CHAPTER 7:
Combined
Distance and
Angular
Measurement
Systems

FIGURE 7.7
Total station instrument, Sokkia Set 2C and data collector SDR 33. (*Courtesy of Sokkia Corporation.*)

replaced by the electronics needed for incremental digital reading of the vertical and horizontal circles and for measurement of distance by EDM.

The illustrated instruments also have vertical axis compensation, and many currently available systems are equipped with dual-axis compensation (Section 6.30). Another feature found on some systems is an onboard memory module or card for data storage, exchange, and downloading. Some of these cards contain software that permits extensive onboard processing of the data. On the system shown in Figure 7.7, the memory card is contained in the receptacle on the left-hand standard.

Because distance and direction are obtained from these systems, short sights are taken to a reflector, mounted on either a tripod or a prism pole. In this case, the center of the reflector serves as the target for horizontal and vertical circle readings. For longer sights, a separate target is needed for accurate horizontal and vertical alignment. If the target is mounted at a height above the station sighted (h.i.), different from that of the reflector, care must be exercised to record and enter both h.i.'s, so that reduction of the slope distance to the horizontal distance and difference in elevation is done properly (Sections 4.41 and 5.55). Preferably, the target would consist of a combined reflector and target similar to the one shown in Figure 7.9. This assembly has a target mounted symmetrically about the reflector that can be tilted to accommodate steeply inclined lines of sight, thus eliminating errors in distance due to lack of perpendicularity of the ray path to the face of the reflector. The reflector also has a sighting collimator that permits accurate alignment of the reflector on the total station system.

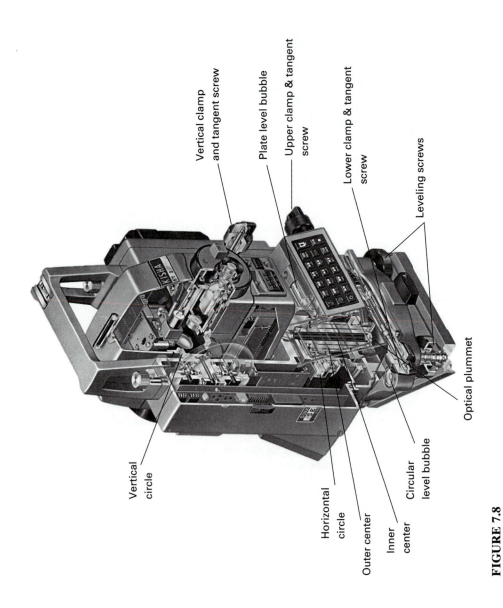

FIGURE 7.8
Sectional view of a total station instrument, Sokkia Set 2. (*Courtesy of Sokkia Corporation.*)

Vertical clamp
and tangent screw

Plate level bubble

Upper clamp & tangent
screw

Lower clamp & tangent
screw

Leveling screws

Optical plummet

Vertical
circle

Horizontal
circle

Outer center

Inner
center

Circular
level bubble

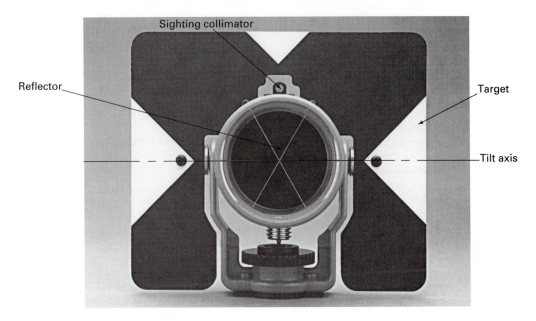

FIGURE 7.9
Combined reflector and target. (*Courtesy of Sokkia Corporation.*)

7.17
FUNCTIONS OF TOTAL STATION SYSTEMS

The simplest and least costly total station systems permit measurement of distance, displayed on the instrument's register as the slope or horizontal distance, and horizontal and zenith or vertical angles. Parameters for atmospheric conditions and system constants also can be entered to correct for the systematic errors in distance measurement by EDM (Sections 4.37 and 4.38). In addition, most such systems have an interface that allows attaching a data collector (Figure 7.7), which permits electronic collection of measurements.

The more sophisticated and, hence, more expensive systems contain a wide array of onboard options and routines to correct various instrumental systematic errors and process collected measurements. Some of the parameters that can be entered through the control panel are (1) units of distance and angular measurement in meters or feet and degrees or gons, respectively; (2) temperature, °F or °C, and atmospheric pressure, millemeters or inches of mercury (some systems have built-in sensors for temperature and atmospheric pressure); (3) prism offset or other system constants; (4) h.i. of instrument and reflector; and (5) point numbers and point incrementation. Examples of typical built-in options for processing measured distances and angles are these:

Distance (horizontal and vertical)

1. A tracking mode that yields distance at 0.4 sec intervals to a moving reflector.
2. A stakeout mode in which the difference between a desired distance, keyed in via the control panel, and the measured distance is displayed, allowing directions to be given to the person at the reflector in a stakeout.

3. An automatic correction of distance and difference in elevation for curvature and refraction (Section 5.2).
4. Averaged distances, where the number of repetitions can be specified.
5. Trigonometric leveling.

Angle measuring

1. Stores collimation and vertical circle index corrections after direct and reversed measurements.
2. Automatically senses alignment errors in the horizontal or trunnion axis and stores this value for correction of subsequently measured horizontal angles.
3. Sensors on opposite sides of the horizontal and vertical circles eliminates eccentricity erros in angles.
4. Displays the average of repeated angles, where the number of repetitions can be specified.

Other functions for measuring, setting out, and recovering points

1. Determination of orthogonal distances from a known reference line. In Figure 7.10a, station A, occupied by a total station system, is a point having known coordinates from which reference line BC is established. When angles and distances are measured from A to points 1, 2, . . . , 4, this routine automatically provides orthogonal offsets d_1, d_2, . . . , d_4.
2. Determination of the components of missing lines between two successive target points such as points 1 and 2 observed from known station G in Figure 7.10b. Given the distances and directions of G-1 and G-2, s_{12}, d_{12}, and $\Delta \text{el.}_{12}$ can be obtained.
3. Determination of the elevation of, or lateral distance to, a remotely located point above, below, or to one side of the reflector, as illustrated in Figure 7.10c. In this example,

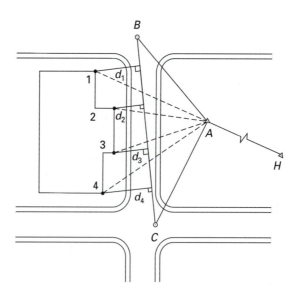

(*a*) Orthogonal distances from a known reference line.

FIGURE 7.10
Other functions for measuring, recovering, and setting out survey points: (*a*) Orthogonal distances from a known reference line.

369

CHAPTER 7:
Combined
Distance and
Angular
Measurement
Systems

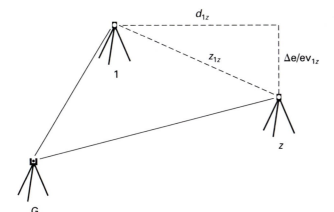

(b)   Length components of a missing line.

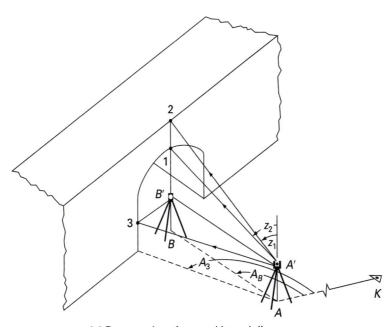

(c) Remote elevations and lateral distances.

FIGURE 7.10 (*continued*)
(b) Length components of a missing line. (c) Remote elevations and
lateral distances.

measurements $A'B'$, z_1, z_2, A_B, A_3, and distance $A'3$ are made from coordinated point A with a backsight on known station K to yield vertical distances $B'1$, $B'2$, and lateral distance $B'3$.

4. Storage of the coordinates for the last station in a traverse as the instrument is moved ahead. Figure 7.10d shows traverse 1-2-3-4 with the instrument on station 3, having taken a backsight on 2 and made a foresight to 4, the last station set. Assuming that

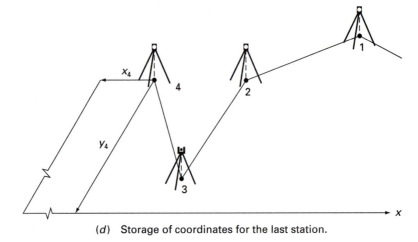

(*d*) Storage of coordinates for the last station.

FIGURE 7.10 (*concluded*)
(*d*) Storage of coordinates for the last station.

known coordinates for station 1 were entered on initiating the traverse, then on completion of the setup at station 3, coordinates for station 4 are automatically stored and retained.

5. Resection. According to the classical definition (Section 9.5), when angles are measured between from three to five points having known coordinates (e.g., points *A*, *B*, and *C* in Figure 7.11a) the position of an unknown, occupied point (*D* in Figure 7.11a) can be calculated (Section 9.6). With total station systems, the unknown position of *D* (Figure 7.11a) also can be found by observing the angle between two points with known coordinates (*A* and *B* in Figure 7.11a) and measuring the distances to those points.

6. Determination of position of an offset point. In Figure 7.11b, the reflector is placed at a known offset that is at a right angle to the line between the reflector and the instrument. The distance and direction to the reflector are measured and entered into the data collector. Then, the direction is observed to the center of the tree. This information is combined to yield coordinates for the center of the tree.

7. Setting stations for construction surveys. In Figure 7.11c, coordinates for known control points *P* and *Q* and for all desired stations on the horizontal curve are precomputed and stored in total station computer memory. In this option, distance and direction to each specified station are calculated and displayed, automatically on command, allowing setting the point by executing the stake-out mode described previously. This procedure greatly facilitates construction layout from random control points (Section 17.7).

The functions just described are present in most total station systems and other functions not mentioned here are found in some of the more exotic systems.

Another feature becoming more common on certain models is a point-guiding system for aiding the rod person to stay on the line. This system consists of two light-emitting diodes (LED) that together emit two beams, one a steady and the other a flashing beam of coherent red light. These beams of light are transmitted through the optical system in such a way that only the flashing beam is visible on one side and only the steady beam is seen when on the other side of the correct alignment. One such system is illustrated in Figure 7.12. When both

371

CHAPTER 7:
Combined
Distance and
Angular
Measurement
Systems

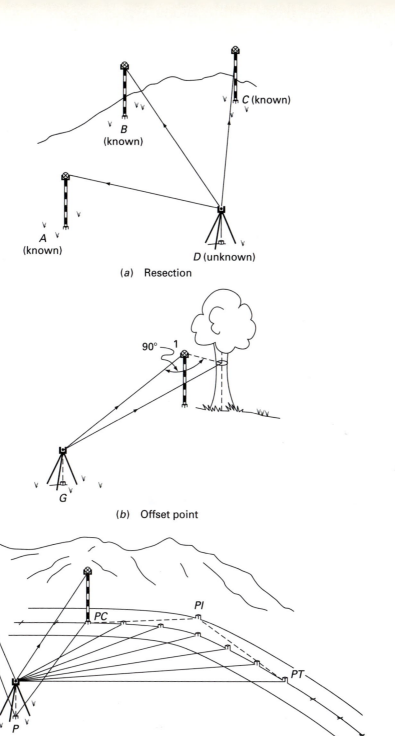

(a) Resection

(b) Offset point

(c) Setting out stations on horizontal curve

FIGURE 7.11
Locating and setting out of point: (a) resection, (b) offset point, (c) setting stations on
a horizontal curve.

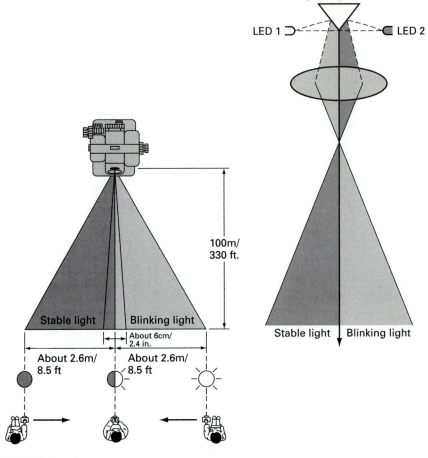

FIGURE 7.12
Point-guiding system. (*Courtesy of Nikon, Inc.*)

the flashing and steady beams are visible, the rod person is within 4 to 6 cm (0.1 to 0.2 ft) of the correct alignment. These lights are visible for distances of up to 100 m (330 ft).

7.18
DATA COLLECTORS

A typical data collector is illustrated in Figure 7.7. These devices are the same size as a hand calculator and communicate with the instrument by way of a cable connected to a serial port found on all total station instruments. Most data collectors are compatible with a wide array of total station systems, and many can be used with all available systems. Although the primary function of the data collector is to permit automatic recording of measured values, which are then downloaded into office computers for further processing, most collectors are progammable and contain software that allows extensive processing of collected data in the

373

CHAPTER 7:
Combined
Distance and
Angular
Measurement
Systems

field. It is also possible to download software from a PC or other computer into the memory of many data collectors. The data collector along with this software constitute the "electronic field book," which is an extremely powerful tool for the surveyor when used in conjunction with a total station system and office-based personal, mini, or mainframe computers. Further details concerning the characteristics and use of data collectors can be found in Section 3.12 under the subsection *Data collectors.*

7.19
USE OF TOTAL STATION SYSTEMS

To operate a total station system, set it over the station, center with the optical plummet, and level using the plate level bubble (sensitivity of 30″ per 2-mm division). The setup procedure for a total station instrument is essentially the same as that for a repeating theodolite, described in Sections 6.31 and 6.32. When the instrument has been leveled and centered over the station, turn on the power and initialize the system, usually by passing the telescope through a zenith angle of 90°. Next, observe the temperature (°C or °F) and atmospheric pressure (mm Hg or in Hg), and enter these values into the onboard microprocessor. As noted previously, some systems contain built-in sensors for temperature and atmospheric pressure so that corrections occur automatically. On the other hand, in some older systems, one uses the observed temperature and atmospheric pressure with a nomogram or chart to obtain an atmospheric correction in parts per million, which then is entered into the system memory. The EDM system constant (Section 4.38) must be known and entered into the system as well. Note that this constant may vary with the reflector being used, so care must be exercised to know which reflectors are being used for a given sight. Because surveying in three dimensions is possible with the total station system, the heights of instrument and reflector above the stations occupied and sighted need to be measured and these values recorded in the notes or entered into the onboard memory or the data collector. When each of these tasks has been accomplished, following a method appropriate to the system being used, surveying in either two or three dimensions can proceed.

When the instrument is set up and initialized with all necessary parameters entered, the telescope is sighted on the reflector and target (see Figure 7.9) at the backsight station and the zero-set key is pressed to set the horizontal circle to 0 or a preselected azimuth is entered via the keyboard. Next, by manipulation of the keys on the control panels of the instrument or data collector, the zenith angle, slope or horizontal distance, and difference in elevation between the horizontal axis of the instrument and the center of the reflector can be displayed on the register and recorded in the field book or stored in the onboard or data-collector memory for future use. If desired, angles can be repeated with the telescope direct and reversed and multiple distances measured. Each measurement is automatically stored and averaged angles and distances are displayed on the register of the instrument and stored in the onboard or data-collector memory. At this stage, if known stations are occupied and sighted, the reflector and target can be moved to another point or station for subsequent sighting from the instrument. Then, any of the functions given for total station systems in Section 7.17 can be performed with the resulting data displayed or stored in the onboard or data-collector memory. When work from this station is finished, the system is moved to another point and similar operations are repeated. On completion of the day's work, the system operator should perform whatever preliminary computations are possible with the collected data in the field, using software in the onboard microprocessor, the memory card or module, and the data collector. This procedure allows detecting mistakes, which can be corrected before leaving the job. On reaching the office, the contents of the onboard memory module or the memory of the data collector should be downloaded into the office computer and a hard-copy listing and diskette made of the field data as backup records. In addition, always keep a minimum amount of written records in a conventional field book (sketches, notes, etc.) as protection against electronic failure or accidental loss of stored records. One cannot overemphasize the importance of this last step.

The routine just described is general and not addressed to specific surveying procedures. The reader should refer to Chapters 8, 15, and 17 on traverse, topographic surveying, and construction surveying for the details of applications of the total station system to these tasks.

As in any survey, certain precautions must be observed to ensure maximum accuracy in total station surveys. The condition and adjustment of support equipment is of particular importance. For example, the adjustment of optical plummets on instruments and tribrachs, circular level bubbles on prism poles, and the fit of tribrachs should be checked on a regular basis. Adjustments of the optical plummet and circular level bubble are covered in Section 6.52, Adjustment 8, and Section 5.63, respectively. In addition, the prism pole ought to be checked on a daily basis to make sure that it is not bent and special care should be exercised to ensure that the system constant for each prism used is known, correct, and stored in the system microprocessor memory.

7.20
REMOTELY CONTROLLED SYSTEMS

A remotely controlled or *robotic* total station system consists of a servo-driven *main station unit* and *remote positioning unit* (RPU). The main station unit is a total station instrument, as described in Section 7.16, in which the horizontal and vertical motions are driven by servo motors controlled by buttons on the instrument or programmed instructions. This unit also contains more than the average amount of internal memory and sophisticated software in the onboard microprocessor. The RPU essentially is a combined prism and remote control unit with its own keyboard and microprocessor. Both units are equipped with a telemetry link so that data can be telemetered from the remote unit to the main station. The main station and RPU for a Geodimeter System 600 are shown in Figures 7.13 and 7.14, respectively.

The main station instrument, illustrated in Figure 7.13 has dual-axis compensation, electronic leveling (sensitivity of 6″), compensation for collimation errors, a point guidance system (Section 7.17; note the tracking light, Figure 7.13b), and can be used as a stand-alone total station system, if desired. The combined keyboard, display register, and data storage unit (Figure 7.13b) is detachable for use on the remote positioning unit.

The RPU, shown in Figure 7.14, consists of a combined reflector and target; keyboard, display register, and storage unit (detachable); and telemetry link, all mounted on a prism pole that can be plumbed using an attached circular level bubble.

The main station can be used simply as a total station system, as in running a traverse (Chapter 8) or turning angles and measuring distances about a point as in triangulation and trilateration operations (Chapter 9). In these cases, the operator needs to locate foresight and backsight targets only once, then the instrument's memory and servo motors automatically locate the directions for repetition of angles a specified number of times. It is necessary only that the operator make the fine adjustment at each station and be sure to register the data in the computer memory.

For remotely controlled operations, first the main station unit is set up and oriented by taking an optical backsight on another known station. With the main station unit thus oriented and set up via the keyboard for remote operation, the operator takes the RPU to the first point to be located, where both the main station unit and RPU are activated by the

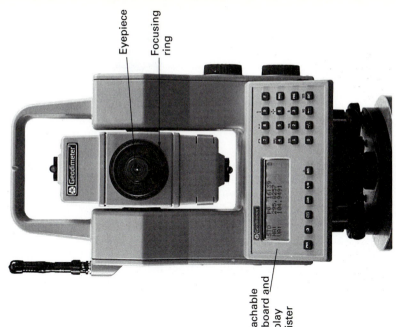

Eyepiece

Focusing ring

Detachable keyboard and display register

(b) Rear (eyepiece) view.

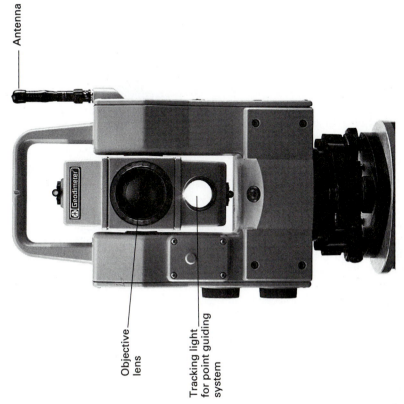

Antenna

Objective lens

Tracking light for point guiding system

(a) Front (objective view).

FIGURE 7.13

Geodimeter System 600: (a) Front (objective view). (b) Rear (eyepiece) view. (*Courtesy of Geotronics of North America, Inc.*)

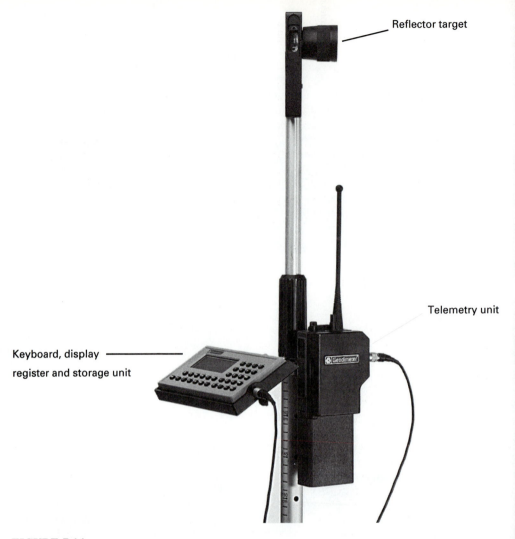

Reflector target

Telemetry unit

Keyboard, display
register and storage unit

FIGURE 7.14
Remote positioning unit, Geodimeter System 600. (*Courtesy of Geotronics of North America, Inc.*)

operator at the RPU. The telescope at the main station then will scan in vertical and horizontal planes (by preprogrammed sectors) until it centers on the reflector of the RPU, where the angle and distance are automatically recorded in the memory of the RPU (capacity for 10,000 points). Note that these operations, which are entirely under the control of the operator at the RPU, occur automatically in a few seconds. The remote operator can add the point number and any necessary attributes via the RPU control panel and proceed to the next point.

For stake-out operations, the coordinates and point numbers of the stations to be set off are stored in the memory of the main station unit. When the point number is entered by the operator at the RPU, the telescope at the main station automatically is oriented in the correct direction. Getting on the line is facilitated, for the operator of the RPU, by the built-in point guiding system of the main station (Section 7.17, Figures 7.12 and 7.13a).

377

CHAPTER 7:
Combined
Distance and
Angular
Measurement
Systems

When the RPU is approximately on the line, the distance to be moved, parallel to and at a right angle to the correct alignment, is indicated on the register of the RPU control panel. When these values are reduced to 0, the location is correct and the station can be set. Data for the point, in the form of a point number and coordinates, can be stored by pressing a single key on the keyboard of the RPU.

Note that, for both stake-out and location of points, the system is under the full control of the operator at the RPU, and, essentially, the entire system can be operated by one person.

The range of such a system is about 500 m (1600 ft), which is more than adequate for most construction and topographic surveys. In addition, the stake-out operation and location of points using the RPU are performed using infrared light and the telemetry link, so that the system can be used after dark. This aspect of the robotic systems would be advantageous for construction layouts, where points always are needed "yesterday," and for topographic surveys of regions congested by traffic during daylight hours.

7.21
APPLICATIONS OF TOTAL STATION SYSTEMS

The systems described in Sections 7.15 to 7.20, which allow surveying in three dimenions, can be applied to practically any surveying task described in this text. Naturally, due regard must be given to the precision of the system versus the specified accuracies required for the application. In general, there are applications for all of these systems in control surveys by traverse (Chapter 8) and by combined triangulation and trilateration (Chapter 9); topographic surveying by the controlling point method or radial survey (Chapter 15); and construction layout (Chapter 17). It should be noted that all of these systems provide data that are very adaptable to forming digital terrain models (DTM; see Chapters 13 and 14). In particular, the total station system provides data that are directly convertible to a DTM by interfacing the data collector with a compatible personal, mini-, or mainframe computer.

Another aspect that needs to be considered relative to applications of total station systems is maintenance of the instruments. Although adjustments for these instruments are about the same as for repeating and direction theodolites (Sections 6.51 and 6.52), these are electronic systems. As such, they are subject to occasional electronic malfunctions which usually require service at the nearest manufacturer's service center. Consequently, one should have a backup system or other equipment to fill the gap if an instrument currently being used needs service. Batteries also constitute a source of difficulty. Always go to a job with two fully charged batteries and have extra ones in the office as well as equipment for recharging batteries. Attention to these details is necessary if these systems are to be used at peak efficiency.

PROBLEMS

7.1. To determine the stadia interval factor, a theodolite is set up at a distance C back of the zero end of a level base line 240.0 m long, the base line being marked by stakes set every 30.00 m. A rod then is held at successive stations along the base line. The stadia interval and each half-interval observed at each location of the rod are in the following table. Compute the lower, upper, and full interval factors for each distance and find the average value for the lower interval, the upper interval, and the full interval.

Distance	Lower interval		Upper interval		Full interval	
$-C$, m	Meters	Factor	Meters	Factor	Meters	Factor
30	0.148		0.150		0.298	
60	0.298		0.293		0.591	
90	0.440		0.445		0.885	
120	0.593		0.592		1.185	
150	0.741		0.739		1.480	
180	0.890		0.885		1.775	
210	1.020		1.000		2.020	
240	1.190		1.180		2.370	
Average						

7.2. A stadia interval of 2.515 m is observed with a theodolite for which the stadia interval factor is 100.0 and C is 0.305 m. The zenith angle is $87°20'30''$ with the middle cross hair set on 1.37 m. If the instrument has a height of instrument (h.i.) of 1.52 m above the point over which it is set and the point has an elevation of 100.05 m, calculate the horizontal distance and elevation of the point sighted by the (*a*) exact stadia equations; (*b*) approximate stadia equations.

7.3. The following observations are taken with a theodolite for which the interval factor is 100.0 and C is 1.00 ft:

Observation	Stadia interval	Zenith angle
	(i)	
(*a*)	10.00 ft	$89°40'$
(*b*)	10.00 ft	$79°50'$
(*c*)	10.00 ft	$64°50'$
	(ii)	
(*d*)	3.000 m	$89°00'$
(*e*)	3.000 m	$79°00'$
(*f*)	3.000 m	$65°00'$

By means of Equations (7.4a) and (7.5a), Section 7.7, compute the horizontal distance and differences in elevation. By means of the approximate Equations (7.8) and (7.9), Section 7.8, determine the same quantities and note the errors introduced by the approximations.

7.4. What would be the amount and sign of error introduced in each computed horizontal distance and difference in elevation if the observations of Problem 7.3, part(ii) were taken (*a*) with a 4-m rod that was unknowingly 0.20 m out of plumb with the top leaning toward the theodolite? (*b*) With the top of the rod leaning 0.20 m away from the theodolite? (*c*) What conclusions may be drawn from these results?

7.5. What error will be introduced in each computed horizontal distance and difference in elevation if in the observations of Problem 7.3, parts (i) or (ii), (*a*) each vertical angle contains an error of $01'$? (*b*) each stadia interval is in error by $1/1000$ of the interval? (*c*) What conclusions may be drawn from these results?

7.6. In determining the elevation of point A and the distance between two points, A and B, a theodolite is set up at A and the following data are obtained: $z = 92°45'$, stadia interval $= 1.311$ m, h.i. $= 1.28$ m, and the line of sight at 2.62 m on rod. The instrument constants are $K = 100.0$ and $C = 0.305$ m. The elevation of B is 38.28 m. Compute the distance AB and the elevation of point A.

7.7. The following are the notes for a line of stadia levels. The elevation of B.M.$_1$ is 637.05 ft. The stadia interval factor is 100.0 and $C = 1.25$ ft. Rod readings are taken at the height of instrument. Determine the elevations of remaining points. Record notes and elevations in a proper note form.

379

CHAPTER 7:
Combined
Distance and
Angular
Measurement
Systems

| | Backsight | | Foresight | |
Station	Stadia interval, ft	Vertical angle	Stadia interval, ft	Vertical angle
B.M.$_1$	4.50	−3°30′		
T.P.$_1$	2.80	−1°41′	3.20	+2°10′
T.P.$_2$	3.30	+1°56′	2.71	+3°35′
T.P.$_3$	3.20	+2°09′	3.35	−0°38′
B.M.$_2$			2.10	+7°27′

7.8. Notes for a line of stadia levels follow:

| | Backsight | | | Foresight | | |
Station	Stadia interval, m	Zenith angle	Rod, m	Stadia interval, m	Zenith angle	Rod, m
B.M.$_{10}$	1.031	95°30′	2.41			
T.P.$_1$	2.041	86°30′	2.06	1.506	81°50′	1.00
T.P.$_2$	2.721	96°30′	1.45	1.432	94°10′	2.50
T.P.$_3$	2.012	100°00′	2.04	0.901	82°30′	0.24
B.M.$_{11}$				1.321	92°40′	2.40

The elevation of B.M.$_{10}$ is 42.852 m. The stadia interval factor is 100.00 and $C = 0.31$ m. Determine elevations for the turning points and B.M.$_{11}$. Record notes and elevations in a proper note form.

7.9. The following are stadia intervals and vertical angles for a theodolite-stadia traverse. The elevation of station A is 418.55 m. The stadia interval factor is 100.0 and $C = 0.31$ m. Rod readings are taken at the height of instrument. Compute the horizontal lengths of the courses and the elevations of the transit stations.

Station	Object	Stadia interval, m	Zenith angle
B	A	2.415	88°10′
	C	1.452	80°35′
C	B	1.455	99°24′
	D	2.459	94°20′
D	C	2.456	85°41′
	E	2.581	93°10

7.10. The following are stadia intervals and vertical angles taken to locate points from a station the elevation of which is 525.5 ft. The height of instrument above the station is 4.0 ft, and rod readings are taken at 4.0 ft except as noted. The stadia interval factor is 100.0 and $C = 0.00$ ft. Compute the horizontal distances and the elevations.

Object	Stadia interval, ft	Vertical angle
43	6.25	+2°30′ on 2.1
44	7.41	−0°58′
45	7.05	−0°44′ on 9.2
46	4.25	−5°36′
47	2.75	+1°20′ on 6.0

7.11. A theodolite is used in locating points from a station the elevation of which is 233.39 m. The instrument constants are $K = 100.0$ and $C = 0$. The height of instrument above the station is 1.37 m. Compute the horizontal distances and the elevations.

Object	Stadia interval, m	Rod reading, m	Zenith angles
114	0.994	1.10	80°31′
115	1.400	1.77	84°10′
116	0.664	1.43	95°24′
117	1.520	1.31	94°35′
118	2.481	1.95	96°30′

7.12. A stadia interval of 2.521 m and vertical angle of 15°30′ are observed between points C and D. The stadia interval factor is 100.0 and $C = 0.31$ m. Calculate the horizontal distance and difference in elevation between points C and D using the exact stadia equations. If $\sigma_s = 0.005$ m, $\sigma_\alpha = 20''$, $\sigma_K = 0.1$, and $\sigma_C = 0.01$ m, determine the estimated standard deviation in the calculated distance and difference in elevation. Comment on the relative importance of uncertainties in s, α, K, and C.

7.13. A theodolite is set over station 10 and stadia observations are taken on point B. The h.i. $= 4.25$ ft, the stadia interval, $s = 3.51$ ft, $\alpha = +9°00'$, $K = 100$, $C = 0.0$, $\sigma_s = 0.02$ ft, $\sigma_{h.i.} = 0.02$ ft, and $\sigma_\alpha = 01'$. The elevation of 10 is 425.50 ft above mean sea level and $K = 100$, assumed errorless. Compute the horizontal distance from 10 to B, the elevation of B, and their respective standard deviations.

7.14. The following data are used to determine the elevation of a photo-control point from a bench mark by stadia: the backsight on B.M.$_{10}$ yields $s = 0.984$ m, $z = 94°50'$, and rod$_B = 2.541$ m; the foresight on point $P1$ yields $s = 1.854$ m. $z = 87°20'$, and rod$_F = 1.524$ m. The constant $K = 100$ and $C = 0.0$. (a) If the elevation of B.M.$_{10} = 30.065$ m, compute the elevation of $P1$. (b) If $\sigma_s = 0.005$ m, $\sigma_z = 20''$, $\sigma_{rod} = 0.005$ m, determine the standard deviation for the elevation of $P1$ calculated in (a).

7.15. For the data given in Figure 7.5 (stadia leveling), the elevation of B.M.$_{42}$ and the constant K are assumed errorless. If $\sigma_s = 0.003$ m, $\sigma_z = 20''$, and $\sigma_{rod} = 0.003$ m, compute the standard deviation for the elevation of T.P.$_3$.

7.16. What are the major components of a total station system? Discuss the function of each of these components.

7.17. Describe the difference between a modular and a self-contained total station system. What are the advantages and disadvantages of the these systems?

7.18. Some total station systems are equipped with dual-axis compensation, one component of which detects and stores misalignment of the horizontal axis with respect to true vertical. Explain how this misalignment can be used in the system to improve the measurements obtained (note, see Section 6.45).

7.19. What basic parameters should be entered into the memory of the microprocessor of a total station system prior to making measurements? Describe how these parameters alter the measurements obtained.

7.20. List some of the onboard functions provided by one of the less expensive total station systems designed for construction layout.

7.21. Describe the functions of a typical data collector when used with a total station system.

7.22. What special considerations are required with respect to the care, maintenance, and adjustment of a total station instrument and the peripheral equipment used with the system?

381

CHAPTER 7:
Combined
Distance and
Angular
Measurement
Systems

REFERENCES

Crawford, W. G. "Shopping for a Total Station?" *P.O.B.*, *Point of Beginning* 22, no. 7 (April 1997), pp. 22–47.

Easa, S. M. "Model of Stadia Surveying with Incomplete Intercepts." *ASCE Journal of Surveying Engineering* 116, no. 3 (August 1990), pp. 139–48.

Easa, S. M. "Maximizing Accuracy in Stadia Surveying." *Surveying and Land Information Systems* 51, no. 3 (September 1991), pp. 149–53.

Kavanagh, B. F., and S. J. G. Bird. *Surveying,* 3rd ed. Englewood Cliffs, NJ: Prentice Hall, 1992.

Pasley, R. M. "P.O.B. Data Collector Survey." *P.O.B., Point of Beginning* 18, no. 4 (August–September 1994), pp. 82–93.

Pepling, A. "The Geodimeter 600 Series and Autolock." *Professional Surveyor* 14, no. 6 (November–December 1994), p. 31.

Reilly, J. P. "P.O.B. 1993 Total Station Survey." *P.O.B., Point of Beginning* 18, no. 4 (April–May 1993), pp. 56–68.

Reilly, J. P. "P.O.B. 1995 Total Station Survey." *P.O.B., Point of Beginning* 20, no. 4 (April–May 1995), pp. 24–40, 98.

Wolf, P. R., and R. C. Brinker. *Elementary Surveying,* 9th ed. New York: HarperCollins Publishers, 1994.

PART III

Survey Operations

CHAPTER 8

Traverse

PART A
ELEMENTARY OPERATIONS

8.1
GENERAL

A *traverse* consists of a series of straight lines connecting successive established points along the route of a survey. The points defining the ends of the traverse lines are called *traverse stations* or *traverse points*. Distances along the line between successive traverse points are determined either by direct measurement using electronic distance measurement (EDM) equipment or a tape or by indirect measurement using tacheometric methods. At each point where the traverse changes direction, an angular measurement is taken using a theodolite. When stand-alone EDM devices and theodolites are used to measure distance and direction, the traverse is performed in two dimensions. If a total station system (Chapter 7) is used, distance and direction are obtained by a single instrument and traverses are possible in two or three dimensions. In Part A of this chapter, emphasis is on the operations, computation, simple adjustment of two-dimensional traverses, and computations with coordinates. Partition of land, error propagation, and rigorous adjustment of traverses by the method of least squares is developed for two-dimensional traverses in Part B.

Traverse by the global positioning system (GPS) also is possible. This operation is a special case, covered in Chapter 12, Section 12.14.

8.2
EQUIPMENT FOR TRAVERSING

Equipment usually consists of a total station system, or theodolite and EDM, or tape. The following are supporting equipment for each system: prism poles, reflectors, targets or combined reflector and target, data collector, range poles and plumb bobs, stakes and hubs, tacks, axe or hammer, field notebook, taping pins, and marking crayon. Also included are devices for marking stations, such as ordinary or special nails, a cold chisel, spray paint, and colored plastic flagging or tape. A two-way radio is invaluable to expedite communication.

Where *forced centering* is to be employed, three tripods compatible with the total station instrument or theodolite, reflectors, and sighting targets are required. In forced centering, after the instrument (total station system or theodolite) has been centered using the optical plummet, the upper part of the instrument (alidade; see Figure 6.22) can be removed from the tribrach by releasing a clamp (Figure 7.6). The centered tribrach remains on the tripod and a reflector or target can be slipped into place and locked into position, where it will be centered automatically and ready for taking a sight or distance measurement.

For a stadia or tacheometric traverse (Section 8.11), distance measuring equipment is omitted and a level or stadia rod is included.

8.3
TRAVERSE STATIONS

Any temporary point of reference over which the instrument is set up is called a *traverse station.* On most surveys, the traverse station is a peg, called a *hub,* driven flush with the ground and having a tack driven into its top to mark the point of reference for the measurements. On pavements, the traverse station may be a driven nail, a cross cut in the pavement or curb, or a tack set in a hole drilled with a star drill and filled with lead wool. In land surveying, the stations often are steel pipes or rebar, stones, or other more or less permanent monuments set at the corners. Pointed metal rods, which are driven by a collar slipped over the top, are available. When driven, the collar is removed and replaced by a special metal cap on which a punch mark defines the point of reference.

The location of a hub, pipe, rod, or whatever marks a station usually is indicated by a flat *guard stake* extending above the ground and driven at a slope so that its top is over the station. This guard stake carries the number or letter of the traverse station over which it stands. Usually the number is marked with keel, lumber crayon, or a felt-tipped pen on the underside of the stake and reads down the stake. Hubs generally are square, say 5 by 5 cm (2 by 2 in.), and guard stakes are flat, perhaps 2 by 8 cm ($\frac{3}{4}$ by 3 in.). Plastic guard stakes are available.

To avoid the necessity for using a rear flagger to give backsights, as far as practical each traverse station is marked by a temporary signal, such as a lath or stick set on the line, with a piece of paper, cloth, or tape attached. When forced centering is employed, reflectors or sighting targets are attached to the tripods at backsight and foresight stations.

8.4
PURPOSE OF THE TRAVERSE

Traversing is a convenient, rapid method for establishing horizontal control. It is particularly useful in densely built-up areas and in heavily forested regions, where the lines of sight are short, so that neither triangulation nor trilateration is suitable, and obstructions overhead or surrounding the points reduce the effectiveness of GPS surveys. Traverses are made for numerous purposes, including

1. Property surveys to locate or establish boundaries.
2. Supplementary horizontal control for topographic mapping.

3. Location and construction layout surveys for highways, railways, and other private and public works.
4. Ground control surveys for photogrammetric mapping.

Frequently, traverse is employed to densify the number of coordinated control points in networks for which the primary controlling points were established by photogrammetric or GPS surveys.

8.5
TYPES OF TRAVERSE

The two general classes of traverse are the open traverse and the closed traverse.

The *open traverse* originates at a point of known position and terminates at a point of unknown position. No computational check is possible to detect errors or blunders in distances or directions in this type of traverse. To minimize errors, distances can be measured twice, angles turned by repetition, magnetic bearings observed on all lines, and astronomic observations made periodically. In spite of these precautions, an open traverse is a risky proposition and should not be used for any of the applications discussed in Section 8.4, because the results are uncertain and always will be subject to question. Open traverses are necessary in certain types of mine surveying. Hence, there are applications of an open traverse when circumstances dictate their use.

Closed traverses originate at a point of known position and close on another point of known horizontal position. A point of known horizontal position is given by either geographic latitude and longitude or X and Y coordinates on a rectangular coordinate grid system. In Figure 8.1, the traverse $MABCDQ$ originates at known point M with a backsight along line MN of known azimuth and closes on known point Q, with a foresight along line QR, also of known azimuth. This type of traverse is preferable to all others because computational checks are possible that allow detection of systematic errors in both distance and direction.

A traverse that originates and terminates on a single point of known horizontal position is called a *closed-loop traverse*. In Figure 8.2, traverse 1-2-3-4-5-6-1, which originates and closes on point 1, is an example of a closed-loop traverse. This type of traverse permits an internal check on the angles, but there is no way to detect systematic errors in distance or errors in the orientation of the traverse. A closed-loop traverse should not be used for major projects. To minimize the chance of errors in a closed-loop traverse, all distance measuring equipment must be calibrated carefully and astronomic observations must be made periodically.

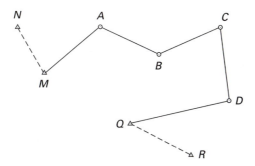

FIGURE 8.1
Closed traverse.

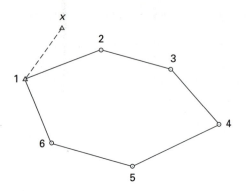

FIGURE 8.2
Closed-loop traverse.

A closed-loop traverse that originates and terminates on a point of assumed horizontal position provides an internal check on angles but no check on systematic errors in distance. An additional disadvantage is that the points are not located on any datum and have no relationship with points in any other survey, although the relative positions of the points are determined. The use of such a method of running a traverse is to be discouraged except under extenuating circumstances, where no feasible alternative is possible.

Each of these types of traverse also can be described according to the method of turning the angles in the traverse. Thus, there are traverses by deflection angles, interior angles, angles to the right, and the azimuth method. Each of these methods for running traverse is discussed in the following sections.

Please note that depending on the type of instrument being used (e.g., repeating optical theodolite, electronic theodolite, Sections 6.22, 6.30, 6.33; or total station system, Section 7.19), the procedure for turning specific angles may vary considerably. In this chapter, instructions are general but the focus is on use of a repeating theodolite and manual recording of the data. The reader should refer to the appropriate sections (or the owner's manual) for detailed instructions for special cases with specific instruments.

8.6
DEFLECTION-ANGLE TRAVERSE

This method of running traverses probably is more commonly employed than any other, especially on open traverses where only a few details are located as the traverse is run. It is used almost entirely for the location surveys for roads, railroads, canals, and pipelines. It is employed to a lesser extent in land surveying and establishing control traverses for topographic and hydrographic surveys.

Successive transit stations are occupied, and at each station a backsight is taken with the horizontal circle set at 0 and the telescope in the reversed position. The telescope is then reversed, the foresight is taken by turning the instrument about the vertical axis on its upper motion, and the deflection angle is observed. The angle is recorded as right (R) or left (L), according to whether the upper motion is turned clockwise or counterclockwise. Usually, it is considered good practice to observe the deflection angle at least twice, once with the telescope direct and once reversed as described in Section 6.35. This process is called turning the angle by *double deflection*.

Figure 8.3 shows a closed deflection angle traverse that originates at point R at one end of line RS having a known azimuth of 170°30′00″ as determined from a previous survey. The

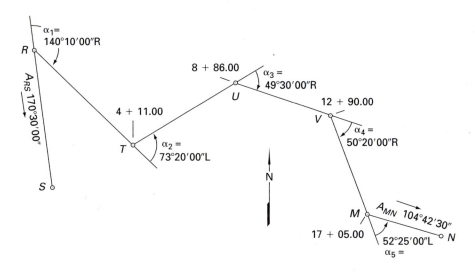

FIGURE 8.3
Deflection-angle traverse between lines of known direction.

traverse closes on point M at one end of line MN, which has an azimuth of $104°42'30''$ also determined from a previous survey.

A portion of the field notes for the traverse is illustrated in Figure 8.4. Angles are recorded on the left with the necessary details sketched on the right. All data are recorded from the bottom to the top of the page, a procedure also followed in preliminary route surveys. Angles are turned by double deflection, and to check the angular closure, azimuths are calculated from known line RS to known line MN.

To compute the azimuth of an unknown line, a deflection angle to the *right* is added to the *forward azimuth* of the previous line. A deflection angle to the *left* is subtracted from the forward azimuth of the previous line. Therefore, the azimuth of line RT is $350°30'00'' + 140°10'00'' - 360°00'00'' = 130°40'00''$. Similarly, the azimuth of line TU is $130°40'00'' - 73°20'00'' = 57°20'00''$. Figure 8.5 illustrates computation of azimuths using deflection angles to the left and right, respectively. The results of these computations for the example traverse in Figure 8.3 are shown in Table 8.1. The process of computing azimuths for unknown lines is continued through lines UV, VM, to MN, where the calculated azimuth fails to agree with the fixed azimuth of $104°42'30''$ by $02'30''$, which is the angular closure for the traverse. The equation by which this error of closure may be computed for the example traverse is

$$A_{SR} + \alpha_1 - \alpha_2 + \alpha_3 + \alpha_4 - \alpha_5 - 360° = A_{MN} \tag{8.1}$$

The general relationship for a traverse that contains m deflection angles to the right and n to the left is

$$A_1 + \sum_{i=1}^{m} \alpha_{R_i} - \sum_{i=1}^{n} \alpha_{L_i} = A_2 + 360° \tag{8.2}$$

in which A_1 is the forward azimuth at station 1, the origin of the traverse, and A_2 is the azimuth of station 2, the closing point for the traverse.

If adjustment of the observed angles to fit the known directions is desired (assuming equal weights for all angles), the closure error may be distributed equally among the five deflection angles. For this example $30''$ will be subtracted from each deflection angle where angles to the right are positive and angles to the left are negative. When azimuths already

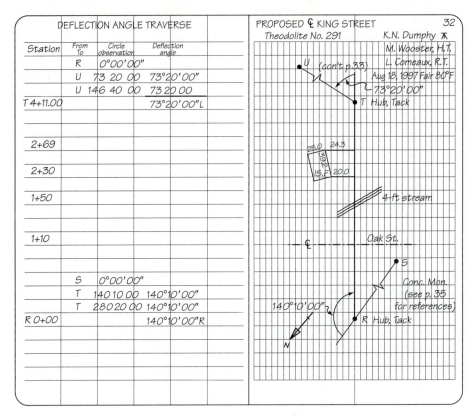

FIGURE 8.4
Closed traverse by deflection angles.

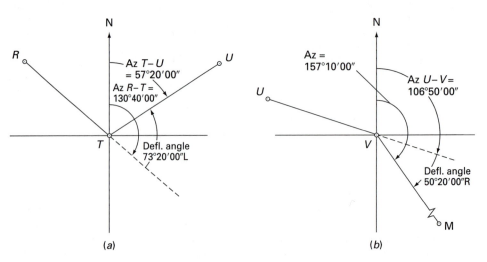

FIGURE 8.5
Azimuth with deflection angle (a) to the left and (b) to the right.

TABLE 8.1

391

CHAPTER 8:
Traverse

Calculation and adjustment of azimuths in a closed traverse by deflection angles

From	To	Azimuth	Correction	Adjusted azimuth	Adjusted bearing
S	R	350°30′00″		350°30′00″	N9°30′00″W
	+∠R	140°10′00″R			
		490°40′00″			
	−	360°00′00″			
R	T	130°40′00″	−30″	130°39′30″	S49°20′30″E
	−∠T	73°20′00″L			
T	U	57°20′00″	−1′00″	57°19′00″	N57°19′00″E
	+∠U	49°30′00″R			
U	V	106°50′00″	−1′30″	106°48′30″	S73°11′30″E
	+∠V	50°20′00″R			
V	M	157°10′00″	−2′00″	157°08′00″	S22°52′00″E
	−∠M	52°25′00″L			
M	N	104°45′00″	−2′30″	104°42′30″	S75°17′30″E
M	N	104°42′30″			

Angular error of closure = +2′30″

have been calculated, as in this case, it is simpler to apply the cumulative correction directly to the unadjusted azimuth. So, the correction to the first azimuth from R to T is $-30''$, to the second from T to U is $-01'00''$, and so forth, as shown in Table 8.1. The same result is achieved using adjusted deflection angles to compute the azimuths.

Figure 8.6 illustrates a closed-loop traverse with deflection angles. This traverse originates and closes on traverse point A. The notes are recorded down the page in the order of occupying the points and a sketch is drawn on the right-hand page. For any closed-loop traverse the initial azimuth, A_1, equals the closing azimuth, A_2. Then, from Equation (8.2), the sum of the deflection angles should equal 360°. The sum of the average values of the angles is recorded at the bottom of the right-hand page, where a closure error of 02′30″ is indicated. If an adjustment is desired and assuming that angles are of equal weight, this error would be distributed equally among the five angles. Because the left deflection angles have a sum that is too large, 30″ is subtracted from each left deflection angle and 30″ is added to the right deflection angle.

The next step is to calculate directions for each of the traverse lines. This is a closed-loop traverse and no angle was observed to a line of known direction, so line AB will be assumed to have an azimuth of 228°00′00″ from true north for determination of azimuths for the balance of the lines. The procedure is similar to that employed for the previous example except that the closure check is on the original line AB, as shown in Table 8.2, which contains the calculation of the directions. Note that the closure is exact, as it should be, because the angles had been balanced previously. The directions, as determined for this example, are not absolute because they are based on an assumed azimuth. In practice, an angle should be turned to a line of known azimuth or an astronomic direction should be determined.

The arithmetical check of the sum of the angles should be performed immediately on completing the traverse to detect blunders or excessively large errors before leaving the field. The angular error of closure should not exceed the estimated standard deviation for observing an angle from a single setup times the square root of the number of instrument stations. In practice, this estimate usually is taken as the least count of the instrument used to turn the angles.

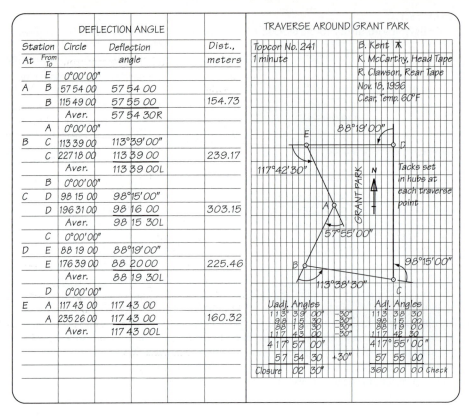

FIGURE 8.6
Deflection-angle traverse, closed-loop.

TABLE 8.2
Azimuths and bearings in a closed-loop traverse

From	To	Azimuth	Bearing
A	B	228°00′00″	S48°00′00″W
	−∠B	113°38′30″L	
B	C	114°21′30″	S65°38′30″E
	−∠C	98°15′00″L	
C	D	16°06′30″	N16°06′30″E
	−∠D	88°19′00″L	
	+	360°00′00″	
D	E	287°47′30″	N72°12′30″W
	−∠E	117°42′30″L	
E	A	170°05′00″	S9°55′00″E
	+∠B	57°55′00″R	
A	B	228°00′00″ checks	S48°00′00″W

INTERIOR-ANGLE TRAVERSE

Field operations for this method of traversing are not materially different from those used for the deflection-angle traverse. At each station the horizontal circle is set at 0, and a backsight to the preceding station is taken. The horizontal clamp is released (the upper clamp on a repeating theodolite) and the alidade is turned until the advance station is sighted and the interior angle is observed. Except for surveys of a very low order of accuracy, all interior angles should be turned at least twice, once with the telescope direct and once with the telescope reversed. Notes may be kept in a form similar to those shown for deflection angles in Figure 8.6.

A closed-loop traverse run by the method of interior angles is illustrated in Figure 8.7. In this traverse, point 1 was occupied first and an angle was observed between line $1Q$, which has a known azimuth, and traverse line 12. After the clockwise interior angle at 1 was observed, points 2, 3, 4, 5, and 6 were occupied with a clockwise interior angle being observed by repetition (at least twice) at each traverse station.

Directions for the traverse are reckoned using the angle $Q12$ turned from the line of known azimuth $1Q$ to line 12. The azimuth of line 12 is $8°30'00'' + 97°50'00'' = 106°20'00''$. The azimuth of each succeeding traverse line then is calculated by adding the clockwise interior angle to the *back azimuth* of the preceding line. For example, the forward azimuth of line $23 = 106°20'00'' + 180°00'00'' + 131°35'00'' - 360°00'00'' = A_{12} + \alpha_2 - 180° = 106°20'00'' + 131°35'00'' - 180° = 57°55'00''$, as illustrated in Figure 8.8. Similarly, the azimuth of line $34 = A_{23} + 180° + \alpha_3 = 57°55'00'' + 180° + 97°35'00'' = 335°30'00''$, and so forth, as illustrated in Table 8.3.

For the example traverse, the angular error of closure can be checked by taking the sum of the angles α_i in Table 8.3 or

$$\alpha_1 + \alpha_2 + \alpha_3 + \alpha_4 + \alpha_5 + \alpha_6 = 4(180°)$$

In general, for any polygon with n sides, the sum of the interior angles is given by

$$\alpha_1 + \alpha_2 + \cdots + \alpha_n = (n - 2)(180°) \tag{8.3}$$

Therefore, the most rapid and simple field check of observed interior angles is to calculate the sum of the observations and compare it with $(n - 2)180°$.

Table 8.4 shows the sum of the observed interior angles for the example traverse. The angular error of closure is $03'$, the same value as determined previously.

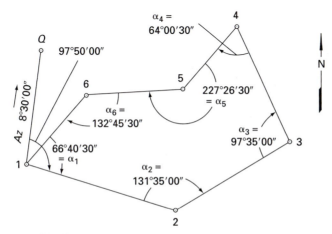

FIGURE 8.7
Closed-loop traverse with interior angles observed.

TABLE 8.3
Azimuths using interior angles

Line	Angle	Unadjusted azimuth		Correction	Adjusted azimuth
12		106°20'00"			106°20'00"
21		286°20'00"			
	$+\alpha_2$	131°35'00"			
23		$\overline{57°55'00''} = A_{12} - 180° + \alpha_2$		−30"	57°54'30"
32		237°55'00"			
	$+\alpha_3$	97°35'00"			
34		$\overline{335°30'00''} = A_{23} + 180° + \alpha_3$		−01'00"	335°29'00"
43		155°30'00"			
	$+\alpha_4$	64°00'30"			
45		$\overline{219°30'30''} = A_{34} - 180° + \alpha_4$		−01'30"	219°29'00"
54		39°30'30"			
	$+\alpha_5$	227°26'30"			
56		$\overline{266°57'00''} = A_{45} - 180° + \alpha_5$		−02'00"	266°55'00"
65		86°57'00"			
	$+\alpha_6$	132°45'30"			
61		$\overline{219°42'30''} = A_{56} - 180° + \alpha_6$		−02'30"	219°40'00"
16		39°42'30"			
	$+\alpha_1$	66°40'30"			
12		$\overline{106°23'00''} = A_{61} - 180° + \alpha_1$		−03'00"	106°20'00"
12		106°20'00"			

Angular error of closure = 03'00"

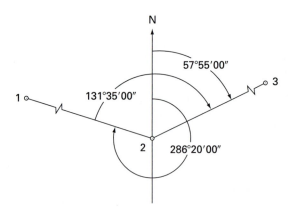

FIGURE 8.8
Calculation of an azimuth using an interior angle.

If adjusted angles are desired at this point, then assuming that all angles were observed with equal precision, the error is distributed equally among the angles and $180''/6 = 30''$ is subtracted from each observed angle, as shown in Table 8.4.

The simplest way to compute adjusted azimuths is to apply a cumulative correction to the unadjusted azimuths in Table 8.3. Therefore, corrections of 30", 01'00", 01'30", 02'00", 02'30", and 03'00" are applied respectively to $A_{23}, A_{34}, \ldots, A_{61}$. These corrections and adjusted azimuths are listed in Table 8.3. An alternative procedure is to use adjusted angles to compute the adjusted azimuths and bearings, as shown in Table 8.5.

TABLE 8.4
Checking and adjustment of interior angles

Station	Observed angles	Correction	Adjusted angles
1	66°40′30″	−30″	66°40′00″
2	131°35′00″	−30″	131°34′30″
3	97°35′00″	−30″	97°34′30″
4	64°00′30″	−30	64°00′00″
5	227°26′30″	−30″	227°26′00″
6	132°45′30″	−30″	132°45′00″
Sum =	720°03′00″		720°00′00″ checks
$(n-2)180° =$	720°00′00″		
Closure =	03′00″		

TABLE 8.5
Azimuth calculation using interior angles

Line	Angle	Azimuth	Bearing
12		106°20′00″	S73°40′00″E
21		286°20′00″	
	$+\alpha_2$	131°34′30″	
23		57°54′30″	N57°54′30″E
32		237°54′30″	
	$+\alpha_3$	97°34′30″	
34		335°29′00″	N24°31′00″W
43		155°29′00″	
	$+\alpha_4$	64°00′00″	
45		219°29′00″	S39°29′00″W
54		39°29′00″	
	$+\alpha_5$	227°26′00″	
56		266°55′00″	S86″55′00″W
65		86°55′00″	
	$+\alpha_6$	132°45′00″	
61		219°40′00″	S39°40′00″W
16		39°40′00″	
	$+\alpha_7$	66°40′00″	
12		106°20′00″ check	

Note that, in this closed-loop traverse, although the calculated azimuths are internally consistent, the absolute orientation is based entirely on one angle observed between lines 1Q and 12. This aspect of a closed-loop traverse is a weakness in this procedure that was mentioned in Section 8.5. To eliminate the weakness, another angle should be observed from some other traverse point to another independent line of known azimuth. If points are not available to permit this second tie, an astronomic observation for azimuth should be made for line 34 or 45 to provide a redundancy in orientation of the traverse. Short traverses, such as the one illustrated in Figure 8.7, may not require such a tie, depending on the nature of the survey. Long closed-loop traverses should always be tied to at least two lines of known direction.

Allowable closures for interior traverses are similar to those for deflection angle traverses. The sum of the interior angles should not deviate from $(n-2)180°$ by more than the square root of the number of instrument setups times the estimated standard deviation in observing the angles, where this estimate usually is taken as the least count of the instrument.

8.8
TRAVERSE BY ANGLES TO THE RIGHT

This method can be used in open, closed, or closed-loop traverses. Repeating, optical, or electronic theodolites or a total station system may be employed to turn the angles. When a repeating instrument is used, the backsight to the preceding station is taken with the horizontal circle set on 0. Then, with the upper clamp loosened, the instrument is turned on the upper motion, a foresight is taken on the next station, and the angle turned to the right is observed on the clockwise circle. If an electronic theodolite or total station instrument is used, first set the angles-right mode by pressing the proper key. Then take a backsight on the preceding station using the horizontal clamp and tangent screw and set the horizontal circle to 0 by pressing the zero-set or equivalent button. With the horizontal clamp loosened, turn to the next station and take a foresight. The clockwise angle will be displayed on the register of the instrument. As in the other methods for observing traverse angles, it is advisable to observe the angles twice, once with the telescope direct and once with the telescope reversed.

A traverse by angles to the right is shown in Figure 8.9. This traverse originates at line YZ for which the azimuth is 250°00′00″ and closes on line TU with a given azimuth of 319°59′45″ from T to U. Figure 8.10 shows the notes from which the angles were determined. Note that angles were observed with the telescope direct and reversed and the average angle was used for computation of the azimuths.

Azimuths are calculated by adding the angle to the right to the back azimuth of the previous line. Computation of these azimuths is shown in Table 8.6. The angular error of closure is −25″. The condition of closure for this example can be expressed by the following equation:

$$A_{YZ} + \alpha_1 + \alpha_2 + \alpha_3 + \alpha_4 + \alpha_5 - (4)(180°) - A_{TU} = 0 \qquad (8.4)$$

which in general becomes

$$A_1 + \alpha_1 + \alpha_2 + \cdots + \alpha_n - (n - 1)(180°) - A_2 = 0 \qquad (8.5)$$

TABLE 8.6
Azimuths calculated by angles to the right

Line	Azimuth	Correction	Corrected azimuth
YZ	250°00′00″		
	210°01′32″		
	460°01′32″		
	−180°00′00″		
AY	280°01′32″	05″	280°01′37″
	140°00′30″		
	420°02′02″		
	−180°00′00″		
RA	240°02′02″	10″	240°02′12″
	290°01′10″		
	530°03′12″		
	−180°00′00″		
SR	350°03′12″	15″	350°03′27″
	90°01′13″		
	440°04′25″		
	−180°00′00″		
TS	260°04′25″	20″	260°04′45″
	59°54′55″		
TU	319°59′20″	25″	319°59′45″
TU	319°59′45″ fixed		
	Angular closure = 25″		

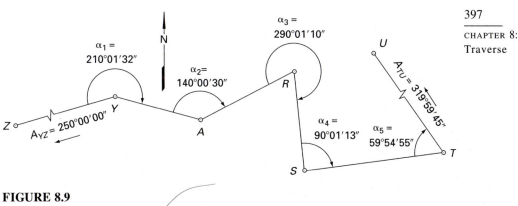

FIGURE 8.9
Traverse by angles to the right with a direction instrument.

in which A_1 and A_2 are fixed azimuths from the north at lines of origin and closing, respectively, and there are n traverse stations (not counting fixed stations Z and U).

If adjusted azimuths are desired at this point, assuming observations of equal precision, the error of closure is distributed equally among the five angles, or $25''/5 = 05''$ per angle. The adjustment is made by applying a cumulative correction to each unadjusted azimuth. The corrections and adjusted azimuths also are listed in Table 8.6.

		GRANT FARM TRAVERSE				Dec. 10, 1997 Clear 40°F 21
Station At	Tel. From To	Circle	Angle to the right	Average angle		R. Hoyt ⚒
						C. Roup
Z		0°00'00"				K. Krone
Y A	D	210 01 30	210 01 30			
A	R	60 03 04	210 01 34			Nikon No. 452
				210° 01' 32"		Electronic theodolite
Y		0°00'00"				
A R	D	140 00 28	140 00 28			Note: Traverse alignment
R	R	280 01 00	140 00 32			and references
				140 00 30		are on p. 20
A		0°00'00"				
R S	D	290 01 05	290 01 05			
S	R	220 02 05	290 01 15			
				290 01 10		
R		0°00'00"				
S T	D	90 01 15	90 01 15			Closure condition
T	R	180 02 26	90 01 11			$A_{YZ} + \sum\limits_{i=1}^{5}\alpha_i + (4)(180°) - A_{TU} = 0$
				90 01 13		250° 00' 00"
						+789° 59' 20"
S		0°00'00"				−720° 00' 00"
T U	D	59 54 53	59 54 53			−319° 59' 45"
U	R	119 49 50	59 54 57			0° 00' 25" = error of closure
				59 54 55		
			Σ	789 59 20		

FIGURE 8.10
Notes for an angles-to-the-right traverse.

8.9
AZIMUTH TRAVERSE

An advantage of the azimuth method is that the simple statement of one angular value, the azimuth, fixes the direction of the line to which it refers. The method is used extensively on topographic and other surveys where a large number of details are located by angular and linear measurements from the traverse stations. Any angular error of closure of a traverse becomes evident by the difference between initial and final observations taken along the first line. The reference meridian may be either true or assumed.

Successive stations are occupied, beginning with the line of known or assumed azimuth. At each station, the theodolite is "oriented" by setting the horizontal circle index (on a repeating optical theodolite) to read the back azimuth (forward azimuth $\pm$ 180°) of the preceding line and then taking a backsight on the preceding station. When an electronic theodolite is used, the azimuth can be set by keying in the value via the control panel or by rotating the alidade about the vertical axis to the desired value, which can be "held" electronically (see instruction manual for instrument) while a backsight is taken. The horizontal clamp (the upper clamp on a repeating instrument) is released, the alidade is turned about the vertical axis, and a foresight on the following traverse station is secured. The reading indicated by the horizontal circle index on a repeating theodolite or displayed on the register of an electronic instrument, as an angle to the right, is the azimuth of the forward line.

Figure 8.11 shows a closed-loop traverse run by the azimuth method using a repeating optical-reading theodolite. The traverse is begun at station 1 with a backsight along line 15, for which the azimuth is known to be 270°28' by a solar observation. With the instrument set over station 1, the horizontal circle index is set to 270°28' on the clockwise circle and a backsight is made on station 5 using the lower motion. The instrument then is turned on the upper motion and a foresight is taken on station 2. Because 0 is oriented toward true north (Figure 8.11), the clockwise circle reading (350°30') is the azimuth from the north of line 12. Note that any other angle observed on the clockwise circle to another point from this setup also is an azimuth from the north. When the instrument is moved to point 2, the back azimuth of line 12 is computed by subtracting 180° from the forward azimuth (350°30' − 180° = 170°30'), and this value is set on the clockwise circle before a backsight, using the lower motion, is taken on station 1. Once again, the clockwise circle is oriented with 0 toward true north, and when the instrument is turned on the upper motion for a foresight on station 3, the angle observed on the clockwise circle (303°05') is the azimuth of line 23 from true north. This operation is repeated for each successive traverse station. When the final setup is made over station 5, the azimuth observed on the foresight along line 51 should equal the azimuth of line 15 minus 180°; in this case, 270°28' − 180° = 90°28'. Any deviation between this last observation and the fixed direction of the closing line is the angular error of closure. Figure 8.12 illustrates the notes recorded for the traverse shown in Figure 8.11. The closure error is 01'. Because angles have been observed only to the nearest minute, this error could be applied to one angle to provide adjusted directions.

8.10
COMPASS TRAVERSE

Use of the compass was described in Section 6.19. When a compass traverse is performed, forward and back bearings are observed from each traverse station and distances are taped

FIGURE 8.11
Azimuth traverse.

399

CHAPTER 8:
Traverse

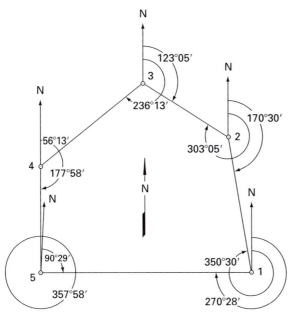

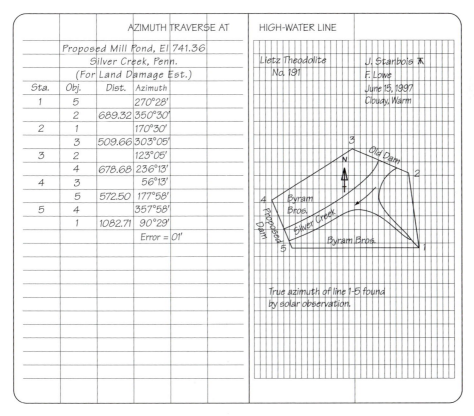

		AZIMUTH TRAVERSE AT		HIGH-WATER LINE
	Proposed Mill Pond, El 741.36			
	Silver Creek, Penn.			Lietz Theodolite J. Starbols
	(For Land Damage Est.)			No. 191 F. Lowe
Sta.	Obj.	Dist.	Azimuth	June 15, 1997
1	5		270°28'	Cloudy, Warm
	2	689.32	350°30'	
2	1		170°30'	
	3	509.66	303°05'	
3	2		123°05'	
	4	678.68	236°13'	
4	3		56°13'	
	5	572.50	177°58'	
5	4		357°58'	
	1	1082.71	90°29'	
			Error = 01'	True azimuth of line 1-5 found
				by solar observation.

FIGURE 8.12
Notes for a short closed azimuth traverse.

or paced. If local attraction exists at any traverse station, both the forward and back bearings are affected equally. Thus, interior angles computed from forward and back bearings are independent of local attraction. Field notes for a closed-loop compass traverse are shown in Figure 8.13. The declination of 20°15'E was set off on the compass (Figure 6.17, Section 6.14) so that bearings are referred to the true meridian.

Computed interior angles are listed in column 4 of the notes shown in Figure 8.13. The sum of these angles reveals a closure error of 25'. Assuming that all bearings are of equal precision and not correlated, this error is distributed equally among the five interior angles so that 05' is added to each unadjusted interior angle to provide an adjusted angle. Adjusted interior angles are listed in column 5 of the notes in Figure 8.13.

Because the forward and back bearings of AB are numerically equal and opposite in direction, the correct forward bearing from A to B can be taken as S30°40'W. Holding the direction of AB fixed, the remainder of the adjusted bearings are computed and listed in column 6 of the notes shown in Figure 8.13.

If the error in the sum of interior angles exceeds 10' to 15' times the square root of the number of angles, it is likely that a blunder in reading the compass has occurred, and the field measurements should be repeated. If the error is within permissible limits but cannot be divided equally among the angles in amounts of 05' or 10', the greater corrections (in multiples of 05') should be applied arbitrarily to those angles for which the conditions of observing were estimated to be least favorable. The precision of most compass measurements does not justify computations with a precision closer than multiples of 05'.

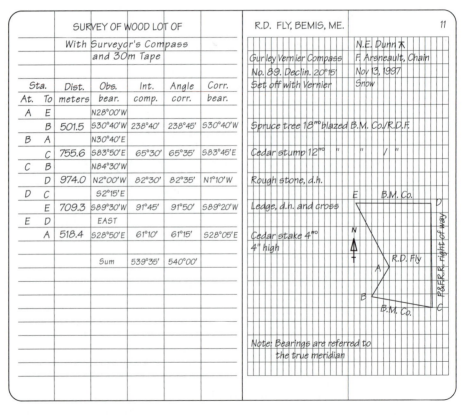

FIGURE 8.13
Notes for a compass survey.

When distances between traverse stations are determined by the stadia method (Chapter 7), the resulting survey is called a *stadia traverse*. This type of traverse is appropriate for certain types of preliminary and reconnaissance surveys, rough surveys for boundary locations, and topographic mapping surveys. Horizontal positions can be determined with relative accuracies of 1 part in 300 to 400 by stadia traverses. The stadia traverse is sufficiently accurate and considerably more rapid and economical than corresponding surveys made with a theodolite and a tape. One advantage of the stadia traverse is that, if desired, elevations may be determined concurrently with horizontal position.

Field notes for a small closed-loop traverse run with an optical-reading repeating theodolite are shown in Figure 8.14. The instrument is set up over traverse station 1, the elevation and location of which are known. The collimation error of the vertical circle is observed and recorded and the height of instrument (h.i.) above the station over which the instrument is set is measured with a rod or tape to the nearest centimeter or nearest tenth of a foot. If the direction of a line in the traverse is not known, a magnetic direction along one line can be observed. In the example traverse, a magnetic bearing first is observed along line 14. This value is converted to an azimuth and corrected for declination. These calculations are shown at the top left of the sample notes in Figure 8.14. This azimuth then is set

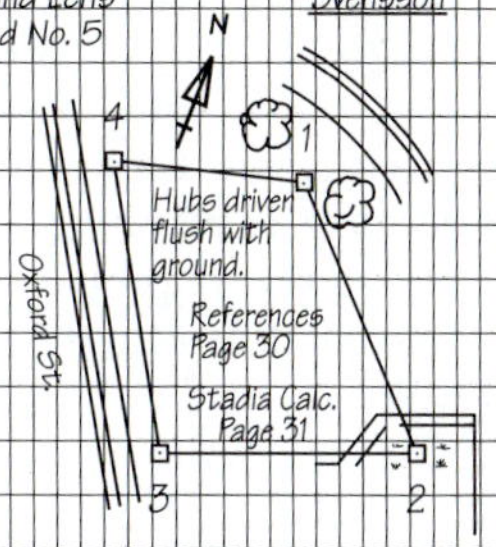

FIGURE 8.14
Stadia traverse for horizontal and vertical control.

off on the clockwise horizontal circle and station 4 is resighted with the lower motion. The stadia interval and zenith angle for the sight on station 4 are observed and recorded in columns 2 and 5, respectively, as shown in the notes in Figure 8.14. In this example, the middle cross hair was set on the value of the h.i. = 1.49 m before reading the zenith angle, so that it was not necessary to record a rod reading. Should it not be possible to set the middle cross hair on the h.i., due to some obstruction on the line of sight, the rod reading sighted is recorded directly above the zenith angle in column 5 of the notes (Figure 8.14, from point 2 to 3). When observations along the backsight to station 4 are completed, the upper clamp is loosened and a foresight is taken on station 2, where the stadia interval, zenith angle, and azimuth are observed and recorded. The azimuth then is observed on the clockwise horizontal circle. This and all subsequent azimuths in the traverse are observed according to the *azimuth method,* as outlined in Section 8.9. If the middle cross hair is set on a rod reading equal to the value of the h.i., it need not be recorded and the difference in elevation is V computed using Equation (7.6a) or (7.6b) (Section 7.7).

When all observations are completed at the first station, the instrument is moved to station 2, where the h.i. is measured (1.64 m) and a backsight is taken on station 1 with the back azimuth of line 12 ($145°19' + 180° = 325°19'$) set on the clockwise horizontal circle. When the stadia interval and zenith angle from 2 to 1 are observed, it is possible to check these readings against the corresponding readings from 1 to 2. The stadia intervals should agree to within 2 or 3 mm and the zenith angles should have a sum that equals 180°. Note that, if a collimation error in the vertical circle is present, the forward and backward zenith or vertical angles will show a systematic difference. Most theodolites now have vertical circle compensators, so the collimation error = 0. A foresight is then taken on station 3, where the stadia interval, zenith angle, and azimuth are observed and recorded.

The procedure just described is repeated for each station in the traverse. When the final station is occupied, the azimuth along the initial line of known direction (line 41, in this example) should correspond to 180° ± the known azimuth or $272°30' - 180°00' = 92°30'$. The difference is the angular error of closure, which is 04' in this example.

Horizontal distances and differences in elevation are computed using the stadia intervals and zenith angles. These values are recorded in columns 3 and 6 in the notes. Horizontal distances are averaged and average differences in elevation are used to check the closure error in elevation. These values are tabulated at the bottom of the left page in the notes. The closure error in elevation for this example is −0.07 m, which is distributed among the traverse stations in proportion to the lengths of the traverse sides (Section 5.42). Because the total length around the traverse is 283.3 m, the correction to the difference in elevation between stations 1 and 2 is $(74/283)(0.07) = +0.02$ m. The remaining corrections are computed in a similar manner and applied to the average differences in elevation, as shown at the bottom of the left page of notes in Figure 8.14. These corrected differences in elevation then are used to calculate adjusted elevations for traverse stations listed in column 7 of the notes.

The error in angular closure of 04' is distributed equally among the four unadjusted azimuths by applying a cumulative correction to each observed direction. Here, the correction per interior angle is 04'/4 = 01' per angle or the corrections are 01', 02', 03', and 04' per azimuth for lines 12, 23, 34, and 41, respectively. These values are tabulated at the bottom of the right page of the sample notes in Figure 8.14.

Larger angular closures can be tolerated in a stadia traverse than one in which distances are determined with a tape or EDM equipment. If the estimated standard deviation of a horizontal distance measured by stadia is assumed to be 0.3 m (Section 7.14), an angular error compatible with this error in a distance of 100 m (Section 3.7) is $\tan^{-1} 0.3/100 \simeq 10'$ of arc. Consequently, angular closure errors of up to 10' are reasonable in a stadia traverse.

Many of the hubs marking the location of highways, railroads, and other public and private works are bound to be uprooted or covered during the progress of construction and must be replaced, often more than once, before construction is completed. Such hubs marking traverse stations usually are tied by angular or linear measurements to temporary wooden hubs, called *reference hubs,* or to other objects that are not likely to be disturbed. A traverse station is said to be *referenced* when it is so tied to nearby objects that it can be replaced readily. The manner of referencing a traverse station is indicated in the notes by an appropriate sketch. The precision of referencing should be comparable to that of the measurements between traverse stations. Particularly on land surveys, the corners should be tied to nearby objects that can be found readily, that are not likely to be moved or obliterated, and that are of a more-or-less permanent character.

Often in land surveying, a corner is incorrectly said to be "witnessed" when angular and linear measurements are taken to nearby objects of the character just mentioned. This unfortunate designation leads to the confusion of reference marks for a corner that has been established with *witness corners* (Section 18.5), which are markers set on one or more of the land lines leading to a corner when the corner falls in a place where it would be either impossible or impractical to establish or to maintain a monument.

Figures 8.15 through 8.18 illustrate several methods of referencing a traverse station. The station shown in Figure 8.15 is tied by linear measurements to three specific points located on semipermanent structures or objects located around the traverse station. Location or relocation of the traverse station is accomplished by intersecting the reference distances. Where possible, reference points should be chosen to be easily identifiable and ought to be located on objects or structures of a permanent nature. Tie lines should intersect at favorable angles (greater than 30° and less than 150°), so that the station can be relocated with certainty.

Figure 8.16 shows a station referenced by setting reference hubs on a line passing through the station and taking linear measurements from these hubs to the station. In this example, four reference hubs are set at the ends of lines AB and CD, which intersect at traverse station Y-1. With the measurements as shown, any two reference hubs may be destroyed and the station may still be relocated.

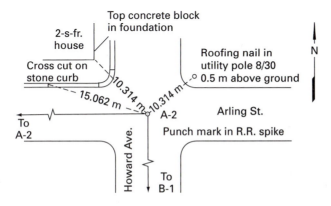

FIGURE 8.15
Reference to a traverse station in an urban area.

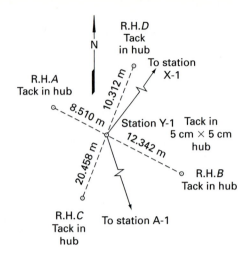

FIGURE 8.16
Referencing by reference hubs.

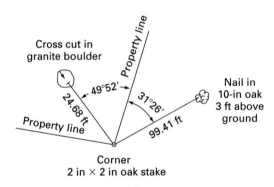

FIGURE 8.17
References to a property corner.

A typical method for referencing property corners in land surveying is illustrated in Figure 8.17. Both angular and linear ties are employed.

8.13
TRAVERSE COMPUTATIONS

Traverse operations in the field yield observed angles or directions and measured distances for a set of lines connecting a series of traverse stations. Angles, or directions converted to angles, can be checked to obtain the angular error of closure that may be distributed among the angles to provide preliminary adjusted values from which preliminary azimuths or bearings are computed. When the observed distances have been corrected for all systematic errors (Sections 4.14 to 4.23, 4.30, 4.36 to 4.39), the preliminary directions and reduced distances are suitable for use in traverse computations performed in a plane rectangular coordinate system.

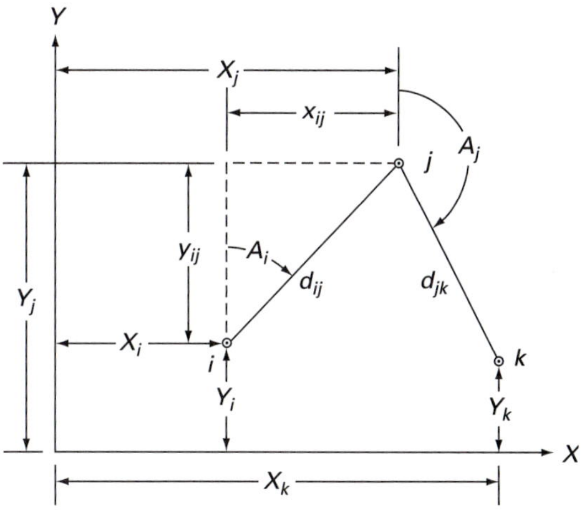

FIGURE 8.18
Departure and latitude of line *ij*.

Computations with plane rectangular coordinates are extremely adaptable for use with pocket calculators and personal, mini-, and mainframe computers. Coordinated points have an unambiguous designation that is ideal for storage and retrieval from spatial information systems (see Sections 1.11, 3.12, and 14.18) and is compatible with many methods for plotting the locations of points, including automatic plotting techniques.

The system within which the coordinates are computed may have a national, state, regional, local, or completely arbitrary origin. Plane coordinate computations in the United States usually are performed in one of the state plane coordinate systems established by the National Ocean Survey for each state. Characteristics of these systems are described in Chapter 11.

Computations with plane coordinates are illustrated by reference to Figure 8.18, where traverse line *ij* has a reduced horizontal distance of d_{ij} and azimuth of A_i. Let x_{ij} and y_{ij} be designated the departure and latitude for line *ij*, so that

$$x_{ij} = d_{ij} \sin A_i$$
$$y_{ij} = d_{ij} \cos A_i$$

(8.6)

The algebraic signs of the departure and latitudes for a traverse line depend on the signs of the sine and cosine of the azimuth of that line. Figure 8.19 shows the algebraic signs of the sine and cosine functions for each of the four quadrants. Pocket calculators and personal, mini-, and mainframe computers with internal routines for trigonometric functions yield the proper sign automatically, given the azimuth. When bearings are used to describe the direction of the line, the sine and cosine always are considered positive and the algebraic sign of the latitude and departure is taken from the quadrant. Consequently, the azimuth is the logical means of defining direction in traverse calculations when modern calculators or computers are used.

Please observe that the terms *departure* and *latitude,* as well as *X* and *Y,* are specified in that order throughout this text. Note that several handheld calculators with direct rectangular to polar or polar to rectangular keys use the reverse notation of latitude and departure, *Y* and *X*.

For example, in Figure 8.18, let A_i = 43°30′00″ and d_{ij} = 432.182 m. Applying Equation (8.6), the departure = x_{ij} = 432.182 sin 43°30′00″ = 432.182(0.6883546) = 297.494 m and the latitude = y_{ij} = 432.182 cos 43°30′00″ = 432.182(0.7253744) =

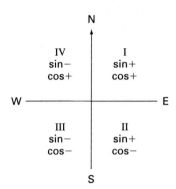

FIGURE 8.19
Signs of azimuth functions.

313.494 m. Continuing, if $A_j = 153°25'40''$, $d_{jk} = 385.151$ m, then the departure $= x_{jk} = 385.151(0.4473255) = 172.288$ m and the latitude $= y_{jk} = 385.151 (-0.8943712) = -344.468$ m.

Given that the coordinates of station i are X_i and Y_i (Figure 8.18), the coordinates of station j are

$$X_j = X_i + x_{ij}$$
$$Y_j = Y_i + y_{ij}$$

(8.7)

If the coordinates of station i in the example are $X_i = 142,482.352$ m, $Y_i = 43,560.805$ m, then the coordinates of station j are

$$X_j = 142,482.352 + 297.494 = 142,779.846 \text{ m}$$

$$Y_j = 43,560.805 + 313.494 = 43,874.299 \text{ m}$$

and the coordinates of station k are

$$X_k = 142,779.846 + 172.288 = 142,952.143 \text{ m}$$

$$Y_k = 43,874.299 - 344.468 = 43,529.831 \text{ m}$$

Consequently, a major portion of traverse computations consists of calculating departures and latitudes for successive traverse lines and cumulating the values to determine the coordinates for consecutive traverse stations.

The inverse solution also is possible, given the coordinates for the two ends of a traverse line. The distance between stations i and j in Figure 8.18 is

$$d_{ij} = [(X_j - X_i)^2 + (Y_j - Y_i)^2]^{1/2}$$

(8.8)

and the azimuth of line ij from the north is

$$A_{Nij} = \arctan \frac{X_j - X_i}{Y_j - Y_i}$$

(8.9a)

The azimuth of line ij from the south is

$$A_{Sij} = \arctan \frac{X_i - X_j}{Y_i - Y_j}$$

(8.9b)

In Equations (8.6) through (8.9b), due regard should be paid to algebraic signs of the trigonometric functions (Figure 8.19) and the differences between coordinates at the two ends of a line.

When coordinates for all the traverse points (or all of the departures and latitudes) for all lines have been computed, a check is necessary on the accuracy of the observations and the validity of the calculations. In a closed traverse, the algebraic sum of the departures should equal the difference between the X coordinates at the beginning and ending stations of the traverse. Similarly, the algebraic sum of the latitudes should equal the difference between the Y coordinates at the beginning and ending stations. If Equation (8.7) is used to calculate coordinates directly, the calculated coordinates should agree with the given values for the final or closing station. It follows that, in a closed-loop traverse, the algebraic sum of the departures and the algebraic sum of the latitudes each must equal 0. For a traverse containing $i = 1, 2, \ldots, n$ stations, starting at station $i = 1$ and terminating at station $i = n$, the foregoing conditions can be expressed as follows:

$$X_n - X_1 = \sum_{i=1}^{n-1} x_{i,i+1} = \sum_{i=1}^{n-1} \text{departures}$$

$$Y_n - Y_1 = \sum_{i=1}^{n-1} y_{i,i+1} = \sum_{i=1}^{n-1} \text{latitudes}$$

(8.10a)

Equations (8.10a) rarely are satisfied exactly in practice, owing to random observational errors, uncorrected systematic errors in observations, and inaccuracies in the given coordinates for a closed traverse. The amounts by which Equations (8.10a) fail to be satisfied are called the *errors in closure in position* or simply the *closures* for a traverse. The closure *corrections* dX and dY, which are of opposite signs to errors, for a traverse defined as for Equations (8.10a) are

$$\mathrm{d}X = (X_n - X_1) - \sum_{i=1}^{n-1} x_{i,i+1}$$

$$\mathrm{d}Y = (Y_n - Y_1) - \sum_{i=1}^{n-1} y_{i,i+1}$$

(8.10b)

When preliminary traverse computations are completed, closure corrections are evaluated and an adjustment is performed. The subject of traverse adjustment is covered in Sections 8.16, 8.17, 8.39, and 9.21. First, consider the complete traverse computations for a closed traverse.

8.14
COMPUTATIONS FOR A CLOSED TRAVERSE

Figure 8.20 shows a traverse that originates at station 1, and closes on station 5. Observed distances, corrected for systematic errors, and angles to the right observed once direct and one reversed with a 30" repeating theodolite, are given in Table 8.7.

The angular error of closure is 01' (Equation (8.5) and Table 8.8). Adjusted azimuths are calculated according to the procedure in Section 8.8. Distribution of the angular error equally among the five angles leads to a correction of −12" per angle. Observation of the angles twice with a 30" instrument does not warrant this precision. Therefore, the corrections to angles are rounded arbitrarily to −10" each to angles 1, 2, and 3 and −15" to angles 4 and 5. Table 8.8 shows azimuth calculations and adjusted azimuths.

Distances, adjusted azimuths, fixed coordinates for the beginning and ending stations, calculated azimuths, and calculated coordinates are listed in Table 8.9. Note that coordinates can be calculated directly using Equation (8.7) and tabulation of departures and latitudes is not necessary unless these values are to be used for some specific purpose. They

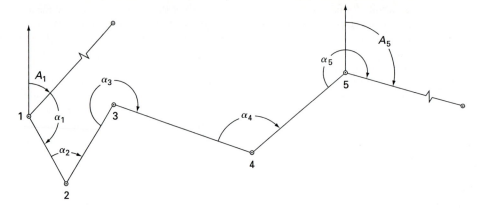

FIGURE 8.20
Closed traverse.

TABLE 8.7
Data for closed traverse

Line	Distance, m	Angle to the right	
12	305.41	$\alpha_1 = 104°00'30''$	$A_{N_1} = 43°31'30''$
23	359.61	$\alpha_2 = 63°20'00''$	$A_{N_5} = 106°17'30''$
34	612.15	$\alpha_3 = 258°11'00''$	
45	485.12	$\alpha_4 = 120°45'00''$	$\sigma_d = 0.01$ m
		$\alpha_5 = 236°30'30''$	$\sigma_\alpha = 30''$

TABLE 8.8
Azimuth calculations

Line	Unadjusted azimuth	Correction	Adjusted azimuth
A_1	43°31'30''		
	α_1 104°00'30''		
12	147°32'00''	$-10''$	147°31'50''
	α_2 63°20'00''		
32	210°52'00''		
23	30°52'00''	$-20''$	30°51'40''
	α_3 258°11'00''		
43	289°03'00''		
34	109°03'00''	$-30''$	109°02'30''
	α_4 120°45'00''		
54	229°48'00''		
45	49°48'00''	$-45''$	49°47'15''
	α_5 236°30'30''		
	286°18'30''		
	$-180°00'00''$		
A_5	106°18'30''	$-01'00''$	106°17'30''
	106°17'30'' fixed		
Angular error	01'00''		

TABLE 8.9
Traverse computations for closed traverse of Figure 8.20

Sta-tion	Distance, m	Azimuth	Departures, x	Latitudes, y	Coordinates, m X	Coordinates, m Y
1					4321.404	6240.562
	305.41	147°31'50"	163.959	−257.668		
2					4485.363	5982.894
	359.61	30°51'40"	184.465	+308.694		
3					4669.828	6291.588
	612.15	109°02'30"	578.654	−199.717		
4					5248.482	6091.871
	485.12	49°47'15"	370.464	+313.205		
5					5618.946	6405.076
	1762.29		$\Sigma = 1297.542$	$\Sigma = 164.514$	[5619.243]	[6405.272]
			$X_5 - X_1 = 1297.839$	$Y_5 - Y_1 = 164.710$		
Closure corrections:			$dX = +0.297$	$dY = +0.196$	$dX = 0.297$	$dY = 0.196$

$$d_c = (dX^2 + dY^2)^{1/2} = [(0.297)^2 + (0.196)^2]^{1/2} = 0.36 \text{ m}$$

are included here for illustrative purposes. The closure corrections in X and Y coordinates at station 5 are 0.297 m and 0.196 m, respectively. These values also may be calculated by applying Equations (8.10b) to yield $dX = (X_5 - X_1) - \Sigma$ dep. $= 1297.839 - 1297.542 = 0.297$ and $dY = (Y_5 - Y_1) - \Sigma$ lat. $= 164.710 - 164.514 = 0.196$ m. The values for closure here are shown graphically in Figure 8.21; the fixed and calculated positions for station 5 are shown to a larger scale. Using Equation (8.8), the resultant closure, $d_c = (dX^2 + dY^2)^{1/2}$, is the distance from the calculated position at 5' to the fixed location 5. In a closed traverse of this type, the resultant closure is caused by random variations in the observations and uncorrected systematic errors in distance and direction. For this example, $d_c = [(0.196)^2 + (0.297)^2]^{1/2} = 0.36$ m.

The ratio of the closure d_c to the distance traversed provides an indication of the goodness of the survey and often is referred to as the *ratio of misclosure* (RoM; Section 2.11) or the *relative accuracy ratio* for the traverse. Surveys frequently are classified according to relative accuracy ratios such as 1/5000, 1/10,000, and so on. When blunders or uncorrected systematic errors in distance or directions are present, the closure and the consequent relative accuracy ratio will be too large. The relative accuracy ratio provides a measure of the relative merits of various traverses but should not be misconstrued as an indication of the absolute accuracy in position for each station in the traverse. For the example traverse, the RoM is 0.36/1762 = 1/4894.

Specifications for various classifications of traverse and corresponding relative accuracies are discussed in Section 8.21.

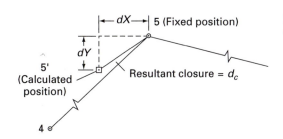

FIGURE 8.21
Closure in a closed traverse.

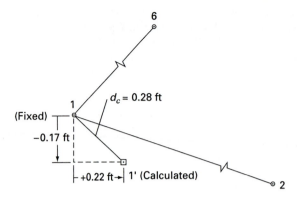

FIGURE 8.22
Closure in a closed-loop traverse.

8.15
COMPUTATIONS FOR A CLOSED-LOOP TRAVERSE

When a traverse originates and closes on the same station, the algebraic sum of the departures and the algebraic sum of the latitudes each must equal 0. As an example, consider the traverse illustrated in Figure 8.7, for which the adjusted azimuths are listed in Table 8.5, and station 1 is held fixed with given coordinates.

Distances, azimuths, calculated departures and latitudes, given coordinates for station 1, and calculated coordinates for the rest of the traverse stations are listed in Table 8.10. The sum of the departures and the sum of the latitudes are 0.22 and -0.17 ft,

TABLE 8.10
Computations for a closed-loop traverse of Figure 8.7

Station	Distance, ft	Azimuth	Departures, x	Latitudes, y	Coordinates, ft X	Y
1					4382.09	6150.82
	405.24	106°20′00″	+388.89	−113.96		
2					4770.98	6036.86
	336.60	57°54′30″	+285.17	+178.83		
3					5056.15	6215.69
	325.13	335°29′00″	−134.92	+295.82		
4					4921.23	6511.51
	212.91	219°29′00″	−135.38	−164.33		
5					4785.85	6347.18
	252.19	266°55′00″	−251.82	−13.56		
6					4534.03	6333.62
	237.69	219°40′00″	−151.72	−182.97		
1					4382.31	6150.65
	1769.76	Closure errors	Σ = 0.22	Σ = −0.17	(4382.09)	(6150.82)
					dX = −0.22	dY = +0.17
	$d_c = [(0.22)^2 + (0.17)^2]^{1/2} = 0.28$ ft				Closure corrections	

respectively. These algebraic sums represent the errors of closures for the loop traverse. According to Equations (8.10b), the closure corrections are $dX = 0.0 - 0.22 = -0.22$ ft, $dY = 0.0 - (-0.17) = +0.17$ ft. Figure 8.22 shows that the resultant closure d_c, calculated by Equation (8.8), is 0.28 ft. For a closed-loop traverse of this type, the resultant closure is a function of random variations in the observations and uncorrected systematic errors in angles or directions. Systematic errors in distance measuring equipment will cancel and are not revealed by the mathematical closure of the traverse. Note also that, although the directions are consistent within the traverse, the entire polygon could be rotated with respect to the grid system with no indication of this rotation in the results from the traverse computations. A second tie to a line of known direction is necessary to reveal this type of error.

8.16
TRAVERSE ADJUSTMENT

Traverse adjustment is for providing a mathematically closed figure and at the same time yielding the best estimates for horizontal positions of all the traverse stations. Methods of adjustment may be classified as approximate or rigorous. Consider the approximate adjustments first.

Traditional methods of approximate traverse adjustment have been developed to accommodate prevailing conditions in certain combinations of angular and linear precision in the observations. In this respect, three combinations still are common.

1. The precision in angles or directions exceeds its equivalent in linear distance observation (Section 3.7). This combination exists in the stadia survey, where azimuths can be determined to within $\pm 30''$ with little difficulty, while the relative precision in distance measurements will never exceed 1 part in 300 to 400.
2. Precision in angles or directions essentially is equal to its equivalent in the precision of distances. This combination is typified by a traverse with an engineer's transit and a steel tape, where the terrain is fairly regular and the distances are corrected for all possible systematic errors.
3. Precision in distances exceeds that in angles and directions. This combination is prevalent in traverses containing long lines, where angles and distances are measured with theodolites and calibrated EDM equipment or by using a total station system.

Approximate adjustments include the transit rule, the compass rule, and the Crandall method. The *transit rule* was developed for the first combination of precisions in observations. Unfortunately, the transit rule is valid only when the traverse lines are parallel with the grid system used for the traverse computations and, therefore, is not considered further in this text. The *compass rule* was developed for the second combination and can be shown to be rigorous when the condition that the angular precision equals the precision in linear distances is rigidly enforced. The compass rule is developed further in Section 8.17. The *Crandall method* is a rather complicated procedure that is more rigorous than either the compass or transit rule but requires substantially more computations. This method is not developed further in this book. The interested reader is referred to Baarda and Albercla (1962), and Richardus (1966) for detailed developments of these and other approximate methods.

The method of least squares provides the most rigorous adjustment, which allows for variation in precision in the observations, minimizes the random variations in the observations, provides the best estimates for positions of all traverse stations, and yields statistics relative to the accuracies of adjusted observations and positions. This method does require more of a computational effort than the approximate adjustments. However, it is well within the capabilities of desktop and even some of the more sophisticated handheld calculators. Theoretical development of the method of least squares is in Sections 2.22 to 2.31. Examples of traverse adjustment by the method of least squares can be found in Section 8.39. The student of surveying and the practicing surveyor should study these examples and become familiar with adjustment of traverse by the method of least squares. The results are well worth the effort.

8.17
ADJUSTMENT OF A TRAVERSE BY THE COMPASS RULE

For any traverse station i, let

δX_i = correction to X_i
δY_i = correction to Y_i
dX_t = total closure correction of the traverse in the X coordinates (by Equation (8.10b))
dY_t = total closure correction of the traverse in the Y coordinates (by Equation (8.10b))
L_i = distance from station i to the initial station
L = total length of traverse

then the corrections are

$$\delta X_i = \left(\frac{L_i}{L}\right) dX_t \quad \text{and} \quad \delta Y_i = \left(\frac{L_i}{L}\right) dY_t \tag{8.11a}$$

In an alternative procedure, corrections may be applied to the departures and latitudes prior to calculating coordinates. In this case,

$$\delta x_{ij} = \left(\frac{d_{ij}}{L}\right) dX_t \quad \text{and} \quad \delta y_{ij} = \left(\frac{d_{ij}}{L}\right) dY_t \tag{8.11b}$$

where δx_{ij} and δy_{ij} are respective corrections to the departure and latitude of line ij, which has a length of d_{ij}, and dX_t, dY_t, and L are as defined previously.

As an example, consider adjustment, by the compass rule, of the coordinates for the traverse shown in Figure 8.20 and calculated in Table 8.9. The closure corrections in departures and latitudes are 0.297 and 0.196 m, respectively. According to Equations (8.11a), $\delta X_2 = (305.41/1762)(0.297) = 0.051$ m, so that X_2 adjusted = 4485.363 + 0.051 = 4485.41 m rounded to two decimal places. Similarly, $\delta Y_2 = (305.41/1762)$ (0.196) = 0.034 m and Y_2 adjusted = 5982.894 + 0.034 = 5982.93 m. To calculate corrections to X_3 and Y_3, the cumulated distance to station 3 is used in Equations (8.11a). Here, $\delta X_3 = (665/1762)(0.297) = 0.112$ m and $\delta Y_3 = (665/1762)(0.196) = 0.074$ m, yielding X_3 adjusted = 4669.828 + 0.112 = 4669.94 m and Y_3 adjusted = 6291.588 + 0.074 = 6291.66 m. Distances, corrections, and adjusted coordinates are listed in Table 8.11.

TABLE 8.11
Adjustment of coordinates by the compass rule (traverse of Figure 8.20, calculated in Table 8.9)

Station	Distance, m	Corrections δX_i	Corrections δY_i	Adjusted coordinates, m X	Adjusted coordinates, m Y
1				4321.404	6240.562
2	305.41	0.051	0.034	4485.41	5982.93
3	665.02	0.112	0.074	4669.94	6291.66
4	1277.17	0.215	0.142	5248.70	6092.01
5	1762.29	0.297	0.196	5619.243	6405.272

4012.99

Traverse adjustment by the compass rule also can be accomplished by adjusting the departures and latitudes using Equations (8.11b) prior to computing the coordinates.

To illustrate this procedure of applying the compass rule adjustment, consider the traverse shown in Figure 8.7 for which calculations are given in Table 8.10. For this traverse, dX and dY, calculated by substituting the sums of departures and latitudes into Equations (8.10b), are -0.22 ft and $+0.17$ ft, respectively. Therefore, applying Equations (8.11b), the correction to the departure of line 1-2, $\delta x_{12} = (405.24/1770)$ $(-0.22) = -0.05$ ft and the correction to the latitude of line 1-2, $\delta y_{12} = (405.24/1770)$ $(+0.17) = +0.04$.

Corrections and adjusted departures and latitudes are listed in Table 8.12. The sums of corrections to latitudes and departures should be computed and checked against their respective closures. Note that the sum of the latitudes was different from the closure in latitudes by 0.01 ft due to a roundoff error. This discrepancy was corrected by examining the third place of the corrections and noting that the correction to the latitude of line 5-6 lay closest to the next highest hundredth, so that 0.01 ft was added to this correction. The sums of the adjusted departures and latitudes also should be checked to ensure against any error in applying the corrections.

Adjusted coordinates computed using the balanced departures and latitudes are given in Table 8.13.

When calculating coordinates with adjusted departures and latitudes, the computations always should be checked by cumulating the adjusted departures and latitudes of the final line to provide a check on the coordinates of the closing station. Regardless of whether coordinates are adjusted directly or departures and latitudes are balanced first, the final adjusted coordinates must be identical.

8.18
ADJUSTED DISTANCES AND AZIMUTHS

When the traverse has been adjusted, the final step in traverse computations is to calculate adjusted (corrected) distances and directions using the adjusted coordinates or adjusted departures and latitudes in Equations (8.8) and (8.9).

For the example traverse (Figure 8.7) calculated in Table 8.10 and using adjusted coordinates from Table 8.13, the corrected distance from 2 to 3 using Equation (8.8) is

$$d_{23} = [(5056.06 - 4770.93)^2 + (6215.76 - 6036.90)^2]^{1/2}$$

$$= 336.59 \text{ m}$$

and the direction of line 23 (from the north) by Equation (8.9) is

$$A_{N23} = \arctan \frac{X_3 - X_2}{Y_3 - Y_2}$$

$$= \arctan \frac{285.13}{178.86} = 57°54'01''$$

Corrected distances and azimuths are listed for all the lines in this example traverse in Table 8.14. These values would be used on the plan or map for which the survey was intended.

TABLE 8.12
Corrections to latitudes and departures (traverse of Figure 8.7, calculated in Table 8.10)

Station	Distance, ft	Departures, ft	Latitudes, ft	Corrections Departures	Corrections Latitudes	Adjusted Departures	Adjusted Latitudes
1	405.24	+388.89	−113.96	−0.05	0.04	+388.84	−113.92
2	336.60	+285.17	+178.83	−0.04	0.03	+285.13	+178.86
3	325.13	−134.92	+295.82	−0.04	0.03	−134.96	+295.85
4	212.91	−135.38	−164.33	−0.03	0.02	−135.41	−164.31
5	252.19	−251.82	−13.56	−0.03	0.03	−251.85	−13.53
6	237.69	−151.72	−182.97	−0.03	0.02	−151.75	−182.95
1	1769.76	0.22	−0.17	−0.22	0.17	$\Sigma = 0.00$	$\Sigma = 0.00$

TABLE 8.13
Adjusted coordinates (traverse of Figure 8.7, calculated in Table 8.10)

Station	Adjusted Departures, ft	Adjusted Latitudes, ft	Adjusted coordinates, ft X	Adjusted coordinates, ft Y
1			4382.09	6150.82
	+388.84	−113.92		
2			4770.93	6036.90
	+285.13	+178.86		
3			5056.06	6215.76
	−134.96	+295.85		
4			4921.10	6511.61
	−135.41	−164.31		
5			4785.69	6347.30
	−251.85	−13.53		
6			4533.84	6333.77
	−151.75	−182.95		
1			4382.09	6150.82

TABLE 8.14
Adjusted distances and azimuths (closed traverse in Figure 8.7)

Station	Adjusted Distance, ft	Azimuth
1		
	405.18	106°19′45″
2		
	336.59	57°54′01″
3		
	325.18	335°28′43″
4		
	212.92	219°29′33″
5		
	252.21	266°55′30″
6		
	237.69	219°40′28″
1		

8.19
COMPUTATIONS WITH RECTANGULAR COORDINATES

The use of rectangular coordinates permits the application of the principles of analytical geometry to solving surveying problems. Some of the more useful relationships are outlined in this section.

Equations (8.8) and (8.9), developed in Section 8.13, allow calculation of distance and azimuth of a line given the coordinates for the two endpoints. Two additional equations

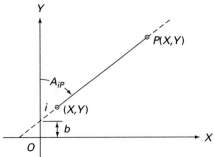

FIGURE 8.23

for the distance of line ij are

$$d_{ij} = \frac{X_j - X_i}{\sin A_{ij}} \tag{8.12a}$$

$$d_{ij} = \frac{Y_j - Y_i}{\cos A_{ij}} \tag{8.12b}$$

When $|X_j - X_i|$ exceeds $|Y_j - Y_i|$, use Equation (8.12a). Conversely, when $|Y_j - Y_i|$ exceeds $|X_j - X_i|$, use Equation (8.12b).

Equations (8.8) to (8.12b) are the best known and most frequently used relationships from analytical geometry involving coordinates that are employed in surveying calculations. Next, consider some equally useful but less used relationships, all of which are derived in Appendix A, Sections A.13 through A.19, A.25, and A.26 but are repeated here for convenience in relating to the given examples.

If the coordinates of point i (Figure 8.23) and the azimuth of line iP are given, the *point slope* equation (Equation (A.46)) for the straight line iP is

$$Y - Y_i = (X - X_i) \cot A_{iP} \tag{8.13}$$

where the $\cot A_{iP}$ represents the slope of the line measured from the positive X axis.

When the azimuth and Y intercept are known, the *slope intercept form* for the equation of a line (Equation (A.44)) is

$$Y - b = X \cot A_{iP} \tag{8.14}$$

If the coordinates for two ends of a line are given, as for line ij in Figure 8.24, then the *two-point form* for the equation of the straight line ij (Equation (A.48)) becomes

$$\frac{Y - Y_i}{Y_j - Y_i} = \frac{X - X_i}{X_j - X_i} \tag{8.15}$$

FIGURE 8.24

Expansion and rearrangement of Equation (8.15) leads to the general equation for a straight line (Equation (A.38)),

$$AX + BY + C = 0 \qquad (8.16a)$$

in which

$$A = Y_j - Y_i, \qquad B = X_i - X_j, \qquad C = X_j Y_i - X_i Y_j \qquad (8.16b)$$

which then leads to

$$X \text{ intercept} = -\frac{C}{A} = \frac{X_j Y_i - X_i Y_j}{Y_i - Y_j} \qquad (8.17)$$

$$Y \text{ intercept} = -\frac{C}{B} = \frac{X_j Y_i - X_i Y_j}{X_j - X_i} \qquad (8.18)$$

$$\tan A_{ij} = -\frac{B}{A} = \frac{X_j - X_i}{Y_j - Y_i} \qquad (8.19)$$

where Equations (8.17) and (8.18) are identical to Equations (A.39) and Equation (8.19) is the same as Equation (A.40), in which $\theta = 90 - A_{ij}$. Note that the symbol A in Equation (8.16a) should not be confused with the notation A_{ij} used to designate azimuth.

Referring to Figure 8.25, the slope of line ij is $m = \cot A_{ij}$, and the slope of the normal $ON = -1/m = \tan \omega$. Therefore, $m = -\cot \omega$. If the length of the normal is p, the Y intercept is $b = p/\sin \omega$ and the *normal form* of the equation of the line ij according to Equation (8.14) is

$$Y - \frac{p}{\sin \omega} = -X \cot \omega$$

or

$$Y \sin \omega - p = -X \cos \omega$$

which becomes

$$X \cos \omega + Y \sin \omega - p = 0 \qquad (8.20)$$

Because $\cot A_{ij} = -\cot \omega = -A/B$ from Equation (8.19),

$$\sin \omega = \frac{B}{\pm (A^2 + B^2)^{1/2}}, \qquad \cos \omega = \frac{A}{\pm (A^2 + B^2)^{1/2}} \qquad (8.21)$$

Because the slope of the line is less than 180°, $\sin \omega$ is positive and the sign of the radical must equal the sign of B. Dividing Equation (8.16) by $\pm (A^2 + B^2)^{1/2}$ yields

$$\frac{A}{\pm \sqrt{A^2 + B^2}} X + \frac{B}{\pm \sqrt{A^2 + B^2}} Y + \frac{C}{\pm \sqrt{A^2 + B^2}} = 0$$

or

$$X \cos \omega + Y \sin \omega = \frac{-C}{\pm \sqrt{A^2 + B^2}} \qquad (8.22)$$

the normal form of the equation for straight line ij (Equation (A.52)) in which A, B, and C are defined by Equation (8.16b).

The normal form of the equation for a line (Equation (8.22)) can be applied to obtain the perpendicular distance from any straight line ij to any coordinated point P, illustrated

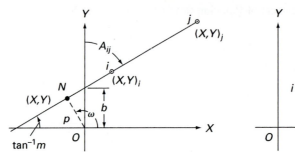

FIGURE 8.25

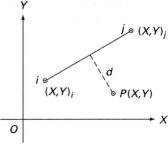

FIGURE 8.26

in Figure 8.26. In this case, the distance d is given by

$$d = \frac{A}{\pm\sqrt{A^2 + B^2}} X_P + \frac{B}{\pm\sqrt{A^2 + B^2}} Y_P + \frac{C}{\pm\sqrt{A^2 + B^2}} \tag{8.23}$$

where X_P and Y_P are coordinates of point P. The coordinates $(XY)_i$ and $(XY)_j$ are used to evaluate A, B, and C with Equation (8.16b).

The preceding equations are related to straight lines and are of the first degree. An example of an equation of the second degree, applicable in surveying calculations, is the equation for a circle.

Let the coordinates for the center of a circle be (H, K), the length of the radius be r, and the coordinates of the point P tracing the circle be (X, Y) (Figure 8.27). By Equation (8.8),

$$r = [(X - H)^2 + (Y - K)^2]^{1/2}$$

or

$$r^2 = (X - H)^2 + (Y - K)^2 \tag{8.24}$$

is the standard equation for a circle. If the center of the circle is taken as the origin of the coordinate system, Equation (8.24) becomes

$$r^2 = X^2 + Y^2 \tag{8.25}$$

Equation (8.24) corresponds to Equation (A.67) in which $X_c = H$ and $Y_c = K$ and Equation (8.25) is identical to Equation (A.68).

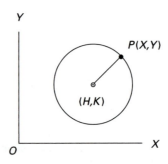

FIGURE 8.27

Consider some applications of the foregoing equations to surveying problems.

EXAMPLE 8.1. Determine the coordinates for the intersection of a line passing from station 1 to 3 with a line going from station 2 to 5 in the closed-loop traverse calculated in Section 8.15 and shown in Figure 8.7.

Solution. The coordinates of stations, 1, 3, 2, and 5 are as follows (Table 8.13):

Station	X, ft	Y, ft	$X' = X - 4300$, ft	$Y' = Y - 6000$, ft
1	4382.09	6150.82	82.09	150.82
3	5056.06	6215.76	756.06	215.76
2	4770.93	6036.90	470.93	36.90
5	4785.69	6347.30	485.69	347.30

Let X and Y be the unknown coordinates for the point of intersection. Using the translated X' and Y' coordinates for stations 1 to 5 to reduce the size of the numbers, and allowing for possible roundoff error, the equation for line 13 by Equation (8.15) is

$$\frac{Y' - 150.82}{215.76 - 150.82} = \frac{X' - 82.09}{756.06 - 82.09} \tag{1}$$

and the equation for line 25 is

$$\frac{Y' - 36.90}{347.30 - 36.90} = \frac{X' - 470.93}{485.69 - 470.93} \tag{2}$$

Expansion and rearrangement of Equations (1) and (2) yield

$$673.97Y' - 64.94X' = 96{,}317.231 \tag{3a}$$

$$14.76Y' - 310.40X' = -145{,}632.028 \tag{3b}$$

Equations (3a) and (3b) can be solved simultaneously for Y' by multiplying Equation (3a) by $310.40/64.94$ and subtracting it from Equation (3b):

$$14.76Y' - 310.40X' = -145{,}632.0280$$

$$-3221.4396Y' + 310.40X' = -460{,}376.7815$$

$$-3206.6796Y' = -606{,}008.8095$$

$$Y' = 188.9833 \text{ ft}$$

Substitution of this value of Y' into Equations (3a) and (3b) as a check gives

$$X' = 478.1618 \text{ ft}$$

The final coordinates of the intersection point are

$$X = 4300.00 + 478.16 = 4778.16 \text{ ft}$$

$$Y = 6000.00 + 188.98 = 6188.98 \text{ ft}$$

EXAMPLE 8.2. Using data from Example 8.1 and the traverse calculated in Section 8.15, compute the shortest distance from station 6 to the straight line connecting stations 1 and 3.

FIGURE 8.28

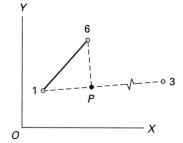

Solution. In Figure 8.28, let the unknown coordinates of P be (X, Y). The coordinates of 6 (Table 8.13) are $X_6 = 4533.84$ ft and $Y_6 = 6333.77$ ft, which may be reduced to $X_6' = 233.84$ ft and $Y_6' = 333.77$ ft as was done in Example 8.1. The distance $6P$ can be found by evaluating the normal form (Equation (8.22)) of Equation (1) for line 13 as developed in Example 8.1, with the coordinates of station 6:

$$A = +64.94, \qquad B = -673.97, \qquad C = 96,317.23$$

and $\sqrt{A^2 + B^2} = 677.09$, so that the normal form of Equation (1) is

$$\frac{-64.94}{677.09}X + \frac{673.97}{677.09}Y + \frac{96,317.23}{677.09} = 0$$

Because B is negative, $\sqrt{A^2 + B^2}$ also is negative. According to Equation (8.23), substitution of X_6' and Y_6' into the normal form of the equation yields the normal distance d, or

$$d = (-0.095910)(233.84) + (0.995390)(333.77) - 142.25 = 167.55 \text{ ft}$$

that is, the distance $6P$.

An alternative solution is possible by forming the point-slope equation for the line passing through 6 and perpendicular to line 13. The equation for line 13 (from Equation (8.13)) is

$$(Y_3 - Y_1) = (X_3 - X_1) \cot A_{13}$$

Line $6P$ is perpendicular to line 1-3; therefore,

$$\tan A_{6P} = -\frac{6215.76 - 6150.82}{5056.06 - 4382.09}$$

$$= -0.096354$$

or

$$\cot A_{6P} = -10.378349$$

so the point-slope equation for line $6P$ is

$$Y - 333.77 = (-10.378349)(X - 233.84)$$

$$Y + 10.378349X = 2760.64$$

This equation and the equation for line 13 (Example 8.1) are

$$Y + 10.378349X = 2760.64 \qquad \text{eq. line } 6P$$

$$673.97Y - 64.94X = 96,317.23 \qquad \text{eq. line } 13$$

Multiplication of the first equation by 673.97 and subtracting the second from the first yields

$$673.97Y + 6994.70X = 1,860,588.54$$

$$-673.97Y + 64.94X = -96,317.23$$

$$\overline{7059.64X = 1,764,271.31}$$

$$X = 249.91 \text{ ft} = X'_P$$

Substitution of X into both equations, as a check, gives

$$Y = 166.99 \text{ ft} = Y'_P$$

The distance $6P$ is evaluated using Equation (8.8):

$$d_{6P} = [(249.91 - 233.84)^2 + (166.99 - 333.77)^2]^{1/2}$$

$$= 167.55 \text{ ft}$$

the same as calculated previously. The coordinates for P in the nontranslated system are $X = 4300.00 + 249.91 = 4549.91$ ft and $Y_P = 6000.00 + 166.99 = 6166.99$ ft.

EXAMPLE 8.3. In route and property surveying, it often is necessary to determine coordinates for the intersection of a straight property line and a curved center line or property line. Figure 8.29 illustrates a curved property line with a radius of $r = 750.000$ m that is intersected by line QR, the center line of a sewer right of way. The azimuth of QR is $30°50'00''$. Calculate the coordinates of point P if the coordinates of Q and O (the center of the circle) are as follows:

Station	X, m	Y, m
Q	2310.638	2560.921
O	2561.051	2110.452

For convenience, these values are translated −2100 m in Y and −2300 m in X to reduce the size of the numbers:

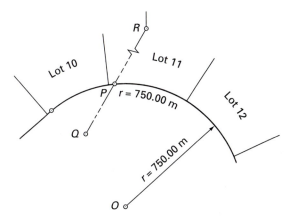

FIGURE 8.29
Intersection of straight and curved lines

Station	X', m	Y', m
Q	10.638	460.921
O	261.051	10.452

Solution. Write the point-slope equation (Equation (8.13)) for line QR and the equation for the circle of radius $r = 750.000$ m. Both equations are formed in terms of the unknown coordinates for $P(X, Y)_P$. Solve these two equations simultaneously for $(X, Y)_P$.

The point-slope equation for line QR is

$$Y'_P - Y'_Q = (X'_P - X'_Q) \cot A_{QR}$$

$$Y'_P - 460.921 = (X'_P - 10.638)(1.675299)$$

$$Y'_P = 1.675299X'_P + 443.0992 \tag{1}$$

The equation for the circle $r = 750.00$ using Equation (8.24) is

$$(X'_P - 261.051)^2 + (Y'_P - 10.452)^2 - (750.00)^2 = 0 \tag{2}$$

Substitution of Y'_P from (1) into (2) yields

$$3.806627\,X'^2_P + 927.5247\,X'_P - 307,168.7757 = 0 \tag{3}$$

Equation (3) is a quadratic equation the roots of which can be computed from

$$x = \frac{-b \pm (b^2 - 4ac)^{1/2}}{2a}$$

where $a = 3.806627$, $b = 927.5247$, and $c = -307,168.7757$, from Equation (3). The two roots of Equation (3) and corresponding values for Y'_P are

$$X'_P = 187.258 \text{ m} \quad \text{or} \quad -430.919 \text{ m}$$

$$Y'_P = 756.813 \text{ m} \quad \text{or} \quad -278.820 \text{ m}$$

Examination of Figure 8.29 reveals that the first set of answers is correct, because the second set yields the intersection point along the extended line RQ to the other side of the circle. To check the solution, $X'_P = 187.258$ and $X'_P = 756.813$ are substituted into Equation (2). The final coordinates for P are

$$X'_P = 187.258 + 2300.000 = 2487.258 \text{ m}$$

$$Y'_P = 756.813 + 2100.000 = 2856.813 \text{ m}$$

8.20
COORDINATE TRANSFORMATIONS

When surveyed points are coordinated, a number of survey problems can be solved conveniently by *coordinate transformations*. Coordinates in a local system may be required in the state plane coordinate system, or it may be desired to convert coordinates in several local systems into one regional framework.

Coordinate transformations are used extensively in the reduction of GPS measurements (Chapter 12). Any time it becomes necessary to convert from one coordinate system to another, a transformation is needed.

Coordinate transformations in two- and three-dimensional space are developed in Appendix C. Applications of these methods to coordinated traverse problems can be found in Section 8.40 of this chapter.

423

CHAPTER 8:
Traverse

8.21
ACCURACY AND SPECIFICATIONS FOR TRAVERSES

Current practice in judging the accuracy of a traverse is to separate the process into two categories involving the evaluation of networks and local surveys. Networks are evaluated by examining rigorously propagated statistics related to the traverse points. This aspect of traverse accuracy falls beyond the scope of elementary traverse computations and is discussed in Section 8.41, Part B. Accuracies for traverses on a local scale continue to be evaluated on the basis of *ratio of misclosure* (RoM; see Section 2.11) or relative positional accuracy, which is the ratio of the resultant error of closure for the traverse (Section 8.14) to the total length of the traverse. For the example traverse in Table 8.9, the resultant closure is 0.36 m for a traverse having a total length of 1762 m. The RoM for this traverse is 0.36/1762 or 1 part in 4894.

Table 8.15 shows the allowable azimuth and position closures for first-, second-, and third-order traverses as recommended by the Federal Geodetic Control Subcommittee (FGCS) in Bossler (1984). Note that second- and third-order surveys are divided into Class I and Class II surveys. The positional closure (RoM) is expressed as a ratio (e.g., 1:10,000 for a third-order, Class I traverse) or as a constant times the total distance traversed, which is designated by K (e.g., $0.4\sqrt{K}$ for a third-order, Class I traverse). The latter criterion is used for longer traverses. It is recommended that the relationship yielding the smallest possible positional closure be used.

Specifications are written to govern the traverse geometry, instrumentation, and field operations so that the surveyor can attempt to meet a standard for a given traverse. Table 8.16 shows a partial listing of specifications recommended by the FGCS. Those interested in further details should obtain the most recent version of Bossler (1994). Rigorous error propagation in traverses is discussed in Section 8.36 of this chapter.

TABLE 8.15
Horizontal control accuracy standards for traverse

Order	First	Second		Third	
Class		I	II	I	II
Azimuth closure At azimuth check point (seconds of arc)	$1.7\sqrt{N}$	$3.0\sqrt{N}$	$4.5\sqrt{N}$	$10.0\sqrt{N}$	$12.0\sqrt{N}$
Position closure	$0.04\sqrt{K}$	$0.08\sqrt{K}$	$0.20\sqrt{K}$	$0.40\sqrt{K}$	$0.80\sqrt{K}$
After azimuth	or	or	or	or	or
adjustment*	1:100,000	1:50,000	1:20,000	1:10,000	1:5,000

Note: N is number of segments, K is route distance in km.
*The expression containing the square root is designed for longer lines, where higher proportional accuracy is required. Use the formula that gives the smallest permissible closure. The closure (e.g., 1:100,000) is obtained by computing the difference between the computed and fixed values, and dividing this difference by K. Do not confuse closure with distance accuracy of the survey.

TABLE 8.16
Horizontal control specifications for traverse

Order	First	Second		Third	
Class		I	II	I	II
Network geometry					
Station spacing not less than (km)	10	4	2	0.5	0.5
Maximum deviation of main traverse from straight line	20°	20°	25°	30°	40°
Minimum number of benchmark ties	2	2	2	2	2
Benchmark tie spacing not more than (segments)	6	8	10	15	20
Astronomic azimuth spacing not more than (segments)	6	12	20	25	40
Minimum number of network control points	4	3	2	2	2
Instrumentation					
Theodolite, least count	0.2″	1.0″	1.0″	1.0″	1.0″
Field procedures					
Directions					
Number of positions	16	8 or 12*	6 or 8**	4	2
Standard deviation of mean not to exceed	0.4″	0.5″	0.8″	1.2″	2.0″
Rejection limit from the mean	4″	5″	5″	5″	5″
Reciprocal vertical angles (along distance sight path)					
Number of independent observations direct/reverse	3	3	2	2	2
Maximum spread	10″	10″	10″	10″	20″
Maximum time interval between reciprocal angles (hr)	1	1	1	1	1
Astronomic azimuths					
Observations per night	16	16	12	8	4
Number of nights	2	2	1	1	1
Standard deviation of mean not to exceed	0.45″	0.45″	0.6″	1.0″	1.7″
Rejection limit from the mean	5″	5″	5″	6″	6″
Infrared distances					
Minimum number of measurements	1	1	1	1	1
Minimum number of concentric observations/measurement	1	1	1	1	1
Minimum number of offset observations/measurement	1	1	1†	—	—
Maximum difference from mean of observations (mm)	10	10	10†	—	—
Minimum number of readings/observation	10	10	10	10	10
Maximum difference from mean of readings (mm)	‡	‡	‡	‡	‡
Microwave distances					
Minimum number of measurements	—	1	1	1	1
Minimum number of concentric observations/measurement	—	2§	1§	1§	1§
Maximum difference from mean of observations (mm)	—	150	150	200	200
Minimum number of readings/observation	—	20	20	10	10
Maximum difference from mean of readings (mm)	—	‡	‡	‡	‡

* 8 if 0.2″, 12 if 1.0″ resolution.
** 6 if 0.2″, 8 if 1.0″ resolution.
† Only if decimal reading near 0 or high 9s.
‡ As specified by manufacturer.
§ Carried out at both ends of the line.

424

When it is impossible or impractical to determine by field observations the length and bearing of every side of a closed traverse, the missing information generally may be calculated, provided no more than two quantities (lengths or azimuths) are omitted. (If only one measurement is omitted, a partial check is obtained on the work.) It must be assumed that the observed values are without error, and hence all errors of measurement are thrown into the computed lengths or directions. Calculated quantities that may be supplied in this manner are

1. Length and direction of one side.
2. Length of one side and direction of another.
3. Lengths of two sides for which the directions have been observed.
4. Directions of two sides for which the lengths have been observed.

There are three general cases: (1) length and direction of one side unknown, (2) omitted measurements in *adjoining* sides of the traverse, and (3) omitted measurements in *nonadjoining* sides. In case 3, the solution involves changing the order of sides in the traverse in such a way as to make the two partly unknown sides adjoin.

When the direction of a side is known but its length is unknown, the solution can be facilitated by assuming that side to lie on the reference meridian.

Methods of subdividing land, which involve the calculation of lengths and directions of unknown sides of a traverse, are described in Sections 8.31 to 8.35.

Length and Direction of One Side Are Unknown

In a closed-loop traverse containing k stations,

$$x_{12} + x_{23} + \cdots + x_{ij} + \cdots + x_{k,1} = 0$$

$$y_{12} + y_{23} + \cdots + y_{ij} + \cdots + y_{k,1} = 0$$

where x and y denote departures and latitudes calculated by Equation (8.6). If the line ij is of unknown distance or direction, then

$$x_{ij} = -(x_{12} + x_{23} + \cdots + x_{k,1}) = -\Sigma D \qquad (8.26)$$

$$y_{ij} = -(y_{12} + y_{23} + \cdots + y_{k,1}) = -\Sigma L$$

and the distance of the unknown line is

$$d_{ij} = [(\Sigma D)^3 + (\Sigma L)^2]^{1/2} \qquad (8.27)$$

where ΣD and ΣL are the sums of departures and latitudes of the known sides of the traverse, respectively.

The direction of an unknown line is

$$\tan A_{ij} = \frac{-\Sigma D}{-\Sigma L} \qquad (8.28)$$

with due regard to signs. Note that Equations (8.27) and (8.28) are similar in form to Equations (8.8) and (8.9).

Length of One Side and Direction of Another Side Are Unknown

Figure 8.30 represents a closed traverse for which the direction of the line $DE = d$ and the length of the line $EA = e$ are not determined by field measurements. Let an imaginary line extend from D to A, cutting off the unknown sides from the reminder of the traverse. Then, $ABCDA$ forms a closed traverse for which the side $DA = f$ is unknown in both direction and length. By Equation (8.28),

$$\tan A_{DA} = \frac{x_{DA}}{y_{DA}} = \frac{-(x_a + x_b + x_c)}{-(y_a + y_b + y_c)}$$

and the distance DA (Equation (8.12)) is

$$d_{DA} = \frac{x_{DA}}{\sin A_{DA}} = \frac{y_{DA}}{\cos A_{DA}}$$

In computing the length of DA it is desirable to use the larger of the two quantities, departure or latitude.

The angle between the lines e and f in triangle ADE is

$$\angle DAE = \text{azimuth of } AE - \text{azimuth of } AD$$

In the triangle ADE, the length of the two sides d and f and one angle DAE are known. By the sine law, in a triangle,

$$\sin DEA = \frac{f}{d} \sin DAE$$

With angle DEA known, angle ADE can be computed, and the remaining unknown length is given by the expression

$$e = f \frac{\sin ADE}{\sin DEA} = d \frac{\sin ADE}{\sin DAE}$$

Also,

$$\text{Azimuth of } DE = \text{azimuth of } DA - \angle ADE$$

FIGURE 8.30

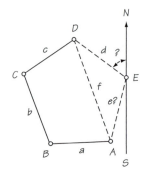

EXAMPLE 8.4. The following table gives the measured lengths and bearings for the courses of a closed traverse a to f (Figure 8.31), together with the departures and latitudes of the known sides. The length of b and the bearing of e are not observed. The general direction of e is southwest. It is desired to compute the unknown length and direction. Quantities in parentheses are derived from following calculations.

Line	Length, ft	Bearing	Departure E (+)	Departure W (−)	Latitude N (+)	Latitude S (−)
a	500.0	N0°00′E	0.0		500.0	
c	854.4	S69°27′E	800.0			299.9
d	1019.8	S11°19′E	200.1			1000.0
f	656.8	N54°06′W		532.0	385.1	
e	1118.0	Unknown (S78°56′30″W)		(1097.2)		(214.4)
b	Unknown (889.8)	N45°00′E	(629.2)		(629.2)	
g	(625.4)	(N48°27′20″W)		(468.1)	(414.8)	

In Figure 8.31, the lines a, c, d, and f are the courses for which the length and bearing are known. The line g is the closing side of the figure formed by the known courses. From the tabulated quantities and using Equation (8.26), the departure for line g, $x_g = -(800.0 + 200.1 - 532.0) = -468.1$ and the latitude $y_g = -(500.0 + 385.1 - 299.9 - 1000.0) = 414.8$. So, by Equation (8.28),

$$\tan A_{HA} = \frac{-468.1}{414.8} = -1.12850$$

so that the azimuth of HA or line g is 311°32′40″, yielding a bearing of N48°27′20″W. The length of g is

$$d_g = \frac{x_g}{\sin A_{HA}} = \frac{-468.1}{-0.74851} = 625.38 = 625.4 \text{ ft}$$

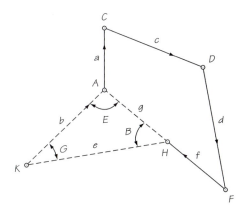

FIGURE 8.31

Because the direction of line b from K to A is N45°00′E,

$$\angle E = N45°00' + 48°27'20'' = 93°27'20''$$

$$\sin G = \frac{g}{e} \sin E = \frac{625.38}{1118.0} \times 0.99818 = 0.55836$$

$$\angle G = 33°56'30''$$

$$\angle B = 180°00' - 93°27'20'' - 33°56'30'' = 52°36'10''$$

$$\text{Length } b = \text{length } e \cdot \frac{\sin B}{\sin E} = 1118 \times \frac{0.79444}{0.99818} = 889.8 \text{ ft}$$

$$\text{Bearing of } e = 180° - 48°27'20'' - 52°36'10'' = S\,78°56'30''\,W$$

As a check on the calculations, the departures and latitudes of the lines b and e are computed and the values are shown in parentheses. The sum of the latitudes and the sum of the departures for courses a, b, c, d, e, and f are found to be approximately 0; hence, the computations for determining unknown length and bearing are correct.

Unknown Courses Are Not Adjoining

The preceding method of solution generally is applicable even though the two partly unknown courses are not adjoining. Obviously, the departure and the latitude of any line of fixed direction and length are the same for one location of the line as for any other. In other words, a line may be moved from one location to a second location parallel with the first, and its departure and latitude will remain unchanged. Because this is the case, it also must be true that the algebraic sum of the latitudes and the algebraic sum of the departures of any system of lines forming a closed figure must be 0, regardless of the order in which the lines are placed. Therefore, the courses shown in the order a, b, c, d, e, in Figure 8.30 are given in the order a, e, b, c, d in Figure 8.32. If now it is assumed that the direction of d and the length of a (Figure 8.30) are unknown, the problem of determining these unknown quantities is seen to be identical with that explained in the preceding section for the case where the partly unknown sides were adjoining.

Lengths of Two Sides Are Unknown

This problem commonly occurs where angular observations are taken from two or more points in the main traverse to some landmark, the measurements being introduced as a check. It occasionally occurs on main traverse lines where there are obstacles to the direct measurement of length but where angles are observed. The solution is nearly identical with that for the case where the direction of one side and the length of another are unknown.

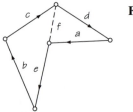

FIGURE 8.32

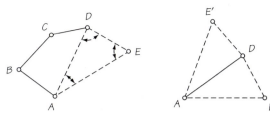

FIGURE 8.33   **FIGURE 8.34**

In Figure 8.33 *ABCD* represents the portion of a closed traverse for which the courses are known in both direction and length, and the lines *DE* and *EA* represent courses for which the direction is known but the length is unknown. From the departures and latitudes of the known sides, the length and bearing of the closing line *DA* are computed; and in the triangle *ADE* the angles *A*, *D*, and *E* are computed from the known directions of the sides. The lengths *DE* and *EA* are determined through the relation

$$\frac{DE}{\sin A} = \frac{EA}{\sin D} = \frac{DA}{\sin E}$$

If the two lines are not adjoining, the problem may be solved as though they were. As the angle between the partly unknown lines approaches 90°, the solution becomes strong, and as the angle approaches 0° or 180°, the solution becomes weak, the problem being indeterminate when the lines are parallel.

Directions of Two Sides Are Unknown

In Figure 8.33 if *DA* is the closing side of the known portion of the traverse, its direction and length are computed; then the lengths of the three sides of the triangle *ADE* are known, and the angles *A*, *D*, and *E* can be computed.

The general direction of at least one of the partly unknown lines must be observed, as the values of the trigonometric functions merely determine the shape of the triangle but do not fix its position. In Figure 8.34, if *DA* is the closing line of the known portion of the traverse forming the base of the triangle of which the courses are of unknown direction but the legs are of known length, then it is evident that the vertex may fall at either *E* or *E'*.

8.23
CALCULATION OF AREAS OF LAND: GENERAL

One of the primary objects of most land surveys is to determine the area of the tract. A closed traverse is run in which the lines of the traverse are made to coincide with property lines where possible. Where the boundaries are irregular or curved or where they are occupied by objects that make direct measurement impossible, they are located with respect to the traverse line by appropriate angular and linear measurements. The lengths and bearings of all straight boundary lines are determined either directly or by computation, the irregular boundaries are located with respect to traverse lines by perpendicular offsets taken at appropriate intervals, and the radii and central angles of circular boundaries are obtained. The following sections explain the several common methods by means of which these data are employed in calculating areas.

In ordinary land surveying, as discussed here, the area of a tract of land is taken as its projection on a horizontal plane, and it is not the actual area of the surface of the land. For precise determinations of the area of a large tract, such as a state or nation, the area is taken as the projection of the tract on the earth's spheroidal surface at the datum.

8.24
METHODS OF DETERMINING AREA

The area of a tract may be determined by any of the following methods:

1. Plot the boundaries to scale. Then, the area of the tract may be found by use of the planimeter or it may be calculated by dividing the tract into triangles and rectangles, scaling the dimensions of these figures, and computing their areas mathematically. This method is useful in roughly determining areas or checking those that have been calculated by more exact methods. Its advantage lies in the rapidity with which calculations can be made.
2. Mathematically compute the areas of individual triangles into which the tract may be divided (Section 8.25). This method is employed when it is not expedient to compute the departures and latitudes of the sides.
3. Calculate the area from the coordinates of the *corners* of the tract (Section 8.26).
4. For tracts having irregular or curved boundaries, the methods of Sections 8.27 to 8.30 are employed.

For computation of the areas of cross sections, see Section 16.30.

8.25
AREA BY TRIANGLES

When the lengths of two sides and the included angle of any plane triangle are known, its area is given by the expression

$$\text{Area} = \tfrac{1}{2}\, ab \sin C \tag{8.29}$$

When the lengths of the three sides of any plane triangle are given, its area is determined by the equation

$$\text{Area} = \sqrt{s(s - a)(s - b)(s - c)} \tag{8.30}$$

in which $s = \tfrac{1}{2}(a + b + c)$.

In surveying small lots, as for a city subdivision, it is common practice to omit the determination of the error of closure of each lot (a practice not condoned in this book), and hence, the computation of departures, latitudes, and coordinates is unnecessary. Under such circumstances the area may be calculated by dividing the lot, usually quadrangular in shape, into triangles, as illustrated by Figure 8.35, for each of which two sides and the included angle have been measured. By Equation 8.29, the areas of *ABD* and *BCD* are computed; the sum of these two areas is the area of the lot. The area thus found can be checked independently by computing the areas of the two triangles *ABC* and *CDA* formed by dividing the quadrilateral by a line from *A* to *C*.

The accuracy of the fieldwork may be investigated by determining the lengths of the diagonals. *BD* can be determined by solving either triangle *ABD* or triangle *BCD*. The field measurements are without error if the length of *BD*, computed by solving one of the triangles, is the same as that computed by solving the other.

Figure 8.36 illustrates a survey made by a single setup of the theodolite or total station system at *O*, such as might be the case for a small lot where the property lines *ABCD* are obstructed or where the theodolite or total station system cannot be set up at the corners. Under these circumstances the angles about *O* and the distances *OA*, *OB*, *OC*, and *OD* are measured in the field. Because in each triangle two sides and the included angle are

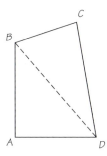

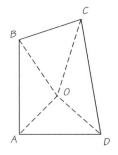

FIGURE 8.35   **FIGURE 8.36**

known, the area can be determined by use of Equation (8.29). If, in addition to the forego-ing measurements, the lengths of the sides *AB*, *BC*, and so forth, are measured, the area of the lot can be checked independently by solving each triangle by Equation (8.30). In general, the area of a triangle computed from two different sets of measurements will not be identical, owing to random measuring errors. Therefore, an alternative would be to use the method of least squares (Sections 2.22, . . . , 2.29) to compute the area using all redundant measurements.

8.26
AREA BY COORDINATES

When the points defining the corners of a tract of land are coordinated with respect to some arbitrarily chosen coordinate axes or are given in a regional system (such as a state plane coordinate system), these coordinates are useful not only in finding the lengths and bearings of the boundaries but also in calculating the area of the tract. Essentially, the calculation involves finding the areas of trapezoids formed by projecting the lines on one of a pair of coordinate axes, usually a true meridian and a parallel at right angles to it.

In Figure 8.37, 12345 represents a tract the area of which is to be determined, where each point 1, 2, . . . , 5 has coordinates $X_1, Y_1, . . . , X_5, Y_5$, as shown in the figure.

The area of the tract can be computed by summing algebraically the areas of the trapezoids formed by projecting the lines on the reference meridian:

Area 12345 = area 23*cb* + area 34*dc* − area 45*fd* − area 15*fa* − area 21*ab*

FIGURE 8.37

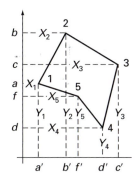

or

$$\text{Area} = \tfrac{1}{2}(X_2 + X_3)(Y_2 - Y_3) + \tfrac{1}{2}(X_3 + X_4)(Y_3 - Y_4) - \tfrac{1}{2}(X_4 + X_5)$$
$$(Y_5 - Y_4) - \tfrac{1}{2}(X_5 + X_1)(Y_1 - Y_5) - \tfrac{1}{2}(X_1 + X_2)(Y_2 - Y_1) \qquad (8.31)$$

By multiplication and a rearrangement of terms in Equation (8.31),

$$2 \times \text{area} = -[Y_1(X_5 - X_2) + Y_2(X_1 - X_3) + Y_3(X_2 - X_4) + Y_4(X_3 - X_5)$$
$$+ Y_5(X_4 - X_1)] \qquad (8.32)$$

and in general for any polygon having n stations,

$$2 \times \text{area} = Y_1(X_2 - X_n) + Y_2(X_3 - X_1) + \cdots$$
$$+ Y_{n-1}(X_n - X_{n-2}) + Y_n(X_1 - X_{n-1}) \qquad (8.33)$$

or, equivalently,

$$2 \times \text{area} = X_1(Y_2 - Y_n) + X_2(Y_3 - Y_1) + \cdots$$
$$+ X_{n-1}(Y_n - Y_{n-2}) + X_n(Y_1 - Y_{n-1}) \qquad (8.34)$$

Note that a negative sign for the final value is of no concern and that the total area always will be positive.

EXAMPLE 8.5. Given the following data, find the required area by applying Equation (8.33).

Corner	1	2	3	4	5
X coordinate, m	300	400	600	1,000	1,200
Y coordinate, m	300	800	1,200	1,000	400

Solution

$$2 \times \text{area} = 300(400 - 1200) + 800(600 - 300) + 1200(1000 - 400)$$
$$+ 1000(1200 - 600) + 400(300 - 1000)$$
$$= -240{,}000 + 240{,}000 + 720{,}000 + 600{,}000 - 280{,}000$$
$$= 1{,}040{,}000 \text{ m}^2$$
$$\text{Area} = \frac{1{,}040{,}000}{2} = 520{,}000 \text{ m}^2$$

Equation (8.33) also can be expressed in the form

$$2 \times \text{area} = X_2 Y_1 + X_3 Y_2 + X_4 Y_3 + \cdots + X_n Y_{n-1} + X_1 Y_n$$
$$- X_1 Y_2 - X_2 Y_3 - \cdots - X_n Y_1 \qquad (8.35)$$

When this form is employed, computations can be made conveniently by tabulating each X coordinate below the corresponding Y coordinate as follows:

$$\frac{Y_1}{X_1} \diagdown \frac{Y_2}{X_2} \diagdown \frac{Y_3}{X_3} \diagdown \cdots \diagdown \frac{Y_n}{X_n} \diagdown \frac{Y_1}{X_1} \qquad (8.36)$$

Then in expression (8.36), the difference between the sum of the products of the coordinates joined by full lines and the sum of the products of the coordinates joined by dashed lines is equal to twice the area of the chart.

EXAMPLE 8.6. Compute the area enclosed by the polygon formed by the closed-loop traverse calculated in Section 8.16 and adjusted in Section 8.17. The adjusted coordinates are listed in Table 8.13. Use Equation (8.34).

Solution

$$2 \times \text{area} = 4382.09(6036.90 - 6333.77) + 4770.93(6215.76 - 6150.82)$$
$$+ 5056.06(6511.61 - 6036.90) + 4921.10(6347.30 - 6215.76)$$
$$+ 4785.69(6333.77 - 6511.61) + 4533.84(6150.82 - 6347.30)$$

Observing the rules of significant figures, the double area is

$$2 \times \text{area} = 314,501 \text{ ft}^2$$

and

$$\text{Area} = 157,250 \text{ ft}^2 = 3.6100 \text{ acres}$$

8.27
AREA OF A TRACT WITH IRREGULAR OR CURVED BOUNDARIES

If the boundary of a tract of land follows some irregular or curved line, such as a stream or road, it is customary to run a traverse in some convenient location near the boundary and to locate the boundary by offsets from the traverse line. Figure 8.38 represents a typical case, AB being one of the traverse lines. The determination of area of the entire tract involves computing the area within the closed traverse, by methods already described, and adding to this the area of the irregular figure between the traverse line AB and the curved boundary. The offset distances are aa', bb', and so on, and the corresponding distances along the traverse line are Aa, Ab, and so on. Where the boundary is irregular, as from a' to f', it is necessary to take offsets at points of change and hence generally at irregular intervals. Where a segment of the boundary is straight, as from f' to g', offsets are taken only at the ends. Where the boundary is a gradual curve, as from g' to m', ordinarily the offsets are taken at regular intervals.

If the offsets are taken sufficiently close together, the error involved in considering the boundary as straight between offsets is small as compared with the inaccuracies of the measured offsets. When this assumption is made, the assumed boundary takes some such form as that illustrated by the dotted lines $g'h'$, $h'k'$, and so forth, in Figure 8.38, and the

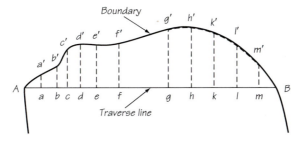

FIGURE 8.38
Irregular boundary.

areas between offsets are of trapezoidal shape. Under such an assumption, irregular areas are said to be calculated by the *trapezoidal rule* (Section 8.28).

Where the curved boundaries are of such definite character as to make it justifiable, the area may be calculated somewhat more accurately by assuming that the boundary is made up of segments of parabolas, as first suggested by Simpson. Under this assumption, irregular areas are said to be computed by *Simpson's one-third rule* (Section 8.29).

8.28
OFFSETS AT REGULAR INTERVALS: THE TRAPEZOIDAL RULE

Let Figure 8.39 represent a portion of a tract lying between a traverse line AB and an irregular boundary CD, offsets $h_1, h_2, \ldots, h_n$ having been taken at the regular intervals d. The summation of the areas of the trapezoids comprising the total area is

$$\text{Area} = \frac{h_1 + h_2}{2} d + \frac{h_2 + h_3}{2} d + \cdots + \frac{h_{n-1} + h_n}{2} d$$

$$= d\left(\frac{h_1 + h_n}{2} + h_2 + h_3 + \cdots + h_{n-1}\right) \qquad (8.37a)$$

Equation (8.37a) may be expressed conveniently in the form of the following rule:

Trapezoidal rule Add the average of the end offsets to the sum of the intermediate offsets. The product of the quantity thus determined and the common interval between offsets is the required area.

> **EXAMPLE 8.7.** By the trapezoidal rule, find the area between a traverse line and a curved boundary, rectangular offsets being taken at intervals of 5 m and the values of the offsets in meters being $h_1 = 3.2$, $h_2 = 10.4$, $h_3 = 12.8$, $h_4 = 11.2$, and $h_5 = 4.4$. By the foregoing rule,
>
> $$\text{Area} = 5\left(\frac{3.2 + 4.4}{2} + 10.4 + 12.8 + 11.2\right) = 191 \text{ m}^2$$

8.29
OFFSETS AT REGULAR INTERVALS: SIMPSON'S
ONE-THIRD RULE

In Figure 8.40, let AB be a portion of a traverse line, DFC a portion of the curved boundary assumed to be the arc of a parabola, and h_1, h_2, and h_3 any three consecutive rectangular offsets from traverse line to boundary taken at the regular interval d.

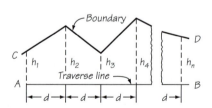

FIGURE 8.39
Area by the trapezoidal rule.

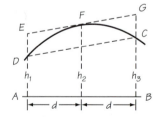

FIGURE 8.40
Area by Simpson's rule.

The area between traverse line and curve may be considered as composed of the trapezoid *ABCD* plus the area of the segment between the parabolic arc *DFC* and the corresponding chord *DC*. One property of a parabola is that the area of a segment (such as *DFC*) is equal to two-thirds the area of the enclosing parallelogram (such as *CDEFG*). Then, the area between the traverse line and curved boundary within the length of $2d$ is

$$\text{Area}_{1,2} = \frac{h_1 + h_3}{2} 2d + \left(h_2 - \frac{h_1 + h_3}{2}\right)2d\left(\frac{2}{3}\right)$$

$$= \frac{d}{3}(h_1 + 4h_2 + h_3)$$

Similarly for the next two intervals,

$$\text{Area}_{3,4} = \frac{d}{3}(h_3 + 4h_4 + h_5)$$

The summation of these partial areas for $(n-1)$ intervals, n being an odd number and representing the number of offsets, is

$$\text{Area} = \frac{d}{3}[h_1 + h_n + 2(h_3 + h_5 + \cdots + h_{(n-2)})$$

$$+ 4(h_2 + h_4 + \cdots + h_{(n-1)})] \qquad (8.37b)$$

Equation (8.37b) may be expressed conveniently in the form of the following rule, which is applicable if the number of offsets is odd.

Simpson's one-third rule Find the sum of the end offsets, plus twice the sum of the odd intermediate offsets, plus four times the sum of the even intermediate offsets. Multiply the quantity thus determined by one-third of the common interval between offsets, and the result is the required area.

EXAMPLE 8.8. By Simpson's one-third rule find the area between the traverse line and the curved boundary of Example 8.7.

Solution

$$\text{Area} = \tfrac{5}{3}[3.2 + 4.4 + 2(12.8) + 4(10.4 + 11.2)] = 199 \text{ m}^2$$

If the total number of offsets is *even,* the partial area at either end of the series of offsets is computed separately, to make n for the remaining area an odd number and thus make Simpson's rule applicable.

Simpson's rule also is useful in other applications, such as finding centers of areas. The prismoidal formula for computing volumes of earthwork (Section 16.31) embodies Simpson's rule for area and a factor for the third dimension.

Results obtained by using Simpson's rule are greater or smaller than those obtained by using the trapezoidal rule, depending on whether the boundary curve is concave or convex toward the traverse line. Some appreciation of the variations between the two methods will be gained by studying the previous examples. It will be seen that the two results differ by more than 4 percent. Under average conditions the difference will be much less than this, but in an extreme case, it may be much larger.

In general, Simpson's rule is always more accurate than the trapezoidal rule. The latter approaches the former in accuracy to the extent that the irregular boundary has curves of contrary flexure, thereby producing the compensative effects mentioned previously.

When the interval along the traverse line is irregular, the trapezoidal rule may be applied by using Equation (8.37a) and the appropriate interval $d_1, d_2, \ldots, d_n$. Simpson's one-third rule is based on an equal interval for each pair of adjacent segments. Therefore, in utilizing Simpson's one-third rule, a regular interval should be used in the field measurements.

Easa (1988) extended the Simpson one-third rule to the case of unequal intervals between offsets and termed it the *generalized Simpson rule*. It is based on selecting parabolic representation over each pair of adjacent intervals. Consequently, it applies directly to situations with an even number of intervals or an odd number of offsets. The formula is given by

$$A = \sum_{i=0, i \neq \text{odd}}^{n=2} \frac{d_i + d_{i+1}}{6} \left[\frac{2d_i - d_{i+1}}{d_i} h_i + \frac{(d_i + d_{i+1})^2}{d_i d_{i+1}} h_{i+1} + \frac{(2d_{i+1} - d_i)}{d_{i+1}} h_{i+2} \right] \qquad (8.38)$$

in which d denotes the length of the interval, h is the magnitude of the offset as before, and n is number of intervals, which is even.

As an example, consider the following data, adapted from Easa (1988):

$$d_0 = 20, \quad d_1 = 10, \quad d_2 = 15, \quad d_3 = 30, \quad d_4 = 10, \quad d_5 = 6$$

$$h_0 = 5.0, \quad h_1 = 7.2, \quad h_2 = 6.3, \quad h_3 = 10.0,$$

$$h_4 = 4.6, \quad h_5 = 7.4, \quad h_6 = 6.8$$

Then,

$$A = \frac{d_0 + d_1}{6} \left[\frac{2d_0 - d_1}{d_0} h_0 + \frac{(d_0 + d_1)^2}{d_0 d_1} h_1 + \frac{(2d_1 - d_0)}{d_1} h_2 \right]$$

$$+ \frac{d_2 + d_3}{6} \left[\frac{2d_2 - d_3}{d_2} h_2 + \frac{(d_2 + d_3)^2}{d_2 d_3} h_3 + \frac{(2d_3 - d_2)}{d_3} h_4 \right]$$

$$+ \frac{d_4 + d_5}{6} \left[\frac{2d_4 - d_5}{d_4} h_4 + \frac{(d_4 + d_5)^2}{d_4 d_5} h_5 + \frac{(2d_5 - d_4)}{d_5} h_6 \right]$$

$$A = \frac{(20 + 10)}{6} \left[\frac{(2 \times 20 - 10)}{20} 5.0 + \frac{(20 + 10)^2}{20 \times 10} 7.2 + \frac{(2 \times 10 - 20)}{10} 6.3 \right]$$

$$+ \frac{(15 + 30)}{6} \left[\frac{(2 \times 15 - 30)}{15} 6.3 + \frac{(15 + 30)^2}{15 \times 30} 10.0 + \frac{(2 \times 30 - 15)}{30} 4.6 \right]$$

$$+ \frac{(10 + 6)}{6} \left[\frac{(2 \times 10 - 6)}{10} 4.6 + \frac{(10 + 6)^2}{10 \times 6} 7.4 + \frac{(2 \times 6 - 10)}{6} 6.8 \right]$$

$$= 696.2 \text{ m}^2$$

When the number of intervals, n, is odd, Equation (8.38) is applied to the first $(n - 1)$ intervals and then the area for the last interval is estimated separately as follows. First, compute the auxiliary quantity (see Easa, 1988),

$$h_{n-1} = \frac{-h_{n-2} d_{n-1}^2 + h_{n-1}(d_{n-1}^2 - d_{n-2}^2) + h_n d_{n-2}^2}{(d_{n-2} + d_{n-1}) d_{n-2} d_{n-1}} \qquad (8.39a)$$

then it also follows that

$$A_l = \frac{d_{n-1}}{6} (4h_{n-1} + 2h_n + d_{n-1} h_{n-1}) \qquad (8.39b)$$

Suppose for the example we had an additional seventh interval, $d_6 = 10$ (note that the first interval is denoted d_0), and an additional post, $h_7 = 3.0$. Then,

$$h_6 = \frac{-7.4 \times (10)^2 + 6.8[(10)^2 - (6)^2] + 3(6)^2}{(6 + 10)\, 6 \times 10} = -0.205$$

and

$$A_l = \tfrac{10}{6}[4 \times 6.8 + 2 \times 3 + 10(-0.205)] = 51.9 \text{ m}^2$$

The total area for all the seven intervals is given by

$$A_t = A + A_l = 696.2 + 51.9 = 748.1 \text{ m}^2$$

8.30
AREA OF SEGMENTS OF CIRCLES

A problem of frequent occurrence in the surveying of city lots and of rural lands adjacent to the curves of highways and railways is that of finding the area where one or more of the lines of the boundary is the arc of a circle.

In Figure 8.41, $ABCDEQF$ may be taken as a boundary of this character, for which it is convenient to run a traverse along the straight portions of the boundary and make the chord EF the closing side of the traverse, the length of the chord $EF = L$ and the middle ordinate $PQ = M$ being measured in the field.

In calculating the area, it is convenient to divide the tract into two parts: (1) that within the polygon formed by the traverse $ABCDEF$, for which the area is found by the coordinate method, and (2) that between the chord EPF and the arc EQF, which is the segment of a circle. The area of this segment is found exactly by subtracting the area of the triangle $OEPF$ from the area of the circular sector $OEQF$. If I is the angle and R is the radius whose arc is

FIGURE 8.41

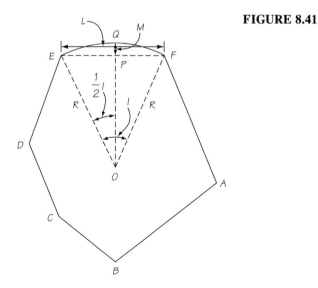

EQF, then by Sections 16.5 and 16.6,

$$\tan \tfrac{1}{4}I = \frac{2M}{L} \tag{8.40}$$

and

$$R = \frac{L}{2 \sin \tfrac{1}{2}I} \tag{8.41}$$

The area of the circular sector *OEQF* is $A_s = \pi R^2 I°/360°$, in which *I* is expressed in degrees. The area of triangle *OEF* is

$$A_t = \frac{LR \cos(I/2)}{2} = \frac{R^2}{2} \sin I$$

The area of the segment is exactly

$$\text{Area} = A_s - A_t = R^2 \left(\frac{\pi I°}{360°} - \frac{\sin I}{2} \right) \tag{8.42}$$

EXAMPLE 8.9. Find the area of a circular segment when the chord length is 80.00 m and the middle ordinate is 10.00 m.

Solution. By Equation (8.40),

$$\tan \tfrac{1}{4}I = 0.2500$$
$$\tfrac{1}{4}I = 14.°036, \qquad I = 56.°145$$

By Equation (8.41),

$$R = \frac{\dfrac{L}{2}}{\left(\sin \dfrac{I}{2} \right)} = \frac{80}{(2)(0.47059)} = 85.00 \text{ m}$$

By Equation (8.42),

$$\text{Area} = (85.000)^2 \left[\frac{(56.145°)\pi}{360°} - \frac{\sin 56.145°}{2} \right]$$
$$= 539.9 \text{ m}^2$$

An alternative method of finding the area of the tract *ABCDEQF* (Figure 8.41) is to divide the area into a rectilinear polygon *ABCDEOF* and the circular sector *OEQF*, and to add the two areas. The polygon has one more side than that just used, but there is no need to compute the area of a circular segment.

Approximation by Parabolic Segment

The area of a parabolic segment is

$$(\text{Area})_p = \tfrac{2}{3}LM \tag{8.43}$$

in which the letters have the same significance as before. This expression may be employed for finding the approximate areas of circular segments, the precision decreasing as the size of the central angle I increases. The following example illustrates the error involved in applying this expression to the conditions of Example 8.9.

EXAMPLE 8.10. By Equation (8.43) find the approximate area of the circular arc of Example 8.9 and determine the percentage of error introduced through using that approximate expression.

Solution. The area, by Equation (8.43), is

$$(\text{Area})_p = \tfrac{2}{3} \times 80 \times 10 = 533.3 \text{ m}^2$$

This value is

$$\frac{539.9 - 533.3}{539.9} \, 100 = 1.2\% \text{ too low}$$

When the central angle is small, the error involved in using Equation (8.43) for circular arcs often is negligible; for example, when $I = 30°$, the error is less than 0.2 percent. But for large values of I, the error introduced is so great as to render the approximate expression of little use; for example, when $I = 90°$, the error is about 3 percent, and when $I = 180°$, the error is about 15 percent.

Alternative Method

When tangents to the curve are property lines, it sometimes is more convenient to establish the traverse, as illustrated by Figure 8.42. Here KA and FC, which are tangent to the curve ABC, are run to an intersection at D, and the distances AD and CD and the angle I are measured. Also E usually is measured as a check.

The work of finding the area is conveniently divided into two parts: (1) that of calculating the area within the polygon $ADCFGHK$ by the coordinate method, and (2) that of calculating the external area between the arc ABC and the tangents AD and CD. The latter area subtracted from the former is the required area.

FIGURE 8.42

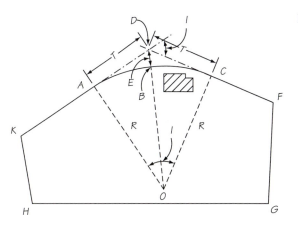

The external area may be found by subtracting the area of the circular sector $OABC = A_s = \pi R^2 I°/360°$ from the area $OADC = TR$, in which T is the tangent distance $AD = CD$, and R is the radius of the curve. If R is unknown, it may be found by the relation

$$R = \frac{T}{\tan \frac{1}{2} I} \qquad \text{(see Section 16.5)} \qquad (8.44)$$

PART B
ADVANCED TOPICS

8.31
PARTITION OF LAND

Four of the simpler cases frequently encountered in the subdivision of irregular tracts of land will be described in the following sections. Methods of subdividing the U.S. public lands are given in Chapter 18.

Where a given tract is to be divided into two or more parts, a resurvey is run; the departures, latitudes, and coordinates are computed; the survey is adjusted; and the area of the entire tract is determined. The corrected departures, latitudes, and coordinates are employed further in the computations of subdivision.

Area Cut Off by a Line between Two Points

In Figure 8.43, let $ABCDEFG$ represent a tract of land to be divided into two parts by a line extending from A to D. A survey of the tract has been made, the adjusted coordinates have been computed for the corners, and the area has been computed.

It is desired to determine the length and direction of the cutoff line AD without additional field measurements and to calculate the area of each of the two parts into which the tract is divided.

When coordinates are available for A and D, the simplest approach is to use Equations (8.8) and (8.9) developed in Section 8.13 to calculate the distance and azimuth of AD. Then the areas of the traverse $ABCD$ and $ADEFG$ can be computed by the coordinate method.

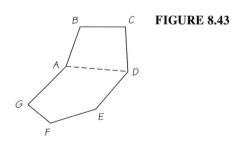

FIGURE 8.43

8.32
AREA CUT OFF BY A LINE RUNNING IN A GIVEN DIRECTION

In Figure 8.44, *ABCDEFG* represents a tract of known dimensions, for which the corrected departures, latitudes, and adjusted coordinates are given; and *DH* represents a line running in a given direction that passes through the point *D* and divides the tract into two parts.

It is desired to calculate from the given data the lengths *DH* and *HA* and the area of each of the two parts into which the tract is divided.

The direction of *DH* is given and the coordinates of *D* are known, so that the point slope Equation (8.13), Section 8.19, can be formed. Because coordinates for *G* and *A* are known, Equation 8.15, Section 8.19, can be written for line *AG*. These two equations, one for line *DH* and one for line *GA*, can be solved simultaneously to obtain the coordinates of *H* (see Example 8.1). The distance *HA* can be calculated with Equation (8.8), using the coordinates of *H* just computed and the known coordinates for *A*. Areas for *ABCDHA* and *HDEFGH* can be computed by the coordinate method (Equation (8.33) or (8.34)).

In the field the length and direction of the side *DH* are laid off from *D*, and a check on fieldwork and computations is obtained if the point *H* thus established lies on the line *GA* and if the computed distance *HA* agrees with the observed distance. The area computations may be checked by observing that the sum of the areas of the two parts, each computed independently, is equal to the area of the entire tract.

8.33
TO CUT OFF A REQUIRED AREA BY A LINE THROUGH A GIVEN POINT

In Figure 8.45 *ABCDEF* represents a tract of land of known dimensions, for which the corrected departures, latitudes, and adjusted coordinates for each point are given; and *G* represents a point on the boundary through which a line is to pass, cutting off a required area from the tract. The area for the total tract can be calculated by the coordinate method and a sketch of the tract has been prepared.

To find the length and direction of the dividing line, the procedure is as follows. A line *GF* is drawn to that corner of the traverse that, from inspection of the sketch, will come nearest to being on the required line of division. Because distance *CG* is specified, the coordinates of *G* can be calculated. Then the distance and azimuth of *GF* are calculated using Equations (8.8) and (8.9). The area enclosed by the traverse and the cutoff line *GF*, *ABCGFA*, is computed by the coordinate method. The difference between this area and the amount specified is found.

In the figure it is assumed that *FABCG* has an area greater than the desired amount, *GH* being the correct position of the dividing line. The triangle *GFH* represents this excess area; and as the angle α may be computed from known directions, in this triangle one side *FG*,

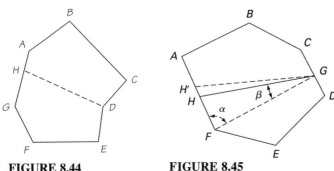

FIGURE 8.44   **FIGURE 8.45**

one angle α, and the area are given. The length HF is computed from the equation for area, area $= \frac{1}{2}(ab \sin C)$, so that, from Figure 8.45,

$$HF = \frac{2 \cdot \text{area } GFH}{FG \sin \alpha} \tag{8.45}$$

The triangle is then solved for angle β and length GH. From the known direction of GF and the angle β, the azimuth of GH is computed. The departures and latitudes of the lines FH, GH, and HA are computed and coordinates are calculated for point H.

In the field, the length GH is laid off in the required direction, and a check on fieldwork and computations is obtained if the point H thus established falls on the line FA and if the computed distance HF or HA agrees with the measured distance.

8.34
TO CUT OFF A REQUIRED AREA BY A LINE RUNNING IN A GIVEN DIRECTION

In Figure 8.46, $ABCDEF$ represents a tract of land of known dimensions and area, which is to be divided into two parts, each of a required area, by a line running in a given direction. The figure is assumed to be drawn at least roughly to scale, and the corrected departures, latitudes, and adjusted coordinates for the corners are known.

Through the corner that seems likely to be nearest the line cutting off the required area, a trial line DG is drawn in the given direction. Then, in the closed traverse $GBCDG$, the departures and latitudes of BC and CD and the directions of DG and GB are known, and the lengths of two sides, DG and GB, are unknown.

Using these data, the coordinates of G can be found by forming the point-slope equation, Equation (8.13), for line DG and Equation (8.15) for line AB and solving these two equations simultaneously for X_G and Y_G. It would also be possible to solve for DG and GB using the methods outlined in Section 8.22. By either procedure, the distance for DG and the distance for AG are calculated.

Next, the area cut off by trial line DG is calculated. The difference between this area and that required is represented in the figure by the trapezoid $DGJH$ in which the side DG is known. The angles at D and G can be computed from the known directions of adjacent

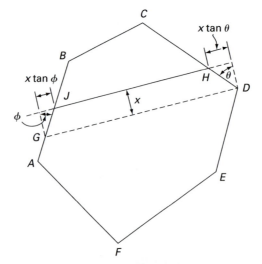

FIGURE 8.46

sides, and in this way θ and ϕ are determined. Tan θ and tan ϕ are positive if they fall inside the trapezoid and negative if they fall outside. With this convention, the

$$\text{Area of trapezoid} = (DG)x + \frac{x^2}{2}(\tan \theta + \tan \phi) \qquad (8.46)$$

in which x is the altitude of the trapezoid. (In the figure, both angles lie outside the trapezoid so that both tangents are negative.) When the known values for DG, θ, and ϕ are substituted into Equation (8.46), it is a quadratic equation in the form $ax^2 + bx + c = 0$, which may be solved for x using $x = [-b \pm (b^2 - 4ac)^{1/2}]/2a$.

In the field, points H and J are established on the lines CD and AB, at the calculated distances from the adjacent corners. The side JH then is measured. If this measured value agrees with the computed value, the fieldwork and portions of the computations are verified. A further check on the computations is introduced by calculating the area $BCHJ$ and comparing it with the required area of this figure.

EXAMPLE 8.11. Determine the length and direction of the line passing through point 5, which divides the area of the tract enclosed by the traverse of Figure 8.7 into two equal areas. Adjusted coordinates are listed in Table 8.13 and the total area within the tract is 157,250 ft^2(see Example 8.6).

Solution. The use of coordinates and analytical geometry provides the most direct solution. To simplify the computations, all coordinates in Table 8.13 are translated so that the Y' axis passes through station 1 and the X' axis through station 2 (Figure 8.47). The X' and Y' translated values are given in Table 8.17.

Assume line 5-2 as a first approximation to the division line. The distance and direction of 5-2 by Equations (8.8) and (8.9) are

$$d_{5\text{-}2} = [(403.60 - 388.84)^2 + (310.40 - 0.0)^2]^{1/2} = 310.751 \text{ ft}$$

$$\tan A_{52} = \frac{388.84 - 403.60}{0 - 310.40}$$

$$A_{52} = 182°43'21''$$

$$\text{Angle } 521 = A_{52} - A_{12} = 76°23'36''$$

TABLE 8.17
Data for computing parting-off land from traverse of Figure 8.7
(see Tables 8.13 and 8.14)

| Station | Corrected | | Translated coordinate | |
	Distance, ft	Azimuths	X', ft	Y', ft
1			0.00	113.92
	405.18	106°19'45"		
2			388.84	0.00
	336.59	57°54'01"		
3			673.97	178.86
	325.18	335°28'43"		
4			539.01	474.71
	212.92	219°29'33"		
5			403.60	310.40
	252.21	266°55'30"		
6			151.75	296.87
	237.69	219°40'28"		
1				

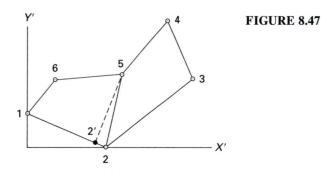

FIGURE 8.47

The total area of the tract, calculated in Example 8.6 by the coordinate method, is 157,250 ft^2, so that area/2 = 78,625 ft^2.

The area enclosed by stations 1256, using expression (8.36), is

$$2(\text{area})_{1256} = \frac{113.92}{0.00} \times \frac{0.00}{388.84} \times \frac{310.40}{403.60} \times \frac{296.87}{151.75} \times \frac{113.92}{0.00}$$

$$(\text{Area})_{1256} = 83,200 \text{ ft}^2$$

$$-\frac{\text{Total area}}{2} = \underline{78,625}$$

$$\text{Error} = 4,575 \text{ ft}^2$$

so that the area enclosed by 1256 is too large by 4575 ft^2. By Equation (8.45) and referring to Figure 8.47, the distance 22' from station 2 to the corrected division line 52' is

$$22' = \frac{(2)(4575)}{(310.751)\sin 76°23'36''} = 30.2951 \text{ ft}$$

The coordinates of 2' by Equations (8.6) and (8.7) are

$$X'_{2'} = 388.84 + (30.2951)(\sin 286°19'45'') = 388.84 + (30.2951)(-0.959662)$$
$$= 359.767 \text{ ft}$$

$$Y'_{2'} = 0.00 + (30.2951)(\cos 286°19'45'') = 0.00 + (30.2951)(0.281155)$$
$$= 8.5176 \text{ ft}$$

The area enclosed by 12'561 is

$$2(\text{area})_{12'561} = \frac{0.00}{113.92} \times \frac{359.767}{8.5176} \times \frac{403.60}{310.40} \times \frac{151.75}{296.87} \times \frac{0.00}{113.92}$$

$$(\text{Area})_{12'561} = 78,625 \text{ ft}^2$$

that is equal to the total area/2. The area enclosed by 2'23452' is computed as a check:

$$2(\text{area})_{2'23452'} = \frac{359.767}{8.5176} \times \frac{388.84}{0.00} \times \frac{673.97}{178.86} \times \frac{539.01}{474.71} \times \frac{403.60}{310.40} \times \frac{359.767}{8.5176}$$

$$(\text{Area})_{2'23452'} = 78,625 \text{ ft}^2$$

By Equations (8.8) and (8.9), $d_{52'} = 305.05$ ft and $A_{52'} = 188°15'42''$.

EXAMPLE 8.12. The tract enclosed by the traverse of Figure 8.7 and also shown in Figure 8.48 is to be divided so that one-fourth of the enclosed area lies south of a line parallel to the X and X' coordinate axes. Compute coordinates for the intersection of this dividing line with the traverse sides. The data listed in Table 8.17 may be used for the computations.

Solution. The initial position for the dividing line is taken passing through station 1, as illustrated in Figure 8.48. An approximate area for triangle 127, made using distances scaled from the sketch, reveals that this line results in an area that is ~20 percent too small, but it does provide a suitable starting line. As in Example 8.11, analytical geometry and coordinates provide the best method of solving the problem. Because the line 17 is parallel to the X' axis,

$$Y'_7 = Y'_1 = 113.92 \tag{1}$$

Equation (8.15) for line 23 is

$$\frac{Y'_7 - Y'_2}{Y'_3 - Y'_2} = \frac{X'_7 - X'_2}{X'_3 - X'_2}$$

$$\frac{Y'_7 - 0.00}{178.86 - 0.00} = \frac{X'_7 - 388.84}{673.97 - 388.84}$$

which yields

$$285.13Y'_7 - 178.86X'_7 = -69,547.92 \tag{2}$$

Substitution of Y'_7 from (1) into (2) gives

$$-178.86X'_7 = -102,029.93$$

$$X'_7 = 570.446 \text{ ft} = d_{17}$$

The area of 127 by area $= \frac{1}{2}(ab)$, where $a = 570.446$ and $b = 113.92$, is

$$\text{area} = \frac{1}{2}(570.446)(113.92) = 32,493 \text{ ft}^2 \tag{3}$$

The total area, calculated in Example 8.6 is 157,250 ft^2, so that (area)$/4 = 39,313$ ft^2. Therefore, the area enclosed by 127 is too small by $39,313 - 32,493 = 6820$ ft^2. Accordingly, the dividing line is moved parallel to itself to position 89, a distance x from the original line. Using Equation (8.46), the trapezoidal area enclosed by 1897 is

$$\text{Area}_{1897} = \frac{x}{2}(d_{17} + d_{17} - x \tan \theta + x \tan \phi) = xd_{17} + x^2\left(\frac{\tan \phi - \tan \theta}{2}\right) \tag{4}$$

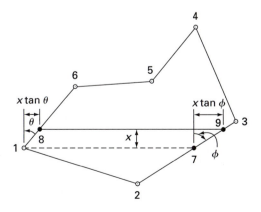

FIGURE 8.48

in which $\theta = 39°40'28''$ and $\phi = 57°54'01''$ are the azimuths of lines 16 and 23 as found in Table 8.17 and shown on Figure 8.48. Substitution of known values into Equation (4) gives

$$6820 = 570.446x + \left(\frac{1.59415 - 0.829463}{2}\right)x^2 = 570.446x + 0.382345x^2$$

which is of the form $ax^2 + bx + c = 0$ and can be solved by

$$x = \frac{-b \pm (b^2 - 4ac)^{1/2}}{2a} = \frac{-570.446 \pm [(570.446)^2 - (4)(0.382345)(-6820)]^{1/2}}{(2)(0.382345)}$$

$$= \frac{-570.446 \pm 579.516}{0.764691} = \frac{9.070}{0.764691} = 11.8613 \text{ ft}$$

Distances d_{18} and d_{79} are calculated using x and the cosines of angles θ and ϕ:

$$d_{18} = \frac{x}{\cos \theta} = \frac{11.8613}{0.769684} = 15.411 \text{ ft}$$

$$d_{79} = \frac{x}{\cos \phi} = \frac{11.8613}{0.531394} = 22.321 \text{ ft}$$

Coordinates for 8 and 9 are computed by Equations (8.6) and (8.7) as follows:

$$X'_8 = X'_1 + d_{18} \sin 39°40'28'' = 0.00 + (15.411)(0.638425) = 9.84 \text{ ft}$$

$$Y'_8 = Y'_1 + d_{18} \cos 39°40'28'' = 113.92 + 11.86 = 125.78$$

$$X'_9 = X'_7 + d_{79} \sin 57°54'01'' = 570.446 + (22.321)(0.847124) = 589.35 \text{ ft}$$

$$Y'_9 = Y'_7 + d_{79} \cos 57°54'01'' = 113.92 + 11.86 = 125.78 \text{ ft}$$

As a final check, the area of the segment enclosed by points 12981 is calculated using the computed coordinates and expression (8.36):

$$2(\text{area}) = \frac{0.00}{113.92} \times \frac{388.84}{0.00} \times \frac{589.35}{125.7813} \times \frac{9.84}{125.7813} \times \frac{0.00}{113.92}$$

$$= 78,625 \text{ ft}^2$$

area $= 39,313$ ft^2 which is equal to one-fourth of the total area. The final coordinates for 8 and 9 are

$$x_8 = 4391.93, \qquad Y_8 = 6162.68, \qquad X_9 = 4971.44, \qquad Y_9 = 6162.68 \text{ ft}$$

8.35
DIRECT METHOD FOR THE PARTITION OF LAND

Computations for the last two cases for partition of land involved trial and error. An alternate method to solve these problems is a general, direct method developed by Easa (1989). Consider subdivision of the tract shown in Figure 8.49, by line AB, into two tracts having areas A_1 and A_2 bounded by n points 1, 2, . . . , $n - 1$, n and m points $n + 1$, $n + 2$, . . . , $n + m$, respectively. The total area $A_1 + A_2$ and the area A_1 are known. It is desired to fix the position of the line AB to satisfy the proportion of A_1 to $A_1 + A_2$.

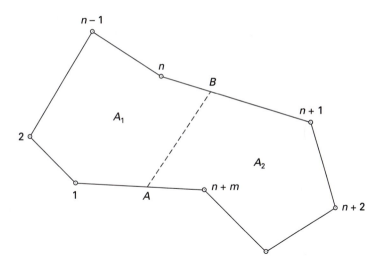

FIGURE 8.49

The problem has four unknowns: X_A, Y_A, X_B, and Y_B. Two equations that relate these four unknown coordinates are

$$Y_A = Y_1 + (X_A - X_1)m_A \tag{8.47}$$

$$Y_B = Y_n + (X_B - X_n)m_B \tag{8.48}$$

in which

$$m_A = \cot A_{1,n+m}$$

$$m_B = \cot A_{n,n+1}$$

where $A_{1,n+m}$ = azimuth from the north of line $1 - n + m$
$A_{n,n+1}$ = azimuth from the north of line $n - n + 1$

The area of A_1 can be computed by the coordinate method using Equation 8.34. Applying this equation to tract A_1 in Figure 8.49 leads to

$$2A_1 = \left\{ [X_B(Y_n - Y_A) + X_A(Y_B - Y_1) + X_1 Y_A - X_n Y_B] \right.$$
$$\left. + \left(\sum_{i=1}^{n-1} X_{i+1} Y_i - X_i Y_{i+1} \right) \right\} \tag{8.49}$$

in which the first group of terms contains coordinates of the subdividing line and the second group has coordinates for the remaining points in the tract.

Equations (8.47) to (8.49) contain four unknowns. To solve for X_A, Y_A, X_B, and Y_B, a fourth constraint, defined by the conditions for subdivision, is needed. For example, consider the case from Section 8.33, where either point A or B is fixed. First suppose that the coordinates of A are known, where A could be one of the given corners of the original tract. Substitution of Y_B from Equation (8.47) into (8.49) and solving for X_B yields

$$X_B = \frac{K_1 - X_A Y_1 + X_1 Y_A + (X_A - X_n)(Y_n - X_n m_B)}{(Y_A - Y_n) - (X_A - X_n)m_B} \tag{8.50}$$

in which

$$K_1 = \left[\sum_{i=1}^{n-1} (X_{i+1} Y_i - X_i Y_{i+1}) - 2A_1 \right] \qquad (8.51)$$

and Y_A can be calculated using Equation (8.47).

If the coordinates of B are known, substitution of Equation (8.47) into Equation (8.49) gives

$$X_A = \frac{-K_1 - X_B Y_n + X_n Y_B + (X_B - X_1)(Y_1 - X_1 m_A)}{(X_B - Y_1) - (X_B - X_1) m_A} \qquad (8.52)$$

in which K_1 is as defined by Equation (8.50). The coordinate, Y_A, can then be found by using Equation (8.47).

EXAMPLE 8.13. Apply the direct method for land partitioning to the problem of Example 8.11.

Solution. Renumber the corners of the tract shown in Figure 8.47 in a clockwise direction, as shown in Figure 8.50. Note the following equivalences between Figure 8.47 and Table 8.17 and Figure 8.50:

Figure 8.50	A	1	2	B	3	4	5
Figure 8.47 and Table 8.17	2'	1	6	5	4	3	2

First solve for K_1 by Equation (8.51) using the coordinates given in Table 8.17 and the preceding renumbering scheme:

$$K_1 = X_2' Y_1' - X_1' Y_2' - 2A_1$$

$$K_1 = (151.75)(113.92) - (0.00)Y_2 - 157,250 = -139,962.64 \text{ ft}^2$$

Next, solve for X_A' using Equation (8.52):

$$X_A' = \frac{139,962.64 - (403.60)(296.87) + (151.75)(310.40) + (403.60)(113.92)}{(310.40 - 113.92) - (403.60 - 0.00)(-0.29297)}$$

$$X_A' = 359.767 \text{ ft}$$

Substitution of this value for X_A' into Equation (8.47) yields

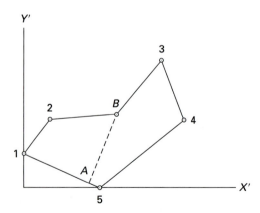

FIGURE 8.50

$$Y'_A = 113.92 + (359.767 - 0)(\cot 106°19'45'') = 8.5179 \text{ ft}$$

By Equations (8.8) and (8.9), $d_{AB} = 305.05$ ft and $A_{AB} = 188°15'41''$, both of which agree with the values obtained by the other solution in Example 8.11.

This concept can be applied to the other cases where (1) the distance between A and B is given, (2) the orientation of the subdividing line is known, and (3) the subdividing line passes through an interior point. For each case, additional equations, which are functions of the conditions involved, must be developed and solved for the coordinates of the ends of the subdivision line in a manner similar to the example shown. Details concerning the methods and equations used for all three cases can be found in Easa (1989).

Note that if $A_{1,n+m}$ or $A_{n,n+1}$ is 0° or 180°, m_A and m_B are undefined. This condition can be handled by rotating the original coordinate system by an arbitrary angle β into an X'', Y'' system using Equation (C.6), Appendix C. The rotated coordinates can be used with the given equations to solve for coordinates of A and B, at the ends of the dividing line, and then rotated back into the original system using the inverse solution of Equation (C.6).

8.36
ERROR PROPAGATION IN TRAVERSES

Although the positional closure (Section 8.21) is an indication of the overall quality of the traverse and is used for traverse classification, it does not yield information on the precision of point locations determined in a traverse. Therefore, in the following paragraphs, the techniques of covariance propagation (see Section 2.19) are employed to determine the covariance matrix for each point in the traverse.

For simplicity, consider the very first point in a traverse, such as point 2, which follows control point 1 in Figure 8.51. The azimuth of line 12, A_{12}, is given by

$$A_{12} = A_r + \alpha_1 - 180°$$

in which A_r is the beginning reference azimuth. If the reasonable assumption is made that A_r is relatively error free and has a standard deviation of 0, then $\sigma_{A_{12}} = \sigma_{\alpha_1}$. Further, let σ_{d_1} represent the standard deviation of the measured distance d_1, and assume that α_1 and d_1 are uncorrelated (i.e., $\sigma_{\alpha_1 d_1} = 0$), which is a very logical assumption. Then,

$$\Sigma_{m1} = \begin{bmatrix} \sigma_{\alpha_1}^2 & 0 \\ 0 & \sigma_{d_1}^2 \end{bmatrix}$$

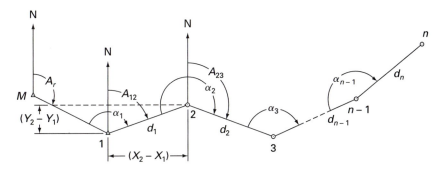

FIGURE 8.51
Open-end traverse.

represents the covariance matrix (actually a variance matrix in this case) for the measurements in the first leg of the traverse. Next, the coordinates of point 2 are given by

$$X_2 = X_1 + d_1 \sin A_{12} = X_1 + d_1 \sin (A_r + \alpha_1 - 180°)$$

$$Y_2 = Y_1 + d_1 \cos A_{12} = Y_1 + d_1 \cos (A_r + \alpha_1 - 180°) \qquad (8.53)$$

in which X_1 and Y_1, the coordinates of the control point 1, are assumed to be errorless. Applying error propagation Equation (2.34a) to Equation (8.53),

$$\Sigma_{C2} = \begin{bmatrix} \sigma^2_{X_2} & \sigma_{X_2 Y_2} \\ \sigma_{X_2 Y_2} & \sigma^2_{Y_2} \end{bmatrix} = J\Sigma_{m1}J' \qquad (8.54)$$

in which Σ_{C2} is the covariance matrix of the *coordinates* of point 2 and J is the Jacobian matrix of partial derivatives, given by

$$J = \begin{bmatrix} \dfrac{\partial X_2}{\partial \alpha_1} & \dfrac{\partial X_2}{\partial d_1} \\ \dfrac{\partial Y_2}{\partial \alpha_1} & \dfrac{\partial Y_2}{\partial d_1} \end{bmatrix} = \begin{bmatrix} d_1 \cos A_{12} & \sin A_{12} \\ -d_1 \sin A_{12} & \cos A_{12} \end{bmatrix}$$

or

$$J = \begin{bmatrix} Y_2 - Y_1 & \dfrac{X_2 - X_1}{d_1} \\ -(X_2 - X_1) & \dfrac{Y_2 - Y_1}{d_1} \end{bmatrix} \qquad (8.55)$$

where we used the substitution $(Y_2 - Y_1)/d_1 = \cos A_{12}$ and $(X_2 - Y_1)/d_1 = \sin A_{12}$.

By substitution of Equation (8.55) into (8.54) and matrix multiplication (see Section B.3.3), the covariance matrix for the coordinates of point 2 becomes

$$\Sigma_{C2} = \begin{bmatrix} (d_1^2 \sigma^2_{\alpha_1} \cos^2 A_{12} + \sigma^2_{d_1} \sin^2 A_{12}) & (\sigma^2_{d_1} - d_1^2 \sigma^2_{\alpha_1}) \sin A_{12} \cos A_{12} \\ (\sigma^2_{d_1} - d_1^2 \sigma^2_{\alpha_1}) \sin A_{12} \cos A_{12} & (d_1^2 \sigma^2_{\alpha_1} \sin^2 A_{12} + \sigma_{d_1} \cos^2 A_{12})] \end{bmatrix}$$

$$= \begin{bmatrix} (Y_2 - Y_1)\sigma^2_{\alpha_1} + \dfrac{(X_2 - X_1)^2}{d_1^2}\alpha^2_{d_1} & -(X_2 - X_1)(Y_2 - Y_1)\sigma^2_{\alpha_1} + \dfrac{(X_2 - X_1)(Y_2 - Y_1)}{d_1^2}\sigma^2_{d_1} \\ \text{symmetric} & (X_2 - X_1)^2 \sigma^2_{\alpha_1} + \dfrac{(Y_2 - Y_1)^2}{d_1^2}\sigma^2_{d_1} \end{bmatrix}$$

$$(8.56)$$

When performing calculations, Equation (8.54) usually is applied without going through the expanded form. In this case, the matrix multiplication is carried out to illustrate two points:

1. Although the original measurements α_1, d_1 may be uncorrelated, the resulting quantities in general are correlated, as shown in Equation (8.56).
2. For the covariance term in Equation (8.56) to be 0,

$$\sigma^2_{d_1} = d_1^2 \sigma^2_{\alpha_1} \qquad \text{or} \qquad \sigma_{d_1} = d_1 \sigma_{\alpha_1}$$

Note that the propagation of error to point 2 in the example is somewhat of a special case, because the beginning point 1 is a control point, assumed to be error free. Now, consider the precision of the coordinates for succeeding point 3 in Figure 8.51. The coordinates of point 3 are given by

$$X_3 = X_2 + d_2 \sin A_{23}$$

$$Y_3 = Y_2 + d_2 \cos A_{23} \tag{8.57}$$

The four values X_2, Y_2, d_2, and α_2 (implicit in A_{23}, see Figure 8.51) are the random variables. Their covariance matrix is given symbolically by

$$\begin{bmatrix} \sigma_{X_2}^2 & \sigma_{X_2 Y_2} & \sigma_{X_2 \alpha_2} & \sigma_{X_2 d_2} \\ \sigma_{X_2 Y_2} & \sigma_{Y_2}^2 & \sigma_{Y_2 \alpha_2} & \sigma_{Y_2 d_2} \\ \sigma_{X_2 \alpha_2} & \sigma_{Y_2 \alpha_2} & \sigma_\alpha^2 & \sigma_{\alpha_2 d_2} \\ \sigma_{X_2 d_2} & \sigma_{Y_2 d_2} & \sigma_{\alpha_2 d_2} & \sigma_{d_2}^2 \end{bmatrix} = \begin{bmatrix} \Sigma_{C2} & \Sigma_{C2m2} \\ \Sigma_{C2m2} & \Sigma_{m2} \end{bmatrix}$$

in which Σ_{C2} is the covariance matrix for the coordinates of point 2 as derived in Equation (8.56) and Σ_{m2} is the covariance matrix for the two measurements, d_2 and α_2, that usually is diagonal with elements $\sigma_{d_2}^2$ and $\sigma_{\alpha_2}^2$. The matrix Σ_{C2m2} represents the correlation between the coordinates of point 2 and the new measurements, d_2 and α_2. At first glance, it may appear that no such correlation exists. However, except for an azimuth traverse (Section 8.9) or a traverse run using a gyroscopic equipped theodolite (Section 10.43), the azimuth A_{23} is computed not only from α_2 but also from A_{12}, which in turn is a function of α_1. Because X_2 and Y_2 were computed in terms of α_1, in general they will be correlated with α_2 and Σ_{C2m2} will not necessarily be 0. To evaluate Σ_{C2m2} requires *cross-covariance* propagation, which is beyond the scope of this textbook (for those interested, see Mikhail, 1976). Instead, Equation (8.57) is extended so that it is expressed in terms of the original observations. Therefore,

$$X_3 = X_1 + d_1 \sin A_{12} + d_2 \sin A_{23}$$

$$Y_3 = Y_1 + d_1 \cos A_{12} + d_2 \cos A_{23}$$

or

$$X_3 = X_1 + d_1 \sin (A_r + \alpha_1 - 180°) + d_2 \sin [A_r + \alpha_1 + \alpha_2 - (2)(180°)]$$

$$Y_3 = Y_1 + d_1 \cos (A_r + \alpha_1 - 180°) + d_2 \cos [A_r + \alpha_1 + \alpha_2 - (2)(180°)]$$

The Jacobian matrix for these equations is (see Section B.10)

$$\begin{aligned} J &= \begin{bmatrix} \dfrac{\partial X_3}{\partial \alpha_1} & \dfrac{\partial X_3}{\partial d_1} & \dfrac{\partial X_3}{\partial \alpha_2} & \dfrac{\partial X_3}{\partial d_2} \\[2mm] \dfrac{\partial Y_3}{\partial \alpha_1} & \dfrac{\partial Y_3}{\partial d_1} & \dfrac{\partial Y_3}{\partial \alpha_2} & \dfrac{\partial Y_3}{\partial d_2} \end{bmatrix} \\[3mm] &= \begin{bmatrix} (d_1 \cos A_{12} + d_2 \cos A_{23}) & \sin A_{12} & d_2 \cos A_{23} & \sin A_{23} \\ -(d_1 \sin A_{12} + d_2 \sin A_{23}) & \cos A_{12} & -d_2 \sin A_{23} & \cos A_{23} \end{bmatrix} \\[3mm] &= \begin{bmatrix} Y_3 - Y_1 & \dfrac{X_2 - X_1}{d_1} & Y_3 - Y_2 & \dfrac{X_3 - X_2}{d_2} \\[2mm] -(X_3 - X_1) & \dfrac{Y_2 - Y_1}{d_1} & -(X_3 - X_2) & \dfrac{Y_3 - Y_2}{d_2} \end{bmatrix} \end{aligned}$$

The four measurements, α_1, d_1, α_2, and d_2, may be assumed to be uncorrelated, and therefore their covariance matrix is diagonal or $\Sigma_{m_{12}} = \text{diag}(\sigma_{\alpha_1}^2 \sigma_{d_1}^2 \sigma_{\alpha_2}^2 \sigma_{d_2}^2)$. Using J just evaluated and this diagonal covariance matrix for the measurements in the error propagation Equation (2.34a), the covariance matrix for the coordinates of point 3 is

$$\Sigma_{C3} = \underset{2,4}{\mathbf{J}} \underset{4,4}{\Sigma_{m12}} \underset{4,2}{\mathbf{J}^t}$$

$$\Sigma_{c3} = \begin{bmatrix} [(Y_3 - Y_1)^2\sigma_{\alpha_1}^2 + (Y_3 - Y_2)^2\sigma_{\alpha_2}^2] + \left[\dfrac{(X_2 - X_1)^2}{d_1^2}\sigma_{d_1}^2 + \dfrac{(X_3 - X_2)^2}{d_2^2}\sigma_{d_2}^2\right] \\ \hline -[(Y_3 - Y_1)(X_3 - X_1)\sigma_{\alpha_1}^2 + (Y_3 - Y_2)(X_3 - X_2)\sigma_{\alpha_2}^2] + \left[\dfrac{(X_2 - X_1)(Y_2 - Y_1)}{d_1^2}\sigma_{d_1}^2 + \dfrac{(X_3 - X_2)(Y_3 - Y_2)}{d_2^2}\sigma_{d_2}^2\right] \\[2em] \text{symmetric} \\ \left[(X_3 - X_1)^2\sigma_{\alpha_1}^2 + (X_3 - X_2)^2\sigma_{\alpha_2}^2 + \dfrac{(Y_2 - Y_1)^2}{d_1^2}\sigma_{d_1}^2 + \dfrac{(Y_3 - Y_2)^2}{d_2^2}\sigma_{d_2}^2\right] \end{bmatrix}$$

which becomes

$$\Sigma_{C3} = \begin{bmatrix} \left[\displaystyle\sum_{i=1}^{2}(Y_3 - Y_i)^2\sigma_{\alpha_i}^2 + \sum_{i=1}^{2}\left(\dfrac{X_{i+1} - X_i}{d_i}\right)\sigma_{d_i}^2\right] & \left[-\displaystyle\sum_{i=1}^{2}(Y_3 - Y_i)(X_3 - X_i)\sigma_{\alpha_i}^2 + \sum_{i=1}^{2}\dfrac{(X_{i+1} - X_i)(Y_{i+1} - Y_i)}{d_i^2}\sigma_{d_i}^2\right] \\[2em] \text{symmetric} & \left[\displaystyle\sum_{i=1}^{2}(X_3 - X_i)^2\sigma_{\alpha_i}^2 + \sum_{i=1}^{2}\left(\dfrac{Y_{i+1} - Y_i}{d_i^2}\right)^2\sigma_{d_i}^2\right] \end{bmatrix}$$

$$\tag{8.58}$$

Equation (8.58) illustrates the error propagation for the special case of two measured lines. For the general case of n traverse stations in an open traverse, starting with point 1 and terminating at point n, the coordinates for point n can be expressed as

$$X_n = X_1 + d_1 \sin(A_r + \alpha_1 - 180°) + d_2 \sin[A_r + \alpha_1 + \alpha_2 - (2)180°] + \cdots$$
$$+ d_n \sin[A_r + \alpha_1 + \alpha_2 + \cdots + \alpha_{n-1} - (n-1)180°]$$
$$Y_n = Y_1 + d_1 \cos(A_r + \alpha_1 - 180°) + d_2 \cos[A_r + \alpha_1 + \alpha_2 - (2)180°] + \cdots$$
$$+ d_n \cos[A_r + \alpha_1 + \alpha_2 + \cdots + \alpha_{n-1} - (n-1)180°]$$

Assuming uncorrelated measurements of angles and distances and errorless coordinates for station 1, application of error propagation Equation (2.34a) yields the covariance matrix for point n as follows:

$$\Sigma_{Cn} = \underset{2n,2}{\mathbf{J}} \underset{2n,2n}{\Sigma_{m(n-1)}} \underset{2n,2}{\mathbf{J}^t} = \begin{bmatrix} \sigma_{X_n}^2 & \sigma_{X_nY_n} \\ \sigma_{X_nY_n} & \sigma_{Y_n}^2 \end{bmatrix} \tag{8.59}$$

in which the elements $\sigma_{X_n}^2, \sigma_{Y_n}^2, \sigma_{X_nY_n}$ are a direct extension to those in Equation (8.58), or

$$\sigma_{X_n}^2 = \sum_{i=1}^{n-1}(Y_n - Y_i)^2\sigma_{\alpha_i}^2 + \sum_{i=1}^{n-1}\left(\frac{X_{i+1} - X_i}{d_i}\right)^2\sigma_{d_i}^2$$

$$\sigma_{Y_n}^2 = \sum_{i=1}^{n-1}(X_n - X_i)^2\sigma_{\alpha_i}^2 + \sum_{i=1}^{n-1}\left(\frac{Y_{i+1} - Y_i}{d_i}\right)^2\sigma_{d_i}^2 \tag{8.60}$$

$$\sigma_{Y_nX_n} = -\sum_{i=1}^{n-1}(Y_n - Y_i)(X_n - X_i)\sigma_{\alpha_i}^2 + \sum_{i=1}^{n-1}\left[\frac{(X_{i+1} - X_i)(Y_{i+1} - Y_i)}{d_i^2}\right]\sigma_{d_i}^2$$

The practical value of this error propagation as developed lies in the application of Equations (8.60) to the data for proposed traverses to estimate the accuracies possible at specific traverse stations using assumed values for σ_{α_i} and σ_{d_i}.

452

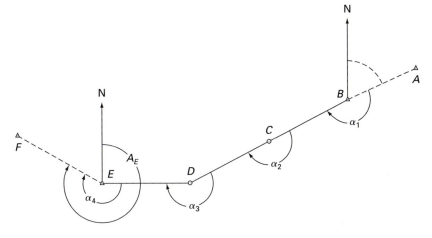

FIGURE 8.52

EXAMPLE 8.14. Consider the traverse used in Examples 8.15 and 8.16 (Section 8.39). Figure 8.52 shows a closed traverse starting at point B and closing on point E, with two intermediate stations C and D. The data follow:

Angle	Value	σ	Distance	Value, m	σ, m
α_1	172°53′34″	2″	d_1	281.832	0.016
α_2	185°22′14″	2″	d_2	271.300	0.016
α_3	208°26′19″	2″	d_3	274.100	0.016
α_4	205°13′51″	2″	**Point**	**X, m**	**Y, m**
A_{Z_B}	68°15′20.7″	0	B	8478.139	2483.826
A_{Z_E}	300°11′30.5″	0	E	7709.336	2263.411

The observations are assumed to be uncorrelated and with the precisions indicated in the table.

It is desired to determine the estimated covariance matrix for point D treating stations B, C, and D as an open traverse and to calculate the parameters of the resulting standard error ellipse at D.

Solution. Distances, azimuths, and unadjusted coordinates for lines BC, CD, and points C and D are as follows:

Station	Distance, m	Azimuth	Departures, x, m	Latitudes, y, m	Coordinates, m X	Coordinates, m Y
B					8478.139	2483.826
	281.832	241°08′57.7″	−246.851	−135.993		
C					8231.288	2347.833
	271.300	246°31′14.6″	−248.838	−108.091		
D					7982.450	2239.742

The standard deviation in angles is $\sigma_{\alpha_i} = (2″)(4.8481368)(10^{-6}) = (9.6962736)(10^{-6})$ rad. Therefore, the variance, $\sigma_{\alpha_i}^2 = (9.401772)10^{-11}$ rad². The variance in each measured distance is $\sigma_{d_i}^2 = (0.016)^2 = (2.56)10^{-4}$ m². Using these estimates, the measured distances and directions, and calculated coordinates in Equation (8.60), the propagated variances and covariances are

$$\sigma_{Xn}^2 = \sum_{i=1}^{2} (Y_3 - Y_i)^2 \sigma_{\alpha_i}^2 + \sum_{i=1}^{2} \left(\frac{X_{i+1} - X_i}{d_i}\right)^2 \sigma_{d_i}^2 = (4.18457)10^{-4} \text{ m}^2$$

$$\sigma_{Yn}^2 = \sum_{i=1}^{2} (X_3 - X_i)^2 \sigma_{\alpha_i}^2 + \sum_{i=1}^{2} \left(\frac{Y_{i+1} - Y_i}{d_i}\right)^2 \sigma_{d_i}^2 = (1.29166)10^{-4} \text{ m}^2$$

$$\sigma_{X_n Y_n} = -\sum_{i=1}^{2} (Y_3 - Y_i)(X_3 - X_i)\sigma_{\alpha_i}^2 + \sum_{i=1}^{2} \left[\frac{(X_{i+1} - X_i)(Y_{i+1} - Y_i)}{d_i^2}\right]\sigma_{d_i}^2$$

$$= (1.87843)10^{-4} \text{ m}^2$$

so that the covariance matrix for point D is

$$\Sigma_{DD} = \begin{bmatrix} (4.18457)10^{-4} & (1.87843)10^{-4} \\ (1.87843)10^{-4} & (1.29166)10^{-4} \end{bmatrix} \text{ m}^2$$

Analysis of this covariance matrix using the methods developed in Section 2.31 and illustrated by Example 2.15 yields the following error ellipse data:

semimajor axis, $a = 0.023$ m
semiminor axis, $b = 0.0006$ m
orientation angle $\theta = 26°12'$ (= angle between semimajor and x-axes)

The procedure outlined in Example 8.14 is particularly useful in the design and planning of traverses. Given approximate distances and directions for a proposed traverse, estimated values for σ_{α_i} and σ_{d_i} are chosen and the covariance matrices for specified traverse points are calculated using Equations (8.59) and (8.60). The values of σ_{α_i} and σ_{d_i} selected are those that produce covariance matrices yielding error ellipses that satisfy specifications for the proposed survey.

It must be emphasized that a covariance matrix propagated for a point using unadjusted data and estimated standard deviations, as illustrated in Example 8.14 does not provide a correct measure for the accuracy of the *adjusted* coordinates for a point. For example, if the method were used for error propagation with coordinates adjusted by the compass rule, the resulting covariance matrix would be incorrect. Consequently, the procedure is of primary value as a design tool for open traverses.

A legitimate measure of the accuracy of an adjusted point can be obtained only by analysis of the covariance matrices propagated as a by-product of a least-squares adjustment of the traverse. Adjustment of traverses by the method of least squares is outlined in Section 8.39 and illustrated by Examples 8.15 and 8.16.

8.37
TRAVERSE IN THREE DIMENSIONS

When elevations also are determined while running a traverse, this method constitutes a traverse in three dimensions. When total station systems (Sections 7.15 to 7.21) are employed to establish horizontal and vertical control, relatively high accuracies in both horizontal and vertical positions are possible. In other words, trigonometric leveling, as described in Sections 5.5 and 5.55, is performed in conjunction with any one of the several methods for running a traverse described in Sections 8.6 through 8.9. Note that the stadia traverse (Section 8.11) also can be performed in three dimensions but is restricted to surveys where accuracy requirements are low (horizontal positions, 1 part in 300 to 400; vertical positions, to a few cm or tenths of a ft). In this section, emphasis is focused on traverse in three dimensions using a total station system.

The equipment required for a three-dimensional traverse consists of a total station system (preferably a self-contained system) with a nominal least count in horizontal and vertical circles of 10″ or smaller; reflectors calibrated for the EDM system of the total station unit; two sighting targets (Section 6.50, Figure 6.53) adaptable to a tripod (combination reflector targets described in Section 7.16, Figure 7.9, are advisable); and three adjustable tripods. Mounting devices or adapters for the total station, reflectors, and targets should be compatible so that forced centering (Section 8.2) can be used. It also is important that the tribrachs supporting the total station system, reflector-target, or reflector and targets be compatible with each device.

Procedure

Assume that a self-contained total station system is to be used to run a three-dimensional traverse from station B to station E, as shown in Figure 8.52. Lines BA and EF are of known directions and X-Y coordinates and elevations are available for stations B and E. The total station system is set over station B and tripods equipped with a combined reflector and targets or just the reflectors are centered over stations A and C. A sample note form is shown in Figure 8.53. The heights of the instrument of the total station system and reflector and targets or just reflectors are measured and recorded at B and C (column 3, Figure 8.53). Environmental and system parameters then are entered into the total station system, and the measuring begins.

Starting with the telescope in the direct or right face position, a backsight is taken on A, the horizontal circle is set to 0 by pressing the zero-set or comparable key (column 6, Figure 8.53), the angle to the right from A to C ($\alpha_1 = 172°53'30''$, column 6, Figure 8.53), and the zenith angle (91°10′00″, column 5, Figure 8.53) and slope distance (281.884, column 7) from A to C are measured and recorded. The telescope then is reversed (left face position), the angle is "held" or the lower motion is used for a backsight and the process is repeated with the zenith and horizontal angles being recorded in columns 5 and 6 in the notes. The direct and reversed zenith and horizontal angles are checked (they should agree to within the least count of the system, 10″ in this example) with the average values being computed and recorded in columns 5 and 6 of the notes, respectively.

At this stage, the tripod and the reflector and target at A are moved ahead to B, where the total station instrument is removed from the tripod and moved toward C. The reflector and target from A is mounted on the tripod at B, the total station instrument is placed on the tripod at C, and the reflector-target combination from C and tripod from A are taken to D to be set up, centered, and aligned, while the instrument at C and reflector-target combination at B are leveled, centered, and aligned as necessary. The process of measuring the h.i.'s, the horizontal angle to the right from B to D, and the zenith angle and slope distance from C to D is repeated, as described for the previous setup, and all readings are recorded, as shown in Figure 8.53 for the instrument set up at station C. This procedure continues until the total station instrument occupies the closing station E, where it is necessary to measure only the horizontal angle to the right from D to F, allowing determination of the horizontal angular error of closure on the line EF, for which the azimuth is known.

Computations

Computations involve the determination of corrected slope distances (column 8, Figure 8.53), horizontal distances (column 9), Δ elevations (column 10), and elevations of

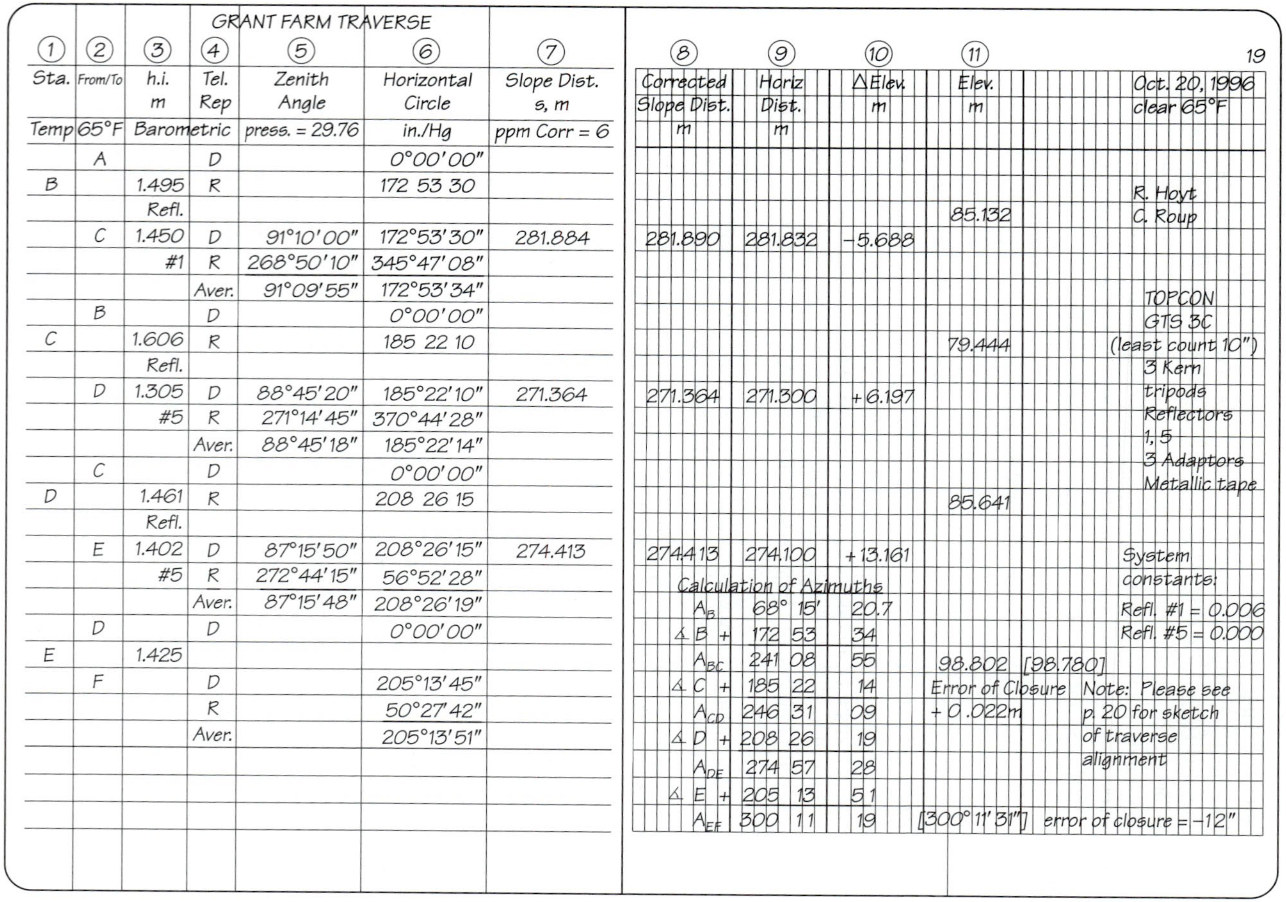

GRANT FARM TRAVERSE

① Sta	② From/To	③ h.i. m	④ Tel. Rep	⑤ Zenith Angle	⑥ Horizontal Circle	⑦ Slope Dist. s, m
Temp 65°F		Barometric press. = 29.76			in./Hg	ppm Corr = 6
	A		D		0°00'00"	
B		1.495	R		172 53 30	
		Refl.				
	C	1.450	D	91°10'00"	172°53'30"	281.884
		#1	R	268°50'10"	345°47'08"	
			Aver.	91°09'55"	172°53'34"	
	B	1.606	D		0°00'00"	
C		Refl.	R		185 22 10	
					208 26 15	
	D	1.305	D	88°45'20"	185°22'10"	271.364
D		#5	R	271°14'45"	370°44'28"	
			Aver.	88°45'18"	185°22'14"	
	C	1.461	D		0°00'00"	
		Refl.	R		208 26 15	
	E	1.402	D	87°15'50"	208°26'15"	274.413
E		#5	R	272°44'15"	56°52'28"	
			Aver.	87°15'48"	208°26'19"	
	D	1.425	D		0°00'00"	
F			R		205°13'45"	
			Aver.		50°27'42"	
	E				205°13'51"	

⑧ Corrected Slope Dist. m	⑨ Horiz. Dist. m	⑩ ΔElev. m	⑪ Elev. m	Remarks
				Oct. 20, 1996
				clear 65°F
281.890	281.832	−5.688	85.132	R. Hoyt / C. Roup
				TOPCON GTS 3C (least count 10")
				3 Km
271.364	271.300	+6.197	79.444	tripods / Reflectors 1, 5
				3 Adaptors / Metallic tape
274.413	274.100	+13.161	85.641	System constants:
	Calculation of Azimuths			Refl. #1 = 0.006
	A_B + 68° 15' 20.7			Refl. #5 = 0.000
	172 53 34			
	A_BC 241 08 55			
	ΔC + 185 22 14		98.802 [98.780]	Error of Closure Note: Please see
	A_CD 246 31 09		Error of Closure	p. 20 for sketch
	ΔD + 208 26 19		+ 0.022m	of traverse
	A_DE 274 57 28			alignment
	ΔE + 205 13 51			
	A_EF 300 11 19	[300°11'51"] error of closure = −12"		

FIGURE 8.53
Sample notes, traverse in three dimensions.

stations at the foresight positions. Most total station systems display the horizontal distance and Δ elevation between the instrumental horizontal axis and the center of the reflector, when the appropriate key on the control panel is pressed. Therefore, if the system constant can be entered into the instrument, the horizontal distance and Δ elevation can be recorded directly, eliminating columns 7 and 8 in the notes. In this case, the h.i.s of the instrument and the reflector at the forward station would have to be applied to the recorded Δ elevation to calculate the correct elevation of the forward station. Elevations of the points are computed, and for this example, the vertical error of closure on known point E is found to be 0.022 m (Figure 8.53).

The azimuths for each line can be computed using averaged angles to the right, as described in Section 8.8, and are tabulated at the bottom of the right page of the notes in Figure 8.53. The angular error of closure on known line EF is found to be $-00°00'12''$ (Figure 8.53). Adjusted azimuths can be computed following the method outlined in Section 8.8. Using these adjusted azimuths and horizontal distances, along with known coordinates for station B, X-Y coordinates and their respective errors of closure can be calculated for stations C and D, according to the methods outlined in Section 8.14. Closure corrections in X and Y coordinates are $dX = -0.041$ m and $dY = 0.033$ m, respectively. For a traverse of this length and configuration, the X-Y coordinates can be adjusted by the compass rule (Section 8.17) and the elevations in proportion to the distance leveled (Section 5.42). Unadjusted coordinates and coordinates adjusted by the compass rule and elevations, both unadjusted and those adjusted in proportion to the distance leveled, for stations C and D, are shown in Table 8.18.

When traverses are longer and more complex, least-squares adjustments of X-Y coordinates and elevations provide more rigorous adjustments for the parameters and also yield propagated errors and statistical measures for the accuracy of the calculated positions. Least-squares adjustment of elevations is described in Section 5.58 and given for X-Y coordinates in Section 8.39 of this chapter.

The best adjustment for this type of traverse is by the method of least squares applied to coordinates and elevations simultaneously or a three-dimensional least squares adjustment, developed in Section 9.21.

Traverse in Three Dimensions Using a Total Station System and a Data Collector

Although the total station instrument provides a powerful tool for a three-dimensional traverse, even when data are manually recorded, its efficiency is immensely enhanced by using a data collector to round out the system. Data collectors were discussed in some detail in Sections 3.12 and 7.18, and a data collector designed for and connected to a specific total station instrument is illustrated in Figure 7.7.

Each brand of total station instrument has its own external or internal data collector. Here, we consider only external data collectors. Each of these collectors has a fairly unique set of software operating instructions, menus, prompts, identifying codes, and the like. Most of these devices also are usable with other total station systems. In addition, "generic" collectors are designed to work with all brands of total station systems. A recent study of these devices showed that 69 external data collectors were on the market in 1996 (Walkowski, 1996). This proliferation of data collectors, each with its own operating system, can be almost overwhelming to the uninitiated student or practitioner.

Moves toward a standard data management system are under way. The American Association of State Highway and Transportation Officials (AASHTO) is developing a

TABLE 8.18
Unadjusted and adjusted coordinates and elevations, three-dimensional traverse

Station	Distance, m	Adjusted azimuth	Unadjusted coordinates X	Unadjusted coordinates Y	Unadjusted elevation, m	Adjusted coordinates X, m	Adjusted coordinates Y, m	Adjusted elevation, m
B			8,478.139	2,483.826	85.132	8,478.139	2,483.826	85.132
	281.832	241°08'58"						
C			8,231.288	2,347.835	79.444	8,231.274	2,347.824	79.437
	271.300	246°31'15"						
D			7,982.450	2,239.744	85.641	7,982.423	2,239.722	85.626
	274.100	274°57'37"						
E			7,709.377 [7,709.336]	2,263.444 [2,263.411]	98.802 [98.780]	7,709.336	2,263.422	98.780

Notes: Closure corrections: $dX = -0.041$, $dY = -0.033$, $dH = +0.022$ m, $d_c = (dX^2 + dX^2)^{1/2} = [(-0.041)^2 + (-0.033)^2]^{1/2} = 0.053$ m.

458

Survey Data Management System (SDMS™) for surveying tasks related to highways.* Until this or a similar system is adopted and in place, the current user of a data collector with a total station system must rely on and master the instruction manual for that specific system.

Figure 8.54 shows the keyboard overlay and display screen of a generic data collector with its main menu displayed. This data collector is a general-purpose pocket calculator (TDS 48GX) modified for use as a data collector; hence, the overlay. The shifted key functions are special routines in the data collector program. These functions are activated by pressing the shift key (⌐) and then the key below the desired function. Note that there is a command line at the bottom of the display screen. The commands given on this line, which change with each screen, are actuated by pressing the keys A, B, and so on, through F directly below the command line. Alphanumeric input is possible by pressing the (α) button and then the letters or numbers desired. Data transmission from the total station instrument to the collector occurs when the key on the data collector, specified for a particular system, is pressed.

In this section, use of a complete total station system, equipped with a data collector, is considered for a three-dimensional traverse. When necessary, reference will be made to a generic collector such as that shown in Figure 8.54. For the most part, comments will be of a general nature, designed to illustrate the overall procedure involved with any type of data collector.

The way in which the total station system and peripheral units (reflector and targets, tripods, etc.) are used in this process is essentially the same as described in the previous section, where the data are recorded manually. The major differences, when using a data collector, are that (1) the bulk of the data, descriptions, and comments are recorded directly or keyed into the data collector; and (2) the preprogrammed menus and display screens generated by the software in the data collector tend to "direct" the survey operation.

As an example, consider the first setup at station $B = 1$ for the traverse shown in Figure 8.52. Preliminary work, using only the data collector, can be done in the field or office, where the following steps are performed with the data collector:

1. Open and create a job identified as a file specified by up to eight alphanumeric characters.

2. Enter plane coordinates and elevations for the initial station and any other known station(s). Identify by point number ($A = 0, B = 1, \ldots, E = 4$) and store in collector memory.

3. Enter total station instrument and model number, such as Topcon GTSB-D, Leitz Set2, or Wild TC2000. If one simply wishes to enter data manually via the keyboard, enter *manual*.

4. Identify and enter operating modes and options, which may include
   (a) Azimuth or bearing: select one.
   (b) Scale factor: Yes or No.
   (c) Distance units: ft or m.
   (d) Angular units: degrees or gons.

5. Set repetition modes and tolerances for angles and distances:
   (a) Horizontal angles: single, directional (1DR), or accumulation (2DR, 3DR).
   (b) Vertical circle: single or multiple.
   (c) Distance: single or multiple.
   (d) Number of repeated sets.
   (e) Tolerances in angles and distances.

---

*American Association of State Highway and Transportation Officials, *Survey Data Management System SDMS™ Data Structure Technical Guide*, 444 N. Capitol St. N.W., Suite 249, Washington, DC. 20001.

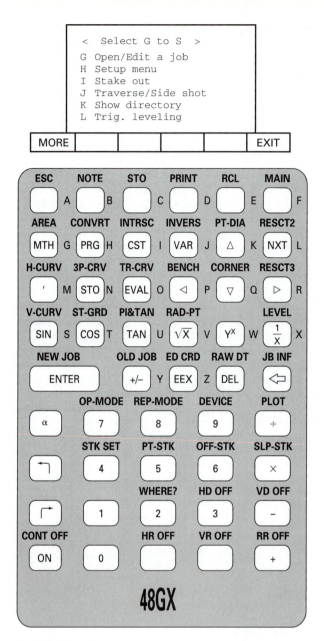

FIGURE 8.54
Keyboard, overlay, and screen, TDS 48GX. (*Courtesy of Tripod Data Systems, Inc.*)

These instructions generally are entered and stored with standard keyboard procedures, following the menus and using the horizontal and vertical cursor keys to display the desired designation or category.

For the example problem of Figures 8.52 and 8.53, starting at point $B = 1$ and terminating at point $E = 4$, a Topcon GTSB-D is designated to turn one set of directional angles or 1DR and single distances, with meters and degrees chosen as the units of measurement.

Tolerances selected were 20″ for angles and no tolerance for distances, because only one measurement is to be made per line. Horizontal angles are to be turned BS and FS direct, BS and FS reversed.

At the first instrument setup ($B = 1$, for the example), the data collector is connected to the system instrument by cable and both collector and system are turned to the "on" position. Following the data collector's menus and displayed prompts, the desired job is chosen. These instructions yield a screen display containing the job number and the starting and ending point numbers, which had been entered in the preliminary steps. Returning to the main menu (Figure 8.54), *traverse/side shot* is selected from the menu. This choice leads to a screen that displays the occupied (OC) point, backsight (BS) point, and foresight (FS) point numbers along with lines for angle right, slope distance, point description (DESC), h.i. of instrument (HI), and h.i. of rod (HR) (see Figure 8.55a). At this stage, if the designation *manual* had been entered when the type of instrument was chosen, then measured angles, distances, and h.i.s simply could be keyed in at the proper lines on the screen and entered so that the data collector would be used as an electronic recording device. For this example, the *automatic* mode with a Topcon GTS3B was chosen, so it is necessary to set up the system for the first backsight.

Most data collectors allow for several options on the initial backsight. For example, one can start with two coordinated points, have a line of known azimuth, make astronomic observations to determine the azimuth, or assume an azimuth for the initial line. For this example, the azimuth from B to A is known, so the first backsight is of a known direction. It is necessary to set up the data collector for the proper backsight (BS) azimuth. To do this for the system shown in Figure 8.54 for which the screen display in Figure 8.55a is given, press key C under the command line instruction *BACK*. This instruction yields the display on the screen shown in Figure 8.55b. On this screen, one can enter the backsight azimuth (68.15207°) and the circle reading (0.0000°), both in decimal degrees, and press the command *SOLVE* (key A), which enters the information into collector memory and yields a display showing this information and the BS bearing (Figure 8.55c). This display is followed by a prompt for the height of instrument, which needs to be measured (1.495, from Figure 8.53) and entered. Exiting from this screen and pressing the command line instruction *TRAV* (key D) leads to a screen prompt that instructs the operator to take a backsight (BS) with the telescope direct and then press the appropriate key to record the data. The steps necessary to complete the setup for this example are as follows:

1. Take a backsight (BS) on point $A = 0$ and press the zero-set or equivalent button on the total station instrument.
2. Press the appropriate key on the controller to transmit data automatically to the data collector. For the system shown in Figure 8.54, press A under *SIDES* on the command line (Figure 8.55d). A screen prompt will appear instructing the operator to take a foresight (FS) with the telescope direct.
3. Take a FS on point $C = 3$ and press key A. The horizontal angle to the right, zenith angle, and slope distance will be transmitted and stored automatically. Then, the operator will be prompted to reverse the telescope and take a backsight (BS) with the telescope reversed.
4. A backsight (BS) is taken on $A = 1$ and again the readings are transmitted to the controller by pressing key A. The prompt on the screen will direct the operator to take a foresight (FS) with the telescope reversed.
5. Sight point $C = 3$ with the telescope reversed and press key A to transmit data to the controller. A prompt will appear on the screen requesting a description of the point sighted and the h.i. of the rod or target. Enter Pt 2 and 1.450 (Figure 8.53) and exit from this screen.

At this stage of the procedure, all measurements have been made from this setup and stored in a raw data file that can be displayed, if desired, by pressing the shift key and the

```
┌─────────────────────────────────────────────┐
│  OC:  1              FS:   2                  │
│  BS pt:              0                        │
│  >Ang right   :          0.0000              │
│  >Zenith ang:            0.0000              │
│   Slope dist  :           0.000              │
│  Desc:  START                                │
│  HI:  0.000     HR:     0.000                │
├───────┬──────┬──────┬──────┬───────┬─────────┤
│ SIDES │ REP  │ BACK │ TRAV │ OFFCT │  EXIT   │
└───────┴──────┴──────┴──────┴───────┴─────────┘
```

(*a*)

```
┌─────────────────────────────────────────────┐
│        Backsight                             │
│  >BS azm:                0.0000              │
│   Circle:                0.0000              │
│                                              │
│   BS Azm:                0.0000              │
│   BS Brg:            N 0.0000 W              │
├───────┬───────┬──────┬──────┬────────┬───────┤
│ SOLVE │ CHECK │      │ FAST │ CIRCLE │ EXIT  │
└───────┴───────┴──────┴──────┴────────┴───────┘
```

(*b*)

```
┌─────────────────────────────────────────────┐
│        Backsight                             │
│  >BS azm:               68.1521              │
│   Circle:                0.0000              │
│                                              │
│   BS Azm:               68.1521              │
│   BS Brg:           N 68.1521 E              │
├───────┬───────┬──────┬──────┬────────┬───────┤
│ SOLVE │ CHECK │      │ FAST │ CIRCLE │ EXIT  │
└───────┴───────┴──────┴──────┴────────┴───────┘
```

(*c*)

```
┌─────────────────────────────────────────────┐
│  OC:  2              FS:   3                  │
│  BS pt:  1                                    │
│  >Ang right   :        172.5334              │
│  >Zenith ang:          91.0955              │
│   Slope dist  :        281.890              │
│  Desc:  PT3                                  │
│  HI:  1.495     HR:    1.450                 │
├───────┬──────┬──────┬──────┬───────┬─────────┤
│ SIDES │ REP  │ BACK │ TRAV │ OFFCT │  EXIT   │
└───────┴──────┴──────┴──────┴───────┴─────────┘
```

(*d*)

FIGURE 8.55
Screen displays, TDS-48GX. (*Modified from
Tripod Data Systems (1991), courtesy Tripod Data
Systems, Inc.*)

have been computed and may be displayed on command by pressing the *ED CRD* or *Z*
key (Figure 8.54). Exiting from the screen, as indicated in step 5, leads to a display that
shows the numbers for the occupied (OC), backsight (BS), and foresight (FS) stations
updated for the next station to 2, 1, and 3, respectively, as shown in Figure 8.55d. Also
displayed on this screen are the average angles, distance, point description, and h.i.s for the
last station.

After step 5 has been performed, the setup is completed. Turn off the system and data
collector, move the instruments ahead one station, and repeat the outlined process for each
station in the traverse. Note that, for each setup after the first, it is *not necessary to set up
for the backsight azimuth* as was described. The coordinates for point *B* = 1 and *C* = 2 are
now stored in the data collector and the back azimuth for line 1-2 is known and stored.
Simply follow steps 1 through 5, for each station occupied, until the closing station has
been sighted.

When all the measurements needed for the traverse are completed, the operator has the
option of doing field calculations for the traverse. Most data collectors also are fairly
powerful computers and have stored programs for traverse computation and adjustment.
Even before the last instrument setup is broken, it is possible to use the data collector menus
to activate the routines required to (1) perform traverse computations; (2) calculate elevations; (3) examine the errors of closure; (4) adjust the coordinates by a simple adjustment,
usually the compass rule; and (5) adjust elevations in proportion to the distance leveled. The
results of these computations can be displayed for examination and stored. Also, one can
plot the traverse points on the screen and scroll through the plot using the horizontal and
vertical cursor keys. If these processes were to be followed for the example problem for
which notes are shown in Figure 8.53, the results displayed would correspond to those given
in Table 8.18 and the sketch in Figure 8.52. These on-site computations and evaluations
allow detection of mistakes that can be corrected immediately, eliminating subsequent
return trips to the job.

Another important supplementary task is to keep minimum conventional notes, in a
field notebook, including a site sketch and references to traverse points, as a backup record
in case of electronic failure or subsequent mistakes in processing files.

After the data have been checked, the contents of the data collector, for the job
specified, can be downloaded to a personal, mini-, or mainframe computer for editing,
computation, and rigorous least-squares adjustment by one of the methods discussed in
Section 8.39.

Side shots. If, at any point along the traverse, a side shot is desired, it is necessary only
to press the appropriate key for this operation and proceed as for a foresight. For the
collector shown in Figure 8.54, press the *SIDES* key (Figure 8.55d) on the command line
to get into the side-shot mode. A screen will appear that directs the operator to take the shot
and then describe the point sighted. This option makes the system particularly applicable for
radial surveys (Section 8.38) and topographic surveys (Chapter 15, Section 15.10).

Point descriptions. Point descriptions or "descriptors" can be entered at the prompts
for such descriptions on the screen or taken from a point description list. Descriptions
entered on a screen usually are restricted to a certain number of alphanumeric characters
(16 for the data collector shown in Figure 8.54). When a point description list is prepared
and stored prior to a survey, the descriptions are preceded by a code. This coded list can be
stored or printed and used in the field, where only the code has to be entered to describe
standard objects such as top of curb, property corner, or hub. A set of typical codes and
descriptions is shown in Table 8.19. For example, if a side shot were on the top of a curb,
TC would be entered for the point description and this would be recorded as "top of curb"

TABLE 8.19
Point list for descriptions

Code	Description
1	Hub with tack
TC	Top curb
POB	Point of Beginning
BF	Barbed wire fence
PC	Property corner
RBC	Rebar with cap

in the stored file. For the collector shown in Figure 8.54, the code may consist of up to seven alphabetic, numeric, or alphanumeric characters and the description is limited to 16 alphanumeric characters, punctuation, symbols, and spaces.

8.38
RADIAL SURVEYS

A traverse performed from a single station is called a *radial traverse*. The station utilized should be a coordinated point in the primary control network for the project and, for three-dimensional surveys, must have a known elevation. A second station is necessary for a backsight and a third known point is useful for checking. Although radial surveys can be performed using separate theodolite and EDM or a modular total station system, a self-contained total station system definitely is preferable and the focus of the discussions that follow.

The method is best used where the terrain is open, with few restrictions to the line of sight. It may be employed to determine locations for traverse stations; fix the position of randomly positioned points, such as in topographic surveys (Chapter 15); or to set off the positions of precalculated points, as in construction surveys (Chapter 17).

Figure 8.56 shows a radial traverse executed from known control point *A* with a backsight on known point *H*. It is desired to locate the positions of traverse points *B*, *C*, *D*, *E*, and *F*. The total station system is set over station *A*, the h.i. is measured and recorded, a backsight is taken on *H*, and horizontal and zenith angles and distances are observed to all stations *B* through *E*, with the number of repetitions being a function of the desired accuracy. The h.i.s for target sights also should be measured for all stations sighted. The procedure followed is similar to that developed for total station instruments in a three-dimensional survey, as described in Section 8.37. The measured data can be recorded manually, in a form similar to that shown in Figure 8.53, or using a data collector. In the latter case, the process is initiated by using the pull-down menus of the data collector software, and the sequence of operations with the total station instrument is controlled by the ensuing prompts on the display screen of the collector. Most data collectors have routines that can be used for radial surveys.

Because lines *AH* and *AK* are of known directions, the azimuths of lines *AB*, *AC*, *AD*, *AE*, *AA'*, and *AF* can be determined. Using the measured slope distances and zenith angles, horizontal distances are computed, allowing the calculation of *X* and *Y* coordinates for all points by Equations (8.6) and (8.7). For three-dimensional surveys, the elevations are determined using the distances, zenith angles, and h.i.s at both ends of the line in Equation (5.11a).

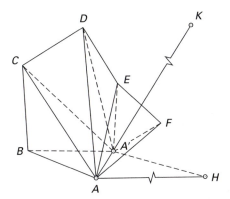

FIGURE 8.56
Radial traverse.

To check the work, the entire operation is repeated from auxiliary station A' with a backsight on station K, yielding a second set of azimuths, horizontal distances, and Δ elevations from A' to all stations A through F. These extra measurements provide a second set of coordinates and elevations for all traverse points. If the values for the two sets of positions are in reasonable agreement, the mean coordinates and elevations are used as the final position for the stations. Alternatively, data from both points A and A' can be used in a least-squares solution for the best estimates of the positions. With the adjusted positions, distances and directions between the coordinated stations can be calculated for lines AB, BC, CD, DE, EF, and FA using Equations (8.8) and (8.9). If distance $A'H$ is observable, the triangle $AA'H$ should be calculated as a check.

Note that the entire procedure could be performed with a backsight on a single line such as AH or AK. However, the extra known point provides an additional check on direction, and if the distance from A' to K can be measured, a redundancy on the position of A' is furnished. In either case, an auxiliary station such as A' is absolutely necessary to provide a check on each point located.

To improve the final positions adopted for the stations, rigorous adjustment methods can be performed to reconcile redundant data. The linearized form of the distance, direction, and Δ elevation Equations (8.8), (8.9), and (5.11a) can be formed for each ray from A and A' to the traverse points. The system of equations so formed then could be solved by the method of least squares in a procedure similar to that developed for a three-dimensional traverse in Section 9.21.

It should be emphasized that there is no standard solution for the application of radial surveys to surveying problems. Each problem is unique and the basic rules as outlined must be used to suit the situation in the field.

8.39
TRAVERSE ADJUSTMENT BY LEAST SQUARES

As explained in Section 8.1, a traverse is composed of consecutive distance and angle measurements. Figure 8.57a shows a traverse between two horizontal control points, 1 and 5, at each of which the azimuths A_1 and A_5 also are known. The observations are five angles, α_1 to α_5, and four distances, l_{12} to l_{45}. A traverse can be adjusted using either of the two techniques of least squares presented in Chapter 2, Sections 2.27 and 2.28. The *technique*

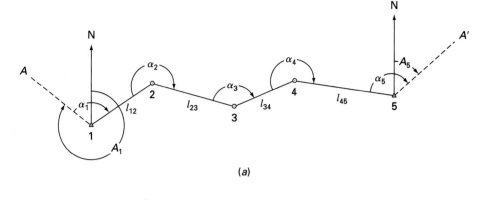

(a)

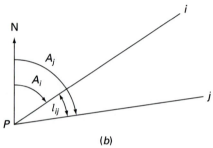

(b)

FIGURE 8.57

of least-squares adjustment of indirect observations is applied more frequently in practice
and therefore is presented first.

There are two types of condition equations: the *angle condition,* and the *distance
condition.* To derive the condition equation for the angle l_{ij} in Figure 8.57b, first write the
two azimuth condition equations (Equation (8.9)) for the azimuths (from the north) A_i
and A_j:

$$A_i = \arctan \frac{X_i - X_P}{Y_i - Y_P} \tag{8.61}$$

$$A_j = \arctan \frac{X_j - X_P}{Y_j - Y_P} \tag{8.62}$$

Then, the *angle condition* equation for the angle l_{ij} is

$$v_{ij} + l_{ij} = A_j - A_i = \arctan \frac{X_j - X_P}{Y_j - Y_P} - \arctan \frac{X_i - X_P}{Y_i - Y_P}$$

or

$$v_{ij} + l_{ij} + \arctan \frac{X_i - X_P}{Y_i - Y_P} - \arctan \frac{X_j - X_P}{Y_j - Y_P} = 0 \tag{8.63}$$

Equation (8.63) obviously is nonlinear in the parameters and must be linearized. Recall that

$$\frac{\partial}{\partial x} \arctan u = \frac{1}{1 + u^2} \frac{\partial u}{\partial x} \tag{8.64}$$

and write Equation (8.63) in a functional form:

$$v_{ij} + l_{ij} + F_{ij}(X_P, Y_P, X_i, Y_i, X_j, Y_j) = 0 \tag{8.65}$$

The linearized form of the angle condition equation is

$$v_{ij} + l_{ij} + F_{ij}(X_P, Y_P, X_i, Y_i, X_j, Y_j,)^0 + \frac{\partial F_{ij}}{\partial X_P} \delta X_P + \frac{\partial F_{ij}}{\partial Y_P} \delta Y_P + \frac{\partial F_{ij}}{\partial X_i} \delta X_i + \frac{\partial F_{ij}}{\partial Y_i} \delta Y_i$$

$$+ \frac{\partial F_{ij}}{\partial X_j} \delta X_j + \frac{\partial F_{ij}}{\partial Y_j} \delta Y_j = 0$$

or

$$v_{ij} + b_1 \delta X_P + b_2 \delta Y_P + b_3 \delta X_i + b_4 \delta Y_i + b_5 \delta X_j + b_6 \delta Y_j$$

$$= -l_{ij} - F_{ij}[(X, Y)_P^0, (X, Y)_i^0, (X, Y)_j^0] = f_{ij} \tag{8.66}$$

Equation (8.64) may be applied to the angle condition given in Equation (8.63) to evaluate the coefficients $b_1, b_2, \ldots, b_6$ in Equation (8.66). As examples, the partial derivatives with respect to X_P and Y_P are (the student should develop $b_3, b_4, \ldots, b_6$ as an exercise)

$$b_1 = \frac{\partial F_{ij}}{\partial X_P} = \frac{-(Y_i - Y_P^0)}{(Y_i - Y_P^0)^2 + (X_i - X_P^0)^2} - \frac{-(Y_j - Y_P^0)}{(Y_j - Y_P^0)^2 + (X_j - X_P^0)^2} \tag{8.67a}$$

$$b_2 = \frac{\partial F_{ij}}{\partial Y_P} = \frac{X_i - X_P^0}{(Y_i - Y_P^0)^2 + (X_i - X_P^0)^2} - \frac{X_j - X_P^0}{(Y_j - Y_P^0)^2 + (X_j - X_P^0)^2} \tag{8.67b}$$

and the right-hand side is given by

$$f_{ij} = -l_{ij} - \arctan \frac{X_i - X_P^0}{Y_i - Y_P^0} + \arctan \frac{X_j - X_P^0}{Y_j - Y_P^0} \tag{8.68}$$

or

$$f_{ij} = (\text{calculated angle}) - (\text{measured angle})$$

As an example, the angle condition for α_2 in Figure 8.57a is

$$\alpha_2 = A_{23} - A_{21} + 360°$$

$$= \arctan \frac{X_3 - X_2}{Y_3 - Y_2} - \arctan \frac{X_1 - X_2}{Y_1 - Y_2} + 360°$$

The distance condition expresses the distance between two points i and j as a function of their coordinates, or

$$l_{ij} + v_{ij} - [(X_i - X_j)^2 + (Y_i - Y_j)^2]^{1/2} = 0 \tag{8.69}$$

Because Equation (8.69) is nonlinear, it must be linearized using a Taylor series by first writing it in the form

$$l_{ij} + v_{ij} + F_{ij}(X_i, Y_i, X_j, Y_j) = 0 \tag{8.70}$$

then applying Equation (B.69) to Equation (8.69):

$$v_{ij} + b_1 \delta X_i + b_2 \delta Y_i + b_3 \delta X_j + b_4 \delta Y_j = -l_{ij} + l_{ij}^0 = f_{ij} \tag{8.71}$$

in which

$$b_1 = \frac{\partial F_{ij}}{\partial X_i} = -\frac{X_i^0 - X_j^0}{l_{ij}^0} \tag{8.72a}$$

$$b_2 = \frac{\partial F_{ij}}{\partial Y_i} = -\frac{Y_i^0 - Y_j^0}{l_{ij}^0} \qquad (8.72b)$$

$$b_3 = \frac{\partial F_{ij}}{\partial X_j} = \frac{X_i^0 - X_j^0}{l_{ij}^0} = -b_1 \qquad (8.72c)$$

$$b_4 = \frac{\partial F_{ij}}{\partial Y_j} = \frac{Y_i^0 - Y_j^0}{l_{ij}^0} = -b_2 \qquad (8.72d)$$

with

$$l_{ij}^0 = [(X_i^0 - X_j^0)^2 + (Y_i^0 - Y_j^0)^2]^{1/2} \qquad (8.72e)$$

The superscript 0 designates an approximate value for the parameters. For the adjustment of the traverse in Figure 8.57a, with the technique of adjustment of indirect observations, five angle and four distance conditions need to be written. The nine condition equations would include as unknown parameters the six coordinates of points 2, 3, and 4. Six normal equations are formed and solved for corrections to the approximate values of the parameters. The corrections are added to the approximations to update their values, and the solution is repeated until the last set of corrections is insignificantly small. The following example demonstrates this technique of adjusting traverses by the method of indirect observations.

EXAMPLE 8.15. Figure 8.52 shows a traverse for which the following data are given:

Angle	Value	σ	Distance	Value, m	σ, m
α_1	172°53'34"	2"	d_1	281.832	0.016
α_2	185°22'14"	2"	d_2	271.300	0.016
α_3	208°26'19"	2"	d_3	274.100	0.016
α_4	205°13'51"	2"	**Point**	**X, m**	**Y, m**
A_B	68°15'20.7"	0	B	8478.139	2483.826
A_E	300°11'30.5"	0	E	7709.336	2263.411

The observations are assumed to be uncorrelated. It is required to compute the coordinates of points C and D using the least-squares technique of adjustment of indirect observations.

Solution. For each of the seven observations given, a condition equation is written in terms of the observation and the coordinates of points C and D carried as parameters and denoted by X_C, Y_C, X_D, and Y_D. Using the regular symbols $l_1, l_2, \ldots, l_7$ to represent the observations $\alpha_1, \alpha_2, \alpha_3, \alpha_4, d_1, d_2$, and d_3, respectively, the condition equations are developed as follows:

$$\alpha_1 + A_{BA} - A_{BC} = 0 \quad \text{or} \quad F_1 = \alpha_1 + A_B - \arctan\frac{X_C - X_B}{Y_C - Y_B} = 0$$

$$\alpha_2 + A_{CB} - A_{CD} = 0 \quad \text{or} \quad F_2 = \alpha_2 + \arctan\frac{X_B - X_C}{Y_B - Y_C} - \arctan\frac{X_D - X_C}{Y_D - Y_C} = 0$$

$$\alpha_3 + A_{DC} - A_{DE} = 0 \quad \text{or} \quad F_3 = \alpha_3 + \arctan\frac{X_C - X_D}{Y_C - Y_D} - \arctan\frac{X_E - X_D}{Y_E - Y_D} = 0$$

$$\alpha_4 + A_{ED} - A_{EF} = 0 \quad \text{or} \quad F_4 = \alpha_4 + \arctan\frac{X_D - X_E}{Y_D - Y_E} - A_E = 0$$

$$F_5 = d_1 - [(X_B - X_C)^2 + (Y_B - Y_C)^2]^{1/2} = 0$$

$$F_6 = d_2 - [(X_C - X_D)^2 + (Y_C - Y_D)^2]^{1/2} = 0$$

$$F_7 = d_3 - [(X_D - X_E)^2 + (Y_D - Y_E)^2]^{1/2} = 0$$

The linearized equations are

$$v_1 + b_{11}\delta X_C + b_{12}\delta Y_C + b_{13}\delta X_D + b_{14}\delta Y_D = f_1$$

$$v_2 + b_{21}\delta X_C + b_{22}\delta Y_C + b_{23}\delta X_D + b_{24}\delta Y_D = f_2$$

$$v_3 + b_{31}\delta X_C + b_{32}\delta Y_C + b_{33}\delta X_D + b_{34}\delta Y_D = f_3$$

$$v_4 + b_{41}\delta X_C + b_{42}\delta Y_C + b_{43}\delta X_D + b_{44}\delta Y_D = f_4$$

$$v_5 + b_{51}\delta X_C + b_{52}\delta Y_C + b_{53}\delta X_D + b_{54}\delta Y_D = f_5$$

$$v_6 + b_{61}\delta X_C + b_{62}\delta Y_C + b_{63}\delta X_D + b_{64}\delta Y_D = f_6$$

$$v_7 + b_{71}\delta X_C + b_{72}\delta Y_C + b_{73}\delta X_D + b_{74}\delta Y_D = f_7$$

which may be expressed in matrix form as

$$\mathbf{v} + \mathbf{B\Delta} = \mathbf{f}$$

in which the coefficients in **B** for angle and distance equations are determined by partial differentiation of the condition equations with respect to the unknown coordinates. The coefficients for this adjustment (refer to Equations (8.67) and (8.72)) are

$$b_{11} = \frac{\partial F_1}{\partial X_C} = -\frac{Y_C^0 - Y_B}{(X_C^0 - X_B)^2 + (Y_C^0 - Y_B)^2} = -\frac{Y_C^0 - Y_B}{L_{BC}^{0^2}}$$

$$b_{12} = \frac{\partial F_1}{\partial Y_C} = \frac{X_C^0 - X_B}{(X_C^0 - X_B)^2 + (Y_C^0 - Y_B)^2} = \frac{X_C^0 - X_B}{L_{BC}^{0^2}}$$

$$b_{13} = \frac{\partial F_1}{\partial X_D} = 0$$

$$b_{14} = \frac{\partial F_1}{\partial Y_D} = 0$$

$$b_{21} = \frac{\partial F_2}{\partial X_C} = -b_{11} + \frac{Y_D^0 - Y_C^0}{(X_D^0 - X_C^0)^2 + (Y_D^0 - Y_C^0)^2} = -b_{11} + \frac{Y_D^0 - Y_C^0}{L_{CD}^{0^2}}$$

$$b_{22} = \frac{\partial F_2}{\partial Y_C} = -b_{12} - \frac{X_D^0 - X_C^0}{(X_D^0 - X_C^0)^2 + (Y_D^0 - Y_C^0)^2} = -b_{12} - \frac{X_D^0 - Y_C^0}{L_{CD}^{0^2}}$$

$$b_{23} = \frac{\partial F_2}{\partial X_D} = -\frac{Y_D^0 - Y_C^0}{(X_D^0 - X_C^0)^2 + (Y_D^0 - Y_C^0)^2} = -(b_{11} + b_{21})$$

$$b_{24} = \frac{\partial F_2}{\partial Y_D} = \frac{X_D^0 - X_C^0}{(X_D^0 - X_C^0)^2 + (Y_D^0 - Y_C^0)^2} = -(b_{12} + b_{22})$$

$$b_{31} = \frac{\partial F_3}{\partial X_C} = b_{23}$$

$$b_{32} = \frac{\partial F_3}{\partial Y_C} = b_{24}$$

$$b_{33} = \frac{\partial F_3}{\partial X_D} = -b_{23} + \frac{Y_E - Y_D^0}{(X_E - X_D^0)^2 + (Y_E - Y_D^0)^2} = -b_{23} + \frac{Y_E - Y_D^0}{L_{DE}^{0^2}}$$

$$b_{34} = \frac{\partial F_3}{\partial Y_D} = -b_{24} - \frac{X_E - X_D^0}{[(X_E - X_D^0)^2 + (Y_E - Y_D^0)^2]} = -b_{24} - \frac{X_E - X_D^0}{L_{DE}^{02}}$$

$$b_{41} = \frac{\partial F_4}{\partial X_C} = 0$$

$$b_{42} = \frac{\partial F_4}{\partial Y_C} = 0$$

$$b_{43} = \frac{\partial F_4}{\partial X_D} = -(b_{23} + b_{33})$$

$$b_{44} = \frac{\partial F_4}{\partial Y_D} = -(b_{24} + b_{34})$$

$$b_{51} = \frac{\partial F_5}{\partial X_C} = \frac{X_B - X_C^0}{L_{BC}^0}$$

$$b_{52} = \frac{\partial F_5}{\partial Y_C} = \frac{Y_B - Y_C^0}{L_{BC}^0}$$

$$b_{53} = \frac{\partial F_5}{\partial X_D} = 0$$

$$b_{54} = \frac{\partial F_5}{\partial Y_D} = 0$$

$$b_{61} = \frac{\partial F_6}{\partial X_C} = -\frac{X_C^0 - X_D^0}{L_{CD}^0}$$

$$b_{62} = \frac{\partial F_6}{\partial Y_C} = -\frac{Y_C^0 - Y_D^0}{L_{CD}^0}$$

$$b_{63} = \frac{\partial F_6}{\partial X_D} = -b_{61}$$

$$b_{64} = \frac{\partial F_6}{\partial Y_D} = -b_{62}$$

$$b_{71} = \frac{\partial F_7}{\partial X_C} = 0$$

$$b_{72} = \frac{\partial F_7}{\partial Y_C} = 0$$

$$b_{73} = \frac{\partial F_7}{\partial X_D} = -\frac{X_D^0 - X_E}{L_{DE}^0}$$

$$b_{74} = \frac{\partial F_7}{\partial Y_D} = -\frac{Y_D^0 - Y_E}{L_{DE}^0}$$

To evaluate the first four elements of **f**, it is necessary to determine the quadrant for each azimuth as implied by the signs associated with the numerator and denominator of the arctan functions. In general, the angles θ computed (in the computer) from the arctan functions in the condition equations are either positive or negative and less than

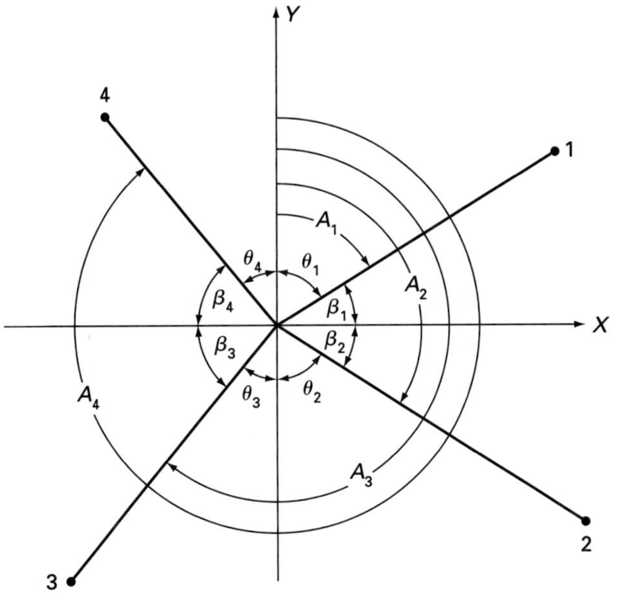

FIGURE 8.58

471

CHAPTER 8:
Traverse

90°. Therefore, the value of the azimuth angle is obtained on the basis of the signs of ΔX and ΔY depicted in Figure 8.58 as follows (azimuth, A, is clockwise from $+Y$ (north), thus consistent with angles turned to the right in the problem):

$$(a) \quad +\Delta X, +\Delta Y \quad A = \theta = \arctan \frac{\Delta X}{\Delta Y}$$

$$(b) \quad +\Delta X, -\Delta Y \quad A = 180° - |\theta| = 180° - \left| \arctan \frac{\Delta X}{\Delta Y} \right|$$

$$(c) \quad -\Delta X, -\Delta Y \quad A = 180° + |\theta| = 180° + \left| \arctan \frac{\Delta X}{\Delta Y} \right|$$

$$(d) \quad -\Delta X, +\Delta Y \quad A = 360° - |\theta| = 360° - \left| \arctan \frac{\Delta X}{\Delta Y} \right|$$

If the computer yields acute positive angles with the signs of ΔX, ΔY as in cases (a) and (c) and acute negative angles in cases (b) and (d) using the arctan function, the following scheme is simpler and more efficient to program (see Figure 8.58):

1. $+\Delta X, A = 90° - \beta = 90° - \arctan (\Delta Y / \Delta X)$; if ΔY is positive, A is less than 90°, and if it is negative, A is between 90 and 180°.
2. $-\Delta X, A = 270° - \beta = 270° - \arctan (\Delta Y / \Delta X)$; again, with $-\Delta Y$, then $180° < A < 270°$; and with $+\Delta Y$, $270° < A < 360°$.

In larger computers, special arctan functions directly evaluate the total angle with a value from $-180°$ to $+180°$. When such functions are available, neither of the foregoing schemes is necessary.

Let us assume the following approximations for the four unknown parameters.

$$X_C^0 = 8200 \text{ m} \qquad Y_C^0 = 2340 \text{ m}$$
$$X_D^0 = 7980 \text{ m} \qquad Y_D^0 = 2230 \text{ m}$$

From Equation (8.68), the rules set forth in (a) to (d), and Equation (8.72e), the constant terms in the $\mathbf{f}$ vector are

$$f_1 = -\alpha_1 + [A_{BC} - A_{BA}]$$

$$= -\alpha_1 + \left[180° + \arctan \frac{\Delta X_{BC}}{\Delta Y_{BC}} - A_{BA} \right] = 0.026318430 \text{ rad}$$

$$f_2 = -\alpha_2 + [A_{CD} - A_{CB}]$$

$$= -\alpha_2 + \left[180° + \arctan \frac{\Delta X_{CD}}{\Delta Y_{CD}} - \arctan \frac{\Delta X_{CB}}{\Delta Y_{CB}} \right] = -0.080146734 \text{ rad}$$

$$f_3 = -\alpha_3 + [A_{DE} - A_{DC}]$$

$$= -\alpha_3 + \left[360° - \arctan \frac{\Delta X_{DE}}{\Delta Y_{DE}} - \arctan \frac{\Delta X_{DC}}{\Delta Y_{DC}} \right] = 0.09011977 \text{ rad}$$

$$f_4 = -\alpha_4 + [A_{EF} - A_{ED}]$$

$$= -\alpha_4 + \left[A_{EF} - \left(180° - \arctan \frac{\Delta X_{ED}}{\Delta Y_{ED}} \right) \right] = -0.036234261 \text{ rad}$$

$$f_5 = -d_1 + [\Delta X_{BC}^2 + \Delta Y_{BC}^2]^{1/2} = -d_1 + L_{BC} = 31.2929 \quad \text{m}$$

$$f_6 = -d_2 + [\Delta X_{CD}^2 + \Delta Y_{CD}^2]^{1/2} = -d_2 + L_{CD} = -25.3325 \text{ m}$$

$$f_7 = -d_3 + [\Delta X_{DE}^2 + \Delta Y_{DE}^2]^{1/2} = -d_3 + L_{DE} = -1.3817 \quad \text{m}$$

The variance of each measured angle is $9.401772217 \times 10^{-11} \text{ rad}^2$. The variance of each measured distance is $2.56 \times 10^{-4} \text{ m}^2$. Therefore, assuming a reference variance of 1.0 and uncorrelated measurements, the covariance matrix (Equation (2.30), Section 2.18) of the measurements is

$$\mathbf{\Sigma} = 2.56 \times 10^{-4} \text{ diag } \{(3.672567)10^{-7} \quad (3.672567)10^{-7} \quad (3.672567)10^{-7}$$
$$(3.672567)10^{-7} \quad 1.0 \quad 1.0 \quad 1.0\}$$

By Equation 2.33, Section 2.18, the weight matrix is $\mathbf{W} = \mathbf{\Sigma}^{-1}$, or

$$\mathbf{W} = 3906.25 \text{ diag } \{2,722,890.898 \quad 2,722,890.898 \quad 2,722,890.898$$
$$2,722,890.898 \quad 1.0 \quad 1.0 \quad 1.0\}$$

With these approximations selected, the $\mathbf{B}$ matrix is

$$\mathbf{B} = 10^{-3} \begin{bmatrix} 1.466905 & -2.836786 & 0 & 0 \\ -3.285087 & 6.473150 & 1.818182 & -3.636364 \\ 1.818182 & -3.636364 & -1.368960 & 7.275529 \\ 0 & 0 & 0.4492217 & -3.639165 \\ 888.2685 & 459.3247 & 0 & 0 \\ -894.4272 & -447.2136 & 894.4272 & 447.2136 \\ 0 & 0 & -992.4672 & 122.5110 \end{bmatrix}$$

The normal equations coefficient matrix is $\mathbf{N} = \mathbf{B'WB}$ (Equation (2.74)), or

$$N = 10^5 \begin{bmatrix} 1.7904 & -3.361 & -0.9313 & 2.6620 \\ & 6.7352 & 1.7657 & -5.3254 \\ & & 0.6421 & -1.5778 \\ \text{symmetric} & & & 8.4536 \end{bmatrix}$$

and its inverse is

$$N^{-1} = 10^{-5} \begin{bmatrix} 11.6976 & 5.1548 & 3.1756 & 0.1566 \\ & 2.8713 & 0.0680 & 0.1983 \\ & & 6.6941 & 0.2923 \\ \text{symmetric} & & & 0.2485 \end{bmatrix}$$

The correction vector is found by solving Equation (2.75) for Δ, or

$$\Delta_1 = N^{-1}B^T Wf = N^{-1}t$$

The vector of corrections after the first iteration is

$$\Delta_1 = [32.5209 \quad 7.8617 \quad 4.8214 \quad 9.1344]^t \text{ (m)}$$

which leads to the following improved approximations: $X_{C_1}^0 = 8232.5209$, $Y_{C_1}^0 = 2347.8617$, $X_{D_1}^0 = 7984.8214$, and $Y_{D_1}^0 = 2239.1344$. Using these values, the solution is iterated. After four iterations, the correction vector is zero to four decimal places, and the final estimates of the coordinates are

$$X_C = 8231.263 \text{ m} \qquad Y_C = 2347.818 \text{ m}$$

$$X_D = 7982.404 \text{ m} \qquad Y_D = 2239.714 \text{ m}$$

According to Equation (2.78), the covariance matrix of the parameters (when the reference variance is unity, as in this example) is equal to the inverse of the normal equations coefficient matrix, or $\Sigma_{\Delta\Delta} = N^{-1}$. At the end of four iterations, the inverse was

$$N^{-1} = \begin{bmatrix} (1.077237)10^{-4} & (5.016788)10^{-5} & (2.456713)10^{-5} & (4.457744)10^{-6} \\ & (2.808001)10^{-5} & (-4.638273)10^{-7} & (3.091250)10^{-6} \\ & & (7.525994)10^{-5} & (4.779247)10^{-6} \\ \text{symmetric} & & & (2.685166)10^{-6} \end{bmatrix}$$

$$= \begin{bmatrix} \Sigma_{CC} & \Sigma_{CD} \\ \Sigma_{CD} & \Sigma_{DD} \end{bmatrix}$$

where Σ_{CC} is the 2×2 covariance matrix for the coordinates of point C, Σ_{DD} is the 2×2 covariance matrix for the coordinates of point D, and Σ_{CD} is a 2×2 cross-covariance matrix between the coordinates of point C and point D. It is used only when the precision of relative information between C and D is desired. The derivation of such precision is beyond the scope of this book.

The covariance matrices Σ_{CC} and Σ_{DD} express the precision with which points C and D are located. Each matrix can be utilized to give the standard deviation for each coordinate. For example, the standard deviation of the X coordinate of point C is $\sigma_{X_c} = [(1.077237)(10^{-4})]^{1/2} = 0.010$ m. Alternatively, each covariance matrix could be utilized to establish a standard error ellipse about the point as explained in Section 2.31 and demonstrated by Example 2.15.

FIGURE 8.59

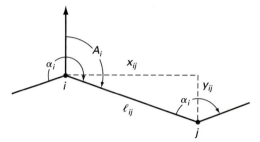

A traverse also may be adjusted using the technique of *least-squares adjustment of observations only* (Section 2.28). Usually, the redundancy is 3 and the following three conditions must be satisfied: (1) the angles must close, (2) the sum of the departures must equal 0, and (3) the sum of the latitudes must equal 0. The angle closure condition is expressed by starting from one end of the traverse and summing the starting azimuth and measured angles through to the other end of the traverse (see Equation (8.5), Section 8.8). The total sum minus multiples of 180° must be equated to the azimuth at the end of the traverse. As an example, the angle closure condition for the traverse in Figure 8.57a is

$$A_1 + (\alpha_1 + v_1) + (\alpha_2 + v_2) + (\alpha_3 + v_3) + (\alpha_4 + v_4) + (\alpha_5 + v_5)$$
$$- (4)(180°) - A_5 = 0$$

To write the departure and latitude conditions, the accumulated azimuth must be evaluated at each station in the traverse. Let x_{ij} and y_{ij} represent the departure and latitude for one traverse leg l_{ij} shown in Figure 8.59. If A_i represents the accumulated azimuth at point i, then

$$x_{ij} = l_{ij} \sin A_i$$
$$y_{ij} = l_{ij} \cos A_i \qquad (8.73)$$

and the coordinates at point j would be

$$X_j = X_i + x_{ij}$$
$$Y_j = Y_i + y_{ij}$$

For the traverse in Figure 8.57a, the departure and latitude conditions are

$$X_1 + x_{12} + x_{23} + x_{34} + x_{45} - X_5 = 0$$
$$Y_1 + y_{12} + y_{23} + y_{34} + y_{45} - Y_5 = 0$$

where (X_1, Y_1) and (X_5, Y_5) are the coordinates of the beginning and ending control points. The following example shows how the technique of least-squares adjustment of observations only is applied to a numerical traverse problem.

EXAMPLE 8.16. Using the data for the traverse in Example 8.15 (Figure 8.52), compute the coordinates of points C and D using the least-squares technique of adjustment of observations only.

Solution. The three conditions that relate the seven observations are

1. Angle closure:

$$F_1 = A_B + (\alpha_1 - \pi) + (\alpha_2 - \pi) + (\alpha_3 - \pi) + \alpha_4 - A_E = 0$$

2. Sum of departures is equal to 0:

$$F_2 = X_B - d_1 \sin(A_B + \alpha_1 - \pi) - d_2 \sin(A_B + \alpha_1 - \pi + \alpha_2 - \pi)$$
$$- d_3 \sin(\alpha_4 + \pi - A_E) - X_E = 0$$

3. Sum of latitudes is equal to 0:

$$F_3 = Y_B - d_1 \cos(A_B + \alpha_1 - \pi) - d_2 \cos(A_B + \alpha_1 - \pi + \alpha_2 - \pi)$$
$$+ d_3 \cos(\alpha_4 + \pi - A_E) - Y_E = 0$$

Linearization yields the following three equations in terms of coefficients, unknown residuals, and constant terms:

$$a_{11}v_1 + a_{12}v_2 + a_{13}v_3 + a_{14}v_4 + a_{15}v_5 + a_{16}v_6 + a_{17}v_7 = f_1$$
$$a_{21}v_1 + a_{22}v_2 + a_{23}v_3 + a_{24}v_4 + a_{25}v_5 + a_{26}v_6 + a_{27}v_7 = f_2$$
$$a_{31}v_1 + a_{32}v_2 + a_{33}v_3 + a_{34}v_4 + a_{35}v_5 + a_{36}v_6 + a_{37}v_7 = f_3$$

which may be expressed in matrix form as

$$\mathbf{Av} = \mathbf{f}$$

where

$$a_{11} = \frac{\partial F_1}{\partial \alpha_1} = 1$$

$$a_{12} = \frac{\partial F_1}{\partial \alpha_2} = 1$$

$$a_{13} = \frac{\partial F_1}{\partial \alpha_3} = 1$$

$$a_{14} = \frac{\partial F_1}{\partial \alpha_4} = 1$$

$$a_{15} = \frac{\partial F_1}{\partial d_1} = 0$$

$$a_{16} = \frac{\partial F_1}{\partial d_2} = 0$$

$$a_{17} = \frac{\partial F_1}{\partial d_3} = 0$$

$$a_{21} = \frac{\partial F_2}{\partial \alpha_1} = -d_1 \cos(A_B + \alpha_1 - \pi) - d_2 \cos(A_B + \alpha_1 + \alpha_2 - 2\pi)$$
$$= d_1 a_{35} + d_2 a_{36} = -244.093 \text{ m}$$

$$a_{22} = \frac{\partial F_2}{\partial \alpha_2} = -d_2 \cos(A_B + \alpha_1 + \alpha_2 - 2\pi) = d_2 a_{36} = -108.098 \text{ m}$$

$$a_{23} = \frac{\partial F_2}{\partial \alpha_3} = 0$$

$$a_{24} = \frac{\partial F_2}{\partial \alpha_4} = -d_3 \cos(\alpha_4 + \pi - A_E) = -d_3 a_{37} = -23.697 \text{ m}$$

$$a_{25} = \frac{\partial F_2}{\partial d_1} = -\sin(A_B + \alpha_1 - \pi) = -0.87583538$$

$$a_{26} = \frac{\partial F_2}{\partial d_2} = -\sin(A_B + \alpha_1 + \alpha_2 - 2\pi) = -0.91719283$$

$$a_{27} = \frac{\partial F_2}{\partial d_3} = -\sin(\alpha_4 + \pi - A_E) = -0.99625383$$

$$a_{31} = \frac{\partial F_3}{\partial \alpha_1} = -d_1 a_{25} - d_2 a_{26} = 495.684 \text{ m}$$

$$a_{32} = \frac{\partial F_3}{\partial \alpha_2} = -d_2 a_{26} = 248.834 \text{ m}$$

$$a_{33} = \frac{\partial F_3}{\partial \alpha_3} = 0$$

$$a_{34} = \frac{\partial F_3}{\partial \alpha_4} = d_3 a_{27} = -273.073 \text{ m}$$

$$a_{35} = \frac{\partial F_3}{\partial d_1} = -\cos(A_B + \alpha_1 - \pi) = -0.48254072$$

$$a_{36} = \frac{\partial F_3}{\partial d_2} = -\cos(A_B + \alpha_1 + \alpha_2 - 2\pi) = -0.39844360$$

$$a_{37} = \frac{\partial F_3}{\partial d_3} = \cos(\alpha_4 + \pi - A_E) = 0.08647715$$

$$f_1 = -A_B - \alpha_1 - \alpha_2 - \alpha_3 - \alpha_4 + 3\pi + A_E = (5.72066569)10^{-5} \text{ rad}$$

$$f_2 = -X_B + d_1 \sin(A_B + \alpha_1 - \pi) + d_2 \sin(A_B + \alpha_1 + \alpha_2 - 2\pi)$$
$$+ d_3 \sin(\alpha_4 + \pi - A_E) + X_E = -0.046217 \text{ m}$$

$$f_3 = -Y_B + d_1 \cos(A_B + \alpha_1 - \pi) + d_2 \cos(A_B + \alpha_1 + \alpha_2 - 2\pi)$$
$$- d_3 \cos(\alpha_4 + \pi - A_E) + Y_E = -0.025221 \text{ m}$$

The covariance matrix Σ for the measurements is given in Example 8.15. Because the reference variance is unity, the cofactor matrix $Q = \Sigma$ and Equations (2.100) and (2.98) in Section 2.28 become $\mathbf{k} = (\mathbf{A}\Sigma\mathbf{A}')^{-1}\mathbf{f}$ and $\mathbf{v} = \Sigma\mathbf{A}'\mathbf{k}$, which lead to $\mathbf{v} = \Sigma\mathbf{A}'(\mathbf{A}\Sigma\mathbf{A}')^{-1}\mathbf{f}$. Substitution of the elements of $\mathbf{A}$ and $\mathbf{f}$, above, and Σ from Example 8.15 into this equation yields the vector of residuals, $\mathbf{v}$, from which adjusted measurements are found by Equation (2.80) or $\hat{\mathbf{l}} = \mathbf{l} + \mathbf{v}$.

From a computer program written to perform these operations, the adjusted observations are

$$\hat{l}_1 = \hat{\alpha}_1 = 3.01755395 \text{ rad} = 172°53'35.2''$$

$$\hat{l}_2 = \hat{\alpha}_2 = 3.23533786 \text{ rad} = 185°22'16.3''$$

$$\hat{l}_3 = \hat{\alpha}_3 = 3.63795702 \text{ rad} = 208°26'22.5''$$

$$\hat{l}_4 = \hat{\alpha}_4 = 3.58197693 \text{ rad} = 205°13'55.8''$$

$$\hat{l}_5 = \hat{d}_1 = 281.862 \text{ m}$$

$$\hat{l}_6 = \hat{d}_2 = 271.325 \text{ m}$$

$$\hat{l}_7 = \hat{d}_3 = 274.095 \text{ m}$$

From these values, the adjusted coordinates of points C and D are computed as follows:

$$\hat{X}_C = X_B - \hat{d}_1 \sin(A_B + \hat{\alpha}_1 - \pi) = 8478.139 - (281.862) \sin(61°08'55.9'')$$
$$= 8231.263 \text{ m}$$

$$\hat{Y}_C = Y_B - \hat{d}_1 \cos(61°08'55.9'') = 2483.826 - 136.008 = 2347.818 \text{ m}$$

$$\hat{X}_D = X_E + \hat{d}_3 \sin(\hat{\alpha}_4 + \pi - A_E) = 7709.336 + (274.095) \sin(85°02'25.3'')$$
$$= 7982.405 \text{ m}$$

$$\hat{Y}_D = Y_E - \hat{d}_3 \cos(85°02'25.3'') = 2263.411 - 23.697 = 2239.714 \text{ m}$$

These adjusted coordinates correspond to within 0.001 m of the values computed in Example 8.15 by the technique of indirect observations. Should error propagation be desired, the propagated covariance matrix for the adjusted coordinates may be calculated by Equation (2.107) in Section 2.28, in which $\mathbf{Q} = \mathbf{\Sigma}$.

8.40
COORDINATE TRANSFORMATIONS: APPLICATIONS

The need for coordinate transformations in surveying computations was previously discussed in Section 8.20. In this section, we consider the use of a linear, two-dimensional, four-parameter, conformal transformation (Equation (C.11), Appendix C) for converting local coordinates into state plane coordinates. As an example, use adjusted coordinates for the traverse in Figure 8.20, which were computed and adjusted in Section 8.14.

EXAMPLE 8.17. Adjusted X and Y coordinates for stations 1, 2, . . . , 5 (Table 8.11) and state plane coordinates referred to as *Eastings* (X coordinates) and *Northings* (Y coordinates) for points 1, 3, and 5 are given in Table 8.20, in which adjusted local coordinates are given to three decimal places for this example and Eastings and Northings are translated so that $E' = E - 1,980,000.000$ m, and $N' = N - 510,000.000$ m. First solve the unique case for the four transformation parameters using points 1 and 3, which are known in both coordinate systems.

TABLE 8.20

Station	Adjusted local coordinates, m		Adjusted state plane coordinates, m	
	X	Y	E'	N'
1	4321.404	6240.562	4275.486	8702.562
2	4485.414	5982.928		
3	4669.940	6291.662	4625.244	8744.479
4	5248.697	6092.013		
5	5619.243	6405.272	5577.207	8833.084

Solution. Equations (C.11) can be written for each point known in both systems. For example, for point 1, we have

$$X_1a + Y_1b + c \quad = E_1'$$

$$Y_1a - X_1b \quad + d = N_1'$$

Due to the four unknown parameters, it is necessary to form equations for at least two points known in both systems. Writing Equations (C.11) for points 1 and 3 and substituting values for the known coordinates, given in Table 8.19, leads to

$$4321.404a + 6240.562b + c = 4275.486$$

$$6240.562a - 4321.404b + d = 8702.562$$

$$4669.940a + 6291.662b + c = 4625.244$$

$$6291.662a - 4669.940b + d = 8744.479$$

Solving these equations simultaneously, for the unknown parameters, yields

$$a = 0.999651, \quad b = 0.026296, \quad c = -208.511, \quad \text{and} \quad d = 2577.816$$

Substitution of these values into Equations (C.11) and solving for E_i' and N_i' for the points unknown in the state plane coordinate system leads to

$$E_2' = (0.999651)(4485.414) + (0.026296)(5982.928) - 208.511 = 4432.664 \text{ m}$$

$$N_2' = (-0.026296)(4485.414) + (0.999651)(5982.928) + 2577.816$$
$$= 8440.708 \text{ m}$$

$$E_4' = (0.999651)(5248.697) + (0.026296)(6092.013) - 208.511 = 5198.550 \text{ m}$$

$$N_4' = (-0.026296)(5248.697) + (0.999651)(6092.013) + 2577.816$$
$$= 8529.683 \text{ m}$$

$$E_5' = (0.999651)(5619.243) + (0.026296)(6405.272) - 208.511 = 5577.204 \text{ m}$$

$$N_5' = (-0.026296)(5619.243) + (0.999651)(6405.272) + 2577.816$$
$$= 8833.089 \text{ m}$$

Next, consider determination of the transformation parameters using points 1, 3, and 5 in a least-squares solution by indirect observations (Section 2.27). For this example, a simplified approach to the least-squares adjustment is used. Using the notation of this example, the condition equations for a point i known in each coordinate system are

$$v_{1i} + X_ia + Y_ib + c \quad = E_i'$$

$$v_{2i} + Y_ia - X_ib \quad + d = N_i'$$

The equations of this form for common points 1, 3, and 5 in this example are

$$v_1 + X_1a + Y_1b + c \quad = E_1'$$

$$v_2 + Y_1a - X_1b \quad + d = N_1'$$

$$v_3 + X_3a + X_3b + c \quad = E_3'$$

$$v_4 + Y_3a + X_3b \quad + d = N_3'$$

$$v_5 + X_5a + Y_5b + c \quad = E_5'$$

$$v_6 + Y_5a - X_5b \quad + d = N_5'$$

which can be expressed as

$$V + B\Delta = f$$

in which

$$V^T = [v_1\, v_2\, v_3\, v_4\, v_5\, v_6], \qquad \Delta^T = [\text{a b c d}],$$

$$B = \begin{bmatrix} 4321.404 & 6240.562 & 1.0 & 0.0 \\ 6240.562 & -4321.404 & 0.0 & 1.0 \\ 4669.940 & 6291.662 & 1.0 & 0.0 \\ 6291.662 & -4669.940 & 0.0 & 1.0 \\ 5619.243 & 6405.272 & 1.0 & 0.0 \\ 6405.272 & -5619.243 & 0.0 & 1.0 \end{bmatrix} \quad \text{and} \quad f = \begin{bmatrix} 4275.486 \\ 8702.562 \\ 4625.244 \\ 8744.479 \\ 5577.207 \\ 8833.084 \end{bmatrix}$$

The normal equations for least-squares adjustment of indirect observations (Section 2.27, Equation (2.70)) are

$$(B^T W B)\Delta = B^T W f$$

Assuming the coordinates in both systems are of equal weight and uncorrelated, $W = I$ and the normal system becomes

$$(B^T B)\Delta = B^T f$$

The normal equation coefficient matrix and constant vector, respectively, are

$$(B^T B) = \begin{bmatrix} 1.9162 \times 10^{10} & -1.9536 \times 10^{-9} & 1.4611 \times 10^5 & 1.8937 \times 10^5 \\ & 1.9162 \times 10^{10} & 1.8937 \times 10^5 & -1.4611 \times 10^5 \\ \text{symmetric} & & 0.3000 \times 10 & 0.0000 \times 10 \\ & & & 0.3000 \times 10 \end{bmatrix}$$

and

$$B^T f = \begin{bmatrix} 2.3732 \times 10^9 \\ -3.6573 \times 10^8 \\ 1.4478 \times 10^5 \\ 2.6280 \times 10^5 \end{bmatrix}$$

Solution of the normal equations yields

$$\Delta = \begin{bmatrix} 0.999654 \\ 0.026298 \\ -208.540 \\ 2577.807 \end{bmatrix}$$

Substitution of these parameters into the original equations and solving for the Northings and Eastings of points 2 and 4 gives

$$\begin{bmatrix} E_2' \\ N_2' \\ E_4' \\ N_4' \end{bmatrix} = \begin{bmatrix} 4432.661 \\ 8440.707 \\ 5198.549 \\ 8529.682 \end{bmatrix} \text{ m}$$

which agree with the answers obtained previously to within a few millimeters.

Adding the original translations, which were applied to the Eastings and Northings to reduce the size of the numbers, yields the following coordinates for points 2 and 4 in the state plane coordinate system:

$$\begin{bmatrix} E_2 \\ N_2 \\ E_4 \\ N_4 \end{bmatrix} = \begin{bmatrix} 1,984,432.661 \\ 518,440.707 \\ 1,985,198.549 \\ 518,529.582 \end{bmatrix} \ m$$

Note that, in the procedure followed for this adjustment, observations were not specified. A residual was attached to each equation, as is done in regression analysis in mathematical statistics. This "resultant residual" in each equation is composed of two parts: one part in the X, Y system; and the other part in the N, E system. So long as the coordinates in both systems are of equal weight and uncorrelated, this is a legitimate method and will give the same answers as if one were to attach a separate residual to each observation and solve the system accordingly. Further details concerning this particular problem can be found in Mikhail (1976).

8.41
STANDARDS FOR TRAVERSE ACCURACY

Specifications and standards for local surveys, based on the 1984 report by the Federal Geodetic Control Subcommittee (FGCS), were discussed in Section 8.21. Currently, the standards and specifications are being reviewed and revised by the FGCS with emphasis being placed on a statistically based standard with a stated probability for regional and local surveys. This approach is designed to accommodate the entire spectrum of users of the highly accurate spatial positions determined by GPS and supplemented by three-dimensional total stations surveys.

The horizontal control standard will be a number that is the radius of a relative error circle with a probability of 0.95. This error circle is computed using error propagation from a least-squares adjustment of the data used to establish the position with respect to the geodetic datum. For the regional accuracy, this datum is defined by the nearest Continuously Operating System (CORS) station. For local accuracy, the error circle is determined by error propagation in a least-squares adjustment between known control points connected by the local survey.

Similar standards are being proposed for ellipsoidal and orthometric heights (Section 12.13, Application 5). Details concerning these standards for both horizontal position and heights can be found in FGCS (1994).

At the time of this writing, these standards are still in the proposal stage and the specifications necessary to maintain them with currently used surveying systems are not yet available. The student of surveying and the practitioner must continuously and carefully monitor the literature in the periodicals and proceedings of the professional societies and associations to stay abreast of new developments and make the transition from the old to the new standards and specifications.

PROBLEMS

8.1. Describe a traverse and its purpose in surveying.

8.2. List four surveying systems that can be used to execute a traverse.

8.3. Describe two general classes of traverses, pointing out the advantages, disadvantages, and applications of each class.

8.4. Describe two specific types of closed traverses, indicating the advantages and disadvantages of each type.

8.5. A traverse originates and closes on Station Roger, which has known X and Y coordinates. What type of traverse is this? Should this traverse be used for a major project? Explain why or why not.

8.6. The interior angles of a five-sided closed traverse are as follows: A, 124°34′; B, 93°30′; C, 139°50′; D, 130°20′. The angle E is not measured. Compute the angle E, assuming the given values to be correct.

8.7. (a) What are the deflection angles of the traverse of Problem 8.6? (b) What are the computed bearings if the bearing of AB is due north?

8.8. The following are the deflection angles of a closed traverse: A, 85°20′R; B, 10°04′L; C, 83°32′R; D, 63°27′R; E, 34°18′R; F, 72°56′R; G, 30°45′R. Compute the error of closure. Adjust the angular values on the assumption that the error is the same for each angle.

8.9. In Problem 8.8, the azimuth of line AB from the north is 120°30′. Compute azimuths for the rest of the lines in the traverse and show a check on the calculations.

8.10. The clockwise interior angles observed with a 01′ theodolite in a traverse that starts and closes on point A are as follows. $A = 83°30′$; $B = 251°50′$; $C = 101°20′$; $D = 83°50′$; $E = 91°40′$; $F = 251°00′$; $G = 89°27′$; and $J = 127°30′$. Compute the error of closure and distribute this error, assuming that all angles are of equal precision. Calculate azimuths for the lines in the traverse assuming that line AJ has a known azimuth from the north of $A_{AJ} = 165°30′$.

8.11. Comment on the relative and absolute angular accuracy of the data in Problem 8.10.

8.12. The following deflection angle traverse was run with a 30″ repeating theodolite from a line of known direction LM to another independent line of known direction TU. Compute the angular error of closure and adjusted azimuths for each line. Comment on the relative and absolute angular accuracy of this traverse.

Station	Deflection angle	Azimuth from north
L		195°30′
M	39°47′L	
N	17°20′L	
O	14°08′L	
P	3°11′L	
Q	49°59′L	
R	32°18′R	
S	18°44′R	
T	7°31′L	
U		114°28′

8.13. Clockwise interior angles in closed-loop traverse $ABCDEFGA$, in the order measured, starting at A and moving counterclockwise around the traverse are 144°56′30″, 92°30′00″, 220°40′00″, 62°19′40″, 58°00′50″, 230°22′00″, and 91°04′00″. Compute (a) the error of closure, (b) the adjusted angles assuming all angles are of equal weight, and (c) the azimuths of all lines. The azimuth of AB is known to be 178°00′00″ from north.

8.14. The azimuth from north of line G-2 $= 180°20′31″$ and the azimuth of another independently surveyed line K-7 $= 14°03′27″$. Deflection angles for closed traverse 2-3-4-5-6-7 run between

G-2 and *K*-7, starting at 2 and terminating at 7, are 89°15′31″L, 85°15′54″R, 97°11′34″L, 74°56′53″L, 124°33′45″R, and 65°17′39″R. Compute the unadjusted azimuths, the error of closure, and the adjusted azimuths for all lines in the traverse. Assume the angles are of equal weight.

8.15. Azimuths from north of two independently surveyed lines, *A-B* and *G-H*, are 181°30′40″ and 195°13′36″. Angles to the right in closed traverse *B-C-D-E-F-G*, initiated at *B* and closing on *G* in the order observed, are 90°44′17″, 265°15′54″, 82°48′26″, 105°03′07″, 304°33′45″, and 245°17′39″. Compute the unadjusted azimuths, the error of closure, and the adjusted azimuths, assuming angles are of equal weight.

8.16. The following are bearings taken on a closed compass traverse. Compute the interior angles and correct them for observational errors. Assuming the observed bearing of the line *AB* to be correct, adjust the bearings of the remaining sides.

Line	Forward bearing	Back bearing
AB	S37°30′E	N37°30′W
BC	S43°15′W	N44°15′E
CD	N73°00′W	S72°15′E
DE	N12°45′E	S13°15′W
EA	N60°00′E	S59°00′W

8.17. The following are stadia intervals and vertical angles for a theodolite-stadia traverse. The elevation of station *A* is 150.485 m. The stadia interval factor is 100 and *C* = 0. Rod readings are taken at the height of instrument.

Station	Azimuth	Interval, m	Zenith angle	Horizontal distance	Differences in elevation	Elevation, m
A						150.485
	85°06′	0.998	85°28′			
B						
	10°18′	1.895	86°08′			
C						
	265°00′	1.551	95°05′			
D						
	173°04′	1.857	92°10′			
A						

(*a*) Calculate the horizontal distance for each line to the nearest decimeter. Determine differences in elevation between stations to the nearest centimeter.

(*b*) Determine elevations of transit stations and distribute the error of closure in proportion to the distance.

8.18. Given the following notes for a closed-loop traverse, the coordinates of which for station *A* are X_A = 1,984,400.612 m, Y_A = 518,430.033 m, compute the error of closure and coordinates for each traverse station adjusted according to the compass rule.

Course	Azimuth	Distance, m
AB	0°42′	372.242
BC	94°03′	164.988
CD	183°04′	242.458
DA	232°51′	197.165

8.19. Given the following adjusted azimuths and measured distances for a closed traverse that starts at C and closes on D, compute

(a) the error of closure;

(b) the coordinates for each traverse station, adjusted by the compass rule; and

(c) adjusted azimuths and distances for each line.

Station	Azimuth	Distance, m	X, m	Y, m
C			16,118.900	31,852.440
	210°43′34″	515.070		
1				
	275°08′30″	766.750		
2				
	283°09′15″	544.412		
3				
	231°23′14″	851.541		
4				
	174°22′28″	665.785		
5				
	252°32′20″	878.724		
D			13,123.601	30,144.626

8.20. The following are the adjusted azimuths and measured distances for a closed traverse that starts on point A and closes on point B, both having known cordinates that also are tabulated.

Station	Azimuth	Distance, m	X, m	Y, m
A			12,272.640	33,442.600
	80°45′15″	978.961		
15				
	108°13′10″	885.808		
16				
	90°08′50″	419.851		
24				
	20°58′13″	984.422		
25				
	95°41′21″	633.005		
B			15,482.321	34,178.324

(a) Compute the error of closure for the traverse and adjusted coordinates for each station, using the compass rule for the adjustment; and

(b) compute adjusted azimuths and distances for each line in the traverse.

8.21. Given the following adjusted azimuths and distances for a closed traverse that starts at station B (X_B = 1,984,625.131 m and Y_B = 518,700.071 m) and closes on station J (X_J = 1,983,419.640 m and Y_J = 516,519.516 m). Compute the error in closure for the traverse and coordinates for each traverse station adjusted by the compass rule. Determine the relative or positional accuracy for this traverse. Discuss the value of this measure of accuracy in terms of absolute position. Describe a better way to describe the absolute positional accuracy of a traverse station for which coordinates are available.

Course	Azimuth	Distance, m
AB	142°08′	
BC	181°37′	349.305
CD	296°13′	158.747
DE	323°46′	248.869
EF	249°51′	221.400
FG	214°03′	567.538
GH	195°45′	852.068
HJ	191°28′	750.930
JK	138°42′	

8.22. Given the following plane rectangular coordinates for traverse stations 3 and 4, compute the distance and azimuth of the line from 3 to 4.

Station	X, m	Y, m
3	525,549.045	148,098.104
4	524,042.132	148,518.135

8.23. Adjusted plane coordinates for stations A and B are as follows:

Station	X, m	Y, m
A	12,275.645	33,444.604
B	15,486.166	34,178.809

Compute the distance and azimuth from A to B.

8.24. Adjusted plane coordinates for stations B and C are as follows:

Station	X, m	Y, m
B	15,486.166	34,178.809
C	16,118.900	31,852.439

Compute the distance and azimuth from B to C.

8.25. Compute adjusted distances and azimuths between all traverse points using adjusted coordinates from Problem 8.18.

8.26. Using the adjusted coordinates in Table 8.13 for the traverse of Figure 8.7, compute the coordinates of the intersection of line 52 with the line from 6 to 3.

8.27. Using the adjusted coordinates in Table 8.13 for the traverse of Figure 8.7, determine the shortest distance from station 5 to a straight line from 6 to 3.

8.28. It is necessary to determine the coordinates for the intersection of a straight property line and a curved line having a radius of 650.000 m, which defines a road right of way. The azimuth of the property line from M to N is 220°20′00″. The coordinates for M and O (the center of the circle) are as follows:

Station	X, m	Y, m
M	5679.533	6072.099
O	5571.166	5120.567

8.29. Using the adjusted coordinates in Table 8.13 for the traverse of Figure 8.7, compute the coordinates of the intersection of line 4-3 with a line passing through 1 and making a clockwise angle of 50°30'00" measured from line 1-6.

8.30. Given the following data for a closed-loop traverse, for which the lengths of *BC* and *DE* have not been measured in the field, compute the unknown lengths.

Course	Bearing	Distance, ft
AB	N9°30'W	685.50
BC	N56°55'W	Unknown
CD	S56°13'W	690.45
DE	S2°02'E	Unknown
EA	S89°31'E	1041.52

8.31. Given the following data for a closed-loop traverse, compute the length and bearing of the unknown side.

Course	Azimuth	Distance, m
AB	278°00'	155.25
BC	Unknown	Unknown
CD	68°30'	242.75
DA	81°45'	130.45

8.32. A square field contains 90.000 acres. What are its dimensions in chains, in rods, in feet, and in meters?

8.33. Rectangular tracts have dimensions of 50.00 × 200.00 m and 2500.0 × 3000.0 m. Compute the areas of these tracts in square meters, hectares, and acres (Section 1.8).

8.34. How many acres are there in a rectangular tract 100 × 200 ft? In a tract 500 × 600 ft? In a tract 3200 × 2000 ft? Express these areas in square meters and in hectares (Section 1.8).

8.35. A triangle has sides 145.50, 310.95, and 250.00 m in length. Calculate the area of the triangle expressed in square meters, hectares, and acres.

8.36. What is the area of a triangle having sides of length 219.0, 350.42, and 301.6 ft? Of a triangle having two sides of length 1167.1 and 392.7 ft and an included angle of 41°25'?

8.37. The mutually bisecting diagonals of a four-sided field are 480 and 360 ft. The angle of intersection between the diagonals is 95°. Find the interior angles and the lengths of the sides.

8.38. The following table gives *X* and *Y* coordinates (in meters) for stations in a closed traverse. Calculate the area enclosed by the traverse using the coordinate method.

Coordinate	Station			
	A	*B*	*C*	*D*
X	4000.0	4100.5	4205.7	4103.4
Y	4000.0	4150.3	4875.0	4870.6

8.39. The following table gives the departures and latitudes of an adjusted closed traverse. Calculate the area by the coordinate method. The coordinates for A are $X_A = Y_A = 1000.00$ ft.

Course	Departures, ft +	Departures, ft −	Latitudes, ft +	Latitudes, ft −
AB		313.6		198.7
BC		274.4	281.1	
CD	189.2		134.1	
DE	110.7		324.9	
EA	288.1			541.4

8.40. (*a*) Find the error of closure of the following traverse and adjust the survey by the compass rule.

Course	Azimuth	Length, m
AB	134°15′	89.733
BC	65°30′	80.284
CD	35°15′	95.585
DE	295°45′	119.482
EF	239°00′	60.107
FA	205°30′	76.261

(*b*) The coordinates for A are $X_A = 520,484.183$ m and $Y_A = 424,323.640$ m. Calculate coordinates for the traverse stations and determine the area in hectares by the coordinate method.

8.41. A traverse *ABCD* is established inside a four-sided field, and the corners of the field are located by angular and linear measurements from the traverse stations, all as indicated by the following data.

(*a*) Compute the departures and latitudes, and adjust the traverse by the compass rule.

(*b*) Compute the coordinates of each traverse point and of each property corner, using *D* as an origin of coordinates.

(*c*) Compute the length and bearing of each side of the field *EFGH*, and tabulate results. Calculate the area of the field by the coordinate method.

Course	Bearing	Length, ft
AB	S88°38′E	296.4
AE	N20°00′W	30.5
BC	S44°20′W	333.9
BF	N35°20′E	16.0
CD	S81°21′W	215.6
CG	S73°00′E	25.2
DA	N28°24′E	314.2
DH	S36°30′W	14.2

8.42. The following table gives offsets from a traverse line to an irregular boundary, measured at points 5 m apart. By the trapezoidal rule (Section 8.28) calculate the area between traverse line and boundary.

Distance, m	Offset, m	Distance, m	Offset, m
0	0.0	25	8.60
5	6.10	30	5.52
10	10.70	35	9.36
15	10.50	40	14.45
20	12.80	45	6.86

8.43. Given the data of Problem 8.42, calculate the required area by Simpson's one-third rule. Note that the number of offsets is even.

8.44. The following are offsets from a traverse line to an irregular boundary, taken at irregular intervals. Calculate the area between traverse line and boundary by means of the trapezoidal rule.

Distance, ft	Offset, ft	Distance, ft	Offset, ft
0	15.0	105	48.9
20	37.7	170	53.9
50	58.2	195	50.0
70	45.2	220	34.2

8.45. The following are offsets from a traverse line to an irregular boundary taken at irregular intervals. Calculate the area between the traverse line and the boundary using
(a) the trapezoidal rule, and
(b) generalized Simpson's rule.

Distance, m	Offset, m	Distance, m	Offset, m
0	5.15	30	13.40
7	12.5	48	18.9
20	16.74	60	14.05
25	10.5	72	8.10

8.46. In Figure 8.41, what is the area of the circular segment *EOF* if the length of the chord *L* is 255.52 m and the middle ordinate *M* is 29.524 m?

8.47. In Figure 8.41, what is the area of the circular segment *EOF* if the chord length *L* is 550 ft and the middle ordinate *M* is 6.54 ft?

8.48. Solve Problems 8.46 and 8.47 using the approximate expression (Equation (8.43) of Section 8.30). Compare the results with those of Problems 8.46 and 8.47, and for each case, compute the percentage of error introduced through use of the approximate expression.

8.49. A curved corner lot is similar in shape to that shown in Figure 8.42. The tangent distances, *T*, are each 20.00 m and the intersection angle *I* is 45°00″. What is the area between the circular curve *ABC* and the tangents *AD* and *CD*? What is the external distance *E*?

8.50. Given the data of Table 8.13, Section 8.18, and illustrated in Figure 8.7, find the area of each of the two parts into which the tract is divided by a meridian line through the point 2.

8.51. Given the data of Problem 8.50, find the length and direction of a line that runs through 1 and divides the tract into two equal parts.

8.52. The data of Problem 8.50 is given. The tract is to be divided into two equal parts by an east-west line. Compute the length of the dividing line, and compute the distances from the ends of the line to adjacent traverse stations.

8.53. Using the data for the traverse in Table 8.7 and shown in Figure 8.20 (Section 8.14), propagate σ_x, σ_y, and σ_{xy} for point 3, treating lines 12 and 23 as an open traverse and assuming that $\sigma_\alpha = 20''$, $\sigma_d = 0.005$ m (see Section 8.36).

8.54. Using the data of Problem 8.20, propagate σ_x, σ_y, and σ_{xy} for station 16, assuming the coordinates of A are errorless and that lines A-15 and 15–16 constitute an open traverse. Assume $\sigma_\alpha = 30''$ and $\sigma_d = 0.010$ m.

8.55. A three-dimensional traverse was run using a total station system from station 49, with a backsight on station 50, to station 12, with a foresight on station 38. The coordinates for the known stations are as follows:

Station	X, m	Y, m	Elevation, m
50	716.464	1564.133	
49	724.638	1500.432	89.733
12	847.531	1312.615	92.821
38	879.185	1238.456	

The observed h.i.'s, in meters, the zenith and horizontal angles, and the slope distances, in meters, are as follows:

At	From/To	h.i., m	Zenith angle	Horizontal angle, right	Distance, m
	50				
49		1.481			
	24	1.296	90°22′45″	203°41′13″	122.270
	49				
24		1.450			
	48	1.271	91°02′10″	152°01′30″	109.913
	24				
48		1.425			
	12	1.448	85°59′48″	86°17′28″	140.587
	48				
12		1.602			
	38			262°11′38″	

Compute the X and Y coordinates and elevations for stations 24 and 48. Determine the error of closure in the coordinates and elevations. Adjust the coordinates by the compass rule and the elevations in proportion to the distance leveled. Assume all slope distances are corrected for system constants and atmospheric conditions.

8.56. A closed-loop traverse, initiated and closing on station 20, was run using a total station system and peripheral equipment. The known state plane coordinates and elevation for station 20 are $X = 1,490,832.529$ m, $Y = 504,294.409$ m, and 68.388 m above the datum. The azimuth from north of line 20-21 $= 80°41'09''$. The observations for the traverse are as follows:

Station			Zenith	Horiz. angle,	Slope
At	From/To	h.i., m	angle	right	distance, m
	20				
21		1.453			
	21A	1.364	84°06'15"	107°11'23"	88.175
	21				
21A		1.518			
	21B	1.383	95°21'15"	45°16'53"	66.495
	21A				
21B		1.357			
	21C	1.486	92°31'34"	187°55'00"	65.046
	21B				
21C		1.526			
	20	1.297	92°01'22"	168°37'00"	60.948
	21C				
20		1.517			
	21	1.418	89°19'18"	31°00'05"	146.220

Compute X and Y coordinates and elevations for stations 21, and 21A through 21C. Determine the error of closure in the coordinates and elevations. Adjust the coordinates by the compass rule and the elevations in proportion to the distance leveled. Assume all slope distances are corrected for system constants and atmospheric conditions.

8.57. In addition to a total station system, describe the items of peripheral equipment required to perform a three-dimensional traverse, assuming a party of two persons.

8.58. Describe how a robotic total station system could be used to perform a traverse in three dimensions (see Section 7.20).

8.59. Using the data from Problem 8.55, adjust the X and Y coordinates by the method of least squares using the techniques of
(*a*) indirect observations and
(*b*) observations only. Assume $\sigma_\alpha = 05''$, $\sigma_d = 0.005$ m, and that measurements are uncorrelated.

8.60. The local adjusted coordinates and state plane adjusted coordinates (for three stations) for the traverse shown in Figure 8.7 and computed in Section 8.15 are

	Adjusted local coordinates, ft		Adjusted state plane coordinates, ft	
Station	X	Y	E	N
1	4,382.09	6,150.82	1,984,951.49	515,645.49
2	4,770.93	6,036.90		
3	5,056.06	6,215.76	1,985,682.56	515,651.44
4	4,921.10	6,511.61		
5	4,785.69	6,347.30		
6	4,533.84	6,333.77	1,985,118.61	515,814.52

Transform the local coordinates of points 2, 4, and 5 into the state plane system using a linear, conformal, four-parameter transformation with

(*a*) two common points 1 and 3, and

(*b*) three common points 1, 3, and 5.

REFERENCES

Baarda, W., and J. Albercla. "The Connection of Geodetic Adjustment Procedures with Methods of Mathematical Statistics." *Bulletin Géodésique* 66 (1962).

Bossler, John D., chairman, Federal Geodetic Control Committee. *Standards and Specifications for Geodetic Control Networks.* Rockville, MD: FGCC, 1984.

Chen, Chun-Sung, and Hung-Chen Lin. "Using the Cubic Spline Rule for Computing the Area Enclosed by an Irregular Boundary." *Surveying and Land Information Systems* 51, no. 2 (June 1991), pp. 113–18.

Easa, S. M. "Area of Irregular Region with Unequal Intervals." *ASCE Journal of Surveying Engineering* 114, no. 2 (May 1988), pp. 50–58.

Easa, S. M. "Direct Method for Land Subdivision." *ASCE Journal of Surveying Engineering* 115, no. 4 (November 1989).

Easa, S. M. "Comment and Discussion re Using the Cubic Spline Rule for Computing the Area Enclosed by an Irregular Boundary." *Surveying and Land Information Systems* 51, no. 4 (December 1991), pp. 251, 252.

Federal Geodetic Control Subcommittee. "Standards for Geodetic Control Networks (Draft)," October 4, 1994.

Hildebrand, F. M. *Introduction to Numerical Analysis,* 2nd ed. New York: Dover Publications, 1987.

Kavanagh, B. F., and S. J. G. Bird. *Surveying,* 3rd ed. Englewood Cliffs, NJ: Prentice Hall, 1992.

Mikhail, E. M. *Observations and Least Squares.* New York: Harper and Row, 1976.

Mikhail, E. M., and G. Gracie. *Analysis and Adjustment of Survey Measurements.* New York: Van Nostrand Reinhold, 1981.

Moffitt, F. H., and H. Bouchard. *Surveying,* 9th ed. New York: HarperCollins Publishers, 1992.

Pasley, R. M. "P.O.B. 1994 Data Collector Survey." *P.O.B., Point of Beginning* 19, no. 6 (August–September 1994), pp. 80–93.

Richardus, P. *Project Surveying.* New York: John Wiley & Sons, 1966.

Târcy-Hornoch, A. "Remarks on Computations for Missing Elements of Closed Traverses." *Surveying and Mapping* 32, no. 4, p. 523.

Tripod Data Systems, Inc. *TDS-48GX Surveying Card Users Manual.* Corvallis, OR, 1991.

Walkowski, W. "Data Collector Survey," *P.O.B., Point of Beginning* 22, no. 3 (December 1996), pp. 33–46.

Wolf, P. R., and R. C. Brinker. *Elementary Surveying,* 9th ed. New York: HarperCollins Publishers, 1994.

Other Methods for Horizontal Positioning

9.1
LOCATION OF POINTS BY INTERSECTION

When coordinates of a point are given and the azimuth and distance to a second point also are known, it is possible to compute the coordinates of the second point (Section 8.13). Similarly, if the coordinates are given for the two ends of a line and directions are observed from each end of this line to a third point not on the line, then coordinates of that third point can be calculated. This procedure is called *location by intersection*.

In Figure 9.1, B and D are points of known coordinates from which angles α and β have been observed so as to locate point C by intersection. The distance c and azimuth A_{BD} from B to D can be found using Equations (8.8) and (8.9):

$$c = [(X_D - X_B)^2 + (Y_D - Y_B)^2]^{1/2}$$

$$\tan A_{BD} = \frac{X_D - X_B}{Y_D - Y_B}$$

Also in triangle DBC, $\gamma = 180° - (\alpha + \beta)$, so that

$$d = \frac{c \sin \beta}{\sin \gamma} \quad \text{and} \quad b = \frac{c \sin \alpha}{\sin \gamma} \tag{9.1}$$

where d is the distance B to C and b is the distance D to C. The azimuths of lines BC and DC can be calculated and the coordinates for C may be computed by Equations (8.6) and (8.7) using data from line BC as a check on calculations with data from line DC.

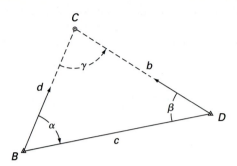

FIGURE 9.1
Location of a point by intersection.

9.2
INTERSECTION BY THE BASE SOLUTION

A more direct solution is possible using the observed angles, α, β, and the base line c computed from the coordinates of B and D. In triangle BCD from Equations (8.6), (8.7), and (9.1),

$$Y_C = Y_D + b \cos A_{DC} = Y_D + \frac{c \sin \alpha}{\sin \gamma} \cos (A_{DB} + \beta)$$

$$Y_C = Y_D + \frac{c \sin \alpha}{\sin (\alpha + \beta)} \cos (A_{DB} + \beta) \tag{9.2}$$

in which

$$\cos (A_{DB} + \beta) = \cos A_{DB} \cos \beta - \sin A_{DB} \sin \beta$$

and the sign of β is determined by careful inspection of the figure. By Equations (8.12a) and (8.12b) (Section 8.19),

$$\sin A_{DB} = \frac{X_B - X_D}{c} \quad \text{and} \quad \cos A_{DB} = \frac{Y_B - Y_D}{c}$$

so that

$$\cos (A_{DB} + \beta) = \frac{Y_B - Y_D}{c} \cos \beta - \frac{X_B - X_D}{c} \sin \beta$$

and by substitution into Equation (9.2),

$$Y_C = Y_D + (Y_B - Y_D) \frac{\sin \alpha \cos \beta}{\sin (\alpha + \beta)} - (X_B - X_D) \frac{\sin \beta \sin \alpha}{\sin (\alpha + \beta)}$$

$$= Y_D + \frac{(Y_B - Y_D) \cot \beta}{\cot \alpha + \cot \beta} - \frac{(X_B - X_D)}{\cot \alpha + \cot \beta} \tag{9.3}$$

$$= \frac{(X_D - X_B) + Y_D \cot \alpha + Y_B \cot \beta}{\cot \alpha + \cot \beta}$$

and in a similar manner,

$$X_C = \frac{(Y_B - Y_D) + X_D \cot \alpha + X_B \cot \beta}{\cot \alpha + \cot \beta} \tag{9.4}$$

Using Equations (9.3) and (9.4), azimuths for the lines to the intersected point need not be determined, and the coordinates can be calculated using adjusted base angles α and β. This method is advantageous when computing through a chain of triangles where adjusted angles may be available but no azimuths have been calculated.

9.3
INTERSECTION WHEN AZIMUTHS ARE GIVEN

When azimuths are given for lines DC and BC, in addition to the coordinates of points D and B, the point-slope equations (Equation (8.13), Section 8.19) for lines DC and BC are

$$Y_C - Y_D = (X_C - X_D) \cot A_{DC}$$
$$Y_C - X_C \cot A_{DC} + X_D \cot A_{DC} - Y_D = 0 \tag{9.5a}$$

and

$$Y_C - Y_B = (X_C - X_B) \cot A_{BC}$$
$$Y_C - X_C \cot A_{BC} + X_B \cot A_{BC} - Y_B = 0 \tag{9.5b}$$

Subtraction of Equation (9.5b) from Equation (9.5a) yields

$$X_C = \frac{(Y_D - Y_B) - X_D \cot A_{DC} + X_B \cot A_{BC}}{\cot A_{BC} - \cot A_{DC}} \tag{9.6}$$

and solution for Y_C in Equations (9.5a) and (9.5b) gives

$$Y_C = Y_D + (X_C - X_D) \cot A_{DC} \tag{9.7}$$

$$Y_C = Y_B + (X_C - X_B) \cot A_{CA} \tag{9.8}$$

Some examples serve to illustrate applications of the different methods.

EXAMPLE 9.1. In Figure 9.2, the coordinates in meters for triangulation stations B and C are

Station	X, m	Y, m
B	3369.287	2890.836
C	3300.259	3082.183

The observed angles are

$$\beta = 64°32'28'', \qquad \alpha = 81°17'38''$$

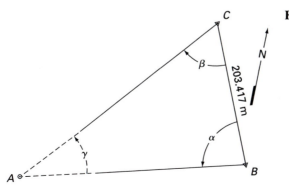

FIGURE 9.2

Determine the X and Y coordinates using the first method (Section 9.1) of solving the intersection problem.

Solution. The distance from C to B is

$$CB = [(3300.259 - 3369.287)^2 + (3082.183 - 2890.836)^2]^{1/2} = 203.41716 \text{ m}$$

The azimuth of BC is

$$\tan A_{BC} = \frac{3300.259 - 3369.287}{3082.183 - 2890.836} = -0.36074775$$

$$A_{BC} = 340°09'47''$$

Compute azimuths for CA and BA:

$$A_{BC} = 340°09'47''$$

$$-\alpha = -81°17'38''$$

$$A_{BA} = 258°52'09''$$

$$A_{CB} = 160°09'47''$$

$$+\beta = 64°32'28'' \qquad\qquad \alpha = 81°17'38''$$

$$A_{CA} = 224°42'15'' \qquad\qquad \beta = 64°32'28''$$

$$\gamma = \angle CAB = 258°52'09'' - 224°42'15'' = 34°09'54''$$

Check sum of angles $\alpha + \beta + \gamma \qquad = 180°00'00''$

In triangle BCA, solve for sides CA and BA using the law of sines:

$$CA = \frac{(203.4172)\sin\alpha}{\sin\gamma} = \frac{(203.4172)(0.9884777)}{0.5615780} = 358.051 \text{ m}$$

$$BA = \frac{(203.4172)\sin\beta}{\sin\gamma} = \frac{(203.4172)(0.9028940)}{0.5615780} = 327.050 \text{ m}$$

The coordinates of A using line CA and Equation (8.7) are

$$X_A = 3300.259 + (358.051)(\sin 224°42'15'') = 3048.389 \text{ m}$$

$$Y_A = 3082.183 + (358.051)(\cos 224°42'15'') = 2827.699 \text{ m}$$

As a check, calculate the coordinates of A using line BA:

$$X_A = 3369.287 + (327.050)(\sin 258°52'09'') = 3048.389 \text{ m}$$

$$Y_A = 2890.836 + (327.050)(\cos 258°52'09'') = 2827.699 \text{ m}$$

For comparison, determine the coordinates from Example 9.1 by Equations (9.3) and (9.4). First solve for Y_A using Equation (9.3), which for this example is

$$Y_A = \frac{(X_C - X_B) + Y_C \cot\alpha + Y_B \cot\beta}{\cot\alpha + \cot\beta} = \frac{U_Y}{V_Y}$$

$$X_C = 3300.259 \qquad \cot\alpha = \cot 81°17'38'' = 0.15313066$$

$$X_B = 3369.287 \qquad \cot\beta = \cot 64°32'28'' = 0.47609507$$

$$X_C - X_B = -69.028 \qquad \cot\alpha + \cot\beta = V_Y = 0.62922573$$

$$Y_C \cot \alpha = \quad 471.9767 \qquad Y_C = 3082.183$$

$$Y_B \cot \beta = \underline{1376.3128} \qquad Y_B = 2890.836$$

$$U_Y = 1779.2615$$

$$Y_A = \frac{U_Y}{V_Y} = 2827.700 \text{ m}$$

Equation (9.3) for X_A is

$$X_A = \frac{(Y_B - Y_C) + X_C \cot \alpha + X_B \cot \beta}{\cot \alpha + \cot \beta} = \frac{U_X}{V_X}$$

$$Y_B = \quad 2890.836 \qquad X_C \cot \alpha = (3300.259)(0.15313066)$$

$$Y_C = \underline{3082.183} \qquad\qquad\quad = 505.3708$$

$$Y_B - Y_C = -191.347 \qquad X_B \cot \beta = (3369.287)(0.47609507)$$

$$X_C \cot \alpha = \quad 505.3708 \qquad\qquad\quad = 1604.1009$$

$$X_B \cot \beta = \underline{1604.1009}$$

$$U_X = 1918.1247 \qquad\qquad V_X = V_Y = 0.62922573$$

$$X_A = \frac{U_X}{V_X} = 3048.389 \text{ m}$$

This method is more compact and efficient computationally but lacks the inherent mathematical check found in the preceding procedure.

EXAMPLE 9.2. In Figure 9.3, point A is intersected from stations C and D, having the following coordinates:

Station	X, m	Y, m
C	3300.259	3082.183
D	3047.954	3048.344

with observed angles

$$\alpha' = \text{angle } DCA = 37°39'28''$$

$$\beta' = \text{angle } ADC = 97°31'31''$$

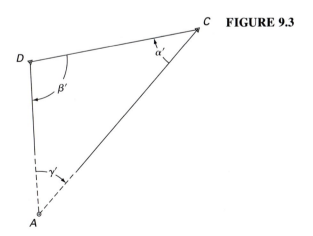

C   FIGURE 9.3

Solution. Use the method developed in Section 9.3, which requires the azimuths to be calculated for lines DA and CA. First compute the azimuth of DC:

$$\text{Azimuth of } DC = A_{DC} = \tan^{-1}\frac{X_C - X_D}{Y_C - Y_D} = \frac{3300.259 - 3047.954}{3082.183 - 3048.344}$$

$$\tan A_{DC} = 7.4560418, \qquad A_{DC} = 82°21'40''$$

Next, compute the azimuths of DA and CA:

Line	Azimuth	
DC	82°21'40''	
+β'	97°31'31''	
DA	179°53'11''	β' 97°31'31''
CD	262°21'40''	α' 37°39'28''
−α'	37°39'28''	γ' 44°49'01''
CA	224°42'12''	180°00'00'' check
−DA	−179°53'11''	
∠DAC = γ'	= 44°49'01''	

The equation for line DA is

$$Y_A - Y_D = (X_A - X_D) \cot A_{DA} \tag{1}$$

and for line CA

$$Y_A - Y_C = (X_A - X_C) \cot A_{CA} \tag{2}$$

Subtraction of Equation (2) from Equation (1) gives

$$X_A = \frac{(Y_D - Y_C) - X_D \cot A_{DA} + X_C \cot A_{CA}}{\cot A_{CA} - \cot A_{DA}} \tag{3}$$

Substitution of the given values into Equation (3) can be tabulated as follows:

$$Y_D - Y_C = 3048.344 - 3082.183 = \quad -33.839$$

$$-X_D \cot A_{DA} = -(3047.954)(-504.31421) = 1{,}537{,}126.5$$

$$X_C \cot A_{CA} = (3300.259)(1.0104096) = \quad 3{,}334.6134$$

$$(Y_D - Y_C) - X_D \cot A_{DA} + X_C \cot A_{CA} = 1{,}540{,}427.3$$

$$\cot A_{CA} - \cot A_{DA} = 505.32462$$

$$X_A = 3048.3916 \text{ m}$$

Calculate Y_A by Equation (9.8):

$$Y_A = Y_C + (X_A - X_C) \cot A_{CA}$$

$$= (3082.183) + (3048.3916 - 3300.259)(1.0104096)$$

$$= 2827.694 \text{ m}$$

As in the method of baseline angles, there is no built-in check on the arithmetic in this solution. A solution using the first method of Section 9.1 reveals that these answers are correct.

Note that, in all three solutions contained in Examples 9.1 and 9.2, one more significant figure was carried than the number of significant digits in the original coordinates.

9.4
INTERSECTION USING REDUNDANT DATA

Frequently, a point can be located by intersection from more than two known points. In Figure 9.4, points B, C, and D are triangulation stations of known positions, from which the angles α, β, α', and β' have been observed so as to locate point A. Therefore, the position of A can be determined using triangle ABC and another separate solution is possible using the data in triangle CDA. Note that there are two extra observations in this example or there is a redundancy of two. Using base line BC and angles α and β, one set of coordinates may be obtained for A. With base line DC and angles α' and β', another set of coordinates can be calculated. The likelihood that these two sets will agree exactly is very small.

The observant reader will have noticed that triangle ABC in Figure 9.4 was used for Example 9.1 and triangle ADC was used in Example 9.2. If the data were perfect, the coordinates in Example 9.1 would have agreed exactly with those computed in Example 9.2. The data are real observations of a high quality but cannot be perfect, because they contain random errors. Note that the values obtained for the coordinates in the two examples differ by 3 mm in X and 5 mm in Y. These observations were made in triangulation for bridge construction, and exceptional care was exercised in the execution of the survey. In an ordinary survey, the difference between the two might be several centimeters or even decimeters and the discrepancy would have to be resolved. One method of resolving the discrepancy is to take the average of the two sets as the final values, particularly if the four angles are of equal precision.

A more rigorous adjustment procedure consists of applying the method of least squares. Such a procedure requires that an equation be formed for each line that intersects at the unknown point. These equations are linearized and solved by the method of least squares to provide a best estimate for the coordinates of the intersected point. Approximate values of coordinates for the unknown point, which are necessary for linearization, can be computed by any of the methods discussed previously, using any pair of angles. Details and an example of this procedure can be found in Example 9.6, Section 9.17.

9.5
RESECTION

When angles between lines to three points of known position are observed from a point of unknown position, the coordinates of the unknown point can be calculated. This procedure is called location by *resection*.

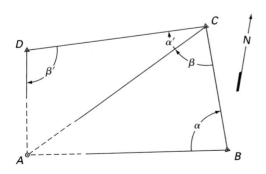

FIGURE 9.4
Intersection with redundant data.

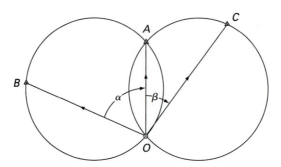

FIGURE 9.5
Three-point resection.

The resection problem is illustrated graphically in Figure 9.5. In the figure, B and A are of known position and O is of unknown position. If the angle α is observed between OB and OA, the position of O is indeterminate because O can be anywhere on the circle circumscribing the triangle OAB. Additional information is needed to make the problem determinate. When the direction of OA or OB is known, the problem is determinate. However, this information usually is not available. It is more convenient to observe angle β, subtended by AC, where C is a third known point. Then O, A, and C lie on the circle that circumscribes triangle OAC. Because O is on the circles that circumscribe both triangles OAB and OAC, it must lie at one of the two intersection points of the two circles. As A is one of these two intersections, O is uniquely determined. This solution presumes that the three known points A, B, C, and O do not fall on the circumference of one circle.

Should the two circumscribing circles tend to merge into one circle, the problem will be less stable and finally becomes indeterminate again when the two circles coincide. Points should be selected in the field to avoid this situation. It can be shown that this condition is present when $\angle BAC + \alpha + \beta = 180°$.

9.6
RESECTION CALCULATION

The method presented here is somewhat similar to the procedure given by the U.S. National Geodetic Survey in Reynolds (1928).

In Figure 9.6, stations B, A, and C are coordinated control points and O is a point of unknown position from which angles α and β are observed. The problem is to determine angles θ and γ so that the distances and directions of AO and BO or CO can be calculated to compute and check the position of point O.

In polygon $OBAC$ (Figure 9.6),

$$\theta + \gamma = 360 - (\phi + \alpha + \beta) = R \tag{9.9a}$$

for the cases represented by Figures 9.6a and 9.6b, and

$$R = \phi - \alpha - \beta \tag{9.9b}$$

for the case shown in Figure 9.6c.

By the law of sines in triangle OAC,

$$AO = \frac{b \sin \gamma}{\sin \beta} = \frac{c \sin \theta}{\sin \alpha} \tag{9.9c}$$

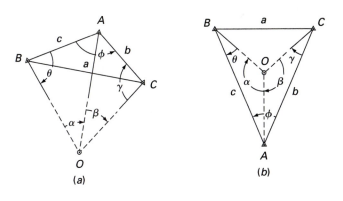

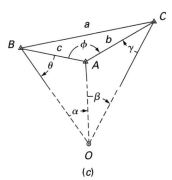

FIGURE 9.6
Three possible cases for three-point resection.

so that

$$\sin \theta = \frac{b \sin \gamma \sin \alpha}{c \sin \beta} \qquad (9.9d)$$

Because $\theta = R - \gamma$ from Equation (9.9a), Equation (9.9d) becomes

$$\sin (R - \gamma) = \frac{b \sin \gamma \sin \alpha}{c \sin \beta} = \sin R \cos \gamma - \cos R \sin \gamma \qquad (9.9e)$$

Division of Equation (9.9e) by $\sin R \sin \gamma$ yields

$$\cot \gamma = \cot R + \frac{b \sin \alpha}{c \sin \beta \sin R} \qquad (9.10)$$

The computational procedure is as follows:

1. Obtain R from Equation (9.9a) or (9.9b).
2. Solve for γ with Equation (9.10).
3. Compute $\theta = R - \gamma$.
4. Solve for the distance AO using b, γ, and β in Equation (9.9c). Check using c, θ, and α.
5. Compute angle $CAO = 180 - (\gamma + \beta)$ to determine the direction of AO. Check by calculating angle BAO.
6. Solve for OC or OB or both using the law of sines.
7. Calculate and check the coordinates of O.

EXAMPLE 9.3. Given the following data for a three-point resection,

Point	X, ft	Y, ft
B	10,000.00	20,000.00
A	16,672.50	20,000.00
C	27,732.76	14,215.24

$c = 6672.5$ ft $\qquad \alpha = 20°05'53''$

$b = 12,481.7$ ft $\qquad \beta = 35°06'08''$

$\phi = 152°23'22''$

$\alpha = 20°05'53'' \qquad R = 360° - (\alpha + \beta + \phi)$

$\beta = 35°06'08'' \qquad\quad = (360°) - 207°35'23''$

$\underline{\phi = 152°23'22''} \qquad R = 152°24'37''$

$\alpha + \beta + \phi = 207°35'23''$

Solution. Determine R, γ, and ϕ using Equations 9.9a and 9.10.

$\cot R \qquad\qquad\qquad\qquad\qquad\qquad\qquad\qquad\qquad = -1.91365958$

$b \sin \alpha = (12,481.7)(0.34362782) \qquad\qquad\qquad = 4289.0594$

$c \sin \beta \sin R = (6672.5)(0.57503698)(0.46313706) = 1777.0265000$

$\cot \gamma = \cot R + \dfrac{b \sin \alpha}{c \sin \beta \sin R} \qquad\qquad\qquad = 0.49995638$

$\gamma \qquad\qquad\qquad\qquad\qquad\qquad\qquad\qquad\qquad = 63°26'13''$

$\theta = R - \gamma = (152°24'37'') - (63°26'13'') \qquad = 88°58'24''$

solve for AO using Equation (9.9c) and calculate the length of OC.

$$AO = \frac{(12,481.7)(0.89444280)}{0.57503698} = 19,414.693 \text{ ft}$$

$$AO = \frac{(6672.5)(0.99983946)}{0.34362782} = 19,414.693 \text{ ft}, \qquad \text{checks}$$

$$\angle CAO = 180° - (\gamma + \beta) = 81°27'39''$$

$$\angle BAO = 180° - (\theta + \alpha) = \frac{70°55'43''}{152°23'22''} = \phi, \qquad \text{checks}$$

To calculate the length of OC,

$$OC = \frac{b \sin CAO}{\sin \beta} = \frac{(12,481.7)(0.98891459)}{0.57503698} = 21,465.289 \text{ ft}$$

To determine the azimuths of lines AO and OC,

Line		Azimuth
AC	117°36′38″	computed from given coordinates
$+\angle CAO$	81°27′39″	
AO	199°04′17″	
OA	19°04′17″	
$+\beta$	35°06′08″	
OC	54°10′25″	
CO	234°10′25″	
$+\gamma$	63°26′13″	
CA	297°36′38″	checks

Finally, to compute coordinates for O,

Point	Distance, ft	Direction	X, ft	Y, ft
A			16,672.50	20,000.00
	19,414.693	199°04′17″		
O			10,328.83	1,650.94
	21,465.289	54°10′25″		
C			27,732.76	14,215.24

$$X_O = 10,328.8 \text{ ft}, \qquad Y_O = 1,650.9 \text{ ft}$$

Many other methods exist for computing the three-point resection problem; El Hassan (1986) and Richardus (1984) provide a sample of the literature on this subject.

9.7
INTRODUCTION TO TRIANGULATION AND TRILATERATION

Triangulation and trilateration formerly were employed extensively to establish horizontal control for topographic mapping; charting lakes, rivers, and ocean coastlines; and for the surveys required for the design and construction of public and private works of large extent.

A *triangulation system* consists of a series of joined or overlapping triangles in which an occasional line is measured and the balance of the sides are calculated from angles measured at the vertices of the triangles. The lines of a triangulation system form a network that ties together all the *triangulation stations* at the vertices of the triangles.

A *trilateration system* also consists of a series of joined or overlapping triangles. However, for trilateration, all the lengths of the triangle's sides are measured and the few directions and angles observed are only those required to establish azimuth. Trilateration became feasible when EDM devices were introduced, making practical the measurement of all lengths with a high order of accuracy under almost all field conditions.

A *combined triangulation and trilateration system* consists of a network of triangles in which all angles and all distances are observed. Such a combined system represents the strongest network for creating horizontal control that can be established by conventional terrestrial methods.

The most notable example of control establishment by triangulation and trilateration in the United States was the transcontinental horizontal system established by the National

Ocean Survey as the basis for the 1927 horizontal datum. This system was designed to provide control for the entire continent, and originally the control network was extended by pure triangulation. With the advent of EDM equipment, trilateration and traverse were used to supplement and strengthen the system. Subsequently, all of the then existing positioning systems such as analytical aerotriangulation (Section 13.22), inertial systems, and satellite doppler systems were employed to supplement the existing triangulation and trilateration data in a readjustment of the entire system to provide a basis for the new horizontal datum of 1983.

Currently, because of its low cost and high precision, almost all of the basic horizontal control in the United States is established by GPS observations (see Chapter 12). Supplementary horizontal control frequently is done by traverse or combined triangulation and trilateration with total station systems.

9.8
TRIANGULATION

A narrow triangulation system employs a chain of figures consisting of *single triangles* (Figure 9.7a), *polygons* (Figure 9.7b), *quadrilaterals* (Figure 9.7c), or combinations of these figures. A triangulation system extending over a wide area likewise is divided into figures irregularly overlapping and intermingling, as illustrated in Figure 9.8. The computations for such a system can be arranged to provide checks on the computed values for most of the sides. As many sides as possible are included in the routes through which the computations are carried from one base line to the next.

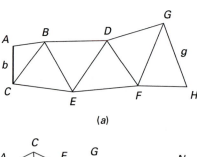

FIGURE 9.7
Chains of (a) single triangles, (b) polygons, and (c) quadrilaterals.

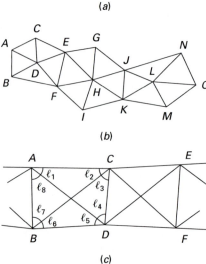

Of the three forms of chains of triangulation figures, the chain of single triangles is the simplest, requiring the measurement of fewer angles than either of the other two. Such a configuration has the obvious weakness that the only check (except between bases) is in the sum of the angles of each triangle considered by itself. This type of chain is not employed in work of high precision but is satisfactory where less precise results are required. For more precise work, quadrilaterals or polygons are used: quadrilaterals are best adapted to long, narrow systems and polygons to wide systems.

The work of triangulation consists of the following steps:

1. Reconnaissance to select the locations of stations.
2. Preanalysis through error propagation (Section 9.16) to evaluate the geometric strength of the proposed network.
3. Setting station marks, erecting signals, and setting towers for elevating signals or instruments where needed.
4. Observations of directions or angles.
5. Measurement of base lines.
6. Astronomic observations at one or more stations to determine the true meridian to which azimuths are referred.
7. Computations, including reduction to the ellipsoid (Section 9.11), calculation of all lengths of triangle sides and coordinates for all triangulation stations, and adjustment of the triangulation network to provide the best estimates of coordinates for all points.

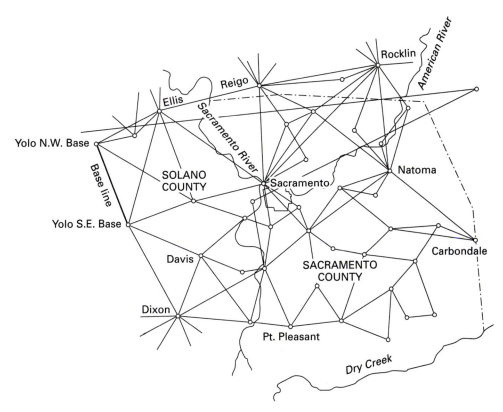

FIGURE 9.8
Area triangulation.

Reconnaissance, an important stage in any survey, consists of the selection of stations and the determination of the size and shape of the resulting triangles, the number of stations to be occupied, and the number of angles or directions to be observed. The intervisibility and accessibility of stations, the usefulness of stations in later work, the cost of necessary signals, and the convenience of base-line measurements must be considered. Information acquired during reconnaissance is utilized later for error propagation and determination of the geometric strength of the network.

After a preliminary study of all maps, survey information, and aerial photographs of the area, the person in charge makes an on-site inspection, choosing the most favorable locations for stations. If the information required cannot be obtained from existing maps, photographs, and the like, angles and distances to other stations are estimated or measured roughly en route, so that the suitability of the system as a whole can be evaluated before detailed work is begun.

The computations of triangulation involve the calculation, by the law of sines, of the lengths of the sides of successive triangles, polygons, or quadrilaterals using the initial measured base line and observed directions or angles. To ensure homogeneous results from these computations, the network must have adequate geometric strength. One important factor that affects geometric strength of a network is the magnitude of the angles measured. Rigorous error propagation can be used to analyze geometric strength, as discussed in Section 9.16. In general, angles should be no less than 30° or greater than 150° to ensure adequate geometry for most triangulation networks.

Directions from which angles are determined for triangulation usually are observed with a direction theodolite (Section 6.29). Precise triangulation requires an electronic or optical-reading direction instrument with a least count of less than 1″. A direction theodolite that has a least count of 1″ is adequate for most ordinary triangulation (Figure 6.30). Use of a direction instrument for measuring directions is covered in Section 6.49.

9.9
BASE-LINE MEASUREMENT USING EDM EQUIPMENT

The primary requirement for measurement of a base line by an EDM instrument is that an unobstructed measuring path exist between the two ends of the line. Selection of the location for the base line can be made entirely on the basis of geometric strength of the network, because measurement of the distance is entirely independent of the type of terrain between the two stations.

Care should be exercised to correct for meteorological conditions, systematic instrumental errors, and the slope of the line of sight, as described in Sections 4.30 and 4.36 to 4.41. With respect to systematic instrumental errors, the EDM system should be calibrated periodically on another base line of known length (Sections 4.38 and 4.39). Zenith or vertical angles or elevations at the two ends of the base line must be obtained to allow conversion of slope distances to horizontal distances (Sections 4.41 and 4.43). At least one ellipsoidal elevation is required to permit reduction of the base line distance to the surface of the ellipsoid (Section 9.11).

9.10
STATION MARKS

For the extensive triangulation systems of the U.S. National Geodetic Survey and the U.S. Geological Survey, every station is permanently marked with a metal disk that is fastened securely in rock or in a concrete monument. These stations are of great value as reference points for local surveys.

Stations for control networks not related to the national systems also should be marked by a punch mark on a metal disk set in a concrete monument or placed securely on a metal rod driven to refusal by a motorized driving device.

Triangulation stations also should be thoroughly referenced and described (Section 8.12) for future recovery and use.

9.11
REDUCTION TO THE ELLIPSOID

Control networks extending over large areas must be reduced to the surface of the ellipsoid (Section 1.3, Figure 1.2). In Figure 9.9, distance d is the horizontal distance from P' to O' at elevation h above the surface of the ellipsoid. If R is the radius of curvature for the surface of the ellipsoid at that section, then by proportion,

$$\frac{b}{d} = \frac{R}{R + h}, \qquad b = d\left(\frac{R}{R + h}\right) \tag{9.11}$$

in which b = the ellipsoid distance and $h = H + N$, where H is the average orthometric elevation of line $P'O'$ and N is the height of the geoid above (Figure 9.9a) or below (Figure 9.9b) the ellipsoid at this section (Section 12.13, Application 5, and Figure 12.12). The value of N is positive when the geoid is above the ellipsoid and is negative when it is below the ellipsoid, which is the case for most areas in the United States. For most ellipsoidal reductions in the United States, a value of R = 6,371,000 m or 20,902,190 ft (based on the U.S. Survey Foot, see Section 1.8) is adequate.

> **EXAMPLE 9.4.** A measured slope distance, corrected for atmospheric conditions and all systematic errors, yielded a horizontal distance of 17,690.817 m along a line having an average orthometric elevation of 773.42 m. The separation between the ellipsoid and geoid in this location is known to be −35.20 m. Determine the ellipsoidal distance for this line.
>
> **Solution.** Calculate the height of the line above the ellipsoid:
> $h = H + N = 773.42 - 35.20 = 738.22$ m. Then, by Equation (9.11) and using the

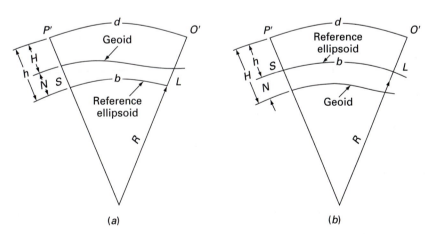

(a)   (b)

FIGURE 9.9
Reduction to the ellipsoid for (a) a positive N and (b) a negative N.

average value of R for the United States,

$$b = \frac{(17,690.819)\,(6,371,000)}{6,371,000 + 738.22} = 17,688.769 \text{ m}$$

Note that the correction in this case is about 2 m, or 1 part in 8000, a significant amount.

9.12
ADJUSTMENT OF TRIANGULATION

When all observed data have been compiled and reduced, the next step in the computations is adjustment of the triangulation network.

Assuming that all systematic errors have been corrected in the measured directions or angles and in the distances, the observed data can be considered randomly distributed so that the method of least squares is the most appropriate adjustment method. Details concerning the least-squares adjustment of triangulation can be found in Section 9.18 of Part B in this chapter.

9.13
TRILATERATION

A trilateration network consists of a system of joined or overlapping triangles or polygons in which all lengths are measured and only enough angles or directions are observed to establish azimuth.

If one considers a pure trilateration network, in which only distances are measured, the number of mathematical conditions in the figure must be formulated in terms of these distances. For example, if a single triangle is trilaterated, as in Figure 9.10, where points A and C are fixed, then two distances, a and c, are measured to fix the position of station B. Because the position of B is defined by two parameters, its X and Y coordinates, no extra or redundant measurements are there to provide a check on the work. Compare this with a triangulation network in which each single triangle has three measured angles and a redundancy of one per triangle to fix the position of the triangulated station. Therefore, trilateration is somewhat less strong, geometrically, than triangulation, and a chain of single triangles obviously is not a suitable configuration for control extension by trilateration.

The number of mathematical conditions for various geometric figures for trilateration is shown in Table 9.1. Note that the number of conditions as determined here for trilateration is synonymous with the redundancy, r, for a least-squares adjustment, as developed in Section 2.24.

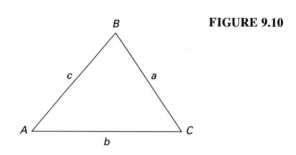

FIGURE 9.10

TABLE 9.1
Number of mathematical conditions for trilateration

Triangle	Braced quadrilateral	Center-point pentagon	Pentagon (all points visible)	Hexagon (all points visible)
0	1	1	3	6

Theoretically, a pentagon or hexagon would be the ideal figure for trilateration. However, from a practical standpoint, establishing even a single pentagon or hexagon in the field would be an almost impossible task.

The U.S. National Geodetic Survey recommends use of the traditional braced quadrilateral (Figure 9.7c), a center-point quadrilateral with an additional measurement (Figure 9.11a), or a center-point pentagon with an additional measurement (Figure 9.11b). It further recommends that angles be measured wherever possible to supplement the trilateration, thus increasing the number of conditions.

The procedure for establishing a trilateration network is similar to that required for triangulation. All available data, such as existing maps, references to existing control, and aerial photographs are assembled for the area in question and preliminary locations for trilateration stations are chosen in the office. A field reconnaissance follows, during which the intervisibility of points, accessibility of stations, and geometric strength of the figure selected are verified.

EDM equipment should be chosen that has the range needed for the job in question and yields a standard deviation in measured lengths suitable for the precision required for the control to be established. Triangles within the figure should contain no angles less than 30°, particularly for the more precise surveys. Because few if any angles will be observed, few signals need to be maintained and the construction of towers and stands for the instruments is generally unnecessary. Reflectors or antenna units for EDM equipment can be extended up to 13 m above the station with no serious effect on the precision of the measurements. In this way, distances of up to 50 km can be measured without the need for towers.

Distances measured by EDM equipment must be corrected for all systematic instrumental errors and for the effects of atmospheric conditions. In addition, EDM systems should be calibrated on a regular basis, using an independent base line of known length (Section 4.38). Slope distances must be reduced to horizontal ellipsoidal distances (Sections 4.41, 4.43, and 9.11). Zenith or vertical angles from both ends of the lines or

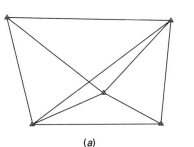
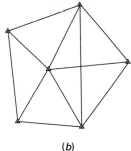

(a)   *(b)*

FIGURE 9.11
(a) Center-point quadrilateral with brace. (b) Center-point pentagon with additional distance.

elevations at the ends of the lines must be obtained. Also, for a given area, the separation between the geoid and ellipsoid must be known to allow reduction of all distances to the ellipsoid.

9.14
ADJUSTMENT OF TRILATERATION

The observed values in a pure trilateration network are the measured distances. As noted in Section 9.13, enough distances must be measured to provide redundant observations. With redundant observed distances, the best procedure for adjustment is the method of least squares using the technique of indirect observations (Section 9.19). Application of this technique to trilateration adjustment requires that the linearized form of Equation (8.8), in Section 8.13, be written for each measured distance. This linearized or so-called *distance condition equation* (Equation (8.71)) is developed in Section 8.39.

The distance condition equation contains measured distances and corrections to the X and Y coordinates for the stations of an unknown position in the network. These corrections have coefficients that are functions of approximate azimuths, distances, and coordinates for all of the measured lines and unknown stations in the network. Consequently, it is necessary to make preliminary calculations to obtain approximate values for these quantities.

For example, Figure 9.12 shows a trilateration network consisting of a center-point quadrilateral 1235 and overlapping braced quadrilaterals 3578 and 3567. An additional distance 4 to 7 was measured to strengthen the figure. Stations 1, 2, and 8 are of known, fixed X and Y coordinates. All distances except 12 have been measured yielding $n = 16$ observations. It is desired to adjust this network by the least-squares technique of indirect observations.

First, it is necessary to solve triangles to obtain approximate azimuths and coordinates for stations 3, 4, 5, 6, and 7. As shown in Figure 9.13, 10 measured lines are required to calculate these coordinates using the triangles labeled (1), (2), (3), (4), and (5). Therefore, the minimum number of measurements for a determinate figure are $n_0 = 10$ and the redundancy is $r = n - n_0 = 16 - 10 = 6$ (Equation (2.43), Section 2.24). Another way to calculate the redundancy is to use Equation (2.44), Section 2.25: $r = c - u$, where $c =$ the number of condition equations written and $u =$ the number of unknown parameters. There are 16 measured lines, requiring 16 distance condition equations, and five stations with unknown X and Y coordinates, so that $u = 10$. Therefore, $r = 16 - 10 = 6$.

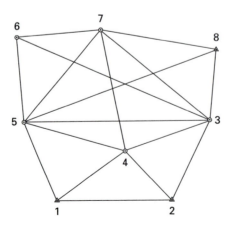

FIGURE 9.12
Trilateration network.

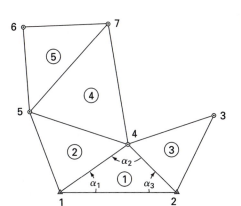

FIGURE 9.13
Trilateration calculations.

Calculation of preliminary approximations proceeds as follows:

1. Calculate the azimuth of line 12 using Equation (8.9), Section 8.13:

$$A_{12} = \arctan \frac{X_2 - X_1}{Y_2 - Y_1}$$

2. Calculate the angles in triangles (1), (2), (3), (4), and (5) using the law of cosines:

$$\cos A = \frac{b^2 + c^2 - a^2}{2bc}$$

Thus, in Figure 9.13 for triangle (1),

$$\cos \alpha_1 = \frac{(L_{14})^2 + (L_{12})^2 - (L_{42})^2}{(2)(L_{14})(L_{12})}$$

$$\cos \alpha_2 = \frac{(L_{14})^2 + (L_{42})^2 - (L_{12})^2}{(2)(L_{14})(L_{42})}$$

$$\cos \alpha_3 = \frac{(L_{12})^2 + (L_{42})^2 - (L_{14})^2}{(2)(L_{12})(L_{24})}$$

To check,

$$\alpha_1 + \alpha_2 + \alpha_3 = 180° \qquad \text{(assume plane triangles)}$$

This procedure is repeated in triangles (2), (3), (4), and (5) yielding angles in all triangles.

3. Using the azimuth of line 12 calculated in step 1, determine azimuths for all lines in triangles (1), (2), . . . , (5).
4. Starting with triangle (1), compute the X and Y coordinates for station 4 using measured distances and computed azimuths. Repeat this step for triangles (2), (3), (4), and (5), obtaining approximate coordinates $[X°, Y°]_i$, $i = 1, 2, 3, 4, 5$, for each unknown station in the network.

At this point all approximations necessary for the least-squares adjustment have been computed. The next step is to form the distance condition equations for each measured line in the network. This procedure is developed in Section 9.19, where Example 9.8 contains details for the adjustment of a braced quadrilateral. Output from the adjustment consists of adjusted X and Y coordinates for each station of unknown position in the network, residuals in measured distances, the reference standard deviation for the adjustment, and covariance matrices from which error ellipses for each adjusted station can be computed.

9.15
COMBINED TRIANGULATION AND TRILATERATION

Acquisition of all distances and directions in a control network constitutes no problem when using a total station system for the survey. Such an operation can be called a *combined control survey* and the subsequent adjustment is designated a *combined adjustment*.

Combined adjustment by the method of least squares using the technique of indirect observations presents no problem at all. For example, if a combined survey of the quadri-lateral *ABCD* in Figure 9.7c is to be performed, all eight angles, $l_1, l_2, \ldots, l_8$, and five distances, $d_{AB}, d_{AD}, d_{BD}, d_{BC}$, and d_{CD}, are measured (assume the coordinates are known for *A* and *C*). For adjustment, eight angle-condition equations (Equation (8.63), Section 8.39, see also Section 9.17), and five distance-condition equations (Equation (8.69), Section 8.39) are formed. There are 13 condition equations, which can be solved by the method of least squares for corrections to four approximate coordinates for unknown stations *A* and *C*. So the redundancy is $r = c - u = 13 - 4 = 9$ (Equation (2.44), Section 2.25), which represents a substantial increase over the redundancy of one for pure trilateration. Details for a combined adjustment by the method of least squares can be found in Section 9.20, Example 9.9.

PART B
ADVANCED TOPICS

9.16
ERROR PROPAGATION IN TRIANGULATION, INTERSECTION, AND RESECTION

Propagation of random errors was presented in Sections 2.15 and 2.19. It deals with unique cases where unknown quantities are calculated from functions of measured quantities, the number of functions being equal to the number of unknowns. Then, the variance, for a single unknown, or the covariance matrix, for multiple unknowns, is computed from Equations (2.16) and (2.34), respectively. When the number of equations exceeds the number of unknowns, redundancy exists and least-squares estimation is applied first, followed by error propagation. For example, when the condition equations are of the form (see Equation (2.53))

$$v + B\Delta = d - l$$

the propagated covariance matrix, $\Sigma_{\Delta\Delta}$, of the unknowns, Δ, is given by (see to Equations (2.74), (2.78), and (2.83), when $\sigma_0 = 1$)

$$\Sigma_{\Delta\Delta} = N^{-1} = (B^T W B)^{-1}$$

or, because $W = \Sigma^{-1}$,

$$\Sigma_{\Delta\Delta} = (B^T \Sigma^{-1} B)^{-1} \tag{9.12}$$

It is important to note that Equation (9.12) applies equally for the unique case when the redundancy, r, is 0. This can be ascertained by showing that Equation (9.12) is identical to direct propagation. When $r = 0$, the number of condition equations becomes equal to the number of parameters, and the residuals, *v*, are 0. Then, **B** is a square, nonsingular matrix, and the condition equations reduce to

$$\mathbf{B}\Delta = d - l$$

or

$$\Delta = \mathbf{B}^{-1}d - \mathbf{B}^{-1}l \tag{9.13}$$

Recalling that Σ is the covariance matrix of the observations, *l*, and *d* is a constant vector, applying the propagation Equation (2.34a) to Eq. (9.13) yields

$$\Sigma_{\Delta\Delta} = \mathbf{J}_{\Delta l}\, \Sigma\, \mathbf{J}_{\Delta l}^{T}$$

$$= (-\mathbf{B}^{-1})\, \Sigma\, (-\mathbf{B}^{-1})^{T}$$

or

$$\Sigma_{\Delta\Delta} = \mathbf{B}^{-1}\Sigma(\mathbf{B}^{-1})^{T} \tag{9.14}$$

With **B** being square and nonsingular, Equation (9.12) reduces to

$$\Sigma_{\Delta\Delta} = (\mathbf{B}^{T}\Sigma^{-1}\mathbf{B})^{-1}$$

$$= \mathbf{B}^{-1}(\Sigma^{-1})^{-1}(\mathbf{B}^{T})^{-1}$$

or

$$\Sigma_{\Delta\Delta} = \mathbf{B}^{-1}\Sigma(\mathbf{B}^{-1})^{T}$$

which is identical to Equation (9.14). Therefore, when there is a least-squares program for a given class of problems, error propagation may be performed by Equation (9.12) (or its equivalent for other least-squares techniques) regardless of whether redundancy exists or not.

Error Propagation in Triangulation

The variance (Section 2.17) of the length of the calculated side in a triangulation net is determined by error propagation through an equation that expresses length in terms of the measured angles and distances. First consider this error propagation in a single triangle as illustrated in Figure 9.10. Assume that side *b* is the measured base line and *A* and *C* are uncorrelated measured angles with respective variances of σ_A^2 and σ_B^2. Then sides *a* and *c* determined by the sine law are

$$a = \frac{b \sin A}{\sin B} = \frac{b \sin A}{\sin (A + C)} \tag{9.15a}$$

$$c = \frac{b \sin C}{\sin B} = \frac{b \sin C}{\sin (A + C)} \tag{9.15b}$$

Propagating the variances in sides *a* and *c* (use Equation (2.16)) yields

$$\sigma_a^2 = \left(\frac{\partial a}{\partial A}\right)^2 \sigma_A^2 + \left(\frac{\partial a}{\partial C}\right)^2 \sigma_C^2 + \left(\frac{\partial a}{\partial b}\right)^2 \sigma_b^2$$

$$\sigma_c^2 = \left(\frac{\partial c}{\partial A}\right)^2 \sigma_A^2 + \left(\frac{\partial c}{\partial C}\right)^2 \sigma_C^2 + \left(\frac{\partial c}{\partial b}\right)^2 \sigma_b^2 \tag{9.16}$$

As an example, the partial derivatives of a with respect to A, C, and b are (the student should evaluate the partials of c as an exercise)

$$\frac{\partial a}{\partial A} = \frac{b \sin (A + C) \cos A - b \sin A \cos (A + C)}{\sin^2 (A + C)} = a \cot A + a \cot B$$

$$\frac{\partial a}{\partial C} = \frac{-b \sin A \cos (A + C)}{\sin^2 (A + C)} = -a \cot (A + C) = a \cot B \qquad (9.17)$$

$$\frac{\partial a}{\partial b} = \frac{\sin A}{\sin (A + C)} = \frac{a}{b}$$

If an errorless base line is assumed, the variances in calculated sides a and c become a minimum when the angles $A + C = 90°$ or $B = 90°$. It can be seen that, for angles less than $45°$ (or more than $135°$), error propagation increases. In practice, angles less than $30°$ or greater than $150°$ are avoided.

The foregoing analysis dealt with the unique case, where the minimum number of measurements of three—two angles and a side—necessary to determine the shape of a single plane triangle, is acquired. When more measurements are given, for one or more triangles, than are required for a unique solution, least squares is used both for estimation and subsequent error propagation.

Error Propagation in Intersection

Equations (9.3) and (9.4) give the coordinates of a point determined by intersection, as functions of the coordinates of two fixed points and two measured angles. The covariance matrix for these two coordinates may be evaluated applying Equation (2.34a) to Equations (9.3) and (9.4). Alternatively, and perhaps more appropriately, least squares may be formulated to accommodate both this unique case and the redundant case, including postadjustment covariance propagation, as discussed in the following section (9.17).

Error Propagation in Resection

In a manner similar to intersection, error propagation may be performed either directly or in conjunction with least-squares adjustment, as presented in Section 9.17.

9.17
LEAST-SQUARES ADJUSTMENT OF RESECTION AND INTERSECTION

Resection Adjustment

Resection is the operation of determining the position of an unknown point from angular measurements made at the point and sighting at known points (refer to Sections 9.5 and 9.6). From the geometric standpoint, a minimum of two angles, such as α and β in Figure 9.14, are necessary for a unique determination of the unknown point. Each additional observation contributes one redundancy to the adjustment. Consequently, it is considerably

FIGURE 9.14

513

CHAPTER 9:
Other Methods
for Horizontal
Positioning

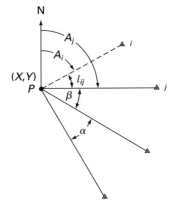

easier to *use the least-squares adjustment of indirect observations for resection problems.*
The form of the condition equation for the angle l_{ij} in Figure 9.14 was derived in Section 8.39 for traverse, where the coordinates of P, i, and j are unknown and given by Equation (8.66), or

$$
\begin{aligned}
v_{ij} + b_1 \delta X_P &+ b_2 \delta Y_P + b_3 \delta X_i + b_4 \delta Y_i + b_5 \delta X_j + b_6 \delta Y_j \\
&= -l_{ij} - F_{ij}[(X, Y)_P^0, (X, Y)_i^0, (X, Y)_j^0] = f_{ij}
\end{aligned}
\tag{8.66}
$$

For the special case of resection, coordinates for points i and j are known so that $b_3 = b_4 = b_5 = b_6 = 0$ and the angle equation for each measured angle contains only two unknown corrections, δX_P and δY_P. Thus, the linearized angle condition equation for resection is

$$
v_{ij} + b_1 \delta X_P + b_2 \delta Y_P = -l_{ij} - F_{ij}(X^0, Y^0)_P = f_{ij}
\tag{9.18a}
$$

in which

$$
b_1 = \frac{\partial F_{ij}}{\partial X_P} = \frac{-(Y_i - Y_P^0)}{(Y_i - Y_P^0)^2 + (X_i - X_P^0)^2} - \frac{-(Y_j - Y_P^0)}{(Y_j - Y_P^0)^2 + (X_j - X_P^0)^2}
\tag{9.18b}
$$

$$
b_2 = \frac{\partial F_{ij}}{\partial Y_P} = \frac{X_i - X_P^0}{(Y_i - Y_P^0)^2 + (X_i - X_P^0)^2} - \frac{X_j - X_P^0}{(Y_j - Y_P^0)^2 + (X_j - X_P^0)^2}
\tag{9.18c}
$$

$$
f_{ij} = -l_{ij} - \arctan \frac{X_i - X_P^0}{Y_i - Y_P^0} + \arctan \frac{X_j - X_P^0}{Y_j - Y_P^0}
\tag{9.18d}
$$

or

$$
f_{ij} = (\text{calculated angle}) - (\text{measured angle})
$$

One condition equation of the type (9.18a) is written for each observed angle. For example, three conditions would be written for the case given in Figure 9.14: one condition for each angle α, β, and l_{ij}. The total set of linearized conditions would be of the form $v + \mathbf{B}\Delta = f$, in which v is 3×1, $\mathbf{B}$ is 3×2, and Δ is 2×1, which contains δX_P and δY_P as parameters. The scheme is demonstrated by the following example, which also shows how the iterations are carried out for the nonlinear problem.

EXAMPLE 9.5. PROBLEM ON RESECTION ADJUSTMENT. Figure 9.15 shows a sketch for the resection of the position of point P from three observed angles l_1, l_2, l_3, to four

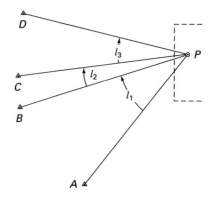

FIGURE 9.15

known control points A, B, C, D, whose coordinates are as follows:

Station	California state plane coordinates, Zone III	
	X, ft	Y, ft
A Flagpole on top of Oakland City Hall	88,237.92	80,232.03
B Top of Durkee tank	82,279.10	97,418.58
C Top of Peet Bros. stack	81,802.35	98,696.21
D Top of Spenger's Fish Grotto tank	80,330.69	102,911.40

The observed directions (mean of two positions), their standard deviations, and the corresponding angles are as follows:

Instrument at point P (mean of two positions): abstract of directions	σ_D	Average angle
A $0°00'00''$	$5''$	
		$l_1 = 44°55'30.5''$
B $44°55'30.5''$	$5''$	
		$l_2 = \ \ 5°56'19.4''$
C $50°51'49.9''$	$5''$	
		$l_3 = 19°56'18.3''$
D $70°48'08.2''$	$5''$	

Compute the coordinates X and Y of the unknown point P.

Solution. Because the angles are computed from directions, which are assumed to be uncorrelated, their covariance matrix must be computed from error propagation.

$$l_1 = D_{PB} - D_{PA}$$
$$l_2 = D_{PC} - D_{PB}$$
$$l_3 = D_{PD} - D_{PC}$$

which in matrix form become

$$
\begin{bmatrix} l_1 \\ l_2 \\ l_3 \end{bmatrix} = \begin{bmatrix} -1 & 1 & 0 & 0 \\ 0 & -1 & 1 & 0 \\ 0 & 0 & -1 & 1 \end{bmatrix} \begin{bmatrix} D_{PA} \\ D_{PB} \\ D_{PC} \\ D_{PD} \end{bmatrix} \qquad \text{or} \qquad l = JD
$$

Because the standard deviation of each direction is $5''$ or the variance is $25 \ \text{sec}^2$ or $5.876107 \times 10^{-10} \ \text{rad}^2$, the variance matrix of the directions is $\Sigma_{DD} = (5.876107)10^{-10}\mathbf{I}$. Then the covariance matrix of the angles (by Equation (2.34)) is

$$
\Sigma_{ll} = J\Sigma_{DD}J^t = (5.876107)10^{-10} \begin{bmatrix} -1 & 1 & 0 & 0 \\ 0 & -1 & 1 & 0 \\ 0 & 0 & -1 & 1 \end{bmatrix} \begin{bmatrix} 1 & 0 & 0 \\ 1 & -1 & 0 \\ 0 & 1 & -1 \\ 0 & 0 & 1 \end{bmatrix} (\text{rad}^2)
$$

$$
= (5.876107)10^{-10} \begin{bmatrix} 2 & -1 & 0 \\ -1 & 2 & -1 \\ 0 & -1 & 2 \end{bmatrix} (\text{rad}^2)
$$

The three angle condition equations (see Equation (8.63)) are

$$
F_1 = l_1 - A_{PB} + A_{PA} = l_1 - \arctan \frac{X_B - X}{Y_B - Y} + \arctan \frac{X_A - X}{Y_A - Y} = 0
$$

$$
F_2 = l_2 - A_{PC} + A_{PB} = l_2 - \arctan \frac{X_C - X}{Y_C - Y} + \arctan \frac{X_B - X}{Y_B - Y} = 0
$$

$$
F_3 = l_3 - A_{PD} + A_{PC} = l_3 - \arctan \frac{X_D - X}{Y_D - Y} + \arctan \frac{X_C - X}{Y_C - Y} = 0
$$

The linearized equations according to the form of Equation (9.18a) for the adjustment are

$$
v_1 + b_{11}\delta X + b_{12}\delta Y = f_1
$$

$$
v_2 + b_{21}\delta X + b_{22}\delta Y = f_2
$$

$$
v_3 + b_{31}\delta X + b_{32}\delta Y = f_3
$$

which may be expressed in matrix form as

$$
v + B\Delta = f
$$

Denoting the approximations for unknown parameters by X^0 and Y^0, the coefficients and constant terms are

$$
b_{11} = \frac{\partial F_1}{\partial X} = -\frac{Y_A - Y^0}{L_{AP}^{0^2}} + \frac{Y_B - Y^0}{L_{BP}^{0^2}}
$$

$$
b_{12} = \frac{\partial F_1}{\partial Y} = \frac{X_A - X^0}{L_{AP}^{0^2}} - \frac{X_B - X^0}{L_{BP}^{0^2}}
$$

$$b_{21} = \frac{\partial F_2}{\partial X} = -\frac{Y_B - Y^0}{L_{BP}^{0^2}} + \frac{Y_C - Y^0}{L_{CP}^{0^2}}$$

$$b_{22} = \frac{\partial F_2}{\partial Y} = \frac{X_B - X^0}{L_{BP}^{0^2}} - \frac{X_C - X^0}{L_{CP}^{0^2}}$$

$$b_{31} = \frac{\partial F_3}{\partial X} = -\frac{Y_C - Y^0}{L_{CP}^{0^2}} + \frac{Y_D - Y^0}{L_{DP}^{0^2}}$$

$$b_{32} = \frac{\partial F_3}{\partial Y} = \frac{X_C - X^0}{L_{CP}^{0^2}} - \frac{X_D - X^0}{L_{DP}^{0^2}}$$

$$f_1 = -l_1 + \arctan \frac{X_B - X^0}{Y_B - Y^0} - \arctan \frac{X_A - X^0}{Y_A - Y^0}$$

$$f_2 = -l_2 + \arctan \frac{X_C - X^0}{Y_C - Y^0} - \arctan \frac{X_B - X^0}{Y_B - Y^0}$$

$$f_3 = -l_3 + \arctan \frac{X_D - X^0}{Y_D - Y^0} - \arctan \frac{X_C - X^0}{Y_C - Y^0}$$

where $L_{AP}^{0^2} = (X_A - X^0)^2 + (Y_A - Y^0)^2$
$\qquad L_{BP}^{0^2} = (X_B - X^0)^2 + (Y_B - Y^0)^2$
$\qquad L_{CP}^{0^2} = (X_C - X^0)^2 + (Y_C - Y^0)^2$
$\qquad L_{DP}^{0^2} = (X_D - X^0)^2 + (Y_D - Y^0)^2$

The value of the azimuth angle computed from the arctan function in evaluating the elements of the f vector will be correct only if that function is properly derived in the computer. If, however, the value given is less than 90° in magnitude, with either a positive or negative sign, the azimuth angle should be determined according to the signs of both the numerator and denominator of the arctan function as explained in Example 8.15, Section 8.39. We repeat, however, that the computer used may yield the azimuth directly as a positive angle between 0° and 360°.

Using the approximations $X^0 = 93,600$ ft, $Y^0 = 104,000$ ft as determined from the geometry of the figure, the numerical values of the B and f matrices are

$$\mathbf{B} = 10^{-6} \begin{bmatrix} 1.50 & 57.06 \\ 6.68 & 4.49 \\ 25.56 & 4.35 \end{bmatrix} \quad \text{and} \quad f = 10^{-3} \begin{bmatrix} 39.12349 \\ 0.45342 \\ -7.35596 \end{bmatrix}$$

Using the covariance matrix $\mathbf{\Sigma}_{ll}$ evaluated previously, the normal equations are formed as $(\mathbf{B}'\mathbf{\Sigma}_{ll}^{-1}\mathbf{B})\mathbf{\Delta}_1 = \mathbf{B}'\mathbf{\Sigma}_{ll}^{-1}f$ and solved to yield

$$\mathbf{\Delta}_1 = \begin{bmatrix} -405.436 \text{ ft} \\ 696.625 \text{ ft} \end{bmatrix}$$

The updated approximations for the parameters become $93,600 - 405.436 = 93,194.564$ and $104,000 + 696.625 = 104,626.625$. Re-forming and solving the normal equations gives the correction vector:

$$\mathbf{\Delta}_2 = \begin{bmatrix} -41.072 \text{ ft} \\ -11.353 \text{ ft} \end{bmatrix}$$

Continuing the process to four iterations gives a $\mathbf{\Delta}_4$, which is essentially 0, and the final least-squares estimates for the coordinates of point P are

$$\hat{X} = 93,153.645 \text{ ft} \quad \text{and} \quad \hat{Y} = 104,685.246 \text{ ft}$$

Intersection Adjustment

Intersection is the operation of determining the position of an unknown point using observed angles taken at known points (refer to Sections 9.1 to 9.4). As shown in Figure 9.16, a unique determination of the unknown point P requires a minimum of two angles, such as α and β. Each additional *measured angle* contributes one redundancy to the adjustment. On the other hand, when one additional control point, such as i in Figure 9.16, is used, two angles l_i and l_i' are added. For each of the measured angles, one condition equation may be written, Equation (8.63). For example, the condition equation for the angle l_i in Figure 9.16 is

$$v_i + l_i = A_{iP} - A_{i2} = \arctan \frac{X_P - X_i}{Y_P - Y_i} - \arctan \frac{X_2 - X_i}{Y_2 - Y_i}$$

or
$$v_i + l_i - \arctan \frac{X_P - X_i}{Y_P - Y_i} + \arctan \frac{X_2 - X_i}{Y_2 - Y_i} = 0 \qquad (9.19)$$

The linearization of Equation (9.19) is very similar to that of Equation (8.63) and is not included here. However, note that, for this special case, the stations occupied and sighted have known coordinates, whereas the station to which the angle is turned is of unknown position. This is the reverse of the resection. In linearizing Equation (9.19), partials are taken with respect to the coordinates of the station to which the angle is turned (X_P and Y_P in Figure 9.16) and the resulting coefficients are not the same as b_1 and b_2 derived in Equations (9.18b) and (9.18c). Coefficients for the intersection adjustment are derived in Example 9.6.

    The angle condition is written as the difference between two azimuths, so the sign convention is very important and must be carefully observed. If the azimuth is defined as *clockwise* from north, the measured angles also must be defined as *clockwise* from the first line encountered from the north to the second line. Figure 9.17 depicts four different cases. In Figure 9.17a, the azimuths A_a and A_b are both in the first quadrant and l easily is seen to be equal to $(A_b - A_a)$. In Figure 9.17b, the azimuth A_a is in the first quadrant but A_b is in the second quadrant, because $A_b = \arctan [(X_b - X_0)/(Y_b - Y_0)]$ and $(X_b - X_0)$ is positive but $(Y_b - Y_0)$ is negative. (Note that the sign of $(X_b - X_0)$ corresponds to the sign of $\sin A_b$, whereas that for $(Y_b - Y_0)$ corresponds to the sign of $\cos A_b$) Here, too, $l = A_b - A_a$. In Figure 9.17c, the azimuth A_b is in the third quadrant and $l = A_b - A_a$. In Figure 9.17d, the azimuth A_b is in the fourth quadrant. There are two possibilities for the observed angle, either l or l' (see Figure 9.17d). If l is given as the clockwise angle from Oa to Ob, it will be directly equal to $(A_b - A_a)$. On the other hand, if l' is given as the clockwise angle from Ob to Oa, then $(A_b - A_a)$ actually will be $(l' - 360°)$, which is a negative value.

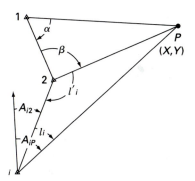

FIGURE 9.16

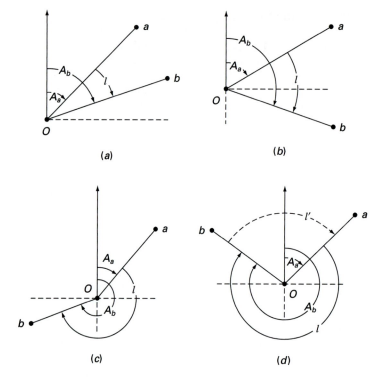

(a)

(b)

(c)

(d)

FIGURE 9.17

Therefore, whenever the difference between two directions is negative, 360° should be added to yield the observed angle.

EXAMPLE 9.6. INTERSECTION ADJUSTMENT PROBLEM. Figure 9.18 shows a sketch for the intersection of the position of point A from angles observed at three horizontal control points B, C, and D, the coordinates of which are as follows:

Point	X, ft	Y, ft
B	11,054.091	9,484.370
C	10,827.622	10,112.150
D	10,000.000	10,000.00

The measured angles, also shown in the figure, are

$$l_1 = \text{angle } ABC = 81°17'37.5''$$

$$l_2 = \text{angle } BCA = 64°32'27.5''$$

$$l_3 = \text{angle } ACD = 37°39'28.2''$$

$$l_4 = \text{angle } CDA = 97°31'31.1''$$

These angles are derived from directions taken with a standard deviation of 2″. Compute the least squares estimates of the coordinates X_A and Y_A of the unknown point A.

Solution. To evaluate the precision of the four given angles, their relationship to the seven directions involved, $d_1, d_2, \ldots, d_7$, shown in Figure 9.18 are

$$l_1 = d_2 - d_1, \qquad l_2 = d_4 - d_3, \qquad l_3 = d_5 - d_4, \qquad l_4 = d_7 - d_6$$

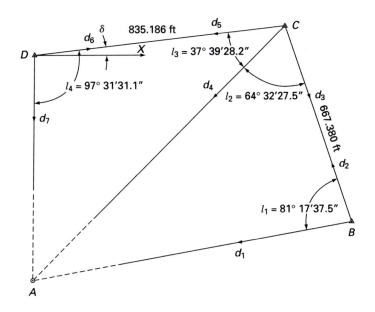

FIGURE 9.18
Position by intersection.

In matrix form, these relations become

$$
\begin{bmatrix} l_1 \\ l_2 \\ l_3 \\ l_4 \end{bmatrix} = \begin{bmatrix} -1 & 1 & 0 & 0 & 0 & 0 & 0 \\ 0 & 0 & -1 & 1 & 0 & 0 & 0 \\ 0 & 0 & 0 & -1 & 1 & 0 & 0 \\ 0 & 0 & 0 & 0 & 0 & -1 & 1 \end{bmatrix} \begin{bmatrix} d_1 \\ d_2 \\ \vdots \\ d_7 \end{bmatrix} = \mathbf{Jd}
$$

The variance matrix for the directions is $\Sigma_{dd} = 4\mathbf{I} \sec^2 = (9.401772)10^{-11}\mathbf{I}$ rad^2. Then, from the rules of error propagation (Equation (2.34)), the covariance matrix for the four angles is

$$
\Sigma_{ll} = \mathbf{J}\Sigma_{dd}\mathbf{J}^t = (9.401772)10^{-11}
$$

$$
\begin{bmatrix} -1 & 1 & 0 & 0 & 0 & 0 & 0 \\ 0 & 0 & -1 & 1 & 0 & 0 & 0 \\ 0 & 0 & 0 & -1 & 1 & 0 & 0 \\ 0 & 0 & 0 & 0 & 0 & -1 & 1 \end{bmatrix}
\begin{bmatrix} -1 & 0 & 0 & 0 \\ 1 & 0 & 0 & 0 \\ 0 & -1 & 0 & 0 \\ 0 & 1 & -1 & 0 \\ 0 & 0 & 1 & 0 \\ 0 & 0 & 0 & -1 \\ 0 & 0 & 0 & 1 \end{bmatrix}
$$

or

$$
\Sigma_{ll} = (9.401772)10^{-11} \begin{bmatrix} 2 & 0 & 0 & 0 \\ 0 & 2 & -1 & 0 \\ 0 & -1 & 2 & 0 \\ 0 & 0 & 0 & 2 \end{bmatrix} \text{(rad}^2)
$$

Before writing the condition equations, compute approximations for the two unknown parameters X_A and Y_A. As shown in Figure 9.18, the line DC makes an angle δ with the X axis, which may be computed from $\delta = \arctan\left[(Y_C - Y_D)/(X_C - X_D)\right] = \arctan$ $112.150/827.622 = 7°7$. With the given value of l_4, the line DA is very nearly parallel with the Y axis. Thus, a good approximation for X_A is to take it equal to X_D or $X_A^0 = 10,000.0000$ ft. To get Y_A^0, the length of the side AD is computed by applying the sine law to the triangle ACD, realizing that the length of CD is

$$[(X_C - X_D)^2 + (Y_C - Y_D)^2]^{1/2} = [(827.622)^2 + (112.150)^2]^{1/2} = 835.186 \text{ ft}$$

Therefore,

$$\overline{AD} = 835.186\frac{\sin l_3}{\sin(180 - l_3 - l_4)} \simeq 700 \text{ ft} \qquad \text{and}$$

$$Y_A^0 = Y_D - \overline{AD} \simeq 10,000 - 700 = 9300 \text{ ft}$$

The condition equations (from Equation (9.19)) are

$$F_1 = l_1 - \arctan\frac{X_C - X_B}{Y_C - Y_B} + \arctan\frac{X_A - X_B}{Y_A - Y_B} = 0$$

$$F_2 = l_2 - \arctan\frac{X_A - X_C}{Y_A - Y_C} + \arctan\frac{X_B - X_C}{Y_B - Y_C} = 0$$

$$F_3 = l_3 - \arctan\frac{X_D - X_C}{Y_D - Y_C} + \arctan\frac{X_A - X_C}{Y_A - Y_C} = 0$$

$$F_4 = l_4 - \arctan\frac{X_A - X_D}{Y_A - Y_D} + \arctan\frac{X_C - X_D}{Y_C - Y_D} = 0$$

Using X_A^0 and Y_A^0 to represent the approximations for the coordinates of point A, linearization of these equations to the form $v + \mathbf{B\Delta} = f$ yields the following coefficients:

$$b_{11} = \frac{\partial F_1}{\partial X_A} = \frac{Y_A^0 - Y_B}{L_{AB}^{0^2}} \qquad\qquad b_{12} = \frac{\partial F_1}{\partial Y_A} = -\frac{X_A^0 - X_B}{L_{AB}^{0^2}}$$

$$b_{21} = \frac{\partial F_2}{\partial X_A} = -\frac{Y_A^0 - Y_C}{L_{AC}^{0^2}} \qquad\qquad b_{22} = \frac{\partial F_2}{\partial Y_A} = \frac{X_A^0 - X_C}{L_{AC}^{0^2}}$$

$$b_{31} = \frac{\partial F_3}{\partial X_A} = \frac{Y_A^0 - Y_C}{L_{AC}^{0^2}} = -b_{21} \qquad b_{32} = \frac{\partial F_3}{\partial Y_A} = -\frac{X_A^0 - X_C}{L_{AC}^{0^2}} = -b_{22}$$

$$b_{41} = \frac{\partial F_4}{\partial X_A} = -\frac{Y_A^0 - Y_D}{L_{AD}^{0^2}} \qquad\quad b_{42} = \frac{\partial F_4}{\partial Y_A} = \frac{X_A^0 - X_D}{L_{AD}^{0^2}}$$

where $L_{AB}^{0^2} = (X_A^0 - X_B)^2 + (Y_A^0 - Y_B)^2$
$L_{AC}^{0^2} = (X_A^0 - X_C)^2 + (Y_A^0 - Y_C)^2$
$L_{AD}^{0^2} = (X_A^0 - X_D)^2 + (Y_A^0 - Y_D)^2$
$L_{BC}^2 = (X_B - X_C)^2 + (Y_B + Y_C)^2 = 445,395.9364 \text{ ft}^2$
$L_{CD}^2 = (X_C - X_D)^2 + (Y_C + Y_D)^2 = 697,535.7974 \text{ ft}^2$

Unless the computer used yields the value of the azimuth in the proper quadrant (i.e., between 0° and 360°) from the arctan function, the quadrant should be determined from the signs of the numerator and denominator as explained in Example 9.5. The elements of the vector f are

$$f_1 = -l_1 + \arctan\frac{X_C - X_B}{Y_C - Y_B} - \arctan\frac{X_A^0 - X_B}{Y_A^0 - Y_B}$$

$$f_2 = -l_2 + \arctan\frac{X_A^0 - X_C}{Y_A^0 - Y_C} - \arctan\frac{X_B - X_C}{Y_B - Y_C}$$

$$f_3 = -l_3 + \arctan\frac{X_D - X_C}{Y_D - Y_C} - \arctan\frac{X_A^0 - X_C}{Y_A^0 - Y_C}$$

$$f_4 = -l_4 + \arctan\frac{X_A^0 - X_D}{Y_A^0 - Y_D} - \arctan\frac{X_C - X_D}{Y_C - Y_D}$$

Using the approximations above, the $\mathbf{B}$ and f matrices are

$$\mathbf{B} = 10^{-4}\begin{bmatrix} -1.6101 & 9.2052 \\ 6.0403 & -6.1554 \\ -6.0403 & 6.1554 \\ 14.2857 & 0 \end{bmatrix}\left(\frac{1}{\text{ft}}\right) \quad \text{and} \quad f = 10^{-2}\begin{bmatrix} -2.110557 \\ 1.459653 \\ -1.597851 \\ 0.334695 \end{bmatrix} \text{ (rad)}$$

With the covariance matrix $\mathbf{\Sigma}$ computed earlier, the normal equations are formed as $(\mathbf{B'\Sigma^{-1}B})\mathbf{\Delta}_1 = \mathbf{B'\Sigma^{-1}}f$ and solved to yield

$$\mathbf{\Delta}_1 = \begin{bmatrix} 2.345 \text{ ft} \\ -22.524 \text{ ft} \end{bmatrix}$$

Adding these corrections to the original approximations gives the new set of approximate coordinates $10,000 + 2.345 = 10,002.345$ and $9300 - 22.524 = 9277.476$. With these new approximations, the normal equations are re-formed and solved for a new vector of corrections. After a total of three iterations, the correction vector becomes essentially 0 and the final least-squares estimates of the coordinates of point A are

$$\hat{X}_A = 10,002.445 \text{ ft} \quad \text{and} \quad \hat{Y}_A = 9277.390 \text{ ft}$$

9.18
LEAST-SQUARES ADJUSTMENT OF TRIANGULATION

As presented in Sections 9.7 through 9.12, triangulation is a means of establishing position by measuring mainly angles and the lengths of a few lines. Each measured line is called a *base line*, which has to be measured with care because its errors will propagate throughout the triangulation network. If the ends of a line are two known horizontal control points, that line can be considered a base line. Consequently, both the resection and intersection subsections of the previous section may be considered special cases of triangulation. In practice, however, the single triangle with measured internal angles is the main unit in triangulation (hence, the name). With this unit, chains of quadrilaterals and central point figures can be constructed. The least-squares adjustment of triangulation may be performed by either the technique of observations only (Section 2.28) for pure triangulation or indirect observations (Section 2.27), which is more appropriate for combined adjustment of triangulation and trilateration. Because pure triangulation rarely is used now, we first consider adjustment of triangulation only, by indirect observations, which is best presented in an example.

EXAMPLE 9.7. Because this is an adjustment by indirect observations, the first step is to perform position computations to determine estimated X and Y coordinates for stations B and C. The quadrilateral $ABCD$ shown in Figure 9.19 has eight interior angles determined from directions observed with a direction theodolite and known X and Y coordinates for stations A and D. The eight interior angles are as follows:

$$l_1 = 22°01'42.51" \qquad l_5 = 86°33'13.45"$$

$$l_2 = 16°44'31.20" \qquad l_6 = 58°46'35.93"$$

$$l_3 = 57°08'57.10" \qquad l_7 = 15°06'52.28"$$

$$l_4 = 19°33'14.13" \qquad l_8 = 84°04'50.66"$$

The given coordinates for stations A and D are

Station	Coordinates, m	
	X	Y
A	15,400.812	10,425.406
D	17,901.905	10,425.406

Therefore, the direction of AD is due east:

$$A_{AD} = 90°00'00.00" \qquad \text{and} \qquad d_{AD} = 2501.093 \text{ m}$$

To avoid using data containing blunders, the sum of angles in each triangle is checked:

$l_1 =$	22°01'42.51"	$l_1 =$	22°01'42.51"	$l_2 =$	16°44'31.20"
$l_6 =$	58°46'35.93"	$l_2 =$	16°44'31.20"	$l_3 =$	57°08'57.10"
$l_7 =$	15°06'52.28"	$l_3 =$	57°08'57.10"	$l_4 =$	19°33'14.13"
$l_8 =$	84°04'50.66"	$l_8 =$	84°04'50.66"	$l_5 =$	86°33'13.45"
$=$	180°00'01.38"	$=$	180°00'01.47"	$=$	179°59'55.88"

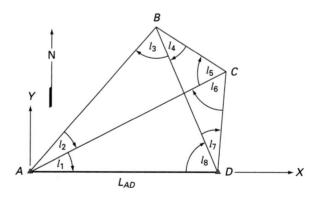

FIGURE 9.19

In triangle *ADC*,

Side	Station	Angles	Distance, m	Sine of angle
AD			2501.093	
	A	l_1 22°01'42.51"		0.37506734
	D	$l_7 + l_8$ 99°11'42.94"		0.98714949
	C	l_6 58°46'35.93"		0.85515305
$DC = (AD \sin l_1)/\sin l_6$			1096.971	
$AC = [AD \sin(l_7 + l_8)]/\sin l_6$			2887.147	

In triangle *ADB*,

Side	Station	Angles	Distance, m	Sine of angle
	D	l_8 84°04'50.66"		0.99466821
	A	$l_1 + l_2$ 38°46'13.71"		0.62620213
	B	l_3 57°08'57.10"		0.84008593
$AB = AD \sin l_8/\sin l_3$			2961.313	
$BD = AD \sin(l_1 + l_2)/\sin l_3$			1864.321	

In triangle *ABC*,

Side	Station	Angles	Distance, m	Sine of angle
	A	l_2 16°44'31.20"		0.28806256
	B	$l_3 + l_4$ 76°42'11.23"		0.97319140
	C	l_5 86°33'13.45"		0.99819161
$AB = AC \sin l_5/\sin(l_3 + l_4)$			2961.315	
$BC = AC \sin l_2/\sin(l_3 + l_4)$			854.589	

Next, azimuths for all lines are computed.
   Triangle *ADC*,

$$A_{AD} = \quad 90°00'00.00" \qquad A_{CA} = \quad 247°58'17.49"$$

$$\text{Angle } l_1 = \underline{-22°01'42.51"} \qquad \text{angle } l_6 = \underline{-58°46'35.93"}$$

$$A_{AC} = \quad 67°58'17.49" \qquad A_{CD} = \quad 189°11'41.56"$$

$$A_{DC} = \quad 9°11'41.56"$$

$$A_{AD} = \underline{\quad 90°00'00.00"}$$

$$l_7 + l_8 = \quad 99°11'42.94" \qquad\qquad 99°11'41.56"$$

   Triangle *ABD*,

$$A_{AD} = \quad 90°00'00.00" \qquad A_{BA} = \quad 231°13'46.29"$$

$$\text{Angle } l_1 + l_2 = \underline{-38°46'13.71"} \qquad \text{angle } l_3 = \underline{-57°08'57.10"}$$

$$A_{AB} = \quad 51°13'46.29" \qquad A_{BD} = \quad 174°04'49.19"$$

$$A_{DB} = \quad 354°04'49.19"$$

$$\text{angle } l_8 = \underline{\quad 84°04'50.66"}$$

$$A_{DA} = \quad 269°59'58.53"$$

Triangle ABC,

$$A_{BA} = 231°13'46.29'' \qquad A_{CB} = 334°31'35.06''$$

$$\text{Angle } l_3 + l_4 = \underline{-76°42'11.23''} \qquad \text{angle } l_5 = \underline{-86°33'13.45''}$$

$$A_{BC} = 154°31'35.06'' \qquad A_{CA} = 247°58'21.61''$$

$$A_{AC} = 67°58'21.61''$$

$$\text{angle } l_2 = \underline{-16°44'31.20''}$$

$$A_{AB} = 51°13'50.41''$$

Finally, coordinates for stations C and B are calculated using Equations (8.6) and (8.7) and following the procedure for computing a closed-loop traverse (Section 8.15) to give

			Coordinates, m		
Station	Distance, m	Azimuth	X	Y	
A			15,400.812	10,425.406	Fixed
	2887.147	67°58'17.49''			
C			18,077.190	11,508.281	Approximate
	1096.971	189°11'41.56''			
D			17,901.902	10.425.405	Fixed
			[17,901.905]	[10,425.406]	
A			15,400.812	10,425.406	
	2961.313	51°13'46.29''			
B			17,709.632	12,279.787	Approximate
	1864.321	174°04'49.19''			
D			17,901.906	10.425.408	
			[17,901.905]	[10,425.406]	

As a check, compute d and A_{BC} by Equations (8.8) and (8.9):

$$d_{CB} = [(18,077.190 - 17,709.632)^2 + (12,279.787 - 11,508.281)^2]^{1/2}$$

$$d_{CB} = 854.588 \text{ m}, \qquad \text{previously calculated as } 854.589$$

$$A_{CB} = \arctan \frac{18,077.190 - 17,709.632}{12,279.787 - 11,508.281} = \arctan 0.47641626$$

$$A_{CB} = 334°31'34.00'', \qquad \text{previously calculated as } 334°31'35.06''$$

The small discrepancies are due to the use of unadjusted observations. The calculated coordinates are more than accurate enough for use as approximations in the least-squares adjustment by indirect observations which follows.

To perform this adjustment, an angle condition equation must be written for each measured angle. Angle condition equation (8.63) is repeated here for convenience:

$$v_{ij} + l_{ij} + \arctan \frac{X_i - X}{Y_i - Y} - \arctan \frac{X_j - X}{Y_j - Y} = 0$$

For the example problem, the following eight angle condition equations of the form of Equation (8.63) are formed:

$$F_1 = l_1 - \arctan \frac{X_D - X_A}{Y_D - Y_A} + \arctan \frac{X_C - X_A}{Y_C - Y_A} = 0$$

$$F_2 = l_2 - \arctan \frac{X_C - X_A}{Y_C - Y_A} + \arctan \frac{X_B - X_A}{Y_B - Y_A} = 0$$

$$F_3 = l_3 - \arctan \frac{X_A - X_B}{Y_A - Y_B} + \arctan \frac{X_D - X_B}{Y_D - Y_B} = 0$$

$$F_4 = l_4 - \arctan \frac{X_D - X_B}{Y_D - Y_B} + \arctan \frac{X_C - X_B}{Y_C - Y_B} = 0$$

$$F_5 = l_5 - \arctan \frac{X_B - X_C}{Y_B - Y_C} + \arctan \frac{X_A - X_C}{Y_A - Y_C} = 0$$

$$F_6 = l_6 - \arctan \frac{X_A - X_C}{Y_A - Y_C} + \arctan \frac{X_D - X_C}{Y_D - Y_C} = 0$$

$$F_7 = l_7 - \arctan \frac{X_C - X_D}{Y_C - Y_D} + \arctan \frac{X_B - X_D}{Y_B - Y_D} = 0$$

$$F_8 = l_8 - \arctan \frac{X_B - X_D}{Y_B - Y_D} + \arctan \frac{X_A - X_D}{Y_A - Y_D} = 0$$

The linearized angle condition equations (see Equation (8.66)) are

$$v_1 + b_{11}\delta X_B + b_{12}\delta Y_B + b_{13}\delta X_C + b_{14}\delta Y_C = f_1$$

$$v_2 + b_{21}\delta X_B + b_{22}\delta Y_B + b_{23}\delta X_C + b_{24}\delta Y_C = f_2$$

$$v_3 + b_{31}\delta X_B + b_{32}\delta Y_B + b_{33}\delta X_C + b_{34}\delta Y_C = f_3$$

$$v_4 + b_{41}\delta X_B + b_{42}\delta Y_B + b_{43}\delta X_C + b_{44}\delta Y_C = f_4$$

$$v_5 + b_{51}\delta X_B + b_{52}\delta Y_B + b_{53}\delta X_C + b_{54}\delta Y_C = f_5$$

$$v_6 + b_{61}\delta X_B + b_{62}\delta Y_B + b_{63}\delta X_C + b_{64}\delta Y_C = f_6$$

$$v_7 + b_{71}\delta X_B + b_{72}\delta Y_B + b_{73}\delta X_C + b_{74}\delta Y_C = f_7$$

$$v_8 + b_{81}\delta X_B + b_{82}\delta Y_B + b_{83}\delta X_C + b_{84}\delta Y_C = f_8$$

which may be expressed in matrix form as

$$v + \mathbf{B}\Delta = f$$

in which the elements in $\mathbf{B}$ are determined by partial differentiation of the condition equations with respect to the unknown coordinates of stations B and C to yield

$$b_{11} = b_{12} = b_{33} = b_{34} = b_{61} = b_{62} = b_{83} = b_{84} = 0$$

$$b_{13} = \frac{\partial F_1}{\partial X_C} = \frac{Y_C^0 - Y_A}{(L_{AC}^0)^2}$$

$$b_{14} = \frac{\partial F_1}{\partial Y_C} = -\frac{X_C^0 - X_A}{(L_{AC}^0)^2}$$

$$b_{21} = \frac{\partial F_2}{\partial X_B} = \frac{Y_B^0 - Y_A}{(L_{AB}^0)^2}$$

$$b_{22} = \frac{\partial F_2}{\partial Y_B} = -\frac{X_B^0 - X_A}{(L_{AB}^0)^2}$$

$$b_{23} = \frac{\partial F_2}{\partial X_C} = -\frac{Y_C^0 - Y_A}{(L_{AC}^0)^2} = -b_{13}$$

$$b_{24} = \frac{\partial F_2}{\partial Y_C} = \frac{X_C^0 - X_A}{(L_{AC}^0)^2} = b_{14}$$

$$b_{31} = \frac{\partial F_3}{\partial X_B} = -\frac{Y_D - Y_B^0}{(L_{BD}^0)^2} + \frac{Y_A - Y_B^0}{(L_{BA}^0)^2}$$

$$b_{32} = \frac{\partial F_3}{\partial Y_B} = \frac{X_D - X_B^0}{(L_{BD}^0)^2} - \frac{X_A - X_B^0}{(L_{BA}^0)^2}$$

$$b_{41} = \frac{\partial F_4}{\partial X_B} = -\frac{Y_C^0 - Y_B^0}{(L_{BC}^0)^2} + \frac{Y_D - Y_B^0}{(L_{BD})^2}$$

$$b_{42} = \frac{\partial F_4}{\partial Y_B} = \frac{X_C^0 - X_B^0}{(L_{BC}^0)^2} - \frac{X_D - X_B^0}{(L_{BD}^0)^2}$$

$$b_{43} = \frac{\partial F_4}{\partial X_C} = \frac{Y_C^0 - Y_B^0}{(L_{BC}^0)^2}$$

$$b_{44} = \frac{\partial F_4}{\partial Y_C} = -\frac{X_C^0 - X_B^0}{(L_{BC}^0)^2}$$

$$b_{51} = \frac{\partial F_5}{\partial X_B} = \frac{Y_B^0 - Y_C^0}{(L_{BC}^0)^2} = -b_{43}$$

$$b_{52} = \frac{\partial F_5}{\partial Y_B} = -\frac{X_B^0 - X_C^0}{(L_{BC}^0)^2} = -b_{44}$$

$$b_{53} = \frac{\partial F_5}{\partial X_C} = -\frac{Y_A - Y_C^0}{(L_{AC}^0)^2} + \frac{Y_B^0 - Y_C^0}{(L_{CB})^2}$$

$$b_{54} = \frac{\partial F_5}{\partial Y_C} = \frac{X_A - X_C^0}{(L_{CA}^0)^2} - \frac{X_B^0 - X_C^0}{(L_{CB}^0)^2}$$

$$b_{63} = \frac{\partial F_6}{\partial X_C} = -\frac{Y_D - Y_C^0}{(L_{DC}^0)^2} + \frac{Y_A - Y_C^0}{(L_{AC}^0)^2}$$

$$b_{64} = \frac{\partial F_6}{\partial Y_C} = \frac{X_D - X_C^0}{(L_{CD}^0)^2} - \frac{X_A - X_C^0}{(L_{AC}^0)^2}$$

$$b_{71} = \frac{\partial F_7}{\partial X_B} = \frac{Y_B^0 - Y_D}{(L_{DB}^0)^2}$$

$$b_{72} = \frac{\partial F_7}{\partial Y_B} = -\frac{X_B^0 - X_D}{(L_{BD}^0)^2}$$

$$b_{73} = \frac{\partial F_7}{\partial X_C} = -\frac{Y_C^0 - Y_D}{(L_{DC}^0)^2}$$

$$b_{74} = \frac{\partial F_7}{\partial Y_C} = \frac{X_C^0 - X_D}{(L_{DC}^0)^2}$$

$$b_{81} = \frac{\partial F_8}{\partial X_B} = -\frac{Y_B^0 - Y_D}{(L_{DB}^0)^2} = -b_{71}$$

$$b_{82} = \frac{\partial F_8}{\partial Y_B} = \frac{X_B^0 - X_D}{(L_{DB}^0)^2} = -b_{72}$$

where $(L_{AC}^0)^2 = (X_C^0 - X_A)^2 + (Y_C^0 - Y_A)^2$
$(L_{AB}^0)^2 = (X_B^0 - X_A)^2 + (Y_B^0 - Y_A)^2$
$$\vdots$$
$(L_{DC}^0)^2 = (X_C^0 - Y_D)^2 + (Y_C^0 - Y_D)^2$

The elements of the matrix f formulated according to Equation (8.68) (Section 8.39) are as follows:

$$f_1 = -l_1 + \arctan \frac{X_D - X_A}{Y_D - Y_A} - \arctan \frac{X_C^0 - X_A}{Y_C^0 - Y_A}$$

$$f_2 = -l_2 + \arctan \frac{X_C^0 - X_A}{Y_C^0 - Y_A} - \arctan \frac{X_B^0 - X_A}{Y_B^0 - Y_A}$$

$$f_3 = -l_3 + \arctan \frac{X_A - X_B^0}{Y_A - Y_B^0} - \arctan \frac{X_D - X_B^0}{Y_D - Y_B^0}$$

$$f_4 = -l_4 + \arctan \frac{X_D - X_B^0}{Y_D - Y_B^0} - \arctan \frac{X_C^0 - X_B^0}{Y_C^0 - Y_B^0}$$

$$f_5 = -l_5 + \arctan \frac{X_B^0 - X_C^0}{Y_B^0 - Y_C^0} - \arctan \frac{X_A - X_C^0}{Y_A - Y_C^0}$$

$$f_6 = -l_6 + \arctan \frac{X_A - X_C^0}{Y_A - Y_C^0} - \arctan \frac{X_D - X_C^0}{Y_D - Y_C^0}$$

$$f_7 = -l_7 + \arctan \frac{X_C^0 - X_D}{Y_C^0 - Y_D} - \arctan \frac{X_B^0 - X_D}{Y_B^0 - Y_D}$$

$$f_8 = -l_8 + \arctan \frac{X_B^0 - X_D}{Y_B^0 - Y_D} - \arctan \frac{X_A - X_D}{Y_A - Y_D}$$

Using the approximations for coordinates and measured angles, the elements for the $\mathbf{B}$ and f matrices, respectively, are

$$\mathbf{B} = 10^{-4} \begin{bmatrix} 0 & 0 & 1.29909 & -3.21077 \\ 2.11461 & -2.63282 & -1.29909 & 3.21077 \\ 2.70246 & 3.18600 & 0 & 0 \\ 5.22866 & 4.47963 & -10.56394 & -5.03283 \\ -10.56394 & -5.03283 & 11.86303 & 1.82206 \\ 0 & 0 & 7.69978 & 1.75412 \\ 5.33528 & 0.55320 & -8.99889 & 1.45665 \\ -5.33528 & -0.55320 & 0 & 0 \end{bmatrix} \left(\frac{1}{\mathrm{m}} \right)$$

$$f = 10^{-7} \begin{bmatrix} 1.77 \\ -1.46 \\ -11.30 \\ 62.43 \\ 150.06 \\ 26.42 \\ -34.81 \\ -60.28 \end{bmatrix} (\mathrm{rad})$$

The normal equations (Equation (2.70)) are

$$(\mathbf{B'WB})\mathbf{\Delta} = \mathbf{B'W}f$$

in which $\mathbf{\Delta}^t = [\Delta X_B \quad \Delta Y_B \quad \Delta X_C \quad \Delta Y_C]^t$ $\mathbf{B}$ and f have been defined, and $\mathbf{W} = \mathbf{Q}^{-1}$ (according to Equation (2.33)), where $\mathbf{Q}$ is determined from the following analysis.

The two angles at each point are derived from three *observed* directions, which are assumed to be uncorrelated and of equal precision ($Q_{directions} = I$). Therefore, every two angles at a point will be correlated. For example, if d_{AB}, d_{AC}, and d_{AD} represent the directions at A, then

$$l_1 = d_{AD} - d_{AC} \quad \text{and} \quad l_2 = d_{AC} - d_{AB}$$

or

$$\begin{bmatrix} l_1 \\ l_2 \end{bmatrix} = \begin{bmatrix} 0 & -1 & 1 \\ -1 & 1 & 0 \end{bmatrix} \begin{bmatrix} d_{AB} \\ d_{AC} \\ d_{AD} \end{bmatrix}$$

The cofactor matrix for l_1 and l_2 can be obtained readily from error propagation (see Equation (2.34b)) using the identity matrix as the cofactor matrix of the directions:

$$Q_{angles} = \begin{bmatrix} 0 & -1 & 1 \\ -1 & 1 & 0 \end{bmatrix} \begin{bmatrix} 0 & -1 \\ -1 & 1 \\ 1 & 0 \end{bmatrix} = \begin{bmatrix} 2 & -1 \\ -1 & 2 \end{bmatrix}$$

The cofactor matrix for all eight angles is the following block diagonal 8×8 matrix:

$$Q = \begin{bmatrix} 2 & -1 & 0 & 0 & 0 & 0 & 0 & 0 \\ -1 & 2 & 0 & 0 & 0 & 0 & 0 & 0 \\ 0 & 0 & 2 & -1 & 0 & 0 & 0 & 0 \\ 0 & 0 & -1 & 2 & 0 & 0 & 0 & 0 \\ 0 & 0 & 0 & 0 & 2 & -1 & 0 & 0 \\ 0 & 0 & 0 & 0 & -1 & 2 & 0 & 0 \\ 0 & 0 & 0 & 0 & 0 & 0 & 2 & -1 \\ 0 & 0 & 0 & 0 & 0 & 0 & -1 & 2 \end{bmatrix}$$

from which $W = Q^{-1}$ and the normal equation coefficient matrix $N = B'WB$ is

$$N = 10^{-6} \begin{bmatrix} 1.28871 & 0.64644 & -1.73919 & -0.36232 \\ & 0.51371 & -0.96007 & -0.31980 \\ & & 3.23750 & 0.58952 \\ \text{symmetric} & & & 0.31569 \end{bmatrix} \left(\frac{1}{m^2} \right)$$

for which the inverse is

$$N^{-1} = 10^6 \begin{bmatrix} 3.73108 & -2.88114 & 1.36624 & -1.18780 \\ & 10.03030 & 0.27060 & 6.34898 \\ & & 1.19345 & -0.38646 \\ \text{symmetric} & & & 8.95786 \end{bmatrix} (m^2)$$

The column matrix of constant terms, $t = B'Wf$, is

$$t = 10^{-8} \begin{bmatrix} -0.87157 \\ -0.33023 \\ 1.80319 \\ 0.05992 \end{bmatrix} \text{(rad/m)}$$

Solution of the normal equations yields the corrections

$$\Delta = \begin{bmatrix} 0.001 \\ 0.001 \\ 0.009 \\ -0.012 \end{bmatrix} (m)$$

Generally, more than one iteration is required for convergence of the solution. In this example, the initial approximations for coordinates are very good, so that one iteration is considered adequate. The very small corrections support this assumption. Application of corrections to the approximate coordinates for stations C and D yields the following adjusted values:

Station	X, m	Y, m
B	17,709.633	12,279.788
C	18,077.199	11,508.269

The residuals are computed using Equation (2.79),

$$v = f - B\Delta = \begin{bmatrix} -1.0'' \\ 1.0'' \\ -0.3'' \\ 1.7'' \\ 1.7'' \\ -0.4'' \\ 1.1'' \\ -1.1'' \end{bmatrix}$$

and the adjusted angles are found by Equation (2.80),

$$\hat{l} = l + v = \begin{bmatrix} 22°01'41.5'' \\ 16°44'32.2'' \\ 57°08'56.8'' \\ 19°33'15.8'' \\ 86°33'15.2'' \\ 58°46'35.5'' \\ 15°06'53.4'' \\ 84°04'49.6'' \end{bmatrix}$$

The estimated reference variance determined using Equation (2.82) is

$$\hat{\sigma}_0^2 = \frac{v'Wv}{r} = \frac{4.84}{4} = 1.21 \text{ sec}^2 = (0.2854)10^{-10} \text{ rad}^2$$

Recall that angles were derived from the original directions, which were assumed noncorrelated and of equal precision, so that $Q = I$. Therefore, σ_0^2 is assumed unknown and $\hat{\sigma}_0^2$ must be used to evaluate the propagated covariance matrix for the adjusted coordinates. Using Equations (2.78) and (2.83), we obtain

$$\Sigma_{\Delta\Delta} = \hat{\sigma}_0^2 \mathbf{N}^{-1}$$

$$= \begin{bmatrix} (1.06485)10^{-4} & (-0.82228)10^{-4} & (0.38993)10^{-4} & -(0.33900)10^{-4} \\ & (0.28627)10^{-3} & (0.77228)10^{-5} & (1.81200)10^{-4} \\ \hline & & (0.34061)10^{-4} & -(1.10296)10^{-5} \\ \text{symmetric} & & & (0.25566)10^{-3} \end{bmatrix}$$

or

$$\Sigma_{\Delta\Delta} = \begin{bmatrix} \Sigma_{BB} & \Sigma_{BC} \\ \Sigma_{BC}^t & \Sigma_{CC} \end{bmatrix}$$

9.19
TRILATERATION ADJUSTMENT

In Section 9.13, trilateration was explained as the procedure for determining horizontal position through the measurement (most frequently by EDM equipment) of only distances between points. A trilatertion network would include several points with a large number of interconnecting lines to provide for redundancy. To fix the orientation of the network, either two horizontal control points or one control point and the azimuth of one line must be known. The least-squares adjustment of trilateration networks is best carried out by the method of indirect observations using the distance condition of Equation (8.69) for each measured line. This adjustment is demonstrated by the following example.

EXAMPLE 9.8. Figure 9.20 shows a quadrilateral $ABCD$ in which A and B are two horizontal control points with the following California Zone III state plane coordinates:

Point	X, ft	Y, ft
A	1,495,316.983	503,991.197
B	1,495,056.547	504,269.054

An EDM instrument was used to measure all five lines between A, B and C, D, and the distances were corrected for reflector and instrument constants and nonlinearity. The distances, reduced to the ellipsoid and plane coordinate grid, are

Line	Distance, ft	Standard deviation, ft
AD	$l_1 = 542.899$	$\sigma_1 = 0.02543 \simeq 0.026$
AC	$l_2 = 678.904$	$\sigma_2 = 0.02678 \simeq 0.026$
BD	$l_3 = 676.289$	$\sigma_3 = 0.02676 \simeq 0.026$
BC	$l_4 = 509.192$	$\sigma_4 = 0.02509 \simeq 0.026$
DC	$l_5 = 479.820$	$\sigma_5 = 0.02479 \simeq 0.026$

The uncertainty in EDM measurements can be assumed as 0.02 ft + 10 ppm. Using the least-squares method of indirect observations, compute the coordinates of points C and D.

Solution. Applying the uncertainty criterion for the measured distances, the standard deviation for each distance is computed. For example, $\sigma_1 = 0.02 + 10(542.899)10^{-6} = 0.02543$ ft. The values of all five standard deviations are given in the third column of the table. It can be seen that all five values are very nearly the same, because the five distances do not vary substantially. Consequently, it is quite reasonable

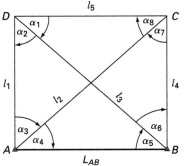

FIGURE 9.20

to consider that all observations are uncorrelated and of equal precision with a standard deviation of 0.026 ft. Hence, the variance matrix is $(0.026)^2\mathbf{I}$ or $(6.76)10^{-4}\mathbf{I}$. So, σ_0^2 is assumed equal to unity.

The five distance condition equations are

$$F_1 = l_1 - [(X_A - X_D)^2 + (Y_A - Y_D)^2]^{1/2} = 0$$

$$F_2 = l_2 - [(X_A - X_C)^2 + (Y_A - Y_C)^2]^{1/2} = 0$$

$$F_3 = l_3 - [(X_B - X_D)^2 + (Y_B - Y_D)^2]^{1/2} = 0$$

$$F_4 = l_4 - [(X_B - X_C)^2 + (Y_B - Y_C)^2]^{1/2} = 0$$

$$F_5 = l_5 - [(X_D - X_C)^2 + (Y_D - Y_C)^2]^{1/2} = 0$$

The linearized distance equations are

$$v_1 + b_{11}\Delta X_C + b_{12}\Delta Y_C + b_{13}\Delta X_D + b_{14}\Delta Y_D = f_1$$

$$v_2 + b_{21}\Delta X_C + b_{22}\Delta Y_C + b_{23}\Delta X_D + b_{24}\Delta Y_D = f_2$$

$$v_3 + b_{31}\Delta X_C + b_{32}\Delta Y_C + b_{33}\Delta X_D + b_{34}\Delta Y_D = f_3$$

$$v_4 + b_{41}\Delta X_C + b_{42}\Delta Y_C + b_{43}\Delta X_D + b_{44}\Delta Y_D = f_4$$

$$v_5 + b_{51}\Delta X_C + b_{52}\Delta Y_C + b_{53}\Delta X_D + b_{54}\Delta Y_D = f_5$$

in which the coefficients in matrix $\mathbf{B}$ and elements in the f matrix are defined in Equations (8.72a) to (8.72e). For this example, the coefficients in $\mathbf{B}$ and elements of the f matrix are

$$b_{11} = \frac{\partial F_1}{\partial X_C} = 0 \qquad\qquad b_{12} = \frac{\partial F_1}{\partial Y_C} = 0$$

$$b_{13} = \frac{\partial F_1}{\partial X_D} = \frac{X_A - X_D^0}{L_{AD}^0} \qquad b_{14} = \frac{\partial F_1}{\partial Y_D} = \frac{Y_A - Y_D^0}{L_{AD}^0}$$

$$b_{21} = \frac{\partial F_2}{\partial X_C} = \frac{X_A - X_C^0}{L_{AC}^0} \qquad b_{22} = \frac{\partial F_2}{\partial Y_C} = \frac{Y_A - Y_C^0}{L_{AC}^0}$$

$$b_{23} = \frac{\partial F_2}{\partial X_D} = 0 \qquad\qquad b_{24} = \frac{\partial F_2}{\partial Y_D} = 0$$

$$b_{31} = \frac{\partial F_3}{\partial X_C} = 0 \qquad\qquad b_{32} = \frac{\partial F_3}{\partial Y_C} = 0$$

$$b_{33} = \frac{\partial F_3}{\partial X_D} = \frac{X_B - X_D^0}{L_{BD}^0} \qquad\qquad b_{34} = \frac{\partial F_3}{\partial Y_D} = \frac{Y_B - Y_D^0}{L_{BD}^0}$$

$$b_{41} = \frac{\partial F_4}{\partial X_C} = \frac{X_B - X_C^0}{L_{BC}^0} \qquad\qquad b_{42} = \frac{\partial F_4}{\partial Y_C} = \frac{Y_B - Y_C^0}{L_{BC}^0}$$

$$b_{43} = \frac{\partial F_4}{\partial X_D} = 0 \qquad\qquad b_{44} = \frac{\partial F_4}{\partial Y_D} = 0$$

$$b_{51} = \frac{\partial F_5}{\partial X_C} = \frac{X_D^0 - X_C^0}{L_{DC}^0} \qquad\qquad b_{52} = \frac{\partial F_5}{\partial Y_C} = \frac{Y_D^0 - Y_C^0}{L_{DC}^0}$$

$$b_{53} = \frac{\partial F_5}{\partial X_D} = -b_{51} \qquad\qquad b_{54} = \frac{\partial F_5}{\partial Y_D} = -b_{52}$$

$$f_1 = -l_1 + L_{AD}^0 \qquad\qquad f_2 = -l_2 + L_{AC}^0$$

$$f_3 = -l_3 + L_{BD}^0 \qquad\qquad f_4 = -l_4 + L_{BC}^0$$

$$f_5 = -l_5 + L_{DC}^0$$

where
$$L_{AD}^0 = [(X_A - X_D^0)^2 + (Y_A - Y_D^0)^2]^{1/2}$$
$$L_{AC}^0 = [(X_A - X_C^0)^2 + (Y_A - Y_C^0)^2]^{1/2}$$
$$L_{BD}^0 = [(X_B^0 - X_D^0)^2 + (Y_B - Y_D^0)^2]^{1/2}$$
$$L_{BC}^0 = [(X_B - X_C^0)^2 + (Y_B - Y_C^0)^2]^{1/2}$$
$$L_{DC}^0 = [(X_D^0 - X_C^0)^2 + (Y_D^0 - Y_C^0)^2]^{1/2}$$

When computations are performed by a computer, coarse values for the coordinates of unknown stations scaled from an accurate drawing of the quadrilateral are adequate to use for estimates. If a handheld calculator is to be used to form coefficients, fairly good esimates are desirable. In this example, the procedure described in Section 9.14 is followed and the two plane triangles ABD and ABC are solved by the law of cosines to obtain angles used to determine preliminary azimuths of the lines. These azimuths then are employed with the measured distances to calculate approximate coordinates for stations C and D. In triangle ABD,

$$\alpha_5^0 = \cos^{-1}\frac{(380.830)^2 + (676.289)^2 - (542.899)^2}{(2)(380.830)(676.289)} = 53°19'29''$$

$$\alpha_3^0 + \alpha_4^0 = \cos^{-1}\frac{(380.830)^2 + (542.899)^2 - (676.289)^2}{(2)(380.830)(542.899)} = 92°26'20''$$

$$\alpha_2^0 = \cos^{-1}\frac{(542.899)^2 + (676.289)^2 - (380.830)^2}{(2)(542.899)(676.289)} = 34°14'11''$$

$$\overline{ 180°00'00''}$$

and in triangle ABC,

$$\alpha_4^0 = \cos^{-1}\frac{(380.830)^2 + (678.904)^2 - (509.192)^2}{(2)(380.830)(678.904)} = 47°54'04''$$

$$\alpha_5^0 + \alpha_6^0 = \cos^{-1}\frac{(380.830)^2 + (509.192)^2 - (678.904)^2}{(2)(380.830)(509.192)} = 98°23'32''$$

$$\alpha_7^0 = \cos^{-1}\frac{(509.192)^2 + (678.904)^2 - (380.830)^2}{(2)(509.192)(678.904)} = \underline{33°42'24''}$$
$$180°00'00''$$

Preliminary azimuths from the north for all lines in the quadrilateral are calculated using the fixed azimuth of line AB determined from the coordinates of these stations:

$$A_{AB} = \arctan\frac{X_B - X_A}{Y_B - Y_A} = \frac{-260.436}{277.857} = \arctan(-0.937302281)$$

$$= 316°51'13.1''$$

Computations for azimuths are as follows:

Triangle ABD		**Triangle ABC**	
A_{AB}	316°51'13.1"	A_{BA}	136°51'13.1"
$-(\alpha_3^0 + \alpha_4^0)$	92°26'20.0"	$+(\alpha_5^0 + \alpha_6^0)$	98°23'31.6"
A_{AD}^0	224°24'53.1"	A_{BC}^0	235°14'44.7"
$-\alpha_2^0$	34°14'11.0"	A_{CB}^0	55°14'44.7"
A_{BD}^0	190°10'42.1"	$+\alpha_7^0$	33°42'24.2"
$-\alpha_5^0$	53°19'29.0"	A_{CA}^0	88°57'08.9"
A_{BA}	136°51'13.1", checks	A_{AC}^0	268°57'08.7"
		$+\alpha_4^0$	47°54'04.3"
		Check	316°51'13.2"

The preliminary azimuths just calculated and measured lengths of lines are next used to calculate approximate coordinates for the adjustment. The results of these computations are given in Table 9.2.

TABLE 9.2
Computation of approximate coordinates

Station	Distance, ft	Azimuth	Coordinates, ft X	Y
Triangle *ADB*				
A			1,495,316.983	503,991.197
	542.899	224°24'53.1"		
D			1,494,937.037	503,603.408
	676.289	10°10'42.1"		
B			1,495,056.546	504,269.054
Triangle *BCA*				
B			1,495,056.547	504,269.054
	509.192	235°14'44.7"		
C			1,494,638.192	503,978.785
	678.904	88°57'08.9"		
A			1,495,316.983	503,991.197

The length and azimuth of line DC are

$$L_{DC}^0 = [(375.377)^2 + (298.845)^2]^{1/2} = 479.808 \text{ ft}$$

$$A_{DC}^0 = \arctan \frac{298.845}{-375.377} = 321°28'34''$$

When computations are programmed for a computer, the approximate coordinates can be used directly to form coefficients for the condition equations. If a handheld calculator is used, it is more convenient to calculate the coefficients with the sine and cosine of the approximate azimuths. Preliminary approximate values for distances and azimuths are listed in Table 9.3.

Using these estimates, measured angles, and distances, the numerical values for the elements of the **B** and f matrices are

$$\mathbf{B} = \begin{bmatrix} 0 & 0 & 0.699847248 & 0.714292538 \\ 0.999832875 & 0.018282758 & 0 & 0 \\ 0 & 0 & 0.176714458 & 0.984262161 \\ 0.821604656 & 0.570057707 & 0 & 0 \\ 0.622840883 & -0.782348537 & -0.622840883 & 0.782348537 \end{bmatrix}$$

$$f^t = \begin{bmatrix} 0 & -0.001 & 0 & 0 & -0.012 \end{bmatrix} \text{ ft}$$

As noted previously, the covariance matrix for measured distances is

$$\Sigma_{tt} = (6.76)10^{-4}\mathbf{I} \quad (\text{ft}^2)$$

and the weight matrix is

$$\mathbf{w} = \Sigma_{tt}^{-1} \quad \left(\frac{1}{\text{ft}^2}\right)$$

The normal equations are

$$(\mathbf{B}'\mathbf{W}\mathbf{B})\Delta = \mathbf{B}'\mathbf{W}f$$

or

$$\mathbf{N}\Delta = t$$

TABLE 9.3
Preliminary values for lines

Line	Calculated lengths, L^0	Approximate azimuth, A^0
AD	542.899	224°24'53.1''
AC	678.903	268°57'08.7''
BD	676.289	190°10'42.4''
BC	509.192	235°14'44.7''
DC	479.808	321°28'34.0''

from which the inverse of the coefficient matrix and vector of constant terms are

$$\mathbf{N}^{-1} = 10^{-4} \begin{bmatrix} 4.96982 & -4.67851 & 5.24397 & -2.99539 \\ & 20.15700 & -14.50300 & 8.28421 \\ & & 18.91870 & -7.15497 \\ \text{symmetric} & & & 6.99401 \end{bmatrix} (\text{ft}^2)$$

$$\mathbf{t}' = [-12.5354 \quad 13.8608 \quad 11.5603 \quad -13.8878]\left(\frac{1}{\text{ft}}\right)$$

Solution of the normal equations yields corrections

$$\mathbf{\Delta}' = [\delta X_C \quad \delta Y_C \quad \delta X_D \quad \delta Y_D]$$

as follows:

$$\mathbf{\Delta}' = [-0.0027 \quad 0.0063 \quad 0.0042 \quad -0.0024](\text{ft})$$

Application of these corrections to approximate values gives the following adjusted coordinates for stations C and D:

	X, ft	Y, ft
C	1,494,638.190	503,978.791
D	1,494,937.041	503,603.406

Since the reference variance σ_0^2 initially was assumed equal to unity, the propagated covariance matrix $\mathbf{\Sigma}_{\Delta\Delta}$ for the adjusted coordinates equals $\mathbf{N}^{-1}$ according to Equations (2.78) and (2.83) and as described for the traverse adjustment in Example 8.15.

9.20
ADJUSTMENT OF COMBINED TRIANGULATION AND TRILATERATION

The need for this adjustment (described in Section 9.15) arises when both angles and distances are measured in a horizontal control survey. Equations (8.63) for angles and (8.69) for distances are formed to yield one set of normal equations that are solved in a least-squares adjustment by indirect observations. The procedures required for this adjustment are illustrated by the following example.

EXAMPLE 9.9. Quadrilateral $ABCD$ (Figure 9.20), for which the five measured distances with respective estimated standard deviations are given in Example 9.8, also has the eight angles $\alpha_1, \alpha_2, \ldots, \alpha_8$ derived from observed directions. These directions are of equal precision and uncorrelated with an estimated standard deviation of $10''$. The angles determined from the directions are listed in Table 9.4.

As in the previous adjustments, approximate values are necessary for the coordinates of stations C and D and the distances. Approximate angles, azimuths, distances, and coordinates previously calculated in Example 9.8 are used for this adjustment. Angles calculated in Example 9.8 also are tabulated in Table 9.4. The differences between the calculated and measured angles constitute the elements of the f matrix of the angle condition equations and are listed in column 3 of Table 9.4.

TABLE 9.4
Measured and approximate angles

	Measured angles α_i	Calculated angles (from Example 9.8) α_i^0	$\alpha_i^0 - \alpha_i$ Seconds	$\alpha_i^0 - \alpha_i$ Radians
1	48°42'06.0"	48°42'08.4"	2.4	0.000011636
2	34°14'03.6"	34°14'10.5"	6.9	0.000033452
3	44°32'06.0"	44°32'15.6"	9.6	0.000046542
4	47°54'00.6"	47°54'04.4"	3.8	0.000018423
5	53°19'16.4"	53°19'29.5"	13.1	0.000063511
6	45°04'00.8"	45°04'02.1"	1.3	0.000006303
7	33°42'29.4"	33°42'24.0"	−5.4	−0.000026180
8	52°31'13.1"	52°31'25.5"	12.4	0.000060117

Coefficients for the eight angle condition equations (8.63) are calculated as described in Example 9.7 using approximate azimuths and distances from Table 9.3 (Section 9.19) or given and approximate coordinates for the triangulation stations as provided in Example 9.8, Table 9.2.

Coefficients for the five distance equations (8.69) and elements of the f matrix as determined in Example 9.8 are used for this adjustment. Therefore, 8 angle and 5 distance equations, in that order, or a total of 13 condition equations, are formed for this adjustment by indirect observations ($v + B\Delta = f$) to form the following B and f matrices:

$$
B = 10^{-3}
\begin{bmatrix}
1.630545 & 1.298104 & -0.175158 & -1.559405 \\
0 & 0 & -0.139685 & -1.027791 \\
0.026931 & -1.472718 & -1.315702 & 1.289092 \\
-0.026931 & 1.472718 & 0 & 0 \\
0 & 0 & 1.455387 & -0.261300 \\
1.119535 & -1.613545 & -1.455387 & 0.261300 \\
-1.092604 & 0.140827 & 0 & 0 \\
-1.65748 & 0.174614 & 1.603545 & 1.298104 \\
0 & 0 & 699.8466 & 714.2932 \\
999.8329 & 18.2827 & 0 & 0 \\
0 & 0 & 176.7146 & 984.2622 \\
821.6046 & 570.0577 & 0 & 0 \\
622.8409 & -782.3485 & -622.8409 & 782.3485
\end{bmatrix}
$$

$$f^t = 10^5[1.1636 \quad 3.3452 \quad 4.6542 \quad 1.8423 \quad 6.3511 \quad 0.6303 \quad -2.6180$$
$$6.0117 \quad 0 \quad 100 \quad 0 \quad 0 \quad -1200]$$

Assuming a variance in measured directions of $100 \sec^2$ or $2.3504(10)^{-9}$ rad^2 the propagated covariance matrix for angles from a single station (see Example 9.7 and propagate with Equation (2.34a)) is

$$\Sigma_{\text{angles}} = \begin{bmatrix} 0 & -1 & 1 \\ -1 & 1 & 0 \end{bmatrix} (10)^{-9} \begin{bmatrix} 2.35044 & 0 & 0 \\ 0 & 2.35044 & 0 \\ 0 & 0 & 2.35044 \end{bmatrix} \begin{bmatrix} 0 & -1 \\ -1 & 1 \\ 1 & 0 \end{bmatrix}$$

$$= (2.35044)10^{-9} \begin{bmatrix} 2 & -1 \\ -1 & 2 \end{bmatrix}$$

Measured distances are of equal precision and noncorrelated with respective variances of 0.000676 ft^2/line. Assuming a reference variance, α_0^2, of unity, the covariance matrix for the eight measured angles and five distances is

$$\Sigma_{ll} = (2.3504)10^{-9} \times$$

$$\begin{bmatrix}
2 & -1 & 0 & & & & & & & & & & & 0 \\
-1 & 2 & 0 & 0 & & & & & & & & & & \\
0 & 0 & 2 & -1 & 0 & & & & & & & & & \\
& 0 & -1 & 2 & 0 & 0 & & & & & & & & \\
& & 0 & 2 & -1 & 0 & & & & & & & \\
& & 0 & -1 & 2 & 0 & 0 & & & & & \\
& & & & 0 & 2 & -1 & & & & & \\
& & & & 0 & -1 & 2 & & 0 & & & & \\
& & & & & 0 & & (2.876)10^5 & 0 & & & \\
& & & & & & 0 & & (2.876)10^5 & 0 & & \\
& & & & & & & 0 & & (2.876)10^5 & 0 & \\
& & & & & & & & 0 & & (2.876)10^5 & 0 \\
0 & & & & & & & & & 0 & & (2.876)10^5
\end{bmatrix}$$

and the weight matrix becomes $W = \Sigma_{ll}^{-1}$. The normal equations are

$$N\Delta = t$$

in which $N = B^t WB$, $\Delta^t = [\delta X_C \quad \delta Y_C \quad \delta X_D \quad \delta Y_D]$, and $t = B^t Wf$. The inverse of the normal equation coefficient matrix and the vector t are

$$N^{-1} = 10^{-4} \begin{bmatrix} 2.25526 & -0.51482 & 1.56597 & 0.10381 \\ & 5.50016 & -2.93210 & 2.05263 \\ & & 5.49383 & -1.18173 \\ \text{symmetric} & & & 2.64349 \end{bmatrix} (\text{ft}^2)$$

$$t^t = [-10.593 \quad 3.4984 \quad 20.4850 \quad 2.3002] \left(\frac{1}{\text{ft}}\right)$$

Solution of the normal equations gives corrections

$$\Delta^t = [0.001 \quad -0.004 \quad 0.008 \quad -0.002] \text{ (ft)}$$

These corrections, applied to the approximate coordinates for stations C and D, yield adjusted coordinates as follows:

$$\hat{X}_C = X_C^0 + \delta X_C = 1,494,638.192 + 0.001 = 1,494,638.193$$
$$\hat{Y}_C = Y_C^0 + \delta Y_C = 503,978.785 \quad - 0.004 = \quad 503,978.781$$
$$\hat{X}_D = X_D^0 + \delta X_D = 1,494,937.037 + 0.008 = 1,494,937.045$$
$$\hat{Y}_D = Y_D^0 + \delta Y_D = 503,603.408 \quad - 0.002 = \quad 503,603.406$$

9.21
ADJUSTMENT OF A THREE-DIMENSIONAL SURVEY BY THE METHOD OF LEAST SQUARES

In a three-dimensional survey each target or survey station is represented by all three coordinates, which locates it in the three-dimensional reference coordinate system. The observed quantities include all four types of measurements: azimuth, horizontal angles, vertical angles, and spatial distances. For each of these a corresponding condition equation is written in terms of some or all of the three-dimensional coordinates of the survey stations involved. The azimuth condition is given by Equation (8.61). The horizontal angle condition is given by Equation (8.63). The vertical angle condition is given by

$$v_1 + l_i - \arctan(Z_i - Z_p)/[(X_i - X_p)^2 + (Y_i - Y_p)^2]^{1/2} = 0 \qquad (9.20)$$

in which l_i is the vertical angle at station i sighting to station P.
The spatial distance condition equation is an extension of Equation (8.69), given by

$$v_{ij} + l_{ij} - [(X_i - X_j)^2 + (Y_i - Y_j)^2 + (Z_i - Z_j)^2]^{1/2} \qquad (9.21)$$

Least-squares adjustment of a three-dimensional survey is accomplished by writing all the necessary condition equations and carrying the three coordinates of all the unknown survey stations as parameters. The condition equations are the linearized, normal equations formed, and the solution is iterated until the final parameter corrections are insignificantly small. The best way to illustrate this procedure is through a numerical example.

EXAMPLE 9.10. ADJUSTMENT OF A CLOSED THREE-DIMENSIONAL TRAVERSE. Figure 9.21 shows a single triangle traverse in which station G is fixed as the origin of a three-dimensional local space rectangular coordinate system. It is required to use least-squares adjustment to calculate the X, Y, and Z coordinates of stations C and P. All possible measurements are available: the azimuth α_{GC}, three horizontal angles, six vertical angles, and six spatial distances. Their values and their standard deviations are as follows (all observations are uncorrelated).
    The azimuth is

$$\alpha_{GC} = 14.9139351628°$$

$$= 14°54'50.166586''$$

$$\sigma = 2'' = 0.00055556° \text{ for all angles}$$

The distances are

$$\sigma = 2 \text{ cm} = 0.020 \text{ m for all distances}$$

$$d_{GP} = 1568.513 \text{ m}, \qquad d_{PG} = 1558.491 \text{ m}$$

$$d_{PC} = 2548.978 \text{ m}, \qquad d_{CP} = 2548.993 \text{ m}$$

$$d_{CG} = 2194.792 \text{ m}, \qquad d_{GC} = 2194.758 \text{ m}$$

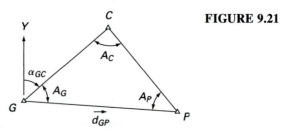

FIGURE 9.21

The horizontal angles are

$$\sigma = 2'' = 0.00055556° \text{ for all angles}$$

$$A_G = 83°29'54.2466''$$

$$A_P = 58°48'48.1475''$$

$$A_C = 37°41'19.9746''$$

The vertical angles are

$$\sigma = 2'' = 0.00055556° \text{ for all angles}$$

$$V_{GP} = 0°20'14.2380'', \qquad V_{PG} = -0°20'12.2811''$$

$$V_{PC} = 0°05'11.6044'', \qquad V_{CP} = -0°05'04.6180''$$

$$V_{CG} = -0°20'23.0624'', \qquad V_{GC} = 0°20'23.4034''$$

Solution. The total number of observations is $n = 1 + 3 + 6 + 6 = 16$. The minimum required to fix three stations in a three-dimensional space is $3 \times 3 = 9$, but because station G is fixed a priori, $n_0 = 6$. Therefore, the redundancy is $r = n - n_0 = 10$. Carrying the six unknown coordinates as parameters, the number of condition equations becomes $c = r + u = 10 + 6 = 16 = n$. Therefore, it is appropriate to use the method of least-squares adjustment of indirect observations. Denoting azimuth by α, horizontal angles by A, vertical angles by V, and spatial distances by d, the 16 condition equations are

$$F_1 = \alpha_{GC} - \arctan \frac{X_C - X_G}{Y_C - Y_G} = 0$$

Because $X_G = Y_G = 0$, these values will be dropped in all remaining relations.

$$F_2 = A_G - \arctan\left(\frac{X_p}{Y_p}\right) + \arctan\left(\frac{X_c}{Y_c}\right) = 0$$

$$F_3 = A_C + \arctan\left(\frac{X_p - X_c}{Y_p - Y_c}\right) - \arctan\left(\frac{X_c}{Y_c}\right) = 0$$

$$F_4 = A_P - \arctan\left(\frac{X_p - X_c}{Y_p - Y_c}\right) + \arctan\left(\frac{X_p}{Y_p}\right) = 0$$

$$F_5 = d_{GP} - \sqrt{X_p^2 + Y_p^2 + Z_p^2} = 0$$

$$F_6 = d_{PG} - \sqrt{X_p^2 + Y_p^2 + Z_p^2} = 0$$

$$F_7 = d_{CP} - \sqrt{(X_p - X_c)^2 + (Y_p - Y_c)^2 + (Z_p - Z_c)^2} = 0$$

$$F_8 = d_{pc} - \sqrt{(X_c - X_p)^2 + (Y_c - Y_p)^2 + (Z_c - Z_p)^2} = 0$$

$$F_9 = d_{GC} - \sqrt{X_c^2 + Y_c^2 + Z_c^2} = 0$$

$$F_{10} = d_{CG} - \sqrt{X_c^2 + Y_c^2 + Z_c^2} = 0$$

$$F_{11} = V_{GP} - \arctan\left(\frac{Z_p}{\sqrt{X_p^2 + Y_p^2}}\right) = 0$$

$$F_{12} = V_{PG} - \arctan\left(\frac{-Z_p}{\sqrt{X_p^2 + Y_p^2}}\right) = 0$$

$$F_{13} = V_{PC} - \arctan\left(\frac{Z_c - Z_p}{\sqrt{(X_c - X_p)^2 + (Y_c - Y_p)^2}}\right) = 0$$

$$F_{14} = V_{CP} - \arctan\left(\frac{Z_p - Z_c}{\sqrt{(X_p - X_c)^2 + (Y_p - Y_c)^2}}\right) = 0$$

$$F_{15} = V_{CG} - \arctan\left(\frac{-Z_c}{\sqrt{X_c^2 + Y_c^2}}\right) = 0$$

$$F_{16} = V_{GC} - \arctan\left(\frac{Z_c}{\sqrt{X_c^2 + Y_c^2}}\right) = 0$$

The linearized form is: $\underset{16,1}{v} + \underset{16,6}{\mathbf{B}} \; \underset{6,1}{\Delta} = \underset{16,1}{f}$ in which:

$$\underset{16,6}{\mathbf{B}} = \begin{bmatrix} \dfrac{\partial F_1}{\partial X_c} & \dfrac{\partial F_1}{\partial Y_c} & \dfrac{\partial F_1}{\partial Z_c} & \dfrac{\partial F_1}{\partial X_p} & \dfrac{\partial F_1}{\partial Y_p} & \dfrac{\partial F_1}{\partial Z_p} \\ \dfrac{\partial F_2}{\partial X_c} & \cdots & \cdots & \cdots & \cdots & \dfrac{\partial F_2}{\partial Z_p} \\ \vdots & \cdots & \cdots & \cdots & \cdots & \vdots \\ \vdots & & & & & \vdots \\ \vdots & \cdots & \cdots & \cdots & \cdots & \vdots \\ \dfrac{\partial F_{16}}{\partial X_c} & \cdots & \cdots & \cdots & \cdots & \dfrac{\partial F_{16}}{\partial Z_p} \end{bmatrix}, \quad \begin{bmatrix} -F_1 \\ -F_2 \\ \vdots \\ \vdots \\ \vdots \\ -F_{16} \end{bmatrix}, \quad f = \begin{bmatrix} f_1 \\ f_2 \\ \vdots \\ \vdots \\ \vdots \\ f_{16} \end{bmatrix}, \quad \Delta = \begin{bmatrix} \delta X_c \\ \delta Y_c \\ \delta Z_c \\ \delta X_p \\ \delta Y_p \\ \delta Z_p \end{bmatrix}$$

The elements of the **B** matrix are

$$b_{11} = \frac{-Y_c}{(X_c^2 + Y_c^2)}$$

$$b_{12} = \frac{X_c}{(X_c^2 + Y_c^2)}$$

$$b_{13} = b_{14} = b_{15} = b_{16} = 0$$

$$b_{21} = -b_{11}$$

$$b_{22} = -b_{12}$$

$$b_{23} = 0$$

$$b_{24} = \frac{-Y_p}{(X_p^2 + Y_p^2)}$$

$$b_{25} = \frac{X_p}{(X_p^2 + Y_p^2)}$$

$$b_{26} = 0$$

$$b_{31} = \frac{-Y_c}{(X_c^2 + Y_c^2)} - \frac{(Y_p - Y_c)}{(X_p - X_c)^2 + (Y_p - Y_c)^2}$$

$$b_{32} = \frac{X_c}{(X_c^2 + Y_c^2)} - \frac{(X_p - X_c)}{(X_p - X_c)^2 + (Y_p - Y_c)^2}$$

$$b_{33} = b_{36} = 0$$

$$b_{34} = \frac{(Y_p - Y_c)}{(X_p - X_c)^2 + (Y_p - Y_c)^2}$$

$$b_{35} = \frac{-(X_p - X_c)}{(X_p - X_c)^2 + (Y_p - Y_c)^2}$$

$$b_{41} = b_{34}$$

$$b_{42} = b_{35}$$

$$b_{43} = b_{46} = 0$$

$$b_{44} = \frac{Y_c - Y_p}{(X_c - X_p)^2 + (Y_c - Y_p)^2} + \frac{Y_p}{(X_p^2 + Y_p^2)}$$

$$b_{45} = \frac{-(X_c - X_p)}{(X_c - X_p)^2 + (Y_c - Y_p)^2} - \frac{X_p}{(X_p^2 + Y_p^2)}$$

$$b_{51} = b_{52} = b_{53} = 0$$

$$b_{54} = \frac{-X_p}{\sqrt{X_p^2 + Y_p^2 + Z_p^2}}$$

$$b_{55} = \frac{-Y_p}{\sqrt{X_p^2 + Y_p^2 + Z_p^2}}$$

$$b_{56} = \frac{-Z_p}{\sqrt{X_p^2 + Y_p^2 + Z_p^2}}$$

$$b_{61} = b_{62} = b_{63} = 0$$

$$b_{64} = b_{54}$$

$$b_{65} = b_{55}$$

$$b_{66} = b_{56}$$

$$b_{71} = \frac{(X_p - X_c)}{\sqrt{(X_p - X_c)^2 + (Y_p - Y_c)^2 + (Z_p - Z_c)^2}}$$

$$b_{72} = \frac{(Y_p - Y_c)}{\sqrt{(X_p - X_c)^2 + (Y_p - Y_c)^2 + (Z_p - Z_c)^2}}$$

$$b_{73} = \frac{(Z_p - Z_c)}{\sqrt{(X_p - X_c)^2 + (Y_p - Y_c)^2 + (Z_p - Z_c)^2}}$$

$$b_{74} = -b_{71}$$

$$b_{75} = -b_{72}$$

$$b_{76} = -b_{73}$$

$$b_{81} = b_{71}$$

$$b_{82} = b_{72}$$

$$b_{83} = b_{73}$$

$$b_{84} = -b_{71}$$

$$b_{85} = -b_{72}$$

$$b_{86} = -b_{73}$$

$$b_{91} = \frac{-X_c}{\sqrt{X_c^2 + Y_c^2 + Z_c^2}}$$

$$b_{92} = \frac{-Y_c}{\sqrt{X_c^2 + Y_c^2 + Z_c^2}}$$

$$b_{93} = \frac{-Z_c}{\sqrt{X_c^2 + Y_c^2 + Z_c^2}}$$

$$b_{94} = b_{95} = b_{96} = 0$$

$$b_{10,1} = b_{91}$$

$$b_{10,2} = b_{92}$$

$$b_{10,3} = b_{93}$$

$$b_{10,4} = b_{10,5} = b_{10,6} = 0$$

$$b_{11,1} = b_{11,2} = b_{11,3} = 0$$

$$b_{11,4} = \frac{X_p Z_p}{(X_p^2 + Y_p^2 + Z_p^2)\sqrt{X_p^2 + Y_p^2}}$$

$$b_{11,5} = \frac{Y_p Z_p}{(X_p^2 + Y_p^2 + Z_p^2)\sqrt{X_p^2 + Y_p^2}}$$

$$b_{11,6} = \frac{-\sqrt{X_p^2 + Y_p^2}}{(X_p^2 + Y_p^2 + Z_p^2)}$$

$$b_{12,1} = b_{12,2} = b_{12,3} = 0$$

$$b_{12,4} = -b_{11,4}$$

$$b_{12,5} = -b_{11,5}$$

$$b_{12,6} = -b_{11,6}$$

$$b_{13,1} = \frac{(X_c - X_p)(Z_c - Z_p)}{[(X_c - X_p)^2 + (Y_c - Y_p)^2 + (Z_c - Z_p)^2]\sqrt{(X_c - X_p)^2 + (Y_c - Y_p)^2}}$$

$$b_{13,2} = \frac{(Y_c - Y_p)(Z_c - Z_p)}{[(X_c - X_p)^2 + (Y_c - Y_p)^2 + (Z_c - Z_p)^2]\sqrt{(X_c - X_p)^2 + (Y_c - Y_p)^2}}$$

$$b_{13,3} = \frac{-\sqrt{(X_c - X_p)^2 + (Y_c - Y_p)^2}}{[(X_c - X_p)^2 + (Y_c - Y_p)^2 + (Z_c - Z_p)^2]}$$

$$b_{13,4} = -b_{13,1}$$

$$b_{13,5} = -b_{13,2}$$

$$b_{13,6} = -b_{13,3}$$

$$b_{14,1} = -b_{13,1}$$

$$b_{14,2} = -b_{13,2}$$

$$b_{14,3} = -b_{13,3}$$

$$b_{14,4} = b_{13,1}$$

$$b_{14,5} = b_{13,2}$$

$$b_{14,6} = b_{13,3}$$

$$b_{15,1} = \frac{-X_c Z_c}{(X_c^2 + Y_c^2 + Z_c^2)\sqrt{X_c^2 + Y_c^2}}$$

$$b_{15,2} = \frac{-Y_c Z_c}{(X_c^2 + Y_c^2 + Z_c^2)\sqrt{X_c^2 + Y_c^2}}$$

$$b_{15,3} = \frac{\sqrt{X_c^2 + Y_c^2}}{(X_c^2 + Y_c^2 + Z_c^2)}$$

$$b_{15,4} = b_{15,5} = b_{15,6} = 0$$

$$b_{16,1} = -b_{15,1}$$

$$b_{16,2} = -b_{15,2}$$

$$b_{16,3} = -b_{15,3}$$

$$b_{16,4} = b_{16,5} = b_{16,6} = 0$$

The approximations for the six parameters are computed as follows:

$$X_c^0 = d_{GC} \cos (V_{GC}) \sin (\alpha_{GC}) = 564.85 \text{ m}$$

$$Y_c^0 = d_{GC} \cos (V_{GC}) \cos (\alpha_{GC}) = 2120.80 \text{ m}$$

$$Z_c^0 = d_{GC} \sin (V_{GC}) = 13.00 \text{ m}$$

$$\alpha_{GP} = \alpha_{GC} + A_G = 98.41233700$$

$$X_p^0 = d_{GP} \cos (V_{GP}) \sin (\alpha_{GP}) = 1551.60 \text{ m}$$

$$V_p^0 = d_{GP} \cos (V_{GP}) \cos (\alpha_{GP}) = -229.46 \text{ m}$$

$$Z_p^0 = d_{GP} \sin (V_{GP}) = 9.23 \text{ m}$$

With these estimates for parameters, coefficients $b_{11}, b_{12}, \ldots, b_{16,4}$ and constant terms $f_1, f_2, \ldots, f_{16}$ are calculated to form the **B** and f matrices. Then estimated standard deviations, σ_i, given for the measurements, allow forming the weight matrix $\mathbf{W} = \sigma_0^2 \Sigma^{-1}$, where $\sigma_0^2 = 1$ and Σ is a diagonal matrix composed of individual variances σ_i^2 for the measurements, which are assumed uncorrelated. Matrices **B**, **W**, and f then are used in a least-squares adjustment of indirect observations to form normal equations from which corrections to the estimates are found.

Using a computer program and after iterating the solution, the final estimates and their standard deviations, obtained from $\mathbf{N}^{-1}$, are

$X_c = 564.854$ m,	$\sigma = 0.021$ m,	$X_p = 1551.596$ m,	$\sigma = 0.012$ m
$Y_c = 2120.800$ m,	$\sigma = 0.012$ m,	$Y_p = -229.451$ m,	$\sigma = 0.018$ m
$Z_c = 13.022$ m,	$\sigma = 0.012$ m,	$Z_p = 9.223$ m,	$\sigma = 0.010$ m

EXAMPLE 9.11. ADJUSTMENT WITH DERIVED OBSERVATIONS. The adjustment of the three-dimensional traverse in Example 9.10 can be simplified by reducing the number of original observations. Replace each pair of lengths by a single measurement and each pair of vertical angles (elevation and depression) by one measurement then repeat the adjustment, making sure that the results are the same as those obtained in Example 9.10.

Solution. Two measurements of the same physical entity, such as a distance, can be replaced properly only if two conditions are met: the two measurements are uncorrelated, and the standard deviation of the resulting *derived measurement* must be computed by rigorous error propagation. Because each pair of measurements has the same standard deviation, σ, that is, the two measurements are of equal precision, the derived observation is their arithmetic mean and its standard deviation, $\sigma' = \sigma/\sqrt{2}$:

$$d'_{GP} = (1568.513 + 1568.491)/2 = 1568.502 \text{ m}, \qquad \sigma' = 0.014 \text{ m}$$

$$d'_{PC} = (2548.978 + 2548.993)/2 = 2548.986 \text{ m}, \qquad \sigma' = 0.014 \text{ m}$$

$$d'_{CG} = (2194.792 + 2194.758)/2 = 2194.775 \text{ m}, \qquad \sigma' = 0.014 \text{ m}$$

$$V'_{GP} = (20'14.2380'' + 20'12.2811'')/2 = 20'13.2596'', \qquad \sigma' = 1.4142''$$

$$V'_{PC} = (5'11.6044'' + 5'04.6180'')/2 = 5'8.1112'', \qquad \sigma' = 1.4142''$$

$$V'_{CG} = (20'23.0624'' + 20'23.4034'')/2 = 20'23.2329'', \qquad \sigma' = 1.4142''$$

The observations are one azimuth, three distances, three horizontal angles, and three vertical angles, with a total of $n = 10$ while $n_0 = 6$ remains the same. The redundancy, therefore, is $r = 10 - 6 = 4$ and with $u = 6$ unknown parameters (coordinates); the number of condition equations becomes $c = r + u = 4 + 6 = 10 = n$. These 10 equations are a subset of the 16 equations written in Example 9.10. In particular, denoting the present equations by F' we have

F'_1, F'_2, F'_3, F'_4 are the same as F_1, F_2, F_3, F_4.
F'_5 is the same form as F_5, replacing d_{GP} by d'_{GP}.
F'_6 is the same as F_8, replacing d_{PC} by d'_{PC}.
F'_7 is the same as F_8, replacing d_{CG} by d'_{CG}.
F'_8 is the same as F_{11}, replacing V_{GP} by V'_{GP}.
F'_9 is the same as F_{13}, replacing V_{PC} by V'_{PC}.
F'_{10} is the same as F_{15}, replacing V_{CG} by V'_{CG}.

As before, the linearized condition equations take the form

$$\underset{10,1}{v} + \underset{10,6}{\mathbf{B}}\ \underset{6,1}{\mathbf{\Delta}} = \underset{6,1}{f}$$

with the elements of $\mathbf{B}$ appropriately selected from those given in Example 9.10. Using the same approximations as in that example, but more important, using the new (proper) standard deviations, the results from the adjustment program are

$$X_c = 564.854 \text{ m}, \qquad \sigma = 0.021 \text{ m}, \qquad X_p = 1551.596 \text{ m}, \qquad \sigma = 0.012 \text{ m}$$

$$Y_c = 2120.801 \text{ m}, \qquad \sigma = 0.012 \text{ m}, \qquad Y_p = -229.451 \text{ m}, \qquad \sigma = 0.018 \text{ m}$$

$$Z_c = 13.022 \text{ m}, \qquad \sigma = 0.012 \text{ m}, \qquad Z_p = 9.223 \text{ m}, \qquad \sigma = 0.010 \text{ m}$$

These results are essentially identical to those obtained in Example 9.10, with no difference exceeding 0.001 m. This demonstrates the importance of using the proper stochastic model and the uniqueness of the least-squares adjustment technique.

9.22
POSTADJUSTMENT ANALYSIS

It is important to assess the quality of the results from least-squares adjustment. This assessment, referred to as *postadjustment analysis,* was discussed in detail in Section 2.30 of Chapter 2. In this section, we apply several of the analysis techniques through numerical examples.

EXAMPLE 9.12. TEST ON REFERENCE VARIANCE. Evaluate the consistency of the posterior estimate of the reference variance $\hat{\sigma}_0^2$, against its a priori value, σ_0^2, for the adjustment problem in Example 9.10. The adjustment program produced a value for the a posteriori reference variance of $\hat{\sigma}_0^2 = 1.0357$.

Solution. Because the variances of the observations were used directly in the adjustment, the prior value of the reference variance is $\sigma_0^2 = 1$. Recalling that $r = 10$ and choosing $\alpha = 0.05$, the 95 percent confidence region, according to Equation (2.112), is between the following two values:

$$10 \times 1.0357/\chi_{10,0.975}^2 = 0.505$$

$$10 \times 1.0357/\chi_{10,0.025}^2 = 3.19$$

Because $\sigma_0^2 = 1$ it falls within the confidence region and $\hat{\sigma}_0^2$ is consistent with σ_0^2.

EXAMPLE 9.13. Apply the same statistical test on the reference variance from Example 9.11 to show the consistency of the results, given $\hat{\sigma}_0^2 = 0.3514$.

Solution. The redundancy in the case of 10 observations is $r = 4$. Therefore,

$$r\hat{\sigma}_0^2/\chi_{r,\alpha/2}^2 = 4 \times 0.3514/\chi_{4,0.025}^2 = 2.904$$

$$r\hat{\sigma}_0^2/\chi_{r,1-\alpha/2}^2 = 4 \times 0.3514/\chi_{4,0.975}^2 = 0.127$$

So σ_0^2 is within the confidence region. This test, therefore, is consistent with the corresponding test in Example 9.12.

EXAMPLE 9.14. ERROR ELLIPSES. Refer to Example 9.9 for the adjustment of combined triangulation and trilateration of the quadrilateral $ABCD$ in Figure 9.20. Use the propagated postadjustment covariance matrix for the coordinates of points C and D to establish their standard deviation, 90 percent confidence region, and error ellipses. Additionally, evaluate the corresponding single precision measures.

Solution. The variances of the observations were used directly in the adjustment, so N^{-1} resulting from the adjustment is the 4×4 covariance matrix of the four coordinates. The pair of 2×2 submatrices on the main diagonal are the covariance matrices for points C and D, respectively. From Example 9.9, the values of these matrices are

$$\Sigma_{CC} = 10^{-4} \begin{bmatrix} 2.25526 & -0.51482 \\ \text{symmetric} & 5.50016 \end{bmatrix} \text{ft}^2$$

$$\Sigma_{DD} = 10^{-4} \begin{bmatrix} 5.49383 & -1.18173 \\ \text{symmetric} & 2.64349 \end{bmatrix} \text{ft}^2$$

Using the standard computation of the eigenvalues (see Section 2.31), the standard error ellipses have the following parameters and are plotted in Figure 9.22, where C and D refer to point C and D, respectively:

$$a_C = \sqrt{\lambda_1} = \sqrt{0.00055799} = 0.024 \text{ ft}$$

$$b_C = \sqrt{\lambda_2} = \sqrt{0.00021755} = 0.015 \text{ ft}$$

$$a_D = \sqrt{\lambda_1} = \sqrt{0.00059200} = 0.024 \text{ ft}$$

$$b_D = \sqrt{\lambda_2} = \sqrt{0.00022173} = 0.015 \text{ ft}$$

The 90 percent error ellipses are evaluated by multiplying the semiaxes of the standard error ellipse by 2.146, which is the value of k corresponding to the probability $P = 0.90$ in Table 2.4. Section 2.30. As also plotted in Figure 9.25, the semiaxes are

$$a_C' = 2.146a_C = 0.051 \text{ ft}, \qquad b_C' = 2.146b_C = 0.032 \text{ ft}$$

$$a_D' = 2.146a_D = 0.052 \text{ ft}, \qquad b_D' = 2.146b_D = 0.032 \text{ ft}$$

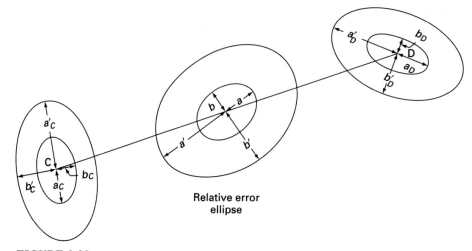

FIGURE 9.22
Error ellipses and relative error ellipse.

As discussed in Section 2.30, when a single accuracy measure is desired, the mean of the coordinate variances is used as an approximate overall indication. Therefore, according to Equation (2.120),

$$\hat{\sigma}_x^2 = \frac{10^{-4}}{4}(2.25526 + 5.50016 + 5.49383 + 2.64349) = 3.97319 \times 10^{-4} \text{ ft}^2$$

or
$$\hat{\sigma}_x = 0.020 \text{ ft}$$

This overall value is used for both points C and D and roughly represents the horizontal accuracy of the adjustment. We could use the circular standard deviation for each point, however, replacing each standard error ellipse by a circle, as developed in Section 2.31. For point C, the ratio of $\sigma_{min}/\sigma_{max}$ is

$$b_C/a_C = 0.625$$

Therefore, according to Equation (2.126)

$$\sigma_c = 0.5(0.024 + 0.015) = 0.020 \text{ ft}$$

For point D,

$$b_D/a_D = 0.625$$

and

$$\sigma_c = 0.5(0.024 + 0.015) = 0.020 \text{ ft}$$

EXAMPLE 9.15. RELATIVE ERROR ELLIPSE. To assess the relative accuracy between points C and D in Example 9.9 (see also Example 9.14), compute and plot the standard deviation and 95 percent confidence level relative error ellipses.

Solution. For Example 9.9, the total 4×4 covariance is

$$\boldsymbol{\Sigma} = \mathbf{N}^{-1} = \begin{bmatrix} \boldsymbol{\Sigma}_{CC} & \boldsymbol{\Sigma}_{CD} \\ \boldsymbol{\Sigma}_{DC} & \boldsymbol{\Sigma}_{DD} \end{bmatrix} = 10^{-4} \begin{bmatrix} 2.25526 & -0.51482 & 1.56597 & 0.10381 \\ & 5.50016 & -2.93210 & 2.05263 \\ & & 5.49383 & -1.18173 \\ \text{symmetric} & & & 2.64349 \end{bmatrix} \text{ ft}^2$$

The elements of the covariance matrix for the differences between the coordinates of points C and D are computed according to Equation (2.124):

$$\sigma_{dx}^2 = 10^{-4}\,[2.25526 - 2(1.56597) + 5.49383] = 0.00046172 \text{ ft}^2$$

$$\sigma_{dy}^2 = 10^{-4}\,[5.50016 - 2(2.05263) + 2.64349] = 0.00040384 \text{ ft}^2$$

$$\sigma_{dxdy} = 10^{-4}\,[-0.51482 - 0.10381 + 2.93210 - 1.18173] = 0.00011317 \text{ ft}^2$$

The corresponding semiaxes of the standard relative error ellipse are

$$a = \sqrt{\lambda_1} = \sqrt{0.00054959} = 0.023 \text{ ft}$$

$$b = \sqrt{\lambda_2} = \sqrt{0.00031596} = 0.018 \text{ ft}$$

From Table 2.4, the multiplier is 2.447 for the 95 percent ellipse, and therefore, its semiaxes are

$$a' = 2.447a = 0.057 \text{ ft}$$

$$b' = 2.447b = 0.045 \text{ ft}$$

Both ellipses are plotted in Figure 9.22 in the middle of the line CD.

EXAMPLE 9.16. ERROR ELLIPSOID. The three-dimensional adjustment in Example 9.10 yields the three coordinates for each of two survey stations C and P. The 6×6 postadjustment covariance matrix associated with these coordinates follows. Compute the semiaxes of the standard deviation and the 90 percent confidence level error ellipsoids for each of the points. Compare the standard spherical, σ_s to the mean of the trace $\bar{\sigma}_c$ for each point:

$$\Sigma_{6,6} = N^{-1} = 10^{-3} \begin{bmatrix} \begin{matrix} 0.4306 & -0.0836 \\ & 0.1389 \end{matrix} & \begin{matrix} 0 & -0.0427 & -0.3012 & 0 \\ -0.0001 & 0.0237 & 0.1132 & 0 \end{matrix} \\ \Sigma_{cc} & \begin{matrix} 0.1473 & 0.0001 & 0 & 0.0404 \end{matrix} \\ & \begin{matrix} 0.1439 & 0.0639 & 0.0003 \\ & 0.3315 & 0 \end{matrix} \\ \text{symmetric} & \Sigma_{pp} \quad 0.0950 \end{bmatrix}$$

Solution. For point C, Σ = the matrix Σ_{cc} in the preceding equation, from which (referring to Table 2.5, Section 2.30, for the 90 percent multiplier)

$$a = \sqrt{\lambda_1} = \sqrt{.4529 \times 10^{-3}} = 0.021 \text{ m}, \qquad a' = 2.5a = 0.053 \text{ m}$$

$$b = \sqrt{\lambda_2} = \sqrt{.1473 \times 10^{-3}} = 0.012 \text{ m}, \qquad b' = 2.5b = 0.030 \text{ m}$$

$$c = \sqrt{\lambda_3} = \sqrt{.1166 \times 10^{-3}} = 0.011 \text{ m}, \qquad c' = 2.5c = 0.027 \text{ m}$$

According to Equations (2.128) and (2.129),

$$\sigma_s = \tfrac{1}{3}(a + b + c) = 0.015 \text{ m}$$

$$\bar{\sigma}_s = [\tfrac{1}{3}(\sigma_x^2 + \sigma_y^2 + \sigma_z^2)]^{\frac{1}{2}} = 0.015 \text{ m}$$

For point P, Σ = the matrix Σ_{pp} in the earlier equation, from which

$$a = \sqrt{\lambda_1} = \sqrt{.3543 \times 10^{-3}} = 0.019 \text{ m}, \qquad a' = 2.5a = 0.048 \text{ m}$$

$$b = \sqrt{\lambda_2} = \sqrt{.1210 \times 10^{-3}} = 0.011 \text{ m}, \qquad b' = 2.5b = 0.028 \text{ m}$$

$$c = \sqrt{\lambda_3} = \sqrt{.0950 \times 10^{-3}} = 0.010 \text{ m}, \qquad c' = 2.5c = 0.025 \text{ m}$$

$$\sigma_s = \tfrac{1}{3}(a + b + c) = 0.013 \text{ m}$$

$$\bar{\sigma}_c = [\tfrac{1}{3}(\sigma_x^2 + \sigma_y^2 + \sigma_z^2)]^{\frac{1}{2}} = 0.014 \text{ m}$$

9.23
STANDARDS OF ACCURACY FOR CONTROL SURVEYS

1984 Standards

The accuracy required for horizontal control depends on the type of survey and ultimate use of the control points. In the United States, standards of accuracy for geodetic control surveys are prepared by the Federal Geodetic Control Subcommittee (FGCS) and reviewed by the American Society of Civil Engineers, the American Congress on Surveying and Mapping, and the American Geophysical Union.

As currently published in the 1984 FGCS Report (Bossler, 1984), these standards provided for three orders of accuracy (first, second, and third), the latter two of which are subdivided into Classes I and II. First-order or primary horizontal control provides the principal framework for the national control network. It is used for earth crustal movement studies in areas of seismic and tectonic activity, for testing defense and scientific equipment, for studying the performance of space vehicles, for engineering projects of high precision and extending over long distances, and for surveys used in metropolitan expansion.

Second-order, Class I or secondary control consists of networks between first-order arcs and detailed surveys where land values are high. Surveys of this class include the basic framework for densification for control. Secondary horizontal control strengthens the entire network and is adjusted as part of the national network.

Second-order, Class II surveys are used to establish control for inland waterways, the interstate highway system, and extensive land subdivision and construction. This class contributes to and is published as part of the national network.

Third-order, Class I and Class II or supplementary surveys are used to establish control for local improvements and development, topographic and hydrographic surveys, or other such projects for which they provide sufficient accuracy. Third-order control may or may not be adjusted to the national network. The surveying engineer should know that third-order, Class I surveys constitute the lowest order permissible for specifying points on the state plane coordinate systems (Chapter 11).

A specific order and class indicates that the control point has a particular relationship with respect to all other points in the network. This relationship is expressed as a relative accuracy or ratio of the relative positional error to the distance between the separation of the points. Control classifications with corresponding relative accuracies or *distance accuracy standards* are shown in Table 9.5.

TABLE 9.5
Distance accuracy standards

Classification	Minimum distance accuracy
First order	1 : 100,000
Second order, Class I	1 : 50,000
Second order, Class II	1 : 20,000
Third order, Class I	1 : 10,000
Third order, Class II	1 : 5,000

The relative or distance accuracy, $1 : a$, is determined from a least-squares adjustment using

$$a = d/s$$

where a is the inverse of the distance accuracy ratio, s is the propagated standard deviation of the distance between survey points obtained from the least-squares adjustment, and d is the distance between survey points.

Theoretically, the distance accuracy is with respect to all pairs of points, but in practice only a sample of pairs is computed. The worst distance accuracy is used as the provisional accuracy. Consider an example to illustrate the process.

EXAMPLE 9.17. In Example 9.7, the covariance matrix was propagated for stations B and C. This matrix is repeated here for convenience and is as follows:

$$\Sigma_{\Delta\Delta} = \begin{bmatrix} (1.06485)10^{-4} & (-0.82228)10^{-4} & (0.38993)10^{-4} & (-0.33900)10^{-4} \\ & (0.28627)10^{-3} & (0.77228)10^{-5} & (1.81200)10^{-4} \\ \text{symmetric} & & (0.34061)10^{-4} & (-1.10296)10^{-5} \\ & & & (0.25566)10^{-3} \end{bmatrix}.$$

$$\Sigma_{\Delta\Delta} = \begin{bmatrix} \sigma^2_{X_B} & \sigma_{X_B Y_B} & \sigma_{X_B X_C} & \sigma_{X_B Y_C} \\ & \sigma^2_{Y_B} & \sigma_{Y_B X_C} & \sigma_{Y_B Y_C} \\ \text{symmetric} & & \sigma^2_{X_C} & \sigma_{X_C Y_C} \\ & & & \sigma^2_{Y_C} \end{bmatrix}$$

Solution. Classify this survey according to the 1984 standards. Using the appropriate elements of the covariance matrix and Equations (2.122) to (2.124) in Section 2.30, the major and minor axes of the relative error ellipse for line BC are found. First, use Equations (2.124) to evaluate

$$\sigma^2_{X_{BC}} = \sigma^2_{X_B} - 2\sigma_{X_B X_C} + \sigma^2_{X_C} = (6.0760)10^{-5}$$

$$\sigma^2_{Y_{BC}} = \sigma^2_{Y_B} - 2\sigma_{Y_B Y_C} + \sigma^2_{Y_C} = (1.79530)10^{-4}$$

$$\sigma_{X_{BC} Y_{BC}} = \sigma_{X_B Y_B} - \sigma_{X_B Y_C} - \sigma_{X_C Y_B} + \sigma_{X_C Y_C} = -(6.2560)10^{-5}$$

The relative covariance matrix for line BC is

$$\Sigma_{D_{BC}} = \begin{bmatrix} (6.2560)10^{-5} & (-6.7080)10^{-5} \\ \text{symmetric} & (1.79530)10^{-4} \end{bmatrix} = \begin{bmatrix} \sigma^2_{X_{BC}} & \sigma_{X_{BC} Y_{BC}} \\ \text{symmetric} & \sigma^2_{Y_{BC}} \end{bmatrix}$$

from which the characteristic polynomial, using Equation (2.122), is

$$\lambda^2 - (\sigma^2_{X_{BC}} + \sigma^2_{Y_{BC}})\lambda + (\sigma^2_{X_{BC}}\sigma^2_{Y_{BC}} - \sigma^2_{X_{BC}Y_{BC}}) = 0$$

$$\sigma^2 - (6.2560 \times 10^{-5} + 1.79530 \times 10^{-4})\lambda + [(6.2560 \times 10^{-5})$$
$$(1.79530 \times 10^{-4}) - (-6.7080 \times 10^{-5})^2] = 0$$

$$\lambda^2 - 2.4029 \times 10^{-4}\lambda + 6.4085 \times 10^{-9} = 0$$

and for which the roots are

$$\lambda_1 = 2.1004 \times 10^{-4} \quad \text{and} \quad \lambda_2 = 3.2050 \times 10^{-5}$$

Here, λ_1 and λ_2, the eigenvalues of the covariance matrix, are the squares of the semimajor axis, a, and semiminor axis, b, of the standard relative error ellipse (with probability of 0.394) for line BC so that

$$a_{BC} = \sqrt{\lambda_1} = 0.014 \text{ m} \quad \text{and} \quad b_{BC} = \sqrt{\lambda_2} = 0.006 \text{ m}$$

In the ratio s/d from the standards, s is the standard deviation, so it is necessary to scale a_{BC} to a probability of 0.632 (see Mikhail and Gracie, 1981, p. 230) using the constant 1.414 to give $a_{BC} = (1.414)(0.014) = 0.020$ m. Therefore, the ratio $d/s = 855/0.020 = 42{,}750$. According to Table 9.5, this value corresponds to a second-order, Class II survey.

Proposed Revised Standards

As previously noted in Section 8.41, the 1984 standards currently are being reviewed and revised by the FGCS and the Federal Geographic Data Committee (FGDC); draft reports were released in December 1994 (FGCS, 1994) and January 1997 (FGDC, 1997). The emphasis in these documents is to propose use of a statistically based standard, consisting of a number with a stated probability, for *network* and *local* surveys.

The horizontal network accuracy of a control point is a number corresponding to the radius of an absolute error circle (Section 2.31), with a confidence level of 95 percent, relative to the geodetic datum. For this case, the datum is defined by the nearest *Continuously Operating Reference Station* (CORS) in the National Spatial Reference System (NSRS), supported by the National Geodetic Survey (NGS).

Local accuracy of a control point is shown by a number that represents the relative error circle (Section 2.31), with a 95 percent confidence level, of that point with respect to known control points connected by the local survey. The local accuracy reported is the average of all the individual values of local accuracy along the lines involved in connecting the point to the local survey.

Horizontal accuracy standards (that are also used for ellipsoid and orthometric heights) are shown in Table 9.6 where classification is by groups of 1-millimeter, 2-millimeter, 5-millimeter, . . . , 5-meter, and 10-meter accuracies.

To illustrate the use of these standards, consider an example.

EXAMPLE 9.18. Evaluate the network and local accuracies for stations B and C in Example 9.7. To assess these accuracies, two assumptions must be made for this example are: The network was adjusted in a minimal constraint adjustment, and stations A and B are CORS points. First, evaluate the network accuracy for these points.

Solution. Determine the semimajor and semiminor axes of the error ellipses for points B and C (see Example 9.17 for elements of these matrices). For point B, the

TABLE 9.6

551

CHAPTER 9:
Other Methods
for Horizontal
Positioning

**Accuracy standards for horizontal position,
ellipsoid height, and orthometric height
(modified courtesy of FGDC, 1997)**

Accuracy classification	95% Confidence, m (less than or equal to)
1 millimeter	0.001
2 millimeters	0.002
5 millimeters	0.005
1 centimeter	0.010
2 centimeters	0.020
5 centimeters	0.050
1 decimeter	0.100
2 decimeters	0.200
5 decimeters	0.500
1 meter	1.000
2 meters	2.000
5 meters	5.000
10 meters	10.000

characteristic polynomial (Equation (2.122)) is

$$\lambda^2 - (\sigma_{X_B}^2 + \sigma_{Y_B}^2)\lambda + (\sigma_{X_B}^2 \sigma_{Y_B}^2 - \sigma_{X_B Y_B}^2) = 0$$

$$\lambda^2 - (1.06485 \times 10^{-4} + 0.28627 \times 10^{-3})\lambda$$
$$+ [(1.06485)10^{-4}(0.28627)10^{-3}$$
$$- (-0.82228 \times 10^{-4})^2] = 0$$

$$\lambda^2 - 3.9275 \times 10^{-4}\lambda + 3.2722 \times 10^{-8} = 0$$

The roots λ_1 and λ_2 are $\lambda_1 = 3.18206 \times 10^{-4}$ and $\lambda_2 = 0.74549 \times 10^{-4}$, so that $a = \sqrt{\lambda_1} = 1.7838 \times 10^{-2}$ m and $b = \sqrt{\lambda_2} = 0.86342 \times 10^{-2}$ m, which are the semimajor and semiminor axes, respectively, of the standard error ellipse (probability of 0.394) for point B.

Similarly, for point C, we have

$$\lambda^2 - (0.34061 \times 10^{-4} + 0.25566 \times 10^{-3})\lambda$$
$$+ [(0.34061 \times 10^{-4})(0.25566 \times 10^{-3}) - (-1.10296 \times 10^{-5})^2] = 0$$

or

$$\lambda^2 - 2.89721 \times 10^{-4}\lambda + 8.5864 \times 10^{-9} = 0$$

The roots of this polynomial are $\lambda_1 = 2.56208 \times 10^{-4}$ and $\lambda_2 = 3.3513 \times 10^{-5}$, so that $a = \sqrt{\lambda_1} = 1.6006 \times 10^{-2}$ m and $b = \sqrt{\lambda_2} = 5.7891 \times 10^{-3}$ m.

Next, evaluate the standard error circles for these points (Section 2.30, Equation (2.124a) or (2.125)). To determine which equation to use, calculate the ratios b/a for each point as follows. For point B, $b/a = 0.5$ and for point C, $b/a = 0.4$. So, conservative estimates for the standard error circle (with probability of 0.394) are

$$\sigma_{cB} \approx 0.5 \, (a + b)_B \approx 0.013 \text{ m} \qquad \text{and} \qquad \sigma_{cC} \approx 0.5(a + b)_C \approx 0.011 \text{ m}$$

These parameters, with a probability of 0.95, are (Table 2.4)

$$\sigma_{cB} \approx (2.447)(0.014) \approx 0.032 \text{ m} \qquad \text{and} \qquad \sigma_{cC} \approx (2.447)(0.011) \approx 0.027 \text{ m}$$

Referring to Table 9.6, the network accuracy classification is 5 centimeters.

To determine the local accuracy, evaluate the standard error circle for the relative standard error ellipse calculated for line BC in Example 9.17. First, calculate the ratio, $\sigma_{min}/\sigma_{max}$ or b/a for this line using results from Example 9.17 to yield $b/a = 0.00566/0.0145 = 0.4$, so that Equation (2.125) is appropriate to calculate σ_c. Now, $\sigma_c \approx 0.5(a + b) = 0.010$ m = the radius of the standard error circle with a probability of 0.394. The radius of this error circle with a probability of 0.95 (see Table 2.4, Section 2.30) is

$$(\sigma_c)_{0.95} \approx (2.447)(0.010) \approx 0.024 \text{ m}$$

Referring to Table 9.6, this value indicates a 5-centimeter survey.

Note that the local accuracy according to the revised standards is approximately the same as the old standard, because both are distance dependent. On the other hand, there is no way to compare network accuracy from the proposed revised standards with the old standards, because it is not distance dependent but entirely a function of the propagated error circle with a probability of 0.95. The revised standards should provide a much more reliable indication of the actual accuracy, especially for the networks.

The standards, as outlined, may continue to be revised and specifications necessary to enable achieving a given accuracy have not yet been developed. Therefore, the practitioner must continually monitor the current professional literature and announcements from professional societies to stay aware of further modified standards and new specifications when they become available and are adopted.

PROBLEMS

9.1. The coordinates for stations C and D are as follows:

Station	X, m	Y, m
C	323,480.540	124,250.503
D	323,980.650	124,850.235

Horizontal angles, measured with a $01''$ theodolite from stations C and D, are angle $BCD = 60°35'25.5''$ and angle $CDB = 79°50'45.3''$. The azimuth from the north of line CD is $A_{CD} = 35°00'00''$. (a) Compute coordinates for station B by intersection, using the base solution. (b) Compute coordinates for station B by intersection, using azimuths calculated for lines DB and CB.

9.2. Station C is located by intersection from triangulation points B and A for which the coordinates are $X_A = 433,191.050$ m and $Y_A = 158,893.500$ m. The distance and azimuth from the north of the line from A to B are 895.425 m and $235°20'32''$, respectively. The measured horizontal angles are angle $BAC = 80°27'35.8''$ and angle $CBA = 54°14'37.8''$. Compute the coordinates of point C by intersection.

9.3. Horizontal clockwise angles are measured from station Ajax to three stations of known position having the following coordinates

Station	X, m	Y, m
A	25,078.670	29,693.183
B	24,484.794	31,367.395
C	24,003.771	32,945.768

The measured angles are

Station	From	To	Angle
	A	B	25°52'07.3"
Ajax			
	B	C	21°07'36.8"

Calculate the coordinates of station Ajax by resection.

9.4. The coordinates for three stations of known horizontal position are

Station	X, ft	Y, ft
H	88,237.92	80,132.03
J	82,279.10	97,418.58
K	80,330.69	102,911.40

Clockwise horizontal angles, measured from station Campbell are

Station	From	To	Angle
	H	J	44°51'58.9"
Campbell			
	J	K	25°52'17.6"

Compute the coordinates of station Campbell by resection.

9.5. The coordinates for three stations of known horizontal position are, for H, $X = 30,071.483$ m, $Y = 24,424,243$ m; for L, $X = 24,933.356$ m, $Y = 30,082.605$ m; for K, $X = 24,484.794$ m, $Y = 31,367.697$ m. The horizontal clockwise angles measured from station R to these stations are, from H to L, $50°51'49.8"$, and from L to K, $19°56'18.3"$. Determine the position of station R by resection.

9.6. An instrument is set at station P, within triangle ABC (Figure 9.23). Given the data that follow, solve the three-point problem (Figure 9.6b) for the angles γ and θ:

$$\angle BAC = 101°40'30", \qquad \alpha = 88°24'45"$$
$$b = 3000.055 \text{ m}, \qquad \beta = 130°10'24"$$
$$c = 2045.235 \text{ m}$$

9.7. Given the data from Problem 9.6 and the additional information that follows, if the instrument is at P' (Figure 9.23), outside the triangle ABC, solve the three-point problem (Figure 9.6c) for the angles γ and θ:

$$\alpha' = 24°30'50" \qquad \text{and} \qquad \beta' = 45°25'35"$$

9.8. Describe the steps involved in the work of triangulation.

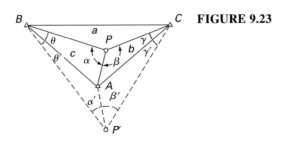

FIGURE 9.23

9.9. Set forth the procedure required in reconnaissance for a control survey network by combined triangulation and trilateration.

9.10. If measurements are needed for a trilateration horizontal control network, what precautions must be observed when measuring the distances with EDM equipment?

9.11. A line in a trilateration network has a horizontal distance of 21,500.540 m (corrected for all systematic EDM instrument errors) measured at an elevation of 235.55 m above the datum in an area where the separation between the geoid and the ellipsoid is -35.55 m. Determine this distance reduced to the surface of the ellipsoid.

9.12. A base line measured by an EDM instrument from station 10 at elevation 240.65 m above the datum to station 11 at elevation 310.45 m above the datum was found to have a slope distance of 19,154.255 m, corrected for atmospheric conditions and EDM system constants. If the average separation between ellipsoid and geoid is known to be -54.25 m for the area covered by this line, determine the ellipsoid distance for the line 10 to 11.

9.13. In triangle ABC, shown in Figure 9.10, the measured base line AC has an estimated standard deviation of 0.008 m. If the angles A, B, and C are observed with estimated standard deviations of 15″, calculate the propagated standard deviation in the computed length of side BC. Assume $\angle A = \angle B = \angle C$.

9.14. In Figure 9.2, line AC or b is the measured base line having an estimated standard deviation of 0.005 m. All of the angles in this chain of triangles are measured, with each angle having an estimated standard deviation of 10″. Determine the standard deviation of line $GH = g$ as calculated through the chain of triangles by the law of sines assuming equilateral triangles and $b = 2000.000$ m. State the assumptions that must be made in doing this error propagation.

9.15. In Figure 9.2 and using the data of Example 9.1, propagate the error in side AB as computed by the law of sines. Assume errorless coordinates for B and C and that angles α and β have respective standard deviations of 10″.

9.16. Supplement the data from Problem 9.4 with one additional clockwise angle of 21°05′34.6″, turned at station Campbell from point K to point S, which has coordinates of $X = 78,752.53$ ft and $Y = 108,069.79$ ft. Using the coordinates computed in Problem 9.4, compute coordinates for Campbell by a least-squares adjustment of resection measurements. Assume the angles are uncorrelated, of equal weight, and $\sigma_{\text{angle}} = 10″$.

9.17. Point C is located by intersection from stations B, A, and D (Figure 9.24), which have the following coordinates:

Station	X, m	Y, m
A	455,773.317	153,616.457
B	455,553.184	154,277.538
C	456,968.552	153,616.457

The clockwise horizontal angles, derived from observed directions $d_1, d_2, \ldots, d_7$ are

$$l_1 = \angle CBA = 114°29′58.2″, \qquad l_3 = \angle CAD = 66°01′16.4″$$
$$l_2 = \angle BAC = 42°19′04.5″, \qquad l_4 = \angle ADC = 69°46′24.9″$$

The azimuth of line AD is due east. (a) Calculate the coordinates of point C by intersection using triangles ABC and ACD. Note the difference between the two sets of values and take the mean as the best estimate. (b) Using the coordinates calculated for C in part (a) as estimates, compute coordinates for C adjusted by a least-squares adjustment of intersection. Assume that observed directions, d_i, are uncorrelated and of equal weight with $\sigma_d = 05″$.

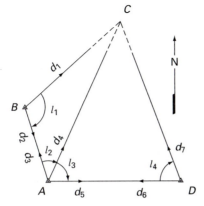

FIGURE 9.24

9.18. If all distances are measured and the coordinates for two stations are known, determine the number of conditions required in a trilateration adjustment by the method of least squares for (a) the center-point quadrilateral with one additional measurement shown in Figure 9.11a, and (b) the center-point pentagon illustrated in Figure 9.11b.

9.19. In Figure 9.12, assume that stations 1, 2, 3, and 7 have known plane coordinates and that all distances have been measured. Determine the number of conditions (or the statistical degrees of freedom) that exist in a least-squares adjustment of trilateration for this network.

9.20. The lengths of the sides of triangle ABC, as measured by and EDM device, are $AB = 1,877.762$ m, $BC = 1,176.903$ m, and $CA = 2,055.764$ m. The azimuth from the north of side AC is $275°44'06.2''$ and the coordinates of station A are $X = 11,997.700$ m and $Y = 9,995.200$ m. Assuming that the given lengths are free of systematic errors, compute preliminary estimates for the azimuths of AC and BC and preliminary coordinates for points B and C to be used in a subsequent adjustment of trilateration by the method of least squares.

9.21. Using the data and answers from Problem 9.20, and measured angles $A = 34°26'42''$, $B = 81°05'15''$, $C = 64°28'10''$, compute adjusted coordinates for points B and C adjusted by the method of least squares. Assume that lengths of lines are uncorrelated, of equal weight, $\sigma_{AB} = \sigma_{BC} = 0.005 + 5$ ppm, and $\sigma_A = \sigma_B = \sigma_C = 10''$. Assume A is east of B.

9.22. An EDM instrument was used to measure the lengths of the sides of triangle BCD to yield $BC = 1,191.09$ ft, $CD = 1,576.62$ ft, $DB = 1,562.14$ ft. These distances are corrected for all systematic errors and are of equal weight, with $\sigma = 0.006 + 10$ ppm. The azimuth from the north of line BC is due east. If the coordinates of B are $X = 20,000.000$ ft and $Y = 40,000.000$ ft, calculate preliminary azimuths for sides CD and DB and preliminary coordinates for points C and D. Assume D is south of C.

9.23. Using the data and answers to Problem 9.22, and measured angles $B = 68°20'45''$, $C = 67°03'25''$, $D = 44°36'00''$ with $\sigma_B = \sigma_C = \sigma_D = 10''$ compute coordinates for points C and D adjusted by the method of least squares.

9.24. The lengths of the sides for quadrilateral $ABCD$, shown in Figure 9.25, are as follows: $AB = 1,931.413$, $BC = 1,201.271$, $BD = 2,113.387$, $AC = 2,114.451$, $AD = 1,303.031$, and $DC = 1,503.335$, all distances being in meters. The azimuth from the north of line $AB = 78°52'10.0''$. The coordinates of A are $X_A = 10,002.564$ m and $Y_A = 9,275.937$ m. Compute preliminary azimuths for the sides and estimates for the coordinates of points C and D, all to be used in a subsequent trilateration adjustment by the method of least squares.

9.25. Using the data and estimates computed in Problem 9.24, determine coordinates for points C and D adjusted by the method of least squares. Assume distances are uncorrelated and that $\sigma_{distance} = 0.005$ m $+ 5$ ppm.

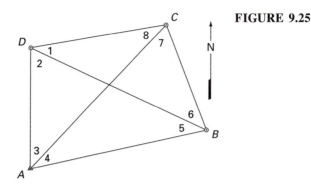

FIGURE 9.25

9.26. Unadjusted angles, arranged and numbered as shown in quadrilateral *ABCD* in Figure 9.25 are as follows: $1 = 33°45'03.6''$, $2 = 63°46'27.5''$, $3 = 44°49'03.8''$, $4 = 34°09'54.9''$, $5 = 37°14'37.6''$, $6 = 44°02'59.4''$, $7 = 64°32'28.2''$, and $8 = 37°39'28.2''$. Using the length and azimuth for *AB* and coordinates for *A* in Problem 9.24, compute lengths of remaining sides and coordinates for *C* and *D* (see Example 9.7). Assume length and azimuth for *AB* are errorless.

9.27. Using the data and answers to Problem 9.26, compute coordinates for points *C* and *D* in a least-squares adjustment of triangulation by the method of least squares. Assume the angles are determined from uncorrelated directions of equal weight, where $\sigma_{d_i} = 05''$.

9.28. Using the data and answers for Problems 9.24 and 9.26, perform a combined adjustment of trilateration and triangulation by the method of least squares to determine the adjusted coordinates for points *C* and *D*.

REFERENCES

Bossler, J. D. "Standards and Specifications for Geodetic Control Networks." Rockville, MD: Federal Geodetic Control Committee, 1984.

El Hassan, I. M. "An Analytical Solution of the Resection Problem." *ASCE Journal of Surveying Engineering* 112, no. 1 (June 1986), pp. 30–35.

Federal Geodetic Control Subcommittee. "Standards for Geodetic Control Networks," draft report. Silver Spring, MD: National Geodetic Survey, October 1994.

Federal Geographic Data Committee. "Draft Geospatial Positioning Accuracy Standards." Reston, VA: Federal Geographic Data Committee Secretariat, U.S. Geological Survey, January 1997.

Kuang, S. "Second-Order Design: Shooting for Maximum Reliability." *ASCE Journal of Surveying Engineering* 119, no. 3 (August 1993), pp. 103–10.

Kuang, S. "Optimization of Horizontal Observations in Triangulation or Triangulation Networks." *ASCE Journal of Surveying Engineering* 119, no. 4 (November 1993), pp. 156–73.

Leick, A. "Minimal Constraints in Two-Dimensional Networks." *ASCE Journal of the Surveying and Mapping Division* 108, no. 2 (August 1982), pp. 53–68.

Leick, A. "Accuracy Standards for Modern Three-Dimensional Geodetic Networks." *Surveying and Land Information Systems* 53, no. 2 (June 1993), pp. 111–16.

Mikhail, E. M. *Observations and Least Squares.* New York: Harper and Row, 1976.

Mikhail, E. M., and G. Gracie. *Survey Computations and Adjustment.* New York: Van Nostrand Reinhold Company, 1981.

Reynolds, W. F. *Manual of Triangulation Computation and Adjustment.* C&GS Special Publication 138, NOAA. Silver Spring, MD: National Geodetic Survey, 1928.

Richardus, P. *Project Surveying: General Adjustment and Optimization Techniques with Applications to Engineering Surveys.* 2nd ed. Boston: A. A. Balkema, 1984.

Tyler, D. A. "Position Tolerance in Land Surveying." *ASCE Journal of Land Surveying* 113, no. 3 (October 1987), pp. 152–63.

Introduction to Astronomy

10.1
GENERAL

The surveyor needs to be familiar with the astronomical and trigonometric principles on which the observations and computations of field astronomy are based. In this chapter, certain fundamentals are given that are applicable to all astronomical observations. However, the discussions are intended to be applied only to surveys of moderate precision and the emphasis will be on the determination of the azimuth of a line.

The azimuth of a line is established by angular observations on some celestial body, most commonly on the sun or Polaris, the North Star or polestar. For computing the azimuth from a celestial observation, the latitude of the place must be known. Also, for certain observations, it is necessary that the longitude be approximately determined. If the survey is through a territory for which a reliable map is available, the latitude and longitude ordinarily may be determined with sufficient precision by scaling from the map.

The requirements of *plane* surveying are met if the true or astronomic azimuth of the survey lines is established with a precision at least equal to that with which the angles between survey lines are measured. Therefore, for plane surveying of ordinary precision, the use of an engineer's theodolite or a total station system and the methods described in this chapter will yield sufficiently precise results.

10.2
THE CELESTIAL SPHERE

In making observations on the sun and stars, the surveyor is not interested in the distance of these celestial bodies from the earth but merely in their angular position. It is convenient to imagine them being attached to the inner surface of a hollow sphere of infinite radius of which the earth is the center. This imaginary globe is called the *celestial sphere*. It also is helpful to imagine the earth as being fixed and to consider the celestial sphere as rotating from east to west, its axis being the prolongation of the axis of the earth. Thus, to the naked eye the polestar appears to remain stationary, but the sun (and similarly the stars near the equator) appears above the horizon in the general direction of east, follows a curved path (convex southward) across the heavens, and disappears below the horizon in the general direction of west.

The portion of the celestial sphere seen by the observer is the hemisphere above the plane of his or her own horizon. The reference plane passes through the center of the earth parallel with the observer's horizon plane, but the radius of the earth is so small in relation to the distances to the stars that the error in vertical angle to a star is negligible. In the case of the sun, the error produced by this assumption is much larger than for any of the stars, amounting under certain conditions to about 9 seconds of arc and requiring an appropriate correction to the observed vertical angle (see Section 10.27). In any case, a refraction correction to observed vertical angles is necessary (Section 10.28).

A vertical line at the location of the observer coincides with the plumb line and is normal to the observer's horizon plane. The point where this vertical line pierces the celestial sphere above the head of the observer is called the *zenith,* and the corresponding point in the opposite hemisphere, directly below the observer, is called the *nadir.*

The *celestial poles* are the points where the earth's axis prolonged pierces the celestial sphere.

The *celestial equator* is the great circle formed by the intersection of the earth's equatorial plane with the surface of the celestial sphere.

Figure 10.1 represents the celestial sphere, the point O being the earth and $NES'W$ being the horizon of an observer, with letters standing for the points of the compass. Figure 10.2 may be taken as an enlarged view of the earth in the same position as that assumed in Figure 10.1. A is an observer in the northern hemisphere, the line $N_a S_a$ being in the horizon plane at that point. Evidently, this observer views everything above the horizon plane or that portion of the celestial sphere (Figure 10.1), which is shown by full lines. B is an observer in the southern hemisphere, at a point on the earth diametrically opposite A; the portion of the celestial sphere that this observer views above horizon plane $N_b S_b$ will be the opposite hemisphere to that seen by A, or that portion of Figure 10.1 which is shown by dash lines. Because the size of the earth is negligible compared with that of the celestial sphere, it may be considered that either $N_a S_a$ or $N_b S_b$ in Figure 10.2 coincides with NS' in Figure 10.1.

Assuming the observer to be in the northern hemisphere (Figure 10.1), Z is the zenith; P and P' are the celestial poles, P being the visible or *elevated* pole; and $EQWQ'$ is the celestial equator, of which the portion EOW is visible to the observer.

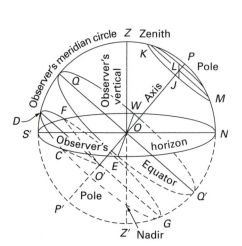

FIGURE 10.1
Celestial sphere.

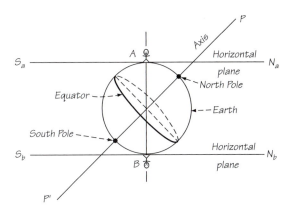

FIGURE 10.2
Observer's horizon.

559

CHAPTER 10:
Introduction to
Astronomy

Because, for the sake of simplicity, we are assuming that the celestial sphere is rotating and the earth remains stationary, N, E, S', W, and Z are regarded as fixed points with respect to any given station on the surface of the earth. If $S'N$ is a line in the plane of the horizon and also lies in the plane passing through the observer's station, then a vertical plane of which this line is an element cuts the celestial sphere in the great circle $S'ZPNZ'P'$, which is called the *meridian circle* or, more often, simply the meridian. At a given instant, the meridian for one station does not occupy the same position in the celestial sphere as the meridian for another station, unless the two stations are at the same longitude.

Any star that is below or south of the equator will follow some path, such as $CDFG$. It will become visible at C, will pass over the meridian at D, and will disappear from view at F. It will be above the horizon for a shorter period of time than it will be below, or the angle whose arc is CDF (angle $CO'F$) is less than 180°. From the figure, it is evident that, if any star is sufficiently far below the equator, it will never appear above the observer's horizon.

Similarly, any star that is above or north of the equator will be above the horizon for a greater length of time than it is below. If it is far enough above the equator, it will be continuously visible to an observer in a northern latitude and, during the course of a single revolution of the celestial sphere, will follow same path as $JKLM$. When it is at the highest point of its apparent path, at K, it is said to be at *upper culmination;* when it is at the lowest point, at M, it is said to be at *lower culmination.*

10.3
OBSERVER'S LOCATION ON EARTH

The location, or *position,* of any point on the surface of a sphere may be fixed by angular measurement from two planes of reference at right angles to each other passing through the center of the sphere; these measurements are called the *spherical coordinates* of the point. The spherical coordinates of any station on the surface of the earth are designated as the *latitude* and *longitude* of the station. Figure 10.3 represents the earth, PP' being the axis and $QUVQ'$ being the equator. Let S be the station of an observer. Then $PSUP'$ is a *meridian circle* through the station. Also, RSR' is a *parallel* passing through the station, the plane RSR' being parallel to that of the equator.

The latitude of a place, for all practical purposes, may be defined as the angular distance of the place above or below the equator. When the station is above the equator, the latitude is north and its sign is positive; when below the equator, the latitude is south and

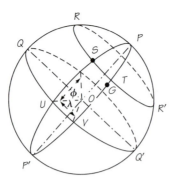

FIGURE 10.3
Observer's location on the earth.

its sign is negative. Hence, in the figure, the latitude of S is given by the angle ϕ or by the corresponding angular distance, measured along any meridian circle, between the equator and the parallel passing through S, such as US, VT, or QR, and the latitude is north or positive. The latitude of a place is stated in degrees. The latitude of the equator is $0°$ and that of the North Pole is $+90°$, or $90°$N.

The longitude of a place is defined as the angular distance measured along the arc of the equator between a reference meridian and the meridian circle passing through the station. The reference meridian is called the *primary meridian.* The primary meridian most generally used is that of Greenwich, England. (Prior to 1912, a Washington, DC, meridian was extensively used in the United States.) In the figure, if the point G represents Greenwich, PGP' is the primary meridian, and the longitude of S is given by the angle λ or by the angular distance VU. Longitudes are expressed either in degrees of arc or in hours of time ($15° = 1$ hr) and are measured either east or west of the Greenwich meridian.

In general, the discussions here are intended to apply in the northern hemisphere and for longitudes west of Greenwich.

10.4
RIGHT-ASCENSION EQUATOR SYSTEM

Figure 10.4 shows the celestial sphere in a position similar to that of the earth in Figure 10.3, S being a celestial body whose position is to be fixed by spherical coordinates. Comparable with the meridian circles or meridians of longitude of the earth are the *hour circles* of the celestial sphere, all of which converge at the celestial poles. The arc PSU is a portion of the hour circle passing through S. Comparable with the parallels of latitude of the earth are the *parallels of declination* of the celestial sphere. RSR' is the parallel of declination passing through S. And comparable with the prime meridian through Greenwich is the *equinoctial colure* of the celestial sphere, which passes through the *vernal equinox,* an imaginary point among the stars where the sun apparently crosses the equator on about March 21 of each year. In the figure, V represents the vernal equinox and PTV is the equinoctial colure.

The *right ascension* of the sun or any star is the angular distance measured along the celestial equator between the vernal equinox and the hour circle through the body. It is comparable with the longitude of a station on the earth. Right ascensions are measured *eastward* from the vernal equinox and may be expressed either in degrees of arc ($0°$ to $360°$) or in hours of time (0^h to 24^h). In the figure, the right ascension of S is given by the angle α in the plane of the equator or by the arc VU.

The *declination* of any celestial body is the angular distance of the body above or below the celestial equator. It is comparable with the latitude of a station on the earth. If the body

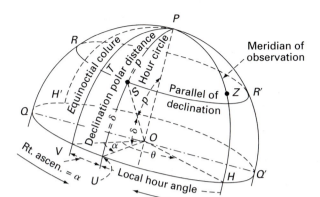

FIGURE 10.4
Equator systems of spherical coordinates.

is above the equator, its declination is said to be north and is considered positive; if it is below the equator, its declination is said to be south and is considered negative. Declinations are expressed in degrees and cannot exceed 90° in magnitude. In the figure, the declination of S is given by the angle δ or by the arc of any hour circle between the equator and the parallel of declination RSR', such as US, VT, or QR.

The *polar distance* or *codeclination* of any celestial body is $p = 90° - \delta$ with due regard to the sign of the declination. In the figure, it is given by the angle p or by the arc PS. Polar distances always are positive. For computations referred to the North Pole, when the declination is north, the polar distance is the complement of the declination; but when the declination is south, as in the case of the sun during the winter months, the polar distance is greater than 90°. In defining the position of a star near either pole, often the polar distance is given instead of the declination.

For present purposes, it may be considered that the vernal equinox is a fixed point on the celestial equator, just as Greenwich is a fixed point on the earth. But, although stations on the earth maintain practically an unvarying location with respect to the equator and the meridian of Greenwich, the coordinates of celestial bodies with respect to the celestial equator and the equinoctial colure change more or less with the passage of time. The fixed stars, or those outside the solar system, alter their positions in the celestial sphere only slightly from month to month and from year to year, the annual change being less than a minute of arc in either right ascension or declination. These variations are due to (1) *precession,* or the slow change in the direction of the earth's axis due to attraction of the sun, moon, and planets; or (2) *nutation,* or small inequalities in the motion of precession, similar to the oscillation of a spinning top.

As the earth actually travels around the sun but not around the stars, the sun appears to move more slowly than the stars, in one year making 365 apparent revolutions (approximately) whereas the stars make 366 apparent revolutions (approximately); therefore the sun apparently makes a complete circuit of the heavens once each year, its right ascension changing from 0^h (or 0°) on March 21 to 12^h (or 180°) on September 22 and continuing to 24^h (or 360°) on the following March 21, when a new cycle begins. Further, as the axis of rotation of the earth is not normal to the plane of the earth's orbit, the path traced by the sun among the stars on the celestial sphere, called the *ecliptic,* is a continuous curved line; each year the sun crosses the equator northward about March 21, reaches a maximum positive declination ($\sim$N23$\frac{1}{2}$°) about June 21, crosses the equator southward about September 22, and reaches a maximum negative declination ($\sim$S23$\frac{1}{2}$°) about December 21.

10.5
HOUR-ANGLE EQUATOR SYSTEM

In many of the problems of field astronomy it is necessary to know not only a star's position in the celestial sphere but also to determine its position with respect to the meridian through a given station on the surface of the earth. In Figure 10.4, let the arc PZH represent the meridian of the observer stationed on the earth and let S be some celestial body whose position is desired with respect to the meridian PZH and the equator $Q'HUVQ$. The spherical coordinates of the star are given by (1) the angular distance of the star above or below the equator, given in Figure 10.4 by the arc US defined in Section 10.4 as the declination δ; and (2) the angular distance HU measured along the equator between the meridian of the observer and the hour angle through the star. This angular measurement is called an *hour angle*. In general, hour angles are measured in a positive sense from east to west (clockwise as viewed from the North Pole) from the reference meridian to the meridian of the celestial body. In Figure 10.4, the reference meridian is that of the observer and the positive hour angle is the angular distance HU. Hour angles are expressed either in hours of time or degrees of arc. When no qualification is stated, it is understood that the hour angle is measured from the upper branch of the meridian; that is, the branch above the station or above the observer's head (meridian PZH in Figure 10.4).

Three hour angles are important in astronomic calculations. To visualize these hour angles, observe the spherical system of Figure 10.4 from the North Pole (i.e., along line PO), as shown by the view from above the North Pole in Figure 10.5. The local hour angle (L.H.A.) is measured clockwise (westward) from the meridian of observation to the meridian of the celestial body and is angle $HPU = t$ in Figure 10.5. When the celestial body is west of the observer's meridian (as in Figure 10.5), the L.H.A. is less than 180° and equals angle t. When the celestial body is east of the observer's meridian, the L.H.A. exceeds 180° and

$$t = \text{L.H.A.} - 360° \tag{10.1a}$$

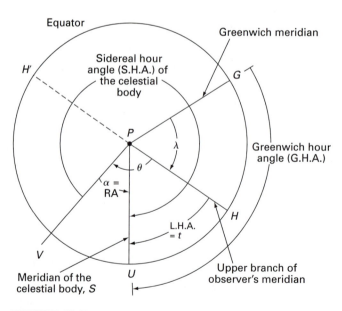

FIGURE 10.5

Local, Greenwich, and sidereal hour angles.

Here, t represents a time interval before or after upper culmination of the sun or moon. The *sidereal hour angle* (S.H.A.) is measured clockwise (westward) from the meridian of the vernal equinox, V, to the meridian of the celestial body and is the angular distance VGU in Figure 10.5. It is obvious from the figure that the right ascension

$$\text{RA} = \alpha = 360° - \text{S.H.A.} \qquad (10.1b)$$

The *Greenwich hour angle* (G.H.A.) is measured clockwise (westward) from the Greenwich meridian to the meridian of the celestial body. In Figure 10.5 the G.H.A. of the celestial body is the angular distance GHU. Also from Figure 10.5 it can be seen that

$$\text{L.H.A.} = \text{G.H.A.} - \text{observer's west longitude} \qquad (10.1c)$$

$$\text{L.H.A.} = \text{G.H.A.} + \text{observer's east longitude} \qquad (10.1d)$$

In connection with the definition of civil time, hour angles are reckoned from the lower branch of the meridian. In Figure 10.5, if the hour angle were reckoned from the lower branch, it would be defined by the angular distance $H'GHU$ and would be 12^h more than that given by the arc HU, which is the hour angle reckoned from the upper branch.

Sometimes, the hour angles of celestial bodies east of the meridian are reckoned eastward from the upper branch of the meridian rather than westward. When an hour angle is expressed in this way, it is preceded by a minus sign. So, if the hour angle of S (Figure 10.5) were reckoned eastward, it would be given by the angular distance $HGH'U$.

10.6
EQUATOR SYSTEMS COMPARED

The system of coordinates described in Section 10.5 is similar to that described in Section 10.4 with this difference: In the hour-angle system the angular distance along the equator is measured (westward) from a *fixed meridian,* whereas in the right-ascension system the angular distance along the equator is measured (eastward) from the *vernal equinox,* which is a point on the celestial equator that rotates with the celestial sphere. Therefore, although right ascensions of fixed stars have annual variations of but a few seconds, hour angles of the stars change rapidly, because the celestial sphere apparently rotates (24^h or $360°$ for each 23^h 56^m of our civil time) and hour angles of the sun change approximately 24^h or $360°$ for each 24^h of our civil time.

The two systems are called the *equator systems of coordinates,* because in each case the primary plane of reference is the celestial equator. The position of a celestial body above or below the equator is given by the declination δ, which is the same in one system as in the other.

Let θ be the hour angle of the vernal equinox represented in Figure 10.5 by the angular distance HUV measured along the equator. At any instant of time, if the hour angle of the vernal equinox with respect to a given meridian is known and the right ascension α of a heavenly body S is known, the hour angle t of the body can be computed because, from the figure,

$$t = \theta - \alpha \qquad \text{or} \qquad \theta = t + \alpha$$

This equation, therefore, is an expression by means of which the coordinates of one system may be transformed to those of the other.

10.7
ASTRONOMICAL TABLES USED BY THE SURVEYOR

By means of astronomical observations and calculations, the positions of many of the celestial bodies are predicted, and values of their right ascensions and declinations for various dates are available in various publications. The position of a celestial body at any time can be obtained by interpolation.

The publication most widely used by astronomers in the United States is the *Astronomical Almanac,* published one or two years in advance by the Nautical Almanac Office, U.S. Naval Observatory. This document is not convenient for use by surveyors.

Most of the astronomical data used by surveyors are presented in *The Nautical Almanac,* which also is published annually in advance by the Nautical Almanac Office, U.S. Naval Observatory, primarily for use in navigation.

Each of these ephemerides is sold by the Superintendent of Documents, U.S. Government Printing Office, Washington, DC 20402.

In Europe, *Apparent Places of Fundamental Stars* is issued annually by the Astronomiches Rechen-Institut, Heidelberg, Germany, and is available from Verlag G. Braun, Karl-Friedriche Strasse 14–18, Karlsruhe, Germany. The Royal Greenwich Observatory (annual) produces a *Star Almanac for Surveyors,* which also is available in the United States.

The most useful publication currently available for the surveyor is the *Sokkia Celestial Observation Handbook and Ephemeris* by Elgin, Knowles, and Senne (annual), published by the Sokkia Corporation. Most of the examples in this chapter are solved using this ephemeris.

Computer programs containing routines that produce ephemeris data, given the date, time, desired celestial body, and so forth, also are available. Specifically, the U.S. Naval Observatory produces Mica for DOS 1990–1999, an interactive astronomical almanac that can be run on a personal computer (PC) and uses the DOS operating system. This software is designed primarily for astronomers and is of limited use to surveyors. Also, Elgin, Knowles, and Senne (e.g., 1986) have several software packages for performing astronomical calculations and computing ephemeris data. These programs are adaptable to handheld calculators and personal computers.

10.8
HORIZON SYSTEM OF SPHERICAL COORDINATES

In the ordinary operations of surveying, angles are measured in the horizontal and vertical planes; to use other planes would be inconvenient. Likewise, in astronomical fieldwork, the angular location of a celestial body at a given instant is determined by measuring its vertical angle (referred to the horizon plane) and its horizontal angle (referred to a given line on the ground).

Figure 10.6 shows a portion of the celestial sphere in which O represents both the earth and the location of the observer, $NES'W$ is the observer's horizon, and $S'ZN$ is the meridian plane passing through the observer's location. The point Z on the celestial sphere directly above the observer is called the *zenith.* The point S represents a celestial body, and BSZ is part of a great circle, called a *vertical circle,* through the body and the zenith. In this *horizon system* of spherical coordinates, the angular location of a celestial body is defined by its azimuth and altitude.

The *azimuth* of a celestial body is the angular distance measured along the horizon in a clockwise direction from the meridian to the vertical circle through the body. Azimuths may be reckoned from either the south point or the north point of the meridian; in astronomical work, azimuths usually are reckoned from south through 360°, except that, for circumpolar stars, they often are reckoned from north. In trigonometric computations, the azimuths of stars west of north or east of south often are expressed as counterclockwise angles from the meridian and considered negative values. In Figure 10.6 the azimuth of S reckoned in the customary manner is given by the angle A or by the angular distance $S'NB$,

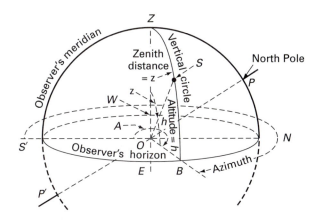

FIGURE 10.6
Horizon system of spherical coordinates.

an arc of the horizon. If the azimuth of *S* were reckoned from north, it would be given by
the angle (*A* − 180°) or by the angular distance *NB*. The negative azimuth reckoned from
south is given by the arc *S'EB*.

The *altitude* of a celestial body is the angular distance measured along a vertical circle,
from the horizon to the body; it corresponds to the vertical angle of ordinary surveying and
is expressed in degrees of arc. The altitude of *S* (Figure 10.6) is given by the vertical angle
h or by the angular distance *BS,* the arc of a vertical circle passing through the zenith.
Except in rare instances, celestial objects are observed when above the true horizon, when
the sign of the altitude is positive. It is seen that positive altitudes may vary between 0°
and 90°.

The complement of the altitude, called the *zenith distance* or *coaltitude,* is the angular
distance from the zenith to the celestial body measured along a vertical circle. In the figure,
the zenith distance is given by the zenith angle, *z*, or by the angular distance, *ZS*. Here,
z = 90° − *h*. Zenith distances always are positive; they correspond to the zenith angle in
ordinary surveying.

Because the celestial sphere apparently is rotating about its axis, whereas the meridian,
horizon, and zenith are imagined as remaining fixed in position, it is clear that in general
both the azimuth and the altitude of a star are changing continuously.

10.9
RELATIONS AMONG LATITUDE, ALTITUDE, AND DECLINATION

Figure 10.7 represents a section of the earth through the poles and the station of an
observer. Because the latitude of a place is its angular distance from the equator measured
along a meridian of longitude, the latitude of the observer is given by the angle ϕ between
the equator and a vertical line through the observer's station, the angle being measured in
the plane of the meridian. Also, from similar triangles, the latitude is given by the angle ϕ
between the axis and the horizon, likewise measured in the plane of the meridian; this angle
is the altitude of the elevated pole.

Similarly Figure 10.8 represents a section of the celestial sphere through the celestial
poles and the observer's zenith. For reasons just explained,

$$\angle QOZ = \angle NOP = \phi = \text{latitude of observer's place}$$

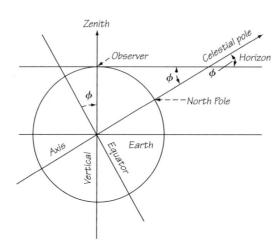

FIGURE 10.7
Latitude of observer.

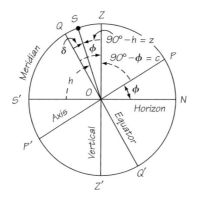

FIGURE 10.8
Relation among latitude, altitude,
and declination.

Hence, the latitude of a place is given by the angular distance NP, which is the altitude of the pole, or by the angular distance QZ, which is the zenith distance or coaltitude of the equator. The angular distance from the pole to the zenith is $90° − \phi = c$, which is the colatitude of the place.

In a northern latitude, if any heavenly body S having declination δ is on the meridian and south of the zenith, then from Figure 10.8,

$$\phi = (90° − h) + \delta = z + \delta$$

Similarly, for any star north of the zenith,

$$\phi = h ± (90° − \delta) = h ± p$$

in which the sign preceding p, the polar distance, is positive or negative according to whether the star is below or above the pole.

10.10
HORIZON AND HOUR-ANGLE EQUATOR SYSTEMS COMBINED

The relation between the coordinates of the horizon system and those of the hour-angle equator system described in Section 10.5 is shown, for a star S not on the meridian, by

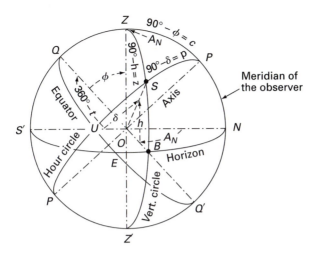

FIGURE 10.9
Horizon and hour-angle equator systems combined.

Figure 10.9. The meridians of the two systems coincide. The place of observation is assumed to be north of the equator at a latitude ϕ, as given either by the angle between the equator and the zenith or by the angular distance NP between the horizon plane and the celestial axis. The star is at a position east of the meridian and above the celestial equator.

In the horizon system the coordinates of S are A_N, the azimuth measured from the north point of the horizon, and h, the altitude. In the equator system the coordinates are t, the hour angle (also called the *local hour angle,* L.H.A.) measured westward from the upper branch of the meridian ($360° - t$ is shown in the figure; see also Figure 10.5), and δ, the declination. The colatitude ($90° - \phi = c$), the zenith distance ($90° - h = z$), and the polar distance ($90° - \delta = p$) define a spherical triangle the vertices of which are the pole, P, the zenith, Z, and the celestial body, S. This triangle is called the *PZS triangle* or the *astronomical triangle.* Most of the problems of field astronomy involve transforming from one system of spherical coordinates to the other and solving the *PZS* triangle for unknown coordinates, having certain coordinates in one or both systems known or observed.

In the figure, the celestial body is shown above the horizon and above the equator. If the body were below the horizon or below the equator, the sides of the *PZS* triangle would be defined in a manner similar to that just described but account would be taken of the algebraic sign of the altitude and the declination.

In the figure, the celestial body is shown as east of the observer's meridian; the angle Z of the spherical *PZS* triangle, therefore, is its azimuth from the north. Also, the angle P of the *PZS* triangle is equal to $360° - t$. When the body is west of the meridian, $Z = 360° -$ azimuth from north and $P = t$.

10.11
SOLUTION OF THE *PZS* TRIANGLE

In surveying, the astronomical triangle is solved in connection with determinations of azimuth. Observations are made on the sun or on some star that can be readily identified. The altitude of the celestial body is measured, its declination at the instant of observation

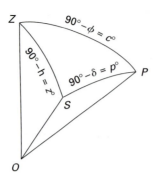

FIGURE 10.10
The astronomical triangle.

is determined from published tables, and the latitude of the place of observation is known or determined by separate observation. In Figure 10.10, PZS is the astronomical triangle having the same orientation as the spherical triangle described in Figure 10.9, where OP is the polar axis and OZ is the observer's zenith. The three known sides of this triangle are $90 - \phi$, the colatitude; $90 - h$, the coaltitude or zenith distance; and $90 - \delta$, the codeclination or polar distance. Determination of the azimuth of the celestial body involves computation of the angle at Z; and determinations of longitude or time involve the computation of the angle at P as a measure of the hour angle.

By applying the law of cosines for an oblique spherical triangle (the rules of spherical trigonometry are similar to those for plane triangles), there results

$$\cos Z = \frac{\cos (90 - \delta) - \cos (90 - \phi) \cos (90 - h)}{\sin (90 - \phi) \sin (90 - h)}$$

$$= \frac{\sin \delta - \sin \phi \sin h}{\cos \phi \cos h} \tag{10.2a}$$

or

$$\cos Z = \frac{\sin \delta}{\cos h \cos \phi} - \tan h \tan \phi \tag{10.2b}$$

The azimuth from the *north* of the celestial body is obtained from Z, as calculated using either Equation (10.2a) or (10.2b), as follows:

1. When the body is *east* of the observer's meridian, Z is *clockwise from the north* and the azimuth from the north is equal to Z.
2. When the body is west of the observer's meridian, the calculated value of Z is *counterclockwise from the north* and the azimuth is equal to $360°$ minus Z.

When azimuths are reckoned from the *south*, Equation (10.2b) takes the following form:

$$\cos Z_s = \tan h \tan \phi - \frac{\sin \delta}{\cos h \cos \phi} \tag{10.3}$$

and the relationships between the azimuth and Z_s are opposite to those above for the azimuth reckoned from north. For example, when the body is west of the observer's meridian, Z_s is *clockwise* from south and equal to the azimuth (also clockwise from south). When the body is *east* of the observer's meridian, Z_s is counterclockwise from south and the azimuth (which is clockwise from south) is equal to $(360° - Z_s)$.

Applying the law of cosines for an oblique spherical triangle to the astronomical triangle shown in Figure 10.10 and solving for the angle $P = t$ yields

$$\cos t = \frac{\sin h - \sin \phi \sin \delta}{\cos \phi \cos \delta} = \frac{\sin h}{\cos \phi \cos \delta} - \tan \phi \tan \delta \qquad (10.4)$$

which is a general expression for determining the hour angle, t, for any celestial body when the three sides of the astronomical triangle are known. The angle t is defined in Section 10.5 and illustrated in Figures 10.4 and 10.5. When measured westward from the meridian of the observer (clockwise when viewed from the North Pole), t is positive; when measured eastward, t is negative. Equation (10.4) is used to determine time by the altitude of the sun and by the altitude of a star. When used to determine time by altitude of the sun, angle t by Equation (10.4) is the time before local noon for a morning observation and the time after local noon for an afternoon observation.

When the unknown angle in Equations (10.1) to (10.4) is either small or near 180°, a relatively small error in the computed value of the cosine will produce a relatively large error in the angle itself, because the magnitude of the cosine is changing slowly. For this reason, *as far as errors in computations are involved,* the foregoing equations are not suitable for precisely computing azimuth and hour angle when the observed celestial body is near the meridian. On the other hand, when the unknown azimuth or hour angle is near 90° or 270°, its cosine is changing rapidly and Equations (10.1) to (10.4) are quite adequate for precise computation of azimuth or hour angle.

By applying a series of substitutions that can be found in any textbook on spherical trigonometry, Equation (10.2b) can be changed to the form

$$\tan^2 \tfrac{1}{2} Z = \frac{\sin (s - h) \sin (s - \phi)}{\cos s \cos (s - p)} \qquad (10.5)$$

and Equation (10.4) can be changed to the form

$$\tan^2 \tfrac{1}{2} t = \frac{\cos s \sin (s - h)}{\cos (s - p) \sin (s - \phi)} \qquad (10.6)$$

In these two equations $p = 90° - \delta =$ polar distance $s = \tfrac{1}{2}(h + \phi + p)$, and the remaining letters have the same significance as in Equations (10.1) to (10.4). In some cases $(s - p)$ will be negative, but the result will not be affected, because the cosine of a negative angle has the same value and the same sign as the cosine of a positive angle of equal size.

For a given angular value, the tangent changes more rapidly than the cosine. Therefore, for a given error of computation of the trigonometric function, Equations (10.5) and (10.6) generally will render a closer determination of azimuth and hour angle than Equations (10.2) and (10.4). For angles near 90° and 270°, the difference between the rates of change of the tangent and the cosine is not large, and results from the two sets of equations are comparable. However, when the object is near the meridian (azimuth near 0° and 180°), for the given computational errors, Equations (10.5) and (10.6) will provide better determinations of angles than it is possible to obtain by Equations (10.2) and (10.4).

Azimuths from the South

When azimuths are reckoned from the south, Equation (10.5) takes the following form, Z_s being the angle measured either clockwise or counterclockwise from south:

$$\cot^2 \tfrac{1}{2} Z_s = \frac{\sin (s - h) \sin (s - \phi)}{\cos s \cos (s - p)} \qquad (10.7)$$

When $\cot \frac{1}{2} Z_s$ has been determined, the computations for hour angle are somewhat reduced if Equation (10.6) is modified as follows:

$$\tan \tfrac{1}{2} t = \frac{\sin (s - h)}{\cot \frac{1}{2} Z_s \cos (s - p)} \tag{10.8}$$

10.12
AZIMUTH AND HOUR ANGLE AT ELONGATION

The most favorable position for determining azimuth by observation on any star that crosses the upper branch of the meridian north of the zenith occurs when it is farthest east or farthest west of the pole, when the star appears to be traveling vertically for some time. In this position, it is said to be at eastern or western *elongation,* according to whether it is east or west of the meridian. At the instant of elongation, because the star appears to be traveling vertically, its apparent path in the celestial sphere is tangent to the vertical circle through the observer's zenith, as illustrated at S in Figure 10.11. Therefore, the angle S between the plane of the hour circle and the plane of the vertical circle is a right angle. For azimuth determinations of this sort, the latitude of the place of observation is known, and either the declination or the polar distance of the star for the given date is obtained from published tables. At the instant of elongation, then, there are known in the astronomical triangle the side $ZP = 90° - \phi$, the side $PS = 90° - \delta$, and the angle $S = 90°$.

In right spherical triangle PZS in Figure 10.11, by the law of sines for a spherical triangle,

$$\sin Z = \frac{\sin (90° - \delta)}{\sin (90° - \phi)}$$

This becomes

$$\sin Z = \frac{\sin p}{\cos \phi} \tag{10.9}$$

which is the general expression employed for determining the azimuth of a circumpolar star when at elongation, Z being the azimuth from the north when the star is at an eastern elongation and $Z = 360° -$ azimuth from the north when the star is at a western elongation.

By applying the law of cosines to the right spherical triangle PZS (Figure 10.11), where $S = 90°$ and $t = $ the angle at P, the following equation can be derived:

$$\cos t = \tan \phi \tan p \tag{10.10}$$

which is an expression for finding the hour angle of a star at the instant of elongation, the hour angle t being reckoned east or west of the upper branch of the meridian, depending on

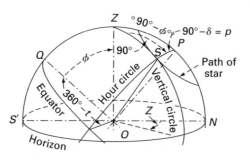

FIGURE 10.11
Star at elongation.

the position of the star. The equation is useful in determining the time at which elongation will occur on any given data.

10.13
AZIMUTH OF THE SUN OR A STAR AT ANY POSITION

In spherical triangle PZS as shown in Figure 10.9, by the law of sines,

$$\frac{\sin (360° - t)}{\sin (90° - h)} = \frac{\sin Z}{\sin (90° - \delta)}$$

or

$$\sin Z = -\frac{\sin t \cos \delta}{\cos h} \tag{10.11}$$

Also, in the same spherical triangle PZS, by the law of cosines,

$$\cos h \cos Z = \sin \delta \cos \phi - \cos \delta \sin \phi \cos t \tag{10.12}$$

Division of Equation (10.11) by Equation (10.12) yields

$$\tan Z = \frac{-\sin t}{\tan \delta \cos \phi - \sin \phi \cos t} \tag{10.13}$$

Equation (10.13) can be used to calculate the azimuth of a star or the sun at any time when the latitude and longitude are known, the declination, δ, is obtained from an ephemeris, and the hour angle, t (as defined in Section 10.5), is determined from the observed time of measurement. This procedure, called *azimuth by the hour-angle method,* is described in Section 10.31 for azimuth of the sun and in Section 10.38 for azimuth of Polaris.

10.14
ALTITUDE OF A STAR

When a star cannot be readily identified through the theodolite's telescope, the process of bringing it into the field of view is considerably expedited if its approximate altitude is computed prior to the observation and laid off on the vertical circle of the theodolite. Also, a check on the correctness of observations and computations for azimuth and hour angle is obtained if the computed value of the altitude agrees with the observed value.

Again, referring to the spherical triangle PZS in Figure 10.9 and applying the law of cosines,

$$\sin h = \sin \phi \sin \delta + \cos \phi \cos \delta \cos t \tag{10.14}$$

in which t is the hour angle at a given time and h is the altitude at the same instant.

TIME

10.15
SOLAR AND SIDEREAL DAY

As the earth rotates about its axis in its travel through space, all celestial bodies apparently rotate about the earth (or about its axis) from east to west. Because the earth in its orbit travels about the sun but does not travel about the fixed stars, which are far outside its orbit, once each year the sun apparently encircles the celestial sphere along a path called the *ecliptic,* which twice cuts the celestial equator during this interval (Figure 10.12). The point among the stars where the sun in its apparent travel northward cuts the celestial equator on about March 21 of each year is called the *vernal equinox,* a point of reference whose position on the celestial sphere is unchanging. No star lies at that point, but it is helpful to imagine that the vernal equinox is an invisible celestial body rigidly fastened in its position on the celestial sphere, while each of the so-called fixed stars slowly moves along a path of extremely small periphery on the surface of the sphere, and the sun travels rapidly along the ecliptic in a direction opposite to that of the rotation of the celestial sphere.

The vernal equinox is referred to in ephemerides as *The First Point of Aries* or simply *Aries;* it is often represented by the zodiacal symbol ♈ as shown in Figure 10.12.

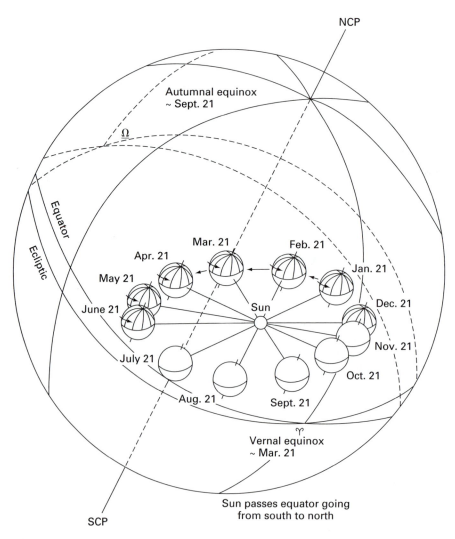

FIGURE 10.12
Plane of the ecliptic and celestial sphere. (*Adapted from Kissam, 1981.*)

Because the sun apparently is traveling from west to east among the stars but the rotation of the celestial sphere about the earth apparently is from east to west, the angular velocity of the sun about the axis of the celestial sphere is less than that of the fixed stars or the vernal equinox, just as the angular velocity of a passenger walking toward the rear of a train on a circular track is less than that of the train. At a given meridian, the hour angle of the sun and that of the vernal equinox will agree at some instant about March 21, but thereafter it will be less for the sun than for the vernal equinox. Six months later, about September 21, when the sun has covered one-half of its annual journey, the hour angle of the sun will be $180°$ or 12^h less than that of the vernal equinox, one year later the hour angle of the sun will be $360°$ or 24^h less than that of the vernal equinox; hence, the hour angles again will agree.

In the course of a tropical year as measured by the time taken by the sun apparently to make a complete circuit of the ecliptic, 366.2422 revolutions of the earth actually occur, or apparently a like number of revolutions of the vernal equinox about the earth. For the reasons just explained, during this interval the sun will have traveled through a total hour angle of $360°$ or 24^h less than that traversed by the vernal equinox; hence, during a tropical year the sun apparently revolves about the earth 365.2422 times.

The interval of time occupied by one apparent revolution of the sun about the earth is called a *solar day,* the unit with which we are all familiar. The interval of time occupied by one apparent revolution of the vernal equinox is called a *sidereal day,* a unit much used by astronomers. Because 366.2422 sidereal days occupy the same period of time as 365.2422 solar days, the sidereal day is a shorter interval of time than the solar day.

When any celestial body, real or imaginary, apparently crosses the upper branch of a meridian, it is said to be at the *upper transit* or *upper culmination;* when any celestial body crosses the lower branch of the meridian, it is said to be at the *lower transit* or *lower culmination.*

The beginning of a sidereal day at a given place occurs at the instant the vernal equinox is at the upper transit.

The solar day is considered to begin at the instant of the lower transit of the sun (midnight), as does the civil day. (Prior to 1925, for astronomical purposes the solar day was considered to begin at noon.)

Both sidereal and solar days are divided into 24 hours each of 60-minute duration. For surveying purposes, the hours are reckoned consecutively from 0 to 24.

10.16
CIVIL (MEAN SOLAR) TIME

Because of the elliptical shape of the earth's orbit, the apparent angular velocity of the sun that we see, called the *true sun,* is not constant; during four periods of each year it is greater and during four intervening periods less than the average velocity. Hence, the days, as indicated by the apparent travel of the true sun about the earth, are not of uniform length. To make our solar days of uniform length, astronomers have invented the *mean sun,* a fictitious body imagined to move at a uniform rate along the celestial equator, making a complete circuit from west to east in one year. The time interval as measured by one daily revolution of the mean sun is called a *mean solar day,* which is the same as the civil day. The mean solar day begins at midnight, as does the civil day, and the *mean solar time* at any place is given by the hour angle of the mean sun plus 12^h. Thus, if the hour angle of the mean sun is $-15° = -1^h$, the mean solar time is $-1^h + 12^h = 11^h$. With regard to time, the terms *mean* and *civil* are interchangeable.

Civil time, has the same meaning as *mean solar time* or *mean time* or *universal time* and, in the form of *standard time* (Section 10.21), is the time in general use by the public. *Local civil time* is that for the meridian of the observer. Civil time for any other meridian is designated by name; for example, *Greenwich civil time.* Civil time for any meridian can be converted into terms of civil time for any other meridian by computations involving the longitude of the two meridians, as described in Section 10.20; 1^h civil time corresponds to 1^h or $15°$ of longitude.

10.17
APPARENT (TRUE) SOLAR TIME

The time interval as measured by one apparent revolution of the true sun about the earth is called an *apparent* or *true solar day.* The true solar day begins at midnight and the true solar time at any place is given by the hour angle of the true sun plus 12^h. Therefore, if the hour angle of the true sun is $45° = 3^h$ the true solar time is $3^h + 12^h = 15^h$. The terms *true* and *apparent* are used interchangeably. *True* is a more accurate designation, but because *apparent* is used in certain ephemerides it must be defined and will be used where necessary. Local true time is the time for the meridian of the observer. True time for any other meridian is designated by name, for example, Greenwich true or apparent time. True time for any meridian can be converted into terms of true time for any other meridian by computations identical with those for civil time, as described in Section 10.20; 1^h true time corresponds to 1^h or $15°$ of longitude.

10.18
EQUATION OF TIME

When the true sun is ahead of the mean sun, true time is faster than mean (civil) time; when behind, slower. The difference between true time and civil time at any instant is called the *equation of time.* It is used to convert civil time at any instant into true time and vice versa.

To formulate the equation of time, it is convenient to utilize the concept of Greenwich hour angle, G.H.A. (Section 10.5) and Greenwich civil time, G.C.T. (Section 10.16). Using these terms, the equation of time is

$$\text{Equation of time} = \text{G.H.A. of the true sun} - \text{G.H.A. of noon at local civil time, or L.C.T.} \quad (10.15a)$$

The equation of time is the elapsed time between an upper transit of the true sun over the observer's meridian and noon L.C.T., or

$$\text{Equation of time} = \text{true solar time} - \text{local civil time} \quad (10.15b)$$

When the equation of time is positive, the true sun precedes noon L.C.T., and it is negative when the true sun follows noon L.C.T.

The maximum value of the equation of time is only about 16^m; hence, for work in which its only use is for the determination of change in declination, it is sometimes neglected.

The equation of time may be obtained from an ephemeris that gives values at given instants of civil time, as follows:

1. In the *Nautical Almanac,* the equation of time, to 01^s, is given for each day at 0^h and 12^h Greenwich civil time. The civil time of meridian passage of the apparent (true) sun is also given.

2. In the *Sokkia Celestial Observation Handbook and Ephemeris* and in the Mica interactive software–generated ephemeris, the equation of time is given to 0.01^s and 0.1^s, respectively, for each day at 0^h universal time with the sign according to Equation 10.15b.

To find the equation of time at any instant other than that for which a value is tabulated, it is necessary to interpolate, adding or subtracting from the tabulated value of the equation of time the change in the equation of time since the instant to which the tabulated value applies.

EXAMPLE 10.1. It is desired to determine by use of the *Nautical Almanac* the equation of time at the instant of $3^h30^m45^3$ P.M. Greenwich civil time on December 15, 1995. Greenwich civil time $= 12^h + 3^h30^m45^s = 15.51^h$. Also, calculate the true (apparent) solar time.

Solution. From the *Nautical Almanac,* the equation of time at 12^h G.C.T. is 05^m04^s. The change in the equation of time in 12 h (0^h December 16) is determined as follows. From the 1995 *Nautical Almanac,*

$$\text{Equation of time at } 0^h \text{ December } 16 = 04^m49^s$$

$$\text{Equation of time at } 12^h \text{ December } 15 = \underline{05^m04^s}$$

$$\text{Change in equation of time} = -15^s$$

The change in the equation of time up to the given instant is

$$\frac{15.51 - 12}{12}(-15) = -4.4^s$$

so that the equation of time at the given instant is

$$05^m04^s - 4.4^s = 5^m00^s$$

Using Equation (10.15b), we obtain

$$05^m00^s = \text{true solar time} - 15^h30^m45^s$$

or

$$\text{True solar time} = 15^h30^m45^s + 05^m00^s = 15^h35^m45^s$$

EXAMPLE 10.2. It is desired to determine by use of the Sokkia ephemeris the Greenwich civil time at the instant of $9^h00^m15^s$ Greenwich apparent (true) time (G.A.T.) on October 10, 1995.

Solution. From the ephemeris the equation of time at 0^h universal time noon is

$$\text{Equation of time at } 0^h \text{ universal time October } 10 = 12^m45.67^s$$

$$\text{Equation of time at } 0^h \text{ universal time October } 11 = 13^m01.81^s$$

$$\text{Change in } 24^h \text{ from October } 10 \text{ to } 11 = +16.14^s$$

The change that will occur in the $9^h00^m15^s$ of elapsed time before G.A.T. noon is

$$\frac{9.0042^h}{24^h}16.14^s = 6.05^s$$

and the equation of time for $9^h00^m15^s$ G.A.T. October 10, 1995, is

$$12^m45.67^s + 6.05^s = 12^m51.72^s$$

Then, by Equation (10.15b),

$$\text{G.C.T. at the instant of } 9^h00^m15^s\text{G.A.T.} = 9^h00^m15^s - 0^h12^m51.7^s = 8^h47^m23.3^s$$

By inspecting the tabulated values of the equation of time as given in the ephemerides, it will be seen that in February the true sun is as much as 14^m behind the mean sun and that in November the true sun is more than 16^m ahead of the mean sun, but on about the dates April 15, June 15, September 1, and December 25, the equation of time is 0 and hence the hour angle of the true sun is for an instant the same as that of the mean sun.

10.19
SIDEREAL TIME

The *sidereal time* at any place is the hour angle of the vernal equinox at that place; and the beginning of the sidereal day, occurring when the vernal equinox crosses the upper branch of the meridian, is called *sidereal noon*. The sidereal hour angle is defined in Section 10.5. Twenty-four-hour clocks regulated to keep sidereal time are called *sidereal clocks*. The vernal equinox, also called *Aries,* is an imaginary point and cannot be observed like the sun; but the right ascensions of stars are referred to the vernal equinox, and therefore the sidereal time can be obtained by determining the hour angle of any star the right ascension of which is known. Then, if θ is the sidereal time, $\theta = t + \alpha$, as explained in Section 10.6.

The sidereal day is shorter than the mean solar day by $3^m55.909^s$ mean solar time, or $3^m56.555^s$ sidereal time. The sidereal hour is shorter than the mean solar hour by $3^m55.909^s/24 = 9.830^s$ mean solar time, or $(3^m56.555^s)/24 = 9.856^s$ sidereal time.

The *Nautical Almanac* is useful for converting sidereal time into civil (mean solar) time and vice versa. The following example illustrates one method of determining the Greenwich sidereal time (G.S.T.) corresponding to a given instant for which the G.C.T. is known.

EXAMPLE 10.3. It is desired to know the Greenwich sidereal time corresponding to $15^h30^m15^s$ G.C.T. on August 1, 1995. The mean solar time interval since 0^h G.C.T. $= 15.5042^h$.

Solution. By the *Nautical Almanac* the Greenwich hour angle of the vernal equinox (Aries) is

G.H.A. at 15^hG.C.T.	$174°45.1'$
G.H.A. at 16^hG.C.T.	$189°47.5'$
Difference in 1^h	$\overline{15°02.4'} = 902.4'$
Difference in $0.5042^h = (0.5042^h)(902.4') =$	$455.0' = 7°35.0'$
G.H.A. at $15^h30^m15^s$ G.C.T. $= 174°45.1' + 7°35.0' = 182°20.1'$	

The sidereal time expressed in hour units is

$$\frac{182°20.1'}{15} = \frac{182.3350°}{15} = 12.1557^h = 12^h09^m20.4^s$$

The conversion from sidereal to civil time or vice versa can be accomplished using the known differences between the times per day and per hour as given here. Consider Example 10.3 using an ordinary ephemeris in which the Greenwich hour angle for the vernal equinox, G.H.A.$_\Upsilon$, is given only for 0^h Greenwich civil time:

Mean solar interval of $15^h30^m15^s$ converted to arc measure

$\qquad = (15.5042^h)(15°/h) = 232.5630° = \qquad\qquad\qquad\qquad$ +232°33.78'

G.H.A. at $15^h30^m15^s$ $\qquad\qquad\qquad\qquad\qquad\qquad\qquad\qquad\qquad$ $\overline{181°41.88'}$

Correction to sidereal interval $= [(9.856^s/h)(15.5042^h)(15''/s)]/(60''/') =$ +38.2'

Sidereal time in angular units $\qquad\qquad\qquad\qquad\qquad\qquad\qquad$ $\overline{182°20.08'}$

Sidereal time converted to hours $= (182.3347°)(15°/h) = 12.1556^h = 12^h09^m20.3^s$

Therefore, to convert a mean solar interval into a sidereal interval, 9.856^s per hour of mean solar interval are added to the solar interval. To convert a sidereal interval into a mean solar interval, 9.830^s per hour of sidereal time are subtracted from the sidereal interval.

10.20
RELATION BETWEEN LONGITUDE AND TIME

As the sun apparently makes a complete revolution (360°) about the earth in one solar day (24 h), and as the longitudes of the earth range from 0° to 360°, it follows that in 1 h the sun apparently traverses $360/24 = 15°$ of longitude. The same statement applies equally well to the sidereal day and the vernal equinox. It follows that, at any instant, the *difference in local time* between two places, whether the time under consideration be sidereal, mean solar, or apparent solar, is equal to the *difference in longitude* between the two places, expressed in hours. This relation is used to determine the difference in time when the difference in longitude between two places is known or vice versa.

Some solar ephemerides are for the meridian of Greenwich, and a frequent problem is to find the local time corresponding to a given instant Greenwich time or vice versa. The local time (L.T.) of a place at a given instant is obtained by adding to or subtracting from the Greenwich time (G.T.) the difference in longitude ($\Delta\lambda$), expressed in hours, between the two places. If the place is east of Greenwich, the difference in longitude is added; if the place is west, the difference in longitude is subtracted.

In these problems concerning time intervals and longitude, conversion from angular units to units of time or the reverse frequently is necessary. These conversions can be accomplished by the following relationships:

Time	Arc		Arc	Time
24^h	$= 360°$		$360°$	$= 24^h$
1^h	$= 15°$		$1°$	$= 4^m$
1^m	$= 15'$		$1'$	$= 4^s$
1^s	$= 15''$		$1''$	$= 0.067^s$

With a pocket calculator, the simplest procedure is to convert to decimal parts of the given units and then divide or multiply by 15. Examples 10.3 and 10.6 furnish examples of this type of conversion.

EXAMPLE 10.4. An observation of the sun is taken at $9^h52^m56^s$ local apparent time (L.A.T.). The longitude of the place is $7^h12^m36^s$ west of Greenwich. What is the Greenwich apparent time (G.A.T.)?

Solution

$$\text{G.A.T.} = \text{L.A.T.} + \Delta\lambda = 9^h52^m56^s + 7^h12^m36^s$$

$$= 17^h05^m32^s$$

EXAMPLE 10.5. On a given date, the mean sun crosses the lower branch of the Greenwich meridian at $3^h52^m48.6^s$, Greenwich sidereal time. At that instant, it is desired to find the local sidereal time (L.S.T.) at a place whose longitude is $5^h12^m24.2^s$ west of Greenwich.

Solution

$$\text{L.S.T.} = \text{G.S.T.} - \Delta\lambda = 3^h52^m48.6^s - 5^h12^m24.2^s + 24^h$$

$$= 22^h40^m24.4^s \text{ of the previous day}$$

EXAMPLE 10.6. At the instant of $18^h48^m15^s$ Greenwich civil time, the local civil time of a place is $10^h37^m45^s$. It is desired to determine the longitude of the place with respect to Greenwich.

Solution

$$\Delta\lambda = 18^h48^m15^s - 10^h37^m42^s = 8^h10^m33^s = 8.17583^h$$

$$= (8.17583)(15) = 122.63750°$$

$$= 122°38'15'' \text{ west of Greenwich}$$

EXAMPLE 10.7. It is desired to find the local civil time at longitude $122°38'15''$W, at the instant of $18^h48^m15^s$ Greenwich civil time.

Solution. The difference in longitude, in hours, is equal to the difference in longitude, in degrees, divided by 15:

$$\text{Local civil time} = 18^h48^m15^s - \frac{122°38'15''}{15} = 10^h37^m42^s$$

Sketches are a valuable aid in the solution of problems involving longitude and time, as they enable the surveyor to visualize the relations. A simple "straight-line" sketch is shown in Figure 10.13, for the instant of 9:00 A.M. Pacific standard time. For clarity, values are given only to 01^m; the actual computations of a surveying problem would be more precise.

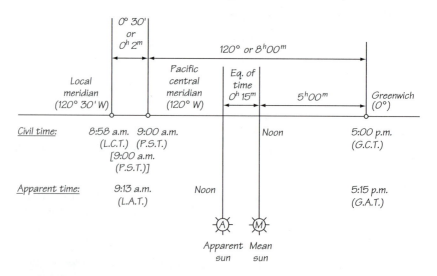

FIGURE 10.13
Relation between longitude and time at a given instant.

10.21

STANDARD TIME, TIME SCALES, AND BROADCAST SIGNALS

579

CHAPTER 10:
Introduction to
Astronomy

To eliminate the confusion resulting from the use of local time by the public, the United States has been divided into belts, each of which occupies a width of approximately 15° or 1^h of longitude. In each belt the watches and clocks that control civil affairs all keep the same time, called *standard time,* which is the local civil time for a meridian near the center of the belt. The time in any belt is a whole number of hours slower than Greenwich civil time, as follows:

Standard time	Abbre- viation	Hours slower than Greenwich civil time	Central meridian	Where used
Atlantic	A.S.T.	4	60°W	Maritime provinces of Canada
Eastern	E.S.T.	5	75°W	Maine through Indiana
Central	C.S.T.	6	90°W	Illinois through central Nebraska
Mountain	M.S.T.	7	105°W	Central Nebraska through western Utah
Pacific	P.S.T.	8	120°W	West of Utah
Yukon	Y.S.T.	9	135°W	Eastern Alaska
Alaska	A.H.S.T.	10	150°W	Central Alaska, Hawaii
Bering Sea	B.S.T.	11	165°W	Western Alaska

The exact boundaries of the time belts are irregular and can be determined only from a map.

In certain localities, "daylight saving time" is employed during the summer months. Daylight saving time is 1^h faster than standard time.

Computations

The Greenwich civil (mean) solar time is found by adding to the standard time the longitude (expressed in hours) of the meridian for which standard time is also local mean time.

EXAMPLE 10.8. At a given instant the Central standard time is 9^h00^m A.M. It is desired to find the Greenwich civil time. The longitude of the meridian to which Central standard time is referred is 90° or 6^h west of Greenwich.

Solution. The Greenwich civil time is $9^h00^m + 6^h00^m = 15^h00^m$, or 3^h00^m P.M.

If the longitude of a place is known, the standard time of the belt in which the place is situated can be determined by adding algebraically to the local mean time the difference in longitude (expressed in hours) between the given place and the meridian for which standard time is also local mean time.

EXAMPLE 10.9. By observation on a star, the local mean time (L.M.T.) at a given instant is found to be $18^h37^m46^s$. The longitude of the place is $\lambda_l = 89°49'30''W = 5^h59^m18^s$. The standard time at the given instant is to be found. The place evidently is in the Central time belt for which the standard time (C.S.T.) is local time for the 90th meridian. The longitude of this meridian expressed in hours is $\lambda_s = 90°/15 = 6^h$.

Solution

$$\lambda_l = 5^h59^m18^s$$

$$\lambda_s = 6^h00^m00^s$$

$$\Delta\lambda = -0^m42^s$$

C.S.T. = L.M.T. + $\Delta\lambda = 18^h37^m46^s + (-0^m42^s) = 18^h37^m04^s = 6^h37^m04^s$ P.M.

Time Scales

The most precise time scale now available is *international atomic time* (TAI, Temps Atomique Internationale), based on analysis of atomic clocks in many countries and independent of the earth's rotation. *Terrestrial dynamic time* (TDT) is the time from an ideal clock on the surface of the earth, related to TAI by the equation TDT = TAI + 32.184 sec (see U.S. Naval Observatory, *User's Guide*; and Elgin et al., annual, p. 21, 1995). TDT is used for the precise calculation of predicted astronomic events and is an option available in the Mica software (see Section 10.7) but is not required for celestial observations by the surveyor.

As noted in Section 10.16, civil or mean time also is called *universal time* (UT), referred to the Greenwich meridian. In astronomy, UT determined from observations on the stars is referred to as UT0, which is somewhat dependent on the place of observation. Because the earth does not rotate at a constant rate, UT and UT0 are not sufficiently precise for all astronomical purposes. When UT0 is corrected for the shift in longitude at the station of observation caused by polar motion, then one has what is known as the *UT1 time scale*. UT1 is used for determining positions of celestial bodies in most ephemerides and in practice should be used by the surveyor.

Coordinated universal time (UTC) is another time scale for general use, where UTC equals TAI plus an integral number of seconds and with leap seconds applied to keep it within 0.9 of a second of UT1 (U.S. Naval Observatory, *User's Guide*, p. 3–6).

Broadcast Signals

Time signals are broadcast in the United States by the National Institute of Standards and Technology (NIST), from station WWV in Fort Collins, Colorado, and WWVH in Hawaii on frequencies of 2.5, 5, 10, 15 MHz. In Canada, Eastern standard time (E.S.T.) is broadcast by radio station CHU on 3.33, 7.35 and 14.67 MHz. All these time signals can be received on inexpensive, pretuned shortwave radio receivers or one can dial (303)499-7111 to get time signals from WWV and WWVH.

The time broadcast by these stations is coordinated universal time. The signals, as broadcast, are adjusted to be within 0.9 sec of UT1 by the application of periodic 1-sec corrections (leap seconds) generally applied at the end of June or December.

To convert UTC to UT1, a correction term labeled *DUT* is applied to UTC, where the relationship is UT1 = UTC + DUT. This correction term is broadcast as a series of double ticks that immediately follow each minute tone of the broadcast signal. Each double tick equals a correction of one-tenth of a second and is positive for the first seven ticks. Starting with the ninth second each double tick is negative. The maximum correction is 0.7 sec and changes about every nine weeks. A detailed description of this correction and how it can be applied using a time module and clock in a pocket calculator can be found in Elgin et al. (1992).

AZIMUTH, LATITUDE, AND LONGITUDE

10.22
GENERAL

The following sections describe the rough methods commonly used in the United States on surveys of ordinary precision where the total station system or repeating theodolite is employed for angular measurements. For more precise methods, such as those necessary on precise geodetic surveys, the reader is referred to texts on geodesy and engineering astronomy (see the references at the end of the chapter). The methods discussed here are based on the relations given in Sections 10.1 to 10.21.

Most observations are taken on the sun and Polaris, so the discussion is concerned chiefly with these two bodies, but the principles involved are the same for any star.

Measurements to the sun cannot be made with the same degree of precision as to a star; hence, the error in computed values is larger than when a fixed star is chosen. However, the sun may be viewed at convenient times, and solar observations are suitable for determination of azimuth, latitude, and longitude with sufficient precision for most ordinary surveys.

Polaris, being near the pole, changes its position slowly. It is the most favorably located of all bright stars for precise determinations of latitude and azimuth, but owing to its slow change in azimuth, it is not suitable for longitude or time observations.

10.23
MEASUREMENT OF ANGLES

Whenever observations are made to determine azimuth, a part of the fieldwork consists of measuring the horizontal angle between the celestial body and a reference mark on the earth's surface. In addition, altitudes quite frequently are observed to these celestial bodies. As the sights to the celestial body in general are steeply inclined, it is highly important that the horizontal axis of the transit or theodolite be in adjustment with respect to the vertical axis and that the instrument be very carefully leveled. Even though the horizontal axis is in perfect adjustment, it will be inclined unless the vertical axis is truly vertical; and the error due to such inclination in general will not be eliminated by a reversal of the telescope between sights. Most theodolites and total station instruments now being used are equipped with vertical circle compensators with precisions of from 0.5″ to 6″ of arc. Such instruments should be leveled precisely, using the vertical circle compensator as described in Section 6.40. Also, sights should be taken with the telescope in both the direct and the reversed positions so that the mean of horizontal angles may be free from other instrumental errors.

The instrument should be supported firmly; if the setup is on soft ground, pegs should be driven to support the tripod legs. Setups on asphalt pavement when the temperature is high and the sun is hot should be avoided.

SOLAR OBSERVATIONS

10.24
OBSERVATIONS ON THE SUN

To observe the sun directly through the telescopic eyepiece may result in serious injury to the eye. *Repeating optical theodolites must be equipped with an eyepiece filter designed to protect the eye. When a total station system is used, the telescope must be equipped with an objective filter so that the direct rays of the sun do not damage the operator's eye or the infrared sensor in the EDM system.* A suitable objective filter should be designed to protect the eye; have optically flat, parallel surfaces; and be made of optical-quality glass.

When using a theodolite, good observations can be made by bringing the sun's image to a focus on a white card held 3 to 4 in., or 7 to 10 cm, to the rear of the telescopic eyepiece.

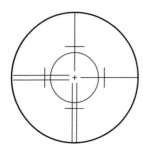

FIGURE 10.14
Reticule with a solar circle.

A rough pointing on the sun is made by sighting over the telescope. The eyepiece, and the objective then are focused until the sun's image and the cross hairs are seen clearly on the card. If the eyepiece of the telescope is erecting, the image on the card will be inverted; if the eyepiece is inverting, the image will be erect. The cross hairs are visible only on the image of the sun.

Because an objective filter must be used on a total station system to protect the EDM sensors, *do not use this indirect* method with a white card with such instruments.

Some theodolites and total station instruments are equipped with a *solar circle*. For this arrangement, the cross-hair reticule, in addition to the usual vertical, horizontal, and stadia cross hairs, has a solar circle inscribed symmetrically about the intersection of the vertical and horizontal cross hairs, as shown in Figure 10.14. This circle has an angular radius of 15′45″, which is slightly smaller than the sun's semidiameter (Section 10.25), so that very accurate pointings on the sun are possible.

10.25
SEMIDIAMETER CORRECTION

Because the sun is large (apparent angular diameter about 32′), its center cannot be sighted precisely with the ordinary theodolite and it is customary to bring the cross hairs tangent to the sun's image. When a sight is taken directly through an erecting telescope and the horizontal cross hair is brought tangent to the lower edge of the sun, the sight is said to be taken to the sun's *lower limb*. This is indicated in the notes by the symbol $\underline{\odot}$. Similarly, the symbol $\overline{\odot}$ indicates a sight to the sun's *upper limb,* $\odot|$ a sight with the vertical cross hair to the sun's *right* or *western limb,* and $|\odot$ a sight to the sun's *left, eastern,* or *trailing limb.*

When a vertical angle is measured to the sun's upper limb, it is necessary to subtract the sun's semidiameter to obtain the observed altitude of the sun's center; when to the lower limb, add the semidiameter. For zenith angles to the sun's upper and lower limbs, subtract and add, respectively, the sun's semidiameter to obtain the zenith distance to the sun's center. The solar ephemerides give values of the semidiameter of the sun for each day of the year. The semidiameter varies from about 15′46″ in July to about 16′18″ in January; for rough calculations, it may be taken as 16′.

When a horizontal angle is measured to the sun's right or left limb, a correction equal to the sun's semidiameter divided by the cosine of the altitude is applied. Therefore, if the altitude h is 60° and the semidiameter is 16′, the correction to a horizontal angle is $16′/\cos h = 32′$. As the sun approaches the zenith, the correction becomes very large and readings should not be taken to one limb only.

The semidiameter correction to a horizontal angle for the sun at any altitude is illustrated in Figure 10.15, in which A is the station of an observer on the earth, S is the sun at the horizon, S' is the sun at some altitude h above the horizon, r is the radius of the sun, and

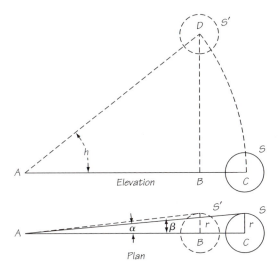

FIGURE 10.15
Semidiameter correction to horizontal angle.

α and β are the small horizontal angles (semidiameter corrections) subtended by the radius of the sun at S and S', respectively. In the plan view, $\alpha \cdot AC = r$ and $\beta \cdot AB = r$; hence, $\alpha \cdot AC = \beta \cdot AB$, or $\beta = \alpha \cdot AC/AB$. But in the elevation view $AC = AD$ and $AD/AB = 1/\cos h$. Therefore,

$$\beta = \frac{\alpha}{\cos h} \tag{10.16}$$

For certain solar observational procedures, an equal number of sights is taken to opposite limbs of the sun; the mean of the horizontal angles and the mean of the vertical angles at the mean of the times are taken, and no corrections for semidiameter are necessary.

10.26
PROCEDURES FOR SIGHTING THE SUN

When the instrument to be used for sighting the sun is not equipped with a solar reticule, it is necessary to take observations on the sun by making the horizontal and vertical cross hairs tangent to the circumference of the sun. The two techniques for accomplishing this task are the center tangent method and the quadrant tangent method. The procedure for each method will be described, assuming a directly observed image as seen through an erecting telescope with either an eyepiece or objective filter, *which must be in place to protect the operator's eyes.*

The *center tangent method,* as performed to determine azimuth by the altitude method (Section 10.31), will be described first. When observations are made in the morning, the telescope is pointed on the sun so that the image of the sun appears as shown in Figure 10.16a, where the vertical cross hair cuts off a segment of the trailing or eastern limb of the sun (dashed circle) and the horizontal cross hair approximately bisects this segment. In the morning, the image of the sun will be moving upward and to the right. Therefore, the segment will get smaller and smaller as the sun moves in its orbit. The horizontal upper

Morning

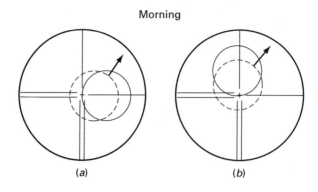

(a)   (b)

Afternoon

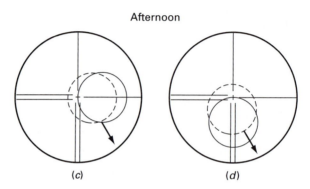

(c)   (d)

FIGURE 10.16
Image of sun as viewed through eyepiece of an erecting
telescope for the center-tangent method. (a) Vertical
cross hair is stationary. (b) Horizontal cross hair is
stationary. (c) Vertical cross hair is stationary.
(d) Horizontal cross hair is stationary.

clamp is locked and the vertical tangent screw is used to adjust the horizontal cross hair to
bisect the disappearing segment of the sun's trailing limb.

When this segment becomes tangent to the vertical cross hair (the solid circle in
Figure 10.16a), the time is recorded and the horizontal and vertical circle readings are
observed. The vertical cross hair is tangent to the trailing limb of the sun so the semidiameter of the sun (Section 10.25) must be added to the horizontal circle reading. Because the
horizontal cross hair bisects the sun, the vertical circle reading corrected for parallax and
refraction (Sections 10.27 to 10.29) requires no correction for the semidiameter of the sun.

Next, the telescope is adjusted so that the vertical cross hair approximately bisects the
disappearing segment of the lower limb of the sun, as shown by the dashed circle in
Figure 10.16b. This time the vertical clamp is locked and the disappearing segment is
bisected by adjusting the upper horizontal tangent motion.

When the lower limb of the sun is tangent to the horizontal cross hair, as shown by the
solid circle in Figure 10.16b, the time and horizontal circle and vertical circle readings are
observed and recorded. The vertical cross hair bisects the sun, so the horizontal circle
reading requires no correction. However, the horizontal cross hair is tangent to the lower
limb of the sun, so that the semidiameter of the sun must be subtracted from a zenith angle
or added to a vertical angle, which first is corrected for parallax and refraction. These

observations on the sun's trailing and lower limbs then are repeated with the telescope in the reversed position. This constitutes one set of readings.

Observations in the afternoon are performed in a similar manner, but the procedure is modified to accommodate the downward movement of the sun in its orbit, which produces an image of the sun moving down and to the right. Therefore, when the vertical cross hair becomes tangent to the trailing limb of the sun, as shown in Figure 10.16c, time and vertical and horizontal circle readings are recorded. The semidiameter of the sun must be added to the horizontal circle reading, and the vertical circle reading must be corrected for parallax and refraction. Next, the horizontal cross hair is made tangent to the upper limb of the sun, as shown in Figure 10.16d. This time the semidiameter of the sun is added to a zenith angle or subtracted from a vertical angle, which first is corrected for parallax and refraction, and the horizontal circle reading requires no correction. As before, these observations should be repeated with the telescope in the reversed position to provide a complete set of readings.

The preceding instructions are equally valid for a solar observation by the hour-angle method (Section 10.31, Procedure 1), except that the image is viewed only as shown in Figures 10.16a (morning) and 10.16c (afternoon), then it is necessary to record only the horizontal angle and instant of time. Observations with the telescope direct and reversed should be performed to provide complete sets in each case.

In the *quadrant tangent method*, both vertical and horizontal cross hairs are brought tangent to the limbs of the sun. The sequence of images seen for morning and afternoon observations by this method is illustrated in Figure 10.17 for an image of the sun viewed directly through the eyepiece of an erecting telescope equipped with a sun filter. Note that the quadrant tangent method is feasible for determining azimuth of the sun by the altitude method but should not be used for the hour-angle procedure.

For observations in the morning, the horizontal cross hair is sighted a short distance above the lower limb of the sun, as shown by the dotted circle is Figure 10.17a. Because the altitude of the sun is increasing, the horizontal cross hair approaches tangency, due to the sun's movement in its orbit. At the same time, the vertical cross hair is kept continuously on the sun's western limb by means of the horizontal upper tangent screw. At the instant when the horizontal and vertical cross hairs are simultaneously tangent to the sun's disk, the motion of the telescope is stopped, the time is observed, and the horizontal and vertical circles are read. A second observation then is taken with the sun in the lower right-hand quadrant, as shown in Figure 10.17b. The procedure is as follows. The vertical cross hair is set a short distance to the right of the sun's eastern limb, as shown in Figure 10.17b by the dashed circle. Because the sun is traveling westward, the vertical cross hair approaches tangency, due to the sun's movement. At the same time, the horizontal cross hair is kept continuously on the sun's upper limb by means of the vertical tangent motion. As before, observations are taken when the horizontal and vertical cross hairs are simultaneously tangent to the sun's disk. The procedure is such that the final setting for either observation requires manipulation of only one tangent screw and the cross hair that is approaching tangency is visible on the sun's disk. The observations just described should be repeated with the telescope in the reversed position to yield one set of measurements. Because the horizontal and vertical circle readings are observed in opposite quadrants, the average values represent measurements to the center of the sun. Of course, the vertical circle readings need to be corrected for parallax and refraction, but corrections for the semidiameter of the sun are not required. The disadvantage of the quadrant tangent method is that the two cross hairs must be made simultaneously tangent to the disk of the sun, a difficult task under the best of conditions.

For afternoon observations in northern latitudes, the procedure is the same as that just described, except that the sun is sighted first in the upper right-hand quadrant (Figure 10.17c) and then in the lower left-hand quadrant (Figure 10.17d).

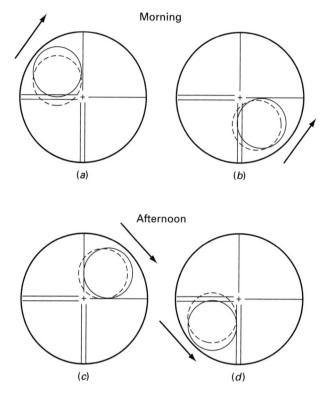

FIGURE 10.17
Image of the sun as viewed through the eyepiece of an
erecting telescope for the quadrant-tangent method:
(a) Horizontal cross hair is fixed. (b) Vertical cross hair
is fixed. (c) Horizontal cross hair is fixed. (d) Vertical
cross hair is fixed.

In both the center tangent and quadrant methods, it is suggested that several sets of
observations be made in rapid succession. When the appropriate corrections have been
applied to the horizontal and vertical circle readings, these corrected values can be plotted
as a function of the time for each measurement. If the data are consistent, the corrected
vertical circle readings and corrected horizontal circle readings will plot as a straight line
versus time, assuming that the observations were taken within a 15- or 20-min period of
time. In this way, blunders can be eliminated before calculations proceed. The procedure for
this operation is described in Section 10.31, Procedure 2.

10.27
PARALLAX CORRECTION

In previous discussions, it was assumed that the celestial sphere is of infinite radius and that
a vertical or zenith angle measured from a station on the surface of the earth is the same
as it would be if measured from a station at the center of the earth. For the fixed stars, this
assumption yields results that are sufficiently accurate for the work described here; but the
distance between the sun and the earth is relatively small, and for solar observations a

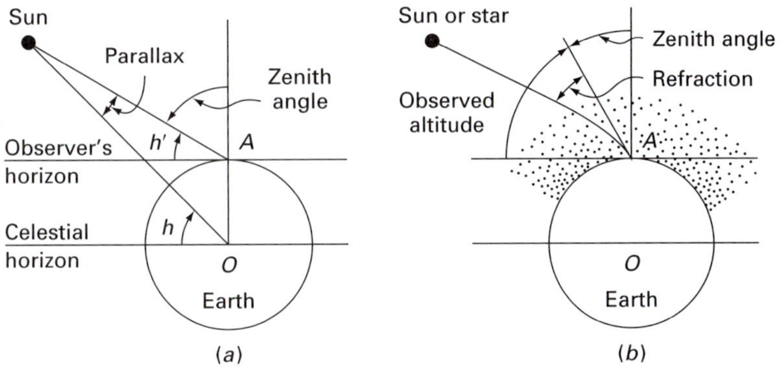

Sun

Parallax

Zenith angle

Observer's horizon

Celestial horizon

h' A

h

O

Earth

(a)

Sun or star

Zenith angle

Refraction

Observed altitude

A

O

Earth

(b)

FIGURE 10.18
(a) Parallax and (b) refraction.

parallax correction is added to the observed altitude to obtain the altitude of the sun from the center of the earth. The parallax correction is subtracted from the zenith angle to obtain the true zenith distance.

In Figure 10.18a, h' is the altitude of the sun above the horizon of an observer at A, and h is the altitude of the sun above the celestial horizon. The parallax correction is equal to the difference between these two angles. As h always is larger than h', the correction must be added to the observed altitude but subtracted from the measured zenith distance.

The parallax correction can be computed, because the distance to the sun is known. The magnitude of the correction depends on the altitude, being 0 when the sun is directly overhead and a maximum when the sun is on the observer's horizon. When the sun is on the observer's horizon, the correction C_h is called the *horizontal parallax*. It can be demonstrated readily that the parallax correction C_p for any observed altitude h' or zenith angle, z', is

$$C_p = C_h \cos h' = 8.94 \cos h' \tag{10.17a}$$

$$C_p = 8.94 \sin z' \tag{10.17b}$$

in which C_p is in seconds.

Corrections for parallax and refraction are usually made together (see Section 10.29).

10.28
REFRACTION CORRECTION

When a ray of light emanating from a celestial body passes through the atmosphere of the earth, the ray is bent downward, as illustrated in Figure 10.18b. Hence, the sun and stars appear to be higher above the observer's horizon than they actually are. The angle of deviation of the ray from its direction on entering the earth's atmosphere to its direction at the surface of the earth is called the *refraction* of the ray. A *refraction correction, C_r,* of an amount equal to the refraction is subtracted from the observed altitude to determine the actual altitude h' above the observer's horizon. When the zenith angle, z', is observed, the refraction correction, C_r, is added to the observation to determine the true zenith angle, z.

The magnitude of the refraction correction depends on the temperature and barometric pressure of the atmosphere and the altitude of the ray, varying as the cotangent of the altitude. It does not depend on the distance to the body from which the ray emanates. The refraction correction can be calculated (U.S. Naval Observatory, 1995, p. B62) using

$$C_r = \frac{0.00452\ p\ \tan z}{(273 + T)} \tag{10.18a}$$

$$C_r = \frac{0.00452\ p\ \cot h}{(273 + T)} \tag{10.18b}$$

in which C_r = refraction correction, deg
p = atmospheric pressure, mbar
z = zenith angle
h = vertical angle to the celestial body
T = temperature, deg. Celsius

or

$$C_r = \frac{0.27306\ p\ \tan z}{(460 + T)} \tag{10.19a}$$

$$C_r = \frac{0.27306\ p\ \cot h}{(460 + T)} \tag{10.19b}$$

where p = atmospheric pressure, in. Hg
T = temperature, deg. Fahrenheit

Owing to uncertainties of the refraction correction for low altitudes, observations for precise determinations are never taken on a celestial body that is near the horizon.

10.29
COMBINED CORRECTION

For solar observations, refraction and parallax corrections usually are made together. The refraction correction, which is additive to zenith angles, is many times larger than the parallax correction, which is subtractive from zenith angles; hence, the combined correction is additive to zenith angles. Conversely, this correction should be subtracted from an observed vertical angle to obtain the true vertical angle.

10.30
DECLINATION OF THE SUN

To determine the azimuth, latitude, or longitude by solar observations, it is necessary that the declination of the sun at the instant of sighting be known. The declination is obtained by interpolating between values given in a solar ephemeris for the current year.

The *Nautical Almanac* gives the apparent declination of the sun for each day and every hour (Greenwich mean or universal time), for every day of the year. The Sokkia ephemeris provides the apparent declination at 0^h universal time for every day of the year. In the Mica interactive software almanac, the user can specify the number of repetitions and tabulation interval. So, it would be possible to get the apparent declination listed for the desired hours of UT1 time on a specified day for a given year.

The following examples illustrate the use of the Sokkia ephemeris and the *Nautical Almanac* to determine declination.

EXAMPLE 10.10. An observation is taken on the sun at 10^h00^m Eastern standard time, on December 15, 1995. It is desired to determine the declination at the given instant.

Solution. The Greenwich civil time at the instant of observation is $10^h00^m + 5^h = 15^h00^m = 15.00^h$. By the Sokkia ephemeris, for 0^h Greenwich civil or universal time, the declinations, δ, for December 15 and 16 are

$$\delta_{\text{Dec. 16}} = -23°13'34.5''$$

$$\delta_{\text{Dec. 17}} = -23°16'46.8''$$

Change $= -03'12.3''$ and the rate of change $= (-03'12.3'')/24 = -8.013''/h$. The change in declination since 0^h G.C.T. $= (-8.013'')(15) = -120.2'' = -02'00.2''$. Therefore, the declination at the instant of observation $= -23°13'34.5'' - 02'00.2''$ or δ at instant of observation $= -23°15'34.7''$.

The *Nautical Almanac* gives $-23°15.6'$ directly for 15.00^h G.C.T. on December 16, 1995.

EXAMPLE 10.11. An observation is taken on the sun as it crosses the meridian on November 16, 1995, at a place where the longitude is $87°49'30''$ west of Greenwich. Determine the declination at the given instant.

Solution

$$\text{G.A.T.} = (87°49'30'')/15 = 5^h51^m18^s = 5.855^h \text{ P.M.}$$

or

$$\text{G.A.T.} = 17.855^h = 17^h51^m18^s$$

It is necessary to compute the change from 0^h G.C.T. to 17.855^h G.A.T. The equations of time from the Sokkia ephemeris for November 16 and 17 are

November 16, equation of time $=$ apparent $-$ mean $= 15^m23.58^s$

November 17, equation of time $=$ apparent $-$ mean $= 15^m12.83^s$

Change in the equation of time $= -10.75^s$ or $(-10.75^s)/24^h = -0.45^s$ per hour. Therefore, at 17.8550^h G.A.T., we have

$$15^m23.58^s + (17.8550^h)(-0.45^s/h) + \text{G.C.T.} = \text{G.A.T.}$$

$$\text{G.C.T.} = 17^h51^m18^s - 15^m15.55^s = 17^h36^m02^s = 17.6007^h$$

or the G.C.T. of local apparent noon $= 17.6007^h$. From the ephemeris,

$$\delta_{\text{Nov. 16}} = -18°33'27.9''$$

$$\delta_{\text{Nov. 17}} = -18°48'30.8''$$

The change $= -15'02.9''$ or $= (-15'02.9'')/24 = -37.62''/h$ of elapsed time. Consequently, the declination at local apparent noon on November 16, 1995 is

$$\delta_{\text{local app. noon}} = (-18°33'27.9'') + (-37.62''/h)(17.6007^h)$$

$$= -18°44'30.0''$$

In the *Nautical Almanac*, computation of G.C.T. of local apparent noon is the same as just shown. Then, because the declination of the sun is given for each hour of every day, we have the following declinations for November 16, 1995 at 0^h G.C.T.:

$$\delta \text{ at } 17^h00^m = -18°44.1'$$

$$\delta \text{ at } 18^h00^m = -18°44.8'$$

The change is $-0.7'$ per hour and, at 17.6006^h G.C.T. of local apparent noon, the change would be $(-0.7'/h)(0.6^h) = -0.4'$ so that δ at 17.6006^h or local apparent noon $= -18°44.5'$. Note that, in the *Nautical Almanac,* declination is given only to tenths of minutes.

The preceding calculations use a linear interpolation to obtain the declination for the sun, which actually moves in a nonlinear orbit. Consequently a very small error, usually insignificant (about 2–$3''$ of arc), is introduced. To ensure optimum accuracy, the following two-point interpolation equation can be used (see Elgin et al., annual):

$$\delta_{obs} = \delta_{0h} + (\delta_{24h} - \delta_{0h})(\text{G.C.T.}_{\cdot obs}/24) + (0.0000395)(\delta_{0h}) \sin [(7.5)(\text{G.C.T.}_{\cdot obs})]$$

in which the G.C.T. is expressed in decimal hours. For this example,

$$\delta_{obs} = -18°44'29.9'' + (0.0000395)(-18.55775) \sin [(7.5)(17.6006)]$$

$$\delta_{obs} = -18°44'29.9'' -2.0'' = -18°44'31.9''$$

The Mica software package for determining ephemeris data was not used in the previous examples, so a few comments in this respect are appropriate. The Mica Interactive Astronomical Almanac is a very powerful tool for generating tables and data for celestial observations. In Example 10.10, for instance, one could specify, by way of pull-down menus, the year, date, and time for which the declination was desired and, by proper use of the programmed instructions, acquire the declination, sidereal hour angle, and equation of time for that instant of time directly. It is beyond the scope of this textbook to go into further detail concerning this or other software packages for generating ephemerides. The interested user needs a suitable handheld calculator or personal computer, the software package, and the user's guide (see U.S. Naval Observatory, for the Mica *User's Guide*; and Elgin et al., 1986, for other software ephemerides).

The balance of the examples in this chapter will be referenced to the Sokkia ephemeris and *Nautical Almanac.*

10.31
AZIMUTH BY DIRECT SOLAR OBSERVATION

Azimuth from direct solar observations can be determined by two procedures: the hour-angle method and the altitude method.

Procedure 1: Hour-Angle Method

The hour-angle method requires very accurate determination of time and no zenith or vertical angle. In the past, the hour-angle method was used little because of the difficulty of obtaining highly accurate time. With pretuned receivers for getting accurate time signals, digital watches that permit recording split times, and time modules usable in pocket calculators, acquisition of accurate time no longer is such a problem.

To determine azimuth of a given line by the hour-angle method, a horizontal angle from the line to the sun is measured and the time of observation is recorded. It also is necessary to know the latitude and longitude of the station of observation, which can be scaled from a map, found by GPS observations, or determined by the methods outlined in Sections 10.33 and 10.34. Then, using the declination of the sun at the given instant, as determined from data in the ephemeris (Section 10.30, Examples 10.10 and 10.11), the *PZS* triangle can be solved by Equation 10.13, which is repeated here for convenience,

$$\tan Z = \frac{-\sin t}{\tan \delta \cos \phi - \sin \phi \cos t} \qquad (10.13)$$

in which t = L.H.A. of the sun (Section 10.5)
  δ = declination of the sun at instant of observation
  ϕ = latitude of the place of observation
  Z = clockwise or counterclockwise angle from north in the astronomical triangle

The azimuth of the sun from the north, A_N, is determined from Z as follows:

1. When the sun is *east* of the meridian (L.H.A. is 180°–360°) and $\tan Z$ is positive, $A_N = Z$. When $\tan Z$ is negative, $A_N = Z + 180°$.
2. When the sun is *west* of the meridian (L.H.A. is 0°–180°) and $\tan Z$ is positive, $A_N = 180° + Z$. If $\tan Z$ is negative, $A_N = 360° + Z$.

The azimuth of the line is computed from the azimuth of the sun and the observed horizontal angle. The usual procedure is as follows. Prior to going into the field, the timing device to be used should be checked very carefully and set using the time signals from WWV or WWVH, received either from a shortwave radio or by calling (303) 499–7111. In each case, the DUT correction (Section 10.21) used to correct UT to UT1 time should be determined. It is recommended that a pocket calculator equipped with a time module and stopwatch be used as a timing device. In this way, the instantaneous time of centering on the sun can be recorded and stored by the instrument operator simply by pressing a single key. This operation reduces the total time required to complete the set of angles and yields the timing accuracy necessary for hour-angle method.

Next, the theodolite or total station system is set up and leveled very carefully over the station at one end of the line. Use of an instrument with a vertical circle compensator is recommended, in which case the final leveling should be performed using the vertical circle compensator (Section 6.40). Sight along the given line with the telescope direct and the horizontal circle set to 0 or, if a direction instrument is being used, observe the horizontal circle. At this stage, before sighting on the sun, *be absolutely certain the objective or eyepiece sun filter is in place or severe damage to the operator's eye may result*. Loosen the horizontal clamp (upper clamp if a repeating instrument is being used) and take a series of sights to the sun, first with the telescope direct and then reversed, according to the directions given in Section 10.24. Record time and horizontal circle readings for each observation, using either a data collector or field notebook.

When the center tangent method is utilized, at least four pointings on the sun, as illustrated in Figure 10.16a, should be made for morning observations. In the afternoon, the pointings would correspond to Figure 10.16c. No zenith or vertical angle is being measured, so it is essential only that the trailing edge of the sun be tangent to the vertical cross hair at a point near the cross-hair intersection. At this instant, the key on the timing device is pressed to record time for the measurement. Next, the telescope is reversed, and the entire group of pointings is repeated. When observations are completed, the telescope, now in reversed position, again is sighted along the line and the circle reading is observed for checking.

When the instrument is equipped with a solar circle, the required number of repetitions with the telescope direct and reversed are taken, centering directly on the sun. As in the other methods, the horizontal angle and time are recorded for each pointing.

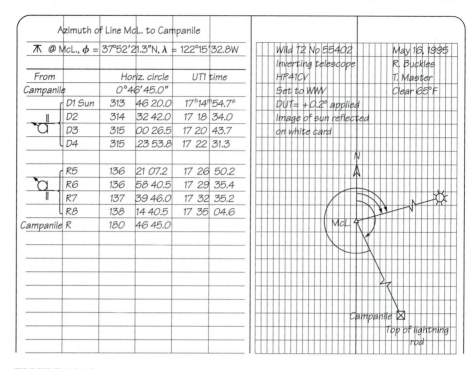

FIGURE 10.19

Field notes for a direct solar observation, hour-angle method.

If sighting on the sun is performed indirectly, using a white card held behind the eyepiece of the telescope (Section 10.24), the party should consist of at least two people: one holding the white card behind the telescope and the other operating the theodolite and manipulating the horizontal tangent screw and the key on the timing device. In this case, motions of the sun are the reverse of those shown in Figure 10.16, with the image of the sun moving down and to the right in the morning and up and to the right in the afternoon.

Figure 10.19 shows notes for a solar observation made from an instrument setup at station McLaughlin (McL.) with a backsight on station Campanile. The instrument used was a direction theodolite, so the recorded measurements of the horizontal circle are directions. Recorded times were determined with a time module in a hand calculator set directly to UT1 time from WWV with the DUT correction applied (Section 10.21). Because the theodolite used was not equipped with either a solar reticule or objective filter, pointing on the trailing limb of the sun was done using the image of the sun reflected on a card (Section 10.26). A correction for the semidiameter of the sun, β, found by Equation (10.16), must be added to the horizontal angles. Because the altitude of the sun is not observed in this method, it must be calculated using Equation (10.14).

Reduction of the data and computation of the azimuth of the line McL. to Campanile, using the second observation with the telescope direct, are shown in Table 10.1 to illustrate the procedure. The geometry of the spherical triangle is shown in Figure 10.20. First, angles are computed from the directions and listed along with UT1 times of observation. The latitude and longitude are known for the point (scaled from a U.S.G.S. topographic quadrangle map) and values for the equation of time and declination of 0^h, May 16 and 17, from the ephemeris permit determining the G.H.A. (UT1 time) and declination at the instant

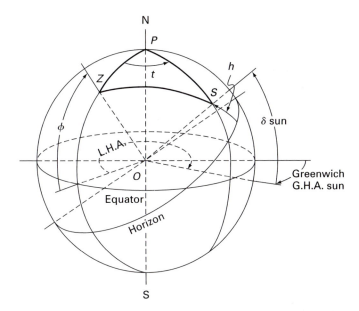

FIGURE 10.20
Spherical triangle for solar observation in Table 10.1.

of observation. Note that the G.H.A. also could be determined by interpolation between tabulated values for G.H.A. of the sun at 0^h on May 16 and 17, as given in the ephemeris and using the following equation:

$$\text{G.H.A.}_{\text{obs}} = \text{G.H.A.}_{\text{oh}} + (\text{G.H.A.}^{24h} - \text{G.H.A.}^{0h} + 360°)(\text{UT1}/24) \qquad (10.20)$$

so that using the given data

$$\text{G.H.A.}_{\text{obs}} = 180°55'02.3'' + (180°54'48.2'' - 180°55'02.3'' + 360°)(17.309444^h/24^h)$$
$$= 80°33'22.2''$$

which corresponds to the value calculated in Table 10.1. The G.H.A.$_{\text{obs}}$ and west longitude next are used in Equation (10.1c) to get the angle t and the L.H.A. Then, the declination, latitude, and t are substituted into Equation (10.13) to yield the angle Z. Because the sun is east of the meridian, the L.H.A. is between 180° and 360°, and Z is negative, A_N of the sun $= Z + 180°$.

The azimuth of line McL. to Campanile is found by applying the observed horizontal angle from the mark to the sun, corrected for the semidiameter of the sun to the azimuth of the sun. The azimuths for the remaining measurements then are calculated and listed. Examination of these azimuths reveals no blunders. However, the sixth value, 151°21'32.3'', deviates from the mean by 17'', so this value and D1 are deleted. The remaining azimuths are in good agreement with $\sigma = 3.1''$ and a mean of 151°21'45.1'' is taken as the final value for the azimuth.

Note that, prior to examining these calculated azimuths, no check on the quality of observed data had been performed. To check for blunders in the observed times and horizontal angles, the angles could have been plotted versus time, as done in Procedure 2 of this section, on azimuth by the altitude method. However, for the hour-angle method, most observations are made with either a total station system equipped with data collector or a theodolite and hand calculator containing a stored program for astronomical reduction

TABLE 10.1
Computation of azimuth by solar observations, hour-angle method

Telescope	Horizontal angle	UT1 time
D1	312°59′35.0″	$17^h14^m54.7^s$
D2	313°45′57.0″	$17^h18^m34.0^s$
D3	314°13′41.5″	$17^h20^m43.7^s$
D4	314°37′08.8″	$17^h22^m31.3^s$
R5	315°34′22.2″	$17^h26^m50.2^s$
R6	316°11″55.5″	$17^h29^m35.4^s$
R7	316°53′01.0″	$17^h32^m35.2^s$
R8	317°27′55.5″	$17^h35^m04.6^s$

Notes:

Date: May 16, 1995; Instrument at Station McL. (McLaughlin)
For the second observation, D2, the horizontal clockwise angle from McL. to sun = 314°45′57.0″ and UT1
time of observation = $17^h18^m34.0^s$ = 17.309444^h, $\phi = 37°52′21.3″$, and $\lambda = 122°15′32.8″$.
From the ephemeris,

Equation of time 0^h, 5-16-95 = apparent − mean   = $00^h03^m40.16^s$

Equation of time 24^h, 5-16-95   = $00^h03^m39.21^s$

Difference in equation of time   = -0.95^s

Correction to time of observation = $(-0.95)(17.3/24)$ = -0.68^s

Equation of time at 17.309444^h = $00^h03^m40.16^s − 0.68^s$ = $00^h03^m39.48^s$

UT1 at observation   = $17^h18^m34.0^s$

G.A.T. at observation   = $17^h22^m13.48^s$

  $− 12^h$

G.H.A. of sun at instant of observation   = $5^h22^m13.48^s$

G.H.A. converted to degrees   =   80°33′22.2″

West longitude   = $-122°15′32.8″$

t = G.H.A. − west longitude (Equation (10.1c))   = $-41°42′10.6″$

L.H.A. = 360° + t (Equation (10.1a))   = 318°17′49.4″

From ephemeris,

$$\text{Declination of sun } 0^h = \delta_{0h} = 18°56′17.0″$$

$$\text{Declination of sun } 24^h = \delta_{24h} = 19°10′11.0″$$

$$\text{Declination}_{obs} = \delta_{0h} + (\delta_{24h} − \delta_{0h})(UT1/24) + (0.0000395(\delta_{0h})) \sin[(7.5)(UT1)]$$

$$\text{Declination}_{obs} = 19°06′20.4″ = 19.105665°$$

Then, by Equation (10.13),

$$Z = \tan^{-1}[-(\sin t)/(\cos \phi \tan \delta − \sin \phi \cos t)] = -74.466700°$$

The L.H.A. is 180–360°, the sun is east of the meridian, and Z is negative, so

$$\text{Azimuth of sun} = 180° + Z = 105.53330 = 105°31′59.8″$$

From the ephemeris,

$$\text{Semidiameter of sun at } 0^h \text{ UT} = 15′50.7″$$

$$\text{Semidiameter of sun at } 24^h \text{ UT} = 15′50.5″$$

Correction at $17^h18^m34.0^s$ = 15′50.7″ + (15′50.5″ − 15′50.7″)(17.3094/24) = 15′50.6″

By Equation (10.16),

$$\beta = (15′50.6″)/ \cos h$$

$$h = \sin^{-1}(\sin \phi \sin \delta + \cos \phi \cos \delta \cos t) = 49.273095°$$

(Equation (10.14))

(*continued*)

TABLE 10.1 (*concluded*)

595

CHAPTER 10:
Introduction to
Astronomy

$$\beta = (15'50.6)/\cos 49.273095 = 0.4047112° = 0°24'16.9''$$

$$\text{Horizontal angle to center of sun} = 313°45'57.0'' + 0°24'16.9''$$

$$= 314°10'13.9''$$

$$\text{Azimuth of line McL. to Campanile} = A_2 = [360° - (314°10'13.9'') + (105°31'59.8'')]$$

$$= 151°21'45.8''$$

Similarly, the balance of the azimuths are

$$\text{Azimuth line McL. to Campanile} = A_1 = 151°21'53.5''$$

$$A_3 = 151°21'50.8''$$

$$A_4 = 151°21'44.1''$$

$$A_5 = 151°21'42.7''$$

$$A_6 = 151°21'27.3''$$

$$A_7 = 151°21'42.1''$$

$$A_8 = 151°21'45.2''$$

of the data. These devices can compute the azimuth of the observed lines in real time. One needs only to load ephemeris data (declination, G.H.A. or equation of time, and semidiameter of sun) before going into the field. Some astronomical modules (e.g., Elgin et al., 1986) generate these parameters internally. Thus, azimuths of the lines observed can be calculated as soon as measurements are completed. Bad data then are deleted and observations repeated before leaving the station. When deleting bad measurements, be sure to remove both a direct and a reversed observation, so that the final mean value contains an equal number of direct and reversed measurements.

Procedure 2: Altitude Method

Finding the azimuth by the altitude method is not as accurate as by the hour-angle method and should be used only if determination of accurate time using a time module set to UT1 time by WWV is not available.

In this method, at a known instant of time, the sun is observed and the zenith angle or altitude of the sun and the horizontal angle from a given reference point to the sun are measured. The declination of the sun at the given instant is found from a solar ephemeris. With the declination, δ, latitude, ϕ, and altitude, h, known, the *PZS* triangle can be solved as described in Section 10.11, using Equation (10.2b), repeated here for convenience:

$$\cos Z = \frac{\sin \delta}{\cos h \cos \phi} - \tan h \tan \delta \qquad (10.2b)$$

in which Z is the clockwise or counterclockwise angle from north used to calculate the azimuth from the north of the sun, as detailed in Section 10.11. The azimuth of the given line then is computed from the azimuth of the sun and the observed horizontal angle.

The observational procedure is similar to that outlined previously for the hour-angle method, except a vertical circle reading is required for each pointing on the sun and time is not so critical. First, assume that the instrument used has an erecting telescope equipped with an objective or eyepiece filter so that the sun can be observed directly through the telescope.

When the center tangent method is utilized, four pointings on the sun, as illustrated in Figure 10.16, should be made with the telescope in the direct position. In the morning, these measurements would be made pointing at the sun as shown in Figure 10.16a, b, a, b. In the afternoon, the pointings would correspond to those shown in Figure 10.16c, d, c, d. Next, the telescope is reversed and the entire group of four pointings is repeated. When the observations are completed, the telescope, now in the reversed position, again is sighted along the line and the circle reading is observed for checking.

When the quadrant tangent method is employed, four pointings, as shown in Figure 10.17a, b, a, b, are made in the morning with the telescope direct. In the afternoon, the pointings would correspond to Figure 10.17c, d, c, d. As in the center tangent method, these four pointings then are repeated with the telescope reversed and finally a sight is taken along the original line.

When the instrument is equipped with a solar circle, the required number of repetitions with the telescope direct and reversed are taken pointing directly at the center of the sun. As in the other methods, time, vertical circle, and horizontal circle readings are recorded for each pointing.

Figure 10.21 shows notes for a solar observation made from an instrument set up on triangulation station 46 with a sight at triangulation station 48. The point sighted and position of the telescope are recorded in columns 1 and 2. Note that a sketch is made to indicate the position of the sun with respect to the cross hairs for each observation and the order of the readings corresponds to Figure 10.17b, a, b, a and d, c, d, c. The procedure followed for this example is the center tangent method using the image of the sun reflected on a card (Section 10.26), so that the sketches represent the reflected image. The image of the sun moves up and to the right for afternoon observations when the card is viewed

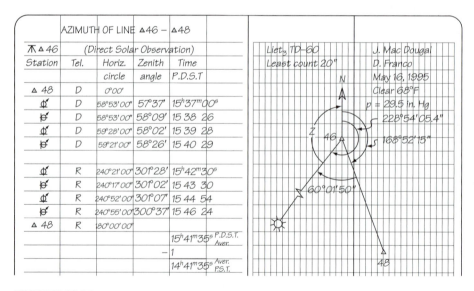

FIGURE 10.21
Field notes for a direct solar observation, altitude method.

from the objective end of the telescope. The horizontal and vertical circle readings for each pointing are recorded in columns 3 and 4, and the time (P.D.S.T.) is in column 5.

On completion of the observations, it is necessary to correct the horizontal and vertical circle readings for refraction and parallax and for the semidiameter of the sun, where necessary. The calculations needed for these corrections are summarized in Table 10.2. Each zenith angle is corrected for parallax and refraction (Sections 10.27 to 10.29) using Equations (10.17) and (10.18) or (10.19). In addition, each zenith angle to the upper limb of the sun (observations were in the afternoon) is corrected by subtracting the semidiameter of the sun, which is found from the ephemeris. Similarly, each horizontal circle reading taken on the left or trailing limb of the sun must be corrected by adding a correction for the semidiameter of the sun to the observation (Section 10.25). Equation (10.16) is used to calculate these corrections. Corrected horizontal- and zenith-angle readings are listed in Table 10.2.

To detect inconsistent data, the corrected horizontal- and zenith-angle readings are plotted on a graph, where the ordinate is the time of observation and the corrected horizontal and zenith angles are plotted on the abscissa. If the observations have been taken over a short period of time of, say, 10 min, the corrected values for horizontal and zenith angles should plot as straight lines versus time. Any substantial deviation from a straight line indicates a mistake in the observations.

Figure 10.22 shows such a graph plotted using the corrected zenith and horizontal angles in Table 10.2 and corresponding times recorded in Figure 10.21. Direct and reversed zenith angles plot as a single straight line, indicating close to no instrumental error

TABLE 10.2
Corrections to zenith and horizontal circle readings for solar observation

Measured vertical angle, z'	Refraction and parallax	Semidiameter of sun	Total correction	Corrected zenith angle, z
57°37′	+1.32′	+15′50.5″	+17′10″	57°54′10″
58°09′	+1.35′		+01′21″	58°10′21″
58°02′	+1.34′	+15′50.5″	+17′11″	58°19′11″
58°26′	+1.36′		+01′22″	58°27′22″
58°32′	+1.37′	+15′50.5″	+17′13′	58°49′13″
58°58′	+1.39′		+01′24″	58°59′24″
58°53′	+1.39′	+15′50.5″	+17′14″	59°10′14″
59°23′	+1.42′		+01′25″	59°24′25″

Average corrected zenith angle = 58°39′18″

Measured horizontal angle	Correction $\beta = \alpha'/\cos h$	Corrected horizontal angle
		58°53′00″
58°53′00″	15′50.5″/sin 58°10′21″ = +18′39″	59°11′39″
		59°28′00″
59°21′00″	15′50.5″/sin 58°27′22″ = +18′35″	59°39′35″
		60°21′00″
60°17′00″	15′50.5″/sin 58°59′24″ = +18′29″	60°35′29″
		60°52′00″
60°55′30″	15′50.5″/sin 59°24′25″ = +18′24″	61°13′54″

Average corrected horizontal angle = 60°01′50″

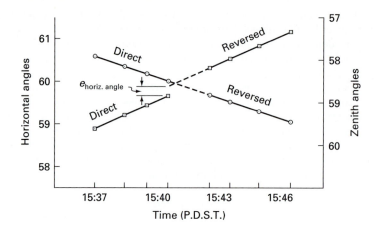

FIGURE 10.22
Zenith and horizontal angles plotted versus time for a solar
observation.

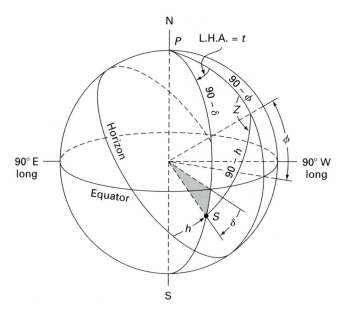

FIGURE 10.23
Spherical triangle for the solar observation in Table 10.3.

present. Direct and reversed horizontal angles plot as two straight parallel lines separated
by an amount $e_{horiz.\ angle}$ that represents the collimation error in the theodolite (Section 6.45). For both zenith and horizontal angles, the averages of direct and reversed
horizontal angles are free from collimation errors. The results of Figure 10.22 indicate no
blunders in the measurements.

Computations of azimuth using the data from Figure 10.21 and Table 10.2 are given
in Table 10.3. Figure 10.23 shows the astronomic triangle used in this computation. The average corrected horizontal and zenith angles and average time are used for this computation.

TABLE 10.3

599

CHAPTER 10:
Introduction to
Astronomy

Calculation of azimuth from solar observations

Date: October 11, 1995
Latitude = 37°52'20"N
Instrument at △ 46, sight on △ 48
Average corrected zenith angle to center of sun = 58°39'18"
Average corrected horizontal angle (clockwise) to the
   center of the sun = 60°01'50"

Average instant of observation	= $14^h41^m35^s$P.S.T.
Longitude of the 120th meridian	= 8^h
Greenwich civil (UT) time of observation	= $\overline{22^h41^m35^s}$ = 22.6931^h

From the ephemeris
δ_{sun} at 0^h, 10/11/95 = −6°45'21.7"
δ_{sun} at 0^h, 10/12/95 = −7°08'00.2" = δ_{24h}
$\delta_{inst.\,obs} = \delta_{0h} + [(\delta_{24h} - \delta_{0h})\,(UT/24)] + [0.0000395\,(\delta_{0h})]\sin[(7.5)(UT)]$
$\delta_{inst\,obs}$ = − 7°06'46.4"
Calculate Z using Equation (10.2a):

$$\cos Z = \frac{\sin\delta - \sin\phi\sin h}{\cos\phi\cos h}, \text{ where } h = 90° - Z = 31°20'42.0"$$

$$= \frac{\sin(-07°06'46.4") - \sin(37°52'20")\sin(31°20'42.0")}{\cos(37°52'20")\cos(31°20'42.0")}$$

$$= -0.65735559$$

So, $Z = 131°05'54.6"$ counterclockwise from the north.

Azimuth of the sun from north	= 360° − 131°05'54.6"
	= 228°54'05" clockwise from north
Clockwise angle backsight on 48 to sun	= 60°01'50" (see Table 10.2)
Azimuth of 46 to 48	= $\overline{168°52'15"}$

The average Pacific daylight saving time (P.D.S.T.) is reduced to Pacific standard time (P.S.T.; Figure 10.21), which in turn is converted to Greenwich civil time or UTC. The declinations for 0^h and 24^h UTC for May 16 and 17, 1995, are obtained from the ephemeris and used to compute a declination corrected for the elapsed time to the instant of observation. Then, the declination, δ, the latitude of the place, ϕ, and true altitude, $h = 90° - z$, are used in Equation (10.2a) to calculate cos Z. The observations were made in the afternoon, so the sun is west of the meridian and Z is the counterclockwise angle from north.

Hence, the azimuth of the sun from the north is $360° - Z$ (Figure 10.23; see the explanation following Equation (10.2b), in Section 10.11). The desired azimuth of the line is calculated by subtracting the clockwise angle from line 46–48 to the sun from the azimuth of the sun as given in Table 10.2 and illustrated in Figure 10.21.

Use of the average zenith angle, horizontal angle, and time implies that the sun apparently is traveling in a straight line, which, of course, is not true. Within a period of 10^m, however, the error introduced is so small as to be of little or no consequence.

The watch time does not have to be of absolute accuracy because it is used only to check the validity of the measurements and to determine declination.

10.32
PRECISION IN AZIMUTH BY SOLAR OBSERVATIONS

Procedure 1: Hour-Angle Method

The error in the computed azimuth, determined by solar observation for a given line, is a function of the error in the horizontal angle turned from the backsight to the sun and the errors propagated in the calculation of the azimuth of the sun.

Error propagation in horizontal angles turned by repetition using repeating and direction theodolites was covered in Section 6.50. For example, an angle turned two times with the telesope direct and two time with the telescope reversed (2DR, Section 6.35), with a repeating theodolite having a least count of 20″ would have an estimated standard deviation of about 7″ (Table 6.1). Another error that affects the horizontal angle is caused by the vertical axis of the instrument not being truly vertical. This condition can cause an error in the horizontal angle that varies as the tangent of the angle of inclination of the telescope when the error is at right angles to the line of sight. Most total station instruments have a plate level bubble with a sensitivity of of 30″/2 mm division. If this bubble is centered to within a quarter of a division, then the vertical axis of the instrument could be in error by 7.5″ or, say, 10″ to be on the safe side. A discrepancy of this size could result in an error of about 8″ in a horizontal angle turned to point with a zenith angle of 50° or at an altitude of 40° (Section 6.45, Part 1 and Figures 6.47 and 6.48).

Errors in the calculated azimuth result from small errors in the time, latitude, longitude, and declination, all of which propagate in Equation (10.13). This error propagation depends not only on the precision of the field observations but also on the shape of the astronomical triangle. The PZS triangle becomes weak as the sun approaches the meridian and becomes indeterminate at the instant of apparent noon.

The effects of errors in the sides of the astronomical triangle on the precision of the azimuth at the instant of observation are given in Table 10.4, which shows changes in azimuth for specified variations in time, latitude, longitude, declination, and misleveling, all at a latitude of 40°N, which is about typical for United States. The changes have been computed by Equation (10.13) and are primarily for comparison but also are useful for determining an error budget for the final azimuth of a given line. In other words, one can assume that the changes in Table 10.4 represent valid estimates for the errors in azimuth due to related parameters (t, ϕ, etc.), so that the square root of the sum of the squares of the changes provides a rough approximation of the total error in the azimuth. The following examples illustrate how Table 10.4 can be used to estimate the accuracy of azimuths obtained from solar observations.

EXAMPLE 10.12. A solar observation by the hour-angle method is to be performed at midafternoon in June at a latitude of about 40°N, using a repeating theodolite with a least count of 20″ to make four direct and four reversed pointings on the sun. The

TABLE 10.4
Changes in azimuth of sun at 40°N and calculated by the hour-angle method, with specified changes in time, latitude, longitude, declination, and leveling

Months	Declina-tion	Hour angle, t	Alti-tude, h	Change in azimuth for changes of				
				0.4^s, in time	4″, in ϕ	4″, in λ	4″, in δ	10″, due to misleveling
November December January	$-20°$	3^h15^m	15°	5″	1″	3″	3″	3″
March September	0°	4^h40^m	15°	4″	1″	3″	3″	3″
	0°	3^h20^m	30°	5″	2″	3″	3″	6″
	0°	1^h30^m	45°	8″	3″	5″	2″	10″
May June July	$+20°$	5^h45^m	15°	4″	1″	2″	3″	3″
	$+20°$	4^h30^m	30°	4″	2″	3″	4″	6″
	$+20°$	3^h10^m	45°	5″	4″	3″	5″	10″
	$+20°$	1^h45^m	60°	8″	6″	6″	5″	17″

estimated standard deviation in the horizontal angle to the sun is 7″ (Section 6.50, Table 6.1) and estimated errors in azimuth due to errors in t, ϕ, λ, δ, and for an error in misleveling of 10″ are approximated by the changes given in Table 10.4 for an hour angle of about 3^h.

Solution. The estimated total error in azimuth, $E_{azimuth}$, is

$$E_{azimuth} = [(5)^2 + (4)^2 + (3)^2 + (5)^2 + (10)^2 + (7)^2]^{1/2} = 15″$$

EXAMPLE 10.13. Assume the same conditions as given for Example 10.12 except the observations are to be made in late afternoon in June.

Solution. The approximate total error in azimuth is

$$E_{azimuth} = [(4)^2 + (1)^2 + (2)^2 + (3)^2 + (3)^2 + (7)^2]^{1/2} = 9″$$

Note that, by simply altering the time of observation, a substantial increase in the estimated accuracy is achieved. In addition, if the theodolite were precisely leveled by using the vertical circle compensator (Section 6.40), if it is so equipped, the error due to misleveling could be reduced significantly.

The preceding analysis is restricted to the conditions of Table 10.4. Similar tables that cover a wider range of latitudes can be found in Elgin, Knowles, and Senne (1992).

Procedure 2: Altitude Method

In the azimuth by the altitude method, errors in ϕ, h (or z), and δ are propagated through Equation (10.2b) and then combined with the error in the horizontal angle caused in the repetition process and misleveling. A table similar to Table 10.4 could be prepared to analyze specific cases. In general, if care is exercised in the instrument setup (e.g., precise leveling) and in sighting the sun, azimuths with a standard deviation in the 10–15″ range should be possible. However, certain precautions are necessary. Because refraction errors are large when the sun is close to the horizon, observational times for this method should be restricted to 8–10 A.M. and 2–4 P.M. Also, a high degree of expertise is required to obtain horizontal and vertical angles of sufficient precision to reach accuracy by this method comparable to that possible by the hour-angle procedure, which is simpler from the operational standpoint.

With respect to both methods, it is important to note that, although the approach taken in Procedure 1 with Table 10.4 is good for comparative purposes and approximate evaluation of the resulting total errors, rigorous error propagation may be applied to Equation (10.13) or (10.2b) when a specific set of data is given. Rewrite Equation (10.13), Section 10.13 as

$$\tan Z = f(t, \phi, \delta)$$

from which

$$Z = \tan^{-1}[f(t, \phi, \delta)]$$

If the three quantities t, ϕ, and δ are *uncorrelated* measurements, then the variance of Z, σ_z^2, can be determined by using Equation (2.38), Section 2.19.

Errors in the horizontal angle are included by propagating through the equation to compute the azimuth of the given line by using Z. For the example computed in Table 10.1, it would be necessary to propagate through the equation

$$A_{Mcl.-Campanile} = [360° - (Z + \alpha)]$$

where α is the corrected horizontal angle from the backsight to the sun. The standard deviation in Z then is

$$\sigma_A = (\sigma_Z^2 + \sigma_\alpha^2)^{1/2}$$

in which σ_z is determined as just outlined and σ_α is estimated on the basis of the equipment and procedures employed for the solar observation.

10.33
LATITUDE BY SOLAR OBSERVATION

Latitude can be determined by observing the altitude or zenith angle to the sun at upper culmination and recording the time of observation. Using this time, the declination, δ, is found in the ephemeris and the latitude, ϕ (Section 10.9, Figure 10.8) is

$$\phi = 90° - h + \delta = z + \delta \qquad (10.21)$$

in which h is the true altitude and z is the true zenith distance.

The accuracy obtainable depends on the precision of the instrument. As the maximum rate of change of declination is only about $01'$ per hour, considerable error in time will affect the declination only slightly. With a total station system having a least count of $5''$ on the vertical circle, the latitude should be attainable with an error no greater than $10''$.

The procedure is as follows. The instrument is set up and very carefully leveled, preferably by using the vertical circle compensator (Section 6.40). The horizontal cross hair is sighted on either the lower or upper limb of the sun until the sun reaches its maximum altitude and begins its apparent descent. At that instant the time is noted and recorded along with the zenith or vertical angle. With the telescope still in the plane of the meridian, the collimation error of the vertical circle is determined, preferably by the method of double centering as described in Section 6.39. The Greenwich civil time (UTC) of observation is calculated, and the declination is found in a solar ephemeris as illustrated by Example 10.10, Section 10.30. The true altitude is determined by applying to the observed zenith distance or altitude the corrections for vertical-circle collimation error, semidiameter (Section 10.25), and refraction and parallax (Sections 10.27 and 10.28). The latitude then is determined by Equation (10.21).

When desired, a second sight may be taken on the opposite limb of the sun with the telescope inverted. The mean of the two vertical circle readings is taken as the altitude of the sun's center at the mean of the two times of observation, and no correction for vertical-circle collimation error is necessary. If the time between sights does not exceed 3 or 4 minutes, the mean vertical circle readings may be considered the zenith distance or altitude at apparent noon. The latitude then is calculated as described in the previous paragraph.

Table 10.5 shows calculations for latitude when direct and reversed zenith angles are measured to the upper and lower limbs of the sun.

10.34
LONGITUDE BY OBSERVATION ON THE SUN AT NOON

If the standard time is known precisely and the meridian has been established, the longitude of a place may be determined by an observation on the sun at local apparent noon. The field procedure is as follows. The theodolite is set up and carefully leveled over the north end of the meridian line, and a sight is taken along the meridian about one-half hour before noon.

TABLE 10.5
Latitude by observation of the sun

Average P.S.T. of observation on March 22, 1995 = $11^h35^m38^s$; watch correction (from WWV) = 29^s; temperature = 55°F; atmospheric pressure = 29.5 in. Hg; average of direct and reversed zenith angles to the upper and lower limbs of the sun = 41°51′30″.

Watch time, P.S.T.	=	$11^h35^m38^s$
Watch correction	=	$+29^s$
Corrected time of observation	=	$11^h36^m07^s$
	=	11.6019^h
G.C.T. (UTC) = P.S.T. + 8^h	=	$19^h36^m07^s$
δ_{0h} G.C.T., March 22 = 0°21′29.4″ (ephemeris)		
δ_{24h} G.C.T., March 22 = 0°45′10.3″ (ephemeris)		
$\delta_{obs} = \delta_{0h} + (\delta_{24h} - \delta_{0h})$ UTC/24	=	+0°40′40.9″
Average zenith angle to center of sun	=	41°51′30″
Correction for parallax (Equation (10.17b))		
$C_p = 8.94 \sin z = 8.94 (\sin 41°51′30″)$	=	−06″
Correction for refraction (Equation (10.19a))		
$C_r = [0.27306 (p \tan z')]/(460° + T)$	=	51″
True zenith angle	=	41°52′15″
δ_{obs}	=	+0°40′50″
$\phi = z + \delta_{obs}$	=	42°33′05″N

The line of sight is elevated to intercept the path of the sun and, at the instant of tangency between the west limb and the vertical cross hair, the time is noted. The telescope is quickly reversed, and a second sight is taken along the meridian. The line of sight again is elevated to intercept the path of the sun, and the time of tangency between the vertical cross hair and the east limb of the sun is observed. The mean of the two times thus observed is the watch time of upper transit of the sun's center, which is local apparent noon.

With the standard time of passage of the center of the true sun (local apparent noon) known, the Greenwich civil time of local apparent noon can be computed, and the equation of time at the instant of local apparent noon can be found readily from an ephemeris for 0^h Greenwich civil time. The standard time of local mean noon (meridian passage of the mean sun) differs from the standard time of local apparent noon by an amount equal to the equation of time, being greater if the mean sun is behind the true sun and less if ahead of it. The difference between the standard time of local mean noon and 12^h standard time is the difference in longitude $\Delta\lambda$ (in time units) between the meridian of the place and the standard-time meridian; if local mean noon occurs before 12^h standard time, the place is east of the standard meridian; if after, west.

EXAMPLE 10.14. The sun's center is observed to pass the meridian at a given place at $11^h30^m12.2^s$ A.M. P.S.T. on December 2, 1995. Determine the longitude of the place.

Solution

P.S.T. December 2, 1995, of local apparent noon	$11^h30^m12.2^s$	
Longitude to P.S.T. (120th meridian)	8^h	
G.C.T. of observation	$19^h30^m12.2^s = 19.50339^h$	
(Equation of time)$_{0h}$, December 2 = $10^m55.48^s$	} ephemeris	
(Equation of time)$_{24h}$, December 2 = $10^m32.66^s$		

(Equation of time observed) = (equation of time)$_{0h}$

$$+ \text{[(equation of time)}_{24h} - \text{(equation of time)}_{0h}](19.5034)/24 = + 10^m36.9^s$$

$$\text{G.A.T. of local apparent noon} = 19^h30^m12.2^s + 10^m36.9^s = \overline{19^h40^m49.1^s}$$

$$-12^h$$

G.H.A. of meridian of observation $\qquad\qquad \overline{7^h40^m49.1^s}$

$$\text{Longitude of meridian of observation} = (7.680306^h)(15) = 115°12'16.5''$$

OBSERVATIONS ON STARS

10.35
GENERAL

In general, the methods of determining azimuth by direct solar observations, with slight modifications, are applicable to observations of the stars.

The measurement of horizontal and zenith or vertical angles to celestial bodies is discussed in Section 10.23. For observations on stars, no correction is required for parallax or semidiameter.

As there are fixed stars in all parts of the heavens, it is an easy matter to select a star in a celestial region favorable to the precise determination of the quantity sought. For determination of azimuth by the hour-angle method with declination and latitude known, observations should be made on a circumpolar star—the nearer to the pole the better—because the azimuth of the star changes more slowly in a given length of time than the azimuth of a star near the equator; hence, any error in time will have less effect. When the azimuth is to be found by measured altitude, declination, and known latitude, observations should be taken on a star in the east or west far enough above the horizon to eliminate uncertain refraction but not so near the meridian as to produce a weak astronomical triangle.

The Greenwich hour angle and declination for Polaris and certain selected stars are given in the *Sokkia Celestial Observation Handbook and Ephemeris,* which constitutes the most convenient source of information on stars for the surveying engineer. The sidereal hour angle and declination for many stars (not for Polaris) and the G.H.A. of the vernal equinox (Aries) are given in the *Nautical Almanac.* Because the positions for fixed stars change very slowly, it is not necessary to determine the values for the hour of observation, as with the sun. The sidereal time corresponding to any given solar time can be found, and the hour angle can be computed by the expression $t = \theta - \alpha$, as explained in Section 10.6.

Stars can be identified by means of charts that show the various constellations. For many stellar observations, however, the published direction and altitude of the star can be set off on the theodolite with sufficient precision that the star will be brought into the field of view at a given time, and it is not necessary to distinguish the star from among its neighbors. In fact, observations often are taken on stars during daylight hours near the hour of darkness, even when the stars are invisible to the naked eye.

In sighting on a star, the objective should be focused until the star appears as a fine, brilliant point of light. Before looking for a star just before sunset or after sunrise, the objective may be focused approximately by sighting at a distant object in the landscape.

Artificial illumination is required to make the cross hairs of the instrument visible in the darkness. Most total station systems and optical-repeating and direction theodolites are equipped with built-in illumination for the reading scales and cross hairs, thus simplifying night-time operations. Some older instruments are equipped with a reflector sleeve that slips over the objective like a sun shade. When a flashlight is held to one side of the reflector, the field of view is illuminated faintly and both cross hairs and the star can be seen. On instruments not so equipped, the cross hairs can be illuminated by holding the flashlight several inches in front of the objective and a little to one side of the telescope barrel, thus causing the rays to enter the telescope diagonally. After some experimenting, a position of the light will be found where both cross hairs and star are visible.

The location of any terrestrial mark used in observing is indicated by a light. The mark may be a strongly illuminated target, the source of illumination being shielded from the observer. Targets adaptable to instrument tripods, such as the one illustrated in Figure 6.53, can be equipped with lights for nighttime observation.

10.36
POLARIS

The polestar, Polaris (α Ursa Minor), is the star predominantly used for observations for latitude and azimuth in the latitudes of the United States. Its distance from the pole is approximately 1°. Its annual change in polar distance (or in declination) is less than 01′, and its maximum daily change in polar distance is less than $\frac{1}{2}$″. It is a second-magnitude (or quite bright) star, the position of which is readily identified by the neighboring constellations of Ursa Major and Cassiopeia. Figure 10.24 shows the position of Polaris with respect to the pole and to these constellations. The seven most brilliant stars in the constellation of Ursa Major are known as the *Great Dipper;* and the two stars forming the part of the bowl farthest from the handle are called the *pointers,* because a line through these stars

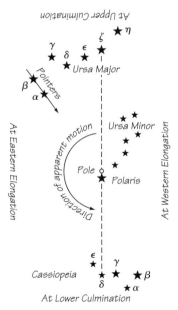

FIGURE 10.24
Positions of constellations near the North Pole when Polaris is at culmination and elongation.

points very nearly to the celestial north pole. It will be noted that the constellation of Cassiopeia is on the same side of the pole as Polaris, so that when Cassiopeia is above the pole, Polaris is near upper culmination; when Cassiopeia is west of the pole, Polaris is near western elongation; and so on. The position of Polaris relative to the pole may be quite closely estimated by noting the positions of δ Cassiopeia and ζ Ursa Major. A line joining these two stars passes nearly through the pole and Polaris. The line is nearly vertical when the star is at either culmination and nearly horizontal when the star is at either elongation.

Culmination of Polaris

Because Polaris, in common with other fixed stars, travels at an angular rate more rapid than that of the sun, it follows that, at a given meridian, it arrives at culmination a little earlier by mean solar (civil) time each day than it did the day before, the amount earlier being approximately equal to the gain of sidereal time on mean solar time for a 24^h interval, or $3^m 56.6^s$ per day or 9.86^s per hour of solar time (Section 10.19). Furthermore, the time of culmination depends on the longitude of the place. To determine the local mean time of the upper culmination of Polaris at any given meridian on any given date, the value for the meridian of Greenwich is taken from the table and reduced to the longitude of the place by means of the variation per hour.

> **EXAMPLE 10.15.** Find the Eastern standard time of upper culmination (U.C.) of Polaris on December 9, 1995, at a place where the longitude is $78°56'15''W = 5^h 15^m 45^s = 5.2625^h$. Figure 10.25 illustrates the time relationships.

Solution

G.C.T. of U.C., December 9, 1995 (ephemeris, under Greenwich transit)

$$= 21^h 16^m 52^s$$

Longitude west of Greenwich to E.S.T. $= -5^h$

E.S.T. of U.C. at Greenwich meridian $= 16^h 16^m 52^s$

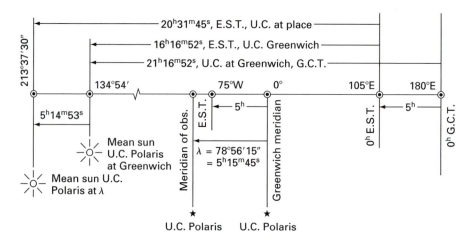

FIGURE 10.25
Upper culmination of Polaris (Example 10.15).

The longitude to the place of observation is 5.2625 sidereal hours. Polaris must move through this interval to cross the meridian at the place of observation. To convert a sidereal interval into a mean solar interval, 9.83^s/sidereal hour are subtracted from the sidereal interval (Section 10.19). The correction is

$$(9.83^s/h)(5.2625^h) = -0^m51.7^s$$

The longitude corrected to yield the solar interval between upper culmination at Greenwich and U.C. at place is

$$\text{Corrected longitude} = (5^h15^m45^s) - 51.7^s = 5^h14^m53^s$$

so that

$$\text{E.S.T. of U.C. at Greenwich (from previously)} = 16^h16^m52^s$$
$$\text{Solar interval U.C. at Greenwich to U.C. at place} = 5^h14^m53^s$$
$$\text{E.S.T of U.C. at place of observation} = 21^h31^m45^s$$

To find the time of lower culmination the quantity 12^h minus one-half of the variation per day (one-half of 3^{m}56^s or 1^{m}58^s) is added to or subtracted from the time of upper culmination. For the data in this example,

E.S.T. of U.C. at place of observation	21^{h}31^{m}45^s
Interval between U.C. and L.C.	$-11^h58^m02^s$
E.S.T of lower culmination on December 9, 1995	9^{h}31^{m}43^s

Elongation of Polaris

The hour angle of Polaris when at elongation can be determined, precisely as explained in Section 10.12, by Equation (10.10), in which cos t = tan ϕ tan p. The hour angle, t, expressed in units of time is the sidereal time interval between upper culmination and eastern or western elongation. The corresponding mean solar time interval is found by deducting, from the computed hour angle, a correction of 9.83^s per hour, which, as explained in Section 10.19, is the difference in solar time between the sidereal hour and the mean solar hour. The standard time of upper culmination is found as shown in Example 10.15. The standard time of eastern or western elongation then is determined by adding or subtracting, from the time of culmination, the mean time interval between upper culmination and elongation. A numerical example follows.

EXAMPLE 10.16. It is desired to find the Eastern standard time of western elongation of Polaris occurring in the early morning hours of November 10, 1995, at a location where the longitude is 5^{h}15^{m}45^s west of Greenwich and the latitude is 50°00'00" north.

Solution. From the ephemeris, the declination δ = 89°14'46.19". Therefore, by Equation (10.10),

$$\cos t = \tan \phi \tan p = \tan (50°00'00") \tan [90° - (89°14'46.19")] = 0.015680888$$

$$t = 89°06'05.85" = 89.101515^h$$

Local sidereal hour angle at time of observation

$$= (89.101522)/15° = 5.9401010^h = 5^h56^m24.4^s$$

Reduction, sidereal to mean solar interval (Section 10.19)

$$= (5.940102^h) (9.83s/h) = - \quad 58.4^s$$

Mean time interval from upper culmination (U.C.) $\quad = \quad 5^h55^m26.0^s$

From Example 10.15, E.S.T. of U.C. at place of observation

on December 9, 1995 $= 21^h31^m45.0^s$

E.S.T. of western elongation, December 10, 1995 $\quad = \quad 3^h27^m11.0^s$

10.37
LATITUDE BY OBSERVATION OF POLARIS

As described in Section 10.9, the latitude of a place is equal to the altitude of the elevated pole. Therefore, if h is the true altitude (vertical angle) or z is the true zenith angle of any circumpolar star as it crosses the meridian, then at upper culmination the latitude ϕ is

$$\phi = h + (90° - \delta) = 180° - z - \delta \qquad (10.22)$$

and at lower culmination

$$\phi = h - (90° - \delta) = \delta - z \qquad (10.23)$$

By this method, the latitude of a station is determined by measuring the altitude or zenith angle of Polaris when it is at either its upper or lower culmination and by applying to this altitude or zenith angle, corrected for refraction, the star's declination as given in the ephemeris. Because the star apparently is traveling in a horizontal line when at either of these two positions, it is not essential to the precision of the latitude determination that the time of culmination be found precisely, but in any case, it facilitates the work of observing if the approximate time is known.

Further it is not essential that the altitude of the star be observed at the instant it crosses the meridian. Some minutes before and after culmination, the star travels in so nearly a horizontal line that, with an ordinary theodolite, vertical movement cannot be detected. Within the period 6^m before to 6^m after culmination, the maximum change in altitude is only $01''$, and within 12^m before to 12^m after culmination the maximum change is only $06''$.

The procedure to be employed in making an observation depends on the precision with which the latitude is to be determined and the time is known. For an observation with a theodolite having a vertical circle reading to $20''$, when the watch time or longitude of the place may be in doubt by a few minutes, the standard time of culmination at the given station (Example 10.16) and the declination of Polaris are determined using the ephemeris. A few minutes before the estimated time of culmination, the theodolite is set up and leveled very carefully, preferably using the vertical circle compensator, if the instrument is so equipped (Section 6.40). The star is found with the naked eye by noting its position with respect to the neighboring constellations, as shown in Figure 10.24. The telescope is focused for a star. If the latitude is known approximately, its estimated value and the star's declination are used in either Equation (10.12) or (10.23) to give h or z, which is set off on the vertical circle to facilitate finding Polaris. The telescope is sighted on Polaris. When the star has been brought within the field of view, the cross hairs are illuminated, if necessary, and the star is continuously bisected with the horizontal cross hair. When, during a period of 3–4 min, Polaris no longer appears to move away from the cross hair but moves horizontally along with it, the star will be practically at culmination. The vertical circle reading is observed and recorded, the theodolite is carefully releveled, the telescope is reversed, and a second observation on the star is taken with the telescope inverted. Usually the instrument then is releveled and a second pair of observations is made. The mean of the observed vertical angles or zenith angles, corrected for refraction (Equation (10.18) or (10.19)), is taken as the true altitude or zenith distance of the star. Note that temperature and atmospheric pressure also must be observed and recorded when Equation (10.18) or (10.19) is to be used. The declination can be found in the ephemeris, which gives the declinations for Polaris at 0^h UT for the days of the current year. Finally, the true altitude, or zenith distance and declination are used in either Equation (10.22) or (10.23) to determine the latitude.

Precise Determination

When it is desired to determine the latitude within a few seconds and the standard time and longitude of the place are known within a minute or so, the watch time of culmination may be precisely computed, as illustrated in Example 10.14, and a series of observations may be taken on the star when it is near culmination. The observing program usually is arranged so that an equal number of observations will be taken before and after culmination. The observations are begun at a given time interval (usually no more than 10^m) before the calculated time of culmination, and at each sighting of the star, the watch time and the altitude or zenith angle are observed. Half of the observations are taken with the telescope direct and half with it reversed; and between pairs of observations, the instrument is carefully releveled. The observed altitudes or zenith angles of the star for positions other than culmination are reduced to the altitude at culmination by applying a correction, which, for altitudes within the United States, is given approximately in Table 10.6.

When the star is near lower culmination, the correction is subtracted from vertical angles and added to zenith angles; when near upper culmination, the correction is added to vertical angles and subtracted from zenith angles. The mean of the altitudes reduced to culmination is corrected for refraction (Equation (10.18) or (10.19)), the declination is found, and the latitude is computed, as for the case described in the preceding section.

When the culmination of Polaris occurs during the daylight hours, latitude by observation of Polaris at any time can be determined by the following equation:

$$\phi = h - p \cos t + \tfrac{1}{2}p^2 \sin^2 t \tan h \sin 1'' \tag{10.24}$$

in which h ($h = 90° - z$) is the true altitude of Polaris, p ($p = 90° - \delta$) is the polar distance of Polaris expressed in seconds, and t is the local hour angle of Polaris. It is

TABLE 10.6
Corrections to be applied to altitudes of Polaris near culmination to give altitude at culmination

Interval from culmination, minutes of time	Change in altitude from culmination, seconds of arc
3	00
6	01
9	03
12	06
15	09
18	12
21	17
24	22
30	34

necessary to measure the vertical or zenith angle to Polaris and record the watch time of each observation. The angle should be observed at least once with the telescope direct and once with the telescope reversed. The average angle then is corrected for refraction to obtain the true altitude. The average watch time is then used in conjunction with the hour angle of Polaris, obtained from the ephemeris, to determine the local hour, t. These data are substituted into Equation (10.24) to solve for the latitude. Consider an example to illustrate the procedure.

EXAMPLE 10.17. The true altitude of Polaris $= 39°40'00''$ observed at $20^h30^m15^s$ E.S.T., July 20, 1995. The longitude of the place is $76°20'00''$W. Determine the latitude for the place of observation.

Solution. The ephemeris used gives the G.H.A. of Polaris at 0^h G.C.T. for every day of the year. To compute the L.H.A. $= t$, determine the G.C.T. of observation and correct the G.H.A. of Polaris for the elapsed *sidereal* interval of time. Then, $t =$ G.H.A. at the instant of observation minus the west longitude of the place of observation. Figure 10.26 shows hour angles at 0^h G.C.T. and at time of observation, respectively.

$$\text{E.S.T. of observation, July 20, 1995} \quad 22^h30^m15^s$$

$$\text{Longitude west from Greenwich} \quad 5^h$$

$$\text{G.C.T. of observation, July 21, 1995} \quad \overline{3^h30^m15^s} = 3.5042^h$$

The local hour angle (L.H.A.) is equal to the G.H.A. of Polaris minus the longitude of the place. However, since 0^h G.C.T. (Figure 10.26a), 3.5042^h of solar time have elapsed (Figure 10.26b). During this period of time, Polaris traveled faster than the sun by 3.943^m per day or $(3.943^m)(60^s)/24^h = 9.86^s/h$, which is the difference between sidereal and solar time (Section 10.19). Consequently, the elapsed time interval since 0^h G.C.T. must be increased by $(3.5042^h)(9.86^s)$. So the sidereal interval, since 0^h G.C.T. equals $(3^h30^m15^s) + (3.5042^h)(9.86^s) = 3^h30^m49.6^s = 3.51378^h$. Converting this interval to degrees, we obtain

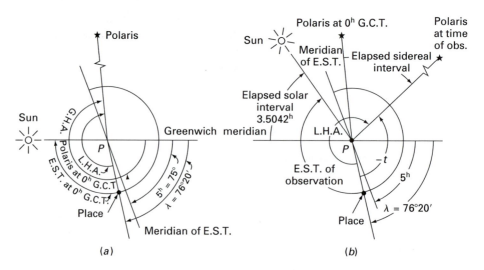

FIGURE 10.26
Latitude from a sight on Polaris at any time (Example 10.17): (a) hour angles at 0^h G.C.T., (b) hour angles at time of observation.

Interval since 0^h G.C.T. $= (3.51378^h)(15°/h) = \quad 52°42'24.1''$

G.H.A. Polaris 0^h G.C.T., July 21 (ephemeris) $= 261°26'32.1''$

G.H.A. Polaris at instant of observation $\quad = 314°08'56.2''$

West longitude $\quad = -76°20'00''$

L.H.A. $\quad = 237°48'56.2''$

$t =$ L.H.A. $-360°$ (Section 10.6) $\quad = -122°11'04''$

The latitude then is calculated by Equation (10.24) using the polar distance, $p = 90° - \delta$, for Polaris, where δ is from the ephemeris and

$$p = 90° - 89°14'18.69'' = 2741.3''$$

$$\phi = h - p \cos t + \tfrac{1}{2}p^2 \sin^2 t \tan h \sin 1''$$

$$= 39°40'00'' - (2741.3'') \cos (237°48'56'')$$

$$\quad + \tfrac{1}{2}(2741.3'')^2 \sin^2 (237°48'56'') \tan (39°40')(0.000004848)$$

$$= 39°40'00'' + 1460'' + 11'' = 39°40'00'' + 24'31''$$

$$= 40°04'31''$$

10.38
AZIMUTH BY OBSERVATIONS ON POLARIS

The azimuth can be determined from observations on Polaris at any time using the *hour-angle* method and by sighting on Polaris at *eastern or western elongation.* The hour-angle method involves more calculations but, with current peripheral equipment (e.g., pocket calculator and time module), is easier to implement, because Polaris can be sighted at any time and acquisition of accurate time and performing the computations present no real problems.

Hour-Angle Method

If the standard time, latitude, and longitude of the place are precisely known, the hour angle of the star at any instant can be found and the azimuth of the star at any instant can be determined, as described in Section 10.13, by Equation (10.13), which is repeated here for convenience:

$$\tan Z = \frac{-\sin t}{\cos \phi \tan \delta - \sin \phi \cos t} \tag{10.13}$$

in which Z is the clockwise or counterclockwise angle from the north (Section 10.11), t is the hour angle of Polaris determined from the observed times, ϕ is the latitude of the place scaled from a map or determined from observations (Section 10.35), and δ is the declination of the star obtained from the ephemeris.

Field observations consist of measuring the horizontal angle between a terrestrial mark and the star. For azimuth determination to within $\pm 30''$, a single set of two observations, one with the telescope direct and the other with it reversed, should be taken with the time (UTC) at which the star passes across the vertical cross hair being recorded at each setting. The L.H.A. is computed using the average time of observation and the G.H.A. of Polaris obtained from the ephemeris (Example 10.17, Section 10.37). The declination is found by interpolation between the values given in the ephemeris. Using these data, the azimuth of Polaris is computed by Equation (10.13). This azimuth combined with the average horizontal angle yields the azimuth of the reference line.

When a higher degree of precision is desired, a series of measurements is taken. At least six observations of the horizontal angle from the reference line to the star should be made. The recommended procedure is to (1) sight the reference mark with the telescope direct, (2) sight on Polaris with the telescope direct, (3) reverse the telescope and sight Polaris, and (4) finally resight the reference mark with the telescope reversed as a check. The horizontal circle readings for each sighting are recorded or keyed into the data collector. Also, the time of each sighting of the star is recorded to the nearest tenth of a second. These observations constitute one set of measurements. At least three and preferably four sets should be observed. For the best results, a pocket calculator with a time module or a digital stopwatch or an inexpensive GPS receiver (Chapter 12), which also provides latitude and longitude good to 100 m, should be used to determine the time. In each case, the timepiece should be checked against WWV (Section 10.21).

Figure 10.27 contains notes for four sets of direct and reversed measurements to Polaris. These measurements were made with a repeating total station system used as a direction theodolite, so that each set was initiated with a different reading on the back-sight to the reference point in order to distribute the angles uniformly around the entire

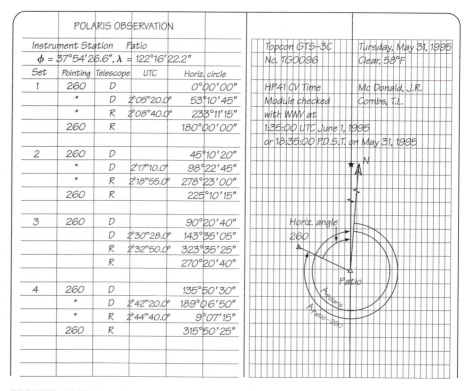

FIGURE 10.27
Notes for Polaris observations by the hour-angle method.

TABLE 10.7

613

CHAPTER 10:
Introduction to
Astronomy

Azimuth by observation of Polaris at any time using the hour-angle method

Date: May 31, 1995, at about $19:00^h$ P.D.S.T; $\phi = 37°54'26.6''$; $\lambda = 122°16'22.2''$

Calculations for set 1

Average horizontal angle $= 53°11'00''$

Average UTC time $\quad = 2^h07^m00.0^s = 2.116667^h$

From Table A, Sokkia ephemeris for June 1, 1995. Note, that since UTC time is less than P.D.S.T., enter the table for the next day, June 1.

G.H.A.$_{.0h}$ Polaris on June 1 $= 212°28'14.8''$

G.H.A.$_{.24h}$ Polaris on June 1 $= 213°27'06.5''$

Declination, δ_{0h} June 1 $\quad = 89°14'24.30''$

Declination, δ_{24h} June 1 $\quad = 89°14'24.10''$

Calculate G.H.A.$_{obs}$ at instant of observation:

G.H.A.$_{obs}$ = G.H.A.$_{.0h}$ + (G.H.A.$_{.24h}$ − G.H.A.$_{.0h}$ + 360°)(UTC/24)

$\quad = 212°28'14.8'' + [(213°27'06.5'') − (212°28'14.8'') + 360°](2.11667/24)$

$\quad = 212°28'14.8'' + 31°50'11.5'' = 244°18'26.3''$

Determine L.H.A. by Equation (10.1c), noting that Polaris is west of observer's meridian so L.H.A. = t (Section 10.5)

L.H.A. = t = G.H.A.$_{obs}$ − λW = $244°18'26.3'' − 122°16'22.2'' = 122°02'04.1''$

Calculate declination at instant of observation:

$\delta_{obs} = \delta_{0h} + (\delta_{24h} − \delta_{0h})$ UTC/24 = 89°14'24.3''

Compute azimuth of Polaris by Equation (10.13):

$Z = \tan^{-1}[(−\sin t)/(\cos \phi \tan \delta − \sin \phi \cos t)]$

$\quad = \tan^{-1}[−\sin 122°02'04.1'']/[\cos 37°54'26.6'' \tan 89°14'24.3'' − \sin 37°54'26.6''$

$\qquad \cos 122°02'04.1'']$

$\quad = \tan^{-1}(−0.0141733) = −0°48'43.3''$

$A_N = 360° + Z = 360° + (−0°48'43.3'') = 359°11'16.7''$

The azimuth from north of the reference line Patio-260 for set 1 is

$A_1 = 359°11'16.7'' − 53°11'00''$

$A_1 = 306°00'16.7''$

Similarly, azimuths are computed for sets 2, 3, and 4 to give

$A_2 = 306°00'14.4''$

$A_3 = 306°00'17.8''$

$A_4 = 306°00'13.0''$

for an average azimuth of 306°00'15.5'' with $\sigma = 2.2''$.

As an additional check, azimuths are computed for each pointing within each set to yield

Set 1: $\quad A_{11} = 306°00'18.2''$

$\qquad A_{12} = 306°00'15.3''$

Set 2: $\quad A_{21} = 306°00'16.8''$

$\qquad A_{22} = 306°00'12.0''$

Set 3: $\quad A_{31} = 306°00'16.7''$

$\qquad A_{32} = 306°00'19.0''$

Set 4: $\quad A_{41} = 306°00'16.3''$

$\qquad A_{42} = 306°00'09.6''$

Average $\quad = 306°00'15.5''$

horizontal circle (Section 6.49). The times at the instant of each observation are from a time module set to WWV and checked immediately before the session.

Table 10.7 shows the calculations for the azimuth of the reference line, using the average time and average horizontal angle for set 1 from Figure 10.27. When direct and reversed observations are obtained within a 3–4 min time span and hand calculations are made, as in this case, use of the average time and horizontal angle is satisfactory. For this example, the G.H.A.$_{obs}$ (at instant of observation) is found by interpolation between the given values in the ephemeris. The G.H.A.$_{obs}$ also could be determined by converting the solar interval since 0^h G.C.T. to a sidereal interval then adding this value converted to degrees, to the G.H.A.$_{obs}$ from the ephemeris, as was illustrated in Example 10.17. This value for G.H.A.$_{obs}$ and the given longitude are used to obtain the L.H.A., and the declination is found by interpolation from given values in the ephemeris with the date and time as arguments. Then, using the given latitude and calculated values for local hour angle and declination, the azimuth of Polaris is computed with Equation (10.13). The azimuth of the reference line is determined by applying the average horizontal angle to the calculated azimuth of Polaris. Azimuths for the remaining sets 2, 3 and 4 also are calculated and show no evidence of mistakes. The average of the four azimuths represents the best estimate for the direction of the line.

As an additional check of the data, an azimuth for each pointing of the star within each set was performed and the values are listed in Table 10.7. This procedure eliminates errors due to curvature of the path of Polaris. The data appear consistent with an average value that is the same as before and $\sigma = 3.2''$. This last method for determining the best estimate of an azimuth from multiple measurements is recommended but is feasible only if a computer program is available for the computations.

Precision

The precision obtained depends on the position of the star, precision of the observations of time and determination of latitude and longitude, number of observations and quality of instrument, and the care and skill of the observer. When the star is near its upper or lower culmination, the azimuth changes at a relatively rapid rate, this change amounting to about $01'$ of arc in about 3^m of time for latitude 40°N. For this reason, the method should not be expected to give good results when Polaris is observed near culmination unless the time is known very precisely. Assuming adequate equipment, well-trained personnel, and optimum positioning of Polaris, the major factors affecting the calculated azimuth of Polaris are errors in time, latitude, longitude, and declination as propagated through Equation (10.13).

Table 10.8 shows the effects of assumed maximum errors of 1^s in time (15" in λ), 10" in latitude, and 0.1" in declination, on azimuths calculated by Equation (10.13) for Polaris and, for comparison, solar observations using the hour-angle method at latitude 40°N. Note that errors in azimuth by solar observations are from 5 to 19 arc seconds due to errors in time and latitude, compared to tenths of seconds for azimuths determined by Polaris measurements. Therefore, observation of Polaris by the hour-angle method provides

TABLE 10.8

Changes in azimuth of Polaris and the sun calculated by the hour-angle method at latitude 40°N for March 1995

Celestial body	Declina-tion, δ	Hour angle, t	Alti-tude, h	Change in azimuth for changes of		
				1^s in time (15" in λ)	10" in ϕ	0.1" in δ
Polaris	89°14'50"	4^h40^m	40°25'	0.1"	0.1"	0.1"
Sun	0°	4^h40^m	15°	10"	7"	0.1"
Polaris	89°14'50"	3^h20^m	40°28'	0.2"	0.0"	0.1"
Sun	0°	3^h20^m	30°	13"	5"	0.1"
Polaris	89°14'50"	1^h30^m	40°40'	0.2"	0.0"	0.1"
Sun	0°	1^h30^m	45°	19"	5"	0.1"

azimuth of a higher order of precision than observing the sun. The major source of error in determining the azimuth of a given line by Polaris measurements is in the horizontal angle from the star to the terrestrial point.

615

CHAPTER 10:
Introduction to
Astronomy

EXAMPLE 10.18. Determine the approximate total error in one set of measurements (1DR) with a total station system (least count 05″) made on Polaris on March 1, 1995, by the hour-angle method, when the estimated hour angle was about 4^h at a place having a latitude of 40°N. The total station system, used as a direction instrument, was leveled with the vertical circle compensator (Section 6.40), resulting in an estimated error due to misleveling of 05″. Estimated errors in t, λ, ϕ, and δ were 1^s, 15″ 10″, and 0.1″, respectively.

Solution. An approximate altitude for Polaris can be found using $\phi = 40°$, $t = 120°$, and $\delta = 89°14'50''$ (from the ephemeris) in Equation (10.14) and $h = \sin^{-1}$ $(\sin \phi \sin \delta + \cos \phi \cos \delta \cos t) = 39°37'$. So, the error in the horizontal angle due to misleveling is

$$e_{ha1} = 05'' \tan (39°37') = 04''$$

The standard deviation in the horizontal angle σ_{ha}, can be evaluated using Equation (6.13) in which the standard deviation in pointing and reading the micrometer are $\sigma_p = 02''$ and $\sigma_m = 0$ (digitized circle) so that the error caused by the repetition process is

$$e_{ha2} = 2[(\sigma_p)^2/n]^{1/2} = [2(02)^2/2]^{1/2} = 02''$$

From Table 10.8, $e_t \approx e_\lambda \approx e_\phi \approx e_\delta \approx 0.1''$ and the total approximate error is

$$E_A = [4(0.01)^2 + (4)^2 + (2)^2]^{1/2} = 4.5''$$

If four sets are observed and the average azimuth is computed, then

$$E_{Aaver} = (E_A)/\sqrt{4} = 2.3''$$

Example 10.18 represents a "worst case" situation. With care, the latitude and longitude can be scaled from a U.S.G.S. topographic map to within 4–5″, and it is possible to register time to within a few tenths of a second when using a time module checked against WWV.* Note that misleveling causes the largest component of the total error, which is in the horizontal angle. This emphasizes the need for a very stable instrument setup, extremely careful leveling, and releveling between sets. The example also shows a substantial decrease in the error by increasing the redundancy, stressing the importance of multiple sets of direct and reversed observations to optimize the precision of the azimuth.

Observation on Polaris at Elongation

The azimuth of a line can be determined by an observation on Polaris at its eastern or western elongation, provided the latitude of the place is known. As shown in Section 10.12, the angle Z in the spherical triangle containing any star at elongation is given by Equation (10.9), repeated here for convenience:

$$\sin Z = \frac{\sin p}{\cos \phi} \tag{10.9}$$

---

*An inexpensive GPS receiver can provide latitude and longitude to within 3–4″ and give time to microseconds.

in which Z is the angle east or west of north depending on whether the star is at its eastern or western elongation (Section 10.12), p is the star's polar distance, and ϕ is the latitude of the place. Figure 10.11 shows the astronomical triangle for a Polaris observation at elongation.

The direction of Polaris from the observer's station at the time of elongation is established by projecting a vertical plane from the star to the earth. The terrestrial line thus established has the same azimuth as the star at elongation; hence, the azimuth of any connecting line can be found if the horizontal angle between the two lines is measured.

The star's polar distance is found approximately from the ephemeris giving values of the declination of Polaris for the days of the year in which the observation is made. The latitude is determined by observation, as explained in Section 10.36 or by scaling from topographic map. The time of elongation is determined as described in Section 10.36, Example 10.16.

It is not essential that the direction to Polaris be observed at the exact instant of elongation. For some minutes before and after elongation, the star travels in so nearly a vertical line that, with a theolodite, horizontal movement cannot be detected. For the latitudes of the United States, within the period 4^m before elongation to 4^m after elongation, the maximum change in the azimuth of Polaris is less than $01''$, and within 10^m of elongation the maximum change in azimuth is only $0.1'$.

A few minutes before the estimated time of elongation, the theolodite is set up over a given station and leveled very carefully. The telescope is focused for a star, the latitude of the place is laid off on the vertical circle to facilitate finding the star, and the theodolite is revolved about the vertical axis until Polaris comes within the field of view. The horizontal and vertical motions then are clamped, the cross hairs are illuminated if necessary, and the star is continuously bisected with the vertical cross hair. When, during a period of 2 or 3 min, Polaris no longer appears to move away from the cross hair but moves vertically along it, the star is practically at its elongation. The telescope is depressed, and a point on the line of sight is marked on a stake or other reference monument 100 m or 300 ft or more away. The telescope then is reversed and another sight is taken on Polaris. The line of sight again is depressed, and a second point is set on the stake alongside the first. Usually, the theolodite is releveled and a second pair of observations is made.

Later, the mean of the points is found and marked on the stake. The line joining the occupied station with the established mean point defines the direction of Polaris at its elongation. Its azimuth is computed by Equation (10.9) using δ from the ephemeris and ϕ that is known. The azimuth of any other line through the station can be determined by measuring the horizontal angle between the two lines, by the method of repetition (Section 6.35).

The precision of azimuth determination by this method necessarily depends on the quality of the instrument, the care and skill of the observer, and the number of observations. For the procedure described, under ordinary conditions, the error should not exceed $10''$.

It should be noted that, for this method, a given error in latitude produces a relatively small error in azimuth. For latitudes of the northern part of the United States, an error of $01'$ in latitude produces an error of about $02''$ in the azimuth, and less for lower latitudes. Because the latitude easily can be scaled from a topographic map or determined by observation using an engineer's theodolite to within $20''$, it is evident that the principal error in azimuth is likely to be due not to errors in the computed azimuth in Polaris, but to the field operations of projecting the direction of the star to the earth. The procedure, if followed, practically eliminates all systematic errors except that due to the vertical axis not being truly vertical, so the need arises for extreme care in leveling the instrument when precise results are to be achieved.

EXAMPLE 10.19. Using the data of Example 10.16, compute the azimuth of Polaris for the time and place of observation. From Example 10.16, $\delta = 89°14'46.19''$ so that $p = 90° - \delta = 0°45'13.81''$ and the latitude $\phi = 50°00'00''$.

Solution. By Equation (10.9),

$$\sin Z = (\sin p)/(\cos \phi) = \sin (0°45'13.81'')/\cos (50°00'00'')$$

$$= 0.02046795$$

$$Z = 1°10'22.1''$$

where Z is the direction of Polaris to the west of north, so the azimuth of Polaris is

$$A_{\text{Polaris}} = 360° - 1°10'22.1'' = 358°49'37.9''$$

The advantage of observing Polaris at either its eastern or western elongation is the simplification of the observational process and computation of the azimuth. The disadvantage is that a given elongation (eastern or western) occurs only once a day, during the dark and frequently at a very inconvenient time. With the current common availability of powerful handheld calculators or data collectors equipped with software for reducing Polaris observations by the hour-angle method, the extent of the computations no longer is a serious problem. Therefore, most surveying engineers prefer to observe Polaris at any time by the hour-angle method, and sighting Polaris at elongation is a less-used method now.

10.39
OBSERVATIONS ON OTHER STARS

Latitude, azimuth, and longitude can be determined by observation on stars distant from the pole by methods similar to those described for the sun.

In the Sokkia ephemeris for each year are tables giving the G.H.A. and declination for eight selected stars of the first order. Knowing the longitude of the place of observation and G.H.A. of the star, the L.H.A. of the star can be computed. With the L.H.A. known, the sidereal interval between 0^h UTC time and the upper transit of the star over the meridian of the place can be calculated. Conversion of this sidereal interval to mean solar time yields the time UTC of upper transit of the star.

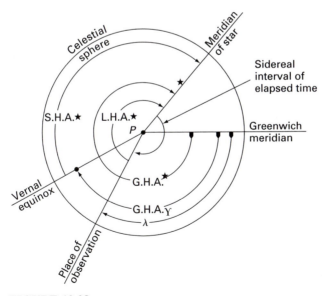

FIGURE 10.28
Hour angles used to determine the transit of a star.

The *Nautical Almanac* contains tables that give the positions of many stars by sidereal hour angle (S.H.A.; Section 10.5) and declination as well as the G.H.A. of the vernal equinox. The G.H.A. plus the S.H.A. of the star yields the G.H.A. of the star from which one can determine the time (UTC) of the star's upper transit, as described in the previous paragraph. Figure 10.28 shows the relationships of these various hour angles. Consider examples for each type of ephemeris.

EXAMPLE 10.20. Determine the Pacific standard time of the upper transit of the star Spica on February 27, 1995, at a place having longitude of $122°15'30.0''$.

Solution. From the Sokkia ephemeris,

G.H.A. of Spica of 0^h UTC, February 27, 1995	$=$	$315°07'08.5''$
Longitude of place	$=$	$-122°15'30.0''$
L.H.A. of Spica west of place	$=$	$192°51'38.5''$
$360° -$ L.H.A. $=$ sidereal interval between 0^h UTC and upper transit at place	$=$	$167°08'21.5''$
	$=$	$11^h08^m33.4^s$
Correction to mean solar time $= (11.14262^h)(9.83)^s$	$=$	$-1^m49.5^s$
UTC time of upper transit at place	$=$	$11^h06^m43.9^s$
Longitude to Pacific standard time	$=$	-8^h
P.S.T. of upper transit at place	$=$	$3^h06^m43.9^s$

EXAMPLE 10.21. What is the Pacific standard time of upper transit of the star Arcturus on February 27, 1995, at a place for which the longitude is $122°15'30''$?

Solution. From the *Nautical Almanac,*

S.H.A of Arcturus at 0^h, February 27, 1995	$=$	$146°08.1'$
G.H.A.Y0^h, February 27, 1995	$=$	$156°21.6'$
G.H.A. Arcturus at 0^h, February 27, 1995	$=$	$302°29.7'$
Longitude of place	$=$	$-122°15.5'$
L.H.A. Arcturus east of place	$=$	$180°14.2'$
$360° -$ L.H.A. $=$ sidereal interval between 0^h UTC and upper transit at place	$=$	$179°45.8'$
	$=$	$11^h59.1^m$
Longitude to P.S.T.	$=$	-8^h
P.S.T. of upper transit at place	$=$	$3^h59.1^m$

When the azimuth, latitude, or longtitude is to be determined from observations on other stars, prior to observing, it is helpful to determine the approximate altitude and azimuth of the star to simplify bringing the star into the field of view of the telescope. If the latitude, longitude, and time of observation are known along with the G.H.A. of the star from the ephemeris, the local hour angle can be calculated. Then, with the declination of the star from the ephemeris, Equations (10.14) and (10.11) are used to compute the altitude and azimuth of the star.

EXAMPLE 10.22. The star Arcturus is to be observed from a place having a latitude and longitude of 42°17′30″ and 72°30′00″, respectively (scaled from a U.S.G.S. quadrangle sheet of the area), on April 1, 1995, at 10 P.M. Eastern standard time. Determine approximate values for altitude and azimuth to assist locating the star for subsequent observations.

Solution. $10^h00^m00^s$ P.M. $= 22^h00^m00^s$ E.S.T., $\delta = $ N19°12′15.9″ (Sokkia ephemeris) UTC of potential observation $= 5^h + 22^h = 27^h = 3^h00^m00^s$ on April 2, 1995 (refer to Figure 10.29a).

G.H.A. Arcturus, April 2, 1995, at 0^h UTC $\quad = 336°00′14.8″$

Longitude to place of observation $\qquad\qquad -72°30′00″$

L.H.A. at 0^h G.C.T. (west of meridian) $\quad = 263°30′14.8″$

t (east of meridian) $=$ L.H.A. $- 360°$ $\qquad = -96°29′45.2″$

Mean solar interval between 0^h UTC and time of observation at 3^h UTC is 3^h.
Therefore, the Sidereal interval $= 3^h + (3^h)(9.86^s/h) = 3^h00^m30^s \quad = \quad 45°07′30″$

L.H.A. at 3^h UTC, April 2 $= t + $ interval $\qquad = -51°22′15.2″$

By Equation (10.14):

$$\sin h = \sin\phi\sin\delta + \cos\phi\cos\delta\cos t$$
$$= \sin(42°17′30″)\sin(19°12′16″) + $$
$$\cos(42°17′30″)\cos(19°12′16″)\cos(-51°22′15″) = 0.65744277$$
$$h = 41°06′18″$$

$\underline{+01′04″}$ refraction correction (assume $p = 29.5$ in. Hg,

$T = 60°F$; use Equation (10.19b))

Vertical angle $= 41°07′22″$

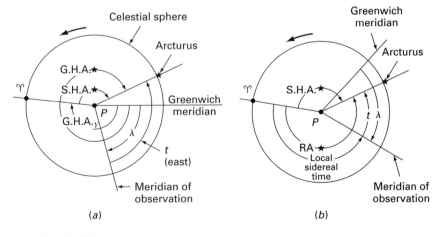

FIGURE 10.29
Hour angles for Example 10.22: (a) hour angles at 0^h UTC, (b) hour angles at time of observation.

By Equation (10.11),

$$\sin Z = -\frac{\sin t \cos \delta}{\cos h}$$

$$= -\frac{[\sin(-51°22'15'') \cos(19°12'16'')]}{\cos(41°07'22'')}$$

$$= (0.97932771)$$

$$Z = 78°19'47'' = \text{azimuth from the north}$$

Had the *Nautical Almanac* been used for Example 10.22, the position of Arcturus would be given by the S.H.A. of the star and the G.H.A. of the vernal equinox, which are combined to give the G.H.A. of Arcturus. The rest of the solution would be the same.

10.40
DETERMINATION OF AZIMUTH BY OBSERVATION ON OTHER STARS

To determine azimuth by observation on any star other than a circumpolar star, one should determine when the star will be above the horizon and then learn to identify the star in the star field. Star charts and diagrams, which can be found in the Sokkia ephemeris and the *Nautical Almanac,* help in this procedure. Once a star and an observational time have been chosen, the approximate position of the star at a specified time is determined as described in Section 10.39, Example 10.22.

The observational procedure is essentially the same as that followed for solar observations described in Section 10.31 by the hour-angle method. The same rules for sighting the star apply as for Polaris, except that other stars, which are at lower declinations than Polaris, move in a manner similar to the sun. Therefore, the rules for night operations apply but the time must be recorded with a higher degree of precision than for Polaris observations. The time module or stopwatch must be set to WWV and DUT correction should be noted and applied to yield UT1 time (Section 10.21). When the star is brought within the field of view, the vertical cross hair is sighted on the star, the time is registered, and the horizontal angle from a backsight to a terrestrial point to the star is read and recorded. Where a high degree of precision is required, several repetitions of the horizontal angle with the telescope in direct and reversed positions should be observed, recording a time with each measurement to the star. The procedure for reducing the data and calculating the azimuth is similar to that outlined for solar observations (Section 10.31) and for observation of Polaris (Section 10.38) by the hour-angle method.

Most of the stars recommended for azimuth determination have declinations about the same as the sun. Consequently, azimuths calculated from observations on these stars have accuracies comparable to the accuracies for solar azimuths by the hour-angle method as discussed in Section 10.32, Procedure 2. In general, the accuracy of azimuths by star shots should be somewhat higher than for solar shots, because stars offer an excellent target on which to center and the atmosphere through which the sight is taken is much less turbulent than for sights on the sun.

10.41
LATITUDE BY OBSERVATION ON OTHER STARS

To determine latitude by an observation on a star at an upper transit, the star's declination and Greenwich hour angle or right ascension and sidereal hour angle are found in an ephemeris and the approximate standard time of the upper transit at the given place is determined, as shown in Examples 10.20 and 10.21. About 10 min before this time, the theodolite or total station system is set up and the estimated altitude or coaltitude (zenith distance) of the star is set off on the vertical circle (the latitude will be known roughly, the declination will be known; hence, the altitude can be estimated with sufficient precision to bring the star within the field of view). The telescope is pointed approximately along the meridian, and the instrument is revolved about the vertical axis back and forth through a small angle until the star is sighted. The star is followed with the horizontal cross hair until its maximum altitude is reached, then the vertical circle is read and recorded. The latitude is determined as described in Section 10.37.

10.42
DETERMINATION OF LONGITUDE

To determine the longitude by observing the transit on any star, the direction of the meridian and the standard time of the place being known, the standard time of the upper transit is calculated for a longitude estimated to be that of the place. The star is found as described in Section 10.39, and the standard time of its upper transit is observed. The interval between the calculated time and the observed time of the upper transit, changed to sidereal time, is the difference between the estimated longitude and the true longitude of the place.

10.43
AZIMUTH BY GYRO ATTACHMENT

A gyro attachment consists of a gyro motor suspended by a thin metal tape from one end of a sealed tube that can be mounted vertically on the standards of a theodolite or a total station system. Figure 10.30 shows the gyro attachment mounted on a total station system. The tube, enclosure, gyro motor, and suspension are so designed that, when the unit is attached to the horizontal axis of a leveled theodolite, the metal tape holding the rotating gyro coincides with the vertical axis of the instrument and also with the direction of gravity. Figure 10.31 illustrates a cross-sectional view of a typical gyro attachment. Note the suspension strip that supports the gyro motor has an axis of rotation perpendicular to the vertical or plumb line. The gyro motor spinning at 22,000 revolutions per minute about this horizontal axis tries to maintain in space its initial random spinning plane created by its moment of inertia. The gyro, fixed to the instrument that is earthbound, is pulled out of its original spinning plane by the earth's rotation. This interference causes the gyro to oscillate around the plumb line until the spin axis is oriented in the north-south plane and the rotation of the gyro corresponds to the rotation of the earth. The gyro does not stabilize immediately in the north-south direction but oscillates about the meridian plane. The oscillation can be observed optically through an eyepiece attached to the tube containing the gyro (Figure 10.31). These oscillations can be observed as a moving light mark projected on a scale that has a V-shaped central index and can be observed through the viewing eyepiece. The midposition of these oscillations can be located by measuring the size of the swing period of the oscillations. Figure 10.32a shows the gyro mark centered and Figure 10.32b shows the mark at one elongation, which is one-half the swing period from the central index. The gyro attachment is oriented on the instrument so that the gyro spin axis and the telescopic line of sight fall in the same vertical plane when the light mark is centered on the index. When this condition is achieved, the telescope is oriented toward true or astronomic north.

The telescope can be oriented approximately toward true north by following the swing of the gyro mark with the upper tangent motion of the theodolite. When the gyro mark

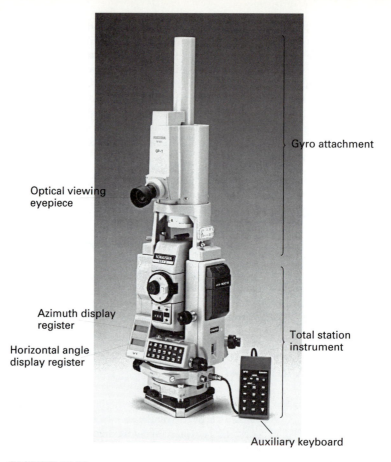

Gyro attachment

Optical viewing
eyepiece

Azimuth display
register

Horizontal angle
display register

Total station
instrument

Auxiliary keyboard

FIGURE 10.30
Sokkia GPI gyro attachment and SF 10 keyboard attached to a
Set 3B11 total station instrument. (*Courtesy of Sokkia
Corporation.*)

reaches an elongation point and reverses direction (a reversal point), the horizontal circle reading is recorded. Next, the gyro mark again is followed, using the upper tangent motion, until a second reversal point on the opposite side of the index is reached. At this time, a second horizontal circle reading is taken. When the average of these two circle readings is set on the horizontal circle, the telescope line of sight is directed approximately toward true north to within ±3–4′. To obtain a more precise direction, measure the time intervals of at least three transits of the gyro mark across the central index and amplitudes of elongation points on both sides are measured. An angular correction to the approximate north direction then can be determined as a function of these observed times and half-swings east and west of the central index.

On the system shown in Figure 10.30, amplitudes of elongation or reversal points (Figure 10.32b) are entered using the auxiliary keyboard (Figure 10.30) for a specified number of transits. For the last three successive transits, as the floating mark passes 0 (Figure 10.32a), the times are entered by pressing one key on the auxiliary keyboard. With these data entered, the angular correction is calculated automatically by onboard software and the azimuth is displayed on the azimuth display register (Figure 10.30). The entire operation requires from 20 to 30 min to complete and permits determination of the azimuth with a standard deviation of 20″.

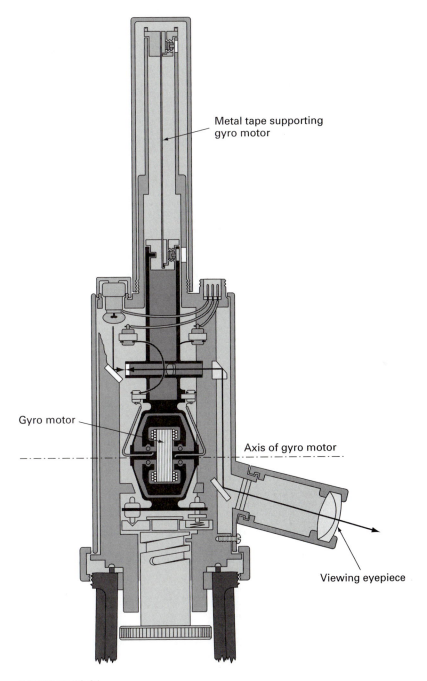

Metal tape supporting
gyro motor

Axis of gyro motor

Gyro motor

Viewing eyepiece

FIGURE 10.31
Cross-sectional view of Wild GAK1 gyro attachment. (*Courtesy of Leica, Inc.*)

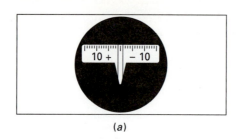
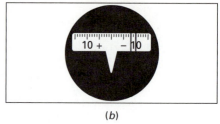

(a)   (b)

FIGURE 10.32
(a) Swing of the gyro mark in the middle of the scale. (b) Gyro mark at elongation—transit method. (*Courtesy of Leica, Inc.*)

The gyro method of establishing an azimuth is applicable to any survey where the attainable precision is adequate. It is particularly useful for the transfer of azimuth in the surveys needed for underground tunneling and mining operations.

PROBLEMS

10.1. Define the term *celestial sphere* as used in astronomy.

10.2. Describe how an observer's position on the surface of the earth is defined in the earth's equatorial system.

10.3. Define *(a)* vernal equinox, *(b)* hour angle, and *(c)* right ascension.

10.4. Explain the relationship between the horizon system of spherical coordinates and the equatorial system.

10.5. On a given day, 0^h Greenwich civil time occurs at $5^h17^m30^s$ Greenwich sidereal time. At that instant, what is the local sidereal time at a place whose longitude is $8^h10^m40^s$W?

10.6. When the local apparent time is $9^h10^m30^s$ at a place where the longitude is $96°10'30''$W, what is the Greenwich apparent time?

10.7. What is the Greenwich civil time when it is 10^h20^m A.M. Pacific daylight savings time?

10.8. The local civil time at a given place is $9^h10^m20^s$ when the Greenwich civil time is $14^h31^m15^s$. What is the longitude of the place?

10.9. The local civil time at a place is $15^h40^m25^s$ and the longitude of the place is $78°35'15''$. Determine the Eastern standard time.

10.10. From an ephemeris find the equation of time for the instant of $2^h10^m10^s$ P.M. Pacific standard time on June 10 of the current year. If the longitude of the place is $7^h46^m03^s$W, calculate the local civil and local apparent times.

10.11. From an ephemeris, find the equation of time for the instant of $3^h19^m30^s$ P.M. apparent time on September 10 of the current year, at a place whose longitude is $8^h11^m38^s$W. Compute the corresponding local civil time.

10.12. At a given place the hour angle of the true sun at 7^h15^m P.M. Greenwich civil time on January 1 of the current year is $38°10'30''$. What is the local sidereal time?

10.13. The mean radius of the earth is 3,956 miles (6367 kilometers), and the mean distance to the sun is 92,900,000 miles (149,508,100 kilometers). What is the sun's mean horizontal parallax? What is the parallax correction when the altitude is $25°$?

10.14. The observed altitude of a star is $26°10'30''$. The temperature is $90°F$ and atmospheric pressure $= 28.70$ in. Hg. Find the refraction correction and compute the true altitude of the star.

10.15. The observed altitude of the sun's center is $15°07'30''$. The temperature is $-10°C$ and atmospheric pressure $= 29.42$ in. Hg. Find the correction for parallax and refraction and compute the true altitude of the sun.

10.16. Find the apparent declination of the sun for the instant 10^h30m A.M. Central standard time on April 5 of the current year, using a current ephermeris.

10.17. Find the apparent declination of the sun for the instant of local apparent noon at a place whose longitude is $7^h40^m30^sW$ for the date of May 15 of the current year, using a current ephemeris.

10.18. The observed coaltitude of the upper limb of the sun is $74°52'30''$. The temperature is $12°F$ and the atmospheric pressure is 27.45 in. Hg. Calculate the true altitude to the sun's center.

10.19. The observed zenith angle to a star is $66°44'40''$ when the temperature is $60°F$ and atmospheric pressure is 28.42 in. Hg. Determine the true coaltitude to the star.

10.20. The observed clockwise horizontal angle from a reference line to the trailing limb of the sun is $312°16'37''$. The angle is measured on December 10 of the current year at $10^h00^m30^s$ Central standard time. Determine the true horizontal angle to the center of the sun at that time.

10.21. On October 9 of the current year at a place where the latitude $= 37°52'21''$ and longitude $= 122°15'33''W$, the following data were recorded for a set of solar observations by the hour-angle method:

Instrument at point number 24, sight taken on point number 49
Instrument: Wild T-2 direction theodolite No. 68730
Sights taken to center of sun using a solar circle attachment

From	To	Telescope	Horizontal circle	UT1 time
49		D	$0°00'42.0''$	
	O	R	$222°13'59.0''$	$23^h15^m33.9^s$
	O	R	$222°24'02.0''$	$23^h16^m23.5^s$
	O	D	$42°44'24.0''$	$23^h18^m03.5^s$
	O	D	$42°56'33.0''$	$23^h19^m03.5^s$
49		R	$180°00'38.0''$	

Using the average UT1 time and average horizontal angle, compute the azimuth of the sun and the azimuth of line 24 to 49 from the north.

10.22. Using the data from Problem 10.21, calculate the azimuths from the north of the sun and of the line 24 to 49 for the first reversed reading.

10.23. On October 8 of the current year, the following data sets were recorded for a solar observation using a direction theodolite with all sights on the center of the sun using a solar attachment.

The instrument was set on station 16 with a backsight on station 38. Latitude and longitude for station 16 are 37°52′21″N and 122°15′33″W, respectively. B.S. 16–38 = 0°45′00.0″

From	To	Telescope	Horizontal circle	UT1 time
16	o	D	222°42′25.9″	23^{h}13^{m}01.9^s
	o	D	223°07′44.7″	23^{h}15^{m}03.2^s
	o	D	223°27′57.3″	23^{h}16^{m}43.1^s
	o	R	44°00′04.8″	23^{h}19^{m}19.5^s
	o	R	44°17′49.8″	23^{h}20^{m}47.4^s
	o	R	44°29′57.7″	23^{h}21^{m}47.4^s
	38	R	180°45′09.2″	

Using the average UT1 time and average horizontal angle, compute the azimuths from the north for the sun and for line 16 to 38 by the hour-angle method.

10.24. Using the data for Problem 10.23, calculate the azimuths from the north for each pointing and compute the mean value. Examine these azimuths and reject any value that varies from the mean by 10″ or more, being sure to reject an equal number of direct and reversed measurements. Determine the mean azimuth from the north for line 16 to 38 using the remaining measurements.

10.25. Solar observations by the altitude method were made on October 20 of the current year, at a place where the latitude = 37°52′20″, the temperature = 74°F, and atmospheric pressure = 29.96 in. Hg. From the instrument setup at station A and a backsight on $A9$, the average horizontal (clockwise from $A9$) and zenith angles to the center of the sun are 56°45′51″ and 60°28′16″, respectively. The average time of observation is 14^{h}31^{m}03^s Pacific standard time. Compute the azimuths from the north of the sun and of line A to $A9$. Use Equation (10.2b).

10.26. On August 1 of the current year the observed altitude of the sun at a given place is 30°51′45″ at 7^{h}42^{m}20^s A.M. local apparent time. The latitude of the place is 37°18′20″ and the longitude is 102°17′30″W. The horizontal angle (measured clockwise) from the reference line to the sun is 89°39′15″. Compute the azimuths from the north of the sun and of the reference line. Use Equation (10.2b). Temperature = 75°F and p = 29.96 in Hg.

10.27. The true altitude of the sun's center from a given station is 24°28′44″ at the instant of 4^{h}13^{m}12^s P.M. Mountain standard time on April 15 of the current year. The clockwise horizontal angle from reference line to the sun is 312°16′37″. The latitude of the place is 39°10′40″. Compute the azimuth and hour angle of the sun at the given instant. Use Equation (10.5) and (10.6), Section 10.11. Determine (a) the azimuth of the reference line from the south, and (b) the longitude of the place.

10.28. The observed zenith angle of the lower limb of the sun, as it crosses the meridian at a given place, is 34°28′30″. The observation is made at 11^{h}34^{m}20^s A.M. Eastern daylight savings time on May 16 of the current year. The temperature is 50°F and atmospheric pressure is 29.48 in. Hg. Calculate the latitude of the place.

10.29. At a given place, the center of the true sun crosses the meridian at 11^{h}41^{m}30^s A.M. Mountain standard time on September 5 of the current year. Calculate the longitude of the place.

10.30. On June 3 of the current year, Polaris was observed from station Patio having latitude = 37°54′26.6″N and longitude = 122°16′22.2″W, where the temperature and atmospheric pressure were 58°F and 28.70 in. Hg, respectively. Two sets of direct and reversed measurements (clockwise horizontal, angles) were made using a Topcon total station system, recorded as follows:

Set	Point	Telescope	Time, UTC	Horizontal circle
1	Kani	D		0°00'00"
	*	D	$3^h51^m28^s$	352°52'20"
	*	R	$3^h56^m38^s$	172°53'20"
2	Kani	D		0°00'00"
	*	D	$3^h59^m56^s$	352°54'15"
	*	R	$4^h05^m30^s$	172°55'30"

Using the average horizontal angle and average time, compute the azimuth of Polaris and the azimuth from north of the line from Patio to Kani by the hour-angle method.

10.31. For the data in Problem 10.30, compute azimuths from the north for the line Patio to Kani for each pointing on Polaris and then determine the average azimuth of this line. Which azimuth would you assume to be more accurate, that for Problem 10.30 or for this problem? Explain why.

10.32. On June 9 of the current year, the average clockwise angle from reference line to Polaris and the average UTC time of observations are 352°55'55" and $3^h42^m45.5^s$, respectively. The latitude and longitude of the place of observation are 37°54'27"N and 122°16'22"W, respectively. Determine the azimuth from the north of the reference line.

10.33. At a given place on August 31 of the current year, the observed altitude of Polaris at upper culmination is 44°30'20". The temperature and atmospheric pressure are 80°F and 29.32 in. Hg, respectively. What is the latitude of the place?

10.34. The zenith angle to Polaris 15^m after the time of upper culmination is 41°27'40" when the declination is 89°10'35". If the temperature and atmospheric pressure are 55°F and 28.95 in. Hg, respectively, at the time of observation, what is the latitude of the place?

10.35. What is the time of western elongation of Polaris at a given place when upper culmination occurs at $5^h15^m20^s$ Pacific standard time, the declination of Polaris is 89°10'30", and the latitude is 42°22'47"N?

10.36. On September 7 of the current year, upper culmination of Polaris at the Greenwich meridian occurs at $2^h35^m29^s$ Greenwich civil time. What is the Eastern standard time of upper culmination on September 10 of the same year at a place where the longitude is 78°30'15"W?

10.37. Find the Pacific standard time of eastern elongation of Polaris on September 20 of the current year for the latitude of 37°54'27"N and longitude $8^h09^m05^s$W.

10.38. Compute the azimuth and hour angle of Polaris when at elongation, the declination being 89°14"22" and the latitude of the place being 43°30'55"N.

10.39. Find the azimuth of Polaris when at elongation on January 3 of the current year for a place where the latitude is 37°52'24"N.

10.40. At a given place, the upper culmination of Polaris occurs at $2^h10^m30^s$ P.M. Central standard time on a given date. On the same date an azimuth observation is made at $8^h20^m05^s$ P.M. Central standard time. The latitude of the place is 42°24'35"N and the declination is of the star is 89°14'24". Compute the hour angle and azimuth of the star.

10.41. On June 11 of the current year, at a place where the latitude and longitude are 37°54'27"N and 122°16'22"W, respectively, observations are to be made on the star Arcturus beginning at 9 P.M. Pacific daylight savings time. Determine the approximate zenith angle and azimuth to be set off on the vertical and horizontal circles of a total station system in order to locate Arcturus.

10.42. The star Spica is to be observed from the same place described in Problem 10.41 on May 15 of the current year at 10 P.M. Pacific daylight savings time. Determine the approximate altitude and azimuth to be set off to locate Spica in the telescope at this time.

REFERENCES

Buckner, R. B. *A Manual on Astronomic and Grid Azimuth.* Rancho Cordova, CA: Landmark Enterprises, 1984.

Elgin, R. L.; D. R. Knowles; and J. H. Senne. *Celestial Observation Handbook and Ephemeris 1995.* Overland Park, KS: Sokkia Corporation, annual (in advance).

Elgin, R. L.; D. R. Knowles; and J. H. Senne. *Practical Surveying Guide to Celestial Observations.* Canton, MI: P. O. B. Publishing Co., 1992.

Elgin, R. L.; D. R. Knowles; and J. H. Senne. *ASTRO*ROM User Manual.* Rolla, MO: Elgin, Knowles & Senne, Inc., 1986.

Kissam, P. *Surveying for Civil Engineers.* 2nd ed. New York: McGraw-Hill Book Co., 1981.

Mackie, J. B. *Elements of Astronomy for Surveyors.* 9th ed. High Wycombe, England: Charles Griffin, 1985.

Mattson, D. F. "Double Sun Shots Without Latitude or Longitude." *Surveying and Land Information Systems* 51, no. 3 (September 1991), pp. 171–77.

Moffitt, F. H., and H. Bouchard. *Surveying.* 9th ed. New York: HarperCollins Publishers, 1992.

Nassau, J. J. *Textbook of Practical Astronomy.* 2nd ed. New York: McGraw-Hill Book Co., 1948.

Royal Greenwich Observatory. *The Star Almanac for Land Surveyors.* Lanham, MD: Bernan-UNIPUB, annual (in advance).

U.S. Naval Observatory. *The Nautical Almanac 1995.* Washington DC: Superintendent of Documents, U.S. Government Printing Office, annual (in advance).

U.S. Naval Observatory. *The Astronomical Almanac 1995.* Washington DC: Superintendent of Documents, U.S. Government Printing Office, annual (in advance).

U.S. Naval Observatory. *Mica and Interactive Astronomical Almanac for 1990–1999, User's Guide.* Washington DC: Astronomical Applications Department.

Wolf, P. R., and R. C. Brinker. *Elementary Surveying.* 9th ed. New York: HarperCollins College Publishers, 1994.

CHAPTER 11

Map Projections

11.1
INTRODUCTION

The location of a point on the surface of the earth is determined by three coordinates. These coordinates are referred to as either a curvilinear system, such as geodetic latitude, ϕ; longitude, λ; and height, h; or a Cartesian (rectangular) system such as geocentric X, Y, and Z or local x, y, and z (see Section C.4, Appendix C). A point on a map is located by only two coordinates.

In plotting a map of a small area, the curvature of the earth need not be considered. A level surface on the earth is assumed to be a plane, and points are plotted on the map in terms of rectangular coordinates from two orthogonal axes, representing the east-west and north-south directions.

For maps of larger areas this simple method is not satisfactory because the curvature of the earth can no longer be ignored. The geometric shape of the earth is an ellipsoid with a polar diameter about one-third of 1 percent shorter than the equatorial diameter. Therefore, a plane passing through the equator would cut the earth in a circle, while a plane through the poles (*meridional* plane) would intersect it in an ellipse. However, with that slight difference between polar and equatorial dimensions, the ellipse is very nearly a circle. Consequently, for many purposes, the earth is assumed to be a *sphere*. Initially, this will make it easier for the reader to visualize the various projections to be discussed. Eventually, the ellipsoidal shape must be considered, especially in connection with the development of equations for various map projections.

Regardless of whether the earth is considered a sphere or an ellipsoid, it is not possible to develop its *surface exactly* onto a plane, just as it is impossible to flatten a section of orange peel without tearing it. It follows, then, that whatever procedure is used to represent a large area on a map, there *always* will be some distortion. To minimize the distortion, points on the map are represented in terms of parallels of latitude and meridians of longitude. A position on the earth in terms of latitude and longitude (e.g., New York City is at 40°45'N latitude and 74°00'W longitude) is transformed into scaled linear dimensions on the map. This is accomplished by using the dimensions of the earth and a selected set of criteria for representing the earth on the map. Such a transformation from latitude ϕ and longitude λ to a map's X and Y coordinates is the function of *map projections.*

Many view the map projection as a mathematical operation in which the map coordinates X and Y are written as a pair of parametric functions of latitude and longitude, ϕ and λ, or

$$
\begin{aligned}
X &= f_x(\phi, \lambda) \\
Y &= f_y(\phi, \lambda)
\end{aligned}
\tag{11.1}
$$

On an *ideal* map *without distortion,* Equation (11.1) must satisfy the following conditions: (1) all distances and areas on the map would have correct relative magnitude, (2) all azimuths and angles would be shown correctly on the map, (3) all great circles on the earth would appear as straight lines on the map, and (4) geodetic latitudes and longitudes of all points would be shown correctly on the map.

As noted earlier, because of the shape of the earth, it is impossible to satisfy *all* of these conditions in the same map. It is possible, however, to satisfy one or more of the four conditions by imposing them on the transformation Equation (11.1). Therefore, several classes of map projection result:

1. *Conformal* or *orthomorphic* projection results in a map showing the correct angle between any pair of short intersecting lines, making small areas appear in correct *shape.* As the scale varies from point to point, the shapes of larger areas are incorrect.
2. An *equal-area* projection results in a map showing all areas in proper relative *size,* although these areas may be much out of shape and the map may have other defects.
3. In an *equidistant* projection, distances are correctly represented from one central point to other points on the map.
4. In an *azimuthal* projection, the map shows the correct *direction* or azimuth of any point relative to one central point.

11.2
TYPES OF MAP PROJECTIONS

Although all map projections are carried out by computing X and Y for each pair of ϕ and λ using different forms of Equation (11.1), there are two methods of projection: geometric and mathematical. In a *geometric* projection, a surface that can be developed into a plane (such as a plane, a cone, or a cylinder) is selected such that it either cuts or is tangent to the earth. A point then is selected as the projection center, from which straight lines are connected to points on the earth and extended until they intersect the selected mapping surface. In a *mathematical* projection, there is no one particular projection point; instead, a form of Equation (11.1) is used to compute the location X, Y of the point on the map from its position ϕ, λ on the earth. In the following sections, the various projections are introduced in groups, beginning with those on a plane tangent to the earth sphere.

11.3
MAP PROJECTION TO A PLANE

The easiest geometric map projection to visualize is that in which the projection surface is a plane tangent to the sphere at any point. The point of tangency becomes the central point in the map. Three possibilities for the point are used as projection center: (1) when the center of the sphere is used (Figure 11.1), the projection is called *gnomonic;* (2) when the

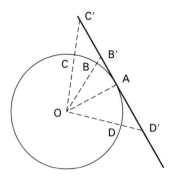

FIGURE 11.1
Gnomonic projection.

FIGURE 11.2
Stereographic projection.

end of the diameter opposite to the point of tangency is used (Figure 11.2), the projection is called *stereographic;* and (3) when the projection center is at infinity, in which case the projection lines become parallel, the projection is called *orthographic.*

11.4
GNOMONIC PROJECTIONS

This, considered to be the oldest true map projection, is assumed to have been devised by the great abstract geometrist Thales in the sixth century B.C. It is a geometric projection to a plane tangent to the sphere at any point such as A in Figure 11.1, with the projection center at the center of the sphere. Lines from the earth's center O through all points to be mapped, A, B, and so forth, are extended until they intersect the tangent plane in the map points A, B', C', and so forth. The new points A, B', C', and so forth, in the plane represent the gnomonic projection of the original points. This projection has the following properties:

1. Great circles show as straight lines and therefore meridians are straight lines.
2. Shapes and sizes undergo extreme distortion as we go farther from the central tangent point.
3. Azimuths of lines drawn from the tangent point to other map points are correct, because such lines are great circles. Therefore, this is an *azimuthal* projection.
4. Because the point of projection is at the center of the sphere, it is not possible to map an entire hemisphere.

This projection is used mostly for navigational purposes. For other purposes, only a limited area around the central point may be used.

11.5
STEREOGRAPHIC PROJECTIONS

This projection is credited to the Greek astronomer Hipparchus dating back to the second century B.C. It is a geometric projection to a tangent plane with the point of projection P diametrically opposite to the point of tangency T (Figure 11.2). Rays from P to surface points B, C, and so on, are extended until they intersect the tangent plane. The points of intersection B'', C'', and so on, are the map points on the stereographic projection on the tangent plane. On any other intersecting plane, such as AE in Figure 11.2, an equivalent projection is possible, with only a reduction scale factor from the projection on the tangent plane.

This projection is both azimuthal and conformal. It has the following properties:

1. It is the only projection on which circles on the earth still appear as circles on the map.
2. Like the gnomonic projection, azimuths of lines from the central point are correct because such lines are great circles.
3. The scale increases as we move away from the tangent point and therefore its main defect is that areas are not correctly shown.

This is an excellent projection for general maps showing a hemisphere. It is also used for plotting ranges from central objects and for navigational purposes at higher latitudes, above 80°, when the plane is tangent at the pole.

11.6
ORTHOGRAPHIC PROJECTIONS

This is another ancient projection, which was used during the Renaissance for artistic representation of the globe. Recent space applications have revived its use. It is again a geometric projection on a tangent plane, with the projection lines parallel to each other and normal to the tangent plane (Figure 11.3). If the tangent point is one of the poles, each parallel of latitude will be shown correctly to scale as a circle, but the distance between successive parallels becomes rapidly smaller as we move farther from the center of the map. This means that unlike the gnomonic and stereographic projections, the scale in the orthographic projection *decreases* away from the central tangent point. On the other hand, this is an azimuthal projection just like the gnomonic and stereographic, so that the

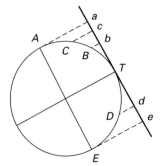

FIGURE 11.3
Orthographic projection.

azimuths from the tangent point are correct because the lines through it are great circles. The representation of the globe on this projection appears as it would in a photograph taken from deep space. Therefore, the orthographic projection is also used for map representations of the moon and planets.

11.7
CONICAL PROJECTIONS

Unlike a sphere, both a cone and a cylinder can be developed into a plane without distortion, and therefore both are used for map projection. A cone fits over a sphere, touching it in a small circle of latitude, called the *standard parallel,* when the cone apex lies on the polar axis (Figure 11.4a). As the height of the cone increases, the standard parallel gets closer to the equator (Figure 11.4b). Finally, when the standard parallel reaches the equator, the elements of the cone become parallel and the cone becomes a cylinder (Figure 11.4c). When the cone height decreases, the standard parallel moves to higher latitudes (Figure 11.4d), and finally the cone becomes a plane tangent to the sphere at one point, the pole (Figure 11.4e). Thus, the projection of a sphere to a plane and a cylinder actually are limiting cases (at least geometrically) to the conical projection.

Because a tangent cone has one parallel of latitude common to the sphere, its representation on the map will be at true scale, and scale distortions will increase as the distance is extended to the north and the south of the parallel. For this reason, many conical projections use a cone that intersects the sphere in *two standard parallels* to minimize scale distortions. Although *geometric* projection on the cone is possible, the more important and useful conical projections are based on *mathematical* projections.

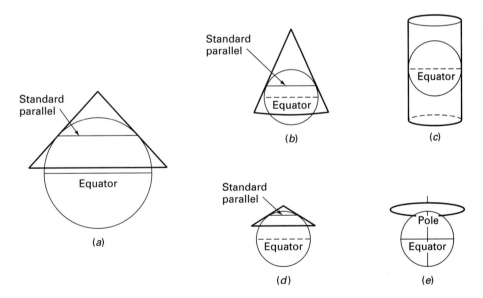

FIGURE 11.4
(a) Cone is tangent to a sphere. (b) Cone is tangent to a sphere with the standard tangent closer to the equator. (c) Cylinder is tangent to a sphere at the equator. (d) Cone has a standard parallel at higher latitudes. (e) Plane is tangent to a sphere at the pole. (*Courtesy of the National Geographic Society.*)

11.8
ALBERS EQUAL-AREA PROJECTIONS

This is a mathematical projection on a cone with two standard parallels, developed by H. C. Albers in 1805. Its main characteristic is that small areas on the map are equal at the scale of the map, to corresponding small areas on the earth. Other properties of this projection include

1. The scale is true along the standard parallels, is small between them, and is large beyond them along *other parallels. Meridional* scale is large between and is small beyond standard parallels.
2. The distances and directions measured on the map are reasonably accurate.
3. The meridians are straight lines meeting at a point that is the center of the concentric circular arcs representing the parallels. Therefore, the meridians and parallels meet at right angles.
4. The parallel spacing increases away from the standard parallels.

This projection is useful for maps requiring equal-area representation and is used for areas at midlatitudes with extensive east-west dimensions.

11.9
POLYCONIC PROJECTIONS

Instead of a single cone, a series of conical surfaces may be used, points on the surface of the earth being considered as projected to a series of frustums of cones that are fitted together. These conical surfaces are developed each way from a central meridian. Owing to differences in radii, the resulting strips would not fit exactly together when laid flat; spaces would appear between them, such spaces increasing in width as the distance from the central meridian increases (Figure 11.5b). To avoid such spaces, the north-south scale must be modified along the various meridians. In such a system of lines, points are plotted by latitude and longitude.

In Figure 11.5, each parallel of latitude appears on the map as the arc of a circle having as radius the corresponding tangent distance; the parallel through A has a radius Aa, that

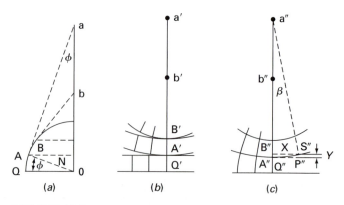

FIGURE 11.5
Polyconic projection: (a) section, (b) development, (c) map.

through B has a radius Bb, and so on. The centers of these circles all lie on the central meridian of the map. The length of each tangent distance Aa, and the like, is $N \cot \phi$, in which N is the length of the normal or vertical at latitude ϕ extended to its intersection with the earth's axis (N equals the radius of curvature in the prime vertical; see Section C.4). More precisely,

$$N = \frac{a}{\sqrt{1 - \varepsilon^2 \sin^2 \phi}} \qquad (11.2)$$

in which a is the earth's equatorial radius and ε is the eccentricity of the ellipse in the meridian section ($\varepsilon^2 = 0.006694380023$). The length of the tangent distance $N \cot \phi$ varies with the latitude. For the assumption that the earth is a sphere, N is equal to the radius of the sphere.

The distances exemplified by $Q'A', A'B'$, along the central meridian on the map (Figure 11.5b) are true scale representations of the corresponding arc distances QA and AB, in the meridian section (Figure 11.5a). The parallels drawn on the map may be selected with as small a difference in latitude as may be desired, and each one is drawn with its own particular radius as shown, with the center of the arc on the central meridian at the proper tangent distance (to scale) above the point where the parallel cuts the central meridian.

It should be observed that the method of drawing these arcs of the parallels of latitude on the map is such that each parallel is developed separately as the circumference of the base of its own distinct cone and that the spacing between them increases with increasing differences of longitude from the central meridian, thereby changing the north-south scale of the map from place to place as the longitude difference increases.

The arc distance $A''S''$ in the map (Figure 11.5c) represents to true scale the difference in longitude between the points A'' and S''. The angle $A''a''S''$ is designated β and is given by

$$\beta = \lambda \sin \phi$$

in which λ is the arc $A''S''$. The rectangular coordinates of the point S'' referred to A'' as its origin are

$$X = A''P'' = a''S'' \sin \beta = N \cot \phi \sin \beta \qquad (11.3)$$

$$Y = P''S'' = a''S''(1 - \cos \beta) = N \cot \phi \cos^{-1} \beta \qquad (11.4)$$

and if the chord $A''S''$ is drawn in the triangle $A''S''P''$, $S''P'' = A''P'' \tan P''A''S''$, or $Y = X \tan(\beta/2)$.

As many points as desired along the parallels on the map, such as point S'', are calculated and plotted. Then the meridians are drawn through such points. These meridians are curved, concave toward the central meridian, but if the parallels are drawn close enough together, each meridian may be drawn as a series of straight lines from parallel to parallel. In the network of parallels and meridians so prepared, points are plotted by latitude and longitude. Near the central meridian there is little error in such a map, but the error increases in proportion to the square of the difference in longitude along any one parallel. The variation with difference in latitude is not in direct proportion.

Note that, along the central meridian and along every parallel, the map is true to scale; along the other meridians, the scale is somewhat changed; near the central meridian, the parallels and meridians intersect nearly at right angles; and areas of great extent north and south may be mapped with a very small distortion.

Although better adapted to mapping an area of great extent in latitude than for an area of great extent east and west, the polyconic projection is sufficiently accurate for maps of considerable areas and was used widely by the U.S. Geological Survey and the U.S. National Geodetic Survey.

11.10
CONFORMAL MAPPING

Conformal projection was defined in the introduction as that which preserves the angle between any pair of intersecting short line segments. In this section, the mathematical condition for conformal mapping is given. However, instead of expressing the condition as a relation between the map coordinates X and Y and geodetic coordinates ϕ and λ an intermediate surface is introduced. This surface is called the *isometric plane,* where a new *isometric latitude, q* (Krakiwsky, 1973; Thomas, 1952), is given by

$$q = \ln\left[\tan\left(\frac{\pi}{4} + \frac{\phi}{2}\right)\left(\frac{1 - \varepsilon \sin \phi}{1 + \varepsilon \sin \phi}\right)^{\varepsilon/2}\right] \tag{11.5a}$$

or also (Stem, 1989a) by

$$q = \tfrac{1}{2}\left(\ln \frac{1 + \sin \phi}{1 - \sin \phi} - \varepsilon \ln \frac{1 + \varepsilon \sin \phi}{1 - \varepsilon \sin \phi}\right) \tag{11.5b}$$

where ε is the eccentricity; $\varepsilon^2 = (a^2 - b^2)/a^2$, with a and b being the semimajor and semiminor axes of the earth ellipsoid. The map coordinates X and Y are then expressed as functions of λ and q:

$$X = f_1(\lambda, q)$$
$$Y = f_2(\lambda, q) \tag{11.6}$$

For the relations in Equation (11.6) to represent a conformal transformation, the following so-called Cauchy-Riemann equations must be satisfied:

$$\frac{\partial X}{\partial \lambda} = \frac{\partial Y}{\partial q}$$

$$\frac{\partial X}{\partial q} = -\frac{\partial Y}{\partial \lambda} \tag{11.7}$$

Equation (11.7) can be used to either derive a conformal mapping or verify that a mapping is conformal.

11.11
LAMBERT CONFORMAL CONIC PROJECTIONS

This projection was developed by Johann Heinrich Lambert in 1772, the same year in which he invented the transverse Mercator projection. The Lambert conformal conic projection is one of the most widely used projections in the United States and even worldwide. It is used for the state plane coordinate systems of states (or their zones) of greater east-west than north-south extent. It is a conic projection with two standard parallels, as shown in Figure 11.6. Therefore, meridians are straight lines meeting in a point outside the map limits; parallels are arcs of concentric circles; and both sets of lines meet at right angles. Scale is true only along the two standard parallels; it is compressed between and expanded beyond them.

The parametric equations for this projection, which must satisfy the conformality conditions of Equation (11.7), are

$$X = Ke^{-lq} \cos l\lambda$$
$$Y = Ke^{-lq} \sin l\lambda \tag{11.8}$$

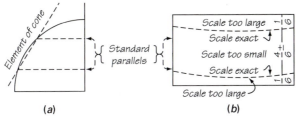

(a)   (b)

FIGURE 11.6
Lambert conformal conic projection: (a) section, (b) map
of state coordinate zone.

in which l and K are constants computed in terms of the two selected standard parallels. The radius of curvature in the prime vertical direction, N, of the ellipsoid at a particular latitude, ϕ, is given by Equation (11.2). Denoting by N_1 and N_2 the radii at the two standard parallels ϕ_1 and ϕ_2, the constant l is given by

$$l = \frac{\ln N_1 - \ln N_2 + \ln \cos \phi_1 - \ln \cos \phi_2}{q_2 - q_1} \qquad (11.9a)$$

An alternative expression in terms of ϕ_0, the latitude of the central parallel or latitude of the grid origin, is given by

$$l = \sin \phi_0 \qquad (11.9b)$$

The second constant, K, is given by

$$K = \frac{N_1 \cos \phi_1}{le^{-lq_1}} = \frac{N_2 \cos \phi_2}{le^{-lq_2}} \qquad (11.10)$$

The radius of any circular arc representing a parallel is computed from

$$R^2 = X^2 + Y^2$$

which from Equation (11.8) becomes

$$R^2 = K^2(e^{-2lq})(\cos^2 l\lambda + \sin^2 l\lambda)$$
$$= K^2 e^{-2lq}$$

Then,

$$R = Ke^{-lq} \qquad (11.11)$$

Equations (11.8) may be rewritten in terms of R as

$$X = R \cos l\lambda$$
$$Y = R \sin l\lambda \qquad (11.12)$$

These equations can be represented graphically as shown in Figure 11.7. The equivalent of the apex of the cone is taken as the origin of coordinates with the X axis in the direction of the central meridian. It becomes the center of the concentric circles representing the parallels and the point in which all the meridians intersect. Unlike the Albers projection, the spacing of parallels increases away from the standard parallels. For a map of the United States on this projection, scale errors at any point need not exceed 2 percent. Computation of state plane coordinates in the Lambert conformal conic projection is explained in detail in Section 11.17.

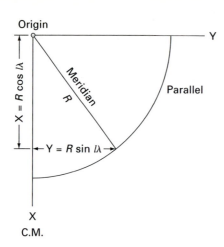

FIGURE 11.7
Graphical representation of coordinates
on the Lambert conformal projection.

11.12
MERCATOR PROJECTIONS

The Mercator projection is a projection on a cylinder tangent to the earth at the equator
(Figure 11.4c). It is a *mathematical* projection, however, and not geometric in order to
enforce the conformality conditions. The projection was created by Mercator in 1569 as a
result of his efforts to have the *rhumb* line, or *loxodrome,* or the line of constant bearing on
the globe, appear as a straight line on the map. With true scale at the equator and taking
the equator as the zero Y value in the map, the mapping equations are

$$X = a\lambda$$
$$Y = aq$$
(11.13)

where λ is the longitude and q the isometric latitude given by Equation (11.5). With these,
the conformality conditions (Equation (11.7)) are satisfied, the meridians are equally
spaced vertical straight lines, and parallels are unequally spaced horizontal lines. The
parallel spacing increases toward the poles, which in this projection are at infinity on the
map. Therefore, although the scale at any one point of intersection of a meridian and a
parallel is *equal in all directions around the intersection* (which is the basis of conformality), the scale expands rapidly at high latitude. If s is the scale number at the equator and
s_ϕ is the scale number at any latitude ϕ, then

$$s_\phi = s \cos \phi$$
(11.14)

As an example, if the scale at the equator is 1 : 20,000, it would be 1 : 10,000 at a latitude
$\phi = 60°$ because $s_\phi = s \cos 60° = (20,000)(0.5) = 10,000$.

Aside from conformality, the particular feature of the Mercator projection is that the
rhumb line on the earth is plotted as a straight line on the map, a property that renders it
invaluable for navigation. The shortest course between two points is determined by drawing
on a gnomonic chart a great circle, which appears there as a straight line. Selected points,
at convenient distances apart, of this great circle then are plotted on the Mercator chart,
after making any necessary corrections on account of shoals, wind, currents, and the like.
The rhumb line connecting any two adjacent points indicates the true bearing of the course,
which is read by means of a protractor. This true bearing, corrected for magnetic declination, gives the compass bearing to be used in steering.

Owing to the rapid variation of scale, maps constructed on the Mercator projection give very inaccurate information as to relative sizes of areas in widely different latitudes. For example, on the map Greenland appears larger than South America, whereas in fact South America is nine times as large as Greenland. Consequently, such a map is not suited to general use, although because of its many other advantages it is widely published.

11.13
TRANSVERSE MERCATOR PROJECTIONS

The transverse Mercator projection was developed by Lambert in 1772, analytically derived by Gauss 50 years later, and then formulas more suitable for calculations were devised by Kruger in 1912. This is perhaps the reason that it is one of the most widely used conformal map projections.

A transverse Mercator projection is the ordinary Mercator projection turned through a 90° angle so that it is related to a central meridian in the same way that the ordinary Mercator projection is related to the equator (Figure 11.8a). Because the cylinder is tangent to the globe at a meridian, the scale is true along that meridian, which is called the *central*

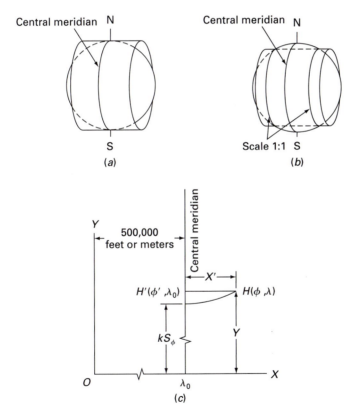

FIGURE 11.8
Transverse Mercator projection: (a) cylinder with one standard line, (b) cylinder with two standard lines, (c) coordinates on the transverse Mercator projection.

meridian and used as the origin of the map X coordinate. The origin of the map Y coordinate is the equator. Although, like the regular Mercator, this projection is conformal, it does not retain the straight-rhumb-line property of the Mercator. Other properties of this projection include (1) both the central meridian and the normal to it are represented by straight lines, (2) other meridians are complex curves that are concave toward the central meridian, (3) parallels are concave curves toward the pole, and (4) the scale is true only along the central meridian.

This projection is used for the state plane coordinate systems of states (or their zones) of greater north-south than east-west extent. For the state systems, the Mercator projection cylinder is made to cut the surface of the sphere along two standard lines parallel to the central meridian instead of being tangent to the sphere as in the ordinary Mercator projection (Figure 11.8b). The projection formulas (Thomas, 1952) are

$$\frac{X}{N} = \lambda' \cos \phi + \frac{\lambda'^3 \cos^3 \phi}{6} (1 - t^2 + \eta^2)$$

$$+ \frac{\lambda'^5 \cos^5 \phi}{120} (5 - 18t^2 + t^4 + 14\eta^2 - 58t^2\eta^2 + 13\eta^4) + \cdots$$

$$\frac{Y}{N} = \frac{S_\phi}{N} + \frac{\lambda'^2}{2} \sin \phi \cos \phi + \frac{\lambda'^4}{24} \sin \phi \cos^3 \phi (5 - t^2 + 9\eta^2 + 4\eta^4) \qquad (11.15)$$

$$+ \frac{\lambda'^6}{720} \sin \phi \cos^5 \phi (61 - 58t^2 + t^4 + 270\eta^2 - 330t^2\eta^2 + 445\eta^4) + \cdots$$

where

$N =$ is given by Equation (11.2)
$t = \tan \phi$

$$\eta^2 = \varepsilon'^2 \cos^2 \phi = \frac{\varepsilon^2 \cos^2 \phi}{1 - \varepsilon^2} = \frac{(a^2 - b^2) \cos^2 \phi}{b^2} \qquad (11.15a)$$

$\lambda' = \lambda - \lambda_0 =$ longitude difference from central meridian, λ_0 in *radians*
$S_\phi =$ length of the meridian arc from the equator to latitude ϕ is given by (Bomford, 1962)

$$S_\phi = \int_0^\phi \frac{a(1 - \varepsilon^2)}{(1 - \varepsilon^2 \sin^2 \phi)^{3/2}} d\phi$$

in which ε is the eccentricity, where $\varepsilon^2 = (a^2 - b^2)/a^2$ as defined in Section 11.10, or

$$S_\phi = a(A_0\phi - A_1 \sin 2\phi + A_2 \sin 4\phi - A_3 \sin 6\phi + \cdots) \qquad (11.16a)$$

in which

$a =$ semimajor axis of the ellipsoid
$A_0 = 1 - \frac{1}{4}\varepsilon^2 - \frac{3}{64}\varepsilon^4 - \frac{5}{256}\varepsilon^6 - \cdots$
$A_1 = \frac{3}{8}\varepsilon^2 + \frac{3}{32}\varepsilon^4 + \frac{45}{1024}\varepsilon^6 + \cdots$
$A_2 = \frac{15}{256}\varepsilon^4 + \frac{45}{1024}\varepsilon^6 + \cdots$
$A_3 = \frac{35}{3072}\varepsilon^6 + \cdots$

The terms used in Equation (11.15) are sufficient to yield X and Y values that are accurate to 0.01 m for zones within $\pm 3°$ of longitude about the central meridian. Computation of state plane coordinates on a transverse Mercator projection is explained in Section 11.19.

The convergence of the meridians (geodetic north-grid north) is

$$\gamma = \lambda' \sin \phi \left[1 + \frac{(\lambda')^2 \cos^2 \phi (1 + 3\eta^2)}{3} + \frac{(\lambda')^4 \cos^4 \phi (2 - t^2)}{15} \right] \quad (11.16b)$$

in which λ' and γ are in radians.

The scale determined from geographic coordinates (latitude and longitude) is

$$k = 1 + \frac{(\lambda')^2 \cos^2 \phi}{2} (1 + \eta^2)$$

$$+ \frac{(\lambda')^4 \cos^4 \phi}{24} (5 - 4t^2 + 14\eta^2 + 13\eta^4 - 28t^2\eta^2 + 4\eta^6 - 48t^2\eta^4 - 24t^2\eta^6)$$

$$+ \frac{(\lambda')^6 \cos^6 \phi}{720} (61 - 148t^2 + 16t^4) \quad (11.17)$$

Latitude and longitude are obtained from the X and Y rectangular map coordinates (Figure 11.8c) by the following equations (Thomas, 1952, for projections 2° on either side of the central meridian):

$$\Delta\phi = \phi - \phi' = t_1 \left[-\frac{X^2}{2R_1 N_1} + \frac{X^4}{24R_1 N_1^3} (5 + 3t_1^2) \right]$$

$$\Delta\lambda = \lambda - \lambda_0 = \sec \phi' \left[\frac{X}{N_1} - \frac{1}{6} \left(\frac{X}{N_1} \right)^3 (1 + 2t_1^2 + \eta^2) \right.$$

$$\left. + \frac{1}{120} \left(\frac{X}{N_1} \right)^5 (5 + 28t_1^2 + 24t_1^4) \right] \quad (11.17a)$$

where

ϕ' = footpoint latitude (Figure 11.8c)

$$R_1 = \frac{a(1 - \varepsilon^2)}{(1 - \varepsilon^2 \sin^2 \phi')^{3/2}}$$

N_1 is given by Equation (11.2) for latitude ϕ'

$t_1 = \tan \phi'$

$$\eta_1^2 = \frac{\varepsilon^2}{1 - \varepsilon^2} \cos^2 \phi'$$

The meridian convergence from rectangular coordinates is

$$\gamma = t_1 \left[\frac{X'}{N_1} - \frac{1}{3} \left(\frac{X'}{N_1} \right)^3 (1 + t_1^2 - \eta_1^2) + \frac{1}{15} \left(\frac{X'}{N_1} \right)^5 (2 + 5t_1^2 + 3t_1^4) \right] \quad (11.17b)$$

where $X' = X - 500,000$ (Figure 11.8c). The scale is given by

$$k = 1 + \frac{1}{2} \left(\frac{X'}{N_1} \right)^2 (1 + \eta_1^2) + \frac{1}{24} \left(\frac{X'}{N_1} \right)^4 (1 + 6\eta_1^2) \quad (11.17c)$$

Computation of state plane coordinates on the transverse Mercator projection is described in Section 11.19.

11.14
UNIVERSAL TRANSVERSE MERCATOR

The universal transverse Mercator (UTM) projection is based entirely on the transverse Mercator projection discussed in the preceding section. Its specifications are

1. Transverse Mercator projection is in zones that are 6° wide.
2. The reference ellipsoid is GRS 80 in North America.

3. The origin of longitude is at the central meridian.
4. The origin of latitude is at the equator.
5. The unit of measure is the meter.
6. False northing is 0 (for the southern hemisphere, a false northing of 10,000,000 m is used).
7. A false easting of 500,000 m is used for the central meridian of each zone.
8. The scale factor at the central meridian is 0.9996.
9. The zones are numbered beginning with 1 for the zone between 180°W and 174°W meridians and increasing to 60 for the zone between meridians 174°E and 180°E (Figure 11.9).
10. The latitude for the system varies from 80°N to 80°S.

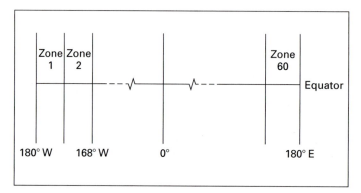

FIGURE 11.9
Universal transverse Mercator zones.

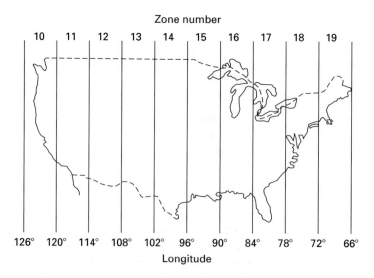

FIGURE 11.10
UTM zones in the United States.

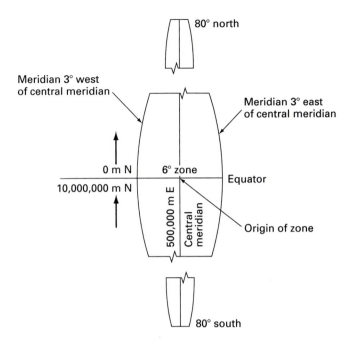

FIGURE 11.11
$X = E$ and $Y = N$ coordinates of the origin of a UTM grid zone.
(*From U.S. Dept. of the Army*, Field Manual, Map Reading, *FM 21–26, 1973.*)

The UTM zones in the United States are shown in Figure 11.10. An individual zone is illustrated in Figure 11.11.

11.15
TRANSVERSE MERCATOR IN 3° ZONES

The UTM projection system is the transverse Mercator projection in 6° zones. It also is possible to have a transverse Mercator projection in 3° zones. Because the zone width is only 3° and the scale factor of 0.9999 is assigned to the central meridian, a smaller scale error can be expected throughout the zone. This scale error is 1/10,000 on the central meridian as compared to 1/2500 for the UTM.

11.16
STATE PLANE COORDINATE SYSTEMS (SPCS)

Each state in the United States has a plane coordinate system based on one or more zones of either the Lambert conic conformal projection (Section 11.11) or the transverse Mercator projection (Section 11.13), except for Alaska, in which zone 1 is the oblique Mercator projection. The Lambert projection is used in states having a large east-west dimension; the transverse Mercator projection is employed for states with a longer north-south dimension.

If the north-south dimension of the Lambert projection and the east-west dimension of the transverse Mercator projection are held to 254 km or 158 mi, the maximum distortion in the projection is 1 part in 10,000. In states that exceed these dimensions, more than one zone of the same projection or perhaps two different projections are used. For example, New York has three transverse Mercator zones and one Lambert zone, which covers Long Island; two transverse Mercator zones cover the peninsula of Florida and a Lambert zone covers the western portion of the state; California has six Lambert zones; and Vermont is covered by one transverse Mercator zone. There are a total of 123 zones (including Alaska, Hawaii, Puerto Rico, and the Virgin Islands) in the state plane coordinate systems in the United States (SPCS 83).

The state plane coordinate systems, introduced in the early 1930s, originally were tied to the 1927 North American datum (NAD 27), utilized the U.S. survey foot (Section 1.8) as the basic unit, and are referred to as SPCS 27.

When the North American datum was redefined in 1983 (NAD 83), two plane coordinate systems were published: the universal transverse Mercator (UTM, as described in Section 11.14), and the state plane coordinate system (SPCS 83). Both systems are referred to the NAD 83 and are in metric units. Thus, conceptually, the previous state coordinate systems have been retained, with shifts in respective origins and changes in constants and units.

Use of state plane coordinates in the individual states becomes official only by legislative act. Prior to the 1983 adjustment, 37 states had enacted enabling legislation for SPCS 27. Legislation to adopt SPCS 83 currently is in process. One aspect that must be dealt with in this legislation is the unit of measurement to be used. Legislation to adopt SPCS 83 must specify whether the unit is to be the meter, U.S. survey foot, or international foot. States are not unanimous in this choice of unit. For example, California specifies the U.S. survey foot; Arizona, the international foot; and Nevada, the meter. Consequently, the survey engineer in a given state must be familiar with the legislative act for that state and the unit of measurement specified to avoid any subsequent conversion discrepancies, which could be significant for long lines.

The primary focus in this chapter is on SPCS 83 and the units considered are in meters.

Note that the state plane coordinate values, formerly referred to as X and Y for SPCS 27, now are defined as eastings $= E$ and northings $= N$, respectively, for the SPCS 83 and are so designated in all examples that follow development of the basic equations in Sections 11.17 and 11.19.

11.17
STATE PLANE COORDINATES USING LAMBERT CONFORMAL CONIC PROJECTIONS (SPCS 83)

The coordinate system illustrated in Figure 11.7 is not suitable for a state plane coordinate system. Instead, Figure 11.12 shows that the origin of coordinates is shifted by an N value of $R_b + N_b$ and the central meridian is given an X value designated by E_0. It follows from the figure that the state plane coordinates are given by

$$X = E = E_0 + R \sin l\lambda = E_0 + R \sin \gamma$$
$$Y = N = R_b + N_b - R \cos l\lambda = R_b + N_b - R \cos \gamma \qquad (11.18)$$

The values of E_0, $l = \sin \phi_0$, N_b, and R_b are given for each state plane coordinate system or its zone. In addition, values for R are tabulated in terms of the latitude ϕ, and γ can be calculated as will be shown. Of course, it is possible to compute the quantity R directly,

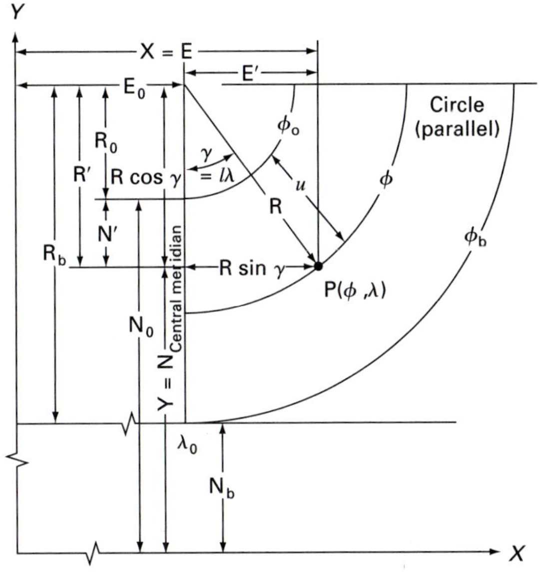

FIGURE 11.12
State plane coordinates on the Lambert projection.

using the relationships given in Section 11.11 and the latitudes for the two standard parallels, ϕ_1 and ϕ_2. For example, the procedure is as follows:

1. Compute N_1 and N_2 with Equation (11.2), using $\phi_1 = \phi_s$ and $\phi_2 = \phi_n$ for the projection.
2. Compute the isometric latitudes q_1 and q_2 for the two standard parallels, using Equation (11.5a) or (11.5b).
3. Compute l, a constant for the projection, using Equation (11.9a).
4. Using Equation (11.10) compute K, another constant for the projection.
5. The isometric latitude, q, for the point of interest is computed by Equation (11.5a) or (11.5b).
6. Equation (11.11) is used to compute R.

Then

$$\gamma = l(\lambda_0 - \lambda) \tag{11.19a}$$

where λ_0 is the longitude of the central meridian, which is given for the projection. Finally, Equations (11.18) are used to calculate the projection coordinates.

The scale of the projection (Thomas, 1952) is given by

$$k = \frac{Ke^{-lq}l}{N \cos \phi}$$

which, by substitution from Equations (11.2) and (11.11) and noting that $l = \sin \phi_0$ (Equation (11.9b)), the expression for k becomes

$$k = \frac{\sqrt{(1 - \varepsilon^2 \sin^2 \phi)}\, R \sin \phi_0}{a \cos \phi} \tag{11.19b}$$

where $\phi_0 =$ the latitude of the central parallel.

If the latitude, ϕ, and the longitude, λ, are required from given coordinates E and N (called the *inverse transformation*), they may be obtained by rearranging Equations (11.18) or

$$R \sin \gamma = X - E_0$$

$$R \cos \gamma = R_b + N_b - Y$$

Division of the first by the second equation yields

$$\tan \gamma = \frac{X - E_0}{R_b + N_b - Y}$$

or

$$\gamma = \arctan\left(\frac{X - E_0}{R_b + N_b - Y}\right) = \frac{E'}{R'} \tag{11.20}$$

Also,

$$R = \frac{R_b + N_b - Y}{\cos \gamma} \tag{11.21}$$

and from Equation (11.19a) compute

$$\lambda = \lambda_0 - \frac{\gamma}{l} \tag{11.22}$$

To compute the latitude ϕ, recall Equation (11.11):

$$R = Ke^{-lq}$$

or

$$\ln R = \ln K - lq$$

from which

$$q = \frac{\ln K - \ln R}{l} \tag{11.23}$$

With the value of the isometric latitude q known, the latitude ϕ can be evaluated from Equation (11.5). Because ϕ cannot be found directly from this equation, an iterative procedure (such as Newton-Raphson; see Scarborough, 1966) must be applied to Equation (11.5) to solve for ϕ.

Of course, if state plane coordinate tables are available, the value of ϕ is obtained directly as a function of R except for linearly interpolating for the seconds portion of the angle.

The United States National Geodetic Survey will provide tables for each state projection. These tables can be obtained from the National Geodetic Survey, N/CG112, 1315 East-West Highway, Station 9202, Silver Spring, MD 20910. A portion of the plane coordinate projection tables for California zone 1 and the constants for this zone are illustrated in Table 11.1. Shown in this table are the ellipsoidal, defining, and derived constants for zone 1, followed by the values of R for every minute of latitude within the zone (only a small part of the tabulation is given in Table 11.1). The tabular difference for 01″ of latitude is provided in the fourth column to obtain values by interpolation. Therefore, given ϕ to some decimal part of a second, R can be found to the nearest tenth of a millimeter. Conversely, given R, the latitude can be determined. The scale factors k, given as ratios, are listed in the last column.

Consider an example to illustrate use of the tables and the basic equations.

Datum: NAD 83
The projection is California 1, zone no. 0401
Ellipsoidal constants
$a = 6378137$ m
$f = 1/298.25722210$

Defining constants

ϕ_b	$= 39°20'$	(latitude of grid origin)
λ_{CM}	$= 122°00'$	(longitude of origin and central meridian, CM)
ϕ_s	$= 40°00'$	(southern standard parallel)
ϕ_n	$= 41°40'$	(northern standard parallel)
E_0	$= 2000000.0000$ m	(easting coordinate of origin)
N_b	$= 500000.0000$ m	(northing coordinate of origin)

Derived constants

l	$= 0.653884305400 = \sin(\phi_0)$	
K	$= 12287826.3052$ m	(mapping radius at the equator)
R_b	$= 7556554.6408$ m	(mapping radius at grid origin)
ϕ_0	$= 40.8351061249$	

Degrees, min	R (meter)	Tab diff (per sec)	Scale factor, k
41°20′	7,334,476.3033	30.84810	0.99993239
41°21′	7,332,625.4175	30.84827	0.99993496
41°22′	7,330,774.5214	30.84844	0.99993762
41°23′	7,328,923.6149	30.84862	0.99994036
41°24′	7,327,072.6980	30.84879	0.99994319
41°25′	7,325,221.7703	30.84898	0.99994610
41°26′	7,323,370.8317	30.84916	0.99994910
41°27′	7,321,519.8822	30.84934	0.99995218

EXAMPLE 11.1. The latitude and longitude for control point 9 in northern California zone 1 are

$$\phi = 41°20'10.403''$$

$$\lambda = 121°05'20.541''W$$

Compute E and N plane coordinates on California Lambert projection zone 1, using the data from Table 11.1 then the rigorous equations developed in Section 11.13.

Solution. Obtain R by interpolation from Table 11.1:

$$R \text{ for } 41°20' = 7,334,476.3033$$

$$(30.84810)(10.403'') = \underline{\quad -320.9128}$$

$$R \text{ for } 41°20'10.430'' = 7,334,155.3905 \text{ m}$$

$$\lambda_{CM} = \lambda_0 = 122°00'00''$$

$$\gamma = (\lambda_0 - \lambda) \sin \phi_0 = (\lambda_0 - \lambda) l$$

$$= [(122°00'00'') - (121°05'20.541'')](0.6538843054)$$

$$= 0.595662991533° = 0°35'44.38676952''$$

$$\sin \gamma = 0.01039609316 \qquad \cos \gamma = 0.999945959163$$

From Table 11.1,

$$R_b = 7,556,554.6408 \text{ m}, \qquad N_b = 500,000.0000 \text{ m}$$

Then, by Equation (11.18),

$$R \sin \gamma = \quad 76,246.5627, \qquad -R \cos \gamma = -7,333,759.0466$$

$$E_0 \quad = 2,000,000.0000, \qquad R_b \quad = \quad 7,556,554.6408$$

$$N_b \quad = \quad 500,000.0000$$

$$E_9 \quad = \quad 2,076,246.563 \text{ m}, \qquad N_9 \quad = \quad 722,795.5942 \text{ m}$$

Also from Table 11.1, using ϕ as the argument, the scale factor = 0.99993284.

Next, compute the E and N coordinates using the basic equations. For this solution, the parameters of GRS 80 and the latitudes for the south and north standard parallels for California zone 1 are

$$a = 6,378,137. \text{ m} \qquad = \text{semimajor axis of the ellipsoid}$$

$$b = 6,356,752.31414 \text{ m} = \text{semiminor axis of the ellipsoid}$$

$$\varepsilon^2 = 0.0066943800229$$

$$\phi_1 = 40°00' = \text{latitude of south standard parallel}$$

$$\phi_2 = 41°40' = \text{latitude of north standard parallel}$$

Solution. Using Equation (11.2), compute the radii of curvature in the prime verticals, N_1 and N_2:

$$N_1 = \frac{6,378,137}{\sqrt{1 - (0.00669438002290) \sin^2 40°}} = 6,386,976.16575 \text{ m}$$

$$N_2 = \frac{6,378,137}{\sqrt{1 - (0.00669438002290) \sin^2 41°40'}} = 6,387,593.18192 \text{ m}$$

Compute q_1 and q_2, the isometric latitudes for the north and south standard parallels. Use Equation (11.5b):

$$q_1 = \frac{1}{2}\left(\ln \frac{1 + \sin 40°}{1 - \sin 40°} - \varepsilon \ln \frac{1 + \varepsilon \sin 40°}{1 - \varepsilon \sin 40°} \right)$$

$$q_1 = \frac{1}{2}(1.52581930414 - 0.00861407698765) = 0.758602613555$$

$$q_2 = \frac{1}{2}\left(\ln \frac{1 + \sin 41°40'}{1 - \sin 41°40'} - \varepsilon \ln \frac{1 + \varepsilon \sin 41°40'}{1 - \varepsilon \sin 41°40'} \right)$$

$$q_2 = \frac{1}{2}(1.60271808904 - 0.0089095859154) = 0.79690425156$$

Compute l using N_1, N_2, q_1, and q_2 as just computed in Equation (11.9):

$$l = \frac{\ln N_1 - \ln N_2 + \ln \cos \phi_1 - \ln \cos \phi_2}{q_2 - q_1}$$

$$l = \frac{\ln 6,386,976.16575 - \ln 6,387,593.18192 + \ln \cos 40° - \ln \cos 41°40'}{0.79690425156 - 0.758602613555}$$

$$l = (0.025044839954)/(0.038301638005) = 0.653884305176$$

Now, calculate K with Equation (11.10), using l as computed previously:

$$K = \frac{N_1 \cos \phi_1}{l} e^{-lq_1} = \frac{N_2 \cos \phi_2}{l} e^{-lq_2}$$

$$K = \frac{(6,387,593.18192) \cos 41°40'}{0.653884305176} e^{(0.653884305176)(0.79690425156)}$$

$$K = (8,034,816.76696)/(0.653884305176) = 12,287,826.3071 \text{ m}$$

Next, compute the isometric latitude for point 9 using Equation (11.5b):

$$q_9 = \frac{1}{2}\left(\ln \frac{1 + \sin 41°20'10.403''}{1 - \sin 41°20'10.403''} - \varepsilon \ln \frac{1 + \varepsilon \sin 41°20'10.403''}{1 - \varepsilon \sin 41°20'10.403''} \right)$$

$$q_9 = \frac{1}{2}(1.58731670274 - 0.00885158439355)$$

$$q_9 = 0.789232559175 \text{ rad}$$

Now, compute R for point 9 using the previously calculated l, K, and q_9 in Equation (11.11):

$$R = \frac{K}{e^{lq_9}} = \frac{12,287,826.3071}{e^{(0.653884305176)(0.789232559175)}} = \frac{12,287,826.3071}{e^{(0.516066783578)}}$$

$$R_9 = (12,287,826.3071)/(1.6754248650)$$

$$R_9 = 7,334,155.3938 \text{ m}$$

Calculate γ with Equation (11.19a)

$$\gamma = l(\lambda_0 - \lambda) = (0.653884305176)[122° - (121°05'20.541'')]$$

$$\gamma = (0.653884305176)(0.910960833°) = 0.595662991329°$$

$$\gamma = 0°35'44.386769878''$$

The E and N coordinates are computed using Equations (11.18) with R_9 and γ just calculated and R_b and N_b from Table 11.1. Note that R_b could be determined using ϕ_b, given for the projection, in Equations (11.5a), (11.5b), (11.10), and (11.11), following the method employed previously to get R_9. For this example, use the tabulated value for R_b:

$$E = 2,000,000.000 + (7,334,155.3938) \sin(0.595662991329°)$$

$$E = 2,076,246.563 \text{ m}$$

$$N = 7,556,554.6408 + 500,000.0000 - (7,334,155.3938) \cos (0.595662991329°)$$

$$N = 722,795.591 \text{ m}$$

Finally, the scale factor is determined with Equation (11.19b):

$$k = \frac{\sqrt{1 - \varepsilon^2 \sin^2 \phi} \, R \sin \phi_0}{a \cos \phi}$$

$$k = \frac{\sqrt{1 - \varepsilon^2 \sin^2 (41°20'10.403'')} \, (7,334,155.3938) \sin (40.8351061249)}{(6,378,137) \cos (41°20'10.403'')}$$

$$k = 0.999932832$$

The coordinates and derived constants computed with the rigorous equations, using a handheld calculator with 12-place accuracy, vary from those determined from the published tables by a maximum amount of 3 mm. Because the tabulated constants and tables for R and the scale factor were produced with an electronic computer using double precision arithmetic, small differences due to round-off errors can be expected in the hand calculations.

Direct use of the rigorous equations is of primary value as an educational device to furnish an example of a theoretical basis for the projection equations and to illustrate the origin of the tabulated constants. If the rigorous equations are used, they ought to be programmed for a personal, mini-, or mainframe computer using double precision arithmetic to avoid what unquestionably are very small discrepancies that may be significant on some projects.

EXAMPLE 11.2. The California state plane coordinates for control point 10 are $E = 2,082,141.812$ m and $N = 737,346.276$ m. Compute the latitude and longitude for the point. For the initial solution, determine these values using Table 11.1.

Solution. First calculate the longitude. From Table 11.1, $l = 0.653884305400$, $R_b = 7,556,554.6408$ m, $E_0 = 2,000,000.000$ m, and $N_b = 500,000.000$ m. Form the following auxiliaries from Equation (11.20):

$$E' = E - E_0 \quad = 2,082,141.812 - 2,000,000.0000 = 82,141.812$$

$$R' = R_b + N_b - N = 7,556,554.6408 + 500,000. - 737,346.276$$

$$= 7,319,208.3648$$

Substituting E' and R' into Equation (11.20), we have

$$\gamma = \arctan \left(\frac{X - E_0}{R_b + N_b - Y} \right) = \arctan \left(\frac{E'}{R'} \right) = \arctan 0.0112227727243$$

$$\gamma = 0.642990517424° = 0°38'34.76586''$$

With this value of γ and tabulated l, compute λ with Equation (11.19a):

$$\lambda_0 - \lambda = \gamma/l = 0.642990517424/0.653884305400 = 0.983339884616°$$

$$\lambda = (122°) - 0.983339884616° = 121.016660115$$

$$\lambda = 121°00'59.9764''$$

Next, compute the latitude. From the preceding computation,

$$R_b + N_b - N = 7,319,208.3648 \text{ m}$$

$$\cos \gamma = 0.999937030634$$

Solve for R using Equation (11.21):

$$R = (R_b + N_b - N)/(\cos \gamma) = 7,319,669.27973 \text{ m}$$

Now, determine the latitude by interpolation from Table 11.1 as follows:

$$R \text{ for } 41°27' = 7,321,519.8822$$

$$R \text{ for } \phi \quad = 7,319,669.27973$$

$$\text{Difference} \quad = \quad 1,850.60247$$

From Table 11.1, column 4, the tabular difference for $01''$ is 30.84934, so that the difference in seconds of arc $= (1850.6025)/(30.8493) = 59.9884''$ and the latitude is $\phi = 41°27'59.9884''$.

Next, consider the solution to this inverse problem without the tables and using only Equations (11.18) through (11.23). The solution for the longitude is exactly the same as developed previously using Equations (11.20) and (11.19a) with constants for the projection and given E and N coordinates. This solution is not repeated here.

To determine latitude, compute R_{10}, where

$$R_{10} = \sqrt{R'^2 + E'^2} = \sqrt{(7{,}319{,}208.3648)^2 + (82{,}141.812)^2}$$

$$R_{10} = 7{,}319{,}669.27974 \text{ m}$$

With these values of R_{10} and K for the projection (Table 11.1), solve for q_{10} by Equation (11.23) that can be modified as follows:

$$q_{10} = [\ln K - \ln R]/l = [\ln(K/R)]/l$$

$$q_{10} = [\ln(12{,}287{,}826.3052/7{,}319{,}669.27974)]/(0.653884305400)$$

$$q_{10} = (0.518043894395)/(0.653884305400) = 0.792256199020 \text{ rad}$$

$$2q = 1.58451239804$$

Then, using the iterative procedure suggested by Stem (1989), a first approximation for ϕ (from Equation (11.5b)) is

$$\sin \phi = \left(\frac{\exp(2q) + 1}{\exp(2q) - 1}\right) = \frac{3.87691280642}{5.87691280642} = 0.659685269141$$

$$\sin \phi = 0.659686269141 \quad \text{and} \quad \phi = 41°16'33.1465''$$

With this value for $\sin \phi$ as an initial approximation, solve for ϕ iteratively, using the following procedure to compute corrections (recall $\varepsilon^2 = 0.00669438002290$ and $\varepsilon = 0.818191910428$):

$$f_1 = \tfrac{1}{2}\left(\ln \frac{1 + \sin \phi}{1 - \sin \phi} - \varepsilon \ln \frac{1 + \varepsilon \sin \phi}{1 - \varepsilon \sin \phi}\right) - q$$

$$f_1 = (\tfrac{1}{2})\left[\ln\left(\frac{1.65968526914}{0.340314730859}\right) - \varepsilon \ln\left(\frac{1.053974191506}{0.946025084936}\right)\right] - q$$

$$f_1 = \tfrac{1}{2}(1.58451239804 - 0.0088409598849) - 0.792256199028$$

$$= -0.00442047994$$

$$f_2 = \frac{1}{1 - \sin^2 \phi} - \frac{\varepsilon^2}{1 - \varepsilon^2 \sin^2 \phi}$$

$$f_2 = 1.770490139 - 0.006713940 = 1.763776199$$

The correction, $\Delta\phi_1 = (-f_1/f_2) = 0.00250625898$, which when applied yields an improved estimate for ϕ:

$$\sin \phi_1 = \sin \phi + \Delta\phi_1 = 0.662191528122, \quad \phi_1 = 41°28'02.0140''$$

The second iteration is

$$f_1 = (\tfrac{1}{2})(1.593413124 - 0.008874614) - q = 0.000013056$$

$$f_2 = 1.780936351 - 0.006714089 = 1.774222262$$

$$\Delta\phi_2 = -(f_1/f_2) = -0.000007359$$

$$\sin \phi_2 = \sin \phi_1 + \Delta\phi_2 = 0.662184169, \quad \phi_2 = 41°27'59.9884''$$

The third iteration is

$$f_1 = (\tfrac{1}{2})(1.593386914 - 0.008874515) - q = 5.00 \times 10^{-10}$$

$$f_2 = (1.780905441 - 0.006714089) = 1.774191352$$

$$\Delta\phi_3 = (5 \times 10^{-10})/(1.774191352) = 2.818 \times 10^{-10}$$

$$\sin \phi_3 = 0.662184169$$

$$\phi_3 = 41°27'59.9884''$$

the solution has converged and this value for the latitude of point 10 is seen to be identical with that obtained in the first solution.

Lambert Projection by Polynomial Coefficients

A relatively simple, accurate method for computing both the direct and inverse solutions on the Lambert projection involves use of polynomial coefficients. These coefficients, determined by least-squares fitting to 10 known points on each map projection, are listed for each zone of every projection in Stem (1989a). Accuracy to the nearest millimeter is provided by these coefficients when using a 10-place handheld calculator, and a programmed solution with the coefficients is more efficient than a program utilizing the rigorous equations. The procedure that follows is based on work by Vincenty (1986) and further developed by Stem (1989a), which contains additional details concerning determination of the coefficients.

The two basic equations used in the polynomial solution are

$$u = L_1 \Delta\phi + L_2 \Delta\phi^2 + L_3 \Delta\phi^3 + L_4 \Delta\phi^4 + L_5 \Delta\phi^5 \tag{11.24a}$$

$$\Delta\phi = \phi - \phi_0 = G_1 u + G_2 u^2 + G_3 u^3 + G_4 u^4 + G_5 u^5 \tag{11.24b}$$

in which

ϕ = latitude of the point
ϕ_0 = latitude of the central parallel
u = radial distance on projection between ϕ and ϕ_0
$L_1, L_2, \ldots, L_5$ are precomputed coefficients
$G_1, G_2, \ldots, G_5$ for the projection

Referring to Figure 11.12, E and N are plane coordinates for the point having latitude ϕ and longitude λ, λ_0 is the longitude of the central meridian of the zone, E_0 is the false easting (X coordinate) of the central meridian, R_0 is the mapping radius at ϕ_0, and γ the convergence angle at the point. Equations used for the computation of (1) plane coordinates from latitude and longitude (forward solution) and (2) latitude and longitude from plane coordinates (inverse solution) follow. For (1), the forward solution,

$$u = \Delta\phi[L_1 + \Delta\phi[L_2 + \Delta\phi(L_3 + \Delta\phi(L_4 + \Delta\phi L_5))]] \tag{11.25}$$

$$R = R_0 - u \tag{11.26}$$

$$\gamma = (\lambda_0 - \lambda) \sin \phi_0, \text{ repeated for convenience} \tag{11.19a}$$

$$E' = R \sin \gamma \tag{11.27}$$

$$N' = u + E' \tan(\gamma/2) \tag{11.28}$$

$$E = E' + E_0 \tag{11.29}$$

$$N = N' + N_0 \tag{11.30}$$

For (2), the inverse solution,

$$N' = N - N_0 \tag{11.31}$$

$$E' = E - E_0 \tag{11.32}$$

$$R' = R_0 - N' = R_b + N_b - N \tag{11.33}$$

$$\gamma = \tan^{-1}(E'/R'), \qquad \text{repeated for convenience} \tag{11.20}$$

$$\lambda = \lambda_0 - \gamma/\sin \phi_0, \qquad \text{repeated for convenience} \tag{11.19a}$$

$$u = N' - E' \tan(\gamma/2) \tag{11.34}$$

$$\Delta\phi = u[G_1 + u[G_2 + u(G_3 + u(G_4 + uG_5))]] \tag{11.35}$$

$$\phi = \phi_0 + \Delta\phi \tag{11.36}$$

The scale factor is

$$k = F_1 + F_2 u^2 + F_3 u^3 \tag{11.37}$$

where F_1, F_2, and F_3 also are listed in Stem (1989a) for the projection.

EXAMPLE 11.3. Using the data from Example 11.1, compute plane coordinates from the given ϕ and λ using the polynomial method. The necessary defining constants and coefficients for California zone 1 follow. The defining constants are

$$\phi_0 = 40.8351061249° \qquad R_0 = 7,389,802.0597 \text{ m}$$

$$\sin \phi_0 = 0.653884305400 \qquad \lambda_0 = 122°00'$$

$$N_0 = 666,752.5811 \text{ m} \qquad E_0 = 2,000,000.0000 \text{ m}$$

The coefficients are

$$L_1 = 111,039.0203 \qquad G_1 = 9.005843038 \times 10^{-6}$$

$$L_2 = 9.65524 \qquad G_2 = -7.05240 \times 10^{-15}$$

$$L_3 = 5.63491 \qquad G_3 = -3.70393 \times 10^{-20}$$

$$L_4 = 0.021275 \qquad G_4 = -1.1142 \times 10^{-27}$$

$$F_1 = 0.999894636561, \qquad F_2 = 1.23062 \times 10^{-14}, \qquad F_3 = 5.47 \times 10^{-22}$$

$$\phi_9 = 41°20'10.403'' = 41.33622306°$$

$$\lambda_9 = 121°05'20.541'' = 121.089039167°$$

Solution. Determine $\Delta\phi$ and u using Equations (11.24b) and (11.25):

$$\Delta\phi = \phi - \phi_0 = 41.33622305° - 40.83510612° = 0.501116930°$$

$$u = \Delta\phi[L_1 + \Delta\phi[L_2 + \Delta\phi(L_3 + \Delta\phi(L_4 + \Delta\phi L_5))]]$$

$$u = \Delta\phi[111,039.0203 + \Delta\phi[9.65524 + \Delta\phi(5.63491$$

$$+ \Delta\phi(0.021275 + \Delta\phi L_5))]]$$

$$u = 55,646.66911 \text{ m}$$

Calculate R using (11.26):

$$R = R_0 - u = 7,389,802.0597 - 55,646.66911$$

$$R = 7,334,155.3906 \text{ m}$$

Compute γ by Equation (11.19a):

$$\gamma = (\lambda_0 - \lambda) \sin \phi_0 = [(122°00') - (121.089039167)](0.653884305400)$$

$$\gamma = 0.595662991533$$

$$\sin \gamma = 0.01039609316$$

so that, using Equation (11.27),

$$E' = R \sin \gamma = 76,246.563 \text{ m}$$

and, using Equation (11.28),

$$N' = u + E' \tan (\gamma/2) = 55,646.66911 + 396.34390$$

$$N' = 56,043.01301 \text{ m}$$

Then, the coordinates E and N by Equations (11.29) and (11.30) are

$$E = E' + E_0 = 76,246.563 + 2,000,000.0000$$

$$E = 2,076,246.563 \text{ m}$$

$$N = N' + N_0 = 56,043.01301 + 666,752.5811$$

$$N = 722,795.594 \text{ m}$$

The scale factor by Equation (11.37) is

$$k = F_1 + F_2 u^2 + F_3 u^3$$

$$= 0.999894636561 + 1.23062 \times 10^{-14} u^2 + 5.47 \times 10^{-22} u^3$$

$$= 0.999894636561 + 3.8107 \times 10^{-5} + 9.42551 \times 10^{-8}$$

$$k = 0.99993284$$

Note that E, N, and k correspond to the values obtained in Example 11.1 using the table. Next, consider the inverse solution using polynomials.

EXAMPLE 11.4. Using the data from Example 11.2, where $E_{10} = 2,082,141.812$ m and $N_{10} = 737,346.276$ m, compute the latitude and longitude for the point using the polynomial approach.

Solution. Solve for N', E', and R' using Equations (11.31), (11.32), and (11.33):

$$N' = N - N_0 = 737,346.276 - 666,752.5811 = 70,593.6949 \text{ m}$$

$$E' = E - E_0 = 2,082,141.812 - 2,000,000.0000 = 82,141.812 \text{ m}$$

$$R' = R_0 - N' = 7,389,802.365 - 70,593.6949 = 7,319,208.365 \text{ m}$$

Use Equation (11.20) to find γ:

$$\gamma = \tan^{-1}(E'/R') = \tan^{-1}[82,141.812)/(7,319,208.365)] = \tan^{-1}0.01222773$$

$$\gamma = 0.642990517°$$

Then, using this value of γ in Equation (11.19a), solve for λ:

$$\lambda = \lambda_0 - \gamma/\sin \phi_0 = (122°00') - (0.642990517)/(0.6538843054)$$

$$= (122°00') - (0.983339885) = 121.0166601°$$

$$\lambda = 121°00'59.976''$$

Note that this solution for longitude corresponds to the one in Example 11.2 except that $R' = R_0 - N'$ is used instead of $R' = R_b + N_b - N$.

To determine the latitude, first compute u by Equation (11.34):

$$u = N' - E' \tan(\gamma/2) = 70{,}593.6949 - (82{,}141.812)/\tan(0.321495259°)$$

$$u = 70{,}593.6949 - 460.9149 = 70{,}132.77997 \text{ m}$$

The change in latitude from the standard parallel, $\Delta\phi$, then is found with Equation (11.35), as follows:

$$\Delta\phi = u[G_1 + u[G_2 + u(G_3 + u(G_4 + uG_5))]]$$

$$= u[9.005843038 \times 10^{-6} + u[-7.0524 \times 10^{-15} + u(-3.70393 \times 10^{-20}$$

$$+ u(-1.1142 \times 10^{-27}))]]$$

$$\Delta\phi = 0.631557316°$$

$$\phi = \phi_0 + \Delta\phi = 40.8351061249° + 0.631557316° = 41.46666344°$$

$$\phi = 41°27'59.9884''$$

Both λ and ϕ correspond to the values obtained in Example 11.2.

11.18
GRID AND GEODETIC AZIMUTHS ON LAMBERT CONFORMAL PROJECTIONS (SPCS 83)

In Figure 11.13, the line AP from the apex of the cone to point P is a meridian and all lines parallel to the central meridian are grid-north lines. The azimuth of a line between two points, such as J and K (Figure 11.13), can be referred to a grid-north line by *grid azimuth, t*, or to a meridian by *geodetic azimuth, α*.

For lines of up to 8 km or 5 mi in length, grid and geodetic azimuths differ by the angle of convergence, γ, as shown in Figure 11.13. As demonstrated in Section 11.17, γ can be

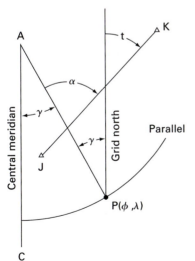

FIGURE 11.13
Grid and geodetic azimuths on the Lambert projection.

calculated with Equation (11.19a), Example 11.1, or computed by Equation (11.20), when the X and Y coordinates or the easting and northing for the point are given (Example 11.2). The difference between geodetic and grid azimuth expressed as an equation is

$$\alpha - t = \gamma - \delta \tag{11.38}$$

in which the sign of γ is determined by Equation (11.19a) or (11.20) and δ is a correction term to be applied when lines exceed 8 km or 5 mi in length.

This correction term is calculated by the following equation:

$$\delta\ ('') = \left(\frac{E_2 - E_1}{2r_0^2}\right)\rho\left(N_1 - N_0 + \frac{N_2 - N_1}{3}\right) \tag{11.39}$$

for the azimuth from point 1 to point 2. In Equation (11.39) (E_1, N_1) and (E_2, N_2) are state plane coordinates, respectively, for points 1 and 2; r_0 is the mean radius of curvature at latitude ϕ_0, scaled to the grid and is a given constant for any zone; and $\rho = 206{,}265''/\text{radian}$.

Frequently a single value for $[1/(2r_0^2)]\rho$ is computed for the average radius of the earth to yield a modification of Equation (11.39), which is

$$\delta('') = 25.4(E_2 - E_1)\left(N_1 - N_0 + \frac{N_2 - N_1}{3}\right)10^{-10} \tag{11.40}$$

where the coordinates are in meters. Equations (11.39) and (11.40) are adequate for most geodetic applications. For geodetic computations requiring the highest precision in the largest Lambert zones, the reader should consult Stem (1989a) for additional refinements in computing the second term or arc-to-chord correction.

Figure 11.14 illustrates the geometry involved in long geodetic lines. The line labeled ϕ_0 is parallel of latitude slightly north of the midpoint between the two standard parallels. The intersection of this parallel with the central meridian has a Y coordinate or northing of N_0, a constant given for the projection. Note that the geodetic line GH curves outward (northward) from the parallel ϕ_0 when projected on the grid. The geodetic line LK also curves away from the parallel ϕ_0. The second term, δ, in Equation (11.38) and as formulated in Equation (11.39) allows for this curvature.

Note that the north-south lines GL and HK project onto the map projection plane as straight lines. When a line is close to or in a grid north or south direction, $E_2 - E_1$ in Equation (11.39) is either very small or 0 so that the correction is negligible or vanishes.

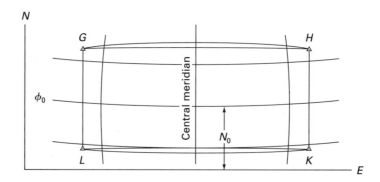

FIGURE 11.14
Geodetic lines on the Lambert projection. (*Adapted from Moffitt and Bouchard, 1992.*)

Grid distances or distances on the map projection are calculated using the given $X = E$ and $Y = N$ coordinates in Equation (8.8) of Section 8.13. To obtain the geodetic distance on the ellipsoid, the scale factors k are determined from the projection tables (Table 11.1) or calculated by Equation (11.19b) or (11.37) for the endpoints of the line. The relationship between grid and geodetic distance on the ellipsoid is then

$$\text{grid distance} = (\text{geodetic distance})(\text{scale factor})$$

As a rule, the average of the two scale factors at the two ends of the line, k_{av}, is used in this computation.

When it is necessary to convert grid distances to geodetic distances on the terrain or reduce measured distances to grid distances, the orthometric elevation of the line must be known and a correction made (Equation (9.11), Section 9.11).

A combined scale and ellipsoidal correction factor may also be calculated, where

$$\text{combined factor} = (\text{scale factor})(\text{ellipsoidal correction factor})$$

or

$$\text{combined factor} = k_{\text{aver}}\left(\frac{R}{R + h}\right) \tag{11.41}$$

where, in the second term, the ellipsoidal correction factor, R, is the average radius of the earth (6,371,000 m or 20,902,190 ft) and h is the average height above the ellipsoid of the line.

EXAMPLE 11.5. The coordinates for points 9 and 10 as calculated in Example 11.1 (using Table 11.1) and given in Example 11.2 are as follows:

Point	E, m	N, m
9	2,076,246.563	722,795.594
10	2,082,141.812	737,346.276

Determine the grid and the geodetic azimuth from both ends of the line. Also, compute the grid distance and geodetic sea-level distance between points 9 and 10.

Solution. Because these are plane coordinates, the grid azimuth is given by Equation (8.6). The grid azimuth from the south of the line from 9 to 10 is

$$\tan A = \frac{E_9 - E_{10}}{N_9 - N_{10}} = 0.407956455$$

so that the azimuth from south is

$$A_{9-10} = 202°11'36.0''$$

The grid azimuth from south of the line from 10 to 9 is exactly 180° different from this value:

$$A_{10-9} = 22°11'36.0''$$

Next, compute the geodetic azimuth of the line from 9 to 10 from the south by Equation (11.38). From Example 11.1, $\gamma = 0°35'44.3868''$. From Stem 1989a, Appendix C,

$$r_0 = 6,374,328. \text{ m} \quad \text{and} \quad N_0 = 666,752.5811 \text{ m}$$

The arc-to-chord second correction term in Equation (11.39) is

$$= \left(\frac{5,895.249}{(2)(6,374,328)^2} \right) \rho \left(722,795.594 - 666,752.581 + \frac{14,550.682}{3} \right) = 0.91''$$

and, by Equation (11.38),

$$\text{geodetic azimuth} = (202°11'36.0'') + (0°35'44.39'') - (0.91'')$$

$$= 202°47'19.5''$$

Similarly, the geodetic azimuth of line 10 to 9 from the south is

$$\text{geodetic azimuth} = (22°11'36.0'') + (0°38'34.7659'')$$

$$- \left(\frac{-5,895.249}{(2)(6,374,328)^2} \right) \rho \left(737,346.276 - 666,752.581 - \frac{14,550.682}{3} \right)$$

$$= 22°11'36.0''$$

$$+0°38'34.77''$$

$$+0°00'00.98''$$

geodetic azimuth 10−9 = 22°50'11.8''

Note that the geodetic azimuth from 9 to 10 does not differ from the geodetic azimuth from 10 to 9 by 180° as do corresponding grid azimuths. Figure 11.15 illustrates the grid and geodetic azimuths from 9 to 10 for Example 11.5. The reader can make a similar sketch for the geodetic and grid azimuths from 10 to 9 as an exercise to verify, graphically, the signs of γ and the correction term in Equation (11.39) for this part of Example 11.5.

The grid distance from 9 to 10 calculated by Equation (8.8) is

$$d_{9-10} = [(E_{10} - E_9)^2 + (N_{10} - N_9)^2]^{1/2} = 15,699.564 \text{ m}$$

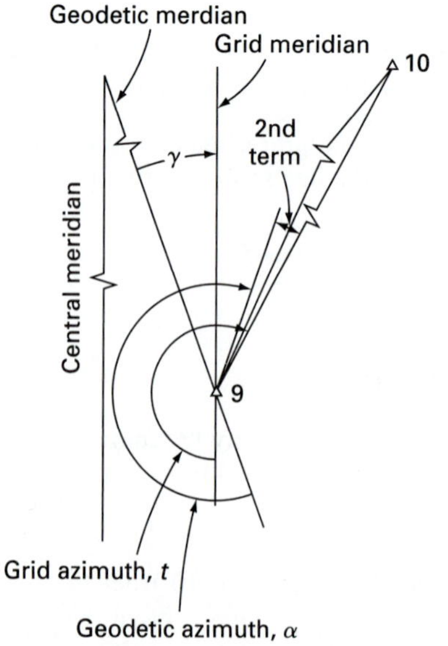

FIGURE 11.15
Geodetic azimuth with the arc-to-chord correction term.

The scale factor for station 9 calculated by Equation (11.19b) in Example 11.1 was found to be 0.999932832. The scale factor for 10 also calculated by Equation (11.19b) using R and ϕ for 10 from Example 11.2 is

$$k_{10} = \frac{(\sin 40°50'06.3820'')(7,319,669.2797)[1 - 0.0066943800229 \sin^2(41°27'59.9885'')]^{1/2}}{(6,378,137)\cos(41°27'59.9885'')}$$

$$= 0.999955347$$

The average scale factor, $k_{av} = (0.999932832 + 0.999955347)/2 = 0.999944090$ and the geodetic distance on the ellipsoid is

$$d_{9-10}(\text{geoid}) = \frac{15,669.564}{0.999944090} = 15,700.442 \text{ m}$$

Assuming an average orthometric elevation of 731.50 m and a separation between the ellipsoid and geoid of -30.50 m for the line, a correction factor can be calculated using Equation (9.11):

$$\text{correction factor} = \frac{R}{R + h} = \frac{6,371,000}{6,371,000 + 731.50 + (-30.50)} = 0.999889982$$

and the geodetic distance at average terrain elevation is

$$d_{9-10}(\text{terrain}) = \frac{15,700.442}{0.999889982} = 15,702.170 \text{ m}$$

A combined factor could have been calculated as follows:

$$\text{combined factor} = (0.999944090)(0.999889982) = 0.999834078$$

and applied directly to the grid distance to yield

$$d_{9-10}(\text{geoid}) = \frac{15,699.564}{0.999834078} = 15,702.170 \text{ m}$$

Note that for a line of length cited in Example 11.5 (in excess of 8 km or 5 mi), a single average scale factor is not adequate. The line should be subdivided into shorter sections with a separate scale factor being calculated for each section (see Section 11.23 and U.S. Dept. of the Army, 1951).

11.19
TRANSVERSE MERCATOR STATE PLANE COORDINATE PROJECTIONS (SPCS 83)

The theory of the transverse Mercator projection is developed in Section 11.13. Because the basic projection formulas presented in that section (Equations 11.15) usually are not used directly in the computation of state plane coordinates, some additional developments are required here. This permits showing the formulation actually used to determine coordinates in a computer program and clarifying use of the projection tables provided for each state plane system by the National Geodetic Survey. For the state systems, the Mercator projection cylinder cuts the surface of the sphere at two standard lines, parallel to and equally spaced from the central meridian as shown in Figure 11.8b. The projection distance between points along these standard lines equals the distance between corresponding points on the sphere. Between the two standard lines, the projection distance is less than the corresponding distance on the sphere and the scale of the projection is too small. Beyond the two standard lines, the projection distance exceeds the corresponding distance on the sphere and the scale of the projection is too large. Therefore, the scale on an east-west line varies from point to point whereas the scale on a north-south line is constant throughout even though it may not be $1:1$. If the two standard circles are set so the distance between them is about 96 km (60 mi) and the total width of the projection is about 254 km (158 mi), then errors due to scale change throughout the projection will not exceed $1:10,000$. Because the scale is constant along the meridian, this type of projection is used in states that have a long north-south dimension.

For the state plane transverse Mercator projections, the central meridian is taken as the Y or northing axis, which has an arbitrarily assigned X coordinate or easting, which is specified for each zone of each state. In this manner there will be no negative X coordinates or eastings within the system. The X coordinate or easting axis is along a line perpendicular to the central meridian at a latitude selected such that there will be no negative northings within the system. Still assuming a sphere, the Y coordinate or northing for a point H (refer to Figure 11.16) would be the length of meridian from the origin to a perpendicular from the central meridian passing through H. The X coordinate or easting would be the length of this perpendicular from the central meridian to the point.

Each state plane transverse Mercator projection has defining constants, which are as follows:

λ_0 = longitude of the central meridian
E_0 = false easting or X coordinate of the central meridian
ϕ_0 = latitude of the grid origin
N_0 = northing or Y coordinate of the grid origin
k_0 = scale factor for the central meridian

Numerical values for these constants are published in Stem (1989a) for each zone of every state that has the transverse Mercator projection. With these constants, two methods currently are used to compute the direct and inverse solutions on the transverse Mercator projection: (1) an approach employing the theoretical equations modified for efficient computer programming; and (2) a procedure that utilizes plane coordinate projection tables available from the NGS for each zone in every state with the transverse Mercator projection.

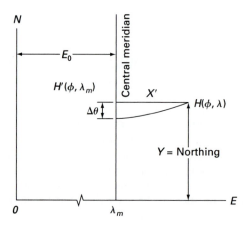

FIGURE 11.16
Coordinates on the transverse Mercator projection.

Theoretical Equations (SPCS 83)

The procedure developed here is based entirely on the method given in Stem (1989a), where the following nine ellipsoid constants for GRS 80 are given (with equations for those who wish to compute them for other ellipsoids):

$$r = \quad 6{,}367{,}449.14577 \text{ m} = \text{radius of the rectifying sphere}$$

$$U_0 = -0.00504\ 82507\ 76 \qquad V_0 = 0.00502\ 28939\ 48$$

$$U_2 = \quad 0.00002\ 12592\ 04 \qquad V_2 = 0.00002\ 93706\ 25$$

$$U_4 = -0.00000\ 01114\ 23 \qquad V_4 = 0.00000\ 02350\ 59$$

$$U_6 = \quad 0.00000\ 00006\ 26 \qquad V_6 = 0.00000\ 00021\ 81$$

from which one can calculate the rectifying latitude

$$\omega_0 = \phi_0 + \sin \phi_0 \cos \phi_0 (U_0 + U_2 \cos^2 \phi_0 + U_4 \cos^4 \phi_0 + U_6 \cos^6 \phi_0) \qquad (11.42a)$$

and

$$S_0 = k_0 \omega_0 r \qquad (11.42b)$$

where S_0 is the distance along the meridian from the equator to latitude ϕ_0. S_0 is a zone-specific constant for the projection, listed in Stem (1989a) for each state. Note that this quantity, S_0, is the same as S_ϕ as defined by Equation (11.16a).

In the direct solution, with ϕ and λ known for a point for which the grid coordinates (E, N) are sought, the following equations are used where all angles are in radians. First, compute the rectifying latitude using Equation (11.42a) in nested form so that

$$\omega = \phi + (\sin \phi \cos \phi)\{U_0 + \cos^2 \phi[U_2 + \cos^2 \phi(U_4 + U_6 \cos^2 \phi)]\} \qquad (11.43)$$

Then the meridional distance for the point by Equation (11.42b) is

$$S = k_0 \omega r \qquad (11.44)$$

and the radius of curvature at latitude ϕ is found by Equation (11.2):

$$R = (k_0 a)/(1 - \varepsilon^2 \sin^2 \phi)^{1/2} \qquad (11.2)$$

Next, define the terms:

$$A_1 = -R \qquad (11.45a)$$

$$A_2 = (1/2)R \tan \phi \qquad (11.45b)$$

$$A_3 = (1/6)(1 - \tan^2 \phi + \eta^2) \qquad (11.45c)$$

$$A_4 = (1/12)[5 - \tan^2 \phi + \eta^2(9 + 4\eta^2)] \qquad (11.45d)$$

$$A_5 = (1/120)[5 - 18 \tan^2 \phi + \tan^4 \phi + \eta^2(14 - 58 \tan^2 \phi)] \qquad (11.45e)$$

$$A_6 = (1/360)[61 - 58 \tan^2 \phi + \tan^4 \phi + \eta^2(270 - 330 \tan^2 \phi)] \qquad (11.45f)$$

$$A_7 = (1/5040)(61 - 479 \tan^2 \phi + 179 \tan^4 \phi - \tan^6 \phi) \qquad (11.45g)$$

in which η^2 is defined and used within Equation (11.15a)

$$\eta^2 = [(a^2 - b^2) \cos^2]/b^2 = [\varepsilon^2/(1 - \varepsilon^2)] \cos^2 \phi = \varepsilon'^2 \cos^2 \phi$$

The last auxiliary quantity is

$$L = (\lambda - \lambda_0) \cos \phi \qquad (11.46)$$

With these terms evaluated and using defining and zone-specific constants, the coordinates E and N are

$$E = E_0 + A_1 L[1 + L^2(A_3 + L^2(A_5 + A_7 L^2))] \qquad (11.47a)$$

$$N = S - S_0 + N_0 + A_2 L^2[1 + L^2(A_4 + A_6 L^2)] \qquad (11.47b)$$

To compute the angle of convergence, γ, and scale factor, k, the following terms are defined:

$$C_1 = -\tan \phi \qquad (11.48a)$$

$$C_3 = (1/3)(1 + 3\eta^2 + 2\eta^4) \qquad (11.48b)$$

$$C_5 = (1/15)(2 - \tan^2 \phi) \qquad (11.48c)$$

$$F_2 = (1/2)(1 + \eta^2) \qquad (11.48d)$$

$$F_4 = (1/12)[5 - 4 \tan^2 \phi + \eta^2(9 - 24 \tan^2 \phi)] \qquad (11.48e)$$

to yield

$$\gamma = C_1 L[1 + L^2(C_3 + C_5 L^2)] \qquad (11.49a)$$

$$k = k_0[1 + F_2 L^2(1 + F_4 L^2)] \qquad (11.49b)$$

The terms involving A_6, A_7, C_5, and F_4 may be ignored when computing within the SPCS 83 boundaries. For computations beyond these boundaries, the terms should be included.

For the inverse solution, grid coordinates (E, N) are known and latitude and longitude (ϕ, λ), convergence angle, and scale factor are sought with the following equations. Calculate the rectifying latitude, ω,

$$\omega = (N - N_0 + S_0)/k_0 r \qquad (11.50)$$

in which N_0, S_0, k_0, and r are given constants for the projection. Then, using the ellipsoidal constants, V_i, determine the foot-point latitude, $\phi_f = \phi'$ (see Figure 11.8c) using Equation (11.42a) in nested form:

$$\phi_f = \omega + (\sin \omega \cos \omega)\{V_0 + \cos^2 \omega[V_2 + \cos^2 \omega(V_4 + V_6 \cos^2 \omega)]\} \quad (11.51)$$

The radius of curvature in the prime vertical at ϕ_f (by Equation (11.2)) is

$$R_f = (k_0 a)/[1 - \varepsilon^2 \sin^2 \phi_f]^{1/2}$$

from which the auxiliary Q is determined by

$$Q = (E - E_0)/R_f = E'/R_f \qquad (11.52)$$

Define the following terms:

$$B_2 = -(1/2) \tan \phi_f(1 + \eta_f^2) \qquad (11.53a)$$

$$B_3 = -(1/6)(1 + 2 \tan^2 \phi_f + \eta_f^2) \qquad (11.53b)$$

$$B_4 = (-1/12)[5 + 3 \tan^2 \phi_f + \eta_f^2(1 - 9 \tan^2 \phi_f) - 4\eta_f^2] \qquad (11.53c)$$

$$B_5 = (1/120)[5 + 28 \tan^2 \phi_f + 24 \tan^4 \phi_f + \eta_f^2(6 + 8 \tan^2 \phi_f)] \qquad (11.53d)$$

$$B_6 = (1/360)[61 + 90 \tan^2 \phi_f + 45 \tan^4 \phi_f$$
$$+ \eta_f^2(46 - 252 \tan^2 \phi_f - 90 \tan^4 \phi_f)] \qquad (11.53e)$$

$$B_7 = (-1/5040)(61 + 662 \tan^2 \phi_f + 1320 \tan^4 \phi_f + 720 \tan^6 \phi_f) \quad (11.53f)$$

$\cdots$

in which η_f^2 is calculated as developed within Equation (11.15a).

Using these evaluated terms, the latitude and longitude are

$$\phi = \phi_f + B_2 Q^2 [1 + Q^2 (B_4 + B_6 Q^2)] \tag{11.54}$$

$$\lambda = \lambda_0 - L/\cos \phi_f \tag{11.55}$$

in which

$$L = Q\{1 + Q^2 [B_3 + Q^2 (B_5 + B_7 Q^2)]\} \tag{11.56}$$

To determine convergence angle and scale factor, first define the terms

$$D_1 = \tan \phi_f \tag{11.57a}$$

$$D_3 = (-1/3)(1 + \tan^2 \phi_f - \eta_f^2 - 2\eta_f^4) \tag{11.57b}$$

$$D_5 = (1/15)(2 + 5 \tan^2 \phi_f + 3 \tan^4 \phi_f) \tag{11.57c}$$

$$G_2 = (1/2)(1 + \eta_f^2) \tag{11.57d}$$

$$G_4 = (1/12)(1 + 5\eta_f^2) \tag{11.57e}$$

then

$$\gamma = D_1 Q[1 + Q^2 (D_3 + D_5 Q^2)] \tag{11.58a}$$

$$k = k_0 [1 + G_2 Q^2 (1 + G_4 Q^2)] \tag{11.58b}$$

When computing within the SPCS 83 boundaries of the projection, terms B_6, B_7, D_5, and G_4 are not significant. For computations outside these boundaries, these terms should be evaluated.

EXAMPLE 11.6. Compute E and N coordinates, the convergence angle, and the scale on the transverse Mercator projection, Arizona eastern zone number 0201, for station Orbit, where

$$\phi = \quad 32°10'31.505'' = \quad 32.17541806° = 0.5615669833 \text{ rad}$$

$$\lambda = 109°40'10.230'' = 109.66950833° = 1.9140940095 \text{ rad}$$

From NOS NGS5 (Stem 1989a), the defining constants for the projection are

$$\lambda_0 = 110°10' = 110.1666667° = 1.922771060 \text{ rad}$$

$$E_0 = 213,360. \text{ m} \qquad k_0 = 0.9999$$

$$N_0 = 0 \qquad\qquad \phi_0 = 31°00' = 0.541052068 \text{ rad}$$

Ellipsoid constants r, U_0, U_2, and U_4 have been listed but are repeated here for convenience, along with the semimajor axis and eccentricity of the ellipsoid, as follows:

$$r = 6,367,449.14577 \text{ m} \qquad S_0 = 3,430,631.2260 \text{ m}$$

$$U_0 = -0.00504\ 82507\ 76 \qquad a = 6,378,137 \text{ m}$$

$$U_2 = \quad 0.00002\ 12592\ 04 \qquad \varepsilon^2 = 0.006694380023$$

$$U_4 = -0.00000\ 01114\ 23$$

$$U_6 = \quad 0.00000\ 00006\ 26$$

Solution. Calculate the rectifying latitude, ω, by Equation (11.43) in nested form:

$$\omega = \phi + \sin \phi \cos \phi \{-0.00504\ 82507\ 76$$
$$+ \cos^2 \phi [2.1259204 \times 10^{-5} + \cos^2 \phi (-1.11423 \times 10^{-7})$$
$$+ 6.26 \times 10^{-10} \cos^2 \phi)]\} = 0.5615669834 - 0.002268563$$

$$\omega = 0.5592984200 \text{ rad}$$

Compute the meridional distance, S, using Equation (11.44):

$$S = k_0 \omega r = (0.9999)(0.5592984200)(6,367,449.146)$$

$$S = 3,560,948.119 \text{ m}$$

Next, determine R by Equation (11.2), where the semimajor axis is scaled to the projection:

$$R = (k_0 a)/(1 - \varepsilon^2 \sin^2 \phi)^{1/2}$$

$$= (0.9999)(6,378,137)/[1 - (0.006694380023) \sin^2 \phi]^{1/2}$$

$$R = 6,383,561.1086 \text{ m}$$

and η^2; where $\eta^2 = [\varepsilon^2/(1 - \varepsilon^2)] \cos^2 \phi = 0.004828375742$

Now, calculate the terms $A_1, A_2, \ldots, A_5$ using Equations (11.45):

$$A_1 = -R = -6,383,561.10859 \text{ m}$$

$$A_2 = (1/2)R \tan \phi = 2,008,059.56405 \text{ m}$$

$$A_3 = (1/6)(1 - \tan^2 \phi + \eta^2) = 0.101503044675$$

$$A_4 = (1/12)[5 - \tan^2 \phi + \eta^2(9 + 4\eta^2)] = 0.3873115247$$

$$A_5 = (1/120)[5 - 18 \tan^2 \phi + \tan^4 \phi + \eta^2(14 - 58 \tan^2 \phi)]$$

$$= -0.01675973344$$

Finally, calculate L with Equation (11.46):

$$L = (\lambda - \lambda_0) \cos \phi = (1.914094009 - 1.922771060) \cos \phi$$

$$L = -0.007344443848 \text{ rad}$$

Then, the coordinates by Equations (11.47) are

$$E = E_0 + A_1 L[1 + L^2(A_3 + L^2(A_5 + A_7 L^2))]$$

in which the term A_7 may be neglected for this solution, so that

$$E = 213,360 + (-6,383,561.1086)(L)[1 + L^2(A_3 + L^2 A_5)]$$

$$E = 260,243.966 \text{ m}$$

$$N = S - S_0 + N_0 + A_2 L^2[1 + L^2(A_4 + A_6 L^2)]$$

where A_6 may be ignored so that

$$N = 3,560,948.119 - 3,430,631.2260 + 0 + A_2 L^2(1 + L^2 A_4)$$

$$N = 130,316.893 + 108.3187$$

$$N = 130,425.212 \text{ m}$$

To determine the convergence angle, γ, and the scale factor, k, the terms $C_1, C_3, F_2,$ and F_4 must be calculated:

$$C_1 = -\tan \phi = -0.629134594$$

$$C_3 = (1/3)(1 + 3\eta^2 + 2\eta^4) = 0.338177251$$

$$F_2 = (1/2)(1 + \eta^2) = 0.502414188$$

Then, by Equations (11.49a) and (11.49b)

$$\gamma = C_1 L(1 + L^2 C_3) = 0.004620729 \text{ rad} = 0.2647483° = 0°15'53.09''$$

$$k = k_0(1 + F_2 L^2) = 0.999927098$$

EXAMPLE 11.7. Compute the latitude and longitude for station Roger where

$$E = 272,013.298 \text{ m}$$

$$N = 107,720.186 \text{ m}$$

on the transverse Mercator projection for Arizona eastern zone number 0201. Projection constants N_0, S_0, k_0, and r are as given in Example 11.6. Ellipsoidal constants, V_i, previously given, are repeated here for convenience:

$$V_0 = 0.00502\ 28939\ 48 \qquad V_4 = 0.00000\ 02350\ 59$$

$$V_2 = 0.00002\ 93706\ 25 \qquad V_6 = 0.00000\ 00021\ 81$$

Solution. Calculate the rectifying latitude, ω, by Equation (11.50):

$$\omega = (N - N_0 + S_0)/k_0 r$$

$$= (107,720.186 - 0 + 3,430,631.226)/[(0.9999)(6,367,449.146)]$$

$$\omega = 0.5557492807 \text{ rad}$$

Determine the foot-point latitude using Equation (11.51):

$$\phi_f = \omega + \sin \omega \cos \omega \{0.00502\ 28939\ 48$$

$$+ \cos^2 \omega[2.9370625 \times 10^{-5}$$

$$+ \cos^2 \omega(2.35059 \times 10^{-7} + 2.181 \times 10^{-9} \cos^2 \omega)]\}$$

$$= 0.555749281 + 0.00226072545$$

$$\phi_f = 0.5580100062 \text{ rad}$$

The radius of curvature at latitude ϕ_f (Equation (11.2)) is

$$R_f = (k_0)(a)/[1 - \varepsilon^2 \sin^2 \phi_f]^{1/2}$$

$$= [(0.9999)(6,378,137)]/[1 - (0.006694380023) \sin^2 \phi_f]^{1/2}$$

$$R_f = 6,383,492.58442 \text{ m}$$

The auxiliary quantity, Q (Equation (11.52)), is

$$Q = (E - E_0)/R_f = (272,013.298 - 213,260)/(6,383,492.584)$$

$$Q = 0.00918827698542$$

Next calculate η^2, where

$$\eta_f^2 = [(\varepsilon^2)/(1 - \varepsilon^2)] \cos^2 \phi_f$$

$$\eta_f^2 = 0.004849948695$$

To determine latitude, calculate the terms B_2 and B_4 in Equations (11.53a) and (11.53c). Because this point is within the boundaries of SPCS 83 projection, B_6 is not required:

$$B_2 = -(1/2) \tan \phi_f (1 + \eta_f^2) = -0.3136040193$$

$$B_4 = -(1/12)[5 + 3 \tan^2 \phi_f + \eta_f^2(1 - 9 \tan^2 \phi_f) - 4\eta_f^4] = -0.5130462431$$

Then, by Equation (11.54),

$$\phi = \phi_f + B_2 Q^2 (1 + Q^2 B_4) = 0.5580100060 - 0.00002647470$$

$$= 0.55798353 \text{ rad}$$

$$\phi = 31.97010139° = 31°58'12.365''$$

The terms B_3 and B_5 are required to compute the longitude by Equations (11.53b) and (11.53d)):

$$B_3 = -(1/6)(1 + 2 \tan^2 \phi_f + \eta_f^2) = -0.2973422125$$

$$B_5 = (1/120)[5 + 28 \tan^2 \phi_f + 24 \tan^4 \phi_f + \eta_f^2(6 + 8 \tan^2 \phi_f)] = 0.1633000798$$

that are used in Equation (11.56) to compute L (B_7 is neglected):

$$L = Q[1 + Q^2(B_3 + Q^2 B_5)] = 0.009188046347$$

and then the longitude (Equation (11.55)) is

$$\lambda = \lambda_0 - (L/\cos \phi_f) = 1.922771060 - 0.01083099461 = 1.911940065 \text{ rad}$$

$$\lambda = 109.5460964° = 109°32'45.947''$$

The terms D_1, D_3, G_2, and G_4 must be evaluated using Equations (11.57a), (11.57b), (11.57d), and (11.57e) to compute the convergence angle and scale at the point:

$$D_1 = \tan \phi_f = 0.624180794$$

$$D_3 = -(1/3)(1 + \tan^2 \phi_f - \eta_f^2 - 2\eta_f^4) = -0.461568223$$

$$G_2 = (1/2)(1 + \eta_f^2) = 0.5024249745$$

$$G_4 = (1/12)(1 + 5\eta_f^2) = 0.08535414529$$

$$\gamma = D_1 Q(1 + Q^2 D_3) = 0.005734923 \text{ rad} = 0.328586884° = 0°19'42.91''$$

$$k = k_0(1 + G_2 Q^2(1 + G_4 Q^2)) = (0.9999)(1.000042417) = 0.99994241$$

Examples 11.6 and 11.7 were computed using a handheld calculator with 10-place accuracy, to verify the theoretical development. The procedure is long, and extreme care must be exercised to avoid time-consuming mistakes. Equations (11.41) through (11.58) were designed to be programmed on personal, mini-, or mainframe computers as well as on programmable handheld calculators with adequate storage capacity. Software packages utilizing these equations for computing SPCS 83 conversions are available from the National Geodetic Survey (NGS). For those surveyors who wish to use calculator computations, use of the projection tables is a viable option. The SPCS 83 projection tables and software are obtainable from the NGS Information Center (see the address cited in Section 11.17).

Transverse Mercator Using Tables (SPCS 83)

The defining constants are as given in the previous section and provided in NOS NGS5 (Stem, 1989a). The tables are based on Equations (11.42) through (11.58) and will provide identical results. The additional equations used in this process for direct conversion from geodetic latitude and longitude to grid coordinates E and N are as follows:

$$\Delta\lambda = (\lambda_0 - \lambda)3600''/\circ \tag{11.59a}$$

$$p = 10^{-4}\Delta\lambda \tag{11.59b}$$

$$N = (\text{I}) + (\text{II})p^2 + (\text{III})p^4 \tag{11.59c}$$

$$E' = (\text{IV})p + (\text{V})p^3 + (\text{VI})p^5 \tag{11.59d}$$

$$E = E_0 + E' \tag{11.59e}$$

$$\gamma = (\text{XII})p + (\text{XIII})p^3 \tag{11.59f}$$

$$q = 10^{-6}E' \tag{11.59g}$$

$$k = k_0[1 + (\text{XVI})q^2 + 0.00003q^4] \tag{11.59h}$$

in which λ_0 and E_0 are defining constants for the projection, λ is the longitude for the point, and the roman numerals I to XVI can be taken from the projection tables using the latitude, ϕ, as the argument. Extracts from these tables are given in Tables 11.2, 11.3, and 11.4.

TABLE 11.2
Transverse Mercator projection tables for Arizona eastern zone number 0201:
Grid coordinates from geodetic coordinates

Latitude	(I) (IV)	Difference 1″	(II) (V)	Difference 1″	(III) (VI)
31°55′	101,627.4403	30.798447	3,366.5774	0.016079	2.212
	262,691.8011	−0.789591	45.7403	−0.001037	−0.012
31°56′	103,475.3471	30.798528	3,367.5421	0.016060	2.212
	262,644.4257	−0.789962	45.6781	−0.001037	−0.012
31°57′	105,323.2588	30.798609	3,368.5056	0.016041	2.211
	262,597.0280	−0.790332	45.6159	−0.001037	−0.012
31°58′	107,171.1753	30.798690	3,369.4681	0.016022	2.211
	262,549.6080	−0.790703	45.5536	−0.001037	−0.012
31°59′	109,019.0967	30.798771	3,370.4294	0.016003	2.210
	262,502.1658	−0.791074	45.4914	−0.001037	−0.012
32° 0′	110,867.0229	30.798852	3,371.3895	0.015984	2.210
	262,454.7014	−0.791444	45.4291	−0.001037	−0.012
⋮	⋮	⋮	⋮	⋮	⋮
32° 8′	125,650.6080	30.799501	3,379.0298	0.015832	2.206
	262,074.1863	−0.794405	44.9312	−0.001037	−0.012
32° 9′	127,498.5781	30.799582	3,379.9797	0.015813	2.206
	262,026.5220	−0.794775	44.8690	−0.001037	−0.012
32°10′	129,346.5530	30.799663	3,380.9285	0.015794	2.206
	261,978.8355	−0.795145	44.8068	−0.001037	−0.012
32°11′	131,194.5328	30.799744	3,381.8761	0.015775	2.205
	261,931.1268	−0.795514	44.7445	−0.001037	−0.012
32°12′	133,042.5174	30.799826	3,382.8226	0.015756	2.205
	261,883.3960	−0.795884	44.6823	−0.001037	−0.013

Second difference corrections:

00″	10″	20″	30″	00″	10″	20″	30″
60″	50″	40″	30″	60″	50″	40″	30″
(I) 0.0000	−0.0003	−0.0005	−0.0006	(IV) 0.0000	0.0015	0.0025	0.0028

TABLE 11.3

Transverse Mercator projection tables for Arizona eastern zone number 0201: Geodetic coordinates from grid coordinates

Latitude	(I) (IX)	Difference 1″	(VII) (X)	Difference 1″	(VIII) (XI)
31°55′	101,627.4403	30.798447	1,584.0443	0.017024	19.928
	38,067.4233	0.114443	277.2604	0.003433	3.729
31°56′	103,475.3471	30.798528	1,585.0657	0.017030	19.946
	38,074.2899	0.114538	277.4664	0.003437	3.734
31°57′	105,323.2588	30.798609	1,586.0875	0.017036	19.963
	38,081.1621	0.114633	277.6726	0.003441	3.739
31°58′	107,171.1753	30.798690	1,587.1096	0.017042	19.981
	38,088.0401	0.114728	277.8791	0.003446	3.744
31°59′	109,019.0967	30.798771	1,588.1321	0.017048	19.999
	38,094.9238	0.114823	278.0858	0.003450	3.750
32° 0′	110,867.0229	30.798852	1,589.1550	0.017054	20.016
	38,101.8132	0.114919	278.2928	0.003454	3.755
⋮	⋮	⋮	⋮	⋮	⋮
32° 8′	125,650.6080	30.799501	1,597.3513	0.017103	20.159
	38,157.1346	0.115684	279.9581	0.003489	3.797
32° 9′	127,498.5781	30.799582	1,598.3774	0.017109	20.176
	38,164.0756	0.115780	280.1674	0.003493	3.802
32°10′	129,346.5530	30.799663	1,599.4040	0.017115	20.194
	38,171.0224	0.115876	280.3770	0.003497	3.807
32°11′	131,194.5328	30.799744	1,600.4309	0.017121	20.212
	38,177.9750	0.115972	280.5868	0.003502	3.812
32°12′	133,042.5174	30.799826	1,601.4582	0.017128	20.230
	38,184.9333	0.116068	280.7969	0.003506	3.818

Second difference corrections:

00″	10″	20″	30″	00″	10″	20″	30″
60″	50″	40″	30″	60″	50″	40″	30″
(I) 0.0000	−0.0003	−0.0005	−0.0006	(IX) 0.0000	−0.0004	−0.0006	−0.0007

The inverse conversion from grid coordinates E and N to geodetic latitude, ϕ, and longitude, λ, is performed similarly, using the equations

$$\phi = \phi' - (\text{VII})q^2 + (\text{VIII})q^4 \tag{11.60a}$$

$$\lambda = \lambda_0 - (\text{IX})q + (\text{X})q^3 + (\text{XI})q^5 \tag{11.60b}$$

$$\gamma = (\text{XIV})q - (\text{XV})q^3 \tag{11.60c}$$

$$k = k_0[1 + (\text{XVI})q^2 + 0.00003q^4] \tag{11.60d}$$

in which ϕ' is obtained from the projection tables by interpolation in term I using N as the argument and E' and q are found with Equations (11.59d) and (11.59g), respectively. The numerals IX, X, XI, XIV, XV, and XVI are taken from the projection tables using ϕ' as the argument. Note that $\phi' = \phi_f$, the foot point latitude calculated in the previous section by Equation (11.51). To demonstrate the procedure, Examples 11.6 and 11.7 are reworked in Examples 11.8 and 11.9, respectively, using the tables.

TABLE 11.4
Transverse Mercator projection for Arizona eastern zone number 0201:
Convergence and scale factor $(t - T)$

Latitude	(I)	Difference 1″	(XII) (XIV)	Difference 1″	(XIII) (XV)	(XVI)
31°55′	101,627.4403	30.798447	5,286.8527	0.041148	3.028	
			20,125.6859	0.217173	227.693	0.012330
31°56′	103,475.3471	30.798528	5,289.3216	0.041141	3.028	
			20,138.7163	0.217251	227.923	0.012330
31°57′	105,323.2588	30.798609	5,291.7900	0.041133	3.029	
			20,151.7513	0.217330	228.153	0.012330
31°58′	107,171.1753	30.798690	5,294.2580	0.041126	3.029	
			20,164.7911	0.217408	228.384	0.012330
31°59′	109,019.0967	30.798771	5,296.7255	0.041118	3.029	
			20,177.8356	0.217487	228.615	0.012330
32° 0′	110,867.0229	30.798852	5,299.1926	0.041111	3.029	
			20,190.8848	0.217565	228.846	0.012330
⋮	⋮	⋮	⋮	⋮	⋮	⋮
32° 8′	125,650.6080	30.799501	5,318.9133	0.041051	3.032	
			20,295.4489	0.218198	230.705	0.012329
32° 9′	127,498.5781	30.799582	5,321.3763	0.041043	3.032	
			20,308.5408	0.218278	230.938	0.012329
32°10′	129,346.5530	30.799663	5,323.8389	0.041036	3.032	
			20,321.6374	0.218357	231.172	0.012329
32°11′	131,194.5328	30.799744	5,326.3011	0.041028	3.033	
			20,334.7388	0.218437	231.406	0.012329
32°12′	133,042.5174	30.799826	5,328.7628	0.041021	3.033	
			20,347.8450	0.218516	231.640	0.012329

Second difference corrections:

00″	10″	20″	30″	00″	10″	20″	30″
60″	50″	40″	30″	60″	50″	40″	30″
(I) 0.0000	−0.0003	−0.0005	−0.0006	(XIV) 0.0000	−0.0003	−0.0005	−0.0006

EXAMPLE 11.8. Using the data from Example 11.6, determine grid coordinates for station Orbit on the Arizona zone 0201 using the tabular approach:

$$\phi = 32°10′31.505″$$

$$\lambda = 109°40′10.230″$$

For constant terms related to the projection, refer to Example 11.6.

Solution. Compute $\Delta\lambda$ (Equation (11.59a)) and p (Equation (11.59b)):

$$\Delta\lambda = (\lambda_0 - \lambda)3600 = [(110°10′) - (109°40′10.230″)3600″$$

$$= (0.4971584°)3600″/₀ = 1789.77024″$$

$$p = 10^{-4}\,\Delta\lambda = 0.178977024″$$

Referring to Table 11.2,

Tabular (I) for 32°10′	=	129,346.5530
(31.505″)(30.799663 difference/″)	=	970.3434
Second difference correction	=	−0.0006
(I)	=	130,316.8958
Tabular (II) for 32°10′	=	3,380.9285
(31.505″)(0.015794 difference/″)	=	0.4976
(II)	=	3,381.4261
Tabular (III)	=	2.206

The Northing, N, then (Equation (11.59c)) is

$$N = (I) + (II)p^2 + (III)p^4$$

$$N = 130,316.8958 + (3,381.4261)(0.178977024)^2 + (2.206)(0.178977024)^4$$

$$N = 130,425.214 \text{ m}$$

Note that this value varies from that computed in Example 11.6 by +0.002 m. To determine the easting, E, obtain IV, V, and VI from Table 11.2:

Tabular (IV) for 32°10″	=	261,978.8355
(31.505″)(−0.795145 difference/″)	=	−25.0510
Second difference correction	=	0.0028
(IV)	=	261,953.7873
Tabular (V) for 31°10′	=	44.8068
(31.505″)(−0.001037 difference/″)	=	−0.0367
(V)	=	44.7741
Tabular (VI)	=	−0.0120

Compute E' and E using Equations (11.59d) and (11.59e):

$$E' = (IV)p + (V)p^3 + (VI)p^5 = 46,883.7093 + 0.256883243 - 0.000002204$$

$$E' = 46,883.9662$$

$$E = E_0 + E' = 213,360. + 46,883.9662$$

$$E = 260,243.966 \text{ m}$$

which is identical to the easting determined in Example 11.6.

The terms XII, XIII, and XVI are needed from Table 11.4 to calculate the convergence angle and scale factor:

Tabular (XII) for 32°10′	=	5,323.8389
(31.505″)(0.041036 difference/″)	=	1.2928
(XII)	=	5,325.1317
Tabular (XIII)	=	3.032
Tabular (XVI)	=	0.012329

The convergence angle, γ, is determined by Equation (11.59f):

$$\gamma = (XII)p + (XIII)p^3 = (5{,}325.1317)(0.178977024) + (3.032)(.178977024)^3$$

$$\gamma = 953.094'' = 0.2647482° = 0°15'53.093''$$

To determine the scale factor, calculate q by Equation (11.59g) and use this value and term (XVI) in Equation (11.59h) to get k:

$$q = 10^{-6}E' = 46{,}883.9662 \times 10^{-6} = 0.0468839662$$

$$k = k_0[1 + (XVI)q^2 + 0.00003q^4] = (0.9999)[1 + (0.012329)q^2 + 0.00003)q^4]$$

$$k = (0.9999)[1 + 0.000027100 + 1.449501 \times 10^{-10}]$$

$$k = 0.99992710$$

Both γ and k correspond to the values found in Example 11.6.

EXAMPLE 11.9. Given the grid coordinates for station Roger in Example 11.7, compute geodetic latitude and longitude using Tables 11.2, 11.4, and 11.5:

$$E = 272{,}013.298 \text{ m}$$

$$N = 107{,}720.186 \text{ m}$$

For the constant terms related to this projection, refer to Example 11.6.

Solution. Calculate E' using Equation (11.59e):

$$E' = E - E_0 = 272{,}013.298 - 213{,}360. = 58{,}653.298 \text{ m}$$

Then q' (Equation (11.59g)) is

$$q' = 10^{-6}E' = 0.058653298$$

Now obtain ϕ' from Table 11.2 by interpolating under (I) using N (given) as the argument:

$$N \qquad = 107{,}720.1860$$

$$\text{Tabular (I)} = 107{,}171.1753$$

$$\text{Difference} = \qquad 549.0107$$

$$\text{Difference/Difference per second} = 549.0107/30.798690 = 17.82578''$$

so that

$$\phi' = 31°58'17.82578'' = 0.558010006 \text{ rad}$$

This value for $\phi' = \phi_f$, the foot point latitude, is identical to ϕ_f evaluated in Example 11.7.

Evaluate terms (VII) and (VIII) from Table 11.3 using ϕ' as the argument (note that ϕ' will be used as the argument for the rest of the table evaluations in this example):

Tabular (VII) for 31°58'	= 1,587.1096
(17.82578'')(0.017042 difference/'') =	0.3038
(VII)	= 1,587.4134
Tabular (VIII) for 31°58'	= 19.981
(17.82578/60)(0.018)	= 0.0053
(VIII)	= 19.9863

The latitude is determined by Equation (11.60a):

$$\phi = \phi' - (VII)q^2 + (VIII)q^4$$

$$= (31°58'17.8258'') - (1{,}587.4134)q^2 + (19.9863)q^4$$

$$= (31°58'17.8258'') - 5.461035'' + 0.00023654''$$

$$\phi = 31°58'12.3650''$$

To determine the longitude, evaluate terms (IX), (X), and (XI) from Table 11.3:

Tabular (IX) for 31°58'	= 38,088.0401
(17.8258'')(0.114728 difference/'') =	2.0451
Second difference	= −0.0004
(IX)	= 38,090.0848
Tabular (X) for 31°58'	= 277.8791
(17.8258'')(0.003446 difference/'') =	0.0614
(X)	= 277.9405
Tabular (XI) for 31°58'	= 3.744
(17.8258''/60'')(0.006)	= 0.00178
(XI)	= 3.7458

Using Eq. (11.60b), the longitude is

$$\lambda = \lambda_0 - (IX)q + (X)q^3 + (XI)q^5 = (38{,}090.0848)q$$
$$+ (277.9405)\,q^3 + (3.7458)q^5$$

$$= (110°10') - 2{,}234.109095'' + 0.056082736'' + 0.0000026002''$$

$$= (110°10') - 2{,}234.053001'' = 110.1666667° - 0.620570280°$$

$$= 109.5460964°$$

$$\lambda = 109°32'45.947''$$

The convergence angle and scale factor require the terms XIV, XV, and XVI from Table 11.4:

Tabular (XIV) for 31°58'	= 20,164.7911
(17.8258'')(0.217408 difference/'') =	3.8755
Second difference	= −0.0005
(XIV)	= 20,168.6661
Tabular (XV) for 31°58'	= 228.384
(17.8258''/60'')(0.231)	= 0.0686
(XV)	= 228.4526
Tabular (XVI)	= 0.012330

The convergence angle (Equation (11.60c)) is

$$\gamma = (XIV)q - (XV)q^3 = (20,168.6661)(0.058653298)$$
$$- (228.4526)(0.058653298)^3$$

$$= 1,182.958783'' - 0.04609708'' = 1,182.912686'' = 0.32858686°$$

$$\gamma = 0°19'42.91''$$

The scale factor (Equation (11.60d)) is

$$k = k_0[1 + (XVI)q^2 + 0.00003q^4] = 0.9999[1 + 0.012330q^2 + 0.00003q^4]$$

$$= 0.9999(1 + 4.2418 \times 10^{-5} + 3.5505 \times 10^{-10}) = 0.9999(1.000042418)$$

$$k = 0.999942414$$

Note that ϕ, λ, γ, and k are the same as the values for these quantities calculated in Example 11.7, verifying the accuracies of both the computational procedure using the equations and the tables.

11.20
GRID AND GEODETIC AZIMUTHS ON TRANSVERSE MERCATOR PROJECTIONS

Grid and geodetic azimuths are illustrated for the transverse Mercator projection in Figure 11.17. Station H has a latitude and longitude of ϕ and λ. The projection of H on the central meridian is at H', which has latitude of ϕ' and longitude λ_m. The parallel of latitude passing through H strikes the central meridian at J with geographic coordinates of ϕ and λ_m. The straight line passing through H and parallel to the central meridian is a grid-north line, from which grid azimuths are measured. The curved line passing through H concave toward the central meridian is a true meridian, from which geodetic azimuths are reckoned. A line tangent to the true meridian at H makes an angle with the grid meridian of γ, which is calculated by Equation (11.49a), (11.58a), (11.59f), or (11.60c), using the constants given in Publication NOS NGS5 (see Stem, 1989a) for the projection. The factor γ, frequently designated the *convergence* of the meridians for this projection, is used in the conversion from grid to geodetic and from geodetic to grid azimuths.

The equation for azimuth on the transverse Mercator projection is

$$\alpha = t + \gamma + \delta \tag{11.61}$$

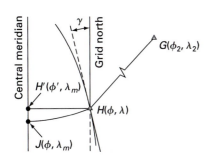

FIGURE 11.17
Geodetic and grid azimuths on the transverse Mercator projection.

in which α is the geodetic azimuth, t is the grid azimuth, γ is the convergence angle, and δ, known as the $t - T$ correction (grid azimuth − geodetic azimuth projected on a plane), is computed for a line between stations 1 and 2 by the equation

$$\delta = \left(\frac{(N_2 - N_1)(2E'_1 + E'_2)}{6r_f^2} \right) \rho \tag{11.62}$$

In this equation, E'_1, E'_2 (referred to the projection central meridian, Figure 11.16) and N_1, N_2 are coordinates of the origin and terminus of the line; $\rho = 206{,}265''/\text{rad}$, and $r_f^2 = R_m N_f$. The value of N_f or radius of the prime vertical for latitude ϕ_f, is calculated by Equation (11.2) and R_m, the radius of curvature the meridian direction, is given by

$$R_m = \frac{a(1 - \varepsilon^2)}{(1 - \varepsilon^2 \sin^2 \phi_f)^{3/2}} \tag{11.63}$$

where a and ε^2 are parameters of the projection and ellipsoid, previously defined; and ϕ_f is the foot point latitude for the average of the N coordinates for points 1 and 2.

Combining Equations (11.2) and (11.63) to solve for r_f^2 leads to

$$r_f^2 = R_m N_f = \frac{(k_0 a)^2 (1 - \varepsilon^2)}{(1 - \varepsilon^2 \sin^2 \phi_f)^2} \tag{11.64}$$

The foot point latitude is found using Equations (11.50) and (11.51) to compute the rectifying latitude and ϕ_f. In this process, set $N = (N_1 + N_2)/2$ in Equation (11.50) and neglect the last three terms in Equation (11.51), leading to

$$\omega_f = [(N_1 + N_2)/2 - N_0 + S_0]/(k_0 r) \tag{11.65}$$

$$\phi_f = \omega_f + V_0 \sin \omega_f \cos \omega_f \tag{11.66}$$

in which N_0, S_0, k_0, r, and V_0 are constants for the projection provided in NOS NGS5 (Stem, 1989a) and also given in Example 11.6.

Frequently, a single value is precomputed for the term $[1/(2r_f^2)]\rho$ in Equation (11.62), using the average radius of the earth to give

$$\delta = \left(\frac{(25.4)(N_2 - N_1)(2E'_1 + E'_2)}{3} \right) 10^{-10} \text{ sec} \tag{11.67}$$

where the coordinates are in meters. Equation (11.67) is adequate for geodetic surveys within the approximate boundaries of SPCS 83 zones (Stem, 1989a).

The third term, or $t - T$ correction, of Equation (11.61) is insignificant for lines that do not exceed 8 km or 5 mi in length. It reaches a maximum for long lines that are predominantly of a north-south direction. The third term is 0 for an east-west line ($N_2 - N_1 = 0$) and very small even for long lines where the line is predominantly east-west.

EXAMPLE 11.10. Determine grid and geodetic azimuths (from the south) of the line from station Orbit (Example 11.8) to station Roger (Example 11.9), which have state plane coordinates on the transverse Mercator projection of Arizona east zone, as follows:

Station	E, m	N, m
Orbit	260,243.966	130,425.214
Roger	272,013.298	107,720.186

Also, calculate the grid and geodetic distances between the two stations.

Solution. Orbit is the origin (station 1) and Roger the terminus of the line (station 2). Therefore, the grid azimuth from the south of the line Orbit-Roger is

$$A_{12}(\text{south}) = \arctan\frac{E_1 - E_2}{N_1 - N_2} = \frac{260{,}243.966 - 272{,}013.298}{130{,}425.214 - 107{,}720.186}$$

$$= \arctan(-0.5183579602) = 332°35'58.830''$$

The convergence γ was calculated in Example 11.8 using Equation (11.59f) and found to be $0°15'53.093''$. The length of the line exceeds 5 mi (it is 25.6 km or 15.9 mi), so that it is necessary to evaluate the third term of Equation (11.61) to calculate the correct geodetic azimuth. Evaluation of this term using Equation (11.67) follows:

$$N_2 - N_1 = (107{,}720.186 - 130{,}425.214) = -22{,}705.028$$

$$2E_1' + E_2' = (2)(260{,}243.966 - 213{,}360) + (272{,}013.298 - 213{,}360)$$

$$= 152{,}421.230$$

and, by Equation (11.67),

$$\delta = \left[\frac{(25.4)(-22{,}705.028)(152{,}421.230)}{3}\right]10^{-10} \text{ sec}$$

$$\delta = -2.930''$$

By Equation (11.61), the geodetic azimuth of the line Orbit-Roger from the south then is

$$\text{geodetic azimuth} = 332°35'58.830'' + 0°15'53.093'' - 0°00'02.930''$$

$$= 332°51'48.993''$$

Figure 11.18 illustrates the relationship between grid and geodetic azimuths for the line from Orbit to Roger. The grid line on the projection is represented by the straight

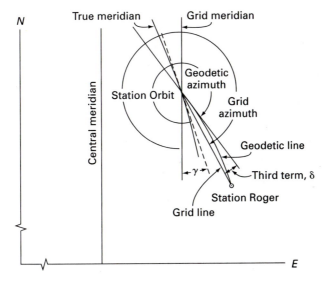

FIGURE 11.18
Grid and geodetic azimuths of the line from Orbit to
Roger on the transverse Mercator projection.

line from Orbit to Roger. The grid azimuth is measured from grid south to the grid line. The geodetic line is a curve concave toward the central meridian. The geodetic azimuth is measured from the geodetic meridian (south) to a tangent to the curving geodetic line at station Orbit. The δ in Equation (11.67) is the angle from the grid line between the points to a tangent to the geodetic line at the station that is the origin for the azimuth.

The grid distance between Orbit and Roger is

$$d_{12} = [(E_1 - E_2)^2 + (N_1 - N_2)^2]^{1/2}$$
$$= [(260,243.966 - 272,013.298)^2 + (130,425.214 - 107,720.186)^2]^{1/2}$$
$$= 25,574.117 \text{ m}$$

To obtain the geodetic distance between the two points, the grid distance must be modified by the average of the scale factors computed for Orbit and Roger in Examples 11.8 and 11.9, respectively. These values are $k_{\text{Orbit}} = 0.99992710$ and $k_{\text{Roger}} = 0.99994241$, so that the average is

$$k_{\text{aver}} = (0.99992710 + 0.99994241)/2 = 0.99993476$$

The grid distance is divided by k_{aver} to yield geodetic distance on the ellipsoid. Therefore, geodetic distance Orbit to Roger = 25,574.117/0.999934755 = 25,575.786 m on the ellipsoid.

For a line of this length and orientation, the difference between the grid and geodetic lengths is substantial. When a geodetic distance is reduced to the grid, the scale factor k_{aver} is multiplied by the geodetic distance on the ellipsoid.

11.21
COMPUTATION OF COORDINATES ON UNIVERSAL TRANSVERSE MERCATOR PROJECTIONS

The basic equations for the transverse Mercator projection are given in Section 11.13, and the universal transverse Mercator projection is described in Section 11.14. Coordinates using the UTM projection can be calculated on any one of the ellipsoids, so this system is applicable anywhere in the world. In the United States, the GRS 80 ellipsoid almost always is specified.

The conversion from latitude and longitude to rectangular coordinates and the inverse solution can be determined by Equations (11.15) and (11.17a). A better option is to use Equations (11.42a) through (11.58b), derived from the basic formulas and designed for applications in computer programs as well as for computation by handheld calculator. When using these equations to determine coordinates on the UTM projection, the only change in the procedure from that given in Section 11.19 for computing state plane coordinates by the transverse Mercator projection is to use the zone-specific constants for UTM (Section 11.14) in the equations. All other constants and terms in the equations are the same for a given ellipsoid.

First, the desired ellipsoid is selected. Example problems in this section are computed on the GRS 80 ellipsoid.

Next, the zone number and longitude of the central meridian are chosen from a list of zone numbers, central, and bounding meridians, as shown in Table 11.5. For example, a point with longitude 77°25′25.452″W is in zone 18, where the longitude of the central meridian, λ_0, is 75°W. In case this table is not available, the zone number and λ_0 (longitude of the central meridian) can be computed using

$$\text{zone number} = \frac{180° - \lambda}{6} \tag{11.68}$$

TABLE 11.5
UTM zone numbers with bounding and central meridians (adapted from _U.S. Dept. of the Army_, 1958)

Zone	Central meridian	Bounding meridians	Zone	Central meridian	Bounding meridians	Zone	Central meridian	Bounding meridians
		180°			60°W			60°E
1	177°W	174°W	21	57°W	54°W	41	63°E	66°E
		174°W			54°W			66°E
2	171°W	168°W	22	51°W	48°W	42	69°E	72°E
		168°W			48°W			72°E
3	165°W	162°W	23	45°W	42°W	43	75°E	78°E
		162°W			42°W			78°E
4	159°W	156°W	24	39°W	36°W	44	81°E	84°E
		156°W			36°W			84°E
5	153°W	150°W	25	33°W	30°W	45	87°E	90°E
		150°W			30°W			90°E
6	147°W	144°W	26	27°W	24°W	46	93°E	96°E
		144°W			24°W			96°E
7	141°W	138°W	27	21°W	18°W	47	99°E	102°E
		138°W			18°W			102°E
8	135°W	132°W	28	15°W	12°W	48	105°E	108°E
		132°W			12°W			108°E
9	129°W	126°W	29	09°W	06°W	49	111°E	114°E
		126°W			06°W			114°E
10	123°W	120°W	30	03°W	00°	50	117°E	120°E
		120°W			00°			120°E
11	117°W	114°W	31	03°W	06°E	51	123°E	126°E
		114°W			06°E			126°E
12	111°W	108°W	32	09°E	12°E	52	129°E	132°E
		108°W			12°E			132°E
13	105°W	102°W	33	15°E	18°E	53	135°E	138°E
		102°W			18°E			138°E
14	99°W	96°W	34	21°E	24°E	54	141°E	144°E
		96°W			24°E			144°E
15	93°W	90°W	35	27°E	30°E	55	147°E	150°E
		90°W			30°E			150°E
16	87°W	84°W	36	33°E	36°E	56	153°E	156°E
		84°W			36°E			156°E
17	81°W	78°W	37	39°E	42°E	57	159°E	162°E
		78°W			42°E			162°E
18	75°W	72°W	38	45°E	48°E	59	165°E	168°E
		72°W			48°E			168°E
19	69°W	66°W	39	51°E	54°E	59	171°E	174°E
		66°W			54°E			174°E
20	63°W	60°W	40	57°E	60°E	60	177°E	180°

For the given example,

$$\text{zone number} = \frac{180° - 77°25'}{6} = \frac{102°35'}{6} = 17+$$

So, zone 18 must be used. Zone 18 is $(6°)(18)$, or $108°$ east of $180°$ longitude and ranges from $72°$ to $78°$ (UTM zones are $6°$ wide). Consequently, the central meridian is 75°W.

To illustrate the use of Equations (11.42a) through (11.58b) and also Equation (11.15) for conversion of geodetic to rectangular coordinates on the UTM projection (GRS 80), consider an example problem.

EXAMPLE 11.11. Compute E and N coordinates for station Orbit, from Example 11.6, where

$$\phi = 32°10'31.505''$$

$$\lambda = 109°40'10.230''$$

From Table 11.5, the zone number is 12. From this table and Section 11.14, the defining constants are

$$\lambda_0 = 111°00' \qquad k_0 = 0.9996$$

$$E_0 = 500,000 \qquad \phi_0 = 0$$

$$N_0 = \qquad 0$$

and parameters for the GRS 80 ellipsoid are

$$a = 6,378,137 \text{ m}, \qquad \varepsilon^2 = 0.006694380023, \qquad r = 6,367,449.14577$$

Because $\phi_0 = 0$, zone-specific constants ω_0 and S_0, by Equations (11.42a) and (11.42b), also are equal to 0.

Solution. The rectifying latitude, ω, a function of ϕ, is the same as in Example 11.6. Therefore,

$$\omega = 0.5592984200 \text{ rad}$$

The meridional distance S, by Equation (11.44), is

$$S = k_0 \omega r = (0.9996)(0.5592984200)(6,367,449.146)$$

$$S = 3,559,879.725 \text{ m}$$

Using the notation of Section 11.19 and NOS NGS5 (Stem, 1989a), the radius of the prime vertical also is affected by k_0 and, with Equation (11.2),

$$R = (k_0 a)/(1 - \varepsilon^2 \sin^2 \phi)^{1/2} = 6,381,645.848 \text{ m}$$

From Example 11.6,

$$\eta^2 = 0.004828375742$$

Terms A_1 and A_2, Equations (11.45a) and (11.45b), are

$$A_1 = -R = -6,381,645.848 \text{ m}$$

$$A_2 = (\tfrac{1}{2}) R \tan \phi = 2,007,457.086 \text{ m}$$

and, from Example 11.6,

$$A_3 = 0.1015030447, \qquad A_4 = 0.3873115247, \qquad A_5 = -0.01675973344$$

Terms A_6 and A_7, by Equations (11.45f) and (11.45g), are

$$A_6 = (1/360)[61 - 58 \tan^2 \phi + \tan^4 \phi + \eta^2(270 - 330 \tan^2 \phi)] = 0.1079796048$$

$$A_7 = (1/5040)(61 - 479 \tan^2 \phi + 179 \tan^4 \phi - \tan^6 \phi) = -0.01996269433$$

Calculate L using Equation (11.46):

$$L = (\lambda - \lambda_0) \cos \phi = (109.6695083° - 111°)(\cos \phi)$$

$$= (-1.3304917)(0.846421711)$$

$$L = -1.126157061° = -0.01965514862 \text{ rad}$$

The $E = X$ and $N = Y$ coordinates, by Equations (11.47a) and (11.47b) are

$$E = E_0 + A_1 L[1 + L^2(A_3 + L^2(A_5 + L^2(A_5 + A_7 L^2)))]$$

$$E = 500,000. + (-6,381,645.848)(-0.01965514862)$$
$$\times [1 + L^2(A_3 + L^2(A_5 + A_7 L^2))] = 500,000. + 125,437.1168$$

$$E = 625,437.117 \text{ m}$$

$$N = S - S_0 + N_0 + A_2 L^2[1 + L^2(A_4 + A_6 L^2)]$$
$$= 3,559,879.725 - 0 + 0 + (2,007,457.086)(-0.01965514862)^2$$
$$\times [1 + L^2(A_4 + A_6 L^2)]$$
$$= 3,559,879.725 + 775.650$$

$$N = 3,560,655.375 \text{ m}$$

As a check on these calculations, compute the coordinates $X = E$ and $Y = N$ using Equation (11.15), Section 11.13. The equations are repeated here for convenience

$$\frac{X}{N} = \lambda' \cos \phi + \frac{\lambda'^3 \cos^3 \phi}{6}(1 - t^2 + \eta^2)$$
$$+ \frac{\lambda'^5 \cos^5 \phi}{120}(5 - 18t^2 + t^4 + 14\eta^2 - 58t^2\eta^2 + 13\eta^4) + \cdots$$

$$\frac{Y}{N} = \frac{S_\phi}{N} + \frac{\lambda'^2}{2} \sin \phi \cos \phi + \frac{\lambda'^4}{24} \sin \phi \cos^3 \phi(5 - t^2 + 9\eta^2 + 4\eta^4)$$
$$+ \frac{\lambda'^6}{720} \sin \phi \cos^5 \phi(61 - 58t^2 + t^4 + 270\eta^2 - 330t^2\eta^2 + 445\eta^4) + \cdots$$

By Equation (11.2),

$$N = \frac{a}{(1 - \varepsilon^2 \sin^2 \phi)^{1/2}}, \qquad \phi = 32°10'31.505''$$

$$a = 6,378,137 \text{ m}$$

$$\varepsilon^2 = 0.006694380023 \qquad \text{for GRS 80}$$

$$N = \frac{6,378,137}{[1 - 0.006694380023(\sin^2 \phi)]^{1/2}} = 6,384,199.528 \text{ m}$$

$$\lambda' = (\lambda_0 - \lambda) = (111° - 109°40'10.230'') = 01°19'49.770'' = 1.33049170°$$

$$\lambda' = 0.023221461 \text{ rad}$$

Note that the sign of $\Delta\lambda = \lambda'$ is the reverse of that used in the transverse Mercator projection for SPCS 83. Because the quantity X (X' in Figure 11.16) in Equation (11.15) is measured relative to the central meridian, this step is necessary to ensure the correct sign for points east and west of the central meridian. Also note that this equation for $\lambda' = \Delta\lambda = (\lambda_0 - \lambda)$ corresponds to Equation (11.59a) for $\Delta\lambda$ in the tabular approach to computing coordinates on the transverse Mercator projection.

$$t = \tan \phi = 0.6291345943 \qquad t^2 = 0.395810330$$

$$\eta^2 = \frac{\varepsilon^2 \cos^2 \phi}{1 - \varepsilon^2} = \frac{(0.006694380023) \cos^2 \phi}{1 - 0.006694380023} = 0.004828375742$$

By definition, the scale of the UTM projection is fixed at $k_0 = 0.9996$ on the central meridian. Therefore, each term in Equation (11.15) must be multiplied by k_0. First, solve for X' (See Figure 11.16):

$$N\lambda' \cos \phi k_0 \qquad\qquad = 125{,}432.1984$$

$$\frac{N(\lambda')^3 \cos^3 \phi}{6}(1 - t^2 + \eta^2)k_0 \qquad = 4.9185$$

$$5 - 18t^2 + t^4 + 14\eta^2 - 58t^2\eta^2 + 13\eta^4 = -2.16753$$

$$\frac{N(\lambda')^5 \cos^5 \phi}{120}(-2.16753)k_0 \qquad = \underline{\quad -0.00034}$$

$$X' \qquad\qquad = 125{,}437.1166$$

The value X' is measured from the central meridian (see Figure 11.16) and is the equivalent of the right side terms, $A_1 L(1 + \cdots)$, in Equation (11.47a). Consequently, the easting $= E = X$ for the point $= E_0 + X'$ so that

$$E = X = 500{,}000. + 125{,}437.117 = 625{,}437.117 \text{ m}$$

In the conversion to the Y coordinate, it is necessary to evaluate S_ϕ, where

$$S_\phi = a(A_0 \phi - A_1 \sin 2\phi + A_2 \sin 4\phi - A_3 \sin 6\phi + \cdots)$$

in which

$$A_0 = 1 - \frac{\varepsilon^2}{4} - \frac{3}{64}\varepsilon^4 - \frac{5}{256}\varepsilon^6 \qquad = 0.9983242984$$

$$A_1 = \frac{3}{8}\varepsilon^2 + \frac{3}{32}\varepsilon^4 + \frac{45}{1024}\varepsilon^6 \qquad = 0.0025146071$$

$$A_2 = \frac{15}{256}\varepsilon^4 + \frac{45}{1024}\varepsilon^6 \qquad = 0.0000026390$$

$$A_3 = \frac{35}{3072}\varepsilon^6 \qquad = 0.0000000003418$$

$$\phi = 0.561566983 \text{ rad}$$

$$S_\phi = 6{,}378{,}137(0.5583612029)(0.9996) = 3{,}559{,}879.726$$

$$N\frac{(\lambda')^2}{2} \sin \phi \cos \phi(0.9996) = 755.5306$$

$$5 - t^2 + 9\eta^4 + 4\eta^4 = 4.647738297$$

$$N\frac{(\lambda')^4}{24} \sin \phi \cos^3 \phi(4.153921350)(0.9996) = (\text{III})\, p^4 = 0.11604$$

$$N = Y = 3{,}560{,}655.373 \text{ m}$$

The answers obtained by both methods are the same to within 1 mm. Equation (11.15) contains sufficient terms to ensure accuracies to a hundredth of a meter (see Thomas, 1952). In general, Equations (11.42a) through (11.58b) provide the most reliable results and are more convenient to use.

Conversion of rectangular coordinates to latitude and longitude on the UTM projection can be accomplished using Equations (11.50) through (11.56) and following the procedure for the inverse solution using state plane coordinates on the transverse Mercator projection. In fact, determination of latitude and longitude using Equations (11.50) through (11.56) is simply a repetition of the process shown in Example 11.7, except that the projection constants for the desired zone in the UTM system and parameters for the ellipsoid used are substituted into the equations. This process is not repeated here. The student should perform this conversion using the UTM coordinates from Example 11.11 and following the method outlined in Example 11.7.

Transformation from rectangular coordinates to latitude and longitude on the UTM projection also is possible by direct solution of Equation (11.17a). A value for ϕ' can be obtained by using the first term in the second set of Equation (11.15), where

$$Y = S_\phi k_0$$

in which Y is the given Y or N coordinate for the station, k_0 is 0.9996, and S_ϕ is defined by Equation (11.16a). Solve for $S_\phi = Y/k_0$ and substitute this value into Equation (11.16a) to determine ϕ'. Because direct solution of Equation (11.16a) is not possible, an iterative solution, such as the Newton-Raphson (Scarborough, 1966) can be used to obtain ϕ'. A simpler approach involves calculating the rectifying latitude ω by Equation (11.50) and $\phi_f = \phi'$ (the foot point latitude) with Equation (11.51), using ω and constants V_i for the ellipsoid. When ϕ' has been determined, Equation (11.17a) is used to compute $\Delta\phi$ and $\Delta\lambda$, from which ϕ and λ can be computed. To demonstrate this procedure, the coordinates calculated in Example 11.11 are transformed to latitude and longitude by Equation (11.17a).

EXAMPLE 11.12. Determine ϕ and λ given $N = X$ and $E = Y$ for station Orbit calculated in the first part of Example 11.11, where

$$E = X = \quad 625,437.117 \text{ m}$$

$$N = Y = 3,560,655.375 \text{ m}$$

The projection constants and parameters and terms for the ellipsoid are

$$\lambda_0 = 111° \qquad k_0 = 0.9996$$

$$E_0 = 500,000 \qquad \phi_0 = 0$$

$$N_0 = 0 \qquad\qquad S_0 = 0$$

$$a = 6,378,137. \text{ m}, \quad \varepsilon^2 = 0.006694380023, \quad r = 6,367,449.14577 \text{ m}$$

$$V_0 = 0.00502\ 28939\ 48 \qquad V_4 = 2.35059 \times 10^{-7}$$

$$V_2 = 0.00002\ 93706\ 25 \qquad V_6 = 2.181 \times 10^{-9}$$

Solution. Calculate the rectifying latitude ω by Equation (11.50):

$$\omega = (N - N_0 + S_0)/(k_0 r) = (3,560,655.375)/[(0.9996)(6,367,449.14577)]$$

$$\omega = 0.5594202836 \text{ rad}$$

Determine the foot point latitude, $\phi_f = \phi'$, using Equation (11.51):

$$\phi' = \omega + \sin\omega\cos\omega\{V_0 + \cos^2\omega[V_2 + \cos^2\omega(V_4 + V_6\cos^2\omega)]\}$$

$$= 0.5594202836 + \sin\omega\cos\omega\{0.00502\ 28939\ 48 + \cos^2\omega[2.9370625 \times 10^{-5}$$

$$+ \cos^2\omega(2.35059 \times 10^{-7} + 2.181 \times 10^{-9}\cos^2\omega)]\}$$

$$\phi' = 0.5594202836 + 0.002268829500 = 0.5616891131 \text{ rad} = 32°10'56.6961''$$

Next, compute $\Delta\phi$ using the first set of Equation (11.17a):

$$\Delta\phi = t_1 \left[\frac{-(X')^2}{2R_1 N_1} + \frac{(X')^4}{24R_1 N_1^3}(5 + 3t_1^2) \right]$$

in which the variables are defined in Section 11.13. Compute R_1, N_1, t_1, η^2, and X':

$$R_1 = \frac{a(1 - \varepsilon^2)}{(1 - \varepsilon^2 \sin^2 \phi')^{3/2}} = \frac{(6{,}378{,}137)(1 - 0.006694380023)}{(1 - 0.006694380023 \sin^2 \phi')^{3/2}}$$

$$= 6{,}353{,}529.373 \text{ m}$$

$$N_1 = \frac{a}{(1 - \varepsilon^2 \sin^2 \phi')^{1/2}} = \frac{6{,}378{,}137}{(1 - 0.006694380023 \sin^2 \phi')^{1/2}}$$

$$= 6{,}384{,}201.886 \text{ m}$$

$$t_1 = \tan \phi' = 0.6293050773$$

$$\eta_1^2 = \frac{\varepsilon^2}{(1 - \varepsilon^2)} \cos^2\phi' = \frac{0.006694380023}{1 - 0.006694380023} \cos^2 \phi'$$

$$= 0.004827633709$$

$$X' = 625{,}437.117 - 500{,}000 = 125{,}437.117 = \text{grid distance}$$

$$\frac{X'}{k_0} = 125{,}487.3119 \text{ m}$$

$$\frac{-t_1(X'/k_0)^2}{2R_1 N_1} = -0.0001221544291 \text{ rad}$$

$$\frac{t_1(X'/k_0)^4}{24R_1 N_1^3}(5 + 3t_1^2) = 0.0000000243372 \text{ rad}$$

$$\Delta\phi = -0.000122130 \text{ rad}$$

$$\Delta\phi = -0°00'25.1911''$$

$$\phi = \phi' + \Delta\phi = (32°10'56.6961'') - (0°00'25.1911'')$$

$$= 32°10'31.5050''$$

which corresponds to the given value to the fourth decimal place in seconds.

The longitude is calculated by the second set of Equation (11.17a) using R_1, N_1, t_1, η_1, and X' computed in determining ϕ'.

$$\Delta\lambda = \sec \phi' \left[\frac{X'}{N_1} - \frac{1}{6}\left(\frac{X'}{N_1}\right)^3(1 + 2t_1^2 + \eta_1^2) + \frac{1}{120}\left(\frac{X'}{N_1}\right)^5(5 + 28t_1^2 + 24t_1^4) \right]$$

$$\frac{X'/k_0}{\cos \phi' N_1} = -0.02322414761 \text{ rad}$$

$$\frac{-1}{(6 \cos \phi')}\left(\frac{X'/k_0}{N_1}\right)^3(1 + 2t_1^2 + \eta_1^2) = 2.687159136 \times 10^{-6} \text{ rad}$$

$$\frac{1}{120 \cos \phi}\left(\frac{X'/k_0}{N_1}\right)^5(5 + 28t_1^2 + 24t_1^4) = -5.73525 \times 10^{-10} \text{ rad}$$

$$\Delta\lambda = -0.02322146102 \text{ rad}$$

$$\Delta\lambda = -1°19'49.7702''$$

$$\lambda = \lambda_0 + \Delta\lambda = 111°00'00'' - 1°19'49.7002'' = 109°40'10.2298''$$

$$\text{original } \lambda \ (10.230'')$$

Equation (11.17a) is for projections with a 4° bandwidth. The UTM has a 6° bandwidth. Consequently, the results obtained by computation with Equation (11.17a) may not correspond to the third or fourth decimal place of seconds with those calculated using Equations (11.50) through (11.56). Additional terms are required to obtain identical results to four places. Equations with the additional terms adequate for projection bandwidths of up to 10 to 12° of arc can be found in Thomas (1952).

11.22
AZIMUTHS ON UTM PROJECTIONS

When *geographic coordinates* (latitude and longitude) are known, the convergence angle can be computed by Equation (11.49a) or (11.16b).

Convergence of the meridians when the *rectangular coordinates* are given is calculated by Equation (11.58a) or (11.17b).

EXAMPLE 11.13. Compute the convergence angle for station Orbit, for which $\phi = 32°10'31.505''$ and $\lambda = 109°40'10.230''$. Defining constants for the projection and ellipsoid parameters are given in Example 11.11.

Solution. First, compute the convergence using Equation (11.49a):

$$\gamma = C_1 L[1 + L^2(C_3 + C_5 L^2]$$

in which C_1 and C_3, computed in Example 11.6, were found to be

$$C_1 = -\tan\phi = -0.629134594$$

$$C_3 = (1/3)(1 + 3\eta^2 + 2\eta^4) = 0.338177251$$

C_5 is found by Equation (11.48c),

$$C_5 = (1/15)(2 - \tan^2\phi) = 0.106945978$$

and L by Equation (11.46) and from Example 11.11,

$$L = (\lambda - \lambda_0)\cos\phi = -0.01965514862 \text{ rad}$$

The convergence angle is

$$\lambda = (-0.629134594)(-0.01965514862)$$
$$\times [1 + (-0.01965514862)^2[0.338177251$$
$$+ 0.106945978(-0.01965514862)^2]$$

$$\lambda = 0.01236736502 \text{ rad} = 0.708597819° = 0°42'30.95''$$

Note that γ is positive because the station is east of the central meridian. As a check on this result, compute the convergence angle using Equation (11.16b), which is

$$\gamma = \lambda' \sin\phi \left[1 + \frac{(\lambda')^2 \cos^2\phi}{3}(1 + 3\eta^2) + \frac{(\lambda')^4 \cos^4\phi}{15}(2 - t^2)\right]$$

where γ and λ' are in radians. From Example 11.11,

$$\lambda' = -0.023221461 \text{ rad}, \qquad \eta^2 = 0.004828375742$$

$$t^2 = \tan^2 \phi = 0.395810338$$

$$\lambda' \sin \phi = 0.01236573404 \text{ rad}$$

$$\frac{(\lambda)^3 \sin \phi \cos^2 \phi (1 + 3\eta^2)}{3} = 1.61546 \times 10^{-6} \text{ rad}$$

$$\frac{(\lambda')^5 \sin \phi \cos^4 \phi (2 - t^2)}{15} = 8.49963 \times 10^{-9} \text{ rad}$$

$$\gamma = 0.01236735800 \text{ rad}$$

$$\gamma = 0°42'30.95''$$

EXAMPLE 11.14. From Example 11.12, the plane coordinates for station Orbit are

$$E = X = 625,437.117 \text{ m}$$

$$N = Y = 3,560,655.375 \text{ m}$$

Also from Example 11.12,

$$\phi' = \phi_f = 0.5616891131 \text{ rad} = 32°10'56.6961'', \qquad \eta_1 = \eta_f = 0.004827633709$$

$$t_1 = \tan \phi' = 0.6293050773, \qquad N_1 = 6,384,201.866 \text{ m}, \qquad E_0 = 500,000. \text{ m}$$

Determine the convergence angle at station Orbit.

Solution. First, use Equation (11.58a):

$$\gamma = D_1 Q[1 + Q^2(D_3 + D_5 Q^2)]$$

in which (Equations (11.52) and (11.2))

$$Q = (E - E_0)/R_f = (E - E_0)/(N_1)k_0$$

$$= (625,437.117 - 500,000)/(6,384,201.866)(0.9996) = 0.1965591223$$

and the terms D_1, D_3, and D_5, by Equations (11.57a) to (11.57c), are

$$D_1 = \tan \phi_f = 0.6293050773$$

$$D_3 = -(1/3)(1 + \tan^2 \phi' - \eta_f^2 - 2\eta_f^4) = -0.4637168783$$

$$D_5 = (1/15)(2 + 5 \tan^2 \phi' + 3 \tan^4 \phi') = 0.2967087680$$

Substitution of these terms and Q into Equation (11.58a) yields

$$\gamma = 0.01236734979 \text{ rad} = 0.7085969467° = 0°42'30.95''$$

Next, calculate the convergence angle at Orbit using Equation (11.17b), where

$$\gamma = t_1 \left[\frac{X'/k_0}{N_1} - \frac{1}{3}\left(\frac{X'/k_0}{N_1}\right)^3 (1 + t_1^2 - \eta_1^2) + \frac{1}{15}\left(\frac{X'/k_0}{N_1}\right)^5 (2 + 5t_1^2 + 3t_1^4) \right]$$

in which the terms are defined in Section 11.13 and $k_0 = 0.9996$ is the scale factor needed to convert grid to arc distance. From Example 11.12,

$$\frac{X'}{0.9996} = 125,487.3119 \text{ m}$$

N_1, t_1, and η_1^2 are as given previously;

$$t_1 \frac{X'/k_0}{N_1} = 0.01236956537 \text{ rad}$$

$$-\frac{1}{3} t_1 \left(\frac{X'/k_0}{N_1}\right)^3 (1 + t_1^2 - \eta_1^2) = -2.21620 \times 10^{-6} \text{ rad}$$

$$\frac{1}{15} t_1 \left(\frac{X'/k_0}{N_1}\right)^5 (2 + 5t_1^2 + 3t_1^4) = 5.47845 \times 10^{-10} \text{ rad}$$

$$\gamma = 0.01236734972 \text{ rad}$$

$$\gamma = 0.708596947°$$

$$\gamma = 0°42'30.95''$$

If the grid azimuth from the south from station Orbit to station Astro has been determined from UTM plane coordinates as $135°45'20.59''$, then by Equation (11.61),

geodetic azimuth $\alpha = t + \gamma + \delta$

$$= (135°45'20.59'') + (0°42'30.95'')$$

$$= 136°27'51.54''$$

where the term δ has been neglected for this example.

For lines greater than 8 km or 5 mi in length, the term δ should be evaluated using either Equation (11.62) or (11.67) and applied when determining the geodetic azimuth with Equation (11.61). The procedure for calculating and applying this correction can be found in Example 11.10, Section 11.20 on azimuth determination for the transverse Mercator projection for state plane coordinates. The same process is applicable to UTM coordinates.

11.23
COMPUTATION OF THE SCALE FACTOR ON UTM PROJECTIONS

Measured distances corrected for systematic errors and reduced to the ellipsoid must be multiplied by a scale factor before being used in plane coordinate computations on the UTM projection.

The scale factor can be computed by Equation (11.49b) or (11.17) when latitude and longitude are given or by Equation (11.58b) or (11.17c) for rectangular coordinates.

EXAMPLE 11.15. Given station Orbit from Example 11.12, for which the rectangular coordinates are

$$E = X = \quad 625,437.117 \text{ m}$$

$$N = Y = 3,560,655.375 \text{ m}$$

compute the scale factor for this station. Additional data from Example 11.12 and 11.14 are

$\phi_f = 0.5616891131$, $\quad \eta_f^2 = 0.004827633709$, $\quad \tan \phi_f = 0.6293050773$,

$E_0 = 500,000$ m

$Q = 0.01965591223$, $\quad N_1 = 6,384,201.866$ m

Solution. First, solve for terms G_2 and G_4 using Equations (11.57d) and (11.57e):

$$G_2 = (\tfrac{1}{2})(1 + \eta_f^2) = (\tfrac{1}{2})(1 + 0.004827633709) = 0.5024138170$$

$$G_4 = (\tfrac{1}{12})(1 + 5\eta_f^2) = 0.08534484742$$

Then, the scale factor k is determined by Equation (11.58b)

$$k = k_0[1 + G_2Q^2(1 + G_4Q^2)] = 0.9996[1 + 0.5024138170(0.01965591223)^2$$
$$\times (1 + (0.08534484742)(0.01965591223)^2)]$$

$$k = 0.999794038$$

For surveys that do not exceed 8 km or 5 mi between the extremities, it is usually adequate to calculate a scale factor based on an average of the E and N coordinates for the area.

For surveys in which the extremities of the region cover distances in excess of 8 km, considerable judgment must be exercised in computing scale factors to achieve specified accuracies. Lines running north and south at $E = 320,000$ m or $E = 680,000$ m have constant scale factors of unity. On the other hand, lines running east and west in these portions of the projection have rapidly changing scale factors. In the first case, a single scale factor would be adequate for lines predominately in the north-south directions. In the latter case, a line may have to be separated into sections with a scale factor calculated for each section.

11.24
STATE PLANE COORDINATE SYSTEMS ON THE 1927
DATUM (SPCS 27)

Prior to the adoption of SPCS 83, all of the states in the United States had state plane coordinate systems based on the 1927 datum (SPCS 27). These systems included the Lambert and transverse Mercator projections adopted in each state by legislative action (Alaska, zone 1 has an oblique Mercator projection). The assignment of projections was approximately the same as in SPCS 83 with some minor changes. All control in the continental network had been coordinated and adjusted using SPCS 27 and not all of these points were included in the readjustment for the horizontal datum of 1983, so many control points still exist with only coordinates on the 1927 datum. Therefore, it is likely that the survey engineer might have to integrate points with SPCS 27 coordinates into a control network or perhaps work with control entirely on the 1927 datum.

Computation of rectangular and geographic coordinates on the SPCS 27 using both the Lambert and transverse Mercator projections generally is the same as for calculating positions on the SPCS 83. However, determination of SPCS 27 positions was based on the Clarke ellipsoid of 1866, so all of the ellipsoidal parameters are different and some of the defining constants for the projections have been changed. Consequently, positions for the same points computed on the two datums do not coincide. The shifts vary from about -20 to 40 m in latitude and 40 to 100 m in longitude, as illustrated in Figures 11.19 and 11.20.

When the SPCS 27 systems were developed, computational power was lacking, so that a tabular approach was taken and tables were published for each state. The published constants were determined using logarithms, the best available method for performing the

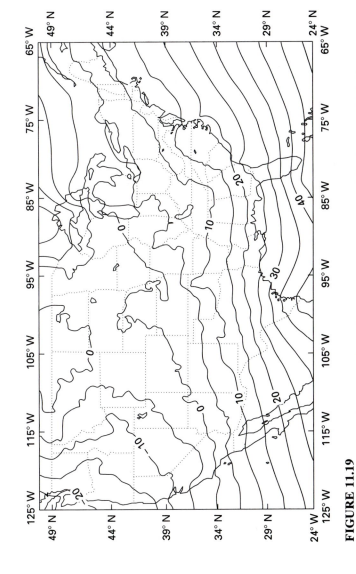

FIGURE 11.19
Latitude datum shift in the conterminous United States in meters (NAD 83 minus NAD 27).
(Adapted with permission of the National Geodetic Survey.)

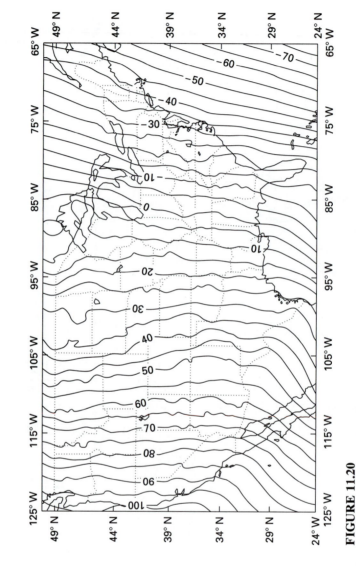

FIGURE 11.20
Longitude datum shift in the conterminous United States in meters (NAD 83 minus NAD 27). (*Adapted with permission of the National Geodetic Survey.*)

task at that time. These tabulated constants were consistent throughout the tables for a given projection and provided *relative positions* for points adequate for the users of the state plane coordinate systems at that time. Use of these tables to compute coordinates on the SPCS 27 is quite similar to the tabular approaches developed in this chapter for the Lambert and transverse Mercator projections.

Note that, if the rigorous equations for either the Lambert or transverse Mercator projections (Sections 11.11 and 11.13) were used to calculate coordinates on the SPCS 27, the answers obtained *would not correspond* to those determined by the tabular method. The differences could be as much as 1.5 m or 5 ft. Consequently, when working with the SPCS 27, use the tabular approach or the equations developed to simulate the tabular method. Tables for each projection in every state (NGS, 1950–1965) and a publication on state plane coordinates by automatic data processing (Claire, 1976) are available from the National Geodetic Survey Information Center, Silver Spring, MD 20910.

11.25
CONVERSION OF SPCS 27 COORDINATES TO
SPCS 83 COORDINATES

When coordinates computed and adjusted on the 1927 datum are to be converted to the 1983 datum, two major methods are appropriate. In the first method, the surveyor should select the local network of control points on the SPCS 27 that provides the strongest control survey around the area under consideration. It is very important that the most reliable points be chosen and that control points previously adjusted in the National Geodetic Survey network of control on the 1927 datum be included. Then the *original measurements* (angles, directions, distances, etc.) should be submitted to the NGS in the Bluebook format as specified in FGCS (1994). The measurements should satisfy first-order horizontal accuracy standards (Section 9.23) and second-order, Class II vertical accuracy standards (Section 5.46, Table 5.2). With these data, the NGS will adjust the control points on the 1983 datum and return to the user, adjusted SPCS 83 coordinates that are consistent with the national network. The surveyor then can use these basic control points to adjust the entire local network by the method of least squares (Sections 2.22–2.30, 8.39, and 9.18–9.21). Another option in this procedure is to include points in the network that have coordinates in a *high-accuracy reference network* (HARN). A HARN consists of control points in the national network in a given state where state agencies and private surveyors have densified the network under the direction of the NGS, which then performs the final adjustment to produce coordinates in this high-precision horizontal control network. The surveyor uses these HARN points as control points to adjust the local network. The resulting adjusted coordinates, in the SPCS 83 system, then are submitted by the surveyor to the NGS. When the surveys performed for this task satisfy horizontal and vertical accuracy standards, as just outlined, the NGS will accept the points and include them in the database for the national network.

The second method consists of using the North American datum conversion (NADCON) program that was developed and is distributed by the NGS. NADCON employs a model of shift values in latitude and longitude for a large area, such as the conterminous United States. These values are used to get estimates for a local model. Local modeling then is performed to determine shift corrections to NAD 27 coordinates. Minimization of curvature (Dewhurst, 1990) is used to calculate the actual shifts. Input to and output from the program are in latitude and longitude and conversion is possible from NAD 27 to NAD 83 or from NAD 83 to NAD 27. For certain specified states, NADCON can convert coordinates to the high-accuracy reference network.

CORPSCON is a software package that allows transformation of coordinates between NAD 27 and NAD 83. This program, a modification of NADCON by the U.S. Corps of Engineers, converts rectangular coordinates and provides output in rectangular coordinates. CORPSCON cannot convert coordinates of points to a HARN. The user may specify units of meters, U.S. survey foot, or the international foot. CORPSCON also is available from the NGS.

PROBLEMS

11.1. "Every map projection has some distortions." Is this statement true or false, and why?

11.2. What four conditions must an ideal distortion-free map satisfy?

11.3. Describe briefly four classes of map projections.

11.4. Explain the basic differences between geometric map projection and mathematical map projection.

11.5. The Lambert conformal conic projection and the transverse Mercator projection are two of the most commonly used map projections; discuss the principal differences between them.

11.6. Enumerate the specifications of the universal transverse Mercator projection.

11.7. "The shape of a state, or its north-south extent relative to its east-west extent, influences the type of map projection used for its state plane coordinate system." Discuss this statement fully.

11.8. Describe the major differences between the state plane coordinate systems on the 1927 horizontal datum and the 1983 horizontal datum.

11.9. List and describe the defining constants for (a) the Lambert projection for SPCS 83 and (b) the transverse Mercator projection for SPCS 83.

11.10. List and describe the derived constants for the Lambert state plane coordinate projections on the 1983 datum.

11.11. The latitude and longitude for station 1 are 41°20′45.618″N and 121°05′50.623″W. Compute on the Lambert projection, California zone 1: (a) the convergence angle γ, for station 1; (b) E and N state plane coordinates for the station; (c) the scale factor at station 1. Use projection Table 11.1.

11.12. Solve Problem 11.11 using the basic equations (Section 11.17 Example 11.1) for Lambert projection.

11.13. The state plane coordinates for station 2 on the Lambert projection for California zone 1 are $E_2 = 2,091,437.726$ m and $N_2 = 734,829.029$ m. The average elevation of the terrain between point 2 and point 10 of Example 11.2 is 1525 m above the ellipsoid. Calculate (a) the grid azimuth from station 2 to station 10; (b) the geodetic azimuth from station 2 to station 10 and from station 10 to station 2; (c) the grid distance, geodetic distance on the ellipsoid, and geodetic distance at average terrain elevation between stations 2 and 10. Assume an average scale factor between 2 and 10 of 0.99995315.

11.14. In Problem 11.13, compute the geographic coordinates for station 2, using projection Table 11.1.

11.15. In Problem 11.14, calculate geographic coordinates for station 2 using polynomial coefficients and the constants for the Lambert projection for California zone 1 as shown in Example 11.4.

11.16. The geographic coordinates for station Glenn are $\phi = 31°55'20.3246''$ and $\lambda = 109°33'50.2415''$. Using Equations (11.59) and the constants of Tables 11.2 and 11.4, compute (*a*) the state plane coordinates for station Glenn on the transverse Mercator projection, Arizona eastern zone; (*b*) the convergence of the meridian and the scale factor for station Glenn.

11.17. The state plane coordinates for station Gail on the transverse Mercator projection, Arizona eastern zone, are $E = 246,986.309$ m and $N = 112,367.528$ m. (*a*) Compute the grid distance and grid azimuth from station Orbit (Example 11.6) to Gail; (*b*) determine the geodetic azimuth from the south of the line from Orbit to Gail.

11.18. Compute geographic coordinates for station Gail in Problem 11.17.

11.19. Calculate the convergence of the meridians and the scale factor for station Gail in Problem 11.17. Also compute the geodetic distance on the ellipsoid and the geodetic azimuth from the south of the line from Gail to Orbit.

11.20. The distance between two stations calculated from their state plane coordinates is 10,500.825 m. The scale factor and average orthometric height are 1.0004 and 1000 m, respectively. If the separation between the geoid and ellipsoid is -30.00 m ($h = H + N$), what is the horizontal distance between these two stations as it would be measured on the ground?

11.21. In the area described for Problem 11.20, the state plane coordinates for two control points A and B are $E_A = 625,571.353$ m, $N_A = 195,158.518$ m and $E_B = 652,612.407$ m, $N_B = 195,223.260$ m. The coordinates for a construction control point C. which is to be established from A, are $E_c = 652,237.131$ m and $N_c = 194,913.531$ m. Compute (*a*) the clockwise angle to be turned from an instrument set up at A with a backsight on B to direct the line of sight toward point C, which is to be established; (*b*) the distance that must be measured over level terrain from A to set point C.

11.22. Control point Arch has geographic coordinates of $\phi = 43°10'30.205''$N and $\lambda = 121°30'45.638''$W. Determine the UTM zone in which this point falls. Compute the plane coordinates on the UTM projection for this point. Use Equations (11.42a) through (11.47b).

11.23. Compute the convergence of the meridian and the scale factor at point Arch for which data are given in Problem 11.22.

11.24. Determine the plane coordinates on the UTM projection for station 1 in Problem 11.11. Use Equations (11.15, a and b), Section 11.13.

11.25. Calculate the grid and geodetic azimuths on the UTM projection for a line from station 1 in problem 11.11 to station Arch in Problem 11.22.

11.26. The plane coordinates on the UTM projection (zone 14) for station Able are $E = 379,272.805$ m and $N = 4,765,501.354$ m. Compute the latitude and longitude for station Able. Use Equations (11.50) through (11.55).

11.27. The plane coordinates for station Plum on the UTM projection (zone 14) are $E = 401,560.354$ m and $N = 4,782,465.550$ m. Compute the geographic coordinates for Plum. Use Equations (11.50) through (11.55).

11.28. Calculate the scale factor and convergence of the meridian on the UTM for station 1 in Problem 11.11. Use Equations (11.16b) and (11.17).

11.29. Calculate the convergence of the meridian and the scale factor for station Able in Problem 11.26.

11.30. Compute the convergence of the meridian and the scale factor for station Plum in Problem 11.27. Use the rigorous equations for this solution.

11.31. Compute the grid and geodetic azimuths from the south on the UTM projection for the line from station Able (Problem 11.26) to station Plum (Problem 11.27).

11.32. Compute the geodetic azimuth from the south on the UTM projection for the line from Plum (Problem 11.27) to Able (Problem 11.26).

11.33. A microwave transmission antenna is to be erected at station Able in Problem 11.26 so that the axis of the antenna is directed toward station Plum in Problem 11.27. From an instrument set up over station Able, a backsight is possible on control point Z, which has plane coordinates of $E_z = 379,071.621$ m and $N_z = 4,765,100.052$ m. Compute the clockwise horizontal angle to be turned from an instrument set up over Able with a backsight on point Z so as to place the instrument line of sight in the proper direction for aligning the antenna.

REFERENCES

Adams, O. S., and C. N. Claire. *Manual of Plane Coordinate Computation.* Special Publication no. 193. Silver Spring, MD: U.S. Department of Commerce, National Geodetic Survey, 1935, reprinted 1971.

Alexander, I. H., and R. J. Alexander. *Projection Tables California Coordinate System.* Santa Rosa, CA: California Land Surveyor's Association, 1989.

Bomford, G. *Geodesy.* London: Oxford University Press, 1962.

Chamberlin, W. *Round Earth on Flat Paper.* National Geographic Society. Washington D.C., March 1950.

Claire, C. N. *State Plane Coordinates by Automatic Data Processing.* Publication 62–4. Silver Spring, MD: U.S. Department of Commerce, National Geodetic Survey, 1976.

Dewhurst, W. T. "The Application of Minimum Curvature-Derived Surfaces in the Transformation of Positional Data from the North American Datum of 1927 to the North American Datum of 1983." NOAA Technical Memorandum NOS NGS50. Silver Spring, MD: National Geodetic Center, 1990.

Ewing, C. E., and M. M. Mitchell. *Introduction to Geodesy.* New York: American Elsivier Publishing Company, 1970.

Federal Geodetic Control Subcommittee (FGCS). *Input Formats and Specifications of the National Geodetic Survey Data Base,* volumes I, (*Horizontal Data*), II (*Vertical Data*), and III (*Gravity Data*). Silver Spring, MD: National Geodetic Survey, September 1994.

Hosmer, G. L. *Geodesy.* 2nd ed. New York: John Wiley and Sons, 1930.

Krakiwsky, E. J. "Conformal Map Projections in Geodesy." University of New Brunswick Lecture Notes no. 37, September 1973.

Meade, B. K. "Program for Computing Universal Transverse Mercator (UTM) Coordinates for Latitudes North or South and Longitudes East or West." *Surveying. and Mapping* 47, no. 1 (March 1987), pp. 37–49.

Moffitt, F. H., and H. Bouchard. *Surveying.* 9th ed. New York: HarperCollins Publishers, 1992.

National Geodetic Survey (NGS), NOAA. *Plane Coordinate Projection Tables.* C&GS Special Publications Series. Silver Spring, MD: National Geodetic Information Branch, NGS, NOAA, 1950–1965.

Scarborough, J. B. *Numerical Mathematical Analysis.* 6th ed. Baltimore: Johns Hopkins Press, 1966.

Schwarz, C. R. *North American Datum of 1983.* NOAA Professional Paper NOS2. Silver Spring, MD: National Geodetic Survey, December 1989.

Snyder, J. P. *MPA Projections—A Working Manual.* U.S. Geological Survey Professional Paper 1395. Washington, DC: U.S. Government Printing Office, 1987.

Stem, J. E. *State Plane Coordinate System of 1983.* NOAA Manual NOS NGS5. Silver Spring, MD: National Geodetic Survey, January 1989a.

Stem, J. E. "User Participation and Impact." *North American Datum 1983,* pp. 237–48. NOAA Professional Paper NOS2. Silver Spring, MD: National Geodetic Survey, 1989b.

Thomas, P. D. *Conformal Projections in Geodesy and Cartography.* Special Publication 251. Silver Spring, MD: U.S. Department of Commerce, National Geodetic Survey, 1952.

U.S. Department of the Army. *Field Manual Map Reading.* Technical Manual FM 21–26. Washington, DC, 1973.

U.S. Department of the Army. *Universal Transverse Mercator Grid.* Technical Manual TM5-241–8. Document no. AOA 176624. Washington, DC: National Technical Information Service, 1958.

U.S. Department of the Army. *The Universal Grid Systems,* Technical Manual TM5-241, T0116-1–233. Washington, DC: 1951.

Vanicek, P., and E. Krakiwsky. *Geodesy, the Concepts,* 2nd ed. New York: North-Holland Publishing Company, 1986.

Vincenty, T. "Precise Determination of the Scale Factor from Lambert Conical Projection Coordinates." *Surveying and Mapping* 45, no. 4 (December 1985), pp. 315–18.

Vincenty, T. "Use of Polynomial Coefficients in Conversion of Coordinates on the Lambert Conformal Projection." *Surveying and Mapping* 46, no. 1 (March 1986), pp. 15–18.

PART IV

Modern Surveying and Mapping

Global Positioning System*

12.1
GENERAL

The Global Positioning System (GPS) is the culmination of developmental work on extraterrestrial positioning systems initiated in the late 1950s. Originally known as the Navigation Satellite Time and Ranging (NAVSTAR) System, GPS is capable of providing positioning and velocity information on a worldwide basis, in all weather conditions, 24 hours a day. The GPS provides the surveyor, engineer, planner, and geographer with timely and accurate positional, timing, and velocity information. The applications of GPS are limited only by the imagination and creativity of the user. Although acquisition of data is simple and existing software packages make data processing straightforward, the users of this very sophisticated tool for surveyors and navigators should have a comprehensive background in the theoretical and practical aspects of this system to obtain consistent and accurate results.

12.2
SYSTEM COMPONENTS

The GPS is composed of three major parts: the satellite system, the control segment, and a user segment. The satellite system consists of 21 satellites or *space vehicles* (sv) and three operational spares. The satellites are distributed in six orbital planes, inclined at 55° and at altitudes of 20,200 km. This orbital configuration was selected to ensure that at least four satellites would be visible, worldwide, 24 hours a day.

The control segment consists of ground stations around the world that are responsible for monitoring the health of each satellite and uploading orbital parameters to the satellites. The master control station (MCS) is located at Falcon Air Force Base, Colorado Springs,

---

* This chapter written by Bryn Fosburgh, Trimble Navigation Limited.

Colorado, with a backup MCS at Onizuka Air Force Base in Sunnyvale, California. Other control stations are found at Hawaii, Kwajalein, Diego Garcia, and Ascension Island.

The user segment is composed of a worldwide community of civilian and military users, equipped with appropriate receivers, who use the GPS for positioning, navigation, and timing.

12.3
SIGNAL STRUCTURE

Each GPS satellite transmits two *carrier frequencies,* labeled L1 and L2. In addition, two *pseudo-random noise* (PRN) codes are modulated onto the carrier frequencies. The *coarse acquisition* (C/A) code is modulated onto the L1 carrier and the *precise* (P) code is modulated onto both the L1 and L2 frequencies (Figure 12.1, Table 12.1). The L1 and L2 carriers and the L1 C/A, L1 P, and L2 P codes are the five primary signals from which *observables* are derived and then used by surveyors and navigators to determine position. The word *observable* describes a *measurable parameter* in a system, to distinguish between the quantity derived from a measurement and the actual measurement. For example, time difference is measured using the C/A and P codes and is the *measurement* used to find the *range* or *pseudorange*. This pseudorange is referred to as an *observable*.

In addition to transmitting the carrier waves and PRN codes, each satellite transmits a broadcast message at 50 bits per second. This message consists of five subframes that contain information on correction parameters (satellite, clock, ionosphere), satellite positions (ephemeris), almanac, and satellite health. Reception of the entire message takes 12.5 min. The almanac, which contains fewer and less precise ephemeris parameters than the broadcast ephemeris for all satellites, is used by the receiver to determine the approximate location of all the other satellites within the constellation and the estimated time that satellites will "rise" and "set" at the receiver's location.

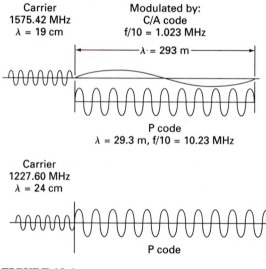

FIGURE 12.1
GPS signal structure.

TABLE 12.1
GPS signal structure

Carrier	Frequency (MHz)	Wavelength (m)	Code	Chipping rate, Mbps	Code length (m)
L1	1575.42	0.19	L1 C/A	1.023	300
L2	1227.60	0.24	L1 P	10.23	30
			L2 P	10.23	30

Notes: Wavelength is determined by dividing the speed of light (c) by the transmitted frequency (f) or $\lambda = c/f$. For example,

$$\lambda_{L1} = (3.0 \times 10^8 \text{ m/sec})/(1575.42 \text{ MHz}) = 0.19 \text{ m} = 19 \text{ cm}$$
$$1 \text{ MHz} = 10^6; \text{ Hz} = 10^6 \text{ cycles/sec}; 1 \text{ Mbs} = 10^6 \text{ bits or chips/sec}$$

12.4
GOVERNMENT DEGRADATION POLICY

Positional and navigational data from the GPS are provided to the user community through either the *precise positioning service* (PPS) or the *standard positioning service* (SPS). Users of PPS consist of the U.S. military, allied forces, authorized government agencies, and authorized contractors. With PPS data, these groups are able to obtain instantaneous horizontal positions having an uncertainty of 25 m with a probability of 95 percent. Commercial surveyors, geodesists, and navigators not authorized to receive the PPS coded signal, must use the SPS. These groups are unable to achieve the same point positioning accuracy as their PPS counterparts, due to lack of access to PPS data and certain governmental policies designed to deliberately perturb all GPS data. On the other hand SPS users, through relative or differential positioning, are able to obtain positional accuracies that equal or exceed those achieved by the PPS user. Differential techniques will be covered later in this chapter (Section 12.7). First, consider government perturbation policies.

Selective Availability

Selective availability (SA) is the deliberate degradation of GPS satellite signals. These perturbations are designed to deny hostile forces the opportunity to capitalize on the accurate point positioning capability provided by the GPS. These system errors degrade the horizontal point positioning accuracy of the SPS user to an uncertainty of 100 m with a probability of 95 percent. In addition, SPS users should encounter no errors greater than 300 m 99 percent of the time. SA errors degrade point positioning by perturbing the satellite clock and the ephemeris data.

Antispoofing

Antispoofing (A-S) is the term applied to a policy of code denial that interchanges the P code with the encrypted Y code. Users of PPS can tolerate this interchange as they are able to receive and interpret the Y code. A-S affects any GPS receiver that tracks the P code for surveying or navigation.

Impact of SA and A-S on the Commercial User

The effects of SA are minimized by employing relative positioning techniques where at least two or preferably several GPS receivers simultaneously collect data (Section 12.7,

699

Method 2). However, for point positioning where a single GPS receiver is used, the accuracy of an SPS user is reduced to 100 m with a 95 percent confidence level. Some GPS receivers designed to receive the P code have hardware and firmware (nonchangeable software imprinted on a chip) which recognize that A-S has been activated. The receiver then squares the L2 signal to remove the unknown Y code. Unfortunately, signal squaring increases the signal noise, eliminates the C/A and P codes, and is a technique sensitive to one-half wavelengths ($\lambda \approx 0.12$ m). This step, in turn, complicates ambiguity resolution (determination of the integral number of wavelengths in a given range), which is essential in obtaining geodetic accuracy in relative positioning. Through signal processing techniques, some manufacturers have developed algorithms that allow recovery of full wavelength L2 observations during periods of A-S.

Thus, even though the effects of SA and A-S can be lessened by clever hardware design and software development, these perturbations cause substantial problems for SPS users.

12.5
RECEIVER COMPONENTS AND DESIGN

The GPS receiver system consists of the antenna, the receiver itself, a command entry and display unit, and power supply. Figure 12.2 illustrates a GPS system in which the receiver and antenna are contained in a compact unit mounted on the tripod. The command entry and display unit or controller is shown attached to one leg of the tripod, but may be removed and held by the operator for more convenient use. The power-supply pack is clamped to another leg of the tripod. The antenna receives the signal from the satellites and converts it into electrical energy usable in the receiver. The receiver, under control of a microprocessor, processes the signal, converts it to a pseudorange, and computes approximate coordinates for the receiver. A data storage unit that is internal or has an output connection that allows interface with another computer also is found on most systems.

A GPS receiver usually has one or more channels, where a channel consists of the hardware and software needed to track the signal from one satellite at one of the two carrier frequencies (L1 or L2). A receiver may have four or more *dedicated channels,* each tracking a different satellite simultaneously. In such a receiver, ranges and phase data from four or more satellites can be obtained, allowing instantaneous determination of the receiver's position and receiver clock error. To simplify the receiver and reduce costs, some receivers have a single channel, which is either *sequencing* or *multiplexing.* Both techniques permit the channel to sample rapidly through a number of satellite signals to get the number of ranges required for position determination.

Receivers that require knowledge of the codes are called *code-correlating* or *code-phase* instruments. These receivers can be single (L1) or dual frequency receivers (L1 and L2) that require knowledge of the C/A and/or P code (when available) to obtain pseudoranges and the ephemeris data. These receivers measure code and carrier phase data for navigation and precise positioning.

In the past, the so-called codeless or carrier-wave receivers were used for high-precision geodetic applications. The *codeless receivers,* as the name implies, require no knowledge of the PRN codes to make precise phase measurements. These receivers use a squaring technique that multiplies the received signal by itself to obtain a second harmonic of the carrier wave that does not contain the code modulation (the C/A and P codes are lost.)

Some receivers are a hybrid between the code-correlating and codeless receivers. These hybrid receivers use code-correlating techniques on the L1 and L2 frequency if A-S

Receiver and antenna

Command entry
and display unit
or controller

Power
supply

FIGURE 12.2
Leica GPS System 200. (*Courtesy of Leica, Inc.*)

is not activated. During times of A-S, the receiver will use the codeless technique (squaring) for the L2 frequency or implement proprietary signal processing techniques that enable the receiver to obtain full wavelength L2 measurements (unsquared). Receiver technology has evolved into receivers with 9 or more dedicated channels that accurately collect L1, L2, and C/A code data during A-S.

GPS receivers also may be designed to collect single or dual frequencies (L1 and L2). Dual frequency receivers have the advantage that ionospheric errors can be accommodated, but they are more expensive and, as described earlier, the P code is subject to A-S.

Most commercially available surveying receivers have multiple channels, single or dual frequency, carrier-phase, code-phase devices. Single frequency receivers are affected more by ionospheric errors than dual frequency receivers, but they are less expensive, making them adaptable for certain surveying applications. Dual frequency, multiple channel receivers can compensate better for ionospheric errors and now are available with the capability of electronic compensation for A-S perturbations. These receivers are ideal for geodetic applications but carry a substantially higher price tag.

12.6
ERRORS IN THE GPS

Systematic errors that affect analysis of the pseudoranges directly include those due to errors in satellite position, satellite timing errors, effects of atmospheric conditions on radio signal transmission, receiver timing and other errors, and site-dependent factors. The magnitude of these errors can be reduced through appropriately chosen mathematical models, equipment design, and site reconnaissance.

Errors in Satellite Position

The positions of the satellites given by the broadcast ephemeris, contained within subframes 2 and 3 of the data message, are accurate to approximately 10 m. However, the government policy of SA can degrade these positions by an amount that causes positional uncertainties of 100 m.

The error in a base line due to orbital errors, Δr, can be estimated by multiplying the ephemeris error, Δe, by the ratio of the base line length, d, to the satellite altitude, $D = 20,200$ km. In GPS, a *base line* is defined as the vector between two stations (azimuth, distance, height, or three-dimensional coordinate differences). Thus, the approximate base-line error is

$$\Delta r = \Delta e (d/D) \qquad (12.1)$$

For example, an ephemeris error of 40 m would cause a corresponding error of approximately 0.020 m in a base line 10 km in length. When measuring longer lines, a precise ephemeris should be used. One precise ephemeris that provides a postfitted adjustment of the orbital parameters can be obtained from the National Geodetic Survey within two or three days to a week after completing the observations.

Satellite Timing Errors

Accurate timing is a critical requirement in the determination of positions using the GPS. The satellites have extremely precise atomic clocks that are very stable, but even so, some residual timing error exists. This error can be modeled using correction parameters contained within subframe 1 of the data message.

In *point positioning,* where a single receiver is used to get position from at least four ranges, a differential clock correction, included as an unknown parameter, is determined in the solution for position.

When several receivers are used for relative positioning, with all stations collecting data simultaneously, the satellite clock error can be eliminated by taking the differences between the equations formed to define the ranges to two different satellites. This process of taking differences between range equations is referred to as forming *linear combinations of observables,* where the observables in this case are the pseudoranges (see Section 12.7).

Effects of Atmospheric Conditions on Signal Transmission

As the GPS signal travels through the atmosphere, ionospheric and tropospheric effects alter the velocity of the signal. In the upper atmosphere (heights of from 80 to 1000 km), the ionospheric error is proportional to the ion content of the medium through which the signal travels and inversely proportional to the square of the signal frequency. By making measurements of both the L1 and L2 carrier frequencies, the constant of proportionality can be calculated, allowing the ionospheric error to be modeled and the measurements corrected. Thus, dual frequency receivers have a definite advantage. For single frequency receivers, ionospheric correction terms are contained within subframe 1 of the data message. However, because the ion content is quite variable, with daily and seasonal variations, modeling this error is difficult.

In the lower atmosphere, tropospheric errors caused by water vapor pressure, temperature, and the pressure of the lower atmosphere are independent of the signal frequency. Tropospheric errors are corrected by using mathematical models. Not tracking satellites at low elevation angles also will help avoid major problems.

In relative positioning, ionospheric and tropospheric errors are minimized by forming linear combinations of the range or phase observables and taking the differences between appropriate equations (see Section 12.7). In addition, mathematical models can be used to estimate the ionospheric and tropospheric errors.

Receiver Timing and Other Errors

Receiver clocks often are less accurate or stable than the atomic clocks in the satellites. The receiver-clock errors are difficult to model and usually are included as unknowns in the equations for the position estimation process, where they are found in the solution. In relative surveying, receiver-clock errors can be eliminated by forming the differences between linear combinations of the range and phase observables (see Section 12.7).

Antenna phase center variations result when the electrical center of the antenna varies for signals arriving from different azimuths and elevation angles. Effects can be minimized by orienting all of the antennas in the same direction during the observation session. This causes the error to be minimized.

Site-Dependent Errors

Site-dependent errors can be minimized through careful procedures and proper site reconnaissance.

Height of instrument (h.i.) The incorrect measurement of the distance between the survey point and the antenna phase center (in the antenna) is a common error source. This distance often is measured as a slant height. To compute the vertical height, the radius from the phase center to the outside edge of the antenna or antenna ground plane (Section 12.5) must be available (Figure 12.3). The slant height should be measured at several locations around the antenna to avoid errors. If these height of instrument (h.i.) measurements do not agree, the operator should determine if the antenna is level or if the antenna ground plane has been warped. Measurements of the h.i. should be taken before and after each observing session to reduce the possibility of mistakes. Measuring h.i.'s in different units (m, ft) provides a good check for mistakes such as transposed digits or misplaced decimals.

Satellite geometry. Observation sessions that take place during times of "poor" satellite geometry can significantly affect the results. The dilution of precision (DOP) is a measure of satellite geometry. A favorable or low DOP exists if the satellites are distributed throughout the horizon and at different elevation angles. A large DOP may exist when the satellites are poorly distributed throughout the horizon or when only a few satellites are at the same elevation angles. Figure 12.4 illustrates satellite constellations with high and low DOPs. Output from position computations is used to evaluate DOP factors. A value for the DOP can be computed by forming the observation equations in Section 12.7 and solving for

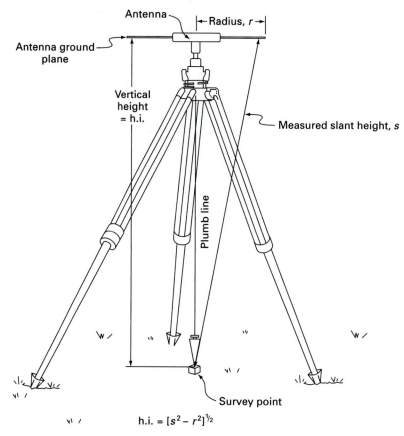

FIGURE 12.3
Measurement of slant height to determine antenna h.i.

the unknown receiver position (ϕ_A, λ_A, h_A) and receiver-clock error (δt_A). The DOP is the square root of the trace (Appendix B, Section B.3) of the cofactor matrix (Section 2.18, Equation (2.30)) of adjusted coordinates and clock error, (Q_{xx}):

$$DOP = [\text{Tr}(Q_{xx})]^{1/2} = [\text{Tr}(1/\sigma_0^2 \Sigma_{xx})]^{1/2} \tag{12.2}$$

in which

σ_0^2 = estimated reference variance
Σ_{xx} = covariance matrix for $(\phi_A, \lambda_A, h_A, \delta t_A)$ (Section 2.18)

DOP factors also can be determined and are given as functions of position, height, and time as follows:

$$\text{Geometric dilution of precision (GDOP)} = \frac{\sqrt{\sigma_\phi^2 + \sigma_\lambda^2 + \sigma_h^2 + c^2 \sigma_{\delta tA}^2}}{\sigma_0}$$

$$\text{Position dilution of precision (PDOP)} = \frac{\sqrt{\sigma_\phi^2 + \sigma_\lambda^2 + \sigma_h^2}}{\sigma_0}$$

$$\text{Horizontal dilution of precision (HDOP)} = \frac{\sqrt{\sigma_\phi^2 + \sigma_\lambda^2}}{\sigma_0}$$

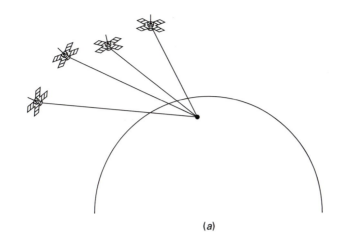

(a)

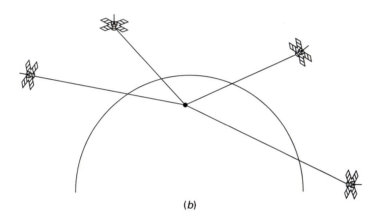

(b)

FIGURE 12.4
(a) Poor (high) PDOP. (b) Good (low) PDOP.

$$\text{Vertical dilution of precision (VDOP)} = \frac{\sigma_h}{\sigma_0}$$

$$\text{Time dilution of precision (TDOP)} = \frac{\sigma_{\delta tA}}{\sigma_0} \qquad (12.3)$$

where (ϕ, λ, h) = latitude, longitude, and ellipsoid height for the point (Appendix C, Figure C.5)

c = speed of light in a vacuum

$\sigma_\phi, \sigma_\lambda, \sigma_h$ = standard deviations in latitude, longitude, and ellipsoid height

$\sigma_{\delta tA}$ = standard deviation of the receiver-clock error

σ_0 = a priori standard deviation of the ranges

The individual DOP values (GDOP, PDOP, etc.) are evaluated and used in planning for GPS surveys, as outlined in Section 12.14. The positional error can be computed by multiplying the individual DOP value by the a priori covariance matrix for the ranges ($\Sigma_{xx} = \sigma_0 Q_{xx}$).

Obstructions. Obstructions from cultural (buildings, bridges, and the like) and natural (trees, hills) features affect the reception and quality of the incoming GPS signal.

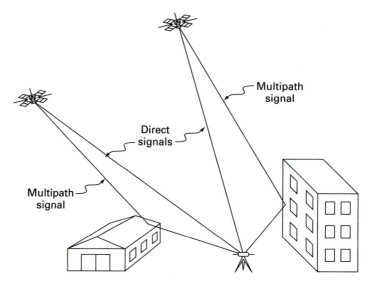

FIGURE 12.5
Multipath errors.

Cycle slips are the temporary loss of the GPS signal during the observation session. Cycle slips may exist when sites are heavily obstructed. *Multipath* errors occur when the GPS signal is reflected from structures and the reflected signal is received at the antenna source (Figure 12.5). Multipath errors can be avoided by selecting a site "free" of obstructions and reflective surfaces. Observing satellites at low elevation angles may increase the incidence of multipath errors.

12.7
POSITIONING METHODS

The positional accuracy that the SPS user obtains depends on the GPS frequencies or codes utilized (L1 and L2 carriers, C/A, and P codes) and the surveying technique employed (point positioning or relative positioning). First, consider the simplest case of point positioning.

Method 1: Point Positioning

One receiver is used in point positioning to determine the three-dimensional position of an unknown location by making measurements of the PRN (C/A or P) code phases. The GPS receiver correlates the incoming PRN code with a similar signal generated in the receiver, yielding the time of transmission of the signal. The difference between the received and transmitted time multiplied by the vacuum speed of light is the pseudorange between the transmitting satellite and the receiver. This quantity is referred to as a *pseudorange* because it is affected by timing errors. The accuracy of the pseudorange also is affected by ionospheric and tropospheric errors, multipath errors, and receiver noise. The timing errors are due to a very small residual satellite-clock error (an atomic clock) and a considerably larger clock offset in the less accurate clock in the receiver. The satellite and receiver timing errors (δt^i, δt_A) combine to form a resultant timing error dt. Figure 12.6 illustrates three pseudoranges that fail to intersect at a single point P due to timing errors.

To determine a point position, satellite orbital parameters are required. Table 12.2 lists the ephemeris parameters from subframes 2 and 3 contained within the data message. The

TABLE 12.2
Ephemeris data definitions (*adapted from NATO Navstar*, 1991)

Parameter	Definition
IODE	Issue of data, ephemeris
C_{rs}	Amplitude of the sine harmonic term to the orbit radius
Δ_n	Mean motion difference from computed value
M_0	Mean anomaly at reference time
C_{uc}	Amplitude of the cosine harmonic correction term to the argument of latitude
e	Eccentricity
C_{us}	Amplitude of the sine harmonic correction term to the argument of latitude
$(A)^{1/2}$	Square root of the semimajor axis
t_{0e}	Reference time ephemeris
C_{ic}	Amplitude of the cosine harmonic correction term to the angle of inclination
Ω_0	Longitude of ascending node of orbit plane at weekly orbit
C_{is}	Amplitude of the sine harmonic correction term to the angle of inclination
i_0	Inclination angle at reference time
C_{rc}	Amplitude of the cosine harmonic correction term to the orbit radius
ω	Argument of perigee
$\dot{\Omega}$	Rate of right ascension
IDOT	Rate of inclination angle

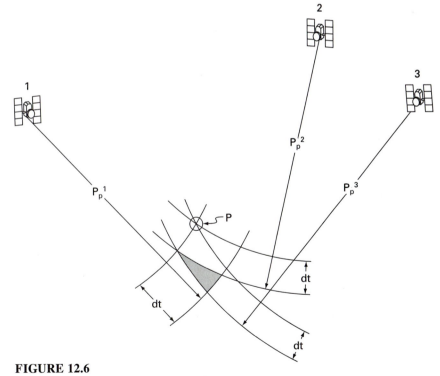

FIGURE 12.6
Three pseudoranges from three satellites.

TABLE 12.3
Elements of coordinate systems (*adapted from NATO Navstar, 1991*)

$\mu = 3.986005 \times 10^{14}\,\text{m}^3/\text{sec}^2$	WGS 84 value of the earth's universal gravitational constant
$\Omega_e = 7.2921151467 \times 10^{-5}$ rad/sec	WGS 84 (World Geodetic System, Chapter 14, Section 14.2) value of the earth's rotation rate
$A = (\sqrt{A})^2$	Semimajor axis
$n_0 = \sqrt{\mu/A^3}$	Computed mean motion
$t_k = t - t_{0e}\,*$	Time from ephemeris epoch
$n = n_0 + \Delta n$	Corrected mean motion
$M_k = M_0 + n t_k$	Mean anomaly
$M_k = E_k - e \sin E_k$	Kepler's equation for eccentric anomaly (may be solved by iteration)
$V_k = \tan^{-1}\left(\dfrac{\sin V_k}{\cos V_k}\right) = \tan^{-1}\left(\dfrac{\sqrt{(1 - e^2}\,\sin E_k)/(1 - e \cos E_k)}{(\cos E_k - e)/(1 - e \cos E_k)}\right)$	True anomaly
$E_k = \cos^{-1}\left(\dfrac{e + \cos V_k}{1 + \cos V_k}\right)$	Eccentric anomaly
$\phi_k = V_k + \omega$	Argument of latitude
$\delta u_k = C_{us} \sin 2\phi_k + C_{uc} \cos 2\phi_k$	Argument of latitude correction ⎫ Second
$\delta r_k = C_{rc} \cos 2\phi_k + C_{rs} \sin 2\phi_k$	Radius correction $\Bigr\}$ harmonic
$\delta i_k = C_{ic} \cos 2\phi_k + C_{is} \sin 2\phi_k$	Correction to inclination ⎭ perturbations
$u_k = \phi_k + \delta u_k$	Corrected argument of latitude
$r_k = A(1 - e \cos E_k) + \delta r_k$	Corrected radius
$i_k = i_0 + \delta i_k + (\text{IDOT})t_k$	Corrected inclination
$\left.\begin{array}{l} x_k' = r_k \cos u_k \\ y_k' = r_k \sin u_k \end{array}\right\}$	Positions in orbital plane
$\Omega_k = \Omega_0 + (\dot{\Omega} - \Omega_e)\,t_k - \Omega_e t_{0e}$	Corrected longitude of the ascending node
$\left.\begin{array}{l} x_k = x_k' \cos \Omega_k - y_k' \cos i_k \sin \Omega_k \\ y_k = x_k' \sin \Omega_k + y_k' \cos i_k \cos \Omega_k \\ z_k = y_k' \sin i_k \end{array}\right\}$	Earth-fixed coordinates

*t is the GPS system time at time of transmission; that is, the GPS time corrected for transit time (range/speed of light). Furthermore, t_k will be the actual total time difference between the time t and the epoch time t_{0e} and must account for beginning or end of week crossovers. That is, if t_k is greater than 302,400 sec, subtract 604,800 sec from t_k. If t_k is less than 302,400 sec, add 604,800 sec to t_k.

ephemeris parameters must be converted to Cartesian coordinates before the unknown receiver location is computed. The relations and algorithms in Table 12.3 (bottom of Table 12.3, Equations for x_k, y_k, and z_k) can be used to transform the ephemeris parameters to Cartesian coordinates.

Observational model for point positioning. The unknown receiver location (X, Y, Z or ϕ, λ, h) is computed by distance resection using pseudoranges from a minimum of four satellites (Figure 12.7). A minimum of four satellites is required to compute the unknown three-dimensional receiver coordinates of the antenna center and receiver-clock error ($X_A, Y_A, Z_A, \delta t_A$). The pseudorange observable is defined as

$$P_A^i = \rho_A^i + c\,(\delta t^i - \delta t_A) + \delta_{A\,\text{iono}}^i + \delta_{A\,\text{trop}}^i + \varepsilon \tag{12.4}$$

in which the superscripts designate the satellites and the subscripts denote the receiver and

$$P_A^i = c\,(t_A - t^i) = \text{measured pseudorange from satellite } i \text{ to receiver } A \text{ (m)}$$
$$t_A = \text{time of reception of signal (sec)}$$
$$t^i = \text{time of transmitting signal (sec)}$$

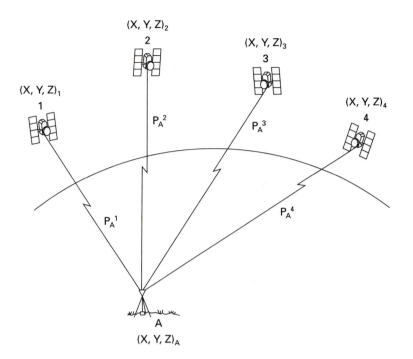

FIGURE 12.7
Four pseudoranges from four satellites.

$$c = \text{vacuum speed of light (m/sec)}$$
$$\rho_A^i = \text{geometric range (m)} = [(X^i - X_A)^2 + (Y^i - Y_A)^2 + (Z^i - Z_A)^2]^{1/2} \text{ in which}$$
$$X^i, Y^i, Z^i = \text{satellite location and } X_A, Y_A, Z_A = \text{receiver location}$$
$$\delta t^i = \text{satellite-clock error (sec)}$$
$$\delta t_A = \text{receiver-clock error (sec)}$$
$$\delta_{A\,iono}^i = \text{ionospheric error (m)}$$
$$\delta_{A\,trop}^i = \text{tropospheric error (delay) (m)}$$
$$\varepsilon = \text{receiver noise (m)}$$

The ionospheric error can be reduced using corrections transmitted within subframe 1. Also, given a dual frequency receiver, frequencies L1 and L2 can be used to compensate for the ionospheric error. The effects of tropospheric and ionospheric delays can be reduced by not observing satellites at low elevation angles. A correction for the tropospheric error is possible using a mathematical model. The satellite-clock error can be corrected using parameters from subframe 1 of the data message. After making these corrections, the equation for the range is

$$P_A^i = [(X^i - X_A)^2 + (Y^i - Y_A)^2 + (Z^i - Z_A)^2]^{1/2} - c\delta t_A \qquad (12.5)$$

When the ranges from four satellites are observed simultaneously, four Equations (12.5) are formed and the unknown receiver location (X_A, Y_A, Z_A) and clock error, δt_A, for that epoch can be computed. If ranges are measured to more than four satellites, a least-squares adjustment is possible.

Accuracy of point positioning. SA restricts the instantaneous point positioning accuracy of the SPS user to 100 m with a probability of 95 percent. If accuracies of less than 100 m are required, the user is advised to utilize relative positioning.

Method 2: Relative Positioning

Two or more GPS receivers simultaneously receiving signals from the same satellites are needed for relative positioning, as shown in Figure 12.8. This method is the most accurate procedure available to the surveyor, engineer, geodesist, and geographer for establishing spatial position. Because observational precision increases with a decrease in wavelength, the accuracy achieved depends on whether the carrier waves or the PRN codes are used as the primary signal measured. The L1 and L2 carriers, with respective wavelengths of 0.19 and 0.24 m, are the "precise tapes" capable of providing centimeter- to millimeter-level positioning. The C/A and P codes, with respective wavelengths of 300 and 30 m, are the "yardsticks" capable of providing decimeter- to meter-level relative positioning.

> ***Relative positioning by carrier-phase measurements.*** The drawback to using carrier-phase measurements is that the signal is ambiguous. The receiver accurately measures the phase difference (fractional phase) between the incoming signal from the satellite and a similar signal generated by the oscillator within the receiver. However, the receiver is unable to determine the whole number of wavelengths between the satellite and the receiver, a factor called *integer ambiguity* or simply *ambiguity*. If ambiguity is resolved correctly, relative positioning using carrier phases provides the user with millimeter to centimeter results.

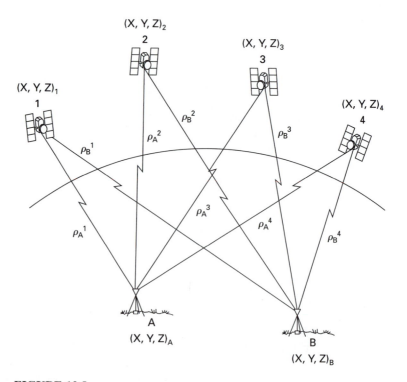

FIGURE 12.8
Eight pseudoranges from four satellites to two stations for relative positioning.

Observational model for carrier-phase positioning. The equation for carrier-phase measurement used in relative positioning and scaled into cycles is

$$\phi_A^i = f\delta^i - (1/\lambda)(\rho_A^i) - f\delta_A + N_A^i - f\delta_{A\,iono}^i + f\delta_{A\,trop}^i + \varepsilon \qquad (12.6)$$

in which ϕ_A^i = carrier phase measurement, in cycles, for satellite i and receiver A
ρ_A^i = geometric range, in m, as defined for Equation (12.4)
δ^i = satellite-clock error (sec)
δ_A = receiver-clock error (sec)
f = the fundamental satellite frequency (Hz)
λ = carrier wavelength, in m (L1 or L2)
c = vacuum speed of light (m/sec)
N_A^i = integer ambiguity (cycles)
$\delta_{A\,iono}^i$ = ionospheric correction (cycles)
$\delta_{A\,trop}^i$ = tropospheric delay (cycles)
ε = receiver noise (cycles)

The mathematical sign on the ionospheric term and the incorporation of the integer ambiguity are the major differences between the carrier-phase (Equation (12.6)) and pseudo-range equations (Equation (12.4)). The reason for the change in the sign of the ionospheric correction is that the phase of the carrier wave is advanced by a higher density of ions in the upper atmosphere. Conversely, the pseudo-random noise signal modulated onto the carrier wave is delayed by the ionosphere. Linear combinations of the carrier-phase equation are formed to eliminate some of the unknowns and facilitate the estimation process. The sections that follow formulate the single, double, and triple difference linear combinations. To simplify these equations, terms for the ionospheric and tropospheric corrections and receiver noise ($\delta_{A\,iono}^i$, $\delta_{A\,trop}^i$, ε) are omitted.

Single difference. Consider two receivers (A and B) simultaneously tracking satellite i (Figure 12.9). The (between station) single difference is formed by subtracting the carrier-phase Equation (12.6) for the range between receiver A and satellite i from the carrier-phase Equation (12.6) for the range between receiver B and satellite i. This single difference equation is

$$\Delta\phi_{AB}^i = \phi_B^i - \phi_A^i$$

$$\Delta\phi_{AB}^i = \left(f\delta^i - \frac{\rho_B^i}{\lambda} - f\delta_B + N_B^i\right) - \left(f\delta^i - \frac{\rho_A^i}{\lambda} - f\delta_A + N_A^i\right)$$

$$\Delta\phi_{AB}^i = \left(\frac{\rho_A^i - \rho_B^i}{\lambda}\right) - f(\delta_B - \delta_A) + \Delta N_{AB}^i \qquad (12.7)$$

Note that the single difference eliminates the satellite-clock offset (δ^i). In addition, the ephemeris error contained within each range (ρ_A, ρ_ε) is eliminated in the single difference.

Double difference. The double difference consists of taking differences between satellites and receivers (Figure 12.10). The double difference operator is defined by the symbol $\nabla\Delta\phi_{AB}^{ij}$ where the superscript j denotes the reference satellite. Normally, the same reference satellite is used in the formulation of each double difference observable. The reference satellite is selected on the basis of signal strength, elevation angle, and length of time the satellite is visible during the observation session. The double difference equation, as formed

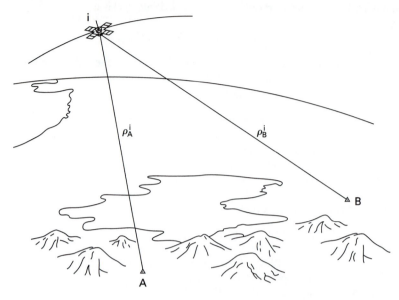

FIGURE 12.9
Single difference, satellite-clock offset eliminated.

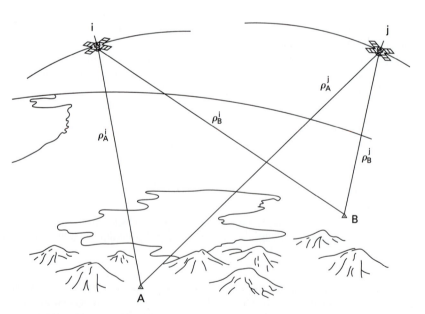

FIGURE 12.10
Double difference, receiver and clock offsets cancel.

for two receivers at A and B receiving signals from satellites i and j, as illustrated in Figure 12.10, is

$$\nabla\Delta\phi_{AB}^{ij} = \Delta\phi_{AB}^{j} - \Delta\phi_{AB}^{i}$$

$$\nabla\Delta\phi_{AB}^{ij} = -\left[\frac{(\rho_B^j - \rho_A^j)}{\lambda} - f(\delta_B - \delta_A) + \Delta N_{AB}^j\right]$$

$$-\left[-\frac{(\rho_B^i - \rho_A^i)}{\lambda} - f(\delta_B - \delta_A) + \Delta N_{AB}^i\right]$$

$$\nabla\Delta\phi_{AB}^{ij} = -\left[\frac{(\rho_B^j - \rho_B^j - \rho_B^i + \rho_A^i)}{\lambda} + \nabla\Delta N_{AB}^{ij}\right] \tag{12.8}$$

The double difference eliminates the receiver-clock error (δ_A, δ_B).

Triple difference. The conditions for a triple difference equation are shown in Figure 12.11, where receivers at stations A and B receive signals first from satellites i_1 and j_1 (the first epoch) and then from the same satellites at i_2 and j_2 (the second epoch). Subtraction of the double difference Equation (12.8) for the first epoch $(\nabla\Delta\phi_{AB_1}^{ij})$, from Equation (12.8) for the second epoch $(\nabla\Delta\phi_{AB_2}^{ij})$ yields a triple difference equation defined by the symbol $\delta\nabla\Delta\phi_{AB}^{ij}$. If a cycle slip has not occurred, the integer ambiguity between the satellite and receiver will remain constant throughout the observing session. The triple difference uses this premise to eliminate the unknown integer ambiguity (N_{AB}^{ij}). The triple difference equation is

$$\delta\nabla\Delta\phi_{AB}^{ij} = \nabla\Delta\phi_{AB_2}^{ij} - \nabla\Delta\phi_{AB_1}^{ij}$$

$$\delta\nabla\Delta\phi_{AB}^{ij} = -\left[\frac{(\rho_B^{j_2} - \rho_A^{j_2} - \rho_B^{i_2} + \rho_A^{i_2})}{\lambda} + \nabla\Delta N_{AB}^{ij}\right]$$

$$-\left[-\frac{(\rho_B^{j_1} - \rho_A^{j_1} - \rho_B^{i_1} + \rho_A^{i_1})}{\lambda} + \nabla\Delta N_{AB}^{ij}\right]$$

$$\delta\nabla\Delta\phi_{AB}^{ij} = -\left[\frac{(\rho_B^{j_2} - \rho_A^{j_2} - \rho_B^{i_2} + \rho_A^{i_2} - \rho_B^{j_1} + \rho_A^{j_1} + \rho_B^{i_1} - \rho_A^{i_1})}{\lambda}\right] \tag{12.9}$$

Note that the ambiguity term cancels. The unknown parameters in Equation (12.9) are either the coordinates of the stations $(X_A, Y_A, Z_A,)$ and (X_B, Y_B, Z_B) or the base-line components $(dx_{AB}, dy_{AB}, dz_{AB})$.

Data processing for carrier-phase measurements. Systems of single difference, triple difference, and double difference equations are formed and solved for base-line vectors using an estimation technique such as least squares or Kalman filtering. In one technique used in GPS processing, a combination of triple differences and double differences is formed to compute the unknown base line. Figure 12.12 outlines a flowchart that illustrates the processing algorithm. First, triple difference equations are formed and solved for the base-line components. Examination of the residuals from this solution allows determination of the presence of cycle slips. When identified, cycle slips are corrected and the data reprocessed. The estimated base lines from this second solution of triple difference equations provide initial approximations for the double difference processing in which cycle slips also have been corrected. The unknown integer ambiguities can be found in this first solution (sometimes called a *float solution*) of double difference equations. If the ambiguities are "close" to integers, they are fixed and the double differencing processing is repeated

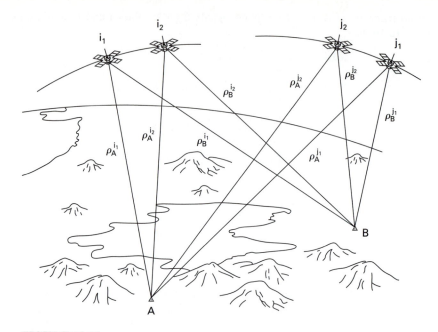

FIGURE 12.11
Triple difference, phase ambiguity eliminated.

to determine improved estimates for the unknown base line. This last process is called the *fixed ambiguity solution.* Output from this sequence of solutions consists of base-line vectors for a line from A to B expressed as the components, dx_{AB}, dy_{AB}, dz_{AB}.

Relative positioning using the carrier phase provides millimeter to centimeter accuracy in positioning. This accuracy depends to a great extent on the correct determination of the integer ambiguities.

Code-phase differential. Differential GPS (DGPS), also known as *pseudorange* or *code-phase differential,* signifies relative positioning using the C/A or P code-phase observables (Equation (12.4)). DGPS can provide the user postprocessed or real-time, three-dimensional positions. Antispoofing (A-S) limits the availability of P code observations; therefore, P code DGPS seldom is available for civilian applications.

Observational model for code-phase differential. Using Equation (12.4), assuming the tropospheric and ionospheric errors have been corrected and no receiver noise, the simplified pseudorange observable is

$$P_A^i = \rho_A^i + c(\delta t^i - \delta t_A) \tag{12.10}$$

from which the single difference equation is

$$\Delta P_{AB}^i = [\rho_B^i + c(\delta t^i - \delta t_B) - (\rho_A^i + c(\delta t^i - \delta t_A))]$$
$$= (\rho_B^i - \rho_A^i) - c(\delta t_B - \delta t_A) \tag{12.11}$$

and the double difference becomes

$$\nabla\Delta P_{AB}^{ij} = [(\rho_B^j - \rho_A^j) - c(\delta t_B - \delta t_A)] - [(\rho_B^i - \rho_A^i) - c(\delta t_B - \delta t_A)]$$
$$\nabla\Delta P_{AB}^{ij} = (\rho_B^j - \rho_A^j) - (\rho_B^i - \rho_A^i) \tag{12.12}$$

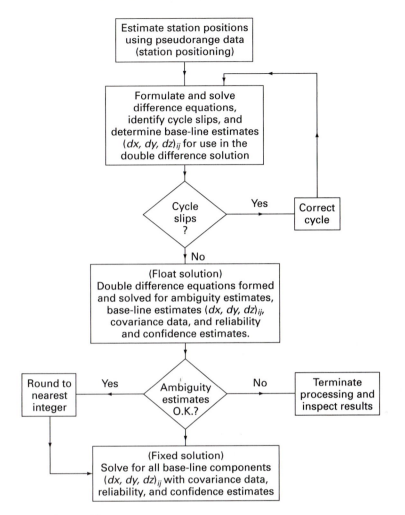

FIGURE 12.12
Data processing flowchart.

where the conditions governing Equations (12.11) and (12.12) are illustrated in Figures 12.9 and 12.10.

Data processing for differential GPS. To survey by DGPS using postprocessing, a reference receiver is placed over a known survey control point and the remote receivers occupy known or unknown "points of interest." Each receiver simultaneously collects data from a minimum of four satellites. After the data have been collected, data files are downloaded to a personal computer, and single and double differenced equations are formed and processed using an estimation algorithm (least-squares or Kalman filtering). Components dx_{AB}, dy_{AB}, dz_{AB}, for the base lines from the known reference station A to remote station B are computed and applied to the known coordinates of the reference station to provide positions for all remote stations relative to the known point.

If real-time positions are desired, the differences between known and computed coordinates at the reference station are used to determine range and range-rate corrections.

These corrections are formatted and transmitted to the remote receiver via a communications link. The range and range-rate corrections are received at the remote receiver and applied to measured ranges. In real-time operations, the reference station must have a processor capable of computing and assessing the quality of the range corrections and formatting the range corrections for transmission to the remote receivers. The Radio Technical Commission for Maritime (RTCM) Users Special Committee (SC 104) has developed a format for the transmission of real-time corrections.

Relative positioning using the code-phase differential is capable of providing the user with positions having accuracies in the decimeter to meter range.

12.8
STATIC SURVEYING

Static surveying is the most reliable relative surveying method. In this procedure, two or more GPS receivers at two or more stations, simultaneously receive signals from a minimum of four common satellites. Occupation time depends on the type of GPS receiver, the separation distance between receivers, the ionospheric activity, number of satellites, and their geometry. These times usually are at least 15 min for dual frequency receivers and 30 min for single frequency receivers.

As the GPS signal travels through the earth's atmosphere (ionosphere and troposphere), it is delayed and bent (Section 12.6). The ionospheric effect is removed by a mathematical model or postprocessing using data collected from dual frequency receivers where an observable, using the basic phase equation (Equation (12.6)) for the L1 and L2 frequencies, is free of ionospheric effects (see Gilbert, 1944). The use of dual frequency receivers is the most effective method of removing ionospheric errors, especially for base lines of 100 km or longer. In static relative positioning with single frequency receivers, modeling the ionospheric error is difficult due to the variable ion content of the ionosphere. However, during times of mild ionospheric activity, the ionospheric errors will be the same for both ranges and when single, double, or triple differences are taken for relative positioning, ionospheric effects will cancel if base lines are relatively short. Thus, single frequency GPS receivers are capable of obtaining geodetic quality data for base-line distances of <25 km.

The tropospheric error is partially removed by using a mathematical model and not observing satellites at low elevation angles.

When data collection is completed, the data files are downloaded to a personal computer and processed using an estimation algorithm. The result is a series of base-line components, $dx_{AB}, dy_{AB}, dz_{AB}$, between stations. After all of the vectors are processed, they are combined and adjusted utilizing a network least-squares adjustment program (Section 12.15, Step 3).

12.9
KINEMATIC SURVEYING

Kinematic surveying provides the user a productive and accurate method to establish relative survey control. In kinematic surveying, one or more reference receivers remain fixed during the observation period and one or more remote receivers are *rovers* that occupy *points of interest* for several minutes at each point. During this survey, the static and remote receivers must continuously track a minimum of the same four satellites. As an example, consider one fixed receiver and antenna and one roving receiver and antenna placed over two stations of known relative position, preferably at the ends of a short base line. For users who do not have software capable of *on-the-fly* (OTF) (Section 12.12) ambiguity resolution, the ambiguity initialization is accomplished by an *antenna swap* (Remondi, 1991), in which the antennas and receiver assemblies are interchanged between the two stations. Another option is to occupy a known base line and perform a static GPS survey or fixed base-line initialization. After initialization, the roving receiver and antenna are moved to the stations to be located and the fixed unit remains at its initial location. When all desired points have been occupied by the roving receiver, this unit and its antenna should then occupy the initial station or another station of known position to provide a closure error.

The drawback to kinematic surveying is that the receivers must maintain a lock to a minimum of four satellites. If loss of lock occurs when tracking four or fewer satellites and if the user lacks the software capable for OTF initialization, the remote station must return to one of the previously surveyed points or occupy a known survey point. This procedure is required to reestablish the ambiguities. Kinematic surveying requires integer ambiguity initialization prior to data processing. Because loss of lock is not tolerated when tracking four or fewer satellites, kinematic surveying is restricted to areas that have a clear view of the horizon (airports, parking lots, open fields).

When the survey is completed, the data files from the remote and reference receivers are downloaded to a personal computer and single, double, or triple difference measurements are processed using an estimation algorithm. It is recommended that the remote receivers occupy the points of interest several times to ensure redundancy and, therefore, reliability. Finally, a network adjustment is performed with all of the processed components for base lines.

12.10
PSEUDOKINEMATIC SURVEYING

Pseudokinematic surveying exploits the fact that a change in satellite configuration facilitates ambiguity resolution. Like kinematic surveying, pseudokinematic surveying consists of one or more reference receivers, which remain static during the observation period, and one or more remote receivers. The remote systems occupy points of interest for approximately 5 min and reoccupy these same points of interest after a substantial change in time (about 1 h). This time interval, which occurs between the first and second occupation, allows for a significant change in the satellite configuration. The disadvantages of pseudokinematic surveying are that the operator must reoccupy each station and losses of lock cannot be tolerated. Advances in initialization processes have made this technique less viable than kinematic (Section 12.9) or rapid static surveying (Section 12.11). However, a combination of kinematic and pseudokinematic surveying is an efficient procedure to provide accurate data on surveys where a large number of points need to be located.

12.11
RAPID STATIC SURVEYING

Rapid static surveying involves one or more reference receivers that remain fixed during the observation period and one or more roving receivers. The roving receivers occupy points of interest from 1 to 20 min per station. The unique characteristic of rapid static is a greatly reduced site occupation time, due to much faster ambiguity resolution through the utilization of dual code and carrier-phase data. These data allow for the formulation of additional linear combinations to determine the correct integer ambiguity. A dual frequency receiver capable of overcoming A-S (P code encryption) is the desired receiver architecture for rapid static applications. The occupation time and relative accuracy depend on the separation distance between the reference and remote receivers, number of satellites tracked, and the receiver architecture. In general, rapid static is effective at distances less than 25 km. The advantage of employing rapid static surveying is that loss of lock is tolerated and the remote receiver need not reoccupy each station. A combination of static, kinematic, and rapid static surveying has the best chance of providing the most efficient and economical execution of a given survey. Data processing is similar to that suggested for static, kinematic, and pseudokinematic surveying.

12.12
ON-THE-FLY AMBIGUITY RESOLUTION

Kinematic surveying with on-the-fly ambiguity resolution is the relative surveying technique that permits very rapid estimation of the integer ambiguities (Abidin, 1994). If loss of satellite lock occurs, the operator continues to survey and need not reoccupy a known location or previously located point. As in kinematic, pseudokinematic, and rapid static surveying, OTF consists of one or more reference and remote receivers.

The ambiguities are resolved by an algorithm that uses initial approximations for the position and ambiguity of the roving receivers by using the triple difference observable or high-quality code-phase measurements. Properties of the code and carrier combinations are exploited to determine the correct integer ambiguity. Various search techniques are used to determine the correct integers, and statistical tests are performed to ensure reliability of the estimated integers. OTF ambiguity resolution requires at least five visible satellites and can be accomplished with dual or single frequency receivers (Abidin, 1994, p. 49). The use of receivers with the capability of obtaining dual frequency code-phase and carrier-phase data greatly facilitates OTF methods. Consequently, implementation of antispoofing has a severe effect on those users not authorized to use the P code or lacking receivers able to overcome A-S through innovative signal processing techniques. Users who must work with L2 squared or single frequency receivers should use as many satellites as possible and have two or three receivers at known stations to get reliable results (Abidin, 1994, p. 49).

The OTF method lends itself to real-time applications since it does not require the user to be stationary or located over a known point. In real-time applications, the entire observation at the base station is transmitted via a communications link to the remote receiver. The base line is computed in real time at the remote receiver using the transmitted phase information from the reference receiver and the phase data collected at the remote receiver.

Under favorable conditions, OTF kinematic methods can yield accuracies to within centimeters.

12.13
APPLICATIONS

GPS techniques have innumerable applications in the fields of surveying, geodesy, and mapping. Before discussing these possibilities, consider the potential of a permanent reference station.

DGPS, static, kinematic, pseudokinematic, and kinematic surveying with OTF initialization require at least one stationary reference receiver and one or more remote receivers.

This commonality in conjunction with the proliferation of GPS receivers throughout the surveying and mapping community has aroused interest in the establishment of fixed reference stations. Fixed reference stations consist of a GPS receiver, antenna assembly, personal computer, and communications link (real-time applications only). They are called *continuously operating reference stations* (CORS). The GPS antenna assembly is set up over a presurveyed location that is clear of obstructions. The GPS receiver collects data 24 hours a day or at predefined time intervals (i.e., 8:00 A.M.–5:00 P.M.). The type of GPS receiver utilized depends on the positional accuracy desired (meter or centimeter level) and the distances between the reference and remote receivers. At predefined intervals (hourly, daily) the data from the reference station are converted from the manufacturers proprietary binary format to the receiver-independent exchange (RINEX) format. RINEX is an ASCII format supported by most receiver manufacturers. The RINEX data are archived and distributed to various users in an on-demand basis (daily, weekly, or monthly). The data collected at the reference station also could be accessed via modem using a bulletin board service (BBS). The remote receivers convert the reference-station derived RINEX files to the necessary file formats required by the remote receiver software. Most GPS manufacturers provide software routines capable of converting RINEX-derived files to their own proprietary formats. Depending on the hardware architecture (single versus dual frequency or code versus carrier tracking) of the reference and remote receivers, the base line can be processed utilizing carrier- or code-phase measurements. If real-time applications are desired, a communications link capable of transmitting digital data is required. A fixed reference station network would reduce the number of GPS receivers a government agency or contractor must procure. This type of network also would provide the surveyor, navigator, or geographer a common geodetic framework, therefore, reducing the use of local coordinate systems.

Geodetic Surveying

GPS has been utilized quite frequently to establish geodetic control to strengthen the National Geodetic Reference System, statewide high-accuracy geodetic networks, photogrammetric control (Section 13.22), and control to determine the relative movement of structures (e.g., dams, buildings). Static, kinematic, pseudokinematic, and OTF positioning techniques can be used for these operations.

Photogrammetry

Techniques such as rapid static and kinematic GPS are appropriate for establishing ground control for the mapping process (Section 14.13). For photogrammetric triangulation (Section 13.22), a GPS receiver can be mounted in an aircraft and used to determine the coordinates of the exposure station. DGPS or kinematic positioning techniques are used to position the aircraft, thus reducing or even eliminating the amount of ground control required for the task. The technique used would depend on the scale of the photography. For small-scale mapping ($<$1:24,000), DGPS with accuracies at the decimeter to meter level would be adequate. For larger scales ($>$1:24,000), relative positioning techniques with OTF initialization are more appropriate.

Land Surveying

Relative GPS positioning techniques provide the surveying community an efficient and accurate method of surveying land parcels. When performing boundary surveys, most land surveyors establish coordinates on land corners to derive the direction (bearing or azimuth) and the distance between these corners. The land surveying community often uses conventional surveying equipment (theodolites, transits, tapes, EDM equipment, and total station systems) and techniques (traversing) to establish land boundaries. In some cases, GPS could provide the land surveyor a more efficient and accurate method of establishing land corners. If a fixed reference station network were available, the land surveyor could procure or lease one GPS receiver that would be used as a remote receiver. The remote receiver could be placed on those land corners that are suitable GPS sites. The data from the reference station and the remote receiver would be processed and adjusted at the land surveyor's office. Each station occupied by the remote receiver would be positioned relative to the fixed reference station. A fixed reference station network would provide the land surveyor a method of establishing reliable coordinates on boundary corners for land information system (LIS) applications.

Location of Controls for Geographic Databases

The accuracy and versatility of GPS has spurred the curiosity of nontraditional geodetic users, such as the geographer, environmentalist, and land surveyor. Members of these disciplines will constitute the majority of the future GPS users.

The GPS is capable of providing the users of geographic information systems (GIS) timely and accurate horizontal positioning data. DGPS is capable of providing decimeter- to meter-level positions relative to the reference station. These data can be provided through postprocessing or in real time via a communications link. The user can assign attribute data to these positions for database compilation, feature identification (vegetation, wildlife habitat, geologic features), and the delineation of ecosystems (wetland boundaries). Many GPS manufacturers provide software routines capable of converting the GPS positional data to line or point data that can be accepted by various GIS manufacturers.

Orthometric Heights

GPS-derived coordinates and base lines are referenced to the WGS 84 reference ellipsoid (Section 14.2), if the broadcast orbits are used. In relative positioning, the lengths of these base lines can be determined to within millimeters. However, GPS-derived heights are referenced to the ellipsoid and the majority of the surveying community is interested in orthometric heights ($\simeq$ mean sea level), which are a function of ellipsoid and geoid heights. Figure 12.13 illustrates the relationship between the reference ellipsoid, geoid, and topographic surface. The separation between the geoid and the ellipsoid (N) is required to transform the ellipsoidal heights (h) to orthometric heights (H). Geoidal heights (N) are computed from gravity observations and interpolated from mathematically derived models (Schwartz, 1994). To obtain orthometric elevations from GPS-derived heights, one must know this separation N from which is derived the orthometric elevation, $H = h - N$ (the angle between these lines is very small, $<1°$, and therefore its effect is negligible) and heights referred to the NAUD 83; it is possible to achieve relative orthometric elevations from GPS data in the 3–5 cm accuracy range (Reilly, 1997).

In the United States, the NGS provides a geoid grid in digital* or graphic form. Using this grid, one can determine the geoidal undulation for a given location and apply the

---

*A computer program on a diskette can be obtained from NOAA, NGS, N/CG174, 1315 East-West Highway, Station 9202, Silver Spring, MD 20910-3282.

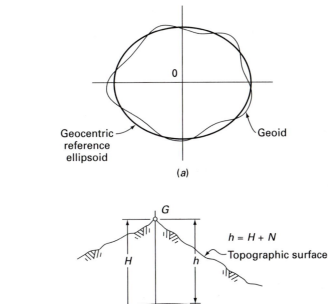

$$h = H + N$$

FIGURE 12.13
Horizontal and vertical datums.

correction to the height determined by GPS to yield an orthometric elevation for the point. Using currently available geoidal undulations, supplemented by regional and local gravity data and heights referred to the NAVD 83, it is possible to achieve relative orthometric elevations from GPS data in the 3–5 cm accuracy range (Reilly, 1997).

Topographic, Route, and Construction Surveys

Relative positioning using static and rapid static GPS techniques is very adaptable to the primary and secondary control networks needed for topographic, route, and construction surveys, which are described in Chapters 15, 16, and 17.

Kinematic, pseudokinematic, and OTF kinematic GPS techniques provide excellent tools for the data collection phases of topographic and route surveys. Problems could occur in areas having obstacles that cause loss of lock in the purely kinematic approach or prevent an adequate number of visible satellites for the OTF kinematic method. Careful planning to exploit the power of GPS in clear, open areas would be necessary.

Kinematic surveying with OTF ambiguity resolution has potential applications in construction layout, where topographical (obstructions, electromagnetic noise, and economic) conditions permit. Some GPS receivers are integrated with data collectors to facilitate data collection (topography) for the entry of feature codes and attributes. Figure 12.14 shows a real-time surveying system with data collector, GPS receiver, and

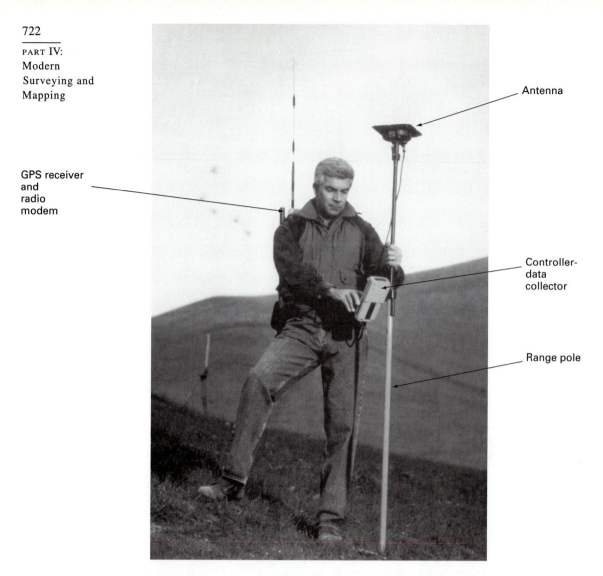

GPS receiver
and
radio
modem

Antenna

Controller-
data
collector

Range pole

FIGURE 12.14
Real-time GPS surveying system. (*Courtesy of Trimble Navigation Limited.*)

antenna assembly. Data collection can be done through postprocessing the GPS data then transferring the positions, feature codes, and attributes to the user's automated mapping program.

Location or the perpetuation of points (stake-out) requires real-time kinematic techniques. In this case, a handheld graphic or text display navigates the user by bearing and distance or station and offset to the point to be staked out or perpetuated. The economic advantage of using GPS for data collection and stake-out is that a line of sight need not be maintained between the GPS base station and the remote receiver, unlike conventional surveying. However, one does need to ensure adequate radio communication with the base station to get real-time corrections.

Use of GPS methods for construction layout is evolving both in the GPS design and manufacturing sector and in the construction industry. Ultimately, GPS techniques will have a major impact on construction layout (Section 17.4).

Deformation Monitoring

723

CHAPTER 12:
Global
Positioning
System

Relative positioning techniques using the GPS are an effective method to determine the movement of natural or human-made structures. GPS base lines are observed from fixed stations to the locations on the structure (levee, dam) or topographic surface where the monitoring is being observed. These fixed stations should be located in stable areas not affected by the movement exhibited at the monitoring station. The base lines could be observed on a daily, weekly, or seasonal basis to determine the magnitude, direction, and rate of movement. Because relative movement is desired, it is not necessary that the fixed stations have known coordinate values.

12.14
PLANNING AND DATA ACQUISITION

The planning and data acquisition process consists of equipment selection, point selection, determination of satellite availability, site reconnaissance, and selection of the survey scheme.

Equipment Selection

Prior to the survey, selection of the proper equipment is essential, with emphasis placed on the special requirements of the project. The required accuracy, number of stations, distance between stations, primary use of the receivers(s), and the intended use of the positional data all are factors to be considered. For example, for short base lines, of up to about 25 km, single frequency receivers are adequate, and if kinematic surveys are to be performed, a system having a high sampling rate with a separate antenna is advisable. For very precise geodetic or deformation surveys, a dual frequency receiver with P code capability is needed, bearing in mind the problems in this type of receiver when antispoofing is activated.

Most of the geodetic- and surveying-quality systems marketed provide very good results, so ultimately the overriding factors governing selection of a system will be cost and ease of use. Detailed characteristics and costs of receivers are given annually in *GPS World* and *P.O.B., Point of Beginning* (e.g., GPS World, 1997 and Reilly, 1996, respectively).

Point Selection

Acquire the best available map of the area under study (e.g., the U.S.G.S. quadrangle maps at 1:24,000, see Chapter 14) and information concerning all existing control points (Section 3.4). Plot the known points and proposed locations for desired points on this map. For the proposed points, keep in mind that an unobstructed view of the sky is advantageous and accessibility to a location not easily disturbed is desirable if not mandatory. Figure 12.15 illustrates a portion of a typical planning map on which the circles (*GG, GH*, etc.) are proposed stations and the triangles at 08 and 06 represent known control points. Written descriptions of how to reach the known and proposed locations also should be prepared to help in the reconnaissance when locations are verified on the ground.

Satellite Availability and Geometry

At this stage, one should use the planning software either provided by the manufacturer of the GPS equipment or programmed in-house by the surveying organization. Input to these programs consists of the latitude and longitude of the proposed site and estimated date of

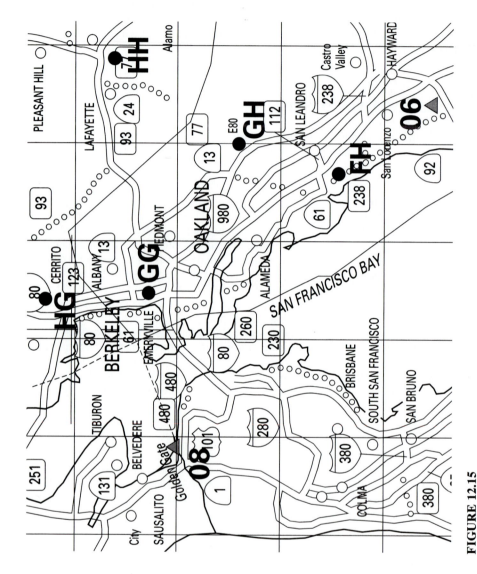

FIGURE 12.15
Location map for a GPS survey. (*Courtesy of California Department of Transportation.*)

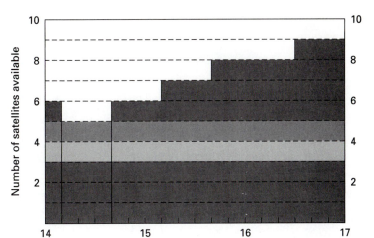

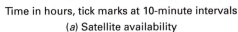

(a) Satellite availability

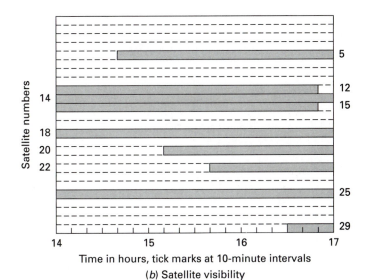

(b) Satellite visibility

FIGURE 12.16
Location Berkeley, CA (latitude 37°52′N, longitude 122°15′W) for a
window of 14:00 to 17:00 PST, December 19, 1994, with a cutoff
vertical angle = 15°, no obstacles, and satellites 5, 12, 14, 15, 18, 20,
22, 25, 29 used (plots were produced using the Leica SKI software).

the survey. The program will provide charts, diagrams, or tables that furnish information on
satellite availability and visibility. Charts of availability and visibility for satellites at a specified date and window of time are shown in Figure 12.16. Note that, in Figure 12.16a, at
least five satellites are visible for the entire time period and six to eight are available from
14:40 to 17:00 Pacific standard time. Figure 12.16b shows that these satellites are numbered 5, 12, 14, 15, 18, 20, 22, 25, and 29. Although the entire period can be used to collect
data, 14:40 to 17:00 is the optimum observation interval at this time span, date, and location.

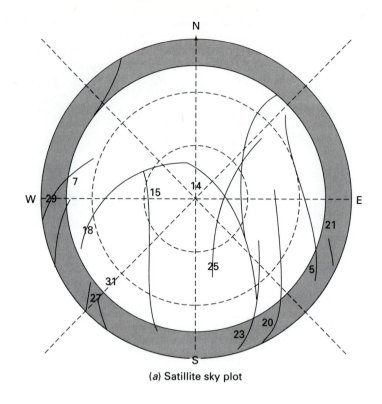

(a) Satillite sky plot

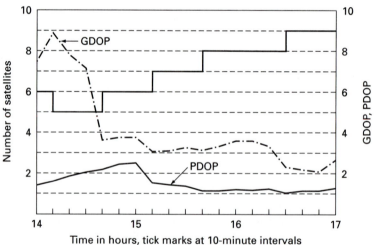

(b) PDOP and GDOP superimposed on satellite availability chart

FIGURE 12.17

Location Berkeley, CA (latitude 37°52′N, longitude 122°15′W) for a window of 14:00 to 17:00 PST, December 19, 1994, with a cutoff vertical angle = 15° and using satellites 5, 12, 14, 15, 18, 20, 22, 25, 29 (plots were produced using the Leica SKI software).

An evaluation of the geometry of the satellite configuration also is useful. A polar satellite sky plot and plots of values for PDOP and GDOP (Section 12.6), superimposed on a satellite availability chart, are shown in Figure 12.17. The sky plot shows the orbital position of the satellites and provides an indication of their altitudes and azimuths. The plots of PDOP and GDOP (Figure 12.17b) show that they are at acceptable levels between 14:40 and 17:00 but at 14:10 the GDOP goes to approximately 9. Recall that low-values for PDOP and GDOP are desirable.

The charts and diagrams cited should be generated prior to site reconnaissance, because they provide a great deal of information useful in that stage of the planning process.

Site Reconnaissance

Using the map with plotted locations of all points, "how to reach" descriptions, and output from the software planning, the survey engineer performs a *site reconnaissance,* which consists of a field verification of all selected locations.

The following guidelines should be used in determining the optimum site locations for GPS observations:

- At tripod ("eye") level, a clear view of the horizon (360°) at an angle of 15° above the horizon is desirable. A clear horizon will help reduce the amount of cycle slips.
- Avoid high transmission lines, radio towers, or other devices that produce electromagnetic frequencies. These sources might cause cycle slips and multipath errors (Section 12.6, Site-Dependent Errors, Obstructions).
- Avoid nearby natural and human-made reflective surfaces that may cause multipath errors.
- Avoid locations that may be affected by subsidence or slope failures, unless stability of surface is being sought.

At each location for a proposed station that has potential obstructions, an *obstruction survey* should be performed. This survey can be made using a magnetic compass and Abney hand level (Sections 6.13 and 5.22), an engineer's transit or theodolite (Chapter 6), or a total station system (Chapter 7) to locate each obstacle by zenith or vertical angle and azimuth. Magnetic azimuths (Section 6.8) are adequate. The outer limits of each obstacle are located and plotted on a polar obstruction diagram (see Figure 12.18), which contains the azimuths numbered around the perimeter, which represents the horizon, and altitudes numbered along the line from the pole to the east point and shown as vertical angles above the horizon. Showing the altitude of satellites by vertical angles is a standard practice in the GPS industry. Figure 12.18 shows for proposed station 22, the obstacles, which vary from the cut-off angle of 15° to a maximum of 30° above the horizon at a magnetic azimuths of 27° and 77°. These data can be used as input to the GPS planning software to provide modified diagrams that reflect the effects of the obstacles.

Figure 12.19a shows the satellite sky plot with an overlay of the obstructions. Note that portions of the orbits of satellites 5, 12, 15, and 29 are affected. A modified version of the satellite availability chart, with plots of PDOP and GDOP values superimposed, illustrated in Figure 12.19b gives a better indication of how the obstacles affect the number of satellites visible and the geometry. For this case, the obstacles around station 22 decrease the number of satellites available but not seriously. From the standpoint of geometric strength, the GDOP becomes excessively high at about 16:30, so that, for a high-precision survey, the optimum window would be from 14:40 until about 16:30.

Site reconnaissance, as outlined previously, should be performed at each station in the proposed survey, with points having excessive obstacles being shifted to better positions

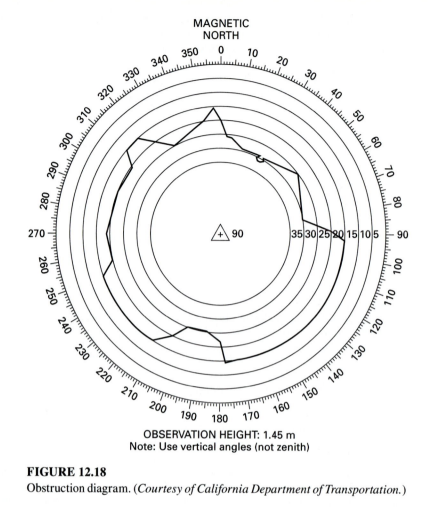

FIGURE 12.18

Obstruction diagram. (*Courtesy of California Department of Transportation.*)

if possible. New data concerning obstacles and revised locations can be entered into the planning software to yield revised charts (similar to those in Figure 12.19) and combined polar sky and obstruction diagrams. These charts and diagrams will assist in formalizing the final plan and choosing the optimum observation times.

Once the plan has crystallized, the stations should be monumented. The type of permanent or semipermanent point set is a function of the project. Photocontrol points could be marked with a steel rod having a metal cap. Surveys of high precision should have concrete monuments that go down to bedrock or drivable, jointed metal rods driven to refusal. Motorized drivers are available for this purpose.

Surveying Scheme

Geodetic or engineering control can be extended through networks, traverses, or radial surveying techniques. The first step in choosing a surveying method is to study the standards and specifications for GPS surveys.

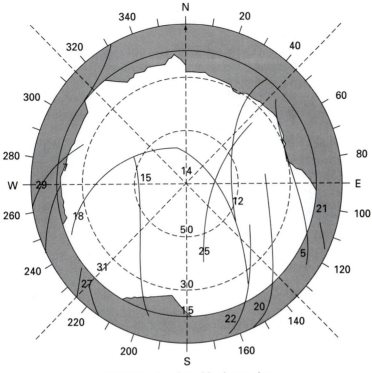

(a) Satillite sky plot with obstructions

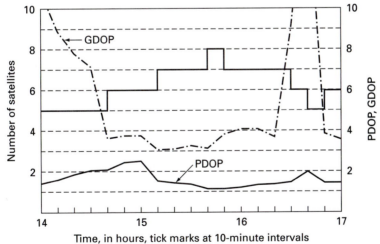

(b) Satellite availability with PDOP and GDOP superimposed,
with obstacles

FIGURE 12.19
Location Berkeley, CA, station 22 (latitude 37°52'N, longitude 122°15'W)
for a window of 14:00 to 17:00 PST, December 19, 1994, with obstructions
and a cutoff vertical angle = 15°, and using satellites 5, 12, 14, 15, 18, 20,
22, 25, 29 (plots generated using the Leica SKI software).

Standards and specifications. Before planning the survey scheme, review the project with respect to the standards and specifications that will be followed and the amount of equipment and personnel necessary. The Federal Geodetic Control Subcommittee has developed standards and specifications for GPS surveying, which can be found in *Geometric Geodetic Accuracy* (1989).

Network. A network consists of a closed polygon, where the base lines connect known and unknown stations (Figure 12.20a). Ties to a minimum of three known horizontal stations (*H*, *G*, and *K*) should be located in different quadrants relative to the center of the network. In addition, ties should be made to at least four known bench marks (B.M.s 10, 20,

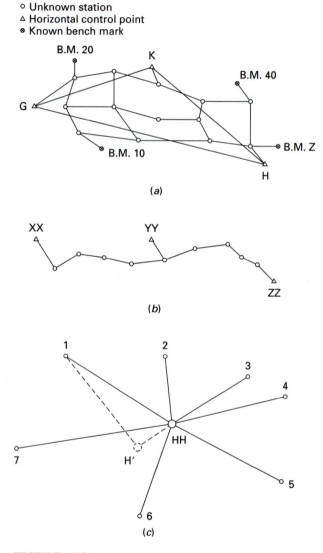

FIGURE 12.20
(a) Network, (b) traverse, (c) radial survey.

40, and Z). Ideally, all control points should be of a higher-order survey than the one being performed.

Traverse. Various types of traverses using conventional equipment are described in Sections 8.6 through 8.11. The traverse is an effective method of establishing horizontal control along corridors for highways, pipelines, railways, canals, and the like. A minimum of three known stations should be used when performing a traverse (Figure 12.20b). Two of the known stations, XX and ZZ, form the beginning and terminus of the traverse line. The third known station, YY, should be located perpendicular to the traverse line and near the center of the project line.

Radial survey. Radial surveys by total stations or by theodolites and EDM are described in Section 8.38. A radial survey by GPS is essentially similar, with the base lines radiating from a central station, continuously occupied by a receiver, to points of interest occupied by a roving receiver as illustrated in Figure 12.20c. Note that the stations being located have no double check. A radial survey is suitable for topographic surveys, determination of azimuth stations, and deformation monitoring. In the latter two cases, a double check can be achieved by setting an auxiliary point, such as at H' in Figure 12.20c, and measuring two base lines to each radial station, where the point is visited by the rover a second time to provide redundancy. For example, in Figure 12.20c, point 1 is located by lines HH-1 AND H'-1 measured from base line HH-H'.

In any GPS survey where stations are marked by semipermanent or permanent monuments, to assist in locating errors in h.i.'s of antennas and centering, 10 to 100 percent of the known and unknown stations should be double or triple occupied, depending on the accuracy of the survey see *Geometric Geodetic Accuracy,* 1989, Table 4.

12.15
DATA PROCESSING AND ANALYSIS

Data analysis depends on the algorithms and output formats used by the manufacturer in the processing software supplied with the receiver. Most software packages have documentation and provide output parameters that assist the user in detecting bad data and evaluating data quality. Processing starts with examining base-line data and proceeds to the network analysis.

Step 1: Base-Line Processing

As indicated in Section 12.7, Method 2, processing GPS data to form base lines consists of formulating linear combinations of the phase observables and their subsequent adjustments. From the computations and adjustments, the a posteriori reference variance, estimated integer ambiguities, covariance matrix, residual plots, and percentage of rejected data are indicators of the quality of the adjusted base line. Consider each of these parameters individually:

- *A posteriori reference variance.* The chi square statistic (Appendix D.5) is used to test the a posteriori reference variance. The results of the chi square test provide the user insight on the possibility of blunders in the measurements or incorrect determination of the a priori weights (optimistic or pessimistic).
- *Integer ambiguities.*The estimated integer ambiguities for each satellite pair should be close to integers. The quality of the estimated base line will be affected if the ambiguities

are determined incorrectly. Assume that the ambiguities from the adjusted double difference combinations were estimated at the following values:

$$\nabla\Delta\phi_{AB}^{0702} = -2{,}558{,}859.990 \text{ (cycles)}$$

$$\nabla\Delta\phi_{AB}^{1502} = 2{,}077{,}856.500 \text{ (cycles)}$$

$$\nabla\Delta\phi_{AB}^{1902} = -8{,}655{,}615.997 \text{ (cycles)}$$

$$\nabla\Delta\phi_{AB}^{2702} = -9{,}301{,}626.123 \text{ (cycles)}$$

$$\nabla\Delta\phi_{AB}^{3102} = -13{,}209{,}164.000 \text{ (cycles)}$$

In this example, six satellites (02, 07, 15, 19, 27, 31) were tracked and used in the base-line computation for stations A and B. Satellite 02 was selected as the reference station. Examination of the given ambiguities reveals that each value for satellites 07, 19, 27, and 31 definitely rounds to an integer value; for example, $-2{,}558{,}859.990$ rounds to $-2{,}558{,}860.00$. On the other hand, the ambiguity for satellite pair $\nabla\Delta\phi_{AB}^{1502}$ could be mathematically rounded to $2{,}077{,}857.00$ or $2{,}077{,}856.00$ (up or down). Therefore, the user may raise the elevation angle for processing or disable satellite 15 and reprocess this base line.

- *Covariance matrix.* The covariance matrix for the base-line components (dx_{AB}, dy_{AB}, dz_{AB}) for the line from A to B is defined as follows:

$$\Sigma_{xyz} = \begin{bmatrix} \sigma_{dx_{AB}}^2 & \sigma_{dx_{AB}dy_{AB}} & \sigma_{dx_{AB}dz_{AB}} \\ \sigma_{dy_{AB}dx_{AB}} & \sigma_{dy_{AB}}^2 & \sigma_{dy_{AB}dz_{AB}} \\ \sigma_{dz_{AB}dx_{AB}} & \sigma_{dz_{AB}dy_{AB}} & \sigma_{dz_{AB}}^2 \end{bmatrix} \tag{12.13}$$

The covariance matrix is where the variances ($\sigma_{dx_{AB}}^2$, $\sigma_{dy_{AB}}^2$, $\sigma_{dz_{AB}}^2$) for the components of the base line can be inspected to determine the quality of the estimated base line. The full covariance matrix also is used as the a priori estimate of quality in the network adjustment.
- *Residual plots.* Plots of the triple and double difference residuals can be used to determine the variability of the data or the presence of cycle slips.
- *Percentage of rejected data.* Rejection of a large amount of data may indicate blunders or inconsistent data. This may result from ionospheric activity, signal obstructions, or multipath errors. A particular satellite that has 5 to 10 percent rejected data should be removed and the base line reprocessed (*Geometric Geodetic Accuracy,* 1989).

Step 2: Preanalysis

Loop closures, differences between repeated base lines, and differences between known and observed base lines are used to determine the internal consistency of the observed base lines. These preanalysis operations should be performed prior to the network adjustment.

Loop closures. Like the closures in the two-dimensional components of a traverse (Section 8.15), loop closures are used to determine misclosures in the base-line components, $dx_{k,k+1}$, $dy_{k,k+1}$, $dz_{k,k+1}$, of a polygon. Loop closures are computed by summing the vector components within the polygon so that, in a polygon having n stations, we have

$$dX = \sum_{k=1}^{n} dx_{k,k+1}, \qquad dY = \sum_{k=1}^{n} dy_{k,k+1}, \qquad dZ = \sum_{k=1}^{n} dz_{k,k+1} \tag{12.14}$$

in which dX, dY, and dZ represent the misclosures in the components of the lines in the loop. A minimum of one base line within the loop should be from a different observing session.

TABLE 12.4
Station vectors

Session	Base lines	$dx_{k,k+1}$	$dy_{k,k+1}$	$dz_{k,k+1}$
A	0001–0002	36,677.885	−12,832.913	−12,874.674
	0002–0003	−126.195	−2,253.214	−2,193.390
	0003–0001	−36,551.683	15,086.138	15,068.069
B	0001–0002	36,677.890	−12,832.900	−12,874.660
	0002–0003	−126.197	−2,253.216	−2,193.392
	0003–0001	−36,551.686	15,086.140	15,068.070

In Figure 12.21, the vectors between stations 0001, 0002, and 0003 were observed during two independent observing sessions A and B. The results are shown in Table 12.4.

The loop closures for this polygon using base lines 0001–0002 and 0002–0003 from session A and base line 0003–0001 from session B, applying Equation (12.14) are $dX = \sum_{k=1}^{n} dx_{k,k+1} = 36,677.885$ m $+ (-126.195$ m$) + (-36,551.686$ m$) = 0.004$ m; $dY = \sum_{k=1}^{n} dy_{k,k+1} = (-12,832.913$ m$) + (-2,253.214$ m$) + (15,086.140$ m$) = 0.013$ m; $dZ = \sum_{k=1}^{n} dz_{k,k+1} = (-12,874.674$ m$) + (-2,193.390$ m$) + (15,068.070$ m$) = 0.006$ m.

The maximum misclosure in each component should not exceed the accuracy of the instrumentation or the required accuracy for the application. Poor misclosures may indicate errors in determining the h.i., inclusion of a range with an incorrect integer ambiguity in a base-line calculation, or station-centering errors.

Loop closures can also be expressed in parts per million (ppm) by

$$dX_{\text{ppm}} = (dX/D)10^6, \qquad dY_{\text{ppm}} = (dY/D)10^6, \qquad dZ_{\text{ppm}} = (dZ/D)10^6 \qquad (12.15)$$

in which D is the perimeter distance of the loop.

For this example, $D = \Sigma(dx_{0001-0002} + dy_{0002-0003} + dz_{0003-0001})$, $D = 40,935.414$ m $+ 3,147.040$ m $+ 42,316.236$ m $= 86,398.690$ m, and $dX_{\text{ppm}} = 0.05$, $dY_{\text{ppm}} = 0.15$, and $dZ_{\text{ppm}} = 0.07$. When the loop closures are presented in terms of parts per million, the user must be careful when evaluating the results, because long distances may mask the presence of an outlier.

Differences in repeated base lines. When the same base line is measured repeatedly, the differences between the components of the respective measurements provide an indication of the precision of the observed base lines. Table 12.5 illustrates the repeated base-line differences for base line 0001–0002.

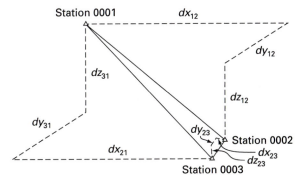

FIGURE 12.21
Base lines 0001–0002, 0001–0003, and 0002–0003.

TABLE 12.5
Repeated base-line differences

Session	Base line	dx (m)	dy (m)	dz (m)
A	0001–0002	36,677.885	−12,832.913	−12,874.674
B	0001–0002	36,677.890	−12,832.900	−12,874.660
(dx, dy, dz)		0.005	0.013	0.014
ppm		0.14	1.01	1.09

Differences between known and observed base lines. Differences between the components of known and observed base lines indicate whether the GPS measurements conform to the known control. Components for the known and observed base lines must be computed from geocentric coordinates. If the known coordinates are presented in terms of latitude (ϕ), longitude (λ), and ellipsoid height (h), these values must be converted to X, Y, Z geocentric, Cartesian coordinates using Equations (C.32), Section C.4, in Appendix C. The inverse solution between these coordinated points then is performed to yield components for each base line. Large differences may be the result of the measurement process (errors in centering or in the instrumental h.i. values) or could reflect on the integrity of the known survey control.

Step 3: Network Adjustment

After all the base lines have been processed and their reliabilies verified, the base lines are adjusted in the network, using a least-squares adjustment. The network adjustment consists of minimally constrained ("free") and constrained adjustments. In a minimally constrained adjustment, the coordinates of one station (ϕ, λ, h) are fixed and all other stations are adjusted relative to that station. This adjustment is used to locate blunders not identified by the base-line processor, visual inspection, loop closures, or repeat base-line analysis. The indicators commonly inspected to determine the presence of blunders are (a) the Chi square test (Section D.5) and (b) the normalized or standardized residuals (Section 2.30).

After all blunders have been removed, the results from the minimally constrained adjustment can be used also to determine the integrity of the other known control stations within the network. Because only one station was held fixed, the minimally constrained adjustment will produce estimated coordinates for all other stations within the network, relative to the station that was fixed. These estimated coordinates can be compared to the published coordinates of the other known stations within the network. If the coordinate differences are greater than the published uncertainties, potential problems exist. The following items are possible causes for large coordinate misclosures:

- *Incorrect coordinates of the station held fixed in minimally constrained adjustment.* Recompute the minimally constrained adjustment with a different station held fixed in the adjustment.
- *Stations with large misclosures may have been derived from different agencies.* The U.S. Geologic Survey (U.S.G.S.), U.S. Army Corps of Engineers (USACE), National Geodetic Survey (NGS), Department of Transportation (DOT), and others may have determined positions using conventional equipment or different specifications.
- *Observational blunders.* All blunders may not have been removed from the adjustment. The minimally constrained adjustment should be inspected to ensure that all blunders were removed.
- *Incorrect datum.* Ensure that the adjustment parameters are consistent with the datum of the known coordinates (e.g., NAD 27, NAD 83, Section 14.2).

The constrained adjustment is performed after all the blunders have been removed and the integrity of the known horizontal and vertical control has been verified. The constrained adjustment consists of fixing the coordinates of all the known horizontal and vertical control within the network. A minimum of two horizontal control stations and three vertical control stations may be held fixed in the constrained adjustment. This amount of control allows for the determination of scale and rotation between the local coordinate system and the GPS derived vectors. If the known local control is in error or has poor geometry, solving for the scale and rotation is not advisable, because the accuracy of the entire network then may be affected. After computing the constrained adjustment, positional error ellipses (semimajor and semiminor ellipse axes, Section 2.31), and relative line accuracies, σ_{AB}/S_{AB} (Section 2.11, Equation (2.11)) should be inspected to determine the quality of the final results.

Most surveyors prefer to fix horizontal and vertical control stations from the national network in a constrained adjustment. These stations consist of control points in the national network from a high-precision GPS static survey, observed and adjusted under the direction of the NGS. Such stations have accuracies consistent with a current GPS static survey. On the other hand, control points from the NAD 83 adjustment lack this inherent accuracy, so that a new GPS survey well could be more accurate than the control points being held in a constrained adjustment. A detailed discussion of this topic can be found in Schwartz (1994).

Adjustment model

The minimum number of observations required to define the model (n_0), in this case, equals the total number of unknown coordinates. This, in turn equals three (coordinates) times the total number of stations (X_s) minus the number of known stations (X_k):

$$n_0 = 3[(X_s) - (X_k)] \tag{12.16}$$

The total number of observations (n) is three times the number of *independently* measured base lines, because each GPS base line is composed of three components (dx, dy, and dz). Consider an example where the GPS base lines between three stations are observed. The coordinates of stations B and C are to be determined relative to station A, which has known coordinates (ϕ_A, λ_A, and h_A). The unknown coordinates can be computed using the following approach.

1. *Elements of the model* (Section 2.24, Equation (2.43)). Because the base lines are derived from the same phase measurements, there are only two *independent* base lines or $n = 2 \times 3 = 6$. The unknowns are six coordinates, and therefore no redundancy would exist in this case. In practice, at least by some GPS suppliers, all three base lines are used in a least-squares adjustment:

$$n = (3 \text{ base lines}) \times 3 = 9$$

$$n_0 = (3 \text{ stations} - 1 \text{ known station}) \times 3 = 6$$

$$r = n - n_0 = 3$$

The example following this discussion presents both approaches for comparison. It is advisable in principle not to use linearly dependent observations, otherwise singularities may be encountered. When selecting independent base lines, the shortest subset should be chosen, as shown in the example. However, the choice of linearly independent measurements in larger networks is rather complex and beyond this introductory treatment of the subject. The reader should consult Goad and Mueller (1988) for a more detailed discussion of the topic.

2. *Convert geodetic coordinates to earth-centered earth-fixed (ECEF) Cartesian coordinates also called geocentric coordinates (Section C.4).*
3. *Develop observation equations.*

$$F_1 = dx_{AB} - X_B + X_A = 0$$

$$F_2 = dy_{AB} - Y_B + Y_A = 0$$

$$F_3 = dz_{AB} - Z_B + Z_A = 0$$

$$F_4 = dx_{AC} - X_C + X_A = 0$$

$$F_5 = dy_{AC} - Y_C + Y_A = 0$$

$$F_6 = dz_{AC} - Z_C + Z_A = 0$$

$$F_7 = dx_{CB} - X_B + X_C = 0$$

$$F_8 = dy_{CB} - Y_B + Y_C = 0$$

$$F_9 = dz_{CB} - Z_B + Z_C = 0$$

where $dx_{AB}, dy_{AB}, dz_{AB}, dx_{AC}, dy_{AC}, dz_{AC}, dx_{CB}, dy_{CB}$, and dz_{CB} are the *measured* GPS vector components; X_A, Y_A, and Z_A are the known ECEF Cartesian coordinates of station A; and X_B, Y_B, Z_B and X_C, Y_C, and Z_C are the unknown ECEF Cartesian coordinates of stations B and C.

4. *Formulate the coefficient matrix* **(B)** *and right-hand vector (f) of the condition (or observations) equations.* The coefficient matrix **B**, which is sometimes called the *design matrix* in the literature, is the Jacobian of the condition equations, with respect to the parameters:

$$\mathbf{B} = \frac{\partial F_r}{\partial P_s}$$

where F_r = condition or observation equations $(r = 1, 2, \ldots, 9)$
P_s = parameters or the unknowns $(s = 1, 2, \ldots, 6)$

The constant term vector, which is sometimes called the *misclosure vector,* is given by the negative of evaluating the equations at the numerical values for the observations:

$$\boldsymbol{f} = (-F_r)$$

The nine condition equations in this case are directly linear and therefore can be rearranged to

$$v_1 + (-1)X_B \qquad = -(dx_{AB} + X_A)$$

$$v_2 + (-1)Y_B \qquad = -(dy_{AB} + Y_A)$$

$$v_3 + (-1)Z_B \qquad = -(dz_{AB} + Z_A)$$

$$v_4 + (-1)X_C \qquad = -(dx_{AC} + X_A)$$

$$v_5 + (-1)Y_C \qquad = -(dy_{AC} + Y_A)$$

$$v_6 + (-1)Z_C \qquad = -(dz_{AC} + Z_A)$$

$$v_7 + (-1)X_B + X_C = -dx_{CB}$$

$$v_8 + (-1)Y_B + Y_C = -dy_{CB}$$

$$v_9 + (-1)Z_B + Z_C = -dz_{CB}$$

or in matrix form

$$\begin{bmatrix} v_1 \\ v_2 \\ v_3 \\ v_4 \\ v_5 \\ v_6 \\ v_7 \\ v_8 \\ v_9 \end{bmatrix} + \begin{bmatrix} -1 & 0 & 0 & 0 & 0 & 0 \\ 0 & -1 & 0 & 0 & 0 & 0 \\ 0 & 0 & -1 & 0 & 0 & 0 \\ 0 & 0 & 0 & -1 & 0 & 0 \\ 0 & 0 & 0 & 0 & -1 & 0 \\ 0 & 0 & 0 & 0 & 0 & -1 \\ -1 & 0 & 0 & 1 & 0 & 0 \\ 0 & -1 & 0 & 0 & 1 & 0 \\ 0 & 0 & -1 & 0 & 0 & 1 \end{bmatrix} \begin{bmatrix} X_B \\ Y_B \\ Z_B \\ X_C \\ Y_C \\ Z_C \end{bmatrix} = \begin{bmatrix} -(dx_{AB} + X_A) \\ -(dy_{AB} + Y_A) \\ -(dz_{AB} + Z_A) \\ -(dx_{AC} + X_A) \\ -(dy_{AC} + Y_A) \\ -(dz_{AC} + Z_A) \\ -dx_{CB} \\ -dy_{CB} \\ -dz_{CB} \end{bmatrix}$$

These equations are in the form of adjustment of indirect observations, for which the general matrix form (Section 2.26, Equation (2.53)) is

$$\underset{9,1}{v} + \underset{9,6}{\mathbf{B}} \underset{6,1}{\boldsymbol{\Delta}} = \underset{9,1}{f}$$

5. *Compute the adjusted coordinates and stochastic parameters using the equations outlined in Section 2.26.*

EXAMPLE 12.1. Differential GPS measurements were made at stations G, C, and P, a three-station network illustrated in Figure 12.22. The given data are as follows. Known coordinates for station G:

$$X_G = 260,115.912 \text{ m}, \qquad Y_G = -4,854,898.970 \text{ m}$$

$$Z_G = 4,114,778.678 \text{ m}$$

Measured base lines:

From/to	dx, m	dy, m	dz, m	Length, m
G–P	1,557.704	−72.590	−168.661	1,568.489
P–C	−1,066.704	1,466.389	1,791.420	2,548.992
G–C	490.991	1,393.800	1,622.756	2,194.786

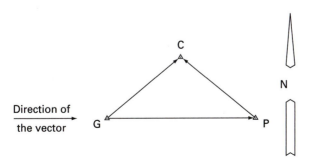

FIGURE 12.22
Network diagram.

Determine coordinates for stations P and C adjusted by the method of least squares, using the technique of indirect observations (see preceding section and Sections 2.26 and 2.27).

Solution. The unknowns are (X_C, Y_C, Z_C) and (X_P, Y_P, Z_P). Assume equal weights for measurements and let $\sigma_0^2 = 1$ so that the stochastic model (Section 2.21) is $\Sigma_{11} = \sigma_0^2 I = I$ and the weight matrix for the measurements (Equation (2.33)) becomes $W = I$.

Begin by considering all three base lines as measurements. Analysis of the adjustment or functional model (Section 2.21), by Equations (12.16) and (2.43), gives

$$n = (3 \text{ base lines}) (3) = 9$$

$$n_0 = 3(3 \text{ stations} - 1 \text{ known station}) = 6$$

$$r = n - n_0 = 3$$

Develop the condition or observation equations

$$F_1 = dx_{G-P} - X_P + X_G = 0$$

$$F_2 = dy_{G-P} - Y_P + Y_G = 0$$

$$F_3 = dz_{G-P} - Z_P + Z_G = 0$$

$$F_4 = dx_{G-C} - X_C + X_G = 0$$

$$F_5 = dy_{G-C} - Y_C + Y_G = 0$$

$$F_6 = dz_{G-C} - Z_C + Z_G = 0$$

$$\text{---} \quad \text{---} \quad \text{---} \quad \text{---}$$

$$F_7 = dx_{C-P} - X_P + X_C = 0$$

$$F_8 = dy_{C-P} - Y_P + Y_C = 0$$

$$F_9 = dz_{C-P} - Z_P + Z_C = 0$$

These equations can be put in matrix form (see the preceding section and Equations (2.49) through (2.53)):

$$v + \mathbf{B}\Delta = f = d - l$$

in which

$$v^T = [v_1 \; v_2 \; v_3 \; v_4 \; v_5 \; v_6 \; v_7 \; v_8 \; v_9]$$

$$
\mathbf{B} =
\begin{bmatrix}
-1 & 0 & 0 & 0 & 0 & 0 \\
0 & -1 & 0 & 0 & 0 & 0 \\
0 & 0 & -1 & 0 & 0 & 0 \\
0 & 0 & 0 & -1 & 0 & 0 \\
0 & 0 & 0 & 0 & -1 & 0 \\
0 & 0 & 0 & 0 & 0 & -1 \\
-1 & 0 & 0 & 1 & 0 & 0 \\
0 & -1 & 0 & 0 & 1 & 0 \\
0 & 0 & -1 & 0 & 0 & 1
\end{bmatrix},
\quad
\Delta =
\begin{bmatrix}
X_P \\ Y_P \\ Z_P \\ X_C \\ Y_C \\ Z_C
\end{bmatrix},
\quad
d =
\begin{bmatrix}
-X_G \\ -Y_G \\ -Z_G \\ -X_G \\ -Y_G \\ -Z_G \\ 0 \\ 0 \\ 0
\end{bmatrix},
\quad
l =
\begin{bmatrix}
dx_{G-P} \\ dy_{G-P} \\ dz_{G-P} \\ dx_{G-C} \\ dy_{G-C} \\ dz_{G-C} \\ dx_{G-P} \\ dy_{G-C} \\ dz_{G-C}
\end{bmatrix}
$$

The column matrix $f = d - l$ is

$$f = \begin{bmatrix} -260,115.912 - & 1,557.704 \\ 4,854,898.970 - & (-72.590) \\ -4,114,778.678 - & (-168.66) \\ -260,115.912 - & 490.991 \\ 4,854,898.970 - & 1,393.800 \\ -4,114,778.678 - & 1,622.756 \\ \hdashline 0.000 - & 1,066.714 \\ 0.000 - & (-1,466.389) \\ 0.000 - & (-1,791.420) \end{bmatrix} = \begin{bmatrix} -261,673.616 \\ 4,854,971.560 \\ -4,114,610.017 \\ -260,606.903 \\ 4,853,505.170 \\ -4,116,401.434 \\ \hdashline -1,066.714 \\ 1,466.389 \\ 1,791.420 \end{bmatrix}$$

The normal equations for least-squares adjustment by indirect observations (Section 2.27, Equation (2.70)) are

$$(\mathbf{B}^T\mathbf{W}\mathbf{B})\, \mathbf{\Delta} = \mathbf{B}^T\mathbf{W}f$$

from which

$$\mathbf{\Delta} = (\mathbf{B}^T\mathbf{W}\mathbf{B})^{-1}\mathbf{B}^T\mathbf{W}f$$

is the solution for the adjusted parameters. Substitution of $\mathbf{W} = \mathbf{I}$ and the matrices just defined yields

$$\mathbf{\Delta} = \begin{bmatrix} X_P \\ Y_P \\ Z_P \\ X_C \\ Y_C \\ Z_C \end{bmatrix} m = \begin{bmatrix} 261,673.616 \\ -4,854,971.559 \\ 4,114,610.016 \\ 260,606.902 \\ -4,853,505.170 \\ 4,116,401.435 \end{bmatrix} m$$

which are the adjusted coordinates for stations P and C.

Next, consider adjustment using any two independent base lines, choosing the shortest two lines $G-P$ and $G-C$ (see the table in the problem statement). In this case, only six condition equations, F_1 to F_6, apply. All matrices used in the previous solution will be truncated by eliminating the bottom three rows, below the dashed lines. Using these new matrices to solve for the six unknown coordinates yields

$$X_P = \quad 261,673.616 \text{ m} \quad (0.000)$$

$$Y_P = -4,854,971.560 \text{ m} \quad (0.001)$$

$$Z_P = \quad 4,114,610.017 \text{ m} \quad (0.001)$$

$$X_C = \quad 260,606.903 \text{ m} \quad (0.001)$$

$$Y_C = -4,853,505.170 \text{ m} \quad (0.000)$$

$$Z_C = \quad 4,116,401.434 \text{ m} \quad (0.001)$$

The quantities in parantheses are differences between the two solutions, which are seen to be extremely small for this example.

PROBLEMS

12.1. In electronic distance measurement, distance can be determined by two-way ranging or one-way ranging. Explain the differences between these two methods of measuring distance and give applications for each method.

12.2. What are the three major segments of the global positioning system (GPS)?

12.3. When determining position by GPS, what measured values are used and what systematic errors affect these measurements?

12.4. Signals must be received from how many satellites to determine the three-dimensional position of a station occupied by a single GPS receiver? Support your answer with a sketch.

12.5. Explain why elevation determined by GPS surveys may not coincide with orthometric elevations determined by differential leveling.

12.6. Explain the basic concept involved in determining positions by GPS.

12.7. Your firm is responsible for establishing basic horizontal and vertical control for a 25-mi segment of highway and also for running supplementary surveys for subsequent construction layout. The firm for which you work has both total station systems and GPS receivers with the necessary software. Which system would you recommend for establishing (a) the basic horizontal control for the entire project and (b) the supplementary surveys? How would you recommend that the basic vertical control be established? In each case, explain why.

12.8. From GPS observations, the elevation of a control point is 240.891 m. The separation between the ellipsoid and geoid is -34.95 m. Determine the orthometric elevation for the control point. Illustrate this relationship using a sketch.

12.9. Because the point position calculation is nonlinear, determine a method to compute the initial approximations for the station position for the unknown receiver coordinates.

12.10. Compute loop closures for the following stations:

From/to	dx, m	dy, m	dz, m
G–P	1557.714	−72.595	−168.668
P–C	−1066.718	1466.430	1791.422
G–C	490.993	1393.801	1622.759

12.11. In Problem 12.10, what could be the cause for the large misclosure in the Y component?

12.12. Perform the example adjustment illustrated in Example 12.1 using least-squares adjustment of observations only (Sections 2.26 and 2.28, Equations (2.57) through (2.64) and (2.98) through (2.108)), where the observation equations are of the form $\mathbf{Av} = \mathbf{f}$.

REFERENCES

Abidin, H. Z. "On-the-Fly Ambiguity Resolution." *GPS World* 5, no. 4 (April 1994), pp. 40–50.
Fiedler, J. "Orthometric Heights from Global Positioning System." *ASCE Journal of Surveying Engineering* 118, no. 3 (August 1992), pp. 70–79.

Geometric Geodetic Accuracy Standards and Specifications for Using GPS Relative Positioning Techniques. Rockville, MD: Federal Geodetic Control Committee, National Geodetic Survey, NOAA, August 1, 1989.

Gilbert, C. "Accuracy Specifications of GPS Data Collection Systems, Code-Based vs. Carrier-Based Systems." *Earth Observations Magazine*, April 1994, p. 57.

Goad, C. C., and A. Mueller. "An Automated Procedure for Generating an Optimum Set of Independent Double Difference Observables Using the Global Positioning Carrier Phaser Measurements." *Manuscripta Geodetica* 13 (1988), pp. 365–69.

"GPS World Receiver Survey." *GPS World* 8, no. 1 (January 1997), pp. 42–59.

Hofmann-Wellenhof, B.; H. Lichtenegger; and J. Collins. *GPS Theory and Practice.* 2nd ed. New York: Springer-Verlag, 1993.

King, R. W.; E. G. Masters; C. Rizos; A. Stolz; and J. Collins. *Surveying with GPS.* Monograph N. 9. Kensington NSW, Australia, University of New South Wales, November 1985.

Kleusberg, A. "Precise Differential Positioning and Surveying." *GPS World* 3, no. 7 (July 1992), pp. 50–52.

Langley, R. "The GPS Observable." *GPS World* 4, no. 4 (April 1993), pp. 52–59.

Leick, A. *GPS Satellite Surveying.* New York: John Wiley and Sons, 1990.

NATO Navstar GPS Project Steering Committe. "Technical Characteristics of the NAVSTAR GPS," 1991.

Reilly, J. P. "1996 GPS Equipment Survey." *P.O.B., Point of Beginning* 21, no. 7 (June 1996), pp. 32–47.

Reilly, J. P. "Elevations from GPS." *P.O.B. Point of Beginning* 22, no. 8 (May 1997), pp. 24–25.

Remondi, R. "Using the Global Positioning System (GPS) Phase Observable for Relative Geodesy: Modeling, Processing, and Results." Doctoral dissertation, University of Texas at Austin, Center for Space Research, Austin, Texas, 1984.

Remondi, R. "Kinematic GPS Results without Static Initialization." NOAA Technical Memorandum NOS NGS-55. Silver Spring, MD: National Information Center, 1991.

Schwarz, C. R. "The Trouble with Constrained Adjustments." *Surveying and Land Information Systems* 54, no. 4 (1994), pp. 202–29.

Schwarz, K. P., and Sideris. "Heights and GPS." *GPS World* 4, no. 2 (February 1993), pp. 50–56.

Townsend, S. "GPS and the Total Station Go Head-to-Head." *Professional Surveyor* 11, no. 4 (July–August 1991), pp. 49–51.

Wells, D. E; N. Beck; D. Delikaraoglu; A. Kleusberg; E. J. Krakiwsky; G. Lachapelle; R. B. Langley; M. Nakiboglu; K. P. Schwarz; J. M. Tranquilla; and P. Vaniček. *Guide to GPS Positioning,* Fredericton, N. B., Canada: Canadian GPS Associates, 1986.

CHAPTER 13

Photogrammetric Surveying and Mapping

13.1
INTRODUCTION

The discipline of photogrammetry involves obtaining information about an object *indirectly*, by measuring photographs taken of the object. Therefore, unlike surveying procedures, where measurements are usually made *directly* on the object in the field, in photogrammetry the object first is recorded on an intermediate medium, such as photographs and images, and the measurements are carried out later in the office. Therefore, the photogrammetric mapping project requires the following operations: (1) planning and acquisition of photography or imagery, (2) processing and preparation of photography or preparation of imagery for use in the photogrammetric restitution systems, and (3) photogrammetric restitution of the imagery to produce results such as point coordinates and other image and digital products.

Two broad categories are involved in photogrammetry: metrical or quantitative activities and interpretive and qualitative work. *Metrical photogrammetry* involves all quantitative work, such as the determination of ground positions, distances, differences in elevations, areas, volumes, various types of maps and many other image and digital products, such as orthophotos (Section 13.25) and digital elevation models. In the second category, classically called *photo interpretation*, photographs are analyzed qualitatively for identifying objects and assessing their significance. Photo interpretation relies on the human ability to assimilate and correlate such photographic elements as sizes, shapes, patterns, tones, textures, colors, contrasts, and relative location. Accurate and reliable photo interpretation requires extensive training and experience. The photo interpreter is called upon in many fields of application, such as ecology, environmental analysis, forestry, geology, engineering site selection, resource inventory, planning, and of course military intelligence. At present, records from other imaging systems, both passive and active, have come into common use for interpretive purposes, and therefore the more general name *remote sensing* is used. Civilian applications of remote sensing, however, still employ aerial photography in addition to other remote sensing imagery.

In photo interpretation, some measurements are necessary; and in metric photogrammetry, some interpretation is required. Therefore, photogrammetry combines both activities and can be broadly defined to include the acquisition, measurement, interpretation, and evaluation of photographs, images, and other remotely sensed data. Because of this broad definition, photogrammetry has found application in a large number of fields, ranging from biomedicine and dentistry to aerospace engineering and astronomy, from cloud-chamber measurements in physics to all types of engineering, from tailoring to transportation and urban planning, from accident analysis to forestry, and so on. But, by far the most common application is in surveying and mapping. Almost all of the topographic maps produced by federal, state, and private organizations are compiled from aerial photographs, and some are beginning to be derived from satellite imagery. Photogrammetric procedures also are used in deriving supplemental ground control, both horizontal and vertical. Therefore, surveying and engineering students must be aware of the various facets of this field and the ways photogrammetry can assist them in their future professional activities.

Metric photogrammetry classically has been divided into terrestrial and aerial types. In *terrestrial photogrammetry,* the photographs are taken from fixed, often known, points on or near the ground. In *aerial photogrammetry,* a high-precision camera is mounted in an aircraft and photographs are taken in an organized manner as the aircraft flies over the terrain. A more recent branch is *space photogrammetry,* which deals with extraterrestrial photography and imagery where the camera may be fixed on earth, mounted on board a satellite, or located on a planet. *Close-range photogrammetry* involves a camera relatively close to the object photographed. Such applications exist in many fields where direct measurement of the object is either impractical, uneconomical, or simply impossible. Examples are found in archaeology, architecture, medicine, and many aspects of experimental engineering laboratory investigations.

This chapter will be devoted to the principles of aerial photogrammetry, its applications in surveying and mapping, with some consideration given to satellite imagery. An important fact should be established at the outset. An aerial photograph is *not* equivalent to a map except in very unusual circumstances. If the terrain is flat and level, if the aerial photograph is exposed with the camera pointing perfectly downward (i.e., its optical axis is truly vertical), and assuming no image aberrations, the resulting photograph will be a map with a constant scale. When any of the foregoing restrictions are not met, the photograph will then only *approximate* a map, the degree of approximation depending on the amount of distortion in the photograph. For example, when the terrain contains relief, the photographic scale at high points will be larger than that at lower points because the higher points will be closer to the camera. To derive a proper map of the terrain, it is necessary that the ground be imaged in at least two successive aerial photographs. Photogrammetric techniques then may be used to reconstruct, to scale, a faithful replication of the ground in all three dimensions from the overlapping photographs. Once this is accomplished, horizontal and vertical ground point positions, planimetric maps, topographic maps, cross sections, and a variety of other image and digital products can be extracted. The objective of this chapter is to introduce the various operations and products of photogrammetry.

Photogrammetry has been affected significantly by technological advances in imaging and computer systems. The photogrammetric plotter, where both the input (photographs) and output (maps) are in *hard-copy* form, progressively is being supplemented and supplanted by the *photogrammetric workstation,* where both input and output are digital. Regular photographs are digitized using high-precision scanners and used in the workstations. Alternatively, imagery may be acquired directly by digital sensors, which may be printed on film and used in a plotter or introduced directly as a digital image into the workstation.

Photogrammetric mapping requires some ground control, which usually has been established by surveying techniques, to extract absolute positional information. The new technology of global positioning system (GPS) discussed in Chapter 12, has had significant impact on that requirement. GPS receivers now are deployed successfully on the aircraft and other platforms used to acquire the photography and imagery. This advancement in technology has resulted in marked reduction in the number of control points established by ground surveying. Finally, current photogrammetric mapping has become almost entirely digital, instead of the graphical plotting that used to be the production norm. The conversion to digital mapping and the introduction of high-accuracy images makes these products an important component of *geographic information systems* (GIS). Such systems, which are spatially based, are rapidly becoming a most universal product as discussed in Chapter 14.

13.2
SENSORS AND PLATFORMS

Photogrammetric imaging is a method of recording a two-dimensional view of a three-dimensional object scene. For many years, this was done using high-quality metric cameras flown in an aircraft and recording the image on a photographic film. Recently, substantial technical advances have taken place in electronic sensing and the ability to carry such sensors in a spacecraft, so that many new sources of imagery emerged to supplement photography. Photogrammetric activities have expanded to include the production of both metric and thematic information from a variety of image sources.

Imaging sensors are classified into passive and active. *Passive* systems rely on sun energy reflected or emitted from the scene. By contrast, *active* systems generate their own energy source, send it to the scene, and record the returned signal. Good examples of active systems are *radar* for earth imaging and *sonar* for imaging under water. The advantage of radar imagery is that it employs long-wavelength energy, which has the capability of penetrating clouds and other hindering particles that would render regular photography useless. Consequently, radar often is used at night and in all types of weather conditions. Passive systems are composed of cameras using film for recording and electro-optical sensors that record data on some electronic media such as tape, diskette, or solid-state memory.

13.3
CAMERAS AND ACCESSORIES

The camera obviously is the first component in the total photogrammetric system, because it is used to obtain the photography from which the data can be extracted. The camera is similar in function to the surveying instruments and is used to gather information about the object in the field. However, it differs greatly with regard to the amount of data it gathers. One sighting by a transit or theodolite yields *one* direction, whereas one photograph makes possible the determination of an essentially limitless number of directions or as many directions as the number of points identified in the photograph. Consequently, the design of the photogrammetric camera is critical to guarantee that the photographs obtained are of good metric quality.

Cameras used for terrestrial photogrammetry differ substantially from those used for aerial photography. In terrestrial photogrammetry, the camera usually is stationary, so its design is much simpler than the fast-moving aerial camera. The terrestrial camera generally is pointed horizontally during exposure.

Aerial cameras, being in motion during exposure, require a fast lens, a reliable shutter system for a very short exposure time, and a high-speed emulsion on the film. All these requirements are necessary to guarantee high-quality aerial photographs, particularly by minimizing the blur due to motion. Cameras used to gather photographs for metric work, as opposed to those for interpretation, are called *cartographic* cameras and are characterized by geometric stability and lenses that are corrected for geometric distortion. Figure 13.1a is a photograph of a modern cartographic camera system; Figure 13.1b is a schematic of a

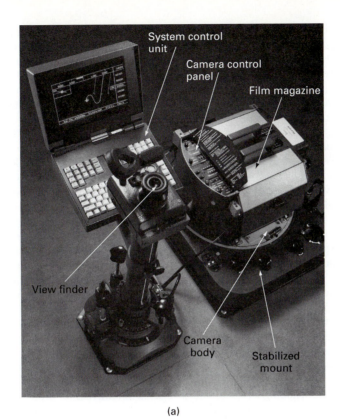

(a)

Magazine

Take-up spool

Supply spool

Supporting plate

Focal plane

Camera body

Lens cone

Optical axis
of the camera

Shutter

Lens

Diaphragm

Filter

(b)

FIGURE 13.1
(a) Aerial camera system, Leica RC30, viewfinder, and Ascot control
unit and (b) components of a typical aerial camera. (*Photograph courtesy
of Leica, Inc.*)

typical camera showing its components. The camera is composed of three main parts: lens cone, camera body, and film magazine. The *lens cone* contains the *lens assembly,* which is composed of a multielement lens, shutter, and diaphragm, and supports the frame that defines the focal plane. The lens is made of several elements, provides for a large angular coverage (about 90° for wide angle and 120° for super wide angle), and is corrected for aberrations and distortions. Because the light from the terrain travels through the atmosphere before reaching the lens, it ends up containing a disproportionate amount of blue light. Consequently, aerial cameras usually are provided with filters that prevent some of that blue light from reaching the film. The *shutter* controls the time interval during which the film is exposed, and the *diaphragm* determines the size of the bundle of light allowed to pass through the lens. The *camera body* houses the camera drive motor and mechanism and a recording chamber. The motor and mechanism operate the shutter and film flattening device and advance the film between exposures. The recording chamber allows for printing on each frame pertinent data such as level bubble, altimeter reading, clock, date, photo and mission numbers, and camera number. As shown in Figure 13.1b the cone and camera body upper surface define the *focal plane,* where a sharp image of the terrain is formed on the emulsion of the film. In this plane, a number of so-called fiducial marks, usually four, are registered on each frame, as shown in Figure 13.2a, which illustrates *side fiducial marks.* These marks are important because they define the coordinate system of the photograph. Furthermore, in well-adjusted cameras, the lines connecting opposite pairs of fiducials intersect in a point that falls on the optical axis on the lens. In this way, the photo system can be tied to the terrain system once the camera position and orientation (i.e., the direction of its optical axis) are determined. Although there are other sizes and shapes (e.g., rectangular), the most common type of aerial photographic format is square and measures 9 in. (23 cm) on each side.

The light entering an aerial camera lens is considered, for all practical purposes, to come from infinity, owing to the distance between the camera and terrain. Consequently, the image of the terrain will be formed at the focal plane of the camera lens. This is why the emulsion plane of the film is located at a distance from the lens equal to the focal length. Focal lengths of aerial camera lenses are nominally 3.5 in. (88 mm), 6 in. (153 mm), 8.25 in. (210 mm), and 12 in. (305 mm), although the 6-in. lens is by far the most common in photogrammetric mapping.

The light-tight *magazine* holds the unexposed and exposed film spools. Magazines vary in capacity from 180 ft (55 m) to 500 ft (150 m) of $9\frac{1}{2}$-in. (24 cm)-wide thin-base film.

Vertical aerial photography is exposed with the intention that the camera optical axis be as truly vertical as possible during exposure. Therefore, it is necessary that the camera be supported in a *stabilized mount* so that it is isolated from aircraft vibrations and perturbations. These mounts employ torquer motors controlled by gyroscopes. Other auxiliary systems used with aerial cameras include viewfinder, intervalometer, and *V/H* computer. The *viewfinder* provides the photographer a clear unobstructed view of the terrain below and ahead of the airplane. The *intervalometer* is a timing device that can be set to trigger the camera shutter at a specified time interval between exposures. The *V/H* (velocity/height) computer develops a voltage directly proportional to the aircraft velocity, *V*, and inversely proportional to its altitude, *H*. It may be used to determine the exposure interval as well as the film shift in the direction of flight during exposure to minimize image blur.

Aerial mapping film usually has panchromatic (i.e., black and white) emulsion, although color film is gaining in use. For thematic mapping, especially of vegetation, *color infrared film* sometimes is preferred. This is still film-based sensing and should not be confused with *infrared thermal* imaging, which requires a particular sensing system different from film. Electro-optical image sensors exist in square format, which could replace film in a frame camera. However, the format size is severely limited compared to film size. For example, a 2048 × 2048 element square CCD (charge coupled device) array,

(a) (b)

(c)

FIGURE 13.2
(a) Vertical aerial photograph with four side fiducial marks. (*Courtesy Pacific Aerial Surveys, Oakland, California.*) (b) Low oblique aerial photograph. (*Courtesy Leica, Inc.*) (c) High oblique aerial photograph. (*Courtesy Carl Zeiss, Inc.*)

with each element being 10 μm on the side, is a square only 20.48 $\times$ 20.48 mm. This hardly qualifies it for aerial camera imaging, but it can be used for close-range applications, where on-line processing may be an advantage. However, other forms of sensing using CCD arrays are becoming more and more common.

13.4
SCANNERS AND LINEAR SENSORS

When acquiring imagery from space, it is not practical to use film-based cameras on board a satellite. Instead, electro-optical sensors are used, and the digital image data are transmitted to earth by telemetry. To cover a scene of suitable extent on the earth surface, various

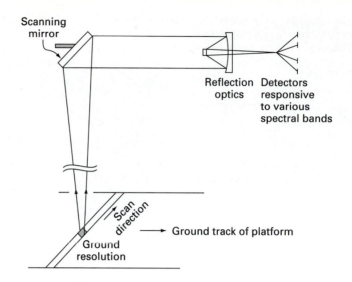

FIGURE 13.3
A multispectral scanner. (Adapted with permission from
CRC Civil Engineering Handbook, Editor-in-Chief, W. F.
Chen, Copyright CRC Press, Boca Raton, FL, © 1995.)

scanning systems have been devised. One of the earliest type is *multispectral scanner,* MSS for short, which has been used on the LANDSAT satellites. Figure 13.3 shows a schematic of the scanner, where the rotating mirror sweeps across the ground track of the satellite. In effect, it images one "line" on the ground, and successive lines are imaged to form a continuous record due to the travel of the satellite along its track.

The light gathered by the mirror is directed toward detectors, such as photomultiplier tubes, PMT, or photodiodes. This light usually is split into several wavelength spectral

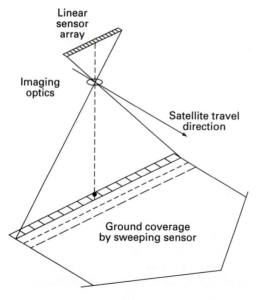

FIGURE 13.4
Push-broom scanner. (Adapted with permission from *CRC Civil Engineering Handbook*, Editor-in-Chief, W. F. Chen, Copyright CRC Press, Boca Raton, FL, © 1995.)

bands; hence, the term *multispectral*. The *instantaneous field of view* (IFOV) of the optics determines the size of the "element" on the ground that is imaged on a single detector. This is referred to as a *pixel* (for picture element), and for LANDSAT MSS it represents 83 m along the track and 68 m across the track. The more recent sensor, the thematic mapper (TM), increased the spatial resolution of a pixel to 30 m on the ground.

The successor to optical-mechanical scanning is the *push-broom* scanning mode. A linear CCD is used to instantaneously image a "line" on the ground. Successive image lines are acquired as the satellite travels, as shown in Figure 13.4. The linear array is oriented perpendicular to the satellite track, and telescopic optics are used to image a swath along the track. Arrays vary in length from 2000 to 8000 elements, each element measuring 10 to 15 μm. The French SPOT system uses 6000 elements with ground resolutions of 10 m at nadir for panchromatic images and 20 m for multispectral images.

13.5
PLATFORMS: IMAGE ACQUISITION CONFIGURATION

From the foregoing sections, it is clear that imagery can be collected from three types of platforms: (1) essentially stationary, as in terrestrial, industrial, or in general *close-range* applications; (2) airborne, which can range from low-flying to rather high-altitude jet aircraft yielding from very-large-scale to small-scale imagery; and (3) space-borne, where the sensors are carried by earth orbiting satellites. Originally, satellite imagery was essentially single coverage with minimal overlap, only to avoid gaps in coverage; no *stereo* coverage was planned. However, recent designs, such as SPOT and other commercial systems planned for the near future in the United States and other countries, consider stereo acquisition an essential function of the sensing systems. Stereo coverage in SPOT is effected by pointing the sensor to the side with an angle up to 23° off the vertical. Thus, two overlapping image swaths are obtained from two different orbital tracks.

Aerial frame photography remains the most common form of imagery for photogrammetric mapping, and therefore the following section is devoted to its acquisition.

13.6
ACQUISITION AND PROCESSING OF AERIAL PHOTOGRAPHS

Once a project area has been specified, the first step in the photogrammetric project is to plan the acquisition of the photography in a manner suitable for the purpose. For topographic mapping and the production of mosaics or orthophotos, vertical aerial photography is most common. A *vertical* photograph is taken with the camera axis vertical, although unavoidable aircraft motion may cause it to tilt a few degrees from the vertical (usually a maximum of 5°, although the average often is 1° or less). An example of a vertical aerial photograph is shown in Figure 13.2a. In small-scale mapping, the camera axis may be intentionally tilted to increase the area covered by one photograph. In this case, if the tilt angle is, say, 20°, the photograph is called a *low oblique*, an example of which is shown in Figure 13.2b. When the tilt angle is large enough (50° or 60°) that the horizon appears in the photograph, it is called a *high oblique*, as shown in Figure 13.2c. The horizon line in the photograph can be used to advantage when deriving metric information from the high-oblique photograph.

In vertical aerial photography with a square format (usually 9 × 9 in. or 228.6 × 228.6 mm), the ground-area coverage of a single photograph is square. As the airplane flies over the ground, successive photographs are exposed in such a way that each two adjacent photographs cover a common area that is more than half the single photo coverage. This common area, called *forward overlap* or simply *overlap,* usually is 60 percent, as shown in Figure 13.5. Thus, if photographs along one pass from one end of the project area to the other are laid down on a table, each two successive photographs would overlap by about 60 percent and each three successive photographs by about 20 percent. This type of coverage is necessary to ensure that each area on the ground is covered at least twice so that its three-dimensional geometry may be recovered. The 20 percent *triple overlap* (see Figure 13.5) is provided to assure the proper connection from one pair of photographs to the next. In some special situations, such as a mosaic of rugged terrain, overlap may be increased to say 80 percent. The nominal line passing through the middle of successive photographs is called a *flight line,* and the set of photographs in one line is often referred to as a *strip.*

Once the airplane has traversed the length of the project area, it turns around and returns in the opposite direction as shown in Figure 13.5. The second strip of photography is exposed in such a way that there is at least 20 percent *sidelap* with the first. This again assures no gaps and allows for the tie-in between strips so that the required maps or image products can be continuously produced.

After an aerial photography mission, the film is brought to the photo lab for processing. The images exposed on the film are called *latent* images, because they are not visible until they have been processed. The degree of complexity in processing the film depends on its type. In general, four types of emulsions are used in aerial photography: (1) *panchromatic black and white* (B&W), which is the most widely used type in metric photogrammetry and interpretation; (2) *color,* composed of three emulsion layers to render the different color hues originally in the object, is used more for interpretation than in mapping; (3) *infrared black and white* (B&W IR) is sensitive to near-infrared light (longer wavelengths), thus penetrates haze, and is used mainly for interpretation and intelligence work as it permits detection of camouflage; and (4) *infrared color* (false color IR), in which different colors are arbitrarily used to code images from different portions of the light spectrum (thus, the name *false color*), is employed in various facets of interpretation, such as crop disease detection or pollution monitoring, as well as intelligence.

After processing in the photo lab, the result is either a film negative or positive, depending on the processing procedure used. On a *film negative,* light portions of the original scene appear dark, while dark scene portions appear light, so the scene illumination is reversed on a film negative. On a *film positive,* the original scene light and dark distribution

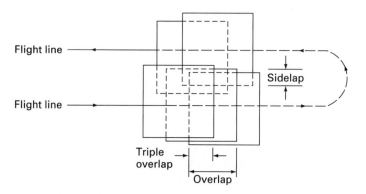

FIGURE 13.5
Forward overlap and sidelap.

is preserved. It is much more common for B&W film (both panchromatic and IR) to be processed to negatives, and both color and false color film are processed either to positives or negatives. When the result of film processing is a negative, positive-tone prints usually are made for initial inspection as well as for planning and execution of the various photogrammetric operations. It is a common and good practice to lay all the photographs in a project together to check for gaps. Also, the tilts of the different photographs are checked in case some are unacceptable. In either case, a reflight may be required to obtain more suitable photographs. Once the photographs are judged as meeting the original specified requirements, the photogrammetric project may proceed.

13.7
GEOMETRIC CONCEPTS

The basic principles of photogrammetry—particularly when using the most common source of imagery, the frame photograph—rely substantially on geometric concepts, both in two dimensions (the photograph) and three dimensions (the object space photographed). Many of the elementary algebraic relationships between these two spaces are derived using relatively simple geometric relations such as parallel lines, proportions, and similar triangles. In addressing these relationships, it is easier to discuss aspects of a single frame photograph and follow with those concepts defined only for overlapping photographs.

13.8
ANALYSIS OF THE SINGLE PHOTOGRAPH

As noted earlier, an aerial photograph will be the same as a map only in the very special situation when the terrain is flat and level and the photograph is truly vertical. In general, neither is the photograph taken precisely vertical nor is it common that the terrain be flat and level. Figure 13.6 shows schematically the analysis of both factors. In Figure 13.6a, the terrain is assumed to be flat and level between points A and B and between points D and E and of variable relief in between. Therefore, $h_A = h_B$ and $h_D = h_E$. The photograph is *assumed* to be truly vertical. The scale at any point in the photograph, such as a, is the ratio of the distance from the camera location, C, to the image point, divided by the distance from C to the object point, A, or $S_a = Ca/CA$.

From the two similar triangles, Cao and CAO, $S_a = Ca/CA = Co/CO$. Because $Co = f$ = focal length and $CO = H - h_A$ = the flying height above the object point of interest, the scale at any point on a truly vertical photograph is

$$S = \frac{f}{H - h} \tag{13.1}$$

in which H, the flying height, and h, the elevation of the point, are both referred to the same datum.

Figure 13.6a shows that the scale is the same for the area between points a and b and is equal to $f/(H - h_A)$. Similarly, the scale is constant between points d and e and is equal to $f/(H - h_D)$. In a region with a constant scale on a vertical photograph, the scale can be derived by dividing an image distance by the corresponding ground distance. Hence, $S_a = ab/AB$ and $S_d = de/DE$. Because h_D is larger than h_A, DE is closer to the camera than AB and the scale along de is larger than along ab. In fact, the two ground distances AB and

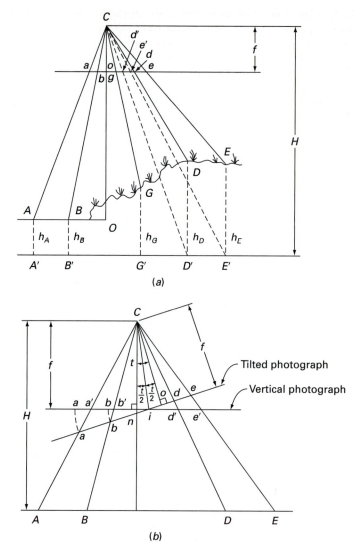

FIGURE 13.6
(a) Geometry of a single vertical aerial photograph and
(b) geometry of a tilted photograph over level terrain.

DE are equal, and Figure 13.6a shows that the image distance *de* is longer than *ab*. It is important, then, to note that, unlike a map whose scale *must* by definition be constant, the scale on an aerial photograph in general is variable. Therefore, it is better to speak of a photographic scale at a *point* unless a nominal, an average, or an overall scale is specified.

These latter scales are approximate and do not apply necessarily to a particular point or area on the photograph. They usually are obtained from a relation similar to Equation (13.1), using an average value h_{aver} for h:

$$S_{\text{aver}} = \frac{f}{H - h_{\text{aver}}} \qquad (13.2)$$

The variability of the scale on an aerial photograph, even when truly vertical, is due to the topographic relief. It is clear from Figure 13.6a that, if the object points D and E were on the datum at D' and E', they would have been imaged at points e' and d'. The ratio between $d'e'$ and $D'E'$ is equal to f/H, often called the *datum scale* of the photograph. Usually, this is the smallest scale unless there are points below the datum. The distance $d'e'$, in fact, is smaller than the image distance de. The effect of relief on the location of image points, called *relief displacement,* is discussed in Section 13.9.

The ground coverage of a 23 cm × 23 cm photograph is obtained by converting this to a square on the ground through applying the scale. Thus, at a scale of 1:10,000, the coverage is 2300 m × 2300 m. Note that, if the scale is reduced (i.e., the *scale number* is increased), the area of coverage is increased. In fact, the increase is proportional to the *square* of the factor. For example, reducing the scale by 2 (i.e., to 1:20,000) yields a coverage of 4.6 km × 4.6 km.

EXAMPLE 13.1. Referring to Figure 13.6a, the elevation of point A is $h_A = 123.2$ m and the scale at the point on the aerial photograph is 1:32,000. The elevation of point E is 275.6 m. Compute the scale at the image for point E and the average scale of the photograph if the focal length is $f = 152.4$ mm.

Solution. The scale at point A, according to Equation (13.1), is

$$S_A = \frac{f}{H - h_A} \quad \text{or} \quad \frac{1}{32,000} = \frac{152.4}{(H - 123.2)\,1000}$$

or
$$H - 123.2 = 152.4 \times 32 = 4876.8 \text{ m}$$

Then,
$$H = 4876.8 + 123.2 = 5000 \text{ m}$$

The scale at point E is

$$S_E = \frac{f}{H - h_E}$$

or
$$S_E = \frac{152.4}{(5000 - 275.6)\,1000} = \frac{1}{31,000} \quad \text{or} \quad 1{:}31{,}000$$

Because of the shape of the terrain in Figure 13.6a, the mean terrain will be taken as the average of the elevations of points A and E. That is, $h_{aver} = \frac{1}{2}(123.2 + 275.6) = 199.4$ m and the average scale, according to Equation (13.2), is

$$S_{aver} = \frac{152.4}{(5000 - 199.4)\,1000} = \frac{1}{31,500} \quad \text{or} \quad 1{:}31{,}500$$

Figure 13.6b shows the effect of the second factor, the tilt in the photograph. To isolate and dramatize this effect, the terrain is assumed to be flat and level, with the distances AB and DE being equal. If a truly vertical photograph is exposed from C, it will have a constant scale equal to f/H. This is borne out by observing that the image distance $a'b'$ and $d'e'$ are equal. An aerial photograph exposed from the same point, C, but with its optical axis tilted an angle t from the vertical contains images of the four object points at $a, b, d,$ and e. It is clear that the scale is variable along the tilted photograph. The image distance ab is larger than *de*, even though both represent the same distance on the ground. The effect of tilt on image location on a tilted photograph, called *tilt displacement,* is discussed Section 13.10.

13.9
RELIEF DISPLACEMENT ON A VERTICAL PHOTOGRAPH

Figure 13.7a shows a section of a vertical plane passing through the camera station, *C*, of a truly vertical photograph and an object point, *A*. The datum position of point *A* is *A′*, and their image points are *a* and *a′*, respectively. Of course, point *a* is what actually appears on the photograph, whereas *a′* is an imaginary point representing the image of *A′* if it were possible for *A′* to be seen by the camera. The distance *d* between *a′* and *a* is the *relief displacement on the photograph.* The extension of line *CaA* intersects the datum in point *A″*, which represents the datum point that would have been imaged at *a*. The distance *D* between *A′* and *A″* is the *relief displacement in the datum plane.* The ratio between *d* and *D* is the datum scale of the photograph, or *f/H*:

$$d = \frac{Df}{H} \tag{13.3}$$

The *principal point, o,* of a truly vertical photograph is the only point that undergoes no relief displacement. Figure 13.7a shows that, no matter how high the ground point, *O*, is above the datum, both *O* and its datum position *O′* are imaged at *o* because both lie on the vertical line through *C*. Consequently, all other points are displaced due to relief radially from *o*, as depicted in Figure 13.7b. The amount of relief displacement depends on the radial distance, *r*, between the image point and the principal point, *o*. The farther the image point from *o*, the larger is the relief displacement, as shown from the relation to be derived.

The two triangles *AA′A″* and *Coa* (Figure 13.7a) are similar and therefore

$$\frac{D}{r} = \frac{h}{f} \quad \text{or} \quad D = \frac{rh}{f}$$

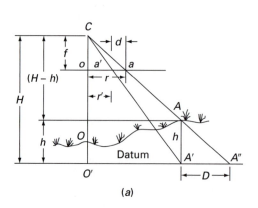
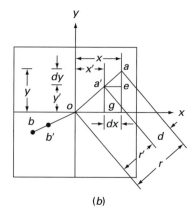

(a) (b)

FIGURE 13.7
Relief displacement in a vertical photograph.

which, when substituting from Equation (13.3), leads to

$$d = \frac{rhf}{fH}$$

or
$$d = \frac{rh}{H} \qquad (13.4)$$

From Equation (13.4), the relief displacement is seen to be directly proportional to r and the elevation of the point, h, and inversely proportional to camera altitude, H; both h and H must be measured above the same datum. One must watch for units in applying Equation (13.4). In general, d and r may be in the same units, say millimeters, while h and H are in meters or feet.

Equation (13.4) gives the total relief displacement for a given image point computed when a scale is used to measure directly the distance r from o to the image. Sometimes, instead of r, the coordinates x and y of the image are measured, as shown in Figure 13.7b. Here, opposite fiducial marks are connected by straight lines to intersect in the principal point, o. One line, usually that in the general direction of the flight line, is taken as the x axis and the other as the y axis, with o as the origin of the coordinate system. One way is to compute r from $r = \sqrt{x^2 + y^2}$ and apply Equation (13.4) to compute d and scale it from or toward o to get a'. Another way is to compute the two components, d_x and d_y, of the relief displacement. In Figure 13.7b, the two triangles $aa'e$ and aog are similar; hence,

$$\frac{aa'}{ao} = \frac{a'e}{og} = \frac{ae}{ag} \qquad \text{or} \qquad \frac{d}{r} = \frac{d_x}{x} = \frac{d_y}{y}$$

But, from Equation (13.4), $d/r = h/H$, so that

$$d_x = \frac{xh}{H}$$

$$\qquad (13.5)$$

$$d_y = \frac{yh}{H}$$

Therefore, if x and y are given, d_x and d_y can be evaluated directly from Equation (13.5). Then, the coordinates of a' are simply $x' = x - d_x$ and $y' = y - d_y$. If one is interested in the total relief displacement, $d = \sqrt{d_x^2 + d_y^2}$. These relations are demonstrated in the following numerical example.

EXAMPLE 13.2. A vertical photograph is taken with a camera having $f = 151.52$ mm. Two images, a and b, appear on the photograph, so that a falls on the x axis with $x_a = +78.70$ mm and b falls on the y axis with $y_b = +91.30$ mm. The two points, A and B, on the ground have the same elevation, 98 m above datum. Compute the image coordinates of the datum point images, a' and b', if the datum scale is 1:16,500.

Solution. The datum scale is f/H; therefore,

$$H = \frac{(16,500)(151.52)}{1000} = 2500 \text{ m}$$

Because point a lies on the photo x axis, point a' also will be on the x axis and the radial distance will be equal to x_a. Therefore, the relief displacement (Equation (13.5)) is

$$d_a = d_{xa} = \frac{x_a h}{H} = \frac{(78.70)(98)}{2500} = 3.09 \text{ mm}$$

and

$$x_{a'} = x_a - d_{xa} = 78.70 - 3.09 = 75.61 \text{ mm}$$

Similarly, because b lies on the y axis, then (Equation (13.5))

$$d_b = d_{yb} = \frac{y_b h}{H} = \frac{(-91.30)(98)}{2500} = -3.58 \text{ mm}$$

and the coordinate of b' is

$$y_{b'} = y_b - d_{yb} = -91.30 - (-3.58) = -87.72 \text{ mm}$$

13.10
TILT DISPLACEMENT

Figure 13.6b shows the relation between a truly vertical photograph and a tilted photograph, both taken from the same camera station. The plane of the figure, in which the tilt angle is measured, called the *principal plane,* is the vertical plane through the optical axis when the tilted photograph is exposed. The trace of that plane on the tilted photograph is a line that passes through the principal point and is called the *principal line,* as shown in Figure 13.8.

 Only one line is common to both the vertical and tilted photographs. This line is represented by point i in Figure 13.6b and by ii' in Figure 13.8. It is called the *isoline* or sometimes the *axis of tilt,* and its intersection with the principal line at i is called the *isocenter.* As Figure 13.6b shows, point i lies on the bisector of the tilt angle t. Therefore, the distance along the principal line from principal point o to the isocenter i is

$$oi = f \tan \frac{t}{2} \tag{13.6}$$

The vertical or plumb line through C pierces the tilted photograph at point n, which is called the *nadir point* and also lies on the principal line. The distance on from the principal point to the nadir point is given by (from Figure 13.6b)

$$on = f \tan t \tag{13.7}$$

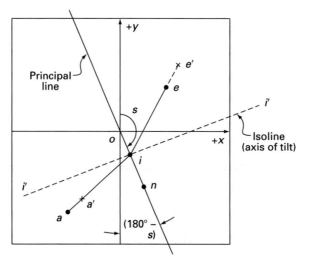

FIGURE 13.8
Principal line and tilt axis on a tilted photograph.

The principal line is located on the photograph by an angle s, called the *swing angle* and defined as the clockwise angle from the photographic $+y$ axis to the line *on* in that direction, as shown in Figure 13.8.

Imagine rotating the tilted photograph about its common line with the truly vertical photograph until they coincide. When this is done, images of the same object point do not match, the difference being the *tilt displacement*. Points a and b, being on the downside of the isoline, are displaced radially *away* from the isocenter, i, while points d and e are shifted radially *toward* i, because they are above the isoline.

Tilt displacement therefore is more complex to calculate than relief displacement. And, of course, when the tilted photograph is taken in the more practical case, where terrain relief exists, both relief and tilt displacements compound each other and geometric treatment becomes impractical.

13.11
ANALYSIS OF OVERLAPPING PHOTOGRAPHS

The basic concept of photogrammetry rests on the fact that, when a three-dimensional object is photographed from two (or more) points of view, its geometry can be reconstructed from these photographs. Therefore, as was discussed in previous sections and depicted in Figure 13.5, to recover the shape of the terrain (for mapping, for example) it is covered by at least two photographs. Each photograph is a two-dimensional representation of the three-dimensional scene. The third dimension (usually the elevation) is lost in a given photograph because the object is projected on the *plane* of the photograph. However, with two different such projections, and knowing the proper relationships between the photographs, the object can be accurately recovered in all three dimensions. This is the fundamental basis of all methods and systems used in photogrammetry, from the simplest, where approximations are made, to the most rigorous and sophisticated approach, as covered in the succeeding sections.

13.12
STEREOSCOPY AND PARALLAX

Stereoscopy refers to the ability of the individual to perceive the object space in three dimensions through using *both* eyes. Each human eye represents a single camera; therefore, *monocular viewing* or viewing with one eye results in *flat perspective* and the person's ability to perceive depth is hampered. *Binocular viewing,* on the other hand, allows the person to view an object from two different locations due to the separation between the eyes. The student can see the similarity between the function of the human eyes and the pair of camera stations when taking overlapping photographs. This is the reason for studying the principles and techniques of stereoscopy in support of photogrammetry.

Viewing with one eye fixes only *one* direction from the eye to the object, which is insufficient for fixing the object's distance from the viewer. When the other eye is utilized, a second direction is fixed and its intersection with the first locates the point. The closer the point to the eyes, the larger is the convergence angle between the two directions, as depicted in Figure 13.9a. Because the angle ϕ_1 is larger than ϕ_2, the observer will perceive point P_1 as closer than point P_2. In fact, the difference in distance from the observer, $H_2 - H_1 = h$, is a function of the difference in convergence angles, $\phi_1 - \phi_2$. The closest distance for distinct, comfortable vision (without aid) is 25 cm (10 in.), for which the convergence angle

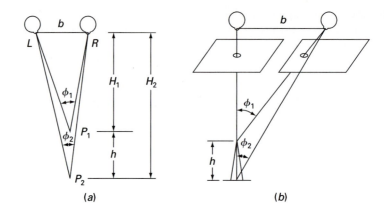

FIGURE 13.9
(a) Convergence angle in stereoscopic vision. (b) Stereoscopic
viewing with a pair of overlapping photographs.

is about $15°$. This can be computed on the basis of an average eye separation of $b = 65$ mm. The lower limit of the convergence angle for unaided vision is 10 to 20 seconds of arc.

Figure 13.9b, as compared to Figure 13.9a, shows the correspondence between natural binocular vision and stereoscopic viewing of a pair of overlapping photographs. An idealized tower is assumed to have been photographed with the camera first directly overhead and second after having traveled a distance past the tower. If it were possible to view the first photograph with the left eye only and the second photograph with the right eye only, the observer would perceive the tower in three dimensions, as shown schematically in Figure 13.9b. The lines joining the eyes and the two images of the top of the tower will intersect at an angle, ϕ_1, whereas those passing through the images of the tower base make an angle ϕ_2. Because ϕ_1 is larger than ϕ_2 the top of the tower will appear closer to the viewer than its base, and the tower will then be perceived in three dimensions. Stereoscopic viewing of objects and terrain from overlapping photographs is very important in almost all operations of both interpretation and metric photogrammetry, as will be shown in subsequent sections of this chapter.

Stereoscopic viewing is affected through the use of various techniques that allow each of the two overlapping photographs to be viewed with a different eye. These techniques are as follows:

1. *Stereoscopes.* These are simple devices, each containing two magnifying lenses for the viewer to look through. There are two types, a pocket stereoscope and a mirror stereoscope. They are used to view either paper prints or photographic transparencies mounted on a light table.
2. *Anaglyphic viewing.* Spectacles made of two complementary colors, red and blue-green (popularly called 3-D glasses), are used to view the images, which are either printed each in the corresponding colors or projected through filters of these two colors.
3. *Polarizing filters.* Anaglyph cannot be used with color imagery. Therefore, image separation for viewing can be accomplished using filters that are polarized in two perpendicular planes. Spectacles are used in conjunction with the light projected through the filters.
4. *Stereo image alternator (mechanical shutter).* A rotating mechanical shutter that is open for one quarter-cycle and closed for a quarter-cycle. The two shutters are out of phase by a quarter-cycle ($90°$). A similarly rotating double viewing shutter allows alternating viewing by the left eye and right eye.

5. *Stereo-optical trains.* Two separate viewing microscopes are used, one for each eye, each composed of lenses, prisms, mirrors, beam splitters, and filters if needed. These are more sophisticated versions of the stereoscope.
6. *Stereo-video display.* This is used in conjunction with a cathode-ray tube (CRT) video display to allow soft-copy stereo viewing. It employs a liquid crystal display (LCD) filter that operates conceptually in a manner similar to either the mechanical shutter or polarized light and filters, as in technique 3. The LCD allows alternate viewing of left and right images when viewed by matching spectacle lenses.

The angle ϕ is called *parallactic angle,* and its linear equivalent is called *parallax.* The quantitative definition of *parallax* depends on the simplifying assumptions that the photographs have *no tilts* and that both photographs are exposed from the same altitude. Any rotation of the camera about its (vertical) axis from one photograph to the other is accommodated by replacing one fiducial system of axes by another. The new system (Figure 13.10) has its abscissa along the projection of the air base on the planes of the photographs. Then the x parallax (or simply parallax) is defined by

$$p = x - x' \qquad (13.8)$$

where x and x' are the coordinates on the left and right photographs, respectively, with respect to the new axis-of-flight, or base-line, system; and p is the parallax. The algebraic values (i.e., with the proper sign) of both x and x' must be used in Equation (13.8). The value of p can be used to determine the elevation of terrain points appearing in the overlap area of two aerial photographs.

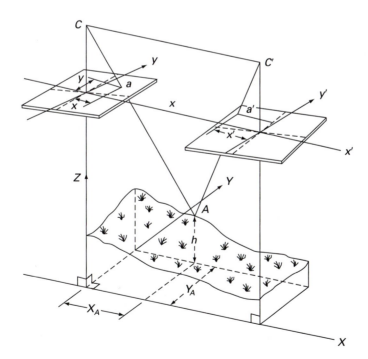

FIGURE 13.10
Parallax in stereophotography.

Figure 13.11 shows a schematic of two photographs and a terrain point drawn in a plane through the flight line. The altitude of each photograph is H above datum, and the elevation of point A is h above the same datum. The two coordinates x and x' are shown for the images a and a' on the left and right photographs, respectively. A line Eb is drawn parallel to $E'A$ by construction; therefore, c_1b is equal to x'. Since from Equation (13.8), the parallax is $(x - x')$, the line segment ab is the parallax p. Because the two triangles abE and $EE'A$ are similar,

$$\frac{H - h}{f} = \frac{B}{p} \quad \text{or} \quad H - h = \frac{Bf}{p}$$

so that

$$h = H - \frac{Bf}{p} \tag{13.9}$$

Therefore, to determine the elevation of a point h from a pair of aerial photographs, it is necessary to have the flying height above datum H, the value of the air base B, and the focal length, f. Usually, f is known from camera calibration data, B is determined from measurements on the photographs, and H is either given from prior computation or determined on the basis of a known *vertical* control point in the overlap area. In the latter case, h would be known and Equation (13.9) would be rearranged to give H:

$$H = h + \frac{Bf}{p} \tag{13.10}$$

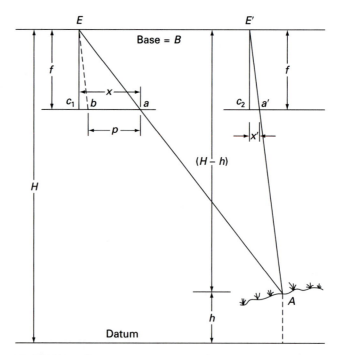

FIGURE 13.11
Relation between parallax and point elevation.

Equation (13.10) can be applied twice to determine the difference in elevation, Δh, between two points, leading to

$$\Delta h = \frac{H'}{b} \Delta p \qquad (13.11)$$

H' is the flying height above mean terrain (e.g., from the altimeter), b is the airbase measured on the photograph between the principal point and the *conjugate principal point* (transferred from the other photo stereoscopically), and Δp is the parallax difference.

Coordinates in the ground system also can be obtained in terms of parallax. Recall from Equation (13.1) that the scale, S, at the point of elevation, h, is

$$S = \frac{f}{H - h}$$

Substitution of H from Equation (13.10) yields

$$S = \frac{p}{B} \qquad (13.12)$$

in which p is the parallax of the point in question and B the air base. If, as shown in Figure 13.10, a coordinate system is established in the terrain with the X axis parallel to the photographic axis-of-flight x axis, then

$$X = \frac{x}{S} = \frac{xB}{p}$$

$$\qquad (13.13)$$

$$Y = \frac{y}{S} = \frac{yB}{p}$$

in which X and Y are the ground coordinates in the specific system shown in Figure 13.10 and x and y are the coordinates of the image point in the *left* photograph with respect to the axis-of-flight coordinate system. On the basis of Equation (13.13), the distance between two points could be determined if their elevations are known, or at least the elevation of one point is known. In this case, the measured parallax values are used together with image coordinates from Equation (13.13). Then, the distance is

$$D = [(X_1 - X_2)^2 + (Y_1 - Y_2)^2]^{1/2} \qquad (13.14)$$

where the subscripts 1 and 2 refer to the point numbers. Directions between points could be determined in a like manner. The use of Equation (13.13) is not so much for determining absolute location as for extracting relative information between points.

13.13
EPIPOLAR PLANES AND LINES

Figure 13.12 shows two photographs, a ground point, G, and its images, g_1 and g_1'. The camera locations are shown as C and C' with the base between them designated B. The three points C, C', and G form a plane in three-dimensional space, called an *epipolar plane*. There obviously are many epipolar planes, one for each object point. Each of these planes intersects the two planes of the photographs in two straight lines, each called an *epipolar line*. Each epipolar line in a pair of photographs is *conjugate* to the other. This means that,

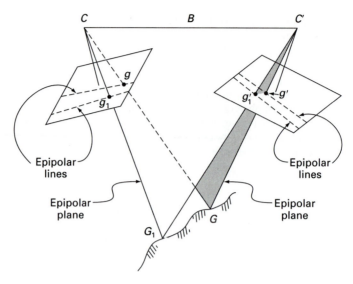

FIGURE 13.12
Epipolar lines and epipolar planes.

when selecting two corresponding images on a pair of conjugate epipolar lines and connecting the rays from C and C', they would intersect in an object point within the epipolar plane. In fact, the conjugate epipolar lines represent all conjugate images of the object points where the corresponding epipolar plane intersects the terrain surface. This concept is particularly useful in digital photogrammetry. For example, when digital imagery originally contains tilt and other angular deviations, its stereo viewing on a screen (CRT) is much facilitated if they are *resampled* along epipolar lines. In this way, the *y parallax* would be eliminated and the overlapping imagery would be properly *registered*. Another example concerns *automated image matching,* which is considerably simpler if it is performed in one dimension along epipolar lines.

13.14
MATHEMATICAL CONCEPTS FOR PHOTOGRAMMETRY

Unlike geometric relationships, which are based on some simplifications, *mathematical condition equations* are based on rigorous concepts. These are the ones implemented in most of the systems currently used in image restitution, such as the *analytical plotter* and the *soft-copy photogrammetric workstation*. The first important concept concerns representing the orientation of a photograph by an orthogonal matrix, M, which is composed of the three sequential rotations ω, ϕ, and κ, as given by Equations (C.21) and (C.22), in Appendix C. These three angles, together with the three coordinates, X_c, Y_c, and Z_c of the camera location, C, are referred to as the *six elements of exterior orientation* of the photograph. The *three elements of interior orientation* are designated x_o and y_o, which represent the image coordinates of the *principal point,* defined as the foot of the perpendicular from C to the plane of the photograph, and f, the calibrated focal length of the camera.

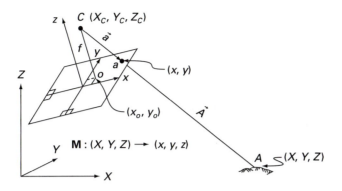

FIGURE 13.13
Photogrammetric collinearity condition.

With reference to Figure 13.13,

The image vector is $a = [(x - x_o)(y - y_o)(-f)]^T$

in an image space coordinate system

The object vector is $A = [(X - X_c)(Y - Y_c)(Z - Z_c)]^T$

in an object space coordinate system

Because a and A are collinear,

$$a = kA \tag{13.15}$$

When expressing these vectors in component form they must be in the same space coordinate system:

$$\begin{bmatrix} x - x_o \\ y - y_o \\ -f \end{bmatrix} = k\mathbf{M} \begin{bmatrix} X - X_c \\ Y - Y_c \\ Z - Z_c \end{bmatrix} \tag{13.16}$$

in which $\mathbf{M}$ is the orientation matrix defined to rotate the object system to become parallel to the image system, and k is a scale factor, which represents the image scale at the point for which the condition is written; that is, whose object and image coordinates are X, Y, Z and x, y, respectively. Because k usually is an unknown variable for each point under consideration, it usually is eliminated from Equation (13.16) by dividing the first and second rows by the third, leading to the most common form of the *collinearity equations:*

$$x - x_o + f\left[\frac{m_{11}(X - X_c) + m_{12}(Y - Y_c) + m_{13}(Z - Z_c)}{m_{31}(X - X_c) + m_{32}(Y - Y_c) + m_{33}(Z - Z_c)}\right] = 0$$

$$y - y_o + f\left[\frac{m_{21}(X - X_c) + m_{22}(Y - Y_c) + m_{23}(Z - Z_c)}{m_{31}(X - X_c) + m_{32}(Y - Y_c) + m_{33}(Z - Z_c)}\right] = 0 \tag{13.17}$$

These are the most fundamental equations used in photogrammetry. They apply to single as well as multiple photographs. They represent the projection from the three-dimensional object (represented by X, Y, Z) to the two-dimensional photograph (represented by x, y). Several other photogrammetric conditions apply equations, derived from these collinearity equations.

13.15
CONCEPT OF STEREOMODEL

Exterior orientation elements, when determined, specify the location and attitude of the bundle of light rays that entered the aerial camera lens at the time of exposure. The shape of the bundle, on the other hand, is determined by the interior orientation elements. These elements express the geometric relationship between the exposure station and the photographic plane. In aerial photography, the interior orientation elements usually are known from camera calibration.

Given a pair of photographs, their 12 exterior orientation elements (assuming interior orientation is known) must be recovered before the geometry of the photographed terrain can be accurately determined. A number of different techniques can be used to accomplish this, the most common being the use of instruments generally called *stereoplotters.* These employ optical-mechanical, electronic, or digital components to simulate and help recover the geometry of the photographs, rigorously.

Figure 13.14 shows what is termed a *stereomodel,* which is a scaled replica of the photographed terrain. The illustration depicts schematically one of the earliest photogrammetric instruments, where each photograph is placed in a projector. When the two projectors are illuminated, each pair of corresponding rays intersects at a point, and an observer would perceive the model in three dimensions. Three orientation operations need to be properly performed for the stereomodel to be suitable for the extraction of absolute

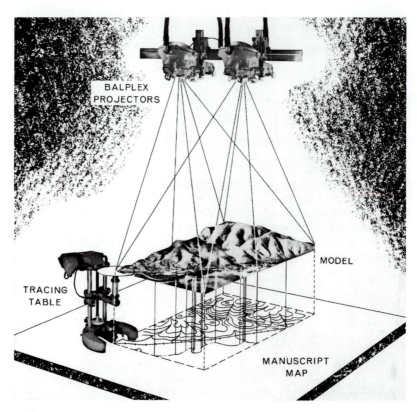

FIGURE 13.14
Model representation.

metric information about the photographed object: interior orientation, relative orientation, and absolute orientation.

Interior orientation is accomplished by the proper placement of the photograph in the projector. After interior orientation is performed, each bundle of rays emerging from a projector lens will duplicate the bundle that entered the aerial camera during the photographic exposure. However, the relative position and orientation will not duplicate the original situation during photography because the 12 exterior orientation elements have yet to be recovered. Each projector has six degrees of freedom, including three translations along X, Y, and Z, and three rotations about the X, Y, and Z axes. The recovery of these 12 elements is divided into two operations: relative orientation and absolute orientation. In *relative orientation*, a stereomodel is established at an arbitrary and unknown scale, location, and orientation relative to the ground coordinate system. This can be visualized by fixing the left projector (i.e., setting arbitrary values for its six elements) and the separation between the projectors, which establishes a stereomodel scale that is unknown at this point. This leaves the remaining five elements of the right projector to accomplish relative orientation. Therefore, relative orientation of a stereopair is a five-parameter operation. In general, these can be selected as any 5 from the 12 elements of exterior orientation.

The remaining seven parameters are determined in *absolute orientation*. They consist of the recovery of the unknown scale of the stereomodel, three translations to locate the model properly relative to the object space, and three rotations to bring the model coordinate system parallel to the ground coordinate system. Mathematically, this is accomplished by the seven-parameter transformation described in Appendix C, Equation (C.23).

13.16
PHOTOGRAMMETRIC SYSTEMS

Although photographs are useful products, particularly for an overall view of the object (terrain) and for general thematic information, they need processing before metric information can be extracted. In general, photogrammetric systems used for the restitution of imagery and the extraction of information necessary to yield the required products are divided into two groups: single image systems and stereo image systems. The first system processes one image or photograph at a time, often in conjunction with other data or information. In stereo systems, usually two overlapping images are used, a stereomodel is established, and the information is extracted by an operator viewing the model in stereo or, more recently, by an automated tool.

13.17
SINGLE-IMAGE SYSTEMS

The input to these systems is a single photograph at a time. The system includes a single-plate or monocomparator, rectifier, and image scanner. The *monocomparator* is a relatively simple instrument composed of a two-axis stage with a measuring microscope and coordinate readout usually with a 1 μm least count. Viewing is either monocular or, more commonly, binocular. The output is a list of two-dimensional Cartesian coordinates for all measured image points, in the comparator coordinate system. These generally are transformed to another system attached to the photograph, and called an image coordinate system, by applying one of the several two-dimensional transformations discussed in Appendix C, Section C.2.

A *rectifier* is an instrument used to produce a *rectified photograph*, in which the tilt has been eliminated. The rectified photograph may be produced at the same average scale as a vertical photograph taken from the same exposure station location as the original tilted photograph or as an enlargement at a larger scale. In the past, rectified prints were produced to construct a *mosaic*. Its use has been significantly reduced since the introduction of the *orthophotograph*, which is more accurate and more readily produced (see Section 13.25).

An *image scanner*, a recently introduced instrument, is used to convert a hard-copy photograph (film) to a digital image. With the onset of digital and soft-copy photogrammetry and because aerial film remains a very efficient recorder of image information, the image scanner will play a major role in photogrammetric activities for many years to come. Scanners, particularly those that scan at fine pixel sizes (5 or 10 μm), employ either linear or areal sensor arrays. Each pixel is usually quantized at 256 gray levels (8 bits), with three of these for color imagery, scanned either sequentially or simultaneously using color filters. It is important to note that image file sizes are quite large; for example, 530 Mb of memory are required for a single digitized 23 $\times$ 23 cm black and white (monchromatic) photograph scanned at 10 μm. Clearly, then, efficient *digital image compression* techniques with a minimum loss of information content will be needed, if digital or soft-copy photogrammetry eventually is to replace hard-copy systems.

13.18
STEREO-IMAGE SYSTEMS

As described in Section 13.15, a stereomodel can be established from an overlapping pair of photographs. When such a model is absolutely oriented, one can extract any required information, such as point coordinates, profiles, contours, and linear and areal features such as roads or lakes. If the imagery is in hard copy, then photogrammetric stereoplotters are used. If, on the other hand, images are in digital form, a soft-copy photogrammetric workstation is employed. In either case, the basic concepts and operations are quite similar and based on the same theory. Because photographs (transparencies) still are in much more common use than digital images, we will first discuss model orientation and follow with a description of the plotters.

13.19
MODEL ORIENTATION

The preceding sections show that, in a photogrammetric stereoplotter, both interior and exterior orientation must be properly recovered for the pair of photographs involved before beginning information extraction, such as plotting a map from the stereomodel. Although the procedure for interior orientation was explained, the method for recovering the 12 elements of exterior orientation was not discussed. Referring to Figure 13.14, note that the plane of the table top where the map manuscript rests represents the XY plane of the plotter; the line normal to it (i.e., the vertical direction of travel of the platen) represents the Z axis. It is possible to bring the two bundles of rays in proper relative registration so that each pair of conjugate rays intersects in a point and one can perceive a model in stereo. The resulting stereomodel, however, may be at an arbitrarily unknown scale and also may be tipped and tilted relative to the plotter's coordinate system. Because, at the time of map plotting, the plotter's system represents the object coordinate system, the model must be rotated, translated, and its scale changed until it sits properly into the instrument system just as the terrain was relative to the object coordinate system and the model is at the scale of the map. This operation of manipulating an established (or *restituted*) stereomodel is called *absolute orientation,* which relies on having an adequate number of survey control points.

The five-parameter operation of *relative orientation* causes each pair of corresponding rays to intersect at a point by recovering the relative tilts of two projectors with respect to each other. A number of specific procedures of relative orientation can be done, depending on the type of plotter available and, to some extent, on the nature of the model (for example, the existence of water covering parts of the model). Interested students should consult the references at the end of the chapter for more details.

Absolute orientation usually is accomplished in two separate operations: scaling and leveling the model. In *scaling,* two horizontal control points are plotted on the map manuscript at the scale at which the map is to be compiled. These control points then are used to make the horizontal distance between the corresponding model points equal to the map distance by changing the model base in the instrument. For *leveling,* an absolute minimum of three vertical control points is needed, because three points determine a plane. In practice, however, the elevations of four points at the four corners of the stereomodel are used and frequently a fifth point in the center is considered desirable. Usually, two rotations, one front to back and the other side to side, are effected to make the stereomodel level. Because leveling affects scaling and vice versa, these two operations often are repeated alternately to refine the results. In modern systems, the operation of absolute orientation is performed analytically by computing the seven parameters in the seven-parameter transformation (refer to Appendix C). For every ground control point of a known X, Y, and Z, three equations (in Equation (C.23)) result. Usually, more equations than the seven unknowns are used and therefore least squares is applied.

13.20
HARD-COPY SYSTEMS: PHOTOGRAMMETRIC STEREOPLOTTERS

A photogrammetric stereoplotter is an instrument of high precision that rigorously reconstructs the geometry of overlapping photographs so that accurate three-dimensional information about the photographed object may be obtained. In the case of aerial photography, planimetric and topographic maps may be compiled or coordinate listings, both horizontal and vertical, may be derived. Although there are many classifications of stereoplotters, it is sufficient for the purpose of this chapter to consider two broad classes: analog plotters and analytical plotters. Analog plotters are easier to understand and therefore will be presented first.

Figure 13.14 is a schematic of a *direct-projection analog stereoplotter.* The aerial camera is replaced by a *projector* similar in function to a slide projector. Two such projectors are needed, so that the pair of overlapping photographs may be projected simultaneously. Each bundle of rays emerging from the projector lens should duplicate, as nearly as possible, the bundle that entered the aerial camera during the photographic exposure. This duplication is accomplished by properly recovering the three elements of interior orientation. The projector accepts photographic transparencies, usually called *diapositives,* that are the same size as the original format used in the aerial camera, and therefore the distance between the photograph and the projector lens (i.e., is principal distance) is set equal to the aerial-camera calibrated focal length. The remaining elements of interior orientation are recovered by placing the photograph such that its center (principal point) is directly over the optical axis of the projector. This is usually accomplished by matching the fiducial marks to reference marks in the diapositive holder of each projector. Relative and absolute orientation then follow, as explained in Section 13.19.

When exterior orientation is properly recovered, a stereoscopic model will result, as shown in Figure 13.14. For the operator to perceive the model in three dimensions, the image from the left projector should be viewed by the left eye and that from the right projector by the right eye. One common procedure, called *anaglyph viewing* (see Section 13.12), is to use two glass filters, one *blue-green* and the other *red,* placed in the path of the projecting light rays. The operator then views the model through a pair of spectacles with corresponding blue-green and red lenses. The resulting stereoscopic model is measured by using a *tracing table* (Figure 13.14), which moves through the model space in all three directions, *X, Y,* and *Z.* It slides freely over the map table (the *X* and *Y* motions) and its top surface, called the *platen,* is moved up and down by a knob (the *Z* motion). All three motions can be monitored and recorded through the use of automatic devices. The reference measuring point is a small hole in the center of the platen illuminated by a small bulb directly underneath it; the point is called the *floating dot* or *floating mark.* Thus, when viewing the model stereoscopically, the operator sees within the model a sharp circular white point that can be moved about and brought in apparent contact with the model surface at any desired spot. Directly underneath the measuring mark is a pencil, which can be lowered to contact the map sheet when plotting. Alternatively, the movements of the floating dot are converted to digital values, which are stored for later use.

Figure 13.15a is a schematic of the old direct-viewing *optical analog stereoplotter.* Figure 13.15b is a schematic of another type of a classical stereoplotter, which is *mechanical,* in that the geometry of the photography is represented by mechanical components such as rods. Both the optical and the mechanical plotters are analog in that they emulate the taking of the photography by direct means. Therefore, each has a limited range of the types of photography (i.e., a specific focal length) that it can process. On the other hand, the *analytical plotter,* shown in Figure 13.16, extends the range and can accomplish all the functions of the analog plotter and additional ones as well. Inclusion of a digital computer provides added flexibility.

Briefly, the analytical plotter is composed of two stages, the motions of which are controlled by a computer, and a stereo viewing microscope. The operator usually controls the motion in the three-dimensional space of the model by two handwheels and a foot disk, as shown in Figure 13.16. The instrument operates primarily through the help of a set of computer software modules, with varying degrees of automation. For example, the measuring

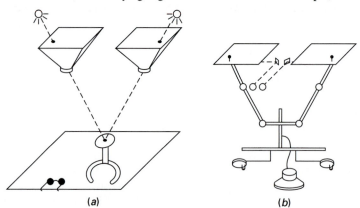

(a) (b)

FIGURE 13.15
Analog stereoplotters: (a) optical and (b) mechanical. (Reprinted with permission from *CRC Civil Engineering Handbook,* Editor-in-Chief, W. F. Chen, Copyright CRC Press, Boca Raton, FL, © 1995.)

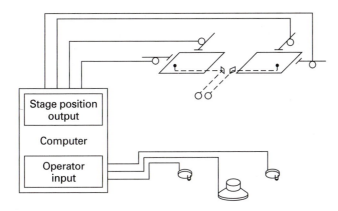

FIGURE 13.16
Analytical stereoplotter. (Reprinted with permission from
CRC Civil Engineering Handbook, Editor-in-Chief, W. F.
Chen, Copyright CRC Press, Boca Raton, FL, © 1995.)

mark is brought automatically to the vicinity of each fiducial mark, but the operator
normally measures each mark. The interior orientation is *computed* and its parameters
stored in the computer. Similarly, relative orientation is accomplished by the operator
clearing or eliminating parallax (actually, measuring the image coordinates on both
photographs at the parallax locations, thus essentially functioning as a stereocomparator)
and a least-squares module estimates the five parameters. Likewise, the model coordinates
of all the control points are measured and the seven parameters of absolute orientation are
estimated by another least-squares module. Unlike analog instruments, where relative and
absolute orientations have to be performed separately, in the analytical plotter they can be
combined into a general step in which all 12 exterior orientation parameters may be
computed simultaneously using the collinearity equations in Equation (13.17). Analytical
plotters can provide all the regular map compilation functions of analog plotters plus the
following: digital elevation models, digital profiles and cross sections, measurements for
aerial triangulation (see Section 13.22), and several other numerical products. These products usually are obtained through an on-line CAD package to allow for direct digitizing. The
resulting digital features are stored in databases either for GIS or cartographic uses (see
Chapter 14).

13.21
SOFT-COPY SYSTEMS

The most modern version of the photogrammetric stereoplotter is the *workstation,* which is
often referred to as a *soft-copy* stereo system (Figure 13.17). In plotters discussed so far, the
input was a hard-copy transparency and the output was either a hard-copy map sheet (the
most classical) or the more current digital data base. The soft-copy "plotter" is totally
digital, in that both the input and output are in digital form. The input may be a directly
digital image, such as that from SPOT, or a digitized frame photograph. The great advantage
of the workstation is the availability of *digital image-processing* tools and the potential
power of automation through the application of *image-understanding* and *pattern-
recognition* technology. Furthermore, the vector graphics, which already may exist in the
data base or have been extracted from the current model, easily can be superimposed on the
stereomodel. Stereo viewing is accomplished on the video monitor by a procedure similar
to using polarized glasses. Extensive research is progressing to provide automated tools for
feature extraction and derivation of robust elevation models.

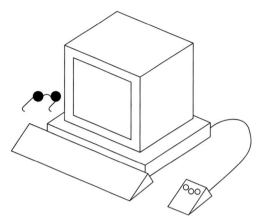

FIGURE 13.17
Digital or soft-copy system.
(Reprinted with permission from
CRC Civil Engineering Handbook,
Editor-in-Chief, W. F. Chen,
Copyright CRC Press, Boca Raton,
FL, © 1995.)

13.22
PHOTOGRAMMETRIC TRIANGULATION

It was explained previously that a stereomodel can be established so that it becomes a faithful reproduction of a segment of the earth's surface at a desired scale. Consequently, the horizontal and vertical coordinates of any terrain point can be extracted from the stereoplotter. This means that, given two horizontal and three (but usually four) vertical control points to absolutely orient a stereomodel, the ground coordinates of any desired number of points in the model can be determined. This analysis can be extended by visualizing a stereoplotter with *three* projectors. Three overlapping photographs can be used and a stereomodel composed of the total overlap (which is equal to the area of the middle photograph) can be relatively oriented. To absolutely orient this large model (composed of two adjacent stereo pairs) still involves seven parameters and requires two horizontal and three (preferably four) vertical control points.

Once this is accomplished, supplemental control, both horizontal and vertical, can be obtained. When map compilation is required, each stereo pair is used separately on a plotter or workstation. For each stereo pair, enough control will be required to perform absolute orientation. Part of this control would have been field determined, but the other part is determined photogrammetrically through restitution of the three photographs together. This concept obviously can be extended to more photographs than just three. It is called *photogrammetric triangulation* and can be defined as the procedure of establishing the geometric relationships among overlapping and sidelapping photographs for determining the positions for supplemental horizontal and vertical control points. In general, the supplemental control thus obtained is used for map compilation. Photogrammetric triangulation is a very powerful tool, because it reduces substantially the control required by field methods, markedly improving the economic aspects of photogrammetric mapping. Photogrammetric triangulation can be performed by one of two techniques: semianalytical or independent model triangulation and analytical or bundle block triangulation. Each technique will be discussed briefly.

Semianalytical or Independent Model Triangulation

In semianalytical triangulation, the model relative orientation may be performed instrumentally on a plotter or analytically using a computer program but model connection,

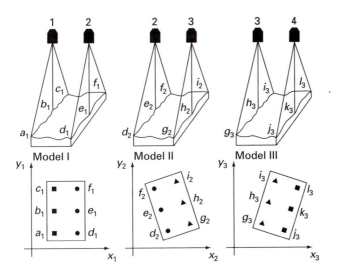

FIGURE 13.18
Three consecutive independent models and their coordinate
systems. (*Courtesy of California Department of
Transportation.*)

including scale transfer between successive models, is accomplished analytically; hence, the
semianalytical designation (except when relative orientation is performed analytically).
The method also is called *triangulation by independent models,* because each model,
composed of photographs $\frac{1}{2}$, $\frac{2}{3}$, $\frac{3}{4}$, and so forth, is relatively oriented completely independently of other models. After relative orientation, each model will have its own coordinate
system, as schematically depicted for three consecutive models in Figure 13.18. Also, each
model will be represented by a list of the coordinates of all model points, such as a_1, b_1, c_1,
d_1, e_1, and f_1 for model I in Figure 13.18. Each two successive models in general will have
at least three points in common, as for example d_1, e_1, f_1 and d_2, e_2, f_2 between models I and
II in the figure. Because the extent of terrain relief usually is considerably less than the
dimension of the model in the y direction, the use of such model points may lead to weak
geometric ties between models. In fact, the determination of the angle about the y axis may
be so weak that the model tie becomes unstable. For this reason, the coordinates of the
perspective centers of both projectors must be determined in the same coordinate system as
the model points in each model. These perspective center coordinates then can be used in
the model connection to provide depth and guarantee a strong geometric tie. This tie is
accomplished by a computer program that implements a transformation involving seven
parameters; three translations, three rotations, and one scale change (see Equation (C.23),
Appendix C). All independent models in a block are adjusted simultaneously to connect to
each other properly and to fit the available ground control. Plotters used for independent
model triangulation also are used for map compilation when the triangulation task is
finished.

Analytical or Bundle Block Triangulation

In this method of analytical triangulation, either a purely image point-measuring instrument such as a *comparator,* an analytical plotter, or a workstation is used to measure the
x–y plate coordinates of each pass point and control point in each photograph. The ideal

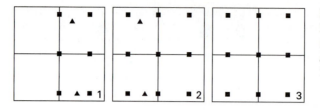

FIGURE 13.19
Ideal pass point locations.
(*Courtesy of California
Department of
Transportation.*)

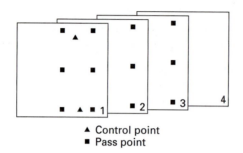

▲ Control point
■ Pass point

locations of the pass points are the same as those used in relative orientation, nine points forming approximately a 3×3 square grid on each photograph, as schematically depicted in Figure 13.19. These photographic coordinates then are used with other camera parameters and available ground control in a relatively extensive program to enforce, analytically, the intersection of conjugate rays and compute the ground coordinates of all pass points. Thus, unlike the semianalytical or independent model method, in analytical triangulation no physical model is constructed. What makes this procedure potentially the most accurate is the ability to correct computationally for all possible systematic errors, such as film shrinkage, lens distortion, and atmospheric refraction.

The effectiveness of triangulation is due to the large number of supplemental ground control points derived photogrammetrically on the basis of a relatively few field-determined control points. The control requirements for a block of photographs call for horizontal control points around its perimeter and for vertical control to be distributed throughout the block. If the global positioning system (see Chapter 12) is used aboard the aircraft to determine the location of the camera at the time of each exposure, requirements for control will be reduced dramatically. The reason is that each exposure station becomes, in effect, a complete ground control point (i.e., of known horizontal and vertical position). Although some proponents advocate that GPS supported triangulation requires *no* ground control, it is advisable that some minimum number, such as four control points at the corners of the block, be used. This is particularly important for vertical control.

13.23
PHOTOGRAMMETRIC PRODUCTS

Photogrammetrically derived products may be classified into two groups: image products and digital products. Classically, most projects were either in photographic or graphic format, with the majority being hard-copy graphics in the form of various types of map sheets. With the introduction of digital technology, most products, even some of those in image form, are digitally based. Image products include single photographs, photographic indexes, rectified prints, photo mosaics, orthophotographs, orthophoto mosaics, and

orthophoto maps (Section 13.25). Digital products include point coordinates such as supplementary control and target locations, digital representation of profiles, cross sections,
contour lines, terrain surface such as digital elevation models, and various terrain features
such as planimetric and spatial vectors.

773

CHAPTER 13:
Photogrammetric
Surveying and
Mapping

13.24
RECTIFIED IMAGERY

A truly vertical photograph of flat terrain essentially is a map of that terrain. A regular
vertical photograph contains tilt and, therefore, its scale varies due to the tilt displacement.
So, if an equivalent truly vertical photograph can be produced from a tilted photograph, tilt
displacements are eliminated. The resulting photograph is called a *rectified photograph* that
approximates a map, with the remaining distortions being due to the relief displacement.
The lower the amount of relief, the less is the distortion. Rectified imagery was used
extensively in the early applications of photogrammetry.

The *rectifier* (Section 13.17) is the instrument used to produce rectified photographs.
It recovers, through analog means, the eight parameters of the projective transformation
between two planes, the original and rectified photographs. A minimum of four control
points is needed for such rectification. Similarly, rectification may be accomplished numerically, where each control point yields two projective transformation equations (see
Equation (C.24), Appendix C) for a total of eight equations (for the four control points) to
solve for the eight transformation parameters. Overlapping rectified photographs may be
assembled carefully to form what is known as *photo mosaic,* usually at a common scale and
in which join lines are eliminated as much as possible.

Plane-to-plane rectification is one form of rectification; another is differential
rectification or orthorectification, which is introduced in the following section.

13.25
ORTHOPHOTOGRAPHY

A recent product of photogrammetry, which is being used increasingly, is the orthophoto.
It can be defined as a pictorial depiction of the terrain derived from aerial photography in
such a way that there are no relief or tilt displacements. Therefore, an orthophoto is
equivalent to a planimetric map except that, instead of lines and symbols, image tonal
variations convey the information. If the photographed terrain is flat and level, a *rectified*
aerial photograph would be the same as an orthophoto.

Orthophotos were produced in the past by optical or electronic means. At present,
almost all orthophotos are produced by digital systems. This requires a digital elevation
model (DEM; see Section 13.28) covering the area for which an orthophoto is to be
produced. Given the interior and exterior orientation elements of the original photograph,
each elemental area, or pixel, is differentially rectified to remove the effects of tilt and
orthographically projected using information from the DEM to remove the effects of relief.
The result is a pictorial equivalent of a map at a constant scale.

An orthophoto offers several advantages over a planimetric map. The details are
reproduced in pictorial form, so the orthophoto carries a wealth of information, whereas in
a line map the operator already has selected only those features deemed important. Being
an accurate map and rendering an abundance of detail makes an orthophoto useful for more
applications than either a planimetric map or a regular aerial photograph. Then, if contour
lines are added to the orthophoto, it becomes even more useful than a topographic map.

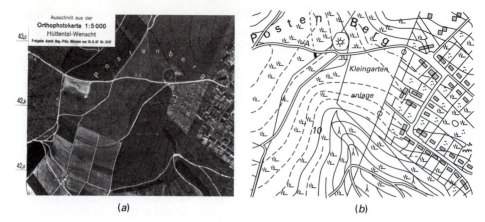

(a)   (b)

FIGURE 13.20
(a) Orthophoto with contours added (scale 1:5000) (b) Topographic map of a portion of
the same area as shown in part (a) at a larger scale. (*Courtesy of Carl Zeiss, Inc.*)

Figure 13.20a shows an example of an orthophoto with added contours. For comparison
purposes, Figure 13.20b illustrates a conventional topographic map of a portion of the same
area at a larger scale. Several orthophotos can be merged digitally to form an *orthophoto
mosaic,* with no seams.

13.26
TOPOGRAPHIC MAPS

Topographic maps are the classic product of photogrammetric mapping. After a photogrammetric stereomodel is absolutely oriented relative to a map manuscript, map compilation
can begin. The planimetry is plotted separately from, and usually before, the contour lines
are drawn. The planimetry is drawn totally before contours are added to make sure that the
locations of the contour lines fall in the proper relationship to the horizontal features (such
as streams and roads). A planimetric feature, for example, a road, is plotted by keeping the
floating dot always in contact with the terrain. If the road is ascending or descending, the
elevation of the floating dot should be continuously changed to keep it in contact with the
apparent terrain surface in the model. It is advisable that the operator compile like features
at the same time: all roads first, then all streams, then structures, and so forth. Whenever
the operator encounters difficulty, the manuscript should be annotated with notes to be used
later in the completion stage.

Contour plotting is somewhat more difficult than planimetry, particularly for an
inexperienced operator. The floating dot is *fixed* at the elevation representing the contour
value then moved in the XY plane *only,* keeping the dot in contact with the terrain surface
in the model. Here lies the difficulty, and often a contour line drawn by an operator with
little experience will zigzag. Therefore, the operator is advised to scan the entire model first
to become familiar with its various areas and begin with those details that are simplest to
compile. After these areas are finished and the operator gains more familiarity with the
model terrain, the plotting may proceed with the remaining, more difficult areas, such as
those that are relatively flat or have ground cover. In fact, in such areas photogrammetric
mapping is at a disadvantage compared to field mapping. Therefore, when the plotting is
finished on the instrument, the compiled manuscript must be checked. Wherever difficulties

are encountered, the respective areas are field completed. Then, the manuscript is checked for mistakes, names are added, and its accuracy is determined by comparing photogrammetric measurements to those correspondingly determined in the field. At this point the manuscript is ready for cartographic and reproduction operations, as explained in Section 14.25.

This classical compilation of map sheets essentially is replaced by a digital approach, where the information is recorded directly into a CAD or GIS system (see Chapter 14).

13.27
DIGITAL PRODUCTS

All products other than those in image form (profiles, contours, and the like) currently are produced in digital form. The simplest product is a list of coordinates for single points, such as supplemental control or targets, where the observer, while viewing in stereo, places the floating mark on the model surface and records the X, Y, and Z coordinates. Another product is the profile of the terrain in either X, or Y, or any specified direction in the model. The analytical plotter or workstation may be programmed so that the reference measuring mark travels in the designated direction. The operator's task is to keep the mark in touch with the terrain surface. The stream of X, Y, and Z coordinates of points is recorded automatically, either at constant intervals of time of travel or a constant travel distance, a choice the operator makes. A contour line also may be collected in the same fashion, except that here the reference floating mark is at a fixed elevation and the operator manipulates it in X and Y. In a profile, the operator has only to control the movement of the mark in Z. The profiling technique can be generalized to collect digital elevation models, as described in the following section.

Planimetric features can be collected digitally in the same manner as profiles and contours. The operator's task is very similar to that described in Section 13.26 for manuscript graphical plotting, except that the features, such as roads or streams, are represented by a string of X, Y, Z coordinate sets. In all cases, the digital data may be used as input to various types of plotters to produce the required graphics.

13.28
DIGITAL ELEVATION MODELS

The DEM is a relatively recent type of product, in which the terrain surface is represented numerically by a very dense network of points of known X, Y, and Z coordinates. This product has been given several names: digital terrain model (DTM), digital elevation model (DEM), digital height model (DHM), or digital terrain elevation data (DTED). The DTM may be produced in a photogrammetric analytical plotter, where the data are collected in profile form with equal spatial interval or time interval during digitization, or selectively at terrain break points as designated. The data also may be obtained as Z values for an X, Y grid with equal X and Y spacings. Another possibility is to digitize along contour lines, where a large set of X and Y values are developed for each given Z value. The DTM may be produced in a digital photogrammetric workstation as well. In fact, it is the first digital product from a workstation in which substantial automation has been implemented. A hierarchical procedure of image matching often is used to produce automatically the elevations at designated grid points. This, however, usually is followed by manual editing.

The various forms of DTM are produced for a number of different purposes. One of the areas where DTM has been successfully applied is in the automation of road location.

Highway engineers have created a DTM, stored it into a computer, and used it over and over to try out different alignments and grades to arrive at optimum design factors (Sections 16.27 and 16.28). The increased significance of terrain data bases makes the use of DTM all the more important. During the next decade or two, a considerable increase in DTM production, processing, and applications will take place. In this regard, photogrammetric instrumentation and methodology will play an important role, particularly in the production of DTM by fast and accurate automated methods applied to digital or digitized imagery.

13.29
GEOGRAPHIC INFORMATION SYSTEMS

As already shown, photogrammetric stereomodels are used to collect various types of map data. Photography and imagery also are a general source of a variety of data. All such mapping data were collected in the past at different times, for different purposes, and stored separately, frequently in different forms, although they often pertain to the same geographic location. When an attempt is made to use data from different sources, which usually were not registered with each other, significant errors often resulted. At present, the approach has changed significantly. All collected data over a geographic region—locational, thematic, and attribute data—are coregistered into what is known as a *geographic information system,* or GIS. This important concept is presented in detail in Chapter 14. Photogrammetric mapping plays a critical role in creating the *land base* layer to which all other GIS layers are registered. Photogrammetric technology also is used for other GIS tasks, such as updates to reflect change and development, for example, subdivision and construction, and collection of all relevant features to be stored in the GIS data base, which are quite economical to compile photogrammetrically.

13.30
PHOTOGRAMMETRIC PROJECT PLANNING

Three interrelated phases are involved in the planning for a photogrammetric survey: (1) the design of a flight plan, which must be followed by the pilot of the airplane while taking the aerial photography; (2) deciding on the amount and location of necessary ground control and performing the field surveys to obtain such control to the designated accuracy standards; and (3) estimation of costs to determine the most economical combination without compromising the accuracy requirements. Analysis of all three phases must be made in relation to the particular end product of the project, which may include topographic maps of a wide range of scales, orthophoto maps, mosaics, digital elevation models, and lists of supplemental control. For conventional topographic mapping, the contour interval that can be derived from vertical aerial photography plays an important role in determining the flight parameters. In the United States, the so-called *C*-factor, where

$$\text{Flying height} = C\text{-factor} \times \text{contour interval}$$

is used for vertical wide-angle photography with nominal focal length of 150 mm or 6 in. Conventional topographic maps, particularly at large and medium scales, are produced at *enlargements* of the original photographic scale. Table 13.1 lists a wide range of these scales as used in both Europe (metric system) and the United States. Note that the *C*-factor is higher for the U.S. mapping; the Europeans do not use it.

TABLE 13.1
Map and photo scales and contour intervals for metric and U.S. maps

Map scale	Photo scale	Enlargement factor	Flying height	Contour interval	C-factor
1 : 500	1 : 3000	6x	450 m	0.5 m	(900)
1 : 600	1 : 4200	7x	2,100 ft	1 ft	2100
(1″ : 50′)					
1 : 1000	1 : 5000	5x	750 m	1 m	(750)
1 : 1200	1 : 7800	6.5x	3,900 ft	2 ft	1950
(1″ : 100′)					
1 : 2400	1 : 12,000	5x	6,000 ft	5 ft	1200
(1″ : 200′)					
1 : 2500	1 : 10,000	4x	1,500 m	2 m	(750)
1 : 4800	1 : 16,800	3.5x	8,400 ft	5 ft	1680
(1″ : 400′)					
1 : 10,000	1 : 25,000	2.5x	3,750 m	5 m	(750)
1 : 12,000	1 : 30,000	2.5x	15,000 ft	10 ft	1500
(1″ : 1000′)					
1 : 24,000	1 : 38,400	1.6x	19,200 ft	10 ft	1920
(1″ : 2000′)					
1 : 48,000	1 : 57,600	1.2x	28,800 ft	20 ft	1440
(1″ : 4000′)					
1 : 50,000	1 : 50,000	1x	7,500 m	10 m	(750)
1 : 100,000	1 : 80,000	0.8x	12,000 m	20 m	(600)

Usually, the factors affecting the acquisition of photography for one type of product may be substantially different from those affecting photography for another. In general, the most important factors to be considered in planning aerial photography include the following:

1. The intended use of the photography (quantitative or interpretive).
2. The desired product (map, orthophoto, mosaic, numerical data).
3. The specified accuracy.
4. The size and shape of the area to be covered by photography.
5. The amount and disposition of relief in the area to be photographed.
6. The scale of photography, which in turn depends on the scale of the final product, the camera to be used, the photogrammetric instruments available, and so forth.
7. The aircraft tilt and crab angles, and drift as they influence the effective overlap and sidelap between adjacent photographs.

When these factors are carefully analyzed for a given project, the basic elements of the flight plan can be resolved. These elements include the flying height above datum, the ground distance between successive photo exposures, and the ground spacing between flight lines. Except for applications requiring only one flight line (such as for highway design), the resulting flight lines usually are laid out on the best available map of the area to be photographed. The general scheme for photographic coverage has been discussed and is shown in Figure 13.5. Although overlap is most commonly taken at 60 percent, it may be increased for some applications (e.g., mosaic). Also, sidelap can be high as 60 percent, particularly for block triangulation to produce supplemental control. The actual details of the steps involved in designing a flight plan and the computations associated with it are outside the scope of this chapter.

PROBLEMS

13.1. Describe concisely the following: (*a*) photo interpretation; (*b*) remote sensing; (*c*) aerial photogrammetry; (*d*) terrestrial photogrammetry; (*e*) close-range photogrammetry.

13.2. Draw a simple sketch showing the main components of an aerial camera and briefly discuss each component.

13.3. "Overlap and sidelap are necessary for proper aerial photographic coverage." Discuss this statement using a sketch to illustrate the two cases.

13.4. Explain the correspondence between stereo perception by the pair of human eyes and the geometry of overlapping vertical photographs.

13.5. Describe fully, with the aid of sketches, the two types of displacements that make an aerial photograph of the terrain different from a map of that terrain.

13.6. Enumerate five different photogrammetric products. Explain each briefly.

13.7. Explain the following operations for a stereo pair of dispositives to be used in a photogrammetric plotter: (*a*) interior orientation; (*b*) exterior orientation; (*c*) relative orientation; (*d*) absolute orientation.

13.8. Describe briefly three procedures for photogrammetric triangulation.

13.9. Explain the principal differences between two photogrammetric products: the digital terrain model and the orthophotograph.

13.10. List the most important factors to be considered in planning vertical aerial photography.

13.11. What is the height of a tower above terrain that appears on a truly vertical photograph with the following data:

Flying height above base of tower = 3200 m

Distance between principal point and the image of the tower base = 75.11 mm

Distance between principal point and the image of the top of the tower = 82.54 mm

13.12. The distance between two well-defined points on a vertical aerial photograph is 72.05 mm. The corresponding distance measured on a U.S. quadrangle map is 55.17 mm. If the map scale is 1:24,000, calculate the approximate scale of the photograph. Calculate the flying height above terrain if the focal length of the camera used is 150.00 mm.

13.13. A vertical photograph is taken with a lens having a focal length of 152.40 mm from a flying height of 2670 m above an airport where the elevation is 380 m above the datum. Determine the representative fraction, expressing the scale of the photograph at a point at which the elevation is 221 m above the datum.

13.14. How high above the datum must an aircraft fly so that photographs at a scale of 1:9600 may be obtained, if the focal length is 8.25 in. and the average elevation of the terrain is 800 ft?

13.15. A distance measured on a vertical photograph between two points, both lying at ground elevation of 366 m, scales 82.677 mm. The focal length of the camera is 152.908 mm. The distance between the same pair of points measures 25.451 mm on a 1:24,000 quadrangle sheet. Compute the flying height above the datum at which the photograph was taken. Calculate the datum scale of the photograph.

13.16. Photographs are to be taken for preparing a highway design map. The lowest elevation in the area to be photographed is 144 m and the highest elevation is 282 m. The minimum photographic scale is to be 1:6000. What must be the flying height above the datum if the camera to be used contains a lens with a focal length of 134.63 mm? What will be the maximum scale?

13.17. The distance between two section corners is assumed to be exactly 1 mi. Both corners lie at 450 ft above the datum. If the distance between the images of these two corners scales 2 : 342 in. on a vertical photograph taken with a lens having an 8.262-in. focal length, what is the flying height at which the photograph was taken?

13.18. The datum scale of a vertical photograph taken with a lens having a focal length of 152.400 mm is 1:8000. A hilltop lies at an elevation of 800 m above the datum and the image of the hilltop is 71.679 mm from the principal point of the photograph. Compute the relief displacement of the hilltop. If this photo should happen to be tilted $2°$ such that the hilltop is now on the "up side" of the tilted photo, explain the effects of tilt on the displacement caused by relief.

13.19. Two overlapping vertical photographs are taken with a camera having a focal length of 152.00 mm. It is assumed that both photographs are exposed at a flying height of 3500 m above mean terrain and that the overlap is 60 percent. A control point with elevation of 100.22 m is used as a basis for calculating the elevations of two other points. If the differences in parallax between the control point and the two points are 4.11 mm and -1.73 mm, respectively, calculate the elevations of the two points.

REFERENCES

Ackerman, F., and H. Shade. "Application of GPS for Aerial Triangulation." *Photogrammetric Engineering and Remote Sensing* 59, no. 11 (November 1993).

American Society for Photogrammetry and Remote Sensing. *Manual of Photogrammetry,* 4th ed. Bethesda, MD, 1980.

Chamard, R. R. "Photogrammetric Mapping for Highways: Western U.S." *ASCE Journal of the Surveying Engineering Division* 109, no. 1 (1983), p. 1.

Cowden, R. W., and R. F. Brinkman. "Digital Orthophotos: A New Alternative for Creating Base Maps." *Geo Info Systems* 3, no. 5 (1993), p. 53.

Karara, H. *Non-Topographic Photogrammetry,* 2nd ed. Bethesda, MD: American Society for Photogrammetry and Remote Sensing, 1989.

Kraus, K. *Photogrammetry,* vol. 1, *Fundamentals and Standard Processes.* Bonn, Germany: Ferd. Dummler Verlag, 1993.

Lapine, L. A. "Practical Photogrammetric Control by Kinematic GPS." *GPS World* 1, no. 3 (1990), p. 44.

LaPrade, G. L. "Stereoscopy—A More General Theory." *Photogrammetric Engineering and Remote Sensing* 38, no. 12 (1972).

Light, D. "The New Camera Calibration System at the U.S. Geological Survey." *Photogrammetric Engineering and Remote Sensing* 43, no. 2 (1992), p. 185.

Moffitt, F. H., and E. M. Mikhail. *Photogrammetry,* 3rd ed. New York: Harper and Row, Publishers, 1980.

Morain, S., et al. "U.S. National Report—Status of Photogrammetry, Remote Sensing and Geographic Information Systems in the United States." *Photogrammetric Engineering and Remote Sensing* 43, no. 8 (1992), p. 1073.

Thorpe, J. "Aerial Photogrammetry: State of the Industry in the U.S." *Photogrammetric Engineering and Remote Sensing* 59, no. 11 (November 1993).

Wolf, P. R. *Elements of Photogrammetry,* 2nd ed. New York: McGraw-Hill, 1983.

Periodicals

Geomatica. Published by the Canadian Institute of Geomatics, Ottawa, Canada.

ISPRS Journal of Photogrammetry and Remote Sensing. Published by Elsevier, Amsterdam, the Netherlands.

Photogrammetric Engineering and Remote Sensing. Published by the American Society of Photogrammetry, Bethesda, MD.

The Photogrammetric Record. Published by the British Society of Photogrammetry, London.

Surveying and Land Information Systems. Published by the American Congress on Surveying and Mapping, Washington, DC.

Mapping, Digital Mapping, and Spatial Information Systems*

14.1
GENERAL

The various classes, types, and general categories of maps are defined in Section 1.11, which noted two main types of maps: line maps and orthophotographic image maps. Rapidly evolving technology now permits most of these two types of maps to be generated digitally, with minimal manual drafting. This evolution represents a significant change in the production of maps and allows use of spatial information systems for storage, retrieval, manipulation, and evaluation of maps and map data.

In this chapter, emphasis is placed on the characteristics and construction of topographic maps in conventional line and digital formats, methods for representing relief, use of digital terrain models in contour generation and mapping, plotting of maps by computer-aided drafting (CAD) systems; integration of the mapping process into spatial information systems such as geographic and land information systems (GIS and LIS), map content and symbology, and standards of accuracy for maps, map data, and the exchange of digital data. A brief overview of modern map production also is presented.

14.2
DATUMS FOR MAPPING

In surveying and mapping, a datum is necessary to relate positions, directions, and elevations measured in the field to global reference systems. Linking observations to coordinates on a particular datum ensures that surveys and adjustments over a wide geographic area will be consistent with each other. A country or a group of countries generally has maintained its own datum. Similar, but not identical, this multiplicity of regional datums requires transformations when survey data cross such regional boundaries. Datum types include horizontal or geodetic and vertical.

---

* This chapter was written by Dr. James Bethel and Dr. Steven R. Lambert, Purdue University, West Lafayette, IN, and Dr. Jolyon D. Thurgood, Space Imaging, Thornton, CO.

A *horizontal* or *geodetic datum* consists of an ellipsoid of revolution approximating the figure of the earth and a set of constants or constraints that specify the size, position, and orientation of the ellipsoid. Two of the constants define the semimajor axis and the flattening of the ellipsoid. The position is often given indirectly by three constants ξ_o, η_o, and N_o at an origin point, O, where ξ_o represents the deflection of the vertical in the meridian plane, η_o represents the the deflection of the vertical in the prime vertical plane, and N_o represents the difference between the geoid and the ellipsoid, which is also called the *geoid undulation*. In addition to these three constants, an observed astronomic latitude and either an observed astronomic longitude or an astronomic azimuth would define the geodetic latitude and longitude of the origin point. More recent datums define the ellipsoid position by specifying that it coincide with the center of mass of the earth. An example of the earlier type of horizontal datum is the North American datum of 1927 (NAD 27), which was based on an origin point at Meades Ranch, Kansas, and used the Clarke 1866 ellipsoid with $a = 6,378,206.4$ meters and $b = 6,356,583.8$. This was the basis for most mapping done in the United States until recently. A general readjustment including satellite observations and very long base-line interferometry has resulted in the North American datum of 1983 (NAD 83), which is an example of the latter type of geodetic or horizontal datum. It is based on the geodetic reference system of 1980 (GRS 80) ellipsoid with $a = 6,378,137.0$ and $1/f = 1/298.257522101$. The adoption of GRS 80 for the NAD 83 is a favorable circumstance for users of the GPS system. The world geodetic system of 1984 (WGS 84), as used in GPS computations, essentially is the same as GRS 80.

A vertical datum is the surface to which elevations or depths are referred. Prior to 1991, all surveys and mapping in the conterminous United States were based on the national geodetic vertical datum of 1929, formerly called the sea-level datum of 1929. This datum was based on a least-squares adjustment to 26 mean sea-level tide stations (Zilkosky, Richards, and Young, 1992) in the United States and Canada and was the result of a general adjustment of level networks in the United States and Canada in 1929. In 1991, the National Geodetic Survey completed incorporating new data into the existing survey network, resulting in the North American vertical datum of 1988 (NAVD 88). Elevations used in topographic mapping, geodetic surveys, engineering studies, engineering construction surveys, and geographic information systems should be referred to the current national vertical datum, NAVD 88. The NAVD 88 is not a mean sea-level datum and should not be confused with local mean sea-level datums. The national datum was determined by a minimal constraint adjustment, holding fixed the height of a primary tidal bench mark at Father Point (Pointe au Pere), Rimouski, Quebec, Canada. Canadian-Mexican-U.S. leveling observations were included in this adjustment. Consequently, small differences exist between the national datum and local mean sea-level for a specific location.

Elevations used in boundary surveys are often referred to a *tidal datum* in tidal waters or the *lake level* in the Great Lakes regions. Tidal datums are defined by the phase of the tide, described as mean high water, mean low water, and mean lower low water. Caution always should be exercised to ensure that bench marks and control points refer to the same datum, particularly in areas where multiple datums are known to exist.

Other horizontal and vertical datums are employed for Alaska (vertical datum), Puerto Rico, the Virgin Islands, Guam, and other oceanic islands. Specifications for these datums can be obtained by requesting information from the National Geodetic Survey.

14.3
REFERENCE COORDINATE SYSTEMS FOR MAPPING

Given a common reference as described by the datum definitions in the previous section, it is still usually necessary to project ellipsoidal data (ϕ, λ) into map projection coordinates

(*X, Y*) or eastings and northings (E, N). This is necessary both for preparing hard-copy maps and for representing map data digitally in a geographic information system. Eventually GIS systems may accommodate latitude and longitude, but at present they generally require Cartesian coordinates. As described in Chapter 11, map projections include the Lambert conformal conic and the transverse Mercator. The newer U.S.G.S. quadrangle maps are on the universal transverse Mercator (UTM) projection (Section 11.14). Older quadrangle maps employed the polyconic projection (Section 11.9). The state plane coordinate systems (SPCS), based on the Lambert conformal conic or the transverse Mercator projections, are widely used as reference coordinates for mapping and GIS. For regional or statewide projects that may cover multiple state plane zones, the UTM projection often is used. For special circumstances, a unique projection can be defined for a particular project, although data interchange is facilitated by using standard projections wherever possible.

783

CHAPTER 14:
Mapping, Digital
Mapping, and
Spatial
Information
Systems

14.4
TOPOGRAPHIC MAPS

A topographic map shows, through the use of suitable symbols, (1) the spatial characteristics of the earth's surface, with such natural features as hills and valleys, vegetation and rivers; and (2) constructed features such as buildings, roads, canals, and cultivation. The distinguishing characteristics of a topographic map, as compared with other maps, is the representation of the terrain relief.

Topographic maps are used in a variety of ways. They are necessary in the design of any engineering project that requires the consideration of elevations for gradients. They also are used for delineating the extent of a flood plain, planning for economic development, and managing natural resources.

The preparation of general topographic maps traditionally has been a function of governmental agencies. However, the rapid development of computer-based tools for terrain modeling and the increasing availability of high-quality field data enable nearly any user of topographic information to create detailed, specialized topographic maps using a desktop computer.

The principal source of topographic data in the United States remains the national map series of medium (1:24,000) to small (1:1 million) scale topographic maps prepared by the National Mapping Division of the U.S. Geological Survey (Figure 14.1). These maps cover the entire United States and its territories in quadrangle tiles from $7\frac{1}{2}' \times 7\frac{1}{2}'$ in latitude and longitude at a map scale of 1:24,000 to tiles $4° \times 12°$ in extent at a scale of 1:1 million. These topographic maps have been compiled using field survey data and photogrammetric compilation techniques. Recently, the U.S.G.S. also provided the same data contained in the printed map series in digital form as digital line graph (DLG) files. Although the printed maps are regularly photo revised, it is common for the latest quadrangle maps not to have been revised in several years. For engineering applications in rapidly developing areas, this time lag may not be acceptable.

The U.S.G.S. currently is in the process of developing a new, large-scale digital orthophoto quadrangle (DOQ) and quarter-quad (DOQQ) map series for the entire country. These new orthophoto maps will be generated at scales of 1:24,000 and 1:12,000, respectively. The DOQQ maps will have a 1-m ground sample (pixel) resolution and will be provided in a UTM system based on NAD 83. The status of the DOQ mapping program is shown graphically in Figure 14.2. The new map series combines the information content of a photograph and the geometric qualities of a standard map and will be uniquely suited for use in spatial information systems. More information on these products is available from the Earth Sciences Information Centers of the U.S.G.S. or by calling 1-800-USA-MAPS.

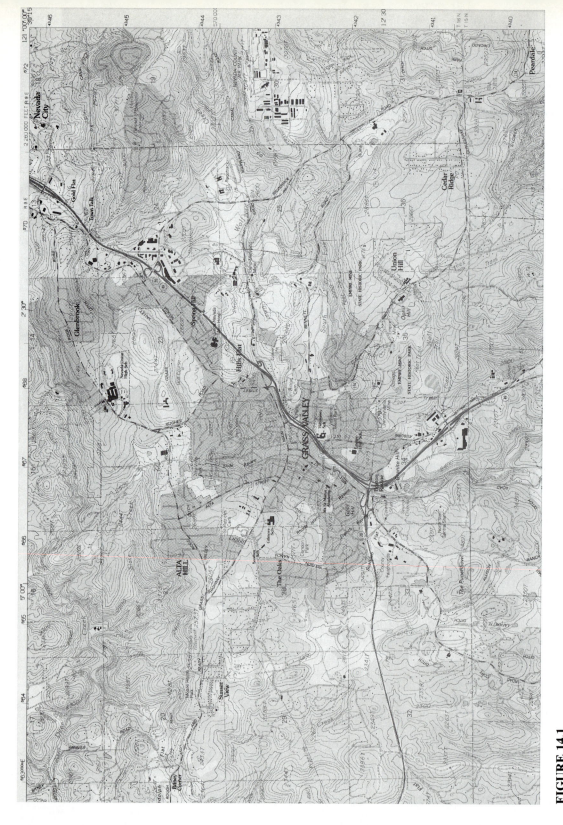

FIGURE 14.1
Typical topographic map of the U.S. Geological Survey. Scale is 1:24,000 (2000 ft/in.). Contour Interval 20 ft. (*U.S. Geological Survey.*)

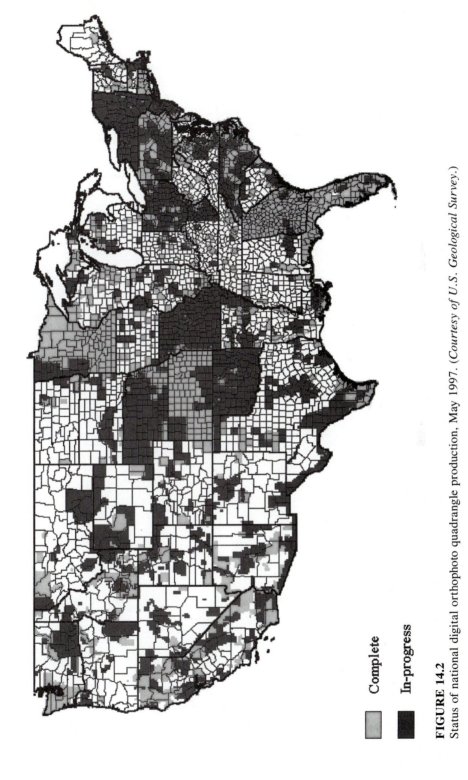

FIGURE 14.2
Status of national digital orthophoto quadrangle production, May 1997. (*Courtesy of U.S. Geological Survey.*)

Complete

In-progress

14.5
REPRESENTATION OF RELIEF

Relief may be represented by relief models, three-dimensional (3D) perspective views with wire-frame or shaded rendering or draped imagery, color coding of elevation or slope, contour lines, cross sections, profiles, hachures, or form lines. Examples of some of these representations are shown in Figure 14.3. For engineering design, the preference historically has been in favor of contour-line representations or cross sections and profiles, because these methods indicated the elevations directly and quantitatively. These techniques now

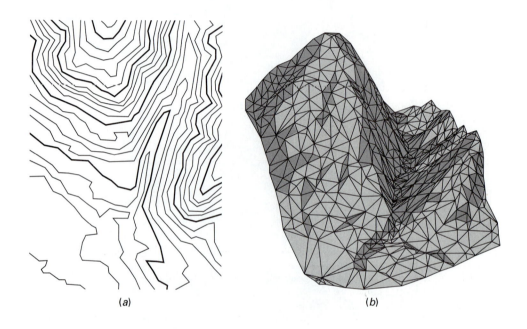

(a)

(b)

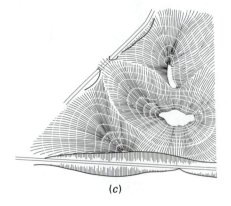

(c)

FIGURE 14.3
Representation of relief: (a) automatically plotted contours, (b) three-dimensional perspective terrain model, (c) hachures.

are being augmented by direct computer evaluation of a *digital terrain model* without necessarily having to produce an intermediate contour plot. This aspect will be addressed more fully in Sections 14.8 through 14.10. For visualization and the rapidly expanding needs of computer-based simulation, the 3D perspective views are becoming more important. These trends suggest a strategy of storing the terrain data in as general a form as possible and producing specific graphic depictions as needed for particular applications. Geographers refer to maps depicting elevation as *hypsographic maps*. Engineering topographic maps generally would fall into this category.

787

CHAPTER 14:
Mapping, Digital
Mapping, and
Spatial
Information
Systems

14.6
CONTOURS AND CONTOUR LINES

A *contour* is an imaginary line of constant elevation on the ground surface. It may be thought of as the trace formed by the intersection of a level surface with the ground surface; for example, the shoreline of a still body of water.

If the locations of several closely spaced ground points of equal elevation are plotted on a drawing, a line joining these points is called a *contour line*. Thus, contours on the ground may be represented by contour lines on the map. The terms *contour* and *contour line* often are used interchangeably. On a given map, successive contour lines represent elevations differing by a fixed vertical distance, called the *contour interval*. The choice of contour interval depends on the map scale, the degree of relief, and the needs of the user for whom the map was prepared.

As mentioned in the previous section, contour-line representations are particularly well suited for quantitative visual interpretation. For drainage analysis, site evaluation, or route planning, the contour-line representation is preferred for visual interpretation over perspective views or hachures, for example. In the computer evaluation of terrain data, contour lines are usually *not* the representation of choice. In that case, a digital terrain model (DTM), consisting of rectangular grid points, or random points organized as a triangulated irregular network (TIN) usually is preferred. Under this scenario, the contour representation may be produced as a by-product of archival DTM data. With the growing acceptance of digital mapping and geographic information systems (GIS), digital terrain model methods often are preferred over an exclusive reliance on contour lines.

Contour lines can represent the terrain effectively when a single terrain point is associated with each planimetric position. As mentioned later, vertical discontinuities and overhangs are not well represented by contour lines. For this reason, contour-line representations (and similar representations) often are referred to in mapping parlance as *2.5D maps*. This indicates that they fall between 2D planimetric maps and full 3D representations, where surface elements cover arbitrarily complex objects.

The principal characteristics of contour lines can be illustrated by reference to Figure 14.4. Figure 14.4 shows that the slope of the river can be disregarded. The height of the river at the time of the survey was at an elevation of 510 m; hence, the shoreline on the map marks the position of the 510-m contour line. For this map, the contour interval is 5 m. If the river were to rise to the 5-m stage, the shoreline would be represented by the 515-m contour line; similarly, the successive contour lines at 520 m, 525 m, and so forth, represent shorelines that the river would have should it rise by 5-m stages.

The principal characteristics of contour lines are as follows:

1. The horizontal distance between contour lines is inversely proportional to slope. Hence, on steep slopes (such as at the railroad and the riverbanks in Figure 14.4), the contour lines are spaced closely.
2. On uniform slopes, the contour lines are spaced uniformly.

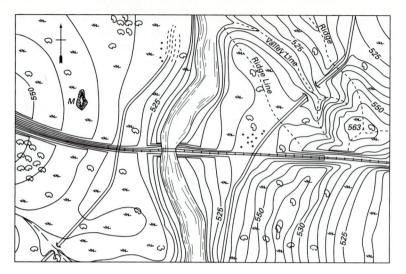

FIGURE 14.4
Contour lines.

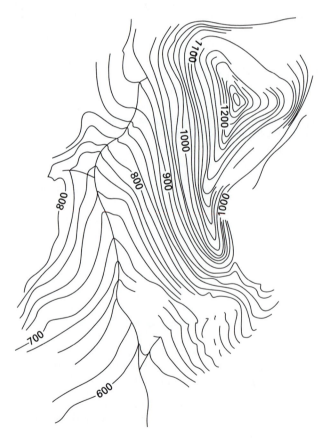

FIGURE 14.5
Mountainous terrain showing a ravine and a ridge.

3. Along plane surfaces (such as those of the railroad cuts and fills in Figure 14.4), the contour lines are straight and parallel to one another.

4. As contour lines represent level lines, they are perpendicular to the lines of steepest slope. They are perpendicular to ridge and valley lines where they cross such lines.

5. As all land areas may be regarded as summits or islands above sea level, evidently all contour lines must close on themselves either within or outside the borders of the map. It follows that a closed contour line on a map always indicates either a local summit or local depression. Depression contours generally are represented by downward pointing hachure lines as shown at *M* in Figure 14.4.

6. As contour lines represent lines of constant elevation on the ground, they cannot merge or cross one another on the map, except in the rare cases of vertical surfaces (see bridge abutments in Figure 14.4) or overhanging ground surfaces, such as at a cliff or a cave. As mentioned previously, this is because a contour map is inherently a 2.5D representation of the terrain surface.

7. A single contour line cannot lie between two contour lines of higher or lower elevation.

Other special features of contours are shown in Figures 14.5 and 14.6. Figure 14.5 shows contours with a 20-m interval for mountainous terrain. On the right is a well-defined ridge. Note that the V or U shape of the contours points down the slope on a ridge. On the left is a ravine with several small streams leading into it. In contrast to the ridge, the V- or U-shaped contours point up the slope to indicate a ravine, a stream, or a ditch. Figure 14.6 shows how contours cross artificial features. In Fig. 14.6a, the contours cross the paved road built with a cut on the upper side and a fill on the lower side. On the upper side is a ditch, indicated by the V-shaped contours pointing up the slope. The contours cross the pavement as a smooth line (the pavement is relatively smooth), which is slightly convex down the slope, because most paved roads are crowned and slightly higher in the center than on the

789

CHAPTER 14:
Mapping, Digital
Mapping, and
Spatial
Information
Systems

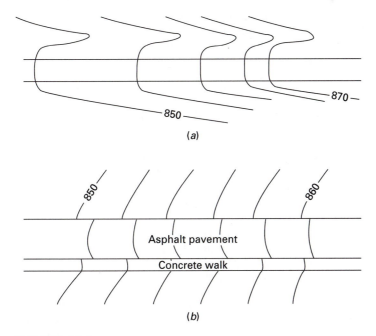

FIGURE 14.6
(a) Contours crossing a paved road. (b) Contours crossing a paved road with a curb.

two edges. In Figure 14.6b, the contours cross a paved road with curbs. Most curbs are 0.15 m high, so the contour crosses the curb at right angles and runs along the face of the curb until it reaches the contour elevation in the gutter line. Therefore, if the grade (Section 3.8) of the road is 5 percent, the distance from where the contour crosses the top of the curb to the point where it meets the contour elevation in the gutter is $0.15/0.05 = 3$ m. Then the contour crosses the pavement slightly convex down the slope to the gutter line on the other side, follows the face of the curb until the elevation at the top of the curb is reached, and crosses the curb and sidewalk at right angles (assuming that the sidewalk is horizontal). The contour interval in Figure 14.6a is 5 ft and that in Figure 14.6b is 2 m.

14.7
CHOICE OF MAP SCALE AND CONTOUR INTERVAL

Formerly, when a hard-copy map was the archival data record used to present the results of a topographic survey, the choice of map scale and contour interval indeed was critical. Current engineering practice often dictates that map data be recorded digitally, usually in ground coordinate units. Thus, within limits, the hard-copy representation of the map can be plotted at any scale desired. This capability is well known to users of CAD systems and other computer graphics programs. Of course, such things as text annotation size and placement, point and line symbology, and feature generalization are scale specific and do not change with plot scale. So, for a map plotted at much smaller scale than intended, the text would be unreadable. A portion of the same map plotted at much larger scale than intended would be graphically unpleasing with the text too large. Likewise, the choice of contour interval was once a fixed parameter of the data collection process. Currently, the parameter that limits the fidelity of the representation of the ground surface often is the density of points in the digital terrain model (DTM, see Section 14.8). This might include the spacing of the points of known elevation or elevation "posts" in a grid layout, the density of points of known spatial position or mass points for construction of a triangular irregular network (TIN) (See Section 14.8), and the density and completeness of breaklines (Section 14.9) and other subordinate features. Having selected the sampling criteria of the terrain surface, a contour representation, again within limits, can be interpolated and plotted at any desired interval. Computer mapping programs, quite readily, and *inappropriately*, will produce 1-dm contours from data points 30-m apart and accurate to only 1 m. Therefore, sound judgment should be used in selecting the sampling density of the terrain points and a contour interval for interpolation and plotting. Hard-copy contour maps will continue to be used for the foreseeable future, and it therefore is useful to look at some of the design parameters used in their construction, in particular scale and contour interval.

Accuracy standards for engineering maps, presented later in this chapter (Section 14.26), generally dictate that well-defined, plotted features should be located within 0.5 mm of their true position at the plotted map scale. Therefore, if a user needs point positions accurate to 2.5 m, a map scale of 1:5000 is indicated. With regard to elevations, accuracy standards generally require that interpolated heights be accurate to one-half contour interval. Therefore, if one needs 0.25-m elevation accuracy, the contour interval should be no greater than 0.5 m. Along with a given map scale, an implied level of detail is expected by the user. Other factors that influence the choice of map scale are (1) the clarity with which features can be shown, (2) the cost (the larger the scale, the higher is the cost), (3) the consistency of the map with other adjoining or overlapping maps, (4) the number and character of the features to be plotted, the nature of the terrain, and the contour interval. Typical map scales, map uses, and corresponding contour intervals are shown in Table 14.1.

791

CHAPTER 14:
Mapping, Digital
Mapping, and
Spatial
Information
Systems

TABLE 14.1
Typical map scales

Map scale	Typical uses	Contour interval for nonmountainous terrain
1 : 1,000	Design	0.25 m
1 : 2,000	Design	0.5 m
1 : 5,000	Planning	1 m
1 : 10,000	Planning	2 m
1 : 25,000	Regional planning	2.5 m
1 : 50,000	Regional planning	5 m
1 : 100,000	Regional planning	5 or 10 m
1 : 250,000	Regional planning	10 m
1 : 500,000	State planning	20 m
1 : 500,000	National planning	20 or 50 m

The choice of a proper contour interval for a topographic survey and map is based on four principal considerations: (1) the desired accuracy of elevations to be determined from the map, (2) the characteristic features of the terrain, (3) the legibility of the map, and (4) the cost. Assuming that heights can be interpolated to one-half of the contour interval, then a map with a 1-m contour interval should yield more accurate heights than a map with a 2-m contour interval. Terrain areas with a fine-textured surface may require a smaller contour interval to represent it than otherwise would be necessary. Contours should not be so close together as to obscure other important map features, although cartographic design and selection of colors, saturations, and line weights can significantly influence the map legibility. Smaller contour intervals, in general, cost more assuming that the conventional accuracy levels are maintained.

As shown in Table 14.1, smaller map scales generally are associated with larger contour intervals. Traditionally, maps have been classified according to scale as large, medium, and small, with these categories generally as in Table 14.2.

The American Society of Civil Engineers' (ASCE) Surveying and Mapping Division has a more detailed classification for map scales and contour intervals.

1. *Design maps.* These maps are used in the design and construction of specific engineering work of all kinds. Scales vary from 1:100 to 1:2000, with contour intervals from 0.1 to 1 m, depending on the type of project, land use, and terrain characteristics. Two subcategories are given within this group. *Critical design maps* are used on projects having critical space, orientation, position, or elevation restrictions; for example, a highway interchange in an urban area. *General design maps* are prepared for projects that have no such rigid restrictions with respect to location; for example, a map prepared for a rural water distribution system.

2. *Planning maps.* These maps include a large group of maps used in planning engineering work or in overall planning at the urban, regional, national, and international levels. These maps may be used as a foundation for GIS, geological studies, land use, agricultural production, population studies, public service planning, and atlases. These maps generally fall in the medium- to small-scale range.

TABLE 14.2
Map-scale categories

Large scale	1 : 20,000 and larger
Medium scale	1 : 20,000–1 : 50,000
Small scale	1 : 50,000 and smaller

TABLE 14.3

Map scales and contour intervals for various topographic maps to be used for designing and planning paved highways in a suburban area (adapted from ASCE, 1972, *courtesy of American Society of Civil Engineers*).

Topographic map subcategory	Scale (representative fraction)	Terrain characteristics	Contour interval in m (ft)
Design critical	1 : 500	Flat	0.1 (0.5–1)
		Rolling	0.5 (2)
		Hilly	1 (5)
General	1 : 500 to 1 : 2000	Flat	0.1 (1)
		Rolling	0.5–1 (2–5)
		Hilly	1–2 (5–10)
Local planning	1 : 4000 to 1 : 6000	Flat	0.5–1 (2–5)
		Rolling	2–5 (10–20)
		Hilly	5–10 (20–50)
Regional	1 : 25,000	Flat	0.5–1 (2–5)
		Rolling	1–5 (10–20)
		Hilly	5–10 (20–50)
National	1 : 100,000 to 1 : 3 million		2–200

An ASCE Task Committee report (1972) addresses the map scales and contour intervals necessary for particular engineering projects in different terrain types. Table 14.3 shows map scales and contour intervals for various maps used in highway design. The reader interested in specific recommendations for other types of engineering projects should consult ASCE (1972).

14.8
DIGITAL TERRAIN MODELS

The pervasiveness of computers has contributed to a significant shift in the way survey and map data are collected, processed, presented, and stored. Well into the 1970s, photogrammetric compilers manually traced contour lines from stereo imagery. This contour representation of the terrain, plotted on stable base material such as Mylar, became the archival medium from which subsequent terrain analysis and engineering design were done. Computer capabilities have introduced two fundamental changes into this scenario. First, terrain data now are collected mostly as a sequence of discrete (X, Y, Z) data points. These *posts,* together with other supplementary data, such as abrupt changes in terrain slope or *breaklines,* constitute a discrete sampling of the continuous terrain surface that should suffice for its mathematical reconstruction. Breaklines are necessary where the first derivative of the terrain surface is not continuous; for example, along a ridge line or a drain line. These data are not tied to a particular graphic representation. Implicitly associated with the terrain data in a DTM is the interpolation algorithm used to reconstruct the terrain surface. Second, the archival record is the digital coordinate file itself rather than a particular graphic depiction. Such graphical depictions as contours, profiles, or wire frame perspective views can be generated as needed, but only the original terrain points and features are considered to be archival data.

793

CHAPTER 14:
Mapping, Digital
Mapping, and
Spatial
Information
Systems

DTMs generally are organized such that the "mass points" lie in a regular grid pattern or they represent vertices of local triangular patches in an array referred to as a *triangulated irregular network*. Companies and government agencies seem to have developed institutional preferences for one method or the other. Just as there are engineering design criteria for selecting a contour interval to represent terrain for a given application, so too similar criteria are used to select a point spacing so that the DTM adequately represents the terrain. These criteria depend on the potential uses for the data, accuracy requirements, the terrain character, and other factors. The advantages of a regular grid layout are a simplified data collection routine, and ease of data access by subsequent programs. The disadvantages are related mostly to the necessity to select a single grid interval, sufficient to define the terrain in the roughest areas although likely to be oversampled in regions where the terrain is smooth and featureless. Conversely, the merits of the irregular point approach are the mirror image of those for the regular grid. The sampling interval can change to match the local terrain character, introducing a kind of stratified sampling, optimizing the quantity of data necessary to define the terrain. Data access for subsequent software analysis is considerably more involved than when using the simple grid structure.

Ideally, during the design of a DTM database, a quantitative analysis is done to determine the magnitude of the errors expected during reconstruction of the terrain surface. The magnitude of these errors should be within the error budget of potential user or client applications. Interpolation methods for generating intermediate points can include patch-wise polynomials, b-splines, moving surface methods, linear prediction with trend surfaces and covariance functions (sometimes called *summation of surfaces*), bilinear methods, plane triangle methods, and ideal reconstruction functions from signal theory. Given a DTM database and an interpolation function, one should be able to construct a profile or cross section along any arbitrary path within the area covered by the DTM. Likewise, one should be able to interpolate heights at arbitrary points within the area covered. This capability would permit one to interpolate heights at regular grid points from an irregular grid as well as interpolate irregular points from a regular grid. Thus, with possibly some cost to accuracy, one could convert between these two popular storage conventions. Converting from irregular points to regular points is straightforward. Converting from regular gridded points to irregular points often is more difficult, especially when there is a desire to reduce drastically the number of points.

Another issue to consider when constructing or using DTM databases is the matter of consistency between the terrain, as defined by the DTM, and the feature data, such as roads, streams, and buildings. On a hard-copy map, these two classes of data always are implicitly consistent. However, if they are collected and stored independently as 3D or 2.5D, then they may be inconsistent or conflicting. This could occur if, for example, a road were above the terrain surface or if a stream were placed so as apparently to flow uphill.

14.9
BREAKLINES AND OTHER TERRAIN FEATURES

Terrain data points, whether in a grid pattern or an arbitrary pattern, always will fail to represent terrain fully in areas where there are sharp breaks or, more properly, discontinuities in slope. Such discontinuities occur along ridge lines, at the upper and lower edges of

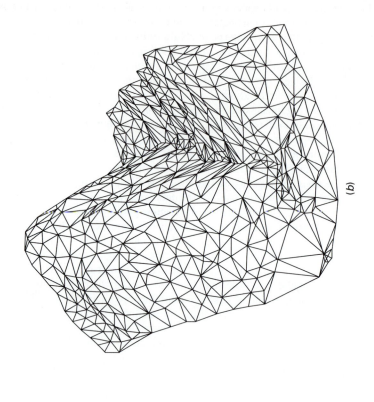

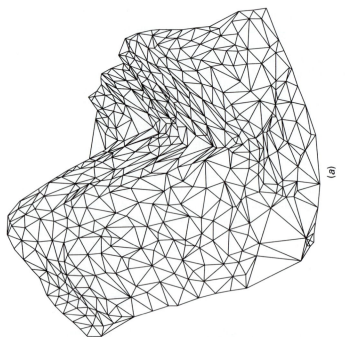

FIGURE 14.7
DTM: (a) without breaklines, (b) with breaklines.

795

CHAPTER 14:
Mapping, Digital
Mapping, and
Spatial
Information
Systems

a steep embankment, along drain lines, and in the vicinity of constructed cuts and fills. In these areas the mass points must be supplemented by breaklines that indicate a sharp change in slope. Other related features, which can help the mathematical terrain reconstruction process, would include form lines (soft breaklines) and spot heights to represent the high point on a hilltop or the low point in a depression. These supplementary features should be used whenever interpolating or computing the height of new points that fall in between the originally sampled data points. They likewise should be used in any contour generation to ensure that a realistic terrain depiction is given by the generated contours. Examples are shown in Figure 14.7 of a DTM with and without needed breaklines.

14.10
CONTOUR GENERATION BASED ON DIGITAL TERRAIN MODELS

Most mapping projects in the future, requiring contours, will employ computer interpolated contours rather than hand-traced contours. The underlying DTM may come from field surveying for small projects or from photogrammetric compilation for more extensive projects. The photogrammetric data could be manually observed in a stereoplotter or derived automatically by digital image-matching techniques. The DTM will represent the terrain via some sampling strategy and an interpolation algorithm. To generate contour lines at some contour interval, a common strategy is to interpolate a fine grid of elevation points over the area in which contours are required. Then, the contour-line interpolation can be done easily in the fine grid. Figure 14.8 illustrates the process of contour interpolation within a rectangular grid cell. Individual segments are generated within each cell and collected to form continuous line strings for a given contour line. Note that an ambiguity can be present when there is a saddle occurring inside a cell. Two valid contour representations are used for this condition, and if the differences are significant, a finer grid interval is indicated.

Contours drafted manually by a skilled cartographer can be very pleasing to the eye. Machine-generated contours, on the other hand, easily can appear "machine generated,"

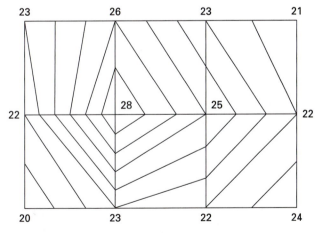

FIGURE 14.8
Contour interpolation within a rectangular cell.

particularly if poor guiding parameters are chosen. Machine-generated contours often look very angular whereas a skilled cartographer always would represent smooth terrain with smooth contour lines. One answer to this situation is to perform some smoothing on the generated contour line. This often is accomplished using piecewise polynomials with slope and continuity constraints. A special case of this technique is the cubic spline (see DeBoor, 1978). To accommodate arbitrary plane curves, one can use a parametric approach where both X and Y become a function of some linear distance or distance approximation. Another characteristic of finely crafted manual contours is the fine detail where a contour crosses a drain line. A U- or V-shaped segment pointing uphill usually will be present on hand-drafted contour maps. This may be missing on computer-generated contours, but here the presence of additional data such as breaklines and form lines can play a significant role in helping to produce pleasing and accurate contour lines.

14.11
USE OF SYMBOLS, COLOR, AND SHADING
TO DEPICT RELIEF

Several techniques have evolved to represent three-dimensional relief on a two-dimensional map. Many of these techniques are used primarily in the manual drafting of topographic maps, including the use of shading and hachures. With the availability of computer-based mapping tools, additional methods, including digital shading, surface textures application, artificial lighting, and perspective views, also can be used to portray relief.

Shading to represent relief portrays the surface viewed from directly above (orthographically) with a light source set at a particular location and angle to flood the surface with parallel light rays. This pattern of lighting creates shadows, enhancing the appearance of terrain relief. This method works particularly well in high-relief areas.

Hachures are short, nearly parallel lines plotted parallel to the steepest slopes (Figure 14.3c). The length and weight of the hachure lines indicate the steepness of the terrain at that point. Hachures give a slightly more visually appealing and intuitive feel for the variation in terrain relief than shading. Both of these methods can be improved on using color density shading or surface textures that change as a function of terrain slope. Both techniques are based on visual rendering of a map using image-processing techniques. The great advantage of these approaches is that the final map can be rendered easily, using different color schemes and textures depending on the needs of the user. In contrast to the traditional shading approach, the artificial lighting of a digital map can be used to enhance the terrain appearance and produce topographic maps that are visually striking.

14.12
SPECIALIZED DEPICTIONS OF RELIEF

In addition to the straightforward graphic approaches to representing relief given in the previous section is another group of depictions that can be extremely useful. These include perspective views, wire-frame representations, ray-traced images, and draped images.

Perspective views are appealing because they approximate the view obtained by the human eye. Perspective images are a function of the location of the viewer's eye, the location of the projection, and the location of the projected rays. To support perspective viewing of terrain relief, a 3D data set is required, where the third dimension is the terrain relief. The terrain relief most often is depicted using a regularly gridded digital elevation model. For example, a perspective view of the terrain in the Cope, Indiana, 1:24,000 U.S.G.S. quadrangle can be seen in Figure 14.9a. To provide further definition in the terrain

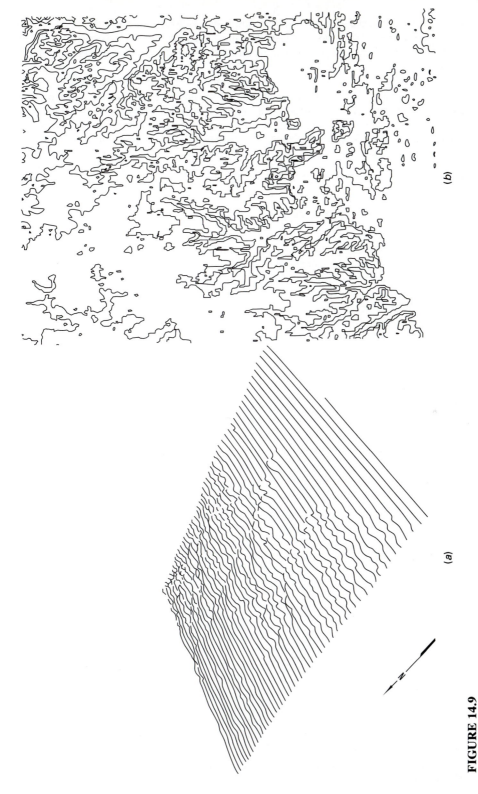

(b)

(a)

FIGURE 14.9
Cope, Indiana, 1:24,000 quadrangle topography looking northeast: (a) perspective view, vertical exaggeration factor = 3: (b) planimetric view of 20-m contour interval.

relief, the vertical scale of the perspective view can be exaggerated where the range of relief is small. This normally is done to provide a more effective image of the terrain relief. A planimetric view of the Cope, Indiana, quadrangle with a 20-m contour interval is shown in Figure 14.9b.

Perspective views also can be generated from a combination of traditional line maps, satellite images, and scanned aerial photos, by draping the planimetric maps over an independently derived terrain model. This approach provides a very useful visual image of the relationship of line map or planimetric features with the terrain relief. Such a method often is used in the analysis and viewing of remotely sensed satellite images, for animation or for "fly-through" simulations.

14.13
TOPOGRAPHIC MAP CONSTRUCTION

Most of large-scale topographic maps are compiled from aerial photographs, using photogrammetric compilation techniques. Photogrammetry has been long recognized as a cost-effective and accurate means of compiling large-scale base maps and developing contour data. The process for using photogrammetric techniques for large-scale base mapping is quite straightforward and can be summarized by the diagram shown in Figure 14.10.

The nine steps illustrated in Figure 14.10 represent the major elements of all photogrammetric mapping projects and indicate their sequence. For very large projects, the last

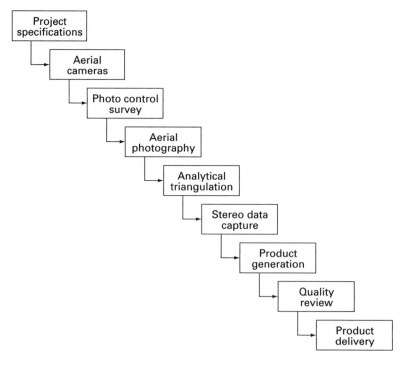

FIGURE 14.10
Major elements of photogrammetric mapping projects.

five steps may be completed in parallel, where phased delivery of mapping products takes place. To provide the proper context for the sections that follow, it is important that each process be understood in general terms; therefore, each will be described briefly.

799

CHAPTER 14:
Mapping, Digital
Mapping, and
Spatial
Information
Systems

Step 1: Project Specifications

Prior to undertaking any photogrammetric project, it is important to define the goals of the project, the products to be generated, potential users' needs, required accuracies for the resulting data and maps, the content of the maps or the features to be mapped, the appropriate map scale, review procedures and acceptance criteria, delivery formats for the digital data, delivery schedules, and approximate cost. Having a clear understanding of the project goals and the procedures required to meet those goals well in advance of the start of the project may prevent costly mistakes and project delays.

Step 2: Aerial Cameras

Modern aerial cameras used for large-scale photogrammetric mapping are very precise instruments and must be regularly calibrated in an optical laboratory. This calibration ensures the proper operation and adjustment of the camera lenses, shutter, filters, platen, and film magazines. The proper calibration of the aerial camera is critical to the mapping, because all information is derived from measurements made on the photographs.

Step 3: Photo Control Survey

To extract the planimetric and topographic data from the aerial photographs, a link must be made between the photographs and the earth's surface. This link is accomplished by surveying and marking points on the earth's surface, preferably prior to the photography. These points, called *control points*, usually are marked with a painted cross or propeller-shaped target so that they can be clearly and unambiguously identified in the photographs. Note that establishing control after the photography sometimes is possible and necessary but in general is not recommended. By determining the coordinates of these control points from the field survey and measuring the photographic coordinates of the images of the control points, a mathematical model based on protective geometry can be defined to compute the ground coordinates of *any* point by measuring the points' photographic coordinates (Section 13.22). Photographic control points are required on the perimeter of the block of photographs and at selected points within the block. Even when the camera system is equipped with a GPS positioning system, a sufficient number of control points is required.

Step 4: Aerial Photography

After the photographic control points are established, the aerial photography can be acquired. When obtaining aerial photographs for use in planimetric and topographic mapping, particular care must be taken to ensure that the proper weather and lighting conditions exist. Aerial photography should not be done during unstable wind conditions, when the sun angle is below 30°, or in the presence of smoke, haze, air pollution, or excessive cloud cover.

A single photograph can support the extraction of two-dimensional data alone. To permit the extraction of three-dimensional data, images of ground points must appear in at least two successive aerial photographs. To accomplish this, the aerial photographs are acquired with a design overlap between successive frames in a strip of photos. The standard overlap for map compilation from aerial photographs is 60 percent. By measuring the photographic coordinates of the same ground points on different photographs, a three-dimensional view of the ground can be created and the elevation of the ground point determined. If aerial photography is to be used to generate orthophoto maps, then the standard overlap between successive frames is 80 percent. The increase is to minimize the effects of relief displacement in the images and improve the quality of the differential rectification of the images. Most mapping projects will require multiple strips of photographs. The standard overlap, or sidelap, between adjacent strips of photographs forming a block is 30 percent.

Step 5: Analytical Aerotriangulation

After the photographs are taken and processed, the photographic coordinates of the images of the premarked photographic control points are measured on all frames. The coordinates of additional points, called *pass points,* used to link successive photographs in a strip, are measured. Finally, the coordinates of tie points, used to link adjacent strips of photography, are measured. Using the measured photographic coordinates and the known locations for photographic control points, the position and orientation of the camera at the instant of each exposure can be computed.

After the camera position and tilt elements are computed, they are substituted back into the projective equations, allowing the computation of the ground coordinates of *any* point on the photographs for which the measured photo coordinates are available. This process is called a *simultaneous intersection.* The entire process of identifying the pass and tie points, measuring the photographic coordinates of these points, and computating the camera position and tilt is called *aerotriangulation.*

Step 6: Stereo Data Capture

The collection of the planimetric and topographic data is completed using an instrument called an *analytical stereoplotter.* In some cases, the data can be captured by tracing directly from digitized orthophotographs. Most large-scale mapping still is completed using the analytical stereoplotter. The stereoplotter permits the operator to view a pair of aerial photographs simultaneously, re-creating the relative positions of the photos as they were taken. The operator's left eye views the image of a point on one photograph, and the right eye views the image of the same point on the other photograph (recall that there is a nominal overlap between successive photos). This permits the operator to create a three-dimensional view of the point and to measure directly not only the planimetric coordinates of the point, but also its elevation. In this fashion, features can be viewed and the coordinates recorded directly. Most stereoplotters allow interface with a CAD program to support the graphic data capture, and features compiled by the operator are stored as points, lines, or polygons in the CAD file. Some systems permit the operator to simultaneously view the CAD graphics overlaid onto the photographic image. Called *superimposition,* this is a valuable quality-control tool that can be used to ensure that all features are captured as well as a means to evaluate the skill of the stereoplotter operator. Normally, planimetric (2D) and topographic (3D) data are captured independently. The planimetric data might include such features as buildings, parking lots, streets, rivers, ponds, and lakes. The topographic data might include spot heights, mass points, profiles, and breaklines. In most cases, contours no longer are traced directly because of the excessive time required. More commonly, a set of points and lines that define the locations where changes in slope occur are captured, and a set of contours are interpolated from this set of points and lines.

Each stereomodel is set in turn, and the data compilation is completed. Rarely will the limits of the stereomodel coverage coincide with the limits of a map or map tile. In this case, the data from several stereomodels must be merged to create a single digital data file. The features crossing model boundaries must be edge matched across the model boundary to ensure that the features are continuous in the final map.

801

CHAPTER 14:
Mapping, Digital
Mapping, and
Spatial
Information
Systems

Step 7: Product Generation

The next step in the production process involves the processing of the captured digital data into the deliverable products. This might require merging several models of data, creating fixed dimension maps or map tiles, merging of several layers of planimetric and topographic data onto a single map, or creating a single, seamless digital data file. In addition, photomaps, orthophotos, hard-copy plotted or printed maps, and perhaps database files containing attributes or qualitative information about the graphic features may be generated.

The digital data sets may require projection into a particular coordinate system, such as state plane coordinates or UTM coordinates, prior to delivery. This process usually is completed after the digital data have been captured and edited. Finally, map collars or border information, including title blocks, legends, scales, coordinate grids, and ticks, are merged with the map tiles to complete the composition of the maps.

Step 8: Quality Review

Prior to accepting the maps and digital data, it is important to conduct a quality review of all or a sampling of the products. This is done to ensure adherence to the specifications for map content and symbols and the consistency and accuracy of the digital data and printed maps. Formal procedures, including checklists, specific acceptance/rejection criteria, and procedures for rectification of errors should be defined during the assembly of the specifications for the project. When the product is finished, it is delivered to the user either in hard copy or digital form accompanied by a report.

This process would be substantially the same for field surveys conducted for engineering work such as construction surveys for route alignment and building construction. In these cases, the aerial photograph and analytical aerial triangulation steps may not apply. The stereo data capture might be replaced by field data capture, perhaps using a data collector and a total station or GPS surveys. The data capture process would include the transfer of the collected data to an appropriate CAD system for the composition of the maps.

14.14
COMPUTER-AIDED DRAFTING AND DESIGN

Computers have revolutionized the process of completing technical drawings and maps. Manual drafting operations, which may have taken hours of painstaking effort, can be completed in minutes using powerful computer-based drafting tools. These tools include drawing simple figures, text placement, complex feature drafting, line fitting, simple and complex curve placement, dimensioning, and coordinate geometry (COGO) for planimetric feature compilation. These CAD systems also perform other drafting tasks, such as fixed grid placement, feature placement, and definition of point symbol libraries, line style libraries, and fill patterns. Many CAD systems support a command macro language or command scripting language of some sort, permitting programming of specific data capture tools, construction of complex figures, and plotting of quality assurance.

The typical CAD drafting system consists of a computer, the CAD programs, a graphics display monitor, a mouse or digitizing pen pointing device for command selection and drafting, an optional printer or plotter, and a digitizing tablet. The choice of a computer host for the CAD software system offers great flexibility. A large number of CAD software systems are available. The choice of one system over another will likely be based on the available computer systems on which to run the software, the required functionality of the CAD system, software support from the publisher, interoperability with other systems in use by a client, and cost. Most CAD publishers provide task-specific layered products that operate with the basic CAD drafting tools. For example, a CAD system for use in civil engineering design may include commands for COGO data entry, transfer of data to and from total station data collectors, profile and cross-section computation, basic geometric design, and contouring. A CAD system for use in utility applications may provide other tools such as engineering work order generation, network tracing, bill of materials generation, or links to billing and maintenance databases.

The capture of drawing data from existing maps can be completed through the use of a digitizing tablet. This tablet allows the tracing and capture of map features using a digitizing cursor or mouse with the CAD drafting tools. To digitize map features, a map is attached to the tablet surface, and based on the known positions of at least two points within the map, a transformation from the tablet digitizer coordinate system to the map coordinate system is determined. This transformation must be done each time the map is placed on the digitizing tablet. Then, the map positions of each feature can be traced and converted to digital form. As the features are captured, the transformation from digitizer to map coordinates is applied so that the features are stored in the drawing file in map coordinates.

More elaborate systems also support the use of raster-scanned map data (see Section 14.17) as input to the data capture process. In these systems, a map is scanned digitally and stored as a raster file. Then, whether using "heads-up" or direct digitizing techniques on a CRT, which simply permit the user to visually trace over the raster file to capture the features in vector form (Section 14.17), or automated vectorization tools, the raster file is converted to a vector file and stored.

14.15
DIGITAL MAPPING (COMPUTER-AIDED MAPPING)

The CAD systems, as just described, are used primarily to capture information for planimetric data or for 2D and 3D design of objects or assemblies. They can be used to obtain planimetric and topographic data from existing maps or through the input of field data to create simplified maps. These systems are not mapping systems until they can be extended to permit direct, first-generation capture of planimetric and topographic data from aerial photos, satellite imagery, or other data sources. Integrated systems, combining traditional CAD system tools and interfaces to stereoplotters or stereo image displays constitute a digital mapping system. Most CAD systems can be configured to use analytical stereoplotters as input devices for data capture, much as a digitizing tablet would be used. However, digital-mapping systems must provide additional functionality for processing stereoplotter input and converting stereo photo coordinates to 2D or 3D coordinates. These additional software components must permit determining the orientation of the photographs on the stereoplotter stages, re-creating the relative positions of the photographs at the instant of the exposure, and real-time computing of image ray intersections from observed photographic coordinate pairs as the operator traces the features. This requirement for continuous real-time computation also is the differentiating factor between digital mapping systems and simple CAD systems. Some digital mapping systems also permit the injection of the graphic images into the stereoplotter optical path so that the technician can view directly the captured graphics superimposed on the images. This is a valuable tool for editing, quality control, and accuracy assessment.

803

CHAPTER 14:
Mapping, Digital
Mapping, and
Spatial
Information
Systems

These digital mapping systems now have completed the transition from the use of a CAD program with a stereo viewing device to fully digital mapping systems, using high-resolution raster scans of aerial photos and stereo displays for new map compilation. These systems use a high-resolution stereo display to view a stereomodel in 3D, just as it would have appeared in a stereoplotter. With such a system, planimetric and topographic features can be compiled directly, simply by tracing the feature from the stereo display screen using the cursor. Most systems permit the extraction of terrain data as well through the use of automatic image correlation of one photo of a stereo pair to the other. By matching pixels digitally from one photograph to the other, the parallax equations can be solved to determine automatically the elevation of each pixel in the stereomodel. Unfortunately, the autocorrelation technique extracts elevation data at places other than the ground surface, such as roof tops, treetops, or water surfaces. Autocorrelated topographic data must be carefully reviewed to ensure that accurate ground-level contours are portrayed. These fully digital systems, because of the inherent software required, also can produce differentially rectified orthophoto maps. With the introduction of the these systems in the last few years, interest in the use of orthophoto maps has seen a sharp increase. This is due, in part, to the time lag between U.S.G.S. quadrangle map updates, quadrangle map scales, and the superior information content of a photomap. Spatial information systems easily can incorporate these orthophotomaps as well. The U.S.G.S. currently is compiling a complete set of digital orthophoto quarter-quads for the conterminous United States (see Section 14.4).

14.16
DATA COLLECTION

Data collection for mapping and GIS (Section 14.18) can include a wide variety of techniques and data sources. Often trade-offs must be made between accuracy, currency, and completeness against cost and level of effort. A major problem with maps is that they start to become obsolete the moment they are finished. Therefore, to produce a map that is unquestionably current and consistent, data collection should come from a new field survey or should be taken from recent imagery, usually in the form of aerial photographs or satellite images. The choice between ground survey and photogrammetric techniques usually is one of efficiency. For areas smaller than about 8 to 12 hectares (~20–30 acres), it usually is less expensive to execute the topographic survey on the ground with total station or GPS equipment (Sections 15.9 through 15.11). For areas larger than this, it usually is less expensive to obtain the topographic data by photogrammetry. In photogrammetry, the cost of mobilizing the aircraft and the crew and the problems involved in scheduling around weather and foliage conditions present obstacles for its application on very small projects. Commercial high-resolution satellites may have an impact on these decisions but that will have to await accuracy, interpretability, and cost evaluations for these systems.

Many municipal officials charged with the task of creating and overseeing the operation of a city- or countywide GIS, are tempted to avoid undertaking new mapping and convert existing maps into digital form via manual digitizing or scanning and semiautomatic digitizing. The arguments in favor of this approach all relate to the cost savings compared to new mapping surveys. The arguments against this approach relate to the issue of starting a new project with an already outdated mapping data base. It would be easier to justify if the existing mapping had been carefully maintained and revised since its compilation, which usually is not the case. An interesting and increasingly popular approach to this problem is presented by the concept of the orthophotograph (Section 13.25). Orthophotographs are differentially rectified images free from displacements due to tilt and relief. Slope discontinuities and occlusions may or may not be handled correctly, depending on the underlying DTM and the robustness of the image resampling strategy. In brief, they provide a current, planimetrically correct base map. Furthermore, in nonurban areas often it is possible to generate the DTM that drives the rectification process via automated image matching. This savings in labor expense can make the digital orthophoto an attractive option for the base-map layer of a GIS. Naturally, the system would have to accommodate and permit registration between both raster and vector data layers.

A special class of data relating to the cadastral or land ownership information in an area often occupies a separate layer in GIS systems. Such data usually originate from the deed descriptions of the individual parcels and subdivisions. These data, unfortunately, often display significant positional discrepancies with the base-map layer. Different users adopt different strategies to handle these discrepancies. Some choose to warp the cadastral data to fit the base map, others leave them as is.

In the digital age, a final source of map information cannot be neglected. Digital cartographic data assembled by the government and made available to the public sometimes can be used in digital mapping systems and GIS. For example, the digital TIGER data from the U.S. Bureau of the Census contain street network information of limited accuracy and useful attributes such as address ranges for individual street segments. This trend of increasing availability of digital map data is certain to continue in the future.

14.17
DATA STORAGE AND MANAGEMENT

Modern map data must be compatible with any number of other applications, from complex design to economic analysis and planning. It therefore is most important that the data composing a map series be structured such that they easily can be used for other processes. A later section in this chapter (14.28) presents current efforts in data transfer standards; the basics of data representation and storage are given here.

Map features can be broadly characterized as point, line, area, or text features. For example, power poles, telephone booths, and manholes often are represented by a specific symbol. The position of the feature is determined by the coordinates of the symbol. Streams, roads, and fences are represented by lines of varying styles, colors, and widths. Forests, lakes, and ponds are represented by closed polygons showing the extent of the particular feature. Street and city names often are placed as simple text labels in several locations on a map. Data of this nature are said to be in *vector* format. Some maps generated from processed satellite images or created by scanning documents with a scanning device represent map features using a grid of fixed-size squares called *pixels*. The color or brightness of individual pixels represent different features on a map. This form of representing map features is called a *raster* format. CAD programs create data sets in vector format; satellite image data and scanned documents normally are represented in raster format.

Vector data are coordinate based, with polygons being composed of a set of lines forming a closed region and the lines being a set of connected points, with points having specific coordinates. This hierarchy of graphic elements (i.e., polygons, lines, and points) is important when devising compact methods of storing data. Also, because only the actual features shown on a map are represented in the vector format, a vector data file is relatively

805

CHAPTER 14:
Mapping, Digital
Mapping, and
Spatial
Information
Systems

compact. On the other hand, capturing image data or scanning a map document requires that even blank areas of a map be scanned and maintained in the data set to ensure that the raster grid is consistent and complete. This characteristic of storing so-called empty data explains why raster data files can be quite large. Moreover, the size of a raster data file also is a function of the resolution with which the scan is completed. For example, a single standard letter-size sheet being printed by a simple black and white (binary image $0 =$ white, $1 =$ black) laser printer using a standard 300 dot per inch resolution would require nearly 1 megabyte of storage ($8.5 \times 11 \times 300^2/8$).

Spatial information systems sometimes are used to store data in addition to the simple graphic representation of map features. This capability differentiates CAD systems from spatial information systems. These types of data management programs often use nongraphic attributes such a road width, road surface type, or street name to complete complex analyses. Further, for some types of map analysis, information as to how features are connected or the direction in which a line is drawn from point to point is important. These characteristics of connectivity of lines at points, incidence of lines at nodes, and inclusion and exclusion of polygons are called the *topology* of a data set. The topology is simply a record of how the various features in a map are connected. Topologic data are not inherent in the collection of map features and must be created and maintained to ensure that they remain consistent. The topology is used when determining shortest, least-cost paths through a network and when modeling flow along streets or rivers.

14.18
SPATIAL INFORMATION SYSTEMS

Spatial information systems is a term for a broad family of computer-based systems for the capture, management, and analysis of spatial data. These systems can be used to facilitate problem solving, assist in decision making, and permit the analysis of complex problems involving spatial data. A modern classification for spatial information systems can be seen in Figure 14.11. Spatial information systems are the most general instance of several information systems. Geographic information systems are a subset of spatial information systems, focusing on geographic, real-world data. Land information systems are a further refinement of both systems, where the primary focus of LIS is in the description of the spatial and legal characteristics of land.

Spatial information systems are contrasted to computer-aided drafting systems, where the primary function is simple drafting or engineering design, by the ability to

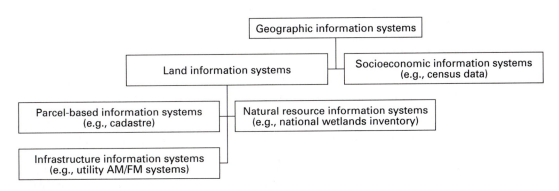

FIGURE 14.11
Classification of spatial information systems.

simultaneously manage and analyze nonspatial data. CAD systems also can be seen as a subset of the spatial information system, providing the ability to manage the graphic representation of spatial data. As such, CAD tools are an important component of spatial information systems.

Surveying and mapping and engineering applications are concerned primarily with real-world data such as are required for contours, volumes, geometric design, and property parcels; therefore, GIS is the best tool to use in processing, analyzing, and managing these data. Additional information on spatial information systems can be found in Laurini and Thompson (1992), Star and Estes (1990), Tomlin (1990), Abler (1971), and Unwin (1981).

A GIS is most clearly differentiated from a CAD system by its ability to manipulate nonspatial data as well as spatial data and to perform spatial analyses and queries. In addition, the development of GIS has advanced the concept of generating maps as indications of solutions to complex problems rather than simple textual information. A GIS also is a visualization tool, permitting the graphic display of complex interactions and alternatives. Detailed numerical analysis can be supported using the stored attribute and graphic data. This makes GIS an ideal modeling and simulation tool for planning purposes. GIS has made enormous inroads in traditional civil engineering practice, surveying and mapping, transportation, and resource management. GIS development has directly benefitted from its multidisciplinary roots in remote sensing, geodesy, cartography, surveying, statistics, computer graphics, mathematics, geography, and operations research.

The basic components of a GIS, after Laurini and Thompson (1992), are shown in Figure 14.12. The graphics management system is concerned with the collection of spatial data, its conversion to digital form, and the production of final maps. This system would consist of a CAD system, digitizing tablets, optical scanners, and printing and plotting devices.

The database component deals with the physical storage of the spatial data in the computer system, the organization of the data, and models for the representation of spatial features. In addition, this component provides the powerful ability to query the spatial data and derive new information from that existing in the database. A spatial query is a question whose answer can be described in terms of the spatial data contained in the database files and the graphic features. For example, a query might request a map showing the location of all private property parcels having a specific zoning classification with areas greater than 10 acres. To respond appropriately, the system must either contain the information explicitly or, using spatial analysis tools, be able to derive the information from the available data. This capability, unique to spatial information systems, is the greatest difference between spatial information systems and CAD systems.

The user interface provides both command line and graphical user input for data analysis and the display of results in a GIS. Most modern GIS are equipped with the graphical user interface (GUI) technology to help make use of the system as intuitive as possible. The GUI enables easier training for new users and permits analysis and query processing without having to deal with underlying abstract algorithms directly.

The operating system is the lowest level of computer instructions used to operate the host computer systems. Examples of operating systems in common use in spatial information systems include UNIX for personal computers, graphics workstations, and mainframe computers, and DOS or Windows for personal computers. In most GIS implementations, the user is removed from having to issue operating system commands directly by the use of the GUI. The operating system also manages communication between programs, other local computers, and computers accessible via a network connection. This set of components taken as a whole characterizes modern GIS.

Applications of GIS have grown dramatically in the last 10 years and closely paralleled the exciting advances in computer technology. Several agencies of the federal government now routinely use GIS to manage resources, develop comprehensive plans, monitor and enforce environmental regulations, gather and analyze census data, and support con-

807

CHAPTER 14:
Mapping, Digital
Mapping, and
Spatial
Information
Systems

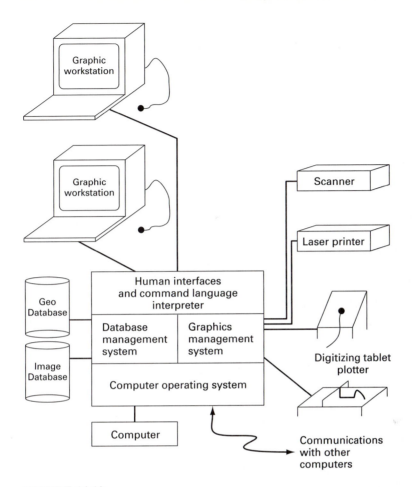

FIGURE 14.12
The components of a spatial information system. (*Adapted from Laurini and Thompson, 1992.*)

gressional redistricting resulting from the decennial census. State and county governments use GIS to model growth, plan for economic development, describe land use, support property tax administration, and assist in the delivery of emergency services. Private corporations and individual interest groups also are using GIS in such diverse applications as fleet management for transportation, marketing analysis, real estate marketing, and presentation of data on important public issues. The availability of GIS and high-quality spatial data and their use in diverse applications in management and problem solving will continue to grow.

14.19
AUTOMATED MAPPING/FACILITIES MANAGEMENT

Automated mapping/facilities management (AM/FM) concerns the application of spatial information systems technology to the management and design of engineering functions. The primary users of AM/FM systems are public and private utility companies, large

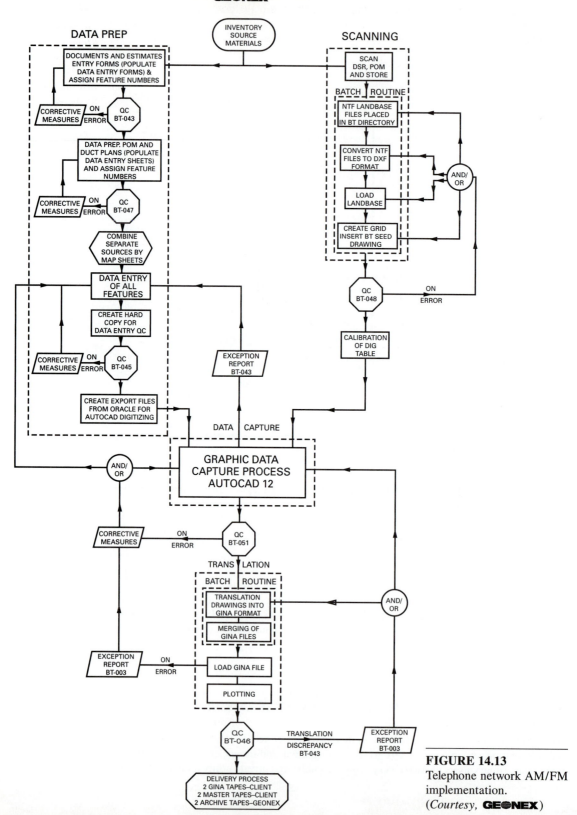

**BRITISH TELECOM PROJECT
PROCESS FLOW DIAGRAM**

GE⊙NEX

FIGURE 14.13
Telephone network AM/FM
implementation.
(*Courtesy,* **GE⊙NEX**)

engineering design firms, and engineering management groups. For example, AM/FM systems are used to collect, manage, model, and design electrical transmission and distribution networks, water and wastewater systems, television and telephone systems, and pipeline systems.

Automated mapping extends the concept of CAD-based engineering drawings to "live," on-line interactive drawings in a computer system. These dynamic drawings are used by all levels of design engineers, and system managers, as well as in customer service applications. Facilities management activities include the maintenance, management, and expansion of the system infrastructure based on the use of the on-line design drawings.

AM/FM systems usually are characterized by vast number of drawings, schematic diagrams, and a dense set of attribute data describing the physical system. The digital drawings also are likely to be at a variety of scales ranging from 1:250 to 1:1000. Other unique traits of the AM/FM systems in contrast to other GIS systems is the demand for concurrent transaction processing, engineering work order processing, engineering work order posting, and the routine generation of new working drawings. System management can also be made part of the AM/FM system to support outage rerouting and repair crew dispatch. For example, a small local telephone wire center may process several thousand system engineering work orders that modify the system configuration, and thus the working drawings, over the course of a year. Electric utilities also routinely process thousands of work orders. In both cases, the changes to the networks must be made to the current engineering drawings. The process of developing and maintaining such an AM/FM system for a telephone system application might proceed as shown in Figure 14.13.

To implement these systems, AM/FM users must make a substantial investment in computer equipment, software systems, and most important, in training for the technical staff. Depending on the scale of the networks, automation of the engineering functions can take place over several months or even several years. Only with the advent of these AM/FM systems have the management and maintenance of these large databases become practical.

14.20
LAND INFORMATION SYSTEMS

As described in Section 14.18, land information systems can be thought of as a subset of spatial information systems, dealing primarily with the description of the geometry and legal interests of real property. This concept of LIS has evolved from the historical roots of the European cadastre. A cadastre, in former times, was a simple listing of the real property parcel ownership, area, value, and land use. Early cadastres were developed to collect taxes on land. The modern view of the cadastre has become that of a multipurpose, multiuse information system for land. In 1979, the National Research Council undertook a study to examine the needs for a national database of land information, potential users, cost, and methods for implementation. The council issued a report, "Need for a Multipurpose Cadastre." This document and the "Procedures and Standards for a Multipurpose Cadastre" have been the stepping stones in the movement to establish modern land information systems in the United States. These two documents clearly identified the social and economic benefits to improving our system of collection and management of land information. Modern developments in computer technology and spatial information systems make the implementation of an LIS feasible.

At this writing, one of the most ambitious projects in LIS implementation is the Bureau of Land Management's Automated Land and Mineral Records System (ALMRS) program. The goal of the project is to modernize the complete holdings of mineral and cadastral records for the public lands administered by the bureau. To get a sense of the scale of the

809

CHAPTER 14:
Mapping, Digital
Mapping, and
Spatial
Information
Systems

project, the bureau is responsible for the management of over *1.8 billion* acres of land in the United States and its territories. The ALMRS system will collect all information related to cadastral surveys, mineral leases, land grants, rangeland leases, easements, and in short create a true multipurpose cadastre for the public lands.

Currently, there are very few examples of true land information systems for private property. Many parcel-based GIS and LIS are implemented at the county level, containing parcel data, owner, area, value, zoning, and other information. However, most of these parcel GIS systems are based on inexact geometric descriptions of the parcels and bear no direct relation to specific deed or plat descriptions. Without the inclusion of the legal interests as conveyed by deed or plat, these systems fall short of being true legal cadastres but still are very useful in the support of county engineering operations, property taxation, land-use management, zoning, and comprehensive planning. Extensive work remains to identify appropriate and efficient methods to link the legal descriptions to these systems. To be a complete legal cadastre, the LIS should provide the complete legal record of real property interests, be self-proving, provide mechanisms for correction of errors, and be easily accessible. As can be imagined, with the volume of transactions in real property being what it is, the implementation of LIS for private property may prove daunting and surely will require substantial changes in the administrative procedures affecting the transfer of interest in land, methods of recording these interests, as well as changes in the social concepts of land tenure and stewardship.

14.21
DATA PROCESSING

After the underlying data for a planimetric or topographic map have been captured, a series of data processing steps are required to convert these data into a map or map series. The steps in this process might include, for example,

1. Design of the map layout.
2. Selection of the map projection, coordinate system, datum, and scale.
3. Manipulation of the map data—analysis, query, extraction, and aggregation.
4. Selection of the map features.
5. Preparation of the map collar information—border, titles, scales, legends, and graticules.
6. Labels of the map features.
7. Preparation of the manuscript.

14.22
INFORMATION SHOWN ON MAPS

Maps are classified according to their specific use or type, but in general they become either a part of the records of land division, a tool for engineering planning and design, or a component of a geographic information system, where they can be used for many purposes. In general, the following information should appear on any map:

1. The direction of the meridian and basis for directions (astronomic, grid, magnetic, etc.).
2. A graphical scale of the map and a corresponding note stating the scale at which the map was produced.

811

CHAPTER 14:
Mapping, Digital
Mapping, and
Spatial
Information
Systems

3. A legend or key to symbols other than the conventional ones.
4. An appropriate title.
5. On topographic maps, a statement of the contour interval.
6. A statement giving the datums to which horizontal and vertical control are referenced.
7. A statement giving the map projection, together with the projection parameters in cases where they are not widely known.
8. A statement giving the coordinate system(s) for which grid lines or grid ticks are shown on the map; 1D ticks along the border should be discouraged in favor of 2D ticks, which can be used for digitizer registration.
9. The date on which the map was produced, and the dates of any revisions.
10. A statement about the source of data used to compile the map, and the date of acquisition of such source data (for example, "compiled by photogrammetric methods from 1:4000 scale photography taken in 1997").
11. A sketch showing the location of the map sheet in a series of sheets covering a project area or municipal region.

In addition, a map that is to become a part of a public record of land division should contain the following information:

1. The length of each line.
2. The azimuth or bearing of each line or the angle between intersecting lines.
3. The location of the tract with reference to established coordinate axes.
4. The number of each formal subdivision, such as section, block, or lot.
5. The location and kind of each monument set, with distances to reference marks.
6. The location and name of each road, stream, landmark, and the like.
7. The names of all property owners, including owners of property adjacent to the tract mapped.
8. A full and continuous description of the boundaries of the tract by bearing and length of sides, and the area of the tract.
9. The witnessed signatures of those possessing title to the tract mapped; if the tract is to be an addition to a town or city, a dedication of all streets and alleys for the use of the public.
10. A certification by the surveyor that the map is correct to the best of his or her knowledge.
11. Surveyor's certification and registration number.

The popular habit of using assumed bearings and arbitrary origin points for coordinates should be discouraged. Such a map may serve a particular purpose well but is difficult or impossible to merge or integrate into a geographic information system. Additional explanatory notes necessary to permit users to make the best use of the map information also are useful, as for example, the precision of the control points used in the map compilation.

14.23
MAP CONTENTS

Scales

The map scale should be indicated on the map sheet numerically and graphically. Numerical designations are best given by the ratio or representative fraction; for example, 1/25,000 or 1:25,000. This method is independent of any particular units. Graphical scales can be given as shown in Figure 14.14. Be aware that digital mapping and GIS

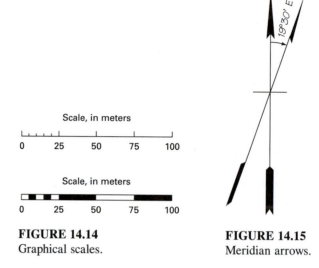

Scale, in meters

| 0 | 25 | 50 | 75 | 100 |

Scale, in meters

| 0 | 25 | 50 | 75 | 100 |

FIGURE 14.14
Graphical scales.

FIGURE 14.15
Meridian arrows.

technology allow maps or portions of maps to be plotted back at any scale desired. A numerical scale therefore would be rendered inaccurate whereas a graphical scale would expand or shrink in proportion to the map itself and thus remain valid.

Meridian Arrows or Grid Lines

Directions on the map should be denoted by grid lines or ticks indicating the projection coordinates (i.e., state plane, UTM, etc.) or lines or ticks of constant latitude and longitude. A graphical north arrow may be used to indicate the relationship between various reference directions (i.e., true north, grid north, or magnetic north). In the northern hemisphere, it is conventional to have the $+Y$ axis correspond to the north direction. An example of a meridian arrow is shown in Figure 14.15.

Text Annotation

The cartographer's skill in lettering has been substantially replaced by the typography capabilities in modern CAD and GIS systems. Text placement and pleasing font selections, however, remain map design decisions that can have a great effect on the user's ability to read the map properly and quickly. Fonts, in general, should be consistent and not overly ornate so as to distract the user from the map data. *Reinhardt* or *Leroy* style lettering is very effective for engineering maps. Examples of Reinhardt letters in vertical and slope forms are illustrated in Fig. 14.16.

Titles

Titles should be constructed so as to readily catch the eye. The best position for the title is the lower right corner of the sheet, except where the shape of the map makes it advantageous to locate the title elsewhere. The space occupied by the title block should be in proportion

813

CHAPTER 14:
Mapping, Digital
Mapping, and
Spatial
Information
Systems

ABCDEFGHI
JKLMNOPQR
STUVWXYZ
abcdefghijklmn
opqrstuvwxyz
1234567890
NORMAL
Excavation 23 cu. yd.
COMPRESSED
RICHARDSON ESTATE 300 Ac.
EXTENDED
RED RIVER
(a)

ABCDEFGHIJ
KLMNOPQ
RSTUVWXYZ
abcdefghijklmn
opqrstuvwxyz
1234567890
NORMAL
Hickory Tree 10 ft.
COMPRESSED
WASHINGTON Sta. 71 + 438
EXTENDED
NEVADA
(b)

FIGURE 14.16
Reinhardt lettering: (a) vertical form, (b) slope form.

| DEPARTMENT OF THE INTERIOR BUREAU OF RECLAMATION |
| GRAND COULEE DAM – WASHINGTON |
| TOPOGRAPHIC MAP OF EAST SIDE GRAVEL PIT |

DRAWN:_____ TRACED:_____

CHECKED:_____ APPROVED:_____

| OCT. 10, 1980 | DENVER, COLO. | 222 – D – 539 |

FIGURE 14.17
Title for a map.

to the size of the map; the general tendency is to make the title too large. Organizations often will have a standard title block required on all maps and drawings with standard contents. The title block should include the type or purpose of the map, the name of the tract, location of the tract, the scale and contour interval, name of the engineer or surveyor, the date, and other pertinent information not shown elsewhere on the map. A sample title block is shown in Figure 14.17.

Notes and Legends

Explanatory notes or legends often are of assistance in interpreting a drawing. They should be as brief as circumstances will allow but should include sufficient information to leave no doubt in the mind of the person using the drawing. A key to the symbols representing various

CONTROL DATA AND MONUMENTS

Aerial photograph roll and frame number*   3-20

Horizontal control

Third order or better, permanent mark	Neace △	Neace △
With third order or better elevation	BM △ 45.1	Pike BM 45.1
Checked spot elevation	△ 19.5	
Coincident with section corner	Cactus	Cactus
Unmonumented*	+	

Vertical control

Third order or better, with tablet	BM × 16.3
Third order or better, recoverable mark	× 120.0
Bench mark at found section corner	BM + 18.6
Spot elevation	× 5.3

Boundary monument

With tablet	BM 21.6	BM 71
Without tablet	171.3	
With number and elevation	67 301.1	

U.S. mineral or location monument   ▲

CONTOURS

Topographic

Intermediate	
Index	
Supplementary	
Depression	
Cut; fill	

Bathymetric

Intermediate	
Index	
Primary	
Index Primary	
Supplementary	

BOUNDARIES

National	— —
State or territorial	— —
County or equivalent	— —
Civil township or equivalent	— —
Incorporated city or equivalent	— —
Park, reservation, or monument	— · —
Small park	— — —

LAND SURVEY SYSTEMS

U.S. Public Land Survey System

Township or range line	———
Location doubtful	— — —
Section line	———
Location doubtful	— — —
Found section corner; found closing corner	+
Witness corner; meander corner	WC MC

Other land surveys

Township or range line	··········
Section line	··········
Land grant or mining claim; monument	— · — □
Fence line	— — —

SURFACE FEATURES

Levee	Levee
Sand or mud area, dunes, or shifting sand	Sand
Intricate surface area	Strip mine
Gravel beach or glacial moraine	Gravel
Tailings pond	Tailings Pond

MINES AND CAVES

Quarry or open pit mine	✕
Gravel, sand, clay, or borrow pit	✕
Mine tunnel or cave entrance	◄
Prospect; mine shaft	X ▪
Mine dump	Mine dump.
Tailings	Tailings

VEGETATION

Woods	
Scrub	
Orchard	
Vineyard	
Mangrove	Mangrove

GLACIERS AND PERMANENT SNOWFIELDS

Contours and limits	
Form lines	

MARINE SHORELINE

Topographic maps

Approximate mean high water	
Indefinite or unsurveyed	— — —

Topographic-bathymetric maps

Mean high water	
Apparent (edge of vegetation)	

*Provisional Edition maps only
Provisional Edition maps were established to expedite completion of the remaining large scale topographic quadrangles of the conterminous United States. They contain essentially the same level of information as the standard series maps. This series can be easily recognized by the title "Provisional Edition" in the lower right hand corner.

FIGURE 14.18
Topographic map symbols. (*Courtesy U.S. Geographical Survey.*)

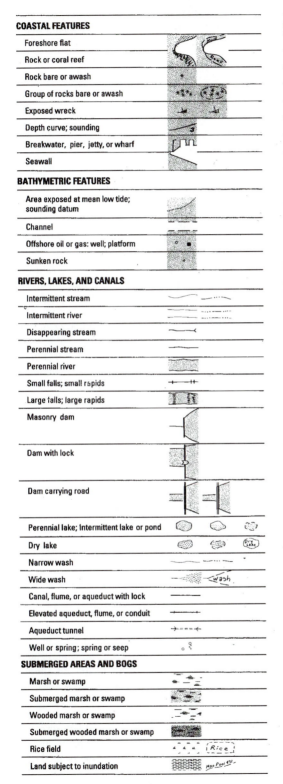

COASTAL FEATURES

Foreshore flat	
Rock or coral reef	
Rock bare or awash	
Group of rocks bare or awash	
Exposed wreck	
Depth curve; sounding	
Breakwater, pier, jetty, or wharf	
Seawall	

BATHYMETRIC FEATURES

Area exposed at mean low tide; sounding datum	
Channel	
Offshore oil or gas: well; platform	
Sunken rock	

RIVERS, LAKES, AND CANALS

Intermittent stream	
Intermittent river	
Disappearing stream	
Perennial stream	
Perennial river	
Small falls; small rapids	
Large falls; large rapids	
Masonry dam	
Dam with lock	
Dam carrying road	
Perennial lake; Intermittent lake or pond	
Dry lake	
Narrow wash	
Wide wash	
Canal, flume, or aqueduct with lock	
Elevated aqueduct, flume, or conduit	
Aqueduct tunnel	
Well or spring; spring or seep	

SUBMERGED AREAS AND BOGS

Marsh or swamp	
Submerged marsh or swamp	
Wooded marsh or swamp	
Submerged wooded marsh or swamp	
Rice field	
Land subject to inundation	

BUILDINGS AND RELATED FEATURES

Building	
School; church	
Built-up Area	
Racetrack	
Airport	
Landing strip	
Well (other than water); windmill	
Tanks	
Covered reservoir	
Gaging station	
Landmark object (feature as labeled)	
Campground; picnic area	
Cemetery: small; large	

ROADS AND RELATED FEATURES

Roads on Provisional edition maps are not classified as primary, secondary, or light duty. They are all symbolized as light duty roads.

Primary highway	
Secondary highway	
Light duty road	
Unimproved road	
Trail	
Dual highway	
Dual highway with median strip	
Road under construction	
Underpass; overpass	
Bridge	
Drawbridge	
Tunnel	

RAILROADS AND RELATED FEATURES

Standard gauge single track; station	
Standard gauge multiple track	
Abandoned	
Under construction	
Narrow gauge single track	
Narrow gauge multiple track	
Railroad in street	
Juxtaposition	
Roundhouse and turntable	

TRANSMISSION LINES AND PIPELINES

Power transmission line: pole; tower	
Telephone line	
Aboveground oil or gas pipeline	
Underground oil or gas pipeline	

FIGURE 14.18 (*concluded*)

815

details ought to be shown unless these are conventional. The nature and source of data on which the map is based should be made known. For example, the data for a map may be obtained from several sources, perhaps partly from old maps, old survey notes, and new surveys. The precision of the surveys, coordinate axes or meridian directions, and the reference datum also should be shown.

14.24
MAP SYMBOLS

Objects are represented on a map by *symbols,* many of which are conventional. Some topographic map symbols published by the U.S. Geological Survey (U.S.G.S.) are shown in Figure 14.18. A chart showing these symbols is available from the U.S.G.S. without charge. On standard U.S.G.S. quadrangle sheets, color is used to identify features. Black is used for constructed or cultural features such as roads, buildings, names, and boundaries. Blue is used for water or hydrographic features such as lakes, rivers, and canals. Brown is used to show the relief or configuration of the ground surface as portrayed by contours. Green is used for wooded or other vegetative cover, with typical patterns to show such features as scrub, vineyards, or orchards. Red emphasizes important roads and public-land subdivision lines and shows built-up urban areas. Purple is reserved for "photo revision," which takes place periodically to keep the maps current. Grid ticks for geographic coordinates, state plane coordinates, and UTM coordinates are shown in black and blue. When color is not available, all map features may be rendered in black. Mapping organizations often define their own standard map symbology, which may be promulgated through a symbol "library" in the CAD system used to produce the drawings. Symbols are necessarily different for large-scale maps, medium-scale maps, and small-scale maps. The U.S. Army Corps of Engineers uses a set of map symbols, some of which are shown in Figure 14.19.

14.25
MAP PRODUCTION

The physical production of final maps has evolved rapidly with the widespread availability of computer-based tools for map compilation, analysis, and publishing. Publishing maps from digital data now increasingly is left to the users of the map data. This permits more flexibility in the types of map products generated, the selection of maps to be published, and the frequency of map update. The concept of "publishing" maps also has taken on new meaning in an environment of Internet communication, network file sharing, and the use of maps in digital form.

Several options are available for the preparation and printing of high-quality hard-copy maps. Output devices include pen plotters, electrostatic plotters, direct raster plotters, and color printing systems using digitally rendered color separates as input. The choice of an output device is a function of the types of maps desired (color, black and white) and the media desired (stable-base media such as polyester, various paper media for thermal, electrostatic, or thermal wax printing, or film). These types of print processes are used primarily for small-volume printing.

Large-volume printing of maps is done by a letterpress or lithographic process. Just as mapping has been greatly affected by the use of computers, the traditional letterpress and lithographic print processes now can accept digital "plates." Digital color separations, each representing a specific map color layer can be combined to generate the final printed maps.

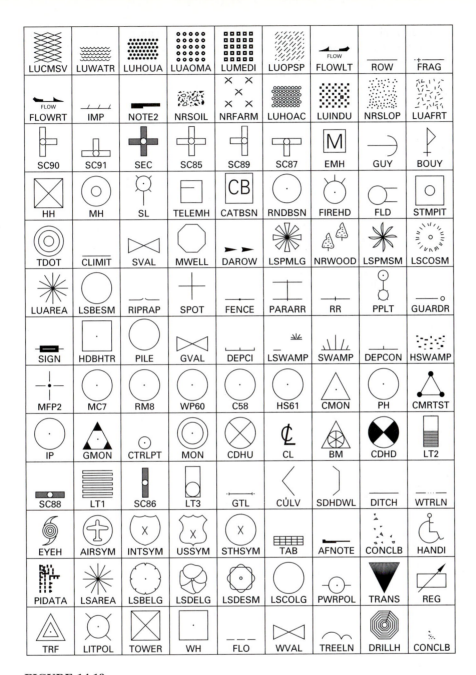

FIGURE 14.19
Symbols, U.S. Corps of Engineers. (*Courtesy of U.S. Corps of Engineers.*)

Map publishing in a computer network environment is quite different. Using digital map formats, simple or complex maps can be "published"; that is, placed in computer data banks or archives or even provided interactively on distributed computer network servers. This approach permits not only users to access and print hard copies of digital maps but also to combine digital maps with other types of data or to revise and update the maps. Maps can be published and distributed in other media as well, including CD-ROM and laser videodisc, which do not permit direct modification on the distribution media.

14.26
STANDARDS FOR ACCURACY OF MAPS AND MAP DATA

National map accuracy standards (NMAS; Bureau of the Budget, 1947) in their current state have been developed for medium- to small-scale maps (1:20,000 and smaller). A great deal of effort is being made to develop and adopt accuracy standards for larger-scale maps. Realizing that maps can be archived, accessed, and used in their digital form, accuracy standards also must address issues other than simple position.

With regard to horizontal accuracy, current NMAS require that no more than 10 percent of *well-defined points* be in error by more than 0.8 mm (1/30th of an inch) at published map scale. Well-defined points are those that are easily visible or recoverable on the ground; in general, those that can be plotted at the scale of the map to within 0.25 mm (1/100th in.). Thus, for a map at a scale of 1:20,000, the horizontal accuracy tolerance is 16.00 m and for the U.S.G.S. standard quadrangle maps at a scale of 1:24,000 (2,000 ft per in.), it is 66.67 ft, both in units on the ground. With regard to vertical accuracy, NMAS require that not more than 10 percent of the elevations tested shall be in error by more than one-half the published contour interval.

A contention is that these standards are not applicable to large-scale maps due to the techniques employed to compile the maps and the uses to which these large-scale maps are applied. Several groups including committees from the American Society of Civil Engineers (ASCE, 1972), American Society for Photogrammetry and Remote Sensing (ASPRS, 1989), the National Committee for Digital Cartographic Data Standards (NCDCDS, 1985) and the Federal Geographic Data Committee (FGDC, 1997) have proposed standards for maps. Proposed standards from the ASPRS Specifications and Standards Committee (1989) are shown for maps in metric and English units in Tables 14.4 and 14.5, respectively. In these tables, horizontal accuracy is expressed in terms of the *root mean square* (RMS) *error* in either the X or Y coordinate of the point. As used for the standards, RMS error corresponds to the square root of the *mean square error* ($\sqrt{\text{MSE}}$), where MSE is defined in Section 2.11 by Equation (2.12) and repeated here for convenience

$$\text{MSE}_x = [\Sigma(x - \tau)^2]/n$$
$$\text{MSE}_y = [\Sigma(y - \tau)^2]/n$$

(2.12)

in which x and y are measured coordinates for the point, τ is the known value of respective x and y coordinates of checkpoints, and n is the number of points tested. Vertical accuracy is defined as the RMS error in elevation for well-defined checkpoints. For the maps shown in Tables 14.4 and 14.5, limiting values for vertical accuracy are set at *one-third* the indicated contour interval for well-defined points and *one-sixth* the indicated contour interval for spot heights.

These recommended standards for horizontal map accuracy can be related to the NMAS by assuming that (1) discrepancies are distributed about a zero mean, (2) the X and Y standard deviations are equal, and (3) a sufficient number of checkpoints are used. When

819

CHAPTER 14:
Mapping, Digital
Mapping, and
Spatial
Information
Systems

TABLE 14.4
Planimetric coordinate accuracy for
well-defined points; class I maps

Planimetric (X or Y) limiting RMS error (m)	Typical map scale
0.0125	1:50
0.025	1:100
0.050	1:200*
0.125	1:500
0.25	1:1,000
0.50	1:2,000
1.00	1:4,000
1.25	1:5,000
2.50	1:10,000
5.00	1:20,000

* Practical limits for aerial mapping methods,
larger scales mapped using ground methods.
(*Adapted from ASPRS, 1989.*)

TABLE 14.5
Planimetric coordinate accuracy requirement
(ground X or Y, in feet) for well-defined points;
class I, maps

Planimetric (X or Y) accuracy[3] (limiting RMS error, ft)	Typical map scale
0.05	1:60
0.1	1:120
0.2	1:240
	------------------------------*
0.3	1:360
0.4	1:480
0.5	1:600
1.0	1:1,200
2.0	1:2,400
4.0	1:4,800
5.0	1:6,000
8.0	1:9,600
10.0	1:12,000
16.7	1:20,000

* Indicates the practical limit for aerial methods—for scales above
this line, ground methods are normally used. (*Adapted from
ASPRS, 1989.*)

these assumptions are satisfied, the circular standard deviation by Equation (2.125) in Section 2.31 corresponds approximately to the *circular map accuracy standard* (CMAS) at the 90 percent level as given by

$$\text{CMAS} = 2.146 \, \sigma_x = 2.146 \, \sigma_y \qquad (14.1)$$

For limiting RMS errors and related scales listed in Tables 14.4 (metric) and 14.5 (in feet), the values for CMAS are 0.54 mm and $\frac{1}{47}$ in., respectively. Note that the recommended accuracies are stated and evaluated at full ground scale.

The CMAS, as realized by Equation (14.1), provides an approximate value useful primarily to provide a comparison with the NMAS of $\frac{1}{30}$ in. of map scale for horizontal accuracy. Currently, the NMAS, still the official standards for map accuracy, are in the process of being revised by the U.S. Department of the Interior (1995). The new standards, with a proposed title of *National Spatial Data Accuracy Standards* (NSDAS), are expected to resemble the ASPRS standards cited previously in this section. Until the new standards are made available to the public, those involved in the production and use of maps should monitor the literature and proceedings of their professional societies to be aware of the adoption and the content of these new standards.

The proposed standards attempt to quantify positional accuracy limits alone. With the increased sophistication of digital mapping systems and fully integrated spatial information systems, it also is important to consider additional accuracy measures. For example, attribute accuracy, logical consistency of graphic features, completeness, and a summary of the lineage of a digital map might be as important as the positional accuracy (NCDCDS, 1985). Are attributes correct (close to their true values)? Are attribute values within acceptable domains for the attribute? What level of attribute accuracy is acceptable? These questions are being actively addressed as digital mapping standards evolve. In digital maps and to a lesser extent in manually drafted maps, are map features logically consistent? Consistency checks can be completed to verify that map polygons in fact are closed, that intersection lines create nodes in a digital cartographic database, and that continuous lines (e.g., streets) are logically connected end to end. These types of accuracy checks formerly were completed by visual inspection of drafted maps, but logical consistency is more critical in the era of digital data sets, because the digital data may be used in other applications requiring this consistency. Also, interest is growing in the maintenance of a map's lineage. From which component data sets was a map compiled? What were the component accuracies? What accuracy statements can one make, based on the propagated cumulative error of the map components? Access to the lineage of a map, which may include source data sets and their accuracies, conversions made on or transformations applied to a map data set, and a summary of the algorithms used would provide valuable information.

14.27
TESTS FOR ACCURACY OF MAPS AND MAP DATA

Maps can be tested for accuracy by using several techniques. The ASPRS (1989) accuracy standards recommend that the conventional rectangular topographic maps be field checked with a minimum of 20 checkpoints, where 20 percent of these points, located in each quadrant of the map sheet, are spaced at intervals of at least 10 percent of the map sheet diagonal. In general, the spatial distribution of checkpoints should not be specified for large-scale, special purpose engineering maps. For these maps, checkpoints ought to be concentrated in critical areas containing structures and drainage facilities, with a less dense concentration of points in areas where no construction is to occur. Check surveys should be conducted according to the Federal Geodetic Control Subcommittee (FGCS) standards for vertical and horizontal control networks (Sections 5.59 and 9.23; and Federal Geodetic Control Subcommittee, 1984, 1994).

Field checks for the planimetric accuracy of a map can be conducted using differential GPS surveys for single-point coordinate observations or by using the results of a field survey completed as noted earlier. Vertical accuracy also can be assessed by running levels or profiles to determine heights. The use of GPS for height determination must recognize the limitations of the precision and accuracy of the geoid model used to convert observed ellipsoidal heights to orthometric heights (Section 12.13, Part 5). The use of GPS to assess vertical accuracy must be completed carefully so that errors or inaccuracies in the geoid model do not falsely indicate vertical accuracy errors of the compiled data.

Another approach to the verification of map accuracy is to compare map features of one map to a map of the same area but at a larger scale.

821

CHAPTER 14:
Mapping, Digital
Mapping, and
Spatial
Information
Systems

14.28
STANDARDS FOR EXCHANGE OF DIGITAL DATA

The rapid expansion of computer-based mapping and drafting systems has spawned a seemingly endless set of digital data formats for the storage of spatial data. These formats usually are proprietary to a specific software system or system publisher. The data content, organization, and format of these various data files follow no given standard. Because of their long presence in the marketplace, the format of some of the CAD drafting systems have become defacto standards. For example, the ASCII DXF file format used by Autodesk's Autocad program is supported by most other software systems. Other proprietary systems provide translations to and from their internal file formats to DXF. Many of these translations may lead to problems because not all of the information contained in the native file format is preserved in the translation file. The DXF format is used primarily to support CAD graphics, and richer or more complex spatial data sets may not be well served by this format.

It is very important that interchange formats be standardized, so that the data survival rate when translating from one form to another is clearly understood. Currently, an effort is underway at the federal level to provide leadership in the development of standards and specifications for the capture, description of (metadata), and exchange of spatial data. This effort is called the national spatial data infrastructure (NSDI).

The NSDI is conceived to be an umbrella of policies, standards, and procedures under which organizations and technologies interact to foster more efficient use, management, and production of spatial data. The NSDI requires and will facilitate cooperation and inter-action among various levels of government, the private sector, and academia. The major components of the NSDI, currently under development, include a basic framework of digital spatial data to act as a foundation for numerous other data collection activities; thematic data sets of critical national importance, with a known quality; standards to facilitate data collection, documentation, access, and transfer; and the means to search, query, find, access, and use spatial data. Strategies to build the NSDI include establishing forums for communication, facilitating access to data, building framework and thematic data sets, developing educational and training programs, and fostering partnerships for data sharing. Key actions underway are developing and implementing standards for framework and thematic data; producing framework and thematic data; implementing standards for spatial data documentation and transfer; establishing procedures to use electronic networks to search for, access, and use spatial data; and cooperating in the development of state and regional councils and private sector agreements to accomplish these actions. In 1995, the Executive Branch of the government charged the Federal Geographic Data Committee (FGDC) with the task of providing the federal leadership for involving the NSDI in co-operation with state and local governments and the private sector.

The second step in developing the infrastructure, involves recognizing that data sharing can be facilitated through the use of common procedures and standards. Metadata, or data about data, are critical, first, to determine that specific data exist and, second, to understand the nature and quality of the data to determine whether they are useful in a specific application. Consistency in how, when, and where to document data will assist others in using the data. A metadata standard for consistency in documentation, the content standards for digital geo-spatial metadata, recently was approved. If members of the spatial data network can agree to use standards such as this or the spatial data transfer standard (SDTS) for data transfer, building another tier in the NSDI becomes possible and everyone moves one step closer to finding and retrieving data. A network-based clearinghouse for geo-spatial data is being developed to provide metadata and spatial data to the community. Instead of centralizing all information in one place, the Internet is used to link all the sites where data are produced or maintained. With this approach, data producers can control and maintain information provided about their data. Users can access this network of information and find what data exist, the quality and condition of the data, and the terms for obtaining them.

PROBLEMS

14.1. What are the two major types of maps and how are these maps currently being generated for the most part?

14.2. Define (*a*) geodetic horizontal datum, and (*b*) vertical datum.

14.3. Describe the reference coordinate systems used in the United States.

14.4. Describe the characteristics of a topographic map.

14.5. In what different ways can relief be represented?

14.6. Define the term *contour* as used with respect to a topographic map.

14.7. Discuss the principal characteristics of contour lines.

14.8. Successive "V"-shaped contours, closely spaced, indicate what type of topography when (*a*) the *V*s point downhill, and (*b*) the *V*s point uphill.

14.9. On a map at a scale of 1:10,000 and a contour interval of 2 m, the distance scaled between two adjacent contours is 15.0 mm. What is the slope of the ground in percent?

14.10. On a map at a scale of 1:2000 with a contour interval of 0.5 m, the distance scaled between two adjacent contours is 15.0 mm. What is the slope of the ground in percent?

14.11. On a map at a scale of 1:5000 with a contour interval of 1 m, the distance scaled along a curb line between the 500 and 499 m contour intervals is 25.0 mm. If the distance scaled in the same direction and along the same curb line between the 500 m contour line and an inlet is 10.0 mm, determine the elevation of the inlet. Assume the curb is 0.15 m high.

14.12. It is possible to plot and scale map details to within 0.5 mm. What scales of maps are required to permit scaling distances between well-defined points to within (*a*) 0.5 m, (*b*) 1.0 m, (*c*) 5 m?

14.13. In Problem 14.12, plotting accuracy was used to determine map scale. What other factors affect the choice of map scale?

14.14. What four main factors ought to be considered when choosing a contour interval for a topographic map?

14.15. What is a DTM and how is it used in topographic mapping?

823

CHAPTER 14:
Mapping, Digital
Mapping, and
Spatial
Information
Systems

14.16. Describe the differences between a rectangular gridded network and a TIN. Discuss the advantages and disadvantages of each type of DTM representation.

14.17. What is a breakline, and why are breaklines necessary to supplement a DTM based on a rectangular grid network of points, for example?

14.18. What techniques other than contours can be used to portray relief on topographic maps?

14.19. Describe the major steps in topographic map compilation and production by photogrammetric methods.

14.20. Discuss the features that distinguish a CAD system from a digital mapping system.

14.21. When collecting data for digital mapping, what sources of information are available and how reliable is the information in these sources?

14.22. What is an orthophotograph, and how does it fit into the spatial information system?

14.23. Describe the differences between data storage in vector and raster formats.

14.24. Give a general definition of a spatial information system and define several of its subsystems.

14.25. What is the difference between a GIS and an LIS?

14.26. Describe the basic components of a GIS.

14.27. Give some examples of the application of a GIS to practical problems.

14.28. Describe the charactersitics of (a) AM/FM systems and (b) land information systems (LIS).

14.29. What information should appear on any map?

14.30. What information should appear on a map of a subdivision that is recorded and becomes a part of the public record?

14.31. What is an advantage of the graphical scale as opposed to a numerical scale on a map?

14.32. Discuss the methods for production of hard-copy maps when (a) a low volume of copies is required and (b) a high volume of copies is needed.

14.33. Describe the national map accuracy standards.

14.34. Are the national map accuracy standards adequate for *all* types of maps? Explain why or why not.

14.35. A map at scale of 1:10,000 has a contour interval of 2 m. Compute the (a) national map accuracy standards for horizontal position and elevation and (b) ASPRS standards for horizontal position and elevation.

14.36. A map with a scale of 1:1200 has a contour interval of 1 ft. Perform the computations of Problem 14.35 (a) and (b) for this map.

14.37. The limiting standard deviations for a map are $\sigma_x = 1$ m and $\sigma_y = 1$ m. Compute the corresponding national map accuracy standards for a horizontal position.

REFERENCES

Abler, R. F. *International Journal of Geographical Information Systems,* The National Science Foundation for Geographic Information and Analysis 1, no. 4 (1971), pp. 303–26.

American Society of Civil Engineers (ASCE), Task Committee for Preparation of a Manual on Selection of Map Types, Scales, and Accuracies for Engineering and Planning. "Selection of Maps for Engineering and Planning." *Journal of the Surveying and Mapping Division, ASCE,* no. SU1 (July 1972).

American Society for Photogrammetry and Remote Sensing (ASPRS), Specifications and Standards Committee. "ASPRS Interim Accuracy Standards for Large-Scale Maps." *Photogrammetric Engineering and Remote Sensing* 55, no. 7 (July 1989), pp. 1038–40.

Bureau of the Budget. "United States National Map Accuracy Standards." Washington, DC: June 17, 1947.

DeBoor, C. *A Practical Guide to Splines.* New York: Springer-Verlag, 1978.

Federal Geodetic Control Subcommittee. *Standards and Specifications for Geodetic Control Surveys.* Silver Spring, MD: September 1984.

Federal Geodetic Control Subcommittee. "Standards for Geodetic Control Networks." Draft Report. Silver Spring, MD: National Geodetic Survey, October 1994.

Federal Geographic Data Committee. "Draft Geospatial Positioning Accuracy Standards." Federal Geographic Data Committee Secretariat. Reston, VA: U.S. Geological Survey, January 1997.

Laurini, R., and D. Thompson. *Fundamentals of Spatial Information Systems.* San Diego, CA: Academic Press, 1992.

Merchant, D. C. "Spatial Accuracy Specifications for Large-Scale Topographic Maps." *Photogrammetric Engineering and Remote Sensing* 52, no. 7 (July 1987), pp. 958–61.

National Committee for Digital Cartographic Data Standards (NCDCDS), H. Moellering, editor. *Digital Cartographic Data Standards: An Interim Proposal.* Report 6. Bethesda, MD: American Congress on Surveying and Mapping (ACSM), 1985.

Star, J. L., and J. E. Estes. *Geographic Information Systems: An Introduction.* Englewood Cliffs, NJ: Prentice Hall, 1990.

Tomlin, D. *Geographic Information Systems and Cartographic Modelling.* Englewood Cliffs, NJ: Prentice Hall, 1990.

Unwin, D. *Introductory Spatial Analysis.* London: Methuen, 1981.

U.S. Department of the Interior. *National Data Accuracy Standard (NSDAS) Draft Report—1995.* Reston, VA: U.S.G.S., 1995.

Zilkosky, D. B.; J. H. Richards; and G. M. Young. "Results of the General Adjustment of the North American Datum of 1988." *Surveying and Land Information Systems* 52, no. 3 (1992), pp. 133–49.

PART V

Types of Surveys

Control and Topographic Surveying

15.1
GENERAL

By definition, a control survey consists of determining the horizontal and vertical or spatial positions of arbitrarily located points (Chapter 1, Section 1.6). The surveying methods for establishing control are described in Chapters 5 (vertical), 8 (horizontal by traversing), 9 (horizontal by intersection, resection, trilateration, and triangulation), and 12 (horizontal and vertical by the global positioning system). Traditionally, horizontal and vertical controls have been established separately, but with the advent of GPS, which provides both horizontal and vertical control in the same operation, this separation is rapidly disappearing.

A geodetic control survey, as described in Section 1.4, takes into account the shape of the earth and generally is used for primary control networks of large extent and high precision, such as those surveys established for continents, states, and counties. The bulk of the geodetic surveys performed currently are done with GPS for the horizontal positions but geodetic leveling (Section 5.52) still is used for precise vertical control. By virtue of the characteristics of the system and the reduction process, differential GPS automatically yields a geodetic horizontal survey.

An engineering control survey provides the horizontal and vertical control for the design and construction of private and public works. Depending on the size and scope of the project, such a survey may be geodetic but often is simply a plane survey for horizontal control with precise or differential leveling for vertical control. Ideally, the engineering survey should originate and close on horizontal and vertical control points in the national or state geodetic network. Naturally, GPS surveying methods also are applicable to engineering surveys.

The distinguishing feature of a topographic survey is the determination of the location, both in plan and elevation, of selected ground points that are necessary for plotting contour lines and the planimetric location of features on the topographic map. A topographic survey consists of (1) establishing, over the area to be mapped, a system of horizontal and vertical controls, which consists of key stations connected by measurements of high precision; and

(2) locating the details, including selected ground points, by measurements of lower precision from the control stations.

Topographic surveys fall roughly into three classes, according to the map scale employed as follows:

> *Large scale*   1:1200 (1 in. to 100 ft) or larger
> *Intermediate scale*   1:1200 to 1:12,000 (1 in. to 100 ft to 1 in. to 1000 ft)
> *Small scale*   1:12,000 (1 in. to 1000 ft) or smaller

Because of the range in uses of topographic maps and variations in the nature of the areas mapped, topographic surveys vary widely in character.

Topographic surveys can be performed by aerial photogrammetric methods, ground survey methods, or some combination of these two procedures. The largest portion of almost all of the small- and intermediate-scale as well as some large-scale topographic mapping now is performed by photogrammetric methods (Section 13.26). This photogrammetric operation includes establishing portions of the horizontal control in addition to compilation of the topographic map. However, ground survey methods still are applicable for large-scale topographic mapping of small areas and for field completion surveys, which usually are needed for photogrammetrically compiled topographic maps. The discussions in this chapter are directed primarily toward the various procedures for topographic surveys by ground survey methods.

15.2
PLANNING THE SURVEY

The choice of field methods for topographic surveying is governed by (1) the intended use of the map, (2) the area of the tract, (3) the map scale, and (4) the contour interval.

1. *Intended use of the map.* Surveys for detailed maps should be made by more refined methods than surveys for maps of a general character. For example, the earthwork estimates to be made from a topographic map by a landscape architect must be determined from a map that represents the ground surface much more accurately in both the horizontal and vertical dimensions than one to be used in estimating the storage capacity of a reservoir. Also, a survey for a bridge site should be more detailed and more accurate in the immediate vicinity of the river crossing than in areas remote therefrom.
2. *Area of the tract.* It is more difficult to maintain a desired precision in the relative location of points over a large area than over a small area. Control measurements for a large area should be more precise than those for a small area.
3. *Scale of the map.* It sometimes is considered that, if the errors in the field measurements are no greater than the errors in plotting, the former are unimportant. But, because these errors may not compensate each other, the errors in the field measurements should be considerably less than the errors in plotting at the given scale. The ratio between field errors and plotting errors should be perhaps one to three.

   The ease with which precision may be increased in plotting, as compared with a corresponding increase in the precision of the field measurements, points to the desirability of reducing the total cost of a survey by giving proper attention to the excellence of the work of plotting points, interpolation, and interpretation in drawing the map.

   The choice of a suitable map scale is discussed in Section 14.7.
4. *Contour interval.* The smaller the contour interval, the more refined should be the field methods. The choice of a suitable contour interval is discussed in Section 14.7.

15.3
ESTABLISHMENT OF CONTROL

Control consists of two parts: (1) *horizontal control,* in which the planimetric positions of specific control points are located, and (2) *vertical control,* in which elevations are established on specified bench marks located throughout the area to be mapped. This control provides the skeleton, which later is clothed with the *details,* or locations of such objects as roads, houses, trees, streams, ground points of known elevation, and contours.

On surveys of wide extent, relatively few stations distributed over the tract are connected by more precise measurements forming the *primary control;* within this system, other control stations are located by less precise measurements, forming the *secondary control.* For small areas, only one control system is necessary, corresponding in precision to the secondary control used for large areas.

15.4
HORIZONTAL CONTROL

Horizontal control can be established by GPS survey, total station system traverse, aerial photogrammetric methods, ordinary traverse, or trilateration and triangulation. Frequently, a combination of certain of these methods is used.

GPS surveys using static (Section 12.8) and rapid static (Section 12.11) methods and total station system traverse (Section 8.37) can be used to establish primary and secondary control for relatively large topographic surveys. These methods also are utilized in areas of lesser extent when field conditions are appropriate (hilly, urban, or mountainous regions). Specifications and standards for these types of surveys can be found in Sections 8.21 and 8.41 for traverse, and in Federal Geodetic Control Committee (1989) for GPS surveys.

Horizontal control determination by aerial photogrammetric methods (Section 13.22) is feasible and particularly applicable to small-scale mapping of large areas. Note that traditional photogrammetric control surveys require a basic framework of horizontal control points established by GPS or total station traverse. However, if a GPS receiver is used aboard the aircraft procuring the aerial photography, the number of ground control points can be substantially reduced (Section 13.22), although as yet, not eliminated.

15.5
VERTICAL CONTROL

The purpose of vertical control is to establish bench marks at convenient intervals over the area to serve (1) as points of departure and closure for operations of topographic parties when locating details and (2) as reference marks during subsequent construction work.

Vertical control usually is accomplished by direct differential leveling (Section 5.33), but for small areas or in rough country the vertical control is frequently established by trigonometric leveling (Sections 5.5, 5.55).

All elevations for topographic mapping should be tied to bench marks that are referred to the North American Vertical Datum of 1988 (NAVD 88; see Section 14.2).

Specifications for first-, second-, and third-order differential levels are given in Table 5.2, Section 5.47 and Table 5.3, Section 5.59. These specifications may be relaxed somewhat depending on map scale, character of the terrain to be mapped, the contour interval desired, and ultimate use of the survey. Table 15.1 gives the ranges of approximate

TABLE 15.1
Topographic survey vertical control specifications

Scale of map	Type of control	Length of circuit		Maximum error of closure	
		mi	km	ft	mm
Intermediate	Primary	1–20	2–30	0.05–0.3 $\times \sqrt{\text{mi}}$	12–72 $\times \sqrt{\text{km}}$
	Secondary	1–5	2–8	0.1–0.5 $\times \sqrt{\text{mi}}$	24–120 $\times \sqrt{\text{km}}$
Large	Primary	1–5	2–8	0.05–0.1 $\times \sqrt{\text{mi}}$	12–24 $\times \sqrt{\text{km}}$
	Secondary	$\frac{1}{2}$–3	1–5	0.05–0.1 $\times \sqrt{\text{mi}}$	12–24 $\times \sqrt{\text{km}}$

closures applicable to intermediate- and large-scale topographic mapping surveys. The smaller error of closure for a given map and type of control is used for very flat regions, where a contour interval of 0.5 m (1 ft) or less is required, and on surveys that are to be used to determine gradients of streams or to establish the grades of proposed drainage or irrigation systems. The higher errors of closure apply to surveys in which no more exact use is made of the results other than to determine the elevations of ground points for contours having 0.5-, 2-, and 3-m or 2-, 5-, and 10-ft intervals.

Revised accuracy standards, such as those currently being developed by the Federal Geographic Data Committee (FGDC, 1997; also see Section 14.26), may be adopted in the near future. New standards such as these, if adopted, would supersede the values given in Table 15.1. Consequently, a constant surveillance of the publications of the surveyor's professional organizations is required to stay abreast of developments in the continuing evolution of accuracy standards.

When an adequate number of points, having known elevations referred to the datum and a reliable mathematical function to model the geoid in the region, are available, elevations by differential GPS survey may be used (Section 12.13, Part 5). Bear in mind that elevations so determined with current methods may have errors in the 3- to 5-cm range. Specifications and standards governing GPS elevations can be found in FGCC (1989). When accuracies required for elevations permit, GPS can provide elevations with substantial improvement in speed of acquisition and economy.

15.6
HORIZONTAL AND VERTICAL CONTROL
BY THREE-DIMENSIONAL TRAVERSE

A three-dimensional, total-station system traverse (Section 8.37) can be used for establishing control for intermediate- and large-scale topographic mapping jobs. Care should be exercised in the trigonometric leveling aspect, with changes in elevations being determined in both directions for each traverse line and corrections for earth curvature and refraction applied for all long lines (Section 5.55).

Differential GPS surveys (Sections 12.8 and 12.9) automatically provide the third dimension. Such surveys are satisfactory for horizontal and vertical control establishment for topographic surveys, assuming that the qualifications with respect to elevations, as detailed in the previous section, are present.

15.7
LOCATION OF DETAILS

In the following sections, the horizontal and vertical controls are assumed to have been established and the field party is concerned only with the location of details.

The adequacy with which the resulting map sheet meets the purposes of the survey depends largely on the task of locating the details. Therefore, the topographer should be informed as to the uses of the map so that proper emphasis is placed on each part of the work.

The instruments currently most used for the location of details are the total station system and GPS equipment. The engineer's optical reading theodolite and engineer's level with level rod or stadia rod, although still used on certain types of topographic surveys, rapidly are being displaced by the previously mentioned electronic systems.

The principal procedure for acquiring topographic detail in the field, using current equipment (e.g., total station system, GPS, or theodolite stadia), is the controlling point method. Other classical procedures, using theodolite, tape, rod, and the like, are the cross-profile, checkerboard, and trace-contour methods. The cross-profile or cross-section method, which is applied primarily in route surveying, is discussed in Chapter 16, Section 16.24. With the prevalence of total station systems and GPS equipment in current surveying operations, much less emphasis now is placed on the checkerboard and trace contour methods for topographic surveying. Therefore, in this chapter, the major focus is on use of the controlling point method for locating topographic details.

15.8
PRECISION

The precision required in locating such definite objects as buildings, bridges, curbs, inlets, and boundary lines should be consistent with the precision of plotting, which may be assumed to be a map distance of about 0.5 mm or $\frac{1}{50}$ in. Such less definite objects as shorelines, streams, and the edges of wooded areas are located with a precision corresponding to a map distance of perhaps 0.9 to 1.3 mm or $\frac{1}{30}$ to $\frac{1}{20}$ in. For use in maps of the same relative precision, more located points are required for a given area on large-scale surveys than on intermediate surveys; hence, the location of details is relatively more important on large-scale surveys.

Contours

The accuracy with which contour lines represent the terrain depends on (1) the accuracy and precision of the observations, (2) the number of observations, and (3) the distribution of the points located. Although ground points are definite, contour lines must necessarily be generalized to some extent. The error of field measurement in a plan should be consistent with the error in elevation, which in general should not exceed one-fifth of the horizontal distance between contours. The error in elevation should not exceed one-fifth of the vertical distance between contours. The purpose of a topographic survey will be better served by locating a greater number of points with less precision, within reasonable limits, than by locating fewer points with greater precision. Therefore, if for a given survey the contour interval is 5 ft, a better map will be secured by locating with respect to each instrument station perhaps 50 points whose standard deviation in elevation is 1 ft than by locating 25 points whose standard deviation is only 0.5 ft. Similarly, if the contour interval is 2 m, it is better to have 50 points with a standard deviation of 0.4 m than 25 points with a standard deviation of 0.2 m.

A general principle that should serve as a guide in the selection of ground points may be noted. As an example, let it be supposed that a given survey is to provide a map that shall be accurate to the extent that if a number of well-distributed points is chosen at random on it, the average difference between the map elevations and ground elevations of identical points shall not exceed one-half of a contour interval (Section 14.26). Under this requirement, an attempt is made in the field to choose ground points such that a straight line between any two adjacent points in no case will pass above or below the ground by more than one contour interval. Therefore, in Figure 15.1, if the ground points were taken only at a, b, c, d, and e, as shown, the resulting map would indicate the straight slopes cd and de; the consequent errors in elevation of mn and op on the profile amount to two contour intervals and show that additional readings should have been taken at the points n and o. The corresponding displacement of the contours on the map is shown by dashed and full lines in Figure 15.2.

Angles

The precision needed in the field measurements of angles to details may be readily determined by relating it to the required precision of corresponding vertical and horizontal distances. For a sight at a distance of 300 m (1000 ft), a permissible error of 0.09 m (0.3 ft) in elevation corresponds to a permissible error of $01'$ in the vertical angle; likewise, a permissible error of 0.09 m (0.3 ft) in azimuth (measured along the arc from the point

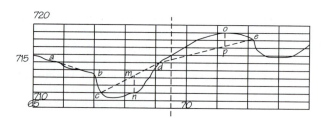

FIGURE 15.1
Effect of omission of significant ground points.

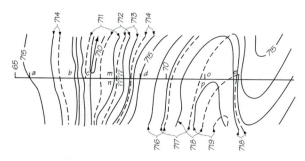

FIGURE 15.2
Errors in contour lines due to insufficient ground points.

sighted) corresponds to a permissible error of 01′ in the horizontal angle. Values for other lengths of sight or degrees of precision are obtained in a similar manner; therefore, if it is desired to locate a point to the nearest 2 ft in azimuth (or elevation) and if the length of the sight is 500 ft, the corresponding permissible error in the angle is 2/500 = 0.004 rad = 14′.

15.9
DETAILS BY THE CONTROLLING POINT METHOD

Details may be located with the controlling point method by employing a total station system (Sections 7.15–7.21) in radial surveys (Section 8.38), GPS kinematic surveys (Sections 12.9 and 12.10), and theodolite stadia (Sections 7.2–7.14). This method is applicable to practically every type of terrain and condition encountered in topographic mapping. Because only the controlling points that govern the configuration of the land and the location of planimetric features are located, a measure of economy is achieved not possible in the other techniques. However, the topographer must be experienced, because the success of the method depends on the selection of key controlling points. Procedures for use of this method are outlined in the following sections.

15.10
RADIAL SURVEYS BY THE TOTAL STATION SYSTEM

The personnel of the topographic party using a total station system usually consists of an instrument operator and one or two people operating reflectors mounted on prism poles.

The procedure is as follows. First, consider the case where the data are recorded manually in a form similar to that shown in Figure 15.3. The total station instrument is set up on either a primary or secondary control station, the elevation and position of which are known. The height of instrument (h.i.) is measured with a rod, tape, or observed on the plumbing rod if the tripod is so equipped. The temperature and atmospheric pressure also are measured and recorded. The total station instrument power is switched on, the environmental data and reflector constant are entered, and a backsight is taken on another station in the primary or secondary network of control. The horizontal circle may be set either on

	TOPOGRAPHIC DETAILS				OAK PARK LOT 10			30
Inst.	at Sta. 10; hi = 1.254 m; Elev. 125.14 m				TOPCON GTS-3C No. T 61034		June 26, 1995	
					Reflector No. 5; Prism Pole 2		Overcast 65°	
	B.S. on Sta. 9; Az 10-9 = 135°25′				p = 29.12 in Hg; Corr 16 ppm		J. Court	
	Horiz Circle, B.S. = 0°00′00″						H. Paley	
(1)	(2)	(3)	(4)	(5)	(6)	(7)	(8)	(9)
To Point	h.i	Horiz Angle Right	Zenith Angle	Slope Dist.	Horiz Dist.	Diff in Elev.	Elev.	Description
	m			m	m	m	m	
101	1.254	123°13′30″	95°46′	49.75	49.50	− 5.00	120.14	SW Corner Mulford Hall
102	1.254	134°12′30″	91°13′	73.82	73.80	− 1.57	123.57	Edge Sidewalk
103	1.254	245°43′00″	88°45′	85.12	85.10	+ 1.86	127.00	3 dm Live Oak
104	1.850	203°40′00″	86°32′	90.45	90.28	+ 4.87	130.01	6 dm Monterey Pine

FIGURE 15.3
Manually recorded note form, topographic details using total station system.

0 or to the azimuth of the line along which the sight is taken. In the example illustrated in Figure 15.3, the circle is set to 0. The horizontal clamp (upper clamp on a repeating instrument) is loosened and sights are taken to the desired points. Point number, h.i. of the reflector on the prism pole, horizontal angle, zenith angle, slope distance, and description of the point are observed and recorded in columns 1 through 5 and 9, respectively (Figure 15.3). Horizontal distance, difference in elevation, and elevation of the point sighted are calculated and entered in columns 6, 7, and 8. Note that most total station systems will show horizontal distance and difference in elevation between the horizontal axis of the instrument and the reflector, directly, when the appropriate control keys are pressed. Therefore, column 5, slope distance, could be eliminated and it would be necessary to compute only the elevation of the point sighted.

Usually, the height of the reflector on the prism pole is set to a height equal to the h.i. of the instrument. In this way, the difference in elevation between the center of the instrument and the reflector is the same as the difference in elevation between the station over which the instrument is set and the point on which the prism pole is held. When the line of sight is obstructed so that another h.i. of the prism pole must be sighted, this difference in h.i.'s must be taken into account in calculating the elevation of the point, as shown for point 104 in Figure 15.3. For this point, the elevation of $104 = 125.14 + 1.254 + (90.45) \cos(86°32') - 1.850 = 130.01$ m. Because the data for this example are only for plotting a topographic map, distances and elevations are rounded to decimeters (dm).

The operators of the reflectors on the prism poles choose points along valley and ridge lines and at summits, depressions, important changes in slope, and definite details. The selection of points is important, so these people should be instructed and trained for their work. They should follow a systematic arrangement of routes so that the entire area is covered and no important objects are overlooked. They should examine the terrain carefully (often with the aid of a hand level) and report any features that cannot be seen from the total station setup.

The example shown in Figure 15.3 illustrates the procedure and the flow of data acquisition. In practice, a data collector normally would be used for obtaining and recording the information. Data collectors (general discussion, Section 3.12; Section 8.37, for 3-D traverse) can be used in topographic surveys as an electronic field book in manual and automatic modes. First, consider the procedure for manual use of a data collector, where the collector is not connected to the total station system.

A sample set of data for a topographic survey is shown in Figure 15.4, tabulated in the approximate order of keyboard entry. The example problem is based on coordinates for the three-dimensional survey used in Section 8.37 and for which the data are given in Table 8.18. The instrument is set up over station 3 with a backsight on station 2. Initial entries via the collector keyboard consist of the job name, date, time, X and Y coordinates and elevation of point 3, backsight on point 2, the azimuth of line 3-2, and horizontal circle setting, which is 0 for this problem. These data could be entered before the station is occupied; in the office, for example.

In the field, after point 3 is occupied by the total station instrument, the collector and instrument are turned on, and environmental and instrumental constants entered, the traverse/side shot (Figure 8.54) or similar option is selected, a backsight is taken on point 2 with 0 being set on the horizontal circle, and the survey commences. The height of the reflector can be set to either the value of the h.i. of the instrument or some other known value. The instrument operator then selects the side shot or similar option from the menu on the data collector, which yields a screen requesting the foresight point number, the h.i. of the instrument, the h.i. of the reflector (rod), the horizontal angle to the right, the zenith angle, the slope distance, and the description of the point. These values, which have been measured and are observed on the instrument registers, are entered at the appropriate

Job Name: TOPOG, Date: 6-27-1995, Time 3:48 POST								
Inst @ 3; X = 7982.423 m, Y = 2239.722 m; Elev = 85.626 m; B.S. on 2; Az 3-2 = 66°31'5"								
Horiz. Circle on 0°00'00"								
Back Sight	Occupied Pt.	Fore Sight	Height Inst.	Height Rod	Horiz Angle	Zenith	Slope	Description
BS	OC	FS	h.i. m	h.i. m	Right	Angle	Dist. m	
2	3	4	1.341	1.341	194°07'00"	93°11'00"	219.76	Edge Stream
		5	1.341	1.341	187°00'00"	93°17'00"	213.36	Edge Stream
		6	1.341	1.341	176°31'00"	94°11'00"	171.00	Edge Stream
		7	1.341	1.341	172°56'00"	92°33'00"	218.20	Top Slope
		8	1.341	1.341	139°06'00"	92°41'00"	239.00	Top Slope
		9	1.341	0.975	208°46'00"	90°00'00"	87.17	Top Slope

FIGURE 15.4
Tabulation of data entries, data collector for topographic details.

location on the screen in the approximate order shown in Figure 15.4, column by column for a single point and line by line for additional points. When the desired number of points have been acquired from a given station, a check sight is made to the backsight station to ensure that the instrument has not shifted. The system then is moved to another control point, and the process is repeated using as input the identification and location of occupied and backsight locations. On completion of the entire job, the entire set of files can be downloaded into a personal, micro-, or mainframe computer for editing, computing, and plotting.

The raw data file TOPOG for the example illustrated in Figure 15.4, as downloaded into a personal computer, is shown in Figure 15.5a. The first five lines contain information about the file name, date, time, occupied point location, and backsight point number and azimuth. The next eight lines show the side-shot data in about the same order as given in Figure 15.4. These data can be edited, if necessary, and then by appropriate use of the software commands, used to compute X and Y coordinates and elevations for each side shot. Figure 15.5b is a tabulation of the coordinate file that results from this computation. The coordinate file, in turn, could be employed with plotting software (e.g., Auto CAD) on a personal computer to plot the points at a specified scale. This plot and the coordinate file could then be stored in a database for a geographic information system (Chapter 14, Section 14.18).

Most data collectors also include an option that permits field personnel to enter feature and command codes in the point description slot. Feature codes describe the objects located such as top of curb = TC, tree = TR, center line = CL, and so forth and usually are user defined. Command codes designate how to connect points; for example, beginning of line = BEG, end of line = END. Feature codes permit the points to be grouped, and command codes tell the plotter when the pen should be up or down and what sort of line and feature to plot. Use of these codes by field personnel, on location, greatly simplifies subsequent plotting of topographic data, but both feature and command codes must be compatible with the plotting software. So, the surveyor who wishes to exploit this opportunity to simplify the field-to-office routine for acquisition of topographic details and mapping, must become familiar with the codes required for the particular system with which the work is to be performed.

Another simplification in the process is to connect the data collector directly to the total station instrument and use the automatic mode to enter observed data (see Section 8.37 for automatic recording of data for a traverse). The procedure is the same as previously outlined

JOB: Name TOPOG, Date 6-27-1995 , Time 15:48:10.62
Mode Setup: North Azimuth, Dist meter, Scale 1.0000, Earth crv OFF, Angle Deg
Store: Pt 3, N 2239.7220, E 7982.4230, Elv 85.6260, START
Occupy: Occ 3, N 2239.7220, E 7982.4230, Elv 85.6260, START
Backsight: Occ 3, BS Pt 2, BS azm 66.3115, Back circle 0.0000
HI / HR : Inst H 1.3410, Rod H 1.3400
Side Shot: 3–4, Ang–Rt 194.0700, Zenith 93.1100, Slp Dst 219.7600, EDGE STREAM
Side Shot: 3–5, Ang–Rt 187.0000, Zenith 93.1700, Slp Dst 213.3600, EDGE STREAM
Side Shot: 3–6, Ang–Rt 176.3100, Zenith 94.1100, Slp Dst 171.0000, EDGE STREAM
Side Shot: 3–7, Ang–Rt 172.5600, Zenith 92.3300, Slp Dst 218.2000, TOP SLOPE
Side Shot: 3–8, Ang–Rt 139.0600, Zenith 92.4100, Slp Dst 239.0000, TOP SLOPE
HI / HR : Inst H 1.3410, Rod H 0.9800
Side Shot: 3–9, Ang–Rt 208.4600, Zenith 90.0000, Slp Dst 87.1700, TOP SLOPE

(a)

Point	Northing	Easting	Elevation	Note
3	2239.722000	7982.423000	85.626	START
4	2204.026560	7765.925038	73.423	EDGE STREAM
5	2179.298223	7778.163041	73.407	EDGE STREAM
6	2162.395929	7830.416174	73.153	EDGE STREAM
7	2128.936581	7794.690249	75.919	TOP SLOPE
8	2024.457878	7879.189457	74.438	TOP SLOPE
9	2247.755008	7895.623923	85.987	TOP SLOPE

(b)

FIGURE 15.5
Files downloaded from data collector to personal computer:
(a) raw data, (b) coordinate file.

for manual entry of data, except that, when sights are taken to topographic control points, the appropriate key for automatic recording must be pressed. The data acquisition flow is similar to that shown in Figure 15.4. The only manually entered data are the h.i. of the instrument, the h.i. of the reflector, point descriptions, and feature and command codes, if the automatic plotting option is available and chosen. The resulting data file is similar to the one shown in Figure 15.5. As a backup for the electronic records, it is advisable to keep minimum manually recorded notes, including points occupied and used for back-sights, references to control stations, and a sketch to show the general location of the area surveyed.

When a remotely controlled or robotic total station system (Section 7.20) is used to acquire topographic detail, the procedure is similar to that just outlined. First, the total station master unit is set over the known control point and oriented by a backsight on another known point. Next, the operator or topographer, in this case, takes the remote positioning unit (RPU; Figure 7.14) to the key topographic control points. From each point, the topographer can exercise complete control of the system by manipulating the data collector that is an integral part of the RPU. In this way, the file of located points is developed from the reflector end of the lines in the total station radial survey. The primary advantage of this method is that the topographer, who is the most knowledgeable person of what constitutes a desirable topographic controlling point, is in the field, choosing the point as well as pushing the buttons to record the data needed for positioning and to describe the point. A secondary advantage is that this is essentially a one-person operation. However, one must recognize that, in certain situations, a second party member would be required to transport equipment and attend the total station master unit to guard against theft or damage.

837

CHAPTER 15:
Control and
Topographic
Surveying

15.11
GPS KINEMATIC SURVEYS FOR TOPOGRAPHIC DETAILS BY THE CONTROLLING POINT METHOD

GPS systems provide very powerful tools for obtaining topographic details by the controlling point method. Kinematic (Section 12.9), pseudo-kinematic (Section 12.10), and kinematic with on-the-fly ambiguity resolution (Section 12.12) are the GPS techniques most useful for topographic surveys. Application of these methods to topographic surveys are discussed in some detail in Section 12.13, Part 6. A brief review of these applications follows.

The equipment and personnel required consist of two or more GPS receivers operated by one person per receiver. One or more known control points are needed. When two receivers and two operators are available, the known control point is occupied by one antenna and receiver and the second antenna with receiver attached is placed over another arbitrarily located point at about the same elevation and a few meters away from the known station. Both receivers are started and data are collected for about 45 sec to 1 min, followed by an antenna swap. In this procedure, which permits ambiguity resolution, the two antennas are switched from one tribrach to the other and data again are collected for about 45 sec. At the end of this time, one antenna and receiver are left occupying the known control point and the second system becomes the rover. The operator of the rover will have the second antenna mounted on a prism pole, the receiver in a backpack or other convenient carrying case, and also will carry the controller or data collector (see Figure 12.14). The rover operator will move from one topographic controlling point to another, occupying each point for about 45 sec. The data collector-controller not only provides storage for point positioning data but also furnishes the capability of entering point attributes and feature codes via keyboard entries by the operator. When all the desired points have been occupied, the rover returns to the initial location where the antenna again is placed over the arbitrarily located point for about 45 sec to furnish a closure error for the survey.

Data from both receivers are downloaded into a laptop, personal, or other computer for postprocessing, which yields the base lines to all occupied points. Note that the elevations so determined are relative to the spheroid and must be converted to the desired datum (Section 12.13, Orthometric Heights) before being used for topographic mapping.

The process just outlined is the minimum case in terms of equipment and personnel. For optimum efficiency and production, four receivers and three operators are necessary. Two receivers occupy known stations continuously and two receivers are rovers.

A disadvantage of this method is that at least four satellites must be continuously tracked by all the receivers. If, for example, a roving receiver loses the signal from one of these four satellites, loss of lock occurs and ambiguity resolution also is lost. In this case, the rover must return to a previously located point and reinitialize the process. Loss of lock can happen when a receiver and antenna go under an overpass, behind a building, or in back of a hill, for example. Therefore, careful preliminary planning of the operation is necessary to avoid this difficulty.

To avoid or minimize the problems caused by loss of lock, pseudokinematic or kinematic resolution with OTF ambiguity resolution can be used for topographic surveying. Pseudokinematic GPS compensates for loss of lock by the procedure used; kinematic resolution with OTF ambiguity depends on very sophisticated software. These procedures are described in more detail in Sections 12.10 and 12.12, respectively. The acquisition of topographic details follows the same general method as described for kinematic GPS surveying.

15.12
THEODOLITE STADIA

For small jobs or special projects, stadia using an optical reading or electronic theodolite and a stadia or level rod is a simple, direct method for obtaining topographic details by the controlling point method. Details concerning the basic concepts for the stadia method are found in Sections 7.1 through 7.14. Consider the procedure to be followed.

The theolodite is set up on either a primary or secondary control station, the elevation and position of which are known and the h.i. of the horizontal axis of the instrument above the station over which it is set is measured with a rod or tape. The vertical circle is checked for vertical collimation error and the theodolite is oriented by backsighting along a line the azimuth of which is known, this azimuth having been set off on the clockwise horizontal circle. The upper motion is unclamped and sights to the desired points are taken.

The notekeeper records all stadia intervals, zenith angles, and azimuths, and describes all points by remarks or sketches so that the draftsperson can interpret the data correctly and draw all features properly on the map. Figure 15.6 shows notes of observations taken

	TOPOGRAPHIC					DETAILS BLACK ESTATE				
Inst. at C;	El. = 127.86 m; h.i. = 1.34 m					Lietz Theodolite No. 1532			G. Burke	
Object	Az.	Rod Int. m	Zenith Angle	Horiz. Dist. m	El. Diff. m	Elev. m	k = 100; ftc = 0		M.D. Rand	Notes
							Description		F.J. & K.D.	Rods
B	176°14'						Tack in Hub		Apr. 4,	1995
1	10°21'	2.204	93°11'	219.7	−12.22	115.64	Water's edge		Cloudy	60°F
2	3°14'	2.140	93°17'	213.3	−12.24	115.62	"	"	Note: All zenith angles	
3	352°45'	1.719	94°11'	171.0	−12.51	115.35	"	"	measured to a rod	
4	7°18'	1.756	94°04'	174.7	−12.42	115.44	"	"	reading = instrumental	
5	349°10'	2.185	92°33'	218.1	−9.71	118.15	Fence line		h.i., unless otherwise	
6	16°55'	1.676	92°50'	167.2	−8.27	119.59	"	"	noted.	
7	315°20'	2.396	92°41'	239.1	−11.20	116.66	"	"		
8	349°15'	1.259	95°46'	124.6	−12.59	115.27	Water's edge			
9	339°30'	1.646	94°22'	163.6	−12.50	115.36	Bank brook 2 m wide			
10	0°05'	1.131	94°12'	112.5	−8.26	119.60				
11	344°40'	1.478	94°54'	146.7	−12.58	115.28	Bank brook 6 m wide			
12	25°00'	0.872	90° on 0.975	87.2	+0.37	128.23	Direct levels			
13	307°45'	1.487	94°56'	147.6	−12.74	115.12	Water's edge			
14	319°10'	1.225	95°56'	121.2	−12.60	115.26	"	"		
15	309°45'	1.768	93°00'	176.3	−9.24	118.26				
16	318°25'	0.997	94°36'	99.1	−7.97	119.89				
17	340°00'	1.932	93°08'	192.6	−10.54	117.32				
18	278°35'	0.765	95°43'	75.7	−7.58	120.28				
19	276°20'	0.936	97°56'	91.8	−12.80	115.07	Water's edge			
20	277°40'	1.292	95°40'	127.9	−12.70	115.16	"	"		
B	176°15'	Check								

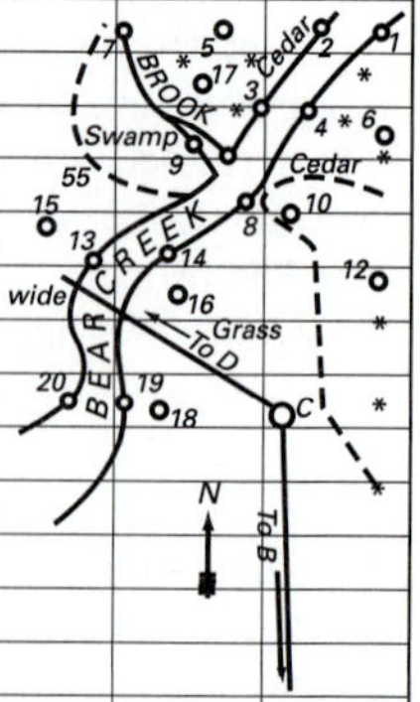

FIGURE 15.6
Stadia notes for location of details.

from station *C* of traverse, the elevation of the station having been previously determined as 127.86 m above the datum (NAVD 88). The h.i. of the instrument is 1.34 m. The theodolite is oriented by sighting to *B*, the azimuth of the line *CB* being set off on the clockwise horizontal circle prior to taking the sight. The first column lists the numbers of the side shots, their locations being shown on the sketch. The second column has the azimuths of the several points sighted; the third column has the rod intervals. The measured zenith angles are recorded in the fourth column, and the following columns show, respectively, the computed horizontal distances, differences in elevations, and elevations. Methods for reducing stadia data to horizontal distances and differences in elevation are outlined in Section 7.9. With one exception (object point No. 12 in the table), each zenith angle was measured to a rod reading equal to the instrumental h.i., thus simplifying computation of elevations. The usual procedure for observing the stadia interval is covered in detail in Section 7.10.

Where measurements are made solely for plotting a map, usually horizontal angles for directions are estimated to 05′. Zenith angles usually are measured to minutes, and differences in elevation are computed to centimeters or tenths of feet. Where elevations to the nearest 3 dm or nearest foot are sufficiently precise, the zenith angles (except for long shots) may be read to the nearest 05′.

When numerous observations are to be taken from a single station, sights to some object the azimuth of which is known are taken at intervals to make sure that no movement of the lower motion of the theodolite is undetected. The notes of Figure 15.4 show a check measurement of this kind taken to station *B* after observations to object 20 had been completed.

15.13
TOPOGRAPHIC SURVEYS AND DIGITAL TERRAIN MODELS

The digital terrain model (DTM) or digital elevation model (DEM) is described in Section 14.8. Because the DTM is a numerical (*X* and *Y* coordinates and elevations) representation of the configuration of the terrain, it can be stored in the memory of a computer and processed by a computer program that permits plotting planimetric details and contours (Section 14.10). Thus, the DTM used in conjunction with a geographic information system (Section 14.18), provides an extremely powerful and flexible tool for storing and processing data that describe terrain configuration.

These comments are inserted at this time to emphasize that data acquired in the process of obtaining topographic details constitute a DTM in the most basic form. At the highest level, data obtained in a kinematic GPS survey for topographic detail and stored in a data collector (Section 15.11) can be downloaded into a computer where the collected information is converted to *X* and *Y* coordinates and elevations and stored in a GIS as a DTM. At the lowest level, data from a theodolite-stadia survey to locate topographic details (Section 15.12) and recorded in a field book by hand may subsequently be keyed into a computer, where it is processed and stored in a GIS as a DTM. In either case, this DTM becomes a layer or specific part of the GIS, in which it may be manipulated to plot a topographic map or be used for other purposes.

The methods for acquiring topographic detail described in this chapter are designed primarily for the small or special purpose topographic survey or map completion survey. The surveyor or engineer involved with these special mapping problems needs to be aware of the potential of the data being collected. In this way, field methods and data recording can be designed for optimum efficiency.

PROBLEMS

15.1. Describe the major steps involved in performing a topographic survey.

15.2. Differentiate among geodetic, engineering, and topographic control surveys.

15.3. What method of topographic surveying is used for the largest portion of small-scale topographic mapping?

15.4. What factors govern the choice of surveying method used for topographic surveying?

15.5. Discuss the various surveying methods appropriate for establishing horizontal control for topographic mapping and the conditions that favor each method.

15.6. What methods are used to establish vertical control for topograhpic mapping? Discuss the advantages and disadvantages of these methods.

15.7. What systems would be most useful for running a three-dimensional traverse or control network for a topographic survey?

15.8. What factor governs the precision of locating horizontal details for topographic mapping?

15.9. What are the advantages of acquiring details for topographic mapping by the controlling point method?

15.10. List three surveying systems or instruments that are ideally adapted for topographic surveying by the controlling point method.

15.11. What is the function of a data collector when obtaining topographic details for mapping?

15.12. When using a data collector to record data for a topographic survey, should any hand-recorded notes be kept in a field book? If so, explain why.

15.13. When a GPS survey is employed to acquire topographic details, what is a major disadvantage and how are the effects of this disadvantage minimized?

15.14. What is an a priori requirement for any topographic survey to acquire details with any system?

15.15. Discuss the relationship between surveys to obtain topographic details and digital terrain models (DTM).

REFERENCES

Canon, M. E. "Integrated GPS-INS for High Accuracy Road Positioning." *A.S.C.E. Journal of Surveying Engineering* 118, no. 4 (November 1992), pp. 103–17.

Federal Geodetic Control Committee. *Geometric Geodetic Accuracy Standards and Specifications for Using GPS Relative Positioning Techniques.* Rockville, MD: National Geodetic Survey, NOAA, August 1, 1989.

Federal Geodetic Subcommittee. "Standards for Geodetic Control Networks (Draft), DGM." Rockville, MD: National Geodetic Survey, NOAA, 1994.

Federal Geographic Data Committee. "Draft Geospatial Accuracy Standards." Federal Geographic Data Committee Secretariat. Reston, VA: U.S. Geological Survey, January 1997.

Fiedler, J. "Orthometric Heights from Global Positioning System." *A.S.C.E. Journal of Surveying and Mapping* 118, no. 3 (August 1992), pp. 70–79.

Helmer, G. "Which Way Is Up in Earth Orbit?" *Professional Surveyor* 11, no. 4 (July–August 1991), pp. 39–44.

Joint Committee of A.S.C.E., A.C.S.M., and A.S.P.R.S., S.R. DeLoach, chairman. *Glossary of the Mapping Sciences.* Bethesda, MD; New York: ASPRS/ACSM and ASCE, 1994.

Roth, A. W. "Effectively Utilizing Your Contouring Program." *P.O.B., Point of Beginning* 17, no. 2 (December 1991–January 1992), pp. 32, 71–73.

Shmutter, B., and Y. Doytsher. "Contouring-Simulation of Cartographer's Procedure." *Journal of Surveying Engineering* 116, no. 4 (November 1990), pp. 193–201.

CHAPTER 16

Route Surveying

16.1
PLANNING THE ROUTE ALIGNMENT: GENERAL

The expression *route surveying* used in a very general sense can be applied to the surveys required to establish the horizontal and vertical alignment for transportation facilities. In the most general case, the transportation facilities are assumed to form a network that includes the transport of people or goods on or by way of highways, railways, rapid transit guideways, canals, pipelines, and transmission lines. For the past four decades in the United States, highways have been the most highly developed form of transportation facility in the overall network. As a result, route surveys for highways are well defined and widely practiced. Most of the methods developed for highway surveys are equally applicable to the other specified means of transport. Consequently, the emphasis in this chapter is on route surveys for highway alignment.

Surveys of some type are required for practically all phases of route alignment planning, design, and construction work. For small projects involving widening or minor improvement of an existing facility, the survey may be relatively simple and require obtaining only sufficient information for the design engineer to prepare plans and specifications defining the work to be done. For more complex projects involving multilane highways on new locations, the survey may require a myriad of details, including data from specialists in related fields to determine the best location; to prepare plans, specifications, and estimates for construction; and to prepare deed descriptions and maps for appraisal and acquisition of the necessary rights of way.

A description of all aspects and the various stages involved for the planning of route alignments is beyond the scope of this book. Details concerning these processes can be found in the references at the end of the chapter.

However, it is the function of the survey or project engineer to plan the surveys and gather all survey data that may be needed to execute the design of a route alignment for a particular project. This process includes obtaining the necessary information regarding terrain and land use, making surveys to determine detailed topography, and establishing horizontal and vertical control required for construction layout.

To plan and perform the surveys needed to acquire these types of data, the survey engineer must be familiar with (1) the geometry of horizontal and vertical curves and how they are used in the route alignment procedure, (2) the methods of acquiring terrain data utilized in the route design procedure, (3) the procedures followed in processing terrain data to obtain earthwork volumes, and (4) the earthwork distribution processes. These topics are covered in the sections that follow.

16.2
CONTROL

An important aspect for any route alignment project is the establishment of the horizontal and vertical control required for the planning, design, construction, and final as-built surveys for the proposed route.

Horizontal control should consist of an adequate number of conveniently placed existing or newly established monuments of sufficient mass and durability to last through the design and construction phases of the project. These monuments must be part of a closed system, having one order of accuracy higher than the project standards and coordinated in the state plane coordinate system on NAD 83 (see Chapter 11, Sections 11.17 through 11.20). Standards of accuracy for horizontal control are discussed in Section 9.23.

Vertical control can consist of elevations on the same monuments used for horizontal control, but usually more bench marks are required, so additional stations must be set. Orthometric heights must be known to one order of accuracy better than the project standards (see Section 5.46) and should be referred to NAVD 88. Establishing a local datum definitely is discouraged. If a GPS is to be used, both orthometric and ellipsoidal heights need to be known (see Chapter 12, Section 12.13, Part 5.)

In both horizontal and vertical control operations, care must be exercised to ensure that all phases of the surveys are performed on the same horizontal and vertical datums. Use of NAD 83 and NAVD 88 are strongly recommended. However, numerous stations in the field still are located on NAD 27 and NAVD 29. If stations on old datums are included in the route survey control network, be sure to convert them properly to account for the datum shift (for horizontal control, see Section 11.25).

16.3
ROUTE CURVES FOR HORIZONTAL AND VERTICAL ALIGNMENT: GENERAL

In highway, railway, canal, and pipeline locations, the horizontal curves employed at points of change in direction are arcs of circles. The straight lines connecting these *circular arcs* are tangent to them and therefore called *tangents*. For the completed line, the transition from tangent to circular curve and from circular curve to tangent may be accomplished gradually by means of a segment in the form of a *spiral* (Sections 16.14 to 16.18).

Vertical curves usually are arcs of parabolas. Horizontal parabolic curves occasionally are employed in route surveying and landscaping; they are similar to vertical curves and will not be discussed in this book.

The subject of route curves is extensive, as indicated by the list of references at the end of the chapter. Here only some of the simpler relationships are discussed.

16.4
CIRCULAR CURVES: GENERAL

The stationing of a route progresses around a curve in the same manner as along a tangent, as indicated in Figure 16.1. The point where a circular curve begins is commonly called the *point of curve* (P.C.), that where the curve ends is called the *point of tangent* (P.T.), and that where two tangents produced intersect is called the *point of intersection* or *vertex* (P.I. or V). A point on the curve is written P.O.C. Other notations also are used; for example, the point of curve may be written T.C., signifying the route changes from tangent to circular curve; the point of tangent is C.T. Or, the beginning of the curve may be written B.C. and the end of the curve E.C. The point of change from tangent to spiral is written T.S. and the point of change from spiral to circular curve S.C.

In the field, the distances from station to station necessarily are measured in straight lines, so that essentially the curve consists of a succession of chords. In this text, a full station is a function of the system of measurement units employed.

Traditionally, the station always was 100 units: In the metric system, one full station is 100 m; when using feet, the full station is 100 ft. When using 100 m as a full station, a point 10,152.159 m beyond station 0 + 00 is designated 101 + 52.159. The numbers to the left of the plus are thousands and hundreds of meters and to the right of the plus are tens and units of meters to the nearest 0.001 m or 1 mm. The intervals used on roads, highways, and marked on maps are 10 m and 20 m or $\frac{1}{10}$ or $\frac{1}{5}$ stations. When the measuring unit is the foot, stationing is similar except the units are different and the distances are expressed to two decimal places or 0.01 ft. For example, a point 1,032.45 ft beyond 0 + 00 has a station of 10 + 32.45. The intervals most commonly used are 100 ft, 50 ft, and 25 ft, or full, $\frac{1}{2}$, and $\frac{1}{4}$ station points.

As the United States moves toward adoption of the metric system (especially for federally supported projects), the American Association of State Highway and Transportation Officials (AASHTO) recently issued new standards (1993a, 1993b) regulating use of the metric system for highway projects. In these regulations, the full station is defined as 1 km or 1000 m. Thus, in the AASHTO metric system, a point 10,152.159 m from the origin, or 0 + 000.000, is designated 10 + 152.159. Numbers to the left of the plus are kilometers and numbers to the right are meters and decimal parts of a meter to the nearest 0.001 m or 1 mm. The interval to be used on most road and highway center lines will be 10 m (urban) and 20 m (rural) or $\frac{1}{100}$ and $\frac{1}{50}$ station points, respectively.

In this chapter, each of the systems just mentioned will be used with emphasis on the metric system. In Figure 16.1, stations are metric with one full station equaling 1 km and the tick marks are at 20 m or $\frac{1}{50}$ station intervals.

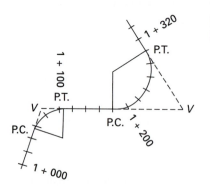

FIGURE 16.1
Route stationing.

Where curves are of long radius, as in railroad practice, the distances along the arc of the curve are considered to be the same as along the chords. In highway practice and along curved property boundaries, the distances usually are considered to be along the *arcs,* and the corresponding chord lengths are computed for measurement in the field.

The sharpness of curvature may be expressed in any of three ways:

1. *Radius.* The curvature is defined by stating the length of radius. This method usually is found in subdivision surveys and frequently in highway work. The AASHTO recommends use of radius, rounded to the nearest 5 m, to designate curvature. In design practice, the radius is chosen on the basis of design speed, allowable superelevation, and friction factor then is rounded *up* to the nearest 5 m (to be conservative).
2. *Degree of curve, arc definition.* Here the curvature is expressed by stating the "degree of curve," D_a, which traditionally has been defined as the angle subtended at the center of the curve by an arc of 100 units. In the metric system, D_a is defined as the angle subtended by a 100-m arc. When the foot is the unit, D_a is defined as the angle subtended by a 100-ft arc. The arc definition is the method most frequently followed in highway practice. In Figure 16.2, D_a is the degree of curve by the arc definition and R is the radius, so that

$$\frac{D_a}{100} = \frac{360°}{2\pi R}$$

or

$$D_a = \frac{36{,}000}{2\pi R} = \frac{5729.578}{R} \qquad (16.1)$$

Equation (16.1) is applicable in both systems of measurement. For example, the radius of a 1° curve is 5729.578 m in the metric system and 5729.578 ft in the foot system, and a 5° curve would have a radius of 1145.916 units in the respective systems.

Because 1 m equals 3.28084 ft (international ft), degrees of curve in the two systems differ by the same proportion. Therefore, a 1° curve in the metric system is the same curve as a 0.3048° curve in the foot system. Conversely, a 1° curve in the foot system is equivalent to a 3.28084° curve in the metric system. Comparable values in both systems, rounded to the nearest tenth of a degree, are shown in Table 16.1 (see Pryor, 1975).

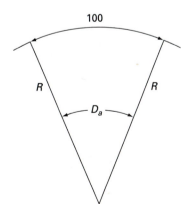

FIGURE 16.2
Degree of curve D_a, arc definition.

TABLE 16.1
Degree-of-curve conversions

Metric	Foot	Foot	Metric
1	0.3	0.5	1.6
5	1.5	1.0	3.3
10	3.0	1.5	4.9
15	4.6	2.0	6.6
20	6.1	2.5	8.2
25	7.6	3.0	9.8

In design practice, D_a usually is selected on the basis of design speed, allowable su-perelevation, and friction factor and is rounded *down* to the nearest integral number. The radius then is calculated using Equation (16.1) for subsequent curve calculations.

3. *Degree of curve, chord definition.* Degree of curve according to the chord definition, D_c, is defined as the angle subtended by a *chord* having a length of one full station or 100 ft in the foot system. In the metric system, D_c would be the angle subtended by a chord of 100 m. This definition in the foot system has been followed almost invariably in railroad practice. From Figure 16.3,

$$\sin\left(\frac{D_c}{2}\right) = \frac{50}{R} \tag{16.2}$$

Note that the radius of curvature varies inversely as the degree of curve; for example, according to the arc definition, the radius of a 1° curve is 5729.578 ft or m and the radius of a 10° curve is 572.958 ft or m. According to the chord definition, the radius of a 1° curve is 5729.651 ft or m and the radius of a 10° curve is 573.686 ft or m.

Field measurements of the curve with the tape, of course, must be made along the chords and not along the arc. When the arc basis is used, either a correction is applied for the difference between arc length and chord length or the chords are made short so as to reduce the error to a negligible amount. In the latter case, usually 100-ft chords are used for curves up to about 3°, 50-ft chords from 3° to 8°, 25-ft chords from 8° to 25°, and 10-ft chords for curves sharper than 25°. In the metric system, 100-m chords can be used for curves up to 1°, 50-m chords from 1° to 3°, 25-m chords from 3° to 5°, and 10-m chords for curves sharper than 5°. Note that curve layout using 100-m or 50-m chords could prove impractical under field conditions where terrain configuration and vegetation present obstacles.

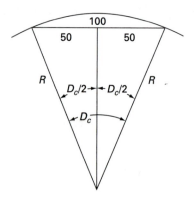

FIGURE 16.3
Degree of curve D_c, chord definition.

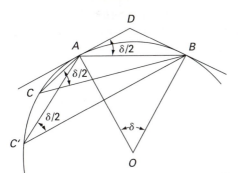

FIGURE 16.4
Geometry of circular curve.

16.5
GEOMETRY OF THE CIRCULAR CURVE

In discussing circular curves, the following geometrical facts are employed:

1. An inscribed angle is measured by one-half its intercepted arc, and inscribed angles having the same or equal intercepted arcs are equal. Therefore, in Figure 16.4, the angle ACB (at any point C on the circumference) subtending an arc AB is one-half the central angle AOB subtending the same arc AB; and the angles at the points C and C' are equal.

2. An angle formed by a tangent and a chord is measured by one-half its intercepted arc. In Figure 16.4, the angle at the point A between AD, is tangent to the curve at that point, and the chord AB, is one-half the central angle AOB subtending the same arc AB. This is a special case of the preceding proposition, when the point C moves to A.

3. The two tangent distances to a circular curve, from the point of intersection of the tangents to the points of tangency, are equal. In Figure 16.4, lines AD and DB are tangent to the curve at points A and B, respectively. Therefore, distance $AD =$ distance DB.

16.6
CURVE FORMULAS

Figure 16.5 represents a circular curve joining two tangents. In the field the intersection angle I between the two tangents is measured. The radius of the curve is selected to fit the topography and the proposed operating conditions on the line. The line OV bisects the angles at V and at O, bisects the chord AB and the arc ADB, and is perpendicular to the chord AB at F. From the figure, $\angle AOB = I$ and

$$\angle AOV = \angle VOB = \angle VAB = \angle VBA = \frac{I}{2}$$

The chord $AB = C$ from the beginning to the end of curve is called the *long chord*. The distance $AV = BV = T$ from the vertex to P.C. or P.T. is called the *tangent distance*. The distance $DF = M$ from the midpoint of arc to the midpoint of chord is called the *middle ordinate*. The distance $DV = E$ from the midpoint of arc to the vertex is called the *external distance*.

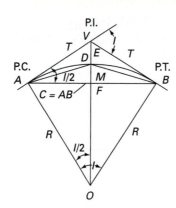

FIGURE 16.5
Basis for curve formulas.

Given the radius of the curve $OA = OB = R$ and the intersection angle I, then in the triangle OAV,

$$\frac{T}{R} = \tan \frac{I}{2}$$

$$T = R \tan \frac{I}{2} = \text{tangent distance} \tag{16.3}$$

$$E = R \sec \frac{I}{2} - R = R\left[\frac{1}{\cos (I/2)} - 1\right] \tag{16.4}$$

$$E = R \operatorname{exsec} \frac{I}{2} = \text{external distance}$$

From the triangle AOF, in which $AF = C/2$,

$$C = 2R \sin \frac{I}{2} = \text{long chord} \tag{16.5}$$

$$M = R - R \cos \frac{I}{2} = R\left(1 - \cos \frac{I}{2}\right)$$

$$M = R \operatorname{vers} \frac{I}{2} = \text{middle ordinate} \tag{16.6}$$

From the triangle AVF, in which $\angle VAF = I/2$ and $AF = C/2$,

$$\frac{C}{2} = T \cos \frac{I}{2}$$

$$C = 2T \cos \frac{I}{2} \tag{16.7}$$

From the triangle ADF, in which $\angle DAF = I/4$,

$$M = \frac{C}{2} \tan \frac{I}{4} \tag{16.8}$$

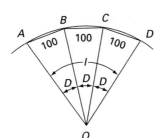

16.7
LENGTH OF THE CURVE

The length of the circumference of a circle is $2\pi R$; this is the arc length for a full circle or 360°. As the arc length corresponding to a given radius varies in direct proportion to the central angle subtended by the arc, the length of arc for any central angle I is

$$\text{arc} = \left(\frac{I°}{360°}\right)2\pi R \tag{16.9}$$

in which angle $I°$ is expressed in degrees.

If the degree of curvature is expressed on the *arc* basis, from Equations (16.1) and (16.9) the length of curve L_a is

$$L_a = 100\frac{I}{D_a} \tag{16.10}$$

Note that, if only the radius is given, Equation (16.9) can be used to calculate L. If the degree of curvature is expressed on the *chord* basis, the length of the curve is considered to be the sum of the lengths of the chords, normally each 100 ft long. In the metric system, the chord lengths would be 100 m. For these cases, the length of curve (on the chords) is

$$L_c = 100\frac{I}{D_c} \tag{16.11}$$

which is somewhat less than the actual arc length. Therefore, if the central angle I of the curve AD (Figure 16.6) is equal to three times the degree of curve D, as shown, then there are three 100-m chords between A and D, and the length of the "curve" on this basis is 300 m. Similarly, in the foot system, there would be three 100-ft chords between A and D, and L_c would be 300 ft.

16.8
LAYING OUT A CURVE BY DEFLECTION ANGLES

Curves can be staked out by the use of deflection angles turned at the P.C. from the tangent to stations along the curve together with the use of chords measured from station to station along the curve. The method is illustrated in Figure 16.7, in which ABC represents the curve, AX is the tangent to the curve at A, and angles XAB and XAC are the deflection angles from the tangent to the chords AB and AC.

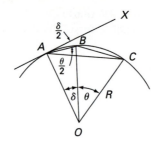

FIGURE 16.7
Curve layout by
deflection angles.

Assume the theodolite to be set up at A. Given R, δ, and θ it is required to locate points B and C. Considering point B,

$$\angle XAB = \frac{\delta}{2} \tag{16.12}$$

and by Equation (16.5)

$$AB = 2R \sin \frac{\delta}{2} \tag{16.13}$$

In the field, point B is located as follows. The deflection angle $XAB = \delta/2$ is set off from the tangent, the distance AB is measured from A, and the forward end of the tape at B is lined in with the theodolite.

Considering point C,

$$\angle BAC = \frac{\theta}{2} \tag{16.14}$$

$$BC = 2R \sin \frac{\theta}{2} \tag{16.15}$$

$$\angle XAC = \frac{\delta + \theta}{2} \tag{16.16}$$

In Figure 16.8, let points a, b, c, and d represent station points on a simple curve. Point a is an odd distance from the P.C. and the distance dB also is an odd increment. The deflection angles are

$$VAa = \frac{d_1}{2}$$

$$VAb = \frac{d_1}{2} + \frac{D}{2}$$

$$VAc = \frac{d_1}{2} + \frac{D}{2} + \frac{D}{2} = \frac{d_1}{2} + D$$

$$VAd = \frac{d_1}{2} + D + \frac{D}{2} = \frac{d_1}{2} + 3\frac{D}{2}$$

$$VAB = \frac{d_1}{2} + \frac{3D}{2} + \frac{d_2}{2} = \frac{I}{2}$$

Note that the sum of all deflection angles must equal $I/2$, a check on the calculations.

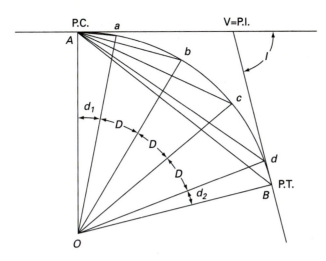

FIGURE 16.8
Deflection angles on a simple curve.

For a curve having degree of curve D (arc or chord) deflection angles per station, half-station, quarter-station, and 1/10th station are $D/2$, $D/4$, $D/8$, and $D/20$, respectively.

For odd lengths of arc or odd stations, it is convenient to calculate the deflection of the arc in minutes per unit of arc. If $D = 9°30'$, then $D/2 = 4°45' = 285'$ per station. Thus, the deflection equals $285'/100$ units $= 2.85'$/unit of arc or station. If the radius is given, then

$$\frac{d}{L} = \frac{360}{2\pi R}$$

so that

$$\frac{d}{2} \text{ (deflection in minutes/unit)} = \frac{1718.873}{R} \qquad (16.17)$$

For $D = 9°30'$, $R = 5729.578/9.5 = 603.113$ units and

$$\frac{d}{2} = \frac{1718.873}{R} = 2.85'\text{/unit of arc or station}$$

When only the radius is given, such as in organizations that have adopted the AASHTO specifications, Equation (16.17) can be used to calculate deflection angles directly.

A curve is normally located in the field as follows. The P.C. and P.T. are marked on the ground. The deflection angle from the P.C. is computed for each full station or substation (e.g., $\frac{1}{5}$, $\frac{1}{10}$ or $\frac{1}{50}$ and $\frac{1}{100}$ stations) on the curve and for any intermediate stations that are to be located. The theodolite is set up at the P.C., a sight is taken along the tangent, and each point on the curve is located by the deflection angle and distance measured from the preceding station. The following examples illustrate the procedure and give the usual form of field notes.

EXAMPLE 16.1. In Figure 16.9, assume that stations have been set as far as $88 + 00$ at C. The directions of tangents CV and MV have been fixed by stakes, but the tangent distances have not been measured. The degree of curve D_a is to be $38°$ according to the

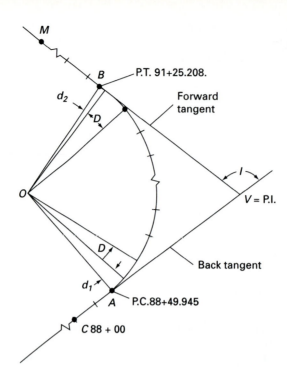

FIGURE 16.9
Laying out a simple curve.

arc definition. It is desired to stake $\frac{1}{5}$ stations on the curve using meters as the unit of measurement and taking one full station $= 100$ m.

Solution. The tangents CA and MB are produced to an intersection at V and the distance from C to V is found to be 245.029 m. Then the theodolite is set at V and the angle I is observed by double deflection to be $104°36'$. Because D_a is given, the radius of the curve, tangent distance, and length of curve using Equations (16.1), (16.3) and (16.10) can be computed as follows:

$$R = \frac{5729.578}{38°} = 150.778 \text{ m}$$

$$T = R \tan \frac{I}{2} = 195.084 \text{ m}$$

$$L = \frac{100I}{D} = 275.263 \text{ m}$$

Therefore, the stations at the P. C. and P. T. are

Station at P.I. $= 88 + 00 + 245.029 =$	$90 +$	45.029
Tangent distance $=$	$- 1$	95.084
Station at the P.C.	$88 +$	49.945
Length of curve	$+ 2$	75.263
Station at the P.T.	$91 +$	25.208

The deflection angle for $\frac{1}{5}$ station is $D/10 = 3°48'$. The distance from the P.C. to the first full station is $(88 + 60) - (88 + 49.945) = 10.055$ m, and the distance from the last full station on the curve to the P.T. is $(91 + 25.208) - (91 + 20) = 5.208$ m. The deflection in minutes per unit of arc equals $(38°)(60'/°)/200 = 11.4'/m$, so that deflection angles for the odd increments at the beginning and end of the curve are

$$\frac{d_1}{2} = (10.055)(11.4) = 1°54'38''$$

$$\frac{d_2}{2} = (5.208)(11.4) = 0°09'22''$$

Chord distances for the initial and final odd increments of arc and the full stations are

$$c_1 = 2R \sin \frac{d_1}{2} = (2)(150.778) \sin 1°54'38'' = 10.054 \text{ m}$$

$$c_2 = (2)(150.778) \sin 0°59'22'' = 5.207 \text{ m}$$

$$c_{100} = (2)(150.778) \sin 3°48'00'' = 19.985 \text{ m}$$

Deflection angles are found by cumulating the individual deflections to full stations from the P.C. to P.T. as follows:

Station	Deflection angles	Chord, m	Curve data
P.T. 91 + 25.208	52°18'00"		
+ 20	51°18'38"	5.207	
91 + 00	47°30'38"	19.985	
+ 80	43°42'38"	19.985	$\Delta = 104°36'$
+ 60	39°54'38"	19.985	$D_a = 38°00'$
+ 40	36°05'38"	19.985	$R = 150.778$ m
+ 20	32°18'38"	19.985	$T = 195.084$ m
90 + 00	28°30'38"	19.985	$L = 275.263$ m
+ 80	24°42'38"	19.985	Sta. at P.I. = 90 + 45.029
+ 60	20°54'38"	19.985	
+ 40	17°06'38"	10.054	
+ 20	13°18'38"	19.985	
89 + 00	9°30'38"	19.985	
+ 80	5°42'38"	19.985	
+ 60	1°54'38"	19.985	
P.C. 88 + 49.945	0°00'00"	19.985	

Note that the total deflection angle should equal $I/2$, which provides a positive check on the cumulated deflection angles.

EXAMPLE 16.2. The directions of two tangents on an existing road are established from the curb lines. The deflection angle to the right is $44°00'00''$. Because the existing curbs are to be retained, a curve must be determined for which the tangent distance does not exceed 50 m but is equal to or greater than 45 m. Determine a radius rounded to 5 m to satisfy this specification and calculate the parameters and deflection angles needed to set stations at 10-m intervals on this curve. The station of the P.I. is $1 + 843.892$, and one full station = 1 km for this example.

Solution. A trial value for R can be found using Equation (16.3):

$$R = \frac{T}{\tan (I/2)} = \frac{45}{\tan 22°} = 111.379 \text{ m}$$

Round this value of R up to 115.000 m and calculate T:

$$T = (115.000) \tan 22° = 46.463 \text{ m}$$

This satisfies the stated specification.

Now solve for L by Equation (16.9):

$$L = \left(\frac{I}{360°}\right) 2\pi R$$

$$L = 88.314 \text{ m}$$

Deflection angles can be calculated using Equation (16.17). The balance of the curve computations are included in the sample note form in Figure 16.10.

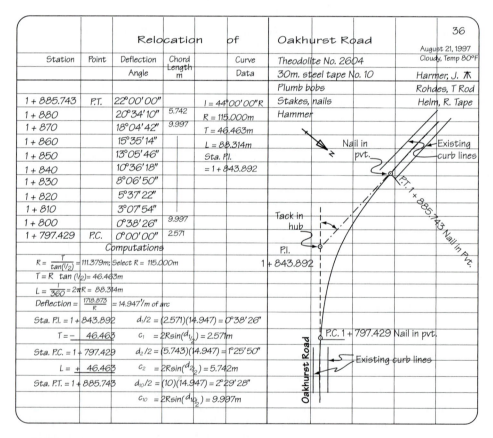

FIGURE 16.10

Notes for the layout of a simple curve.

EXAMPLE 16.3. Using the alignment shown for tangents CV and VM in Figure 16.9 and assuming that station C at $88 + 00$ has been set, determine curve parameters and deflection angles to full stations where $D_a = 12°$, according to the arc definition, the unit of measurement is the foot, and one full station equals 100 ft.

Solution. As in Example 16.1, tangents CV and MB are produced to an intersection at V, the distance from C to V is measured as 803.90 ft, and the angle at V is found to be $104°36'$. D_a is known, so that the radius of the curve, tangent distance, and length of curve can be calculated using Equations (16.1), (16.3), and (16.10) as follows:

$$R = \frac{5729.578}{12} = 477.46 \text{ ft}$$

$$T = R \tan \frac{I}{2} = 617.77 \text{ ft}$$

$$L = \frac{100I}{D} = 871.67 \text{ ft}$$

So, the stations at the P.C. and P.T. are

Station at P.I. $= 88 + 00 + 803.90 =$	$96 +$	03.90
Tangent distance $=$	$- 6$	17.77
Station at the P.C.	$89 +$	86.13
Length of curve	$+ 8$	71.67
Station at the P.T.	$98 +$	57.80

The deflection angle for one full 100-ft station is $D/2 = 6°00'$. The distance from the P.C. to the first full station is $(90 + 00) - (89 + 86.13) = 13.87$ ft, and the distance from the last full station on the curve to the P.T. is $(98 + 57.80) - (98 + 00) = 57.80$ ft. The deflection in minutes per unit of arc equals $(12°)(60'/°)/200 = 3.6'/\text{ft}$, so that deflection angles for the odd increments at the beginning and end of the curve are

$$\frac{d_1}{2} = (13.87)(3.6) = 0°49'56'' = 0°49.93'$$

$$\frac{d_2}{2} = (57.80)(3.6) = 208.08 = 3°28'05'' = 3°28.08'$$

Chord distances for the initial and final odd increments of arc and the full stations are

$$c_1 = 2R \sin \frac{d_1}{2} = (2)(477.46) \sin 0°49'56'' = 13.87 \text{ ft}$$

$$c_2 = (2)(477.46) \sin 3°28'05'' = 57.76 \text{ ft}$$

$$c_{100} = (2)(477.46) \sin 6°00'00'' = 99.82 \text{ ft}$$

Deflection angles are found by cumulating the individual deflections to full stations from the P.C. to P.T. as follows:

Station	Deflection angles	Chord, ft	Curve data
P.T. 98 + 57.80	52°18.0′		
98	48°49.9′	57.76	
97	42°49.9′	99.82	
96	36°49.9′	99.82	$\Delta = 104°36′$
95	30°49.9′	99.82	$D_a = 12°00′$
94	24°49.9′	99.82	$R = 477.46$ ft
93	18°49.9′	99.82	$T = 617.77$ ft
92	12°49.9′	99.82	$L = 871.67$ ft
91	6°49.9′	99.82	Sta. at P.I. = 96 + 03.90
90	0°49.9′	99.82	
P.C. 89 + 86.13	0°00.0′	13.87 ft	

Note that the total deflection angle should equal $I/2$, which provides a positive check on the cumulated deflection angles.

16.9
MOVING THE INSTRUMENT AHEAD ON THE CURVE

Because of obstacles, great lengths of curve, and so on, often it is impractical or impossible to run all of a given curve with the theodolite at the P.C. In such cases, one or more setups are required along the curve between the P.C. and the P.T.

Figure 16.11 illustrates the case where the instrument is set up at an intermediate point A. The curve is begun at the P.C. and is located as far as A, where a hub is to be set. The instrument is moved ahead and set up at A. A backsight (with the horizontal circle set to 0 and the telescope direct) is taken on the preceding station at which the theodolite was set up, in this case the P.C. The telescope then is turned toward B with the angle $(\alpha + \beta)/2$ being set on the horizontal circle. The line of sight thus is directed along chord AB. The angle $(\alpha + \beta)/2$, turned from the line P.C. to A extended, is equal to the deflection angle at the P.C. for point B; therefore, the horizontal circle setting to locate point B from theodolite station A is the same as that which would have been used had the theodolite remained at the P.C. According to this method, the following procedure may be used to orient the theodolite at any point on the curve:

1. Compute deflection angles as for use at the P.C.

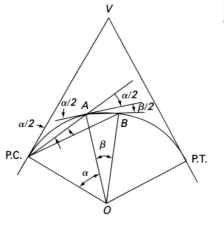

FIGURE 16.11
Theodolite setup on a simple curve.

2. When set at any point on the curve, take a backsight on any preceding instrument station with the telescope direct and the horizontal circle on $180°$ + the deflection angle to the station sighted for curves deflecting to the right, and $180°$ − the deflection angle to the station sighted for curves deflecting to the left.
3. To locate stations in a forward direction along the curve, turn the instrument through $180°$, and set to the previously determined deflection angles.

An alternative procedure for orienting the horizontal circle is possible if the instrument is in good adjustment. Take a backsight on the last station occupied with the telescope reversed and the horizontal circle set on the deflection angle for the station sighted. If the backsight is the P.C., the horizontal circle is set to 0. Then reverse the telescope to the direct position and set the previously determined deflection angles.

16.10
LAYOUT OF A CURVE USING A TOTAL STATION SYSTEM

When a total station system (Sections 7.15 to 7.17 and 7.19) is employed for curve layout, the procedures and calculations change somewhat. First consider a method, similar to the procedure described in Section 16.8, where the layout occurs from the P.C.

Layout from the P.C.

Assume that sufficient information is provided so that curve parameters and deflection angles can be calculated as in Example 16.3. In addition, the total chords from the P.C. to each station on the curve are calculated. In Figure 16.12, the chords A-90 + 00, A-91 + 00, through A-98 + 00, and A-B are calculated. For example, the chord A-90 = $2R \sin (0°49.9') = 13.87$ ft, A-91 = $2R \sin (6°49.9') = 113.59$ ft, and so on to A-98 = $2R \sin (48°49.9') = 718.66$ ft, A-B = $2R \sin (52°18.0) = 755.55$ ft.

The field layout would proceed as in Example 16.3 except that the EDM of the total station instrument, in the scanning mode, and a prism on a prism pole would be used to set the P.C. and P.T. from an instrument set up on the P.I. Next, the total station is set on the P.C., a sight taken on the P.I. with the horizontal circle set to 0, and the first deflection angle is turned to the left ($0°49'54''$) to set station 90 + 00 by measuring the chord of 13.87 ft. Then, each station is set by deflection angle and total chord from the P.C., successively. When chord A-B is measured, a check is provided on the accuracy of setting of the P.C. and P.T. As an added precaution, the curve should be inspected visually for consistency and smoothness, and chords between selected stations ought to be checked using a tape. Obviously, this procedure depends on clear lines of sight from the P.C. to all stations. If obstacles are present, the total station instrument can be moved ahead and the procedure described in Section 16.9 followed. For this case, it would be necessary to compute the total chords from the intermediate station to the remaining stations on the curve.

Layout from a Random Control Point

Construction layout from random control points is discussed in Chapter 17, Section 17.7. Setting a horizontal curve from a random control point is a special case of that general procedure.

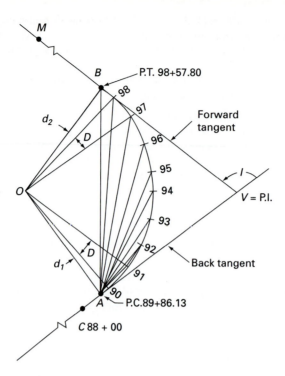

FIGURE 16.12
Layout of a curve by total
station system at the P.C.

For this procedure, it is assumed that the azimuth of either the back tangent or forward tangent of the curve is known and the P.I. and at least two horizontal control points in the control network for the highway, conveniently located with respect to the curve to be located, are coordinated in the state plane coordinate system of the region in which the highway falls.

The method consists of occupying a horizontal control point of known position with the total station instrument, taking a backsight along a line of known azimuth to another horizontal control point, and then locating the control points and all stations on the curve by direction and distance. Each point on the curve must be coordinated so that the inverse solution between the random control point and each station yields distance and direction. This procedure is ideally suited for the capabilities of a total station system equipped with a data collector.

To illustrate the method, consider an example.

EXAMPLE 16.4. The curve calculated in Example 16.2 is to be located from random control point CP-4. State plane coordinates for CP-4 and the P.I. of the curve are

Point	E, m	N, m
CP-4	454,621.541	153,734.552
P.I.	454,636.508	153,751.813

The grid azimuth from north along the back tangent from the P.I. to the P.C. is $58°30'00''$. The azimuth from CP-4 to control point Campanile is $94°00'00''$. Compute the distances and directions necessary to set the P.I., P.C., P.T., and $\frac{1}{10}$ stations for the curve calculated in Example 16.2.

Solution. Using given coordinates for the P.I. and curve data from Example 16.2, calculate coordinates for the P.C., $\frac{1}{10}$ stations on the curve, and the P.T. First, determine directions for the lines in a closed-loop traverse that runs from the P.I. to the P.C., then to each $\frac{1}{10}$ station on the curve to the P.T., closing back on the P.I. (see Figure 16.13). This is a deflection-angle traverse so that azimuths can be calculated by following the instructions given in Section 8.6. Computation of the directions follows, where deflection angles for the curve are from Example 16.2 and the geometry involved at each station is illustrated in Figures 16.4 and 16.7:

P.I.-P.C.	58°30′00″ (given)
P.C.-P.I.	238°30′00″
$+ d_1/2$	+ 38′26″
P.C. to 1 + 800	239°08′26″
$+ d_1/2 + d_{10}/2$	+ 3°07′54″
1 + 800 to 1 + 810	242°16′20″
$+ d_{10}$	+ 4°58′56″
1 + 810 to 1 + 820	247°15′16″
$+ d_{10}$	+ 4°58′56″
1 + 820 to 1 + 830	252°14′12″
$+ d_{10}$	+ 4°58′56″
1 + 830 to 1 + 840	257°13′08″
$+ d_{10}$	+ 4°58′56″
1 + 840 to 1 + 850	262°12′04″
$+ d_{10}$	+ 4°58′56″
1 + 850 to 1 + 860	267°11′00″
$+ d_{10}$	+ 4°58′56″

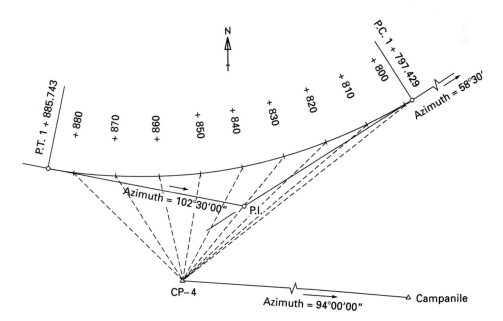

FIGURE 16.13
Layout of a simple curve from a random control point.

1 + 860 to 1 + 870	272°09′56″
+ d_{10}	+ 4°58′56″
1 + 870 to 1 + 880	277°08′52″
+ $d_{10} + d_2/2$	+ 3°55′18″
1 + 880 to P.C. 1 + 885.743	281°04′10″
+ $d_1/2$	+ 1°25′50″
	−180°00′00″
P.C. 1 + 885.743 to P.I.	102°30′00″
− I	−44°00″00″
P.I. to P.C.	58°30′00″ checks

The distances on the traverse consist of the tangent distances for the curve and chords between successive stations. Distances, directions, given coordinates, and coordinates for control and station points on the curve are listed in Table 16.2. Coordinates are computed directly as described in Section 8.15. Note that the closure errors equal 0 and no adjustment is needed.

With coordinates for each point on the curve known, the next step is to perform the inverse solution between known control point CP-4 and each station on the curve using Equations (8.8) and (8.9) to obtain distance and direction, respectively. For

TABLE 16.2
Coordinate computations, layout of horizontal curve from random control point

Station	Grid distance, m	Grid azimuth	E, m	N, m	Distance, m Grid	Distance, m Terrain	Azimuth
CP-4			454,621.541	153,734.552			
PI			454,636.508	153,751.813	22.846	22.848	40°55′43″
	46.463	58°30′00″					
PC 1 + 797.429			454,676.124	153,776.090	68.591	68.597	52°43′43″
	2.571	239°08′26″					
+ 800			454,673.917	153,774.771	66.036	66.041	52°48′47″
	9.997	242°16′20″					
+ 810			454,665.068	153,770.120	56.211	56.216	50°44′46″
	9.997	247°15′16″					
+ 820			454,655.849	153,766.254	46.713	46.717	47°15′36″
	9.997	252°14′12″					
+ 830			454,646.328	153,763.205	37.886	37.889	40°51′44″
	9.997	273°13′08″					
+ 840			454,636.579	153,760.993	30.418	30.420	29°37′43″
	9.997	262°12′04″					
+ 850			454,626.674	153,759.636	25.604	25.606	11°33′54″
	9.997	267°11′00″					
+ 860			454,616.690	153,759.145	25.067	25.069	348°50′30″
	9.997	272°09′56″					
+ 870			454,606.700	153,759.523	29.048	29.050	329°16′33″
	9.997	277°08′52″					
+ 880			454,596.780	153,760.767	36.060	36.063	316°38′02″
	5.742	281°04′10″					
PT 1 + 885.743			454,591.145	153,761.869	40.867	40.870	311°56′46″
	46.463	102°30′00″					
PI			454,636.508	153,751.813			

example, the grid distance from CP-4 to the P.I. by Equation (8.8) is

$$d_{\text{CP4-PI}} = [(E_{\text{PI}} - E_{\text{CP4}})^2 + (N_{\text{PI}} - N_{\text{CP4}})^2]^{1/2} = [(14.967)^2 + (17.261)^2]^{1/2}$$

$$d_{\text{CP4-PI}} = 22.846 \text{ m}$$

and the grid azimuth of CP-4 to the P.I. by Equation (8.9) is

$$A_{\text{CP4-PI}} = \arctan(E_{\text{PI}} - E_{\text{CP4}})/(N_{\text{PI}} - N_{\text{CP4}}) = \arctan[(14.967)/(17.261)]$$

$$A_{\text{CP4-PI}} = 40°55'43''$$

Grid distances and directions from CP-4 to the remaining stations and the P.C. are listed in columns 6 and 8 of Table 16.2. Column 7 contains *terrain distances*, which are to be used for the actual layout of the curve. The reason for this additional computation is that the given values of E and N are defined on the map projection surface (Lambert projection for this example, see Section 11.17) and not on the surface where the measurements are to be made. Therefore, the calculated distances in column 6, Table 16.2, need to be adjusted by the map projection scale factor (Section 11.17, Equation (11.19b) or (11.37)) and the ellipsoidal reduction factor (Section 9.11, Equation (9.11)). Using Equation (11.41) (Section 11.18) for the location and elevation of the example project, the combined scale and ellipsoidal reduction factor is 1.0000817. Multiplication of this combined factor by each calculated grid distance yields the terrain distances in column 7, Table 16.2.

Note that the differences for this example are small and vary from 2 to 6 mm. However, depending on the location and elevation of the project, the differences between grid and terrain distances could be significant, so that scale and ellipsoidal reduction factors always should be taken into account when using state plane coordinates for field layout and location problems.

The layout procedure in the field, for Example 16.4, is as follows. Control point CP-4 is occupied by the total station system, a backsight is taken on the Campanile with an azimuth of $94°00'00''$ set in a clockwise direction. Then, the P.I., P.C., P.T., and $\frac{1}{10}$ stations on the curve are set off by angle and distance from CP-4 using the values shown in Table 16.2 and measuring distances and directions with the total station system and a reflector mounted on a prism pole. When all points have been set, the curve is visually inspected for consistency and several chords between stations are checked using a tape. A field note form for this procedure is shown in Figure 16.14.

This operation, as discussed, involves extensive calculations if all computations are done on a hand calculator. Generally, when this method is applied, the total station system is equipped with a data collector. Also, the coordinates for control points and stations on the center line are natural by-products of the route alignment procedure performed at a workstation using a GIS (Section 16.27). This list of coordinates is in the form of an electronic file that can be downloaded to a PC and then again downloaded from the PC to a data collector. This operation usually is done in the office. The data collector and total station system are taken to the site where existing coordinated control points, conveniently located with respect to the layout site, are found. Known coordinates for these points are entered into the data collector, which contains software that enables real-time calculation of distance and direction to each of the coordinated points in the file stored in the data collector. This problem provides a good example of the use of electronic records from design to layout. It still is advisable to keep minimum hand-lettered field notebook records, as illustrated in Figure 16.14, as backup for the electronic files.

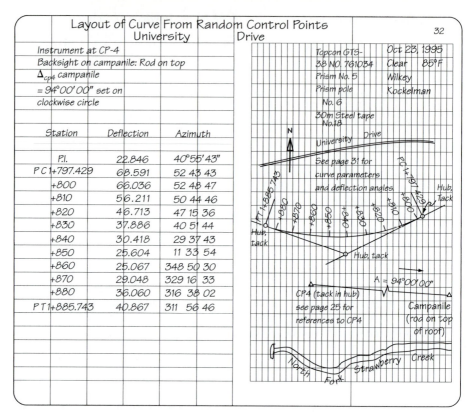

FIGURE 16.14
Note form curve layout from random control points.

16.11
COMPOUND CURVES

A compound curve consists of two or more simple curves that deflect in the same direction, are tangent to one another, and have two or more centers on the same side of the curve.

Use of compound curves permits better fitting of highway and railroad center lines to difficult topographic conditions. Compound curves should not be used where a simple curve satisfies alignment conditions.

Figure 16.15 illustrates a two-center compound curve. Of the several variables, I_1, I_2, T_1, T_2, R_1, R_2, and I, six are independent because $I = I_1 + I_2$. It is beyond the scope of this treatment to deal with all solutions to compound curves. Two methods are discussed, the vertex triangle procedure and the traverse method.

Vertex Triangle Method

In Figure 16.15, assume that R_1, R_2, I_1, I_2, I, and the station of the P.I. at V are known. The line $A'B'$ is tangent to the curve at the point of compound curve (P.C.C.) and forms the vertex triangle $A'B'V$. The distance $A'D = T_1 = R_1 \tan(I_1/2)$ and $DB' = T_2 = R_2 \tan(I_2/2)$, yielding $A'B' = T_1 + T_2$. The triangle $A'B'V$ is solved by the law of sines for p and q. Then, $T_L = T_1 + p$ and $T_R = T_2 + q$.

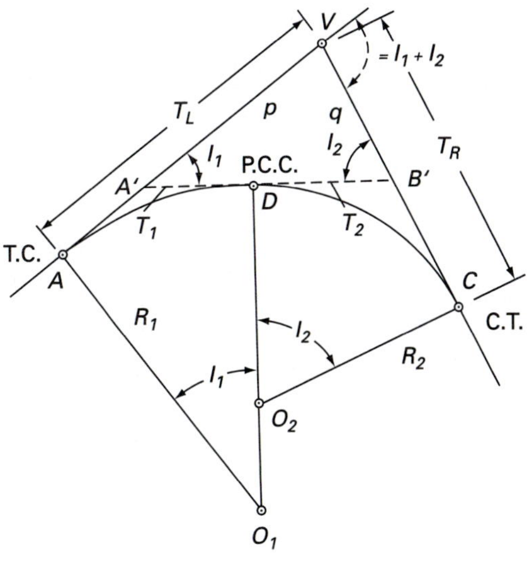

FIGURE 16.15
Compound curve.

863

CHAPTER 16:
Route Surveying

Traverse Method

The procedure for applying this method to compound curves is as follows:

1. Draw a sketch of the curve to be computed. Form a closed traverse (Section 8.15), including all independent variables and label variables. In Figure 16.15, polygon $AVCO_2O_1A$ forms the closed traverse.
2. Assume a direction of zero azimuth to be parallel or perpendicular to one unknown line. Therefore, in Figure 16.15, O_1A can be assumed due north.
3. Calculate azimuths for each line, including unknown directions expressed as functions of unknown angles. For example, in Figure 16.15, the azimuth of O_1O_2 is I_1^o, assuming that O_1A is due north.
4. Tabulate data as for traverse calculations and compute departures and latitudes (Section 8.13), expressing those for unknown lines as functions of unknowns.
5. Take Σ latitudes $= 0$ and the Σ departures $= 0$ to form two equations that can be solved for the unknowns.

The tabular form used to set up the traverse solution of the compound curve in Figure 16.15 is illustrated in Table 16.3. Taking Σ latitudes $= 0$ yields

$$R_1 - (R_1 - R_2) \cos (180° + I_1) - T_R \cos (90° + I) + R_2 \cos (180° + I) = 0$$

TABLE 16.3
Solution of a compound survey by traverse

Side	Azimuth	Length	Departures		Latitudes	
			E	W	N	S
O_1A	$0°$	R_1			R_1	
AV	$90°$	T_L	T_L			
VC	$90° + I$	T_R	$T_R \sin (90° + I)$			$T_R \cos (90° + I)$
CO_2	$180° + I$	R_2	$R_2 \sin (180° + I)$		$R_2 \cos (180° + I)$	
C_2O_1	$180° + I_1$	$R_1 - R_2$	$(R_1 - R_2) \sin (180° + I_1)$			$(R_1 - R_2) \cos (180° + I_1)$

which can be solved for T_R:

$$T_R = \frac{R_1 - (R_1 - R_2)\cos(180° + I_1) + R_2\cos(180° + I)}{\cos(90° + I)} \tag{16.18}$$

Setting Σ departures $= 0$ and substituting T_R from Equation (16.18) into this equation yields a value for T_L:

$$T_L = R_2\sin(180° + I) + (R_1 - R_2)\sin(180° + I_1) + T_R\sin(90° + I) \tag{16.19}$$

Many solutions to various compound curve problems are possible, depending on the combinations of known and sought parameters. Reference to a route surveying text is desirable to become familiar with the many possibilities (e.g., Meyer and Gibson, 1980).

16.12
INTERSECTION OF A CURVE AND A STRAIGHT LINE

A problem that frequently occurs in route alignment is that of locating the intersection of a straight line with a curve. In Figure 16.16, the line XP' of known direction intersects the curve at P. It is desired to locate point P so as to fix the corner of property bounded by the arc BP and the line PX. Therefore, the angle θ and distance PP' must be calculated. The angle α, distance AP', I, and R are known. One solution to this problem is as follows.

In triangle NAP',

$$NA = AP'\tan\alpha \qquad \text{and} \qquad NP' = \frac{AP'}{\cos\alpha}$$

so that

$$ON = R - NA = R - AP'\tan\alpha$$

Because $\beta = 90 - \alpha$, $\sin\beta = \cos\alpha$. In triangle ONP,

$$\frac{ON}{\sin\phi} = \frac{R}{\sin\beta} \qquad \text{or} \qquad \frac{\sin\phi}{\cos\alpha} = \frac{ON}{R} = \frac{R - AP'\tan\alpha}{R}$$

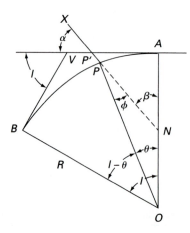

FIGURE 16.16
Intersection of a curve by a straight line.

so that

$$\sin \phi = \cos \alpha - \frac{AP'}{R} \sin \alpha$$

from which ϕ may be determined and used to calculate θ, where

$$\theta = \beta - \phi = 90 - \alpha - \phi$$

In triangle ONP, by the law of sines,

$$NP = \frac{R \sin \theta}{\cos \alpha}$$

and

$$PP' = NP' - NP = \frac{AP'}{\cos \alpha} - \frac{R \sin \theta}{\cos \alpha}$$

This problem also can be solved by the traverse method, as outlined in Section 16.11. The closed traverse would consist of $PP'AOP$. A bearing of due north could be assumed for line PP'. Traverse data are shown in Table 16.4, from which

$$\Sigma \text{ departures} = P'A \sin \alpha + R \cos (\alpha + \theta) - R \cos \alpha = 0$$

which can be solved for

$$\cos (\alpha + \theta) = \frac{R \cos \alpha - P'A \sin \alpha}{R} \tag{16.20}$$

to yield θ because α is known. Then,

$$\Sigma \text{ latitudes} = PP' + R \sin (\alpha + \theta) - P'A \cos \alpha - R \sin \alpha = 0$$

from which

$$PP' = P'A \cos \alpha + R \sin \alpha - R \sin (\alpha + \theta) \tag{16.21}$$

This completes the solution.

A third solution to this problem consists of using the method of coordinates according to the procedures developed in Section 8.19 and illustrated in Example 8.3. To apply this method the rectangular coordinates (preferably state plane) for the center of the circular curve and for points on the property line must be available.

TABLE 16.4
Intersection of a curve by a straight line

Line	Direction	Distance	Departures E	Departures W	Latitudes N	Latitudes S
PP'	N0°E	PP'			PP'	
$P'A$	S α E	$P'A$	$P'A \sin \alpha$			$P'A \cos \alpha$
AO	S(90° $-\alpha$)W	R		$R \cos \alpha$		$R \sin \alpha$
OP	N[90° $- (\alpha + \theta)$]E	R	$R \cos (\alpha + \theta)$		$R \sin (\alpha + \theta)$	

16.13
SUPERELEVATION

On high-speed highways and railroad curves, the velocity of movement of the vehicle or train develops a horizontal centrifugal force. So that the plane of the pavement or rails may be normal to the resultant of the horizontal and vertical forces acting on the vehicle, the outer edge of pavement or outer rail of the track is *superelevated,* or elevated above the inner edge of pavement or inner rail of track. In railway work, the amount of superelevation is made equal to approximately $0.00067V^2D$ (AREA, 1995) expressed in inches, where V is the train speed in miles per hour and D the degree of curve according to the chord definition (one full station = 100 ft). The amount of the superelevation should not exceed 7 or 8 in. (0.18 to 0.20 m) because of the use of most tracks by slow trains. For a speed of 40 mi (64 km) per hour, the superelevation in inches equals a fraction more than the degree of curve in degrees. The elevation of the inner rail is maintained at grade.

The superelevation of highway curves usually is combined with a friction factor, f, between the wheels of the vehicle and the pavement. In the metric system, the amount of pavement superelevation plus the side friction factor is given approximately (modified on the conservative side for safety) by (Pryor, 1975)

$$e + f = \frac{V^2}{127R} \tag{16.22}$$

where

$\quad\quad e$ = superelevation, m/m
$\quad\quad f$ = side friction factor
$\quad\quad V$ = vehicle design speed, km/hr
$\quad\quad R$ = radius of curve, m

Substituting for R using Equation (16.1) (rounded to one decimal place) and solving the resulting equation for D yields

$$D_a = \frac{730,000(e + f)}{V^2} \tag{16.23}$$

in which D_a is the degree of curve according to the arc definition, where one full station equals 100 m. Maximum values for e and assumption for f depend on climatic conditions, terrain configuration, type of area (urban vs. rural), and frequency of slow-moving vehicles. Depending on these factors, the American Association of State Highway and Transportation Officials recommends values of $e + f$ that vary from 0.20 to 0.28. With a knowledge of local conditions, Equation (16.23) can be used to calculate the maximum safe degree of curve for specified vehicle speeds and values for $(e + f)$. For example, assuming that V = 80 km/hr, e = 0.08 m/m, and f = 0.16, computation with Equations (16.23) and (16.1) yields R = 209.3 m and D = 27.38°. These values could be rounded to 210 m and 27°, respectively, without sacrificing safety.

Superelevation and side friction factor for highway curves, when the foot is the unit of measure, are given (AASHTO, 1994) by

$$e + f = \frac{V^2}{15R} \tag{16.24}$$

or

$$D = \frac{85,660(e + f)}{V^2} \tag{16.25}$$

where
D = angle subtended by 100 ft
e = superelevation, feet/foot
f = side friction factor
V = vehicle design speed, mi per hour

Equation (16.25) can be used to determine maximum degree of curve, given $e + f$ and vehicle design speed. For example, assuming that $e = 0.08, f = 0.14$, and $V = 50$ mi/hr, the maximum $D_a = 7.6°$ and the minimum safe radius is 758 ft. Tables for selecting D_a and R (in feet), given the design speed and $e + f$, can be found in many route surveying textbooks and in the AASHTO's *A Policy on Geometric Design of Highways and Streets* (1994).

When the kilometer is specified as the definition of one station and radius defines the curvature (Section 16.4), Equation (16.22) can be used to determine the minimum safe radius, given the values for $e + f$ and vehicle design speed. As the transition to the metric system occurs, most state departments of transportation have issued interim tables and standards to cover design specifications such as these.

Superelevation on highway pavements generally is introduced by gradually rotating the pavement cross section about the center line. In freeways and interchange design, special problems often require that the rotation occur about the inner edge or some intermediate point on the pavement cross section. Obviously, full superelevation cannot be introduced immediately at the beginning of a curve but must be developed gradually. To provide a smooth transition from normal pavement crown on a tangent to full superelevation on the circular curve, a spiral curve could be used and is introduced in the next section.

16.14
SPIRALS

Transition curves were first utilized by railways as early as 1880 to provide easement between tangents and circular curves. The curve chosen for this purpose was a *clothoid*, in which the curvature varies inversely as the radius and increases linearly from 0, at the tangent to spiral (T.S.), to the degree of curvature of the simple curve at the point where the spiral is tangent to the curve. Spiral curves are also called *easement* or *transition curves*. They are used as easement curves in both railway and highway alignment.

16.15
GEOMETRY OF THE SPIRAL

Figure 16.17 illustrates a spiral with a length of L_s. Because the degree of curvature for a spiral increases from 0 at the T.S. to D_a at the S.C., the rate of change of curvature of a spiral in degrees per station is

$$K = \frac{100D}{L_s} \qquad (16.26)$$

in which L_s is in the units of measurement and D is in degrees. For example, if $L_s = 200$ m and D_a at the S.C. is $20°$, then $K = (100)(20)/200 = 10°$ per station. Therefore, if L_s and

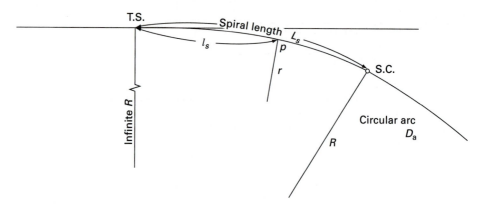

FIGURE 16.17
Spiral curve.

D_a or R are known for the circular curve at the S.C., then D_p or the radius r at any point p a distance l_s from the T.S. (Figure 16.17) can be determined as follows:

$$D_p = \frac{l_s K}{100}$$

$$r = \frac{5729.578}{D_p} = \frac{(5729.578)(100)}{l_s K}$$

(16.27)

The radius at the S.C. is

$$R = \frac{5729.578}{D_a} = \frac{(5729.578)(100)}{L_s K}$$

(16.28)

Division of Equation (16.27) by Equation (16.28) yields

$$\frac{r}{R} = \frac{L_s}{l_s}$$

(16.29)

which indicates that the radius of a spiral varies inversely as the length of the spiral times a constant.

To provide room for the spiral, the original circular curve is shifted inward from the main tangent, such as to position $KCC'K''$ in Figure 16.18. The portion CC' is retained and spirals are introduced from A to C and from C' to B.

The central angle of a spiral is a function of the average degree of curvature of the spiral or $D_a/2$, given by

$$\Delta = \frac{L_s D_a}{200}$$

(16.30)

Note that, from here on, Δ refers to the central angle of a spiral and I is used to designate the central angle of a simple curve. So, if a simple curve having a central angle I has equal spirals with central angles of Δ introduced at both ends of the curve, as shown in Figure 16.18, the central angle of the remaining simple curve from C to C' is $I - 2\Delta$.

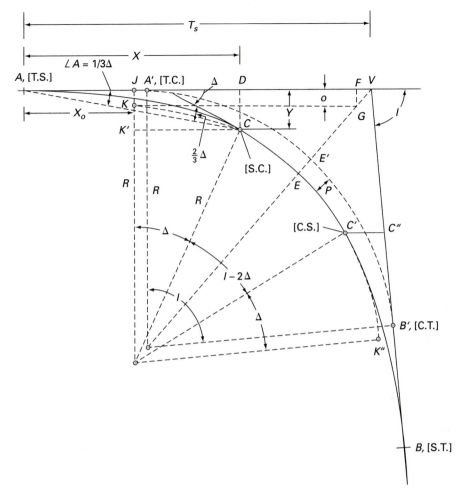

FIGURE 16.18
Field layout of a spiraled curve.

In Figure 16.19, consider the equations required to evaluate deflection angles to points on spirals. For a segment of the spiral dl at point p located a distance l_s from the T.S.,

$$d\delta = \frac{dl}{r} \qquad (16.31)$$

From Equation (16.29),

$$r = \frac{RL_s}{l_s} \qquad (16.32)$$

Substitution of r from Equation (16.32) into Equation (16.31) gives

$$d\delta = \frac{dl\, l_s}{RL_s} \qquad (16.33)$$

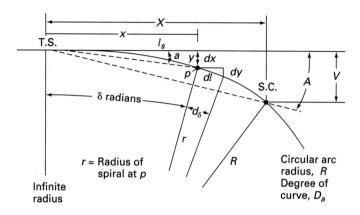

FIGURE 16.19
Mathematical development of a spiral curve.

Integration of Equation (16.33) from 0 to l_s yields δ, or

$$\delta_p = \frac{1}{RL_s} \int_0^{l_s} l\,dl = \frac{l_s^2}{2RL_s} \tag{16.34}$$

At the S.C., Δ in radians equals $\delta_{S.C.}$, so that

$$\Delta = \frac{L_s}{2R} \tag{16.35}$$

Division of Equation (16.34) by Equation (16.35) leads to

$$\delta_p = \frac{l_s^2}{L_s^2}\Delta \tag{16.36}$$

Therefore, the angle from the T.S. to any point on the spiral is proportional to the square of the distance of the point from the T.S.

Using the equations derived, several approximations useful for field layout of spirals can be developed. Referring to Figure 16.19, we obtain

$$\sin \delta = \frac{dy}{dl} = \delta \quad \text{(approximate)} \tag{16.37}$$

and

$$dy = \delta dl \quad \text{(approximate)} \tag{16.38}$$

Substituting Equation (16.34) into Equation (16.38) and integrating from 0 to l yields

$$y = \frac{l^3}{6RL_s} \quad \text{(approximate)} \tag{16.39}$$

or y, which is equal to the tangent offset to any point on the spiral, varies closely as the cube of the distance from the T.S. to the point.

Next, let a be the deflection angle to any point p on the spiral, where

$$\sin \alpha = \frac{y}{l_s} = a \quad \text{(approximate)} \tag{16.40}$$

Substitution of y from Equation (16.39) into Equation (16.40) yields

$$a = \frac{l_s^2}{6RL_s} \quad \text{(approximate)} \tag{16.41}$$

Applying similar logic to the total deflection angle, A, from the T.S. to the S.C., results in

$$A = \frac{L_s}{6R} \quad \text{(approximate)} \tag{16.42}$$

From Equation (16.35), $R = L_s/2\Delta$, so that

$$A = \frac{\Delta}{3} \quad \text{(approximate)} \tag{16.43}$$

Division of Equation (16.41) by Equation (16.42) yields

$$a = \frac{l_s^2}{L_s^2} A \quad \text{(approximate)} \tag{16.44}$$

Also, it can be shown by using Equations (16.43), (16.36), and (16.44) that

$$a = \frac{\delta}{3} \quad \text{(approximate)} \tag{16.45}$$

The approximations given by Equations (16.37) to (16.45) are sufficiently accurate for fieldwork in most practical situations.

Several parameters remain to be determined before field layout of a spiral is possible. To locate the S.C. from the T.S. or the C.S. from the S.T., X and Y must be calculated for the spiral (Figure 16.18). Also, to locate the T.S. and S.T. with respect to the P.I. or point V, the distance $o = FG = JK$ must be determined.

Using Equations (16.31) and (16.26), it can be shown that

$$\delta = \frac{Kl_s^2}{20,000} \tag{16.46}$$

in which δ is in radians, K in rad/station, and l_s in the units of the curve. From Figure 16.19,

$$dx = dl \cos \delta \tag{16.47a}$$

$$dy = dl \sin \delta \tag{16.47b}$$

Substitution of δ from Equation (16.46) into Equations (16.47a) and (16.47b) gives

$$dx = \cos (\alpha l^2) \, dl \tag{16.47c}$$

$$dy = \sin (\alpha l^2) \, dl \tag{16.47d}$$

in which $\alpha = K/20,000 = \delta/(l^2)$. Integration of Equations (16.47c) and (16.47d) from 0 to l_s gives x and y or

$$x = \int_0^{l_s} \cos (\alpha l^2) \, dl \tag{16.47e}$$

$$y = \int_0^{l_s} \sin (\alpha l^2) \, dl \tag{16.47f}$$

Given δ and l_s, Equations (16.47e) and (16.47f) can be integrated numerically using a programmable pocket calculator or PC program to yield x and y directly. Should it be necessary to compute x and y manually with a pocket calculator (given δ and l_s), the cosine and sine in Equations (16.47a) and (16.47b) can be found in a series expansion to give

$$dx = dl\left(1 - \frac{\delta^2}{2!} + \frac{\delta^4}{4!} - \frac{\delta^6}{6!} + \cdots\right) \tag{16.48a}$$

$$dy = dl\left(\delta - \frac{\delta^3}{3!} + \frac{\delta^5}{5!} - \frac{\delta^7}{7!} + \cdots\right) \tag{16.48b}$$

Substitution of δ from Equation (16.46) into Equations (16.48) and integration of the resulting modified Equations (16.48a) and (16.48b) from 0 to l_s yields

$$x = l_s\left[1 - \frac{\delta^2}{(5)(2!)} + \frac{\delta^4}{(9)(4!)} - \frac{\delta^6}{(13)(6!)} + \cdots\right] \tag{16.49a}$$

$$y = l_s\left[\delta/3 - \frac{\delta^3}{(7)(3!)} + \frac{\delta^5}{(11)(5!)} - \frac{\delta^7}{(15)(7!)} + \cdots\right] \tag{16.49b}$$

For the S.C., $\Delta = \delta$ and $L_s = l_s$. Given Δ in radians and L_s, exact values for X and Y can be calculated either by numerical integration on a calculator or PC, using Equations (16.47e) or (16.47f) or manually with Equations (16.49a) and (16.49b).

With a value for Y calculated, the "throw" or distance $o = JK$ (see Figure 16.18) is

$$o = Y - KK' = Y - R(1 - \cos \Delta) \tag{16.50}$$

and the external distance EV is

$$EV = EG + GV = R\left[\frac{1}{\cos (I/2)} - 1\right] + \frac{o}{\cos (I/2)} \tag{16.51a}$$

16.16
CALCULATION AND FIELD LAYOUT OF A SPIRAL

In the usual case, the station of the P.I. (point V), direction of the tangent(s), I, and D_a are known. A brief review of the computational details involved follows.

1. Select L_s to fit the existing conditions (refer to the AASHTO, 1994).
2. Calculate Δ using Equation (16.30).
3. Calculate $(I - 2\Delta)$, R, and L_a (length of circular arc), using Equations (16.1) and (16.10) for the last two.
4. Compute K and δ by Equations (16.26) and (16.46), then X and Y using Equations (16.47e) and (16.47f) or (16.49a) and (16.49b).
5. Calculate o using Equation (16.50).
6. Calculate T_s, where $T_s = AJ + KG + FV$ (Figure 16.18), in which $AJ = X - R \sin \Delta = X_0$, $KG = R \tan (I/2)$, and $FV = o \tan (I/2)$ to yield

$$T_s = X - R \sin \Delta + (R + o) \tan \frac{I}{2} \tag{16.51b}$$

7. Calculate the stations of control points as follows:

$$\text{Station T.S.} = \text{station P.I.} - T_s$$

$$\text{Station S.C.} = \text{station T.S.} + L_s$$

$$\text{Station C.S.} = \text{station S.C.} + L_a$$

$$\text{Station S.T.} = \text{station of C.S.} + L_s$$

8. Calculate spiral deflection angles using Equations (16.43) and (16.44).
9. Calculate deflection angles to desired stations from the S.C. to the C.S. on the simple curve.

Field layout of the equally spiraled curve is as follows:

1. Occupy the P.I. and sight along the back tangent setting the T.S. and point D (Figure 16.18).
2. Establish the direction of the forward tangent by turning I by double deflection (Section 6.36). Measure to set C'' and the S.T.
3. Occupy points D and C'', setting the S.C. and C.S. by right-angle offset, where $DC = C'C'' = Y$.
4. Occupy the T.S. and set stations on the approach spiral to the S.C. by deflection angles and chords. Usually, the spiral is divided into an equal number of chords. Chord lengths can be taken equal to the nominal difference in stationing for relatively flat spirals.
5. Occupy the S.C., backsight the T.S. with $180 \pm 2A$ set on the horizontal circle. Turn through $180°$ and set stations on the simple curve by deflection angles and chords. Check the closure on the C.S.
6. Occupy the S.T. and set stations on the leaving spiral from the S.T. to the C.S. On an equally spiraled curve, the same station interval and deflection angles calculated for the approach spiral can be used. When even stations are being set, new deflection angles must be calculated for the leaving spiral.
7. Stations on the leaving spiral can be set from the C.S. with a backsight on the S.C. where $180° \pm (I - 2\Delta)/2$ is set on the horizontal circle. In this case, deflection angles from the C.S. to the T.S. may be calculated using the osculating-circle principle (Section 16.17).

At this point, an example problem is useful.

EXAMPLE 16.5. It is desired to place spirals on an existing curve in meters for which the station of the P.I. is $54 + 61.460$ and I is $40°00'$ right. The design speed is 80km/hr and $e + f = 0.08 + 0.15 = 0.23$. Using Equation (16.25), $D_a = (730,000) (0.23)/(80)^2 = 26.2°$. This value is rounded down to $26°$, yielding a slightly longer radius, to be on the safe side. The length of spiral L_s generally is selected on the basis of design speed and superelevation. Using tables from the AASHTO(1994), the length of spiral required for superelevation runoff for $e = 0.08$ and a design speed of $50 \text{ mi/hr}(\sim 80 \text{ km/hr})$ for a two-lane highway is 190 ft, so that L_s is chosen as 60 m.

By Equation (16.30),

$$\Delta = \frac{L_s D_a}{200} = \frac{(60)(26)}{200} = 7.8° = 7°48'$$

By Equation (16.1),

$$R = \frac{5729.578}{26°} = 220.368 \text{ m}$$

The central angle of the spiraled circular curve is

$$I - 2\Delta = 40° - (2)(7.8°) = 24.4° = 24°24'$$

From Equation (16.10), the length of circular curve is

$$L_a = \frac{100(I - 2\Delta)}{D_a} = \frac{(100)(24.4)}{26} = 93.846 \text{ m}$$

Calculate X and Y using Equations (16.47e) and (16.47f) or (16.49a) and (16.49b) (using only the first two terms), where $\delta = 0.136136$ rad and $l_s = L_s = 60.000$ to give

$$X = 59.889 \text{ m}$$

$$Y = 2.719 \text{ m}$$

The throw, o, is calculated by Equation (16.50):

$$o = Y - R(1 - \cos \Delta) = 2.719 - (220.368)(1 - 0.990748)$$

$$= 0.680 \text{ m}$$

Using Equation (16.51b), T_s is computed:

$$T_s = X - R \sin \Delta + (R + o) \tan \frac{I}{2}$$

$$= 59.889 - (220.368) \sin 7°48' + (220.368 + 0.680) \tan 20°$$

$$= 110.437 \text{ m}$$

The stations at T.S., S.C., C.S., and T.S. are

Station at P.I.	54 + 61.460
$-T_s$	1 10.437
Station at T.S.	53 + 51.023
L_s	60.000
Station at S.C.	54 + 11.023
L_a	93.846
Station at C.S.	55 + 04.869
L_s	60.000
Station at S.T.	55 + 64.869

The total deflection angle from the T.S. to the S.C. is calculated by Equation (16.43):

$$A = \frac{\Delta}{3} = 2.6° = 2°36' = 156'$$

Deflection angles are determined using Equation (16.44). Computations for spiral deflection angles on the approach are accomplished most easily by arranging them in tabular form, as shown in Table 16.5.

Deflection angles to 20-m stations on the simple curve are computed in the usual way and established from an instrument setup at the S.C. A backsight is taken on the

TABLE 16.5
Computation of spiral deflection angles

Station	$l/10$	$l^2/100$	$L_s^2/100$	l^2/L_s^2	$a_i = (l_i^2/L_s^2)A$ $i = (1, 2, \ldots, 6)$
S.C. 54 + 11.023	6	36	36	1	2°36.0'
+ 01.023	5	25	36	25/36	1°48.3'
+ 91.023	4	16	36	4/9	1°09.3'
+ 81.023	3	9	36	1/4	0°39.0'
+ 71.023	2	4	36	1/9	0°17.3'
+ 61.023	1	1	36	1/36	0°04.3'
T.S. 53 + 51.023					0°00.0'

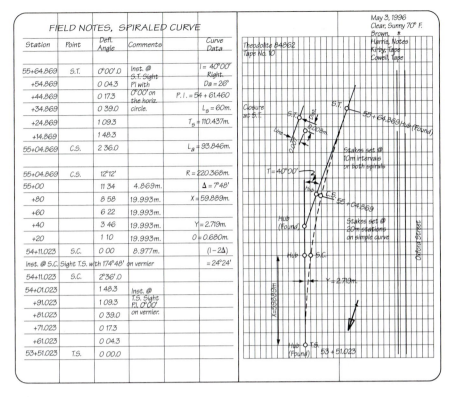

FIGURE 16.20
Field notes for a spiraled curve.

T.S. with the clockwise horizontal circle set on $180° - \frac{2}{3}\Delta = 180° - 2A = 174°48'$. The leaving spiral is set from the S.T. using the same uniformly spaced chord intervals and the same deflection angles as were calculated from the T.S. to the S.C.

Figure 16.20 illustrates the field notes for the spiraled curve. Note that the chord lengths on the spiral can be taken equal to the length of spiral between stations without introducing significant errors because the station interval is small and the spiral is relatively short and flat.

16.17
PRINCIPLE OF THE OSCULATING CIRCLE

The osculating circle is utilized for intermediate instrument setups on a spiral, calculating deflection angles to points on a spiral from the C.S. to the S.T., and the calculations necessary to place a spiral between two simple curves.

At any point p on the spiral illustrated in Figure 16.21, the radius r is defined by Equation (16.1), in which D_a is the degree of curve at that point. The circular arc having radius r and tangent to the spiral at p is called the *osculating circle*. The following significant properties govern the relationship between the osculating circle and the spiral:

1. The osculating circle departs inside the spiral toward the T.S. and outside the spiral toward the S.C.

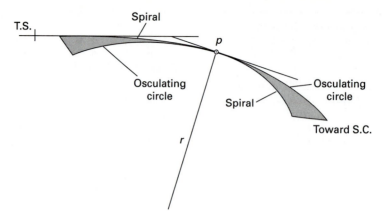

FIGURE 16.21
Osculating circle.

2. The rate of departure between the spiral and the osculating circle in both directions is the same as the rate of departure of the same spiral from the tangent to the spiral at the T.S.

The latter statement pertains only to rate of departure of degree of curve and is not exact with respect to rate of departure of deflection angles and offsets. However, the approximations that result from applying the theory to deflection angles and offsets are sufficiently close for the relatively flat spirals employed in highway and railway alignment. Consider how the osculating circle can be used to calculate deflection angles from the C.S. to the S.T. of a spiral.

EXAMPLE 16.6. Calculate deflection angles from the C.S. to the S.T., using the data of Example 16.5.

Solution. As illustrated in Figure 16.22, the osculating circle at the C.S. ($R = 220.368$ m and $D_a = 26°$) departs inside the spiral as stations increase from the C.S.

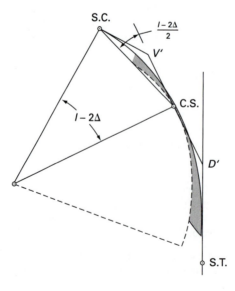

FIGURE 16.22
Deflection angles from C.S. to S.T.

TABLE 16.6

877

CHAPTER 16:
Route Surveying

Deflection angles C.S. to the T.S.

Station	Deflection to osculating circle, $b_i(i = 1, 2, \ldots, 6)$	Deflection between osculating circle and spiral, $a_i(i = 1, 2, \ldots, 6)$	Deflection angles local tangent to spiral, $c_i = b_i - a_i$
S.T. 55 + 64.869(6)	7°48′	2°36.0′	5°12.0′
+ 54.869(5)	6°30′	1°48.3′	4°41.7′
+ 44.869(4)	5°12′	1°09.3′	4°02.7′
+ 34.869(3)	3°54′	0°39.0′	3°15.0′
+ 24.869(2)	2°36′	0°17.3′	2°18.7′
+ 14.869(1)	1°18′	0°04.3′	1°13.7′
C.S. 55 + 04.869			0°00.0′

toward the S.T. With the theodolite set at the C.S., the telescope can be oriented to a 0 setting along the local tangent $V'D'$ by setting $180° - [(I - 2\Delta)/2] = 167°48′$ on the clockwise horizontal circle and taking a sight with the telescope direct on the S.C. Deflection angles can be calculated from the local tangent to stations at 10-m intervals on the osculating circle (a simple curve $D_a = 26°$). These angles are tabulated in column 2 of Table 16.6. Now, the rate of departure between the osculating circle and spiral is the same as that between the tangent and the spiral from the T.S. Consequently, the deflection angles to 10-m station intervals from the T.S. to the spiral calculated in Example 16.5 and listed in Table 16.6 are equal to the deflections between the spiral and osculating circle at corresponding intervals from the C.S. toward the S.T. Therefore, the deflection from the local tangent at the C.S. to 55 + 14.869 equals the deflection from the local tangent to the osculating circle ($b = 1°18′$) minus the deflection between spiral and osculating circle ($a = 0°04.3′$) equals 1°13.7′. Calculations to all the equal 10-m stations on the leaving spiral are shown in Table 16.6. For convenience, stations are labeled 1, 2, . . . , 6 so that the final angle can be expressed as $c_i = b_i - a_i$.

The osculating circle also is used when an intermediate setup is necessary on a spiral.

EXAMPLE 16.7. Assume that, in setting stations for the spiral of Example 16.5, it is necessary to move ahead to station 53 + 71.023 to complete laying out the spiral. In this case, the spiral departs inside the osculating curve as stationing progresses toward the S.C. (Figure 16.23 illustrates the relationships involved). Therefore, the deflection angles from the local tangent at 53 + 71.023 to subsequent stations will be the *sum* of the deflection angle to the osculating curve, b_i, and the deflection angle from the osculating curve to the spiral, a_i. First, it is necessary to calculate the degree of curve of the spiral at 54 + 71.023. Again, for convenience in notation, label the stations on the spiral 1, 2, 3, 4, 5, 6, as shown in Table 16.7. Then,

$$D_2 = \left(\frac{20}{60}\right)26° = 8.6667°$$

or the deflection per meter = 2.6′/meter and the deflection per one 10-m station of osculating curve is 0°26′. Table 16.7 contains the deflection angles from 53 + 71.023 to subsequent points on the spiral. With the instrument set at 53 + 71.023, a backsight is taken on the T.S. with the telescope direct and the horizontal circle set on $180° - 2a_2$, or $180° - (2)(0°17.3′) = 179°25.4′$.

Should another intermediate setup be required to complete the spiral layout, a backsight must be made to a station on the spiral. In this case, the osculating circle must

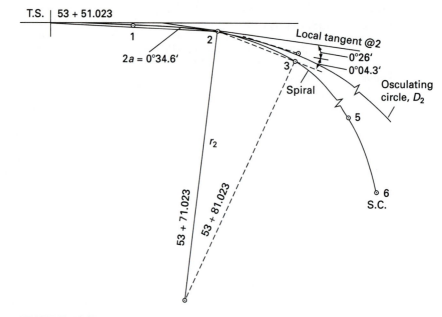

FIGURE 16.23
Intermediate setup on a spiral.

be used to calculate the backsight setting as well as the deflection angles to subsequent stations. Suppose, in the previous example, that an instrument setup is necessary at $53 + 91.923$ or chord point (4) and a backsight is made on chord point (2) (Figure 16.24). The degree of curve of the osculating circle at (4), D_4, is

$$D_4 = (\tfrac{4}{6})26° = 17.333°$$

yielding a deflection of $0°52'$ per 10-m station. As illustrated in Figure 16.24, the osculating circle departs inside the spiral toward the T.S. For two 10-m stations, the deflection between the local tangent and the osculating circle is $1°44'$. The deflection between the osculating circle and the spiral for two stations is $0°17.3'$ (a_2, Table 16.6). Therefore, the angle between local tangent and spiral is $(1°44') - (0°17.3') = 1°26.7'$ and the backsight from 4 to 2 is made with the telescope direct and the horizontal circle set on $180° - (1°26.7') = 178°33.3'$. Deflection angles to stations 5 and 6 (the S.C.) are calculated as in Example 16.7, using $D_4 = 17.333°$ for the osculating circle.

TABLE 16.7
Intermediate setup on a spiral

Station	Deflection to osculating circle, $b_i(i = 3, 4, 5, 6)$	Deflection between osculating circle and spiral, $a_i(i = 3, 4, 5, 6)$	Deflection angle local tangent to spiral, $c_i = a_i + b_i$
S.C 54 + 11.023 (6)	1°44′	1°09.3′	2°53.3′
54 + 01.023 (5)	1°18′	0°39.0′	1°57.0′
+ 91.023 (4)	0°52′	0°17.3′	1°09.3′
+ 81.023 (3)	0°26′	0°04.3′	0°30.3′
53 + 71.023 (2)	0°00′		0°00.0′

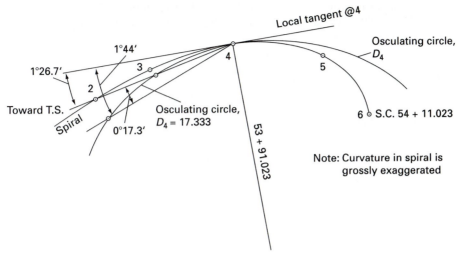

FIGURE 16.24
Backsight from an intermediate station on a spiral.

In the previous examples, the assumption was made that the total length of spiral L_s, equals the sum of the chords used to set the stations. This assumption is valid for relatively flat spirals of moderate length. Chord corrections are necessary only for long chords near the end of a long sharp spiral. When necessary, the chord correction can be calculated using the methods employed for simple curves, as outlined in Section 16.6.

16.18
CALCULATION AND LAYOUT OF LONG SPIRALS

The approximate equations derived in Section 16.15 for deflection angles are valid for relatively flat spirals of moderate length. For long, sharp spirals, use of the rigorous equations throughout is advisable. First consider the computations involved.

Computations

One approach is to use Equations (16.47) or (16.49) to compute the x and y coordinates to points on the curve. The coordinates are in a local system having an origin at the T.S. and an x axis that coincides with the tangent at the T.S. and is positive toward the P.I. (Figure 16.19). For a curve deflecting to the right, this x-y system is *left-handed* (Figure 16.19) and for a curve that deflects to the left (not shown), the system is right-handed. These coordinates permit computation of exact deflection angles to each station on the spiral and chords between each pair of consecutive stations. Given L_s and D_a for the spiral, K is computed by Equation (16.26), and for any point p on the spiral at a distance l_s from the T.S. (Figure 16.19), Equation (16.46) is used to calculate δ_p in radians. Then, δ_p (in radians) is substituted into Equations (16.47) or (16.49) and x and y are calculated for point p. The exact value for the deflection angle a_p (Figure 16.19) to station p is

$$a_p = \arctan (y/x) \qquad (16.52)$$

Chords between successive stations and the total chord from the T.S. to any station are found by applying the inverse solution with the coordinates and using Equation (8.8) (Section 8.13) to yield

$$c_{i,i+1} = [(x_{i+1} - x_i)^2 + (y_{i+1} - y_i)^2]^{1/2} \tag{16.53}$$

for the chord between stations i and $i + 1$ and

$$c_i = (x_i^2 - y_i^2)^{1/2} \tag{16.54}$$

for the total chord between from the T.S. ($x_i = y_i = 0$) to station i. An example problem illustrates the process.

EXAMPLE 16.8. Compute exact deflection angles from the T.S. to $\frac{1}{10}$ stations for the spiral given in Example 16.5, where

$$L_s = 60.0 \text{ m}, \qquad I = 40°00'00 \text{ right}, \qquad D_a = 26°00'00''$$

$$X = 59.889 \text{ m}, \qquad Y = 2.719 \text{ m by the rigorous equations,}$$

The station at the T.S. = 53 + 51.023, and the station at the S.C. = 54 + 11.023.

Solution. Compute K by Equation (16.26):

$$K = (100)(26)/60. = 43.3333°/\text{station} = 0.75630934 \text{ rad}$$

Let $\alpha = K/20{,}000$ from Equation (16.46), so that

$$\delta = \alpha l^2 = 3.718544 \times 10^{-5} l^2$$

for this problem. This equation can be used to compute δ and the x and y coordinates for each point on the spiral. For station 51 + 81.023,

$$\delta_{+81.023} = (3.718544 \times 10^5)(30)^2 = 0.03403389 \text{ rad}$$

Next, x and y coordinates for the each station are calculated for each point using Equations (16.49), which are repeated here for convenience:

$$x = l_s[1 - \delta^2/10 + \delta^4/216 - \delta^6/9360 + \cdots]$$

$$y = l_s[\delta/3 - \delta^3/42 + \delta^5/1320 - \delta^7/75{,}600 + \cdots]$$

Using these equations for station 51 + 81.023,

$$x = (30.0)(0.9998842) = 29.996 \text{ m}$$

$$y = (30.0)(0.01134369) = 0.34031 \text{ m}$$

With these values and Equation (16.52),

$$a_{+81.023} = \arctan(y/x) = (0.34031)/(29.996) = 0.011345$$

$$a_{+81.023} = 0°39'00''$$

The total chord to the station by Equation (16.54) is

$$c_{+81.023} = [(29.996)^2 + (0.34031)^2]^{1/2} = 29.998 \text{ m}$$

The results of calculating x and y and deflection angles to each station, chords between each pair of consecutive stations, and total chords to each station from the T.S. are given in Table 16.8.

Chords between stations are calculated using Equation (16.53). Thus, the chord between 51 + 81.023 and 51 + 91.023 is

$$c_{+81.023 \text{ to } +91.023} = [(39.9854 - 29.9965)^2 + (0.806724 - 0.340339)^2]^{1/2}$$

$$c_{+81.023 \text{ to } +91.023} = 9.9998 = 10.000 \text{ m}$$

Note that the exact deflection angles in Example 16.8 agree to within a few seconds of the those computed by the approximate equations and the chords are the same except for the last stations, which differ by only 1 mm. Obviously, use of the approximate equations is satisfactory for short, flat spirals.

Next consider an example with a long, sharp spiral.

EXAMPLE 16.9. A spiral having $L_s = 900$ ft and with $D_a = 10°$ is to be staked with 15 60-ft chords. Use the rigorous equations to compute exact deflection angles to these stations and chords between each pair of consecutive stations. Compare the deflection angles with values calculated by the approximate equations.

Solution. Calculate K by Equation (16.26):

$$K = (100)(10)/(900) = 1.11111°/\text{station} = 0.019392547 \text{ rad/station}$$

Compute δ using Equation (16.46), in which $\alpha = K/20{,}000 = 9.69627 \times 10^{-7}$, so that

$$\delta = \alpha l_s^2 = 9.69627 \times 10^{-7} l^2$$

Designating the stations 1, 2, ..., 15, the calculations for station 4 ($l_s = 240$ ft) are as follows:

$$\delta_4 = 9.69627 \times 10^{-7}(240)^2 = 0.0558505 \text{ rad}$$

Using Equations (16.49),

$$x = (240)(0.9998688) = 239.9251 \text{ m}$$

$$y = (240)(0.01861269) = 4.467045 \text{ m}$$

The deflection angle by Equation (16.52) is

$$a_4 = \arctan(y/x) = (4.467045)/(239.9251) = \arctan(0.01861850)$$

$$a_4 = 1°04'00''$$

Similarly, the coordinates and deflection angles are computed for the remainder of the stations on the spiral. The x and y coordinates, exact and approximate deflection

TABLE 16.8
Spiral deflection angles and chords using rigorous equations, short spiral

Station	l, m	$\delta = \alpha l_i^2$, rad	x, m	y, m	Deflection angle	Deflection angle (approximate)	Chord, station to station, m	Total chord, m
TS.51 + 51.023							10.000	
	10							
+ 61.023		3.781544×10^{-3}	10.000	0.0126	0°04'20"	0°04'18"		10.000
	20						10.000	
+ 71.023		0.01512618	19.9995	0.100839	0°17'20"	0°17'18"		20.000
	30						10.000	
+ 81.023		0.034033891	29.9965	0.340339	0°39'00"	0°39'00"		29.998
	40						10.000	
+ 91.023		0.060504700	39.9854	0.806724	1°09'21"	1°09'18"		39.993
	50						9.999	
52 + 01.023		0.094538595	49.9553	1.574637	1°48'20"	1°48'18"		49.980
	60						9.999	
S.C. 52 + 11.023		0.136135577	59.8889	2.719128	2°35'59"	2°36'00"		59.951

TABLE 16.9
Spiral deflection angles and chords using rigorous equations, long curve

Station	l	$\delta = al_s^2,$ rad	x	y	Deflection angle = $\tan^{-1}(y/x)$ (exact)	Approximate angle = $(\Delta/3)l_s^2/L_s^2$	Chord (exact)
T.S. 0							
1	60	0.0034907	59.9999	0.069810	0°04'00"	0°04'	
2	120	0.0139626	119.9977	0.558498	016 00	0 16	60.000
3	180	0.0314159	179.9822	1.884883	036 00	0 36	59.999
4	240	0.0558505	239.9251	4.467047	104 00	1 04	59.998
5	300	0.0872664	299.7716	8.721899	14000	1 40	59.998
6	360	0.1256637	359.4319	15.06264	223 59	2 24	59.996
7	420	0.1710423	418.7729	23.89592	315 57	3 16	59.995
8	480	0.2234021	477.6099	35.61712	415 54	4 16	59.993
9	540	0.2827433	535.6989	50.60392	523 47	5 24	59.991
10	600	0.3490658	592.7303	69.20791	639 35	6 40	59.989
11	660	0.4223697	648.3266	91.74397	803 16	8 04	59.986
12	720	0.5026548	702.0199	118.4774	9 34 46	9 36	59.984
13	780	0.5899213	753.2893	149.6088	11 13 59	11 16	59.981
14	840	0.6841690	801.5235	185.2569	13 00 52	13 04	59.978
S.C. 15	900	0.7853982	846.0463	225.4394	14 55 14	15 00	59.974

angles, and chords between stations (calculated by Equation (16.53)) are listed in
Table 16.9.

For this example, the differences between exact and approximate deflection angles vary
from 03" at station 7 to 04'46" at the S.C. (station 15). Chords differ from the nominal
length of curve between stations, from 2 mm at station 4 to 0.026 m at the S.C. Therefore,
for long, sharp spirals, use of the rigorous equations is necessary when setting alignment for
curbs, pavement, retaining walls, and the like (see Chapter 17).

The same principles can be applied to calculating the deflection angles from the S.C.
to the T.S. or from the C.S. to the S.T. Referring to Figure 16.25, consider determination

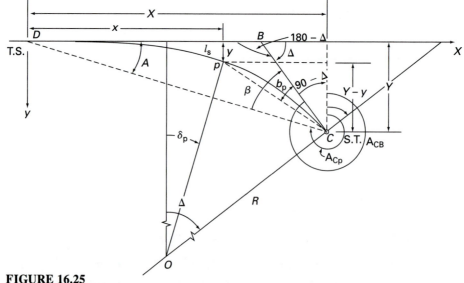

FIGURE 16.25
Exact angles from the S.T.

882

of the deflection angle to point p from a setup at the S.T., at C. Calculate X and Y for the spiral using Equations (16.49). The exact value for angle A is

$$A = \arctan (Y/X) \tag{16.55}$$

and in triangle DBC

$$A + 180° - \Delta + \beta = 180°$$

so that

$$\beta = \Delta - A \tag{16.56}$$

Next, compute x and y to any point p on the spiral using Equations (16.49). Let the azimuth of line CP be A_{CP}, where

$$A_{CP} = \arctan (Y - y)/(X - x) \tag{16.57}$$

The deflection angle b_p to point p from local tangent CB is

$$b_p = A_{CB} - A_{CP} \tag{16.58}$$

in which

$$A_{CB} = 360° - (90° - \Delta) = 270° - \Delta$$

In general,

$$b_i = A_{CB} - \arctan [(Y - y_i)/(X - x_i)] \tag{16.59}$$

When coordinates are computed for all stations, following the procedure outlined previously in this section, chords are calculated using the coordinates in Equation (16.53), as shown in Example 16.8.

EXAMPLE 16.10. Compute exact angles from the S.C. to the T.S (or from the C.S. to the S.C. on the forward tangent) using the curve data from Example 16.9, where $L_s = 900$ ft, $D_a = 10°$, and stations are to be set at 60-ft intervals along the spiral.

Solution. Compute Δ using Equation (16.30):

$$\Delta = (L_s D_a)/200 = (900)(10)/200 = 45°00'00''$$

Solve for β with Equation (16.56) and angle $A = 14°55'14''$ from Example 16.9:

$$\beta = \Delta - A = (45°00'00'') - (14°55'14'') = 30°04'46''$$

The azimuth of line CB (see Figure 16.25) is

$$A_{CB} = 270° + \Delta = 315°00'00''$$

From Example 16.9, $X = 846.0643$ m, $Y = 225.4394$ m, $x_{14} = 801.5235$ m, and $y_{14} = 185.2569$ m. Substitution of these values into Equation (16.59) yields

$$b_{14} = A_{CB} - \tan^{-1}(Y - y_{14})/(X - x_{14})$$

$$= 315°00'00 - \tan^{-1}(225.4394 - 185.2569)/(846.0463 - 801.5235)$$

$$= 315°00'00'' - \tan^{-1}(0.90251512) = 315°00'00'' - 312°04'00.2''$$

$$b_{14} = 2°55'59.8''$$

In similar fashion, $b_{13}, b_{12}, \ldots, b_0$ are calculated. Approximate and exact deflection angles for stations 14, 13, . . . , 0 are given in Table 16.10. Chords are the same as those calculated for stations from the T.S. to the S.C. and shown in Table 16.9.

Finally consider computation of the long tangent $= LT = DB$ and the short tangent $= ST = BC$ (Figure 16.25) of a spiral for which the length of spiral and

TABLE 16.10

Deflection angles (exact) from S.C. to T.S, long spiral

Station	Approximate deflection angle, local tan to spiral	Azimuth $C - i$, $A_{c-i} = \tan^{-1} \dfrac{(Y - y_i)}{X - x_i}$ $i = 14, 13, \ldots, 1$	Exact deflection angle $= A_{CB} - A_{ci}$
S.C. 14	2°56′	312°04′00.4″	2°55′59.6″
13	5°44′	309°15′59.9″	5°44′00.1″
12	8°24′	306°35′58.7″	8°24′01.3″
11	10°56′	304°03′55.8″	10°56′04.1″
10	13°20′	301°39′50.5″	13°20′09.5″
9	15°36′	299°23′41.7″	15°36′18.3″
8	17°44′	297°15′28.8″	17°44′31.2″
7	19°44′	295°15′11.2″	19°44′48.8″
6	21°36′	293°22′48.6″	21°37′11.4″
5	23°20′	291°38′21.0″	23°21′39.0″
4	24°56′	290°01′48.8″	24°58′11.1″
3	26°24′	288°33′12.8″	26°26′47.1″
2	27°44′	287°12′34.0″	27°47′25.9″
1	28°56′	285°59′53.8″	29°00′06.2″
T.S. 0	30°00′	284°55′13.7″	30°04′46.2″

degree of curve at the S.C. are known. First, compute Δ by Equation (16.30) and use this value to calculate X and Y with Equations (16.49). Next, determine angle A at the T.S. using Equation (16.52), so that

$$A = (Y)/(X)$$

The chord from D to C is

$$C_{D-C} = [X^2 + Y^2]^{1/2}$$

In triangle DBC (Figure 16.25), the $LT = DB$ is

$$LT = C_{DC}[\sin \beta]/(\sin \Delta) \qquad (16.60a)$$

where β is found by Equation 16.56.

The $ST = BC$ is

$$ST = C_{DB}(\sin A)/(\sin \Delta) \qquad (16.60b)$$

Layout of the Spiral Using a Total Station System

When coordinates for stations on a spiral have been determined, the spiral can be set using a total station system set on a single horizontal control point where there also is a backsight along a line of known azimuth. Several methods can be used to accomplish this task.

In the first option, where it is assumed that the control points for the spiral (T.S., S.C., P.I.) are established and data corresponding to that in Table 16.8 are available, occupy the T.S. with the total station instrument, sight the P.I., and set points on the spiral by deflection angle and total chord from the T.S. If data are given as shown in Tables 16.9 and 16.10, occupy the C.S. with the total station instrument, sight the T.S. with the horizontal circle set on $360° - \beta$ to put $0°$ on the local tangent CB and again set stations by the deflection angle and total chords from the S.C. at point C (Figure 16.25). Another possibility would be to compute local coordinates (system as shown in Figures 16.19 and 16.25) for the P.I. of the spiraled curve. Then, calculate the inverse solution from the P.I. to each coordinated station on the spiral. Occupy the P.I. with the total station instrument, sight along the back tangent with 180° set on the horizontal circle, and set the stations by angle and distance from the P.I. This procedure would be attractive if center-line control were established and the T.S. and S.C. could not be occupied.

Usually, the center-line control and construction control points are coordinated in the state plane coordinate system and the most efficient way to set the spiral is from a random control point, as described in Section 16.10. Because coordinates for the stations on the spiral are in a local system, a coordinate transformation is required to convert local coordinates for the spiral into state plane coordinates. To make this transformation at least two points must be known in both systems. For example, the T.S. and P.I. for the curve could easily be determined in the local and state plane systems. With these two common points, the four equations expressed by Equation (C.12) (Appendix C, Section C.2) can be formed to solve for the four transformation parameters of scale, rotation, and two translations (see Section 8.40, Example 8.18). These parameters are used to convert local spiral coordinates to state plane coordinates using Equations (C.14). Then, the inverse solution can be performed between a given coordinated random control point and the stations on the spiral to yield direction and distances from the random control point to each station. With the total station system set on the random control point and a backsight on another horizontal control point, these distances and directions are used to set points on the curve in the same procedure as described for layout of a simple curve from a random control point in Section 16.10, Example 16.4. As in that example, grid distances need to be converted to terrain distances using a combined scale and ellipsoidal reduction factor.

The computations involved in coordinating stations on spirals for layout from random control points are substantial, especially when using a hand calculator. However, programs are available for main frame computers, PCs, and data collectors that minimize the computational effort. Thus, coordinating the curve and layout from random control points becomes very practical.

16.19
VERTICAL CURVES

On highways and railways, to avoid an abrupt change in the vertical direction of moving vehicles, adjacent segments of differing grade are connected by a curve in the vertical plane, called a *vertical curve*. Usually, the vertical curve is the arc of a parabola, as this shape is well adapted to gradual change in direction, and elevations along the curve are computed readily.

Figure 16.26 illustrates crest and sag parabolic and equal tangent vertical curves. Point A is the beginning of the vertical curve or BVC, V is the P.I. or intersection of the tangents, and B is the end of the vertical curve or EVC. The length of the curve AB is designated L and measured in horizontal units. The two grades in the direction of stationing along tangents AV and VB expressed in percent are G_1 and G_2, respectively. When the tangent rises in the direction of stationing, the grade is positive; and when the tangent slopes downward, the grade is negative.

The design of crest and sag vertical curves is a function of the grades of the intersecting tangents, stopping or passing sight distance, which in turn are functions of vehicle design speed and height of the driver's eye above the roadway, and drainage. In addition to these factors, design of a sag vertical curve depends on headlight beam distance, rider comfort, and appearance. Details governing design of vertical curves are beyond the scope of this text and may be found in route surveying textbooks as well as in the AASHTO (1994).

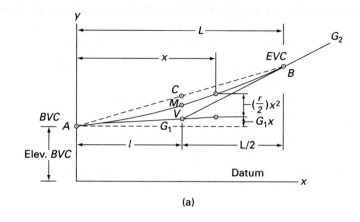

(a)

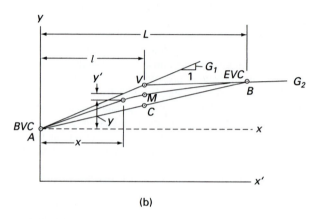

(b)

FIGURE 16.26
Vertical curves: (a) sag, (b) crest.

Elevations on vertical parabolic curves can be computed by using (1) the equation of the parabola directly, (2) the geometric properties of the parabola to calculate vertical offsets from the tangent, and (3) the geometric properties of the parabola as exemplified by the chord gradient method.

16.20
VERTICAL CURVES BY THE EQUATION OF THE PARABOLA (HURDLE, 1974)

Consider the plane parabolic curve, AB, of Figure 16.26a with the y axis passing through the BVC and the x axis corresponding to the datum for elevations. Let L be the length of the vertical curve in stations. The slope of the curve at BVC is G_1 and the slope at EVC is G_2. Because the rate of change in slope of a parabola is constant, the second derivative of y with respect to x is a constant:

$$\frac{d^2y}{dx^2} = \text{constant} = r \tag{16.61}$$

Integration of Equation (16.61) yields the first derivative or the slope of the parabola:

$$\frac{dy}{dx} = rx + H \qquad (16.62)$$

When $x = 0$, the slope $= G_1$; and when $x = L$, the slope $= G_2$. Therefore,

$$G_1 = 0 + H \qquad \text{and} \qquad G_2 = rL + H$$

from which

$$r = \frac{G_2 - G_1}{L} \qquad (16.63)$$

The value of r is the rate of change of the slope in percent per station. Substitution of Equation (16.63) into Equation (16.62) gives

$$\frac{dy}{dx} = \left(\frac{G_2 - G_1}{L}\right)x + G_1 \qquad (16.64)$$

To obtain y, integrate Equation (16.64) to yield

$$y = \left(\frac{G_2 - G_1}{L}\right)\frac{x^2}{2} + G_1 x + c \qquad (16.65)$$

which becomes

$$y = \left(\frac{r}{2}\right)x^2 + xG_1 + (\text{elevation of } BVC) \qquad (16.66)$$

that is the equation of the parabolic curve AB and can be used to calculate elevations on the curve given G_1, G_2, L, and the elevation of BVC. Note that the first term, $(r/2)x^2$, is the vertical offset between the curve and a point on the tangent to the curve at a distance x from BVC. The second term, $G_1 x$, represents the elevation on the tangent at a distance x from BVC, and the third term is the elevation of BVC above the datum (see Figure 16.26a). Any system of units can be employed in Equation (16.66). The grades G_1 and G_2 are dimensionless ratios and expressed as percentages, but r (Equation (16.63)) must be determined in units compatible with the units of x and y. If distances are measured in meters, L in Equation (16.63) is in metric stations where 100 m = one station. If distances are measured in feet, L must be in stations where 100 ft = one station. When one full station is defined as 1 km (Section 16.4), Equations (16.63) and (16.65) become

$$r = \frac{G_2 - G_1}{L'} \qquad (16.66a)$$

and

$$y = \left(\frac{r}{2}\right)x'^2 + x'G_1 + (\text{elevation of } BVC) \qquad (16.66b)$$

in which $L' = 10(L)$, $x' = 10(x)$, and L and x are substations of the 1000-m station.

The high or low point of a vertical curve frequently is of interest for drainage purposes. At the high or low point, the tangent to the vertical curve is 0. Equating the first derivative of y with respect to x (Equation (16.64)) to 0 gives

$$xr + G_1 = 0 \qquad (16.67)$$

or

$$x = -\frac{G_1}{r} \qquad (16.68)$$

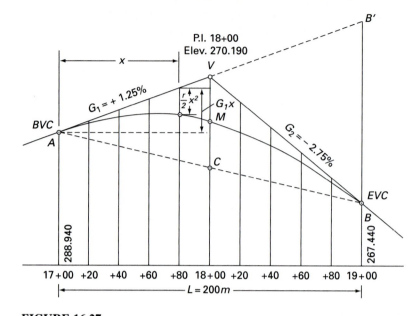

FIGURE 16.27
Vertical curve by equation of the parabola and by vertical offsets.

EXAMPLE 16.11. A 200-m equal tangent parabolic vertical curve is to be placed between grades of $G_1 = 1.25$ percent, and $G_2 = -2.75$ percent intersecting at station $18 + 00$, which has an elevation of 270.190 m above the datum. Calculate elevations at even 20-m stations on the vertical curve and determine the station and elevation of the high point on the vertical curve. Figure 16.27 shows the geometry involved.

Solution. The elevation of $BVC = 270.190 - (0.0125)(100) = 268.940$ m and the elevation of $EVC = 270.190 - (0.0275)(100) = 267.440$ m above the datum. Next, compute the rate of change in grade by Equation (16.63), where $L = 2$ stations:

$$r = \frac{-2.75 - 1.25}{2} = -2.000$$

so that

$$\frac{r}{2} = -1.000$$

From Equation (16.66), the equation of the parabola for this vertical curve is

$$y = -1.000x^2 + 1.25x + 268.940$$

in which y equals the elevation of a point on the vertical curve x stations from the BVC. Calculation of elevations on the vertical curve by this equation is best illustrated by Table 16.11. Note that elevations on the curve for a given station are the sum of columns 3, 4, and 6 of the table.

By Equation (16.68), the station of the high point is

$$x = -\frac{G_1}{r} = \frac{1.25}{2} = 0.625 \text{ station or } 62.500 \text{ m}$$

TABLE 16.11
Elevations on a vertical curve by equation of parabola

Station	x	$(r/2)x^2$	G_1x	$(r/2)x^2 + G_1x$	Elevation BVC (m)	Elevation on curve (m)
BVC 17+00	0.	0.	0.	0.	268.940	268.940
+20	0.2	−0.04	0.250	0.210	268.940	269.150
+40	0.4	−0.16	0.500	0.340	268.940	269.280
+60	0.6	−0.36	0.750	0.390	268.940	269.330
+80	0.8	−0.64	1.000	0.360	268.940	269.300
18+00	1.0	−1.00	1.250	0.250	268.940	269.190
+20	1.2	−1.44	1.500	0.060	268.940	269.000
+40	1.4	−1.96	1.750	−0.210	268.940	268.730
+60	1.6	−2.56	2.000	−0.560	268.940	268.380
+80	1.8	−3.24	2.250	−0.990	268.940	267.950
EVC 19+00	2.0	−4.00	2.500	−1.500	268.940	267.440

so that the

$$\text{station of high point} = 17 + 00 + 62.500 = 17 + 62.500$$

Using the equation of the parabola, the elevation of this station is

$$\text{elevation at } 17 + 62.500 = (-1.0)(0.625)^2 + (1.25)(0.625) + 268.940$$

$$= 269.331 \text{ m above the datum}$$

16.21
VERTICAL CURVES BY VERTICAL OFFSETS
FROM THE TANGENT

In Figure 16.26b, translate the x coordinate axis so that it passes through BVC at A. Then, by similar triangles and noting that y' in Figure 16.26b is negative,

$$\frac{y - y'}{x} = G_1$$

or

$$y = xG_1 + y' \qquad (16.69)$$

Substituting Equation (16.69) into Equation (16.65) ($c = 0$) yields

$$xG_1 + y' = \left(\frac{G_2 - G_1}{2L}\right)x^2 + G_1 x$$

which can be solved for y' to give

$$y' = \left(\frac{G_2 - G_1}{2L}\right)x^2 = \left(\frac{r}{2}\right)x^2 \qquad (16.70)$$

or vertical offsets from the tangent vary as the square of the distance from the point of tangency. Now, at point V, the P.I. of the vertical curve, $y' = VM$ and $x = L/2$, so that

$$y' = VM = \left(\frac{G_2 - G_1}{2L}\right)\frac{L^2}{4} = \frac{(G_2 - G_1)L}{8} \qquad (16.71)$$

Divide Equation (16.70) by Equation (16.71) and solve for y' to give

$$y' = \left(\frac{x^2}{l^2}\right)VM \qquad (16.72)$$

in which $l = L/2$.

In Figure 16.26b, point C bisects the chord AB and, because the midpoint M on the vertical curve is halfway between V and C, $VM = \frac{1}{2}VC$. Given the elevations of BVC and EVC, VM can be calculated. On sag vertical curves, VM is positive; and on crest curves, VM is negative, so that y' is positive on sag (Figure 16.26a) and negative on crest (Figure 16.26b) vertical curves.

To determine elevations on the vertical curve by vertical offsets, calculate the elevations for the desired stations along the tangent to the vertical curve from BVC to V using G_1 and from V to EVC using G_2. Next, calculate the vertical offsets (Equation (16.72)) and apply these to the elevations on the tangent. Because an equal tangent parabolic vertical curve is symmetrical, vertical offsets need to be computed for only one-half the curve when stations are equally spaced and symmetrical about the P.I., as in Example 16.11. Then, these same vertical offsets can be applied to elevations along the tangent at EVC, allowing determination of elevations on the second half of the curve. Table 16.12 shows the calculations necessary to compute the vertical curve of Example 16.11 by vertical offsets from the tangent.

VM is calculated as follows:

Point	Elevation
BVC	268.940
EVC	267.440
	536.380
C	268.190

TABLE 16.12
Elevations on a vertical curve by vertical offsets

Station	Grade along tangent	x, in stations	$(x^2/l^2)VM*$	Elevation on curve	Difference First	Difference Second
17 + 00 BVC	268.940	0.	0.	268.940		
					> 0.210	
+20	269.190	0.2	−0.040	269.150		>0.08
					> 0.130	
+40	269.440	0.4	−0.160	269.280		>0.08
					> 0.050	
+60	269.690	0.6	−0.360	269.330		>0.08
					> −0.030	
+80	269.940	0.8	−0.640	269.300		>0.08
					> −0.110	
18+00	270.190	1.0	−1.000	269.190		>0.08
					> −0.190	
+20	269.640	0.8	−0.640	269.000		>0.08
					> −0.270	
+40	269.090	0.6	−0.360	268.730		>0.08
					> −0.350	
+60	268.540	0.4	−0.160	268.380		>0.08
					> −0.430	
+80	267.990	0.2	−0.040	267.950		>0.08
					> −0.510	
19+00 EVC	267.440	0.	0.	267.440		

*$VM = -1.0$; $l = 1$ station.

$$VM = -\frac{\text{el. } V - \text{el. } C}{2}$$

$$= -\frac{270.190 - 268.190}{2}$$

$$= -1.000$$

Elevations along the tangent are computed from BVC to V using G_1. Elevations along the tangent V to EVC are calculated with G_2. In this case, where points are equally spaced, second differences of the elevations must be equal. This is an important check that can be applied to all vertical curves calculated by the equation of the parabola (as in Example 16.11) or by vertical offsets.

If stations are not equally spaced and symmetrical about the P.I., grades are calculated for the desired stations along the tangent AV extended to B' (Figure 16.27) and vertical offsets are computed with Equation (16.72), where x always is measured from BVC.

16.22
ELEVATIONS ON VERTICAL CURVES BY CHORD GRADIENTS

The rate of change in slope of a tangent to the vertical curve, r, has been defined by Equations (16.63) and (16.66a). Using these equations, the grade of chords can be calculated. Knowing these grades, elevations on the vertical curve can be computed by the *chord gradient* method.

Figure 16.28 illustrates chord gradients for full stations on a 500-ft vertical curve. Because r is defined as the change in grade between tangents to the vertical curve, the change between the tangent at BVC and the first station is $r/2 = a$, the change between chords of successive full stations is $r = 2a$, and the change between the last chord and the tangent at EVC is a. For the example in Figure 16.28, the chord gradients to full stations are as follows: $G_1 + a$, $G_1 + 3a$, $G_1 + 5a$, $G_1 + 7a$, $G_1 + 9a$, and as a check $G_1 + 10a = G_2$. Elevations at full stations are elevation of $BVC + G_1 + a$, elevation of $16 + 00 + G_1 + 3_a$, . . . , elevation of $19 + 00 + G_1 + 9a$. Because elevations are calculated cumulatively, the method is self-checking.

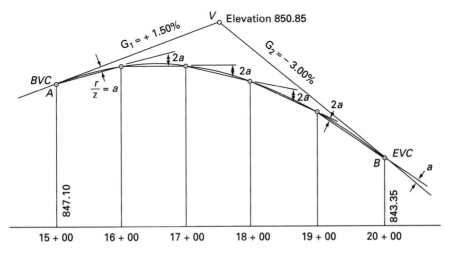

FIGURE 16.28
Vertical curve by chord gradients.

Chord gradients for subchords are determined as follows:

1. The change in gradient between a tangent and a subchord is a times the length of the chord in stations.
2. The change in gradient between two adjacent subchords is a times the sum of the lengths of the subchords in stations.

In the vertical curve of Example 16.11, a 200-m curve with 20-m stations, $r = 2a = -2.000$, so that $a = -1.000$. Consequently, for the first 20-m subchord, $a_{20} = (-1.0)(0.2) = -0.2$ percent and for subsequent consecutive 20-m subchords $2a_{20} = (-1.0)(0.4) = -0.4$ percent. The same logic applies when the full station is defined as 1 km, except that Equation (16.66a) is used to determine r from which the value of a is evaluated. As an exercise, the reader should work Example 16.11 using chord gradients, first with the full station as 100 m and then with the full station as 1000 m = 1 km.

EXAMPLE 16.12. In the vertical curve illustrated by Figure 16.28, $G_1 = 1.50$ percent, $G_2 = -3.00$ percent, and the elevation of the P.I. is 850.85 ft above the datum. Calculate elevations at half-station points on the vertical curve.

Solution. The elevation of $BVC = 850.85 - (250)(0.015) = 847.10$. The elevation at $EVC = 850.85 - (250)(0.03) = 843.35$ ft above the datum. The rate of change in gradient is

$$r = \frac{G_2 - G_1}{L} = \frac{-3.00 - 1.50}{5} = \frac{-4.50}{5} = -0.9\%/\text{station}$$

so that $a = -0.45$ percent. According to rule 1 for subchords, the change in gradient between the tangent at BVC and a chord to the first half-station is equal to $(-0.45)(0.5) = -0.225$ percent, and the change between two successive 50-ft subchords is $(-0.45)(1.0) = -0.45$ percent. Changes in gradients and elevations for half-stations are listed in Table 16.13.

The high point on the curve by Equation (16.68) is

$$x = -\frac{G_1}{r} = \frac{1.50}{0.90} = 1.6667 \text{ stations} = 166.67 \text{ ft}$$

TABLE 16.13
Vertical curve by chord gradients*

From/to station	Chord gradient, % $= G_1 + na$ $n = (0.5, 1.5, \ldots, 9.5)$	$\Delta\text{el.} =$ $(G_1 + na)(0.5)$	Elevation, ft above the datum
BVC 15 + 00			847.10
15 + 00 to 15 + 50	$G_1 + 0.5a = \quad 1.275$	0.638	847.74
15 + 50 to 16 + 00	$G_1 + 1.5a = \quad 0.825$	0.413	848.15
16 + 00 to 16 + 50	$G_1 + 2.5a = \quad 0.375$	0.188	848.34
16 + 50 to 17 + 00	$G_1 + 3.5a = -0.075$	-0.038	848.30
17 + 00 to 17 + 50	$G_1 + 4.5a = -0.525$	-0.263	848.04
17 + 50 to 18 + 00	$G_1 + 5.5a = -0.975$	-0.488	847.55
18 + 00 to 18 + 50	$G_1 + 6.5a = -1.425$	-0.713	846.84
18 + 50 to 19 + 00	$G_1 + 7.5a = -1.875$	-0.938	845.90
19 + 00 to 19 + 50	$G_1 + 8.5a = -2.325$	-1.163	844.74
EVC 19 + 50 to 20 + 00	$G_1 + 9.5a = -2.775$	-1.388	843.35
	$G_1 + 10.a = -3.000$ (check)		

*$r = -0.90$ percent; $a = -0.45$ percent; $G_1 = 1.50$ percent; $G_2 = -3.00$ percent.

so that the station of the high point is $16 + 66.67$. The elevation of the high point by the chord gradient method is as follows:

Chord gradient $16 + 00$ to $16 + 50$	$= \quad 0.375$
Change in gradient at $16 + 50$	
$\quad = (0.5 + 0.16667)(-0.45)$	$= \quad -0.300$
Subchord gradient $16 + 50$ to $16 + 66.67$	$= \quad 0.075$
Elevation at $16 + 50$	$= 848.34$
Δel. $16 + 50$ to $16 + 66.67 = (0.075)(0.16667) =$	$\quad 0.01$
Elevation at $16 + 66.67$	$= 848.35$

To check this elevation into station $17 + 00$, proceed as follows:

Chord gradient $16 + 50$ to $16 + 66.67$	$= \quad 0.075$
Change in gradient at $16 + 66.67$	
$\quad = (0.16667 + 0.3333)(-0.45)$	$= \quad -0.225$
Subchord gradient $16 + 66.67$ to $17 + 00$	$= \quad -0.150$
Elevation $16 + 66.67$	$= 848.35$
Δel. $16 + 66.67$ to $17 + 00 = (-0.150)(0.333) =$	$= -0.050$
Elevation $17 + 00$	$= 848.30 \quad$ checks

Therefore, the chord gradient method, although somewhat more difficult to visualize, has the advantage of being self-checking.

16.23
EARTHWORK OPERATIONS

Earthwork operations consist of the movement of materials to establish a predetermined surface for the construction of public and private works and determination of the volume of materials moved. Fieldwork involves acquisition of terrain data (usually location of terrain breakpoints, profile, and cross sections) and setting points (grade stakes and slope stakes) to guide construction work on the site. Office work involves acquisition of terrain data from maps or by photogrammetric methods, processing terrain data at a workstation using a geographic information system (Section 14.18) to form digital terrain models (Section 14.8), calculating volumes of excavated or embanked materials, and determining the most economic procedure for performing the excavation and embankment. Earthwork operations and computations are key elements in the overall route-location procedure.

For large projects, most of the terrain data can be extracted more accurately and at a lower cost from a DTM formed using data from all sources, which consist of field surveys, large-scale photogrammetrically compiled maps, and data obtained by applying photogrammetric methods to remotely sensed data such as can be obtained from aerial photographs.

Although the main emphasis in this chapter is on earthwork operations for route surveying, the field and office methods developed also are applicable for determining the volumes of stockpiles of materials, reservoir volumes, and volumes of concrete structures.

16.24
TERRAIN DATA FOR ROUTE LOCATION

As noted in the previous section, terrain data can be obtained in a number of ways. Traditionally, one of the most common methods for obtaining and using terrain data for route location is by way of the *terrain or route cross section*. Cross-section data can be obtained from field surveys, topographic maps compiled by photogrammetric methods, photogrammetric stereomodels (Section 13.15), and digital terrain models (Section 14.8).

These methods are listed in increasing order of complexity and current usage. Consider the simplest method first to aid in understanding exactly what constitutes a cross section.

Route Cross Sections: Field Methods

Surveys for highways and streets, railroads, and canals often are made by establishing a center line along the proposed route. This center line may be a line of numerically known positions with respect to a digital terrain model, a location on paper using a topographic map, or points set on the ground. In any case, station points usually are designated or set at some fractional part of a station along the center line. In the metric system, where one full station = 1 km and also one full station = 100 m, station points generally are set at 10- or 20-m intervals. When one full station is defined as 100 ft, station points are set at intervals of 50 and 100 ft. Odd station points are set at breaks in the surface slope configuration (breakpoints). Then, to furnish data for location studies and estimating volumes of earthwork, a profile is run along the center line (Sections 5.43 and 5.44) and cross-profiles are taken along lines passing through each station and at right angles to the center line. The elevations along these cross lines and respective distances to the right and left (as one looks toward the next, numerically higher station) constitute terrain cross-section data. The solid line in Figure 16.29 is the center line, the dashed lines represent cross-section lines, and the tick marks are points where elevations are obtained to the left and right of center line, which is usually designated as ℄.

In the traditional method for obtaining cross-section data, the center line is set in the field and a profile is run over the center-line stations by differential leveling (Section 5.43). The directions of short cross lines are set off by eye; that of long cross lines by theodolite

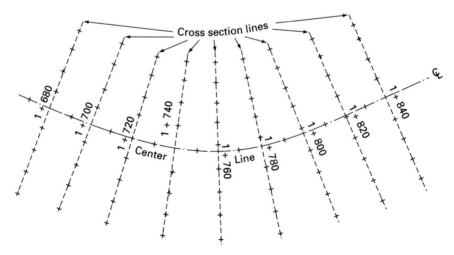

FIGURE 16.29
Cross lines for cross sections by the plus and offset method.

Sta.	B.S.	H.I.	F.S.	Elev.	
					PRELIMINARY CROSS SECTIONS
					HALF MOON BAY
44+00					
TPI	0.815	210.263	3.752	209.448	
43+00					(Dist.)
					(Rod)
					(Elev.)
BM24	2.750	213.200		210.450 meters	

Right-hand notes:

O.H. Ellis ★
C.O. Lord, Rod
J.E. Crum, Tape   June 5, 1996   Fair, Warm 75° F   28

Level No. 421 Rod No. 10

25	15	8	℄	7	15	25
3.42	3.15	2.95	2.25	2.04	1.25	0.95
206.84	207.11	207.31	208.01	208.22	209.01	209.31

206.25 .... 210.01
4.01 .... 0.25
35 35

(Dist.)	25	16	8	℄	8	16	24
(Rod)	3.05	3.60	4.40	2.51	2.05	1.45	1.00
(Elev.)	210.15	209.60	208.80	210.69	211.15	211.75	212.20

209.25 .... 212.90
3.95 .... 0.30
36 36

Brass plate set in concrete. See page 10

FIGURE 16.30
Notes for cross sections taken concurrently with profile leveling (metric units).

or total station. Elevations are determined with an engineer's level and rod in level terrain and with a hand level and rod in rough, irregular country. For each cross section, the height of instrument is established by a backsight on the center-line station. The rod then is held on the cross line at specified distances or at breakpoints, where rod readings are observed and distances measured with a metallic tape. In very steep, irregular terrain, use of a total station system to obtain distance and elevation is advantageous. Figure 16.30 shows a typical set of notes, where the profile and cross sections are run concurrently. The cross section for station 43 + 00, such as might be used for route design and plotted at horizontal and vertical scales of 1:500 and 1:100, respectively, is shown in Figure 16.31.

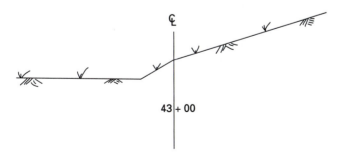

FIGURE 16.31
Route cross section.

The traditional field method, just described, provides a good illustration of what constitutes a cross section, but the procedure is difficult and time consuming in the field and rarely used now, especially for large projects. The method is appropriate for projects of small extent, relatively level terrain, and field conditions where the ground surface is concealed by heavy vegetation.

Terrain Data by the Total Station System

A total station system (Sections 7.15–7.20) can be used to obtain cross-section data in a procedure analogous to that just described. The total station instrument is set on a center-line station for which spatial coordinates (X, Y, and elevation) are known. A sight is taken along the center line (of known azimuth), and the direction of the cross line is established. With the operator of the prism pole and reflector moving left and right along the section line, offset distances and elevations at breakpoints are obtained by EDM, processed in the system software, and stored in the data collector. Next, assuming that other stations with spatial coordinates are visible from the occupied point, distances and directions can be measured from the total station instrument, at the initial station, to points on offset lines at these other stations. The directions of these other cross-section lines would be estimated by the prism pole operator and shots taken at all breakpoints along the lines. This process yields a file of data from which X, Y coordinates and elevations for all breakpoints on each cross line can be calculated. The inverse solutions between the coordinates of breakpoints and the coordinates of the center-line station provide horizontal distances along the cross line from the breakpoints to the center line. These distances, along with the elevations of respective breakpoints, yield a cross section for each station so observed. In this way, data from which cross sections may be derived are observed at all stations visible from a single instrument setup. When no more stations and cross lines are visible, the instrument is moved ahead to another station from which multiple stations can be seen and the procedure is repeated. All data are stored in the data collector (Section 7.18) and subsequently checked and processed in a computer in the field or later in the office. The cross section of the terrain is obtained as a direct by-product of this operation.

A more efficient way to utilize a total station system to acquire terrain data is to use the system to obtain spatial positions of topographic controlling points in the area of interest. In this procedure, horizontal and vertical control must exist in the area to be surveyed but the center line is not set in the field. For example, the total station instrument is set on a control point of known horizontal and vertical position and a sight is taken on another point of known position. Then, distances and directions are measured to break-points and along breaklines on the surface of the surrounding terrain using the total station system, with all the data being stored in the data collector. In general, the procedure corresponds to a topographic survey with a total station system by the controlling point method or the radial method (Sections 15.9 and 15.10). The data can be processed in the field or in the office to yield a file of X, Y coordinates and elevations for each point. These spatial positions can be utilized to form a digital terrain model, from which cross sections can be obtained with respect to a proposed alignment established at a workstation using a GIS (Section 14.18). In this case, the cross section is obtained indirectly from the data measured in the field.

Terrain Data by GPS

Kinematic and kinematic with on-the-fly (OTF) ambiguity resolution (Sections 12.9 and 12.12), provide powerful tools for obtaining terrain data from which cross sections can be derived. These techniques can be used in the same way as the total station systems; that is, to collect a file of spatial positions of breakpoints and breaklines that are processed in the office to furnish data from which the cross sections are obtained. The methods for accomplishing the field survey for this task essentially are the same as those required for topographic surveys by GPS (Section 15.11), where the points controlling the configuration of the terrain (breakpoints and breaklines) are located by the kinematic GPS techniques given previously. Although a reference control point of known position is required, the center line is not staked. Special care must be exercised in the advanced planning to minimize the possibility of loss of lock with satellites and to ensure that final elevations are on the proper datum. Recall that GPS elevations are on the ellipsoid and elevations on the NAVD 88 datum are required for all route locations. Therefore, sufficient orthometric elevations must be available in the area surveyed to allow reliable conversion of ellipsoidal heights to NAVD 88 elevations.

Note that, when either a total station system or GPS is used to obtain terrain data in the field by locating breakpoints and breaklines, the center line is not set in the field. Also, in each case, the cross sections are obtained indirectly in the office at a workstation by manipulating software in a geographic information system (Section 16.27).

16.25
TERRAIN CROSS SECTIONS FROM TOPOGRAPHIC MAPS

During the successive stages of highway mapping for route alignment (reconnaissance, project planning, and design), trial locations for proposed highway center lines are superimposed on video displayed maps or plotted on the conventional topographic maps drawn on paper for evaluation of grades and earthwork. One method of obtaining cross-section data for evaluating earthwork volumes is to plot the cross-section lines on the topographic map at each given station interval and all breaks in the terrain along the center line. Offset distances then are scaled from the center line and elevations are interpolated from the contours. The simplest and most effective illustration of this approach is to plot the center and cross-section lines on a conventional hard-copy topographic map and scale the offset distances to contours with an engineer's scale, recording offset distances and elevations as terrain notes.

A portion of a topographic map at a scale of 1:500 and with a contour interval of 1 m is shown in Figure 16.32 (scale in the figure is reduced). A trial location for a route center line and cross lines at 20-m intervals also are plotted on the map. Cross-section data are obtained by scaling from the center line to the contours that intersect the cross line. If the cross line is parallel to the contour, elevations are interpolated at specified intervals left and right of the center line. Cross-section notes determined from station 51 + 00 on the map in Figure 16.32 are shown in Figure 16.33a. The plotted cross section for station 51 + 00 with a horizontal scale of 1:500 and vertical scale of 1:100 is illustrated in Figure 16.33b (scale in the figure is reduced).

The process just described provides an excellent example implemented with simple tools to illustrate the method. However, this procedure is time consuming and tedious, especially for large projects.

The current practice in highway mapping is to utilize digital topographic maps (Section 14.15), where the map may have been produced directly from a photogrammetric stereoplotter (Section 13.20) and stored as digital data in computer memory. Another option is to scan an existing hard-copy topographic map and store the map as digital data in computer memory. In either case, using an appropriate software package, such as one of the components of a GIS (Section 14.18), the digital map can be displayed on the video screen of a workstation. Working in the same software module, one can introduce the parameters of a proposed highway alignment (e.g., coordinates of P.I.s, P.C.s, P.T.s, curve radii)

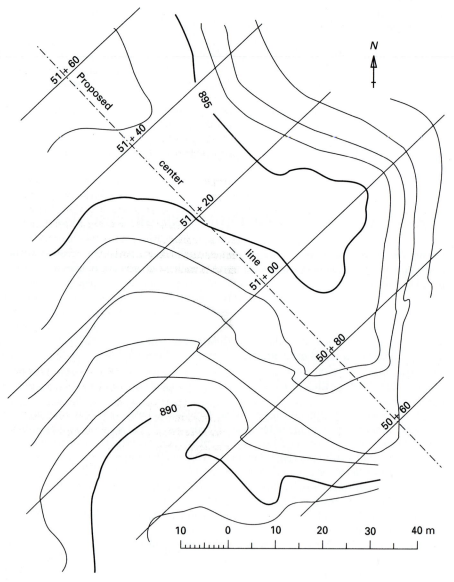

FIGURE 16.32
Cross-section data from topographic maps.

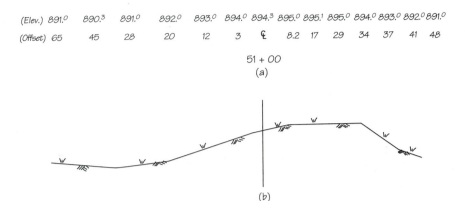

(Elev.)	891.º	890.³	891.º	892.º	893.º	894.º	894.³	895.º	895.¹	895.º	894.º	893.º	892.º	891.º
(Offset)	65	45	28	20	12	3	₵	8.2	17	29	34	37	41	48

51 + 00

(a)

(b)

FIGURE 16.33
Cross section: (a) notes and (b) plotted, both from a topographic map.

and display this alignment superimposed on the topographic map with cross lines at specified station intervals. Then, it is necessary only to place the cursor on the intersection of the cross line and contours and click on the mouse, digitizing the data for the cross section, which can be stored in memory and employed later for design and earthwork considerations.

The degree of automation built into the overall procedure is a function of the software package. What has been discussed here is a very simple example of acquiring cross-section data from a digitized map.

16.26
TERRAIN CROSS-SECTION DATA FROM PHOTOGRAMMETRIC STEREOMODELS

Topographic maps prepared for the planning and design stages of route locations usually are compiled by photogrammetric methods. Consequently, aerial photography of the study area is available almost from the very beginning of a proposed alignment project. To compile a topographic map by photogrammetric methods, overlapping aerial photographs are placed in a stereoplotter or analytical plotter and are oriented relatively and absolutely according to the methods outlined in Section 13.20. The stereomodel that results from these operations provides a scaled, three-dimensional model of the photographed terrain (see Figure 13.14). This model may be in the image space of the stereoplotter or analytical plotter or also can be stored as digital data in computer memory. In any case, given the proper instrumentation, it is possible to obtain cross-section data directly from the stereomodel. Because of the fewer operations involved, cross-section data from a stereomodel are more accurate than those taken from a topographic map compiled from the same stereomodel.

To obtain cross sections from a stereomodel, the proposed alignment must be projected or superimposed onto the model. When a conventional, two-projector, analog stereoplotter is used (Figure 13.14), the simplest procedure is to plot the positions of known horizontal control points and the location of the center line on a base sheet. All the center-line stations and cross lines through each station are plotted as well. Profile and cross-section data may be obtained by centering the measuring mark of the photogrammetric system over the station or offset mark and placing the floating dot on the apparent surface of the stereomodel (Section 13.20). The elevation is read from the counter on the tracing table and the station and elevation are recorded. The offset distance is scaled and also recorded. This procedure is repeated for every station and cross-section line. The process, which is a good way to illustrate the concept, is tedious, time consuming, and requires manual recording of the data. During the years when this type of plotter was common, numerous schemes involving hardware and software were devised that allowed digitizing the output and making the system suitable for practical use. These systems no longer are used and are not discussed here.

The analytical plotter currently is the best-suited system for obtaining cross sections from the stereoplotter. After the model has been oriented in the analytical plotter (Section 13.20), it is possible to project or display the proposed alignment into the image space of the plotter. Then, the operator can view the three-dimensional terrain with the center line and cross-section lines superimposed over the apparent surface of the terrain. At this point, it is necessary only to center the measuring mark over the center-line station or cross-section offset point, place the floating dot (Section 13.20) on the apparent surface of the ground, enter the station number at the keyboard, and press a specified key so that station, elevation, and offset are automatically stored in a digital database for future use in a GIS.

16.27
CROSS-SECTION DATA FROM A DIGITAL TERRAIN MODEL

Digital terrain models are described in some detail in Section 14.8. Briefly, a DTM is a mathematical representation of the earth's surface in which X, Y coordinates and elevations are given for all intersections of all lines in a rectangular or triangular network covering the study area. The data used to form the DTM are collected from various sources, including (1) the photogrammetric stereomodel (Section 16.26), (2) digitized topographic maps (Section 16.25), and (3) GPS and total station surveys to acquire terrain data (Section 16.24). The techniques already described for obtaining terrain data for cross sections (each of which creates an individual DTM) provide data that may be merged to form the final and most complete DTM.

The terrain surface formed from the DTM consists of triangles that have spatial coordinates (X, Y, and elevation) at each vertex. The files that contain this information are called *triangulated irregular networks (TIN)* (Section 14.8). A TIN such as might be used for the alignment of a portion of a highway is shown in Figure 16.34a. When a DTM is formed, it is possible to determine cross sections by linear interpolation between vertices of triangles for any alignment that falls within the DTM.

The process requires a workstation (Figure 16.34b) consisting of a high-resolution video display screen and a high-speed computer with adequate storage capacity to hold the operating system, programs for manipulating the data, and the data files. The terrain data (already preprocessed after field collection or downloaded from the map digitizer or analytical plotter) are entered into the computer and the DTM is formed and displayed (see Figure 16.34c). Following menu-driven prompts, the user can specify parameters for horizontal and vertical alignments for the proposed route. On command at the terminal, this alignment can be superimposed on the DTM and displayed on the video screen. Then, cross sections and profiles may be obtained, displayed, and stored in the database for future evaluation of earthwork. If changes in alignment are made, the profile and cross sections reflect these changes.

The DTM provides a flexible, powerful tool for obtaining terrain data for route location and design. It consolidates data from all sources into one compact package that is easily accessed and manipulated.

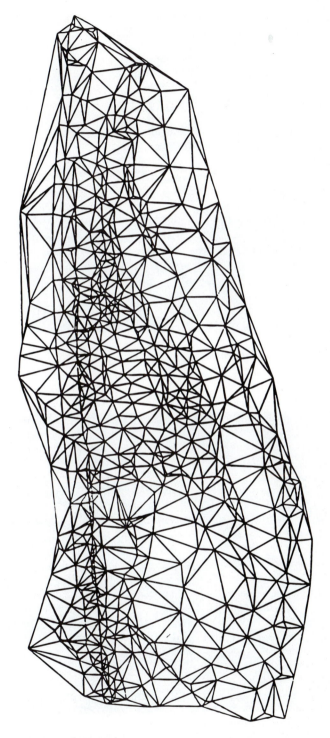

FIGURE 16.34
(a) Sample TIN file. (*Courtesy of CALTRANS.*)

Computer

Video display

Scanner

FIGURE 16.34 (b) Workstation. (*Courtesy of Integraph Corporation.*)

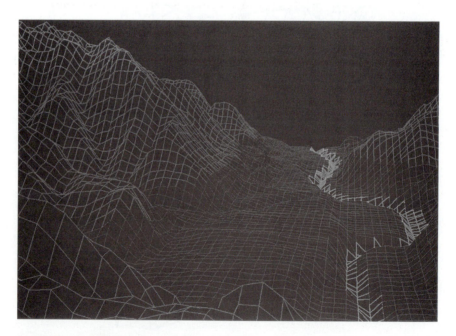

FIGURE 16.34 (c) Perspective view of a DTM projected on a video screen. (*Courtesy of Integraph Corporation.*)

Preliminary terrain cross sections are obtained by any one of the methods described in Sections 16.24 through 16.27, which then are used for studying various alignments and for making preliminary estimates of earthwork.

When a preliminary alignment and grade are being studied, a trial grade is selected and plotted on a profile sheet (Figure 16.35) or displayed on a video screen (Figure 16.36). This trial grade is based on maximum permissible rates of grade and elevations of controlling points such as termini, stream, railway, and roadway crossings. A crude effort also is made at this time to balance the amount of cut and fill on the center line. When a grade line is temporarily fixed, grade elevations are computed for all center-line stations and the finished roadbed templets superimposed on the terrain cross sections are plotted to a scale of from 1:60 to 1:120. Figure 16.37 a to c shows typical cross sections in cut, in fill, and a sidehill cross section for a highway. During the design at a workstation, the cross sections may be displayed for the entire portion of roadway being considered (Figure 16.38) or individually at scales specified by the user of the system.

The roadbed templet consists of a subgrade, which usually is a plane surface, transversely level but on highway curves perhaps superelevated, and side slopes at a given rate of slope. On a given road, the subgrade is of one width in cut (Figure 16.37a), of a second and smaller width in fill (Figure 16.37b), and of still a third width where the section is partly in cut and partly in fill (Figure 16.37c). The finished cross section may be sloped in various ways to provide shoulders, drainage, and rounded corners, as illustrated in Figure 16.37c.

When the metric system is used, the cross slope of pavement and subgrade is expressed in percent. For example the cross slope of the pavement in Figure 16.37c is 2.000 percent from center line to the edge of pavement. The rate of the side slope is stated as a ratio of the *vertical component to the horizontal component,* where the vertical component is unity for slopes of less than 45°. Therefore, a 1:2 slope indicates a vertical rise or fall of 1 m in 2 m horizontal distance. This convention is in keeping with the mathematical definition of slope (Section A.7). For slopes greater than 45°, the horizontal component is unity (e.g., a 5:1 slope). The slope most commonly employed for cuts and fills through ordinary earth is 1:1.5. For coarse gravel, the slope is often made 1:1; for loose gravel, it is 2:1 (greater than 45°); for solid rock, it is 4:1; and for soft sand or clay, it is 1:2 or 1:3.

When the English system (feet) is used, cross-slopes are expressed in inches per foot. For example, pavement usually slopes $\frac{1}{4}$ in./ft from the center line to the edge of pavement. Side slopes are the ratio of the *horizontal component to the vertical component.* A 2:1 slope represents 2 ft horizontal distance for a 1-ft vertical rise or fall, so that for soft clay or sand, the slope would be 2:1 or 3:1.

After study of the tentative profile and cross-section data, areas of cross sections are determined for each station according to the methods given in Section 16.30, and cumulated volumes of earthwork are computed by procedures outlined in Section 16.31.

For major highway alignment projects, the design is performed and earthwork volumes are determined at a workstation such as was described in Section 16.27. Terrain cross-section data to form a DTM and roadbed templet information plus alignment and grade parameters are used as input to the system. The operator then can display the plan and profile (Figure 16.36) and desired cross sections (Figure 16.38) with the roadbed templet superimposed and make adjustments to the proposed horizontal and vertical alignments, if necessary. When the alignment appears satisfactory, earthwork volumes, positions of slope stakes (Section 16.29), and ordinates for the mass diagram (Section 16.40) can be computed on-line. These data then are listed for evaluation, and the mass diagram displayed for subsequent balancing of cut and fill volumes (Section 16.41).

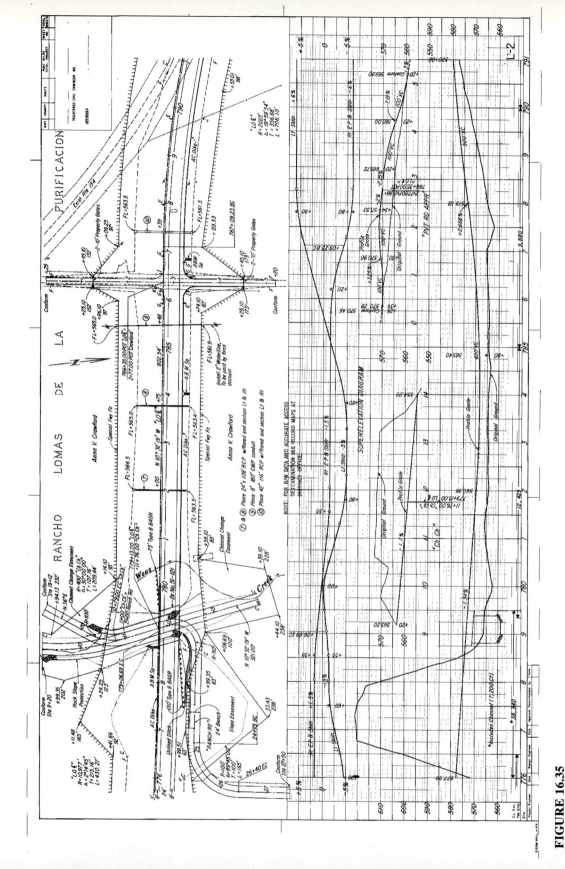

FIGURE 16.35
Typical location plan and profile. *(Courtesy California Department of Transportation.)*

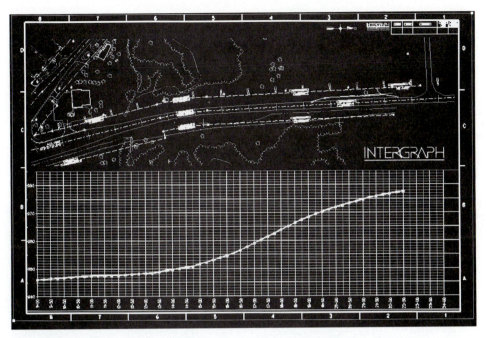

FIGURE 16.36
Plan and profile displayed on a video screen. (*Courtesy of Intergraph Corporation.*)

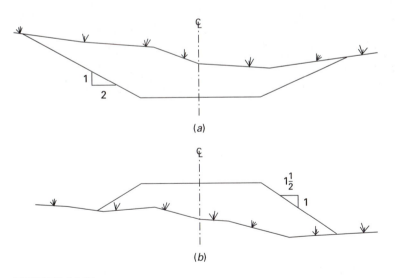

FIGURE 16.37
Preliminary cross sections: (a) cross section in cut, (b) cross section in fill, (c) side-hill cross section (see page 906).

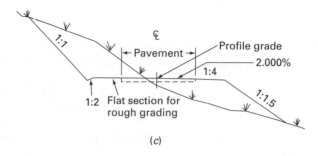

(c)

FIGURE 16.37 (continued)

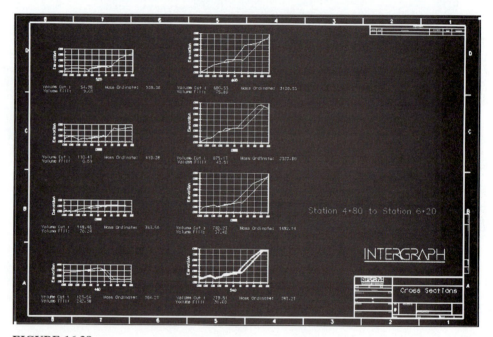

FIGURE 16.38
Cross sections displayed on a video screen. (*Courtesy of Intergraph Corporation.*)

Earthwork volumes and the mass diagram for the tentative alignment are studied, and necessary modifications are made to provide the most economic solution. At the same time, all the standards specified for the horizontal and vertical alignment of the proposed highway must be satisfied.

16.29
FINAL CROSS SECTIONS AND SLOPE STAKES

When a highway project reaches the final design stage, the horizontal and vertical alignments are fairly well fixed. During this stage, final design plans are prepared and construction of the roadway begins.

Prior to the actual construction, final cross sections are determined and slope-stake positions marking the intersections of the side slope with the natural ground surface (also called *catch points*) are determined or calculated and set opposite each center-line station.

Final cross-section data, which are used to calculate earthwork quantities for payment, can be determined by field surveys, photogrammetric methods, or using a DTM. Distances to slope stakes can be determined by trial and error in the field concurrently with running the final cross sections or calculated from the photogrammetric or DTM cross section. If calculated, the slope stakes are set in the field during construction layout. By far the majority of the final cross sections are determined from the DTM. However, to visualize better the concept and process, first consider final cross sections and slope stakes by field surveys.

Final Cross Sections and Slope Stakes by Field Methods

Level readings for final cross sections generally are taken with an engineer's level, and distances to the right or left of the center line are measured with a metallic tape, all to centimeters or tenths of feet. In rough, steep terrain, final cross sections may be taken by measuring slope distances and zenith or vertical angles. Total station systems can be used to advantage in such situations.

Prior to going to the field, the surveyor secures a record of elevations of ground points as obtained from the profile levels and the elevation of the established grade at each station. In the field, the instrument is set up in any convenient location, and the H.I. is obtained by a backsight on a bench mark. At each station, the rod is held on the ground, a foresight is taken, and the ground elevation is checked against that obtained by profile leveling. Next, the computed cut or fill is marked on the back of the center stake, and a cross line through the station is established at right angles to the center line.

Figure 16.39 shows the engineer's level in position for taking rod readings at a section *in cut*. The height of instrument (H.I.) has been determined and the elevation of grade at the particular station is known. The leveler computes the difference between the H.I. and the grade elevation, a difference known as the *grade rod;* that is, H.I. − elevation of grade = grade rod. The rod is held at any point for which the cut is desired, and a reading, called the *ground rod,* is taken. The difference between the grade rod and the ground rod is equal to the cut designated c.

Figure 16.40 is a similar illustration for a cross section *in fill*. If the H.I. is *above* grade (as at A), the fill is the *difference* between the grade rod and the ground rod; if the H.I. is *below* grade (as at B), the fill is the *sum* of the grade rod and the ground rod.

If the ground is level in a direction transverse to the center line, the only rod reading necessary is that at the center stake, and the distance to the slope stake can be calculated

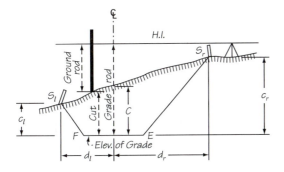

FIGURE 16.39
Road cross section in cut.

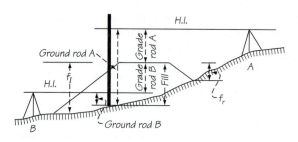

FIGURE 16.40
Road cross section in fill.

after the center cut or fill has been determined; such a cross section is called a *level section*. When rod readings are taken at each slope stake in addition to the reading taken at the center, as will normally be done if the ground is sloping, the cross section is called a *three-level section*. When rod readings are taken at the center stake, the slope stakes, and at points on each side of the center at a distance of half the width of the roadbed, the cross section is called a *five-level section*. A cross section for which observations are taken to points between center and slope stakes at irregular intervals is called an *irregular section* (Figure 16.37b). When the cross section passes from cut to fill, it is called a *side-hill section* (Figure 16.37c), and an additional observation is made to determine the distance from center to the grade point; that is, the point where the subgrade will intersect the natural ground surface. A peg usually is driven to grade at this point, and its position is indicated by a guard stake marked *grade*. In this case, cross sections also are taken at additional plus stations where the center line, the left edge, and the right edge of the roadway pass from cut to fill or from fill to cut, as at stations C, D, and F in Figure 16.41.

The process of setting slope stakes in the field consists of locating, by trial and error, the position of the point where the side slope of the cut or fill intersects the terrain. Therefore, in Figure 16.39, a slope stake would be set at S_r at an offset distance d_r to the right, where the cut equals c_r. On the left, a slope stake is set at S_l at an offset distance d_l to the left with a cut equal to c_l to subgrade. The final cross-section notes would be

$$\frac{c_l}{d_l}, \quad \frac{c}{0.0}, \quad \frac{c_r}{d_r}$$

or for a fill section, as in Figure 16.40,

$$\frac{f_l}{d_l}, \quad \frac{c}{0.0}, \quad \frac{f_r}{d_r}$$

Elevations and offset distances are measured with a level, rod, and tape or with a total station system and reflector mounted on a prism pole. The actual detailed procedure for setting slope stakes requires some additional explanation and is covered in detail in Section 17.7, on construction surveying for highways.

Final Cross Sections by Photogrammetric Methods and Using a DTM

When terrain cross sections are determined by photogrammetric methods (Section 16.26) or using a DTM (Section 16.27), the cross-section data are stored in a file in the database of the GIS used for the highway location and design. To obtain final cross sections from these data, the positions of the slope stakes or catch points must be calculated using design roadbed templet information such as center-line grade elevation, width of pavement, cross slope, and side slopes. Such a roadbed templet is shown in Figure 16.42b superimposed on

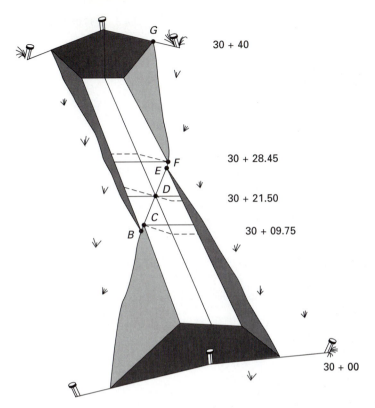

30 + 40

30 + 28.45

30 + 21.50

30 + 09.75

30 + 00

FIGURE 16.41
Cross sections in a transition area.

the terrain cross section of Figure 16.33. The simplest way to determine the position of the catch points S_r and S_l is to use coordinate geometry in a local, right-handed x, y system, where the origin is at the intersection of the center line with the subgrade and the x axis coincides with a line through the subgrade positive to the right side of the cross section when facing toward the higher station (Figure 16.42). The offset distances are x coordinates and terrain elevations minus grade elevation are the y coordinates. Using these coordinates, the intersection of the side slope with the terrain can be calculated.

EXAMPLE 16.13. Determine the offset distance and elevation for the catch points for station 51 + 00 (Figure 16.42) when the roadbed width $w = 13.20$ m, the center line grade $= 890.000$ m above the datum, and side slopes are 1:1.50. Terrain elevations for the portion of cross section used are shown in Figure 16.42a.

Solution. In a purely computational process, no plot of the cross section is available and an estimate of the approximate location for the catch point is needed. This estimate is obtained from the equation $d = w/2 + c's$ or $d = w/2 + f's$, where $c' =$ the average cut to the right or left of center line, $f' =$ the average fill to the right or left of center line, and $s =$ the denominator of the cut or fill side-slope ratio. For this example, on the right side of 51 + 00, $d_{r1} = 13.2/2 + [(894.3 + 895.0 + 895.1 + 894.0)/4 - 890.00)] (1.50) = 6.6 + 6.9 = 13.5$ m, which falls between the offsets of 8.2 and 17 m right (Figure 16.42a). Consequently, the catch point is located at the

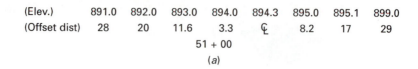

(Elev.)	891.0	892.0	893.0	894.0	894.3	895.0	895.1	899.0
(Offset dist)	28	20	11.6	3.3	₵	8.2	17	29

51 + 00

(a)

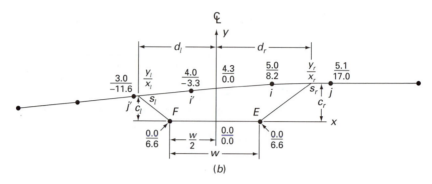

(b)

FIGURE 16.42

(a) Preliminary cross-section notes, (b) final cross section.

intersection of the terrain line i-j and side slope E-S_r (Figure 16.42a). Now, write the two-point Equation (8.15, Section 8.19) for line i-j:

$$(y_{S_r} - y_i)/(y_j - y_i) = (x_{S_r} - x_i)/(x_j - x_i)$$

$$(y_{S_r} - 5.0)/(5.1 - 5.0) = (x_{S_r} - 8.2)/(17.0 - 8.2)$$

$$0.1 x_{S_r} - 8.8\, y_{S_r} = -43.18$$

Next, form the point-slope Equation (8.13) for line E-S_r,

$$y_{S_r} - y_E = (x_{S_r} - x_E) \cot A_{E-S_r}$$

$$y_{S_r} - 0.0 = (x_{S_r} - 6.6)(0.67)$$

$$0.67 x_S - y_S = 4.40$$

Solving the two-point and point-slope equations simultaneously yields

$$0.10 x_{S_r} - 8.8 y_{S_r} = -43.18$$

$$-5.90 x_{S_r} + 8.8 y_{S_r} = -38.89$$

$$-5.80 x_{S_r} = -82.08$$

$$x_{S_r} = 14.15 \text{ m}$$

$$y_{S_r} = (0.67)(14.15) - 4.42 = 5.06 \text{ m}$$

Therefore, $\qquad d_r = 14.2$ m and $c_r = 5.1$ m.

Similarly, on the left side of the cross section, the estimate for d_l is $d_{l_1} = 6.6 +$ [(894.3 + 894.0 + 892.0 + 891.0)/4 − 890.000)](1.50) = 10.8 m which falls between offsets 3.3 and 11.6 to the left of center line or points i' and j'. As on the right, an Equation (8.15) and an Equation (8.13) can be formed for lines i'-j' and F-S_l, respectively, to give

$$x_{S_l} - 8.3 y_{S_l} = -36.50$$

$$-0.67 x_{S_l} - y_{S_l} = 4.42$$

which can be solved for $x_{s_l} = -11.2$ m and $y_{s_l} = 3.1$ m so that $d_\ell = 11.2$ m to the left and $c_l = 3.1$ m = the cut at the left catch point. The final cross-section notes for station $51 + 00$ are

$$\frac{c3.1}{11.2}, \quad \frac{c4.0}{3.3}, \quad \frac{c4.3}{0.0}, \quad \frac{c5.0}{8.2}, \quad \frac{c5.1}{14.2}$$

The readings 3.3 m left and 8.2 m right are included, because the points represent breaks in the terrain.

When the highway is designed and located using programs in a GIS, slope-stake positions are calculated for each station specified and final cross sections are kept in a file that can be listed or downloaded to a data collector for subsequent stakeout of slope stakes in the field. Note that this file contains not only the offset distances and cuts or fills to catch points but also the coordinates for each catch point. This information facilitates setting the slope stakes from a nearby random control point, a procedure described in Section 17.7.

Location of Cross Sections

When measurements are in meters, cross sections are taken at 10- or 20-m intervals and at all intermediate stations where significant breaks in the terrain occur. When the English system is used, stations generally are set at all full or half stations and intermediate stations when needed. If grading is very heavy or rock is present, cross sections should be taken at closer intervals.

Additional cross sections are required for transitions from cut to fill on a side-hill location. Figure 16.41 shows a transition from fill to cut from station $30 + 00$ to station $30 + 40$. Additional cross sections are required at B, C, D, E, and F. Generally, B is so close to C and E is so close to F that these points are omitted. Consequently, three additional cross sections are required: (1) at the fill-base grade point at C, (2) at the center-line grade point at D, and (3) at the cut-base grade point at F.

16.30
AREAS OF CROSS SECTIONS

For regular cross sections, levels are taken at one point on each side of the center line. Level and three-level sections as defined in Section 16.29 are regular cross sections.

The area of a level cross section (Figure 16.43a) is

$$A = f(w + sf) \tag{16.73}$$

in which f is the center-line fill, w is the roadway width, and s is the side-slope ratio. If the section is in cut, the center-line cut, c, is substituted for f in Equation (16.73).

A three-level section is illustrated in Figure 16.43b. As shown in the figure, this section can be divided into four parts, for which the respective areas are

$$A_1 = \frac{1}{2}(f_l)\left(\frac{w}{2}\right), \quad A_2 = \frac{1}{2}(f_r)\left(\frac{w}{2}\right), \quad A_3 = f\left(\frac{d_l}{2}\right), \quad A_4 = f\left(\frac{d_r}{2}\right)$$

in which f_r and f_l are the fills at the right and left catch points and d_r and d_l are the distances from center line to right and left catch points. Taking the sum of the four areas and collecting terms yields

$$A_{3l} = \frac{1}{2}f(d_l + d_r) + \frac{w}{4}(f_l + f_r) \tag{16.74}$$

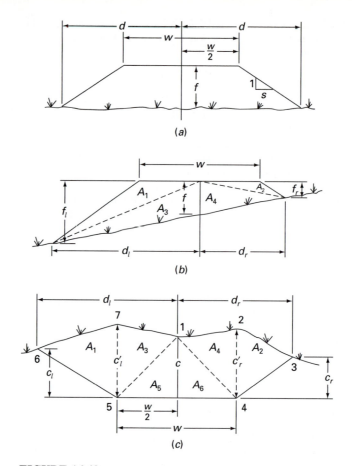

FIGURE 16.43
Cross sections: (a) level, (b) three-level, (c) five-level.

If the section is in cut, substitute $c = f$, $c_l = f$, and $c_r = f$, in Equation (16.74).

A five-level section is shown in Figure 16.43c. This cross section can be divided into five triangles (where $A_5 = A_6$, so $2A_5 = A_5 + A_6$), as indicated in the figure. The area of each triangle then can be determined as a function of c, $w/2$, c_l', c_r', d_l, and d_r. For example, $A_1 = \frac{1}{2}c_l'(d_l - w/2)$, $A_2 = \frac{1}{2}c_r'(d_r - w/2)$, and so forth. Taking the sum of these areas and collecting terms gives

$$A_{5l} = \frac{1}{2}(c_l'\,d_l + c_r'\,d_r + cw) \tag{16.75}$$

An irregular cross section is illustrated in Figure 16.44. The area of this type of section can be determined by dividing the section into trapezoids and triangles. However, the most efficient procedure is to use the method of coordinates as developed to determine the area within a closed traverse in Equation (8.33), Section 8.26. To compute the area of a section by Equation (8.33), let the center line in the plane of the cross section define the plus Y axis and the finished grade line define the X axis. The coordinates of slope stake positions or catch points, breakpoints, and *hinge points* (the angle points between the grade line and side slope) in the cross section are the offset distances from center line (X coordinates) and

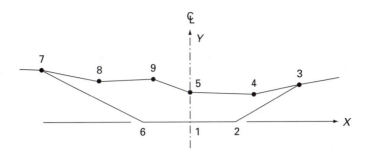

FIGURE 16.44
Irregular cross section.

differences in elevation between the finished grade and the terrain (Y coordinates). The coordinates of these points can be determined from the final cross-section notes (Figure 16.42a, and Example 16.13) with due regard for sign of offsets and differences in elevation between roadbed templet and terrain. The area of a cross section having n sides by Equation (8.33) (Section 8.26) is

$$A = \tfrac{1}{2}[Y_1(X_2 - X_n) + Y_2(X_3 - X_1) + Y_3(X_4 - X_2) + \cdots$$
$$+ Y_{n-1}(X_n - X_{n-2}) + Y_n(X_1 - X_{n-1})]$$

For the cross section shown in Figure 16.43c, with seven sides,

$$A = \tfrac{1}{2}[Y_1(X_2 - X_7) + Y_2(X_3 - X_1) + Y_3(X_4 - X_2)$$
$$+ \cdots + Y_6(X_7 - X_5) + Y_7(X_1 - X_6)]$$

A convenient way to systematize the calculation of cross-section areas by the coordinate method is to divide the section at the center line, taking the intersection of the grade line and center line (point 1 in Figure 16.44) as the origin of the coordinate system. Then applying expression (8.36) (Section 8.26), the area of the right half of the section in Figure 16.44 is

$$A_R = \frac{1}{2}\left[\frac{Y_1}{X_1} \diagdown \frac{Y_2}{X_2} \diagdown \frac{Y_3}{X_3} \quad \cdots \quad \frac{X_5}{X_5} \diagdown \frac{Y_1}{X_1} \right]$$

where the coordinates are listed for each point with Y in the numerator and X in the denominator, starting and ending with the first point and moving in a counterclockwise direction around the loop. Similarly, the area to the left of the center line, beginning with point 1 and moving in a clockwise direction around the loop, is

$$A_L = \frac{1}{2}\left[\frac{Y_1}{X_1} \diagdown \frac{Y_6}{X_6} \diagdown \frac{Y_7}{X_7} \quad \cdots \quad \frac{Y_5}{X_5} \diagdown \frac{Y_1}{X_1} \right]$$

The total area enclosed by the cross section is the sum of the absolute values of A_R and A_L.

EXAMPLE 16.14. Compute the area enclosed by the cross section and roadbed templet of Figure 16.45. Use the three-level equation (16.74) and check by the coordinate method. The final cross-section notes are as follows (units are in meters):

$$\frac{c0.56}{4.62}, \quad \frac{c1.11}{0.00}, \quad \frac{c1.86}{7.22}$$

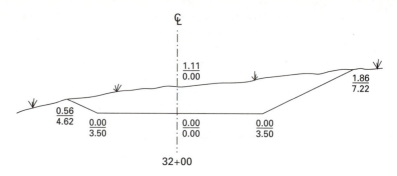

FIGURE 16.45
Final cross section, Example 16.14.

Solution. By Equation (16.74), the area is

$$A_{3l} = \tfrac{1}{2}(1.11)(4.62 + 7.22) + \tfrac{7}{4}(0.56 + 1.86)$$

$$= 10.81 \text{ m}^2$$

Checking by the coordinate method yields

$$A_L = \tfrac{1}{2}\left[\frac{0.00}{0.00} \diagdown\diagup \frac{1.11}{0.00} \diagdown\diagup \frac{0.56}{4.62} \diagdown\diagup \frac{0.00}{3.50} \diagdown\diagup \frac{0.00}{0.00} \right]$$

$$= \tfrac{1}{2}[(0.00)(0.00) + (1.11)(4.62) + (0.56)(3.50) + (0.00)(0.00) - (1.11)(0)$$
$$- (0.56)(0) - (0)(4.62) - (0)(3.50)]$$

$$= 3.54 \text{ m}^2$$

$$A_R = \tfrac{1}{2}\left[\frac{0.00}{0.00} \diagdown\diagup \frac{1.11}{0.00} \diagdown\diagup \frac{1.86}{7.22} \diagdown\diagup \frac{0.00}{3.50} \diagdown\diagup \frac{0.00}{0.00} \right]$$

$$= \tfrac{1}{2}[(0)(0) + (1.11)(7.22) + (1.86)(3.50) + (0)(0) - (1.11)(0) - (1.86)(0)$$
$$- (0)(7.22) - (0)(3.50)]$$

$$= 7.26 \text{ m}^2$$

$$A = A_L + A_R = 3.54 + 7.26 = 10.80 \text{ m}^2$$

16.31
VOLUMES OF EARTHWORK

Practically all volumes of earthwork for extensive projects are calculated on an electronic computer using programs that are an integral part of the GIS with which the highway is being designed. Therefore, after vertical and horizontal alignment have been decided and slope-stake data generated, the next step would be to determine volumes that can be displayed or listed for the record or for subsequent study and evaluation.

Volumes of earthwork are calculated by a variety of methods, depending on the nature of the excavation and the data. If cross sections have been taken along the route, their areas are determined as described in the preceding sections, and the volumes of the prismoids between successive cross sections are calculated by one of the following procedures: method of average end areas (Section 16.32), prismoidal formula (Sections 16.33 and 16.34), or finite element method (Section 16.37).

The same procedure may be followed for borrow pits and similar excavations, or if elevations are observed at the same points before and after excavating, the volume may be computed by dividing it into vertical truncated prisms (Section 16.42). Estimates for grading frequently are based on a topographic map showing the contours for the undisturbed ground and contours for the ground as it will appear when grading has been completed. The volume is determined conveniently by dividing it into prismoids with horizontal bases and sloping sides. Volumes of earthwork may be computed by the use of grading contours as described in Section 16.43. Total volumes are expressed in cubic yards or cubic meters.

Because of the repetitive nature of computations for earthwork, calculation by electronic computer is highly desirable.

16.32
VOLUMES BY AVERAGE END AREA

The common method of determining volumes of excavation along the line of highways, railroads, canals, and similar works is that of *average end areas*. It is assumed that the volume between successive cross sections is the average of their areas multiplied by the distance between them, or

$$V = \frac{l}{2}(A_1 + A_2) \tag{16.76}$$

in which V is the volume (in cubic meters or cubic feet) of the prismoid of length, l(meters or feet) between cross sections having areas (square meters or square feet) A_1 and A_2.

If l is in feet and cross sections at successive stations are in square feet, the volume between the stations in cubic yards, V_y is

$$V_y = \frac{l}{54}(A_1 + A_2) \tag{16.77}$$

Formulas (16.76) and (16.77) are valid only when $A_1 = A_2$ but are approximate for $A_1 \neq A_2$. As one of the areas approaches 0, as on running from cut to fill on side-hill work, a maximum error of 50 percent would occur if the formulas were followed literally. In this case, however, the volume is usually calculated as a pyramid; that is,

$$\text{Volume} = \tfrac{1}{3}(\text{area of base})(\text{length})$$

Considering that cross sections usually are a considerable distance apart and minor inequalities in the surface of the earth between sections are not considered, the method of average end areas is sufficiently precise for ordinary earthwork.

Where heavy cuts or fills occur on sharp curves, the computed volume of earthwork may be corrected for curvature (Section 16.35), but ordinarily the correction is not large enough to be considered.

16.33
PRISMOIDAL FORMULA

The volume of a prismoid can be shown to be

$$V = \frac{l}{6}(A_1 + 4A_m + A_2) \tag{16.78}$$

in which l is the distance between end sections, A_1 and A_2 are the areas of the end sections, and A_m is the middle area or area halfway between the end sections. A_m is determined by averaging the corresponding linear dimensions of the end sections and *not* by averaging the end areas A_1 and A_2. Equation (16.78) is one application of Simpson's rule (Section 8.29).

> **EXAMPLE 16.15.** Table 16.14 shows the three-level cross-section notes for two stations 100 ft apart. The width of the roadbed is 20 ft, and the side-slope ratio is 1.5 to 1.
> The volume of earthwork between the two stations is to be computed by the prismoidal formula. Below the regular cross-section notes are shown those for the midsection obtained by averaging the values given for sections at stations 115 and 116. The column *Area* lists areas of cross sections computed by Equation (16.74) (Section 16.30). By the prismoidal formula just given,
>
> $$V = \frac{100}{6}[212.0 + (4)(154.0) + 103.0] = 15{,}520 \text{ ft}^3 \text{ or } 575 \text{ yd}^3$$

For this example, the volume computed by average end areas is 583 yd^3 and the difference between the results obtained by the two methods is 8 yd^3 or about 1.4 percent.

As far as volumes of earthwork are concerned, the use of the prismoidal formula is justified only if cross sections are taken at short intervals, if small surface deviations are observed, and if the areas of successive cross sections differ widely. Usually, it yields smaller values than those computed from average end areas. For excavation under contract, the basis of computation should be understood in advance; otherwise, the contractor may claim (and obtain) the benefit of the common method of average end areas.

TABLE 16.14
Earthwork data

Station	Cross section ₵	Cross section ₵	Cross section R	Area, ft^2	Volume, yd^3
115	$c4.0$ / 16.0	$c6.0$ / 0	$c12.0$ / 28.0	212	
					575
116	$c2.0$ / 13.0	$c3.0$ / 0	$c8.0$ / 22.0	103	
Midsection	$c3.0$ / 14.5	$c4.5$ / 0	$c10.0$ / 25.0	154	

PRISMOIDAL CORRECTION

The difference between the volume computed by average end areas and that computed by the prismoidal formula for the prismoids defined by three-level sections can be shown to be

$$C_V = \frac{l}{12}(c_0 - c_1)(d_0 - d_1) \qquad (16.79)$$

where C_v = difference, in volume or correction, for a prismoid l meters or feet in
   length, in cubic meters or cubic feet
   c_0 = center height at one end section, meters or feet
   c_1 = center height at the other end section, meters or feet
   d_0 = distance between slope stakes at the end section where the center height is c_0,
      meters or feet
   d_1 = distance between slope stakes at the other end section, meters or feet

C_V is known as the *prismoidal correction;* it is *subtracted algebraically* from the volume as determined by the average end area method to give the more nearly correct volume as determined by the prismoidal formula.

All units in Equation (16.79) must be the same. When all distances are in feet, the prismoidal correction in cubic yards is

$$C_{V_y} = \frac{l}{324}(c_0 - c_1)(d_0 - d_1) \qquad (16.80)$$

EXAMPLE 16.16. For the prismoid of Example 16.15, the prismoidal correction by Equation (16.80) is

$$C_{V_y} = \frac{100}{324}(6.0 - 3.0)(44.0 - 35.0) = 8.33 \text{ yd}^3$$

which is consistent with the difference $(583 - 575) = 8 \text{ yd}^3$ between the volumes obtained by the average end area method and the prismoidal equation, respectively.

16.35
EARTHWORK CURVATURE CORRECTION

The average end area method for calculating volumes is based on parallel end areas. Volumes computed on curves will have discrepancies if this correction is neglected.

According to a theorem of Pappus, a plane area revolved about an axis generates a volume equal to the product of the revolving area and the path generated by the center of gravity of the revolving area. Let

   C_e = curvature correction = corrected volume − volume by the average end area
      method
   R = radius of the highway center line
   $\overline{e}$ = eccentricity of the cross-section area

The corrected volume, V_{corr}, is

$$V_{\text{corr}} = \frac{l}{2}(A_1 + A_2) \pm C_e \qquad (16.81a)$$

where A_1 and A_2 have respective eccentricities of $\bar{e}_1$ and $\bar{e}_2$, and l is the horizontal center-line distance between stations. Then, $V_1 = A_1 l_1$, where l_1 is the length of arc at $R + e_1$, so that by proportion

$$l_1 = l\left(\frac{R + \bar{e}_1}{R}\right)$$

and

$$V_1 = A_1\left(\frac{R + \bar{e}_1}{R}\right)l \tag{16.81b}$$

Similarly,

$$V_2 = A_2\left(\frac{R + \bar{e}_2}{R}\right)l \tag{16.81c}$$

Next, assume that

$$V_{\text{corr}} = \frac{V_1 + V_2}{2} \tag{16.81d}$$

Substitution of Equations (16.81b) and (16.81c) into (16.81d) and the result of this manipulation into Equation (16.81a) yields

$$C_e = \frac{l}{2R}(\bar{e}_1 A_1 + \bar{e}_2 A_2) \tag{16.82}$$

The curvature correction is positive when the excess area is on the outside of the curve and negative when the excess area is on the inside.

The centroid of the cross section is determined by taking moments about the center line. For example, the three-level section shown in Figure 16.46a is divided into four triangles having respective areas of a_1, a_2, a_3, and a_4, each with a centroid at distances of r_1, r_2, r_3, and r_4 from the center line. Taking moments about the center line of the section yields

$$\bar{e} = \frac{r_1 a_1 + r_2 a_2 + r_3 a_3 + r_4 a_4}{a_1 + a_2 + a_3 + a_4} \tag{16.83}$$

For a three-level section such as that illustrated in Figure 16.46a, the areas of the triangles can be calculated using the cuts or fills and offsets taken from the cross-section notes and the dimensions of the roadbed templet. The respective centers of gravity can be scaled from the cross section plotted at a scale of 1:100 [1:120 (10 ft per 1 in.)] and recalling that the centroid of any triangle falls at the intersection of its medians, where the median of any triangle connects any vertex with the midpoint of the side opposite that vertex. The centroid also falls on the median, two-thirds of the horizontal length of the median from the vertex.

Most volumes are determined by a computer program, so it is better to locate the centroid analytically. Using the cuts or fills and offsets from the final cross sections as coordinates defined as for slope-stake and area computations (Section 16.29, Figure 16.42), the coordinates for the endpoint of the median and, by proportion, those for the centroid can be calculated quite easily. The x coordinate is the offset to the centroid of that triangle. The geometry involved is illustrated in Figure 16.46a.

EXAMPLE 16.17. Assume that the center line containing stations 115 and 116 of Example 16.15 is a 10° circular curve (arc definition, see Section 16.4) that turns to the left. Calculate the curvature correction to the earthwork volume between these two stations.

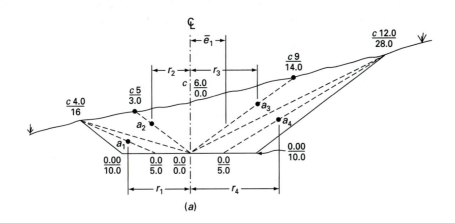

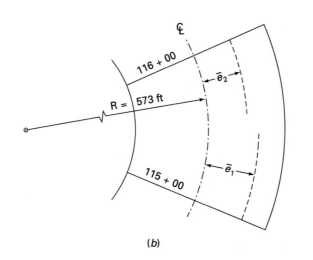

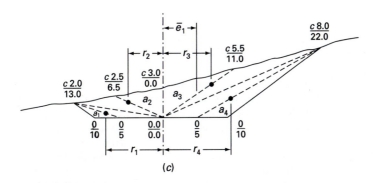

FIGURE 16.46
Center of gravity of cross sections: (a) 115 + 00, (b) plan view,
(c) 116 + 00.

Solution. The cross sections for stations 115 and 116 are plotted in Figure 16.46a and 16.46c. From Figure 16.46a and the cross-section notes of Example 16.15, the cuts and offsets for the midpoints of one side of triangles a_1, a_2, a_3, and a_4 can be found by simple proportion and are annotated on the figure. Areas and individual centroids for each triangle in the cross section for station 115 too are as follows:

$$a_1 = (\tfrac{1}{2})(10.0)(4.0) = 20 \text{ ft}^2, \quad r_1 = 16.0 - (\tfrac{2}{3})(16.0 - 5.0) = 8.7 \text{ ft}$$

$$a_2 = (\tfrac{1}{2})(6.0)(16.0) = 48 \text{ ft}^2, \quad r_2 = (\tfrac{2}{3})(8.0 - 0.0) = 5.3 \text{ ft}$$

$$a_3 = (\tfrac{1}{2})(6.0)(28.0) = 84 \text{ ft}^2, \quad r_3 = (\tfrac{2}{3})(14.0 - 0.0) = 9.3 \text{ ft}$$

$$a_4 = (\tfrac{1}{2})(10.0)(12.0) = 60 \text{ ft}^2, \quad r_4 = 28.0 - (\tfrac{2}{3})(28.0 - 5.0) = 12.7 \text{ ft}$$

$$\Sigma a_i = = 212 \text{ ft}^2$$

$$\bar{e}_1 = \frac{(20)(-8.7) + (48)(-5.3) + (84)(9.3) + (60)(12.7)}{212} = 5.3 \text{ ft}$$

Similarly, for station 116 + 00,

$$\bar{e}_2 = \frac{-(\tfrac{1}{2})(10)(2.0)(7.7) - (\tfrac{1}{2})(3)(13)(5.3) + (\tfrac{1}{2})(3)(22)(7.3) + (\tfrac{1}{2})(10)(8)(10.7)}{103}$$

$$= 4.7 \text{ ft}$$

From Equation (16.1), the radius of the center line is

$$R = \frac{5729.58}{10} = 573 \text{ ft}$$

By Equation (16.82), the curvature correction is

$$C_e = \frac{100}{(2)(573)}[(5.3)(212) + (4.9)(103)] = 142 \text{ ft}^3 = 5 \text{ yd}^3$$

Because the center of gravity falls outside the center line, the correction is positive, so

$$V = 583 \text{ yd}^3 + 5 \text{ yd}^3 = 588 \text{ yd}^3$$

Note that this correction is only 1 percent of the total. Unless side-hill cuts or fills are very large, this correction frequently is neglected.

16.36
VOLUMES IN TRANSITIONAL AREAS

The cross sections required in the transition from cut to fill or fill to cut are discussed in Section 16.29, and illustrated in Figure 16.41. In that illustration, additional sections are necessary at 30 + 09.75 (point C), where the cut runs out; 30 + 21.50 (point D), the grade point at center line; and 30 + 28.45 (point F), where the fill runs out. These cross sections are plotted and shown in Figure 16.47 together with the sections for 30 + 00 and 30 + 40. Note that these sections are plotted assuming a straight line for the finished grade in cut.

From 30 + 00 to 30 + 09.75, the volume is all fill. From 30 + 09.75 to 30 + 21.50, the volume consists of fill on the right and cut on the left. Note that the volume on the left between 30 + 09.75 (where cut runs out) and 30 + 21.50 must be calculated using the equation for volume of a pyramid (Section 16.32). From 30 + 21.50 to 30 + 28.45, the volume on the left is cut and on the right is fill, which also must be calculated using the

FIGURE 16.47

921

CHAPTER 16:
Route Surveying

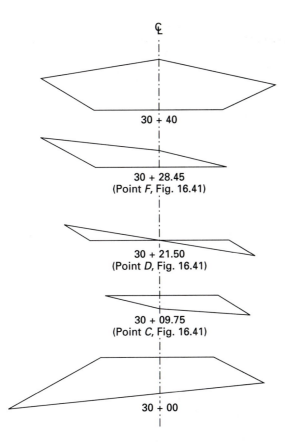

FIGURE 16.47
Cross sections in transition
from cut to fill.

equation for a pyramid. In calculating the volumes as pyramids, the bases are the cross-sectional areas at 30 + 21.50 and the altitudes of the pyramids are the distances from 30 + 21.50 to the stations where fill or cut runs out. From 30 + 28.45 to 30 + 40, the volume is all cut.

EXAMPLE 16.18. Table 16.15 contains cross-section data from which the cross sections of Figure 16.47 were plotted. Compute the volumes of cut and fill in this transitional area from 30 + 00 to 30 + 40 by the end area method and then using the prismoidal correction. The units are meters. The roadbed templet is $w = 21$ m in cut and 18 m in fill with 1:2 side slopes.

For 30 + 00,

$$A = \tfrac{1}{2}(5.00)(27.20 + 16.20) + \tfrac{18}{4}(9.10 + 3.60)$$

$$= 165.65 \text{ m}^2 \qquad \text{fill}$$

For 30 + 09.75,

$$A = \tfrac{1}{2}(2.20)(9 + 13.56) + \tfrac{18}{4}(0 + 28)$$

$$= 35.08 \text{ m}^2 \qquad \text{fill}$$

TABLE 16.15
Cross-section data for transition area

Station	Surface elevation	Grade elevation	Left	₵	Right
30 + 00	149.56	154.565	$\dfrac{f9.10}{27.20}$	$\dfrac{f5.00}{0.00}$	$\dfrac{f3.60}{16.20}$
30 + 09.75	152.87	155.073	$\dfrac{0.00}{9.00}$	$\dfrac{f2.20}{0.00}$	$\dfrac{f2.28}{13.56}$
30 + 21.50	155.69	155.686	$\dfrac{c2.64}{15.78}$	$\dfrac{0.00}{0.00}$	$\dfrac{f2.35}{13.70}$
30 + 28.45	158.20	156.048	$\dfrac{c4.10}{18.70}$	$\dfrac{c2.15}{0.00}$	$\dfrac{0.00}{10.50}$
30 + 40	164.10	156.650	$\dfrac{c4.15}{18.80}$	$\dfrac{c7.45}{0.00}$	$\dfrac{c3.84}{18.18}$

For 30 + 21.50, calculate as triangles, first on the left,

$$A_{\text{left}} = (\tfrac{1}{2})(10.5)(2.64) = 13.86 \text{ m}^3 \qquad \text{cut}$$

$$A_{\text{right}} = \tfrac{1}{2}(9)(2.35) = 10.58 \text{ m}^2 \qquad \text{fill}$$

For 30 + 28.45,

$$A = \tfrac{1}{2}(2.15)(18.7 + 10.5) + \tfrac{21}{4}(4.1 + 0.0)$$

$$= 52.92 \text{ m}^2 \qquad \text{cut}$$

For 30 + 40,

$$A = \tfrac{1}{2}(7.45)(18.80 + 18.18) + \tfrac{21}{4}(4.15 + 3.84)$$

$$= 179.70 \text{ m}^2 \qquad \text{cut}$$

These results and the accompanying volumes are listed in Table 16.16.
Volumes are computed as follows. From 30 + 00 to 30 + 09.75 by Equation (16.76),

$$V = \frac{9.75}{2}(165.65 + 35.08) = 978.6 \text{ m}^3 \qquad \text{fill}$$

From 30 + 09.75 to 30 + 21.50 the volume in fill by Equation (16.76) is

$$V = \frac{11.75}{2}(35.08 + 10.58) = 268.3 \text{ m}^3 \qquad \text{fill}$$

and the pyramid of cut is

$$V = \tfrac{1}{3}(13.86)(11.75) = 54.3 \text{ m}^3 \qquad \text{cut}$$

TABLE 16.16
Summary of areas and volumes

Station	Area, m²		Volume, m³	
	Cut	Fill	Cut	Fill
30 + 00		165.65		
				978.6
30 + 09.75		35.08		
			54.3	268.3
30 + 21.50	13.86	10.58	232.1	24.5
30 + 28.45	52.92			
			1343.4	
30 + 40	179.70		——	——
			1629.8	1271.4

From 30 + 21.50 to 30 + 28.45 there is a pyramid of fill on the right,

$$V = \tfrac{1}{3}(10.58)(6.95) = 24.5 \text{ m}^3 \qquad \text{fill}$$

and the volume of cut is computed by Equation (16.76) for this pair of stations and between 30 + 28.45 and 30 + 40:

$$V = \frac{6.95}{2}(13.86 + 52.92) = 232.1 \text{ m}^3 \qquad \text{cut}$$

and

$$V = \frac{11.55}{2}(52.92 + 179.70) = 1343.4 \text{ m}^3 \qquad \text{cut}$$

The total volumes are 1629.8 m³ of cut and 1271.4 m³ of fill, both by the average end area method.

Next, determine the prismoidal correction for each interval except for the two pyramids. Referring to the data in Table 16.15 and using Equation (16.79) the corrections are as follows. From 30 + 00 to 30 + 09.75,

$$C_{V_1} = (l/12)(c_0 - c_1)(d_0 - d_1) = (9.75/12)(5.00 - 2.20)(43.40 - 22.56)$$
$$= 47.4 \text{ m}^3$$

$$V_{1,\text{corr}} = 978.6 - 47.4 = 931.2 \text{ m}^3 \qquad \text{fill}$$

From 30 + 09.75 to 30 + 21.50,

$$C_{V_2} = [(11.75)/(12)](2.20 - 0.00)(22.56 - 13.71) = 19.1 \text{ m}^3$$

$$V_{2,\text{corr}} = (268.3 - 19.1) = 249.2 \text{ m}^3 \qquad \text{fill}$$

From 30 + 21.50 to 30 + 28.45,

$$C_{V_3} = [(6.95)/(12)](0.00 - 2.15)(15.78 - 29.20) = 16.7 \text{ m}^3$$

$$V_{3,\text{corr}} = (232.1 - 16.7) = 215.4 \text{ m}^3 \qquad \text{cut}$$

From 30 + 28.45 to 30 + 40,

$$C_{V_4} = [(11.55)/(12)](2.15 - 7.45)(29.20 - 36.98) = 39.7 \text{ m}^3$$

$$V_{4,\text{corr}} = (1,343.4) - 39.7) = 1,303.7 \text{ m}^3 \qquad \text{cut}$$

A summary of the corrected volumes follows:

| | Volume, m³ | |
Station	Cut	Fill
30 + 00		
		931.2
30 + 09.75		
	54.3	249.2
30 + 21.50		
	215.4	24.5
30 + 28.50		
	1,303.7	
30 + 40	1,573.4	1,204.9

Note that volumes are reduced by 3.5 and 5.2 percent in cut and fill, respectively, by applying the prismoidal correction. In Example 16.15, the cross sections are relatively close together and some minor irregularities exist in the terrain. In general, if the terrain is highly irregular through a transition area, use of the prismoidal correction is advisable.

16.37
VOLUMES BY FINITE ELEMENTS

An alternative method for computing volumes was reported by Davis (1994). Based on a well-known mathematical concept, called *finite elements,* it is applied to a variety of engineering problems. Davis (1994) presents a detailed description of the method and associated derivations. Here, we restrict ourselves to a summary of the method, his equations, and a numerical example for comparison with the other two methods for volume computation presented in previous sections.

The procedure for computing volumes by finite elements differs from the other two methods as follows. In the average end area and prismoidal methods, the volume is computed as the product of *cross-sectional areas* by a distance in the direction of the center line. By contrast, finite element volume is computed as the sum of volumes of a large number of *elements;* the volume of each element is calculated by multiplying the area of a *longitudinal section* in the direction of the center line by the width of the element, see Figure 16.48a, 16.48b, and 16.48c. The smaller the selected value of the element width, W, the more accurate is the resulting volume.

The relationships used in this method are summarized following the definitions of the symbols used:

L = the length along the center line, $\mathcal{L}$
L_0 = the corresponding length of a line at an offset, e, from $\mathcal{L}$
A_c = cross-sectional area
R = $\mathcal{L}$ radius of curvature
W = width of an element in plan, see Figure 16.48b
H = height of such elements
e = the plan distance from the $\mathcal{L}$ to an offset line in plan

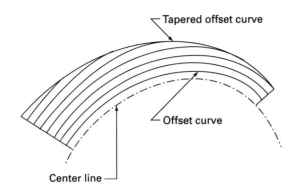

Tapered offset curve

Offset curve

Center line

FIGURE 16.48a
Plan view, tapered offset
curve. (*Adapted. Courtesy
Davis*, 1994.)

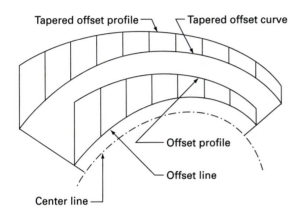

Tapered offset profile

Tapered offset curve

Offset profile

Offset line

Center line

FIGURE 16.48b
Perspective view, offset
profile and longitudinal
element. (*Adapted. Courtesy
Davis*, 1994.)

The equations that follow are used in the calculation of volumes by the finite elements method. For circular curves,

$$L_0 = L\left(1 + \frac{e}{R}\right) \tag{16.84a}$$

For spiral-curve segments,

$$L_0 = L\left[1 + \frac{e}{2}\left(\frac{1}{R_1} + \frac{1}{R_2}\right)\right] \tag{16.84b}$$

For a straight-line or circular-curve segment, the longitudinal element area is given by

$$A_l = \tfrac{1}{2}L_0(H_1 + H_2) \tag{16.84c}$$

where H_1 and H_2 are the initial and terminal element heights, respectively, obtained from cross-section data by interpolation. For spiral segments, the longitudinal element area is

$$A_l = \tfrac{1}{2}L_0(H_1 + H_2) + \tfrac{1}{12}Le(H_2 - H_1)\left(\frac{1}{R_2} - \frac{1}{R_1}\right) \tag{16.84d}$$

Finally, the volume of an element is given by

$$V = A_l W \tag{16.84e}$$

The sum of all positive element volumes is the required fill, and that of all negative element volumes is the cut quantity.

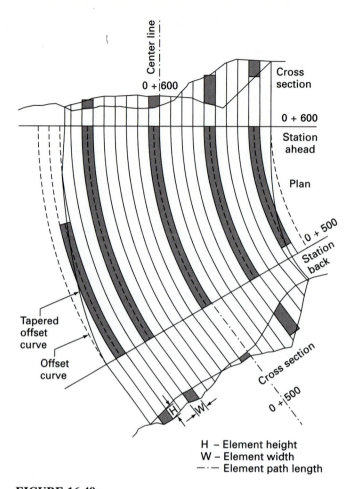

H – Element height
W – Element width
— · — Element path length

FIGURE 16.48c
Plan and transverse section views of elements. (*Adapted. Courtesy Davis, 1994.*)

EXAMPLE 16.19 (from Davis, 1994). Figure 16.48d shows the cross sections at stations 0 + 500 and 0 + 600 for a fill segment. Assume circular curvature with a center-line radius, $R = 200$ m (Figure 16.48e). Compare the volumes as computed by the following three methods: average end area, prismoidal, and finite elements.

Solution. For the average end area method,

$$V_A = \tfrac{1}{2}L(A_{c_1} + A_{c_2}) = \tfrac{1}{2}100(30 + 7.5) = 1875 \text{ m}^3$$

For the prismoidal method, the middle height is

$$H_m = \frac{1 + 2}{2} = 1.5 \text{ m}$$

and the middle area is

$$A_m = \tfrac{1}{2}(22.5)(1.5) = 16.875 \text{ m}^2$$

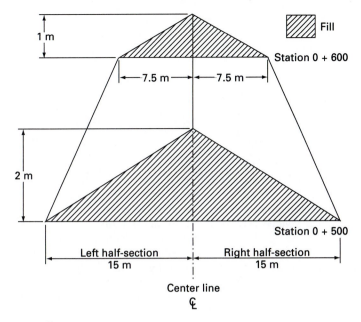

FIGURE 16.48d
Cross section data for Example 16.19. (*Adapted. Courtesy Davis, 1994.*)

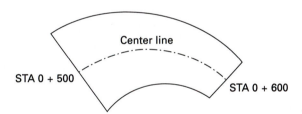

FIGURE 16.48e
Plan view circular curvature, $R = 200$ m, Example 16.19. (*Adapted, Courtesy Davis, 1994.*)

The volume is

$$V_p = \tfrac{1}{6}L(A_1 + 4A_m + A_2)$$
$$= \tfrac{1}{6}\,100[30 + 4(16.875) + 7.5] = 1750 \text{ m}^3$$

Considering each of the three possibilities as a separate case, we first correct the prismoidal volume for curvature and compare it to the finite element volume.

For the case of a circular center line with $R = 200$ m, the curvature correction is of opposite sign for the left and right sides of the center-line alignment: positive for the left side and negative for the right side. The eccentricities for stations $0 + 500$, the mid cross section, and $0 + 600$ are $\bar{e}_1 = 5$ m, $\bar{e}_m = 3.75$, and $\bar{e}_2 = 2.5$ m, respectively. Noting that A represents *half-section areas,* then it can be shown that the curvature correction for the prismoidal case is (Davis, 1994)

$$V_{corr} = \frac{1}{6R}L(A_1 e_1 + 4A_m\bar{e}_m + A_2\bar{e}_2)$$

$$= \frac{100}{6(200)}[15(5) + 4(8.4375)(3.75) + 3.75(2.5)]$$

$$= 17.578 \text{ m}^3$$

Therefore, the left volume is $875 + 17.578 = 892.578$ m^3 and the right volume is $875 - 17.578 = 857.422$ m^3 to give a total of 1750.0 m^3. Note that, because both cross sections involved are symmetrical about the center line, the curvature correction for the total volume between $0 + 500$ and $0 + 600$ is 0: ($+17.58$ on the left) + (-17.58 on the right) $= 0.00$. This value is identical to the answer that would be obtained for C_v using Equation (16.79) (applicable to the average end area volume) for the two cross sections. If the equation just given for V_{corr} is applied to Example 16.17, the answers agree to within a few hundreths of a cubic yard (5.05 yd^3), so the two equations for curvature correction provide essentially the same results for the data in Example 16.17 and this example.

To apply the finite element method, an element width, W, is selected first (see Figure 16.48c). The smaller the value of W, the closer is the volume to the theoretically correct value. Let $W = 0.1$ m. The first element adjacent to the center line has the following data:

$$e = 0.05 \text{ m}$$

$$L_0 = L(1 + e/R) = 100\left[1 + \frac{0.05}{200}\right] = 101.00025 \text{ m}$$

$$H_1 = (2)\frac{14.95}{15} = 1.9933 \text{ m}$$

$$H_2 = (1)\frac{7.45}{7.5} = 0.9933 \text{ m}$$

$$A = \frac{L_0}{2}(H_1 + H_2)$$

$$= \tfrac{1}{2}(101.00025)(1.9933 + 0.9933) = 150.827 \text{ m}^2$$

$$V = WA = 0.05 \times 150.827 = 7.541 \text{ m}^3$$

The data for the next element are

$$e = 0.15 \text{ m}$$

$$L_0 = 100\left[\frac{0.15}{200}\right] = 101.00075 \text{ m}$$

$$H_1 = (2)\frac{14.85}{15} = 1.980 \text{ m}$$

$$H_2 = (1)\frac{7.35}{7.5} = 0.981 \text{ m}$$

$$A = \tfrac{1}{2}(101.00075)(1.980 + 0.981) = 149.532 \text{ m}^2$$

$$V = 0.05 \times 149.532 = 7.477 \text{ m}^3$$

The rest of the elements are processed in a similar fashion. Note, however, that, when the distance e on the front station, $0 + 500$, exceeds 7.5 m, the distance from £ to catch point at $0 + 600$, the elements will terminate on a tapered curve. This will lead to progressively decreasing values for L_0. Such intermediate values are evaluated by interpolation.

When all the elements are all evaluated and summed, the left-side volume will be 893.9 m^3 and the right-side volume will be 858.6 m^3 to yield a total volume of 1752.5 m^3 (see Davis, 1994). If an element width of 0.001 m is used, the total volume by the finite element method is 1750.0 m^3 (Davis, 1994), which is the same as obtained by the prismoidal method with the curvature correction. Thus, the finite element method provides a powerful computational tool that takes into account center-line curvature and disparities between end areas.

The reader should note that the idealized, symmetric conditions used in Example 16.19, although very good to support understanding of the finite element method, rarely occur in actual practice.

16.38
ERROR PROPAGATION IN VOLUME COMPUTATION

It is useful to consider the propagated errors in earthwork volume computation. As an example, consider three-level cross sections, computation of the area by Equation (16.74), (Section 16.30), and calculation of volumes by the average end area method using Equation (16.76) or (16.77). Equation (16.74) for area of a three-level section is repeated here for convenience:

$$A = \tfrac{1}{2}c(d_l + d_r) + \frac{w}{4}(c_l + c_r) \tag{16.74}$$

Measured values are the center-line cut, c; the cut at the left slope stake, c_l; the cut at the right slope stake, c_r; and d_l and d_r, the horizontal distances to the left and right slope stakes, respectively. Applying the error propagation Equation (2.38) (Section 2.19) to Equation (16.74) yields an estimated variance in the area A of

$$\sigma_A^2 = \left(\frac{\partial A}{\partial c}\right)^2 \sigma_c^2 + \left(\frac{\partial A}{\partial d_l}\right)^2 \sigma_{d_l}^2 + \left(\frac{\partial A}{\partial d_r}\right)^2 \sigma_{d_r}^2 + \left(\frac{\partial A}{\partial c_l}\right)^2 \sigma_{c_l}^2 + \left(\frac{\partial A}{\partial c_r}\right)^2 \sigma_{c_r}^2$$

in which $c, d_l, \ldots, c_r$ are assumed to be uncorrelated. Assuming that $\sigma_c^2 = \sigma_{c_r}^2 = \sigma_{c_l}^2$ and $\sigma_{d_l}^2 = \sigma_{d_r}^2 = \sigma_d^2$,

$$\sigma_A^2 = \left[\left(\frac{d_l + d_r}{2}\right)^2 + \frac{w^2}{8}\right]\sigma_c^2 + \left(\frac{c^2}{2}\right)\sigma_d^2 \tag{16.85}$$

Equation (16.76) for volume by the average end area method is

$$V = \frac{l}{2}(A_1 + A_2) \tag{16.76}$$

in which l is the measured horizontal distance between stations and A_1 and A_2 are derived from measured elevations and distances using Equation (16.74). The propagated variance in the volume by applying Equation (2.38) to Equation (16.76) is

$$\sigma_V^2 = \left(\frac{\partial V}{\partial l}\right)^2 \sigma_l^2 + \left(\frac{\partial V}{\partial A_1}\right)^2 \sigma_{A_1}^2 + \left(\frac{\partial V}{\partial A_2}\right)^2 \sigma_{A_2}^2$$

$$\sigma_V^2 = \left(\frac{A_1 + A_2}{2}\right)^2 \sigma_l^2 + \left(\frac{l}{2}\right)^2 (\sigma_{A_1}^2 + \sigma_{A_2}^2) \tag{16.86}$$

in which $\sigma_{A_1}^2$ and $\sigma_{A_2}^2$ are estimated by Equation (16.85) and σ_l^2 is estimated on the basis of the equipment and methods used. Application of error propagation to Equation (16.77), where volumes are expressed in cubic yards, yields

$$\sigma_V^2 = \left(\frac{A_1 + A_2}{54}\right)^2 \sigma_l^2 + \left(\frac{l}{54}\right)^2 (\sigma_{A_1}^2 + \sigma_{A_2}^2) \tag{16.87}$$

When slope stakes are set in the field, a total station system and prism pole or a metallic tape, level, and rod are used. Estimated standard deviations of 0.015 m or 0.05 ft in horizontal distance and 0.015 m or 0.05 ft in elevation are reasonable with these types of equipment.

Propagated errors for areas and volumes computed by Equations (16.74) and (16.77) for several three-level cross sections of varying size are given in Table 16.17. The assumptions used are roadway width $w = 12$ m, $l = 20$ m, slopes are 1 to 2, $\sigma_c = 0.015$ m, and $\sigma_l = \sigma_d = 0.015$ m.

An inspection of the tabulation shows (1) that the percentage of error in the area and in the volume varies inversely with the depth of the cut or fill, (2) that the magnitude of the estimated standard deviation is not important as compared with the errors due to variations over the ground surface, and (3) that the estimated standard deviations indicate an uncertainty of one or more in the decimeter place of the computed quantities. Hence, it will be consistent to carry two decimal places in intermediate computations of areas and volumes,

TABLE 16.17
Propagated errors in volume computations*

Left slope stake	$\mathcal{L}$	Right slope stake	Area, A, m^2	σ_A^2, m^4	σ_A, m^2	σ_A/A, percent	Volume, m^3	σ_V, m^3	σ_V/V, percent
$\dfrac{c0.30}{6.55}$	$\dfrac{c0.70}{0}$	$\dfrac{c0.90}{7.80}$	8.6	0.012	0.11	1.3			
							261	1.75	0.7
$\dfrac{c0.60}{7.20}$	$\dfrac{c1.55}{0}$	$\dfrac{c1.20}{8.4}$	17.5	0.018	0.13	0.8			
							669	2.25	0.3
$\dfrac{c1.50}{9.10}$	$\dfrac{c3.54}{0}$	$\dfrac{c2.74}{11.6}$	49.4	0.030	0.17	0.4			
							1081	2.58	0.2
$\dfrac{c1.83}{9.80}$	$\dfrac{c4.24}{0}$	$\dfrac{c2.74}{11.4}$	58.7	0.030	0.17	0.3			

*$l = 20$ m, $w = 12$ m, s is 1 to 2, $\sigma_l = \sigma_d = 0.015$ m, $\sigma_c = 0.015$ m.

but it is absurd to record values beyond the last one-tenth meter place, either of areas or of volumes.

The procedures in the preceding paragraphs apply equally to volumes using irregular cross sections and borrow-pit excavation. Therefore, if volumes are calculated with irregular cross sections, the variance in areas is propagated by applying Equation (2.38) to Equation (8.33) for calculating area by coordinates. Estimated standard deviations in volumes computed with irregular cross sections obtained by field methods that satisfy $\sigma_c = 0.015$ m and $\sigma_d = 0.015$ m would be approximately the same as those given in Table 16.17. The critical item in the error propagation is in selecting valid estimates for the standard deviations in elevation and horizontal distance.

Suppose that the cross-section data are obtained from a topographic map at a scale of 1:500 with a contour interval of 0.5 m. Map accuracy standards as developed by the American Society of Photogrammetry and Remote Sensing (see Section 14.26, Tables 14.4 and 14.5 and discussion) are suitable for this case. From Table 14.4, the limiting RMS error in X or Y for a map at a scale of 1:500 is 0.125 m, so that $\sigma_d = \sigma_l = 0.125$ m. From the discussion in Section 14.26, the limiting RMS error in elevation is one-third the contour interval or $\sigma_c = 0.167$ m. Propagation of these errors using Equations (16.85) and (16.87) for the four cross sections of Table 16.17 yields standard deviations in the volumes that are 7.8, 3.6, and 2.6 percent of the total volumes between consecutive cross sections, respectively.

When the foot is the unit and a map at a scale of 1:600 (1 in. to 50 ft) with a contour interval of 2 ft is used to acquire cross-section data, a similar analysis leads to values of $\sigma_d = \sigma_l = 0.5$ ft and $\sigma_c = 0.6$ ft. Propagation of errors in computed volumes for a problem similar to Example 16.18 but with units in feet yields errors in volumes that are 9, 4, and 3 percent of the volumes between consecutive cross sections. Therefore, cross-section data obtained from topographic maps, at the given scales and contour intervals, should not be used to compute volumes as a basis for payment, because the errors are likely to be excessive. If such an approach is necessary, map scales and contour intervals of 1:200 and 0.25 m or 1:360 (1 in. to 20 ft) and 0.5 ft would be required.

When cross-section data are determined photogrammetrically, using photographs taken over cleared terrain, $\sigma_c = 0.12$ m (0.4 ft) and $\sigma_l = \sigma_d = 0.06$ m (0.2 ft) are reasonable estimates. Using these values, propagated standard deviations in volumes for the cross sections in Table 16.17 are 4.0, 1.8, and 1.3 percent of the volumes between respective pairs of cross sections.

As indicated previously, most cross sections now are determined from a DTM, where the DTM is constructed using data from photogrammetric stereomodels, topographic maps, and field surveys. In this case, it is reasonable to assume estimated standard deviations in offset distances and elevation and propagated errors in volumes comparable to those for photogrammetric cross sections. In general, the strictest possible control must be exercised over each operation contributing data to the DTM to derive cross-section data of sufficient accuracy to provide volumes for estimated payment quantities of earthwork.

16.39
DISTRIBUTION ANALYSIS OF EARTHWORK

Earthwork contracts usually allow payment for earthwork by a lump sum or on the basis of the volume of materials moved. So, for a given job the contractor needs to know the volumes of cut and fill, where these quantities are located, and the distances materials must be moved. Preparatory to detailed discussions concerning distribution analysis, consider some definitions related to the movement of cut and placing of fill or embankment.

Excavation is a pay quantity consisting of materials in cut that are transported to another location and placed in a fill or embankment. The distance that a cubic unit (cubic yard or cubic meter, where 1 yd^3 = 0.765 m^3 and 1 m^3 = 1.308 yd^3) of material is transported from cut to fill is called a *haul distance* or simply *haul*. The *free-haul distance* (usually specified in the contract) is the distance a contractor can haul a cubic unit of excavated material and place it in fill without extra cost above the cost for excavation. Any haul distance beyond free haul is called *overhaul,* for which there is an extra charge. Material excavated but not used for fill is called *waste*. Material needed for fill but not obtained from the roadway grading is called *borrow*. Some types of soils shrink when placed in fill and others swell. Consequently, *shrinkage* or *swell* of the material must be taken into account when calculating volumes used in the earthwork analysis.

The distribution analysis of earthwork involves determining balance points along the roadway center line between cut and fill. On a simple job, one could make separate subtotals of cuts and fills. The balance points would occur where these subtotals are equal. On larger projects, this method is inadequate and a more comprehensive method is necessary. Three procedures can be used to analyze earthwork operations: (1) the station-to-station method, (2) by study of the mass diagram, and (3) by optimization using linear programming. Details concerning the station-to-station method can be found in Meyer and Gibson (1980). The mass diagram, a general graphical approach that provides an excellent introduction to earthwork operations, is described in the sections that follow. Optimization by linear programming is a numerical method beyond the scope of this book. The interested reader should consult Easa (1987) and Mayer and Stark (1981).

16.40
THE MASS DIAGRAM

The earthwork mass diagram (see Foote, 1955) is a continuous graphical display of cumulated cut and fill volumes plotted as ordinates versus stations along the center line as abscissas. Cut is positive and fill is negative. Usually, the initial ordinate of the diagram is translated so that all ordinates are positive . The mass diagram is plotted above or below the center-line profile so that both can be used to analyze the earthwork operation. The abscissas generally are plotted at the same scale as the profile (say, 1:5000 in the metric system or 1 in. to 400 ft), and ordinates are plotted at as large a scale as possible considering the range in the cumulated cuts and fills. Figure 16.49 illustrates a portion of a profile plotted over the mass diagram for the corresponding stations.

Figure 16.49 shows the following features of the mass diagram: (1) a rising curve indicates a cut, such as from *A* to *C*; (2) a falling curve indicates fill, such as from *C* to *B*; (3) a maximum point on the curve such as at *C* indicates a change from cut to fill in the direction of stationing (note that if there is extensive side-hill cut and fill, this maximum point and the grade point *c* on the profile may not coincide exactly); (4) a minimum point such as at *D* indicates a change from fill to cut in the direction of stationing; and (5) a horizontal line intersects the curve of the mass diagram at balance points between which there is a balance of cut and fill volumes. In Figure 16.49b, the line *AB* is called a *balance line*.

In Figure 16.49a, if *aa′* represents 1 cubic unit of cut that is transported a distance of *x* stations to be placed as a fill of 1 cubic unit at *bb′*, this volume times the distance transported in stations constitutes the number of *station units* of haul represented by the area *AA′B′B* under the mass diagram in Figure 16.49b. In the metric system, haul is expressed in station meters, where one station equals 100 m. In the English system, haul is expressed in station yards, where one station equals 100 ft. The total haul resulting from placing the entire cut *aa′c* in the fill *cbb′* (Figure 16.49a), is the area between the curve of

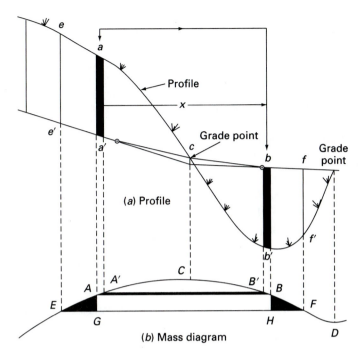

FIGURE 16.49
(a) Profile. (b) Corresponding mass diagram.

the mass diagram ACB and the balance line AB (Figure 16.49b). If the balance line AB equals the free-haul distance at the scale of the abscissa of the mass diagram, material in $aa'c$ can be excavated and placed in fill at cbb' at the unit price for excavation.

Overhaul

Again referring to Figure 16.49, if it is desired to place excavation $eaa'e'$ in fill at $bff'b'$, the volumes so handled and transported constitute *overhaul* in station units. An additional charge is assigned for overhaul. The total amount of overhaul in Figure 16.49b is represented by the areas EAG and BFH.

Limit of Economic Overhaul

Under certain conditions, it may pay to borrow or waste. In other words, the overhaul distance is profitable only to a limit. Let

$$
\begin{aligned}
C_E &= \text{cost to excavate 1 unit volume (includes free haul)} \\
C_{EW} &= \text{cost to excavate and waste 1 unit volume} \\
C_B &= \text{cost to borrow 1 unit volume} \\
C_{OH} &= \text{cost for overhaul per station unit} \\
F &= \text{free-haul distance, stations} \\
\text{L.E.H.} &= \text{limit of economic haul, stations}
\end{aligned}
$$

The cost to excavate and waste plus the cost to borrow equals the cost of excavation plus the cost for overhaul:

$$C_{\text{EW}} + C_B = C_{\text{OH}}(\text{L.E.H.} - F) + C_E \tag{16.88}$$

Next, assume that $C_{\text{EW}} = C_E$, so that

$$\text{L.E.H.} = \frac{C_B}{C_{\text{OH}}} + F \tag{16.89}$$

Equation (16.89) governs the limit to economic overhaul given the free-haul distance and unit costs for borrow and overhaul.

> **EXAMPLE 16.20.** The free-haul distance is specified as 10 stations (1000 ft) and the unit costs are $C_E = \$1.80/\text{yd}^3$, $C_B = \$2.00/\text{yd}^3$, and $C_{\text{OH}} = \$0.40/\text{station yd}$. Compute the limit of economic haul.
>
> *Solution.* By Equation (16.89),
>
> $$\text{L.E.H.} = \frac{2.00}{0.40} + 10 = 15 \text{ stations}$$
>
> Therefore, if AB (Figure 16.48) is taken as 10 stations, it is less expensive to excavate and waste at E and borrow at F than to haul the materials farther than 15 stations (assume that $EF > 15$ stations).

> **EXAMPLE 16.21.** The free-haul distance $F = 240$ m, $C_E = \$3.60/\text{m}^3$, $C_B = \$4.00/\text{m}^3$, and $C_{\text{OH}} = \$0.50/\text{station m}$. Compute the limit of economic haul.
>
> *Solution.* By Equation (16.89),
>
> $$\text{L.E.H.} = \frac{4.00}{0.50} + 2.4 = 10.4 \text{ stations}$$
>
> beyond which it is less expensive to excavate and waste and borrow.

Shrinkage and Swell

Shrinkage is small in granular soils (sands and gravels) but can be substantial (up to 30 percent) in fine-grained silts and clays. Shrinkage occurs when excavated soils are compacted into fills. Thus, 1 cubic unit of excavated material will not provide 1 cubic unit of compacted fill. In contrast to fine-grained soils, rock when excavated will occupy a larger volume in fill than when in place. In this case swell occurs and 1 cubic unit of excavated rock will occupy more than 1 cubic unit of fill. Swell of up to 30 percent is possible in excavated rocky materials.

When excavated materials have uniform shrinkage or swell factors, compensation is accomplished by increasing or decreasing the calculated fill quantities before computing the mass diagram ordinates. For example, if the excavated material shrinks 20 percent when placed in fill, 125 m³ of cut is required to provide 100 m³ of fill. Consequently, all fills must be increased by 20 percent to have a proper balance.

When excavated materials are nonhomogeneous and several different shrinkage factors are involved, the proper factor is applied to each cut volume and actual fills are utilized. Thus, if the shrinkage factor is 20 percent, a measured cut of 100 m³ will occupy a fill of 80 m³ when placed in fill. Compensation is achieved by decreasing all cut volumes by 20 percent prior to computing the cumulated cuts and fills and the ordinate of the mass diagram represents actual volumes of fill and volumes of cut available for fill.

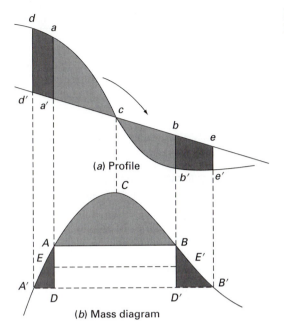

FIGURE 16.50
Free haul and overhaul.

935

CHAPTER 16:
Route Surveying

(a) Profile

(b) Mass diagram

Determination of Overhaul

The amount of overhaul can be determined from the mass diagram. Portions of one loop of a mass diagram and the corresponding profile are shown in Figure 16.50. Assume that balance line AB is the free-haul distance, at the scale of the mass diagram, and balance line $A'B'$ is the limit of economic haul. Material from cut $aa'd'd$, excavated, hauled, and placed in fill at $bb'e'e$, is overhaul, which is represented under the mass diagram by areas $AA'D$ and $BB'D'$. When AA' and BB' are relatively straight, the sum of these two areas equals $AD(EE' - AB)$, where EE' lies midway between AB and $A'B'$. Consequently, the overhaul in station units (station m or station yd) is found by multiplying the difference in ordinates between the free-haul balance line and the overhaul balance line by the difference between the overhaul distance and free-haul distance in stations. This graphical method of determining overhaul quantities is valid when the two overhaul areas ($AA'D$ and $BB'D$ in Figure 16.50) are nearly equal and of similar configuration. When the mass diagram is very curved or irregular, the areas that represent overhaul can be measured by a planimeter, an analog device for measuring area. Another, more rigorous numerical approach consists of approximating the profile and cross slopes by higher-order polynomials and integrating to obtain volumes and respective centroids of the volumes (Easa, 1988a). In this section, attention is focused on graphical analysis of the mass diagram.

16.41
BALANCING PROCEDURES

So far, only the case involving one loop of the mass diagram has been studied. In practice, several or many loops of varying magnitude may occur consecutively, and the mass diagram is very useful in determining the most economical solution by applying certain balancing procedures.

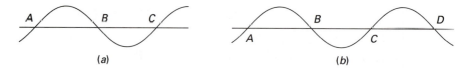

FIGURE 16.51
Balancing procedures: (a) two loops, (b) three loops.

When two loops occur, as in Figure 16.51a, the most economical position for the balance line ABC is that in which $AB = BC$, providing the distances AB and BC are each less than the limit of economic haul.

If three consecutive loops occur, as in Figure 16.51b, the most economical position for balance line $ABCD$ is that in which $(AB + CD) - BC = $ L.E.H., provided that AB, CD, and BC each is less than the limit of economic haul.

In general, the most economical position for a balance line cutting an even number of loops is where the sum of the segments cutting concave loops equals the sum of the segments cutting convex loops. Each segment should be less than the limit of economic haul. The most economical position for a balance line cutting an odd number of loops is where the sum of the segments cutting concave (or convex) loops minus the sum of the segments cutting loops in the opposite direction equals the limit of economic haul. As before, each segment should be less than the limit of economic haul.

EXAMPLE 16.22. Figure 16.52 shows the mass diagram and corresponding center-line profile for a small project. The unit costs for this project are cost of excavation $C_E = \$4.00/\text{m}^3$, cost of borrow $C_B = \$4.50/\text{m}^3$, cost of overhaul $C_{OH} = \$0.45/\text{sta-tion m}$, and the free-haul distance $F = 800\text{ m} = 8$ stations. The total excavation from the volume summary is 24,848 m^3. Because variable shrinkage factors are involved (1.00 and 0.80, see Figure 16.52), excavated volumes adjusted for shrinkage and actual fills are used to compute ordinates for the mass diagram.

Solution. By Equation (16.89), the limit of economic haul is

$$\text{L.E.H.} = \frac{4.50}{0.45} + 8 = 18 \text{ stations}$$

In Figure 16.52, balance lines AB (free-haul distance) and CD (limit of economic haul) are placed on the first large loop. The cut from $A'B$ can be placed in fill at $A'A$ at the cost of excavation. Balance line CD is the limit of economic haul, so the cut DB placed in fill at AC is overhaul and from the origin to C requires borrow.

The balance line EFG for the two consecutive loops is placed such that $EF = FG$. Note that both EF and FG are less than the free-haul distance. Thus, excavation from D to E and from G to H is wasted. The cut from E to E' can be placed in fill from E' to F and the cut from G to G' can be placed in the fill from G' to F at the cost for excavation because $EF = FG < $ the free-haul distance of 8 stations.

From the mass diagram, the distance JK is scaled and found to be 15 stations, so that

$$\text{overhaul} = (15 - 8)(6170 - 3730) = 17,080 \text{ station m}$$

and the total costs are

$$\text{Excavation} = (24,848)(4.00) \qquad = \$99,392$$
$$\text{Borrow} = (10,000 - 6170)(4.50) = \quad 17,235$$
$$\text{Overhaul} = (17,080/0.80)(0.45) \quad = \quad \underline{9,608}$$
$$\text{Total cost} = \$126,235$$

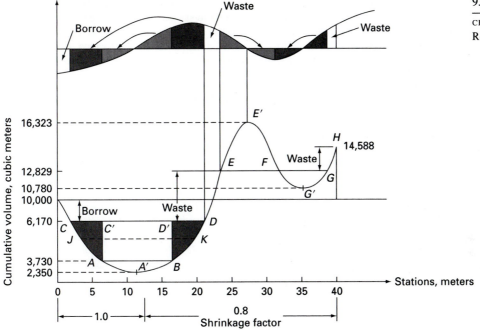

FIGURE 16.52
Mass diagram and profile for a small project (Example 16.22).

Note that the total cost of excavation must be calculated from the earthwork volume summary and the overhaul is adjusted by the shrinkage factor. These steps are necessary for this example, because ordinates on the mass diagram reflect cut volumes reduced by the shrinkage factor.

Certain checks can be applied to the mass-diagram ordinates: (1) the difference between the final and initial ordinates is equal to the difference between the total volume of waste and total volume of borrow, and (2) the final ordinate is equal to the difference between the total volume of cut and the total volume of fill taken from the earthwork volume summary.

Applying the first check to the mass diagram of Figure 16.52, the total volume of borrow from the origin to C is 3830 m³ and the total volume of waste from D to E and from G to H is (6659 + 1759) = 8418 m³. Therefore, the total waste minus the total borrow is 8418 − 3830 = 4588 m³, which equals the difference between the final and initial ordinates, or 14,588 − 10,000 = 4588 m³.

Application of the second check requires data from the volume summary. The total volume of cut is 24,848 m³ and the total volume of fill is 10,260 m³, yielding a difference of 14,588 m³. This value corresponds to the final ordinate of the mass diagram as taken from Figure 16.52.

16.42
BORROW PITS

The traditional method of determining the volume of a borrow pit was to cross section the area before and after excavating. In this procedure, a rectangular grid, with a regular uniform grid interval and referenced to points of known horizontal position, is set out over the area containing the borrow pit. Elevations are obtained at each grid intersection. A theodolite and tape and an engineer's level and rod are used to determine the horizontal and vertical control. These data constitute a DTM that covers the borrow pit region. After excavation, a similar survey, tied to the same control points, is performed to yield a DTM of the bottom and sides of the borrow pit. Comparison of the surfaces corresponding to these two DTMs yields the volume.

Using current technology, the horizontal and vertical positions for points on the original and excavated surfaces can be determined much more easily by field surveys using a total station system or GPS techniques, photogrammetric methods, and digitized topographic maps. Given data obtained by any one or combination of these methods and stored in computer memory, a DTM can be formed for each surface and a rectangular grid, referenced to known control points, superimposed numerically over each surface. This procedure yields elevations at the grid intersections of the rectangular grid on each surface, the original ground level, and the same points after excavation. The differences between elevations at corresponding intersection points, on the original ground and the excavated surface, yield cuts at each intersection of the grid network.

Figure 16.53 illustrates the plan view of a truncated rectangular grid that encompasses a borrow pit and has been superimposed over the DTMs representing the original ground and excavated surfaces. The numbers positioned diagonally are the cuts in meters at each intersection point. This example requires evaluation of volumes for rectangular and triangular prisms.

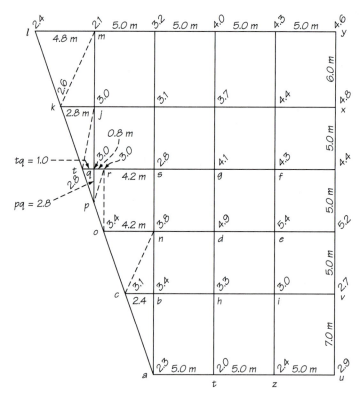

FIGURE 16.53
Borrow pit.

The volume of a triangular truncated prism (such as *abc* in Figure 16.53) is

$$V = \frac{A}{3}(h_1 + h_2 + h_3) \qquad (16.90)$$

in which A is the horizontal sectional area and h_1, h_2, and h_3 are the corner heights at a, b, and c in this figure.

The volume of any rectangular prism (as *defg* in Figure 16.53) can be found using the equation

$$V = A\left(\frac{h_1 + h_2 + h_3 + h_4}{4}\right) \qquad (16.91)$$

in which A is the area of the base of the rectangular prism and h_1, h_2, h_3, and h_4 are the corner depths of the prism from original terrain surface to the finished grade of the borrow pit.

When a uniform grid is used, the base area for all rectangular prisms is the same, and it is possible to derive a single equation to determine the volume within the area included by the uniform grid. This is accomplished by summing the common corner heights. For example, in rectangular prism *ivuz* the height at u is used once, at v and z twice, and at i four times. Let

h_1 = sum of corner heights used once
h_2 = sum of corner heights used twice
h_3 = sum of corner heights used three times
h_4 = sum of corner heights used four times

Then the sum of the volumes of the rectangular prisms all having a base area of A is

$$V = \frac{A}{4}\left(\sum h_1 + 2\sum h_2 + 3\sum h_3 + 4\sum h_4\right) \qquad (16.92)$$

The volumes of marginal prisms must be calculated separately and added to the volume determined by Equation (16.92) to get the total volume of the borrow pit.

EXAMPLE 16.23. Compute the volume of the borrow pit shown in Figure 16.53. First, compute the volume of the rectangular prisms having uniform base areas of $(5)(5) = 25$ m^2 enclosed by *xjqsbv*.

Solution. Use Equation (16.92) to compute the following:

h_1	h_2	h_3	h_4
4.8	4.4	2.8	4.1
3.0	3.7	$\Sigma h_3 = \overline{2.8}$	4.3
3.0	3.1		4.9
3.4	3.8		5.4
2.7	3.3		$\Sigma h_4 = \overline{18.7}$
$\Sigma h_1 = \overline{16.9}$	3.0		
	5.2		
	4.4		
	$\Sigma h_2 = \overline{30.9}$		

By Equation (16.92), the volume of *xjqsbv* is

$$V = \frac{625}{4}[16.9 + (2)(30.9) + (3)(2.8) + (4)(18.7)] = 25{,}297 \text{ ft}^3$$

Next, compute the volumes of *bvua* and *xymj*, both composed of rectangular prisms:

	bvua			xymj	
	h_1	h_2		h_1	h_2
	2.7	3.0		4.8	4.3
	3.4	3.3		4.6	4.0
	2.3	2.0		2.1	3.2
	2.9	2.4		3.0	3.1
$\Sigma h_1 = \overline{11.3}$		$\Sigma h_2 = \overline{10.7}$	$\Sigma h_1 = \overline{14.5}$		3.7
					4.4
				$\Sigma h_2 = \overline{22.7}$	

$$V_{bvua} = \frac{(5)(7)}{4}[11.3 + (2)(10.7)] = 286 \text{ m}^3$$

$$V_{xymj} = \frac{(6)(5)}{4}[14.5 + (2)(22.7)] = 449 \text{ m}^3$$

The volumes of the remaining triangular truncated prisms and the one remaining rectangular prism are calculated by Equations (16.90) and (16.91), respectively. The results follow, together with the volumes already computed.

Prism	Σ Corner heights, m	Area, m²	Times	Volume, m³
abc	8.8	(1.2)(7.0)	$\frac{1}{3}$	24.6
bcn	10.3	(1.2)(5.0)	$\frac{1}{3}$	20.6
con	10.3	(2.1)(5.0)	$\frac{1}{3}$	36.1
por	9.6	(0.4)(5.0)	$\frac{1}{3}$	6.4
pqr	9.2	(0.4)(1.4)	$\frac{1}{3}$	3.4
pqt	9.0	(1.4)(1.0)	$\frac{1}{3}$	4.2
tqj	8.8	(0.5)(5.0)	$\frac{1}{3}$	7.3
kjt	8.4	(1.4)(5.0)	$\frac{1}{3}$	19.6
kjm	7.7	(1.4)(6.0)	$\frac{1}{3}$	21.6
klm	7.1	(2.4)(6.0)	$\frac{1}{3}$	34.1
nors	13.0	(4.2)(5.0)	$\frac{1}{4}$	68.3
xjqsbv				1012
bvua				286
xymj				449
Total				1993 m³

In the method just presented, linearity is assumed between intersection points on the terrain. Because linearity of the actual surface of the ground occurs rarely, this assumption can lead to erroneous results, where the error varies directly with the irregularity of the original terrain.

Several methods have been developed for determining volumes of borrow pits that use the superimposed rectangular grid but approximate the terrain surface with second- and third-degree polynomials or, in another case, a cubic spline formula to compensate for nonlinearity between intersection points. Certain of these methods also can accommodate an unequal grid interval. Details of these procedures can be found in Chambers (1988), Chen and Lin (1991), and Easa (1988b).

16.43
EARTHWORK VOLUMES BY GRADING CONTOURS

Contours plotted on a topographic map to portray the finished surface of a proposed grading operation are called *grading contours*. Because the finished surface usually is smooth and of constant slope, grading contours are generally smooth, equally spaced, parallel lines that may be straight or curved, depending on the character of the final surface. Volumes from grading contours are determined by the method of horizontal planes and the method of equal-depth contours.

Horizontal Planes

For preliminary estimates for grading areas, especially where the graded surface is more or less irregular, the common practice is to use the topographic map directly as a basis for calculation of volume. On the map are shown contours for the natural ground and contours for the proposed graded surface. This method consists of determining the volumes of earth to be moved between the horizontal planes at the elevations of successive contours.

The light solid lines of Figure 16.54 represent contours of the original ground, and the dashed lines represent contours of the proposed graded surface. The heavy solid lines are drawn through points of no cut or fill. Thus, the line *abcdefa* bounds an area that is entirely in fill, and the line *dghjked* bounds an area that is entirely in cut. The "no cut or fill" lines are seen to pass through the points of intersection between full contours and the corresponding dash contours, as at *a, b, d, e, h,* and *j*. The conditions surrounding the problem make it possible to estimate the position of the lines where the cut or fill runs out between contours, such as the lines *bcd, efa,* and *jke*. The shaded portions are the horizontal sections of earth cut or fill at the contour elevations; so, F_1 represents the horizontal section of earth filled at elevation 98 m. The volume of earthwork between the two horizontal planes at the elevations of successive contours is a solid, the altitude of which is the contour interval and the top and bottom bases of which are the horizontal projections of the cut or fill at the contour elevations (such as the fill between the 97- and 98-m contours where the height is 1 m and the bases are F_2 and F_1, the areas of which may be determined by use of a planimeter). Where the cut or fill runs out between contours (such as along line *bcd*), the height of the final volume will be less than the contour interval. This height may be estimated by assuming the slope of the ground to be uniform between contours; therefore, point *c* is estimated to be at elevation 98.6 m and the volume above the 98-m contour is a solid, the base of which is F_1 and the altitude of which is $98.6 - 98 = 0.6$ m. The final volumes may be considered pyramids.

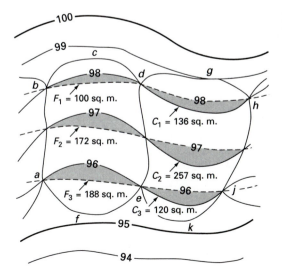

FIGURE 16.54
Volume by horizontal planes.

EXAMPLE 16.24. It is desired to determine the volume of earthwork in fill bounded by the line *abcdefa* (Figure 16.54). The intermediate volumes are to be calculated by the method of average end areas; the end volumes are to be considered as pyramids. The areas of fill at the contours are as shown in the figure. Point *c* is estimated to lie 0.6 m above the 98-m contour, and point *f* is estimated to be 0.8 m below the 96-m contour.

Solution

Elevation	Base area, m²	Altitude, m		Volume, m³
c = 98.6	0			
		0.6	$(\frac{1}{3})(0.6)(100)$	20
98	100			
		1.0	$(\frac{1}{2})(1.0)(272)$	136
97	172			
		1.0	$(\frac{1}{2})(1.0)(360)$	180
96	188			
		0.8	$(\frac{1}{3})(0.8)(188)$	50
f = 95.2	0			

Total   386 m³

Equal-Depth Contours

This method consists of determining volumes between irregularly inclined upper and lower surfaces bounding certain increments of cut or fill. Horizontal projections of the inclined areas are taken from the map, usually with the planimeter, and the volume between any two successive areas is determined by multiplying the average of the two areas by the depth between them.

Figure 16.55 represents the topographic map of a portion of a tract that is to be graded by filling. The light solid lines represent contours of the original ground surface, and the dashed lines represent contours of the proposed graded surface. Above the dashed 102-ft

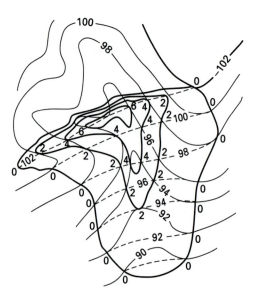

FIGURE 16.55
Volume by equal-depth contours.

contour, the fill drops abruptly to the natural ground. Along the bank thus formed, just above and paralleling the 102-ft contour, actually there would be 100-, 98-, and 96-ft contour lines, which are not shown to avoid confusion. At the intersection of each light solid line with each of the dashed lines the depth of fill (or cut) is recorded.

The heavy solid lines drawn through points of equal fill are called *lines of equal fill* (or *cut*). The heavy outer line passes through points of zero fill and marks the limit of the fill. The next heavy line encloses the area over which the fill is a minimum of one contour interval and passes through points of intersection between a full contour and a dashed contour, the elevation of which is 2 ft greater, and so on. Along the side of the bank above the dashed 102-ft contour, the heavy lines are seen to be close together and nearly parallel.

The fill between the graded surface and the surface 2 ft below is represented by the solid figure the altitude of which is 2 ft and the upper and lower surfaces of which are shown in horizontal projection by the line of zero fill and the line of 2-ft fill, respectively. Similarly, the lines of 2- and 4-ft fill define the volume of fill between the depths of 2 ft and 4 ft from the graded surface. The volume below the innermost line of equal fill may be considered a pyramid, the base area of which is that bounded by the line and the altitude of which is estimated, being always less than the full contour interval. Volumes usually are determined by multiplying the contour interval by the average of the areas of successive surfaces of equal cut or fill. When there is a large difference between successive areas the prismoidal, Equation (16.78) (Section 16.33) or Equation (16.79) for prismoidal correction is sometimes used.

EXAMPLE 16.25. An estimate of volume of earthwork in fill is to be made from a contour map similar to that of Figure 16.55. Lines of equal fill are drawn, and the areas of the horizontal projections of surfaces of equal fill are determined by measurement with a planimeter. The altitude of the pyramid below the innermost surface of equal fill is estimated to be 1 ft.

Solution

Fill, ft	Area, ft^2	Altitude, ft		Volume, ft^3
0	101,000			
		2.0	$(\frac{1}{2})(2)(134,000)$	134,000
2	33,000			
		2.0	$(\frac{1}{2})(2)(50,000)$	50,000
4	17,000			
		2.0	$(\frac{1}{2})(2)(22,000)$	22,000
6	5,000			
		1.0	$(\frac{1}{3})(1)(5000)$	2,000
7	0			
			Total	208,000 ft^3
			or	7700 yd^3

16.44
EARTHWORK FOR A ROADWAY BY GRADING CONTOURS

Figure 16.56 shows the contour lines for a proposed roadway (the grade line of which has been drawn on the profile) drawn dotted over the existing contour map. Above the contour map are shown a profile of the ground along the center line and the grade of the proposed roadway. The side slopes of the earthwork are 1.5 to 1. The width of the roadway is 36 ft in cut and 30 ft in fill. From a study of these two drawings, the following observations may be made:

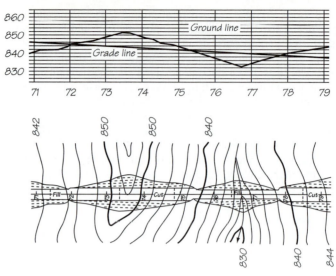

FIGURE 16.56
Earthwork for roadway by grading contours.

1. The 840-ft contour line of the proposed roadway crosses the roadway at a point on the map vertically beneath the point on the profile where the grade line crosses the 840-ft elevation line, and similarly for the other gradient contours.
2. On the side slopes of the earthwork at any station, the distance out from the edge of the roadway to a contour line is given by the difference in elevation (between that which the contour line represents and the elevation of the grade at that station), multiplied by the side-slope ratio. Therefore, at station 76 + 40 the elevation of grade is 840.0 ft and the elevation of the first contour line out from the edge of the fill is 838.0 ft; hence, the distance out is $2 \times 1.5 = 3$ ft (or 18 ft from ℄). (For clearness, in the illustration, the lateral scale is exaggerated.)
3. As the grade is not level, the contour lines on the earthwork slopes are not parallel to the roadway. Therefore, the 844-ft dashed contour line that crosses the roadway at station 73 + 30 is so inclined in direction that, at station 74 + 80, where the elevation of grade is 842 ft, the 844-ft contour line is out from the edge of the roadway a distance of $2 \times 1.5 = 3$ ft.
4. The top or toe of a slope is drawn on the contour map by connecting the points where the dashed lines intersect the corresponding solid lines.

The volume of earthwork is estimated by means of horizontal planes, as described in Section 16.43. Figure 16.57, an enlargement of the portion of Figure 16.56 from stations 72 + 00 to 75 + 00, illustrates the method. As an example, consider the volume contained between the 848 and 850 grading contours. The area *abdc* enclosed by the 850 grade contours and the 850-ft contour line is determined using a planimeter. Next, the area *efgh* enclosed by the 848-ft grade contours and the 848-ft contour also is measured by the planimeter. These two areas constitute end areas separated by the contour interval, or 2 ft in this case. The volume of this solid can be calculated by the average end area equation (16.77) (Section 16.32). Volumes of the remaining segments are calculated in a similar manner. The volume of the segment bounded by the 852 grading contour, the 852 contour, and the top of the slope at elevation 853 (interpolated from the map) is determined as the volume of a pyramid having a base area of *jkl* and an altitude of 1 ft. The total volume is the sum of the volumes of all segments within the area of interest.

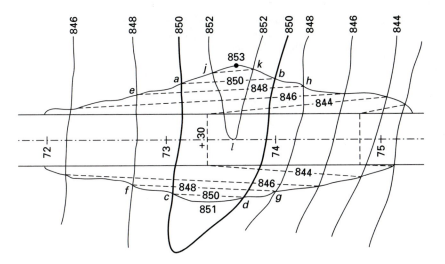

FIGURE 16.57
Roadway earthwork volumes by grading contours.

16.45
RESERVOIR AREAS AND VOLUMES

A contour map can be employed to determine the capacity of a reservoir, the location of the flow line, the area of the reservoir, and the area of the drainage basin. The procedure may be illustrated by reference to the fill across the valley in Figure 16.56, the fill being considered a dam that extends from the 841-ft elevation on the left to about the 838-ft elevation on the right. If water is imagined to stand at the elevation of 834 ft, the water surface is represented by that within the full and dashed 834-ft contour line. If the water were to rise through a 2-ft stage to the elevation of 836 ft, the water surface would be represented by that within the solid and dashed 836-ft contour lines (the solid line being continued until the two parts meet). The volume of water that caused the 2-ft rise is given by the average of the two surface areas multiplied by the vertical distance of 2 ft. Similarly, the volume of water required to cause a rise of the water surface from 836 ft to 838 ft may be found, and so on. By a similar procedure, the volume of any reservoir can be calculated.

The outline of the submerged area of a proposed reservoir is given by the contour line representing the maximum stage of the impounded water. The drainage area may be estimated by sketching the watershed line on the map and measuring the extent of the watershed with a planimeter.

PROBLEMS

16.1. Given $I = 34°00'R$, D_a (arc definition) $= 30°00'$, and P. C. = station $65 + 35.550$ (assume one full station $= 100$ m). Calculate R, L, T, E, and deflection angles to points at 20 m intervals on this curve. Arrange deflection angles in notebook form.

16.2. Given $I = 30°00'L$, D_c (chord definition) $= 5°00'$, and station of the P. I. $= 24 + 10.23$ (assume one full station $= 100$ ft). Compute R, L, T, E, and deflection angles to full station points on the curve. Arrange deflection angles in notebook form.

16.3. Given the data of Problem 16.1, except one full station is defined as 1 km so that the station of the P. C. = 6 + 535.550. Compute deflection angles to points at 10-m intervals on the curve. Tabulate in a proper field notebook form.

16.4. Given $I = 50°30'$ and $E = 38.252$ m. Calculate R, D_a (arc definition), C, T, M, and L.

16.5. The intersection angle, $I = 36°00'R$, $D_a = 25°00'$, and the station of the P. I. = 14 + 80.000. Compute (a) T, E, L, the middle ordinate M for the entire curve, and M for one full station of 100 m; (b) the stations of the P. C. and the P. T. and deflection angles for staking the curve at even 1/10th station points; and (c) the true lengths of 1/10th station chords and the true lengths for subchords at either end of the curve. (d) Write the necessary field notes for staking this curve in proper field notebook form.

16.6. Given $I = 28°30'00''$, $D_a = 28°30'$, the station of the P. I. = 14 + 63.210, and one full station = 100 m. Calculate R, L, T, E, stations of the P.C. and P. T., and the chord lengths for full and one-half stations.

16.7. Given $I = 60°30'$ and $E = 125.450$ ft, compute R, D_a, C, T, M, and L.

16.8. In Problem 16.1, the rectangular coordinates for the P. I. are $E = 82,141.800$ m and $N = 37,345.252$ m and the azimuth of the back tangent from P. I. to P. C. is $190°30'00''$. Perform the computations needed for layout of the curve from a single setup of a total station system at (a) the P. C. with a backsight on the P. I. and (b) the P. I. with a backsight on the P. C. (c) Tabulate the data in a form suitable for field layout.

16.9. The curve in Problem 16.5 is to be staked from a single setup of a total station system on random control point GG1 having rectangular coordinates $E = 82,141.050$ m and $N = 37,454.235$ m. A backsight will be taken on HH1 where the azimuth GG1-HH1 = $84°30'30''$. The coordinates for the P. I. of the curve are $E = 82,000.500$ m, $N = 37,350.290$ m, and the azimuth along the forward tangent from P. I. to P. T. is $45°40'00''$. Make the necessary computations for staking this curve and tabulate in a form suitable for field layout.

16.10. Two tangents AV and BV have an intersection angle of $40°00'L$. A point C is located by coordinates $VH = 82.350$ m and $HC = 49.970$ m, VH being measured along tangent VA and HC being measured perpendicular to it. It is desired to connect the two tangents with a curve passing through point C. Required are R, D_a, T, L, and E. The station of the P. I. = 1 + 740.000. Assume A is the P. C.

16.11. Given the data of Problem 16.1, make the necessary computations for insertion of a 60-m spiral at each end of the curve. Calculate deflection angles to six stations at 10-m intervals on each spiral and to one-fifth station points on the circular curve.

16.12. Given $I = 54°00'$ to the right, $D_a = 28°00'$, and the station of the P. I. = 25 + 10.455. The sharpness of this curve requires a spiral at each end of the circular curve. The rate of change of curvature of the spiral is to be $20°00'$ per full station (100 m). Using the rigorous equations for X and Y, compute (a) the stations for the T.S., S.C., C.S., and S.T. points on the spiraled curve; (b) the deflection angles (use approximate equations) for staking the curve with 20 chords on the spirals and 1/10th station points (10 m) on the circular curve. The spiral from the T.S. to the S.C. is to be laid out from a setup on the T.S. and the second spiral is to be staked from an instrument set at the C.S. Tabulate the results in the form of field notes showing the necessary information for orienting the theodolite when set on the points specified.

16.13. Using the data in Problem 16.11, compute deflection angles and chords, by rigorous equations, for points at 20-m intervals on the entering spiral. Compare the deflection angles to those determined by the approximate equations and the chords to the nominal differences in stationing between points.

16.14. A spiral having $L_s = 300$ m and $D_a = 30°$ is to be staked with 15 20-m chords. The station for the T.S. is $25 + 30.245$. Use rigorous equations to compute exact deflection angles to these stations and chords between each pair of consecutive stations. Compare these deflection angles with angles calculated by the approximate equations.

16.15. The curve in Problem 16.11 is to be staked from a single setup of a total station system set on control point GG1 (Problem 16.9) with a backsight on HH1 (Problem 16.9). The coordinates for the P. I. of the curve in Problem 16.11 are $E = 82,220.451$ m and $N = 37,500.525$ m. Perform the calculations necessary to stake the curve and tabulate the results in a form usable in the field. Assume the azimuth from the P. I. to the T. S. is $270°00'00''$.

16.16. The elevations of two preliminary grade lines that intersect at station $27 + 46.143$ (the P. I. or point V) are as follows:

Station	Elevation, m above the datum
6 + 00	1,900.200
27 + 46.143 (P.I.)	1,990.000
47 + 25.470	1,950.000

An equal tangent, parabolic vertical curve with a total length of 200.000 m and centered about $27 + 46.143$ is to be calculated for construction layout. Compute (a) the grade elevations for BVC and EVC and also at even one-fifth station points (e.g., $26 + 60$, $26 + 80$) throughout the vertical curve; and (b) the station and elevation of the high (or low) point on the vertical curve. Use the equation of the parabola to solve this problem.

16.17. Do the computations for Problem 16.16, parts (a) and (b), using the method of vertical offsets from the tangent.

16.18. In Problem 16.16, do the computations for parts (a) and (b) by the chord gradient method.

16.19. List the various methods for obtaining terrain data for route location. List these methods in decreasing order of current use and complexity.

16.20. Describe what constitutes a route cross section and how it is related to the profile of a route.

16.21. List the various methods for obtaining cross-section data by (a) field surveys and (b) office procedures.

16.22. Following are the notes for cross sections at stations 109 and 110. The width of the roadbed is 24 ft, and the side slopes are 2 to 1. Plot these two cross sections at a scale of 1:120 (1 in. per 10 ft). Compute the areas of the two sections.

Station	Cross section		
109	$c4.1$	$c3.0$	$c1.5$
	20.0	0.0	15.0
110	$c10.2$	$c7.2$	$c5.5$
	32.6	0.0	23.2

16.23. Plot the preliminary cross section for station $43 + 00$ in Figure 16.30 using a horizontal scale of 1:200 and vertical scale of 1:100. Assuming a roadbed width of 10 m, side slopes of 2 to 1, and a center-line grade elevation of 207.300 m, plot the roadbed templet and compute the cross-section area by the coordinate method.

16.24. Using the roadbed templet specifications given in Problem 16.23, compute the cross-section area in square meters for the data of station 44 + 00 in Figure 16.30.

16.25. Compute the volume, in cubic meters between stations 43 + 00 and 44 + 00, of Problems 16.23 and 16.24, by the average end area method.

16.26. The width of roadbed for a highway is 8 m with side slopes of 1 to 2; the grade elevation at station 48 + 00 is 340.4 m, and the center-line elevation is 339.29 m. The H. I. for obtaining the final cross section at this station is 341.940 m. The rod readings at the right and left slope stakes are 4.60 and 7.10 m, respectively. Compute the grade rod, the center-line cut or fill, and the cut or fill and distance to each slope stake.

16.27. Compute the volume in cubic yards between stations 109 and 110 of Problem 16.22. Use both the average end area method and the prismoidal formula. Note the discrepancy expressed as a percentage between volumes as determined by the two methods.

16.28. What error in volume between station 109 and station 110 of Problem 16.22 would be introduced if the recorded cuts at centers and slope stakes were 0.2 ft too great? What is the error in terms of percentage of the volume by average end areas?

16.29. Assume that the center line for the cross-sectional data in Problem 16.22 turns to the right with a simple circular curve having $D_a = 12°$ (arc definition). Compute the curvature correction in cubic yards to the earthwork volume between these two stations and the percentage of the total volume of earthwork this correction constitutes.

16.30. Compute the volume in cut and in fill for the given cross sections between stations 62 + 50 and 63 + 50. The roadbed is 24 ft wide in cut and 20 ft wide in fill, and the side slopes are 1.5 to 1. Tabulate the data in the following form: station, cross section, area, and volume. Use the prismoidal formula.

Station	Cross section			
62 + 50	$\dfrac{c2.6}{15.9}$	$\dfrac{c4.8}{0.0}$	$\dfrac{c6.4}{21.6}$	
63 + 00	$\dfrac{0.0}{12.0}$	$\dfrac{c3.1}{0.0}$	$\dfrac{c4.4}{18.6}$	
63 + 25	$\dfrac{f4.6}{16.9}$	$\dfrac{0.0}{0.0}$	$\dfrac{c2.6}{15.9}$	
63 + 50	$\dfrac{f7.2}{20.8}$	$\dfrac{f4.8}{0.0}$	$\dfrac{0.0}{6.0}$	$\dfrac{c1.8}{14.7}$

16.31. Solve Problem 16.30 by the method of average end areas, computing the volume of pyramids by the relation, volume $= \frac{1}{3}$ (area of base times length).

16.32. In plan, a borrow pit is 75 by 150 ft. Before and after excavation, levels are run and offsets are measured from stations along one of the 150-ft sides. The computed cuts are shown in the following table:

	Cut, ft						
Offset	Station 0	Station 0 + 25	Station 0 + 50	Station 0 + 75	Station 1 + 00	Station 1 + 25	Station 1 + 50
0	0.0	1.5	0.0	4.5	6.2	4.7	0.0
25	1.2	2.9	10.6	9.7	7.9	8.4	2.5
50	2.5	3.7	8.7	8.7	9.4	8.4	3.6
75	0.0	0.0	1.9	7.6	6.8	6.3	0.0

Compute the volume of excavation in cubic yards using the method outlined in Section 16.42.

FIGURE 16.58

949

CHAPTER 16:
Route Surveying

16.33. Compute the volume in cut and fill in cubic meters for the given cross sections (all distances and elevations in meters) between stations $10 + 00$ and $10 + 70$. The roadbed is 8 m wide in cut and 6 m wide in fill and the slopes are 1 to 1.5. Tabulate the results as directed in Problem 16.30. Use the method of average end areas.

Station	Cross section		
$10 + 00$	$\dfrac{c2.00}{7.00}$	$\dfrac{c1.50}{0.00}$	$\dfrac{c0.80}{5.20}$
$10 + 25$	$\dfrac{c1.60}{6.40}$	$\dfrac{c1.00}{0.00}$	$\dfrac{0.00}{4.00}$
$10 + 45$	$\dfrac{c0.85}{5.28}$	$\dfrac{0.00}{0.00}$	$\dfrac{f1.40}{5.10}$
$10 + 70$	$\dfrac{c1.0}{5.50}$	$\dfrac{0.00}{2.00}$ $\dfrac{f1.62}{0.00}$	$\dfrac{f2.54}{6.81}$

16.34. In Figure 16.58, the rectangular grid over borrow pit *ABCDE* has uniform grid spacing of 20 m and the cuts to the finished grade are positioned diagonally at the intersection points. Assuming vertical slopes for all sides of the excavation, compute the volume removed in cubic meters.

REFERENCES

American Association of State Highway and Transportation Officials (AASHTO). *Guide to Metric Conversion.* Washington, DC: AASHTO Task Force on Geometric Design, 1993a.

American Association of State Highway and Transportation Officials. *Interim Selected Metric Values for Geometric Design, an Addendum to a Policy on Geometric Design of Highways and Streets, 1990.* Washington, DC: AASHTO Task Force on Geometric Design, 1993b.

American Association of State Highway and Transportation Officials. *A Policy on Geometric Design of Highways and Streets.* Washington, DC: AAHSTO, 1994.

American Railway Engineering Association (AREA). *Manual for Railway Engineering,* Chapter 5, pp. 53–9. Washington, DC: AREA 1995.

CALTRANS. *User's Guide to Photogrammetric Products and Services.* Sacramento, CA: State of California Department of Transportation, Engineering Service Center, 1994a.

CALTRANS. *Getting into Metrics, a Metric Primer,* 2nd ed. Sacramento, CA: California Department of Transportation, February 1994b.

Chambers, D.W. "Estimating Pit-Excavation Volume Using Unequal Intervals." *ASCE Journal of Surveying Engineering* 114, no. 2 (May 1988), pp. 71–83.

Chen, C.-S., and H.-C. Lin. "Estimating Pit-Excavation Volume Using Cubic-Spline Volume Formula." *ASCE Journal of Surveying Engineering* 117, no. 2 (May 1991), pp. 51–66.

Davis, T. G. "Finite Element Volumes." *ASCE Journal of Surveying Engineering* 120, no. 3 (August 1994).

Easa, S. M. "Earthwork Allocations with Nonconstant Unit Costs." *ASCE Journal of Construction Engineering and Management* 113, no. 1 (March 1987), pp. 34–50.

Easa, S. M. "Improved Method for Locating Centroid of Earthwork." *ASCE Journal of Surveying Engineering* 114, no. 1 (February 1988a), pp. 13–25.

Easa, S. M. "Estimating Pit Excavation Volume Using Nonlinear Ground Profile." *ASCE Journal of Surveying Engineering* 114, no. 2 (May 1988b), pp. 71–83.

Foote, F. S. "Notes on the Mass Diagram as Applied to Earthwork Computations and Operations." *Information Circular*, no. 24, Institute for Transportation and Traffic Engineering, ITTE, University of California, Berkeley, CA, May 1955.

Hurdle, V. F. "Vertical Curves." Unpublished notes, University of California, Institute of Transportation and Traffic Engineering, Berkeley, 1974.

Kavanagh, B. F., and S. J. Bird. *Surveying: Principles and Applications,* 3rd ed. Englewood Cliffs, NJ: Prentice-Hall, 1992.

Mayer, R., and R. Stark. "Earthmoving Logistics." *ASCE Journal of the Construction Division* 107, no. CO2 (1981), pp. 297–312.

Meyer, C. F., and D. W. Gibson. *Route Surveying,* 5th ed. New York: Harper and Row, Publishers, 1980.

Moffitt, F. H., and H. Bouchard. *Surveying,* 9th ed. New York: HarperCollins, Publishers, 1992.

Pryor, W. T. "Metrication." *Surveying and Mapping* 35, no. 3 (September 1975), pp. 229–37.

West, T. W. "Metrication of Federal Highway Systems." ACSM Bulletin, no. 145 (September–October 1993), pp. 27–29.

Construction Surveying

17.1
GENERAL

Surveys for construction involve the following: (1) establishing on the ground a system of stakes or other markers, both in plan and in elevation, from which measurement of earthwork and structures can be taken conveniently by the construction force; (2) giving line and grade as needed either to replace stakes disturbed by construction or to reach additional points on the structure itself; and (3) making measurements necessary to verify the location of completed parts of the structure (the as-built survey) and determine the volume of work actually performed up to a given data (usually each month), as a basis of payment to the contractor.

Prior to construction, a topographic survey (Chapter 15) of the site is performed and maps are prepared to be used in the development of plans for the project. As soon as approval of the project is assured, property line surveys are initiated to be used for acquisition of lands or rights of way. The control network established for these topographic and property surveys contains many of the horizontal and vertical control points that eventually will form the basis for subsequent construction surveys. Consequently, the survey engineer in charge of planning the surveys has the responsibility for organizing the initial basic control surveys to provide the maximum number of horizontal and vertical control points, which are useful not only for topographic and property surveys but also for the construction surveys that follow. At this stage, design of the survey system to be used also should be directed toward integrating all survey data into the geographic information system (Section 14.18) of the overall project. This step is of special importance, not only for small jobs that must be related to the town, city, or county in which they exist but also for large projects, such as airports, bridges, dams, drainage systems, harbors, and the so-called megaprojects, which include combinations of these facilities on a citywide or regional scope.

Equipment, such as total station systems, GPS, and laser-equipped alignment instruments (Section 17.5), provide powerful tools with many applications in construction surveys. Although the application of these devices results in somewhat altered operations, the basic ideas concerning construction surveying as just outlined remain the same.

The detailed methods employed on construction surveys vary greatly with the type, location, and size of the structure and the preference of the engineering and construction organizations. Much depends on the ingenuity of the surveyor so that the correct information is given without confusion or needless effort. Consequently, the discussions that follow are of a general nature, with a minimum number of specific examples and procedures set forth.

17.2
ALIGNMENT

Temporary stakes or other markers usually are set at the corners of the proposed structure, as a rough guide for beginning the excavation. Outside the limits of excavation or probable disturbance but close enough to be convenient are set permanent stations, which are established with the precision required for the measurement of the structure itself. These permanent stations should be well referenced (Section 8.12), with the reference stakes in such number and in such position that the loss of one or two will not invalidate any portion of the location survey. Permanent targets or marks called *foresights* may be erected as convenient means of orienting the theodolite or other alignment instrument on the principal lines of the structure and for sighting along these lines by eye.

Stakes or other markers are set on all important lines to mark clearly the limits of the work. The number of such markers should be sufficient to avoid the necessity for many

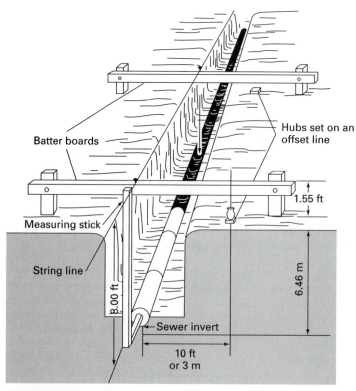

FIGURE 17.1
Batter boards for sewer construction.

measurements by the workers but should not be so great as to cause confusion. A simple and uniform system of designating the various points, satisfactory to the construction superintendent, should be adopted. Also, the exact points, lines, and planes from which and to which measurements are to be made should be well understood.

In many cases, line and grade are given more conveniently by means of batter boards than by stakes. A *batter board* is a board (usually 1 by 6 in. (2.5 by 15 cm)) nailed to two substantial posts (usually 2 by 4 in. (5 by 10 cm)) with the board horizontal and its top edge preferably either at grade or at some whole number of feet above or below grade. The alignment is fixed by a nail driven in the top edge of the board. A stout cord or wire is stretched between two such batter boards to define the line and grade. Batter boards for sewer construction are shown in Figure 17.1. Note that the string line provides the reference line for alignment and grade between the surveyed stations.

Laser-equipped instruments, such as those described in Section 17.5, provide a means of establishing a laser beam, which can be used in the same way as the string line shown in Figure 17.1.

Often, it is impractical to establish permanent markers on the line of the structure. For example, the face of a bridge abutment may be beyond the shoreline and therefore inaccessible. Also, stakes placed at the edge of a concrete pavement would interfere with grading and setting the forms. In such cases, the survey line is established parallel to the structure line, as close as practical and with the offset distance some whole number of meters, decimeters, or feet.

17.3
GRADE

A system of bench marks is established near the structure in convenient locations that probably will not be subject to disturbance, either of the bench mark itself or the supporting ground. From time to time these bench marks should be checked against one another to detect any disturbance. Every care should be taken to preserve existing bench marks of state and federal surveys; if construction necessitates the removal of such marks, the proper organization should be notified and the marks transferred in accordance with its instructions.

The various grades and elevations are defined on the ground by means of stakes or batter boards, as a guide to the workers. The grade stakes may or may not be the same as the stakes used in giving line. When stakes are used, the vertical measurements may be taken from the top of the stake, from a crayon mark or a nail on the side of the stake, or (for excavation) from the ground surface at the stake; to avoid mistakes, only one of these bases for measurement should be used for a given kind of work, and the basis should be made clear at the beginning of construction. When batter boards are used, the vertical measurements are taken from the top edge of the board, which is horizontal. The stake or the batter board may be set either at grade or at a fixed whole number of meters, decimeters, or feet above or below grade.

On some jobs, grade stakes consisting of hubs driven flush with the ground are set, elevations are obtained on the tops of the hubs, and cuts or fills to finished grade are calculated in whole and fractions of units from the top of the hub. These hubs also are referred to as grade stakes and usually are guarded by a flat stake. The construction engineer has the responsibility of setting batter boards or other alignment devices as desired using the given stakes, cuts, and fills.

Use of instruments equipped with lasers to establish grades is discussed in Section 17.5.

17.4
ALIGNMENT AND GRADE BY TOTAL STATION SYSTEMS AND GPS

When construction layout is performed using theodolite and tape, alignment may be established in the field by direct measurement from the highway center line, the exterior dimensions of a structure, or from the property line. This means that center line, side line, or property line must be established prior to setting offset stakes or batter boards. This procedure has definite advantages in that it provides a check on office calculations and furnishes proof that the structure as designed is consistent with basic control points in the field. Also, the process for setting offset stakes or batter boards from center line or building corners is straightforward, involves right-angle offsets or parallel lines, and provides ample visual checks for the validity of the layout. The disadvantages are that the procedure is time consuming, expensive, and during initial stages of construction (e.g., rough excavation) most of the points set will be lost and will have to be replaced for the final layout. Use of a total station system or GPS eliminates these disadvantages and facilitates construction layout.

Total Station Systems

Use of these systems (Sections 7.16 to 7.21), with direct readout of direction and distance and equipped with a tracking mode and data collectors in conjunction with coordinate geometry, make construction layout from randomly located control points feasible. To perform this operation, horizontal and vertical control points are positioned at locations appropriate for layout. These locations can be outside the area of construction but must provide good visibility for construction layout. All random control points must be coordinated and part of a closed survey network. It also is advantageous to have elevations on these control points. The accuracy maintained is a function of the type of work being done.

The points to be staked also must be coordinated. If stakes on right-angle offsets from the corner of a building or offsets from the catch point in slope staking (Section 17.7) are to be set, the offset point is coordinated. Next, the inverse solution between the random control point and coordinated offset point is computed to yield distance and direction (azimuth or bearing). Most surveying organizations have standard computer programs and all data collectors contain menu-driven routines for this computation, given the coordinates for both ends of the line. The minimum amount of data provided for field operations consists of azimuth or bearing and horizontal distance from the random control point to the offset point. When offset stakes are to be set for alignment and grade (as for slope stakes in highway excavation), azimuth and zenith or vertical angle, slope distance, horizontal distance, and elevation of the offset point are given. Data usually are precalculated and furnished to field personnel as a file to be downloaded to a data collector or in the form of a hard-copy listing.

In the field, the operation consists of occupying the designated random control point, taking a backsight on an intervisible random control station, turning off the direction, and measuring the distance to set the specified offset stake. Ideally, a series of offset stakes are set from a single well-located random control point. To verify calculations and fieldwork, the locations of the stakes as set are checked by measuring distance and direction from another random control point or by measurements between offset stakes on the site.

Obviously, the locations for random control stations must be chosen carefully to achieve the maximum benefits from this method. In preparation for layout from random stations, enough inverse data must be generated from several control stations to provide for checking and to guard against lack of visibility along certain lines. In case insufficient inverses are generated to complete the job, data collectors provide enough computing power to allow calculation of the necessary inverses on the site.

Construction Layout by GPS

Kinematic and static GPS are GPS techniques adaptable to construction surveys. In this section, emphasis will be on kinematic GPS with static initialization, the details of which will be discussed later.

As noted in Sections 12.9–12.12, GPS surveys by real-time kinematic methods require one or more receivers at a fixed-base station(s) and one or more roving receivers (rovers). In addition, a communications link is needed between the base receiver and the rover so that data can be transmitted from the fixed receiver to the rover where they are combined with data being received at that station to provide a real-time position. The communications links are designated *spread spectrum* and UHF or VHF. Radios operating on the spread spectrum are of lower power and have a shorter range but require no license. UHF and VHF radios are more powerful and have a longer range. However, one must have a license from the Federal Communication Commission (FCC) to operate UHF and VHF units.

The base station may consist of (1) a local GPS receiver at a control point of known position or (2) a regionally located, continuously operating reference station (CORS), which occupies a point of known position (Section 12.13). The user generally subscribes for this latter service and the central station should be within 10 to 40 km of the area where work is to be performed. Under option (1), at least two GPS receivers (one at the base station and the other used as a rover) and a communications link are required. When the user subscribes to a service from a central station, at least one GPS receiver is needed, to be used as a rover. A roving system on a construction site is shown in Figure 17.2, which illustrates the antenna mounted on a range pole, the controller unit held by the operator, and the receiver and radio modem in the backpack.

As described in Section 12.9, ambiguity resolution or ambiguity initialization is needed in kinematic GPS to determine the integer number of double-differenced wavelengths being received at the GPS receiver. In static initialization, after communications have been established between the base unit and the rover, the roving GPS receiver is moved to the area where it is desired to set points, and the antenna mounted on a range pole (Figure 17.2) is held motionless for up to 2 min. This allows time for initialization, which is indicated on the system control unit.

Assume that a file of the coordinates for points to be set has been stored in the controller memory. The operator, on retrieving a particular point number and its coordinates, follows visual directions (azimuth and distance or offsets) displayed on the controller screen and proceeds to the precise location of the first point, which is set by holding the pole with antenna at the directed location for a few seconds. A typical display on a controller screen is shown in Figure 17.3, where point F212 is being set. The current position of the prism pole is at the cursor, below and to the left of the correct position at the target. According to the display, the operator needs to move 0.195 m right and 0.212 m ahead. These values were determined in real time by comparing rover coordinates with the prestored coordinates for F212. As the operator moves in the designated directions, these values will reduce to 0.000 and the cursor will be centered on the target. At this time the point will be located correctly and the cut or fill to finished grade will be indicated on the screen.

After the point has been set and marked, the operator proceeds to the next point, and so forth until all points in this area have been set or the signal is lost. When signals are lost, the controller indicates that by a beep and an instruction displayed on the controller screen. Then the antenna again must be held motionless for up to 2 min while the software is

GPS receiver and radio modem

Antenna

Controller

Range pole

FIGURE 17.2
GPS system (rover) on a construction site. (*Courtesy of Leica, Inc.*)

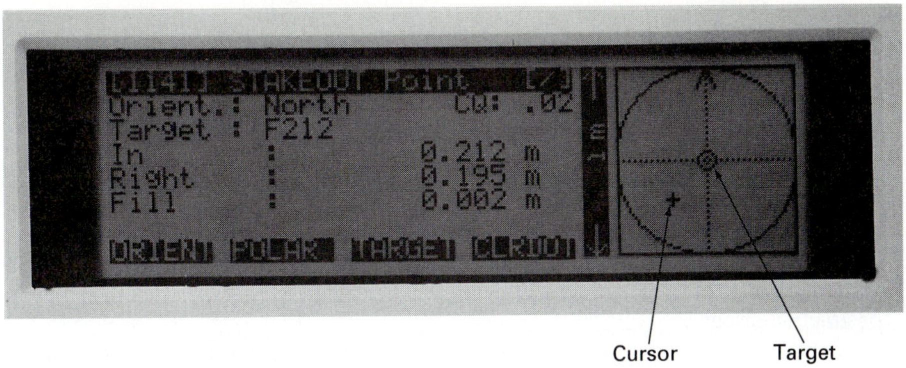

Cursor   Target

FIGURE 17.3
Display on screen of controller for rover unit. (*Courtesy Leica, Inc.*)

initialized, after which setting the points proceeds as before. If difficulties are experienced while trying to regain initialization, one solution is to move a short distance to another area, reestablish initialization and then proceed with the survey. Under favorable conditions (e.g., continuous reception from five satellites by both base and rover units, low noise, and no multipath interference), multiple points can be set from one initialization. When a section of points has been staked, it is good practice to move the base to another control point of known position, initialize the rover again, and repeat setting the same points, thus providing an independent check.

When the kinematic, on-the-fly (OTF) technique is used, the procedure is the same except no 2-min initialization period is required. Initializing and regaining the initialization can be achieved while the operator carrying the system is moving. The capability for doing this type of initialization is built into the optional software package, which increases the price of the system substantially.

At least three points of known orthometric elevation must be observed so that heights are referred to the datum employed for the construction project. In most systems now available for construction layout, the geoid correction model is not used and heights are modeled on an inclined secant plane at the location of the survey. This method is adequate for small projects. For large projects, the data should be downloaded and processed by a geoid correction modeling program* before final cuts or fills are established.

The most serious problem now affecting construction layout by kinematic GPS surveys is the potential for multipath errors (Section 12.6, Site-Dependent Errors, Obstructions) caused by signals being reflected or refracted from constructed and natural obstructions in the area. Preplanning for the project can help alleviate but not eliminate this problem. Alternate stakeout procedures should be recommended for areas containing known sources of multipath errors, such as buildings, walls, power transmission poles, and deep cuts in the soil.

The choice as to whether construction layout should be performed in the traditional manner, from random control points using a total station system, or by GPS depends on many factors related to characteristics of the job and site. These factors generally are unique to each project. In all probability, a combination of all methods will be used on most projects. For example, in highway construction, basic horizontal control may be established by static GPS survey, slope stakes set from random control points using a total station system, and the final grade stakes established from the center line, which is set from random control points or by kinematic GPS surveys. Examples of setting slope stakes from center line and from random control points are given in Section 17.7.

---

* The National Geodetic Survey (NGS) sells a PC program for this purpose. Order from NOAA, NGS, 1315 East-West Highway, Station 9202, Silver Spring, MD 20910-3282, telephone (301)713-3242.

17.5
LASER ALIGNMENT AND LEVELING EQUIPMENT FOR CONSTRUCTION LAYOUT

As noted in Section 4.29, laser beams are used in certain EDM instruments as the carrier beam for modulated light in distance measurement, increasing the range of these instruments substantially. Laser light is of a single color or monochromatic, is coherent (i.e., the light waves are in step with each other), and is highly collimated, so that the beam spreads only very slightly as the distance from generator to target increases. These characteristics make the laser a useful device for surveying instruments to be utilized in various types of construction layout.

Development of *diode laser* technology has greatly simplified the incorporation of lasers into surveying instruments. Using this technology, the laser generator can be built directly into an eyepiece attachment. Figure 17.4 shows a diode laser eyepiece. The diode laser can be quickly adapted to all total station instruments, theodolites, levels, and optical plummets equipped with detachable eyepieces. The laser is controlled by an intensity regulator (Figure 17.5) fastened to a tripod leg. A total station system instrument equipped with a diode laser eyepiece, intensity regulator, and battery pack is shown in Figure 17.5.

Within the telescope, the laser beam is focused on a reticle that masks out a cross. The laser beam passes through a beam-splitting cube that directs the laser along the axis of the telescope. The surface of the beam-splitting cube, visible through the eyepiece, contains a second reticle designed to permit the telescope to be used as usual under good lighting conditions. A filter is installed to absorb any laser light directed toward the eyepiece. Directed toward the telescope objective is a narrow beam containing the projected cross, which may be brought to focus on a target by focusing the telescope on the target.

As already mentioned, diode laser eyepieces also can be attached to levels and the optical plummet. A laser plummet that projects a vertical laser beam up and down is illustrated in Figure 17.6.

The system just described and shown in Figure 17.5 is a *single-beam* laser that projects a visible "string line," which can be seen on targets under all lighting conditions. If a sight

Eyepiece

Diode laser

FIGURE 17.4
Diode laser eyepiece, Leica DL2. (*Courtesy of Leica, Inc.*)

is taken on a distant point in the conventional manner, alignment can be achieved at any intermediate point by moving a special target about until the project cross is centered on the target. This alignment can be established at any intermediate point without instructions from an operator behind the instrument.

The laser beam can be projected at any inclination or may be used as a level beam of light. For leveling, the graduated rod would have a movable target equipped with a vernier that is moved up or down until the beam or projected cross is centered on the target. The operator of the rod then would read the rod to the least count of the scale. The range of the laser attachment shown in Figure 17.5 is from 5 to 150 m, in daylight, or 400 m, in darkness (5 to 500 ft, in daylight, or 1300 ft, in darkness) with a maximum range of approximately 300 m (1000 ft) under moderate dust conditions, as might be common in mining and tunneling operations. Using an output beam of 0.6 mm (0.002 ft) at 5 m (15 ft), the estimated standard deviation in alignment is about 6 to 10 seconds of arc or a displacement of from 6 to 10 mm (0.019 to 0.033 ft) in 200 m ($\approx$ 650 ft). This accuracy can be achieved most easily by utilizing a laser beam with a dark spot or cross at its center for more precise centering on the target. A stable instrument setup is extremely important, and whenever possible permanent stands with built-in centering plates or trivets should be used.

FIGURE 17.5
Total station instrument (Wild TC2002) with diode laser attachment. (*Courtesy of Leica, Inc.*)

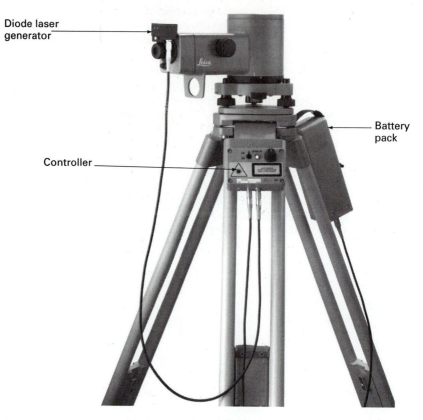

Diode laser
generator

Battery
pack

Controller

FIGURE 17.6
Optical plummet (Leica ZL), equipped with diode laser attachment. (*Courtesy of Leica, Inc.*)

The system described is quite elementary but illustrates the principles involved in application of laser technology to existing equipment.

Experiments have been performed using prototypes of laser-equipped theodolites and levels for high-precision alignment and leveling. The reader should consult Chrzanowski, Jarzymowski, and Kaspar (1976) and Chrzanowski and Janssen (1972) for details.

An impressive array of special-purpose laser-equipped instruments is available to the surveying community. Most of these instruments have been designed for construction layout and can be classified into two general groups:

1. Instruments that project single beams visible on targets under all lighting conditions. These devices usually are employed for horizontal and vertical linear alignment. Included in this group are systems or theodolites equipped with lasers (Figure 17.5), lasers for alignment of tunneling machinery, and lasers for alignment of pipes and drains.
2. Instruments in which the laser beam is rotated by rapidly spinning optics to provide a reference plane in space over open areas and trace reference lines that are visible indoors. The speed of rotation can be varied and at zero revolutions per minute a single beam is available.

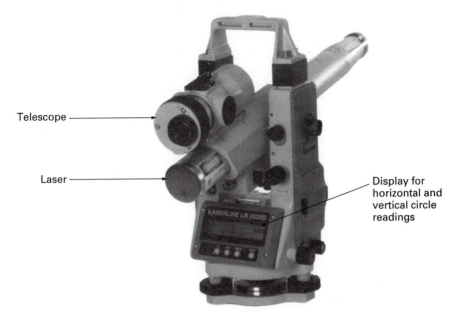

Telescope

Laser

Display for
horizontal and
vertical circle
readings

FIGURE 17.7
Laser alignment instrument, Laserline LR2020D. (*Courtesy of Laserline Mfg., Inc.*)

A single-beam laser alignment instrument is shown in Figure 17.7. This laser is mounted on the framework of an electronic theodolite which has horizontal and vertical motions, a vertical circle, and a digitized horizontal circle. Horizontal and vertical circles have a digital display with a least count of 20 sec. Mounted on top of the laser enclosure is a telescope that allows the operator to sight the location of the transmitted laser spot. The spot diameter of the beam is 0.016 m (0.053 ft), at the laser, and due to spread becomes 0.062 m at 1 km (0.33 ft at 1 mi) and 0.50 m at the maximum range of 8 km (1.67 ft at 5 mi). This range is effective in mild haze and increases somewhat at night.

A fanning lens can be attached to the emission end of the laser housing. This attachment allows the beam spot to be converted into a horizontal or vertical line beam shape (Figure 17.8). The dimensions of this beam are 0.062 m by 161 m at a distance of 1 km (0.33 ft by 850 ft in 1 mi), increasing to 0.50 m by 1288 m at 8 km (1.67 ft by 4250 ft in 5 mi). This instrument can be used for horizontal and vertical control of dredging operations, control of tunneling machines, and azimuth and grade control for pipe-laying operations.

A self-leveling, rotating laser level and sensor are shown in Figure 17.9a and 17.9b. This instrument is equipped with a laser diode to generate the laser beam, has a circular level bubble (bull's eye) for approximate leveling, and a liquid compensator that then takes over to level the system automatically. The range of the compensator is ± 10 minutes of arc. The battery-powered sensor, which can be handheld or attached to a level rod, contains a liquid crystal display that gives a graphical indication of whether the sensor is high, on grade, or low (see Figure 17.9c), also distinct audible tones match the visual signal, providing *audiovisual detection*. The tolerance specified for the position of the reference plane with respect to true level, for the instrument shown in Figure 17.9, is 15 seconds of arc. Therefore, in a distance of 100 m (330 ft) from the instrument, a deviation of 7.3 mm (0.024 ft) is possible. The operating range of this system is approximately 200 m (660 ft).

A rod equipped with a laser sensor (Figure 17.10) is used in conjunction with the rotating laser. A sliding, battery-powered sensor is supported on one side of the rod. This

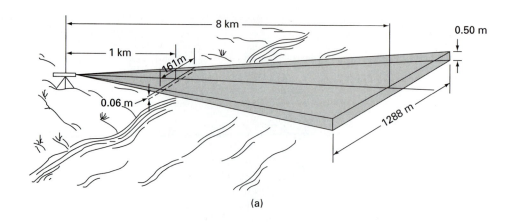

(a)

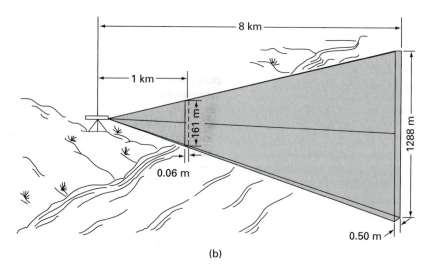

(b)

FIGURE 17.8
Horizontal and vertical line beams produced by fanning a laser beam: (a) horizontal,
(b) vertical. (*Courtesy of Spectra Physics, Laserplane, Inc.*)

sensor has a detection range of 0.050 m (0.164 ft) and a detection precision of 0.5 mm to
1.3 mm (0.003 to 0.005 ft) with audiovisual indication that the sensor is high, on grade, or
low, mentioned previously and illustrated in Figure 17.9c. When within 0.050 m (0.164 ft)
of the beam, the sensor emits a "beep," and the operator of the rod then makes a precise
adjustment until the visual display indicates "on grade" (Fig. 17.9c). The operator reads the
rod at an index on the sensor that allows reading the scale on the front of the rod. If the rod
is held on a bench mark of known elevation, this rod reading establishes the height of
instrument (H.I.) and provides a reference plane from which subsequent stakeout operations can be performed.

Another system utilizing the rotating laser is the combination rotating laser and digital
theodolite shown in Figure 17.11. This system consists of an electronic theodolite (Section 6.30) with a rotating laser supported on the top of the telescope housing in such a
manner that the projected laser plane is parallel to the optical line of sight of the telescope.

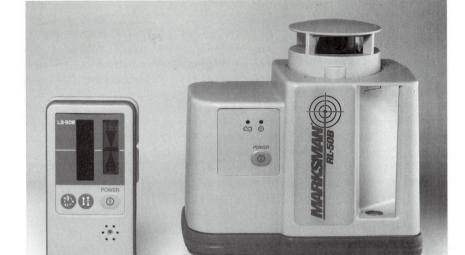

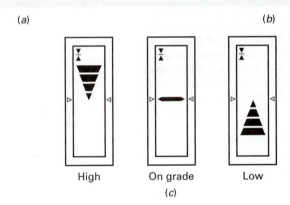

High   On grade   Low

(c)

FIGURE 17.9
Rotating laser level and sensor: (a) sensor, LS-50B; (b) laser level, RL-50B (Both *courtesy of Topcon Laser Systems, Inc.*); (c) typical sensor display.

Thus, the laser plane can be oriented according to a given azimuth (least count $10''$) and inclination (least count $20''$). This system is especially adaptable to setting slopes and inclined grade lines, where the detector is mounted on grading machinery (see the next subsection).

Grade and Alignment by Laser-Equipped Instruments

The basic idea in the application of laser devices to construction layout is that the laser plane replaces the horizontal line of sight of the level and the laser beam replaces the string line. This plane or beam can be located in space at any intermediate point by using a small plastic

FIGURE 17.10
Level rod equipped with laser sensor. (*Courtesy of Spectra Physics, Laserplane, Inc.*)

target or an electronic laser sensor that automatically seeks the laser plane or beam when it is within a specified range. Consequently, no operator need be stationed at the instrument when it is desired to get on a line or obtain a rod reading.

When using a rotating laser beam or laser beam to form a horizontal reference plane or line, the operation of setting grade stakes to a given elevation is the same as described

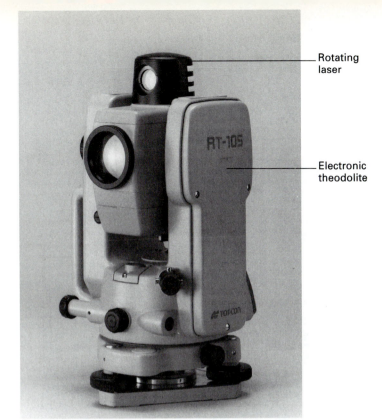

Rotating laser

Electronic theodolite

FIGURE 17.11
Combination electronic theodolite and rotating laser.
(*Courtesy of Topcon Laser Systems, Inc.*)

in Section 17.3 except that there is no need for instructions from the operator of the instrument. The horizontal laser plane is detected by a sliding sensor on the face of the rod, which provides an audiovisual indication (Figure 17.9c) when the sensor index mark coincides with the laser plane. The operator of the rod must know what rod reading is required to set the grade stake, and then can set the stake or form accordingly without assistance from the other end of the line at the instrument. When the laser beam is used as a substitute for the string line, the laser transmitting device is set over a control point at one end of the line and sighted at a control point at the other end of the line. The laser beam serves the same function as the string line described in Section 17.2.

As an example of the layout process, a laser level in an operating environment is shown in Figure 17.12, where the elevation of the top of a form is being checked by bringing the sensor onto the reference plane formed by the rotating beam. In this operation, the laser level is set at a central location, leveled, and a backsight is taken on a level rod equipped with a sensor held on a bench mark of known elevation. The rod reading, thus determined, added to the elevation of the bench mark establishes the H.I. of the instrument or the elevation of the plane formed by the rotating laser beam. The grade elevation of the top of the form subtracted from the H.I. of the level is then set off on the staff to the index mark of the sensor attached to the staff. Afterward, all forms within the range of the laser level are set to this grade by raising or lowering the staff until the sensor is "on grade" by audiovisual detection. Note that this entire process essentially is a *one-person* operation. Another possibility involves attaching the laser sensor to the machine that is grading, paving, tunneling, and so on. The operator of the machine can observe the sensor and, by keeping the sensor "on grade," can maintain the machine on the proper alignment and grade.

FIGURE 17.12
Laser level in a working environment. (*Courtesy of Spectra Physics, Laserplane, Inc.*)

A simple example consists of the grading operations for an area to be graded horizontally at a specified elevation. Assume that a laser level is set up near the area to be graded, leveled, and a reading is taken on a nearby bench mark. This reading establishes the height of the laser level above the datum (H.I.) so that the cut to the given elevation desired for the area can be calculated. Assume that this cut is 1.84 m (6.05 ft), so that the blade of the grader must be adjusted to 1.84 m below the laser plane. Special graduated devices that support laser sensors and can be attached to various types of machinery are available for this purpose. Figure 17.13 shows a laser level mounted on a tripod and a power grader and bulldozer with laser sensors mounted on their respective blades. When the laser sensor is within about 0.050 m (0.15 ft) of the beam, it will automatically energize display lights in the cab that indicate low, on grade, or high positioning of the blade. A pointer display also shows how much adjustment is needed to get the detector on the beam. When the on-grade light is illuminated, the cutting edge of the blade is on grade. The same principles can be used to guide a ditch-digging backhoe (Figure 17.13) and tunneling and paving machinery.

Another application consists of laser control of graded slopes using the combination electronic theodolite and rotating laser shown in Figure 17.11. In Figure 17.14, the combination theodolite and rotating laser beam is set up on a control point of known position at the bottom of the slope on the right. The height of instrument is measured (h.i.) to establish the instrumental H.I., and a backsight is taken on a station of known horizontal position to obtain azimuth. Then, the correct horizontal angle is turned from the backsight to establish the direction normal to the slope, and the desired inclination of the slope (in degrees or percent) is set off on the vertical circle. When the rotating laser is energized, the resulting laser plane will be at the inclination of the desired slope at a known height above the finished grade. Attached to the blade of the bulldozer is an electronic laser sensor, visible to the operator, with audiovisual indicators for high, on-grade, or low operation. By observing this sensor, the operator can adjust the blade and the amount of cut or fill until the slope is at the correct inclination and at grade elevation.

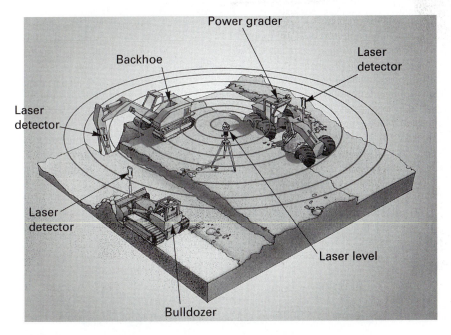

FIGURE 17.13
Various types of grading machinery controlled by a laser level and laser detectors.
(*Courtesy of Spectra Physics, Laserplane, Inc.*)

Sensor

Combined
theodolite
and laser
level

FIGURE 17.14
Combination theodolite and rotating laser beam controlling the grading of a slope.
(*Courtesy of Topcon Laser Systems, Inc.*)

There is a precaution (in addition to the usual steps to ensure safety) which must be observed when using any of the laser alignment devices. Theoretically, an advantage of these types of systems is that, once the laser transmitter is set up and aligned, no operator needs to stand by the instrument. However, laser transmitters are subject to the same disturbances that cause conventional optical alignment telescopes to become misaligned. Settlement, vibration, and accidental bumping are common factors around any construction job that can cause a perfectly aligned laser beam to be deflected. Consequently, periodic checks of the laser-beam alignment are absolutely necessary. Such checks can consist of having a party member check the alignment device at regular intervals or the checking device may be built into the system. Most laser alignment systems are equipped with a warning signal or shut off when the laser beam deflects from horizontal. This constitutes a built-in check on the system.

Laser alignment devices and systems have many applications, particularly in construction layout. However, it should be emphasized that many of these devices are not general-purpose surveying instruments having the flexibility in application of an engineer's theodolite or total station system (with the exception of the lasers attached to a total station system and optical plummet as shown in Figures 17.5 and 17.6). Consequently, laser systems are more likely to be useful to contractors with a large volume of construction layout operations.

Various applications of laser alignment devices are described in the sections that follow.

17.6
PRECISION

For purposes of excavation only, usually elevations are given to the nearest 3 cm (0.1 ft). For points on the structure, usually elevations to 3 mm (0.01 ft) are sufficiently precise. Alignment to the nearest 3 mm (0.01 ft) will serve the purpose of most construction, but greater precision may be required for prefabricated steel structures or members.

It is still desirable to give dimensions to the workers in feet, and decimal parts of feet, inches, and fractions of an inch, since the metric system is still not officially authorized in the United States. Ordinarily, measurements to the nearest 0.02 to 0.01 ft or $\frac{1}{4}$ to $\frac{1}{8}$ in. (6 or 3 mm) are sufficiently precise, but certain of the measurements for the construction of buildings and bridges should be given to the nearest 0.005 ft or $\frac{1}{16}$ in. (1.5 mm). Often, it is convenient to use the relation that $\frac{1}{8}$ in. equals approximately 0.01 ft ($\sim$3 mm).

17.7
HIGHWAYS

As a general rule, numerous horizontal and vertical control points are available along a highway right of way prior to construction. These control points are set during the planning and design stages for the highway and usually provide a good basic network of control points for construction surveys.

The two basic approaches to highway construction surveys consist of layout from (1) center-line control and (2) random control stations. The method chosen depends on terrain configuration, amount of vegetation, equipment available, construction requirements, and other factors unique to the particular project.

Center-Line Control

Just prior to beginning the construction of a section of highway, the center line is rerun as a base line for construction, missing stakes are replaced, and hubs are referenced. Borrow pits (if necessary) are staked out and cross-sectioned. Lines and grades are staked out for bridges, culverts, and other structures. Slope stakes are set, except where clearing is necessary; in that case, they are set when the right of way has been cleared. For clearing, only rough measurements from the center-line stakes are necessary.

Slope stakes. The process of setting slope stakes in the field, which constitutes an important part of highway construction layout, requires some additional explanation beyond the brief description given in connection with final cross sections in Section 16.30. To explain the concept, first consider the procedure when setting slope stakes with the traditional equipment of engineer's level, level rod, and tape.

If w is the width of roadbed, d is the measured distance from center line to slope stake, s is the side-slope ratio (ratio of drop or rise to horizontal distance, in the metric system, ratio of horizontal distance to drop or rise when units are in feet), and c is the cut (or f the fill) at the slope stake, then from Figure 17.15, when the slope stake is at the correct position (at 3) and units are in meters,

$$d = \frac{w}{2} + \frac{c}{s} \tag{17.1}$$

and

$$d = \frac{w}{2} + \frac{f}{s} \tag{17.2}$$

When units of measurement are feet, the second terms in Equations (17.1) and (17.2) become (cs) and (fs), respectively.

The following example for a cross section in fill illustrates the steps involved in establishing the correct location for a slope stake in the field.

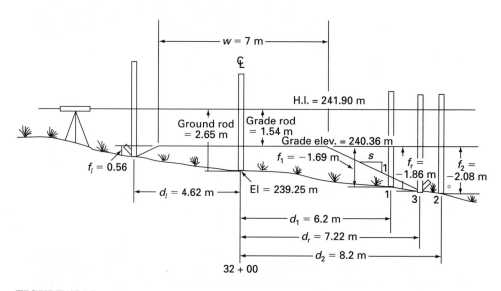

FIGURE 17.15
Slope stakes for a cross section in fill.

EXAMPLE 17.1. Let $w = 7$ m, side slopes $1:2$, grade elevation at $32 + 00 =$ 240.36 m, center line elevation $= 239.25$ m.

A backsight on a B.M. gives the H.I. of 241.90 m. The rod reading on center line is 2.65 m, which verifies the given center-line elevation. The grade rod = H.I. − grade elevation $= 241.90 − 240.36 = 1.54$ m. To set the right slope stake the rod is held, as a first trial, at point 1, where the ground rod = 3.23 m. Therefore, $f_1 =$ grade rod − ground rod $= 1.54 − 3.23 = −1.69$ m, where the negative sign indicates a fill. The computed distance for the absolute value of f_1 by Equation (17.2) is $3.5 + (1.69)(2) = 6.88$ m. The measured value for d_1 is 6.2 m, so that the rod must be moved farther from the center line.

A second trial is made at point 2, where the ground rod is 3.62, $f_2 = 1.54 − 3.62 = −2.08$ m, and the computed distance to f_2 is $3.5 + (2.08)(2) = 7.66$ m. Because the measured distance, d_2, is 8.2 m, the rod is out too far.

Eventually, by trial, the rod is held at 3, where the ground rod = 3.40, $f_r = 1.54 − 3.40 = −1.86$ m, and the computed distance is $3.5 + (1.86)(2) = 7.22$ m. This distance agrees with the measured distance d_r; hence, point 3 is the correct location for the slope stake. The notes for this final cross section are

H.I.
241.90

$$32 + 00 \qquad \frac{f0.56}{4.62} \qquad \frac{f1.11}{0.00} \qquad \frac{f1.86}{7.22}$$

where the left slope stake also is set by trial and error and the symbol f designates a fill.

For cross sections in a cut, a similar procedure is followed, except Equation (17.1) is used to obtain the computed distances.

Slope stakes also can be set in the field by slope distance and zenith or vertical angle and using a total station system. This procedure is particularly appropriate where the cuts are deep and fills or embankment are high, so that setting slope stakes from a single setup of a level is not possible.

In Figure 17.16 the instrument is set over the center-line station with a height above the ground equal to the h.i. A sight is taken on the rod at A (the correct position for the slope stake) such that the rod reading AB equals the h.i. and the vertical angle, α, or zenith angle, z, is observed. The difference in elevation between the center-line station at D and the ground at the slope stake at A is $BC = V$, where

$$V = l \sin \alpha = l \cos z \qquad (17.3)$$

and the horizontal distance from D to A is

$$d = l \cos \alpha = l \sin z \qquad (17.4)$$

If c is the cut to grade elevation at the center line, taken from the profile and grade plans, the difference in elevation between the center line and the ground at the slope stake is $(c + V)$ for an uphill sight and $(c − V)$ for a downhill sight. Therefore, the calculated distance from center line to slope stake is

$$d(\text{calculated}) = \frac{w}{2} + s(c \pm V) \qquad (17.5)$$

When the value $d(\text{calculated})$ from Equation (17.5) equals d obtained by Equation (17.4) using field measurements the rod is in the correct position to set the slope stake. If d is greater or less than $d(\text{calculated})$, the rod must be moved toward or away from the center line and another set of measurements taken. The correct position for the slope stake is found by trial.

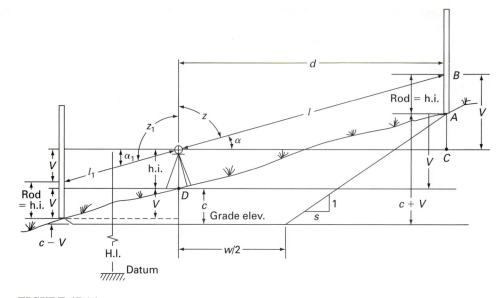

FIGURE 17.16
Setting slope stakes by slope distances and vertical or zenith angles.

For cross sections in fill, the calculated distance is

$$d(\text{calculated}) = \frac{w}{2} + s(f \pm V) \tag{17.6}$$

where f is the fill at the center line (taken as a positive value) and the height of instrument (H.I.) above the datum is assumed to be below the center-line grade elevation. The value of V is negative when the sight is to the uphill side and V is positive when the sight is to the downhill side of the center line. Equation (17.6) also can be used if the H.I. is above the grade elevation at the center line, but it is usually more convenient to set slope stakes with a horizontal line of sight and distance measurement as described in Example 17.1.

Slope stakes are set side to the line, sloping outward in fill and inward in cut. On the back of the stake the station number is marked. On the front (the side nearest the center line) are marked the cut or fill at the stake and the distance from center line to slope stake. The numbers read down the stake.

Some organizations do not set stakes at the top or bottom of the slope (frequently referred to as the *catch point*) but set a reference point at a fixed distance back from the catch point. This distance is from 3 to 7 m (10 to 20 ft) from the catch point. Figure 16.41 shows this method for stations 30 + 00 and 30 + 40. In this way, the stakes are not as likely to be destroyed during roadway construction. Some organizations set stakes at both the catch points and reference points.

The procedures just discussed illustrate the steps involved in setting slope stakes from center-line control using traditional equipment and the total station system. In current practice, the majority of final cross sections are determined numerically from the parameters of the final center-line alignment, using either photogrammetric methods or a DTM as described in Section 16.29. Therefore, all final cross-section data (center-line stations, slope-stake offsets, catch points, cuts or fills, and the like) are computed in a rectangular coordinate system (preferably state plane coordinates) and stored in an electronic file or provided as a hard-copy listing.

Slope stake left		Catch point left	Center line ditch	Center line	Center-line ditch	Catch point right		Slope stake right
	_R.P._ _R.P._			C15.52			_R.P._ _R.P._	
C30.19	El. 251.1 −2.0	249.07	218.88	220.48	218.88	228.95	+1.5 El. 227.4	C10.07
@ 90.57								@ 30.23
s = 0.333	₵ 140.6 15.0	125.57	35.00	0.00	35.00	65.23	15.0 ₵ 80.2	s = 0.333
				46 + 00				
				C23.34				
C36.16		254.22	218.06	219.66	218.06	235.52		C17.46
@ 106.47								@ 52.37
s = 0.333		143.47	35.00	0.00	35.0	87.37		s = 0.333
				45 + 50				

(a)

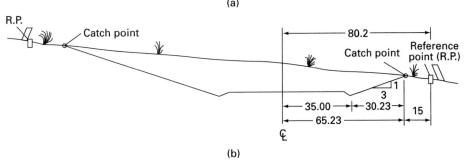

(b)

FIGURE 17.17
(a) Listing of slope-stake notes derived from computer output for field layout. (b) Plotted cross section for station 46 + 00.

A typical set of slope-stake notes derived from a computer listing for a field layout is shown in Figure 17.17a. The data in the computer listing could have been determined from field notes, topographic map, by photogrammetric methods, or from a DTM. In this particular example, the data were derived photogrammetrically. These data also could be stored as an electronic file, stored in the memory, and then displayed on the screen of a data collector for use in setting the slope stakes in the field. Figure 17.17b shows the plotted cross section for station 46 + 00. At this station the grade at center line is 220.48 and the cut at center line is 15.52 ft. The center line of the ditch 35 ft right has a grade elevation of 218.88 ft and the catch point 65.23 ft right has an elevation of 228.95 ft. The slope-stake data are given in the column farthest to the right, where the $C = (30.23)(0.333) = 10.07$ ft is for a point 30.23 ft to the right of the center line of the ditch (see Figure 17.17b). This catch point is calculated by the computer program as the intersection of a 3 to 1 slope (0.333) and the terrain surface from the photogrammetric cross section. Similar data are given for the left side of the cross section. The hand-lettered entries in Figure 17.17a are inserted on the listing or in the data collector during field operations.

In the field, the center-line station is occupied, the direction of the cross-section line is established, and an offset stake or reference point (R.P.) is set 80.23 ft right of the center line or 15 ft from the catch point. In this way, the reference point is outside the area of construction and will not be destroyed. The elevation of the reference point is determined trigonometrically as 227.4 ft. Therefore, the catch point is 1.5 ft higher and 15 ft toward the center line from the R.P. (see Figure 17.17b). These data concerning offset distance and elevation of the R.P. and distance and difference in elevation from R.P. are entered on the listing by hand in the field, as shown in Figure 17.17a, or entered and stored in the appropriate file of the data collector. The difference in elevation and offset distance between catch point and R.P. (+1.50/15) also are marked on the center-line side of the face of a stake set to guard the R.P. hub. Next, the position of the catch point is set, and the elevation is checked against that given on the computer listing. If the elevation of the catch point as determined in the field does not agree with the theoretical location given in the listing, the true position of the catch point is determined by trial according to the methods given previously in this section, and the reference point and slope-stake data are adjusted accordingly. Any changes should be shown directly on the listing or entered into the data collector.

Where the depth of cuts and fills does not average more than about 1 m or 3 ft, the slope stakes may be omitted; in this case, the line and grade for earthwork may be indicated by a line of hubs (with guard stakes) along one side of the road and offset a uniform distance such that they will not be disturbed by the grading operations. Hubs usually are placed on both sides of the road at curves and may be so placed on tangents; when this is done, measurements for grading may be taken conveniently by sighting across the two hubs or by stretching a line or tape between them.

When rough grading has been completed, a second set of grade stakes is set on both sides of the roadway. It should be understood whether the cuts and fills from the tops of these stakes are to subgrade or to finished grade (final grade). If the slopes of cuts are terraced to provide drainage, finishing stakes are set along the terraces.

To give line and grade for the pavement of concrete highways, a line of stakes is set along each side, offset a uniform distance (usually 1 m or 2 ft) from the edge of the pavement. The grade of the top of the pavement, at the edge, is indicated either by the top of the stake or by a nail or line on the side of the stake. The alignment is indicated on one side of the roadway only, by means of a tack in the top of each stake. Laser levels (Section 17.5) can be used to advantage when setting grade for pavement or curbs. The distance between stakes in a given line is usually 30 or 10 m (100 or 50 ft) on tangents at uniform grade and half the normal distance on horizontal or vertical curves. The dimensions of the finished subgrade and the finished pavement are checked by the construction inspector, usually by means of a templet.

Random Control Stations

Construction layout from random control stations is described in Section 17.4. Three procedures are possible: (1) staking directly from random control stations, (2) establishing a random line along or near the work area tied into existing random control points, and (3) establishing the center line or a line offset from the center line tied to existing random control points.

For each of these procedures, the order of establishing offset stakes for grading is the same as the center-line method described in the preceding subsection. However, the method of establishing the positions of the stakes differs.

As an example, consider staking directly from random control points. The coordinates for all control points, reference points, and catch points are calculated. Inverses between conveniently located random control points and the reference points to the slope stakes are calculated using a computer program. A listing from the computer giving inverse data from random control point CM 131 to reference points to slope stakes (RPSS) for stations 130 + 00 through 133 + 00 is shown in Figure 17.18. Given in this listing are the offset distance of the catch point (or slope stake) left or right of the center line, azimuth and zenith angle from the control point to the reference point, slope and horizontal distance from the control point to the reference point in meters and feet, and elevation and coordinates of the reference point. These data also may be provided in the form of an electronic file that can

13 OG 01 01 PAGE 2

TRANSIT AT	PT							ELEVATION	NORTH COORD	EAST COORD
CM 131	02							1,228.944	550,190.755	1,999,254.589
DESCRIPTION (POINT SIGHTED)	PT	AZIMUTH (NORTH)	ZENITH ANGLE	SLOPE DISTANCE (METERS)	SLOPE DISTANCE (FEET)	HORIZONTAL DISTANCE (METERS)	HORIZONTAL DISTANCE (FEET)	ELEVATION	NORTH COORD	EAST COORD
CM 124	01	198 54 19	90 40 51	288.265	945.751	288.245	945.684	1,217.708	549,296.085	1,998,948.184
CM 134	03	333 15 37	92 04 50	221.642	727.169	221.496	726.690	1,202.545	550,839.733	1,998,927.624
130+00 RPSS 140 LT	04	237 22 34	92 01 30	223.793	734.226	223.653	733.767	1,203.000	549,795.165	1,998,636.590
130+00 RPSS 373 RT	05	213 56 58	126 25 28	95.196	312.321	76.598	251.306	1,043.500	549,982.289	1,999,114.244
130+50 RPSS 145 LT	06	240 39 01	91 59 17	223.551	733.433	223.416	732.992	1,203.500	549,831.487	1,998,615.681
130+50 RPSS 368 RT	07	230 05 53	127 31 15	86.250	282.971	68.407	224.433	1,056.600	550,046.787	1,999,082.417
131+00 RPSS 152 LT	08	243 52 27	92 09 55	225.427	739.589	225.266	739.061	1,201.000	549,865.313	1,998,591.039
131+00 RPSS 370 RT	09	248 41 01	127 45 20	84.100	275.919	66.492	218.149	1,060.000	550,111.454	1,999,051.364
131+50 RPSS 154 LT	10	247 00 46	92 35 38	227.253	745.577	227.020	744.813	1,195.200	549,899.885	1,998,568.920
131+50 RPSS 372 RT	11	266 08 25	125 28 43	88.985	291.944	72.463	237.739	1,059.500	550,174.752	1,999,017.389
132+00 RPSS 133 LT	12	250 30 10	91 23 13	223.446	733.088	223.380	732.873	1,211.200	549,946.149	1,998,563.741
132+00 RPSS 359 RT	13	277 17 02	118 11 00	100.124	328.490	88.253	289.544	1,073.800	550,227.465	1,998,967.381
132+50 RPSS 107 LT	14	254 21 59	89 45 27	220.150	722.274	220.148	722.267	1,232.000	549,996.116	1,998,559.042
132+50 RPSS 365 RT	15	286 42 46	115 37 57	114.667	376.201	103.382	339.178	1,066.200	550,288.294	1,998,929.738
133+00 RPSS 82 LT	16	258 24 48	88 08 56	219.440	719.944	219.325	719.568	1,252.200	550,046.229	1,998,549.684
133+00 RPSS 353 RT	17	290 49 05	111 19 15	132.909	436.052	123.813	406.208	1,070.400	550,335.122	1,998,874.901

ZENITH ANGLES ARE CALCULATED TO A ROD READING = (HI + 0.000 FEET) TRANSIT AT CM 131 POINT 02 PAGE 2

FIGURE 17.18
Computer listing for direction and distances from a random control station to reference points to slope stakes. (*Courtesy of California Department of Transportation.*)

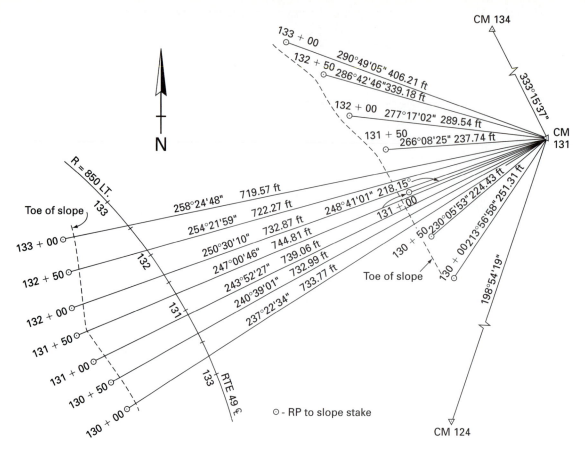

FIGURE 17.19
Slope staking from random control points. (*Courtesy of California Department of Transportation.*)

be downloaded into a data collector for use in the field. Figure 17.19 shows a plan view of the reference-point layout from control station CM 131 to reference points for slope stakes from stations 131 through 133.

Consider, as an example, setting the reference point for station 130 + 00 (140 ft left) using a total station system. The total station instrument is set on station CM 131 and a backsight is taken on station CM 124. An angle of 237°22'34" − 198°54'19" = 38°28'15" (Figure 17.19) is turned to the right, and a zenith angle of 92°01'30" is set on the vertical circle. The reflector is placed on the line at the approximate location of the reference point and with the EDM in the tracking mode, the reflector is moved until the EDM registers the slope distance of 734.226 ft (223.793 m), and a hub is set. The measurements then are repeated, including a backsight on CM 124, and the distance and elevation are determined to the hub just set. The field party also should have the slope-stake listing or file (Figure 17.20) so that the elevation of the catch point can be checked. The catch point is set and checked from the reference point using a tape, hand level, and rod. Note that data for the reference-point location is entered on the computer listing (Figure 17.20), or entered into the memory of the data collector for station 130 + 00 on the left as R.P. El. = 1203.0@140 ℄. The catch point is set by measuring 15 ft toward the center line and found

976

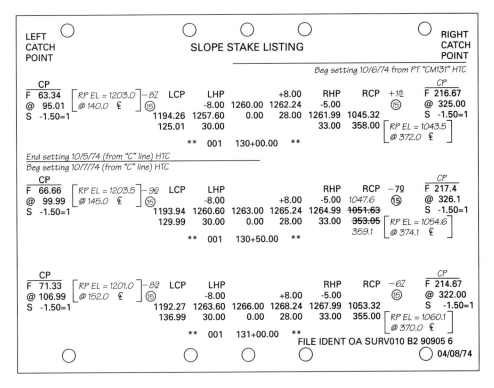

FIGURE 17.20

Slope-stake listing for field layout. (*Courtesy of California Department of Transportation.*)

to be −8.7 ft lower than the reference point. The field elevation of the catch point is 1203.0 − 8.7 = 1194.3, which checks the precomputed elevation for the catch point also given in the listing. As a rule for earthwork operations, this discrepancy should not exceed 5 ft or about 2 m. Reference points to slope stakes can be checked from a second instrument setup, at random control point CM 134 (Figure 17.19), for example. However, if the elevations of reference and catch points as determined in the field check the theoretical values in the data collector or on computer listings (Figure 17.20), these checks are assumed sufficient and a second setup is unnecessary.

In Figure 17.20, at station 130 + 50 the field measurements do not check the elevations and offsets in the listing (see the voided entries on the right). Consequently, the correct position of the catch point is determined in the field and adjusted values are entered in the listing as shown or entered into the data collector.

When a random line is established near the working area, it may be a line connecting two random control points or a series of supplementary points established between existing random control points. These supplementary points are used as angle points on the random line. The random line should be located as close as possible to the points to be set. The relationship between the center line and the random line must be determined and all reference points to slope stakes set from the stations on the random line. Figure 17.21 illustrates an example of a random line established to set reference points to slope stakes for a portion of curved center line for stations 129 through 134. As in the case of staking directly from random points, this procedure is simplified by computer calculations to determine the distances and directions from the random line to the reference line. Slope-

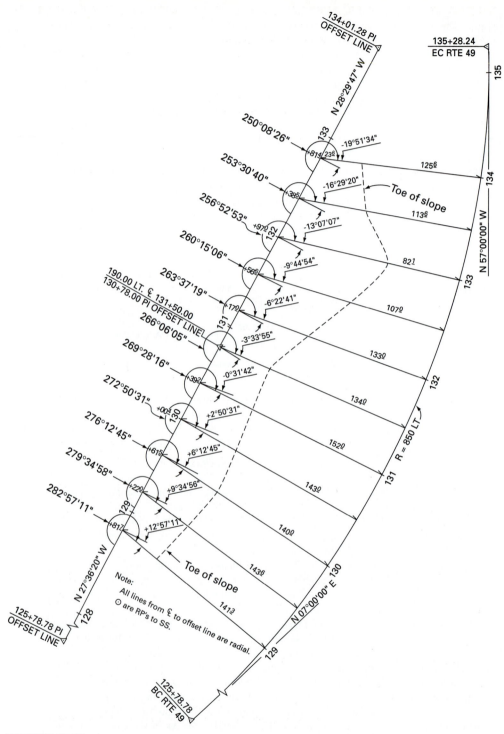

FIGURE 17.21
Slope staking from a random offset line. (*Courtesy of California Department of Transportation.*)

stake and offset data stored in a data collector are used in the field to assist in setting the points. The method is adaptable to the use of a total station system and in general involves measuring shorter distances than when staking directly from the control stations.

When excavation is heavy (cuts greater than 10 m or 30 ft), intermediate slope stakes may be required after grading has been partially completed. These stakes can be set from random control points but are more conveniently established from a random offset line or line of constant offset (offset line parallel to the center line).

The center line of the alignment is established from ties to random control points when rough excavation is completed. The balance of the layout work then becomes a combination of staking from random control points and staking from the center line. Prior to setting final grade stakes, bench-mark levels are run throughout the graded alignment and bench marks are set to provide vertical control for all subsequent operations. These bench marks should be established so as to satisfy third-order specifications (Table 5.2, Section 5.46) and should have a minimum density of two per kilometer (three per mile).

Final grade stakes for subgrade excavation and paving usually are set from the center line. Laser leveling and alignment instruments are useful at this stage of the project. However, grade stakes in the vicinity of an interchange may be set more efficiently from random control points.

As construction proceeds, monthly tallies are made of the work completed to date. A quantity survey is made near the close of each month, and the volumes of earthwork and the like are classified and summarized as a basis for payment.

17.8
STREETS

The procedure of surveying for street construction is similar to that just described for highways. Ordinarily, the curb is built first. The line and grade for the top of each curb are indicated by hubs driven just outside the curb line, usually at 15-m (50-ft) intervals.

On alignments containing horizontal and vertical curves, grade stakes should be set at closer intervals, commensurate with the curvature of the center line and grade line. Differential levels are run, obtaining an elevation on the top of each hub. Cuts or fills are calculated from the top of each hub to finished grade at the top of the curb. A laser level (Figure 17.12) can be used as described in Section 17.5 for this operation. These cuts and fills are marked on guard stakes set next to each hub and entered into the data collector or tabulated in the field notebook. Frequently, a separate summary of cuts and fills is prepared for the construction superintendent. A copy of this summary should be filed by the surveyor.

When the curb is completed, levels should be run over the finished curb to check the accuracy of the construction. The grade for the edge of the pavement then is marked on the face of the completed curb; for a combined curb and gutter, it is indicated by the completed gutter. Hubs are set on the center line of the pavement, either at the grade of the finished subgrade (in which case holes are dug when necessary to place hubs below the ground surface) or with the cut or fill indicated on the hub or on an adjacent stake. Where the street is wide, an intermediate row of hubs may be set between center line and each curb. Usually, the hubs must be reset after the street is graded. Where driving stakes is impractical because of hard or paved ground, nails or spikes may be driven or marks may be cut or painted on the surface.

The surveys for street location and construction should determine the location of all surface and underground utilities that may affect the project; and notification of necessary changes should be given well in advance.

On paving projects where the rate of grade is constant, the possibility of using a laser level, sensor, and rod such as illustrated in Figures 17.9 and 17.10 should be investigated.

The use of laser levels for setting grade stakes is discussed in Section 17.5. A bench mark must be within the work area, and care must be exercised to avoid sights of excessive lengths. The laser level plane is accurate to within ±7 mm or 0.024 ft in a sight of 100 m (330 ft).

The laser alignment equipment described in Section 17.5 can be used to guide graders in establishing subgrade elevations and paving machines utilized for placing concrete to finished grade elevations.

17.9
RAILROADS

Construction surveys for railroads are similar to those described in Section 17.7 for highways. Prior to construction, the located center line is rerun, missing stakes are replaced, control station hubs are referenced, borrow pits are staked out, slope stakes are set, and lines and grades for structures are established on the ground. When rough grading is completed, final grade stakes are set to grade at the outer edges of the roadbed, as a guide in trimming the slopes.

The foregoing operations can also be performed from random control points or using a random offset line as described in Sections 17.4 and 17.7.

When the roadbed has been graded, alignment is established precisely by setting tacked stakes along the center line at full stations on tangents and usually at fractional stations on horizontal and vertical curves. Spiral curves are staked out at this time. An additional line of hubs is set on one side of the track and perhaps 1 m (3 ft) from the proposed line of the rail, with the top of the hub usually at the elevation of the top of the rail. Track usually is laid on the subgrade and is lifted into position after the ballast has been placed and compacted around the ties.

17.10
SEWERS AND PIPELINES

The center line for a proposed sewer is located on the ground with stakes or other marks set usually at 15-m or 50-ft intervals where the grade is uniform and as close as 3 m or 10 ft on vertical curves. At one side of this line, just far enough from it to prevent being disturbed by the excavation, a parallel line of hubs is set, with the hubs at the same intervals as those on the center line. A guard stake is driven beside each hub, with the side to the line; on the side of the guard stake farthest from the line is marked the station number and offset, and on the side nearest the line is marked the cut (to the nearest 3 mm or 0.01 ft). In paved streets or hard roads, where it is impossible to drive stakes and pegs, the line and grade are marked with spikes (driven flush), chisel marks, or paint marks.

When the trench has been excavated, batter boards are set across the trench at the intervals used for stationing. The top of the board is set at a fixed whole number of decimeters or feet above the sewer invert (inside surface of bottom of sewer pipe), and a measuring stick of the same length is prepared. A nail is driven in the top edge of each batter board to define the line. As the sewer is being laid, a cord is stretched tightly between these nails, and the free end of each section of pipe is set at the proper distance below the cord as determined by measuring with the stick (see Figure 17.1).

For pipelines, the procedure is similar to that for sewers, but the interval between grade hubs or batter boards may be greater, and less care need be taken to lay the pipe at the exact grade.

For both sewers and pipelines, the extent of excavation in earth and rock is measured in the trench, and the volumes of each class of excavation are computed as a basis of payment to the contractor.

The records of the survey should include the location of underground utilities crossed by, or adjacent to, the trench.

Laser alignment instruments can be used instead of the batter boards and string line described previously. The simplest approach consists of utilizing a laser-equipped theodolite or total station system (Figure 17.5) to establish a laser beam parallel with the flow line of the sewer. This alignment beam would be used with a rod having an attached sensor (Figure 17.10) first for rough grading to get the ditch excavated and then for placing the pipe on line and at grade in the ditch.

Also, several special-purpose laser alignment devices are designed specifically for sewer construction. One such device, which has horizontal and vertical laser beams, is shown in Figure 17.22. The vertical beam is used to plumb the instrument in the ditch or sewer invert (flow line of sewer) and the percent grade of the horizontal beam is adjusted by manipulating the appropriate keys on the control panel, which also contains a register on which the gradient is displayed. A remote control is available that permits turning the power on and off and adjusting the gradient. Figure 17.23 shows a similar instrument placed in the invert of a manhole. These instruments are self-leveling and transmit a laser beam that can be adjusted from a -10 to a $+40$ percent grade. Instruments such as the laser level (Figure 17.9) and this one are not useful for general surveying purposes but would be appropriate for a general contractor engaged in substantial amounts of excavation and sewer construction.

Note that the laser alignment instruments serve only to establish a line between previously surveyed control points and bench marks. Therefore, the sewer alignment still needs to be established initially as in the traditional approach.

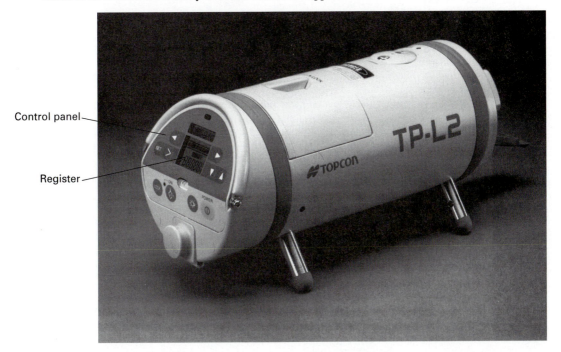

FIGURE 17.22
Pipe laser with horizontal and vertical beams. (*Courtesy of Topcon Laser Systems, Inc.*)

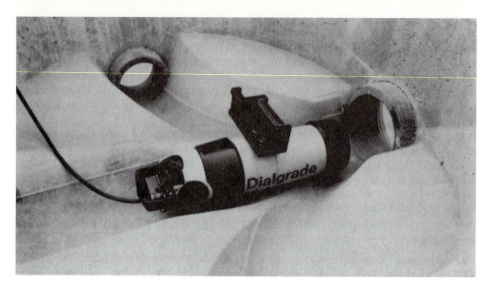

FIGURE 17.23
Laser device for sewer alignment. (*Courtesy of Spectra Physics, Laserplane, Inc.*)

17.11
TUNNELS

Tunnel surveys are run to determine by field measurements and computations the length, direction, and slope of a line connecting given points and to lay off this line by appropriate field measurement. The methods employed naturally vary somewhat with the purpose of the tunnel and the magnitude of the work. A coordinate system is particularly appropriate for tunnel work.

For a short tunnel, such as a highway tunnel through a ridge, a traverse and a line of levels are run between the terminal points, and the length, direction, and grade of the connecting line are computed. Where practical, the surface traverse between the terminals takes the form of a straight line. Outside the tunnel, on the center line at both ends, permanent monuments are established. Additional points are established in convenient surface locations on the center line to fix the direction of the tunnel on each side of the ridge. As construction proceeds, the line at either end is given by setting up at the permanent monument outside the portal, taking a sight at the fixed point on line, and then setting points along the tunnel, usually in the roof. Grade is given by direct levels taken to points in either the roof or the floor, and distances are measured from the permanent monuments to stations along the tunnel. If the survey line is on the floor of the tunnel, it usually is offset from the center line to a location relatively free from traffic and disturbance; from this line a rough temporary line is given as needed by the construction force.

The dimensions of the tunnel usually are checked by some form of templet transverse to the line of the tunnel but may be checked by direct measurement with the tape or by using a specially designed laser scanner that transmits a visible laser beam with range of 0.5 to 30 m, yielding an accuracy of 5 to 10 mm (see Bickel and Kuesel, 1982). Photogrammetric methods also may be used to check the cross-sectional dimensions of the tunnel (see Katibah, 1968).

Railroad and aqueduct tunnels in mountainous country often are several miles in length and not uniform in either slope or direction. Tunnels of this character usually are driven not only from the ends but also from several intermediate points where shafts are sunk to

intersect the center line of the tunnel. The surface surveys for the control of the tunnel work usually consist of a precise GPS static survey (Section 12.8) or total station system traverse tied to monuments at the portals of the main tunnel and at the entrances of shafts and adits, and a precise system of differential levels connecting the same points. With these data as a basis, the length, direction, and slope of each of the several sections of the tunnel are calculated; and construction is controlled by establishing these lines and grades as the work progresses.

Alignment of tunneling machinery was an earlier application of laser-equipped instruments. The procedure for establishing control and the center line of the tunnel is the same as described previously in this section except that a laser beam is projected ahead to control the direction of the tunneling machine (see "Laser Control," 1971, and Peterson and Frobenius, 1973). As the center line of the tunnel is produced, a laser alignment instrument (Figure 17.7, Section 17.5) is suspended from a support at a center-line station on the roof of the tunnel. The line of sight is produced ahead by double centering from a backsight. The laser transmitter is turned on to create a laser beam parallel to the center line of the tunnel. The tunneling machine has two targets placed on the top of its framework. The first target is glass with a center cross, and the second is an opaque target with a center mark. The machine is adjusted until the laser beam passes through the center of the first target and strikes the center of the second target. When these two conditions are met, the tunneling machine is on line and work can progress. The operator of the machine can detect and correct for any deviation from this alignment. The laser beam also is used to control setting of forms for the tunnel lining. Figure 17.24 illustrates in schematic form the arrangement for control of the tunneling operation.

The range of the laser beam is only about 300 m(1000 ft), owing to the dust created by tunneling operations. Therefore, periodically the line is produced ahead by double centering and a target is centered on a new station attached to the tunnel roof and immediately behind the tunnel machine. The target and alignment instrument are interchanged, the instrument is centered by forced centering, a backsight is taken on the target, and the line again is produced by double centering. As the tunneling progresses, the backsight should be checked periodically to ensure that the laser beam does not deviate from the correct position.

When the alignment of the tunnel is curved, the machine must be driven ahead along chords on the curve. To get directed along a chord from the point of curvature, the targets on top of the machine are shifted laterally be a precomputed amount to produce the desired amount of deflection of the cutting heads in the direction of the chord. When the machine

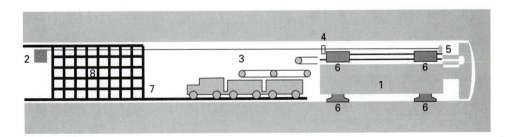

FIGURE 17.24
Laser alignment to control a tunneling machine: (1) tunneling machine, (2) laser theodolite station, (3) laser beam, (4) glass target, (5) second target, (6) hydraulic bracing, (7) tunnel invert (lining), (8) concrete forms. (*Courtesy Leica, Inc.*)

has moved ahead enough to provide clearance, the chord point is set on the roof of the tunnel, the theodolite is moved ahead, a backsight is taken on the previous station, and the deflection angle to the next chord point is turned. This procedure establishes the direction of the tunnel until the cutting heads of the machine advance to the next chord point, where the entire procedure is repeated. The machine is driven forward from chord point to chord point until the end of the curve is reached. Obviously, to maintain correct direction of the tunnel, very careful attention must be given to centering the alignment instrument and producing the laser beam by double centering. Further details concerning use of lasers for tunnel alignment can be found in Peterson and Frobenius (1973) and Robinson et al. (1995).

Electronic distance measuring equipment (EDM) also is very useful in establishing the distances to theodolite station, points of curvature, and chord points as the tunnel is driven.

To establish absolute directions (azimuth), a gyroscopic theodolite can be used. This instrument, designed to determine direction, is described in Section 10.43. Periodic determination of the azimuth of the tunnel center line provides a check on the direction established by producing the line by double centering.

17.12
BRIDGES

Normally, the location survey will provide sufficient information for use in the design of culverts and small bridges; but for long bridges and grade-separation structures, a special topographic survey of the site frequently is necessary. This survey should be made as early as possible to allow time for design and—in the case of grade crossings or navigable streams—permit securing approval from the appropriate government agencies. The site map should show all the data of the location survey, including the line and grade of the roadway and the marking and referencing of all survey stations.

A scale of 1:1200 (1 in./100 ft) with a 2-m (5-ft) contour interval usually is suitable for large bridges in rough terrain. Where the terrain is level and in urban regions, a scale of 1:600 with a contour interval of 1 m (2 ft) is desirable.

Where the bridge crosses rivers or other bodies of water, a hydrographic survey (see Dept. of the Army, 1994) that will provide a continuous profile of the bottom of the river or bay at 3- to 8-m (10- to 25-ft) intervals is required.

When planning the horizontal and vertical control for design mapping, due consideration should be given to establishing control points suitably located for the construction survey. Long crossings over water require elaborate control networks of first-order accuracy (Section 9.23), which must be planned well in advance of the actual construction. For a short bridge with no offshore piers, first the center line of the roadway is established, the stationing of some governing line such as the abutment face is established on the located line, and the angle of intersection of the face of the abutment with the located line is turned off. This governing cross line may be established by two well-referenced control points at each end of the cross line offset beyond the limits of excavation or, if the face of the abutment is in the stream, by a line parallel to the face of the abutment offset on the shore. Similarly, governing lines for each of the wing walls are established on shore beyond the limits of excavation, with two stations on the line prolonged at one, or preferably both, end(s) of the wing-wall line. If the faces are battered, usually one line is established for the bottom of the batter and another for the top. Stakes are set as a guide to the excavation and replaced as necessary. When the foundation concrete has been cast, line is given on the footings for setting the forms and then by sighting with the theodolite or total station system for the top of the forms. As the structure is built, grades are carried up by leveling, with marks on the forms or on the hardened portions of the concrete. Also, the alignment is established on completed portions of the structure. The data are recorded in data collectors and in field books kept especially for the structure, principally by means of sketches.

For long sights or work of high precision, as in the case of offshore piers, various control points are established on shore or on specially built towers by static GPS surveys (Section 12.8) supplemented by combined trilateration-triangulation (Section 9.15), such that favorable intersection angles and checks will be obtained for all parts of the work. Survey-quality GPS systems, theodolites or total station systems with a least count of one second, and carefully calibrated EDM units should be employed in establishing this control network. For ordinary river crossings, a single quadrilateral or two adjacent quadrilaterals with the common side coincident with the center line of the bridge are appropriate figures and provide well-located stations for construction layout. For river and bay crossings, vertical control is established by simultaneous, reciprocal leveling (Section 5.54). For long crossings, special illuminated targets are used on the rod and all measurements are made at night.

To establish the location of offshore structures, at least two coordinated control points that have favorable geometric locations with respect to the point being set should be occupied by total station systems. Thus, the points are located by redundant distances and directions providing a better estimate of the position and a double check on calculations and fieldwork. The survey usually is accomplished in stages, with the pier locations being established with lower precision than those of the fittings subsequently established on top of the pier.

Laser-equipped instruments can be used for establishing the alignment and vertical positions for cofferdams and pile-driving equipment used in the construction of foundations for piers. The laser alignment device shown in Figure 17.7 used in conjunction with EDM equipment provides an excellent system for this purpose.

As soon as the pier foundation is placed, survey control is transferred to the foundation to verify the location and provide control for construction of the rest of the pier. Elevations also are transferred to the pier at this time. When the pier is completed, the precise locations of anchor bolts and other fastenings to which the bridge structure is connected are set out on the top of the pier. All center-line points, anchor-bolt locations, and control monuments are coordinated in a system referred to a common datum. Thus, it is possible to locate positions of anchor bolts and the like by distance and direction from at least two control monuments that have good geometric positions with respect to the points being located. Note that this procedure is an application of location by random control points (Sections 17.4 and 17.7).

When anchor bolts have been placed, their positions should be verified by distance and direction from control monuments. In addition, distances between anchor bolts and other pier connections on adjacent piers should be measured using EDM equipment.

On completion of the structure, permanent survey points are established and referenced for use in future surveys to determine the direction and extent of any movement of the piers and bridge structure.

17.13
CULVERTS

At the intersection of the center line of the culvert with the located survey line of the highway or railway, the angle of intersection is turned off, and a survey line defining the direction of the culvert is projected for a short distance beyond its ends and is well referenced. At (or offset from) each end of the culvert, a line defining the face is turned off and referenced. If excavation is necessary for the channel to and from the culvert, it is staked out in a manner similar to that for a roadway cut. Bench marks are established nearby, and hubs are set for convenient leveling to the culvert. Line and grade are given as required for the particular type of structure.

17.14
BUILDINGS

At the beginning of excavation, the corners of the building are marked by stakes, which of course will be lost as excavation proceeds. Sighting lines are established and referenced on each outside building line and line of columns, preferably on the center line of wall or column. A batter board is set at each end of each outside building line, about 1 m or 3 ft outside the excavation. If the ground permits, the tops of all boards are set at the same elevation; in any event, the boards at opposite ends of a given line (or portion thereof) are set at the same elevation so that the cord stretched between them will be level. The elevations are chosen at some whole number of meters or feet above the bottom of the excavation, usually that for the floor rather than that for the footings, and established by holding the level rod on the side posts for the batter board and marking the grade on the post. When the board has been nailed on the posts, a nail is driven in the tip edge of the board on the building line, which is given by the theolodite. Carpenter's lines stretched between opposite batter boards define both the line and the grade, and measurements can be made conveniently by the workers for excavation, setting forms, and aligning masonry and framing. If the distance between batter boards is great enough that the sag of the carpenter's line is appreciable, the grades must be taken as approximate only. Laser alignment and leveling devices provide an excellent substitute for the string line (Section 17.5) and are adaptable for establishing line and grade. Use of these devices is convenient particularly when obstructions prohibit the erection and use of batter boards.

When excavation is completed, grades for column and wall footings are given by hubs driven in the ground to the elevation of either the top of the footing or top of the floor. Lines for footings are given by batter boards set in the bottom of the excavation. Column bases and wall plates are set to grade directly by a level and level rod. The position of each column or wall is marked in advance on the footing; and when a concrete form, a steel member, or a first course of masonry has been placed on the footing, its alignment and grade are checked directly.

In setting the form for a concrete wall, the bottom is aligned and fixed in place before the top is aligned.

As before, laser alignment and leveling devices also can be utilized at this stage to establish line and grade.

Similarly, at each floor level the governing lines and grade are set and checked, except that, for prefabricated steel framing, the structure as a whole is plumbed by means of a total station instrument, theodolite, or plumbed laser beam at every second- or third-story level. All data are stored in a data collector supplemented by notes kept in a field book used especially for the purpose, principally by means of sketches.

Whenever the elevation of a floor is given, it should be understood clearly whether the value refers to the bottom of the base course, the top of the base course, or the top of the finished floor.

Throughout the construction of large buildings, selected key points are checked by means of total station systems, plumb lines, optical plummet, plumbed laser beam, plumbed laser plane, theodolite, or level to detect settlement, excessive deflection of forms or members, or mistakes. Bench marks are checked to detect any disturbance.

17.15
DAMS

987

CHAPTER 17:
Construction
Surveying

Prior to the design of a dam, a topographic survey is made to determine the feasibility of the project, the approximate size of the reservoir, and the optimum location and height of the dam. To provide information for the design, a topographic survey of the site similar in many respects to that for a bridge (Section 17.12), is made, usually by photogrammetric methods supplemented by field surveys where necessary. Extensive soundings and borings are made, and topography is taken in detail sufficient to define not only the dam itself but also the appurtenant structures, necessary construction plant, roads, and perhaps a branch railroad. A property-line survey is made of the area to be covered by, or directly affected by, the proposed reservoir. As noted previously (Section 17.1), all survey data should be integrated into a GIS to allow the efficient retrieval, storage, and coordination of all surveying and mapping information throughout the duration of the project.

Prior to construction, horizontal control stations, sighting points, and bench marks are permanently established and referenced upstream and downstream from the dam at advantageous locations and elevations for sighting on the various parts of the structure as work proceeds. These basic control and reference points usually are established by static GPS surveys, combined trilateration-triangulation (Section 9.15), and precise traverse with total station systems. All points are referred to a system of rectangular coordinates. In the United States, the state plane coordinate systems should be used.

Precise levels (Sections 5.47 to 5.52) should be run to establish basic vertical control. For major dam projects, first-order specifications should be satisfied in establishing both horizontal and vertical control (Sections 9.23 and 5.46, Table 5.2). A dam is a vast project, in which construction may last for a period of several years. Comprehensive planning of the survey control is essential from the beginning of the project. Careful attention must be given to setting horizontal and vertical control points in locations suitable for subsequent construction layout.

To establish the horizontal location of a point on the dam, as for setting concrete forms or checking the alignment of the dam, the method of staking from random control points as described in Section 17.4 can be used to advantage. Coordinated points on the dam structure are set from at least two coordinated horizontal control points by direction and distance using total station systems (Sections 7.15 to 7.20).

Vertical control usually is established by direct leveling. When warranted, trigonometric leveling (Section 5.5 and 5.55) and GPS systems can be used to establish elevation. Note that sufficient orthometric elevations must be available to permit conversion of GPS elevations on the ellipsoid to elevations on the geoid (Section 12.13).

Wherever possible, laser alignment instruments should be utilized. For example, lasers can be used in controlling tunneling operations in the underground excavation associated with the construction of the dam and also for alignment and setting grades of forms from nearby control points run in from the basic control network.

A traverse is run around the reservoir above the proposed shoreline and monuments are set for use in connection with property-line surveys and for future reference. Similarly, bench marks are established at points above the shoreline. The shoreline may be marked out with stakes set at intervals along the specified contour line. The area to be cleared is defined with reference to these stakes. The area and volume of the reservoir may be computed as described in Section 16.45.

When the dam is completed, permanent horizontal control monuments and bench marks are placed on the dam structure in appropriate locations for subsequent studies related to deformation of the dam. A network of control points suitable for geodetic GPS surveys or combined trilateration-triangulation (Sections 12.8 and 9.15) provides optimum control for monitoring possible movement of the dam. Observations on this control network should be made with geodetic GPS receivers (Section 12.5) or total station systems having horizontal and vertical angular least counts of at least 1 sec of arc (Section 7.16). Elevations should be determined by geodetic leveling (Section 5.52). First-order specifications should govern all horizontal and vertical control surveys (Section 9.23 and Table 5.2, Section 5.47).

17.16
AS-BUILT SURVEYS

The *as-built survey* is performed on completion of a construction project for the purpose of reestablishing the principal horizontal and vertical controlling points and to locate the positions of all structures and improvements. As-built surveys should be performed after the construction of highways, railroads, airports, bridges, buildings (including private homes), dams, and underground facilities.

Essentially, the as-built survey consists of rerunning center lines or property lines, setting permanent monuments at controlling points, and locating all improvements relative to these lines. These data are plotted on a map to provide a permanent record or inventory of the work done and its precise location.

Static and kinematic GPS surveys, total station systems, and photogrammetric surveys can be used to advantage in the establishment of controlling points and the acquisition of data that define the as-built facilities. These systems permit acquisition of data in data collectors and on film or as digitized images for transfer to a GIS that provides the ideal tool for subsequent reduction, processing, retrieval, review, and automatic plotting of the final map, which can be a digital map or the standard line map on paper.

Vehicle-mounted systems that combine GPS, inertial surveying systems (INS), and photogrammetric sensors (solid-state CCD cameras) into one integrated system are now under development (Li, 1994). This type of system, which yields real-time positions and an electronic file of locations of all objects visible as the vehicle travels along a route, has potential for as-built surveys, especially along highways and railroads. Final output would consist of a digital or line map.

REFERENCES

Barry, R. A. *Construction Measurements,* 2nd ed. New York: John Wiley & Sons, 1988.

Bickel, J. O., and T. R. Kuesel. *Tunnel Engineering Handbook.* New York: Van Nostrand Reinhold Company, 1982.

Chrzanowski, A., and H. D. Janssen. "Use of Laser in Precision Leveling." *The Canadian Surveyor* 26, no. 4 (September 1972).

Chrzanowski, A.; A. Jarzymowski; and M. Kaspar. "A Comparison of Precision Alignment Methods." *The Canadian Surveyor* 30, no. 2 (June 1976).

Cooney, A. "Laser Alignment Techniques in Tunneling." *Journal of the Surveying and Mapping Division, ASCE* 94, no. SU2 (September 1968), p. 203.

Department of the Army, U.S. Army Corps of Engineers. *Engineering and Design, Hydrographic Surveying,* Manual No. 1110-2-1003. Washington, DC: October 1994.

Greening, W. J. T. "GPS Surveys and Boston's Central Artery/Third Harbor Tunnel Project." *ASCE Journal of Surveying Engineering* 114, no. 4 (November 1988).

Katibah, G. P. "Photogrammetry in Highway Practice." Unpublished report, California Department of Transportation, 1968.

Kavanagh, B. F. *Surveying with Construction Applications.* Englewood Cliffs, NJ: Prentice-Hall, 1989.

"Laser Control in Tunnel Driving." *Kern Bulletin 15.* Aaru, Switzerland: Kern Instrument, April 1971.

Li, R.; K. R. Schwarz; M.A. Chapman; and M. Gravel. "VISAT: A Real-Time System for Highway Spatial Information Acquisition." *Technical Papers,* vol. 1, pp. 344–349. 1994 ASPRS/ACSM Annual Convention, Reno, Nevada, April 25–28, 1994.

More, N. L., and R. G. Begell. "Control Surveys for Major Bridges." *Journal of the Surveying and Mapping Division, ASCE* 98, no. SU1 (May 1971).

Peterson, E. W., and P. Frobenius. "Tunnel Surveying and Tunneling Machine Control." *Journal of the Surveying and Mapping Division, ASCE* 99, no. SU1 (September 1973), p. 21.

Robinson, G. L.; W. J. Trevor Greening; P. W. DeKrom; A. Chrzanowski; E.C. Silver; G.C. Allen; and M. Falk. "Surface and Underground Geodetic Control for Superconducting Super Collider." *Journal of Surveying Engineering, ASCE* 121, no. 1 (February 1995), pp. 13–34.

Thompson, B. J. "Planning Economical Tunnel Surveys." *Journal of the Surveying and Mapping Division, ASCE* 100, no. SU2 (November 1974), p. 95.

Wand, B.T., and W. Oren. "Accelerator Alignment." *Surveying and Land Information Systems* 53, no. 1 (March 1993), pp. 19–28.

CHAPTER 18

Land Surveys

18.1
GENERAL

Land surveying deals with laying off or measuring the lengths and directions of lines forming the boundaries of real or landed property. Land surveys are made for one or more of the following purposes:

1. To secure the necessary data for writing the legal description and for finding the area of a designated tract of land, the boundaries of the property being defined by visible objects.
2. To reestablish the boundaries of a tract for which a survey previously has been made and for which the description as defined by the previous survey is known.
3. To subdivide a tract into two or more smaller units in accordance with a definite plan that predetermines the size, shape, and location of the units.

Whenever real estate is conveyed from one owner to another, it is important to know and state the location of the boundaries, particularly if there is a possibility of encroachment by structures or roadways.

The functions of the land surveyor are to carry out field surveys, calculate dimensions and areas, prepare maps showing the lengths and directions of boundary lines and areas of lands, and write descriptions by means of which lands may be legally conveyed, by deed, from one party to another.

The land surveyor must be familiar with not only technical procedures but also the legal aspects of real property and boundaries. A baccalaureate degree in surveying or survey engineering should be, and in many states must be, acquired if one is to be registered as a land surveyor. Usually, the land surveyor is required to be licensed by the state, either directly or as a civil engineer.

In this chapter, practices as applied to both rural and urban properties are described and some of the legal aspects of land surveying are discussed. The United States system of subdividing the public lands also is outlined. Methods of calculating and subdividing areas are discussed in Chapter 8.

18.2
KINDS OF LAND SURVEYS

In accordance with the purposes listed in Section 18.1, land surveys may be classified as follows:

1. *Original surveys,* to measure the unknown lengths and directions of boundaries already established and in evidence. Surveys of this character usually are of rural lands. For example, Adams may purchase from Brown a certain parcel of land bounded or defined by features or objects such as fences, roads, or trees. So that the deed may contain a definite description of the tract, a survey is necessary.
2. *Resurveys,* to reestablish the boundaries of a tract for which a survey has previously been made. The surveyor is guided by a description of the property based on the original survey and by evidence on the ground. The description may be in the form of the original survey notes, an old deed, or a map or plat on which are recorded the measured lengths and bearings of sides and other pertinent data. When land is transferred by deed from one party to another, often a resurvey is made.
3. *Subdivision surveys,* to subdivide land into more or less regular tracts in accordance with a prearranged plan. The division of the public lands of the United States into townships, sections, and quarter sections is an example of the subdivision of rural lands into large units. Laying out blocks and lots in a city addition or subdivision is an example of the subdivision of urban lands.

18.3
EQUIPMENT AND METHODS

Land surveys can be run with total station systems (Sections 7.15 to 7.21), static and kinematic GPS surveys (Sections 12.8 to 12.12), theodolite and EDM devices, and the traditional theodolite and tape. In some states, land surveys have been performed using photogrammetric methods to establish the control points from which property corners have been located and set. Ample evidence exists to indicate that photogrammetric methods provide adequate accuracy for property surveys (Brown, Robillard, and Wilson, 1995). Acceptance of photogrammetry as one of the tools useful for performing property surveys is bound to come as more surveyors become familiar with the procedures (Chapter 13) and as the legality of the method is recognized in the courts.

Directions of lines for property surveys usually are referred to the true or astronomic meridian, and angular measurements are transformed into bearings. Distances still are given in feet and decimal parts of the foot, but with metrication on the horizon, it is recommended that distances be given in feet and meters. Angles are measured in degrees, minutes, and seconds. On the U.S. public-land surveys, all distances are in Gunter's chains (66 ft) with 100 links per chain as prescribed by law.

Formerly, the surveyor's compass and 66-ft link chain were used extensively, particularly in rural surveying; and the directions and lengths of lines contained in many old deeds are given in terms of magnetic bearings and Gunter's chains. In retracing old surveys of this character, allowance must be made for change in magnetic bearing since the time of the original survey. Also, it must be kept in mind that the compass and link chain used on old surveys were relatively inaccurate instruments and that great precision was not regarded as necessary because usually the land values were low. Further, for many years the United States public lands were surveyed under contract, at the low price of a few dollars per mile. Many of the lines and corners established by old surveys are not where they theoretically should be; nevertheless, these boundaries legally remain fixed as they were originally established.

This principle was eloquently stated in the following extract from "The Judicial Function of Surveyors," by Justice William Cooley of the Michigan Supreme Court, in a paper delivered to the Michigan Society of Surveyors and Engineers in 1894:

> The corner and quarter stakes were often nothing but green sticks driven into the ground. Stones might be put around or over these if they were handy, but often they were not, and the witness trees must be relied upon after the stake was gone. Too often the first settlers were careless in fixing their lines with accuracy while monuments remained, and an irregular brush fence, or something equally untrustworthy, may have been relied upon to keep in mind where the blazed line once was. A fire running through this might sweep it away, and if nothing were substituted in its place, the adjoining proprietors might in a few years be found disputing over their lines, and perhaps rushing into litigation, as soon as they had occasion to cultivate the land along the boundary.
>
> If now the disputing parties call in a surveyor, it is not likely that anyone summoned would doubt or question that his duty was to find, if possible, the place of the original stakes which determined the boundary line between the proprietors. However erroneous may have been the original survey, the monuments that were set must nevertheless govern, even though the effect be to make one half-quarter section 90 acres and the one adjoining but 70; for the parties buy or are supposed to buy in reference to those monuments, and are entitled to what is within their lines, and no more, be it more or less.

Wherever possible, the field procedure is such that the lengths of boundary lines and the angles between boundaries are determined by direct measurement. Therefore, the land survey in general is a traverse, the stations being at corners of the property and the traverse lines coinciding with property lines. Where obstacles render direct measurement of boundaries impossible, a traverse is run as near the property lines as practical, and measurements are made from the traverse to property corners; the lengths and directions of the property lines then are calculated. Where the boundary is irregular or curved, the traverse is established in a convenient location, and offsets are taken from the traverse line to points on the boundary; the length of the boundary then is calculated. If GPS surveys are used, the corners of the property should be occupied by the system antenna, or when obstacles are present, the control points in the survey should be as close as possible to the corners and boundaries.

In general, the required precision of land surveys depends on the value of the land, being higher in urban than rural areas. The possibility of increase in land values also should be considered.

Wherever possible, the survey utilized to establish property corners should be tied to monuments for which state plane coordinates are available (Chapter 11). Use of state plane coordinates integrates all surveys into a common system, facilitates replacing lost corners, and provides a means of parcel identification that is compatible with filing survey data in electronic computer data banks.

18.4
LEGAL ASPECTS OF LAND SURVEYS: DEFINITIONS

Property lines are established and indicated by acts on the ground or by legal documents. Old boundaries are located on the basis of acts and documents exercised in the past. New boundaries are created by current acts or new documents. Consequently, the present positions of property lines or boundaries depend on work performed on the ground and documents executed in the past as well as in the present.

The positions of boundaries are controlled by two principles of law: (1) the *intent* of the parties involved in originally establishing the boundaries, and (2) *acceptance* of conditions as they have existed over a period of years. Intent is judged on the basis of the acts and documents of the involved parties. If old evidence of intent is obscure and misleading, then conditions as defined by existing evidence are accepted. The longer the period of acceptance, the stronger the evidence becomes.

Two types of law govern the foregoing principles: *common law,* which consists of the body of laws that have been inherited from our predecessors, and *statutory law,* which is established by governing bodies. Much of the common law in the United States is based on laws originating in England. Statutory law *usually takes precedence over common law.*

The evidence used to establish property lines consists of title transfers, transfer of rights such as in easements for right of way, by agreement, by marks on the ground, and by acts leading to adverse possession. Consider each of these types of evidence briefly.

Title Transfer

A title is transferred by legal instruments such as deeds or wills, inheritance without a will, or by adverse possession (Section 18.12). Deeds usually are recorded in some public office (Section 18.6) and contain a description of the property (Section 18.7). This description generally contains references to marks on the ground. The surveyor makes much use of this type of evidence in relocating old boundary lines.

In connection with deeds, it is necessary to define the *senior deed.* When land is partitioned, in most cases the first parcel sold is the senior deed. Subsequent deeds related to partitioning of the remainder of the land have seniority in the order of their execution. In a dispute, the owner of the senior deed has rights over those with less senior rights.

Easements for Rights of Way

The right to use the land of another for some specific purpose constitutes an *easement.* An easement that grants the right to pass over the land of another is a *right of way.* An easement can be established by an owner through deed or by *dedication.* The state has the right of *eminent domain,* which is the right to use a person's property for a specific use (such as a highway) provided that the owner of the property involved receives fair remuneration for the land taken. *Condemnation* proceedings usually are instituted to acquire land by eminent domain. These proceedings are recorded and become part of the public record.

Visual evidence of property-line location consists of marks on the ground. These marks include monuments (Section 18.5), iron pipes and rods, stakes, fences, and structures. Evidence of this type can be extremely valuable when attempting to reestablish old property lines.

Adverse Possession

Land acquired by the open and hostile use of another's land with the full knowledge of the owner is said to have been obtained by adverse possession. Details concerning adverse possession can be found in Section 18.12.

Additional Legal Terms

Additional legal terms, quoted from Bouvier's *Law Dictionary* (1991), include a few of the more common legal terms having to do with the transfer of land.

Color of title. Color of title, for the purposes of adverse possession under the statute of limitations, is that which has the semblance or appearance of title, legal or equitable, but which in fact is no title.

A writing which upon its face professes to pass title but which does not in fact do so, either from a want of title in the person making it or from the defective conveyance used, is also known as color of title. The term is also applied to a title that is imperfect, but not so obviously that it would be apparent to one not skilled in the law.

Fee. The word "fee" signifies that the land or other subject of property belongs to its owner and is transmissible, in the case of an individual, to those whom the law appoints to succeed him under the appellation of heirs.

Fee simple. An estate of inheritance is a fee simple. The word "simple" adds no meaning to the word fee standing by itself, but it excludes all qualifications or restrictions as to the persons who may inherit it as heirs.

Parol. Parol is a term used to distinguish contracts which are made verbally, or in writing not under seal, which are called parol contracts, as distinguished from those which are under seal, which bear the name of deeds or specialties.

Patent. A patent is the title deed by which a government, either state or federal, conveys its lands.

18.5
MONUMENTS

Monuments are classified as natural, artificial, record, or legal.

Examples of *natural monuments* are trees, large stones, or other substantial, naturally occurring objects in place before the survey was made.

Artificial monuments can consist of an iron pipe or bar driven in the ground; a concrete or stone monument with drill hole, cross, or metal plug marking the exact corner; a stone with identifying mark, placed below the ground surface; charcoal placed below the surface; a mound of stones; a mound of earth above a buried stone; and a metal marker set in concrete below the surface, reached through a covered shaft. Monuments for city lots usually are set nearly flush with the ground. Subsurface stones commonly were used for corner monuments in localities where roads followed section lines. On many old government surveys through wooded country where stones were not available, corners were established by building up a mound of earth over a quart of charcoal or a charred stake, or by building a mound around a tree at which the corner fell. The U.S. Bureau of Land Management has adopted as the standard for the monumenting of the public-land surveys a post made of iron pipe filled with concrete, the lower end of the pipe being split and spread to form a base, and the upper end being fitted with a brass cap with identifying marks (Section 18.43).

A record monument exists because of a reference in a deed or description. For example, the phrase in a deed "to the side line of a street" is a call for a record monument (the street) that could be marked by improvements, stakes, a fence, or may not be marked in any way.

A legal monument is controlling in the description. Therefore, the statement in a deed "to a concrete post" is a call to a legal artificial monument. Similarly, the statement "to Johnson's property line" is a call to a legal record monument.

Original, natural monuments control over artificial monuments and record monuments. Original, artificial monuments control over record monuments. Further details concerning the principles that govern conflicts between monuments can be found in Bouvier (1991).

If there is a possibility that a corner monument will become displaced, the corner should be *referenced* (Section 8.12) or connected to nearby objects of more or less permanent character in such manner that it may be replaced readily in case of loss. Usually, the recorded measurement is called a *connection,* and the object is called a *reference mark* or a *corner accessory;* essentially, it is a part of the monument. Examples of corner accessories are trees, mounds, pits, large stones, and buildings. In many large cities, systems of permanent monuments are established and all surveys are referred to them. On public-land surveys, the bearing and distance from a corner to a tree are taken where possible, the tree being blazed and so marked as to identify the section on which the tree stands, the mark terminating with the letters *B.T.*, signifying bearing tree. The Bureau of Land Management specifies that every corner established in the public-land surveys shall be referenced by one or more objects of any of the following classes: (*a*) "bearing trees, or other natural objects . . . ; (*b*) permanent improvements and memorials; (*c*) mound of stone; and (*d*) pits."

If the location of a corner within reasonable accuracy can be determined beyond reasonable doubt, the corner is said to *exist;* otherwise, it is said to be *lost*. If the monument marking an existing corner cannot be found, the corner is said to be *obliterated,* but it is not necessarily lost.

Where a corner falls in such location as to make it impossible or impractical to establish a monument in its true location, it is customary to set a point on one or more of the boundary lines leading to the corner, as near to the true corner as practical. A point thus established is called a *witness corner*. Everything that has been said concerning monuments at the true corners also applies to witness corners. Witness corners are necessary where the true corner falls in a roadway or body of water, within a building, or on a precipitous slope. Under certain circumstances, as when boundaries are in roads, it is impossible to place the witness corner on any of the property lines approaching the true corner, in which case the witness corner is established in any convenient location.

The field notes should give detailed information concerning the character, size, and location of all monuments and reference marks; and the data should be recorded in such manner that there will be no possibility of misinterpretation. As far as possible, all points established in the field should be marked clearly to indicate the object they represent.

18.6
BOUNDARY RECORDS

Descriptions of the boundaries of real property may be found from deeds, official plats or maps, or notes of original surveys. Typical descriptions are given in Sections 18.8 and 18.9.

Records of the transfer of land from one owner to another usually are kept in the office of the county registry of deeds, exact copies of all deeds to transfer being filed in deed books. These files are open to the public and are a frequent source of information for the land surveyor in search of boundary descriptions when it is inconvenient or impossible to secure permission of the owner to examine the original deed.

Originally, all official records were copied by hand into the record books. The first step toward modernization of this system was to make photocopies of each record, which still were filed in the books. As records proliferated, many recording offices switched to making microfilm copies of new and old records to save space. The microfilm copies can be stored on cards (microfiche) and filed in a system that permits manual or automatic retrieval.

No matter what system of filing the records is employed, some method for indexing, usually dictated by statute, is required. The system that has and still is in most general use in the United States consists of an alphabetic index, usually kept by years, giving in one part the names of the *grantors* or persons selling property and in the other the names of the *grantees* or persons buying property. If either or both parties to the transfer and the approximate date of the transfer are known, a given deed can be located. Usually, the preceding transfer of the same property is noted on the margin of the deed.

In most cases the deeds of transfer of city lots give only the lot or block number and the name of the addition or the subdivision. The official plat or map showing the dimensions of all lots and the character and location of permanent monuments is on file either in the office of the city clerk or in the country registry of deeds; copies also are on file in the offices of city and county assessors. Filing land transfer records according to these data is called indexing by *lot, block, and parcel,* or more simply indexing by *land parcel.*

Some organizations, usually called *title companies,* for a fee, will search the records for boundary descriptions and guarantee the title against possible defects in description, legal transfer, and certain types of claims such as those for right of way. Title insurance does not necessarily mean that the property corners are correctly marked on the ground; and if assurance is desired, a survey should be made. The files of title companies usually are based on land parcel indexing.

As U.S. public lands are subdivided, official plats are prepared showing the dimensions of subdivisions and the character of monuments marking the corners. When the surveys within a state have been completed, records are given to the state. States in possession of records have them on file with a designated state agency. Usually, information concerning these records can be secured from the state secretary of state. Photographic copies of the official plats are obtainable at nominal cost. Land transfer records maintained in this fashion are said to be filed according to the federal rectangular survey system.

Several states have passed land registration acts that specify the procedures to be followed in the transfer of land. These acts usually designate the state plane coordinate systems as the reference framework to be used. The state of Massachusetts has a special "land court" where title to land can be confirmed (see Woodbury, 1973).

Current trends are toward storing legal records related to land transfers as part of a geographic or land information system (GIS or LIS) (see Sections 14.18 and 14.20). Such a procedure has the advantage that title searches can be performed more rapidly and at lower cost. Most title companies already have automated or semiautomated search procedures for land records. Electronic computer storage and retrieval methods and interactive computer techniques are utilized in these procedures (Chapter 14). The problems encountered in such systems are related mostly to the fact that the grantor-grantee alphabetic listing is not exactly compatible with efficient computer usage and a unique identifier for each parcel involved in a land transfer is necessary.

One proposal is to pass legislation requiring that all property surveys be tied into the state plane coordinate system. In this way, records for each property transaction would have a parcel identifier consisting of the state plane coordinates for the approximate center of the parcel (Cook, 1977). These coordinates would be preceded by identifier digits that specify the state, county, and other necessary municipal subdivision. Naturally, the success of such a system would hinge on stringent but realistic accuracy requirements for all property surveys to avoid ambiguity in coordinates of the visual centers for adjacent parcels.

18.7
PROPERTY DESCRIPTIONS

As noted in Section 18.4, a description of the property is an essential item in the legal document referring to the transfer of the land. The major purpose of this description is to identify the property for title purposes and describe its size, shape, and location. The description should be precise, clear, and concise. It should be worded with sufficient legal terms to perpetuate the intent of the parties in a legal sense. The dimensions given to describe the property lines should be mathematically correct, and there should be no conflicts in the description of the property or with respect to adjoining areas.

There are two basic methods for describing property: metes and bounds descriptions and subdivision descriptions. Each of these methods for describing land is discussed in more detail in the sections that follow.

In the older portions of the United States, nearly all of the original land grants were of irregular shape, many of the boundaries following stream and ridge lines. Also, in the process of subdivision, the units were taken without much regard for regularity, and it was thought sufficient if lands were specified as bounded by natural or artificial features of the terrain and if the names of adjacent property owners were given. An example of a description of a tract as recorded in a deed reads:

> Bounded on the north by Bog Brook, bounded on the northeast by the irregular line formed by the southwesterly border of Cedar Swamp of land now or formerly belonging to Benjamin Clark, bounded on the east by a stone wall and land now or formerly belonging to Ezra Pennell, bounded on the south and southeast by the turnpike road from Brunswick to Bath, and bounded on the west by the irregular line formed by the easterly fringe of trees of the wood lot now or formerly belonging to Moses Purington.

As the country developed and land became more valuable, and as many boundaries such as those listed in the preceding description ceased to exist, land litigations became numerous. It then became the general practice to determine the lengths and directions of the boundaries of land by measurements with the link chain and surveyor's compass and to fix the locations of corners, permanently by monuments. The lengths ordinarily were given in rods, perches, or chains; and the directions were expressed as bearings usually referred to the magnetic meridian. Surveys of this character usually now are made with theodolite and EDM device or a total station system, distances being recorded in feet, chains, or meters and directions being given in astronomic, geodetic, or grid bearings computed from angular measurements. In describing a tract surveyed in this manner, the lengths and bearings of the several courses are given in order and the objects marking the corners are described; if any boundary follows some prominent feature of the terrain, the fact is stated; and the calculated area of the tract is given. When the bearings and lengths of the sides are thus given, the tract is said to be described by *metes and bounds*. Within the limits of the precision of the original survey, it is possible to relocate the boundaries of a tract if its description by metes and bounds is available, provided that at least one of the original corners can be identified and the true direction of one of the boundaries can be determined.

It should be noted that conveyances of land having senior and junior rights, as discussed in Section 18.4, will come under the classification of metes and bounds descriptions. This is in contrast to the subdivision description, in which all lots within the subdivision have equal rights (Section 18.9).

There are several types of metes and bounds descriptions, the most common of which is the method of *successive bounds,* as discussed previously. This type of metes and bounds description is sometimes regarded as the true metes and bounds description.

EXAMPLE 18.1. Alan Tart has sold a portion of his property to Richard Merlin. Tart's land was obtained in a previous conveyance from Biggert to Tart. There is no recorded plan of lots (subdivision), so that a metes and bounds description is necessary in the deed that is the legal document in which the transaction from Tart to Merlin will be recorded. Prior to the sale, the land to be sold is surveyed and the corners are marked as illustrated in Figure 18.1. The following is one form of a deed description with a metes and bound description by successive bounds:

A parcel of land in the City of _____ , County of _____ , State of _____ , being a part of the land conveyed by John Biggert to Alan Tart by deed

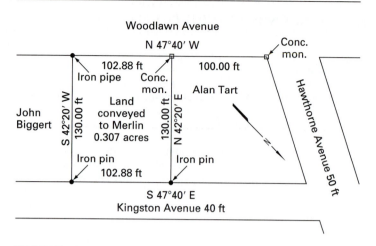

FIGURE 18.1

Plan of property described in Example 18.1.

dated June 14, 1956, and recorded in Book 201, page 131, at the _____ County Recorder's Office, and described as follows:

Beginning at a concrete monument on the northerly line of Woodlawn Avenue, said monument bearing S47°40′E a distance of 100.00 ft (30.480 m) along the northerly line of Woodlawn Avenue from the intersection of the northerly line of Woodlawn Avenue with the easterly line of Hawthorne Avenue and running: (1) thence along the northerly line of Woodlawn Avenue S47°40′E a distance of 102.88 ft (31.358 m) to an iron pipe (2 in. or 51 mm inside diameter) at the line of the land of John Biggert; (2) thence along said land N42°20′E a distance of 130.00 ft (39.624 m) to an iron pin ($\frac{1}{2}$ in. or 13 mm outside diameter) on the southerly line of Kingston Avenue; (3) thence along the southerly line of Kingston Avenue N47°40′W a distance of 102.88 ft (31.358 m) to an iron pin ($\frac{1}{2}$ in. or 13 mm outside diameter) at the line of the land of the grantor; (4) thence along the land of the grantor S42°20′W a distance of 130.00 ft (39.624 m) to the point of beginning.

All directions are based on the stated bearing of the northerly line of Woodlawn Avenue.

This description was written on July 1, 1978, by _____, Professional Land Surveyor, Registration No. _____, State of _____, and is based on data gathered in the field by _____, on June 15, 1978.

The deed contains a general statement that indicates the approximate location of the land. The body of the deed includes a specific description of the land conveyed by the method of successive bounds. Note that the description has a point of beginning that is referenced to the block corner. Such a reference is necessary to fix the position of the land being conveyed. Then, each course is described in turn and the description ends with a return to the point of beginning.

Following the description are given the basis for bearings, the name and registration number of the surveyor, and the dates when the deed was written and fieldwork performed.

The basis for bearings should always be stated clearly. Bearings should be described as astronomic, grid, or referenced to a stated line. The basis for bearings can be stated indirectly. In Example 18.1, the initial statement "said monument bearing S47°40′E along the northerly line of Woodlawn Avenue . . ." fixes the basis for bearings along that line. Hence, the statement on basis for bearings, following the body of the deed, is redundant only in situations such as this example.

Note that the deed conveying land from Biggert to Tart has senior rights over the deed in Example 18.1 conveying land from Tart to Merlin. Therefore, if the monuments along the

survey reveals that the record distance of 202.88 ft is 201.88 ft to the iron pin, Tart must
get his 100.00 ft of frontage along Woodlawn Avenue even at the expense of Merlin, who
has the junior deed. When senior and junior rights are involved, there is no proportioning
of errors in surveys (Sections 18.18 and 18.47).

Several other types of metes and bounds descriptions exist:

1. *Strip conveyances* are employed to describe a road easement or right of way. The form
   used is "a right of way for highway purposes over and across a strip of land lying 25 ft
   (7.620 m) on either side of the following described center line. . . ." Stationing is used
   to denote distance along the center line.
2. *Conveyance by division line:* "All of lot 62 lying south of Rock Creek" is a description
   by division line.
3. *Conveyance by distance:* "The westerly 20 m of lot 32" constitutes the very brief
   description by distance.
4. *Proportional conveyances* describe a fractional part of the whole tract, such as "the
   eastern $\frac{1}{8}$ of lot 11."
5. Land may be conveyed by *exception,* such as "all except the westerly 100 ft of lot 61,"
   or by *acreage,* such as "the west 2 acres of lot 15."

Further details concerning the variations on the true metes and bounds descriptions can
be found in *Black's Law Dictionary* (1979).

Descriptions for Condominiums

In general, the term *condominium* refers to a method of ownership. This term may be
applied to a multiunit building or to a unit or units within the building. Therefore, the owner
of an individual unit or of several units within a multiunit building is called a *condominium
owner.* The limits of ownership of a condominium owner are the top surface of the floor,
the surfaces of the walls, and the bottom surface of the ceilings of the specified unit.
Condominium ownership also includes a fractional interest in the common elements associated with each unit such as swimming pools, tennis courts, and other common areas in the
building. A grant of interest also may be made for items such as garage, patio, and storage
space.

Because the potential condominium owner is buying a volume of space as just defined,
it is necessary to provide a description of the unit in three dimensions. To describe each
condominium unit in three dimensions by a metes and bounds description would be a
challenging task, to say the least. A simpler approach is to prepare a map of the land on
which the building is located and a set of plans of the building referenced to the map. These
plans of the building show each individual unit with an identifying label. In this way, any
specified unit can be described by the identifying label and a reference to the map.

The land on which the condominium building is constructed can be described by a
metes and bounds description, a lot or parcel on a map already recorded, or by a new
subdivision plan of one or more lots. The most convenient instrument of record is a map that
shows the boundaries of the land and the position of the exterior walls of the condominium
building with respect to these boundaries. The condominium property acts of some states
require that this type of map be recorded to establish a condominium. Along with the map
there should be a set of plans of the building. These plans show all horizontal dimensions
of the interior walls and the relationships of these walls with the exterior walls of the
structure. Elevations also are required for the top of floor and bottom of ceiling of each unit.
These elevations can be shown on the plans, indicated on a cross section of the building, or
tabulated for each unit. All elevations should be based on bench marks referred to a
specified datum.

The dimensions shown on the plan should be certified by the surveyor as conforming to the physical building. Therefore, the surveyor needs to perform an as-built survey (Chapter 17) of the structure after it is completed. If this is not possible and measurements in the field are made to the unfinished elements of the structure, the interior dimensions can be based on the architect's plans, which must be indicated on the plans. When the building is finished, the surveyor should resurvey the building, including the interior horizontal and vertical dimensions. Any differences between dimensions determined from this final as-built survey and those projected from the architect's plans should be indicated on the plans. If major differences occur, a new plan should be filed.

The following example description for a condominium unit is one of several forms recommended by the California Title Association.

EXAMPLE 18.2. Assume that the condominium building is on land described by a recorded subdivision plan and that all units are within one lot.

A condominium comprised of:
PARCEL 1: An undivided _____ interest in and to Lot _____ of Tract No. _____, in the City of _____ County of _____, State of _____, as per map recorded in Book _____, of Maps, at Page _____, in the Office of the County Recorder of said County.
PARCEL 2: Unit _____, as shown upon the map referred to in PARCEL 1.

If the condominium building rests on more than one lot, this must be noted in the description of Parcel 1. Where the condominium buildings rest on land described by metes and bounds, the description for Parcel 1 in the example is modified as follows:

An undivided _____ interest in and to the land described as follows: . . . insert the metes and bounds description here and add the description of Parcel 2.

For further details concerning condominiums and their descriptions, consult Laundry (1974) and Vorhies (1977).

Descriptions by Coordinates

As noted in Section 18.6, there is a trend toward requiring all property surveys to be performed in the state plane coordinate systems. In some states, the locations of land corners are legally described by state plane coordinates. Although practice is not as yet uniform in this regard, the following description by the Tennessee Valley Authority illustrates the description of land both by metes and bounds and by coordinates, with further reference to corners and lines of the U.S. public-land survey. The public-land survey is referred to the Huntsville principal meridian. A map of the tract is shown in Figure 18.2.

EXAMPLE 18.3

A tract of land lying in Jackson County, State of Alabama, in the South Half (S $\frac{1}{2}$) of the Northwest Quarter (NW $\frac{1}{4}$) of Section Three (3), Township Six (6) South, Range Five (5) East of the Huntsville principal meridian, and more particularly described as follows:
    Beginning at a fence corner at the southwest corner of the Northwest Quarter (NW $\frac{1}{4}$) of Section Three (3) (coordinates X416,239; Y1,470,588), said corner being North six degrees twenty-four minutes West (N6°24'W) twenty-six hundred (2600) feet from the southwest corner of Section Three (3) (X416,529; Y1,468,004), and a corner to the land of T. E. Morgan; thence with Morgan's line, the west line of Section Three (3), and a fence line, North five degrees thirty-three minutes West (N5°33'W) thirteen hundred four (1304) feet to a fence corner (X416,113; Y1,471,886), a corner of the lands of T. E. Morgan, and the G. T. Cabiness Estate. . . thence with Weeks' line, the south line of the Northwest Quarter (NW $\frac{1}{4}$) of

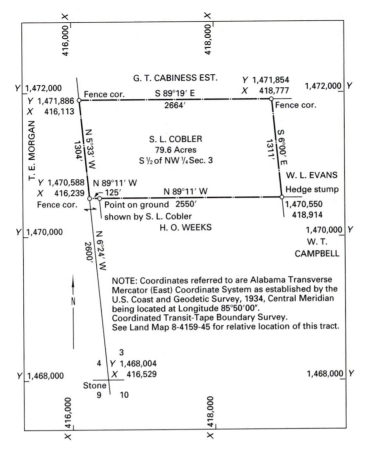

FIGURE 18.2
Land map.

Section Three (3), and a fence line, North eighty-nine degrees eleven minutes West (N89°11′W) two thousand five hundred fifty (2550) feet to a point on the ground shown by S. L. Cobler, a corner of the lands of H. O. Weeks and T. E. Morgan; thence with Morgan's line, the south line of the Northwest Quarter (NW $\frac{1}{4}$) of Section Three (3), and the fence line North eighty-nine degrees eleven minutes West (N89°11′W), one hundred twenty-five (125) feet to the point of beginning.

The above described land contains seventy-nine and six-tenths (79.6) acres more or less, subject to the rights of a county road which affects approximately five-tenths (0.5) acres, and is known as Tract No. GR 275, as shown on Map No. 8-4159-45, prepared by the Engineers of the Tennessee Valley Authority.

The coordinates referred to in the above description are for the Alabama Mercator (East) Coordinate System as established by the U.S. Coast and Geodetic Survey, 1934. The Central Meridian for this coordinate system is Longitude eight-five degrees (85°) fifty minutes (50′) no seconds (00″).

An outstanding advantage of recording the coordinates of the corners in the deed description is that lost corners can be replaced at any subsequent time without having to resurvey the entire property. As discussed in Section 18.6, coordinates for the approximate center of the property can be utilized as an identifier for the parcel when filing the deed. For this example, the approximate center is taken as the average values of the coordinates for the corners, which are $X = 417,510$ ft and $Y = 1,471,219$ ft.

Note that both the deed and map for this example fail to indicate the date of the survey or name of the surveyor. Normally, these items of information are included in both the description and on the map.

In Example 18.3, coordinates are given for each corner. Theoretically, a single coordinated point in a figure that closes mathematically is adequate to position (but not to orient) the property. As a rule, coordinates for at least two points should be given. Consider the deed of Example 18.1 and Figure 18.1 using coordinates for two points.

> **EXAMPLE 18.4.** To illustrate the use of two coordinated points in this description only the necessary portions of the deed are repeated as follows:
>
> A parcel of land in the City of _____ , County of . . . and described as follows:
>   Beginning at a concrete monument on the northerly line of Woodlawn Avenue, having established grid coordinates of $X =$ _____ , $Y =$ _____ , North Zone of the _____ State Coordinate System, said monument bearing S47°40′E a distance of 100.00 ft (30.480 m) along the northerly line of . . . (3) thence along the southerly line of Kingston Avenue N47°40′W a distance of 102.88 ft (31.358 m) to an iron pin ($\frac{1}{2}$ in. or 13 mm outside diameter), having grid coordinates of $X =$ _____ , $Y =$ _____ , of said North Zone, thence along the land of the grantor S42°20′W a distance of 130.00 ft (39.624 m) to the point of beginning.
>   All directions are based on the stated . . . data gathered in the field by _____ on June 15, 1978.

Whenever possible, the coordinates cited in the description should be for control points of a higher-order survey. Note that in each description the state plane coordinate system and zone are designated.

Coordinates provide an excellent method for identifying and reestablishing the location of a point and an unambiguous means for parcel identification compatible with computer technology. As such, coordinates should be employed in all descriptions. However, coordinates alone provide no legal description for a property corner except where laws have been established to recognize coordinated points as the sole criteria for property lines. In most regions, court decisions in the past have firmly established monuments as the best evidence of property ownership.

Properly written legal descriptions provide checks and ties to which property lines must correspond. Coordinates in a given system referred to a known datum certainly provide an invaluable supplement to the description but do not furnish sole evidence of property ownership.

It is the responsibility of the surveyor to be familiar with the state, county, and local laws of the area in which the surveying practice occurs so as to know the legal status of coordinates in a description.

18.9
DESCRIPTIONS BY SUBDIVISIONS

When land is partitioned into lots and a map showing the manner in which the land has been divided and also satisfying state, county, and local laws is placed on record, a subdivision has been created. The significant aspects of land descriptions by subdivision are (1) a plan that shows all dimensions, directions, areas, and monuments is recorded; and (2) all lots within a given subdivision are created simultaneously, so that no senior rights are involved among lots within a subdivision. Sectionalized land surveyed according to the Public Land Surveys in the United States (Sections 18.21 to 18.48) and private subdivisions fall within this definition.

Most states have laws that control subdivision development. In the more densely populated regions, counties and local municipal governments also have laws regulating subdivisions. Details concerning the development of subdivision plans and their subsequent approval and recording are given in Section 18.18.

When the boundaries of the property to be described coincide exactly with a lot within a subdivision for which there is an official recorded map, the lot may be described legally by a statement giving the lot and block numbers and the name, date, and place of filing the map. Most city property and many lots in urban developments are described in this way. The following are descriptions of this type.

EXAMPLE 18.5

Lot 29, Block 0, Map of Berkeley Highlands Addition, filed November 20, 1912, in Map Book 8, page 194, Contra Costa County Records, State of California.

Lot 15 in Block 5 as said lots and blocks are delineated and so designated upon that certain map entitled *Map of Thousand Oaks, Alameda County, California,* filed August 23, 1909, in Liber 25 of Maps, page 2, in the office of the County Recorder of the said County of Alameda.

If the boundaries of a given tract within a subdivision for which there is a recorded map do not conform exactly to boundaries shown on the official map, the tract is described by metes and bounds (Section 18.8), with the point of beginning referred to a corner shown on the official map. Also, the numbers of lots of which the tract is composed are given. The following is an example of a description of this kind.

EXAMPLE 18.6

Beginning at the intersection of the Northern line of Escondido Avenue, with the Eastern boundary line of Lot 16, hereinafter referred to; running thence Northerly along said Eastern boundary line of Lot 16, and the Eastern boundary line of Lot 17, eighty-nine (89) feet; thence at right angles Westerly, fifty-one (51) feet; thence South 12°06′ East, seventy-five (75) feet to the Northern line of Escondido Avenue; thence Easterly along said line of Escondido Avenue; fifty-three and $\frac{13}{100}$ (53.13) feet, more or less, to the point of beginning.

Being a portion of Lots 16 and 17, in Block 5, as said lots and blocks are delineated and so designated upon that certain map entitled *Map of Thousand Oaks, Alameda County, California,* filed August 23, 1909, in Liber 25 of Maps, page 2, in the office of the County Recorder of the said County of Alameda.

Note that if land is divided within a subdivision and metes and bounds descriptions are written into the deed, senior rights must be observed in the order in which the division of the land occurs.

18.10
LEGAL INTERPRETATION OF DEED DESCRIPTIONS

The descriptions of the boundaries of a tract include the objects that fix the corners, the lengths and directions of lines between the corners, and the area of the tract. A deed description may contain errors or mistakes of measurement or mistakes of calculation or record, introducing inconsistencies that cannot be reconciled completely when retracement becomes necessary. In such cases, where uncertainty has arisen as to the location of property lines, it is a universal principle of law that the endeavor is to make the deed effectual rather than void and to execute the intentions of the contracting parties. The following general rules are pursuant to this principle:

1. *Monuments.* It is presumed that the visible objects that marked the corners when a conveyance of ownership was made indicated best the intentions of the parties

concerned; hence, it is agreed that a corner is established by an existing material object or conclusive evidence as to the previous location of the object. A corner thus established will prevail against all other conflicting evidence, provided there is reason to believe that the monument was set in accordance with the original intention and its location has not been disturbed. The kinds of evidence valid in relocating obliterated corners are stated in Section 18.48. Specifically, monuments will control over distance, bearing, and area.

2. *Distance, direction, and area.* In case of conflicts among "calls" in the deed (or dimensions on a recorded subdivision map) for distance, bearings, and area, the following order of importance is observed: distance controls over bearings; and bearings control over area.

3. *Mistakes.* It is a well-established principle that a deed description which taken as a whole, plainly indicates the intentions of the parties concerned will not be invalidated by evident mistakes or omissions. For example, such obvious mistakes as the omission of a full tape length in a dimension or the transposition of the words *northeast* for *northwest* will have no effect on the validity of a description, provided it otherwise is complete and consistent or provided its intention is manifest.

4. *Purchaser favored.* In the case of a description capable of two or more interpretations, that one will prevail which favors the purchaser (the "grantee").

5. *Ownership of highways.* Land described as being bounded by a highway or street often conveys ownership to the center of the highway or street. Any variation from this interpretation must be explicitly stated in the description.

6. *Original government surveys presumed correct.* Errors found in original government surveys do not affect the location of the boundaries established under those surveys, and the boundaries remain fixed as originally established.

18.11
RIPARIAN RIGHTS

An owner of property that borders on a body of water is a *riparian proprietor* and has riparian rights (pertaining to the use of the shore or of the water), which may be valuable. Because of the difficulties arising from the irregularity of such boundaries, it is important that the surveyor be familiar with the general principles relating to riparian rights and the statutes and precedents established in the particular state where work is being performed. This article addresses three major riparian rights issues: ownership of the bed of a body of water, division of submerged lands, and changes to the shoreline.

The reader should bear in mind that the following discussion refers to natural lakes or watercourses. The law at present is unclear on lakes or watercourses that are artificial (Cameron, 1989, p. 68). Before developing the individual topics, some basic definitions are in order.

> *Lake.* "A considerable body of standing water in a depression of land or expanded part of a river" (*Black's Law Dictionary,* 1979, p. 788). "Lakes are distinguished from watercourses by the absence of a single prevailing current" (Cameron, 1989, p. 68).
>
> *Watercourse.* "A running stream of water; a natural stream fed from permanent or natural sources, including rivers, creeks, runs, and rivulets. There must be a stream, usually flowing in a particular direction, though it need not flow continuously" (*Black's Law Dictionary,* 1979, p. 1428).
>
> *Bed.* The bed of a lake or watercourse generally is that portion covered by water for a long enough time to prevent vegetation from growing.

High-water mark. The high-water mark occurs wherever the presence of the water is so common as to mark on the soil a character, in respect to vegetation, distinct from that of the banks; it does not include low lands that, though subject to periodic overflow, are valuable for agricultural purposes. That part of the shore of the sea to which the waves ordinarily reach when the tide is at its highest also is known as the *high-water mark.*

Low-water mark. The low-water mark is that part of the shore of the sea to which the waters recede when the tide is lowest, that is, the line to which the ebb tide usually recedes, or it is the ordinary low-water mark unaffected by drought. It has been said to be the point to which a river recedes at its lowest stage.

Meander lines. "Meander lines were established in the original government survey 'for the purpose of ascertaining the quantity of land remaining after segregation of the water area.' Meander lines should not be confused with boundary lines. Since a lake, river, or stream is a natural monument, the monument, and not the meander line, determines the property boundaries" (Cameron, 1989, p. 136).

Ownership of the Bed of the Body of Water

"Unless there are reservations or exceptions in the pertinent grant from the federal government, the laws of the state determine the extent and nature of the ownership of riparian proprietors" (Cameron, 1989, p. 69). In many states, riparian proprietors own the bed only to the high-water mark or the low-water mark. In other states, riparian proprietors own the bed out to the middle of the lake or watercourse (Cameron, 1989, p. 69). For example, as regards the ownership of the bed of a navigable river, the two states of Iowa and Illinois bordering on the same river have very different laws. Clark (1959) states: "It is a rule of property in Illinois, that the fee of the riparian owner of land in that state bordering on the Mississippi River extends to the middle line of the main channel of the river," whereas Iowa courts hold "that the bed of the Mississippi River and the banks to the high water mark belong to the state, and that the title of the riparian proprietor extends only to that line."

Riparian ownership may vary within a state. "In Michigan, the owner of riparian land that borders on an inland lake owns the land under the water to the middle of the lake . . . an owner of land bordering one of the Great Lakes owns title only to the water line of the lake" (Cameron 1989, pp. 69, 73, and 74).

A distinction should be made between navigable and nonnavigable waterways. *Black's Law Dictionary* defines navigable waters as "Those waters which afford a channel for useful commerce" (1979, p. 926). Cameron states that "The basic legal distinction between them is that the public as well as riparian owners have legal rights to use navigable waters, but the public has no rights to use or go upon nonnavigable waterways" (Cameron 1989, p. 84). The designation of navigable or nonnavigable is a function of state and federal statute, and the professional surveyor would be wise to consult legal counsel when in doubt.

Ownership of the bed and navigability or nonnavigability should be looked on as separate issues. Whether one is a condition of the other will vary from state to state.

Division of Submerged Lands

Lakes. Three general cases govern the division of submerged lands beneath a lake: (1) circular lakes, (2) oblong lakes, and (3) lakes of irregular or undefined shapes.

Ownership of the bed of a circular lake is defined by establishing a point in the center. Riparian lines are projected from the shore to the center point forming a sort of triangle (Figure 18.3).

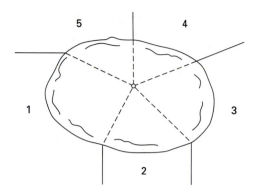

FIGURE 18.3
Riparian boundaries, circular lake.

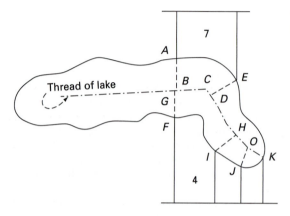

FIGURE 18.4
Riparian boundaries, oblong lake.

In the case of an oblong lake (Figure 18.4), a reference line is established along the center line or along the thread of the lake and riparian lines are established perpendicular to it, again to the boundary points on the shore line. With this method it can be "difficult to determine how far the midline if the lake should be extended to either end before the boundaries of the parcels on the ends of the lake are extended" (Cameron, 1989, p. 72).

If either of these two methods prove impractical, as in the cases of lakes or irregular or undefined shapes, the lake bed is divided in proportion to the shoreline owned (Cameron, 1989, p. 72). For a further discussion of this method, the reader is referred to the discussion of bays and coves covered later in this section.

Streams and rivers. The lines fixing the riparian rights of owners of property bordering on streams and rivers are established by lines perpendicular to the thread of the stream if nonnavigable or to the low-water line (sometimes to the middle of the channel) if navigable. Therefore, the lines for lots 2 and 3 of Figure 18.5 are established perpendicular to the line *BCD*.

Origin of dividing lines. There are two opposing lines of court decisions. One decision holds that a dividing line has its origin at the high-water line of a river or at the shoreline of a lake and not at the meander line. Thus, in Figure 18.5, the dividing line between lots 2 and 3 would be made perpendicular to the thread of the stream, beginning at *e* (on the high-water line) and not at *E* (on the meander line). Under the other line of decisions, the reverse is held.

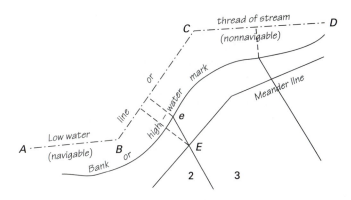

FIGURE 18.5
Origin of riparian boundaries.

Bays and coves. Property lines fixing riparian rights in bays or coves sometimes are established by lines beginning at the extremities of the property lines on shore and having a direction perpendicular to a line connecting the adjacent headlands of the bay or cove. Therefore, the lines *BB'*, *CC'*, and the like for lots 1, 2, and 3 (Figure 18.6) are established perpendicular to the line *AF*, which connects the two headlands *A* and *F*.

Other court decisions have fixed the lines according to the following rule: Divide the straight line joining the headlands (*AF* in Figure 18.6) into parts proportional to the lengths of the shoreline held by each owner; the property line of the inundated land is determined by joining the extremities of the property lines on shore and the corresponding points of subdivision on the line between headlands. These are shown in the figure by lines *BB"*, *CC"*, and so forth.

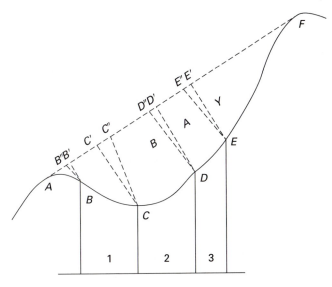

FIGURE 18.6
Riparian boundaries in a bay.

Changes in the Shoreline

Any water boundary is subject to change. Change may come about from the rising or falling of the level of the water, the wearing away of the shoreline, or deposits of sediment. Change also may come about by a shift in the course of a flowing body of water. That increase of earth on a bank of a river or on the shore of the sea by the force of the water, as by a current or waves or by the recession of the water, which is so gradual that no one can judge how much is added in a given interval, is known as *alluvium*. The process by which alluvium is formed is called *accretion* or *alluvion*. The gradual cutting away of the shoreline by the action of the water is termed *erosion*. The gradual withdrawal of water from the shoreline is *reliction*. In general, "a riparian [owner] gains by accretion and reliction and loses by erosion" (Cameron, 1989, p. 93).

The direction of the property lines dividing areas created by alluvium or reliction is determined by the proportional lengths of the old and of the new shorelines. The extremities of these lines are fixed either by definite bends, such as A, F, A', and F' (Figure 18.7) or by the intersections of the old and new lines, such as A and F (Figure 18.8). The general rule is to measure along the old shoreline between the old extremities, such as A and F, and divide the new line ($A'F'$) into parts proportional in length to those of the old line (AF). Thus, for lots 4 and 12 (Figures 18.7 and 18.8, respectively) by proportion $B'C'/BC = A'F'/AF$. The area $BB'C'C$ represents the area of alluvium added by accretion or alluvion to lots 4 and 12.

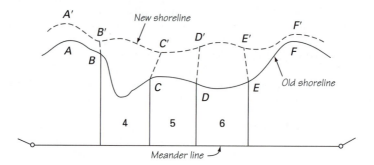

FIGURE 18.7
Riparian boundaries for areas added by alluvion or accretion between definite bends in a shoreline.

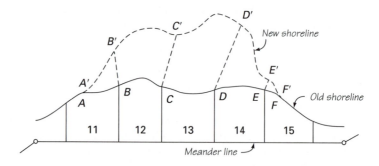

FIGURE 18.8
Riparian boundaries for areas added by alluvion or accretion between intersection of old and new shorelines.

The removal of a considerable quantity of soil from the land of one owner and its deposit on or annexation to the land of another, suddenly and by the perceptible action of water, is *avulsion*. In such a case, the property belongs to the first owner. Avulsion by the Missouri River, the middle of whose channel forms the boundary line between the states of Missouri and Nebraska, works no change in such a boundary but leaves it in the center line of the old channel. Avulsion is the opposite of accretion and is sometimes called *revulsion*.

18.12
ADVERSE POSSESSION

The many legal aspects of adverse possession cannot be treated here, but it is desirable to direct attention to the important fact that property lines may be fixed by continued possession and use of the land (usually for 20 years) as against original survey boundaries. The conditions and the period of time necessary to gain title are fixed by statute in the various states.

Adverse possession, to become effective, must be plainly evident to the owner, without his or her permission, to his or her exclusion, and hostile to his or her interests. Such possession may be evidenced by fencing, cultivation, erection of buildings, and the like.

Right to title by adverse possession may be acquired by individuals, corporations, and even by the state. But the statute does not run *against* the state; that is, property in a street or highway cannot be acquired by adverse possession.

Under this principle, if a person should use the land up to a fence and recognize it as a boundary line, to the exclusion of the owner, for the statutory period, the fence then becomes the legal property line even though it may be shown later that it is not on the true and original line. However, if the possession of the land has not been held adversely, that is, to the exclusion of the owner, and if the fence has merely served the convenience of the persons concerned, both parties recognizing that it was probably not on the true line, title cannot be claimed.

It therefore is clear that the application of the principle of adverse possession is entirely a matter of intention and belief. If land is held openly and notoriously with the intent to acquire title or with the belief that the occupation is proper and right, title will be granted if and when the statutory requirements are fulfilled. But, if by parol agreement or by actions, it is manifest that the parties concerned had no intention to occupy beyond the true line, at the same time knowing that the location of the true line was uncertain, title cannot be gained adversely.

Adverse possession under "color of title" will "ripen into title" under the statute of limitations in some jurisdictions in half the time required without color of title; for example, if title may be gained without color of title in 20 years, it may be gained in 10 years with it.

18.13
LEGAL AUTHORITY OF THE SURVEYOR

A resurvey may be run to settle a controversy between owners of adjoining property. The surveyor should understand that, although it is possible to act as an arbiter in such cases, it is not within the power of the surveyor legally to fix boundaries without the mutual consent and authority of all interested parties. In the event of a dispute involving court action, the surveyor may present evidence and argument as to the proper location of a boundary but has no authority to establish such a boundary against the wishes of either party concerned. A competent surveyor by wise counsel can usually prevent litigation; but if no agreement occurs, the boundaries in dispute become valid and defined only by a decision of the court. In boundary disputes the surveyor is an expert witness, not a judge.

The right to enter on property to make public surveys usually is provided by law, but there is no similar provision regarding private surveys. The surveyor (or employer, whether public or private) is liable for damage caused by cutting trees, destroying crops or fences, and so forth.

Permission should be obtained before entering private property, to avoid a possible charge of trespassing. If permission is not granted, it may be necessary to traverse around the property or to resort to aerial photographs. In certain cases, for example, where a boundary is in doubt or dispute, a court order may be obtained.

18.14
LIABILITY OF THE SURVEYOR

It has been held in court decisions that county surveyors and surveyors in private practice are members of a learned profession and may be held liable for incompetent services rendered. Clark (1959), quoting from court decisions, states: "If a surveyor is notified of the nature of a building to be erected on a lot, he may be held liable for all damages resulting from an erroneous survey; and he may not plead in his defense that the survey was not guaranteed." Similarly, it has been held that, in any case where the surveyor knows the purpose for which the survey is made, he or she is liable for damages resulting from incompetent work.

The general principle invoked in such cases is that the surveyor is bound to exhibit that degree of prudence, judgment, and skill which reasonably may be expected of a member of the profession. Therefore, in the following quotations from Clark, a Connecticut court says "the gist of the plaintiff's cause of action was the negligence of the defendant in his employment as a civil engineer. Having accepted that service from the plaintiff, the defendant . . . was bound to exercise that degree of care which a skilled civil engineer would have exercised under similar circumstances." Also, a Kansas court declared, "reasonable care and skill is the measure of the obligation created by the implied contract of a surgeon, lawyer, or any other professional practitioner." But Ruling Case Law says, "yet a person undertaking to make a survey does not insure the correctness of his work, nor is absolute correctness the test of the amount of skill the law requires. Reasonable care, honesty, and a reasonable amount of skill are all he is bound to bring to the discharge of his duties."

18.15
ORIGINAL SURVEY

The need for an original survey usually arises when one person desires to transfer to another a tract of land that has not been previously surveyed but is defined by certain natural or artificial features of the terrain.

With the desired boundaries of the land given, the surveyor establishes monuments at the corners and runs a survey around the property. This survey can be a closed traverse run by total station system or theodolite and EDM using the methods described in Chapter 8. A static GPS survey also could be employed to locate the corners and establish primary control points around the tract. Where boundaries are not straight, offsets from the traverse line to the curved boundary are measured at known intervals. Where obstructions make direct measurement along the boundaries impossible, the traverse is run or the control points are set as close to the boundary as convenient and measurements are taken from the control points nearest the corners of the tract. Whenever possible, the control survey should be tied to monuments in the state plane coordinate systems established by the National Geodetic Survey (NGS) of the National Oceanic and Atmospheric Administration (NOAA) (see Sections 11.11 to 11.20). Preferably, the corners should be referenced to permanent objects. Also, the direction of the astronomic meridian should be determined, usually by solar observation (see Section 10.31).

The information thus obtained is recorded in a data collector or the surveyor's notebook, the angles and distances of the main traverse being recorded or tabulated, and the remaining data being transcribed in the form of a sketch. The bearings of the sides then are computed with respect to the true or astronomic rather than the magnetic meridian.

A description of the tract, by metes and bounds or by coordinates, is prepared (Sections 18.7 and 18.8). Usually, a plat is drawn, the boundaries being plotted by one or another of the methods described in Chapter 14 and details being shown as suggested by Figure 18.2 and Section 14.22. The area is calculated as described in Chapter 8. In the process of calculation, the error of closure of the traverse is determined and thus a check on the reliability of the survey is obtained. A copy of the description and a tracing or prints of the plat are submitted to the person for whom the survey is made.

When state, county, or other local laws require it, the map or plat must be approved and filed in the county recorder's office. These regulations vary from state to state and from county to county. It is the responsibility of the surveyor to become familiar with the rules governing division of land in the region where the work occurs.

18.16
RESURVEYS

The resurvey of lands is attended with greater difficulty than is usually appreciated by those inexperienced in work of this character. This is particularly true in the older sections of the United States, where the early surveys were not of the rectangular system and were not under the control of the U.S. Bureau of Land Management. The proper relocation of old lines calls for greater ingenuity and broader experience on the part of the surveyor than any other kind of surveying.

The purpose of the resurvey is to reestablish boundaries in their original locations. As a guide the surveyor has available the description contained in the deed or obtained from old records and descriptions of adjoining property.

The first step in a resurvey is to obtain a copy of the deed description from the owner. It should be established that the document is the deed, not a tax statement or other incomplete document. Next, the surveyor should obtain copies of deed descriptions and maps of all adjoining properties, city or county surveyor's records, township plats, utility maps, field notes (if any), and record of surveys. Depending on the region being surveyed and the local regulations in effect, many of these records may not be available or even exist. However, a thorough search should be performed to ensure that no records potentially useful for the survey have been neglected. Copies of deeds generally are found in the county recorder's office, wills at the recorder's office, and maps or plans at the county surveyor's office or in city hall. These arrangements vary from community to community, and it is necessary to become familiar with the situation in the region where the surveying takes place.

When all records are assembled, the surveyor examines the descriptions, maps, and plats for gross errors, computes departures and latitudes for the courses given in the description, determines the closure error, and plots the boundaries of the tract to scale. When a metes and bounds description is involved, the seniority of the deeds must be determined. An abstract of the title gives the order of seniority of the deeds related to the property being surveyed and should be used if available.

Next, data concerning the locations of all marks that show the positions of streets, roads, or other public properties near the tract to be surveyed should be ascertained. The state, county, township, or city engineer may have this information. If the property is near a newly constructed state highway, the right of way for the highway generally is marked by monuments that have state plane coordinates and ties to the nearest property corners. In many states, this information would be filed as a record of survey and is available to the public. Many large cities have established systems of control monuments that have been coordinated in either an arbitrary or the state plane coordinate system. Quite frequently, these monuments are on a line parallel to but offset from the property line and are located in the sidewalks at the corners of the city blocks. The city engineer would have data for these monuments.

Only when all available data have been gathered and analyzed should field surveys commence. The importance of a thorough search for all records followed by an equally thorough analysis of the data cannot be overemphasized.

In rural regions, initial survey operations should be concentrated on locating monuments. When monuments are not available or cannot be found, old fence lines, corner posts, stakes, pipes, iron pins, and hedge lines are some of the features that provide indications of the existing boundary locations. Old aerial photographs are very beneficial in some cases.

In urban regions or in the city, monuments should be sought initially. In the absence of monuments, property corners marked by iron pins, metal survey markers, iron pipes, stakes, fence lines, walls, and hedge rows provide a method for establishing the boundaries. When there are sidewalks and curbs, cross marks or drill holes frequently are chiseled into the walks or on the tops of curbs. These marks generally are on the lot lines produced to a line offset an even number of feet or meters from the side line of the street. In the absence of all visible marks on the ground, the center line of the street can be established approximately by splitting the distance between the curb lines or the edges of the pavement. In the city, buildings frequently are constructed on the property lines or may be offset a few tenths of a foot or a few centimeters inside of the line. By measuring the distances between buildings and between curb lines and taking an average of one-half of these measurements, it is possible to establish an approximate center line for the street. This procedure provides a means of getting started in the search for evidence of old monuments or property corners that are not visible.

When the approximate positions for boundaries of the property to be surveyed have been located and in the absence of positive visible evidence of property corners, a control traverse is run around the tract. The traverse lines should be established approximately parallel and as close as is practical to the estimated positions for the property lines. If possible, traverse lines can coincide with the estimated positions for the property lines when existing survey markers are found and there are no obstacles on the line. Permanent markers should be set at all traverse points and each station should be referenced for subsequent relocation. For city and urban surveys, the control traverse may extend around the entire block, particularly if this is the first survey in the area. In rural surveys, the control traverse generally is restricted to the outline of the property being surveyed plus the necessary ties to control monuments not in the immediate vicinity of the survey. Whenever possible in city, urban, or rural surveys, the property survey control traverse should be tied to control points in the national geodetic network so that the survey can be coordinated on the state plane coordinate system.

As the control traverse is being run, ties should be measured to all details relevant to the boundaries. Locate property corners, fence lines, hedge rows, walls, and all buildings on the lot being surveyed and on adjoining properties.

The data collected in the field are reduced, calculated, and plotted in the office. Coordinates are determined for the traverse stations and potential property corners, and by inverse computations, the distance and bearings between located, possible property corners can be calculated. The control traverse and all evidence of property lines and corners are plotted to scale.

The deed information and the physical location of all evidence in the field then are compared and evaluated and a solution is sought that will best fit all the data, written and physical. This procedure requires substantial skill, experience, and judgment on the part of the surveyor. Frequently, a preliminary examination and computations of all the evidence allows calculation of a bearing and distance from a traverse control station to the possible location of a property corner or monument for which there is a call in the deed. If the called-for corner was not found during the initial survey, a return trip should be made to the field with every effort being expended to locate the corner. The bearing and distance can be set off from the designated traverse station and the area examined very carefully to locate any evidence pertaining to the old corner. If no evidence exists on the surface, the soil should be carefully sliced away in increments in an effort to detect the rotted remains of an old corner stake or post or the remains of an old monument.

When a final solution has been reached, the property corners chosen as those that best fit all the data are coordinated and ties by direction and distance to the nearest traverse control stations are computed. Using these computed ties, the property corners, as determined from the final evaluation, are set off in the field from the traverse control stations.

With the property corners established, structures on the tract are located by perpendicular offsets from the nearest property line, and a map is prepared showing the property corners, property lines, fence lines, and buildings. Figure 18.9 illustrates such a map, sometimes referred to as a *plat* or *plot plan*.

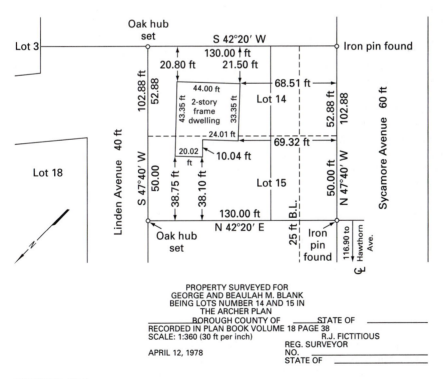

FIGURE 18.9
Plat or plot plan for a property survey.

The map is submitted to the client along with a report stating exactly what evidence was found and what procedure was followed in reestablishing the property corners.

Depending on the state, county, or local regulations, it also may be necessary for the surveyor to file a record of survey in the appropriate government office. Many of the more densely populated counties in several states now have statutes requiring that any division of land be approved by the county surveyor and that the map of the survey be filed with the county recorder.

Obviously, the procedure as outlined is an extensive and expensive project. The cost of such a survey would be more than one could reasonably expect to charge a client for a lot survey. However, the method as described is necessary when doing a first job in an area with which the surveyor is unfamiliar. Subsequent surveys in the same vicinity can be accomplished with much less expenditure of time and money. Thus, the initial investment in a complete search and survey of the area will be returned with dividends. Surveyors frequently must indulge in a certain amount of speculation in order to become familiar with the conditions that govern the locations of boundaries in a given region. In order to exploit the efforts invested in an initial survey, field notes and office records for the survey should be rigorously maintained and all relevant monuments and control traverse stations must be very carefully referenced for future use.

18.17
SUBDIVISION SURVEY OF RURAL LAND

A subdivision survey of rural land implies a survey conducted to subdivide into two or more tracts, in accordance with some prearranged plan, an area whose boundaries already are established. In such cases, a resurvey of the tract is run, new monuments are established on the new boundary lines, and a new plat and description are prepared as in the case of an original survey.

Public Lands

The public lands are divided into townships, sections, and quarter sections by United States land surveyors, in a manner prescribed by law. In general, the United States surveys establish the boundaries of sections and establish quarter-section corners on section lines; and any further subdivision is made after the lands have passed into the hands of private individuals, the work being carried out by surveyors in private practice. Subdivisions of this kind are described in detail in Sections 18.35 through 18.37.

Irregular Subdivisions

Surveys of this class are conducted for a variety of purposes. The following examples illustrate the procedure for certain cases.

EXAMPLE 18.7. A railroad is to traverse the land belonging to Black, and the railroad company desires to secure title to a right of way of a definite width on either side of the center line, which already has been surveyed and marked with stakes. A description of Black's tract has been secured.

The right-of-way surveyor reruns the boundaries of Black's tract that are intersected by the railroad line, establishes the directions of the right-of-way boundaries parallel with the center line, and sets monuments at the intersections of these boundaries with those of Black's tract. The surveyor then makes a survey of the tract thus defined, securing sufficient data that the lengths and directions of the boundaries of the right-of-way tract now within Black's tract are obtained. He or she also ties right-of-way corners that he (or she) has established to the nearest old corners of Black's property.

With these data, the area of the right-of-way tract is calculated and a description of the tract is prepared as for an original survey (see Sections 18.7, 18.8, and 18.15). The point of beginning is referred to one of the old monuments marking the original tract, and not to the center line of the railroad.

EXAMPLE 18.8. It is stipulated in the will of Green that his New England farm is to be divided equally among his three sons, each to have an equal frontage on the highway that forms one of the boundaries of the tract. The farm is of irregular shape and has not been surveyed for many years.

In a case of this kind, the surveyor first makes a resurvey of the entire tract. From the data thus obtained the area is calculated. In connection with the resurvey, subdivision corners are established on the highway.

The simplest division is one for which the subdividing lines are straight, each cutting off the required area from the given tract. With the area of the entire tract known, the area to be received by each son is calculated and the length and bearing of each of the lines rendering the subdivision are computed, as described in Section 8.31 through 8.35.

Finally, from each of the two subdivision corners already established on the highway frontage, the surveyor lays off the computed direction of the subdividing line through that point and establishes the remaining unknown corner at the point where this subdividing line intersects the opposite boundary. The distances from this latter corner to adjacent corners are measured, and the survey is considered checked if these measured distances agree closely with the computed lengths of the same courses. A plat is drawn, as for an original survey, the lengths and bearings of all lines and the area of each subdivision being shown; and a description of each of the three subdivisions is prepared.

18.18
SUBDIVISION SURVEY OF URBAN LAND

As a city or town develops, unimproved lands are subdivided into lots, which are placed on sale as residential or business property. In most instances, such extensions result from the activities of real-estate operators who acquire a tract of rural land of considerable area, develop a plan of subdivision that is approved by the authorities of the municipality to which the tract is to be attached, and cause surveys to be made to establish the boundaries of individual lots. A tract thus divided according to an acceptable plan is known as an *addition* or *subdivision*.

For large and important developments, the work of originating the general plan often is carried out by persons specializing in city planning and landscape architecture, under whose direction the surveyor works. Such developments require a high degree of skill, and usually extensive surveys (particularly in hilly sections) are carried out before the actual plan of subdivision can be decided on. Problems of this character can be adequately discussed only in treatises on city planning. However, it is appropriate to state here that the preliminary studies should consider the probable future character of the district; the probable location of business sections; the probable magnitude, direction, and character of future traffic; the topography of the land; the location, width, grade, and character of paving of the streets; the size and shape of lots and blocks; the location and size of storm and sanitary sewers; and the disposition of electric and telephone wires and cables.

For the ordinary real-estate development, the owner usually calls for the services of an engineer or surveyor who has had experience in such work. The surveyor confers with the owner, and they discuss a general plan. The surveyor makes a resurvey of the entire property; and if the character of the topography is irregular, certain preliminary surveys to find the location and elevation of the governing features of the terrain are made. In some cases a complete topographic survey may be made. With the general plan fixed and having studied the results of the field investigation and considered the items listed in the previous paragraph, the surveyor works out a detailed plan on paper, showing on the drawing the names of all streets and the numbers of all blocks and lots, the dimensions of all lots, the width of streets, the length and bearing of all street tangents, and the radius and length of all street curves. The surveyor also prepares a report that, in addition to a discussion of the plan of subdivision, may consider the cost of subdividing including not only the establishing of boundaries but also the work of grading, paving, constructing sewers, and landscaping.

This detailed plan, when approved by the owner, is submitted to the governing body in the municipality. If it meets with the requirements of this body, it is approved.

On the authority of the owner, the surveyor then proceeds to execute the necessary subdivision surveys, including the laying out of roads, walks, blocks, and lots. Often the lot and block corners are marked with permanent monuments; but in many cases, contrary to what may be considered good practice, the lot corners are marked by wooden stakes. When the surveys are completed, the map of the subdivision is revised to show minor changes made during the survey, together with the location and character of permanent monuments. A tracing is submitted to the municipality, and this, when duly signed by those in authority, becomes the official map of the subdivision. It then becomes a part of the public records and

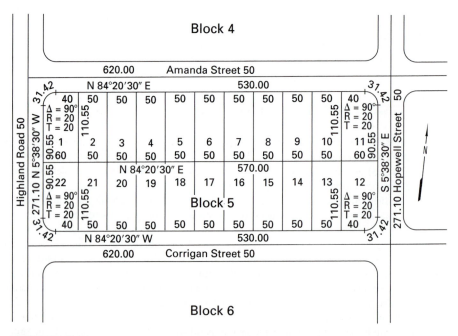

FIGURE 18.10a Portion of a subdivision lot plan plotted by hand.

FIGURE 18.10b
Portion of a subdivision lot plan plotted automatically. (*Courtesy of Raymond Vail and Associates, Sacramento, California.*)

usually is filed in the registry of deeds of the county in which the municipality lies. On this approval, if the subdivision is outside the corporate limits of the municipality, the limits are extended to include it.

Sample portions of typical subdivision plans are shown in Figure 18.10a and b. The line work, distances, and directions on the subdivision plan illustrated in Fig 18.10b were plotted automatically (see Section 14.14). The latter plan is in a form suitable for recording as the final official map of the subdivision.

Resurveys within a Subdivision

When land is bought and sold within a subdivision, often a resurvey is required to relocate lost property corners. These surveys, in general, are performed according to the procedures set forth in Section 18.16. Because an official recorded map contains all distances and directions, which are referenced to monuments and lot corners, many of which exist on the ground, the problems related to subdivision resurveys are not so complex as those found in metes and bounds resurveys. However, problems do occur. Many older subdivision plans are incomplete and contain incorrect information. Furthermore, incorrect information can occur on any such plan, even one that was approved and presumably satisfied all requirements.

Prior to performing the survey, always calculate a mathematical closure using recorded map data not only for the lot to be surveyed but also for the entire block within which the lot falls. When the field survey is performed, always close around the block in which the lot falls. In this way the surveyor can be certain that the lot corners established by the resurvey do not conflict with any other property owner in the block.

Because all lots set off according to a recorded map for a given subdivision are created simultaneously, all property owners have equal rights and there are no senior deeds. Therefore, if the surveyor finds an excess or shortage of distance within the block in which a lot is being surveyed, it is distributed proportionately to all the lots within the block except where there is an obvious mistake in the exterior survey, which is corrected where it occurs.

For example, in Figure 18.10a, assume that the corners for lot 6 in block 5 are to be resurveyed. The center lines of Highland Road and Amanda and Hopewell Streets are established from monuments and original property corners referenced in the original survey notes for the subdivisions. Next, the distance of 285.00 ft (86.868 m) is measured from the center-line intersection of Hopewell and Amanda Streets to the east boundary of lot 6. Temporary points are set on the center line of Amanda Street at the east and west lot lines of lot 6 extended. Then the distance is measured from the west boundary of lot 6 along the center line of Amanda Street to the center line of Highland Road. If this distance measured in the field compares with the map distance to yield an acceptable relative accuracy for the block, the property corners for lot 6 are set from the temporary points on the center line of Amanda Street.

However, suppose that an excess of 0.60 ft in the dimension of block 5 was detected on closing at the center line of Highland Road. After double checking to be sure there are no errors in measurement, such as uncorrected systematic errors or mistakes, that the plan dimensions close for block 6, and that the center-line intersections are correctly established, the excess of 0.60 ft is distributed proportionately among all the lots in block 5. Because the distance corner to corner is 570 ft, lots 1 and 11 receive $(60/570)(0.60) = 0.063$ ft $(0.019$ m) each and the remaining lots receive $(50/570)(0.60) = 0.053$ ft $(0.016$ m) each. Consequently the distance from the center line of Hopewell Street to the east boundary of lot 6 is 285.274 ft (79.332 m), the frontage for lot 6 is 50.053 ft (15.256 m), and the distance from the west boundary of lot 6 to the center line of Highland Road should check with a measurement of 285.274 ft (86.952 m). Distances are carried to the third place to avoid round-off error in cumulating the proportional change in lot dimensions. In practice, all distances would be rounded to the second place using feet or to the third place using meters.

18.19
CITY SURVEYING

The term *city survey* refers to an extensive coordinated survey of the area in and near a city to fix reference monuments, locate property lines and improvements, and determine the configuration and physical features of the land for a city lot. Such a survey is of value for a wide variety of purposes, particularly for planning city improvements. Briefly, the work consists of the following:

1. Selecting an appropriate *geographic or land information system* (GIS or LIS, see Section 14.18) so that all subsequent field and office work can be organized in a manner compatible with the selected information system. If the city already has a GIS or LIS in place, every effort must be made to design the data acquisition and processing methods to accommodate this system.

2. Establishing *horizontal and vertical control,* as described for topographic surveying. The primary horizontal control is usually by static GPS survey, supplemented as desired by precise total station traversing. Secondary horizontal control is established by total station traversing of appropriate precision. Photogrammetric methods for establishing horizontal control are feasible for setting control points in the secondary network. All horizontal control should be tied to the state plane coordinate system. Primary vertical control is by precise leveling and should be referred to the North American vertical datum 1988 (NAVD 88; Section 14.2).

3. Making a *topographic survey* and a *topographic map.* Usually, the scale of the map may range from 1:1200 to 1:2400 (1:1000 to 1:2500 in the metric system) with contour intervals of from 0.5 m to 2 m (1 ft to 5 ft). Photogrammetric methods for map compilation should be utilized.

4. *Monumenting* a system of selected points at suitable locations such as street corners, for reference in subsequent surveys. These monuments are referred to the state plane coordinate system and the national datum.

5. Making a *property map.* The survey for the map consists of collecting recorded information regarding property; determining the location on the ground of street intersections, angle points, and curve points; monumenting the points so located; and determining the coordinates of the monuments. Usually, the scale of the property map is 1:500 in metric units (1:600 or 50 ft/in.). The property map shows the length and bearing of all street lines and boundaries of public property, coordinates of governing points, control, monuments, important structures, natural features of the terrain, and so forth, all with appropriate legends and notes.

6. Making a *small-scale map,* which shows essentially the same information as the topographic map but drawn to a smaller scale; the scale should be no less than 1:25,000 (2000 ft per in. or 1:24,000).

7. Making a *map of underground utilities.* Usually, the scale and the size of the map sheets are the same as those for the property map. The underground map shows street and easement lines, monuments, surface structures and natural features affecting underground construction, and underground structures and utilities (with dimensions), all with appropriate legends and notes.

All map data should be stored in the GIS or LIS in digital format so that a given section or sections of the city with the desired attributes (topography, property lines, underground utilities, and so on) can be retrieved and either displayed on a video screen (individually or as overlays) or automatically plotted as a hard-copy line map.

18.20
CADASTRAL SURVEYING

Cadastral surveying is a general term referring to extensive surveys relating to land boundaries and subdivisions made to create units suitable for transfer or to define limitations of title. The expression is derived from the word *cadastre,* meaning register of the real property of a political subdivision with details of area, ownership, and value (see American Society of Civil Engineers, 1994). The term is applied to the U.S. public-land surveys (Sections 18.21 to 18.48) by the U.S. Bureau of Land Management and also may be used to describe corresponding surveys outside the public lands. However, the term *property, land,* or *boundary surveys* usually is used by preference.

A cadastral map shows individual tracts of land with the corners, length and azimuth or bearing of boundaries, acreage, ownership, and sometimes the cultural and drainage features. The surveying methods are the same as those described for topographic surveying for maps of intermediate and large scale (Chapter 15).

18.21
U.S. PUBLIC-LANDS SURVEYS: GENERAL

The following sections deal with the methods of subdividing the public lands of the United States in accordance with regulations imposed by law. The public lands are subdivided into townships, sections, and quarter sections—in early years by private surveyors under contract, later by the Field Surveying Service of the General Land Office, and currently by civil service employees of the Bureau of Land Management, which succeeded the General Land Office in 1946. Further subdivision of such lands is made after the lands have passed into the hands of private owners, the work being carried out by surveyors in private practice.

The methods described here are those now in force, but with minor differences they have been followed in principle since 1785, when the rectangular system of subdivision was inaugurated. Under this system, the public lands of 30 states have been or are in progress of being surveyed. In general, these methods of subdividing land do not apply in the 13 original states and in Hawaii, Kentucky, Tennessee, Texas, and West Virginia. As the progress of the public-land surveys has been from east to west, the details in states east of the Mississippi River differ somewhat from those of present practice.

The laws regulating the subdivision of public lands and the surveying methods employed are fully described in the *Manual of Instructions for the Survey of the Public Lands of the United States,* published by the Bureau of Land Management (1973). Much of the material for this chapter is drawn from this manual.

Field notes and plats of the public-land surveys may be examined in the regional offices of the bureau, and copies may be procured for a nominal fee.

18.22
GENERAL SCHEME OF SUBDIVISION

The regulations for the subdivision of public lands have been altered from time to time; hence, the methods employed in surveying various regions of the United States show marked differences, depending on the dates when the surveys were made. In general principle, however, the system has remained unchanged, the primary unit being the *township,* bounded by meridional and latitudinal lines and as nearly as may be 6 mi (9.66 km) square. The township is divided into 36 secondary units called *sections,* each as nearly as may be 1 mi (1.61 km) square. Because the meridians converge (Section 18.29), it is impossible to lay out a square township by such lines; and because the township is not square, not all the 36 sections can be 1 mi square even if all measurements are without error.

18.23
STANDARD LINES

Since the time of the earliest surveys, the townships and sections have been located with respect to principal axes passing through an origin called an *initial point;* the north-south axis is a true or astronomic meridian called the *principal meridian,* and the east-west axis is an astronomic parallel of latitude called the *base line.*

The principal meridian is given a name to which all subdivisions are referred. Therefore, the principal meridian that governs the rectangular surveys (wholly or in part) of the states of Ohio and Indiana is called the *First Principal Meridian;* its longitude is 84°48′11″W, and the latitude of the base line is 40°59′22″N. The extent of the surveys

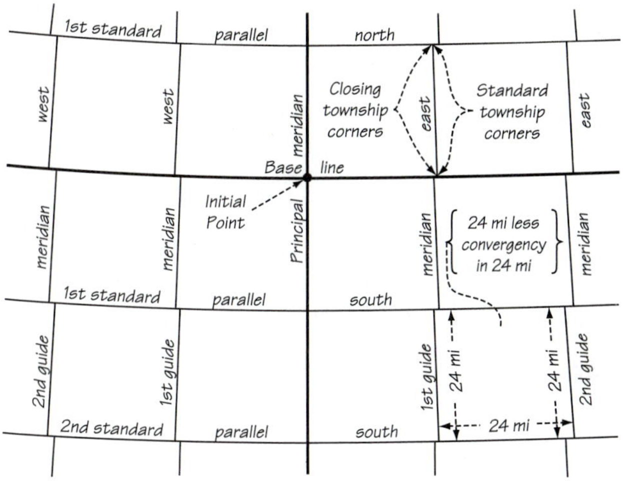

FIGURE 18.11
Standard lines.

referred to a given initial point may be found by consulting a map, published by the Bureau of Land Management, entitled "United States, Showing Principal Meridians, Base Lines, and Areas Governed Thereby," or from the U.S. Bureau of Land Management (1973).

Secondary axes are established at intervals of 24 mi (38.62 km) east or west of the principal meridian and at intervals of 24 mi north or south of the base line, dividing the tract being surveyed into quadrangles bounded by astronomic meridians 24 mi long and by true or astronomic parallels, the south boundary of each quadrangle being 24 mi long, and the north boundary being 24 mi long less the convergence of the meridians in that distance. (In some early surveys, these distances were 30 (48.28 km) or 36 (57.94 km) mi.) The secondary parallels are called *standard parallels* or *correction lines,* and each is continuous throughout its length. The secondary meridians are called *guide meridians,* and each is broken at the base line and at each standard parallel.

The principal meridian, base line, standard parallels, and guide meridians are called *standard lines.*

A typical system of principal and secondary axes is shown in Figure 18.11. The base line and standard parallels, being everywhere perpendicular to the direction of the meridian, are laid out on the ground as curved lines, the rate of curvature depending on the latitude. The principal meridian and guide meridians, being astronomic north and south lines, are laid out as straight lines but converge toward the north, the rate of convergence depending on the latitude.

Standard parallels are counted north or south of the base line; the *second standard parallel south* indicates a parallel 48 mi (77.25 km) south of the base line. Guide meridians are counted east or west of the principal meridian; the *third guide meridian west* is 72 mi (154.50 km) west of the principal meridian.

18.24
TOWNSHIPS

The division of the 24-mi (38.62 km) quadrangles into townships is accomplished by laying off astronomic meridional lines, called *range lines,* at intervals of 6 mi (9.66 km) along each standard parallel, the range line extending north 24 miles to the next standard parallel, and

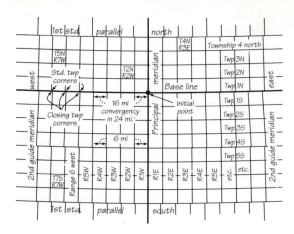

FIGURE 18.12
Township and range lines.

by joining the township corners established at intervals of 6 mi (9.66 km) on the range lines, guide meridians, and principal meridian with latitudinal lines called *township lines*.

The plan of subdivision is illustrated by Figure 18.12. A row of townships extending north and south is called a *range;* and a row extending east and west is called a *tier*. Ranges are counted east or west of the principal meridian and tiers are counted north or south of the base line. Usually, for purposes of description the word *township* is substituted for *tier*. A township is designated by the number of its tier and range and the name of the principal meridian. For example, T7S, R7W (read *township seven south, range seven west*) designates a township in the seventh tier south of the base line and the seventh range west of the principal meridian.

18.25
SECTIONS

The division of townships into sections is performed by establishing, at intervals of 1 mi (1.61 km), "meridional" lines parallel to the east boundary of the township and by joining the section corners established at intervals of 1 mi with straight latitudinal lines. (These lines are not exactly meridional but are parallel to the east boundary of the township, which is a meridional line.) These lines, called *section lines,* divide each township into 36 sections, as shown in Figure 18.13. The sections are numbered consecutively from east to west and from west to east, beginning with no. 1 in the northeast corner of the township and ending with no. 36 in the southeast corner. Thus, section 16 is a section whose center is $3\frac{1}{2}$ mi (5.63 km) north and $3\frac{1}{2}$ mi (5.63 km) west of the southeast corner of a township.

A section is legally described by giving its number, the tier and range of the township, and the name of the principal meridian; for example, Section 16, T7S, R7W, of the Third Principal Meridian.

On account of the convergence of the range lines (true or astronomic meridians) forming the east and west boundaries of townships, the latitudinal lines forming the north and south boundaries of townships are less than 6 mi (9.66 km) in length, except for the south boundary of townships that lie just north of a standard parallel. As the north-south section lines are run parallel to the *east* boundary of the township, it follows that all sections except those adjacent to the west boundary will be 1 mi (1.61 km) square, but that those adjacent to the west boundary will have a latitudinal dimension less than 1 mi by an amount

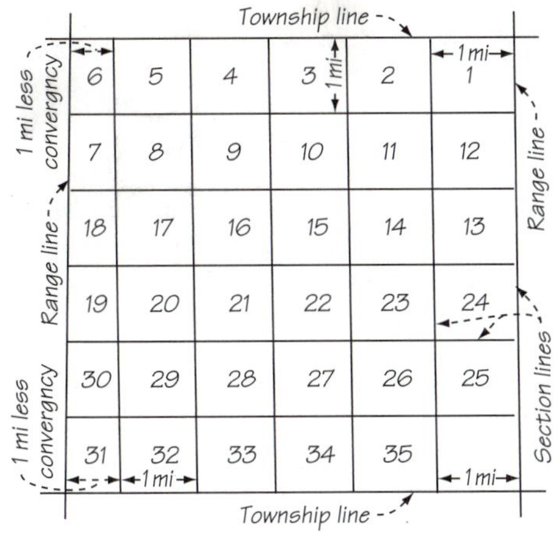

FIGURE 18.13
Numbering of sections.

1023

CHAPTER 18:
Land Surveys

equal to the convergence of the range lines within the distance from the section to the nearest standard parallel to the south.

The subdivision of sections is described in Sections 18.35 through 18.37.

18.26
STANDARD CORNERS

Corners called *standard corners* are established on the base line and standard parallels at intervals of 40 chains or 2640 ft (804.67 m); these standard corners govern the meridional subdivision of the land lying between each standard parallel and the next standard parallel to the north. Other corners, called *correction corners* or *closing corners,* are established later on the base line and standard parallels during the process of subdivision; these corners fall at the intersection of the base line or standard parallel either with the meridional lines projected from the standard township corners of the next standard parallel to the south (see Figure 18.12) or with the intermediate section and quarter-section lines. Standard parallels also are called *correction lines.*

18.27
IRREGULARITIES IN SUBDIVISION

It should be understood that the plan of subdivision just described is the one carried out when conditions allow. Of course, errors of measurement always are present, so that the actual lengths and directions established in the field do not entirely agree with the theoretical values. In addition, conditions met in the field often make it inexpedient or impossible to establish the lines of the survey in exact accordance with the specified plan. There are numerous instances of standard parallels and guide meridians having been originally established at intervals of 30 and 36 mi (48.27 and 57.92 km), under old regulations, and regions having been only partly surveyed. Later, under present regulations, meridians have been established between the old guide meridians; and recent subdivisions, therefore, are referred to standard lines, many of which are less than 24 mi (38.62 km) apart. Also, large bodies of water, mountain ranges, Indian reservations, and such may greatly modify the method of division, many townships and sections being made fractional.

18.28
ESTABLISHING STANDARD LINES

Principal Meridian

The principal meridian is established as an *astronomic meridian* through the initial point, either north or south or in both directions, as conditions require. Permanent quarter-section and section corners are established alternately at intervals of 40 chains ($\frac{1}{2}$ mi or 0.80 km), and regular township corners are placed at intervals of 480 chains (6 mi or 9.66 km).

Two independent linear measurements are taken. When the discrepancy between two sets of measurements taken in the prescribed manner exceeds 20 links (13.02 ft or 4.02 m) per mile, the line must be remeasured to reduce the difference. The use of electronic distance measurement (EDM) equipment or the global positioning system is feasible and simplifies setting corners substantially. The corners are set at the mean distances. When successive independent tests of the alignment, as determined by astronomical observations, indicate that the line has departed from the astronomic meridian by more than 03', the necessary correction must be made to reduce the deviation in azimuth.

Base Line

From the initial point, the base line is extended east and west on a true or astronomic parallel of latitude, standard quarter-section and section corners being established alternately at intervals of 40 chains ($\frac{1}{2}$ mi or 0.80 km) and standard township corners being placed at intervals of 480 chains (6 mi or 9.66 km). The manner of taking the linear measurements of the base line and the required precision of both linear measurements and alignment are the same as for the survey of the principal meridian. Any of the three methods described in Section 18.30, for laying out the astronomic latitude curve, may be used.

Standard Parallels

At intervals of 24 mi (38.62 km) north and south of the base line, true parallels of latitude, called *standard parallels* or *correction lines,* are run east and west from the principal meridian, these lines being established in a manner identical with that prescribed for the survey of the base line.

Guide Meridians

The guide meridians are extended north from the base line and standard parallels at intervals of 24 mi (38.62 km) east and west of the principal meridian. Each guide meridian is established as a true meridian in a manner identical to that employed in laying off the principal meridian. The guide meridians terminate at the points of their intersection with the standard parallels and hence are broken lines, each segment being theoretically 24 mi (38.62 km) long. Errors of measurement are placed in the most northerly half mile (0.80 km) of each 24-mi segment. At the point of intersection of the guide meridian and standard parallel, a township corner, called a *correction corner* or *closing corner,* is established by retracing the standard parallel between the first standard corners to the east and west of the point for the closing corner; and the distance from the closing corner to the nearest standard corner on the standard parallel is measured.

18.29
CONVERGENCE OF MERIDIANS

The angular convergence of two meridians is a function of the distance between the meridians, the latitude, and the dimensions of the reference ellipsoid. In Figure 18.14, d is the distance between the meridians along parallel of latitude AB that lies at latitude ϕ. The angular convergence in seconds is

$$\hat{\theta} = \frac{d \tan \phi (1 - \epsilon^2 \sin^2 \phi)^{1/2}}{a} \rho \qquad (18.1)$$

where a = semimajor axis of the ellipsoid = 6378.137 km = 3963.1906 mi
   ϵ = eccentricity of the ellipsoid (GRS 80)
    = 0.081819191
   ρ = 206,265″/rad

The linear convergence, c, of two meridians having length l and separated by a distance d (Figure 18.14) is given by

$$c = \frac{ld \tan \phi (1 - \epsilon^2 \sin^2 \phi)^{1/2}}{a} \qquad (18.2)$$

in which c, l, d, and a are in the same units.

EXAMPLE 18.9. Find the angular convergence of two guide meridians 24 mi (38.62 km) apart at latitude 43°20′.

Solution. By Equation (18.1),

$$\hat{\theta} = \frac{(24)(\tan 43°20′)[1 - (0.081819191)^2 \sin^2 43°20′]^{1/2}(206,265)}{(3963.2)}$$

$$= 1176.6″$$

$$= 0°19′36.6″$$

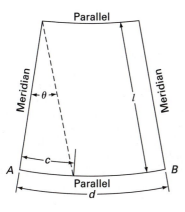

Parallel

Meridian

θ

l

Meridian

c

A

B

Parallel

d

FIGURE 18.14
Convergence between two meridians.

EXAMPLE 18.10. Find the linear convergence for the data in Example 18.9.

Solution. By Equation (18.2),

$$c = \frac{(24)^2(\tan 43°20')[1 - (0.081819191)^2 \sin^2 43°20']^{1/2}}{3963.2} = 0.136902 \text{ mi}$$

$$= 10.952 \text{ chains} = 722.84 \text{ ft} = 220.323 \text{ m}$$

Note in Example 18.9 that the convergence of the two guide meridians is nearly one-third of a degree. In Example 18.10, the linear convergence is nearly 11 chains (726 ft or 220 m) in 24 mi; this amount represents the jog at the correction line in the first guide meridian east or west, or one-half of the jog in the second guide meridian.

Meridional Section Lines

In the subdivision of townships into sections, the establishment of section lines parallel to the east boundary of the township necessitates a correction in azimuth of these section lines on account of the angular convergence of the meridians. While meridional section lines are being run north, they are made to deflect to the left or west of the true meridian by an angle equal to the convergence in the distance to the section line from the east boundary. The deflections from the true meridian for section lines, respectively, 1, 2, 3, 4, and 5 mi (1.61, 3.22, 4.83, 6.44, and 8.05 km) west of the east boundary of the township, can be calculated using Equation (18.1) with d = 1, 2, 3, 4, and 5 mi, respectively, and θ = the latitude of the place.

18.30
LAYING OFF A PARALLEL OF LATITUDE

As the base line, standard parallels, and latitudinal township lines are astronomic parallels of latitude, they are curved lines when established on the surface of the earth. This is evident from the fact that meridians converge and that a parallel of latitude is a line whose direction at any point is perpendicular to the direction of the meridian at that point. It is defined by a plane at right angles to the earth's polar axis cutting the earth's surface on a circle whose radius is less for higher latitudes. The rate of curvature within the latitudes of the United States is so small that two points $\frac{1}{4}$ mi (0.4 km) apart on the same parallel of latitude, for all practical purposes, will define the direction of the curve at either point; but the continuation of a line so defined in either direction would describe a great circle of the earth, gradually departing southerly from the astronomic parallel. The great circle tangent to the parallel at any point along the parallel is called the *tangent to the parallel*, and it coincides with the astronomic latitude curve only at the point of tangency.

Although the tangent to the parallel is a straight line in plan, its bearing is not constant but varies with the distance from the point of tangency, the deflection from true east or true west being equal to the angle of convergence of the meridians within the distance from the point of tangency to the given point. Hence, the angles of convergence given by Equation (18.1) for a specified distance and latitude also represent the deviations in azimuth of the tangent from the parallel.

Within the limits of precision necessary in land surveying, the offset from tangent to parallel at any distance from the point of tangency is one-half of the linear convergence of the meridians within the same distance. In Figure 18.15, let a_i be the tangent offset, where

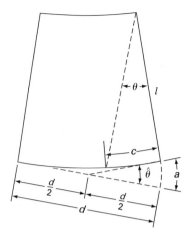

$i = (\frac{1}{2}, 1, 1\frac{1}{2}, \ldots, 5\frac{1}{2}, 6)$; c = the linear convergence calculated by Equation (18.2); $\hat{\theta}$ = the angular convergence calculated by Equation (18.1); and d = the distance along the parallel between the meridians. Then, from Figure 18.15,

$$c = \hat{\theta}d$$

$$a = \hat{\theta}\frac{d}{2} \tag{18.3}$$

$$a = \frac{c}{2} \quad \text{(approximate)}$$

Hence, values one-half as great as the values of the linear convergence in 6 mi (9.66 km) calculated by Equation (18.2) represent the offset from tangent to parallel, measured along the meridian, at a distance of 6 mi from the point of tangency. With small error, a parallel of latitude, within the limits of distance here considered, may be assumed to behave as a parabola. Hence, the offset from tangent to curve at any point, for all practical purposes, may be said to vary as the square of the distance from the point of tangency; and the offsets at $\frac{1}{2}$, 1, $1\frac{1}{2}$, 2, and so on mi from the point of tangency would bear to the offset at 6 mi the ratios $\frac{1}{144}$, $\frac{1}{36}$, $\frac{1}{16}$, $\frac{1}{9}$, and so on, respectively.

Three general methods of establishing a true or astronomic parallel of latitude may be employed independently to arrive at the same result: (1) the solar method, (2) the tangent method, and (3) the secant method. The secant method is most commonly employed in heavily wooded areas because offsets from the secant line to the parallel are small, eliminating a lot of brush cutting.

Solar Method

The azimuth at each 40-chain point is determined by solar observation, according to the methods outlined in Chapter 10, Section 10.31, and the true parallel may be established by turning an angle of 90° in either direction from the meridian. If sights taken with the telescope pointing in the latter direction are no longer than 20 to 40 chains (1320 to 2640 ft or 402.3 to 804.7 m), the line thus defined will not depart appreciably from the true parallel.

Tangent Method

The tangent method consists in determining the true meridian at the point of tangency, from which the tangent to the parallel is established by laying off an angle of 90°. The tangent is extended in a straight line for a distance of 6 mi and as each distance of 40 chains ($\frac{1}{2}$ mi) is laid off along the tangent, the corresponding section or quarter-section corner is established on the parallel by laying off along the meridian the appropriate offset from tangent to parallel.

At the end of 6 mi a new tangent is laid off, and the process just described is repeated. The values of the offsets may be found using Equation (18.2).

EXAMPLE 18.11. Compute tangent offsets for a standard parallel at latitude 44°30′. Figure 18.16 shows the configuration of a tangent to the parallel.

Solution. Calculate the linear and angular convergence at 6-mi increments by Equations (18.2) and (18.1), respectively.

$$c = ld \frac{\tan (44°30′)[1 - (0.0818191910)^2 \sin^2 44°30′]^{1/2}}{3963.2}$$

$$= (6)(6)(0.000247547) = 0.008912 \text{ mi} = 47.05 \text{ ft} = 14.342 \text{ m} = 71.3 \text{ links}$$

$$\hat{\theta}'' = \frac{(6) \tan (44°30′)[1 - (0.0818191910)^2 \sin^2 44°30′]^{1/2}(206,265)}{3963.2}$$

$$= (6)(0.000247547)(206,265) = 306.4''$$

$$= 0°05′06.4''$$

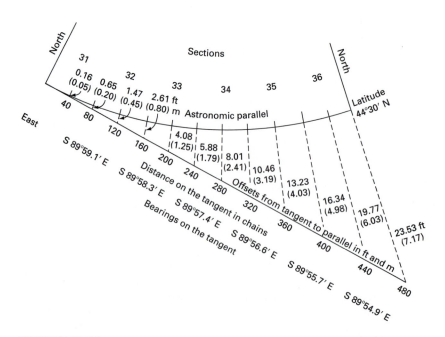

FIGURE 18.16
Tangent method for establishing a parallel of latitude (44°30′N).

The tangent offset at d = 6 mi is approximately $c/2 = 23.53$ ft $= 7.17$ m $= 35.7$ links. Tangent offsets at the $\frac{1}{2}$, 1, $1\frac{1}{2}$, 2, $2\frac{1}{2}$, 3, and so forth, mile points are proportional to the squares of distances from the point of tangency. Therefore, the tangent offset from the $\frac{1}{2}$-mi point $= (\frac{1}{144})(23.53$ ft$) = 0.163$ ft, and so on. Results of these computations for each $\frac{1}{2}$ mi are given in Table 18.1 in links, feet, and meters.

The angles of convergence for each 1-mi point are $\frac{1}{6}$, $\frac{1}{3}$, $\frac{1}{2}$, $\frac{2}{3}$, and $\frac{5}{6}$ of the convergence at the 6-mi point. These values also are listed in Table 18.1, in column 6, to the nearest 10th of a second and are used to compute bearings along the tangent to the nearest 10th of a minute, as shown in Figure 18.16.

Secant Method

The secant method is a modification of the tangent method, in which the secant is a straight line 6 mi in length forming the arc of a great circle, which intersects the true parallel at the end of the first and fifth miles from the point of beginning, as illustrated by Figure 18.17. For the latitude of the given parallel, the offsets (in feet and meters) from secant to parallel are given in the figure, at intervals of $\frac{1}{2}$ mi. From the figure it is clear that the secant is parallel with a tangent to the parallel at the end of the third mile (240 chains); hence, the offset south from the 3-mile point on the secant line to the corner on the true parallel is the same as the offset from the tangent to the parallel in a distance of 2 mi. Also, it is evident that the offset south of the point of beginning to the initial point on the secant and the offset north of the secant to the true parallel at the end of the sixth mile are equal to the difference between the tangent offset in a distance of 3 mi and the tangent offset in a distance of 2 mi.

In general, offsets from the secant line to corners on the parallel at $\frac{1}{2}$-mi intervals are calculated by combining the appropriate tangent offsets computed using Equations (18.2) and (18.3). Let the secant offset be designated $(so)_i$, where $i = (0, \frac{1}{2}, 1, \ldots, 5\frac{1}{2}, 6)$. The determination of the secant offsets for the example illustrated in Figure 18.17, where the latitude of 44°30′N is the same as for Example 18.11, is as follows (note: use the tangent offsets $a_{\frac{1}{2}}, a_1, \ldots, a_6$ from Table 18.1):

$$(so)_0 = a_3 - a_2 = 5.88 - 2.61 = 3.27 \text{ ft } (1.00 \text{ m})$$

$$(so)_0 = (so)_6$$

$$(so)_{\frac{1}{2}} = a_{2\frac{1}{2}} - a_2 = 4.08 - 2.61 = 1.47 \text{ ft } (0.45 \text{ m})$$

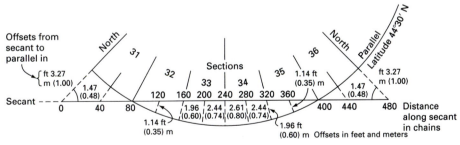

FIGURE 18.17
Secant-line method for establishing parallel of latitude (latitude 44°30′N).

TABLE 18.1
Tangent offsets and angular convergence

Distance along parallel d, mi	$d^2/36$	Tangent offset			Angular convergence
		links	ft	m	
$\frac{1}{2}$	$\frac{1}{144}$	0.2	0.16	0.050	
1	$\frac{1}{36}$	1.0	0.65	0.199	51.1″
$1\frac{1}{2}$	$\frac{1}{16}$	2.2	1.47	0.448	
2	$\frac{1}{9}$	4.0	2.61	0.797	1′42.1″
$2\frac{1}{2}$	$\frac{25}{144}$	6.2	4.08	1.245	
3	$\frac{1}{4}$	8.9	5.88	1.793	2′33.2″
$3\frac{1}{2}$	$\frac{49}{144}$	12.1	8.01	2.440	
4	$\frac{4}{9}$	15.8	10.46	3.187	3′24.3″
$4\frac{1}{2}$	$\frac{9}{16}$	20.0	13.23	4.033	
5	$\frac{25}{36}$	24.8	16.34	4.980	4′15.3″
$5\frac{1}{2}$	$\frac{121}{144}$	30.0	19.77	6.025	
6	1	35.6	23.53	7.160	5′06.4″

$$(so)_{\frac{1}{2}} = (so)_{5\frac{1}{2}}$$

$$(so)_1 = 0 = (so)_5$$

$$(so)_{1\frac{1}{2}} = a_2 - a_{1\frac{1}{2}} = 2.61 - 1.47 = 1.14 \text{ ft } (0.35 \text{ m})$$

$$(so)_{1\frac{1}{2}} = (so)_{4\frac{1}{2}}$$

$$(so)_2 = a_2 - a_1 = 2.61 - 0.65 = 1.96 \text{ ft } (0.60 \text{ m})$$

$$(so)_2 = (so)_4$$

$$(so)_{2\frac{1}{2}} = a_2 - a_{\frac{1}{2}} = 2.61 - 0.16 = 2.45 \text{ ft } (0.75 \text{ m})$$

$$(so)_{2\frac{1}{2}} = (so)_{3\frac{1}{2}}$$

$$(so)_3 = a_2 = 2.61 \text{ ft } (0.80 \text{ m})$$

Owing to the convergence of meridians, the azimuth of the secant—a straight line in plan—varies along its length. If the secant is laid off toward the east, the direction of the secant from the point of beginning to the end of the third mile is north of astronomic east, and beyond the end of the third mile is south of astronomic east, the variation from astronomic east increasing directly with the distance in either direction from the 3-mi point. At the 3-mi point, the secant bears astronomic east; at the point of beginning, the secant bears north of east by an amount equal to the angular convergence of meridians 3 mi apart; and at the end of the sixth mile, the secant bears south of east by the same amount.

For the data of Figure 18.17, where $\phi = 44°30'$N, the values for angular convergence by Equation (18.1) for distances d of 1, 2, and 3 mi are

$$\hat{\theta}_1'' = (1)\frac{\tan(44°30')[1 - (0.081819191)^2 \sin^2 44°30']^{1/2} \, 206,265}{(3963.2)}$$

$$= 51.06''$$

$$\hat{\theta}_2'' = (2)(51.06)$$

$$= 102.12'' = 01'42.12''$$

$$\hat{\theta}_3'' = (3)(51.06) = 153.18'' = 02'33.18''$$

The bearings, B_i, $i = (0, 1, \ldots, 6)$ for segments of the secant line at 1-mi intervals are as follows. First, from the point of beginning to the 3-mi point,

$$B_0 = (90°) - \hat{\theta}_3 = (90°) - (0°02'33.18'') = \text{N}89°57'26.82''\text{E} = \text{N}89°57.4'\text{E}$$

$$B_1 = B_0 + \hat{\theta}_1 = (89°57'26.82'') + (51.06'') = \text{N}89°58'17.88''\text{E} = \text{N}89°58.3'\text{E}$$

$$B_2 = B_0 + \hat{\theta}_2 = B_1 + \hat{\theta}_1 = (89°58'17.88'') + (51.06'') = \text{N}89°59'08.94''\text{E}$$

$$= \text{N}89°59.1'\text{E}$$

$$B_3 = B_0 + \hat{\theta}_3 = B_2 + \hat{\theta}_1 = (89°59'08.94'') + (51.06'') = \text{due east}$$

Next, from the 3-mi point to the 6-mi point,

$$B_4 = (90) - \hat{\theta}_1 = (90°) - (0°00'51.06'') = \text{S}89°59'08.94''\text{E} = \text{S}89°59.1'\text{E}$$

$$B_5 = B_4 - \hat{\theta}_1 = (89°59'08.94'') - (0°00'51.06'') = \text{S}89°58'17.88''\text{E} = \text{S}89°58.3'\text{E}$$

$$B_6 = B_5 - \hat{\theta}_1 = (89°58'17.88'') - (0°00'51.06'') = \text{S}89°57'26.82''\text{E} = \text{S}89°57.4'\text{E}$$

$$B_6 = 90° - \hat{\theta}_3 = (90°) - (0°02'33.18'') = \text{S}89°57'26.82''\text{E} = \text{S}89°57.4'\text{E}$$

as a check.

Bearings are calculated to two places to avoid round-off error. Final values should be rounded to tenths of minutes.

The procedure for establishing a true parallel by the secant method is as follows. The initial point on the secant is located by measuring south of the beginning corner a distance equal to the secant offset for 0 mi determined by calculations, as described previously in this section (3.27 ft = 1.00 m = 4.97 links, Figure 18.17). The theodolite is set up at this point, and the direction of the secant line is established by laying off from astronomic north the azimuth or bearing calculated, as discussed previously in this section; for the conditions illustrated by Figure 18.17, the bearing of the secant that extends east from the point of beginning is N89°57.4'E. The secant then is projected in a straight line for 6 mi (9.66 km); and as each 40 chains ($\frac{1}{2}$ mi or 0.805 km) is laid off along the secant, the proper offset is taken to establish the corresponding section or quarter-section corner on the true parallel.

At the end of 6 mi, if it is not convenient to determine the true meridian, the succeeding secant line may be established by laying off, at the 6-mi point, a deflection angle from the prolongation of the preceding secant to the succeeding secant line, the angle being equal to the convergence of meridians 6 mi apart. The angular convergence can be calculated using Equation (18.1). When the direction of the new secant line has been thus defined, the process of measurement to establish corners on the true parallel is continued as before.

18.31
ESTABLISHING TOWNSHIP EXTERIORS

When practical, the township exteriors are surveyed successively through a 24-mi quadrangle in ranges, beginning each range with the township on the south (Figure 18.18). The range lines or meridional boundaries of the townships take precedence in the order of survey and are run from south to north on astronomic meridians.

For example, in Figure 18.18, the meridional line from a is run due north to b, setting quarter-section and section corners alternately at intervals of 40 chains (2640 ft or 804.67 m). At the end of 6 mi, a temporary township corner is set at b, pending latitudinal measurements necessary to close the township exterior and calculate the error of closure.

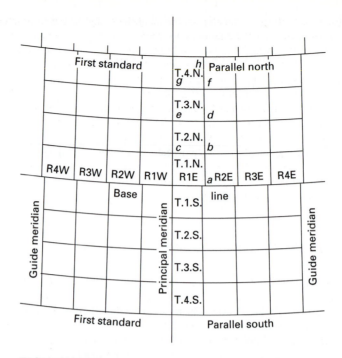

FIGURE 18.18
Four quadrangles of 16 townships, each bounded by
standard lines.

Next, a random line is run from the corner at *b* west to intersect the principal meridian (or guide meridian). Temporary monuments are set at quarter-section and section corners alternately at intervals of 40 chains (2640 ft or 804.67 m) as this line is being run. The closure on the corner at *c* is noted, and this amount (assuming it is within the permissible value) is used to determine the direction of an improved estimate for the true line between *b* and *c*. This correction line is run from *c* to *b*, correcting the temporary monuments at quarter-section and section corners and setting permanent monuments. Owing to the convergence of the meridians, this northern boundary of the township will be less than 6 mi. All errors due to convergence and errors in measurement are thrown into the westernmost $\frac{1}{2}$ mi.

The meridional or range line then is extended from the corner at *b* north to the corner at *d*, setting quarter-section and section corners alternately at 40-chain (2640-ft or 804.67-m) intervals. From the corner at *d*, a random line is run west to the principal meridian or range line, where the closure is noted at corner *e*. A correction line is run from *e* to *d*, setting permanent quarter corners and section corners alternately at 40-chain intervals. Similarly, the range line is extended to *f*, a random line is run west to *g*, and a correction line is run from *g* to *f*. Finally, the range line is run from *f* to *h*, where it is terminated at its intersection with the standard parallel, the excess or deficiency between standard parallels being placed in the most northerly half-mile. At the point of intersection between the range line and the standard parallel, a closing township corner (correction corner) is established. To determine the alignment of the line closed upon, the standard parallel is retraced between the two standard corners adjacent to the closing corner. The distance from the closing corner to the nearest standard corner is measured so that the error of closure may be calculated.

In a similar manner, the boundaries of townships forming the next or second range line to the east are established. The range lines are extended north, setting quarter-section and section corners alternately at 40-chain ($\frac{1}{2}$-mi or 0.805-km) intervals, and random lines are run west to close on the previously established township corners on the first range line.

When the third range line is extended north from the southwest corner of the southeast township, quarter-section corners are set as before, but random latitudinal lines at the 6-, 12-, and 18-mi points are run (1) to the west to connect with corresponding township corners, and (2) to the east to connect with the first, second, and third regular township corners north of the standard parallel on the guide meridian.

As the line run to the east is corrected back to the west, with the initial measurement being made from the guide meridian, quarter-section and section corners are set alternately at 40-chain ($\frac{1}{2}$-mi or 0.805-km) intervals and any error is thrown into the westerly half-mile of the township, as was done for the others.

18.32
ALLOWABLE LIMITS OF ERROR OF CLOSURE

According to the *Manual of Instructions for the Survey of Public Lands of the United States* (U.S. Bureau of Land Management, 1973), the maximum *limit of closure* (ratio of misclosure, Sections 8.21 and 2.11) is 1/905, provided that the limit of closure in either the sum of the departures or sum of the latitudes does not exceed 1/1280. Where a survey qualifies under the latter limit, the first must be satisfied. A cumulative error of $6\frac{1}{4}$ links (4.13 ft or 1.26 m) should not be exceeded in either departure or latitude.

The departures and latitudes of a normal section each should close within 25 links (16.5 ft or 5.03 m); of a normal range or tier of sections, to within 88 links (58.1 ft or 17.70 m); and of a normal township, within 150 links (99.0 ft or 30.18 m). The boundaries of all fractional sections, irregular claim lines or meanders, and all broken or irregular boundaries should have a limit of closure (relative accuracy) of 1/1280 to be determined separately for the sum of the departures and the sum of the latitudes.

Whenever a closure is effected, the departures, latitudes, and error of closure of the lines composing the figure (quadrangle, township, section, meander, etc.) must be calculated, and corrective steps must be taken whenever the test discloses an error in excess of the allowable value.

18.33
RECTANGULAR LIMITS

Before considering further the methods employed in the subdivision of townships, the legal requirement relative to the rectangular surveys of the public lands should be stated. Of the 36 sections in each normal township (Figure 18.19), 25 are returned as containing 640 acres (259 ha) each; 10 sections adjacent to the north and west boundaries (comprising sections 1–5, 7, 18, 19, 30, and 31), each containing regular subdivisions totaling 480 acres (194 ha), and in addition 4 *fractional lots,* each containing 40 acres (16 ha) plus or minus definite differences to be determined in the survey; and 1 section (section 6) in the northwest corner containing regular subdivisions totaling 360 acres (146 ha) and in addition 7 fractional lots each containing 40 acres (16.2 ha), plus or minus certain definite differences to be determined in the survey. The aliquot* parts of 640 acres (259 ha), or regular

---

*A part of the distance which divides the distance without a remainder (American Society of Civil Engineers, 1994).

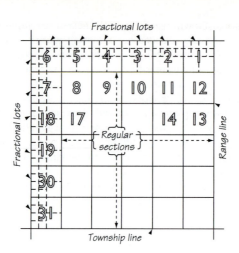

FIGURE 18.19
Subdivision of a township.

subdivisions of a section, are the quarter section ($\frac{1}{2}$ mi or 0.805 km square), the half-quarter or eighth section ($\frac{1}{4}$ by $\frac{1}{2}$ mi or 0.402 by 0.805 km), and the quarter-quarter or 16th section ($\frac{1}{4}$ mi square or 0.402 by 0.402 km), the last containing 40 acres (16.2 ha) and being the legal minimum for purposes of disposal under the general land laws.

18.34
SUBDIVISION OF TOWNSHIPS

Following the normal plan for subdividing townships with regular boundaries, the subdivisional survey is begun on the south boundary of the township at the section corner between sections 35 and 36 (see Figure 18.20). The line between sections 35 and 36 is run in a northerly direction parallel to the east boundary of the township, the quarter-section corner between 35 and 36 being set at 40 chains (2640 ft or 804.67 m) and the section corner common to sections 25, 26, 35, and 36 being set at 80 chains (5280 ft or 1609.34 m). From the latter corner a random line is run eastward on a course calculated to be parallel with the south boundary of section 36, a temporary quarter-section corner being set at 40 chains. If this random line intersects the east boundary of the township at the corner of sections 25

FIGURE 18.20
Order of establishing section lines.

and 36, it is blazed (a blaze is a mark made with an axe on a living tree, see Robillard and Bouman, 1992, and Section 18.40) and established as the true line, and if the linear error of closure is within the allowable limits, the temporary quarter-section corner is made permanent by shifting it to a position midway between adjacent section corners, as determined by field measurements.

If the point of intersection between the random line and east boundary falls to the north or south of the section corner on the township boundary, as will generally be the case, the falling is measured and, from the data thus obtained, the bearing of the true return course is calculated and the true line joining the section corners is blazed and established, the quarter-section corner common to sections 25 and 36 being placed midway between section corners, as described previously.

This process is repeated for the successive meridional and latitudinal lines in the eastern range of sections until the north boundary of section 12 is established, the order in which the lines are surveyed being as indicated by the numbers on the section lines in Figure 18.20.

When the northern boundary of the township is not a base line or standard parallel, the line between sections 1 and 2 is run north as a random line parallel to the east boundary, the distance to its points of intersection with the northern boundary of the township being measured. If the random line intersects the northern boundary at the corner of sections 1 and 2 and the linear error of closure of the tier of sections is within the allowable limit, the random line is blazed back and established as the true line, the fractional measurement being thrown into that portion of the line between the quarter-section corner and the north boundary of the township.

If, as is usually the case, the random line intersects the north boundary to the east or to the west of the corner of sections 1 and 2, the falling is measured, the bearing of the true return course is calculated, and the true line joining the section corners is established, the permanent quarter-section corner common to sections 1 and 2 being placed a full 40 chains (2640 ft or 804.67 m) from the south boundary of these sections. In this way the excess or deficiency in linear measurement, as before, is placed in that portion of the line between the permanent quarter-section corner and the north boundary of the township.

When the north boundary of the township is a base line or a standard parallel, the line between sections 1 and 2 is run as a true line parallel to the east boundary of the township, a permanent quarter-section corner being set at 40 chains (2640 ft or 804.67 m), a closing section corner being established at the point of intersection of the section line and base line or standard parallel, and the distance from this closing corner to the nearest standard corner being measured.

The successive ranges of sections from east to west are surveyed in a manner identical to the procedure described in the preceding paragraphs for the most easterly range until the two most westerly ranges are reached.

The west and north boundaries of section 32 are established as for corresponding sections to the east. A random line parallel to the south boundary of the township then is run west from the corner of sections 29, 30, 31, and 32, and the point of intersection between the random line and the west boundary of the township is determined. The falling of the intersection from the true corner then is measured, the course of the true line is calculated, and the true line is blazed and established, the permanent quarter-section corner being placed on the true line at a full 40 chains (2640 ft or 804.67 m) from the corner of sections 29, 30, 31 and 32. Thus, the deficiency due to convergence of the meridians and the excess or deficiency due to errors in linear measurements are thrown in the most westerly half-mile.

The survey of the other sections comprising the two most westerly ranges is continued in similar manner, the order in which the lines are surveyed being indicated by the numbers shown in Figure 18.20.

18.35
SUBDIVISION OF SECTIONS

The function of the United States surveyor is to establish the official monuments so that the officially surveyed lines may be identified and the subdivision of the section may be controlled as contemplated by law. There the duties of the United States surveyor cease, and those of the surveyor in private practice begin. In the work of subdividing sections into the parts shown on the official plat, the local surveyor cannot properly serve his or her client without being familiar with the land laws regarding the subdivision of sections, nor in the event of the loss of original monuments can the surveyor expect legally to restore them unless the principles employed in the execution of the original survey are thoroughly understood.

18.36
SUBDIVISION BY PROTRACTION

On the official government township plats, the interior boundaries of quarter sections are shown as dashed straight lines connecting opposite quarter-section corners. The sections adjacent to the north and west boundaries of a normal township, except section 6, are further subdivided by protraction into parts containing two regular half-quarter sections and four *lots,* the latter containing the fractional areas resulting from the plan of subdivision of the normal township. This process of plotting the interior, unsurveyed boundaries on the official plat is called *subdivision by protraction.* Figure 18.21 illustrates the plan of the normal subdivision of sections. The regular half-quarter sections are protracted by laying off a full 20 chains (1320 ft or 402.34 m) from the line joining opposite quarter-section corners. The lines subdividing the fractional half-quarter sections into the fractional lots are protracted from midpoints of the opposite boundaries of the fractional quarter sections.

In section 6, the two quarter-quarter-section corners on the south and east boundaries of the fractional northwest quarter are similarly fixed, one at a point 20 chains north and the other at a point 20 chains west of the center of the section, from which points lines are

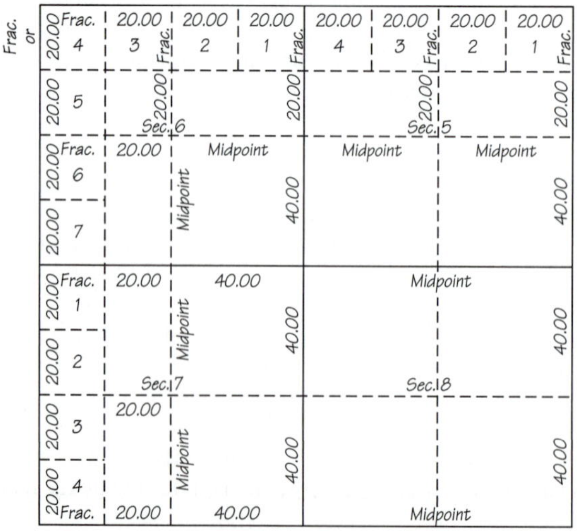

FIGURE 18.21
Subdivision of sections.

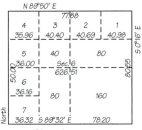

FIGURE 18.22
Subdivisional areas of
section 6 in acres.

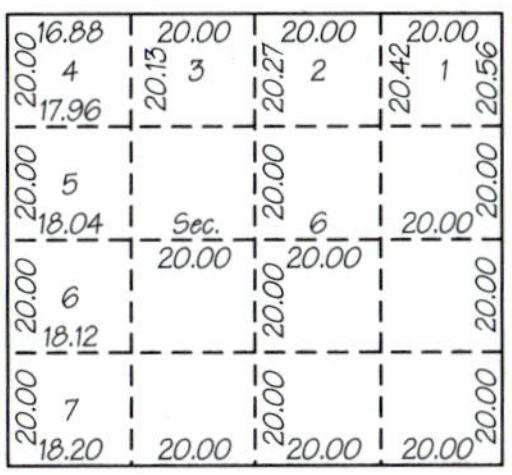

FIGURE 18.23
Subdivisional dimensions of
section 6 in chains.

protracted to corresponding points on the west and north boundaries of the section. Hence, the subdivision of the northwest quarter of section 6 results in one regular quarter-quarter section and three lots.

In all sections bordering on the north boundary the fractional lots are numbered in succession beginning with no. 1 at the east. In all sections bordering on the west boundary the fractional lots are numbered in succession beginning with no. 1 at the north, except section 6, which, being common to both north and west boundaries, has its fractional lots numbered in progression beginning with no. 1 in the northeast corner and ending with no. 7 in the southwest corner, all as illustrated by Figure 18.21.

Figure 18.22 illustrates a typical plat of section 6 on which the protracted areas are shown. Figure 18.23 is a similar section giving the calculated dimensions of the protracted areas.

Fractional Lots

In addition to sections made fractional by reason of being adjacent to the north and west boundaries of a township, sections also are made fractional on account of meanderable bodies of water (Section 18.38), mining claims, Indian reservations, and other segregated areas within their limits. Such sections are subdivided by protraction into such regular and fractional parts as are necessary for the entry of the undisposed public lands and to describe these lands separately from the segregated areas.

18.37
SUBDIVISION BY SURVEY

The rules for the subdivision of sections given in the following paragraphs are based upon the general land laws. When an entryman* has acquired title to a certain legal subdivision, he or she becomes the owner of the ground area represented by the same subdivision on the official plat. Preliminary to subdivision it is necessary to identify the actual boundaries of the section, as it cannot be legally subdivided until the section and exterior quarter-section

---

* One who initiates procedures to acquire title to public lands under public land laws (American Society of Civil Engineers, 1994).

corners have been found or restored to their original locations and the resulting courses and distances have been redetermined in the field. When the opposite quarter-section corners have been located, the legal center of the section, or interior quarter-section corner, may be placed. If the boundaries of quarter-quarter sections or of fractional lots are to be established on the ground, it is necessary to measure the boundaries of the quarter section and fix thereon the quarter-quarter-section corners at distances in proportion to those given upon the official plat; then the interior quarter-quarter-section corner may be placed.

Subdivision of Sections into Quarter Sections

According to law, the procedure to be followed in the subdivision of a section into quarter sections is to run straight lines between the established opposite quarter-section corners. The point of intersection of lines thus run is the quarter-section corner common to each of the four quarter sections into which the section is divided. Called the *interior quarter-section corner,* it is the legal center of the section.

Subdivision of Fractional Sections

Where opposite corresponding quarter-section corners of a section have not been or cannot be fixed, as is frequently the case when sections are made fractional by streams, lakes, and so forth, the lines of sectional subdivisions are run on courses with the mean of the bearings of adjoining established section lines or on courses parallel to the east, south, west, or north boundary of the section where there is no opposite section line.

Subdivision of Quarter Sections into Quarter-Quarter Sections

Preliminary to the subdivision of regular quarter sections, the quarter-quarter-section (16th-section) corners are established at points midway between the section and exterior quarter-section corners and between the exterior quarter-section corners and the center of the section. The quarter-quarter-section corners having been thus established, the center lines of the quarter section are run as straight lines between opposite corresponding quarter-quarter-section corners on the boundaries of the quarter section. The intersection of these lines is common to the four quarter-quarter sections into which the quarter section is divided. Called the *interior quarter-quarter-section corner,* it marks the legal center of the quarter section.

1. *Irregular quarter sections.* This case arises (1) when the quarter section is adjacent to the north or west boundary of a regular township and (2) when the quarter section adjoins any irregular boundary of an irregular township. The procedure is the same as that outlined in the preceding paragraph, except that the quarter-quarter-section corners on the boundaries of the quarter section that are normal to the township exterior are placed at 20 chains (1320 ft or 402.34 m), proportionate measurement (Section 18.47), counting from the regular quarter-section corner.
2. *Fractional quarter sections.* The subdivisional lines of fractional quarter sections are run from properly established quarter-quarter-section corners, with courses governed by the conditions represented on the official plat, to the lake, water course, or reservation that renders such quarter sections fractional.

18.38
MEANDERING

In the process of surveying the public lands, all navigable bodies of water and other important rivers and lakes below the line of mean high water are segregated from the lands open to private ownership. In the process of subdivision, the regular section lines are run to an intersection with the mean high-water mark of such a body of water, at which intersection corners, called *meander corners,* are established. The traverse run between meander corners, approximately following the margin of a permanent body of water, is called a *meander line,* and the process of establishing such lines is called *meandering.* The mean high-water mark is taken as the line along which vegetation ceases. The fact that an irregular line must be run in tracing the boundary of a reservation does not entitle such a line to be called a *meander line* except where it follows closely the shore of a lake or the bank of a stream.

Meander lines are not boundaries but lines run to locate the water boundaries approximately, and although the official plats show fractional lots as bounded in part by meander lines, it is an established principle that ownership does not stop at such boundaries. A Supreme Court decision reads as follows:

> Meander lines are run in surveying fractional portions of the public lands bordering on navigable rivers, not as boundaries of the tract, but for the purpose of defining the sinuosities of the banks of the stream and as the means of ascertaining the quantity of land in the fraction subject to sale, which is to be paid for by the purchaser. In preparing the official plat from the field notes, the meander line is represented as the border line of the stream, and shows to a demonstration that the water-course, and not the meander line as actually run on the land, is the boundary.

In running a meander line, the surveyor begins at a meander corner and follows the bank or shoreline, as closely as convenience permits, to the next meander corner, the traverse being a succession of straight lines. The true length and bearing of each of the courses of the meander line are observed with precision, but for convenience in plotting and computing areas the intermediate courses are laid off to the exact quarter degree and each intermediate occupied station is placed a whole number of chains, or at least a multiple of 10 links (6.60 ft or 2.01 m), from the preceding station. Inasmuch as meander lines are not true boundaries, this procedure defines the sinuosities of the mean high-water line with sufficient accuracy. When a meander line is "closed" on a second meander corner, the departures and latitudes of the courses bounding the fractional lot are computed and the error of closure is determined. If this exceeds the allowable value, the line is rerun until an error in bearing or distance is discovered that will bring the closure within the specified limits (maximum error in either departure or latitude 1/1280).

Rivers

Proceeding downstream, the riverbank on the left hand is termed the *left bank* and that on the right hand, the *right bank.* Navigable rivers and bayous as well as all rivers not embraced in the class denominated "navigable," the right-angle width of which is 3 chains (198 ft or 60.35 m) and upward, are meandered on both banks, at the ordinary mean high-water mark, by taking the general courses and distances of their sinuosities.

Lakes

Regulations provide for the meandering of all lakes having an area of 25 acres (101,171.4 m^2) or greater, the procedure being the same as for the meandering of streams. If the lake lies entirely within a section, there obviously will be no regular meander corners, and a *special meander corner* is established at the intersection of the shore of the lake with a line run from one of the quarter-section corners on a theoretical course to connect with the opposite quarter-section corner, the distance from the quarter-section corner to the special meander corner being measured. The lake then is meandered by a line beginning and ending at the special meander corner. If a meanderable lake is found to lie entirely within a quarter section, an *auxiliary meander corner* is placed at any convenient place on its margin, and this is connected by traverse with one of the regular corners established on the boundary of the section.

Islands

In the progress of the regular surveys, every island of any meanderable body of water, except those islands that have formed in navigable streams since the admission of a state to the union, is located with respect to regular corners on section boundaries and is meandered and shown on the official plat. Also in the survey of lands fronting on any nonnavigable body of water, any island opposite such lands is subject to survey.

18.39
FIELD NOTES

The field notes taken in connection with the survey of public lands are required to be in narrative form and are designed to furnish not only a record of the exact surveying procedure followed in the field but also a report showing the character of the land, soil, and timber traversed by the line of subdivision and a detailed schedule of the topographic features adjacent to the lines, together with reference measurements showing the location of the lines with respect to natural objects, improvements, and the lines of other surveys. In this way, the notes serve three purposes: (1) the field procedure is made a matter of official record, (2) the general characteristics of the territory served by the subdivision surveys are secured, and (3) the reference measurements to objects along the surveyed lines furnish evidence by which the established points and lines become practically unchangeable. These notes may be recorded in a standard field book or cadastral electronic field book consisting of a data collector (Section 3.12) equipped with software that permits collecting measurements and recording descriptive information for the survey in the prescribed narrative form (see Kopach, 1993).

18.40
MARKING LINES BETWEEN CORNERS

A final step in the survey of the public lands is to fix the location of the legal lines of subdivision permanently with reference to objects on the surface of the earth. This aim is accomplished (1) by setting monuments, of a character later to be defined, at the regular corners, (2) by finding the location of the officially surveyed lines with respect to natural features of the terrain, and (3) by indicating the position of the regular lines through living timber by blazing and by hack marks.

The last method of fixing the location of the regular subdivisional lines is required by law just as definitely as is the establishment of monuments at the corners. All legal lines of the public-land surveys through timber are marked in this manner. Those trees that lie on the line, called *line trees,* are marked with two horizontal notches, called *hack marks,* on each side of the tree facing the line; and an appropriate number of trees on either side of the line and within 50 links (33 ft or 10.06 m) are marked by flat axe marks, called *blazes,* a single blaze on each of two sides quartering toward the line.

18.41
CORNERS

In the subdivision of the public lands as described in the preceding sections, it is required that the United States surveyors shall permanently mark the location of the township, section, exterior quarter section, and meander corners, as well as those quarter-quarter-section corners necessary in connection with the subdivision of fractional sections. Monuments of a character specified by regulations of the Bureau of Land Management are employed for this purpose. The location of every such corner monument, in accordance with definite rule, is referred to such nearby objects as are available and suitable for this purpose; and where the corner itself cannot be marked in the ordinary manner, an appropriate witness corner is established (Section 18.42).

At the appropriate place in the field notes of the survey, a record of each established monument is introduced, this record including the character and dimensions of the monument itself, the manner in which it is placed, the significance of its location, its markings, and the nature of the objects to which reference measurements are taken, together with the reference measurements.

18.42
WITNESS CORNERS

Where a true corner point falls within an unmeandered stream or lake, within a marsh, or in an inaccessible place, a witness corner is established in a convenient location nearby, preferably on one of the surveyed lines leading to the location of the regular corner. Also, where the true point falls within the traveled limits of a road, a cross-marked stone is deposited below the road surface, and a witness corner is placed in a suitable location outside the roadway.

The witness corner is placed on any one of the surveyed lines leading to a corner, if a suitable place within a distance of 10 chains (660 ft or 201.168 m) is available; but if no secure place is to be found on a surveyed line within the stated limiting distance, the witness corner may be located in any direction within a distance of 5 chains (330 ft or 100.584 m).

18.43
CORNER MONUMENTS

The Bureau of Land Management has adopted a standard iron post for monumenting the public-land surveys. This post is to be used unless exceptional circumstances warrant the use of other material (U.S. Bureau of Land Management, 1973).

Where the procedure is duly authorized, durable native stone may be substituted for the model iron post described previously, provided the stone is at least 20 in. (51 cm) long and at least 6 in. (15.2 cm) in its least lateral dimension. Stone may not be used as a monument for a corner whose location is among large quantities of loose rock. The required corner markings are cut with a chisel, and usually the stone is set with about three-fourths of its length in the ground.

Where the ground is underlaid with rock close to the surface and it is impractical to complete the excavations for monuments to the regular depth, the monument is placed as deep as practical and supported above the natural ground surface by a mound of stone. Where the solid rock is at the surface, the exact corner point is marked by a cross cut in the rock; and if it is practical to do so, the corner monument is established in its proper location and supported by a mound of stones.

Where the corner point falls within the trunk of a living tree which is too large to be removed readily, the tree becomes the corner monument and, as such, is scribed with the proper marks of identification.

Legal penalties are prescribed for damage to government survey monuments or marked trees.

18.44
MARKING CORNERS

A complete treatment of the system of marking corner monuments established in the survey of the public lands is beyond the scope of this text. However a brief description of the general features of the system is given. For further details the reader is referred to the *Manual of Instructions for the Survey of Public Lands of the United States* (U.S. Bureau of Land Management, 1973).

All classes of monuments are marked in accordance with a system designed to provide a ready identification of the location and character of the monument on which the markings appear. Iron posts and tree corners are marked with capital letters, which are themselves keys to the character of the monument, and with arabic figures, giving the section and township and range numbers of the adjacent subdivisions and the year in which the survey was made.

In the case of stone monuments, certain marks in the form of *notches* and *grooves* are placed on the vertical edges or faces; for an exterior corner, the number of marks is made equal to the distance in miles from the adjoining township corner along the township or range line to the monument, and for an interior corner, the number of marks is made equal to the distance in miles from the adjoining township boundary along section lines to the monument. These marks furnish a means of determining the number of the adjoining sections.

A witness corner and its accessories are constructed and marked similarly to a regular corner for which it stands, with the additional letters *WC* to signify *witness corner* and with an arrow pointing to the true corner.

The following is an index of the ordinary markings common to all classes of corners:

Mark	Meaning	Mark	Meaning
AMC	Auxiliary meander corner	S	Section
BO	Bearing object	S	South
BT	Bearing tree	SC	Standard corner
C	Center	SMC	Special meander corner
CC	Closing corner	T	Township
E	East	W	West
MC	Meander corner	WC	Witness corner
N	North	WP	Witness point
R	Range	$\frac{1}{4}$	Quarter section
RM	Reference monument	$\frac{1}{16}$	Quarter-quarter section

18.45
CORNER ACCESSORIES

When a corner is referred by direction and distance to some other more-or-less permanent object nearby and the operation becomes a matter of record, it is possible to relocate the corner with respect to the object. In land surveying, a recorded measurement of this kind often is called a *connection,* and the object thus located is called a *corner accessory.* It is specified that the United States surveyors in the survey of the public lands shall employ at least one accessory for every corner established. The character of the accessories is to fall within the following groups: (*a*) bearing trees, or other natural objects such as notable cliffs and boulders, permanent improvements, and reference monuments; (*b*) mounds of stone; and (*c*) pits and memorials. Essentially, such an accessory is a part of the monument.

The marks on a bearing tree are made on the side nearest the corner, in the manner already described for tree-corner monuments. The mark includes the section number in which the tree stands and is terminated by the letters *BT.*

Where a bearing object is of rock formation, the point to which measurements are taken is indicated by a cross, and it is marked with the letters *BO* and the section number, all marks being cut with a chisel.

Where it is impossible to make a single connection to a bearing tree or other bearing object and where a mound of stone or a pit is impracticable, a *memorial,* or durable article such as glassware, stoneware, a cross-marked stone, a charred stake, a quart of charcoal, or piece of metal, is deposited alongside the base of the monument.

Where native stone is at hand, a mound of stones of sufficient size to be conspicuous is employed as an accessory.

Where accessories such as those mentioned in the preceding paragraphs are not available, pits may be used if conditions are favorable to their permanence. Where the ground is covered with sod, the soil is firm, and the slope is not steep, the pit will gradually fill with a material different in color or in texture from the original soil; and often a new species of vegetation will spring up. Thus, it may be possible to identify the location of a pit after the lapse of many years.

18.46
RESTORATION OF LOST CORNERS

Although it has been the aim of the Bureau of Land Management in the subdivision of the public lands to monument the established corners so that there always will be physical evidence of their location, it is a matter of common experience that many corner marks become obliterated with the progress of time. An important duty of the local or county surveyor, in the relocation of property lines or the further subdivision of lands, is to examine all available evidence and identify the official corners if they exist. Should a search of this kind result in failure, the surveyor is to employ a process of field measurement that will result in the obliterated corner being restored to its most probable original location.

As here employed, the term *corner* designates a point established by a survey, whereas the term *monument* indicates the object placed to mark the corner point on the surface of the earth.

A corner is said to *exist* when its location, within reasonable accuracy, can be determined beyond reasonable doubt, either by means of the original monument, by means of the accessories to which connections were made at the time of the original survey, by the expert testimony of surveyors who may have identified the original corner and recorded connections to other accessories, or even by landowners who have indisputable knowledge of the exact location of the original monument. If the original location of a corner cannot be determined beyond reasonable doubt, the corner is said to be *lost*. If the monument of an existent corner cannot be found, the corner is said to be *obliterated*, but it is not necessarily lost.

In the absence of an original monument, either a line tree or a definite connection to natural objects or improvements may fix a point of the original survey for both departure and latitude. The mean location of a line marked by blazed trees, when identified as the original line, sometimes may help to fix a meridional line for departure or a latitudinal line for latitude. Other calls of the original field notes in relation to various items of topography may assist materially in the recovery of the locus of the original survey. Such evidence may be developed in infinite variety.

A lost corner is restored to its original location, as nearly as possible, by processes of surveying that involve the retracement of lines leading to the corner. Restoration of a corner does not ensure that it is placed exactly in its original location, and when a corner is restored the record of the survey should so state.

18.47
PROPORTIONATE MEASUREMENTS

It is essential that the laying off of a given distance at the time of a resurvey to restore a lost corner should render the same absolute distance between two points on the ground as was measured during the original survey. For reasons that have been discussed in earlier chapters (Sections 4.14 to 4.26), the measurement of a given known line at the time of a resurvey in general will not agree with the length of the line as recorded in the original survey. Therefore, where linear measurements are necessary to the restoration of a lost corner, the principle of proportionate measurement must be employed. *Proportionate measurement* distributes an excess or deficiency in an overall remeasured distance so that each of the remeasured parts will have the same ratio to the remeasured distance as the corresponding original parts had to the originally measured distance.

Single proportionate measurement consists of first comparing the resurvey measurement with the original measurement between two existing corners on opposite sides of the lost corner, and then laying off a proportionate distance from one of the existing corners to the lost corner. Double proportionate measurement consists of single proportionate measurement on each of two such lines perpendicular and intersecting at the lost corner.

18.48
FIELD PROCESS OF RESTORATION

The following are the field procedures to be followed in a few of the simpler cases of the restoration of lost corners. In any event, the restorative process must be in harmony with the methods employed in originally establishing the lines involved, and the preponderant lines must be given the greater weight in determining whether a corner should be relocated by single or double proportionate measurement or by some other method. Standard parallels are given precedence over township exteriors, the latter are given precedence over subdivisional lines, and quarter-section corners are relocated after adjoining section corners have been restored. Detailed instructions for the relocation of lost corners are given in the U.S. Bureau of Land Management (1973).

1. *Township corner common to four townships.* Where all the connecting lines have been established in the field, retracement is made between the nearest existing corners on the meridional line, one north and one south of the lost corner, and a temporary stake is set at the proportionate distance for the lost corner; this defines the latitude of the lost

corner. Similarly, measurement is made between the nearest existing corners on the latitudinal line through the point, and at the proper proportionate distance a second temporary stake is set; this marks the departure (or longitude) of the lost corner. The location of the lost corner then is found at the intersection of an east-west line through the first stake and a north-south line through the second; the corner thus is relocated by double proportionate measurement.

2. *Section corner common to four sections in the interior of a township.* Where all lines have been run, the section corner common to four sections in the interior of a township is restored by double proportionate measurement, in the manner described in step 1.

3. *Regular corner on a range line but not at the corner of a township.* The range line is straight between township corners. Two original corners on the 6-mi (9.66 km) segment of the range line, one north and one south of the point sought, are identified and a line is run between them. The lost corner is relocated by a single proportionate measurement along this line. This procedure applies either to section or quarter-section corners.

4. *Regular corner on a township line but not at the corner of a township.* The township line originally was run as a parallel of latitude for 6 mi (9.66 km). A parallel is rerun between the nearest existing corners to the east and west of the point sought, and the corner is relocated by single proportionate measurements along this line.

5. *Standard corner.* The standard corner includes any township, section, quarter-section, or meander corner, established on a base line or standard parallel at the time the line was originally run. The corner is relocated by the process explained in step 4; that is, by single proportionate measurement along the parallel reestablished between the nearest existing standard corners on opposite sides of the point sought.

6. *Quarter-section corner on either a meridional or latitudinal section line but not on a range or township line.* The corner is relocated by single proportionate measurement along the straight line joining the adjacent section corners of the same section. If these section corners cannot be identified, they must be restored, as previously explained, before the quarter-section corner can be reestablished.

7. *Quarter-section corner at the center of a section.* The corner is relocated at the intersection of meridional and latitudinal lines between opposite quarter-section corners on the boundaries of the section.

8. *Closing corner on a standard parallel.* The parallel is reestablished between the nearest existing corners on opposite sides of the corner sought. The lost corner is relocated by single proportionate measurement along the parallel from the nearest *standard* corners on opposite sides of the point sought.

9. *Quarter-quarter-section corner on a section and quarter-section lines.* The corner is relocated by single proportionate measurement between quarter-section and section corners on opposite sides of the point sought.

10. *Quarter-quarter-section corner at the center of a quarter section.* The corner is relocated at the intersection of the meridional and latitudinal lines between opposite quarter-quarter-section corners on the exterior of the quarter section.

18.49
PHOTOGRAMMETRIC METHODS APPLIED
TO PROPERTY SURVEYS

Photogrammetric triangulation (Section 13.22) is a method for determining positions of photographed objects. Qualitative interpretation of terrain features also is possible with overlapping aerial photographs. These two aspects of photogrammetry, quantitative and qualitative, make it a powerful tool for performing property surveys.

The procedure involved in preparing and executing a photogrammetric property survey is similar in many respects to the methods described for original surveys, resurveys, and subdivision surveys described in Sections 18.15 to 18.18. The primary differences are in the methods employed to determine horizontal positions of points. Generally, the surveyor and photogrammetrist work as a team on a photogrammetric surveying project. Few individuals possess sufficient experience and expertise to do the job alone. Consider, as an example, a subdivision survey for a large tract of land.

A first priority should be the selection of a suitable geographic or land information system (GIS or LIS; see Section 14.18) so that all subsequent field and office operations can be designed to coordinate data acquisition and processing to be compatible with the selected spatial information system. As in any property survey, all available maps, aerial photographs, descriptions of the property to be surveyed, descriptions of adjacent lands, descriptions and references to all existing horizontal and vertical control points, and all field notes should be assembled and analyzed prior to any field operations or photogrammetric work. Assume, for purposes of this discussion, that a topographic map adequate for planning is available for the area and a tentative plan of streets and lots has been prepared for the subdivision. The objective of the photogrammetric survey is to establish horizontal positions of control points for two purposes: (1) to fix the positions of boundaries around the perimeter of the tract, and (2) to provide horizontal control points around the perimeter and throughout the interior of the area with sufficient density to permit locating subdivision monuments and center-line control points for subsequent layout of all street center lines and property corners in the subdivision.

Assume that analytical photogrammetric triangulation is to be used to establish positions for horizontal control points (Section 13.22). First, photography must be planned to provide complete stereoscopic coverage of the area at a scale compatible with the specified accuracy requirements (Section 13.30). During this stage, ground control points are located around the perimeter to provide control for the block of aerial photographs. These points should be positioned to be usable in a subsequent total station system traverse around the perimeter. In addition, points should be located as closely as possible to existing property corners and other areas where evidence of boundary positions may exist. These latter points may or may not be included in the perimeter traverse. At the same time, control-point locations also are selected throughout the interior of the tract. These interior or *photogrammetric control points* should be in locations suitable for photogrammetric positioning: points must be in the overlap of at least two aerial photographs and ought not to be under trees, next to steep cliffs, and so forth. These points also should be chosen in locations convenient for subsequent layout of subdivision controlling points, such as center-line intersections, points of intersections for curves, or angle points in lot lines. Each interior photogrammetric control point must be placed to be intervisible with at least one other point to provide azimuth in subsequent field surveys. Selection of locations for perimeter ground control points and photogrammetric control points (perimeter and interior) can be expedited by use of existing aerial photography. Stereoscopic examination of this photography allows locating perimeter control points near presumed property corners, fence lines, or other evidence of boundary position as interpreted from studying the three-dimensional model. This type of analysis is helpful in choosing strategic locations for interior photogrammetric control points as well. The selected locations can be marked directly on the photographs for later use in the field. Vertical control points used in the photogrammetric triangulation also must be planned. The highest density of vertical control should be around the perimeter of the tract. Elevations obtained from a static GPS survey (properly corrected from ellipsoid to geoid; Section 12.13) or trigonometric leveling (Sections 5.5 and 5.55), performed concurrently with the perimeter survey, provide adequate accuracy in elevation.

Prior to taking the aerial photography, all ground control points, photogrammetric control points, existing monuments and bench marks and all potential and identified property corners are marked with targets. These targets should be of a size, configuration, and color suitable for the analytical photogrammetric triangulation. In choosing the size of the target, the scale of the photography to be flown must be taken into account.

After all points have been marked, or concurrently with placing the targets, a control survey is run around the tract. This may be performed by static GPS techniques (Section 12.8) or by total station system traverse tied to at least two monuments in the national control system. *No point should remain unchecked.* Ties should be made to all U.S. National Geodetic Survey, U.S. Geological Survey, and county monuments in the area. Computations should be performed in the state plane coordinate system and all elevations should be referred to the North American vertical datum (NAVD 88; Section 14.2).

When marking of all points and field surveys is completed, the data are reduced and adjusted yielding state plane coordinates and elevations for all ground points. After aerial photography has been acquired, the images of all ground control and photogrammetric control points are measured using a comparator. These measurements and the X and Y coordinates and elevations for known ground control points are utilized in an electronic computer program that performs the calculations required for photogrammetric triangulation. This program yields as output the X and Y coordinates and elevations for all unknown perimeter and interior photogrammetric control points included in the adjustment. These data are retained for future computations.

The control point locations around the perimeter of the tract determined by both field survey and photogrammetric methods now are used to make final determination of boundary locations around the property. Evidence located by stereoscopic examination of aerial photographs permits defining the area of search within rather narrow limits to minimize field examination for evidence of property lines. Azimuths and distances are computed from ground and photogrammetric control points to positions of potential property corners and monuments to facilitate the search in the field. When all data from examining photography and field searches are consolidated, analyzed, and correlated with the deed description of the property, the exterior boundaries of the tract are fixed and coordinates are computed for each property corner and angle point. Next, coordinates are computed for all center-line and lot-line controlling points and all lot corners within the subdivision. Finally, the azimuths and distances from photogrammetric control points to the nearest subdivision monument or control point are calculated. A listing or electronic file of these azimuths, distances, and coordinates for points is prepared for use in the field. The positions of photogrammetric control points and the nearest subdivision monuments also can be plotted automatically (Section 14.14) on work sheets, which can be used in the field.

Field layout surveys consist of occupying the photogrammetric control points, backsighting on another photogrammetric control point, and setting the nearest subdivision monument by angle and distance from the electronic files used in data collectors or data sheets prepared in the office. If conditions are favorable, a kinematic GPS survey (as described in Section 17.4) can be used to set monuments and other control points. When a sufficient number of subdivision monuments have been established, the center lines of streets and the balance of the lot corners should be set to provide a check on the layout. No lot corner or subdivision monument should remain unchecked.

As the monumentation and lot survey proceeds, spot checks on the accuracy of the photogrammetric survey can be made. Independent field traverses are run from the perimeter control traverse for this purpose. These traverses should be well distributed throughout the tract to provide an adequate sample of check points.

The photogrammetric survey is completed when the monumentation survey and checking have been completed to the satisfaction of the surveyor, the photogrammetrist, and the county engineer or surveyor. All control points and monuments should be thoroughly referenced for subsequent use. At this stage, all data and mapping (topographic, property, etc.) should be retrievable from the GIS or LIS, allowing automatic plotting of a subdivision plan, which should be submitted to the proper authorities for approval and recording.

The procedure as outlined for a new subdivision can be applied in essentially the same way to the resurvey of an old subdivision, a group of lots located by metes and bounds, and land subdivided according to the U.S. public-land surveys. To summarize, some of the principal factors to be remembered in any photogrammetric property survey are these:

1. The surveyor and photogrammetrist must work as a team. Careful advance planning to correlate field and photogrammetric work is absolutely necessary.
2. Whenever possible, locate all visible evidence of property lines and corners by examination of existing photography and field searches prior to acquiring the aerial photography to be used in the photogrammetric survey.
3. Set targets on all ground control points, photogrammetric control points, and existing monuments and bench marks. Set targets on or near all identified and potential property corners.

Finally, it is necessary to remember that the photogrammetrist can determine the position of a signalized control point with an accuracy more than adequate for most property surveys. However, to locate a property corner properly from this control point requires the services of a surveyor with substantial experience in property surveying and an adequate knowledge of the laws and customs regulating property subdivision in the region where the work is being performed.

Brown (1973) and Lafferty (1971) contain additional details concerning accuracies possible and application of photogrammetric methods to property surveys.

PROBLEMS

18.1. Name and discuss the different types of property or land surveys.

18.2. Define the following terms: senior deed, easement, eminent domain, oral agreement, and adverse possession.

18.3. Define and give examples for natural, artificial, and record monuments.

18.4. Describe three methods by which land transfer records may be filed.

18.5. List the major elements that should be included in a land description.

18.6. What is a metes and bounds description? Describe five different types of metes and bounds descriptions.

18.7. Write a metes and bounds description for the property shown in Figure 18.9. Do not use coordinates in the description.

18.8. Give examples of descriptions by subdivision. What is a significant difference between a description by subdivision and one by metes and bounds?

18.9. Discuss how a description can be written for the property owned in a condominium.

18.10. Assume that you have bought unit 10 in a condominium constructed on the property shown in Figure 18.2 and described in Example 18.3. Write the description for this condominium unit.

18.11. The coordinates (Lambert projection, south zone, Pennsylvania) for the southernmost corner of the property shown in Figure 18.9 are $X = 1,005,431.05$ ft and $Y = 405,110.85$ ft. All directions shown are referred to grid north for the projection. Write a metes and bounds description for this property using coordinates.

18.12. Write a lot and block description for lot 22 in the plan shown in Figure 18.10a. Assume that this subdivision is part of the Artesian Heights Addition recorded in Plan Book 140, page 32, in the Office of the Recorder of _____ County, State of _____ .

18.13. Give the order of precedence for calls in a deed when conflicts exist.

18.14. Define the following terms that are related to riparian rights: thread of a river, meander line, alluvium, alluvion, bed, high-water mark, low-water mark, and reliction.

18.15. List the conditions that must be satisfied to acquire possession of property by adverse possession.

18.16. Outline the steps that should be taken by a registered surveyor in preparing to execute a resurvey of property located in a region unfamiliar to the surveyor.

18.17. Brown owns a farm containing about 60 acres (about 243 ha). In 1965 he sold a 10-acre (40.5-ha) portion to Boyd, and in 1970 sold a 20-acre (80.9-ha) portion to Wilson. Brown continues to live on and farm the remaining land. Among the three property owners, who has senior rights? Explain why.

18.18. In block 5 of the subdivision illustrated in Figure 18.10a, lot 7 was sold in March 1992. Thereafter, lots 20, 8, 13, and 11 were sold in April, July, August, and October 1993. Among these five property owners, which has senior rights? Explain why.

18.19. Joyce, who owns a lot 200 m in width along Route 82, sold the west half, a lot 100 m wide, to Eliot in 1987. Both properties were surveyed at the time and metes and bounds descriptions were recorded. Ten years later both properties were resurveyed and a shortage of 1 m was discovered in the frontage along Route 82. How is this discrepancy proportioned in the resurvey? Support your conclusions.

18.20. A subdivision contains twenty 60-ft lots numbered 1, 2, . . . , 20 from east to west in one block. During a resurvey to set the corners of lot 5, the total distance between block corners (proven to be correct) is shown to be 1200.80 ft. What distance should be measured from the eastern corner of the block to set the eastern corner of lot 5, and how much frontage should be allocated to this lot? Support your reasons for proportioning the discrepancy.

18.21. Describe, briefly, how photogrammetric methods can be used for property surveys.

18.22. In performing a property survey using photogrammetric methods, describe the responsibilities of (*a*) the photogrammetric engineer, (*b*) the registered land surveyor. Assume that both are involved in a joint effort in the task.

18.23. Find the angle of convergence between two meridians 6 mi apart at a mean latitude of $32°20'$. Compute the linear convergence, measured along a parallel of latitude, in a distance of 6 mi.

18.24. Find the length of $1°$ longitude at a latitude of $40°06'20''$.

18.25. Find the offsets between the tangent and the parallel at intervals of $\frac{1}{2}$ mi over a distance of 6 mi at a latitude of $40°06'20''$.

18.26. Find the azimuth of the secant and the offsets from the secant to the parallel at intervals of $\frac{1}{2}$ mi over a distance of 6 mi at a latitude of 40°06′20″.

18.27. Show the dimensions and areas of the protracted subdivisions of section 7, as required by law to be shown on the official plat, when the north, east, south, and west boundaries are, respectively, 76.84, 80.00, 76.64, and 80.00 chains.

18.28. Show the dimensions and areas of the protracted subdivisions of section 6, as required by law to be shown on the official plat, when the north, east, south, and west boundaries are, respectively, 76.36, 80.44, 76.60, and 80.00 chains.

18.29. A lost interior section corner is to be restored by a resurvey. The nearest corners that can be identified are regular section corners 1 mi north, 2 mi east, 3 mi south, and 1 mi west of the point sought. The records show the corresponding original measured distances to be 80.40, 160.56, 240.00, and 78.32 chains. The resurvey measurement between the nearest existing monuments on the meridional line through the lost corner is 320.16 chains and that along the latitudinal line between the nearest existing corners is 238.48 chains. Calculate the proportionate measurements to be used in the relocation of the lost corner and state the procedure to be employed in its reestablishment.

18.30. A lost section corner on a range line is to be restored by a resurvey. One mile to the south the township corner is identified, and $2\frac{1}{2}$ mi to the north the quarter-section corner is found. According to the records the corresponding distances measured at the time of the original survey were 80.00 and 200.00 chains. The resurvey distance between the existing corners is 279.64 chains. State the procedure to be followed in restoring the lost corner, and calculate the proportionate distances to be employed.

REFERENCES

American Society of Civil Engineers. "Proposed Model Standard of Practice for Property Boundary Surveys." *Journal of Surveying Engineering* 119, no. 3 (August 1993), pp. 111–17.

American Society of Civil Engineers. *Glossary of the Mapping Sciences.* New York and Bethesda, MD: Joint Committee of the ASCE, American Congress of Surveying and Mapping, and American Society for Photogrammetry and Remote Sensing, 1994.

Black's Law Dictionary, 5th ed. St. Paul, MN: West Publishing Co., 1979.

Bouvier, J. *Bouvier's Law Dictionary.* New York: Gordon Press, 1991.

Brown, C. M. "Identifying Monuments." *Surveying and Mapping* (September 1976).

Brown, C. M.; W. G. Robillard; and D. A. Wilson. *Brown's Boundary Control and Legal Principles,* 4th ed. New York: John Wiley & Sons, 1995.

Brown, D. C. "Accuracies of Analytical Triangulation in Applications to Cadastral Surveys." *Surveying and Mapping* (September 1973).

Cameron, J. C. *Michigan Real Property Law: Principles and Commentary,* vols. 1 and 2. Ann Arbor, MI: The Institute of Continuing Legal Education, 1989.

Clark, F. F. *Law of Surveying and Boundaries,* 3rd ed. Indianapolis, IN: Bobbs-Merill Company, 1959.

Cook, R. N. "Multi-Purpose Land Data System, the Legal Parcel." *Proceedings 37th Annual Meeting, ACSM,* 1997, pp. 612–17.

Danial, N. F. "Some Guidelines for Laying Out Subdivisions." *Surveying and Mapping* (June 1972).

Dean, D. R., and J. G. McEntyre. "Surveyor's Guide to the Use of a Law Library." *Surveying and Mapping* 35, no. 3 (September 1975).

Greulich, G. "The Use of Metric System in Metes and Bounds Surveys." *Proceedings, Fall Convention, ACSM,* 1974.

Hill, J. M. "Riparian Rights—From Estuary to Tidal Headwaters." *Surveying and Mapping* (June 1973).

Kissam, P. *Surveying for Civil Engineers.* New York: McGraw-Hill Book Company, 1981.

Kopach, S. G. "Cadastral Surveys Automation Related to Surveying and Land Information Systems." *Surveying and Land Information Systems* 53, no. 4 (1993), pp. 238–40.

Lafferty, M. E. "Photometric Control for Subdivision Monumentation." *Proceedings, Fall Convention, ACSM,* 1971.

Laundry, M. E. "The Condominium Survey: Its Legal and Practical Aspects under Illinois Law." *Surveying and Mapping* 34, no. 3 (September 1974).

McIntyre, J. G. *Land Survey Systems.* New York: John Wiley and Sons, 1978.

Moyer, D. D., and P. K. Fisher. "Land Parcel Identifiers for Information Systems." *Proceedings, CLIPP,* sponsored by the American Bar Foundation, Atlanta, GA, 1972.

Onsrud, H. J.; M. C. Swanson; and S. D. Johnson. "Technical Standards for Boundary Surveys Developing Model Law." *Journal of Surveying Engineering* 113, no. 2 (June 1987), pp. 101–15.

Robillard, W. G. "The Surveyor and the Law." *Surveying and Mapping* (June 1978).

Robillard, W. G., and L. J. Bouman. *Clark on Surveying and Boundaries.* Charlottesville, VA: Michie Company, 1992.

U.S. Bureau of Land Management, U.S. Department of the Interior. *Manual of Instructions for the Survey of Public Lands of the United States.* Technical Bulletin No. 6. Washington DC: Government Printing Office, 1973.

U.S. Department of the Interior. *Restoration of Lost or Obliterated Corners and Subdivision of Sections, a Guide for Surveyors.* Washington, DC: Government Printing Office, 1979.

Vorhies, J. H. "A Computerized Land Title Plant in Joint Use." *Proceedings, Annual Meeting, ACSM,* 1977.

Wattles, G. H. *Writing Legal Descriptions in Conjunction with Survey Boundary Control,* 2nd ed. Orange, CA: Gurdon H. Wattles Publications, 1979.

Wattles, W. C. *Land Survey Descriptions,* 10th ed. Orange, CA: Gurdon H. Wattles Publications, 1974.

Woodbury, R. L. "The Surveyor and the Law (Massachusetts Land Court History and Procedure)." *Surveying and Mapping* 33, no. 2 (June 1973).

Elementary Mathematical Concepts

A.1
ANGLES

Angles, such as θ in Figure A.1a, are measured either in degrees, minutes, and seconds or in radians (abbreviated rad). The full circle in Figure A.1a is divided into 360 *degrees* or 360°. The angle θ is 1/12th of the circle and therefore is

$$\theta = \frac{360°}{12} = 30°$$

Each degree is divided into 60 parts called *minutes of arc,* or simply *minutes:*

$$1 \text{ degree} = 1° = 60 \text{ minutes} = 60'$$

Each minute is divided into 60 parts called *seconds of arc,* or simply *seconds:*

$$1 \text{ minute of arc} = 1' = 60 \text{ seconds} = 60''$$

The *radian* measure is a dimensionless quantity. One *radian,* or 1 rad, is an angle whose subtending arc is equal to the radius of that arc, such as the angle ϕ in Figure A.1b. Because π is the ratio of the circumference of a circle to its diameter, d, or to twice its radius, r,

$$\pi = \frac{\text{circumference}}{\text{diameter}}$$

$$\text{circumference} = \pi d = 2\pi r$$

The number of radians in a circle is

$$\frac{\text{circumference}}{r} = 2\pi = 6.28 \text{ rad}$$

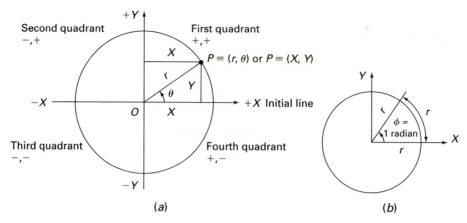

FIGURE A.1

Therefore,

$$360° = 2\pi \text{ rad}$$

$$180° = \pi \text{ rad}$$

$$1° = \frac{\pi}{180°} = 0.0175 \text{ rad}$$

$$1 \text{ rad} = \frac{180°}{\pi} = 57.29578$$

$$= 57°17'44.8''$$

$$= 206,264.8''$$

The last quantity is referred to by ρ. It is used when one needs to convert from radians to seconds and vice versa. Because $\rho = 206,264.8''/\text{rad}$, then

Multiply by ρ when converting radians to seconds.
Divide by ρ or multiply by $1/\rho = 0.00000485$ when converting seconds to radians.

The degrees, minutes, and seconds (also called the *sexagesimal*) measure is used quite frequently in the United States. In Europe, the *centesimal* system is used, where the circle is divided into 400 equal parts, called *gons,* abbreviated[g], each gon having 100 centesimal minutes, abbreviated[c],* and each minute having 100 centesimal seconds, abbreviated[cc]** (see Section 1.8, Chapter 1).

A.2
POLAR AND RECTANGULAR (CARTESIAN) COORDINATES

In a plane, any point, such as P in Figure A.1a, is determined by two quantities, called *coordinates.* For *polar* coordinates, a fixed reference point, such as the origin O, and a fixed line, called the *initial line,* are required. The point P is determined by the distance r from O, and the angle θ from the *initial line.* The radius r *always* is positive, and the angle θ takes any value from 0 to 360° and is counterclockwise from the initial line, as shown in Figure A.1a.

---

* The[c] stands for the centigon, or one one-hundredth of a gon.
** The[cc] stands for centi-centigon; or one ten-thousandth of a gon.

TABLE A.1

Quadrant	Sign of X	Sign of Y	Range of values for θ, deg
First	+	+	0–90
Second	−	+	90–180
Third	−	−	180–270
Fourth	+	−	270–360

A much more common system of coordinates is the *rectangular* or *Cartesian* system. The position of any point P is determined by two distances X and Y from two fixed lines that are perpendicular to each other at the origin (point O), as shown in Figure A.1a. The X coordinate is called the *abscissa,* and the Y coordinate is called the *ordinate.* Each coordinate can be either positive or negative. The signs of the coordinates depend on the quadrant in which the point falls. There are four such quadrants, as shown in Table A.1. The data in Table A.1 also are shown in Figure A.1a.

A.3
TRIGONOMETRIC FUNCTIONS

For an angle, such as θ in Figure A.1a, there are six functions, which are defined as the following ratios:

$$\sin \theta = \frac{Y}{r} \qquad \operatorname{cosec} \theta = \frac{1}{\sin \theta} = \frac{r}{Y}$$

$$\cos \theta = \frac{X}{r} \qquad \sec \theta = \frac{1}{\cos \theta} = \frac{r}{X}$$

$$\tan \theta = \frac{Y}{X} \qquad \cot \theta = \frac{1}{\tan \theta} = \frac{X}{Y}$$

$$\tan \theta = \frac{\sin \theta}{\cos \theta} \qquad \cot \theta = \frac{\cos \theta}{\sin \theta}$$

Each of these trigonometric functions is either positive or negative, depending on the signs of X and Y (recall that r always is positive). For example, $\cos 130°$ is negative because in the second quadrant the X value is negative; and $\tan 210°$ is positive because in the third quadrant both X and Y are negative (from Table A.1) and dividing a negative quantity by another negative quantity yields a positive quantity. Table A.2 lists the signs of all six trigonometric functions for angles in each of the four quadrants. In most computing systems, the quadrant of the angle is determined automatically by checking the signs of the quantities involved. A particularly useful function is the inverse of the tan function, which is referred to as

$$a \tan 2 \, (Y, X) = \theta$$

TABLE A.2

Quadrant	sin	cosec	cos	sec	tan	cot
First	+	+	+	+	+	+
Second	+	+	−	−	−	−
Third	−	−	−	−	+	+
Fourth	−	−	+	+	−	−

TABLE A.3

Angle	sin	cosec	cos	sec	tan	cot
0	0	∞	1	1	0	∞
$\dfrac{\pi}{2} = 90°$	1	1	0	∞	∞	0
$\pi = 180°$	0	∞	-1	-1	0	∞
$\dfrac{3\pi}{2} = 270°$	-1	-1	0	∞	∞	0
$2\pi = 360°$	0	∞	1	1	0	∞

where the signs of Y and X are checked to produce the proper quadrant of θ (also see Equation (A.24)).

For angles such as 0°, 90°, and the like, the functions assume limiting values. These values are listed in Table A.3 for reference.

A.4
USEFUL RELATIONS BETWEEN TRIGONOMETRIC FUNCTIONS

This information is given in the following equations, and in Table A.4.

$$\sin^2 \theta + \cos^2 \theta = 1 \tag{A.1}$$

$$1 + \tan^2 \theta = \sec^2 \theta \tag{A.2}$$

$$1 + \cot^2 \theta = \operatorname{cosec}^2 \theta \tag{A.3}$$

$$\sin (\alpha \pm \beta) = \sin \alpha \cos \beta \pm \cos \alpha \sin \beta \tag{A.4}$$

$$\cos (\alpha \pm \beta) = \cos \alpha \cos \beta \mp \sin \alpha \sin \beta \tag{A.5}$$

$$\tan (\alpha \pm \beta) = \frac{\tan \alpha \pm \tan \beta}{1 \mp \tan \alpha \tan \beta} \tag{A.6}$$

$$\sin 2\theta = 2 \sin \theta \cos \theta \tag{A.7}$$

$$\cos 2\theta = \cos^2 \theta - \sin^2 \theta$$

$$= 1 - 2 \sin^2 \theta$$

$$= 2 \cos^2 \theta - 1 \tag{A.8}$$

$$\sin^2 \theta = \tfrac{1}{2} - \tfrac{1}{2}\cos 2\theta \tag{A.9}$$

$$\cos^2 \theta = \tfrac{1}{2} + \tfrac{1}{2}\cos 2\theta \tag{A.10}$$

$$\tan 2\theta = \frac{2 \tan \theta}{1 - \tan^2 \theta} \tag{A.11}$$

$$\sin 3\theta = 3 \sin \theta - 4 \sin^3 \theta \tag{A.12}$$

$$\cos 3\theta = 4 \cos^3 \theta - 3 \cos \theta \tag{A.13}$$

$$\tan 3\theta = \frac{3 \tan \theta - \tan^3 \theta}{1 - 3 \tan^2 \theta} \tag{A.14}$$

Angle	Quadrant*	sin	cos	tan
$-\theta$	4	$-\sin\theta$	$\cos\theta$	$-\tan\theta$
$+\theta$	1	$\sin\theta$	$\cos\theta$	$\tan\theta$
$90° - \theta = \dfrac{\pi}{2} - \theta$	1	$\cos\theta$	$\sin\theta$	$\cot\theta$
$90° + \theta = \dfrac{\pi}{2} + \theta$	2	$\cos\theta$	$-\sin\theta$	$-\cot\theta$
$180° - \theta = \pi - \theta$	2	$\sin\theta$	$-\cos\theta$	$-\tan\theta$
$180° + \theta = \pi + \theta$	3	$-\sin\theta$	$-\cos\theta$	$\tan\theta$
$270° - \theta = \dfrac{3\pi}{2} - \theta$	3	$-\cos\theta$	$-\sin\theta$	$\cot\theta$
$270° + \theta = \dfrac{3\pi}{2} + \theta$	4	$-\cos\theta$	$\sin\theta$	$-\cot\theta$
$360° - \theta = 2\pi - \theta$	4	$-\sin\theta$	$\cos\theta$	$-\tan\theta$

* The quadrant is given under the assumption that θ is an acute angle.

$$\sin\alpha + \sin\beta = 2\sin\tfrac{1}{2}(\alpha + \beta)\cos\tfrac{1}{2}(\alpha - \beta) \tag{A.15}$$

$$\sin\alpha - \sin\beta = 2\cos\tfrac{1}{2}(\alpha + \beta)\sin\tfrac{1}{2}(\alpha - \beta) \tag{A.16}$$

$$\cos\alpha + \cos\beta = 2\cos\tfrac{1}{2}(\alpha + \beta)\cos\tfrac{1}{2}(\alpha - \beta) \tag{A.17}$$

$$\cos\alpha - \cos\beta = -2\sin\tfrac{1}{2}(\alpha + \beta)\sin\tfrac{1}{2}(\alpha - \beta) \tag{A.18}$$

$$\sin\alpha\cos\beta = \tfrac{1}{2}[\sin(\alpha + \beta) + \sin(\alpha - \beta)] \tag{A.19}$$

$$\cos\alpha\sin\beta = \tfrac{1}{2}[\sin(\alpha + \beta) - \sin(\alpha - \beta)] \tag{A.20}$$

$$\cos\alpha\cos\beta = \tfrac{1}{2}[\cos(\alpha + \beta) + \cos(\alpha - \beta)] \tag{A.21}$$

$$\sin\alpha\sin\beta = -\tfrac{1}{2}[\cos(\alpha + \beta) - \cos(\alpha - \beta)] \tag{A.22}$$

A.5
RELATION BETWEEN POLAR AND CARTESIAN COORDINATES

Figure A.2 shows the two sets of coordinates, polar (r, θ) and rectangular (X, Y), for any point P. Given r and θ, one can calculate X and Y, using

$$X = r\cos\theta$$
$$Y = r\sin\theta \tag{A.23}$$

FIGURE A.2

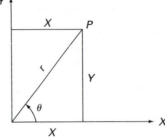

Conversely, from X and Y, the polar coordinates r and θ can be calculated, using

$$r = \sqrt{X^2 + Y^2}$$

$$\tan \theta = \frac{Y}{X} \tag{A.24}$$

or

$$\theta = \arctan \frac{Y}{X} = \tan^{-1} \frac{Y}{X} = a \tan 2 \, (Y, X)$$

As was mentioned earlier, only the positive square root is considered, because r always is positive. The symbols *arctan* and *tan*$^{-1}$ indicate "the angle whose tangent is Y/X." These *inverse* relations exist for all trigonometric functions, such as $\sin^{-1}$ and $\cos^{-1}$. The quadrant in which the calculated angle θ falls is determined by the signs of X, Y, as explained in Section A.2 and given in Table A.1.

EXAMPLE A.1. The Cartesian coordinates of a point are $X = -6.51$ m and $Y = 17.89$ m. Calculate its polar coordinates.

Solution

$$r = \sqrt{(-6.51)^2 + (17.89)^2} = 19.04 \text{ m}$$

$$\theta = \tan^{-1} \frac{17.89}{-6.51} = \tan^{-1}(-2.7481) = 110°$$

Because X is negative and Y is positive, the angle falls in the second quadrant according to Table A.1. No ambiguity exists when one knows the signs of the numerator and denominator.

In some calculators, when a negative number is used for the tan, a negative value for the angle is given: $-70°$ in this case. This is because the tan is negative in the second and fourth quadrants. If, as in this case, the angle is known to be in the second quadrant, then the final value of the angle is $180°$ + the value from the calculator, or $180 - 70 = 110°$.

A.6
DISTANCE BETWEEN TWO POINTS

Although all the relations to be developed in this and following sections are in rectangular coordinates, comparable relations can be derived in polar coordinates. The reason for concentrating on rectangular coordinates is that they are much more common in surveying practice than polar coordinates.

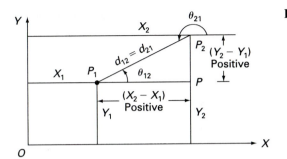

FIGURE A.3

1059

APPENDIX A:
Elementary
Mathematical
Concepts

Figure A.3 shows two points, P_1 and P_2, with coordinates X_1, Y_1 and X_2, Y_2, respectively. The distance between them, d_{12}, is obtained by employing the Pythagorian relationship to the triangle $P_1 P_2 P$ or

$$(P_1 P_2)^2 = (P_1 P)^2 + (P_2 P)^2$$

or

$$d_{12}^2 = (X_2 - X_1)^2 + (Y_2 - Y_1)^2$$

Then,

$$d_{12} = \sqrt{(X_2 - X_1)^2 + (Y_2 - Y_1)^2} \tag{A.25}$$

Note that $d_{12} = d_{21}$ because the square of each component is the same regardless whether positive or negative; for example, $(X_2 - X_1)^2 = (X_1 - X_2)^2$.

A.7
SLOPE OF A LINE BETWEEN TWO POINTS

The slope of any line is equal to the tangent of the angle it makes with the X axis. The direction of the line is not important in regard to the slope, but *is important* for the slope angle. The slope angle is the counterclockwise angle from $+X$ to the line in the specified direction. For the specific case shown in Figure A.3, the slope angle for the line $P_1 P_2$ is θ_{12}, which is $< 90°$. On the other hand, the slope angle for the line $P_2 P_1$ is θ_{21}, which is $> 180°$. In fact,

$$\theta_{21} = \theta_{12} + 180° \tag{A.26}$$

The slope m_{12} of the line $P_1 P_2$ is

$$m_{12} = \tan \theta_{12} = \frac{Y_2 - Y_1}{X_2 - X_1} \tag{A.27}$$

and that for the line $P_2 P_1$ is

$$m_{21} = \tan \theta_{21} = \frac{Y_1 - Y_2}{X_1 - X_2} \tag{A.28}$$

Note that $m_{21} = \tan \theta_{21} = \tan (\theta_{12} + 180°) = \tan \theta_{12}$ according to Table A.4, or $m_{21} = m_{12}$, as noted previously.

1060

APPENDIX A:
Elementary
Mathematical
Concepts

FIGURE A.4

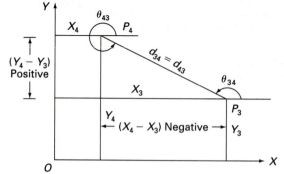

It does not matter which of the two equations, (A.27) or (A.28), is used. Calculating the coordinate differences so that one gets the correct sign of the slope is important. This is done by remembering that the first coordinates in both the numerator and denominator belong to the same point. For example,

$$m_{AB} = \frac{Y_B - Y_A}{X_B - X_A} = \frac{Y_A - Y_B}{X_A - X_B} \qquad (A.29)$$

The slope of the line through points P_1 and P_2, Figure A.3, is *positive* because θ_{12} and θ_{21} are in the first and third quadrants, respectively, and the tangent is positive in these two quadrants (see Table A.2). The slope of the line through points P_3 and P_4, Figure A.4, is negative because θ_{34} and θ_{43} are in the second and fourth quadrants, respectively, where the tangent is negative according to Table A.2. Or,

$$m_{34} = \frac{Y_4 - Y_3}{X_4 - X_3} = \frac{+}{-} = \text{negative quantity}$$

$$m_{43} = \frac{Y_3 - Y_4}{X_3 - X_4} = \frac{-}{+} = \text{negative quantity}$$

(A.30)

In Equations (A.27) through (A.30), one must check to make sure that the denominator is not 0. If it is 0, then the line is parallel to the Y axis, and the slope is infinite.

A.8
AZIMUTH AND BEARING

Azimuth always is a clockwise angle with a magnitude between 0 and 360°. It is measured either from the north or the $+Y$ axis, or from the south or the $-Y$ axis. For any line such as OP_1, in Figure A.5a, α_N is the azimuth from north or $+Y$, and α_S is the azimuth from the south or $-Y$. One azimuth angle is obtained from the other by simply adding 180° and dropping 360° whenever the sum exceeds 360°:

$$\alpha_S = \alpha_N + 180°$$

$$\alpha_N = \alpha_S + 180°$$

(A.31)

Figure A.5 shows four lines, one in each of the four quadrants. Application of Equation (A.31) is demonstrated in Table A.5.

A.5

1061

APPENDIX A:
Elementary
Mathematical
Concepts

	α_N, deg	$\alpha_N + 180°$, deg	α_S, deg	$\alpha_S + 180°$, deg	Figure
	35	215	215	$395 - 360 = 35$	A.5a
OP_2	300	$480 - 360 = 120$	120	300	A.5b
OP_3	247	$427 - 360 = 67$	67	247	A.5c
OP_4	115	295	295	$475 - 360 = 115$	A.5d

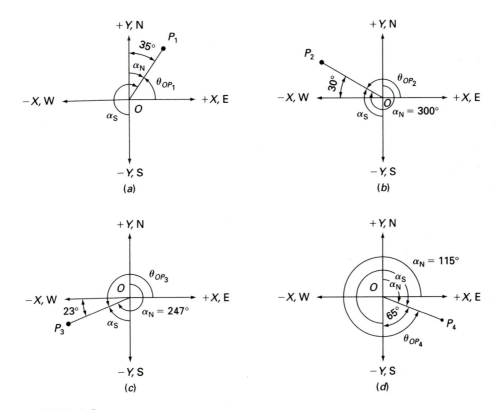

FIGURE A.5

Another way of expressing the direction of a line in surveying is the *bearing,* which always is an *acute* angle, with a magnitude between 0 and 90°. The bearing is the angle the line makes with either the $+Y$ or $-Y$ axis. Instead of using $+Y$ and $-Y$, N (for north) and S (for south) are used. The value of the bearing always is a positive quantity. Therefore, the quadrant is indicated by specifying whether the angle is on the east or west side of the meridian (Y axis). Thus, the bearing angle is preceded by either N or S and succeeded by either E or W. Four examples are given in Figure A.6.

If β designates a bearing, Table A.6 shows how to convert to azimuth. Table A.7 shows how to convert azimuth to bearing.

Although Tables A.5 through A.7 are useful and cover various cases, it always is advisable to draw a clear sketch and make sure that the angles are in their correct relationship and proper quadrants. Nothing is more helpful than a good sketch of the situation. This is particularly true when angles and lines are added consecutively, as in a traverse.

TABLE A.6

Bearing	α_N, deg	α_S, deg
$N\beta°E$	β	$180 + \beta$
$S\beta°E$	$180 - \beta$	$360 - \beta$
$S\beta°W$	$\beta + 180$	β
$N\beta°W$	$360 - \beta$	$180 - \beta$

TABLE A.7

α_N, deg	α_S, deg	β	Bearing
0–90	180–270	α_N or $\alpha_S - 180°$	$N\beta°E$
90–180	270–360	$180° - \alpha_N$ or $360° - \alpha_S$	$S\beta°E$
180–270	0–90	$\alpha_N - 180°$ or α_S	$S\beta°W$
270–360	90–180	$360° - \alpha_N$ or $180° - \alpha_S$	$N\beta°W$

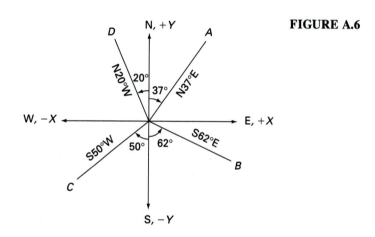

FIGURE A.6

A.9
RELATION BETWEEN AZIMUTH AND SLOPE ANGLE

Because it is relatively easy to convert from one form of azimuth to the other, consider only the more commonly used azimuth from north, α_N. In Section A.7, the well-accepted practice in algebra of defining the slope angle of a line as the counterclockwise angle from the $+X$ axis was used. Given α_N, the slope angle θ is obtained from

$$\theta = 90° - \alpha_N \tag{A.32}$$

Except for $\alpha_N \leq 90°$, the value of θ will be negative. This will cause no problems with trigonometric functions because the proper quadrant will be accommodated automatically. However, if it is preferable to work with positive angles, one simply can add 360°:

$$\theta = 450° - \alpha_N \quad \text{for} \quad \alpha_N > 90° \tag{A.33}$$

TABLE A.8

1063

APPENDIX A:
Elementary
Mathematical
Concepts

α_N, deg	$450° - \alpha_N$, deg	θ, deg	$\alpha_N = 450 - \theta$, deg	Figure
35	415 or 55 = θ_{op_1}	55	395 or 35	A.5a
300	150 = θ_{op_2}	150	300	A.5b
247	203 = θ_{op_3}	203	247	A.5c
115	335 = θ_{op_4}	335	115	A.5d

For the four cases shown in Figure A.5, the slope angles are calculated in Table A.8 as an example (see also Table A.5).

The azimuth can be calculated directly, too, if the line is fixed by the coordinates of its terminal points. The azimuth of any line *from* a point *to* another, α_{ft} (where subscript *f* stands for "from" and *t* for "to"), is given by

$$\alpha_{ft} = \tan^{-1}\frac{X_{to} - X_{from}}{Y_{to} - Y_{from}} \tag{A.34}$$

The four possible cases, one in each quadrant, are depicted in Figure A.7. For these cases, the azimuths are given by the following. For Figure A.7a,

$$\alpha_{12} = \tan^{-1}\frac{X_2 - X_1}{Y_2 - Y_1} = \tan^{-1}\left(\frac{+}{+}\right) \qquad \text{first quadrant}$$

$$\alpha_{21} = \tan^{-1}\frac{X_1 - X_2}{Y_1 - Y_2} = \tan^{-1}\left(\frac{-}{-}\right) \qquad \text{third quadrant}$$

$$\alpha_{34} = \tan^{-1}\frac{X_4 - X_3}{Y_4 - Y_3} = \tan^{-1}\left(\frac{-}{+}\right) \qquad \text{fourth quadrant}$$

$$\alpha_{43} = \tan^{-1}\frac{X_3 - X_4}{Y_3 - Y_4} = \tan^{-1}\left(\frac{+}{-}\right) \qquad \text{second quadrant}$$

Note that the proper quadrant is determined by the signs of the numerator and denominator. Also, the value of the denominator first must be checked against 0. If it is 0, tan α is ∞

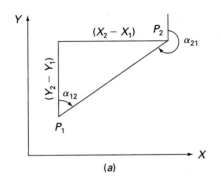

(a)

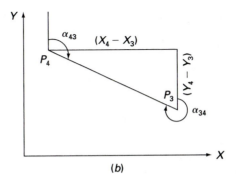

(b)

FIGURE A.7

according to Table A.3. Then, the numerator is checked; if it is positive, $\alpha = 90°$; if it is negative, $\alpha = 270°$.

A.10
INTERNAL AND EXTERNAL DIVISION OF A LINE SEGMENT

Given a line segment between two points of known coordinates, it is useful to find the coordinates of either an internal point, I, or an external point, E (Figure A.8). In Figure A.8a, point I is between P_1 and P_2 and divides the line in the proportion s_1/s_2 or

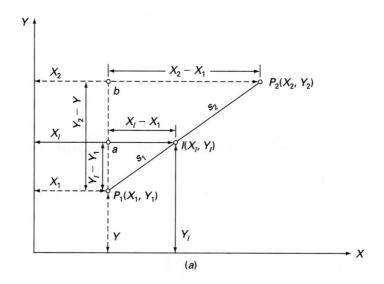

(a)

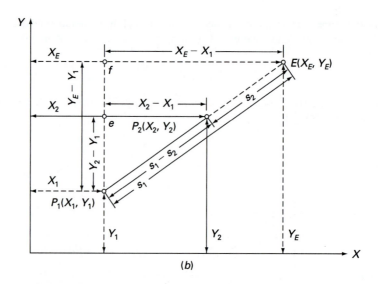

(b)

FIGURE A.8

$(P_1 I)/(IP_2) = s_1/s_2$. By proportion in similar triangles $P_1 Ia$ and $P_1 P_2 b$ (see Figure A.8a), the coordinates of point I are given by

$$X_I = \frac{s_1 X_2 + s_2 X_1}{s_1 + s_2}$$

$$Y_I = \frac{s_1 Y_2 + s_2 Y_1}{s_1 + s_2}$$

(A.35)

In Figure A.8b, the division is external; that is, point E is on the extension of $P_1 P_2$ such that $(P_1 E)/(EP_2) = s_1/s_2$. Again by proportion in similar triangles $P_1 Ef$ and $P_1 P_2 e$ (Figure A.8b), the coordinates of point E are given by

$$X_E = \frac{s_1 X_2 - s_2 X_1}{s_1 - s_2}$$

$$Y_E = \frac{s_1 Y_2 - s_2 Y_1}{s_1 - s_2}$$

(A.36)

A.11
VARIOUS FORMS OF THE EQUATION OF A STRAIGHT LINE

There are many ways to algebraically express a straight line in a plane that are quite useful in survey computations. In the following sections, several of these are given.

A.12
LINE WITH KNOWN X AND Y INTERCEPTS

See Figure A.9:

$$\frac{X}{X_0} + \frac{Y}{Y_0} = 1$$

$$Y_0 X + X_0 Y = X_0 Y_0$$

(A.37)

where X_0 and Y_0 are the X and Y intercepts, respectively. This may be referred to as the *two-intercept form* of the line equation.

FIGURE A.9

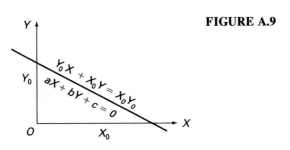

A.13
GENERAL EQUATION OF A LINE

Rearranging Equation (A.37) yields

$$Y_0 X + X_0 Y - X_0 Y_0 = 0$$

or, in general terms,

$$aX + bY + c = 0 \qquad \text{(A.38a)}$$

Because two points define a line in a plane, only two of these three coefficients are independent. We can divide by c to get

$$\frac{a}{c}X + \frac{b}{c}Y + 1 = 0$$

or

$$a'X + b'Y + 1 = 0 \qquad \text{(A.38b)}$$

The two intercepts are obtained by substituting 0 for each of the two variables, X and Y, and calculating the value of the other. Thus, with $Y = 0$ and $X = 0$, Equation (A.38a) gives the X intercept and Y intercept, respectively:

$$X \text{ intercept} = -\frac{c}{a}$$

$$Y \text{ intercept} = -\frac{c}{b} \qquad \text{(A.39)}$$

The slope of the line is given by

$$m = \tan \theta = -\frac{a}{b} \qquad \text{(A.40)}$$

A.14
LINES PARALLEL TO THE AXES

The equation of a line parallel to the X axis is simply

$$Y = k_1 \qquad \text{(A.41)}$$

where k_1 is the distance of the line from the X axis. Similarly, the equation of a line parallel to the Y axis is

$$X = k_2 \qquad \text{(A.42)}$$

where k_2 is the distance between the Y axis and the line.

A.15
LINE WITH A GIVEN SLOPE AND Y INTERCEPT

The line shown in Figure A.10 has a slope $m = \tan \theta$ and a Y intercept $= c$. Let P be any point with coordinates X, Y. Then,

$$m = \tan \theta = \frac{Y - c}{X}$$

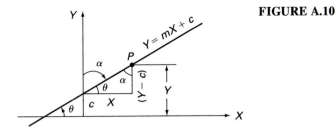

FIGURE A.10

1067

APPENDIX A:
Elementary
Mathematical
Concepts

or

$$Y - c = mX$$

or the equation of the line is

$$Y = mX + c \tag{A.43}$$

In terms of the azimuth, α, this equation becomes

$$Y = X \cot \alpha + c \tag{A.44}$$

This may be referred to as the *slope intercept* form of the equation of a line.

A.16
LINE WITH A GIVEN SLOPE PASSING THROUGH A GIVEN POINT

The line shown in Figure A.11 has a slope $m = \tan \theta$ and passes through point P_1. If P is any point with coordinates X, Y, then

$$m = \tan \theta = \frac{Y - Y_1}{X - X_1}$$

or the equation of the line is

$$Y - Y_1 = m(X - X_1) \tag{A.45}$$

In terms of azimuth, the equation becomes

$$Y - Y_1 = (X - X_1) \cot \alpha \tag{A.46}$$

If the given point is the origin of the coordinate system, then the equations reduce to

$$Y = mX$$
$$Y = X \cot \alpha \tag{A.47}$$

This may be called the *point-slope* form of the equation of a line.

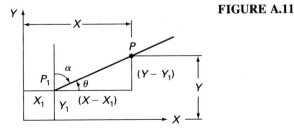

FIGURE A.11

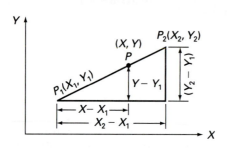

FIGURE A.12

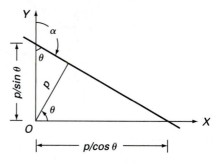

FIGURE A.13

A.17
LINE JOINING TWO POINTS

Figure A.12 shows a line passing through two points, P_1 and P_2, with coordinates X_1, Y_1 and X_2, Y_2, respectively. With $P(X, Y)$ being any point on the line,

$$\frac{Y - Y_1}{X - X_1} = \frac{Y_2 - Y_1}{X_2 - X_1} \tag{A.48}$$

This is called the *two-point* form of the equation of a line.

A.18
LINE WITH A GIVEN LENGTH AND SLOPE OF THE PERPENDICULAR FROM THE ORIGIN

As shown in Figure A.13 the perpendicular from the origin, O, to the line in question has a length p and slope angle θ. As is clear from the figure, the X and Y intercepts are $p/\cos \theta$ and $p/\sin \theta$, respectively. Consequently, according to Equation (A.37), the equation for the straight line is

$$\frac{X}{p/\cos \theta} + \frac{Y}{p/\sin \theta} = 1$$

or

$$X \cos \theta + Y \sin \theta = p \tag{A.49}$$

This is referred to as the *normal form* of the equation of a line.

A.19
PERPENDICULAR DISTANCE FROM THE ORIGIN TO A LINE

The general form of the equation of a line (Equation (A.38a)) is

$$aX + bY + c = 0$$

or

$$aX + bY = -c \tag{A.50}$$

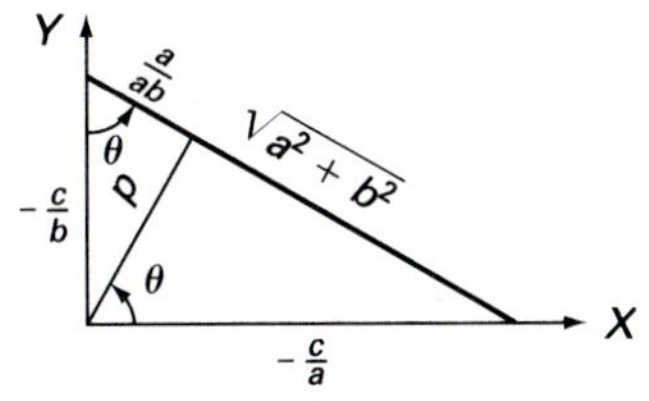

FIGURE A.14

1069

APPENDIX A:
Elementary
Mathematical
Concepts

Comparing Equations (A.49) and (A.50), the coefficients of X and Y relate to trigonometric functions of an angle. Figure A.14 shows that the sine and cosine of the slope angle of the normal, θ, are given by

$$\cos \theta = \frac{-a}{\sqrt{a^2 + b^2}}$$

$$\sin \theta = \frac{-b}{\sqrt{a^2 + b^2}}$$
(A.51)

Dividing both sides of Equation (A.50) by $\sqrt{a^2 + b^2}$ yields

$$\frac{a}{\sqrt{a^2 + b^2}} X + \frac{b}{\sqrt{a^2 + b^2}} Y = \frac{c}{\sqrt{a^2 + b^2}}$$

or, in view of Equations (A.51),

$$X \cos \theta + Y \sin \theta = \frac{c}{\sqrt{a^2 + b^2}}$$
(A.52)

Consequently, from Equations (A.49) and (A.52), the perpendicular from the origin to the line with the general equation is

$$p = \frac{c}{\sqrt{a^2 + b^2}}$$
(A.53)

The radical is a quantity that represents a length and so does p. To avoid confusion, always take the positive sign of the radical. This means that the sign of p will be either positive or negative, depending on whether the sign of c is negative or positive, respectively. Again, to avoid confusion, regardless of the resulting sign for p, always take it as a positive length. In other words, Equation (A.53) may be rewritten in absolute value form:

$$p = \left| \frac{c}{\sqrt{a^2 + b^2}} \right|$$
(A.54)

Note that, when $c = 0$, the distance p also is 0 and the line passes through the origin.

EXAMPLE A.2. Figure A.15 shows four lines with given equations:

$$A: \quad X + Y - 4 = 0$$
$$B: \quad X - Y + 3 = 0$$
$$C: \quad X + Y + 5 = 0$$
$$D: \quad X - Y - 3 = 0$$

Calculate the length of the perpendicular from the origin to each of the lines.

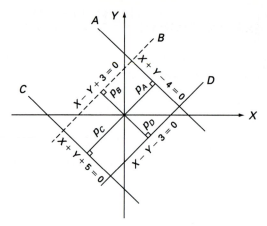

FIGURE A.15

Solution. By Equation (A.54),

$$p_A = \left| \frac{-4}{\sqrt{1+1}} \right| = \frac{4}{\sqrt{2}} = \frac{4\sqrt{2}}{2} = 2\sqrt{2}$$

$$p_B = \left| \frac{3}{\sqrt{2}} \right| = 1.5\sqrt{2}$$

$$p_C = \left| \frac{5}{\sqrt{2}} \right| = 2.5\sqrt{2}$$

$$p_D = \left| -\frac{3}{\sqrt{2}} \right| = 1.5\sqrt{2}$$

EXAMPLE A.3. Given the following length and slope angles of the four perpendiculars,

$$p_A = 2\sqrt{2} \qquad \theta_A = 45°$$
$$p_B = 1.5\sqrt{2} \qquad \theta_B = 135°$$
$$p_C = 2.5\sqrt{2} \qquad \theta_C = 225°$$
$$p_D = 1.5\sqrt{2} \qquad \theta_D = 315°$$

show that the resulting equations of the four lines are the same as those given in Example A.2 and shown in Figure A.15.

Solution. By Equation (A.49),

$A:$ $X\cos 45° + Y\sin 45° = p_A$ or $\dfrac{X}{\sqrt{2}} + \dfrac{Y}{\sqrt{2}} = 2\sqrt{2}$ or $X + Y - 4 = 0$

$B: X\cos 135° + Y\sin 135° = p_B$ or $-\dfrac{X}{\sqrt{2}} + \dfrac{Y}{\sqrt{2}} = 1.5\sqrt{2}$ or $X - Y + 3 = 0$

$C: X\cos 225° + Y\sin 225° = p_C$ or $-\dfrac{X}{\sqrt{2}} - \dfrac{Y}{\sqrt{2}} = 2.5\sqrt{2}$ or $X + Y + 5 = 0$

$D: X\cos 315° + Y\sin 315° = p_D$ or $\dfrac{X}{\sqrt{2}} - \dfrac{Y}{\sqrt{2}} = 1.5\sqrt{2}$ or $X - Y - 3 = 0$

A.20
PERPENDICULAR DISTANCE FROM A POINT TO A LINE

Figure A.16 shows a line, A, with equation $aX + bY + c = 0$ and a point, P, with coordinates X_1, Y_1. It is required to find the perpendicular distance d of point P from the line A. An alternative form of the equation of line A is $X \cos \theta + Y \sin \theta - p = 0$, where (see Equations (A.52) and (A.53))

$$\cos \theta = \frac{-a}{\sqrt{a^2 + b^2}} \qquad \sin \theta = \frac{-b}{\sqrt{a^2 + b^2}}$$

$$p = \frac{c}{\sqrt{a^2 + b^2}}$$

(A.55)

The line B, which is parallel to A and passes through point P, has the equation $X \cos \theta + Y \sin \theta = p + d$. Point P falls on line B, so its coordinates X_1, Y_1 satisfy its equation:

$$X_1 \cos \theta + Y_1 \sin \theta = p + d \qquad \text{(A.56)}$$

Substituting Equation (A.55) into (A.56) and rearranging, one gets

$$d = -\frac{aX_1 + bY_1 + c}{\sqrt{a^2 + b^2}} \qquad \text{(A.57)}$$

The quantity in the numerator can be either positive or negative because point P can fall on either side of the line A (more explanation of this is given in Section A.21). Because the radical always will be taken as positive, it is best to take the absolute value:

$$d = \left| \frac{aX_1 + bY_1 + c}{\sqrt{a^2 + b^2}} \right| \qquad \text{(A.58)}$$

EXAMPLE A.4. Figure A.17 shows a straight line, A, with X intercept $X_0 = 3$ and Y intercept $Y_0 = -2$. Find the distance d_1 and d_2 of the two points $P_1(4, 4)$ and $P_2(9, 2)$ from line A.

Solution. By Equation (A.37), the equation of line A is

$$-2X + 3Y = -6$$

or

$$2X - 3Y - 6 = 0$$

Next,

$$\sqrt{a^2 + b^2} = \sqrt{4 + 9} = 3.61$$

FIGURE A.16

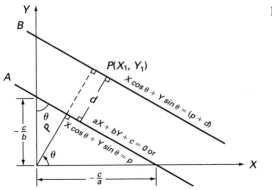

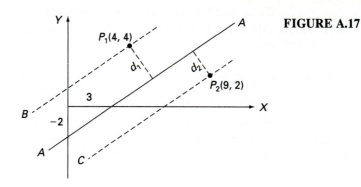

FIGURE A.17

For $P_1(4, 4)$,

$$aX_1 + bY_1 + c = 2(4) - 3(4) - 6 = -10$$

then

$$d_1 = -\left[\frac{-10}{3.61}\right] = 2.77$$

For $P_2(9, 2)$,

$$aX_2 + bY_2 + c = 2(9) - 3(2) - 6 = 6$$

then

$$d_2 = -\left[\frac{6}{3.61}\right] = -1.66$$

A.21
EQUATION OF A LINE THROUGH A POINT AND PARALLEL TO ANOTHER LINE

The equation sought here was basically given in Section A.20. In the normal form it is

$$X \cos \theta + Y \sin \theta = p + d \qquad (A.59)$$

where $X \cos \theta + Y \sin \theta = p$ is the equation of the given line, and d is the perpendicular distance between the new line and the given line. The value of d cannot be taken in absolute value as was the case in Section A.20 where the magnitude was sufficient. Here, the proper sign of d must be determined. Before proceeding, note that the evaluated value of a line equation will always be positive for all points on one side of the line and negative for all points on the other side. (The value is of course zero for points falling on the line.) This fact was demonstrated numerically in Example A.4 (see also Figure A.17). For the line $2X - 3Y - 6 = 0$, the calculated value of d for point P_1 is positive (equal to 2.77), and is negative for point P_2 (equal to -1.66). Point P_1 is on the same side of line A as the origin. As a further demonstration, substitute 0, 0, the coordinates of the origin in Equation (A.57)

$$d_0 = -\frac{a(0) + b(0) + c}{\sqrt{a^2 + b^2}} = 6/\sqrt{13} = 1.66$$

which is also positive as for point P_1.

Using Equations (A.55) and (A.57) substituted into Equation (A.59), one gets

$$\frac{a}{\sqrt{a^2 + b^2}}X + \frac{b}{\sqrt{a^2 + b^2}}Y = \frac{-c}{\sqrt{a^2 + b^2}} + \frac{aX_1 + bY_1 + c}{\sqrt{a^2 + b^2}}$$

which when clearing fractions becomes

$$aX + bY - (aX_1 + bY_1) = 0 \qquad (A.60)$$

that is the equation sought for the line. The coefficients a and b belong to the given line and X_1, Y_1 are the coordinates of the given point.

An alternative procedure is to determine the slope of the given line and use the point-slope form, Section A.16. Thus, by Equation (A.40)

$$m = -\frac{a}{b}$$

and then by Equation (A.45)

$$Y - Y_1 = m(X - X_1) = -\frac{a}{b}(X - X_1)$$

or

$$bY - bY_1 = -aX + aX_1$$

or

$$aX + bY - (aX_1 + bY_1) = 0$$

which is the same as Equation (A.60).

EXAMPLE A.5. Write the equations for lines through points P_1 and P_2 and parallel to line A shown in Figure A.17, using the data in Example A.4.

Solution. Line B, through point $P_1(4, 4)$, will have the equation

$$2X - 3Y - [(2)(4) - (3)(4)] = 0$$

or

$$2X - 3Y + 4 = 0$$

Line C, through point $P_2(9, 2)$, has the equation

$$2X - 3Y - [(2)(9) - (3)(2)] = 0$$

or

$$2X - 3Y - 12 = 0$$

A.22
EQUATION OF A LINE THROUGH A POINT AND PERPENDICULAR TO ANOTHER LINE

In Figure A.18, the given line, A, has the general equation $aX + bY + c = 0$, and the given point, P, has coordinates X_1, Y_1. Required is the equation of line B, which passes through point P and is perpendicular to line A. According to Equation (A.40), the slope of line A is

$$m_A = \tan \theta_A = -\frac{a}{b}$$

And from Figure A.18, the slope angle θ_B of line B is

$$\theta_B = \theta_A + 90°$$

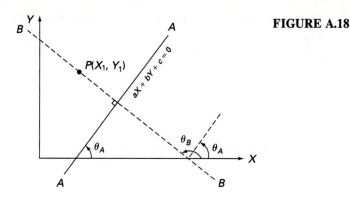

FIGURE A.18

Then the slope of line B is

$$m_B = \tan \theta_B = \tan(\theta_A + 90) = -\cot \theta_A \qquad \text{(see Table A.4)}$$

or

$$m_B = -\frac{1}{m_A} \qquad (A.61)$$

Equation (A.61) indicates that the slope of perpendicular line B is the negative of the reciprocal of the slope of given line A. With m_B and point P, the equation of line B can be written using the point slope from Section A.16 or from Equation (A.45):

$$Y - Y_1 = m_B(X - X_1) = \frac{b}{a}(X - X_1)$$

or

$$aY - aY_1 = bX - bX_1$$

or

$$bX - aY - (bX_1 - aY_1) = 0 \qquad (A.62)$$

A.23
ANGLE BETWEEN TWO LINES

Figure A.19 shows two lines, the angle between them being designated by γ. It is clear from the figure that

$$\gamma = \theta_2 - \theta_1 \qquad (A.63)$$

where θ_1 and θ_2 are the slope angles of line 1 and line 2, respectively. In terms of slope, one gets

$$\tan \gamma = \tan(\theta_2 - \theta_1)$$

which according to Equation (A.6) becomes

$$\tan \gamma = \frac{\tan \theta_2 - \tan \theta_1}{1 + \tan \theta_1 \tan \theta_2}$$

With slopes $m_1 = \tan \theta_1$ and $m_2 = \tan \theta_2$, of lines 1 and 2, one finally gets

$$\tan \gamma = \frac{m_2 - m_1}{1 + m_1 m_2} \qquad (A.64)$$

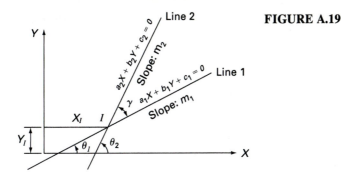

FIGURE A.19

1075

APPENDIX A:
Elementary
Mathematical
Concepts

If $\tan \gamma = 0$, or $m_1 = m_2$, the two lines are parallel. On the other hand, if $\tan \gamma = \infty$, or $m_1 m_2 = -1$ (see also Equation (A.61)), the two lines will be perpendicular to each other. The slopes m_1 and m_2 can be calculated from whatever form is given for the equations of the lines.

A.24
POINT OF INTERSECTION OF TWO LINES

The two lines in Figure A.19 intersect at point I. Because this point lies on both lines, its coordinates X_I, Y_I satisfy their equations. Then, to get X_I, Y_I we simultaneously solve the two equations of the lines:

$$a_1 X_I + b_1 Y_I + c_1 = 0$$

$$a_2 X_I + b_2 Y_I + c_2 = 0$$

Multiply the first equation by b_2, the second equation by b_1, and subtract:

$$a_1 b_2 X_I + b_1 b_2 Y_I + c_1 b_2 = 0$$
$$\underline{a_2 b_1 X_I + b_1 b_2 Y_I + c_2 b_1 = 0}$$
$$(a_1 b_2 - a_2 b_1)X_I = c_2 b_1 - c_1 b_2$$

or

$$X_I = \frac{c_2 b_1 - c_1 b_2}{a_1 b_2 - a_2 b_1} \tag{A.65}$$

In a similar manner, for Y_I, one gets

$$Y_I = \frac{c_2 a_1 - c_1 a_2}{a_2 b_1 - a_1 b_2} \tag{A.66}$$

A.25
EQUATION OF A CIRCLE

Figure A.20 shows a circle of radius R, the center of which is at point C with coordinates X_c, Y_c. From Section A.6, the square of the distance between C and any point on the circle, such as $P(X, Y)$, is given by

$$(X - X_c)^2 + (Y - Y_c)^2 = R^2 \tag{A.67}$$

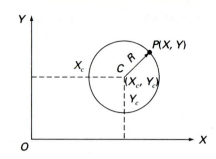

FIGURE A.20

This is the equation of a circle when the coordinates of its center and its radius are given. If $X_c = Y_c = 0$, that is, if the circle's center is the origin of the coordinate system, then the equation simplifies to

$$X^2 + Y^2 = R^2 \qquad (A.68)$$

Equation (A.67) may be expanded to

$$X^2 - 2XX_c + X_c^2 + Y^2 - 2YY_c + Y_c^2 - R^2 \qquad = 0$$

or

$$X^2 + Y^2 - 2X_cX - 2Y_cY + (X_c^2 + Y_c^2 - R^2) = 0$$

and written in the form

$$X^2 + Y^2 + 2dX + 2eY + f = 0 \qquad (A.69)$$

Equation (A.69) represents the general form of the equation of a circle. It contains three coefficients, d, e, and f, which represent three geometric elements such as the radius and the two coordinates of its center.

A.26
INTERSECTION OF A LINE AND A CIRCLE

In general, a straight line intersects a circle in two points, as shown in Figure A.21a. Let the equation of the circle (Equation (A.69)) be

$$X^2 + Y^2 + 2dX + 2eY + f = 0$$

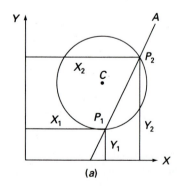
(a)

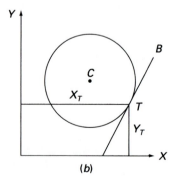
(b)

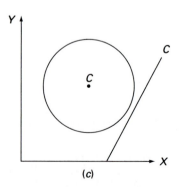
(c)

FIGURE A.21

and that of the straight line (Equation (A.43)) be

$$Y = mX + c$$

Substituting Equation (A.43) into (A.69) yields

$$X^2 + (mX + c)^2 + 2dX + 2e(mX + c) + f = 0$$

which can be reduced to the general *quadratic* equation form:

$$AX^2 + BX + C = 0 \tag{A.70}$$

Whatever form the equations for the circle and straight line are, the quadratic form in Equation (A.70) always will result from the substitution. In general, two values for X are obtainable from Equation (A.70):

$$X_1 = \frac{-B + \sqrt{B^2 - 4AC}}{2A} \quad \text{and} \quad X_2 = \frac{-B - \sqrt{B^2 - 4AC}}{2A} \tag{A.71}$$

and, for each of these, a value for Y is obtained from Equation (A.43). In general, a straight line intersects a circle in two points such as $P_1(X_1, Y_1)$ and $P_2(X_2, Y_2)$ in Figure A.21a. Two other situations also are possible, depending on the quantity under the radical, $\sqrt{B^2 - 4AC}$. If this quantity is 0, or $B^2 = 4AC$, then $X_1 = X_2$ and the line is tangent to the circle at one point, $T(X_T, Y_T)$, as shown in Figure A.21b. In the second case, $B^2 < 4AC$, the quantity is negative, which means that one cannot take the square root. (This usually is referred to as the *imaginary* solution to the quadratic equation.) This case is depicted in Figure A.21c, where the line misses the circle and therefore no intersection takes place.

A.27
EQUATION OF AN ELLIPSE

Figure A.22 shows an ellipse with a *semimajor axis, a*, and *semiminor axis, b*, and its center, c, has the coordinates X_c and Y_c. An ellipse is the locus of a point such that the sum of its distances to two fixed points is constant. The semiaxes of this ellipse are parallel to the coordinate axes as shown. The equation of the ellipse is given by

$$\frac{(X - X_c)^2}{a^2} + \frac{(Y - Y_c)^2}{b^2} = 1 \tag{A.72}$$

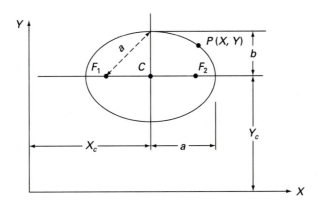

FIGURE A.22

This equation contains four parameters, X_c, Y_c, a, and b; therefore, four points are needed to define it uniquely. The general case is when the ellipse axes are oriented at an angle, θ, relative to the coordinate axes. In such a case, a minimum of five points are required to define a general ellipse in a plane. On the other hand, the simplest case is when the ellipse is centered at the origin of the coordinate system ($X_c = 0$ and $Y_c = 0$), the semimajor axis is along the X axis, and the semiminor axis is along the Y axis. In that case, its equation simplifies to the form

$$\frac{X^2}{a^2} + \frac{Y^2}{b^2} = 1 \tag{A.73}$$

In Figure A.22, F_1 and F_2 are called the *focii* of the ellipse and represent the two fixed points mentioned in the definition of the ellipse. The distance between a *focus* and the center of the ellipse divided by a is the *eccentricity, e*, which is given by

$$e^2 = \frac{a^2 - b^2}{a^2} \tag{A.74}$$

from which, with reference to Figure A.22, the following relations result:

$$a^2 = b^2 + (CF_1)^2$$

$$CF_1 = CF_2 = ae$$

Note that $e < 1$. When $e = 0$ and $a = b$, the ellipse becomes a circle.

A.28
AREAS

Circle

$$A = \pi R^2 = \frac{\pi D^2}{4} \tag{A.75}$$

where R and D are the radius and diameter, respectively.

Sector

$$A = \tfrac{1}{2} Rs = \frac{\pi R^2 \gamma^\circ}{360^\circ} \tag{A.76}$$

where R is the radius, s is the arc length, and γ is the angle at the center in degrees, as shown in Figure A.23.

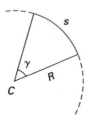

FIGURE A.23

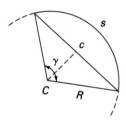

FIGURE A.24

Segment

A segment is less than a semicircle (see Figure A.24):

$$A = \tfrac{1}{2}Rs - \tfrac{1}{2}R^2 \sin \gamma \tag{A.77}$$

or

$$A = \frac{\pi R^2 \gamma^\circ}{360^\circ} - \tfrac{1}{2}R^2 \sin \gamma \tag{A.78}$$

where R, s, and γ are as already defined. If the chord length c is given (Figure A.24) then

$$\sin \frac{\gamma}{2} = \frac{c}{2R} \tag{A.79}$$

Ellipse

$$A = \pi \, ab \tag{A.80}$$

where a and b are the semimajor and semiminor axes, respectively.

Triangle

For a *general* triangle, as shown in Figure A.25, the area K is given by

$$K = \tfrac{1}{2}bh \qquad (h \text{ is perpendicular to } b) \tag{A.81}$$

or

$$K = \tfrac{1}{2}bc \sin A \tag{A.82}$$

or

$$K = \sqrt{s(s-a)(s-b)(s-c)} \tag{A.83}$$

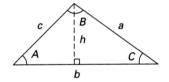

FIGURE A.25

with

$$s = \tfrac{1}{2}(a + b + c) \tag{A.84}$$

where a, b, and c are the lengths of the sides; h is the height of the triangle (which is perpendicular to the base b); and A, B, and C are the interior angles, as shown in Figure A.25. Other useful relations for a plane triangle are

$$A + B + C = 180° \tag{A.85}$$

$$\frac{\sin A}{a} = \frac{\sin B}{b} = \frac{\sin C}{c} \tag{A.86}$$

$$a^2 = b^2 + c^2 - 2bc \cos A \tag{A.87a}$$

$$b^2 = c^2 + a^2 - 2ca \cos B \tag{A.87b}$$

$$c^2 = a^2 + b^2 - 2ab \cos C \tag{A.87c}$$

$$\tan \frac{A}{2} = \frac{1}{s - a} \sqrt{\frac{(s - a)(s - b)(s - c)}{s}} \tag{A.88}$$

$$\cos \frac{A}{2} = \sqrt{\frac{s(s - a)}{bc}} \tag{A.89a}$$

$$\cos \frac{B}{2} = \sqrt{\frac{s(s - b)}{ca}} \tag{A.89b}$$

$$\cos \frac{C}{2} = \sqrt{\frac{s(s - c)}{ab}} \tag{A.89c}$$

$$\frac{a - b}{a + b} = \frac{\tan(A - B)/2}{\tan(A + B)/2} \tag{A.90}$$

$$h = \frac{2}{b} \sqrt{s(s - a)(s - b)(s - c)} \tag{A.91}$$

For an *equilateral* triangle, where sides $a = b = c$ and angles $A = B = C = 60°$, the area K becomes

$$K = \frac{a^2 \sqrt{3}}{4} \quad \left(\text{with } h = \frac{a\sqrt{3}}{2} \right) \tag{A.92}$$

Square

$$A = a^2 \tag{A.93}$$

where a is the side length.

Rectangle

$$A = ab \tag{A.94}$$

where a and b are its width and length.

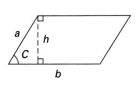

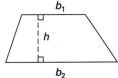

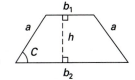

FIGURE A.26   FIGURE A.27   FIGURE A.28

Parallelogram

Let a and b be the sides, h be the altitude on side b, C be the acute angle, and A be the area (see Figure A.26). Then,

$$A = bh = ab \sin C \qquad (A.95)$$

Trapezoid

If b_1 and b_2 are the parallel sides, and h is the altitude between them (see Figure A.27), then the area A is

$$A = \tfrac{1}{2}h(b_1 + b_2) \qquad (A.96)$$

For an *isosceles* trapezoid (Figure A.28), if a is the length of one of the two nonparallel sides and C is the acute angle between a and b_2, then

$$A = \tfrac{1}{2}a\,a(b_1 + b_2) \sin C \qquad (A.97)$$

PROBLEMS

A.1. Find the sine, cosine, tangent, secant, cosecent, and cotangent of each of these angles: $17°$, $135°$, $184°$, and $296°$.

A.2. Refer to Figure A.2; given $X = 6$ ft and $Y = 8$ ft, calculate r, θ in degrees, minutes, and seconds; θ in radians; $\sin\theta$, $\cos\theta$, $\tan\theta$.

A.3. Given $r = 10$ ft and $Y = 5$ ft, in Figure A.29, find X, ϕ, θ, $\sin\theta$, $\cos\theta$, and $\tan\theta$.

A.4. Refer to Figure A.2; given $r = 10$ ft and $\theta = 50°$, find X, Y, $\sin\theta$, $\cos\theta$, and $\tan\theta$.

A.5. Refer to Figure A.29; given $r = 15$ ft and $\theta = 145°$, find X, Y, $\sin\theta$, $\cos\theta$, and $\tan\theta$.

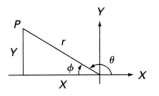

FIGURE A.29

A.6. Find θ and r for each point, given the following tabulated coordinates:

Point	X coordinate, ft	Y coordinate, ft
1	+68	−16
2	−28	+61
3	−39	−30
4	+46	+82

A.7. Find the distance between points 1 and 2 and the slope of the line through points 1 and 2, for the following cases:

	Point 1		Point 2	
	X, ft	Y, ft	X, ft	Y, ft
(a)	6	8	25	14
(b)	3	11	9	4
(c)	−7	12	13	−5
(d)	−10	−17	33	2
(e)	−20	−19	−15	−16
(f)	−18	−1	9	−22

A.8. Convert the following azimuths to bearings: 26°, 106°, 216°, and 306°.

A.9. Convert the following bearings to azimuths: N6°E, S46°W, S76°E, and N16°W.

A.10. Find θ (the slope angle) for the bearings and azimuths in Problems A.8 and A.9.

A.11. Refer to Figure A.8a; find the coordinates of point I if the coordinates of point P_1 are $X_1 = 2$ ft, $Y_1 = 1$ ft, and point P_2 are $X_2 = 8$ ft, $Y_2 = 10$ ft, and if $s_1 = 1$ and $s_2 = 2$.

A.12. Refer to Figure A.8b; find the coordinates of point E if the coordinates of point P_1 are $X_1 = 3$ ft, $Y_1 = 4$ ft, and point P_2 are $X_2 = 9$ ft, $Y_2 = 7$ ft, and if $s_1 = 3$ and $s_2 = 2$.

A.13. Refer to Figure A.9; write the equation of the line if $X_0 = 5$ and $Y_0 = 4$.

A.14. Given the equation of a line $2X - Y + 3 = 0$, find the slope m, the X intercept, and the Y intercept.

A.15. Write the equation of the line that has a Y intercept of 4 and a slope of −2.

A.16. Write the equation of the line that passes through point $P(X = +3, Y = +7)$ with a slope of +1. Find the X and Y intercepts of this line.

A.17. Write the equation of the line that passes through point $P_1(X_1 = -3, Y_1 = 5)$ and $P_2(X_2 = 6, Y_2 = -1)$. Find its slope and the X intercept and the Y intercept.

A.18. Write the equation of the line in Figure A.13 if $p = 4$ and $\theta = 60°$.

A.19. Refer to Figure A.15; calculate p for the equation $X + 2Y - 5 = 0$.

A.20. Given the following length and slope angle for the perpendicular p, write the equation of the line (see Figure A.13): $p = \sqrt{5}$, $\theta = 69°$.

A.21. Find the perpendicular distance from point $P(X = 2, Y = 5)$ to the line $X - Y + 1 = 0$.

A.22. Refer to Figure A.17; if $X_0 = 2$ and $Y_0 = 3$, find the distances d_1 and d_2 of points $P_1(-3, -4)$ and $P_2 (2, -5)$ from the line.

A.23. Write the equations for lines through points $P_1(5, 6)$ and $P_2(8, 1)$ and parallel to line A whose equation is $2X - 3Y - 6 = 0$.

A.24. Referring to Figure A.18, the equation of line A is $2X - Y - 6 = 0$. Find the equation of line B, which passes through point $P(7, 3)$ and is perpendicular to line A.

A.25. Find the coordinates X_I and Y_I of the point of intersection of the two lines $3X - Y + 4 = 0$ and $X - Y + 8 = 0$.

A.26. Write the equation of a circle of radius 2 with its center at point $C(X_c = 4, Y_c = 3)$.

A.27. Write the equation of a circle of radius 3 with its center at point $P(5, 4)$. Refer to Figure A.21 and determine which case applies for the intersection of each of the following lines with the circle. If the line intersects the circle or is tangent to the circle, give the coordinates of the points of intersection or tangency.

Line	Equation
1	$-Y + 3 = 0$
2	$-Y + 1 = 0$
3	$X - 1 = 0$

A.28. Refer to Figure A.23; find the area of the sector if $\gamma = 33°$ and $R = 3$ units.

A.29. Refer to Figure A.24; find the area of the segment if $\gamma = 111°$ and $R = 11$ units. Find the chord length c and the arc s.

A.30. Refer to Figure A.24; find the area of the segment if $c = 8$ units and $R = 5$ units. Find the arc lengths s and the angle α.

A.31. Refer to Figure A.25; find the area of the triangle if the three sides have lengths of 6, 8, and 10 m. Find the value of the angles.

APPENDIX B

Introduction to Vector and Matrix Algebra

B.1
THREE-DIMENSIONAL COORDINATE SYSTEMS

Two-dimensional coordinate systems, both polar and Cartesian or rectangular, and their relationship were introduced in Sections A.2 and A.5 of Appendix A. A coordinate system in three-dimensional (3D) space requires six elements for its definition, three associated with a reference point, and three with orientation. If the linear unit of measurement also is to be fixed, the minimum required becomes seven (see also the seven-parameter transformation in Section C.2 in Appendix C).

Figure B.1 shows two systems of three-dimensional coordinates: one is *spherical* (α, β, r) and the other Cartesian or rectangular (x_1, x_2, x_3). In the spherical system, any point, p, is located by the angle α from the x_1 axis to r' (the projection of r onto the x_1, x_2 plane), the angle β between r and r', and the distance, r, from the origin. The Cartesian coordinate system is composed of three mutually perpendicular axes; x_1, x_2, and x_3. The system depicted in Figure B.1 is *right handed* because a right-threaded screw rotated by an angle less than 90° from $+x_1$ to $+x_2$ would advance in the direction of $+x_3$. In general, any point may be determined in space by three coordinates: x_1, x_2, and x_3.

The relations between these two systems are as follows:

$$r' = r \cos \beta, \qquad x_1 = r' \cos \alpha, \qquad x_2 = r' \sin \alpha; \qquad \text{and}$$

$$x_1 = r \cos \alpha \cos \beta \tag{B.1}$$

$$x_2 = r \sin \alpha \cos \beta$$

$$x_3 = r \sin \beta$$

$$r = (x_1^2 + x_2^2 + x_3^2)^{1/2}$$

$$\alpha = \tan^{-1}(x_2/x_1) \tag{B.2}$$

$$\beta = \sin^{-1}(x_3/r)$$

1085

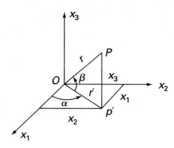

FIGURE B.1
Three-dimensional right-hand
Cartesian coordinate system.

B.2
VECTOR ALGEBRA

A *vector* is any entity that has a magnitude and direction. In two- and three-dimensional space, it is a directed line segment from one point to another. (Later, this will be generalized to the multidimensional case.) It may be represented by a single lower case boldface letter, such as **a**, or by $\overrightarrow{PQ}$ to represent the vector from point P to point Q. An example is shown in Figure B.2. The projections of the vector on the x_1 and x_2 axes are a_1 and a_2, which are called the *vector components*. We represent the vector's components in a column:

$$a = \begin{bmatrix} a_1 \\ a_2 \end{bmatrix}$$

It is clear from Figure B.2 that $a_1 = x_{1Q} - x_{1P}$ and $a_2 = x_{2Q} - x_{2P}$. Therefore, the components of a vector are obtained by subtracting the coordinates of its beginning point P, from the coordinates of its terminal point (arrow), Q. Because the vectors p and q in the figure begin at the origin,

$$p = \begin{bmatrix} x_{1P} \\ x_{2P} \end{bmatrix} \quad \text{and} \quad q = \begin{bmatrix} x_{1Q} \\ x_{2Q} \end{bmatrix}$$

and

$$\begin{bmatrix} a_1 \\ a_2 \end{bmatrix} = \begin{bmatrix} x_{1Q} \\ x_{2Q} \end{bmatrix} - \begin{bmatrix} x_{1P} \\ x_{2P} \end{bmatrix} \tag{B.3}$$

or

$$a = q - p$$

Expanding to three dimensions, we write

$$a = \begin{bmatrix} a_1 \\ a_2 \\ a_3 \end{bmatrix}$$

The *length* of the vector is designated by $|a|$, given by

$$|a| = (a_1^2 + a_2^2 + a_3^2)^{1/2} \tag{B.4}$$

when the components are referred to a rectangular coordinate system.

A vector's *direction* is given by either the angles it makes with the axes (α, β, γ) or by their cosines. The latter, called *direction cosines,* are given by

$$\cos \alpha = \frac{a_1}{|a|}, \quad \cos \beta = \frac{a_2}{|a|}, \quad \cos \gamma = \frac{a_3}{|a|} \tag{B.5}$$

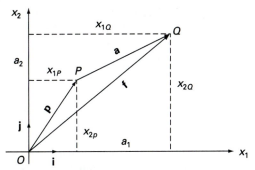

FIGURE B.2
A vector in a plane.

1087

APPENDIX B:
Introduction to
Vector and
Matrix Algebra

Only two of these are independent because a direction in space is completely determined by only two angles; in Figure B.1, OP is fixed by α and β. Consequently, the direction cosines are related by the equation

$$\cos^2 \alpha + \cos^2 \beta + \cos^2 \gamma = 1 \qquad (B.6)$$

which can be readily proven from Equations (B.4) and (B.5).

One now can generalize a vector to n dimensions:

$$a = \begin{bmatrix} a_1 \\ a_2 \\ \cdot \\ \cdot \\ a_n \end{bmatrix}$$

Vector Operations

Equality. $a = b$ when

$$a_1 = b_1, \qquad a_2 = b_2, \qquad \cdots, \qquad a_n = b_n$$

Addition and subtraction:

$$c = a \pm b \qquad \text{or} \qquad c_1 = a_1 \pm b_1, \qquad c_2 = a_2 \pm b_2, \qquad \cdots, \qquad c_n = a_n \pm b_n$$

$$a + b = b + a, \qquad (a + b) + c = a + (b + c)$$

Multiplication by a scalar. A *scalar* is a quantity that has only magnitude but no direction, such as mass, temperature, or time, designated by a lowercase Greek letter:

$$\lambda a = \begin{bmatrix} \lambda a_1 \\ \lambda a_2 \\ \cdot \\ \cdot \\ \lambda a_n \end{bmatrix}$$

$$\lambda(\mu a) = (\lambda \mu) a = \mu(\lambda a) \qquad (B.7)$$

$$(\lambda + \mu) a = \lambda a + \mu a$$

$$\lambda(a + b) = \lambda a + \lambda b$$

$$|\lambda a| = \lambda |a|$$

Any vector a, other than the zero vector, is reduced to a unit vector a° when dividing its components by its length, which is a scalar, or $a^\circ = a/|a|$. The components of a° are the direction cosines of a. Unit vectors along the coordinate axes, called *base vectors,* are given by

$$i = \begin{bmatrix} 1 \\ 0 \\ 0 \end{bmatrix}, \qquad j = \begin{bmatrix} 0 \\ 1 \\ 0 \end{bmatrix}, \qquad k = \begin{bmatrix} 0 \\ 0 \\ 1 \end{bmatrix} \qquad (B.8)$$

(see Figure B.2 for **i** and **j**). Any vector in 3D space is uniquely expressed as

$$a = a_1 i + a_2 j + a_3 k \qquad (B.9)$$

The right-handed system introduced in Section B.1 can be generalized for three vectors: a, b, and c. If they are not coplanar and have the same initial point, they are said to form a *right-handed* system if a right threaded screw rotated through an angle *less than 180°* from **a** to **b** would advance in the direction **c**.

Vector Products

Dot (or scalar) product

$$a \cdot b = \sum_{p=1}^{n} a_p b_p = a_1 b_1 + a_2 b_2 + \cdots + a_n b_n \qquad (B.10)$$

This also is called the *inner product*. It is a scalar and has the following properties:

$$a \cdot b = b \cdot a$$
$$a \cdot (b + c) = a \cdot b + a \cdot c$$
$$\lambda(a \cdot b) = (\lambda a) \cdot b = a \cdot (\lambda b) = (a \cdot b)\lambda \qquad (B.11)$$
$$i \cdot i = j \cdot j = k \cdot k = 1$$
$$i \cdot j = j \cdot k = k \cdot i = 0$$

The dot product of a vector with itself is equal to the square of its length:

$$a \cdot a = a_1^2 + a_2^2 + \cdots + a_n^2 = |a|^2 \qquad (B.12)$$

If θ is the angle between two vectors a and b (in two- or three-dimensional space), it can be shown that

$$a \cdot b = |a||b| \cos \theta \qquad (B.13)$$

It follows that, if a is perpendicular to b, then $a \cdot b = 0$.

Cross (or vector) product.

The term $a \times b$ (read *a cross b*) is another vector c that is perpendicular to both **a** and **b** and in a direction such that a, b, and c (in this order) form a right-handed system. The length of c is given by

$$|c| = |a \times b| = |a||b| \sin \theta \qquad (B.14)$$

where θ is the angle between **a** and **b**. This quantity is the area of the parallelogram determined by a and b. If $a = a_1 i + a_2 j + a_3 k$, and $b = b_1 i + b_2 j + b_3 k$, then c is given by the determinant

$$c = a \times b = \begin{vmatrix} i & j & k \\ a_1 & a_2 & a_2 \\ b_1 & b_2 & b_3 \end{vmatrix} \qquad (B.15)$$

It has the following properties:

$$a \times b = -(b \times a)$$

$$a \times (b + c) = a \times b + a \times c \quad \text{(observing the order)}$$

$$a \cdot (a \times c) = 0$$

$$|a \times b|^2 = |a|^2|b|^2 - (a \cdot b)^2 \tag{B.16}$$

$$i \times i = j \times j = k \times k = 0$$

$$i \times j = k; \quad j \times k = i; \quad k \times i = j$$

For two nonzero vectors, if $a \times b = 0$, then a and b are parallel.

Scalar triple product

$$a \times b \cdot c = \begin{vmatrix} a_1 & a_2 & a_3 \\ b_1 & b_2 & b_3 \\ c_1 & c_2 & c_3 \end{vmatrix}$$

$$= a_1(b_2c_3 - b_3c_2) - a_2(b_1c_3 - b_3c_1) \tag{B.17}$$

$$+ a_3(b_1c_2 - b_2c_1)$$

is a scalar (because it is a 3×3 determinant) equal to the volume of the parallelepiped determined by $a, b,$ and c. If it is 0, then the three vectors are coplanar. It has the following properties:

$$a \times b \cdot c = b \times c \cdot a = c \times a \cdot b$$

$$a \times b \cdot c = a \cdot b \times c$$

Planes and lines. If p_0 is a given point in a plane, n is a nonzero vector normal to the plane, and p is any point in the plane, then the equation of the plane takes the form (see Figure B.3)

$$(p - p_0) \cdot n = 0 \quad \text{or} \quad p \cdot n - p_0 \cdot n = 0 \tag{B.18}$$

Let $n = Ai + Bj + Ck$, $p_0 = X_0i + Y_0j + Z_0k$, and $p = Xi + Yj + Zk$. Then, Equation (B.18) becomes

$$A(X - X_0) + B(Y - Y_0) + C(Z - Z_0) = 0$$

or

$$AX + BY + CZ + D = 0 \tag{B.19}$$

where $D = -(AX_0 + BY_0 + CZ_0)$. Two planes are parallel when they have a common normal vector, n, and are perpendicular when their normals are $n_1 \cdot n_2 = 0$.

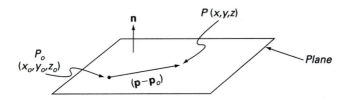

FIGURE B.3
Elements of a vector equation of a plane.

If p_0 represents a given point on a line, p is any other point on the line, and v is a given nonzero vector parallel to the line,

$$p = p_0 + \lambda v \tag{B.20}$$

is an equation of the line. In component form, it yields three scalar equations describing the parametric form (λ is the running parameter):

$$X = X_0 + \lambda v_x$$

$$Y = Y_0 + \lambda v_y \tag{B.21}$$

$$Z = Z_0 + \lambda v_z$$

If λ is eliminated, one gets the usual two-equation form of a straight line in space.

B.3
MATRIX ALGEBRA

Definitions

A matrix is simply a collection of numbers or symbols collected in an array form. The following are examples of matrices:

$$\begin{bmatrix} 1 & 2 & 0 \\ 6 & 4 & 3 \end{bmatrix} \quad \begin{bmatrix} 1 \\ 5 \end{bmatrix} \quad [7 \quad 9 \quad 3] \quad \begin{bmatrix} a & b \\ c & d \end{bmatrix}$$

$$(1) \qquad (2) \qquad (3) \qquad (4)$$

Every matrix has a specified number of rows and columns. The matrix in (1) has two rows and three columns and is said to be a 2×3 matrix. Similarly, (2) is a 2×1 matrix, (3) is a 1×3 matrix, and (4) is a 2×2 matrix. The two numbers representing the rows and columns are referred to as the *matrix dimensions*.

A matrix normally is designated by a boldface capital roman letter. If $\mathbf{A}$ is an $m \times n$ matrix, it can be written symbolically as

$$\mathbf{A} = \begin{bmatrix} a_{11} & a_{12} & \cdots & a_{1n} \\ a_{21} & a_{22} & \cdots & a_{2n} \\ \cdots\cdots\cdots\cdots\cdots\cdots \\ a_{m1} & a_{m2} & \cdots & a_{mn} \end{bmatrix}$$

A lowercase letter with a double subscript designates an element in a matrix, so a_{ij} represents a typical element of the matrix $\mathbf{A}$. The first subscript, i, refers to the number of the row in which a_{ij} lies, starting with 1 at the top and proceeding down to m at the bottom. The second subscript, j, refers to the number of the column containing a_{ij}, starting with 1 on the left and proceeding to n at the right. Therefore, a_{ij} lies at the intersection of the ith row and jth column. As an example, a_{23} in matrix (1) is 3, while a_{12} in matrix (3) is equal to 9. The smallest matrix size is the 1×1, which is then called a *scalar*.

Types of Matrices

Square matrix. A *square matrix* is a matrix with an equal number of rows and columns. In this case $\mathbf{A}_{m,m}$ is a square matrix of order m. The principal (main, or leading) diagonal is that composed of the elements a_{ij} for $i = j$. The following are examples of the square matrix:

$$\mathbf{A} = \begin{bmatrix} 1 & 2 \\ 3 & 4 \end{bmatrix}, \qquad \mathbf{B} = \begin{bmatrix} a & b & c \\ d & e & f \\ g & h & k \end{bmatrix}, \qquad \mathbf{C} = [3]$$

The main diagonal of $\mathbf{A}$ is composed of 1 and 4, while that for $\mathbf{B}$ contains a, e, and k. Two special cases of the square matrix are the *symmetric* and *skew-symmetric* ones. The case of a symmetric matrix will be introduced later.

Row matrix. A *row matrix* or *row vector* is a matrix composed of only one row. It will be designated by a lowercase boldface roman letter; for example,

$$\underset{1,n}{\mathbf{a}} = [a_1 \quad a_2 \cdots \quad a_n] \qquad \text{or} \qquad \underset{1,3}{\mathbf{c}} = [1 \quad 2 \quad 4]$$

Column matrix. A *column matrix* or *column vector* is composed of only one column, such as

$$\underset{m,1}{\mathbf{b}} = \begin{bmatrix} b_1 \\ b_2 \\ \vdots \\ b_m \end{bmatrix} \qquad \text{or} \qquad \mathbf{d} = \begin{bmatrix} -1 \\ 3 \end{bmatrix}$$

This is the same definition of a vector as introduced in Section B.2.

Diagonal matrix. A *diagonal matrix* is a square matrix in which all elements above and below the main diagonal are 0:

$$\mathbf{D} = \begin{bmatrix} d_{11} & & & \mathbf{0} \\ & d_{12} & & \\ & & \ddots & \\ \mathbf{0} & & & d_{mm} \end{bmatrix}$$

where

$$d_{ij} = 0 \qquad \text{for all } i \neq j$$
$$d_{ij} \neq 0 \qquad \text{for some or all } i = j$$

For example,

$$\mathbf{G} = \begin{bmatrix} 1 & 0 & 0 \\ 0 & 0 & 0 \\ 0 & 0 & -3 \end{bmatrix} \qquad \text{and} \qquad \mathbf{H} = \begin{bmatrix} p & 0 & 0 \\ 0 & q & 0 \\ 0 & 0 & r \end{bmatrix}$$

Scalar matrix. A *scalar matrix* is a diagonal matrix whose diagonal elements *all* are equal to the same scalar; hence,

$$\mathbf{A} = \begin{bmatrix} a & & & \mathbf{0} \\ & a & & \\ & & \ddots & \\ \mathbf{0} & & & a \end{bmatrix}, \qquad \begin{matrix} a_{ij} = 0 & \text{for all } i \neq j \\ a_{ij} = a & \text{for all } i = j \end{matrix}$$

and

$$\mathbf{H} = \begin{bmatrix} 2 & 0 & 0 \\ 0 & 2 & 0 \\ 0 & 0 & 2 \end{bmatrix}$$

are scalar matrices.

Unit or identity matrix. A *unit* or *identity matrix* is a diagonal matrix whose diagonal elements all are equal to 1. It will always be referred to by

$$\mathbf{I} = \begin{bmatrix} 1 & 0 & \cdots & 0 \\ 0 & 1 & & \vdots \\ \vdots & & \ddots & \vdots \\ 0 & \cdots & & 1 \end{bmatrix}, \qquad \begin{aligned} a_{ij} &= 0 & \text{for all } i \neq j \\ a_{ij} &= 1 & \text{for all } i = j \end{aligned}$$

Null or zero matrix. A *null* or *zero matrix* is a matrix whose elements *all* are 0. It is denoted by a boldface zero, $\mathbf{0}$.

Triangular matrix. A *triangular matrix* is a square matrix whose elements above (or below), but not including the main diagonal, all are 0. An upper-triangular matrix takes the form

$$\mathbf{A} = \begin{bmatrix} a_{11} & a_{12} & \cdots & a_{1m} \\ 0 & a_{22} & \cdots & a_{2m} \\ \cdots\cdots\cdots\cdots\cdots\cdots \\ 0 & 0 & \cdots & a_{mm} \end{bmatrix}, \qquad \text{with } a_{ij} = 0, \qquad \text{for } i > j$$

The matrix

$$\mathbf{A} = \begin{bmatrix} -1 & 3 & 4 \\ 0 & 1 & 0 \\ 0 & 0 & 7 \end{bmatrix}$$

is an example of an upper-triangular matrix of order 3.

On the other hand, a lower-triangular matrix is of the form

$$\mathbf{A} = \begin{bmatrix} a_{11} & 0 & \cdots & 0 \\ a_{21} & a_{22} & \cdots & 0 \\ \cdots\cdots\cdots\cdots\cdots\cdots \\ a_{m1} & a_{m2} & \cdots & a_{mm} \end{bmatrix}, \qquad \text{where } a_{ij} = 0, \qquad \text{for } i < j$$

For example,

$$\mathbf{B} = \begin{bmatrix} 18 & 0 \\ 2 & -11 \end{bmatrix}$$

is a lower-triangular matrix of order 2.

Matrix Operations

Equality. Two matrices $\mathbf{A}$ and $\mathbf{B}$ of the *same dimensions* are *equal* if each element $a_{ij} = b_{ij}$ for all i and j. Matrices of different dimensions cannot be equated. Some relationships that apply to matrix equality include

$$\text{If } \mathbf{A} = \mathbf{B} \text{ then } \mathbf{B} = \mathbf{A} \text{ for all } \mathbf{A} \text{ and } \mathbf{B} \tag{B.22}$$

$$\text{If } \mathbf{A} = \mathbf{B} \text{ and } \mathbf{B} = \mathbf{C}, \text{ then } \mathbf{A} = \mathbf{C} \text{ for all } \mathbf{A}, \mathbf{B}, \text{ and } \mathbf{C} \tag{B.23}$$

As an example, let

$$\mathbf{A} = \begin{bmatrix} 1 & 2 \\ 3 & 4 \end{bmatrix} \quad \text{and} \quad \mathbf{B} = \begin{bmatrix} b_{11} & b_{12} \\ b_{21} & b_{22} \end{bmatrix}$$

If $\mathbf{A} = \mathbf{B}$ (noting that both are 2×2 matrices), then

$$b_{11} = 1, \quad b_{12} = 2, \quad b_{21} = 3, \quad \text{and} \quad b_{22} = 4$$

Sums. The sum of two matrices $\mathbf{A}$ and $\mathbf{B}$, of the *same* dimensions, is a matrix $\mathbf{C}$ of the same dimensions, the elements of which are given by $c_{ij} = a_{ij} + b_{ij}$ for all i and j. Matrices of different dimensions cannot be added. The following relationships apply to matrix addition:

$$\mathbf{A} + \mathbf{B} = \mathbf{B} + \mathbf{A} \tag{B.24}$$

$$\mathbf{A} + (\mathbf{B} + \mathbf{C}) = (\mathbf{A} + \mathbf{B}) + \mathbf{C} = \mathbf{A} + \mathbf{B} + \mathbf{C} \tag{B.25}$$

With the null or zero matrix, $\mathbf{0}$, we have

$$\mathbf{A} + \mathbf{0} = \mathbf{0} + \mathbf{A} = \mathbf{A} \tag{B.26}$$

$$\mathbf{A} + (-\mathbf{A}) = \mathbf{0} \tag{B.27}$$

where $(-\mathbf{A})$ is the matrix composed of $(-a_{ij})$ as elements. As an example, if

$$\mathbf{A} = \begin{bmatrix} 1 & 2 & 0 \\ 0 & 3 & 5 \end{bmatrix}, \quad \mathbf{B} = \begin{bmatrix} 6 & 4 & 2 \\ 3 & 2 & 7 \end{bmatrix}, \quad \mathbf{C} = \begin{bmatrix} x & y & z \\ u & v & w \end{bmatrix}$$

and $\mathbf{C} = \mathbf{B} - \mathbf{A}$, compute the values of the six variables $x, y, z, u, v,$ and w. First, compute $\mathbf{B} - \mathbf{A}$ as

$$\begin{bmatrix} 6 & 4 & 2 \\ 3 & 2 & 7 \end{bmatrix} - \begin{bmatrix} 1 & 2 & 0 \\ 0 & 3 & 5 \end{bmatrix} = \begin{bmatrix} 5 & 2 & 2 \\ 3 & -1 & 2 \end{bmatrix}$$

and then from

$$\mathbf{C} = \begin{bmatrix} x & y & z \\ u & v & w \end{bmatrix} = \begin{bmatrix} 5 & 2 & 2 \\ 3 & -1 & 2 \end{bmatrix}$$

we get $x = 5$, $y = 2$, $z = 2$, $u = 3$, $v = -1$, and $w = 2$.

Scalar multiplication. Multiplication of a matrix $\mathbf{A}$ by a scalar α is another matrix $\mathbf{B}$ whose elements are $b_{ij} = \alpha a_{ij}$, for all i and j. (A scalar generally is denoted by a lowercase Greek letter.) Therefore, $\mathbf{B} = \alpha \mathbf{A}$. As an example, for

$$\mathbf{A} = \begin{bmatrix} 7 & 2 \\ 3 & -4 \end{bmatrix}, \quad \mathbf{B} = 3\mathbf{A} = \begin{bmatrix} 21 & 6 \\ 9 & -12 \end{bmatrix}$$

The following relations hold for scalar multiplication:

$$\alpha(\mathbf{A} + \mathbf{B}) = \alpha\mathbf{A} + \alpha\mathbf{B} \tag{B.28}$$

$$(\alpha + \beta)\mathbf{A} = \alpha\mathbf{A} + \beta\mathbf{A} \tag{B.29}$$

$$\alpha(\mathbf{AB}) = (a\mathbf{A})\mathbf{B} = \mathbf{A}(\alpha\mathbf{B}) \tag{B.30}$$

$$\alpha(\beta\mathbf{A}) = (\alpha\beta)\mathbf{A} \tag{B.31}$$

Some examples are

$$A = \begin{bmatrix} 1 & 2 \\ 2 & 1 \\ 3 & 0 \end{bmatrix} \quad \text{and} \quad B = \begin{bmatrix} 0 & 3 \\ 2 & 0 \\ 1 & 1 \end{bmatrix}$$

To verify Equation (B.28), compute $2(A + B)$ both ways:

$$A + B = \begin{bmatrix} 1 & 5 \\ 4 & 1 \\ 4 & 1 \end{bmatrix} \quad \text{and} \quad 2(A + B) = \begin{bmatrix} 2 & 10 \\ 8 & 2 \\ 8 & 2 \end{bmatrix}$$

On the other hand,

$$2A = \begin{bmatrix} 2 & 4 \\ 4 & 2 \\ 6 & 0 \end{bmatrix} \quad \text{and} \quad 2B = \begin{bmatrix} 0 & 6 \\ 4 & 0 \\ 2 & 2 \end{bmatrix}$$

and then

$$2A + 2B = \begin{bmatrix} 2 & 10 \\ 8 & 2 \\ 8 & 2 \end{bmatrix}$$

which checks with the previous answer.

To demonstrate Equation (B.29),

$$(2 + 1)A = 3A = \begin{bmatrix} 3 & 6 \\ 6 & 3 \\ 9 & 0 \end{bmatrix}$$

Also,

$$2A + A = \begin{bmatrix} 2 & 4 \\ 4 & 2 \\ 6 & 0 \end{bmatrix} + \begin{bmatrix} 1 & 2 \\ 2 & 1 \\ 3 & 0 \end{bmatrix} = \begin{bmatrix} 3 & 6 \\ 6 & 3 \\ 9 & 0 \end{bmatrix}$$

The relation in Equation (B.30) could not be applied until matrix multiplication is defined, which is given in the next section. Using $\alpha = 2$ and $\beta = 3$, one should be able to verify Equation (B.31) as an exercise.

Matrix multiplication. The product of two matrices in general is another matrix. A relation between the dimensions of both matrices must exist before multiplication can be performed. The relation is that the number of columns of the first matrix must equal the number of rows of the second matrix. If A is $m \times k$ and B is $k \times n$, the product AB *in that order* is another matrix C with m rows (as in A) and n columns (as in B). Each element c_{ij} in C is obtained taking the k elements of the ith row in A and multiplying each by the corresponding element in the jth column in B and adding the results. Algebraically, this is written as

$$c_{ij} = a_{i1}b_{ij} + a_{i2}b_{2j} + \cdots + a_{ik}b_{kj} \tag{B.32}$$

This process may be shown schematically as

$$
\begin{bmatrix} a_{11} & a_{12} & \cdots & a_{1k} \\ \cdots\cdots\cdots\cdots\cdots \\ a_{i1} & a_{i2} & \cdots & a_{ik} \\ \cdots\cdots\cdots\cdots\cdots \\ a_{m1} & a_{m2} & \cdots & a_{mk} \end{bmatrix}
\begin{bmatrix} b_{11} & \cdots & b_{ij} & \cdots & b_{1n} \\ b_{21} & \cdots & b_{2j} & \cdots & b_{2n} \\ \cdots\cdots & \cdots & \cdots\cdots \\ b_{k1} & \cdots & b_{kj} & \cdots & b_{kn} \end{bmatrix}
$$

$$
= \begin{bmatrix} c_{11} & \cdots & & c_{1j} & & \cdots & c_{1n} \\ \cdots\cdots\cdots\cdots\cdots\cdots\cdots\cdots \\ c_{i1} & \cdots & & \boxed{c_{ij} = \sum_{r=1}^{r=k} a_{ir} b_{rj}} & & \cdots & c_{in} \\ \cdots\cdots\cdots\cdots\cdots\cdots\cdots\cdots \\ c_{m1} & \cdots & & c_{mj} & & \cdots & c_{mn} \end{bmatrix}
\qquad (B.33)
$$

From this definition it follows that, if $\mathbf{A}$ is a row matrix and $\mathbf{B}$ is a column matrix, then $\mathbf{AB}$ is a scalar. For example, if

$$
\mathbf{A} = \begin{bmatrix} 1 & 2 & 3 \end{bmatrix} \qquad \text{and} \qquad \mathbf{B} = \begin{bmatrix} 4 \\ 5 \\ 6 \end{bmatrix}
$$

then

$$
\underset{1,3}{\mathbf{A}} \quad \underset{3,1}{\mathbf{B}} = \underset{1,1}{\mathbf{C}} = (1)(4) + (2)(5) + (3)(6) = 32
$$

As examples for multiplication of matrices, let

$$
\mathbf{A} = \begin{bmatrix} 1 & 1 \\ 2 & 0 \end{bmatrix}, \qquad \mathbf{B} = \begin{bmatrix} 1 & 2 \\ 0 & 1 \\ 1 & 0 \end{bmatrix}, \qquad \mathbf{C} = \begin{bmatrix} 1 & 0 & 1 \\ 0 & 1 & 2 \\ 1 & 0 & 1 \end{bmatrix}
$$

Then,

$$
\underset{3,2\ 2,2}{\mathbf{B}\ \mathbf{A}} = \underset{3,2}{\mathbf{D}} = \begin{bmatrix} 1 & 2 \\ 0 & 1 \\ 1 & 0 \end{bmatrix} \begin{bmatrix} 1 & 1 \\ 2 & 0 \end{bmatrix} = \begin{bmatrix} (1)(1) + (2)(2) & (1)(1) + (2)(0) \\ (0)(1) + (1)(2) & (0)(1) + (1)(0) \\ (1)(1) + (0)(2) & (1)(1) + (0)(0) \end{bmatrix} = \begin{bmatrix} 5 & 1 \\ 2 & 0 \\ 1 & 1 \end{bmatrix}
$$

$$
\underset{3,3\ 3,2}{\mathbf{C}\ \mathbf{B}} = \underset{3,2}{\mathbf{E}} = \begin{bmatrix} 1 & 0 & 1 \\ 0 & 1 & 2 \\ 1 & 0 & 1 \end{bmatrix} \begin{bmatrix} 1 & 2 \\ 0 & 1 \\ 1 & 0 \end{bmatrix}
$$

$$
= \begin{bmatrix} (1)(1) + (0)(0) + (1)(1) & (1)(2) + (0)(1) + (1)(0) \\ (0)(1) + (1)(0) + (2)(1) & (0)(2) + (1)(1) + (2)(0) \\ (1)(1) + (0)(0) + (1)(1) & (1)(2) + (0)(1) + (1)(0) \end{bmatrix} = \begin{bmatrix} 2 & 2 \\ 2 & 1 \\ 2 & 2 \end{bmatrix}
$$

Note that, because of the given matrix dimensions, the multiplications $\underset{2,2\ 3,2}{\mathbf{A}\ \mathbf{B}}$, $\underset{3,2\ 3,3}{\mathbf{B}\ \mathbf{C}}$, and so on, are not possible. Therefore, it is a general rule that $\mathbf{TS} \neq \mathbf{ST}$, even if both matrices are square. For example, if

$$
\mathbf{T} = \begin{bmatrix} 1 & 2 \\ 5 & 0 \end{bmatrix} \qquad \text{and} \qquad \mathbf{S} = \begin{bmatrix} 3 & 4 \\ 0 & 2 \end{bmatrix}
$$

then

$$\mathbf{TS} = \mathbf{P} = \begin{bmatrix} 1 & 2 \\ 5 & 0 \end{bmatrix} \begin{bmatrix} 3 & 4 \\ 0 & 2 \end{bmatrix} = \begin{bmatrix} 3 & 8 \\ 15 & 20 \end{bmatrix}$$

whereas

$$\mathbf{ST} = \mathbf{Q} = \begin{bmatrix} 3 & 4 \\ 0 & 2 \end{bmatrix} \begin{bmatrix} 1 & 2 \\ 5 & 0 \end{bmatrix} = \begin{bmatrix} 23 & 6 \\ 10 & 0 \end{bmatrix}$$

with the obvious result that $\mathbf{P} \neq \mathbf{Q}$ and thus $\mathbf{TS} \neq \mathbf{ST}$.

The following relationships regarding matrix multiplication hold:

$$\mathbf{AI} = \mathbf{IA} = \mathbf{A} \qquad \text{with } \mathbf{I} = \text{identity matrix} \qquad (B.34)$$

$$\mathbf{A(BC)} = \mathbf{(AB)C} = \mathbf{ABC} \qquad \text{(associative law)} \qquad (B.35)$$

$$\mathbf{A(B + C)} = \mathbf{AB} + \mathbf{AC} \qquad \text{(distributive law)} \qquad (B.36)$$

$$\mathbf{(A + B)C} = \mathbf{AC} + \mathbf{BC} \qquad \text{(distributive law)} \qquad (B.37)$$

In all these relations, the sequence of the matrices is preserved *strictly,* because if the order is reversed, the results will be different, as has just been shown.

An important property of matrix multiplication, which distinguishes it from scalar multiplication, is that the product of two matrices can be the null or zero matrix, without either matrix being the zero matrix. Three examples follow:

1. $\begin{bmatrix} 1 & 1 \\ 0 & 0 \end{bmatrix} \begin{bmatrix} 2 & 3 \\ -2 & -3 \end{bmatrix} = \begin{bmatrix} 0 & 0 \\ 0 & 0 \end{bmatrix}$

2. $\begin{bmatrix} 1 & 2 & 3 \\ 1 & -1 & 0 \end{bmatrix} \begin{bmatrix} 3 & -5 \\ 3 & -5 \\ -3 & 5 \end{bmatrix} = \begin{bmatrix} 0 & 0 \\ 0 & 0 \end{bmatrix}$

3. $\begin{bmatrix} 2 & 2 & -2 \end{bmatrix} \begin{bmatrix} 1 & 2 & 1 \\ 0 & -2 & 1 \\ 1 & 0 & 2 \end{bmatrix} = \begin{bmatrix} 0 & 0 & 0 \end{bmatrix}$

Matrix transpose. The transpose of a matrix $\mathbf{A}$ of dimensions m by n is an $n \times m$ matrix formed from $\mathbf{A}$ by interchanging rows and columns such that the ith row of $\mathbf{A}$ becomes the ith column of the transposed matrix. We denote the transpose by adding a superscript t to the matrix, or $\mathbf{A}^t$. If $\mathbf{B} = \mathbf{A}^t$, it follows that $b_{ij} = a_{ji}$ for all i and j. For example,

$$\text{if} \quad \mathbf{A} = \begin{bmatrix} 3 & 2 \\ -1 & 1 \end{bmatrix} \qquad \text{then} \qquad \mathbf{A}^t = \begin{bmatrix} 3 & -1 \\ 2 & 1 \end{bmatrix}$$

$$\mathbf{B} = \begin{bmatrix} -1 & 6 \\ 0 & 4 \\ 5 & 0 \end{bmatrix} \qquad \qquad \mathbf{B}^t = \begin{bmatrix} -1 & 0 & 5 \\ 6 & 4 & 0 \end{bmatrix}$$

$$\mathbf{C} = \begin{bmatrix} a & b & c \end{bmatrix} \qquad \qquad \mathbf{C}^t = \begin{bmatrix} a \\ b \\ c \end{bmatrix}$$

The following relationships apply to a matrix transpose:

$$(\mathbf{A} + \mathbf{B})' = \mathbf{A}' + \mathbf{B}' \qquad (B.38)$$

$$(\mathbf{AB})' = \mathbf{B}'\mathbf{A}' \qquad \text{(note reverse order)} \qquad (B.39)$$

$$(\alpha\mathbf{A})' = \alpha\mathbf{A}' \qquad (B.40)$$

$$(\mathbf{A}')' = \mathbf{A} \qquad (B.41)$$

The first relationship (B.38) can be verified readily by recalling that matrix addition is element by element; therefore, whether you sum first and transpose, or transpose first then add, the result will be the same. The second relation (B.39) is rather important, because transposing a matrix product leads to transposing each matrix, then *reversing* the sequence before performing the multiplication. This becomes logical when we recall that, if $\mathbf{A}$ is $m \times k$, then $\mathbf{A}'$ is $k \times m$, and if $\mathbf{B}$ is $k \times n$, then $\mathbf{B}'$ is $n \times k$ and multiplication is defined only for $\mathbf{B}'\mathbf{A}'$. Furthermore, because $\mathbf{AB}$ is m by n, then $(\mathbf{AB})'$ by definition is an $n \times m$ matrix, which is the same as the dimensions of $\underset{n,k}{\mathbf{B}'} \underset{k,m}{\mathbf{A}'}$. As a demonstration of Equation (B.39), let

$$\underset{2,3}{\mathbf{A}} = \begin{bmatrix} 1 & 1 & 0 \\ 0 & 2 & 3 \end{bmatrix} \quad \text{and} \quad \underset{3,1}{\mathbf{B}} = \begin{bmatrix} 1 \\ 1 \\ 2 \end{bmatrix}$$

then

$$\underset{2,3}{\mathbf{A}} \underset{3,1}{\mathbf{B}} = \underset{2,1}{\mathbf{C}} = \begin{bmatrix} 1 & 1 & 0 \\ 0 & 2 & 3 \end{bmatrix}\begin{bmatrix} 1 \\ 1 \\ 2 \end{bmatrix} = \begin{bmatrix} 2 \\ 8 \end{bmatrix} \quad \text{and} \quad \mathbf{C}' = \begin{bmatrix} 2 & 8 \end{bmatrix}$$

$$\mathbf{B}'\mathbf{A}' = \begin{bmatrix} 1 & 1 & 2 \end{bmatrix}\begin{bmatrix} 1 & 0 \\ 1 & 2 \\ 0 & 3 \end{bmatrix} = \begin{bmatrix} 2 & 8 \end{bmatrix}$$

which is equal to $\mathbf{C}'$. The last two relationships (B.40) and (B.41) are rather straightforward and one should verify them by numerical examples.

When the original matrix is square, the operation of transpose does not affect the elements of the main diagonal. For example, for

$$\mathbf{A} = \begin{bmatrix} a & b \\ c & d \end{bmatrix}$$

the transpose is

$$\mathbf{A}' = \begin{bmatrix} a & c \\ b & d \end{bmatrix}$$

It follows that, if the matrix has zero elements above and below the main diagonal, then it is equal to its transpose. We have introduced three matrices with this property: the identity matrix, $\mathbf{I}$; the diagonal matrix, $\mathbf{D}$; and the scalar matrix, $\mathbf{K}$. Hence, $\mathbf{I}' = \mathbf{I}$, $\mathbf{D}' = \mathbf{D}$, and $\mathbf{K}' = \mathbf{K}$. The scalar matrix, $\mathbf{K}$, also can be written $\mathbf{K} = k\mathbf{I}$, where k is the value of each element along the main diagonal. Therefore, $\mathbf{K}' = (k\mathbf{I})' = k\mathbf{I}' = k\mathbf{I} = \mathbf{K}$. If $\mathbf{x}$ is a column matrix, which usually is called a *column vector,* then $\mathbf{x}'\mathbf{x}$ is a positive scalar equal to the sum of the squares of the vector components, or the square of its length. For example,

$$\text{if } \mathbf{x} = \begin{bmatrix} x_1 \\ x_2 \\ x_3 \end{bmatrix}, \quad \text{then} \quad \mathbf{x}'\mathbf{x} = \begin{bmatrix} x_1 & x_2 & x_3 \end{bmatrix}\begin{bmatrix} x_1 \\ x_2 \\ x_3 \end{bmatrix} = x_1^2 + x_2^2 + x_3^2$$

On the other hand, $\mathbf{xx}'$ is a *square-symmetric* matrix, as explained in the following section.

Symmetric matrix. A square matrix is called *symmetric* if it is equal to its transpose; $\mathbf{A}' = \mathbf{A}$. Because transposing a matrix does not change the elements of the main diagonal, the elements above the main diagonal of a symmetric matrix are "mirror images" of those below the diagonal. For example,

$$\begin{bmatrix} 3 & 2 & 1 \\ 2 & 5 & 6 \\ 1 & 6 & 4 \end{bmatrix}, \qquad \begin{bmatrix} a & b \\ b & c \end{bmatrix}, \qquad \text{and} \qquad \begin{bmatrix} 6 & 3 & 0 & 1 \\ 3 & 7 & 2 & 0 \\ 0 & 2 & 4 & 0 \\ 1 & 0 & 0 & 5 \end{bmatrix}$$

are symmetric matrices.

For any matrix $\mathbf{A}$ (not necessarily square), both $\mathbf{AA}'$ and $\mathbf{A}'\mathbf{A}$ are symmetric. The proof is direct. For example, if $\mathbf{C} = \mathbf{AA}'$ then $\mathbf{C}' = [\mathbf{AA}']' = (\mathbf{A}')'(\mathbf{A})' = \mathbf{C}$, which means that $\mathbf{C}$ is symmetric. If $\mathbf{B}$ is a symmetric matrix of suitable dimensions, then for any matrix $\mathbf{A}$, both $\mathbf{ABA}'$ and $\mathbf{A}'\mathbf{BA}$ are symmetric. Let

$$\mathbf{A} = \begin{bmatrix} 1 & 1 & 0 \\ 0 & 2 & 1 \end{bmatrix} \qquad \text{and} \qquad \mathbf{B} = \begin{bmatrix} 3 & 1 \\ 1 & 4 \end{bmatrix}$$

Then,

$$\mathbf{A}'\mathbf{BA} = \begin{bmatrix} 1 & 0 \\ 1 & 2 \\ 0 & 1 \end{bmatrix}\begin{bmatrix} 3 & 1 \\ 1 & 4 \end{bmatrix}\begin{bmatrix} 1 & 1 & 0 \\ 0 & 2 & 1 \end{bmatrix} = \begin{bmatrix} 3 & 1 \\ 5 & 9 \\ 1 & 4 \end{bmatrix}\begin{bmatrix} 1 & 1 & 0 \\ 0 & 2 & 1 \end{bmatrix} = \begin{bmatrix} 3 & 5 & 1 \\ 5 & 23 & 9 \\ 1 & 9 & 4 \end{bmatrix}$$

which is obviously symmetric. The sum and difference of symmetric matrices also are symmetric. The product, however, in general is different.

Matrix inverse. Unlike scalars, *division* of matrices is *not defined*. In fact, the relationship $\mathbf{AB} = \mathbf{AC}$ may exist without having $\mathbf{B} = \mathbf{C}$. This implies that the operation of "dividing" by $\mathbf{A}$, even if $\mathbf{A} \neq \mathbf{0}$, is not possible. As an example, let

$$\mathbf{A} = \begin{bmatrix} 2 & 0 \\ 4 & 0 \end{bmatrix}, \qquad \mathbf{B} = \begin{bmatrix} 2 & -2 \\ 5 & 3 \end{bmatrix}, \qquad \mathbf{C} = \begin{bmatrix} 2 & -2 \\ 1 & 4 \end{bmatrix}$$

where obviously $\mathbf{B} \neq \mathbf{C}$. Computation of both $\mathbf{AB}$ and $\mathbf{AC}$ yields

$$\mathbf{AB} = \begin{bmatrix} 4 & -4 \\ 8 & -8 \end{bmatrix} = \mathbf{AC}$$

In place of division, the concept of a matrix *inverse* is used. It is symbolized by $\mathbf{A}^{-1}$, for a matrix $\mathbf{A}$ (similar to the reciprocal α^{-1} of the scalar α).

The *inverse* of a square matrix $\mathbf{A}$, *if it exists,* is the unique matrix $\mathbf{A}^{-1}$ with the following property:

$$\mathbf{AA}^{-1} = \mathbf{A}^{-1}\mathbf{A} = \mathbf{I} \tag{B.42}$$

where $\mathbf{I}$ is the identity matrix. For

$$\mathbf{A} = \begin{bmatrix} 3 & 1 \\ 2 & 1 \end{bmatrix}$$

the matrix

$$\mathbf{A}^{-1} = \begin{bmatrix} 1 & -1 \\ -2 & 3 \end{bmatrix}$$

is the inverse because

$$\begin{bmatrix} 1 & -1 \\ -2 & 3 \end{bmatrix}\begin{bmatrix} 3 & 1 \\ 2 & 1 \end{bmatrix} = \begin{bmatrix} 1 & 0 \\ 0 & 1 \end{bmatrix} \quad \text{and} \quad \begin{bmatrix} 3 & 1 \\ 2 & 1 \end{bmatrix}\begin{bmatrix} 1 & -1 \\ -2 & 3 \end{bmatrix} = \begin{bmatrix} 1 & 0 \\ 0 & 1 \end{bmatrix}$$

Some of the properties of the inverse are

$$(\mathbf{AB})^{-1} = \mathbf{B}^{-1}\mathbf{A}^{-1} \tag{B.43}$$

$$(\mathbf{A}^{-1})^{-1} = \mathbf{A} \tag{B.44}$$

$$(\mathbf{A}')^{-1} = (\mathbf{A}^{-1})^t \tag{B.45}$$

$$(\alpha\mathbf{A})^{-1} = \frac{1}{\alpha}\mathbf{A}^{-1} \tag{B.46}$$

assuming that all inverses exist. The square matrix for which an inverse exists is called *nonsingular,* and that which does not have an inverse is called *singular.*

As shown previously, the product $\mathbf{AB}$ can equal $\mathbf{0}$ without either $\mathbf{A} = \mathbf{0}$ or $\mathbf{B} = \mathbf{0}$. On the other hand, if either $\mathbf{A}$ or $\mathbf{B}$ is nonsingular, the other matrix must be a null matrix. Hence, the product of two nonsingular matrices cannot be a null matrix.

Several procedures are used to compute the inverse of a square matrix. One such procedure requires introduction of the concept of determinants.

Associated with each *square* matrix $\mathbf{A}$ is a unique scalar value called the *determinant* of $\mathbf{A}$, denoted by either det $\mathbf{A}$ or $|\mathbf{A}|$. For $\mathbf{A} = \begin{bmatrix} 3 & 1 \\ 1 & 2 \end{bmatrix}$, the determinant is expressed as $|\mathbf{A}| = \begin{vmatrix} 3 & 1 \\ 1 & 2 \end{vmatrix}$, the value of which is computed as follows. One should be careful to differentiate between the square brackets used for the matrix and the vertical lines used for the determinant.

The determinant of order n (for an $n \times n$ square matrix) can be defined in terms of determinants of order $n - 1$ and less. To apply this procedure, the determinant of a 1×1 matrix must be defined. Accordingly, for a matrix consisting of a single element, the determinant is defined as the value of the element; that is, for $\mathbf{A} = [a_{11}]$, $|\mathbf{A}| = $ det $\mathbf{A} = a_{11}$.

If $\mathbf{A}$ is an $n \times n$ matrix and one row and one column of $\mathbf{A}$ are deleted, the resulting matrix is an $(n - 1) \times (n - 1)$ *submatrix* of $\mathbf{A}$. The determinant of such a submatrix is called a *minor* of $\mathbf{A}$, designated m_{ij}, where i and j correspond to the deleted row and column, respectively. More specifically, m_{ij} is known as the *minor of the element* a_{ij} in $\mathbf{A}$. For example, consider

$$\mathbf{A} = \begin{bmatrix} a_{11} & a_{12} & a_{13} \\ a_{21} & a_{22} & a_{23} \\ a_{31} & a_{32} & a_{33} \end{bmatrix}$$

Each element of $\mathbf{A}$ has a minor. The minor of a_{11}, for example, is obtained by deleting the first row and first column from $\mathbf{A}$ and taking the determinant of the 2×2 submatrix that remains; that is,

$$m_{11} = \begin{vmatrix} a_{22} & a_{23} \\ a_{32} & a_{33} \end{vmatrix}$$

The *cofactor* c_{ij} of an element a_{ij} is defined as

$$c_{ij} = (-1)^{i+j}m_{ij} \tag{B.47}$$

Obviously, when the sum of the row number i and column number j is even $c_{ij} = m_{ij}$, and when $i + j$ is odd, $c_{ij} = -m_{ij}$.

The determinant of an $n \times n$ matrix $\mathbf{A}$ can now be defined as

$$|\mathbf{A}| = a_{11}c_{11} + a_{12}c_{12} + \cdots + a_{1n}c_{1n} \tag{B.48}$$

which states that the determinant of $\mathbf{A}$ is the sum of the products of the elements of the first row of $\mathbf{A}$ and their corresponding cofactors. (It is equally possible to define $|\mathbf{A}|$ in terms of any other row or column but, for simplicity, the first row is used here.) On the basis of the definition, the 2×2 matrix

$$\mathbf{A} = \begin{bmatrix} a_{11} & a_{12} \\ a_{21} & a_{22} \end{bmatrix}$$

has cofactors $c_{11} = |a_{22}| = a_{22}$, and $c_{12} = -|a_{21}| = -a_{21}$, and the determinant of $\mathbf{A}$ is

$$|\mathbf{A}| = a_{11}c_{11} + a_{12}c_{12} = a_{11}a_{22} - a_{12}a_{21}$$

Therefore,

$$\begin{vmatrix} 3 & -1 \\ 2 & 4 \end{vmatrix} = (3)(4) + (-1)(-2) = 14$$

The *cofactor matrix* $\mathbf{C}$ of a matrix $\mathbf{A}$ is the square matrix of the same order as $\mathbf{A}$ in which each element a_{ij} is replaced by its cofactor c_{ij}. For example, the cofactor matrix of

$$\mathbf{A} = \begin{bmatrix} 1 & 2 \\ -3 & 4 \end{bmatrix}$$

is

$$\mathbf{C} = \begin{bmatrix} 4 & 3 \\ -2 & 1 \end{bmatrix}$$

The *adjoint matrix* of $\mathbf{A}$, denoted adj $\mathbf{A}$, is the transpose of its cofactor matrix; that is,

$$\text{adj } \mathbf{A} = \mathbf{C}^t \tag{B.49}$$

It can be shown that

$$\mathbf{A}(\text{adj } \mathbf{A}) = (\text{adj } \mathbf{A})\,\mathbf{A} = |\mathbf{A}|\,\mathbf{I} \tag{B.50}$$

Comparison of Equations (B.42) and (B.50) leads directly to a procedure for evaluating the inverse from the adjoint matrix; namely,

$$\mathbf{A}^{-1} = \frac{\text{adj } \mathbf{A}}{|\mathbf{A}|} \tag{B.51}$$

As an example,

$$\mathbf{A} = \begin{bmatrix} 3 & 1 \\ 2 & 1 \end{bmatrix}; \quad |\mathbf{A}| = 1; \quad \mathbf{C} = \begin{bmatrix} 1 & -2 \\ -1 & 3 \end{bmatrix}; \quad \text{adj } \mathbf{A} = \begin{bmatrix} 1 & -1 \\ -2 & 3 \end{bmatrix}$$

and

$$\mathbf{A}^{-1} = \frac{\text{adj } \mathbf{A}}{|\mathbf{A}|} = \begin{bmatrix} 1 & -1 \\ -2 & 3 \end{bmatrix}$$

Note that for a 2 × 2 matrix, the adjoint matrix is simply

$$\begin{bmatrix} a_{22} & -a_{12} \\ -a_{21} & a_{11} \end{bmatrix}$$

A square matrix is called *orthogonal* if its inverse is equal to its transpose: $\mathbf{A}^{-1} = \mathbf{A}^t$. A matrix $\mathbf{M}$ is orthogonal when

$$\mathbf{M}'\mathbf{M} = \mathbf{M}\mathbf{M}' = \mathbf{I} \tag{B.52}$$

The columns of an orthogonal matrix are mutually orthogonal vectors of unit length. Also,

$$|\mathbf{M}| = \pm 1 \tag{B.53}$$

When $|\mathbf{M}| = +1$, then $\mathbf{M}$ is called *proper orthogonal*, otherwise, it is termed *improper orthogonal*. The product of two orthogonal matrices also is an orthogonal matrix.

Trace. The *trace* of a *square* matrix is the scalar that is equal to the sum of its main diagonal elements. It is denoted by tr($\mathbf{A}$) and thus tr($\mathbf{A}$) = $a_{11} + a_{22} + a_{33} + \ldots$. As an example,

$$\text{for } \mathbf{A} = \begin{bmatrix} 1 & 4 & 7 \\ 2 & 5 & 8 \\ 3 & 6 & 9 \end{bmatrix}, \quad \text{tr}(\mathbf{A}) = 1 + 5 + 9 = 15$$

The following are properties of the trace:

$$\text{tr}(\mathbf{A}) = \text{tr}(\mathbf{A}')$$

$$\text{tr}(\lambda\mathbf{A}) = \lambda\text{tr}(\mathbf{A})$$

$$\text{tr}(\mathbf{A} + \mathbf{B}) = \text{tr}(\mathbf{A}) + \text{tr}(\mathbf{B})$$

$$\text{tr}(\mathbf{AB}) = \text{tr}(\mathbf{BA})$$

$$\text{tr}(\mathbf{FAF}^{-1}) = \text{tr}(\mathbf{A}) \quad (\mathbf{F} \text{ nonsingular matrix, see the following section})$$

B.4
SOLUTION OF LINEAR EQUATIONS

Linear equations are encountered frequently in surveying and mapping adjustment and computation problems. When the number of equations is equal to the number of unknowns, the system of equations is said to be *unique*. The following is a set of n equations:

$$a_{11}x_1 + a_{12}x_2 + \cdots + a_{1n}x_n = b_1$$

$$a_{21}x_1 + a_{22}x_2 + \cdots + a_{2n}x_n = b_2 \tag{B.54}$$

$$\ldots \ldots \ldots \ldots \ldots \ldots \ldots \ldots \ldots$$

$$a_{n1}x_1 + a_{n2}x_2 + \cdots + a_{nn}x_n = b_n$$

In Equation (B.54), the symbols a_{ij} and b_i are numerical coefficients and constant terms, whereas the x_j are the unknowns. The solution to Equation (B.54) is a unique set of numerical values for x_j that, when substituted into the equations, would be simultaneously satisfied. Equation (B.54) may be expressed in matrix form as

$$\underset{n,n}{\mathbf{A}} \; \underset{n,1}{\mathbf{x}} = \underset{n,1}{\mathbf{b}} \tag{B.55}$$

where $\mathbf{A}$ is a square matrix of order n. Consideration here is limited to the more practical unique case for which $\mathbf{A}$ must be nonsingular. Premultiplying both sides of Equation (B.55) by $\mathbf{A}^{-1}$ yields

$$\mathbf{A}^{-1}\mathbf{A}\mathbf{x} = \mathbf{A}^{-1}\mathbf{b}$$

or

$$\mathbf{x} = \mathbf{A}^{-1}\mathbf{b} \tag{B.56}$$

because $\mathbf{A}^{-1}\mathbf{A} = \mathbf{I}$ by definition. Consequently, the solution to linear equations is equivalent to finding the inverse of the coefficient matrix. If the equations contain only two or three unknowns (i.e., $\mathbf{A}$ is of dimensions 2×2 or 3×3), it may be practical to compute the inverse by a simple procedure such as the adjoint method described previously. However, computing the inverse by the adjoint method becomes quite tedious and inefficient for matrices of order more than three. Instead, other procedures are employed for both finding the inverse and direct solutions of linear equations. Many of these procedures use what are called *primary row or column operations*, which are given in the next section.

B.5
PRIMARY ROW OR COLUMN OPERATIONS

These operations, which do not change the order of the matrix being manipulated, are the following:

1. The interchange of any two rows (or two columns).
2. The multiplication of all the elements of any row (or column) by the same nonzero constant.
3. The addition to any row (or column) of an arbitrary multiple of any other row (or column).

   Two procedures for solving linear equations, which apply these operations, are given in the next two sections.

B.6
THE GAUSS METHOD

In the Gauss method, the elementary operations are used to reduce the coefficient matrix to an upper-triangular form (preferably with diagonal elements of 1). This operation is called the *forward solution*. The subsequent computation of $\mathbf{x}$ is called *backward solution*. As an example, solve the system

$$x_1 + 2x_2 = 3$$

$$-3x_1 + 4x_2 = 1$$

or

$$\begin{bmatrix} 1 & 2 \\ -3 & 4 \end{bmatrix} \begin{bmatrix} x_1 \\ x_2 \end{bmatrix} = \begin{bmatrix} 3 \\ 1 \end{bmatrix}$$

The first equation may be written as

x_1	x_2	b
1	2	3

which needs no modification because the coefficient of x_1 already is 1. The next step is to reduce the coefficient of x_1 in the second equation to 0. This is done by multiplying the first equation by 3 and adding to the second. The new second equation is

$$\begin{array}{ccc} x_1 & x_2 & b \\ \hline 0 & 10 & 10 \end{array}$$

which already completes the forward solution. The backward solution begins with the new second equation, because it has only one unknown, x_2, and therefore $x_2 = 1$. Using this in the first equation, compute $x_1 = 3 - 2x_2 = 3 - 2 = 1$. As another example, consider the following set of three equations:

$$x_1 + \qquad -x_3 = -1$$

$$x_2 + 3x_3 = \quad 7$$

$$-x_1 + \quad + 2x_3 = \quad 3$$

or, in matrix form,

$$\begin{bmatrix} 1 & 0 & -1 \\ 0 & 1 & 3 \\ -1 & 0 & 2 \end{bmatrix} \begin{bmatrix} x_1 \\ x_2 \\ x_3 \end{bmatrix} = \begin{bmatrix} -1 \\ 7 \\ 3 \end{bmatrix}$$

In this system, the initial coefficients of x_1 are 1 in each equation. To achieve the desired upper-triangular form, it is necessary only to add the first and third equations as follows:

	x_1	x_2	x_3	b	Remarks
Eq. 1	1	0	-1	-1	The original data
Eq. 3	-1	0	2	3	
	1	0	-1	-1	First and second equations unaltered because
	0	1	3	7	$a_{11} = a_{22} = 1$ already; add original first equation to
	0	0	1	2	original third equation to get a revised third equation; this completes the forward solution.

The backward solution proceeds as follows:

$$x_3 = \quad 2$$

$$x_2 = \quad 7 - 3x_3 = 1$$

$$x_1 = -1 + \quad x_3 = 1$$

B.7
THE GAUSS-JORDAN METHOD

In the Gauss-Jordan procedure, the primary operations are continued until the coefficient matrix reduces to the identity matrix. At this point, $\mathbf{b}$ becomes the answer vector. If all the operations are performed simultaneously on an identity, it will become $\mathbf{A}^{-1}$ at the end. To show that such is the case, solve the two-equation problem of the preceding section.

Line	x_1	x_2	b	I		Remarks
1	1	2	3	1	0 ⎱	Original data augmented by I
2	−3	4	1	0	1 ⎰	
3	1	2	3	1	0 ⎱	First equation (line 1) unaltered;
4	0	10	10	3	1 ⎰	3 times line 1 added to line 2
5	1	0	1	0.4	−0.2 ⎱	Divide line 4 by 10 to get line 6; subtract
6	0	1	1	0.3	0.1 ⎰	2 times line 6 from line 3 to get line 5
	I		x	A^{-1}		

It can be seen that **A** reduced to **I**, **b** reduced to **x**, and **I** became A^{-1}. Next, work the three-equation problem of the preceding section:

Line	x_1	x_2	x_3	b	I			Remarks
1	1	0	−1	−1	1	0	0 ⎱	
2	0	1	3	7	0	1	0 ⎬	Given data
3	−1	0	2	3	0	0	1 ⎰	
4	1	0	−1	−1	1	0	0	Same as line 1
5	0	1	3	7	0	1	0	Same as line 2
6	0	0	1	2	1	0	1	Line 1 + line 3
7	1	0	0	1	2	0	1	Line 4 + line 6
8	0	1	0	1	−3	1	−3	(Line 5) − (3)(line 6)
9	0	0	1	2	1	0	1	Same as line 6
	I			x	A^{-1}			

An important note concerns the possibility of having to divide by 0. For example, if the coefficient of x_1 in the first equation is 0, the procedure for both the Gauss and Gauss-Jordan methods could not progress unaltered. In such a case the equations are rearranged so that the first equation will have a nonzero coefficient for x_1. This also should be done for x_2 in the second step, for x_3 in the third step, and so on.

B.8
EIGENVALUES AND EIGENVECTORS

For each square matrix **A** there exists a set of scalars λ_i and vectors x_i, one vector for each scalar, such that the following relationship holds:

$$Ax_i = \lambda_i x_i \qquad (B.57)$$

The scalar λ_i is called an *eigenvalue* and x_i is called an *eigenvector*. If the matrix **A** is symmetric, all eigenvalues are real and all eigenvectors are mutually orthogonal to each other. This is the case most often encountered in surveying and mapping, where this technique is used to determine error ellipses from known covariance matrices. This will be shown in conjunction with quadratic forms in the following section.

Equation (B.57) may be rearranged and the subscript i dropped to allow for determining all eigenvalues:

$$Ax - \lambda x = 0$$

or

$$(A - \lambda I)x = 0 \qquad (B.58)$$

Equation (B.58) represents a set of n homogeneous equations (assuming that $\mathbf{A}$ is of order n) that will have a nontrivial solution if the determinant of the coefficient matrix $(\mathbf{A} - \lambda\mathbf{I})$ is 0. Expanding such a determinant leads to an nth-degree polynomial in λ, the n roots of which would be the eigenvalues. From a practical standpoint, consider the two-dimensional case. Therefore,

$$|\mathbf{A} - \lambda\mathbf{I}| = 0$$

or

$$\begin{vmatrix} a_{11} - \lambda & a_{12} \\ a_{21} & a_{22} - \lambda \end{vmatrix} = 0$$

or

$$(a_{11} - \lambda)(a_{22} - \lambda) - a_{12}a_{21} = 0$$
$$\lambda^2 - (a_{11} + a_{22})\lambda + (a_{11}a_{22} - a_{12}a_{21}) = 0 \qquad (B.59)$$
$$\lambda^2 - \text{tr}(\mathbf{A})\lambda + |\mathbf{A}| = 0$$

in which $\text{tr}(\mathbf{A})$ is the *trace* of $\mathbf{A}$ and $|\mathbf{A}|$ is the determinant of $\mathbf{A}$. In general, two values of λ are obtained from Equation (B.59). For each λ, the corresponding eigenvector is derived using Equation (B.58). Note that because Equation (B.58) represents a set of homogeneous equations in the components of the eigenvector, it can be determined only by direction. The concept is illustrated by the numerical computation of the eigenvalues and eigenvectors of the matrix

$$\mathbf{B} = \begin{bmatrix} 2 & 1 \\ 1 & 2 \end{bmatrix}$$

First, $\text{tr}(\mathbf{B}) = 4$ and $|\mathbf{B}| = (2)(2) - (1)(1) = 3$. Then substituting in Equation (B.59) gives $\lambda^2 - 4\lambda + 3 = 0$ or $(\lambda - 3)(\lambda - 1) = 0$, from which $\lambda_1 = 3$ and $\lambda_2 = 1$. For $\lambda_1 = 3$, the eigenvector is evaluated from [see Equation (B.58)]

$$(\mathbf{B} - 3\mathbf{I})\mathbf{x} = 0$$

or

$$\left[\begin{bmatrix} 2 & 1 \\ 1 & 2 \end{bmatrix} - \begin{bmatrix} 3 & 0 \\ 0 & 3 \end{bmatrix} \right] \begin{bmatrix} x_1 \\ x_2 \end{bmatrix} = \begin{bmatrix} 0 \\ 0 \end{bmatrix}$$

or

$$\begin{bmatrix} -1 & 1 \\ 1 & -1 \end{bmatrix} \begin{bmatrix} x_1 \\ x_2 \end{bmatrix} = \begin{bmatrix} 0 \\ 0 \end{bmatrix}$$

for which $x_1 = x_2$. This means that for $\lambda_1 = 3$ the eigenvector makes $45°$ with the x_1 axis and can be taken as $(1, 1)$. For $\lambda_2 = 1$,

$$\begin{bmatrix} 1 & 1 \\ 1 & 1 \end{bmatrix} \begin{bmatrix} x_1 \\ x_2 \end{bmatrix} = \begin{bmatrix} 0 \\ 0 \end{bmatrix}$$

from which $x_1 = -x_2$. This means that for $\lambda = 2$ the eigenvector makes a $135°$ angle with the x_1 axis and can be taken as $(1, -1)$.

B.9
QUADRATIC FORMS

If $\mathbf{A}$ is a symmetric matrix of order n and $\mathbf{x}$ is an $n \times 1$ vector of variables, the scalar

$$u = \mathbf{x}^t \mathbf{A} \mathbf{x} \qquad (B.60)$$

is called a *quadratic form* because it is the sum of second-order terms in the elements of **x**. As a demonstration, the two-dimensional case is given by

$$u = [x_1 \quad x_2] \begin{bmatrix} a_{11} & a_{12} \\ a_{12} & a_{22} \end{bmatrix} \begin{bmatrix} x_1 \\ x_2 \end{bmatrix} = a_{11}x_1^2 + 2a_{12}x_1x_2 + a_{22}x_2^2 \qquad (B.61)$$

A good example of quadratic forms is the sum of weighted residuals squared, which is minimized in least squares. It takes the general form

$$\phi = \mathbf{v}^t \mathbf{W} \mathbf{v}$$

in which **v** is the vector of observational residuals and **W** is the symmetric weight matrix of the observations.

If the matrix **A** in Equation (B.61) represents the inverse of a covariance matrix, that equation will express a family of ellipses the axes of which are inclined with respect to the x_1, x_2 system. If the directions of the eigenvectors of the covariance matrix are designated by y_1, y_2, it can be shown that these coincide with the semimajor and semiminor axes of the ellipses. Furthermore, if $u = 1$ in Equation (B.61), the resulting ellipse is called the *standard error ellipse* with semimajor and semiminor axes which are equal to $\sqrt{\lambda_1}$ and $\sqrt{\lambda_2}$, respectively. The quantities λ_1 and λ_2 are the eigenvalues of the covariance matrix. Consider an example. The covariance matrix is

$$\mathbf{B} = \begin{bmatrix} 2 & 1 \\ 1 & 2 \end{bmatrix} \qquad \text{and} \qquad \mathbf{A} = \mathbf{B}^{-1} = \frac{1}{3} \begin{bmatrix} 2 & -1 \\ -1 & 2 \end{bmatrix}$$

The quadratic form for the standard ellipse is

$$[x_1 \quad x_2] \frac{1}{3} \begin{bmatrix} 2 & -1 \\ -1 & 2 \end{bmatrix} \begin{bmatrix} x_1 \\ x_2 \end{bmatrix} = 1$$

$$\tfrac{2}{3}x_1^2 - \tfrac{2}{3}x_1x_2 + \tfrac{2}{3}x_2^2 = 1 \qquad (B.62)$$

Equation (B.62) represents the equation of the ellipse in Figure B.4 with respect to the x_1, x_2 axis system. If the dimensions of that ellipse are computed then the semimajor axis, $a = \sqrt{3}$, and the semiminor axis, $b = 1$, and the axes are oriented at 45° as shown in the figure. This will agree with the fact that $\lambda_1 = 3$ and $\lambda_2 = 1$ for the covariance matrix **B** as was computed in the preceding example. Furthermore, y_1 and y_2 do, in fact, represent the directions of the eigenvectors of the covariance matrix **B**. The equation of the standard ellipse with respect to the y_1, y_2 system is

$$\frac{y_1^2}{3} + \frac{y_2^2}{1} = 1 \qquad (B.63)$$

The reader should prove this equality by substituting, for example, the x_1, x_2 coordinates for one or two points on the ellipse, such as c and d, and see that they satisfy Equation (B.62) and, correspondingly, the y_1, y_2 coordinates and see that they satisfy Equation (B.63).

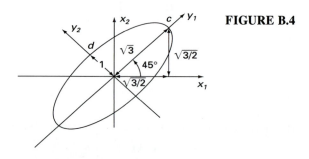

FIGURE B.4

B.10
DIFFERENTIATION OF MATRICES AND QUADRATIC FORMS

The following are useful relations encountered in surveying and mapping computation and adjustment.

1. If the elements of a matrix $\mathbf{A}$ are functions of a (scalar) variable u, then the derivative $d\mathbf{A}/du$ of the matrix with respect to u is another m by n matrix given by

$$\frac{d\mathbf{A}}{du}_{m,n} = \begin{bmatrix} \dfrac{da_{11}}{du} & \cdots & \dfrac{da_{1n}}{du} \\ \cdots\cdots\cdots\cdots\cdots \\ \dfrac{da_{m1}}{du} & \cdots & \dfrac{da_{mn}}{du} \end{bmatrix} \tag{B.64}$$

2. If a (column) vector $\mathbf{y}$ is composed of m functions of some or all of the elements of another variable (column) vector $\mathbf{x}$, then the partial derivative $\partial\mathbf{y}/\partial\mathbf{x}$ of $\mathbf{y}$ with respect to $\mathbf{x}$ is an m by n matrix, called the *Jacobian matrix,* whose elements are

$$\mathbf{J}_{yx} = \frac{\partial\mathbf{y}}{\partial\mathbf{x}} = \begin{bmatrix} \dfrac{\partial y_1}{\partial x_1} & \dfrac{\partial y_1}{\partial x_2} & \cdots & \dfrac{\partial y_1}{\partial x_n} \\ \cdots\cdots\cdots\cdots\cdots \\ \dfrac{\partial y_m}{\partial x_1} & \dfrac{\partial y_m}{\partial x_2} & \cdots & \dfrac{\partial y_m}{\partial x_n} \end{bmatrix} \tag{B.65}$$

For example, if $y_1 = x_1 + 2x_2 + 3x_3^2$ and $y_2 = 7 - 2x_2^2 + 5x_3$, then the Jacobian $\mathbf{J}_{yx}$ is

$$\mathbf{J}_{yx} = \begin{bmatrix} \partial y_1/\partial x_1 & \partial y_1/\partial x_2 & \partial y_1/\partial x_3 \\ \partial y_2/\partial x_1 & \partial y_2/\partial x_2 & \partial y_2/\partial x_3 \end{bmatrix} = \begin{bmatrix} 1 & 2 & 6x_3 \\ 0 & -4x_2 & 5 \end{bmatrix}$$

3. The matrix $\mathbf{A}$ is symmetric in the quadratic form $u = \mathbf{x}^t \mathbf{A}\mathbf{x}$ (see Section B.9) and also is independent of x. If both $\mathbf{A}$ and $\mathbf{x}$ are of order n, then $\partial u/\partial\mathbf{x}$ must be a 1 by n row vector according to relation 2. That partial derivative may be evaluated by visualizing partially differentiating first with respect to $\mathbf{x}$ and then with respect to $\mathbf{x}^t$ and adding the results. Keeping in mind that the result is $1 \times n$, then

$$\frac{\partial u}{\partial\mathbf{x}} = \left(\underset{1,n}{\mathbf{x}^t} \; \underset{n,n}{\mathbf{A}} \right) + \left(\underset{1,n}{\mathbf{x}^t} \; \underset{n,n}{\mathbf{A}^t} \right) = \mathbf{x}^t(\mathbf{A} + \mathbf{A}^t) \tag{B.66}$$

Because $\mathbf{A}$ is symmetric, $\mathbf{A}^t = \mathbf{A}$ and Equation (B.66) becomes

$$\frac{\partial u}{\partial\mathbf{x}} = 2\mathbf{x}^t\mathbf{A} \tag{B.67}$$

We demonstrate this for the two-dimensional case by referring to Equation (B.61):

$$\frac{\partial u}{\partial x_1} = 2a_{11}x_1 + 2a_{12}x_2 = 2[x_1 \quad x_2]\begin{bmatrix} a_{11} \\ a_{12} \end{bmatrix}$$

$$\frac{\partial u}{\partial x_2} = 2a_{12}x_1 + 2a_{22}x_2 = 2[x_1 \quad x_2]\begin{bmatrix} a_{12} \\ a_{22} \end{bmatrix}$$

Therefore,

$$\frac{\partial u}{\partial\mathbf{x}} = 2[x_1 \quad x_2]\begin{bmatrix} a_{11} & a_{12} \\ a_{12} & a_{22} \end{bmatrix} = 2\mathbf{x}^t\mathbf{A}$$

B.11
LINEARIZATION OF NONLINEAR EQUATIONS

In practice, surveying problems yield condition equations that frequently are nonlinear in the observations or the parameters. The direct use of nonlinear equations in least squares is very complex and rarely done. Instead, the equations are linearized using Taylor's series expansion and solving the resulting linear equations, then iterating until the effect of the neglected higher-order terms is minimized.

Let $y = f(x)$ be a function that is nonlinear in x. The Taylor series expansion is

$$y = f(x^0) + \frac{dy}{dx}\bigg|_{x^0} \Delta x + \frac{1}{2!}\frac{d^2y}{dx^2}\bigg|_{x^0} (\Delta x)^2 + \cdots \tag{B.68}$$

in which x^0 is the approximate value of the variable at which the function is evaluated. The first term on the right-hand side of Equation (B.68) is the zero-order term, which is equal to the value of the function evaluated at $x = x^0$; the second term is the first-order term, which contains the first derivative evaluated at $x = x^0$; the third term is the second-order term, which includes the second derivative evaluated at $x = x^0$; and so on. To have a linear form, only the zero- and first-order terms are used from the series expansion. If $y = f(x_1, x_2)$ is a nonlinear function of the two variables x_1 and x_2, the linearized form is

$$y = f(x_1^0, x_2^0) + \frac{\partial y}{\partial x_1}\bigg|_{x_1^0, x_2^0} \Delta x_1 + \frac{\partial y}{\partial x_2}\bigg|_{x_1^0, x_2^0} \Delta x_2$$

$$= f(x_1^0, x_2^0) + j_1\Delta x_1 + j_2\Delta x_2$$

$$= y^0 + [\,j_1 j_2\,]\begin{bmatrix} \Delta x_1 \\ \Delta x_2 \end{bmatrix}$$

or
$$y = y^0 + \mathbf{J}_{yx}\,\Delta\mathbf{x} \tag{B.69}$$

in which $\mathbf{J}_{yx}$ is the Jacobian matrix, containing the partial derivatives of y with respect to x_1 and x_2, respectively, as described in Section B.10. The relation in Equation (B.69) can be generalized to the case of m functions $\mathbf{y}$, each in terms of some or all of n variables $\mathbf{x}$ (i.e., $\mathbf{y} = f(\mathbf{x})$):

$$\mathbf{y} = \mathbf{y}^0 + \mathbf{J}_{yx}\,\Delta\mathbf{x} \tag{B.70}$$

where

$$\mathbf{y}^0 = \begin{bmatrix} y_1^0 \\ y_2^0 \\ \vdots \\ y_m^0 \end{bmatrix} = \begin{bmatrix} f_1(x_1^0, x_2^0, \ldots, x_n^0) \\ f_2(x_1^0, x_2^0, \ldots, x_n^0) \\ \cdots\cdots\cdots\cdots \\ f_m(x_1^0, x_2^0, \ldots, x_n^0) \end{bmatrix}, \quad \Delta\mathbf{x} = \begin{bmatrix} \Delta x_1 \\ \Delta x_2 \\ \vdots \\ \Delta x_n \end{bmatrix}$$

$$\mathbf{J}_{yx} = \frac{\partial \mathbf{y}}{\partial \mathbf{x}} = \begin{bmatrix} \dfrac{\partial y_1}{\partial x_1} & \dfrac{\partial y_1}{\partial x_2} & \cdots & \dfrac{\partial y_1}{\partial x_n} \\ \dfrac{\partial y_2}{\partial x_1} & \dfrac{\partial y_2}{\partial x_2} & \cdots & \dfrac{\partial y_2}{\partial x_n} \\ \cdots\cdots\cdots\cdots\cdots \\ \dfrac{\partial y_m}{\partial x_1} & \dfrac{\partial y_m}{\partial x_2} & \cdots & \dfrac{\partial y_m}{\partial x_n} \end{bmatrix}$$

evaluated at $\mathbf{x} = \mathbf{x}^0$.

PROBLEMS

B.1.

$$A = \begin{bmatrix} 1 & 0 \\ -2 & 1 \end{bmatrix}, \quad B = \begin{bmatrix} -3 & 4 & 0 \\ 4 & -1 & 3 \end{bmatrix}, \quad C = \begin{bmatrix} -1 & 6 \\ 0 & 3 \\ 5 & 2 \end{bmatrix}$$

Calculate the following matrices:

$$E = AB, \quad F = BC, \quad G = EC, \quad H = AC', \quad K = F + G$$

B.2.

$$C = \begin{bmatrix} 1 & 2 & -3 \\ 2 & 0 & 4 \\ -3 & 4 & 1 \end{bmatrix}, \quad D = \begin{bmatrix} 0 & 1 & -2 \\ -1 & 0 & 3 \\ 2 & -3 & 0 \end{bmatrix}$$

Compute $E = CD$ and $F = DC$. Can you detect a relationship between E and F?

B.3.

$$A = \begin{bmatrix} 1 & 2 & 3 \\ 0 & -1 & 2 \\ 0 & 0 & -3 \end{bmatrix}, \quad B = \begin{bmatrix} 1 & 0 & 0 \\ -2 & 3 & 0 \\ 1 & -1 & -4 \end{bmatrix}$$

Compute $A^{-1}, B^{-1}, C = AB$, and C^{-1}. Show that $C^{-1} = B^{-1}A^{-1}$.

B.4. Solve the following system of linear equations by two methods:

$$-X_1 + 2X_2 - X_3 = -5$$
$$3X_1 \quad\quad + 2X_3 = \quad 7$$
$$2X_1 + 5X_2 + X_3 = -1$$

B.5. Find the eigenvalues for the following matrices:

$$A = \begin{bmatrix} 3 & 1 \\ 1 & 3 \end{bmatrix}, \quad B = \begin{bmatrix} 4 & -1 \\ -1 & 4 \end{bmatrix}, \quad C = \begin{bmatrix} 2 & 1 \\ 1 & 3 \end{bmatrix}$$

B.6. Compute the elements of the standard error ellipses (i.e., a, b, and θ) for the three matrices, A, B, and C, in Problem B.5.

REFERENCES

Anderson, J. M., and E. M. Mikhail. *Introduction to Surveying.* New York: McGraw-Hill, 1985.

Golub, G. H. and G. F. Van Loan. *Matrix Computations.* Baltimore: Johns Hopkins University Press, 1983.

Mikhail, E. M. *Observations and Least Squares.* New York: Harper and Row, 1976.

Mikhail, E. M., and G. Gracie. *Analysis and Adjustment of Survey Measurements.* New York: Van Nostrand Reinhold, 1980.

Moffitt, F. H. and E. M. Mikhail. *Photogrammetry,* 3rd ed. New York: Harper and Row, 1980.

APPENDIX C

Coordinate Transformations, Including Transformations between Geodetic Reference Systems

C.1
INTRODUCTION

Transformation of coordinates can be either linear or nonlinear. Also, transformation can occur in two-dimensional space, in three-dimensional space, or from a three-dimensional to a two-dimensional space, as in photogrammetry. We begin the discussion with the most commonly used linear transformations.

C.2
LINEAR TRANSFORMATIONS

A general *linear transformation* of a vector x to another vector y takes the form

$$y = Mx + t \tag{C.1}$$

Each element of the y vector is a linear combination of the elements of x plus a translation or shift represented by an element of the t vector. The matrix M is called the *transformation matrix*, which is in general rectangular, and t is called the translation vector. For our use, we restrict M to being square and nonsingular. Therefore, the inverse relation exists:

$$x = M^{-1}(y - t) \tag{C.2}$$

in which case it is called *affine transformation*. Although both Equations (C.1) and (C.2) apply to higher-dimension vectors, we limit our discussions, without loss of generality, to the more practical two- and three-dimensional spaces, where the elements of the transformations can be depicted geometrically.

Two-Dimensional Linear Transformations

The six *elementary* transformations, each representing a single effect, are represented geometrically in Figure C.1. Initially, four vectors (1, 3) (1, 5), (3, 3) (3, 5), representing

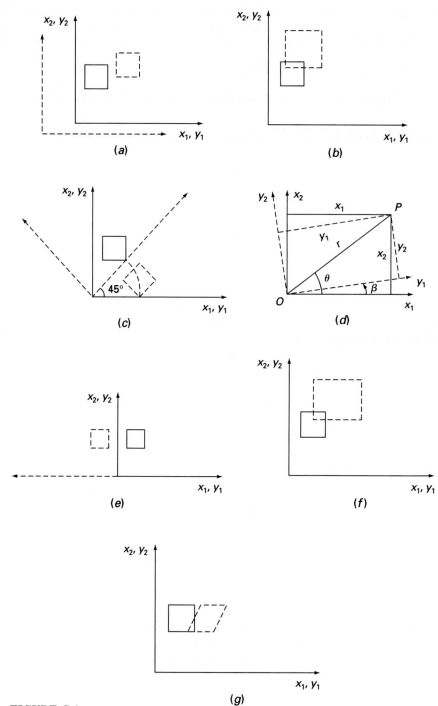

FIGURE C.1
Elementary transformations: (a) translation, (b) uniform scale, (c) rotation,
(d) rotation of a two-dimensional coordinate system, (e) reflection, (f) stretch
(nonuniform scale), (g) skew (nonperpendicularity of axes).

the corners of a square (solid lines in Figure C.1), are referred to the x_1, x_2 coordinate system. Each of the six elementary transformations operates on the square and the resulting y_1, y_2 coordinates plotted to show the effect on the location, orientation, size, and shape of the square after the transformation (dotted lines in Figure C.1). In displaying the effects of the transformations, we either display the new figure in the same coordinate system or change the coordinate system. It is easier for the reader to visualize these transformations if the new figure is drawn without changing the coordinate system, which was done in Figure C.1. However, as each elementary transformation is discussed, we comment on the second interpretation when appropriate.

Translation

$$y = x + t \tag{C.3}$$

where $M = I$. The square is shifted three units in the x_1 direction and one unit in the x_2 direction, as shown in Figure C.1a. Alternatively, the solid square remains and the coordinate axes are shifted (in the opposite direction).

Uniform scale

$$y = Mx, \qquad M = U = \begin{bmatrix} u & 0 \\ 0 & u \end{bmatrix} = uI \tag{C.4}$$

The (dotted) square is enlarged by the uniform scale u ($= 1.5$ in Figure C.1b), which results from all four point coordinate pairs multiplied by u. Alternatively, the square with solid lines is referred to the same coordinate system, except that the units along the axes are now $1/u$ of the original units.

Rotation

$$y = Mx, \qquad M = R = \begin{bmatrix} \cos \beta & \sin \beta \\ -\sin \beta & \cos \beta \end{bmatrix} \tag{C.5}$$

The square retains its shape but is rotated through θ about the origin of the coordinate system. In Figure C.1c, the coordinate system also is rotated. The elements of R are derived from Figure C.1d as follows:

$$y_1 = r \cos(\theta - \beta) = r \cos \theta \cos \beta + r \sin \theta \sin \beta$$

$$y_2 = r \sin(\theta - \beta) = r \sin \theta \cos \beta - r \sin \theta \sin \beta$$

or

$$y_1 = x_1 \cos \beta + x_2 \sin \beta$$

$$y_2 = -x_1 \sin \beta + x_2 \cos \beta$$

or

$$\begin{bmatrix} y_1 \\ y_2 \end{bmatrix} = \begin{bmatrix} \cos \beta & \sin \beta \\ -\sin \beta & \cos \beta \end{bmatrix} \begin{bmatrix} x_1 \\ x_2 \end{bmatrix} \tag{C.6}$$

The matrix R is proper orthogonal (refer to Equation (B.53)), $R^{-1} = R'$ and $|R| = +1$. Rotation matrices do not change the length of the vector; that is, $|x| = |y|$. Considering the square of the vector length

$$y'y = (Mx)' Mx = x'M'Mx = x'x$$

or

$$x'(M'M - I)x = 0$$

which for a nontrivial solution means that $M'M = I$, showing that M is an orthogonal matrix.

Reflection

$$y = \mathbf{M}x, \qquad \mathbf{M} = \mathbf{F} = \begin{bmatrix} -1 & 0 \\ 0 & 1 \end{bmatrix}$$

Figure C.1e shows the reflection of the x_1 axis (i.e., about the x_2 axis). $\mathbf{F}$ is called improper orthogonal, $\mathbf{F}^{-1} = \mathbf{F}$ and $|\mathbf{F}| = -1$.

Stretch (two-scale factors)

$$y = \mathbf{M}x, \qquad \mathbf{M} = \mathbf{S} = \begin{bmatrix} s_1 & \\ & s_2 \end{bmatrix} \tag{C.7}$$

The square is transformed into a rectangle, as shown in Figure C.1f.

Skew (shear)

$$y = \mathbf{M}x, \qquad \mathbf{M} = \mathbf{K} = \begin{bmatrix} 1 & k \\ 0 & 1 \end{bmatrix} \tag{C.8}$$

The square is transformed into a parallelogram, Figure C.1g. From these elementary transformations several affine transformations may be constructed using various sequences. The following are two of the commonly used transformations in surveying and photogrammetry.

Four-parameter transformation

$$\begin{bmatrix} y_1 \\ y_2 \end{bmatrix} = \begin{bmatrix} u & 0 \\ 0 & u \end{bmatrix} \begin{bmatrix} \cos \beta & \sin \beta \\ -\sin \beta & \cos \beta \end{bmatrix} \begin{bmatrix} x_1 \\ x_2 \end{bmatrix} + \begin{bmatrix} t_1 \\ t_2 \end{bmatrix} \tag{C.9}$$

or

$$y_1 = ux_1 \cos \beta + ux_2 \sin \beta + t_1$$
$$y_2 = -ux_1 \sin \beta + ux_2 \cos \beta + t_2 \tag{C.10}$$

or

$$y_1 = ax_1 + bx_2 + c$$
$$y_2 = -bx_1 + ax_2 + d \tag{C.11}$$

or

$$\begin{bmatrix} y_1 \\ y_2 \end{bmatrix} = \begin{bmatrix} a & b \\ -b & a \end{bmatrix} \begin{bmatrix} x_1 \\ x_2 \end{bmatrix} + \begin{bmatrix} c \\ d \end{bmatrix} \tag{C.12}$$

The inverse transformation is given by

$$\begin{bmatrix} x_1 \\ x_2 \end{bmatrix} = \frac{1}{u} \begin{bmatrix} \cos \beta & -\sin \beta \\ \sin \beta & \cos \beta \end{bmatrix} \begin{bmatrix} y_1 - t_1 \\ y_2 - t_2 \end{bmatrix} \tag{C.13}$$

or

$$\begin{bmatrix} x_1 \\ x_2 \end{bmatrix} = \frac{1}{a^2 + b^2} \begin{bmatrix} a & -b \\ b & a \end{bmatrix} \begin{bmatrix} y_1 - c \\ y_2 - d \end{bmatrix} \tag{C.14a}$$

This transformation has four parameters: a uniform scale, a rotation, and two translations. It is a conformal transformation (Section 11.1).

Three-parameter transformation. In many situations, only a change in the reference coordinate system is needed. In this case, the transformation involves only three parameters (and so called a *three-parameter transformation*): one rotation β, which changes the orientation of the coordinate system, and two translations, which reflect the location change in the origin of the coordinate system. In such a case, the scale change in the four-parameter transformation becomes unity, or $u = 1$. The three-parameter transformation equations take the form

$$\begin{bmatrix} y_1 \\ y_2 \end{bmatrix} = \begin{bmatrix} \cos \beta & \sin \beta \\ -\sin \beta & \cos \beta \end{bmatrix} \begin{bmatrix} x_1 \\ x_2 \end{bmatrix} + \begin{bmatrix} t_1 \\ t_2 \end{bmatrix} \tag{C.14b}$$

or

$$y_1 = x_1 \cos \beta + x_2 \sin \beta + t_1$$
$$y_2 = -x_1 \sin \beta + x_2 \cos \beta + t_2 \tag{C.14c}$$

Note that in this case it is not possible to use simpler parameters, such as a, b, and so on, in Equations (C.11) and (C.12).

The inverse three-parameter transformation is obtained from Equation (C.13) by enforcing $u = 1$:

$$\begin{bmatrix} x_1 \\ x_2 \end{bmatrix} = \begin{bmatrix} \cos \beta & -\sin \beta \\ \sin \beta & \cos \beta \end{bmatrix} \begin{bmatrix} y_1 - t_1 \\ y_2 - t_2 \end{bmatrix} \tag{C.14d}$$

Six-parameter transformation in 2D space

$$\begin{bmatrix} y_1 \\ y_2 \end{bmatrix} = \begin{bmatrix} s_1 & 0 \\ 0 & s_2 \end{bmatrix} \begin{bmatrix} 1 & k \\ 0 & 1 \end{bmatrix} \begin{bmatrix} \cos \beta & \sin \beta \\ -\sin \beta & \cos \beta \end{bmatrix} \begin{bmatrix} x_1 \\ x_2 \end{bmatrix} + \begin{bmatrix} t_1 \\ t_2 \end{bmatrix} \tag{C.15}$$

or

$$\begin{bmatrix} y_1 \\ y_2 \end{bmatrix} = \begin{bmatrix} a & b \\ d & e \end{bmatrix} \begin{bmatrix} x_1 \\ x_2 \end{bmatrix} + \begin{bmatrix} c \\ f \end{bmatrix} \tag{C.16}$$

The six parameters of this transformation are two scales, one skew factor (lack of perpendicularity of the axes), one rotation, and two shifts. The inverse transformation is given by

$$\begin{bmatrix} x_1 \\ x_2 \end{bmatrix} = \frac{1}{ae - bd} \begin{bmatrix} e & -b \\ -d & a \end{bmatrix} \begin{bmatrix} y_1 - c \\ y_2 - f \end{bmatrix} \tag{C.17}$$

Three-Dimensional Linear Transformations

Similar to the two-dimensional case, affine transformation in three dimensions can be factored out in several elementary transformations: translation, uniform scale, nonuniform scale, rotations, reflections, are some examples. However, we limit consideration to the seven-parameter transformation used extensively in photogrammetry and geodesy. It is composed of a uniform scale change, three translations, and three rotations. First consider rotations in three-dimensional space.

Rotations of a three-dimensional coordinate system. There are three elementary rotations, one about each of the three axes. They frequently are performed in sequence one after the other. A set of three of these is shown in Figure C.2, where x is the original system, x' is once rotated, and x'' is twice rotated. The convention is as follows:

1. β_1 about the x_1 axis, positive rotation advances $+x_2$ to $+x_3$.
2. β_2 about the x_2' axis, positive rotation advances $+x_3'$ to $+x_1'$.
3. β_3 about the x_3'' axis, positive rotation advances $+x_1''$ to $+x_2''$.

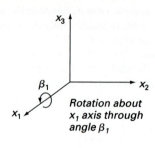

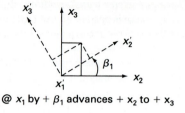

(a)

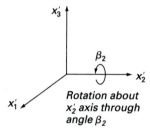

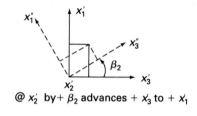

(b)

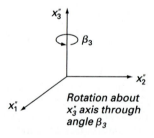

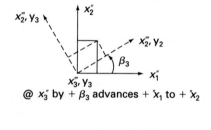

(c)

FIGURE C.2
Rotating a Cartesian system about each axis.

Each of the three elementary rotations is represented in matrix form as follows:

$$\begin{bmatrix} x_1' \\ x_2' \\ x_3' \end{bmatrix} = \begin{bmatrix} 1 & 0 & 0 \\ 0 & \cos\beta_1 & \sin\beta_1 \\ 0 & -\sin\beta_1 & \cos\beta_1 \end{bmatrix} \begin{bmatrix} x_1 \\ x_2 \\ x_3 \end{bmatrix} = \mathbf{M}_{\beta_1} \begin{bmatrix} x_1 \\ x_2 \\ x_3 \end{bmatrix} \qquad (C.18)$$

where x_1, x_2, and x_3 are the coordinates before rotation and x_1', x_2', and x_3' are the coordinates after rotation. Similarly, a rotation of $+\beta_2$ about the x_2' axis and $+\beta_3$ about the x_3'' axis are given by

$$\begin{bmatrix} x_1'' \\ x_2'' \\ x_3'' \end{bmatrix} = \begin{bmatrix} \cos\beta_2 & 0 & -\sin\beta_2 \\ 0 & 1 & 0 \\ \sin\beta_2 & 0 & \cos\beta_2 \end{bmatrix} \begin{bmatrix} x_1' \\ x_2' \\ x_3' \end{bmatrix} = \mathbf{M}_{\beta_2} \begin{bmatrix} x_1' \\ x_2' \\ x_3' \end{bmatrix} \qquad (C.19)$$

$$\begin{bmatrix} y_1 \\ y_2 \\ y_3 \end{bmatrix} = \begin{bmatrix} x_1''' \\ x_2''' \\ x_3''' \end{bmatrix} = \begin{bmatrix} \cos\beta_3 & \sin\beta_3 & 0 \\ -\sin\beta_3 & \cos\beta_3 & 0 \\ 0 & 0 & 1 \end{bmatrix} \begin{bmatrix} x_1'' \\ x_2'' \\ x_3'' \end{bmatrix} = \mathbf{M}_{\beta_3} \begin{bmatrix} x_1'' \\ x_2'' \\ x_3'' \end{bmatrix} \qquad (C.20)$$

The three rotations in Equations (C.18) through (C.20) often are referred to as *elementary* rotations, because they may be used to construct any required set of sequential rotations. By successive substitution, the total rotation matrix is obtained:

$$y = x''' = \mathbf{M}_{\beta_3}\mathbf{M}_{\beta_2}\mathbf{M}_{\beta_1}x = \mathbf{M}x \tag{C.21}$$

in which $\mathbf{M}$ now is a function of the three rotation angles, β_1, β_2, and β_3. The most commonly used set of sequential rotations in photogrammetry is given the symbols ω, ϕ, κ, where $\omega \equiv \beta_1$, $\phi \equiv \beta_2$, $\kappa \equiv \beta_3$. In this case, the matrix $\mathbf{M}$ that rotates the object coordinate system (X, Y, Z) parallel to the photo coordinate system (x, y, z) is given by

$$\mathbf{M} =$$

$$\begin{bmatrix} \cos\phi\cos\kappa & \cos\omega\sin\kappa + \sin\omega\sin\phi\cos\kappa & \sin\omega\sin\kappa - \cos\omega\sin\phi\cos\kappa \\ -\cos\phi\sin\kappa & \cos\omega\cos\kappa - \sin\omega\sin\phi\sin\kappa & \sin\omega\cos\kappa + \cos\omega\sin\phi\sin\kappa \\ \sin\phi & -\sin\omega\cos\phi & \cos\omega\cos\phi \end{bmatrix}$$

$$\tag{C.22}$$

in which ω is about X axis, ϕ is about the once rotated Y axis, and κ is about the twice rotated Z axis. The matrix $\mathbf{M}$ is orthogonal, because $\mathbf{M}_\omega$, $\mathbf{M}_\phi$, and $\mathbf{M}_\kappa$ are all orthogonal (see Appendix B.3, Matrix Operations).

Seven-parameter transformation. This transformation contains seven parameters: a uniform scale change, three rotations (β_1, β_2, and β_3), and three translations (t_1, t_2, and t_3). It takes the general form

$$y = u\mathbf{M}x + t \tag{C.23a}$$

The orthogonal matrix $\mathbf{M}$ is a function of only three independent parameters, in this case the angles β_1, β_2, and β_3.

Six-parameter transformation in 3D space. As in the two-dimensional case, a change in the location and orientation of a three-dimensional reference coordinate system involves only six parameters: three angles (orientation change) and three translations (location change). By imposing a unit scale change, or $u = 1$, in the seven-parameter transformation, Equation (C.23a), we get the following equation for the six-parameter transformation:

$$y = \mathbf{M}x + t \tag{C.23b}$$

The three angles (β_1, β_2, and β_3) are involved in the elements of the matrix $\mathbf{M}$, and the three translations (t_1, t_2, and t_3) are the elements of the shift vector t.

C.3
NONLINEAR TRANSFORMATIONS

In addition to the linear transformations discussed so far, we occasionally use nonlinear transformations, in both two and three dimensions. In two dimensions, we have the following two transformations.

Eight-parameter transformation

$$y_1 = \frac{a_1 x_1 + b_1 x_2 + c_1}{a_0 x_1 + b_0 x_2 + 1}$$

$$y_2 = \frac{a_2 x_1 + b_2 x_2 + c_2}{a_0 x_1 + b_0 x_2 + 1} \tag{C.24}$$

is the projective transformation from the x to the y coordinate systems, with the eight-transformation parameters being $a_0, b_0, a_1, b_1, a_2, b_2, c_1, c_2$. Its inverse is given by

$$x_1 = \frac{(c_1 - y_1)(b_0 y_2 - b_2) - (c_2 - y_2)(b_0 y_1 - b_1)}{(a_0 y_1 - a_1)(b_0 y_2 - b_2) - (a_2 y_2 - a_2)(b_0 y_1 - b_1)}$$

$$x_2 = \frac{(a_0 y_1 - a_1)(c_2 - y_2) - (a_0 y_2 - a_2)(c_1 - y_1)}{(a_0 y_1 - a_1)(b_0 y_2 - b_2) - (a_0 y_2 - a_2)(b_0 y_1 - b_1)} \tag{C.25}$$

These equations describe the central projectivity between two planes.

Two-dimensional general polynomials

$$y_1 = a_0 + a_1 x_1 + a_2 x_2 + a_3 x_1 x_2 + a_4 x_1^2 + a_5 x_2^2 + \cdots$$

$$y_2 = b_0 + b_1 x_1 + b_2 x_2 + b_3 x_1 x_3 + b_4 x_1^2 + b_5 x_2^2 + \cdots \tag{C.26}$$

These polynomials can obviously be extended to higher powers in x_1 and x_2. A special case of these is the conformal form given next.

Two-dimensional conformal polynomials.

The conformal property preserves the angles between intersecting lines after the transformation. If we impose the two conditions

$$\frac{\partial y_1}{\partial x_1} = \frac{\partial y_2}{\partial x_2} \quad \text{and} \quad \frac{\partial y_1}{\partial x_2} = -\frac{\partial y_2}{\partial x_1} \tag{C.27}$$

on the general polynomials in Equations (C.26), we get

$$y_1 = A_0 + A_1 x_1 + A_2 x_2 + A_3(x_1^2 - x_2^2) + A_4(2x_1 x_2) + \cdots$$

$$y_2 = B_0 - A_2 x_1 + A_1 x_2 - A_4(x_1^2 - x_2^2) + A_3(2x_1 x_2) + \cdots \tag{C.28}$$

Note that the first three terms after the equals signs are the same as the four-parameter transformation given in Equation (C.11). Equation (C.28) also can be derived using complex numbers, by writing

$$(y_1 + y_2 i) = (a_0 + b_0 i) + (a_1 + b_1 i)(x_1 + x_2 i) + (a_3 + b_3 i)(x_1 + x_2 i)^2 + \cdots$$

in which $i = \sqrt{-1}$. Expanding and equating y_1 to the real part and y_2 to the imaginary part (multiplier of i) on the right side leads to Equation (C.28).

Three-dimensional general polynomials

$$y_1 = a_0 + a_1 x_1 + a_2 x_2 + a_3 x_3 + a_4 x_1^2 + a_5 x_2^2 + a_6 x_3^2$$
$$+ a_7 x_1 x_2 + a_8 x_2 x_3 + a_9 x_1 x_3 + \cdots$$

$$y_2 = b_0 + b_1 x_1 + b_2 x_2 + b_3 x_3 + b_4 x_1^2 + b_5 x_2^2 + b_6 x_3^2$$
$$+ b_7 x_1 x_2 + b_8 x_2 x_3 + b_9 x_1 x_3 + \cdots \tag{C.29}$$

$$y_3 = c_0 + c_1 x_1 + c_2 x_2 + c_3 x_3 + c_4 x_1^2 + c_5 x_2^2 + c_6 x_3^2$$
$$+ c_7 x_1 x_2 + c_8 x_2 x_3 + c_9 x_1 x_3 + \cdots$$

We can extend these polynomials to a higher order. Unlike the two-dimensional case, conformal transformation does *not* exist in three dimensions beyond the first order (or linear) case given by the seven-parameter transformation, Equation (C.23a). A close approximation, which exists for only second-degree terms, is derived by imposing conditions similar to those in Equation (C.27), on every pair of coordinates in Equation (C.29). This makes the projections of the three-dimensional space onto each of the three planes conformal. Imposing the following on the general polynomials in Equation (C.29)

$$\frac{\partial y_1}{\partial x_1} = \frac{\partial y_2}{\partial x_2} = \frac{\partial y_3}{\partial x_3}$$

$$\frac{\partial y_1}{\partial x_2} = -\frac{\partial y_2}{\partial x_1}, \quad \frac{\partial y_2}{\partial x_3} = -\frac{\partial y_3}{\partial x_2}, \quad \frac{\partial y_1}{\partial x_3} = -\frac{\partial y_3}{\partial x_1}$$

(C.30)

leads to

$$y_1 = A_0 + Ax_1 + Bx_2 - Cx_3 + E(x_1^2 - x_2^2 - x_3^2)$$
$$+ 0 + 2Gx_3x_1 + 2Fx_1x_2 + \cdots$$

$$y_2 = B_0 - Bx_1 + Ax_2 + Dx_3 + F(-x_1^2 + x_2^2 - x_3^2)$$
$$+ 2Gx_2x_3 + 0 + 2Ex_1x_2 + \cdots$$

(C.31)

$$y_3 = C_0 + Cx_1 - Dx_2 + Ax_3 + G(-x_1^2 - x_2^2 + x_3^2)$$
$$+ 2Fx_2x_3 + 2Ex_3x_1 + 0 + \cdots$$

C.4
GOEDETIC COORDINATE SYSTEMS AND TRANSFORMATIONS

Two types of coordinate systems are used in geodesy and control surveys of large extent: Cartesian or space rectangular and curvilinear. In the curvilinear systems, two coordinates are angular and the third is linear. The direction of the third coordinate depends on which of the geodetic surfaces is used. Figure C.3 shows the three surfaces involved: the physical

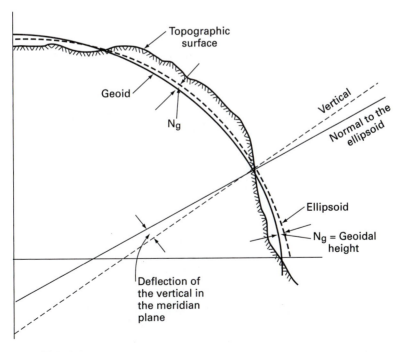

FIGURE C.3
Geodetic surfaces.

surface of the earth representing its *topography,* the *geoid* or level (equipotential) surface, and the purely mathematical or geometric surface of the *ellipsoid.*

Geocentric Coordinate System

This is an earth-centered, earth-rotating Cartesian coordinate system with an origin at the center of mass of the earth (Figure C.4). The Z axis is directed toward the adopted North Pole. The XY plane coincides with equator and the XZ plane passes through Greenwich meridian. The Y axis is selected such that XYZ is a right-handed coordinate system.

Geodetic Coordinate System

This is a curvilinear system with the following characteristics:

1. *Geodetic longitude,* λ, is the angle between the geocentric XZ plane and the meridian plane passing through the point of interest, positive to the east (or positive to the west for NAD27, NAD83 datums, Soler 1988).
2. *Geodetic latitude,* ϕ, is the angle between the normal to the ellipsoid at the point of interest, and the geocentric XY plane; the numerical limits on ϕ are $-90° \leq \phi \leq +90°$.
3. *Geodetic height, h,* is the distance along the normal to the ellipsoid between the point of interest and the surface of this reference ellipsoid. It is often referred to as the *ellipsoidal height,* see also Figure C.5.

A given location on the earth's surface may have different values for λ, ϕ, and h, depending on the reference ellipsoid (geodetic datum) used.

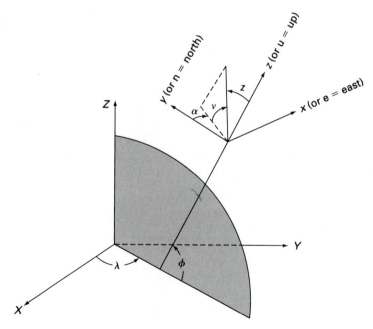

FIGURE C.4
Geocentric and local coordinate systems.

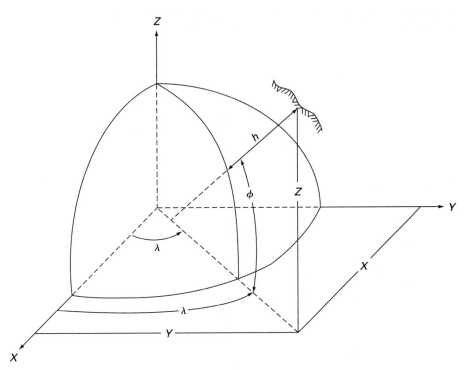

FIGURE C.5
Relation between λ, ϕ, h and x, y, z.

Local Space Rectangular Coordinate System

This is a Cartesian coordinate system with its origin at a selected point on or near the earth's surface (see Figure C.4). When the point is on the surface of the earth, the system is called *topocentric*. The z axis is along the normal to the ellipsoid with the positive direction away from the ellipsoid, corresponding to the "up" direction. (This is why sometimes this system is designated e, n, u, for *east, north,* and *up,* instead of x, y, z.) The x axis is perpendicular to z and to the meridian plane through the selected point of origin. Note that, if the ellipsoidal height of this point is 0, that is, $h = 0$, the x-y plane becomes tangent to the ellipsoidal surface. A positive x is in the direction of increasing longitude, λ, or *east.* The y axis is selected to form a right-handed system with x and z. The z axis lies in the *meridian* plane through the selected point no matter what is the value of h. A positive y is in the direction of increasing latitude, ϕ, or *north.*

Related to the local system just defined are two angles used in surveying, Figure C.4:

1. *Geodetic azimuth,* α, is the clockwise angle from $+y$ (north) measured in the plane of the local geodetic horizon and to the vertical plane containing the point: $0° \le \alpha \le 360°$.
2. *Vertical angle, V,* is the angle in the plane containing the geodetic vertical (i.e., the ellipsoid normal, see Figure C.3), from the local geodetic horizon to the line of sight to the observed point; it is positive above the horizon: $-90° \le V \le 90°$.
3. Zenith angle, $z = 90 - V$.

Transformation from Geodetic (λ, ϕ, h) to Geocentric (X, Y, Z)

In Figure C.5,

$$X = (N + h) \cos \phi \cos \lambda$$
$$Y = (N + h) \cos \phi \sin \lambda \qquad \text{(C.32)}$$
$$Z = [N(1 - \epsilon^2) + h] \sin \phi$$

The curvilinear geodetic coordinates λ, ϕ, and h are with respect to a reference ellipsoid specified by its semimajor axis, a, and f. The auxiliaries in Equation (C.32) are

$$N = a/(1 - \epsilon^2 \sin^2 \phi)^{1/2} \qquad \text{(C.33)}$$

which equals the radius of curvature of ellipsoid in the prime vertical plane and

$$\epsilon^2 = 2f - f^2 = (2 - f)f = \frac{a^2 - b^2}{a^2} \qquad \text{(C.34)}$$

f is called the *flattening* and is given by $f = (a - b)/a$.

Transformation from Geocentric (X, Y, Z) to Geodetic (λ, ϕ, h)

This transformation frequently is presented as the following iterative scheme:

1. Calculate

$$\lambda = \tan^{-1} \frac{Y}{X} \qquad \text{(C.35)}$$

2. Iterate for ϕ; and consequently for h. The initial value for ϕ is the spherical latitude,

$$\phi_0 = \tan^{-1} \left(\frac{Z}{\sqrt{X^2 + Y^2}} \right) \qquad \text{(C.36)}$$

Then, compute

$$N = \frac{a}{(1 - \epsilon^2 \sin^2 \phi)^{1/2}} \qquad \text{(C.37)}$$

$$h = \frac{\sqrt{X^2 + Y^2}}{\cos \phi} - N \qquad \text{(C.38)}$$

giving
$$\phi = \tan^{-1} \left[\frac{Z}{\sqrt{X^2 + Y^2}} \left(1 - \epsilon^2 \frac{N}{N + h} \right)^{-1} \right] \qquad \text{(C.39)}$$

According to Soler and Hothem (1988), the following noniterative alternative method, which is based on work by Bowring, for all practical purposes is rigorous:

$$\tan \lambda = Y/X \qquad \text{(C.40)}$$
$$\tan \phi = (Z + \epsilon^2 a \sin^3 \mu)/(p - \epsilon^2 a \cos^3 \mu) \qquad \text{(C.41)}$$
$$h = p \cos \phi + Z \sin \phi - (a^2/N) \qquad \text{(C.42)}$$

in which

$$p = (X^2 + Y^2)^{1/2} \qquad \text{(C.43)}$$
$$r = (p^2 + Z^2)^{1/2} \qquad \text{(C.44)}$$
$$\tan \mu = \frac{Z(1 - f)}{p} \left(1 + \frac{\epsilon^2 a}{r} \right) \qquad \text{(C.45)}$$

An alternative solution proposed by Professor C. Goad,* as a stable iterative scheme, is as follows.

Refer to Equations (C.32), (C.33), and (C.35), repeated here for convenience,

$$X = (N + h) \cos \phi \cos \lambda$$

$$Y = (N + h) \cos \phi \sin \lambda \qquad \text{(C.32)}$$

$$Z = [N(1 - \epsilon^2) + h] \sin \phi$$

$$N = (a)/(1 - \epsilon^2 \sin^2 \phi)^{1/2} \qquad \text{(C.33)}$$

$$\lambda = \tan^{-1}(Y)/(X) \qquad \text{(C.35)}$$

Then, using Equations (C.43) and (C.32),

$$p = (X^2 + Y^2)^{1/2} = (N + h) \cos \phi$$

from which

$$\begin{aligned} p(\phi, h) &= p(\phi_o + \Delta\phi, h_o + \Delta h) \\ &= p(\phi_o, h_o) + p_o\Delta\phi + p_h \Delta h + \cdots \\ &\approx p(\phi_o, h_o) - (N + h) \sin \phi \, \Delta\phi + \cos \phi \, \Delta h \end{aligned}$$

or

$$\Delta p \approx -(N + h) \sin \phi \, \Delta\phi + \cos \phi \, \Delta h \qquad \text{(C.46a)}$$

where initial values for ϕ_0 and h_0 are

$$\phi_O = \sin^{-1}[Z/r]$$

$$h_O = r - a(1 - f \sin^2 \phi_O)$$

in which

$$r = (p^2 + Z^2)^{1/2}$$

Similarly,

$$\Delta Z \approx (N + h) \cos \phi \, \Delta\phi + \sin \phi \, \Delta h \qquad \text{(C.46b)}$$

Expressing Equations (C.46a) and (C.46b) in matrix form

$$\begin{bmatrix} \Delta Z \\ \Delta p \end{bmatrix} = \begin{bmatrix} \cos \phi & \sin \phi \\ -\sin \phi & \cos \phi \end{bmatrix} \begin{bmatrix} (N + h) \Delta\phi \\ \Delta h \end{bmatrix}$$

and solving for $\Delta\phi$ and Δh,

$$\begin{bmatrix} (N + h) \Delta\phi \\ \Delta h \end{bmatrix} = \begin{bmatrix} \cos \phi & -\sin \phi \\ \sin \phi & \cos \phi \end{bmatrix} \begin{bmatrix} \Delta Z \\ \Delta p \end{bmatrix}$$

or

$$\Delta\phi = [\cos \phi \, \Delta Z - \sin \phi \, \Delta p]/(N + h)$$

$$\Delta h = \sin \phi \, \Delta Z + \cos \phi \, \Delta p$$

Corrections $\Delta\phi$ and Δh are applied to the initial estimates ϕ_O and h_O and the solution is iterated until convergence.

---

* Former head of the Department of Geodetic Sciences, The Ohio State University, Columbus, through personal communication.

Transformation from Geocentric (X, Y, Z) to a Local Space Rectangular (x, y, z)

The origin of the local system needs to be specified in the geocentric coordinate system. If the geodetic location of that origin is λ_O, ϕ_O, and h_O, then from Equation (C.32),

$$
\begin{aligned}
X_O &= (N_O + h_O) \cos \phi_O \cos \lambda_O \\
Y_O &= (N_O + h_O) \cos \phi_O \sin \lambda_O \\
Z_O &= [N_O(1 - \epsilon^2) + h_O] \sin \phi_O
\end{aligned}
\tag{C.47}
$$

The geocentric coordinates first are shifted to this local origin then rotated by an orthogonal matrix $\mathbf{R}$. This matrix is composed of two sequential rotations:

1. $\mathbf{R}_3$ is a rotation about the Z axis by $(\lambda_O - 270°)$ in which λ_O is measured with a value between $0°$ and $360°$, which brings the X axis perpendicular to the meridian plane and in the "east" direction (see Figure C.4); substituting $(\lambda_O - 270°)$ for β_3 in Equation (C.20), we get

$$
\mathbf{R}_3 = \begin{bmatrix} -\sin \lambda_O & \cos \lambda_O & 0 \\ -\cos \lambda_O & -\sin \lambda_O & 0 \\ 0 & 0 & 1 \end{bmatrix}
\tag{C.48}
$$

2. $\mathbf{R}_1$ is a rotation about the once-rotated X axis by $(90° - \phi)$ so that the new Z axis is along the local vertical, which makes the new Y axis point to the north; substituting $(90° - \phi)$ for β_1 in Equation (C.18) yields

$$
\mathbf{R}_1 = \begin{bmatrix} 1 & 0 & 0 \\ 0 & \sin \phi_O & \cos \phi_O \\ 0 & -\cos \phi_O & \sin \phi_O \end{bmatrix}
\tag{C.49}
$$

$$
\mathbf{R} = \mathbf{R}_1 \mathbf{R}_3 = \begin{bmatrix} -\sin \lambda_O & \cos \lambda_O & 0 \\ -\cos \lambda_O \sin \phi_O & -\sin \lambda_O \sin \phi_O & \cos \phi_O \\ \cos \lambda_O \cos \phi_O & \sin \lambda_O \cos \phi_O & \sin \phi_O \end{bmatrix}
\tag{C.50}
$$

The transformation equation then becomes

$$
\begin{bmatrix} x \\ y \\ z \end{bmatrix} = \mathbf{R} \begin{bmatrix} X - X_O \\ Y - Y_O \\ Z - Z_O \end{bmatrix}
\tag{C.51}
$$

Transformation from a Local Space Rectangular (x, y, z) to Geocentric (X, Y, Z)

From Equation (C.51),

$$
\begin{bmatrix} X - X_O \\ Y - Y_O \\ Z - Z_O \end{bmatrix} = \mathbf{R}^{-1} \begin{bmatrix} x \\ y \\ z \end{bmatrix} = \mathbf{R}^t \begin{bmatrix} x \\ y \\ z \end{bmatrix} \qquad (\mathbf{R} \text{ is an orthogonal matrix})
$$

The final form of the transformation is

$$\begin{bmatrix} X \\ Y \\ Z \end{bmatrix} = \begin{bmatrix} -\sin \lambda_O & -\cos \lambda_O \sin \phi_O & \cos \lambda_O \cos \phi_O \\ \cos \lambda_O & -\sin \lambda_O \sin \phi_O & \sin \lambda_O \cos \phi_O \\ 0 & \cos \phi_O & \sin \phi_O \end{bmatrix} \begin{bmatrix} x \\ y \\ z \end{bmatrix} + \begin{bmatrix} X_O \\ Y_O \\ Z_O \end{bmatrix} \quad \text{(C.52)}$$

PROBLEMS

C.1. Given a point $P\,(3, 2)$ in an (x_1, x_2) system, compute its coordinates in a new system (y_1, y_2) that results from rotating (x_1, x_2) by an angle $\beta = -25°$.

C.2. Compute the new coordinates of point P in Problem C.1 If the (y_1, y_2) system is translated by the amounts $(3, -1)$.

C.3. Given the rotations $\beta_1 = 10°$, $\beta_2 = -30°$, and $\beta_3 = 40°$, all defined as in the standards given in Section C.2. Compute $\mathbf{M}_{\beta_1}$, $\mathbf{M}_{\beta_2}$, and $\mathbf{M}_{\beta_3}$.

C.4. The system $(x_1, x_2, x_3,)$ is rotated by β_1 then by β_2 as in Problem C.3 to yield the new system (y_1, y_2, y_3). A point is given by the coordinates $x_1 = -5, x_2 = 7$, and $x_3 = 10$, compute its new (y_1, y_2, y_3) coordinates.

C.5. A new system (z_1, z_2, z_3) is obtained by two successive rotations, β_3 first then β_1 as given in Problem C.3, to the (x_1, x_2, x_3) system. Given the following two points in the X system, $P_1(3, 4, 5)$ and $P_2(-1, 2, 6)$, compute their coordinates in the Z system. Compute further the distances, L_x and L_z, between the two points using their x coordinates and z coordinates, respectively. Determine the relationship between L_x and L_z and explain it.

REFERENCES

Mikhail, E. M. *Observations and Least Squares*. Lanham, MD: University Press of America. 1976.

Mikhail, E. M., and G. Gracie. *Analysis and Adjustment of Survey Measurements*. New York: Van Nostrand Reinhold, 1980.

Moffit, F. H., and E. M. Mikhail. *Photogrammetry*, 3rd ed. New York: Harper and Row, 1980.

Soler, T., and L. D. Hothem. "Coordinate Systems Used in Geodesy: Basic Definitions and Concepts." *ASCE Journal of Surveying Engineering* 114, no. 2 (May 1988), pp. 84–97.

APPENDIX D

Introductory Probability and Statistics

D.1
RANDOM VARIABLES AND PROBABILITY

A *random variable* is defined as a variable that takes on any of *several possible values,* with each of which is associated a probability. Probability is the likelihood associated with a random event. A *random event* is one whose relative frequency of occurrence approaches a stable limit as the number of observations or repetitions of an experiment is increased to infinity. The limit of the relative frequency of occurrence of a random event is known as the *probability* of the event.

The probability of a random event is a number that lies somewhere between 0 and 1. If the probability is neither 0 nor 1, but some number between 0 and 1, the event may or may not occur, its chance of occurring being given by the specific value of the probability. In mathematical language, if E represents a random event, and $P(E)$ the probability of the event occurring, then

$$0 \le P[E] \le 1 \tag{D.1}$$

EXAMPLE D.1. A distance is measured repeatedly using a steel tape. With much diligence and patience, 5000 measurements are made to 0.001 m. The measurements are corrected for systematic error and the resulting corrected values are rounded off to the nearest centimeter. For each resulting value, the relative frequency of occurrence is calculated and recorded (see Table D.1). On the basis of the large number of

TABLE D.1

Rounded off value of distance measurement (m)	No. of measurements occurring, a	Relative frequency of occurrence, $a/5000$
489.51 or less	0	0
489.52	206	0.0412
489.53	3633	0.7266
489.54	1161	0.2322
489.55 or more	0	0
Total	5000	1.0000

measurements made, the calculated relative frequencies are accepted as the limiting values; that is, they are accepted as probabilities.

Let A be the event that the rounded-off value of any one of the distance measurements is 489.51 m or less, B be the event the measurement is 489.52 m, C be the event the measurement is 489.53 m, D be the event the measurement is 489.54 m, and E be the event the measurement is 489.55 m or more. Accepting the relative frequencies of occurrence in Table D.1 as the probabilities, we have

$$P[A] = 0$$

$$P[B] = 0.0412$$

$$P[C] = 0.7266$$

$$P[D] = 0.2322$$

$$P[E] = 0$$

A very important and useful concept in probability theory is that of independence. Two events are said to be *independent* if the occurrence of one has no influence on the occurrence of the other. If two events are independent, the probability of the two events occurring together is equal to the product of the probabilities of the events occurring individually. In mathematical language, a random variable can take on distinct or discrete values, such as the number of dots on the face of a die or the distance measurement in Table D.1 when rounded off to the nearest centimeter. If, on the other hand, we consider measurements and their errors in a more general way, the probability functions will be continuous. The continuous function that describes all possible relative frequencies is termed the *probability density function, $f(x)$.* On the other hand, the function that directly gives probabilities for various values the random variable can take is called the *probability distribution function,* or simply the *distribution function, $F(x)$.* The lowest value of $F(x)$ is 0 (representing no probability), and the highest is 1. The density function is the first derivative of the distribution function:

$$f(x) = F'(x) = \frac{\mathrm{d}}{\mathrm{d}x} F(x) \tag{D.2}$$

Because $F(x)$ is nondecreasing, its slope must be nonnegative. Hence,

$$f(x) \geq 0 \tag{D.3}$$

From the fundamental theorem of integral calculus, we have

$$\int_a^b f(x)\, dx = F(b) - F(a) \tag{D.4}$$

Denoting the random variable by X, then

$$P[a < X \leq b] = F(b) - F(a) = \int_a^b f(x)\, dx \tag{D.5}$$

which provides the means for evaluating the probability that the random variable X takes on a value between a and b.

Note that $(a < X \leq b)$ is a random event and that the probability of this random event is a definite integral of the density function. Hence, the probability of this event is represented by the *area* under the density function between a and b.

It is quite important to recognize that evaluation of a density function does *not* yield probability; that is, the density function, $f(x)$, does not give the probability that X takes on

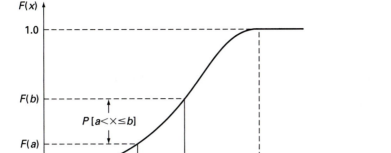

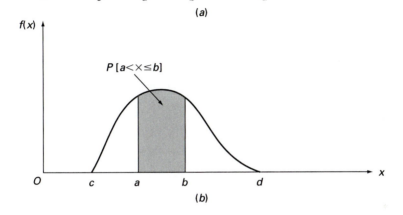

FIGURE D.1
(a) Distribution and (b) density functions.

the value x, as $p(x)$ does for the discrete case. To repeat, probability is represented by the *area* under the density curve *not by the ordinate value*.

Figure D.1 illustrates the relationship between a continuous distribution function and the corresponding density function. Note carefully how $P[a < X \le b]$ is represented in each case. In Figure D.1a, $P[a < X \le b]$ is represented by a difference in ordinate values, $F(b) - F(a)$; Figure D.1b, $P[a < X \le b]$ is represented by the shaded area. Note also that $F(x)$ indeed is continuous (i.e., no jumps) and lies entirely between 0 and 1 and that $f(x)$ is nonnegative.

Two important and very useful relationships can be obtained as special cases of Equation (D.5). First, by making $a = -\infty$ and $b = \infty$, we have

$$\int_{-\infty}^{\infty} f(x)\, dx = P[-\infty < X \le \infty] = F(\infty) - F(-\infty) = 1 \qquad \text{(D.6)}$$

which states that the total area under the density curve must equal unity. Second, by making $a = -\infty$ and $b = x$ and recognizing that the event $(X \le x)$ is equivalent to the event $(-\infty < X \le x)$, we have

$$F(x) = P[X \le x] = P[-\infty < X \le x] = F(x) - F(-\infty) = \int_{-\infty}^{x} f(u)\, du \qquad \text{(D.7)}$$

(It is necessary to change x to u, because x is the upper limit of the integral.)

D.2
NORMAL DISTRIBUTION

Of all existing probability distributions, none is more important than the normal distribution. The *normal distribution* has widespread application in science, technology, and industry; it is used as the basic model for all physical measurements, including measurements in surveying.

The density function of the normal distribution is

$$f(x) = \frac{1}{\sigma\sqrt{2\pi}} \exp\left\{ -\frac{(x - \mu)^2}{2\sigma^2} \right\} \qquad \text{for } -\infty < x < \infty \qquad (D.8)$$

The quantities μ and σ are referred to as the *distribution parameters,* called the *mean* and *standard deviation,* respectively.

The distribution function of the normal distribution is

$$F(x) = \int_{-\infty}^{x} f(u)\, du = \frac{1}{\sigma\sqrt{2\pi}} \int_{-\infty}^{x} \exp\left\{ -\frac{(u - \mu)^2}{2\sigma^2} \right\} du \qquad (D.9)$$

These two functions are plotted in Figure D.2. It is quite clear from Figure D.2a that the normal distribution is symmetric about μ. The density function's points of inflection are located at $x = \mu - \sigma$ and $x = \mu + \sigma$, see Figure D.3. Maximum density occurs at $x = \mu$.

If X is a random variable with a normal distribution, then $P[X \le x]$, the probability that X is less than or equal to x, is represented by the left shaded area in Figure D.2a and by the ordinate, $F(x)$, in Figure D.2b.

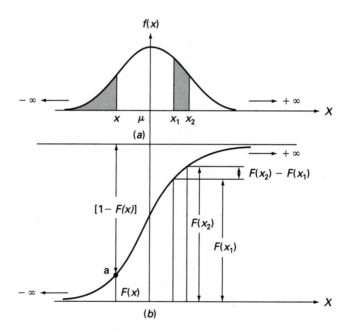

FIGURE D.2
(a) Density distribution and (b) cumulative distribution curves.

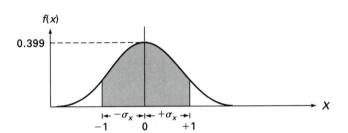

FIGURE D.3
Points of inflection on the density distribution curve. When
$\mu = 0$, $\sigma = 1$, it is called the standard normal.

EXAMPLE D.2. Let X be a normally distributed random variable with parameters
$\mu = 12$ and $\sigma = 2$; that is, the density function of X is

$$f(x) = \frac{1}{2\sqrt{2\pi}} \exp\left\{-\frac{(x-12)^2}{2(2)^2}\right\} = 0.19947 \exp\{-0.1250(x-12)^2\}$$

and the distribution function of X is

$$F(x) = 0.19947 \int_{-\infty}^{x} \exp\{-0.1250(u-12)^2\}\, du*$$

Specific values of these two functions are listed in Table D.2.

Evaluate the probability that (a) X is less than or equal to 10, (b) X lies between
11 and 15, (c) X is greater than 16; (d) X is precisely 10.

Solution.

a. The probability that X is less than or equal to 10 is $P[X \le 10] = F(10) = 0.1587$.
This probability would be represented by the shaded area to the left of the ordinate
$x = 10$ in Figure D.2a.

b. The probability that X lies between 11 and 15 is

$$P[11 < X < 15] = F(15) - F(11) = 0.9332 - 0.3085 = 0.6247$$

c. The probability that X is greater than 16 is

$$P[X > 16] = P[16 < X < \infty] = F(\infty) - F(16) = 1 - 0.9772 = 0.0228$$

d. The probability that X is precisely 10, of course, is 0, because this is represented by
the 0 area under the density curve. If, however, we interpret 10 as the interval
between 9.5 and 10.5, then it is possible to evaluate a probability that is nonzero:

$$P[9.5 < X < 10.5] = F(10.5) - F(9.5) = 0.2266 - 0.1056 = 0.1210$$

A random variable Z is said to have a *standard normal distribution* if its density
function is

$$f(z) = \frac{1}{\sqrt{2\pi}} \exp\left\{-\frac{z^2}{2}\right\} \qquad \text{for } -\infty < z < \infty \tag{D.10}$$

---

*$F(x)$ is left in the integral form because it cannot be integrated directly. To evaluate this function, a series
approximation is used.

TABLE D.2

Evaluation of density and distribution functions for a normally distributed random variable with $\mu = 12$ and $\sigma = 2$

x	Density function, $f(x)$	Distribution function, $F(x)$
4	0.0001	0.0000
5	0.0004	0.0002
6	0.0022	0.0013
7	0.0088	0.0062
8	0.0270	0.0228
9	0.0648	0.0668
10	0.1210	0.1587
11	0.1760	0.3085
12	0.1995	0.5000
13	0.1760	0.6915
14	0.1210	0.8413
15	0.0648	0.9332
16	0.0270	0.9772
17	0.0088	0.9938
18	0.0022	0.9987
19	0.0004	0.9998
20	0.0001	1.0000
9.5	0.0913	0.1056
10	0.1210	0.1587
10.5	0.1506	0.2266

This function is obtained from the normal density function by setting $\mu = 0$ and $\sigma = 1$; it is shown in Figure D.3.

The standard normal distribution is an important special case of the normal distribution because it provides a convenient way of evaluating probabilities associated with any normal distribution. The density function of the normal distribution cannot be integrated directly, so it presents a problem whenever probabilities have to be evaluated for specific values of μ and σ. Fortunately, the problem can be circumvented by first transforming the normal random variable, X, into the standard normal random variable, Z, and then evaluating probabilities for Z.

The transformation, known as *standardization,* is given by

$$Z = \frac{X - \mu}{\sigma} \tag{D.11}$$

Even though it is just as impossible to integrate $f(z)$ directly as it is $f(x)$ for the normal distribution, there is only one $f(z)$ to integrate. The resulting function is

$$P[Z \leq z] = \phi(z) = \frac{1}{\sqrt{2\pi}} \int_{-\infty}^{z} \exp\left\{-\frac{u^2}{2}\right\} du \quad \text{for} \quad -\infty < z < \infty \tag{D.12}$$

The integral can be evaluated by approximate means once and for all, and values of $\phi(z)$ can be tabulated. Such tabulation is given in Table E.2 of Appendix E.

To show how standardization and Table E.2 are applied, consider an example.

EXAMPLE D.3. As in Example D.2, let X be a normally distributed random variable with parameters $\mu = 12$ and $\sigma = 2$. Standardize X and use Table E.2 of Appendix E to evaluate the probability that (a) X is less than or equal to 10, (b) X lies between 11 and 15, (c) X is greater than 16.

Solution. According to Equation (D.11), the standard normal random variable is

$$Z = \frac{X - \mu}{\sigma} = \frac{X - 12}{2}$$

a. The probability that X is less than or equal to 10 is evaluated first noting that if $X \le 10$ then $(X - 12)/2 \le (10 - 12)/2$; that is, $Z \le -1$. Therefore,

$$P[X \le 10] = P\left[\frac{X - 12}{2} \le \frac{10 - 12}{2}\right] = P[Z \le -1]$$

and from Table E.2 of Appendix E, $P[Z \le -1] = \phi(-1) = 0.1587$.

b. The probability that X lies between 11 and 15 is evaluated in similar fashion:

$$P[11 < X < 15] = P\left[\frac{11 - 12}{2} < \frac{X - 12}{2} < \frac{15 - 12}{2}\right] = P[-0.5 < Z < 1.5]$$

$$= \phi(1.5) - \phi(-0.5) = 0.9332 - 0.3085 = 0.6247$$

c. The probability that X is greater than 16 is

$$P[X > 16] = P[16 < X < \infty] = P\left[\frac{16 - 12}{2} < \frac{X - 12}{2} < \infty\right]$$

$$= P[2 < Z < \infty] = \phi(\infty) - \phi(2)$$

$$= 1 - 0.9772 = 0.0228$$

The suitability of the normal distribution as the probability model for measurements is based firmly on the behavior of actual observations. In effect, the model is chosen to fit the observed data. The choice of the normal distribution has a theoretical basis as well. However, if the resulting error in a measurement is the sum of several component errors, each of which has its own probability distribution, it can be shown that the total error has approximately a normal distribution, whether or not the component errors are normally distributed themselves.

D.3
SAMPLING

The probability distribution associated with a random variable, except for a few cases, cannot be constructed theoretically. Instead, it usually is *estimated statistically* from repeated observations. A *random sample* is the collection of *independent* measurements that are repetitions of the same operation under the same conditions. The *number* of measurements in the sample is the *sample size*. By contrast to a sample, the *population* is the total of all possible values from which a sample may be drawn. Whereas in some applications the population may be of finite size, in surveying, the population usually is infinite: it includes all the possible values of the measurement under consideration.

D.4
STATISTICS

Given a sample of n observations, a new quantity, which is called a *statistic,* may be computed from the sample. For example, the arithmetic mean of the n observation, called the *sample mean,* is a statistic. Because the statistic is a function of random variables (the elements of the sample), it also is a random variable with its own probability distribution, referred to as a *sampling* distribution. Under the realistic assumption that survey measurements are normally distributed, the following two sampling distributions, the chi-square and student (t), are used to assess sample statistics.

D.5
CHI-SQUARE DISTRIBUTION

If $Z_1, Z_2, \ldots, Z_n$ are n independent random variables, each with a *standard* normal distribution (i.e., $\mu = 0$, and $\sigma = 1$), it can be shown that the quantity

$$Y = Z_1^2 + Z_2^2 + \cdots + Z_n^2 \tag{D.13}$$

has a *chi-square* (χ^2) *distribution* with n degrees of freedom. It can be shown that the mean and variance of the distribution are

$$\mu_y = n \tag{D.14}$$

$$\sigma_y^2 = 2n \tag{D.15}$$

There is a different density function $f(y)$ for each value of n. For $n = 4$, the values of $f(y)$ are given in Table D.3, and the density function is given in Figure D.4, which also shows plots for several values of n. The mean and standard deviation of Y are

$$\mu_y = n = 4$$

$$\sigma_y = \sqrt{2n} = \sqrt{8} = 2.83$$

For small values of n, the density functions are asymmetric with long tails to the right. For larger values of n, the curves approach the symmetric shape of the density function of

TABLE D.3
Evaluation of the chi-square density function for $n = 4$

f	$f(y)$	y	$f(y)$
0	0.000	5	0.103
0.5	0.097	6	0.075
1.0	0.152	7	0.053
1.5	0.177	8	0.037
2.0	0.184	9	0.025
2.5	0.179	10	0.017
3.0	0.166	11	0.011
3.5	0.152	12	0.007
4.0	0.135	13	0.005
4.5	0.114	14	0.003

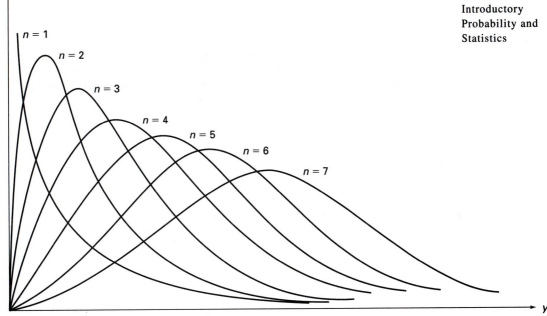

FIGURE D.4 Chi-square density functions for several values of n.

Reproduced with permission of Van Nostrand Reinhold from *Analysis and Adjustmet of Survey Measurements*, by E. M. Mikhail and G. Gracie.

the normal distribution with mean n and variance $2n$. The values of the probability distribution functions, which are the integrals of $f(y)$, are designated by χ and are given in Table E.3 of Appendix E. The subscript p represents the probability.

D.6
THE t (STUDENT)* DISTRIBUTION

If Z is a random variable that has a standard normal distribution and Y is an independent random variable that has a chi-square distribution with n degrees of freedom, then the random variable

$$T = \frac{Z}{\sqrt{Y/n}} \tag{D.16}$$

is said to have a *t distribution with n degrees of freedom.*

The mean and variance of T are

$$\mu_t = 0 \qquad \text{for } n > 1$$
$$\tag{D.17}$$
$$\sigma_t^2 = \frac{n}{n-2} \qquad \text{for } n > 2$$

---

* "Student" was the pseudonym of W.S. Gosset, the statistician who first derived the t distribution.

The distribution is symmetric about 0. Each value of n determines the shape of the density function. For $n = 4$, the values of t are listed in Table D.4, and plotted in Figure D.5. The standard deviation of T is

$$\sigma_t = \sqrt{\frac{n}{n - 2}} = \sqrt{2} = 1.414$$

It is obvious from Figure D.5 that the shape of the t distribution's density function resembles that of normal distribution. However, the two distributions are not the same,

TABLE D.4
Evaluation of the t density
function for $n = 4$

t	$f(t)$
0	0.375
±0.5	0.322
±1.0	0.215
±1.5	0.123
±2.0	0.066
±2.5	0.036
±3.0	0.020
±3.5	0.011
±4.0	0.007

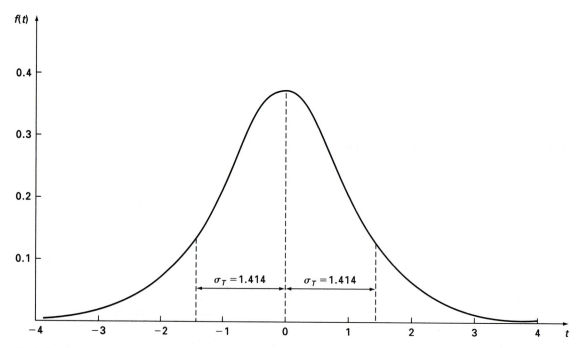

FIGURE D.5 Density function for the t distribution for $n = 4$.

Reproduced with permission of Van Nostrand Reinhold from *Analysis and Adjustment of Survey Measurements*, by E. M. Mikhail and G. Gracie.

particularly for small values of n. As n increases, the t distribution approaches the standard normal distribution. Indeed, for large n, the standard normal distribution can be used to approximate the t distribution.

The probability function of T is the integral of $f(t)$. If it is fixed at some specified value, p (i.e., $F(t) = P[T \le t] = p$), the value of t associated with p is designated $t_{p,n}$, representing the pth percentile of the t distribution with n degrees of freedom. Values of $t_{p,n}$ are given in Table E.4 of Appendix E.

Another distribution, called the *tau distribution* (or the τ distribution) is closely related to the t distribution and used in blunder detection. The τ distribution with r degrees of freedom is given by

$$\tau_r = \frac{\sqrt{r}\, t_{r-1}}{\sqrt{(r-1) + t_{r-1}^2}} \tag{D.18}$$

in which t_{r-1} is the student t distribution with $(r-1)$ degrees of freedom.

D.7
STATISTICAL ANALYSIS

Assessment of the quality of estimates from samples of survey measurement is done through statistical evaluation. Two types exist, confidence interval about the estimate and statistical tests. Each of these will be given in a separate section for each separate sample statistic.

D.8
CONFIDENCE INTERVAL FOR THE MEAN

If a random sample of size n is drawn from a normally distributed population with mean μ and variance σ^2, the sample mean $\bar{X}$ has a normal distribution with mean μ and variance σ^2/n. The quantity

$$Z = \frac{\bar{X} - \mu}{\sigma/\sqrt{n}} \tag{D.19}$$

has a standard normal distribution with zero mean and unit variance, and it follows that

$$P\left[-z < \frac{\bar{X} - \mu}{\sigma/\sqrt{n}} < z\right] = 2\phi(z) - 1 \tag{D.20}$$

where $\phi(z)$ is the value of the standard normal distribution function, obtainable from Table E.2 of Appendix E.

Rearranging the inequality inside the brackets of Equation (D.20), we get

$$P\left[\bar{X} - \frac{z\sigma}{\sqrt{n}} < \mu < \bar{X} + \frac{z\sigma}{\sqrt{n}}\right] = 2\phi(z) - 1 \tag{D.21}$$

which reads as follows. The probability that μ lies between $\bar{X} - (z\sigma/\sqrt{n})$ and $\bar{X} + (z\sigma/\sqrt{n})$ is $2\phi(z) - 1$. When a specific numerical value $\bar{x}$ is provided for $\bar{X}$, the foregoing probability statement becomes a *confidence statement*. The values $\bar{x} - (z\sigma/\sqrt{n})$ and $\bar{x} + (z\sigma/\sqrt{n})$ are known as *confidence limits,* the interval between them is known as a *confidence interval,* and $2\phi(z) - 1$ is known as the *degree of confidence,* or *confidence level,* often stated as a percentage. The construction of a confidence interval for a particular distribution parameter, such as μ is known as *interval estimation.*

In Equations (D.20) and (D.21), σ is assumed to be known. More often, σ is estimated by the sample standard deviation S. Consequently, instead of using Equation (D.19), we use

$$T = \frac{\bar{X} - \mu}{S/\sqrt{n}} \tag{D.22}$$

where T has a t distribution with $n - 1$ degrees of freedom. And, instead of Equation (D.21), we have

$$P\left[\bar{X} - \frac{tS}{\sqrt{n}} < \mu < \bar{X} + \frac{tS}{\sqrt{n}}\right] = 2F(t) - 1 \tag{D.23}$$

When specific numerical values $\bar{x}$ and s are provided for $\bar{X}$ and S, we obtain a confidence interval with confidence limits $\bar{x} - (ts/\sqrt{n})$ and $\bar{x} + (ts/\sqrt{n})$ and degree of confidence $2F(t) - 1$. Recall that, as n increases, the values of t approach the corresponding values of z. Indeed, for a sample size of 30 or larger, the t distribution can be approximated very well by the standard normal distribution.

D.9
CONFIDENCE INTERVAL FOR THE VARIANCE

When a random sample of size n is drawn from a normal population, the fact that $[(n - 1)/\sigma^2]S^2$ has a chi-square distribution with $n - 1$ degrees of freedom can be used to construct a confidence interval for the population variance, σ^2:

$$P\left[\chi^2_{a,n-1} < \frac{(n - 1)S^2}{\sigma^2} < \chi^2_{b,n-1}\right] = b - a \tag{D.24}$$

where $\chi^2_{a,n-1}$ and $\chi^2_{b,n-1}$ are the ath and bth percentiles, respectively, of the chi-square distribution with $n - 1$ degrees of freedom.

From Equation (D.24) it follows that

$$P\left[\frac{(n - 1)S^2}{\chi^2_{b,n-1}} < \sigma^2 < \frac{(n - 1)S^2}{\chi^2_{a,n-1}}\right] = b - a \tag{D.25}$$

and when a specific numerical value s^2 is provided for S^2, we obtain a confidence interval with limits $(n - 1)\, s^2/\chi^2_{b,n-1}$ and $(n - 1)\, s^2/\chi^2_{a,n-1}$, and degree of confidence $b - a$.

In constructing an appropriate confidence interval for σ^2, it is customary to make the two percentiles complementary; that is $a + b = 1$.

If a confidence interval for the standard deviation σ is desired, positive square roots of the confidence limits for σ^2 are taken.

D.10
STATISTICAL TESTING

Statistical testing deals with testing of hypotheses about the probability distribution, based on statistics computed from a sample. The main or *null hypothesis* is designated H_0 and its complementary *alternative hypothesis* H_1.

A hypothesis is tested by drawing a sample from the population, computing the value of a specific sample statistic, then making the decision to accept or reject the hypothesis on the basis of the value of the statistic. The statistic used for making the test is called the *test statistic*.

Testing of a statistical hypothesis H_0 is not infallible, because it is based on a sample drawn from a population rather than on the entire population. Four possible outcomes can occur.

1. H_0 is accepted, when H_0 is true.
2. H_0 is rejected, when H_0 is true.
3. H_0 is accepted, when H_0 is false.
4. H_0 is rejected, when H_0 is false.

When outcomes 1 or 4 occur, no error is made, in that the correct course of action has been taken. Outcome 2 is known as a *type I error,* outcome 3 is known as a *type II error.*

The *size of the type I error,* designated α, is defined as the probability of rejecting H_0 when H_0 is true; that is,

$$\alpha = P[\text{reject } H_0 \text{ when } H_0 \text{ is true}] \tag{D.26}$$

When α is fixed at some level for H_0 and is expressed as a percentage, it is known as the *significance level* of the test. Although the choice of significance level is arbitrary, common practice indicates a significance level of 5 percent as "significant" and 1 percent as "highly significant."

D.11
HYPOTHESIS TEST OF THE MEAN OF A PROBABILITY DISTRIBUTION

Under certain conditions we may expect the mean μ of a probability distribution to have a specific value, μ_0. The hypothesis that $\mu = \mu_0$ can be tested by drawing a sample of size n and using the sample mean $\bar{X}$ as the test statistic. Specifically, we have

$$H_0 : \mu = \mu_0$$

$$H_1 : \mu \neq \mu_0$$

$\bar{X}$ is assumed to be normally distributed, or at least approximately normally distributed. Under the hypothesis that $\mu = \mu_0$, the following probability statement can be derived from Equation (D.20), assuming σ is known:

$$P[(\mu_0 - c) < \bar{X} < (\mu_0 + c)] = 2\phi(z) - 1 \tag{D.27}$$

where $c = z\sigma/\sqrt{n}$.

If σ is unknown, the following probability statement can be derived from Equation (D.23):

$$P[(\mu_0 - c) < \bar{X} < (\mu_0 + c)] = 2F(t) - 1 \tag{D.28}$$

where $c = tS/\sqrt{n}$. H_0 is accepted if $\bar{x}$, the specific value of $\bar{X}$ calculated from the sample, lies between $\mu_0 - c$ and $\mu_0 + c$; otherwise, H_0 is rejected. The regions of acceptance and rejection are shown in Figure D.6. If α is the probability that H_0 is rejected when it is true, then $1 - \alpha$ must be the probability that H_0 is accepted when it is true. It follows, then, that

$$1 - \alpha = 2\phi(z) - 1 \quad \text{for known } \sigma \tag{D.29}$$

$$= 2F(t) - 1 \quad \text{for unknown } \sigma \tag{D.30}$$

Solving for $\phi(z)$ or $F(t)$, we get

$$\phi(z) = 1 - \frac{\alpha}{2} \tag{D.31}$$

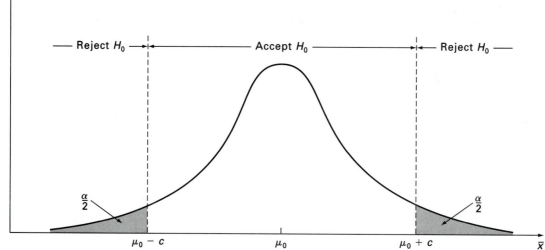

FIGURE D.6 Regions of acceptance and rejection for test of the mean.

Reproduced with permission of Van Nostrand Reinhold from *Analysis and Adjustment of Survey Measurements,* by E. M. Mikhail and G. Gracie.

or

$$F(t) = 1 - \frac{\alpha}{2} \tag{D.32}$$

Therefore, the value of z or t is obtained from the significance level of the test, α. Specifically, Table E.2 of Appendix E is used to evaluate z; Table E.4 of Appendix E is used to evaluate t. (Note that in Table E.4, $t = t_{p,n-1}$, where $p = F(t)$.)

D.12
HYPOTHESIS TEST OF THE VARIANCE OF A PROBABILITY DISTRIBUTION

Under the assumption that a population is normally distributed, we can test the null hypothesis H_0 that the population variance is σ_0^2 against the alternative that it is not σ_0^2, using the sample variance S^2 as the test statistic.

Noting that $(n - 1)S^2/\sigma_0^2$ is distributed as chi-square with $n - 1$ degrees of freedom, we can make the following probability statement:

$$P[\chi^2_{a,n-1} < (n - 1)S^2/\sigma_0^2 < \chi^2_{b,n-1}] = b - a \tag{D.33}$$

from which we get

$$P\left[\frac{\chi^2_{a,n-1}\sigma_0^2}{(n - 1)} < S^2 < \frac{\chi^2_{b,n-1}\sigma_0^2}{(n - 1)}\right] = b - a \tag{D.34}$$

H_0 is accepted if s^2, the specific value of S^2 calculated from the sample, lies between $\chi^2_{a,n-1}\sigma_0^2/(n - 1)$ and $\chi^2_{b,n-1}\sigma_0^2/(n - 1)$; otherwise, H_0 is rejected. The regions of acceptance and rejection are shown in Figure D.7.

Again, $1 - \alpha$ is the probability that H_0 is accepted when it is true. Therefore,

$$1 - \alpha = b - a \tag{D.35}$$

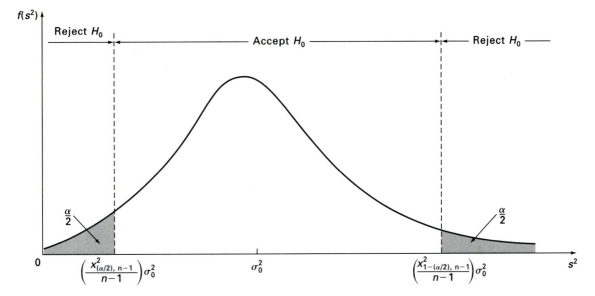

FIGURE D.7 Regions of acceptance and rejection for test of the variance.

Reproduced with permission from Van Nostrand Reinhold from *Analysis and Adjustment of Survey Measurements* by E. M. Mikhail and G. Gracie.

and for $a + b = 1$ (complementary percentiles), we obtain

$$a = \frac{\alpha}{2} \tag{D.36}$$

$$b = 1 - \frac{\alpha}{2} \tag{D.37}$$

This test procedure, of course, can be used to test the standard deviation as well as the variance.

REFERENCE

Pope, A. "The Statistics of Residuals and the Detection of Outliers." International Union of Geodesy and Geophysics IAG, XVI General Assembly, Grenoble, France, 1975.

APPENDIX E

Trigonometric Formulas and Statistical Tables

TABLE E.1
Trigonometric formulas

Right triangles

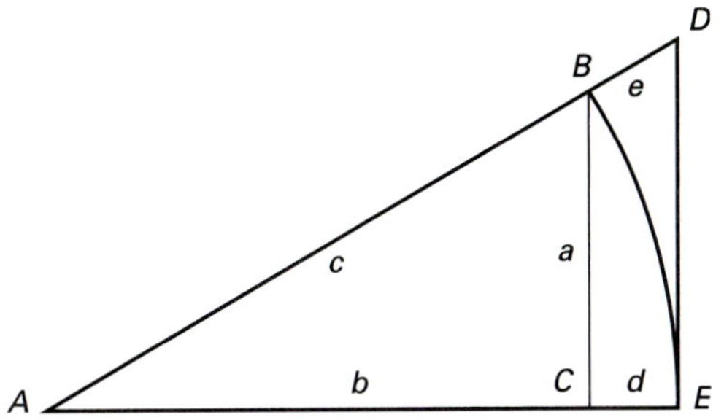

FIGURE E.1
Right triangle.

$\sin A = \dfrac{a}{c} = \cos B$ $\qquad$ $\sec A = \dfrac{c}{b} = \operatorname{cosec} B$

$\cos A = \dfrac{b}{c} = \sin B$ $\qquad$ $\operatorname{cosec} A = \dfrac{c}{a} = \sec B$

$\tan A = \dfrac{a}{b} = \cot B$ $\qquad$ $\operatorname{vers} A = 1 - \cos A = \dfrac{c - b}{c} = \dfrac{d}{c}$

$\cot A = \dfrac{b}{a} = \tan B$ $\qquad$ $\operatorname{exsec} A = \sec A - 1 = \dfrac{e}{c}$

$$a = c \sin A = c \cos B = b \tan A = b \cot B = \sqrt{c^2 - b^2}$$

$$b = c \cos A = c \sin B = a \cot A = a \tan B = \sqrt{c^2 - a^2}$$

$$c = \frac{a}{\sin A} = \frac{a}{\cos B} = \frac{b}{\sin B} = \frac{b}{\cos A} = \frac{d}{\operatorname{vers} A} = \frac{e}{\operatorname{exsec} A} = \sqrt{a^2 + b^2}$$

$d = c \operatorname{vers} A$ $\qquad$ $e = c \operatorname{exsec} A$

(continued)

1144

APPENDIX E:
Trigonometric
Formulas and
Statistical
Tables

TABLE E.1 *(Continued)*

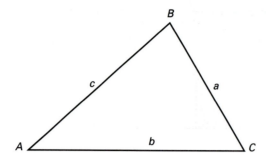

FIGURE E.2
Oblique triangle.

Oblique triangles		
Given	**Sought**	**Formulas**
A, B, a	b, c	$b = \dfrac{a \sin B}{\sin A} \qquad c = \dfrac{a \sin(A + B)}{\sin A}$
A, a, b	B, c	$\sin B = \dfrac{b \sin A}{a} \qquad c = \dfrac{a \sin C}{\sin A}$
C, a, b	$\dfrac{1}{2}(A + B)$	$\dfrac{1}{2}(A + B) = 90° - \dfrac{1}{2}C$
	$\dfrac{1}{2}(A - B)$	$\tan\dfrac{1}{2}(A - B) = \dfrac{a - b}{a + b}\tan\dfrac{1}{2}(A + B)$
a, b, c	A	$\cos A = \dfrac{b^2 + c^2 - a^2}{2bc}$
	Area	If $s = \dfrac{1}{2}(a + b + c)$,
		$\text{Area} = \sqrt{s(s - a)\,(s - b)\,(s - c)}$
C, a, b	Area	$\text{Area} = \dfrac{1}{2}ab \sin C$
A, B, C, a	Area	$\text{Area} = \dfrac{a^2 \sin B \sin C}{2 \sin A}$

TABLE E.2
Values of the standard normal distribution function

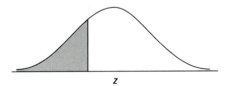

$$\phi(z) = \int_{-\infty}^{z} \frac{1}{\sqrt{2\pi}} e^{-u^2/2} \, du = P[Z \le z]$$

z	0	1	2	3	4	5	6	7	8	9
−3.	.0013	.0010	.0007	.0005	.0003	.0002	.0002	.0001	.0001	1.0000
−2.9	.0019	.0018	.0017	.0017	.0016	.0016	.0015	.0015	.0014	.0014
−2.8	.0026	.0025	.0024	.0023	.0023	.0022	.0021	.0021	.0020	.0019
−2.7	.0035	.0034	.0033	.0032	.0031	.0030	.0029	.0028	.0027	.0026
−2.6	.0047	.0045	.0044	.0043	.0041	.0040	.0039	.0038	.0037	.0036
−2.5	.0062	.0060	.0059	.0057	.0055	.0054	.0052	.0051	.0049	.0048
−2.4	.0082	.0080	.0078	.0075	.0073	.0071	.0069	.0068	.0066	.0064
−2.3	.0107	.0104	.0102	.0099	.0096	.0094	.0091	.0089	.0087	.0084
−2.2	.0139	.0136	.0132	.0129	.0126	.0122	.0119	.0116	.0113	.0110
−2.1	.0179	.0174	.0170	.0166	.0162	.0158	.0154	.0150	.0146	.0143
−2.0	.0228	.0222	.0217	.0212	.0207	.0202	.0197	.0192	.0188	.0183
−1.9	.0287	.0281	.0274	.0268	.0262	.0256	.0250	.0244	.0238	.0233
−1.8	.0359	.0352	.0344	.0336	.0329	.0322	.0314	.0307	.0300	.0294
−1.7	.0446	.0436	.0427	.0418	.0409	.0401	.0392	.0384	.0375	.0367
−1.6	.0548	.0537	.0526	.0516	.0505	.0495	.0485	.0475	.0465	.0455
−1.5	.0668	.0655	.0643	.0630	.0618	.0606	.0594	.0582	.0570	.0559
−1.4	.0808	.0793	.0778	.0764	.0749	.0735	.0722	.0708	.0694	.0681
−1.3	.0968	.0951	.0934	.0918	.0901	.0885	.0869	.0853	.0838	.0823
−1.2	.1151	.1131	.1112	.1093	.1075	.1056	.1038	.1020	.1003	.0985
−1.1	.1357	.1335	.1314	.1292	.1271	.1251	.1230	.1210	.1190	.1170
−1.0	.1587	.1562	.1539	.1515	.1492	.1469	.1446	.1423	.1401	.1379
− .9	.1841	.1814	.1788	.1762	.1736	.1711	.1685	.1660	.1635	.1611
− .8	.2119	.2090	.2061	.2033	.2005	.1977	.1949	.1922	.1894	.1867
− .7	.2420	.2389	.2358	.2327	.2297	.2266	.2236	.2206	.2177	.2148
− .6	.2743	.2709	.2676	.2643	.2611	.2578	.2546	.2514	.2483	.2451
− .5	.3085	.3050	.3015	.2981	.2946	.2912	.2877	.2843	.2810	.2776
− .4	.3446	.3409	.3372	.3336	.3300	.3264	.3228	.3192	.3156	.3121
− .3	.3821	.3783	.3745	.3707	.3669	.3632	.3594	.3557	.3520	.3483
− .2	.4207	.4168	.4129	.4090	.4052	.4013	.3974	.3936	.3897	.3859
− .1	.4602	.4562	.4522	.4483	.4443	.4404	.4364	.4325	.4286	.4247
− .0	.5000	.4960	.4920	.4880	.4840	.4801	.4761	.4721	.4681	.4641

Reprinted with permission of Macmillan Publishing Co., Inc. from *Introduction to Probability and Statistics* by B.W. Lindgren and G.W. McElrath. Copyright © 1969 by B.W. Lindgren and G.W. McElrath.

TABLE E.2 (*concluded*)

z	0	1	2	3	4	5	6	7	8	9
.0	.5000	.5040	.5080	.5120	.5160	.5199	.5239	.5279	.5319	.5359
.1	.5398	.5438	.5478	.5517	.5557	.5596	.5636	.5675	.5714	.5753
.2	.5793	.5832	.5871	.5910	.5948	.5987	.6026	.6064	.6103	.6141
.3	.6179	.6217	.6255	.6293	.6331	.6368	.6406	.6443	.6480	.6517
.4	.6554	.6591	.6628	.6664	.6700	.6736	.6772	.6808	.6844	.6879
.5	.6915	.6950	.6985	.7019	.7054	.7088	.7123	.7157	.7190	.7224
.6	.7257	.7291	.7324	.7357	.7389	.7422	.7454	.7486	.7517	.7549
.7	.7580	.7611	.7642	.7673	.7703	.7734	.7764	.7794	.7823	.7852
.8	.7881	.7910	.7939	.7967	.7995	.8023	.8051	.8078	.8106	.8133
.9	.8159	.8186	.8212	.8238	.8264	.8289	.8315	.8340	.8365	.8389
1.0	.8413	.8438	.8461	.8485	.8508	.8531	.8554	.8577	.8599	.8621
1.1	.8643	.8665	.8686	.8708	.8729	.8749	.8770	.8790	.8810	.8830
1.2	.8849	.8869	.8888	.8907	.8925	.8944	.8962	.8980	.8997	.9015
1.3	.9032	.9049	.9066	.9082	.9099	.9115	.9131	.9147	.9162	.9177
1.4	.9192	.9207	.9222	.9236	.9251	.9265	.9278	.9292	.9306	.9319
1.5	.9332	.9345	.9357	.9370	.9382	.9394	.9406	.9418	.9430	.9441
1.6	.9452	.9463	.9474	.9484	.9495	.9505	.9515	.9525	.9535	.9545
1.7	.9554	.9564	.9573	.9582	.9591	.9599	.9608	.9616	.9625	.9633
1.8	.9641	.9648	.9656	.9664	.9671	.9678	.9686	.9693	.9700	.9706
1.9	.9713	.9719	.9726	.9732	.9738	.9744	.9750	.9756	.9762	.9767
2.0	.9772	.9778	.9783	.9788	.9793	.9798	.9803	.9808	.9812	.9817
2.1	.9821	.9826	.9830	.9834	.9838	.9842	.9846	.9850	.9854	.9857
2.2	.9861	.9864	.9868	.9871	.9874	.9878	.9881	.9884	.9887	.9890
2.3	.9893	.9896	.9898	.9901	.9904	.9906	.9909	.9911	.9913	.9916
2.4	.9918	.9920	.9922	.9925	.9927	.9929	.9931	.9932	.9934	.9936
2.5	.9938	.9940	.9941	.9943	.9945	.9946	.9948	.9949	.9951	.9952
2.6	.9953	.9955	.9956	.9957	.9959	.9960	.9961	.9962	.9963	.9964
2.7	.9965	.9966	.9967	.9968	.9969	.9970	.9971	.9972	.9973	.9974
2.8	.9974	.9975	.9976	.9977	.9977	.9978	.9979	.9979	.9980	.9981
2.9	.9981	.9982	.9982	.9983	.9984	.9984	.9985	.9985	.9986	.9986
3.	.9987	.9990	.9993	.9995	.9997	.9998	.9998	.9999	.9999	1.0000

TABLE E.3
Percentiles of the chi-square distribution

1147

APPENDIX E:
Trigonometric
Formulas and
Statistical
Tables

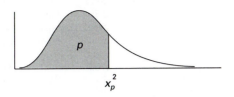

Degrees of freedom	$\chi^2_{.005}$	$\chi^2_{.01}$	$\chi^2_{.025}$	$\chi^2_{.05}$	$\chi^2_{.10}$	$\chi^2_{.20}$	$\chi^2_{.30}$
1	.000	.000	.001	.004	.016	.064	.148
2	.010	.020	.051	.103	.211	.446	.713
3	.072	.115	.216	.352	.584	1.00	1.42
4	.207	.297	.484	.711	1.06	1.65	2.20
5	.412	.554	.831	1.15	1.61	2.34	3.00
6	.676	.872	1.24	1.64	2.20	3.07	3.83
7	.989	1.24	1.69	2.17	2.83	3.82	4.67
8	1.34	1.65	2.18	2.73	3.49	4.59	5.53
9	1.73	2.09	2.70	3.33	4.17	5.38	6.39
10	2.16	2.56	3.25	3.94	4.87	6.18	7.27
11	2.60	3.05	3.82	4.57	5.58	6.99	8.15
12	3.07	3.57	4.40	5.23	6.30	7.81	9.03
13	3.57	4.11	5.01	5.89	7.04	8.63	9.93
14	4.07	4.66	5.63	6.57	7.79	9.47	10.8
15	4.60	5.23	6.26	7.26	8.55	10.3	11.7
16	5.14	5.81	6.91	7.96	9.31	11.2	12.6
17	5.70	6.41	7.56	8.67	10.1	12.0	13.5
18	6.26	7.01	8.23	9.39	10.9	12.9	14.4
19	6.83	7.63	8.91	10.1	11.7	13.7	15.4
20	7.43	8.26	9.59	10.9	12.4	14.6	16.3
21	8.03	8.90	10.3	11.6	13.2	15.4	17.2
22	8.64	9.54	11.0	12.3	14.0	16.3	18.1
23	9.26	10.2	11.7	13.1	14.8	17.2	19.0
24	9.89	10.9	12.4	13.8	15.7	18.1	19.9
25	10.5	11.5	13.1	14.6	16.5	18.9	20.9
26	11.2	12.2	13.8	15.4	17.3	19.8	21.8
27	11.8	12.9	14.6	16.2	18.1	20.7	22.7
28	12.5	13.6	15.3	16.9	18.9	21.6	23.6
29	13.1	14.3	16.0	17.7	19.8	22.5	24.6
30	13.8	15.0	16.8	18.5	20.6	23.4	25.5
40	20.7	22.1	24.4	26.5	29.0	32.3	34.9
50	28.0	29.7	32.3	34.8	37.7	41.4	44.3
60	35.5	37.5	40.5	43.2	46.5	50.6	53.8

1148

APPENDIX E:
Trigonometric
Formulas and
Statistical
Tables

TABLE E.3 (*concluded*)

Degrees of freedom	$\chi^2_{.50}$	$\chi^2_{.70}$	$\chi^2_{.80}$	$\chi^2_{.90}$	$\chi^2_{.95}$	$\chi^2_{.975}$	$\chi^2_{.99}$	$\chi^2_{.995}$
1	.455	1.07	1.64	2.71	3.84	5.02	6.63	7.88
2	1.39	2.41	3.22	4.61	5.99	7.38	9.21	10.6
3	2.37	3.66	4.64	6.25	7.81	9.35	11.3	12.8
4	3.36	4.88	5.99	7.78	9.49	11.1	13.3	14.9
5	4.35	6.06	7.29	9.24	11.1	12.8	15.1	16.7
6	5.35	7.23	8.56	10.6	12.6	14.4	16.8	18.5
7	6.35	8.38	9.80	12.0	14.1	16.0	18.5	20.3
8	7.34	9.52	11.0	13.4	15.5	17.5	20.1	22.0
9	8.34	10.7	12.2	14.7	16.9	19.0	21.7	23.6
10	9.34	11.8	13.4	16.0	18.3	20.5	23.2	25.2
11	10.3	12.9	14.6	17.3	19.7	21.9	24.7	26.8
12	11.3	14.0	15.8	18.5	21.0	23.3	26.2	28.3
13	12.3	15.1	17.0	19.8	22.4	24.7	27.7	29.8
14	13.3	16.2	18.2	21.1	23.7	26.1	29.1	31.3
15	14.3	17.3	19.3	22.3	25.0	27.5	30.6	32.8
16	15.3	18.4	20.5	23.5	26.3	28.8	32.0	34.3
17	16.3	19.5	21.6	24.8	27.6	30.2	33.4	35.7
18	17.3	20.6	22.8	26.0	28.9	31.5	34.8	37.2
19	18.3	21.7	23.9	27.2	30.1	32.9	36.2	38.6
20	19.3	22.8	25.0	28.4	31.4	34.2	37.6	40.0
21	20.3	23.9	26.2	29.6	32.7	35.5	38.9	41.4
22	21.3	24.9	27.3	30.8	33.9	36.8	40.3	42.8
23	22.3	26.0	28.4	32.0	35.2	38.1	41.6	44.2
24	23.3	27.1	29.6	33.2	36.4	39.4	43.0	45.6
25	24.3	28.2	30.7	34.4	37.7	40.6	44.3	46.9
26	25.3	29.2	31.8	35.6	38.9	41.9	45.6	48.3
27	26.3	30.3	32.9	36.7	40.1	43.2	47.0	49.6
28	27.3	31.4	34.0	37.9	41.3	44.5	48.3	51.0
29	28.3	32.5	35.1	39.1	42.6	45.7	49.6	52.3
30	29.3	33.5	36.2	40.3	43.8	47.0	50.9	53.7
40	39.3	44.2	47.3	51.8	55.8	59.3	63.7	66.8
50	49.3	54.7	58.2	63.2	67.5	71.4	76.2	79.5
60	59.3	65.2	69.0	74.4	79.1	83.3	88.4	92.0

TABLE E.4
Percentiles of the *t* distribution

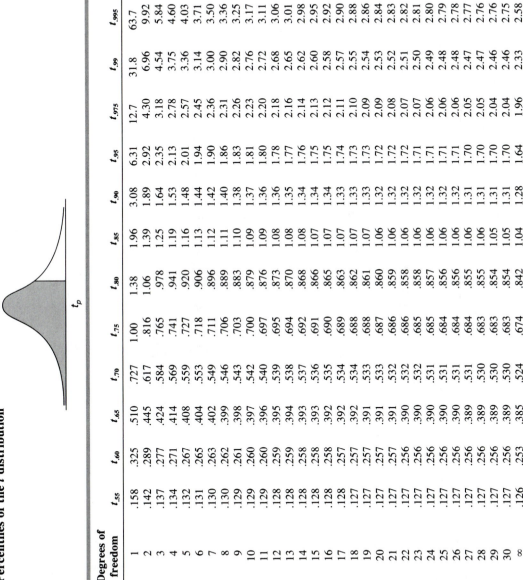

Degrees of freedom	$t_{.55}$	$t_{.60}$	$t_{.65}$	$t_{.70}$	$t_{.75}$	$t_{.80}$	$t_{.85}$	$t_{.90}$	$t_{.95}$	$t_{.975}$	$t_{.99}$	$t_{.995}$	$t_{.9995}$
1	.158	.325	.510	.727	1.00	1.38	1.96	3.08	6.31	12.7	31.8	63.7	637
2	.142	.289	.445	.617	.816	1.06	1.39	1.89	2.92	4.30	6.96	9.92	31.6
3	.137	.277	.424	.584	.765	.978	1.25	1.64	2.35	3.18	4.54	5.84	12.9
4	.134	.271	.414	.569	.741	.941	1.19	1.53	2.13	2.78	3.75	4.60	8.61
5	.132	.267	.408	.559	.727	.920	1.16	1.48	2.01	2.57	3.36	4.03	6.86
6	.131	.265	.404	.553	.718	.906	1.13	1.44	1.94	2.45	3.14	3.71	5.96
7	.130	.263	.402	.549	.711	.896	1.12	1.42	1.90	2.36	3.00	3.50	5.40
8	.130	.262	.399	.546	.706	.889	1.11	1.40	1.86	2.31	2.90	3.36	5.04
9	.129	.261	.398	.543	.703	.883	1.10	1.38	1.83	2.26	2.82	3.25	4.78
10	.129	.260	.397	.542	.700	.879	1.09	1.37	1.81	2.23	2.76	3.17	4.59
11	.129	.260	.396	.540	.697	.876	1.09	1.36	1.80	2.20	2.72	3.11	4.44
12	.128	.259	.395	.539	.695	.873	1.08	1.36	1.78	2.18	2.68	3.06	4.32
13	.128	.259	.394	.538	.694	.870	1.08	1.35	1.77	2.16	2.65	3.01	4.22
14	.128	.258	.393	.537	.692	.868	1.08	1.34	1.76	2.14	2.62	2.98	4.14
15	.128	.258	.393	.536	.691	.866	1.07	1.34	1.75	2.13	2.60	2.95	4.07
16	.128	.258	.392	.535	.690	.865	1.07	1.34	1.75	2.12	2.58	2.92	4.02
17	.128	.257	.392	.534	.689	.863	1.07	1.33	1.74	2.11	2.57	2.90	3.96
18	.127	.257	.392	.534	.688	.862	1.07	1.33	1.73	2.10	2.55	2.88	3.92
19	.127	.257	.391	.533	.688	.861	1.07	1.33	1.73	2.09	2.54	2.86	3.88
20	.127	.257	.391	.533	.687	.860	1.06	1.32	1.72	2.09	2.53	2.84	3.85
21	.127	.257	.391	.532	.686	.859	1.06	1.32	1.72	2.08	2.52	2.83	3.82
22	.127	.256	.390	.532	.686	.858	1.06	1.32	1.72	2.07	2.51	2.82	3.79
23	.127	.256	.390	.532	.685	.858	1.06	1.32	1.71	2.07	2.50	2.81	3.77
24	.127	.256	.390	.531	.685	.857	1.06	1.32	1.71	2.06	2.49	2.80	3.74
25	.127	.256	.390	.531	.684	.856	1.06	1.32	1.71	2.06	2.48	2.79	3.72
26	.127	.256	.390	.531	.684	.856	1.06	1.32	1.71	2.06	2.48	2.78	3.71
27	.127	.256	.389	.531	.684	.855	1.06	1.31	1.70	2.05	2.47	2.77	3.69
28	.127	.256	.389	.530	.683	.855	1.06	1.31	1.70	2.05	2.47	2.76	3.67
29	.127	.256	.389	.530	.683	.854	1.05	1.31	1.70	2.04	2.46	2.76	3.66
30	.127	.256	.389	.530	.683	.854	1.05	1.31	1.70	2.04	2.46	2.75	3.65
∞	.126	.253	.385	.524	.674	.842	1.04	1.28	1.64	1.96	2.33	2.58	3.29

Reprinted with permission of Macmillan Publishing Co., Inc. from *Introduction to Probability and Statistics* by B. W. Lindgren and G. W. McElrath. Copyright © 1969 by B. W. Lindren and G. W. McElrath.

Index

Other Useful Relationships

0.000 004 848	$= \sin 1'' = \tan 1''$
0.000 290 888	$= \sin 1' = \tan 1'$
0.017 452 406	$= \sin 1°$
0.017 455 065	$= \tan 1°$
3.141 592 654	$= \pi$
0.574	$=$ coefficient of combined curvature & refraction (c&r), ft/mi^2
0.0675	$=$ coefficient of combined (c&r), m/km^2
1.15 mi	$= 1'$ of latitude ≈ 1 nautical mile
69.1 mi	$\approx 1°$ latitude
6076.10 ft	$= 1$ nautical mile
101 ft	≈ 1 second (1″) of latitude
6 mi	$=$ length and width of a normal township
*80 chains	$= 1$ Gunter's mile
640 acres	$= 1 \text{ mi}^2 =$ area of 1 normal section
480 chains	$=$ length and width of a normal township
*36	$=$ number of sections in a normal township
*15° longitude	$=$ width of one time zone $= 360°/24$ hr
*24 hr	$= 360°$ of longitude
*20°C = 68°F	$=$ standard temperatures (Celsius & Fahrenheit) in taping
299,792.5 km/sec	$=$ speed of light and other magnetic waves in a vacuum
6,356,752.3 m	$=$ earth's semi-minor axis (GRS80 ellipsoid)
6,378,137.0 m	$=$ earth's semi-major axis (GRS80 ellipsoid)
6,378,206.4 m	$=$ earth's semi-major axis (Clark ellipsoid, 1866)
6,371,000 m	$=$ mean radius of earth $= 20,902,000$ ft
2,000,000 kg/cm^2	$=$ Young's modulus of elasticity for steel $= 29,000,000 \text{ lb/in}^2$

*Denotes exact value. All others correct to figures shown.

Multiplication Factors

Multiple	Prefix	Symbol
$1\ 000\ 000\ 000 = 10^9$	giga	G
$1\ 000\ 000 = 10^6$	mega	M
$1\ 000 = 10^3$	kilo	k
$100 = 10^2$	hecto	h
$10 = 10^1$	deka	da
$0.1 = 10^{-1}$	deci	d
$0.01 = 10^{-2}$	centi	c
$0.001 = 10^{-3}$	milli	m
$0.000\ 001 = 10^{-6}$	micro	μ
$0.000\ 000\ 001 = 10^{-9}$	nano	n

Useful relations between trigonometric functions given in the following equations, and in the table below.

$$\sin^2 \theta + \cos^2 \theta = 1$$
$$1 + \tan^2 \theta = \sec^2 \theta$$
$$1 + \cot^2 \theta = \csc^2 \theta$$
$$\sin(\alpha \pm \beta) = \sin \alpha \cos \beta \pm \cos \alpha \sin \beta$$
$$\cos(\alpha \pm \beta) = \cos \alpha \cos \beta \mp \cos \alpha \sin \beta$$
$$\tan(\alpha \pm \beta) = \frac{\tan \alpha \pm \tan \beta}{1 \mp \tan \alpha \tan \beta}$$

$$\sin 2\theta = 2 \sin \theta \cos \theta$$
$$\cos 2\theta = \cos^2 \theta - \sin^2 \theta$$
$$= 1 - 2 \sin^2 \theta$$
$$= 2 \cos^2 \theta - 1$$
$$\sin^2 \theta = \tfrac{1}{2} - \tfrac{1}{2} \cos 2\theta$$
$$\cos^2 \theta = \tfrac{1}{2} + \tfrac{1}{2} \cos 2\theta$$
$$\tan 2\theta = \frac{2 \tan \theta}{1 - \tan^2 \theta}$$

$$\sin 3\theta = 3 \sin \theta - 4 \sin^3 \theta$$
$$\cos 3\theta = 4 \cos^3 \theta - 3 \cos \theta$$
$$\tan 3\theta = \frac{3 \tan \theta - \tan^3 \theta}{1 - 3 \tan^2 \theta}$$

$$\sin \alpha + \sin \beta = 2 \sin \tfrac{1}{2}(\alpha + \beta) \cos \tfrac{1}{2}(\alpha - \beta)$$
$$\sin \alpha - \sin \beta = 2 \cos \tfrac{1}{2}(\alpha + \beta) \sin \tfrac{1}{2}(\alpha - \beta)$$
$$\cos \alpha + \cos \beta = 2 \cos \tfrac{1}{2}(\alpha + \beta) \cos \tfrac{1}{2}(\alpha + \beta)$$
$$\cos \alpha - \cos \beta = -2 \sin \tfrac{1}{2}(\alpha + \beta) \sin \tfrac{1}{2}(\alpha - \beta)$$

$$\sin \alpha \cos \beta = \tfrac{1}{2}[\sin(\alpha + \beta) + \sin(\alpha - \beta)]$$
$$\cos \alpha \sin \beta = \tfrac{1}{2}[\sin(\alpha + \beta) - \sin(\alpha - \beta)]$$
$$\cos \alpha \cos \beta = \tfrac{1}{2}[\cos(\alpha + \beta) + \cos(\alpha - \beta)]$$
$$\sin \alpha \sin \beta = -\tfrac{1}{2}[\cos(\alpha + \beta) - \cos(\alpha - \beta)]$$

Angle	Quadrant[†]	sin	cos	tan
$-\theta$	4	$-\sin \theta$	$\cos \theta$	$-\tan \theta$
$+\theta$	1	$\sin \theta$	$\cos \theta$	$\tan \theta$
$90° - \theta = \dfrac{\pi}{2} - \theta$	1	$\cos \theta$	$\sin \theta$	$\cot \theta$
$90° + \theta = \dfrac{\pi}{2} + \theta$	2	$\cos \theta$	$-\sin \theta$	$-\cot \theta$
$180° - \theta = \pi - \theta$	2	$\sin \theta$	$-\cos \theta$	$-\tan \theta$
$180° + \theta = \pi + \theta$	3	$-\sin \theta$	$-\cos \theta$	$\tan \theta$
$270° - \theta = \dfrac{3\pi}{2} - \theta$	3	$-\cos \theta$	$-\sin \theta$	$\cot \theta$
$270° + \theta = \dfrac{3\pi}{2} + \theta$	4	$-\cos \theta$	$\sin \theta$	$-\cot \theta$
$360° - \theta = 2\pi - \theta$	4	$-\sin \theta$	$\cos \theta$	$-\tan \theta$

[†] The quadrant is given under the assumption that θ is an acute angle.